Genetics

A Conceptual Approach

FOURTH EDITION

Benjamin A. Pierce

Southwestern University

W. H. Freeman and Company
New York

Publisher: Kate Ahr Parker
Executive Editor: Susan Winslow
Development Editor: Lisa Samols
Senior Project Editor: Georgia Lee Hadler
Manuscript Editor: Patricia Zimmerman
Art Director: Diana Blume
Illustrations: Dragonfly Media Group
Illustration Coordinator: Janice Donnola
Photo Editor: Ted Szczepanski
Photo Researcher: Elyse Rieder
Production Coordinator: Paul Rohloff
Media Editor: Aaron Gass
Supplements Editor: Anna Bristow
Associate Director of Marketing: Debbie Clare
Composition: Preparé
Printing and Binding: RR Donnelly

Library of Congress Control Number: 2010934358

ISBN-13: 978-1-4292-3250-0
ISBN-10: 1-4292-3250-1

Printed in the United States of America

First printing

W. H. Freeman and Company
41 Madison Avenue
New York, NY 10010
Houndsmills, Basingstoke RG21 6XS. England

www.whfreeman.com

To my parents, Rush and Amanda Pierce;

my children, Sarah and Michael Pierce;

and my genetic partner, friend, and soul mate
for 30 years, Marlene Tyrrell

Contents in Brief

Contents

Chapter 10   DNA: The Chemical Nature of the Gene   271

Chapter 11   Chromosome Structure and Transposable Elements   291

Chapter 12   DNA Replication and Recombination   321

Chapter 18 Gene Mutations and DNA Repair   481

Chapter 19 Molecular Genetic Analysis and Biotechnology   513

Chapter 26 Evolutionary Genetics   721

TASTER GENES IN SPITTING APES   721

Letter from the Author

One of my passions in life is teaching. Probably like many of you, I've been inspired by past teachers. Miss Amos, my high-school English teacher, demonstrated the importance of having excitement for the subject that you teach. My organic chemistry professor, Harold Jesky, taught me that students learn best when they are motivated. And my first genetics professor, Ray Canham, taught me that the key to learning genetics is to focus on concepts and problem solving. All are important lessons that I learned from my teachers—lessons that I've tried to incorporate into my own teaching and into my textbook.

Seventeen years ago, I set out to write a new genetics textbook. My vision was to create a book that conveys the excitement of genetics, that motivates students, and that focuses on concepts and problem solving. Those were the original goals of the first edition of *Genetics: A Conceptual Approach* and they remain the core features of this fourth edition of the book.

In this book, I've tried to share some of what I've learned in my 30 years of teaching genetics. I provide advice and encouragement at places where students often have difficulty, and I tell stories of the people, places, and experiments of genetics—past and present—to keep the subject relevant, interesting, and alive. My goal is to help you learn the necessary details, concepts, and problem-solving skills while encouraging you to see the elegance and beauty of the discipline.

At Southwestern University, my office door is always open, and my students often drop by to share their own approaches to learning, things that they have read about genetics, and their experiences, concerns, and triumphs. I learn as much from my students as they learn from me, and I would love to learn from you—by email (pierceb@southwestern.edu), by telephone (512-863-1974), or in person (Southwestern University, Georgetown, Texas).

Ben Pierce
Professor of Biology and
holder of the Lillian Nelson Pratt Chair
Southwestern University

Preface

The title *Genetics: A Conceptual Approach* precisely conveys the major goals of the book: to help students uncover major concepts of genetics and make connections among those concepts so as to have a fuller understanding of genetics. This conceptual and holistic approach to genetics has proved to be effective in the three preceding editions of this book. After taking part in class testing of those editions and of a sample chapter of the current edition, students say that they come away with a deeper and more complete understanding of genetics, thanks to the accessible writing style, simple and instructive illustrations, and useful pedagogical features throughout the book.

Hallmark Features

■ **Key Concepts and Connections** Throughout the book, I've included features to help students focus on the major concepts of each topic.

• *Concept boxes* throughout each chapter summarize the key points of preceding sections. *Concept Checks* ask students to pause for a moment and make sure that they understand the take-home message. Concept Checks are in multiple-choice and short-answer format, and answers are listed at the end of each chapter.

> **CONCEPTS**
>
> The process of transformation indicates that some substance—the transforming principle—is capable of genetically altering bacteria. Avery, MacLeod, and McCarty demonstrated that the transforming principle is DNA, providing the first evidence that DNA is the genetic material.
>
> ✔ **CONCEPT CHECK 3**
>
> If Avery, MacLeod, and McCarty had found that samples of heat-killed bacteria treated with RNase and DNase transformed bacteria, but samples treated with protease did not, what conclusion would they have made?
>
> a. Protease carries out transformation.
> b. RNA and DNA are the genetic materials.
> c. Protein is the genetic material.
> d. RNase and DNase are necessary for transformation.

• *Connecting Concepts* sections draw on concepts presented in several sections or several chapters to help students see how different topics of genetics relate to one another. These sections compare and contrast processes or integrate ideas across chapters to create an overarching, big picture of genetics. All major concepts are listed in the *Concepts Summary* at the end of each chapter.

■ **Accessibility** The welcoming and conversational writing style of this book has long been one of its most successful features for both students and instructors. In addition to carefully walking students through each major concept of genetics, I invite them into the topic with an **introductory story**. These stories include relevant examples of disease or other biological phenomena to give students a sample of what they'll be learning in a chapter. More than a third of the introductory stories in this edition are new.

■ **Clear, Simple Illustration Program** I have worked closely with illustrators to create attractive and instructive illustrations, which have proved to be an effective learning tool for students. Each illlustration was carefully rendered to highlight main points and to step the reader through experiments and processes. Most illustrations include textual content that walks students through the graphical presentation. Illustrations of experiments reinforce the scientific method by first proposing a hypothesis, then pointing out the methods and results, and ending with a conclusion that reinforces concepts explained in the text.

I liked the amount of concept reinforcement that is in the chapter. The concept check questions seem like they would be very useful in helping students understand the material by getting them to stop and think about what they just read. —William Seemer, Student, University of North Florida

The style of writing is easy to follow and is directed at college-level students. I appreciate the use of real-world language, and examples that can be found in daily life. —Shannon Forshee, Student, Eckerd College

The figures and balloon dialogue really make the concepts easier to understand. Visualization is really important to me when I am studying —Jamie Adams, Student, Arkansas State University

■ **Emphasis on Problem Solving** One of the things that I've learned in my 30 years of teaching is that students learn genetics best through problem solving. Working through an example, equation, or experiment helps students see concepts in action and reinforces the ideas explained in the text. In the book, I help students develop problem-solving skills in a number of ways. **Worked Problems** follow the presentation of difficult quantitative concepts. Walking through a problem and solution within the text reinforces what the student has just read. New **Problem Links** spread throughout each chapter point to end-of-chapter problems that students can work to test their understanding of the material that they have just read. I provide a wide range of end-of-chapter problems, organized by chapter section and split into Comprehension, Application Questions and Problems, and Challenge Questions. Some of these questions draw on examples from published papers and are marked by a data analysis icon.

I liked the addition of the yellow tags telling you what problems to do that go along with the section. I also really liked that a worked problem was included in the text and not just at the end. I felt like that helped my problem-solving skills while I was reading. —Alexandra Reynolds, Student, California Polytechnic State University, San Luis Obispo

New to the Fourth Edition

The fourth edition builds on successful features of the preceding editions of *Genetics: A Conceptual Approach* while providing an up-to-date look at the field.

■ **More Help with Problem Solving** I encourage students to practice applying the concepts that they've just learned with new **Problem Links**, spread throughout each chapter. Each link points to an end-of-chapter problem that addresses the concept just discussed, so that students can immediately test their understanding of the concept. These problem links, in addition to the Concept Check questions, provide valuable just-in-time practice, enabling students to monitor their own progress.

■ **Updated Coverage** The fourth edition addresses recent discoveries in the field of genetics, corresponding to our ever-changing understanding of inheritance, the molecular nature of genetic information, and genetic evolution. **Epigenetics**, an exciting new area of genetics, has been given extended and updated coverage in this edition. Information about epigenetics is provided in five different chapters so that students get a glimpse of all aspects of genetics that are affected by this new field. Additional updates include:

- genetic-conflict hypothesis of genomic imprinting (Chapter 5)
- epigenetics and the development of queen bees (Chapter 5)
- interpreting genetic tests (Chapter 6)
- direct-to-consumer genetic tests (Chapter 6)
- genomewide association studies (Chapter 7)
- horizontal gene transfer (Chapter 8)
- rapid evolution of influenza viruses (Chapter 8)
- segmental duplications (Chapter 9)
- copy-number variations (Chapter 9)
- epigenetic changes and chromatin modifications (Chapter 11)
- evolution of transposable elements and their role in genetic variation (Chapter 11)

- RNA world (Chapter 13)
- Piwi-interacting RNAs (Chapter 14)
- epigenetic effects in eukaryotic gene regulation (Chapter 17)
- molecular mechanism of epigenetics (Chapter 17)
- the epigenome (Chapter 17)
- repair of double-strand breaks (Chapter 18)
- next-generation sequencing techniques (Chapter 19)
- metagenomics (Chapter 20)
- synthetic biology (Chapter 20)
- epigenetic changes in development (Chapter 22)
- apoptosis and development (Chapter 22)
- epigenetic changes associated with cancer (Chapter 23)

NEW Problem Link

is the probability of Y^+y ($^1/_2$) multiplied by the probability of C^+c ($^1/_2$), or $^1/_4$. The probability of each progeny genotype resulting from the testcross is:

Progeny genotype	Probability at each locus		Overall probability	Phenotype
$Y^+y\ C^+c$	$^1/_2 \times ^1/_2$	=	$^1/_4$	red peppers
$Y^+y\ cc$	$^1/_2 \times ^1/_2$	=	$^1/_4$	peach peppers
$yy\ C^+c$	$^1/_2 \times ^1/_2$	=	$^1/_4$	orange peppers
$yy\ cc$	$^1/_2 \times ^1/_2$	=	$^1/_4$	cream peppers

When you work problems with gene interaction, it is especially important to determine the probabilities of single-locus genotypes and to multiply the probabilities of *genotypes*, not phenotypes, because the phenotypes cannot be determined without considering the effects of the genotypes at all the contributing loci. **TRY PROBLEM 25 →**

Gene Interaction with Epistasis

Sometimes the effect of gene interaction is that one gene masks (hides) the effect of another gene at a different locus, a phenomenon known as **epistasis**. In the examples of genic

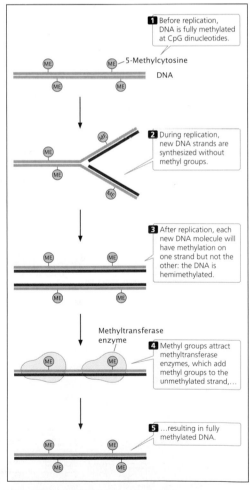

1 Before replication, DNA is fully methylated at CpG dinucleotides.

ME ME —5-Methylcytosine
DNA
ME ME

2 During replication, new DNA strands are synthesized without methyl groups.

3 After replication, each new DNA molecule will have methylation on one strand but not the other: the DNA is hemimethylated.

Methyltransferase enzyme

4 Methyl groups attract methyltransferase enzymes, which add methyl groups to the unmethylated strand,…

5 …resulting in fully methylated DNA.

17.4 DNA methylation is stably maintained through DNA replication.

NEW Introductory Stories

15 The Genetic Code and Translation

HUTTERITES, RIBOSOMES, AND BOWEN–CONRADI SYNDROME

The essential nature of the ribosome—the cell's protein factory—is poignantly illustrated by children with Bowen–Conradi syndrome. Born with a prominent nose, small head, and an unusual curvature of the small finger, these children fail to thrive and gain weight, usually dying within the first year of life.

Almost all children with Bowen–Conradi syndrome are Hutterites, a branch of Anabaptists who originated in the 1500s in the Tyrolean Alps of Austria. After years of persecution, the Hutterites immigrated to South Dakota in the 1870s and subsequently spread to neighboring prairie states and Canadian provinces. Today, the Hutterites in North America number about 40,000 persons. They live on communal farms, are strict pacifists, and rarely marry outside of the Hutterite community.

Bowen–Conradi syndrome is inherited as an autosomal recessive disorder, and the association of Bowen–Conradi syndrome with the Hutterite community is a function of the group's unique genetic history. The gene pool of present-day Hutterites in North America can be traced to fewer than 100 persons who immigrated to South Dakota in the late 1800s. The increased incidence of Bowen–Conradi syndrome in Hutterites today is due to the founder effect—the presence of the gene in one or more of the original founders—and its spread as Hutterites intermarried within their community. Because of the founder effect and inbreeding, many Hutterites today are as closely related as first cousins. This close genetic relationship among the Hutterites increases the probability that a child will inherit two copies of the recessive gene and have Bowen–Conradi syndrome; indeed, almost 1 in 10 Hutterites is a heterozygous carrier of the gene that causes the disease.

The Hutterites are a religious branch of Anabaptists who live on communal farms in the prairie states and provinces of North America. A small number of founders, coupled with a tendency to intermarry, has resulted in a high frequency of the mutation for Bowen–Conradi syndrome among Hutterites. Bowen–Conradi syndrome results from defective ribosome biosynthesis, affecting the process of translation. [Kevin Fleming/Corbis.]

- **More Flexibility** Model genetic organisms are now presented at the end of the book in a **Reference Guide to Model Genetic Organisms**. Each model organism is presented in a two-page spread that summarizes major points about the organism's life cycle, genome, and uses as a model organism. This new organization makes it easier to teach model genetic organisms as a single unit at any time in the course or to cover individual organisms in the context of particular studies throughout the course.

- **More Connections** New **summary tables** throughout the book help students make connections between concepts. I use tables to make side-by-side comparisons of similar processes, to summarize the steps and players in complex biological pathways, and to help visually organize important genes, molecules, or ideas.

- **More Relevance** Each chapter begins with a brief **introductory story** that illustrates the relevance of a genetic concept that students will learn in the chapter. These stories—a favorite feature of past editions—give students a glimpse of what's going on in the field of genetics today and help to draw the reader into the chapter. Among new introductory topics are "The Strange Case of Platypus Sex," "Topoisomerase, Replication, and Cancer," "Death Cap Poisoning," "Hutterites, Ribosomes, and Bowen–Conradi Syndrome," and "Helping the Blind to See." New end-of-chapter problems specifically address concepts discussed in most introductory stories, both old and new.

Media and Supplements

The complete package of media resources and supplements is designed to provide instructors and students with the most innovative tools to aid in a broad variety of teaching and learning approaches—including e-learning. All the available resources are fully integrated with the textbook's style and goals, enabling students to connect concepts in genetics and think like geneticists as well as develop their problem-solving skills.

GENETICS P⊘RTAL

http://courses.bfwpub.com/pierce4e

GeneticsPortal is a dynamic, fully integrated learning environment that brings together all of our teaching and learning resources in one place. It features problem-solving videos, animations of difficult-to-visualize concepts, and our new **problem-solving engine**—an engaging tool that contains all of the end-of-chapter questions and problems in the textbook, converted into multiple-choice problems, including illustrations from the textbook and drop-down menus. Easy-to-use assessment tracking and grading tools enable instructors to assign problems for practice, as homework, quizzes, or tests.

Some of the GeneticsPortal features are as follows:

- Hundreds of self-graded end-of-chapter practice problems allow students to fill in Punnett Squares, construct genetic maps, and calculate probabilities in multiple-choice versions.

- Step-by-step problem-solving videos walk through the specific process to solve select problems.

- Animations and activities to help students visualize genetics.

- A personalized calendar, an announcement center, and communication tools all in one place to help instructors manage the course.

NEW Genetics Portal

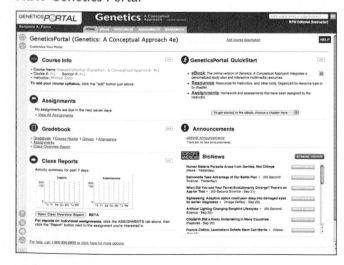

GeneticsPortal also includes the fully interactive **eBook** and all of the Student and Instructor Resources that are on the **Book Companion Website**. The GeneticsPortal is included with all new copies of the fourth edition of *Genetics: A Conceptual Approach*.

eBook
http://ebooks.bfwpub.com/pierce4e

The eBook is a completely integrated electronic version of the textbook—the ultimate hybrid of textbook and media. Problems and resources from the printed textbook are incorporated throughout the eBook along with Check Your Understanding questions that are linked to specific sections of the textbook, to ensure that students can easily review specific concepts. The eBook is available as a stand-alone textbook (sold for approximately 60% of the retail price of the printed textbook) or packaged with the printed textbook at a discounted rate. The eBook enables students to:

- Access the complete book and its electronic study tools from any internet-connected computer by using a standard Web browser;
- Navigate quickly to any section or subsection of the book or any page number of the printed book;
- Add their own bookmarks, notes, and highlighting;
- Access all the fully integrated media resources associated with the book, including the Interactive Animations and the Problem-Solving Videos (described on p. xx);
- Review quizzes and personal notes to help prepare for exams; and
- Search the entire eBook instantly, including the index and spoken glossary.

eBook

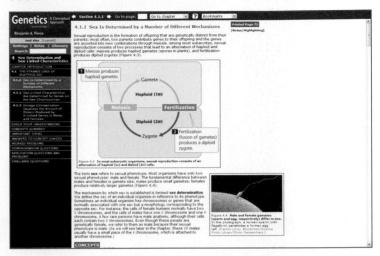

Instructors teaching from the eBook can assign either the entire textbook or a **custom version** that includes only the chapters that correspond to their syllabi. They can choose to add notes to any page of the eBook and share these notes with their students. These notes may include text, Web links, animations, or photographs. Also available is a **CourseSmart eBook**.

Instructor Resources

Instructors are provided with a comprehensive set of teaching tools, carefully developed to support lecture and individual teaching styles.

On the Book Companion Site
www.whfreeman.com/pierce4e

- All **Textbook Images and Tables** are offered as high-resolution JPEG files in PowerPoint. Each image has been fully optimized to increase type sizes and adjust color saturation. These images have been tested in a large lecture hall to ensure maximum clarity and visibility.
- **Layered or Active PowerPoints** deconstruct key concepts, sequences, and processes, step-by-step in a visual format, allowing instructors to present complex ideas in clear, manageable chunks.
- **Clicker Questions** allow instructors to integrate active learning in the classroom and to assess students' understanding of key concepts during lectures. Available in Microsoft Word and Power-Point, numerous questions are based on the Concepts Check questions featured in the textbook.
- The **Test Bank** has been prepared by Brian W. Schwartz, Columbus State University; Bradley Hersh, Allegheny College; Paul K. Small, Eureka College; Gregory Copenhaver, University of North Carolina at Chapel Hill; Rodney Mauricio, University of Georgia; and Ravinshankar Palanivelu, University of Arizona. It contains 50 questions per chapter, including multiple-choice, true-or-false, and short-answer questions, and has been updated for the fourth edition with new problems. The Test Bank is available on the Instructor's Resource DVD as well as on the Book Companion Site as chapter-by-chapter Microsoft Word files that are easy to download, edit, and print. A computerized test bank also is available.

- **Lecture Connection PowerPoint Presentations** for each chapter have been developed to minimize preparation time for new users of the book. These files offer suggested lectures including key illustrations and summaries that instructors can adapt to their teaching styles.

- The **Solution and Problem-Solving Manual** (described below) is available to download as pdf files.

- The **Instructor's Resource DVD** contains all of the resources on the Book Companion Site, including text images in PowerPoint slides and as high-resolution JPEG files, all animations, the Solutions and Problem-Solving Manual, and the Test Bank in Microsoft Word format.

- **Blackboard** and **WebCT** cartridges are available and include the Test Bank and end-of-chapter questions in multiple-choice and fill-in-the-blank format.

The most popular images in the textbook are also available as **Overhead Transparencies**, which have been optimized for maximum visibility in large lecture halls.

Student Resources

Students are provided with media designed to help them fully understand genetic concepts and improve their problem-solving ability.

- **Solutions and Problem-Solving Manual** by Jung Choi, Georgia Institute of Technology, and Mark McCallum, Pfeiffer University, contains complete answers and worked-out solutions to all questions and problems in the textbook. The manual has been found to be an indispensable tool for success by students and has been reviewed extensively by instructors for the fourth edition. (ISBN: 1-4292-3254-4)

On the Book Companion Site

www.whfreeman.com/pierce4e

- **Problem-Solving Videos** developed by Susan Elrod, California Polytechnic State University, San Luis Obispo, offer students valuable help by reviewing basic problem-solving strategies. Students learn first-hand how to deconstruct difficult problems in genetics by using a set of questioning techniques. The problem-solving videos demonstrate in a step-by-step manner how these strategies can be applied to the in-text worked problems and selected end-of-chapter problems in many of the chapters.

- **Interactive Animations/Podcasts** illuminate important concepts in genetics. These tutorials help students understand key processes in genetics by outlining these processes in a step-by-step manner. The tutorials are also available on the eBook. Podcasts adapted from the tutorial presentations in the following list are available for download from the Book Companion Site and from the eBook. Students can review important genetics processes and concepts at their convenience by downloading the animations to their MP3 players. The major animated concepts are:

Interactive Animations

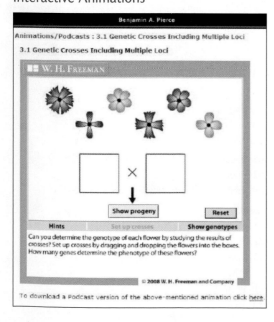

2.1 Cell Cycle and Mitosis
2.2 Meiosis
2.3 Genetic Variation in Meiosis
3.1 Genetic Crosses Including Multiple Loci
4.1 X-Linked Inheritance
7.1 Determining Gene Order by Three-Point Cross

8.1 Bacterial Conjugation
11.1 Levels of Chromatin Structure
12.1 Overview of Replication
12.2 Bidirectional Replication of DNA
12.3 Coordination of Leading- and Lagging-Strand Synthesis

Also available for students is *Genetics: A Conceptual Approach*, Fourth Edition, in looseleaf (ISBN: 1-4292-3251-X) at a reduced cost, and *Transmission and Population Genetics*, Fourth Edition (ISBN: 1-4292-5494-7), for courses focused solely on the transmission and population areas of genetics.

Acknowledgments

I am indebted to the thousands of genetics students who have filled my classes in the past 30 years, first at Connecticut College, then at Baylor University, and now at Southwestern University. The intelligence, enthusiasm, curiosity, and humor of these students have been a source of motivation and pleasure throughout my professional life. From them I have learned much about the art of teaching and the subject of genetics. I am also indebted to my genetics teachers and mentors, Dr. Raymond Canham and Dr. Jeffry Mitton, for introducing me to genetics and encouraging me to be a lifelong learner and scholar.

Five years ago, Southwestern University provided me with the opportunity to return to teaching and research at a small undergraduate university, and I have greatly enjoyed the experience. The small classes, close interaction of students and faculty, and integration of teaching and research have made working at Southwestern fun and rewarding. My colleagues in the Biology Department continually sustain me with friendship and advice. I thank James Hunt, Provost of Southwestern University and Dean of the Brown College, for valued friendship, collegiality, and support.

I have been blessed with an outstanding team of colleagues at W. H. Freeman and Company. Publisher Kate Ahr Parker provided support for this new edition. Executive Editor Susan Winslow expertly shepherded the project, providing coordination, creative ideas, encouragement, and support throughout the project. Developmental Editor Lisa Samols was my daily partner in crafting this edition. Her organizational skills, creative insight, and superior editing were—as always—outstanding. Importantly, Lisa also kept me motivated and on schedule with a positive attitude and good humor.

Patty Zimmerman, an outstanding manuscript editor, kept close watch on details and contributed valuable editorial suggestions. Patty has worked with me on all four editions of *Genetics: A Conceptual Approach*, as well as *Genetics Essentials*; her editorial touch resonates throughout the book. Senior Project Editor Georgia Lee Hadler at W. H. Freeman expertly managed the production of this fourth edition, as well as all preceding editions. I thank Craig Durant at Dragonfly Media Group for creating and revising the book's illustrations and Bill Page and Janice Donnola for coordinating the illustration program. Thanks to Paul Rohloff at W. H. Freeman and Pietro Paolo Adinolfi at Peparé for coordinating the composition and manufacturing phases of production. Diana Blume developed the book's design and the cover for this edition. I thank Ted Szczepanski and Elyse Rieder for photo research. Anna Bristow did an outstanding job of managing the supplements and assisting with the editorial development of the book. Aaron Gass and Ashley Joseph coordinated the excellent multimedia package that accompanies the book. I am grateful to Jung Choi and Mark McCallum for writing solutions to new end-of-chapter problems. Brian W. Schwartz, Bradley Hersh, Paul K. Small, Gregory Copenhaver, Rodney Mauricio, and Ravinshankar Palanivelu developed the Test Bank. Debbie Clare brought energy, creative ideas, and much fun to the marketing of the book.

I am grateful to the W. H. Freeman sales representatives, regional managers, and regional sales specialists, who introduce my book to genetic instructors throughout world. I have greatly enjoyed working with this sales staff, whose expertise, hard work, and good service are responsible for the success of Freeman books.

A number of colleagues served as reviewers of this book, kindly lending me their technical expertise and teaching experience. Their assistance is gratefully acknowledged. Any remaining errors are entirely my own.

Marlene Tyrrell—my spouse and best friend for 30 years—and our children Michael and Sarah provide love, support, and inspiration for everything that I do.

My gratitude goes to the reviewers of this new edition of *Genetics: A Conceptual Approach*.

Ann Aguanno
Marymount Manhattan College

Andrea Bailey
Brookhaven College

Paul W. Bates
University of Minnesota

Bethany V. Bowling
Northern Kentucky University

Natalie Bronstein
Mercy College

Gerald L. Buldak
Loyola University Chicago

J. Aaron Cassill
University of Texas at San Antonio

Maria V. Cattell
University of Colorado

Henry C. Chang
Purdue University

Mary C. Colavito
Santa Monica College

John F. Davis

Kimberley Dej
McMaster University

Steve H. Denison
Eckerd College

Laura E. DeWald
Western Carolina University

Moon Draper
University of Texas at Austin

Richard Duhrkopf
Baylor University

Cheryld L. Emmons
Alfred University

Michele Engel
University of Colorado Denver

Victor Fet
Marshall University

David W. Foltz
Louisiana State University

Wayne Forrester
Indiana University

Robert G. Fowler
San Jose State University

Laura Frost
Point Park University

Maria Gallo
University of Florida

Alex Georgakilas
East Carolina University

William D. Gilliland
DePaul University

Jack R. Girton
Iowa State University

Doreen R. Glodowski
Rutgers University

Elliott S. Goldstein
Arizona State University

Jessica L. Goldstein
Barnard College

Steven W. Gorsich
Central Michigan University

Liu He
University of Denver

Stephen C. Hedman
University of Minnesota Duluth

Bradley Hersh
Allegheny College

Carina Endres Howell
Lock Haven University of Pennsylvania

Colin Hughes
Florida Atlantic University

Dena Johnson
Tarrant County College NW

David H. Kass
Eastern Michigan University

Margaret Kovach
The University of Tennessee at Chattanooga

Brian Kreiser
University of Southern Mississippi

Catherine Kunst
University of Colorado at Denver

Mary Rose Lamb
University of Puget Sound

Haiying Liang
Clemson University

William J. Mackay
Edinboro University of Pennsylvania

Cindy S. Malone
California State University at Northridge

Jessica L. Moore
University of South Florida

Daniel W. Moser
Kansas State University

Mary Rengo Murnik
Ferris State University

Sang-Chul Nam
Baylor University

Ann V. Paterson
Williams Baptist College

Helen Piontkivska
Kent State University

Michael Lee Robinson
Miami University

Katherine T. Schmeidler
Irvine Valley College

Rodney J. Scott
Wheaton College

Barbara B. Sears
Michigan State University

Barkur S. Shastry
Oakland University

Mark Shotwell
Slippery Rock University

Anoop S. Sindhu
University of Saskatchewan

Paul K. Small
Eureka College

Thomas Smith
Ave Maria University

Lara Soowal
University of California at San Diego

Ruth Sporer
Rutgers University

Nanette van Loon
Borough of Manhattan Community College

Sarah Ward
Colorado State University

Lise D. Wilson
Siena College

Kathleen Wood
University of Mary Hardin-Baylor

Malcolm Zellars
Georgia State University

Zhongyuan Zuo
University of Denver

1 Introduction to Genetics

Hopi bowl, early twentieth century. Albinism, a genetic condition, arises with high frequency among the Hopi people and occupies a special place in the Hopi culture. [The Newark Museum/Art Resource, NY.]

ALBINISM IN THE HOPIS

Rising a thousand feet above the desert floor, Black Mesa dominates the horizon of the Enchanted Desert and provides a familiar landmark for travelers passing through northeastern Arizona. Not only is Black Mesa a prominent geological feature, but, more significantly, it is the ancestral home of the Hopi Native Americans. Fingers of the mesa reach out into the desert, and alongside or on top of each finger is a Hopi village. Most of the villages are quite small, having only a few dozen inhabitants, but they are incredibly old. One village, Oraibi, has existed on Black Mesa since 1150 A.D. and is the oldest continuously occupied settlement in North America.

In 1900, Alĕs Hrdlĭĕka, an anthropologist and physician working for the American Museum of Natural History, visited the Hopi villages of Black Mesa and reported a startling discovery. Among the Hopis were 11 white persons—not Caucasians, but actually white Hopi Native Americans. These persons had a genetic condition known as albinism (**Figure 1.1**).

Albinism is caused by a defect in one of the enzymes required to produce melanin, the pigment that darkens our skin, hair, and eyes. People with albinism don't produce melanin or they produce only small amounts of it and, consequently, have white hair, light skin, and no pigment in the irises of their eyes. Melanin normally protects the DNA of skin cells from the damaging effects of ultraviolet radiation in sunlight, and melanin's presence in the developing eye is essential for proper eyesight.

The genetic basis of albinism was first described by the English physician Archibald Garrod, who recognized in 1908 that the condition was inherited as an autosomal recessive trait, meaning that a person must receive two copies of an albino mutation—one from each parent—to have albinism. In recent years, the molecular natures of the mutations that lead to albinism have been elucidated. Albinism in humans is caused by defects in any one of several different genes that control the synthesis and storage of melanin; many different types of mutations can occur at each gene, any one of which may lead to albinism. The form of albinism found in the Hopis is most likely oculocutaneous albinism type 2, due to a defect in the *OCA2* gene on chromosome 15.

The Hopis are not unique in having albinos among the members of their tribe. Albinism is found in almost all human ethnic groups and is described in ancient writings; it has probably been present since humankind's beginnings. What is unique about the

1.1 Albinism among the Hopi Native Americans. In this photograph, taken about 1900, the Hopi girl in the center has albinism. [The Field Museum/Charles Carpenter.]

Hopis is the high frequency of albinism. In most human groups, albinism is rare, present in only about 1 in 20,000 persons. In the villages on Black Mesa, it reaches a frequency of 1 in 200, a hundred times as frequent as in most other populations.

Why is albinism so frequent among the Hopi Native Americans? The answer to this question is not completely known, but geneticists who have studied albinism in the Hopis speculate that the high frequency of the albino gene is related to the special place that albinism occupied in the Hopi culture. For much of their history, the Hopis considered members of their tribe with albinism to be important and special. People with albinism were considered pretty, clean, and intelligent. Having a number of people with albinism in one's village was considered a good sign, a symbol that the people of the village contained particularly pure Hopi blood. Albinos performed in Hopi ceremonies and assumed positions of leadership within the tribe, often becoming chiefs, healers, and religious leaders.

Hopi albinos were also given special treatment in everyday activities. The Hopis farmed small garden plots at the foot of Black Mesa for centuries. Every day throughout the growing season, the men of the tribe trekked to the base of Black Mesa and spent much of the day in the bright southwestern sunlight tending their corn and vegetables. With little or no melanin pigment in their skin, people with albinism are extremely susceptible to sunburn and have increased incidences of skin cancer when exposed to the sun. Furthermore, many don't see well in bright sunlight. But the male Hopis with albinism were excused from this normal male labor and allowed to remain behind in the village with the women of the tribe, performing other duties.

Geneticists have suggested that these special considerations given to albino members of the tribe are partly responsible for the high frequency of albinism among the Hopis. Throughout the growing season, the albino men were the only male members of the tribe in the village during the day with all the women and, thus, they enjoyed a mating advantage, which helped to spread their albino genes. In addition, the special considerations given to albino Hopis allowed them to avoid the detrimental effects of albinism—increased skin cancer and poor eyesight. The small size of the Hopi tribe probably also played a role by allowing chance to increase the frequency of the albino gene. Regardless of the factors that led to the high frequency of albinism, the Hopis clearly respected and valued the members of their tribe who possessed this particular trait. Unfortunately, people with genetic conditions in many societies are often subject to discrimination and prejudice.

TRY PROBLEMS 1 AND 25 →

Genetics is one of the most rapidly advancing fields of science, with important new discoveries reported every month. Pick up almost any major newspaper or news magazine and chances are that you will see articles related to genetics: the completion of another genome, such as that of the platypus; the discovery of genes that affect major diseases, including multiple sclerosis, depression, and cancer; a report of DNA analyzed from long-extinct animals such as the woolly mammoth; and the identification of genes that affect skin pigmentation, height, and learning ability in humans. Even among the advertisements, one is likely to see genetics: ads for genetic testing to determine paternity, one's ancestry, and susceptibility to diseases and disorders. These new findings and applications of genetics often have significant economic and ethical implications, making the study of genetics relevant, timely, and interesting.

This chapter introduces you to genetics and reviews some concepts that you may have encountered briefly in a biology course. We begin by considering the importance of genetics to each of us, to society at large, and to students of biology. We then turn to the history of genetics, how the field as a whole developed. The final part of the chapter presents some fundamental terms and principles of genetics that are used throughout the book.

1.1 Genetics Is Important to Us Individually, to Society, and to the Study of Biology

Albinism among the Hopis illustrates the important role that genes play in our lives. This one genetic defect, among the 20,000 genes that humans possess, completely changes the life of a Hopi who possesses it. It alters his or her occupation, role in Hopi society, and relations with other members of the tribe. We all possess genes that influence our lives in significant ways. Genes affect our height, weight, hair color, and skin pigmentation. They affect our suscepti-

(a)   **(b)**

Laron
dwarfism

Susceptibility
to diphtheria

Low-tone
deafness

Limb–girdle
muscular
dystrophy

**Diastrophic
dysplasia**

Chromosome 5

1.2 Genes influence susceptibility to many diseases and disorders. (a) An X-ray of the hand of a person suffering from diastrophic dysplasia (bottom), a hereditary growth disorder that results in curved bones, short limbs, and hand deformities, compared with an X-ray of a normal hand (top). (b) This disorder is due to a defect in a gene on chromosome 5. Braces indicate regions on chromosome 5 where genes giving rise to other disorders are located. [Part a: (top) Biophoto Associates/Science Source/Photo Researchers; (bottom) courtesy of Eric Lander, Whitehead Institute, MIT.]

1.3 In the Green Revolution, genetic techniques were used to develop new high-yielding strains of crops. (Left) Norman Borlaug, a leader in the development of new strains of wheat that led to the Green Revolution. Borlaug was awarded the Nobel Peace Prize in 1970. (Right) Modern, high-yielding rice plant (left) and traditional rice plant (right). [Left: UPI/Corbis-Bettman. Right: IRRI.]

bility to many diseases and disorders (**Figure 1.2**) and even contribute to our intelligence and personality. Genes are fundamental to who and what we are.

Although the science of genetics is relatively new compared with sciences such as astronomy and chemistry, people have understood the hereditary nature of traits and have practiced genetics for thousands of years. The rise of agriculture began when people started to apply genetic principles to the domestication of plants and animals. Today, the major crops and animals used in agriculture are quite different from their wild progenitors, having undergone extensive genetic alterations that increase their yields and provide many desirable traits, such as disease and pest resistance, special nutritional qualities, and characteristics that facilitate harvest. The Green Revolution, which expanded food production throughout the world in the 1950s and 1960s, relied heavily on the application of genetics (**Figure 1.3**). Today, genetically engineered corn, soybeans, and other crops constitute a significant proportion of all the food produced worldwide.

The pharmaceutical industry is another area in which genetics plays an important role. Numerous drugs and food additives are synthesized by fungi and bacteria that have been genetically manipulated to make them efficient producers of these substances. The biotechnology industry employs molecular genetic techniques to develop and mass-produce substances of commercial value. Growth hormone, insulin, and clotting factor are now produced commercially by genetically engineered bacteria (**Figure 1.4**). Genetics has

also been used to produce bacteria that remove minerals from ore, break down toxic chemicals, and inhibit damaging frost formation on crop plants.

Genetics plays a critical role in medicine. Physicians recognize that many diseases and disorders have a hereditary component, including rare genetic disorders such as sickle-cell anemia and Huntington disease as well as many common diseases such as asthma, diabetes, and hypertension. Advances in genetics have resulted in important insights into the nature of diseases such as cancer and in the development

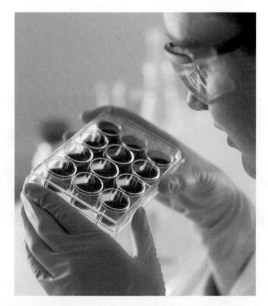

1.4 The biotechnology industry uses molecular genetic methods to produce substances of economic value. [Andrew Brooks/Corbis.]

of diagnostic tests such as those that identify pathogens and defective genes. Gene therapy—the direct alteration of genes to treat human diseases—has now been administered to thousands of patients.

The Role of Genetics in Biology

Although an understanding of genetics is important to all people, it is critical to the student of biology. Genetics provides one of biology's unifying principles: all organisms use genetic systems that have a number of features in common. Genetics also undergirds the study of many other biological disciplines. Evolution, for example, is genetic change taking place through time; so the study of evolution requires an understanding of genetics. Developmental biology relies heavily on genetics: tissues and organs develop through the regulated expression of genes (**Figure 1.5**). Even such fields as taxonomy, ecology, and animal behavior are making increasing use of genetic methods. The study of almost any field of biology or medicine is incomplete without a thorough understanding of genes and genetic methods.

Genetic Diversity and Evolution

Life on Earth exists in a tremendous array of forms and features in almost every conceivable environment. Life is also characterized by adaptation: many organisms are exquisitely suited to the environment in which they are found. The history of life is a chronicle of new forms of life emerging, old forms disappearing, and existing forms changing.

Despite their tremendous diversity, living organisms have an important feature in common: all use similar genetic systems. A complete set of genetic instructions for any organism is its **genome**, and all genomes are encoded in nucleic acids—either DNA or RNA. The coding system for genomic information also is common to all life: genetic instructions are in the same format and, with rare excep-

tions, the code words are identical. Likewise, the process by which genetic information is copied and decoded is remarkably similar for all forms of life. These common features of heredity suggest that all life on Earth evolved from the same primordial ancestor that arose between 3.5 billion and 4 billion years ago. Biologist Richard Dawkins describes life as a river of DNA that runs through time, connecting all organisms past and present.

That all organisms have similar genetic systems means that the study of one organism's genes reveals principles that apply to other organisms. Investigations of how bacterial DNA is copied (replicated), for example, provide information that applies to the replication of human DNA. It also means that genes will function in foreign cells, which makes genetic engineering possible. Unfortunately, these similar genetic systems are also the basis for diseases such as AIDS (acquired immune deficiency syndrome), in which viral genes are able to function—sometimes with alarming efficiency—in human cells.

Life's diversity and adaptation are products of evolution, which is simply genetic change through time. Evolution is a two-step process: first, inherited differences arise randomly and, then, the proportion of individuals with particular differences increases or decreases. Genetic variation is therefore the foundation of all evolutionary change and is ultimately the basis of all life as we know it. Furthermore, techniques of molecular genetics are now routinely used to decipher evolutionary relationships among organisms; for example, recent analysis of DNA isolated from Neanderthal fossils has yielded new information concerning the relationship between Neanderthals and modern humans. Genetics, the study of genetic variation, is critical to understanding the past, present, and future of life. **TRY PROBLEM 17 →**

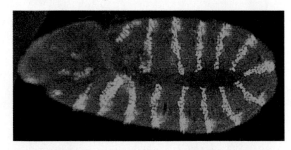

1.5 The key to development lies in the regulation of gene expression. This early fruit-fly embryo illustrates the localized expression of the *engrailed* gene, which helps determine the development of body segments in the adult fly. [Stephen Paddock Digital Image Gallery.]

CONCEPTS

Heredity affects many of our physical features as well as our susceptibility to many diseases and disorders. Genetics contributes to advances in agriculture, pharmaceuticals, and medicine and is fundamental to modern biology. All organisms use similar genetic systems, and genetic variation is the foundation of the diversity of all life.

✔ CONCEPT CHECK 1

What are some of the implications of all organisms having similar genetic systems?

a. That all life forms are genetically related

b. That research findings on one organism's gene function can often be applied to other organisms

c. That genes from one organism can often exist and thrive in another organism

d. All of the above

Divisions of Genetics

The study of genetics consists of three major subdisciplines: transmission genetics, molecular genetics, and population genetics (**Figure 1.6**). Also known as classical genetics, **transmission genetics** encompasses the basic principles of heredity and how traits are passed from one generation to the next. This area addresses the relation between chromosomes and heredity, the arrangement of genes on chromosomes, and gene mapping. Here, the focus is on the individual organism—how an individual organism inherits its genetic makeup and how it passes its genes to the next generation.

Molecular genetics concerns the chemical nature of the gene itself: how genetic information is encoded, replicated, and expressed. It includes the cellular processes of replication, transcription, and translation (by which genetic information is transferred from one molecule to another) and gene regulation (the processes that control the expression of genetic information). The focus in molecular genetics is the gene, its structure, organization, and function.

Population genetics explores the genetic composition of groups of individual members of the same species (populations) and how that composition changes geographically

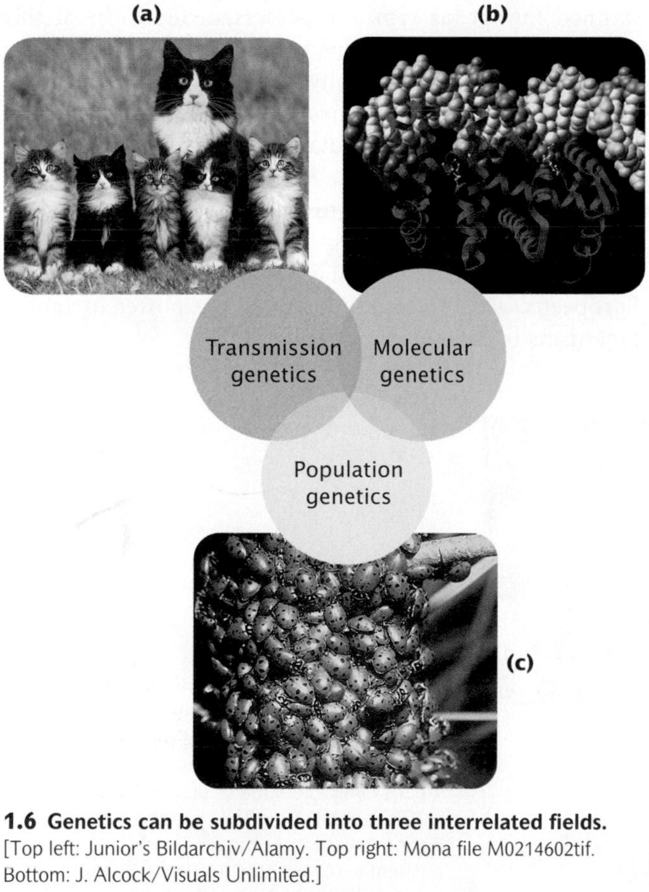

1.6 Genetics can be subdivided into three interrelated fields.
[Top left: Junior's Bildarchiv/Alamy. Top right: Mona file M0214602tif. Bottom: J. Alcock/Visuals Unlimited.]

and with the passage of time. Because evolution is genetic change, population genetics is fundamentally the study of evolution. The focus of population genetics is the group of genes found in a population.

Division of the study of genetics into these three groups is convenient and traditional, but we should recognize that the fields overlap and that each major subdivision can be further divided into a number of more-specialized fields, such as chromosomal genetics, biochemical genetics, quantitative genetics, and so forth. Alternatively, genetics can be subdivided by organism (fruit fly, corn, or bacterial genetics), and each of these organisms can be studied at the level of transmission, molecular, and population genetics. Modern genetics is an extremely broad field, encompassing many interrelated subdisciplines and specializations.
TRY PROBLEM 18 →

Model Genetic Organisms

Through the years, genetic studies have been conducted on thousands of different species, including almost all major groups of bacteria, fungi, protists, plants, and animals. Nevertheless, a few species have emerged as **model genetic organisms**—organisms having characteristics that make them particularly useful for genetic analysis and about which a tremendous amount of genetic information has accumulated. Six model organisms that have been the subject of intensive genetic study are: *Drosophila melanogaster*, the fruit fly; *Escherichia coli*, a bacterium present in the gut of humans and other mammals; *Caenorhabditis elegans*, a nematode worm; *Arabidopsis thaliana*, the thale-cress plant; *Mus musculus*, the house mouse; and *Saccharomyces cerevisiae*, baker's yeast (**Figure 1.7**). These species are the organisms of choice for many genetic researchers, and their genomes were sequenced as a part of the Human Genome Project. The life cyles and genetic characteristics of these model genetic organisms are described in more detail in the Reference Guide to Model Genetic Organisms located at the end of the book.

At first glance, this group of lowly and sometimes despised creatures might seem unlikely candidates for model organisms. However, all possess life cycles and traits that make them particularly suitable for genetic study, including a short generation time, large but manageable numbers of progeny, adaptability to a laboratory environment, and the ability to be housed and propagated inexpensively. Other species that are frequently the subjects of genetic research and considered genetic models include *Neurospora crassa* (bread mold), *Zea mays* (corn), *Danio rerio* (zebrafish), and *Xenopus laevis* (clawed frog). Although not generally considered a genetic model, humans also have been subjected to intensive genetic scrutiny; special techniques for the genetic analysis of humans are discussed in Chapter 6.

(a)

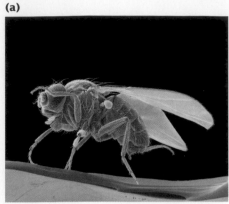

Drosophila melanogaster
Fruit fly

(b)

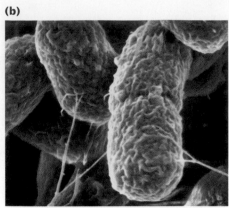

Escherichia coli
Bacterium

(c)

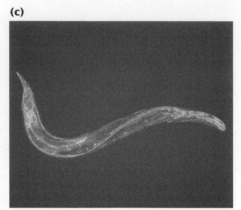

Caenorhabditis elegans
Nematode

1.7 Model genetic organisms are species having features that make them useful for genetic analysis. [Part a: SPL/Photo Researchers. Part b: Gary Gaugler/Visuals Unlimited. Part c: Natalie Pujol/Visuals Unlimited. Part d: Peggy Greb/ARS. Part e: Joel Page/AP. Part f: T. E. Adams/Visuals Unlimited.]

The value of model genetic organisms is illustrated by the use of zebrafish to identify genes that affect skin pigmentation in humans. For many years, geneticists have recognized that differences in pigmentation among human ethnic groups are genetic (**Figure 1.8a**), but the genes causing these differences were largely unknown. The zebrafish has recently become an important model in genetic studies because it is a small vertebrate that produces many offspring and is easy to rear in the laboratory. The mutant zebrafish called *golden* has light pigmentation due to the presence of fewer, smaller, and less-dense pigment-containing structures called melanosomes in its cells (**Figure 1.8b**). Light skin in humans is similarly due to fewer and less-dense melanosomes in pigment-containing cells.

Keith Cheng and his colleagues at Pennsylvania State University College of Medicine hypothesized that light skin in humans might result from a mutation that is similar to the *golden* mutation in zebrafish. Taking advantage of the ease with which zebrafish can be manipulated

in the laboratory, they isolated and sequenced the gene responsible for the *golden* mutation and found that it encodes a protein that takes part in calcium uptake by melanosomes. They then searched a database of all known human genes and found a similar gene called *SLC24A5*, which encodes the same function in human cells. When they examined human populations, they found that light-skinned Europeans typically possessed one form of this gene, whereas darker-skinned Africans, Eastern Asians, and Native Americans usually possessed a different form of the gene. Many other genes also affect pigmentation in humans, as illustrated by mutations in the *OCA2* gene that produce albinism among the Hopi Native Americans (discussed in the introduction to this chapter). Nevertheless, *SLC24A5* appears to be responsible for 24% to 38% of the differences in pigmentation between Africans and Europeans. This example illustrates the power of model organisms in genetic research.

(a)

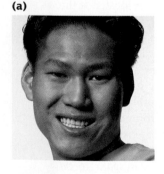

(b)

Normal zebrafish

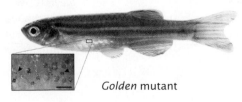

Golden mutant

1.8 The zebrafish, a genetic model organism, has been instrumental in helping to identify genes encoding pigmentation differences among humans. (a) Human ethnic groups differ in degree of skin pigmentation. (b) The zebrafish *golden* mutation is caused by a gene that controls the amount of melanin pigment in melanosomes. [Part a: PhotoDisc. Part b: K. Cheng/J. Gittlen, Cancer Research Foundation, Pennsylvania State College of Medicine.]

(d)

Arabidopsis thaliana
Thale-cress plant

(e)

Mus musculus
House mouse

(f)

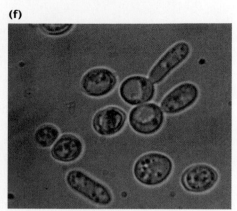

Saccharomyces cerevisiae
Baker's yeast

The three major divisions of genetics are transmission genetics, molecular genetics, and population genetics. Transmission genetics examines the principles of heredity; molecular genetics deals with the gene and the cellular processes by which genetic information is transferred and expressed; population genetics concerns the genetic composition of groups of organisms and how that composition changes geographically and with the passage of time. Model genetic organisms are species that have received special emphasis in genetic research; they have characteristics that make them useful for genetic analysis.

✔ CONCEPT CHECK 2

Would the horse make a good model genetic organism? Why or why not?

1.2 Humans Have Been Using Genetics for Thousands of Years

Although the science of genetics is young—almost entirely a product of the past 100 years or so—people have been using genetic principles for thousands of years.

The Early Use and Understanding of Heredity

The first evidence that people understood and applied the principles of heredity in earlier times is found in the domestication of plants and animals, which began between approximately 10,000 and 12,000 years ago in the Middle East. The first domesticated organisms included wheat, peas, lentils, barley, dogs, goats, and sheep (**Figure 1.9a**). By 4000 years ago, sophisticated genetic techniques were already in

1.9 Ancient peoples practiced genetic techniques in agriculture. (Left) Modern wheat, with larger and more numerous seeds that do not scatter before harvest, was produced by interbreeding at least three different wild species. (Right) Assyrian bas-relief sculpture showing artificial pollination of date palms at the time of King Assurnasirpalli II, who reigned from 883 to 859 B.C. [Left: Scott Bauer/ARS/USDA. Right: The Metropolitan Museum of Art, gift of John D. Rockefeller, Jr., 1932. (32.143.3) Photograph © 1996 Metropolitan Museum of Art.]

use in the Middle East. Assyrians and Babylonians developed several hundred varieties of date palms that differed in fruit size, color, taste, and time of ripening (**Figure 1.9b**). Other crops and domesticated animals were developed by cultures in Asia, Africa, and the Americas in the same period.

Ancient writings demonstrate that early humans were also aware of their own heredity. Hindu sacred writings dating to 2000 years ago attribute many traits to the father and suggest that differences between siblings are produced by the mother. The Talmud, the Jewish book of religious laws based on oral traditions dating back thousands of years, presents an uncannily accurate understanding of the inheritance of hemophilia. It directs that, if a woman bears two sons who die of bleeding after circumcision, any additional sons that she bears should not be circumcised; nor should the sons of her sisters be circumcised. This advice accurately corresponds to the X-linked pattern of inheritance of hemophilia (discussed further in Chapter 6).

The ancient Greeks gave careful consideration to human reproduction and heredity. Greek philosophers developed the concept of **pangenesis**, in which specific particles, later called gemmules, carry information from various parts of the body to the reproductive organs, from which they are passed to the embryo at the moment of conception (**Figure 1.10**). Although incorrect, the concept of pangenesis was highly influential and persisted until the late 1800s.

Pangenesis led the ancient Greeks to propose the notion of the **inheritance of acquired characteristics**, in which traits acquired in a person's lifetime become incorporated into that person's hereditary information and are passed on to offspring; for example, people who developed musical ability through diligent study would produce children who are innately endowed with musical ability. The notion of the inheritance of acquired characteristics also is no longer accepted, but it remained popular through the twentieth century.

Although the ancient Romans contributed little to an understanding of human heredity, they successfully developed a number of techniques for animal and plant breeding; the techniques were based on trial and error rather than any general concept of heredity. Little new information was added to the understanding of genetics in the next 1000 years.

Dutch eyeglass makers began to put together simple microscopes in the late 1500s, enabling Robert Hooke (1635–1703) to discover cells in 1665. Microscopes provided naturalists with new and exciting vistas on life, and perhaps excessive enthusiasm for this new world of the very small gave rise to the idea of **preformationism**. According to preformationism, inside the egg or sperm there exists a fully formed miniature adult, a *homunculus*, which simply enlarges in the course of development (**Figure 1.11**). Preformationism meant that all traits were inherited from

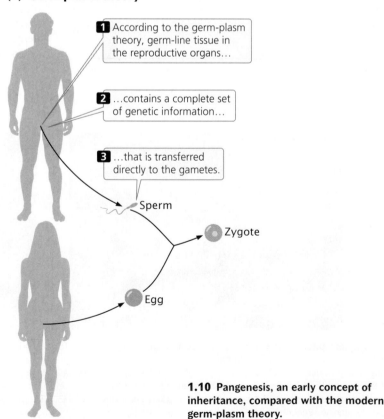

(a) Pangenesis concept

1 According to the pangenesis concept, genetic information from different parts of the body…

2 …travels to the reproductive organs…

3 …where it is transferred to the gametes.

Sperm

Zygote

Egg

(b) Germ-plasm theory

1 According to the germ-plasm theory, germ-line tissue in the reproductive organs…

2 …contains a complete set of genetic information…

3 …that is transferred directly to the gametes.

Sperm

Zygote

Egg

1.10 Pangenesis, an early concept of inheritance, compared with the modern germ-plasm theory.

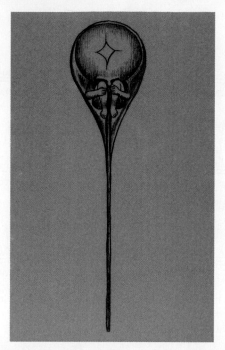

1.11 Preformationists in the seventeenth and eighteenth centuries believed that sperm or eggs contained fully formed humans (the homunculus). Shown here is a drawing of a homunculus inside a sperm. [Science VU/Visuals Unlimited.]

only one parent—from the father if the homunculus was in the sperm or from the mother if it was in the egg. Although many observations suggested that offspring possess a mixture of traits from both parents, preformationism remained a popular concept throughout much of the seventeenth and eighteenth centuries.

Another early notion of heredity was **blending inheritance**, which proposed that offspring are a blend, or mixture, of parental traits. This idea suggested that the genetic material itself blends, much as blue and yellow pigments blend to make green paint. Once blended, genetic differences could not be separated out in future generations, just as green paint cannot be separated out into blue and yellow pigments. Some traits do *appear* to exhibit blending inheritance; however, we realize today that individual genes do not blend.

The Rise of the Science of Genetics

In 1676, Nehemiah Grew (1641–1712) reported that plants reproduce sexually by using pollen from the male sex cells. With this information, a number of botanists began to experiment with crossing plants and creating hybrids, including Gregor Mendel (1822–1884; **Figure 1.12**), who went on to discover the basic principles of heredity. Mendel's conclusions, which were not widely known in the scientific community for 35 years, laid the foundation for our modern understanding of heredity, and he is generally recognized today as the father of genetics.

Developments in cytology (the study of cells) in the 1800s had a strong influence on genetics. Robert Brown (1773–1858) described the cell nucleus in 1833. Building on the work of others, Matthias Jacob Schleiden (1804–1881) and Theodor Schwann (1810–1882) proposed the concept

of the **cell theory** in 1839. According to this theory, all life is composed of cells, cells arise only from preexisting cells, and the cell is the fundamental unit of structure and function in living organisms. Biologists interested in heredity began to examine cells to see what took place in the course of cell reproduction. Walther Flemming (1843–1905) observed the division of chromosomes in 1879 and published a superb description of mitosis. By 1885, it was generally recognized that the nucleus contained the hereditary information.

Charles Darwin (1809–1882), one of the most influential biologists of the nineteenth century, put forth the theory of evolution through natural selection and published his ideas in *On the Origin of Species* in 1859. Darwin recognized that heredity was fundamental to evolution, and he conducted extensive genetic crosses with pigeons and other organisms. However, he never understood the nature of inheritance, and this lack of understanding was a major omission in his theory of evolution.

In the last half of the nineteenth century, cytologists demonstrated that the nucleus had a role in fertilization. Near the close of the nineteenth century, August Weismann (1834–1914) finally laid to rest the notion of the inheritance of acquired characteristics. He cut off the tails of mice for 22 consecutive generations and showed that the tail length in descendants remained stubbornly long. Weismann proposed the **germ-plasm theory**, which holds that the cells in the reproductive organs carry a complete set of genetic information that is passed to the egg and sperm (see Figure 1.10b).

1.12 Gregor Mendel was the founder of modern genetics. Mendel first discovered the principles of heredity by crossing different varieties of pea plants and analyzing the transmission of traits in subsequent generations. [Hulton Archive/Getty Images.]

The year 1900 was a watershed in the history of genetics. Gregor Mendel's pivotal 1866 publication on experiments with pea plants, which revealed the principles of heredity, was rediscovered, as considered in more detail in Chapter 3. The significance of his conclusions was recognized, and other biologists immediately began to conduct similar genetic studies on mice, chickens, and other organisms. The results of these investigations showed that many traits indeed follow Mendel's rules. Some of the early concepts of heredity are summarized in **Table 1.1**.

After the acceptance of Mendel's theory of heredity, Walter Sutton (1877–1916) proposed in 1902 that genes, the units of inheritance, are located on chromosomes. Thomas Hunt Morgan (1866–1945) discovered the first genetic mutant of fruit flies in 1910 and used fruit flies to unravel many details of transmission genetics. Ronald A. Fisher (1890–1962), John B. S. Haldane (1892–1964), and Sewall Wright (1889–1988) laid the foundation for population genetics in the 1930s by synthesizing Mendelian genetics and evolutionary theory.

Geneticists began to use bacteria and viruses in the 1940s; the rapid reproduction and simple genetic systems of these organisms allowed detailed study of the organization and structure of genes. At about this same time, evidence accumulated that DNA was the repository of genetic information.

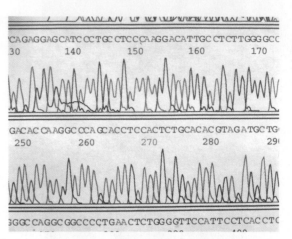

1.13 The human genome was completely sequenced in 2003. A chromatograph of a small portion of the human genome. [Science Museum/SSPL.]

James Watson (b. 1928) and Francis Crick (1916–2004), along with Maurice Wilkins (1916–2004) and Rosalind Franklin (1920–1958), described the three-dimensional structure of DNA in 1953, ushering in the era of molecular genetics.

By 1966, the chemical structure of DNA and the system by which it determines the amino acid sequence of proteins had been worked out. Advances in molecular genetics led to the first recombinant DNA experiments in 1973, which touched off another revolution in genetic research. Walter Gilbert (b. 1932) and Frederick Sanger (b. 1918) developed methods for sequencing DNA in 1977. The polymerase chain reaction, a technique for quickly amplifying tiny amounts of DNA, was developed by Kary Mullis (b. 1944) and others in 1983. In 1990, gene therapy was used for the first time to treat human genetic disease in the United States, and the Human Genome Project was launched. By 1995, the first complete DNA sequence of a free-living organism—the bacterium *Haemophilus influenzae*—was determined, and the first complete sequence of a eukaryotic organism (yeast) was reported a year later. A rough draft of the human genome sequence was reported in 2000, with the sequence essentially completed in 2003, ushering in a new era in genetics (**Figure 1.13**). Today, the genomes of numerous organisms are being sequenced, analyzed, and compared. TRY PROBLEMS 22 AND 23 ➜

The Future of Genetics

Numerous advances in genetics are being made today, and genetics remains at the forefront of biological research. New, rapid methods for sequencing DNA are being used to sequence the genomes of numerous species, from bacteria to elephants, and the information content of genetics is increasing at a rapid pace. New details about gene structure and function are continually expanding our knowledge of how genetic information is encoded and how it specifies traits. These findings are redefining what a gene is.

The power of new methods to identify and analyze genes is illustrated by recent genetic studies of heart attacks in

Table 1.1	Early concepts of heredity	
Concept	Proposed	Correct or Incorrect
Pangenesis	Genetic information travels from different parts of the body to reproductive organs.	Incorrect
Inheritance of acquired characteristics	Acquired traits become incorporated into hereditary information.	Incorrect
Preformationism	Miniature organism resides in sex cells, and all traits are inherited from one parent.	Incorrect
Blending inheritance	Genes blend and mix.	Incorrect
Germ-plasm theory	All cells contain a complete set of genetic information.	Correct
Cell theory	All life is composed of cells, and cells arise only from cells.	Correct
Mendelian inheritance	Traits are inherited in accord with defined principles.	Correct

humans. Physicians have long recognized that heart attacks run in families, but finding specific genes that contribute to an increased risk of a heart attack has, until recently, been difficult. In 2009, an international team of geneticists examined the DNA of 26,000 people in 10 countries for single differences in the DNA (called single-nucleotide polymorphisms, or SNPS) that might be associated with an increased risk of myocardial infarction. This study and other similar studies identified several new genes that affect the risk of coronary artery disease and early heart attacks. These findings may make it possible to identify persons who are predisposed to heart attack, allowing early intervention that might prevent the attacks from occurring. Analyses of SNPS are helping to locate genes that affect all types of traits, from eye color and height to glaucoma and heart attacks.

Information about sequence differences among organisms is also a source of new insights about evolution. For example, recent analysis of DNA sequences at 30 genes has revealed that all living cats can trace their ancestry to a pantherlike cat living in Southeast Asia about 11 million years ago and that all living cats can be divided into eight groups or lineages.

In recent years, our understanding of the role of RNA in many cellular processes has expanded greatly; RNA has a role in many aspects of gene function. The discovery in the late 1990s of tiny RNA molecules called small interfering RNAs and micro RNAs led to the recognition that these molecules play central roles in gene expression and development. Today, recognition of the importance of alterations of DNA and chromosome structure that do not include the base sequence of the DNA is increasing. Many such alterations, called epigenetic changes, are stable and affect the expression of traits. New genetic microchips that simultaneously analyze thousands of RNA molecules are providing information about the activities of thousands of genes in a given cell, allowing a detailed picture of how cells respond to external signals, environmental stresses, and disease states such as cancer. In the emerging field of proteomics, powerful computer programs are being used to model the structure and function of proteins from DNA sequence information. All of this information provides us with a better understanding of numerous biological processes and evolutionary relationships. The flood of new genetic information requires the continuous development of sophisticated computer programs to store, retrieve, compare, and analyze genetic data and has given rise to the field of bioinformatics, a merging of molecular biology and computer science.

A number of companies and researchers are racing to develop the technology for sequencing the entire genome of a single person for less than $1000. As the cost of sequencing decreases, the focus of DNA-sequencing efforts will shift from the genomes of different species to individual differences within species. In the not-too-distant future, each person will likely possess a copy of his or her entire genome sequence, which can be used to assess the risk of acquiring various diseases and to tailor their treatment should they arise. The use of genetics in agriculture will contine to improve the productivity of domestic crops and animals, helping to feed the future world population. This ever-widening scope of genetics will raise significant ethical, social, and economic issues.

This brief overview of the history of genetics is not intended to be comprehensive; rather it is designed to provide a sense of the accelerating pace of advances in genetics. In the chapters to come, we will learn more about the experiments and the scientists who helped shape the discipline of genetics.

CONCEPTS

Humans first applied genetics to the domestication of plants and animals between 10,000 and 12,000 years ago. Developments in plant hybridization and cytology in the eighteenth and nineteenth centuries laid the foundation for the field of genetics today. After Mendel's work was rediscovered in 1900, the science of genetics developed rapidly and today is one of the most active areas of science.

✔ CONCEPT CHECK 3

How did developments in cytology in the nineteenth century contribute to our modern understanding of genetics?

1.3 A Few Fundamental Concepts Are Important for the Start of Our Journey into Genetics

Undoubtedly, you learned some genetic principles in other biology classes. Let's take a few moments to review some fundamental genetic concepts.

Cells are of two basic types: eukaryotic and prokaryotic. Structurally, cells consist of two basic types, although, evolutionarily, the story is more complex (see Chapter 2). Prokaryotic cells lack a nuclear membrane and possess no membrane-bounded cell organelles, whereas eukaryotic cells are more complex, possessing a nucleus and membrane-bounded organelles such as chloroplasts and mitochondria.

The gene is the fundamental unit of heredity. The precise way in which a gene is defined often varies, depending on the biological context. At the simplest level, we can think of a gene as a unit of information that encodes a genetic characteristic. We will enlarge this definition as we learn more about what genes are and how they function.

Genes come in multiple forms called alleles. A gene that specifies a characteristic may exist in several forms, called alleles. For example, a gene for coat color in cats may exist as an allele that encodes black fur or as an allele that encodes orange fur.

Genes confer phenotypes. One of the most important concepts in genetics is the distinction between traits and genes. Traits are not inherited directly. Rather, genes are inherited and, along with environmental factors, determine the expression of traits. The genetic information that an individual organism possesses is its genotype; the trait is its phenotype. For example, albinism seen in some Hopis is a phenotype, the information in *OCA2* genes that causes albinism is the genotype.

Genetic information is carried in DNA and RNA. Genetic information is encoded in the molecular structure of nucleic acids, which come in two types: deoxyribonucleic acid (DNA) and ribonucleic acid (RNA). Nucleic acids are polymers consisting of repeating units called nucleotides; each nucleotide consists of a sugar, a phosphate, and a nitrogenous base. The nitrogenous bases in DNA are of four types: adenine (A), cytosine (C), guanine (G), and thymine (T). The sequence of these bases encodes genetic information. DNA consists of two complementary nucleotide strands. Most organisms carry their genetic information in DNA, but a few viruses carry it in RNA. The four nitrogenous bases of RNA are adenine, cytosine, guanine, and uracil (U).

Genes are located on chromosomes. The vehicles of genetic information within a cell are chromosomes (**Figure 1.14**), which consist of DNA and associated proteins. The cells of each species have a characteristic number of chromosomes; for example, bacterial cells normally possess a single chromosome; human cells possess 46; pigeon cells possess 80. Each chromosome carries a large number of genes.

Chromosomes separate through the processes of mitosis and meiosis. The processes of mitosis and meiosis ensure that a complete set of an organism's chromosomes exists in each cell resulting from cell division. Mitosis is the separation of chromosomes in the division of somatic (nonsex) cells. Meiosis is the pairing and separation of chromosomes in the division of sex cells to produce gametes (reproductive cells).

Genetic information is transferred from DNA to RNA to protein. Many genes encode traits by specifying the structure of proteins. Genetic information is first transcribed from DNA into RNA, and then RNA is translated into the amino acid sequence of a protein.

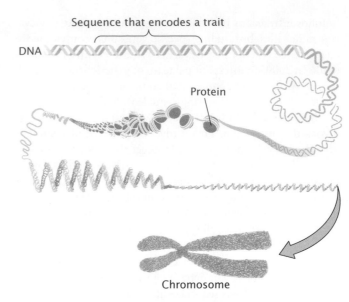

1.14 Genes are carried on chromosomes.

Mutations are permanent changes in genetic information that can be passed from cell to cell or from parent to offspring. Gene mutations affect the genetic information of only a single gene; chromosome mutations alter the number or the structure of chromosomes and therefore usually affect many genes.

Some traits are affected by multiple factors. Some traits are affected by multiple genes that interact in complex ways with environmental factors. Human height, for example, is affected by hundreds of genes as well as environmental factors such as nutrition.

Evolution is genetic change. Evolution can be viewed as a two-step process: first, genetic variation arises and, second, some genetic variants increase in frequency, whereas other variants decrease in frequency. **TRY PROBLEM 24** →

CONCEPTS SUMMARY

- Genetics is central to the life of every person: it influences a person's physical features, susceptibility to numerous diseases, personality, and intelligence.

- Genetics plays important roles in agriculture, the pharmaceutical industry, and medicine. It is central to the study of biology.

- All organisms use similar genetic systems. Genetic variation is the foundation of evolution and is critical to understanding all life.

- The study of genetics can be broadly divided into transmission genetics, molecular genetics, and population genetics.

- Model genetic organisms are species about which much genetic information exists because they have characteristics that make them particularly amenable to genetic analysis.

- The use of genetics by humans began with the domestication of plants and animals.

- Ancient Greeks developed the concepts of pangenesis and the inheritance of acquired characteristics. Ancient Romans developed practical measures for the breeding of plants and animals.

- Preformationism suggested that a person inherits all of his or her traits from one parent. Blending inheritance proposed that offspring possess a mixture of the parental traits.

- By studying the offspring of crosses between varieties of peas, Gregor Mendel discovered the principles of heredity. Developments in cytology in the nineteenth century led to the understanding that the cell nucleus is the site of heredity.

- In 1900, Mendel's principles of heredity were rediscovered. Population genetics was established in the early 1930s, followed closely by biochemical genetics and bacterial and viral genetics. The structure of DNA was discovered in 1953, stimulating the rise of molecular genetics.

- Cells are of two basic types: prokaryotic and eukaryotic.
- The genes that determine a trait are termed the genotype; the trait that they produce is the phenotype.
- Genes are located on chromosomes, which are made up of nucleic acids and proteins and are partitioned into daughter cells through the process of mitosis or meiosis.
- Genetic information is expressed through the transfer of information from DNA to RNA to proteins.
- Evolution requires genetic change in populations.

IMPORTANT TERMS

genome (p. 4)
transmission genetics (p. 5)
molecular genetics (p. 5)
population genetics (p. 5)

model genetic organism (p. 5)
pangenesis (p. 8)
inheritance of acquired
    characteristics (p. 8)

preformationism (p. 8)
blending inheritance (p. 9)
cell theory (p. 9)
germ-plasm theory (p. 9)

ANSWERS TO CONCEPT CHECKS

1. d

2. No, because horses are expensive to house, feed, and propagate, they have too few progeny, and their generation time is too long.

3. Developments in cytology in the 1800s led to the identification of parts of the cell, including the cell nucleus and chromosomes. The cell theory focused the attention of biologists on the cell, eventually leading to the conclusion that the nucleus contains the hereditary information.

COMPREHENSION QUESTIONS

Answers to questions and problems preceded by an asterisk can be found at the end of the book.

Section 1.1

*1. How does the Hopi culture contribute to the high incidence of albinism among members of the Hopi tribe?

2. Outline some of the ways in which genetics is important to each of us.

*3. Give at least three examples of the role of genetics in society today.

4. Briefly explain why genetics is crucial to modern biology.

*5. List the three traditional subdisciplines of genetics and summarize what each covers.

6. What are some characteristics of model genetic organisms that make them useful for genetic studies?

Section 1.2

7. When and where did agriculture first arise? What role did genetics play in the development of the first domesticated plants and animals?

*8. Outline the notion of pangenesis and explain how it differs from the germ-plasm theory.

9. What does the concept of the inheritance of acquired characteristics propose and how is it related to the notion of pangenesis?

*10. What is preformationism? What did it have to say about how traits are inherited?

11. Define blending inheritance and contrast it with preformationism.

12. How did developments in botany in the seventeenth and eighteenth centuries contribute to the rise of modern genetics?

*13. Who first discovered the basic principles that laid the foundation for our modern understanding of heredity?

14. List some advances in genetics made in the twentieth century.

Section 1.3

15. What are the two basic cell types (from a structural perspective) and how do they differ?

*16. Outline the relations between genes, DNA, and chromosomes.

APPLICATION QUESTIONS AND PROBLEMS

Section 1.1

17. What is the relation between genetics and evolution?

*18. For each of the following genetic topics, indicate whether it focuses on transmission genetics, molecular genetics, or population genetics.

a. Analysis of pedigrees to determine the probability of someone inheriting a trait

b. Study of people on a small island to determine why a genetic form of asthma is prevalent on the island

c. Effect of nonrandom mating on the distribution of genotypes among a group of animals

d. Examination of the nucleotide sequences found at the ends of chromosomes

e. Mechanisms that ensure a high degree of accuracy in DNA replication

f. Study of how the inheritance of traits encoded by genes on sex chromosomes (sex-linked traits) differs from the inheritance of traits encoded by genes on nonsex chromosomes (autosomal traits)

19. Describe some of the ways in which your own genetic makeup affects you as a person. Be as specific as you can.

20. Describe at least one trait that appears to run in your family (appears in multiple members of the family). Does this trait run in your family because it is an inherited trait or because it is caused by environmental factors that are common to family members? How might you distinguish between these possibilities?

Section 1.2

*21. Genetics is said to be both a very old science and a very young science. Explain what is meant by this statement.

22. Match the description (*a* through *d*) with the correct theory or concept listed below.

> Preformationism
> Pangenesis
> Germ-plasm theory
> Inheritance of acquired characteristics

a. Each reproductive cell contains a complete set of genetic information.

b. All traits are inherited from one parent.

c. Genetic information may be altered by the use of a characteristic.

d. Cells of different tissues contain different genetic information.

*23. Compare and contrast the following ideas about inheritance.

a. Pangenesis and germ-plasm theory

b. Preformationism and blending inheritance

c. The inheritance of acquired characteristics and our modern theory of heredity

Section 1.3

*24. Compare and contrast the following terms:

a. Eukaryotic and prokaryotic cells

b. Gene and allele

c. Genotype and phenotype

d. DNA and RNA

e. DNA and chromosome

CHALLENGE QUESTIONS

Introduction

25. The type of albinism that arises with high frequency among Hopi Native Americans (discussed in the introduction to this chapter) is most likely oculocutaneous albinism type 2, due to a defect in the *OCA2* gene on chromosome 15. Do some research on the Internet to determine how the phenotype of this type of albinism differs from phenotypes of other forms of albinism in humans and which genes take part. Hint: Visit the Online Mendelian Inheritance in Man Web site (http://www.ncbi.nlm.nih.gov/omim/) and search the database for albinism.

Section 1.1

26. We now know as much or more about the genetics of humans as we know about that of any other organism, and humans are the focus of many genetic studies. Should humans be considered a model genetic organism? Why or why not?

Section 1.3

*27. Suppose that life exists elsewhere in the universe. All life must contain some type of genetic information, but alien genomes might not consist of nucleic acids and have the same features as those found in the genomes of life on Earth. What might be the common features of all genomes, no matter where they exist?

28. Choose one of the ethical or social issues in parts *a* through *e* and give your opinion on the issue. For background information, you might read one of the articles on ethics marked with an asterisk in the Suggested Readings section for Chapter 1 at www.whfreeman.com/pierce4e.

a. Should a person's genetic makeup be used in determining his or her eligibility for life insurance?

b. Should biotechnology companies be able to patent newly sequenced genes?

c. Should gene therapy be used on people?

d. Should genetic testing be made available for inherited conditions for which there is no treatment or cure?

e. Should governments outlaw the cloning of people?

*29. A 45-year old women undergoes genetic testing and discovers that she is at high risk for developing colon cancer and Alzheimer disease. Because her children have 50% of her genes, they also may have increased risk of these diseases. Does she have a moral or legal obligation to tell her children and other close relatives about the results of her genetic testing?

*30. Suppose that you could undergo genetic testing at age 18 for susceptibility to a genetic disease that would not appear until middle age and has no available treatment.

a. What would be some of the possible reasons for having such a genetic test and some of the possible reasons for not having the test?

b. Would you personally want to be tested? Explain your reasoning.

2 Chromosomes and Cellular Reproduction

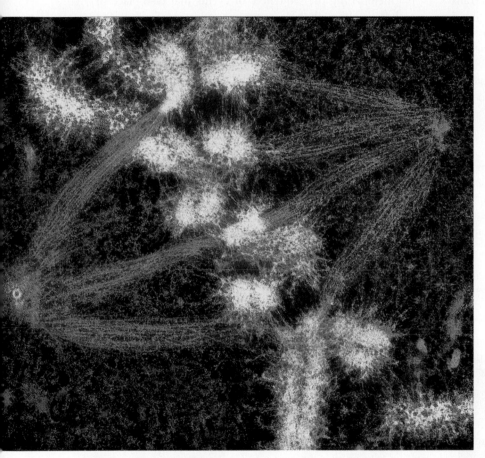

Chromosomes in mitosis, the process through which each new cell receives a complete copy of the genetic material. [Photograph by Conly L. Reider/Biological Photo Service.]

THE BLIND MEN'S RIDDLE

In a well-known riddle, two blind men by chance enter a department store at the same time, go to the same counter, and both order five pairs of socks, each pair a different color. The sales clerk is so befuddled by this strange coincidence that he places all ten pairs (two black pairs, two blue pairs, two gray pairs, two brown pairs, and two green pairs) into a single shopping bag and gives the bag with all ten pairs to one blind man and an empty bag to the other. The two blind men happen to meet on the street outside, where they discover that one of their bags contains all ten pairs of socks. How do the blind men, without seeing and without any outside help, sort out the socks so that each man goes home with exactly five pairs of different colored socks? Can you come up with a solution to the riddle?

By an interesting coincidence, cells have the same dilemma as that of the blind men in the riddle. Most organisms possess two sets of genetic information, one set inherited from each parent. Before cell division, the DNA in each chromosome replicates; after replication, there are two copies—called sister chromatids—of each chromosome. At the end of cell division, it is critical that each new cell receives a complete copy of the genetic material, just as each blind man needed to go home with a complete set of socks.

The solution to the riddle is simple. Socks are sold as pairs; the two socks of a pair are typically connected by a thread. As a pair is removed from the bag, the men each grasp a different sock of the pair and pull in opposite directions. When the socks are pulled tight, it is easy for one of the men to take a pocket knife and cut the thread connecting the pair. Each man then deposits his single sock in his own bag. At the end of the process, each man's bag will contain exactly two black socks, two blue socks, two gray socks, two brown socks, and two green socks.*

Remarkably, cells employ a similar solution for separating their chromosomes into new daughter cells. As we will learn in this chapter, the replicated chromosomes line up at the center of a cell undergoing division and, like the socks in the riddle, the sister chromatids of each chromosome are pulled in opposite directions. Like the thread connecting two socks of a pair, a molecule called cohesin holds the sister chromatids together until severed

*This analogy is adapted from K. Nasmyth. Disseminating the genome: joining, resolving, and separating sister chromatids, during mitosis and meiosis. *Annual Review of Genetics* 35:673–745, 2001.

by a molecular knife called separase. The two resulting chromosomes separate and the cell divides, ensuring that a complete set of chromosomes is deposited in each cell.

In this analogy, the blind men and cells differ in one critical regard: if the blind men make a mistake, one man ends up with an extra sock and the other is a sock short, but no great harm results. The same cannot be said for human cells. Errors in chromosome separation, producing cells with too many or too few chromosomes, are frequently catastrophic, leading to cancer, miscarriage, or—in some cases—a child with severe handicaps.

This chapter explores the process of cell reproduction and how a complete set of genetic information is transmitted to new cells. In prokaryotic cells, reproduction is relatively simple, because prokaryotic cells possess a single chromosome. In eukaryotic cells, multiple chromosomes must be copied and distributed to each of the new cells, and so cell reproduction is more complex. Cell division in eukaryotes takes place through mitosis or meiosis, processes that serve as the foundation for much of genetics.

Grasping mitosis and meiosis requires more than simply memorizing the sequences of events that take place in each stage, although these events are important. The key is to understand how genetic information is apportioned in the course of cell reproduction through a dynamic interplay of DNA synthesis, chromosome movement, and cell division. These processes bring about the transmission of genetic information and are the basis of similarities and differences between parents and progeny.

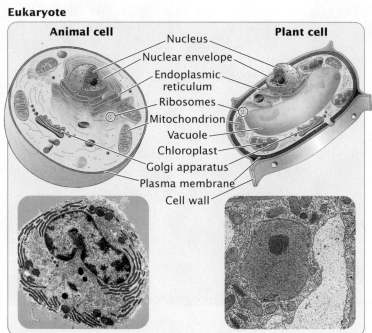

	Prokaryotic cells	Eukaryotic cells
Nucleus	Absent	Present
Cell diameter	Relatively small, from 1 to 10 μm	Relatively large, from 10 to 100 μm
Genome	Usually one circular DNA molecule	Multiple linear DNA molecules
DNA	Not complexed with histones in eubacteria; some histones in archaea	Complexed with histones
Amount of DNA	Relatively small	Relatively large
Membrane-bounded organelles	Absent	Present
Cytoskeleton	Absent	Present

2.1 Prokaryotic and eukaryotic cells differ in structure. [Photographs (left to right) by T. J. Beveridge/ Visuals Unlimited/Getty Images (prokaryotes); W. Baumeister/Science Photo Library/Photo Researchers; G. Murti/Phototake; Biophoto Associates/Photo Researchers.]

2.1 Prokaryotic and Eukaryotic Cells Differ in a Number of Genetic Characteristics

Biologists traditionally classify all living organisms into two major groups, the *prokaryotes* and the *eukaryotes* (**Figure 2.1**). A **prokaryote** is a unicellular organism with a relatively simple cell structure. A **eukaryote** has a compartmentalized cell structure with components bounded by intracellular membranes; eukaryotes are either unicellular or multicellular.

Research indicates that a division of life into two major groups, the prokaryotes and eukaryotes, is not so simple. Although similar in cell structure, prokaryotes include at least two fundamentally distinct types of bacteria: the **eubacteria** (true bacteria) and the **archaea** (ancient bacteria). An examination of equivalent DNA sequences reveals that eubacteria and archaea are as distantly related to one another as they are to the eukaryotes. Although eubacteria and archaea are similar in cell structure, some genetic processes in archaea (such as transcription) are more similar to those in eukaryotes, and the archaea are actually closer evolutionarily to eukaryotes than to eubacteria. Thus, from an evolutionary perspective, there are three major groups of organisms: eubacteria, archaea, and eukaryotes. In this book, the prokaryotic–eukaryotic distinction will be made frequently, but important eubacterial–archaeal differences also will be noted.

From the perspective of genetics, a major difference between prokaryotic and eukaryotic cells is that a eukaryote has a nuclear envelope, which surrounds the genetic material to form a **nucleus** and separates the DNA from the other cellular contents. In prokaryotic cells, the genetic material is in close contact with other components of the cell—a property that has important consequences for the way in which genes are controlled.

Another fundamental difference between prokaryotes and eukaryotes lies in the packaging of their DNA. In eukaryotes, DNA is closely associated with a special class of proteins, the **histones**, to form tightly packed chromosomes. This complex of DNA and histone proteins is termed **chromatin**, which is the stuff of eukaryotic chromosomes (**Figure 2.2**). Histone proteins limit the accessibility of enzymes and other proteins that copy and read the DNA, but they enable the DNA to fit into the nucleus. Eukaryotic DNA must separate from the histones before the genetic information in the DNA can be accessed. Archaea also have some histone proteins that complex with DNA, but the structure of their chromatin is different from that found in eukaryotes. Eubacteria do not possess histones; so their DNA does not exist in the highly ordered, tightly packed arrangement found in eukaryotic cells (**Figure 2.3**). The copying and reading of DNA are therefore simpler processes in eubacteria.

Genes of prokaryotic cells are generally on a single, circular molecule of DNA—the chromosome of a prokaryotic cell. In eukaryotic cells, genes are located on multiple,

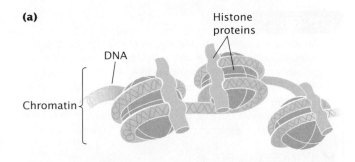

(a)

DNA

Histone proteins

Chromatin

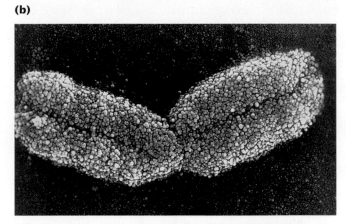

(b)

2.2 Eukaryotic chromosomes consist of DNA and histone proteins. (a) DNA wraps around the histone proteins to form chromatin, the material that makes up chromosomes. (b) A eukaryotic chromosome. [Part b: Biophoto Associates/Photo Researchers.]

usually linear DNA molecules (multiple chromosomes). Eukaryotic cells therefore require mechanisms that ensure that a copy of each chromosome is faithfully transmitted to each new cell. This generalization—a single, circular chromosome in prokaryotes and multiple, linear chromosomes in eukaryotes—is not always true. A few bacteria have more than one chromosome, and important bacterial genes are frequently found on other DNA molecules called *plasmids* (see Chapter 8). Furthermore, in some eukaryotes, a few genes are located on circular DNA molecules found in certain organelles (see Chapter 21).

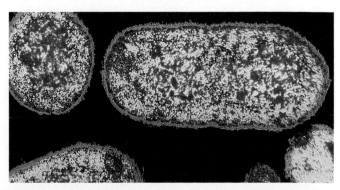

2.3 Prokaryotic DNA (shown in red) is neither surrounded by a nuclear membrane nor complexed with histone proteins.
[A. B. Dowsett/Science Photo Library/Photo Researchers.]

Organisms are classified as prokaryotes or eukaryotes, and prokaryotes consist of archaea and eubacteria. A prokaryote is a unicellular organism that lacks a nucleus, its DNA is not complexed to histone proteins, and its genome is usually a single chromosome. Eukaryotes are either unicellular or multicellular, their cells possess a nucleus, their DNA is complexed to histone proteins, and their genomes consist of multiple chromosomes.

✔ CONCEPT CHECK 1

List several characteristics that eubacteria and archaea have in common and that distinguish them from eukaryotes.

Viruses are neither prokaryotic nor eukaryotic, because they do not possess a cellular structure. Viruses are actually simple structures composed of an outer protein coat surrounding nucleic acid (either DNA or RNA; **Figure 2.4**). Neither are viruses primitive forms of life: they can reproduce only within host cells, which means that they must have evolved after, rather than before, cells evolved. In addition, viruses are not an evolutionarily distinct group but are most

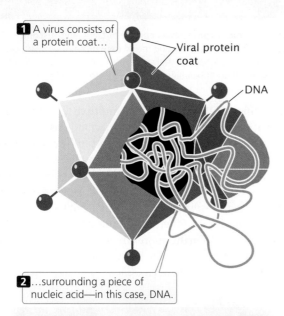

① A virus consists of a protein coat…

Viral protein coat

DNA

② …surrounding a piece of nucleic acid—in this case, DNA.

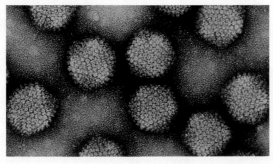

2.4 A virus is a simple replicative structure consisting of protein and nucleic acid. Adenoviruses are shown in the micrograph. [Micrograph by Hans Gelderblom/Visuals Unlimited.]

closely related to their hosts: the genes of a plant virus are more similar to those in a plant cell than to those in animal viruses, which suggests that viruses evolved from their hosts, rather than from other viruses. The close relationship between the genes of virus and host makes viruses useful for studying the genetics of host organisms.

2.2 Cell Reproduction Requires the Copying of the Genetic Material, Separation of the Copies, and Cell Division

For any cell to reproduce successfully, three fundamental events must take place: (1) its genetic information must be copied, (2) the copies of genetic information must be separated from each other, and (3) the cell must divide. All cellular reproduction includes these three events, but the processes that lead to these events differ in prokaryotic and eukaryotic cells because of their structural differences.

Prokaryotic Cell Reproduction

When prokaryotic cells reproduce, the circular chromosome of the bacterium replicates and the cell divides in a process called binary fission (**Figure 2.5**). Replication usually begins at a specific place on the bacterial chromosome, called the origin of replication. In a process that is not fully understood, the origins of the two newly replicated chromosomes move away from each other and toward opposite ends of the cell. In at least some bacteria, proteins bind near the replication origins and anchor the new chromosomes to the plasma membrane at opposite ends of the cell. Finally, a new cell wall forms between the two chromosomes, producing two cells, each with an identical copy of the chromosome. Under optimal conditions, some bacterial cells divide every 20 minutes. At this rate, a single bacterial cell could produce a billion descendants in a mere 10 hours.

Eukaryotic Cell Reproduction

Like prokaryotic cell reproduction, eukaryotic cell reproduction requires the processes of DNA replication, copy separation, and division of the cytoplasm. However, the presence of multiple DNA molecules requires a more-complex mechanism to ensure that exactly one copy of each molecule ends up in each of the new cells.

Eukaryotic chromosomes are separated from the cytoplasm by the nuclear envelope. The nucleus was once thought to be a fluid-filled bag in which the chromosomes floated, but we now know that the nucleus has a highly organized internal scaffolding called the nuclear matrix. This matrix consists of a network of protein fibers that maintains precise spatial relations among the nuclear components and takes part in DNA replication, the expression of genes, and the modification of gene products before they leave the

(a)

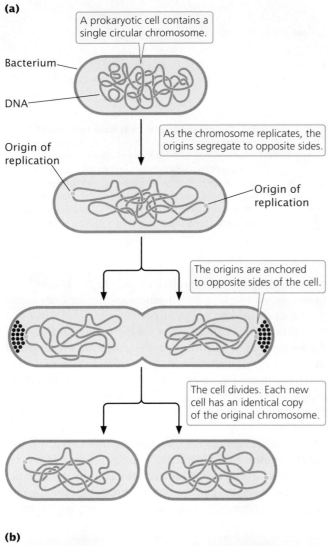

A prokaryotic cell contains a single circular chromosome.

Bacterium

DNA

Origin of replication

As the chromosome replicates, the origins segregate to opposite sides.

Origin of replication

The origins are anchored to opposite sides of the cell.

The cell divides. Each new cell has an identical copy of the original chromosome.

(b)

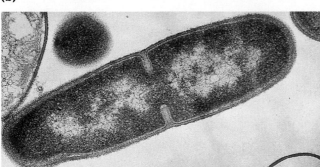

2.5 Prokaryotic cells reproduce by binary fission. (a) Process of binary fission. (b) Micrograph showing a bacterial cell undergoing binary fission. [Part b: Lee D. Simon/Photo Researchers.]

nucleus. We will now take a closer look at the structure of eukaryotic chromosomes.

Eukaryotic chromosomes   Each eukaryotic species has a characteristic number of chromosomes per cell: potatoes have 48 chromosomes, fruit flies have 8, and humans have 46. There appears to be no special relation between the

complexity of an organism and its number of chromosomes per cell.

In most eukaryotic cells, there are two sets of chromosomes. The presence of two sets is a consequence of sexual reproduction: one set is inherited from the male parent and the other from the female parent. Each chromosome in one set has a corresponding chromosome in the other set, together constituting a homologous pair (**Figure 2.6**). Human cells, for example, have 46 chromosomes, constituting 23 homologous pairs.

The two chromosomes of a **homologous pair** are usually alike in structure and size, and each carries genetic information for the same set of hereditary characteristics. (The sex chromosomes are an exception and will be discussed in Chapter 4.) For example, if a gene on a particular chromosome encodes a characteristic such as hair color, another copy of the gene (each copy is called an allele) at the same position on that chromosome's homolog *also* encodes hair color. However, these two alleles need not be identical: one might encode brown hair and the other might encode blond hair. Thus, most cells carry two sets of genetic information; these cells are **diploid**. But not all eukaryotic cells are diploid: reproductive cells (such as eggs, sperm, and spores) and even nonreproductive cells of some organisms may contain a single set of chromosomes. Cells with a single set of chromosomes are **haploid**. A haploid cell has only one copy of each gene.

CONCEPTS

Cells reproduce by copying and separating their genetic information and then dividing. Because eukaryotes possess multiple chromosomes, mechanisms exist to ensure that each new cell receives one copy of each chromosome. Most eukaryotic cells are diploid, and their two chromosome sets can be arranged in homologous pairs. Haploid cells contain a single set of chromosomes.

✔ CONCEPT CHECK 2

Diploid cells have

a. two chromosomes

b. two sets of chromosomes

c. one set of chromosomes

d. two pairs of homologous chromosomes

Chromosome structure   The chromosomes of eukaryotic cells are larger and more complex than those found in prokaryotes, but each unreplicated chromosome nevertheless consists of a single molecule of DNA. Although linear, the DNA molecules in eukaryotic chromosomes are highly folded and condensed; if stretched out, some human chromosomes would be several centimeters long—thousands of

(a)

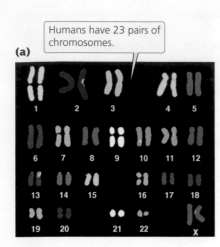

Humans have 23 pairs of chromosomes.

(b)

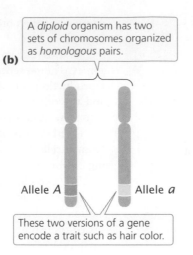

A *diploid* organism has two sets of chromosomes organized as *homologous* pairs.

Allele *A*   Allele *a*

These two versions of a gene encode a trait such as hair color.

2.6 Diploid eukaryotic cells have two sets of chromosomes. (a) A set of chromosomes from a female human cell. Each pair of chromosomes is hybridized to a uniquely colored probe, giving it a distinct color. (b) The chromosomes are present in homologous pairs, which consist of chromosomes that are alike in size and structure and carry information for the same characteristics. [Part a: Courtesy of Dr. Thomas Ried and Dr. Evelin Schrock.]

times as long as the span of a typical nucleus. To package such a tremendous length of DNA into this small volume, each DNA molecule is coiled again and again and tightly packed around histone proteins, forming a rod-shaped chromosome. Most of the time, the chromosomes are thin and difficult to observe but, before cell division, they condense further into thick, readily observed structures; it is at this stage that chromosomes are usually studied.

A functional chromosome has three essential elements: a centromere, a pair of telomeres, and origins of replication. The centromere is the attachment point for spindle microtubules—the filaments responsible for moving chromosomes in cell division (**Figure 2.7**). The centromere appears as a constricted region. Before cell division, a multiprotein complex called the kinetochore assembles on the centro-

mere; later, spindle microtubules attach to the kinetochore. Chromosomes lacking a centromere cannot be drawn into the newly formed nuclei; these chromosomes are lost, often with catastrophic consequences for the cell. On the basis of the location of the centromere, chromosomes are classified into four types: metacentric, submetacentric, acrocentric, and telocentric (**Figure 2.8**). One of the two arms of a chromosome (the short arm of a submetacentric or acrocentric chromosome) is designated by the letter p and the other arm is designated by q.

Telomeres are the natural ends, the tips, of a whole linear chromosome (see Figure 2.7). Just as plastic tips protect the ends of a shoelace, telomeres protect and stabilize the chromosome ends. If a chromosome breaks, producing new ends, the chromosome is degraded at the newly broken ends.

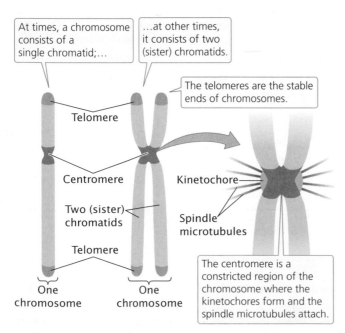

At times, a chromosome consists of a single chromatid;…

…at other times, it consists of two (sister) chromatids.

The telomeres are the stable ends of chromosomes.

Telomere

Centromere

Two (sister) chromatids

Telomere

One chromosome

One chromosome

Kinetochore

Spindle microtubules

The centromere is a constricted region of the chromosome where the kinetochores form and the spindle microtubules attach.

2.7 Each eukaryotic chromosome has a centromere and telomeres.

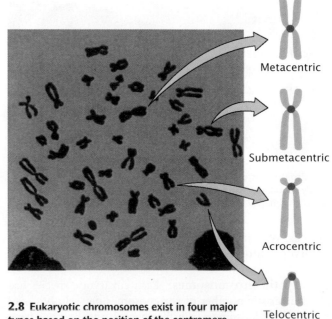

Metacentric

Submetacentric

Acrocentric

Telocentric

2.8 Eukaryotic chromosomes exist in four major types based on the position of the centromere.
[Micrograph by L. Lisco, D. W. Fawcett/Visuals Unlimited.]

Telomeres provide chromosome stability. Research shows that telomeres also participate in limiting cell division and may play important roles in aging and cancer (discussed in Chapter 12).

Origins of replication are the sites where DNA synthesis begins; they are not easily observed by microscopy. Their structure and function will be discussed in more detail in Chapter 12. In preparation for cell division, each chromosome replicates, making a copy of itself, as already mentioned. These two initially identical copies, called **sister chromatids**, are held together at the centromere (see Figure 2.7). Each sister chromatid consists of a single molecule of DNA.

CONCEPTS

Sister chromatids are copies of a chromosome held together at the centromere. Functional chromosomes contain centromeres, telomeres, and origins of replication. The kinetochore is the point of attachment for the spindle microtubules; telomeres are the stabilizing ends of a chromosome; origins of replication are sites where DNA synthesis begins.

✔ CONCEPT CHECK 3

What would be the result if a chromosome did not have a kinetochore?

The Cell Cycle and Mitosis

The **cell cycle** is the life story of a cell, the stages through which it passes from one division to the next (**Figure 2.9**). This process is critical to genetics because, through the cell cycle, the genetic instructions for all characteristics are passed from parent to daughter cells. A new cycle begins after a cell has divided and produced two new cells. Each new cell metabolizes, grows, and develops. At the end of its cycle, the cell divides to produce two cells, which can then undergo additional cell cycles. Progression through the cell cycle is regulated at key transition points called **checkpoints**.

The cell cycle consists of two major phases. The first is **interphase**, the period between cell divisions, in which the cell grows, develops, and functions. In interphase, critical events necessary for cell division also take place. The second major phase is the **M phase** (mitotic phase), the period of active cell division. The M phase includes **mitosis**, the process of nuclear division, and **cytokinesis**, or cytoplasmic division. Let's take a closer look at the details of interphase and the M phase.

Interphase Interphase is the extended period of growth and development between cell divisions. Although little activity can be observed with a light microscope, the cell is quite busy: DNA is being synthesized, RNA and proteins are being produced, and hundreds of biochemical reactions necessary for cellular functions are taking place. In addition to growth and development, interphase includes several checkpoints, which regulate the cell cycle by allowing or prohibiting the cell's division. These checkpoints, like the checkpoints in the M phase, ensure that all cellular components are present and in good working order before the cell proceeds to the next stage. Checkpoints are necessary to

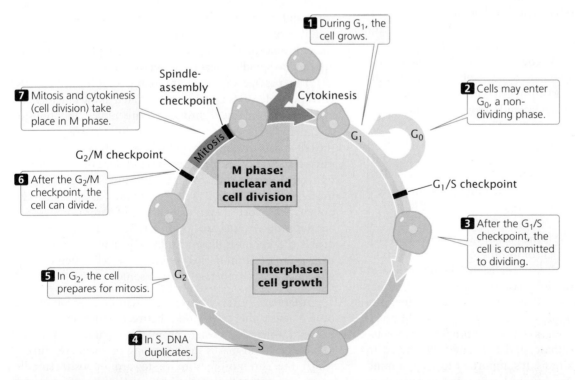

1 During G_1, the cell grows.

7 Mitosis and cytokinesis (cell division) take place in M phase.

Spindle-assembly checkpoint

Cytokinesis

2 Cells may enter G_0, a non-dividing phase.

G_2/M checkpoint

6 After the G_2/M checkpoint, the cell can divide.

Mitosis

M phase: nuclear and cell division

G_1

G_0

G_1/S checkpoint

3 After the G_1/S checkpoint, the cell is committed to dividing.

5 In G_2, the cell prepares for mitosis.

G_2

Interphase: cell growth

4 In S, DNA duplicates.

S

2.9 The cell cycle consists of interphase and M phase.

prevent cells with damaged or missing chromosomes from proliferating. Defects in checkpoints can lead to unregulated cell growth, as is seen in some cancers. The molecular basis of these checkpoints will be discussed in Chapter 23.

By convention, interphase is divided into three subphases: G_1, S, and G_2 (see Figure 2.9). Interphase begins with G_1 (for gap 1). In G_1, the cell grows, and proteins necessary for cell division are synthesized; this phase typically lasts several hours. Near the end of G_1, a critical point termed the G_1/S checkpoint holds the cell in G_1 until the cell has all of the enzymes necessary for the replication of DNA. After this checkpoint has been passed, the cell is committed to divide.

Before reaching the G_1/S checkpoint, cells may exit from the active cell cycle in response to regulatory signals and pass into a nondividing phase called G_0, which is a stable state during which cells usually maintain a constant size. They can remain in G_0 for an extended length of time, even indefinitely, or they can reenter G_1 and the active cell cycle. Many cells never enter G_0; rather, they cycle continuously.

After G_1, the cell enters the S phase (for DNA synthesis), in which each chromosome duplicates. Although the cell is committed to divide after the G_1/S checkpoint has been passed, DNA synthesis must take place before the cell can proceed to mitosis. If DNA synthesis is blocked (by drugs or by a mutation), the cell will not be able to undergo mitosis. Before the S phase, each chromosome is unreplicated; after the S phase, each chromosome is composed of two chromatids (see Figure 2.7).

After the S phase, the cell enters G_2 (gap 2). In this phase, several additional biochemical events necessary for cell division take place. The important G_2/M checkpoint is reached near the end of G_2. This checkpoint is passed only if the cell's DNA is undamaged. Damaged DNA can inhibit the activation of some proteins that are necessary for mitosis to take place. After the G_2/M checkpoint has been passed, the cell is ready to divide and enters the M phase. Although the length of interphase varies from cell type to cell type, a typical dividing mammalian cell spends about 10 hours in G_1, 9 hours in S, and 4 hours in G_2 (see Figure 2.9).

Throughout interphase, the chromosomes are in a relaxed, but by no means uncoiled, state, and individual chromosomes cannot be seen with a microscope. This condition changes dramatically when interphase draws to a close and the cell enters the M phase.

M phase   The M phase is the part of the cell cycle in which the copies of the cell's chromosomes (sister chromatids) separate and the cell undergoes division. The separation of sister chromatids in the M phase is a critical process that results in a complete set of genetic information for each of the resulting cells. Biologists usually divide the M phase into six stages: the five stages of mitosis (prophase, prometaphase, metaphase, anaphase, and telophase), illustrated in **Figure 2.10**, and cytokinesis. It's important to keep in mind that the M phase is a continuous process, and its separation into these six stages is somewhat arbitrary.

Prophase. As a cell enters prophase, the chromosomes become visible under a light microscope. Because the chromosome was duplicated in the preceding S phase, each chromosome possesses two chromatids attached at the centromere. The mitotic spindle, an organized array of microtubules that move the chromosomes in mitosis, forms. In animal cells, the spindle grows out from a pair of centrosomes that migrate to opposite sides of the cell. Within each centrosome is a special organelle, the centriole, which also is composed of microtubules. Some plant cells do not have centrosomes or centrioles, but they do have mitotic spindles.

Prometaphase. Disintegration of the nuclear membrane marks the start of prometaphase. Spindle microtubules, which until now have been outside the nucleus, enter the nuclear region. The ends of certain microtubules make contact with the chromosomes. For each chromosome, a microtubule from one of the centrosomes anchors to the kinetochore of *one* of the sister chromatids; a microtubule from the opposite centrosome then attaches to the other sister chromatid, and so the chromosome is anchored to both of the centrosomes. The microtubules lengthen and shorten, pushing and pulling the chromosomes about. Some microtubules extend from each centrosome toward the center of the spindle but do not attach to a chromosome.

Metaphase. During metaphase, the chromosomes become arranged in a single plane, the metaphase plate, between the two centrosomes. The centrosomes, now at opposite ends of the cell with microtubules radiating outward and meeting in the middle of the cell, center at the spindle poles. A spindle-assembly checkpoint ensures that each chromosome is aligned on the metaphase plate and attached to spindle fibers from opposite poles.

The passage of a cell through the spindle-assembly checkpoint depends on tension generated at the kinetochore as the two conjoined chromatids are pulled in opposite directions by the spindle fibers. This tension is required for the cell to pass through the spindle-assembly checkpoint. If a microtubule attaches to one chromatid but not to the other, no tension is generated and the cell is unable to progress to the next stage of the cell cycle. The spindle-assembly checkpoint is able to detect even a single pair of chromosomes that are not properly attached to microtubules. The importance of this checkpoint is illustrated by cells that are defective in their spindle-assembly checkpoint; these cells often end up with abnormal numbers of chromosomes.

Anaphase. After the spindle-assembly checkpoint has been passed, the connection between sister chromatids breaks down and the sister chromatids separate. This chromatid separation marks the beginning of anaphase, during which the chromosomes move toward opposite spindle poles. The microtubules that connect the chromosomes to the spindle poles are composed of subunits of a protein

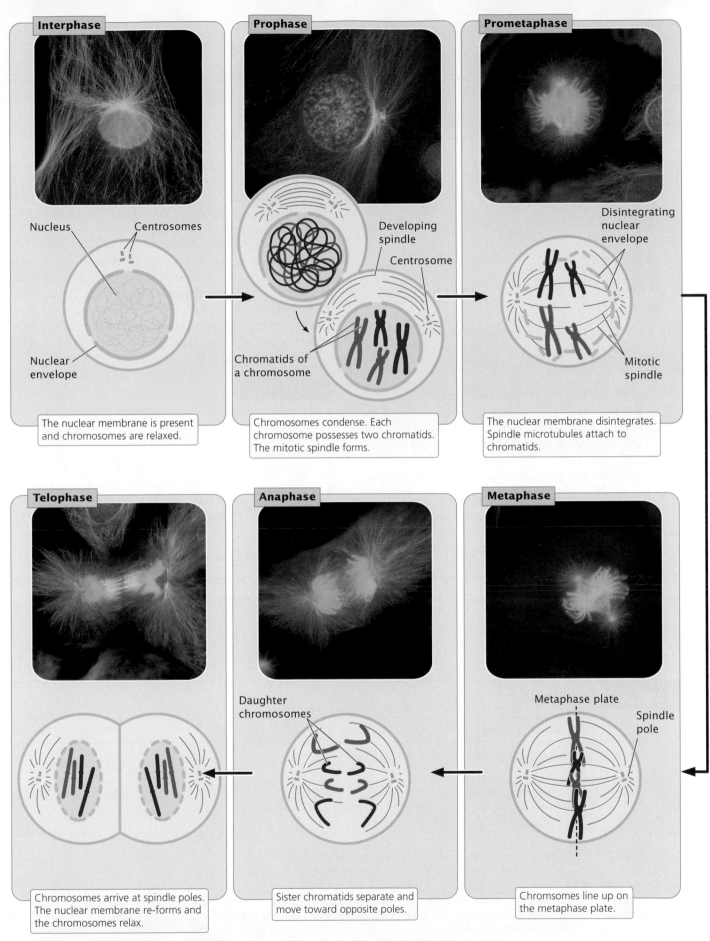

Interphase

Nucleus   Centrosomes

Nuclear envelope

The nuclear membrane is present and chromosomes are relaxed.

Prophase

Developing spindle

Centrosome

Chromatids of a chromosome

Chromosomes condense. Each chromosome possesses two chromatids. The mitotic spindle forms.

Prometaphase

Disintegrating nuclear envelope

Mitotic spindle

The nuclear membrane disintegrates. Spindle microtubules attach to chromatids.

Telophase

Chromosomes arrive at spindle poles. The nuclear membrane re-forms and the chromosomes relax.

Anaphase

Daughter chromosomes

Sister chromatids separate and move toward opposite poles.

Metaphase

Metaphase plate

Spindle pole

Chromsomes line up on the metaphase plate.

2.10 The cell cycle is divided into stages. [Photographs by Conly L. Rieder/Biological Photo Service.]

▶ www.whfreeman.com/pierce4e   Animation 2.1 illustrates events of the cell cycle dynamically. See what happens if different processes in the cycle fail.

23

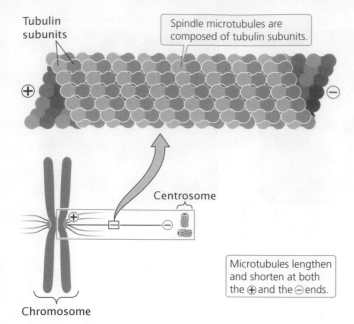

Tubulin subunits

Spindle microtubules are composed of tubulin subunits.

Centrosome

Microtubules lengthen and shorten at both the ⊕ and the ⊖ ends.

Chromosome

2.11 Microtubules are composed of tubulin subunits. Each microtubule has its plus (+) end at the kinetochore and its negative (−) end at the centrosome.

called tubulin (**Figure 2.11**). Chromosome movement is due to the disassembly of tubulin molecules at both the kinetochore end (called the + end) and the spindle end (called the − end) of the spindle fiber. Special proteins called molecular motors disassemble tubulin molecules from the spindle and generate forces that pull the chromosome toward the spindle pole.

Telophase. After the chromatids have separated, each is considered a separate chromosome. Telophase is marked by the arrival of the chromosomes at the spindle poles. The nuclear membrane re-forms around each set of chromosomes, producing two separate nuclei within the cell. The chromosomes relax and lengthen, once again disappearing from view. In many cells, division of the cytoplasm (cytokinesis) is simultaneous with telophase. The major features of the cell cycle are summarized in **Table 2.1**. TRY PROBLEM 22 ➡

Genetic Consequences of the Cell Cycle

What are the genetically important results of the cell cycle? From a single cell, the cell cycle produces two cells that contain the same genetic instructions. The resulting daughter cells are genetically identical with each other and with their parent cell because DNA synthesis in the S phase creates an exact copy of each DNA molecule, giving rise to two genetically identical sister chromatids. Mitosis then ensures that one of the two sister chromatids from each replicated chromosome passes into each new cell.

Another genetically important result of the cell cycle is that each of the cells produced contains a full complement of chromosomes: there is no net reduction or increase in chromosome number. Each cell also contains approximately half the cytoplasm and organelle content of the original parental cell, but no precise mechanism analogous to mitosis ensures that organelles are evenly divided. Consequently, not all cells resulting from the cell cycle are identical in their cytoplasmic content.

CONCEPTS

The active cell cycle phases are interphase and the M phase. Interphase consists of G_1, S, and G_2. In G_1, the cell grows and prepares for cell division; in the S phase, DNA synthesis takes place; in G_2, other biochemical events necessary for cell division take place. Some cells enter a quiescent phase called G_0. The M phase includes mitosis and cytokinesis and is divided into prophase, prometaphase, metaphase, anaphase, and telophase. The cell cycle produces two genetically identical cells each of which possesses a full complement of chromosomes.

✔ CONCEPT CHECK 4

Which is the correct order of stages in the cell cycle?

a. G_1, S, prophase, metaphase, anaphase

b. S, G_1, prophase, metaphase, anaphase

c. Prophase, S, G_1, metaphase, anaphase

d. S, G_1, anaphase, prophase, metaphase

Table 2.1	Features of the cell cycle
Stage	**Major Features**
G_0 phase	Stable, nondividing period of variable length.
Interphase	
G_1 phase	Growth and development of the cell; G_1/S checkpoint.
S phase	Synthesis of DNA.
G_2 phase	Preparation for division; G_2/M checkpoint.
M phase	
Prophase	Chromosomes condense and mitotic spindle forms.
Prometaphase	Nuclear envelope disintegrates, and spindle microtubules anchor to kinetochores.
Metaphase	Chromosomes align on the metaphase plate; spindle-assembly checkpoint.
Anaphase	Sister chromatids separate, becoming individual chromosomes that migrate toward spindle poles.
Telophase	Chromosomes arrive at spindle poles, the nuclear envelope re-forms, and the condensed chromosomes relax.
Cytokinesis	Cytoplasm divides; cell wall forms in plant cells.

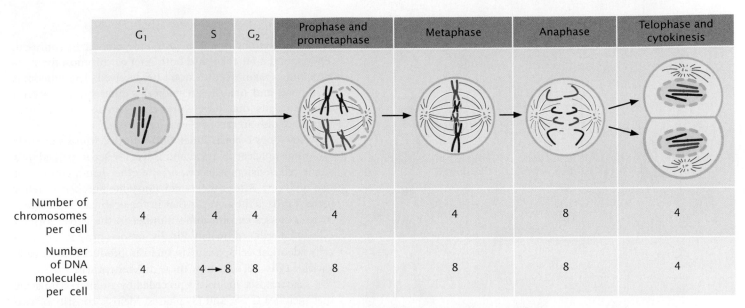

	G₁	S	G₂	Prophase and prometaphase	Metaphase	Anaphase	Telophase and cytokinesis
Number of chromosomes per cell	4	4	4	4	4	8	4
Number of DNA molecules per cell	4	4 → 8	8	8	8	8	4

2.12 **The number of chromosomes and the number of DNA molecules change in the course of the cell cycle.** The number of chromosomes per cell equals the number of functional centromeres. The number of DNA molecules per cell equals the number of chromosomes when the chromosomes are unreplicated (no sister chromatids present) and twice the number of chromosomes when sister chromosomes *are* present.

CONNECTING CONCEPTS

Counting Chromosomes and DNA Molecules

The relations among chromosomes, chromatids, and DNA molecules frequently cause confusion. At certain times, chromosomes are unreplicated; at other times, each possesses two chromatids (see Figure 2.7). Chromosomes sometimes consist of a single DNA molecule; at other times, they consist of two DNA molecules. How can we keep track of the number of these structures in the cell cycle?

There are two simple rules for counting chromosomes and DNA molecules: (1) to determine the number of chromosomes, count the number of functional centromeres; (2) to determine the number of DNA molecules, first determine if sister chromatids are present. If sister chromatids are present, the chromosome has replicated and the number of DNA molecules is twice the number of chromosomes. If sister chromatids are not present, the chromosome has not replicated and the number of DNA molecules is the same as the number of chromosomes.

Let's examine a hypothetical cell as it passes through the cell cycle (**Figure 2.12**). At the beginning of G₁, this diploid cell has two complete sets of chromosomes, inherited from its parent cell. Each chromosome is unreplicated and consists of a single molecule of DNA, and so there are four DNA molecules in the cell during G₁. In the S phase, each DNA molecule is copied. The two resulting DNA molecules combine with histones and other proteins to form sister chromatids. Although the amount of DNA doubles in the S phase, the number of chromosomes remains the same because the sister chromatids are tethered together and share a single functional centromere. At the end of the S phase, this cell still contains four chromosomes, each with two sister chromatids; so $4 \times 2 = 8$ DNA molecules are present.

Through prophase, prometaphase, and metaphase, the cell has four chromosomes and eight DNA molecules. At anaphase, however, the sister chromatids separate. Each now has its own functional centromere, and so each is considered a separate chromosome. Until cytokinesis, the cell contains eight unreplicated chromosomes; thus, there are still eight DNA molecules present. After cytokinesis, the eight chromosomes (eight DNA molecules) are distributed equally between two cells; so each new cell contains four chromosomes and four DNA molecules, the number present at the beginning of the cell cycle.

In summary, the number of chromosomes increases only in anaphase, when the two chromatids of a chromosome separate and become distinct chromosomes. The number of chromosomes decreases only through cytokinesis. The number of DNA molecules increases only in the S phase and decreases only through cytokinesis. **TRY PROBLEM 23** →

2.3 Sexual Reproduction Produces Genetic Variation Through the Process of Meiosis

If all reproduction were accomplished through mitosis, life would be quite dull, because mitosis produces only genetically identical progeny. With only mitosis, you, your children, your parents, your brothers and sisters, your cousins, and many people you don't even know would be clones—copies of one another. Only the occasional mutation would introduce any genetic variability. All organisms reproduced in this way for the first 2 billion years of Earth's existence (and it is the way in which some organisms still reproduce today). Then, some 1.5 billion to 2 billion years ago, something remarkable evolved: cells that produce genetically variable offspring through sexual reproduction.

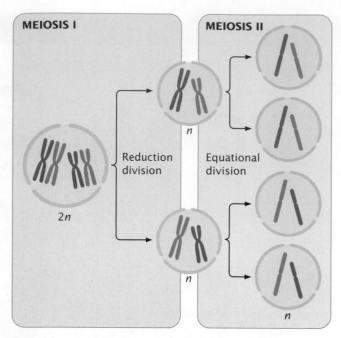

2.13 Meiosis includes two cell divisions. In this illustration, the original cell is $2n = 4$. After two meiotic divisions, each resulting cell is $1n = 2$.

Meiosis

The words *mitosis* and *meiosis* are sometimes confused. They sound a bit alike, and both refer to chromosome division and cytokinesis. But don't be deceived. The outcomes of mitosis and meiosis are radically different, and several unique events that have important genetic consequences take place only in meiosis.

How does meiosis differ from mitosis? Mitosis consists of a single nuclear division and is usually accompanied by a single cell division. Meiosis, on the other hand, consists of two divisions. After mitosis, chromosome number in newly formed cells is the same as that in the original cell, whereas meiosis causes chromosome number in the newly formed cells to be reduced by half. Finally, mitosis produces genetically identical cells, whereas meiosis produces genetically variable cells. Let's see how these differences arise.

Like mitosis, meiosis is preceded by an interphase stage that includes G_1, S, and G_2 phases. Meiosis consists of two distinct processes: meiosis I and meiosis II, each of which includes a cell division. The first division, which comes at the end of meiosis I, is termed the reduction division because the number of chromosomes per cell is reduced by half (**Figure 2.13**). The second division, which comes at the end of meiosis II, is sometimes termed the equational division. The events of meiosis II are similar to those of mitosis. However, meiosis II differs from mitosis in that chromosome number has already been halved in meiosis I, and the cell does not begin with the same number of chromosomes as it does in mitosis (see Figure 2.13).

The evolution of sexual reproduction is one of the most significant events in the history of life. As will be discussed in Chapters 24 and 25, the pace of evolution depends on the amount of genetic variation present. By shuffling the genetic information from two parents, sexual reproduction greatly increases the amount of genetic variation and allows for accelerated evolution. Most of the tremendous diversity of life on Earth is a direct result of sexual reproduction.

Sexual reproduction consists of two processes. The first is **meiosis**, which leads to gametes in which chromosome number is reduced by half. The second process is **fertilization**, in which two haploid gametes fuse and restore chromosome number to its original diploid value.

Meiosis I During interphase, the chromosomes are relaxed and visible as diffuse chromatin. **Prophase I** is a lengthy stage, divided into five substages (**Figure 2.14**). In leptotene, the chromosomes contract and become visible. In zygotene, the chromosomes continue to condense; homologous chromosomes pair up and begin **synapsis**, a very close pairing association. Each homologous pair of

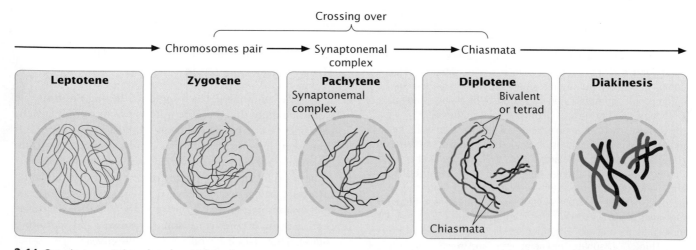

2.14 Crossing over takes place in prophase I.

synapsed chromosomes consists of four chromatids called a **bivalent** or **tetrad**. In pachytene, the chromosomes become shorter and thicker, and a three-part synaptonemal complex develops between homologous chromosomes. The function of the synaptonemal complex is unclear, but the chromosomes of many cells deficient in this complex do not separate properly.

Crossing over takes place in prophase I, in which homologous chromosomes exchange genetic information. Crossing over generates genetic variation (see Sources of Genetic Variation in Meiosis later in this chapter) and is essential for the proper alignment and separation of homologous chromosomes. The centromeres of the paired chromosomes move apart in diplotene; the two homologs remain attached at each chiasma (plural, chiasmata), which is the result of crossing over. In diakinesis, chromosome condensation continues, and the chiasmata move toward the ends of the chromosomes as the strands slip apart; so the homologs remain paired only at the tips. Near the end of prophase I, the nuclear membrane breaks down and the spindle forms, setting the stage for metaphase I. The stages of meiosis are outlined in **Figure 2.15**.

Metaphase I is initiated when homologous pairs of chromosomes align along the metaphase plate (see Figure 2.15). A microtubule from one pole attaches to one chromosome of a homologous pair, and a microtubule from the other pole attaches to the other member of the pair.

Anaphase I is marked by the separation of homologous chromosomes. The two chromosomes of a homologous pair are pulled toward opposite poles. Although the homologous chromosomes separate, the sister chromatids remain attached and travel together. In **telophase I**, the chromosomes arrive at the spindle poles and the cytoplasm divides.

Meiosis II   The period between meiosis I and meiosis II is **interkinesis**, in which the nuclear membrane re-forms around the chromosomes clustered at each pole, the spindle breaks down, and the chromosomes relax. These cells then pass through **prophase II**, in which the events of interkinesis are reversed: the chromosomes recondense, the spindle re-forms, and the nuclear envelope once again breaks down. In interkinesis in some types of cells, the chromosomes remain condensed, and the spindle does not break down. These cells move directly from cytokinesis into **metaphase II**, which is similar to metaphase of mitosis: the individual chromosomes line up on the metaphase plate, with the sister chromatids facing opposite poles.

In **anaphase II**, the kinetochores of the sister chromatids separate and the chromatids are pulled to opposite poles. Each chromatid is now a distinct chromosome. In **telophase II**, the chromosomes arrive at the spindle poles, a nuclear envelope re-forms around the chromosomes, and the cytoplasm divides. The chromosomes relax and are no longer visible. The major events of meiosis are summarized in **Table 2.2**.

Table 2.2	Major events in each stage of meiosis
Stage	**Major Events**
Meiosis I	
Prophase I	Chromosomes condense, homologous chromosomes synapse, crossing over takes place, the nuclear envelope breaks down, and the mitotic spindle forms.
Metaphase I	Homologous pairs of chromosomes line up on the metaphase plate.
Anaphase I	The two chromosomes (each with two chromatids) of each homologous pair separate and move toward opposite poles.
Telophase I	Chromosomes arrive at the spindle poles.
Cytokinesis	The cytoplasm divides to produce two cells, each having half the original number of chromosomes.
Interkinesis	In some types of cells, the spindle breaks down, chromosomes relax, and a nuclear envelope re-forms, but no DNA synthesis takes place.
Meiosis II	
Prophase II*	Chromosomes condense, the spindle forms, and the nuclear envelope disintegrates.
Metaphase II	Individual chromosomes line up on the metaphase plate.
Anaphase II	Sister chromatids separate and move as individual chromosomes toward the spindle poles.
Telophase II	Chromosomes arrive at the spindle poles; the spindle breaks down and a nuclear envelope re-forms.
Cytokinesis	The cytoplasm divides.

*Only in cells in which the spindle has broken down, chromosomes have relaxed, and the nuclear envelope has re-formed in telophase I. Other types of cells proceed directly to metaphase II after cytokinesis.

Meiosis I

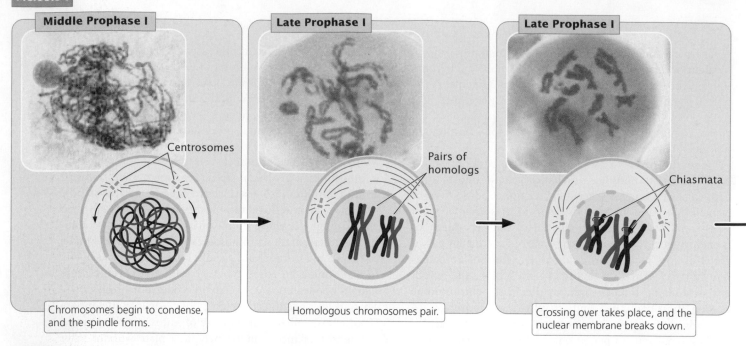

Middle Prophase I

Centrosomes

Chromosomes begin to condense, and the spindle forms.

Late Prophase I

Pairs of homologs

Homologous chromosomes pair.

Late Prophase I

Chiasmata

Crossing over takes place, and the nuclear membrane breaks down.

Meiosis II

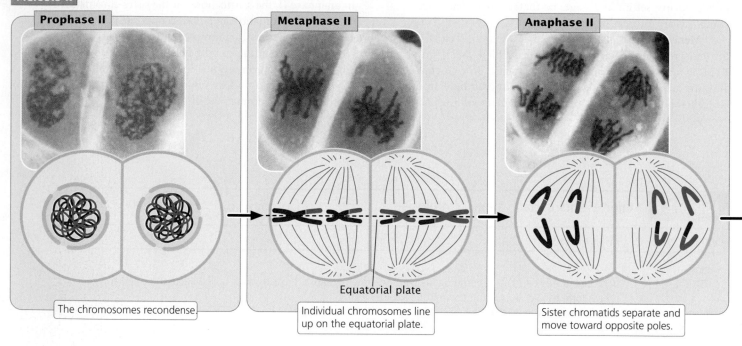

Prophase II

The chromosomes recondense.

Metaphase II

Equatorial plate

Individual chromosomes line up on the equatorial plate.

Anaphase II

Sister chromatids separate and move toward opposite poles.

CONCEPTS

Meiosis consists of two distinct processes: meiosis I and meiosis II. Meiosis I includes the reduction division, in which homologous chromosomes separate and chromosome number is reduced by half. In meiosis II (the equational division) chromatids separate.

✔ CONCEPT CHECK 5

Which of the following events takes place in metaphase I?

a. Crossing over

b. Chromosomes contract.

c. Homologous pairs of chromosomes line up on the metaphase plate.

d. Individual chromosomes line up on the metaphase plate.

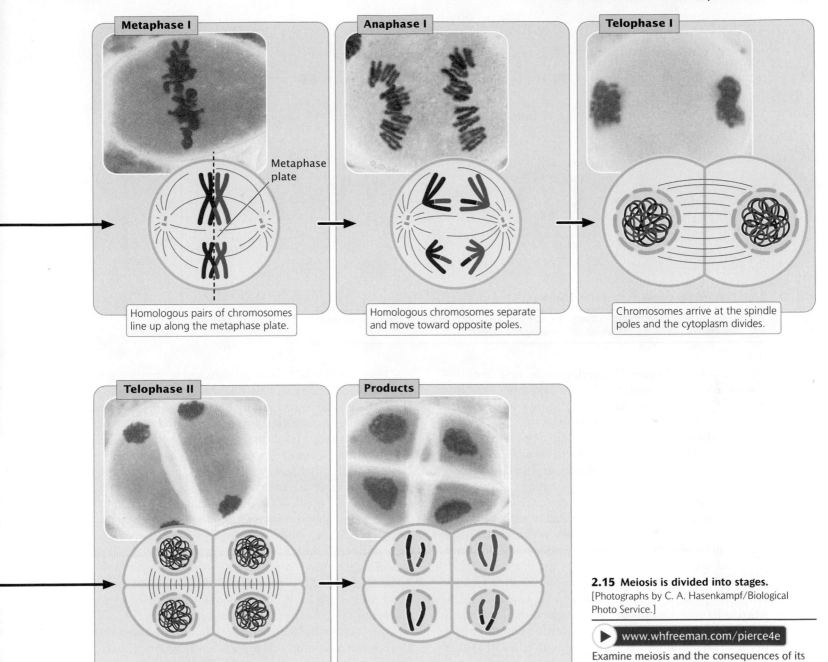

2.15 Meiosis is divided into stages.
[Photographs by C. A. Hasenkampf/Biological Photo Service.]

▶ www.whfreeman.com/pierce4e

Examine meiosis and the consequences of its failure by viewing Animation 2.2.

The labels and captions within the figure:

Metaphase I — Homologous pairs of chromosomes line up along the metaphase plate. / Metaphase plate

Anaphase I — Homologous chromosomes separate and move toward opposite poles.

Telophase I — Chromosomes arrive at the spindle poles and the cytoplasm divides.

Telophase II — Chromosomes arrive at the spindle poles and the cytoplasm divides.

Products

Sources of Genetic Variation in Meiosis

What are the overall consequences of meiosis? First, meiosis comprises two divisions; so each original cell produces four cells (there are exceptions to this generalization, as, for example, in many female animals; see Figure 2.20b). Second, chromosome number is reduced by half; so cells produced by meiosis are haploid. Third, cells produced by meiosis are genetically different from one another and from the parental cell. Genetic differences among cells result from two processes that are unique to meiosis: crossing over and random separation of homologous chromosomes.

Crossing over   Crossing over, which takes place in prophase I, refers to the exchange of genes between nonsister chromatids (chromatids from different homologous chromosomes). Evidence from yeast suggests that crossing over is initiated in zygotene, before the synaptonemal complex develops, and is not completed until near the end of prophase I (see Figure 2.14). In other organisms, recombination is initiated after the formation of the synaptonemal complex and, in yet others, there is no synaptonemal complex.

After crossing over has taken place, the sister chromatids may no longer be identical. Crossing over is the basis for

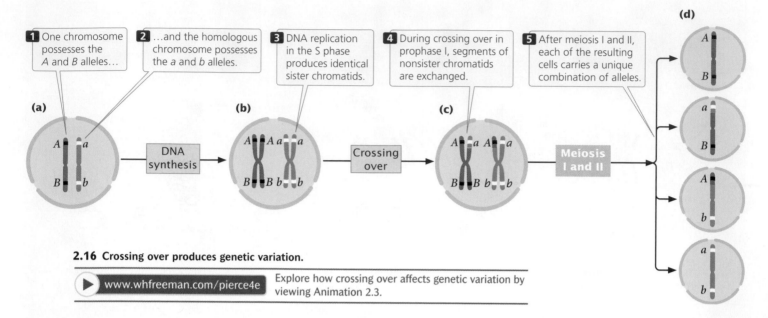

1 One chromosome possesses the *A* and *B* alleles…

2 …and the homologous chromosome possesses the *a* and *b* alleles.

3 DNA replication in the S phase produces identical sister chromatids.

4 During crossing over in prophase I, segments of nonsister chromatids are exchanged.

5 After meiosis I and II, each of the resulting cells carries a unique combination of alleles.

2.16 Crossing over produces genetic variation.

▶ www.whfreeman.com/pierce4e   Explore how crossing over affects genetic variation by viewing Animation 2.3.

intrachromosomal **recombination**, creating new combinations of alleles on a chromatid. To see how crossing over produces genetic variation, consider two pairs of alleles, which we will abbreviate *Aa* and *Bb*. Assume that one chromosome possesses the *A* and *B* alleles and its homolog possesses the *a* and *b* alleles (**Figure 2.16a**). When DNA is replicated in the S phase, each chromosome duplicates, and so the resulting sister chromatids are identical (**Figure 2.16b**).

In the process of crossing over, there are breaks in the DNA strands and the breaks are repaired in such a way that segments of nonsister chromatids are exchanged (**Figure 2.16c**). The molecular basis of this process will be described in more detail in Chapter 12. The important thing here is that, after crossing over has taken place, the two sister chromatids are no longer identical: one chromatid has alleles *A* and *B*, whereas its sister chromatid (the chromatid that underwent crossing over) has alleles *a* and *B*. Likewise, one chromatid of the other chromosome has alleles *a* and *b,* and the other chromatid has alleles *A* and *b*. Each of the four chromatids now carries a unique combination of alleles: *A B* , *a B* ,   *A b* , and   *a b* . Eventually, the two homologous chromosomes separate, each going into a different cell. In meiosis II, the two chromatids of each chromosome separate, and thus each of the four cells resulting from meiosis carries a different combination of alleles (**Figure 2.16d**).

Random Separation of Homologous Chromosomes

The second process of meiosis that contributes to genetic variation is the random distribution of chromosomes in anaphase I after their random alignment in metaphase I. To illustrate this process, consider a cell with three pairs of chromosomes, I, II, and III (**Figure 2.17a**). One chromosome of each pair is maternal in origin (I_m, II_m, and III_m); the other is paternal in origin (I_p, II_p, and III_p). The chromosome pairs line up in the center of the cell in metaphase I and, in anaphase I, the chromosomes of each homologous pair separate.

How each pair of homologs aligns and separates is random and independent of how other pairs of chromosomes align and separate (**Figure 2.17b**). By chance, all the maternal chromosomes might migrate to one side, with all the paternal chromosomes migrating to the other. After division, one cell would contain chromosomes I_m, II_m, and III_m, and the other, I_p, II_p, and III_p. Alternatively, the I_m, II_m, and III_p chromosomes might move to one side, and the I_p, II_p, and III_m chromosomes to the other. The different migrations would produce different combinations of chromosomes in the resulting cells (**Figure 2.17c**). There are four ways in which a diploid cell with three pairs of chromosomes can divide, producing a total of eight different combinations of chromosomes in the gametes. In general, the number of possible combinations is 2^n, where *n* equals the number of homologous pairs. As the number of chromosome pairs increases, the number of combinations quickly becomes very large. In humans, who have 23 pairs of chromosomes, there are 2^{23}, or 8,388,608, different combinations of chromosomes possible from the random separation of homologous chromosomes. The genetic consequences of this process, termed independent assortment, will be explored in more detail in Chapter 3.

In summary, crossing over shuffles alleles on the *same* chromosome into new combinations, whereas the random distribution of maternal and paternal chromosomes shuffles alleles on *different* chromosomes into new combinations. Together, these two processes are capable of producing tremendous amounts of genetic variation among the cells resulting from meiosis. TRY PROBLEMS 30 AND 31 ➡

CONCEPTS

The two mechanisms that produce genetic variation in meiosis are crossing over and the random distribution of maternal and paternal chromosomes.

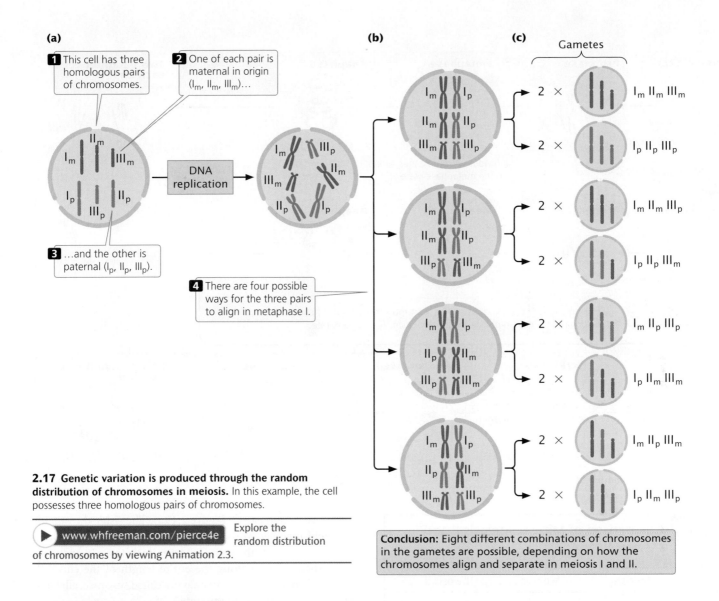

2.17 Genetic variation is produced through the random distribution of chromosomes in meiosis. In this example, the cell possesses three homologous pairs of chromosomes.

> ▶ www.whfreeman.com/pierce4e   Explore the
> random distribution
> of chromosomes by viewing Animation 2.3.

Conclusion: Eight different combinations of chromosomes in the gametes are possible, depending on how the chromosomes align and separate in meiosis I and II.

CONNECTING CONCEPTS

Mitosis and Meiosis Compared

Now that we have examined the details of mitosis and meiosis, let's compare the two processes (**Figure 2.18** and **Table 2.3**). In both mitosis and meiosis, the chromosomes contract and become visible; both processes include the movement of chromosomes toward the spindle poles, and both are accompanied by cell division. Beyond these similarities, the processes are quite different.

Mitosis results in a single cell division and usually produces two daughter cells. Meiosis, in contrast, comprises two cell divisions and usually produces four cells. In diploid cells, homologous chromosomes are present before both meiosis and mitosis, but the pairing of homologs takes place only in meiosis.

Another difference is that, in meiosis, chromosome number is reduced by half as a consequence of the separation of homologous pairs of chromosomes in anaphase I, but no chromosome reduction takes place in mitosis. Furthermore, meiosis is characterized by two processes that produce genetic variation: crossing over (in prophase I) and the random distribution of maternal and paternal chromosomes (in anaphase I). There are normally no equivalent processes in mitosis.

Mitosis and meiosis also differ in the behavior of chromosomes in metaphase and anaphase. In metaphase I of meiosis, *homologous pairs* of chromosomes line up on the metaphase plate, whereas *individual chromosomes* line up on the metaphase plate in metaphase of mitosis (and in metaphase II of meiosis). In anaphase I of meiosis, *paired chromosomes* separate and migrate toward opposite spindle poles, each chromosome possessing two chromatids attached at the centromere. In contrast, in anaphase of mitosis (and in anaphase II of meiosis), *sister chromatids* separate, and each chromosome that moves toward a spindle pole is unreplicated.

TRY PROBLEMS 25 AND 26 →

The Separation of Sister Chromatids and Homologous Chromosomes

In recent years, some of the molecules required for the joining and separation of chromatids and homologous chromosomes have been identified. **Cohesin,** a protein that holds the chromatids together, is key to the behavior of chromosomes

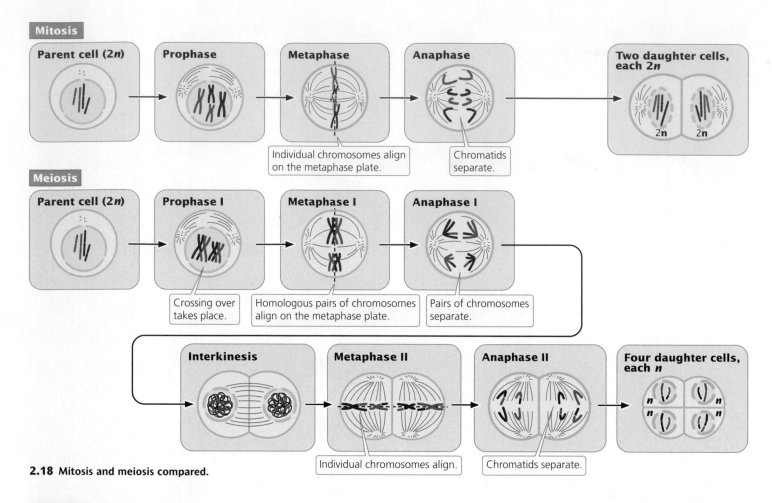

Mitosis

| Parent cell (2*n*) | Prophase | Metaphase | Anaphase | Two daughter cells, each 2*n* |

Individual chromosomes align on the metaphase plate.

Chromatids separate.

Meiosis

| Parent cell (2*n*) | Prophase I | Metaphase I | Anaphase I |

Crossing over takes place.

Homologous pairs of chromosomes align on the metaphase plate.

Pairs of chromosomes separate.

| Interkinesis | Metaphase II | Anaphase II | Four daughter cells, each *n* |

Individual chromosomes align.

Chromatids separate.

2.18 Mitosis and meiosis compared.

Table 2.3	Comparison of Mitosis, Meiosis I, and Meiosis II		
Event	Mitosis	Meiosis I	Meiosis II
Cell division	Yes	Yes	Yes
Chromosome reduction	No	Yes	No
Genetic variation produced	No	Yes	No
Crossing over	No	Yes	No
Random distribution of maternal and paternal chromosomes	No	Yes	No
Metaphase	Individual chromosomes line up	Homologous pairs line up	Individual chromosomes line up
Anaphase	Chromatids separate	Homologous chromosomes separate	Chromatids separate

in mitosis and meiosis (**Figure 2.19a**). The sister chromatids are held together by cohesin, which is established in the S phase and persists through G$_2$ and early mitosis. In anaphase of mitosis, cohesin along the entire length of the chromosome is broken down by an enzyme called separase, allowing the sister chromatids to separate.

As we have seen, mitosis and meiosis differ fundamentally in the behavior of chromosomes in anaphase (see Figure 2.18). Why do homologs separate in anaphase I of meiosis, whereas chromatids separate in anaphase of mitosis and anaphase II of meiosis? It is important to note that the forms of cohesin used in mitosis and meiosis differ. At the beginning of meiosis, the meiosis-specific cohesin is found along the entire length of a chromosome's arms (**Figure 2.19b**). The cohesin also acts on the chromosome arms of homologs at the chiasmata, tethering two homologs together at their ends.

In anaphase I, cohesin along the chromosome arms is broken, allowing the two homologs to separate. However, cohesin at the centromere is protected by a protein called shugoshin, which means "guardian spirit" in Japanese. Because of this protective action by shugoshin, the centromeric cohesin remains intact and prevents the separation of the two sister chromatids during anaphase I of meiosis. Shugoshin is subsequently degraded. At the end of meta-

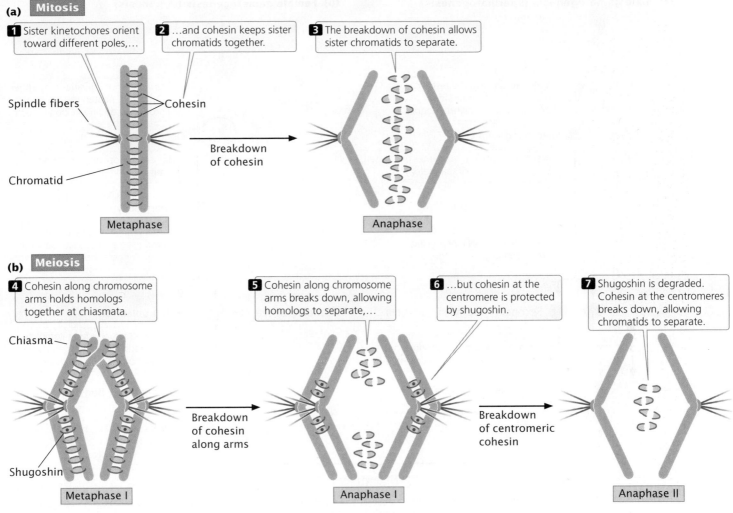

(a) Mitosis

1 Sister kinetochores orient toward different poles,...

2 ...and cohesin keeps sister chromatids together.

3 The breakdown of cohesin allows sister chromatids to separate.

Spindle fibers

Cohesin

Chromatid

Breakdown of cohesin

Metaphase

Anaphase

(b) Meiosis

4 Cohesin along chromosome arms holds homologs together at chiasmata.

5 Cohesin along chromosome arms breaks down, allowing homologs to separate,...

6 ...but cohesin at the centromere is protected by shugoshin.

7 Shugoshin is degraded. Cohesin at the centromeres breaks down, allowing chromatids to separate.

Chiasma

Shugoshin

Breakdown of cohesin along arms

Breakdown of centromeric cohesin

Metaphase I

Anaphase I

Anaphase II

2.19 **Cohesin controls the separation of chromatids and chromosomes in mitosis and meiosis.**

phase II, the centromeric cohesin—no longer protected by shugoshin—breaks down, allowing the sister chromatids to separate in anaphase II, just as they do in mitosis (Figure 2.19b). **TRY PROBLEM 27** →

CONCEPTS

Cohesin holds sister chromatids together during the early part of mitosis. In anaphase, cohesin breaks down, allowing sister chromatids to separate. In meiosis, cohesin is protected at the centromeres during anaphase I, and so homologous chromosomes, but not sister chromatids, separate in meiosis I. The breakdown of centromeric cohesin allows sister chromatids to separate in anaphase II of meiosis.

✔ CONCEPT CHECK 6

How does shugoshin affect sister chromatids in meiosis I and meiosis II?

Meiosis in the Life Cycles of Animals and Plants

The overall result of meiosis is four haploid cells that are genetically variable. Let's now see where meiosis fits into the life cycles of a multicellular animal and a multicellular plant.

Meiosis in animals The production of gametes in a male animal, a process called **spermatogenesis**, takes place in the testes. There, diploid primordial germ cells divide mitotically to produce diploid cells called **spermatogonia** (**Figure 2.20a**). Each spermatogonium can undergo repeated rounds of mitosis, giving rise to numerous additional spermatogonia. Alternatively, a spermatogonium can initiate meiosis and enter into prophase I. Now called a **primary spermatocyte**, the cell is still diploid because the homologous chromosomes have not yet separated. Each primary spermatocyte completes meiosis I, giving rise to two haploid **secondary spermatocytes** that then undergo meiosis II,

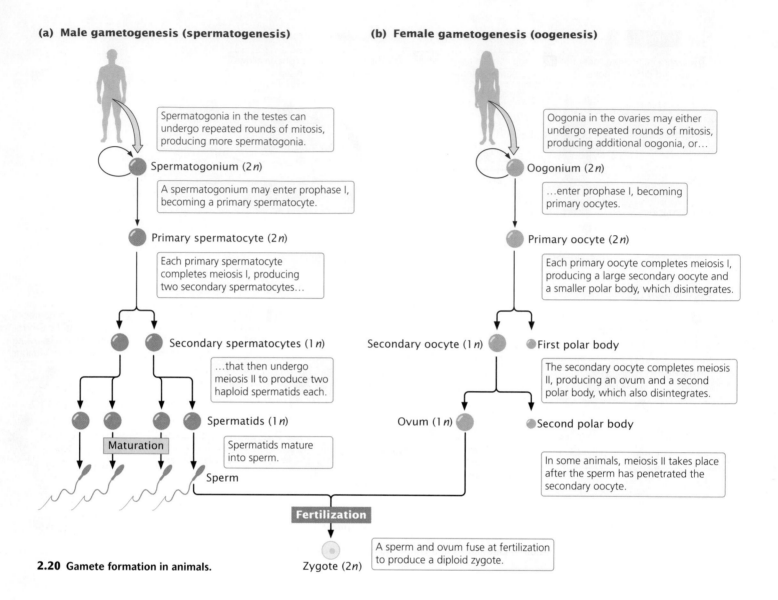

(a) Male gametogenesis (spermatogenesis)

Spermatogonia in the testes can undergo repeated rounds of mitosis, producing more spermatogonia.

Spermatogonium (2*n*)

A spermatogonium may enter prophase I, becoming a primary spermatocyte.

Primary spermatocyte (2*n*)

Each primary spermatocyte completes meiosis I, producing two secondary spermatocytes…

Secondary spermatocytes (1*n*)

…that then undergo meiosis II to produce two haploid spermatids each.

Spermatids (1*n*)

Maturation

Spermatids mature into sperm.

Sperm

(b) Female gametogenesis (oogenesis)

Oogonia in the ovaries may either undergo repeated rounds of mitosis, producing additional oogonia, or…

Oogonium (2*n*)

…enter prophase I, becoming primary oocytes.

Primary oocyte (2*n*)

Each primary oocyte completes meiosis I, producing a large secondary oocyte and a smaller polar body, which disintegrates.

Secondary oocyte (1*n*)   First polar body

The secondary oocyte completes meiosis II, producing an ovum and a second polar body, which also disintegrates.

Ovum (1*n*)   Second polar body

In some animals, meiosis II takes place after the sperm has penetrated the secondary oocyte.

Fertilization

A sperm and ovum fuse at fertilization to produce a diploid zygote.

Zygote (2*n*)

2.20 Gamete formation in animals.

with each producing two haploid **spermatids**. Thus, each primary spermatocyte produces a total of four haploid spermatids, which mature and develop into sperm.

The production of gametes in a female animal, a process called **oogenesis**, begins much as spermatogenesis does. Within the ovaries, diploid primordial germ cells divide mitotically to produce **oogonia** (**Figure 2.20b**). Like spermatogonia, oogonia can undergo repeated rounds of mitosis or they can enter into meiosis. When they enter prophase I, these still-diploid cells are called **primary oocytes**. Each primary oocyte completes meiosis I and divides.

At this point, the process of oogenesis begins to differ from that of spermatogenesis. In oogenesis, cytokinesis is unequal: most of the cytoplasm is allocated to one of the two haploid cells, the **secondary oocyte**. The smaller cell, which contains half of the chromosomes but only a small part of the cytoplasm, is called the **first polar body**; it may or may not divide further. The secondary oocyte completes meiosis II, and, again, cytokinesis is unequal—most of the cytoplasm

passes into one of the cells. The larger cell, which acquires most of the cytoplasm, is the **ovum**, the mature female gamete. The smaller cell is the **second polar body**. Only the ovum is capable of being fertilized, and the polar bodies usually disintegrate. Oogenesis, then, produces a single mature gamete from each primary oocyte.

In mammals, oogenesis differs from spermatogenesis in another way. The formation of sperm takes place continuously in a male throughout his adult reproductive life. The formation of female gametes, however, is often a discontinuous process, and may take place over a period of years. Oogenesis begins before birth; at this time, oogonia initiate meiosis and give rise to primary oocytes. Meiosis is then arrested, stalled in prophase I. A female is born with primary oocytes arrested in prophase I. In humans, this period of suspended animation may last 30 or 40 years. Before ovulation, rising hormone levels stimulate one or more of the primary oocytes to recommence meiosis. The first division of meiosis is completed and a secondary oocyte is ovulated from the ovary. In humans and many other species, the second division of meiosis is then delayed until contact with the sperm. When the sperm penetrates the outer layer of the secondary oocyte, the second meiotic division takes place, the second polar body is extruded from the egg, and the nuclei of the sperm and newly formed ovum fuse, giving rise to the zygote.

CONCEPTS

In the testes, a diploid spermatogonium undergoes meiosis, producing a total of four haploid sperm cells. In the ovary, a diploid oogonium undergoes meiosis to produce a single large ovum and smaller polar bodies that normally disintegrate.

✔ CONCEPT CHECK 7

A secondary spermatocyte has 12 chromosomes. How many chromosomes will be found in the primary spermatocyte that gave rise to it?

a. 6   c. 18
b. 12   d. 24

Meiosis in plants   Most plants have a complex life cycle that includes two distinct generations (stages): the diploid sporophyte and the haploid gametophyte. These two stages alternate; the sporophyte produces haploid spores through meiosis, and the gametophyte produces haploid gametes through mitosis (**Figure 2.21**). This type of life cycle is sometimes called alternation of generations. In this cycle, the immediate products of meiosis are called spores, not gametes; the spores undergo one or more mitotic divisions to produce gametes. Although the terms used for this pro-

cess are somewhat different from those commonly used in regard to animals (and from some of those employed so far in this chapter), the processes in plants and animals are basically the same: in both, meiosis leads to a reduction in chromosome number, producing haploid cells.

In flowering plants, the sporophyte is the obvious, vegetative part of the plant; the gametophyte consists of only a few haploid cells within the sporophyte. The flower, which is part of the sporophyte, contains the reproductive structures. In some plants, both male and female reproductive structures are found in the same flower; in other plants, they exist in different flowers. In either case, the male part of the flower, the stamen, contains diploid reproductive cells called **microsporocytes**, each of which undergoes meiosis to produce four haploid **microspores** (**Figure 2.22a**). Each microspore divides mitotically, producing an immature pollen grain consisting of two haploid nuclei. One of these nuclei, called the tube nucleus, directs the growth of a pollen tube. The other, termed the generative nucleus, divides mitotically to produce two sperm cells. The pollen grain, with its two haploid nuclei, is the male gametophyte.

The female part of the flower, the ovary, contains diploid cells called **megasporocytes**, each of which undergoes meiosis to produce four haploid **megaspores** (**Figure 2.22b**), only one of which survives. The nucleus of the surviving megaspore divides mitotically three times, producing a total of eight haploid nuclei that make up the female gametophyte, the embryo sac. Division of the cytoplasm then produces separate cells, one of which becomes the egg.

When the plant flowers, the stamens open and release pollen grains. Pollen lands on a flower's stigma—a sticky platform that sits on top of a long stalk called the style. At

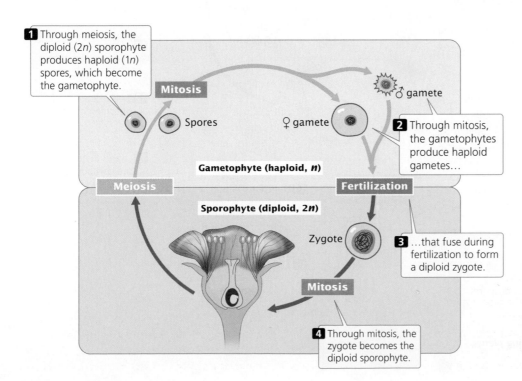

1 Through meiosis, the diploid (2*n*) sporophyte produces haploid (1*n*) spores, which become the gametophyte.

Mitosis

Spores   ♀ gamete   ♂ gamete

2 Through mitosis, the gametophytes produce haploid gametes…

Gametophyte (haploid, *n*)

Meiosis   Fertilization

Sporophyte (diploid, 2*n*)

Zygote

3 …that fuse during fertilization to form a diploid zygote.

Mitosis

4 Through mitosis, the zygote becomes the diploid sporophyte.

2.21 Plants alternate between diploid and haploid life stages (female, ♀; male, ♂).

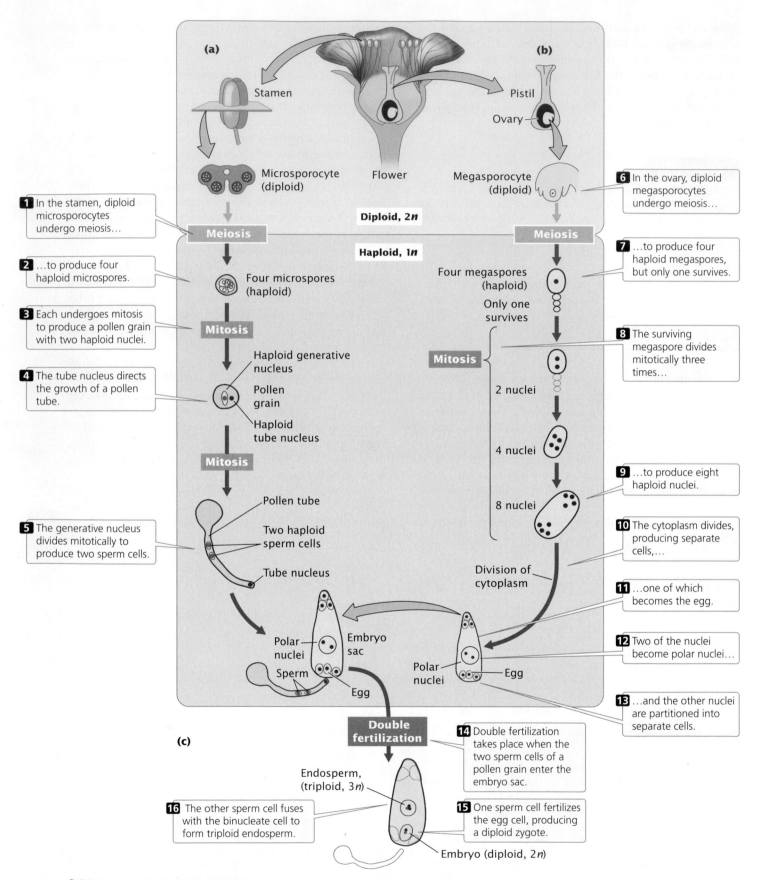

(a)

Stamen

(b)

Pistil

Ovary

Flower

Microsporocyte (diploid)

Megasporocyte (diploid)

1 In the stamen, diploid microsporocytes undergo meiosis…

6 In the ovary, diploid megasporocytes undergo meiosis…

Diploid, 2*n*

Meiosis

Meiosis

Haploid, 1*n*

2 …to produce four haploid microspores.

7 …to produce four haploid megaspores, but only one survives.

Four microspores (haploid)

Four megaspores (haploid)

Only one survives

3 Each undergoes mitosis to produce a pollen grain with two haploid nuclei.

Mitosis

8 The surviving megaspore divides mitotically three times…

Mitosis

Haploid generative nucleus

Pollen grain

2 nuclei

4 The tube nucleus directs the growth of a pollen tube.

Haploid tube nucleus

4 nuclei

Mitosis

9 …to produce eight haploid nuclei.

8 nuclei

Pollen tube

5 The generative nucleus divides mitotically to produce two sperm cells.

Two haploid sperm cells

Tube nucleus

10 The cytoplasm divides, producing separate cells,…

Division of cytoplasm

11 …one of which becomes the egg.

Polar nuclei

Embryo sac

12 Two of the nuclei become polar nuclei…

Sperm

Polar nuclei

Egg

Egg

13 …and the other nuclei are partitioned into separate cells.

(c)

Double fertilization

14 Double fertilization takes place when the two sperm cells of a pollen grain enter the embryo sac.

Endosperm, (triploid, 3*n*)

16 The other sperm cell fuses with the binucleate cell to form triploid endosperm.

15 One sperm cell fertilizes the egg cell, producing a diploid zygote.

Embryo (diploid, 2*n*)

2.22 Sexual reproduction in flowering plants.

the base of the style is the ovary. If a pollen grain germinates, it grows a tube down the style into the ovary. The two sperm cells pass down this tube and enter the embryo sac (**Figure 2.22c**). One of the sperm cells fertilizes the egg cell, producing a diploid zygote, which develops into an embryo. The other sperm cell fuses with two nuclei enclosed in a single cell, giving rise to a 3*n* (triploid) endosperm, which stores food that will be used later by the embryonic plant. These two fertilization events are termed double fertilization.

CONCEPTS

In the stamen of a flowering plant, meiosis produces haploid microspores that divide mitotically to produce haploid sperm in a pollen grain. Within the ovary, meiosis produces four haploid megaspores, only one of which divides mitotically three times to produce eight haploid nuclei. After pollination, one sperm fertilizes the egg cell, producing a diploid zygote; the other fuses with two nuclei to form the endosperm.

✔ CONCEPT CHECK 8

Which structure is diploid?

a. Microspore

b. Egg

c. Megaspore

d. Microsporocyte

We have now examined the place of meiosis in the sexual cycle of two organisms, a typical multicellular animal and a flowering plant. These cycles are just two of the many variations found among eukaryotic organisms. Although the cellular events that produce reproductive cells in plants and animals differ in the number of cell divisions, the number of haploid gametes produced, and the relative size of the final products, the overall result is the same: meiosis gives rise to haploid, genetically variable cells that then fuse during fertilization to produce diploid progeny. **TRY PROBLEMS 33 AND 34** →

CONCEPTS SUMMARY

- A prokaryotic cell possesses a simple structure, with no nuclear envelope and usually a single, circular chromosome. A eukaryotic cell possesses a more-complex structure, with a nucleus and multiple linear chromosomes consisting of DNA complexed to histone proteins.

- Cell reproduction requires the copying of the genetic material, separation of the copies, and cell division.

- In a prokaryotic cell, the single chromosome replicates, each copy moves toward opposite sides of the cell, and the cell divides. In eukaryotic cells, reproduction is more complex than in prokaryotic cells, requiring mitosis and meiosis to ensure that a complete set of genetic information is transferred to each new cell.

- In eukaryotic cells, chromosomes are typically found in homologous pairs. Each functional chromosome consists of a centromere, telomeres, and multiple origins of replication. After a chromosome has been copied, the two copies remain attached at the centromere, forming sister chromatids.

- The cell cycle consists of the stages through which a eukaryotic cell passes between cell divisions. It consists of (1) interphase, in which the cell grows and prepares for division, and (2) the M phase, in which nuclear and cell division take place. The M phase consists of (1) mitosis, the process of nuclear division, and (2) cytokinesis, the division of the cytoplasm.

- Progression through the cell cycle is controlled at checkpoints that regulate the cell cycle by allowing or prohibiting the cell to proceed to the next stage.

- Mitosis usually results in the production of two genetically identical cells.

- Sexual reproduction produces genetically variable progeny and allows for accelerated evolution. It includes meiosis, in which haploid sex cells are produced, and fertilization, the fusion of sex cells. Meiosis includes two cell divisions. In meiosis I, crossing over takes place and homologous chromosomes separate. In meiosis II, chromatids separate.

- The usual result of meiosis is the production of four haploid cells that are genetically variable. Genetic variation in meiosis is produced by crossing over and by the random distribution of maternal and paternal chromosomes.

- Cohesin holds sister chromatids together. In metaphase of mitosis and in metaphase II of meiosis, the breakdown of cohesin allows sister chromatids to separate. In meiosis I, centromeric cohesin remains intact and keeps sister chromatids together so that homologous chromosomes, but not sister chromatids, separate in anaphase I.

- In animals, a diploid spermatogonium undergoes meiosis to produce four haploid sperm cells. A diploid oogonium undergoes meiosis to produce one large haploid ovum and one or more smaller polar bodies.

- In plants, a diploid microsporocyte in the stamen undergoes meiosis to produce four pollen grains, each with two haploid sperm cells. In the ovary, a diploid megasporocyte undergoes meiosis to produce eight haploid nuclei, one of which forms the egg. During pollination, one sperm fertilizes the egg cell and the other fuses with two haploid nuclei to form a 3*n* endosperm.

IMPORTANT TERMS

prokaryote (p. 17)
eukaryote (p. 17)
eubacteria (p. 17)
archaea (p. 17)
nucleus (p. 17)
histone (p. 17)
chromatin (p. 17)
virus (p. 18)
homologous pair (p. 19)
diploid (p. 19)
haploid (p. 19)
telomere (p. 20)
origin of replication (p. 21)
sister chromatid (p. 21)
cell cycle (p. 21)
checkpoint (p. 21)
interphase (p. 21)
M phase (p. 21)
mitosis (p. 21)
cytokinesis (p. 21)

prophase (p. 22)
prometaphase (p. 22)
metaphase (p. 22)
anaphase (p. 22)
telophase (p. 24)
meiosis (p. 26)
fertilization (p. 26)
prophase I (p. 26)
synapsis (p. 26)
bivalent (p. 27)
tetrad (p. 27)
crossing over (p. 27)
metaphase I (p. 27)
anaphase I (p. 27)
telophase I (p. 27)
interkinesis (p. 27)
prophase II (p. 27)
metaphase II (p. 27)
anaphase II (p. 27)
telophase II (p. 27)

recombination (p. 30)
cohesin (p. 31)
spermatogenesis (p. 33)
spermatogonium (p. 33)
primary spermatocyte (p. 33)
secondary spermatocyte (p. 33)
spermatid (p. 34)
oogenesis (p. 34)
oogonium (p. 34)
primary oocyte (p. 34)
secondary oocyte (p. 34)
first polar body (p. 34)
ovum (p. 34)
second polar body (p. 34)
microsporocyte (p. 35)
microspore (p. 35)
megasporocyte (p. 35)
megaspore (p. 35)

ANSWERS TO CONCEPT CHECKS

1. Eubacteria and archaea are prokaryotes. They differ from eukaryotes in possessing no nucleus, a genome that usually consists of a single, circular chromosome, and a small amount of DNA.

2. b

3. The kinetochore is the point at which spindle microtubules attach to the chromosome. If the kinetochore were missing, spindle microtubules would not attach to the chromosome, the chromosome would not be drawn into the nucleus, and the resulting cells would be missing a chromosome.

4. a

5. c

6. During anaphase I, shugoshin protects cohesin at the centromeres from the action of separase; so cohesin remains intact and the sister chromatids remain together. Subsequently, shugoshin breaks down; so centromeric cohesin is cleaved in anaphase II and the chromatids separate.

7. d

8. d

WORKED PROBLEMS

1. A student examines a thin section of an onion-root tip and records the number of cells that are in each stage of the cell cycle. She observes 94 cells in interphase, 14 cells in prophase, 3 cells in prometaphase, 3 cells in metaphase, 5 cells in anaphase, and 1 cell in telophase. If the complete cell cycle in an onion-root tip requires 22 hours, what is the average duration of each stage in the cycle? Assume that all cells are in the active cell cycle (not G_0).

• **Solution**

This problem is solved in two steps. First, we calculate the proportions of cells in each stage of the cell cycle, which correspond to the amount of time that an average cell spends in each stage. For example, if cells spend 90% of their time in interphase, then, at any given moment, 90% of the cells will be in interphase. The second step is to convert the proportions into

lengths of time, which is done by multiplying the proportions by the total time of the cell cycle (22 hours).

Step 1. Calculate the proportion of cells at each stage. The proportion of cells at each stage is equal to the number of cells found in that stage divided by the total number of cells examined:

Interphase	$^{94}/_{120} = 0.783$
Prophase	$^{14}/_{120} = 0.117$
Prometaphase	$^{3}/_{120} = 0.025$
Metaphase	$^{3}/_{120} = 0.025$
Anaphase	$^{5}/_{120} = 0.042$
Telophase	$^{1}/_{120} = 0.08$

We can check our calculations by making sure that the proportions sum to 1.0, which they do.

Step 2. Determine the average duration of each stage. To determine the average duration of each stage, multiply the proportion of cells in each stage by the time required for the entire cell cycle:

Interphase	0.783×22 hours $= 17.23$ hours
Prophase	0.117×22 hours $= 2.57$ hours
Prometaphase	0.025×22 hours $= 0.55$ hour
Metaphase	0.025×22 hours $= 0.55$ hour
Anaphase	0.042×22 hours $= 0.92$ hour
Telophase	0.008×22 hours $= 0.18$ hour

2. A cell in G_1 of interphase has 8 chromosomes. How many chromosomes and how many DNA molecules will be found per cell as this cell progresses through the following stages: G_2, metaphase of mitosis, anaphase of mitosis, after cytokinesis in mitosis, metaphase I of meiosis, metaphase II of meiosis, and after cytokinesis of meiosis II?

• Solution

Remember the rules about counting chromosomes and DNA molecules: (1) to determine the number of chromosomes, count the functional centromeres; (2) to determine the number of DNA molecules, determine whether sister chromatids exist. If sister chromatids are present, the number of DNA molecules is $2 \times$ the number of chromosomes. If the chromosomes are unreplicated (don't contain sister chromatids), the number of DNA molecules equals the number of chromosomes. Think carefully about when and how the numbers of chromosomes and DNA molecules change in the course of mitosis and meiosis.

The number of DNA molecules increases only in the S phase, when DNA replicates; the number of DNA molecules decreases only when the cell divides. Chromosome number increases only when sister chromatids separate in anaphase of mitosis and in anaphase II of meiosis (homologous chromosomes, not chromatids, separate in anaphase I of meiosis). Chromosome number, like the number of DNA molecules, is reduced only by cell division.

Let's now apply these principles to the problem. A cell in G_1 has 8 chromosomes, and sister chromatids are not present; so 8 DNA molecules are present in G_1. DNA replicates in the S phase and now each chromosome consists of two sister chromatids; so, in G_2, $2 \times 8 = 16$ DNA molecules are present per cell. However, the two copies of each DNA molecule remain attached at the centromere; so there are still only 8 chromosomes present. As the cell passes through prophase and metaphase of the cell cycle, the number of chromosomes and the number of DNA molecules remain the same; so, at metaphase, there are 16 DNA molecules and 8 chromosomes. In anaphase, the chromatids separate and each becomes an independent chromosome; at this point, the number of chromosomes increases from 8 to 16. This increase is temporary, lasting only until the cell divides in telophase or subsequent to it. The number of DNA molecules remains at 16 in anaphase. The number of DNA molecules and chromosomes per cell is reduced by cytokinesis after telophase, because the 16 chromosomes and DNA molecules are now distributed between two cells. Therefore, after cytokinesis, each cell has 8 DNA molecules and 8 chromosomes, the same numbers that were present at the beginning of the cell cycle.

Now, let's trace the numbers of DNA molecules and chromosomes through meiosis. At G_1, there are 8 chromosomes and 8 DNA molecules. The number of DNA molecules increases to 16 in the S phase, but the number of chromosomes remains at 8 (each chromosome has two chromatids). The cell therefore enters metaphase I having 16 DNA molecules and 8 chromosomes. In anaphase I of meiosis, homologous chromosomes separate, but the number of chromosomes remains at 8. After cytokinesis, the original 8 chromosomes are distributed between two cells; so the number of chromosomes per cell falls to 4 (each with two chromatids). The original 16 DNA molecules also are distributed between two cells; so the number of DNA molecules per cell is 8. There is no DNA synthesis in interkinesis, and each cell still maintains 4 chromosomes and 8 DNA molecules through metaphase II. In anaphase II, the two sister chromatids of each chromosome separate, temporarily raising the number of chromosomes per cell to 8, whereas the number of DNA molecules per cell remains at 8. After cytokinesis, the chromosomes and DNA molecules are again distributed between two cells, providing 4 chromosomes and 4 DNA molecules per cell. These results are summarized in the following table:

Stage	Number of chromosomes per cell	Number of DNA molecules per cell
G_1	8	8
G_2	8	16
Metaphase of mitosis	8	16
Anaphase of mitosis	16	16
After cytokinesis of mitosis	8	8
Metaphase I of meiosis	8	16
Metaphase II of meiosis	4	8
After cytokinesis of meiosis II	4	4

COMPREHENSION QUESTIONS

Section 2.1

*1. What are some genetic differences between prokaryotic and eukaryotic cells?

2. Why are the viruses that infect mammalian cells useful for studying the genetics of mammals?

Section 2.2

*3. List three fundamental events that must take place in cell reproduction.

4. Outline the process by which prokaryotic cells reproduce.

5. Name three essential structural elements of a functional eukaryotic chromosome and describe their functions.

*6. Sketch and identify four different types of chromosomes based on the position of the centromere.

7. List the stages of interphase and the major events that take place in each stage.

*8. List the stages of mitosis and the major events that take place in each stage.

9. Briefly describe how the chromosomes move toward the spindle poles during anaphase.

*10. What are the genetically important results of the cell cycle?

11. Why are the two cells produced by the cell cycle genetically identical?

12. What are checkpoints? List some of the important checkpoints in the cell cycle.

Section 2.3

13. What are the stages of meiosis and what major events take place in each stage?

*14. What are the major results of meiosis?

15. What two processes unique to meiosis are responsible for genetic variation? At what point in meiosis do these processes take place?

*16. List similarities and differences between mitosis and meiosis. Which differences do you think are most important and why?

17. Briefly explain why sister chromatids remain together in anaphase I but separate in anaphase II of meiosis.

18. Outline the process of spermatogenesis in animals. Outline the process of oogenesis in animals.

19. Outline the process by which male gametes are produced in plants. Outline the process of female gamete formation in plants.

APPLICATION QUESTIONS AND PROBLEMS

Introduction

*20. Answer the following questions regarding the Blind Men's Riddle, presented at the beginning of the chapter.
  a. What do the two socks of a pair represent in the cell cycle?
  b. In the riddle, each blind man buys his own pairs of socks, but the clerk places all pairs into one bag. Thus, there are two pairs of socks of each color in the bag (two black pairs, two blue pairs, two gray pairs, etc.). What do the two pairs (four socks in all) of each color represent?
  c. In the cell cycle, what is the thread that connects the two socks of a pair?
  d. In the cell cycle, what is the molecular knife that cuts the thread holding the two socks of a pair together?
  e. What in the riddle performs the same function as spindle fibers?
  f. What would happen if one man failed to grasp his sock of a particular pair and how does it relate to events in the cell cycle?

Section 2.2

21. A certain species has three pairs of chromosomes: an acrocentric pair, a metacentric pair, and a submetacentric pair. Draw a cell of this species as it would appear in metaphase of mitosis.

22. A biologist examines a series of cells and counts 160 cells in interphase, 20 cells in prophase, 6 cells in prometaphase, 2 cells in metaphase, 7 cells in anaphase, and 5 cells in telophase. If the complete cell cycle requires 24 hours, what is the average duration of the M phase in these cells? Of metaphase?

Section 2.3

*23. A cell in G_1 of interphase has 12 chromosomes. How many chromosomes and DNA molecules will be found per cell when this original cell progresses to the following stages?

  a. G_2 of interphase
  b. Metaphase I of meiosis
  c. Prophase of mitosis
  d. Anaphase I of meiosis
  e. Anaphase II of meiosis
  f. Prophase II of meiosis
  g. After cytokinesis following mitosis
  h. After cytokinesis following meiosis II

24. How are the events that take place in spermatogenesis and oogenesis similar? How are they different?

*25. All of the following cells, shown in various stages of mitosis and meiosis, come from the same rare species of plant. What is the diploid number of chromosomes in this plant? Give the names of each stage of mitosis or meiosis shown.

26. The amount of DNA per cell of a particular species is measured in cells found at various stages of meiosis, and the following amounts are obtained:

Amount of DNA per cell

_____ 3.7 pg   _____ 7.3 pg   _____ 14.6 pg

Match the amounts of DNA above with the corresponding stages of the cell cycle (*a* through *f*). You may use more than one stage for each amount of DNA.

Stage of meiosis

a. G_1
b. Prophase I
c. G_2
d. Following telophase II and cytokinesis
e. Anaphase I
f. Metaphase II

*27. How would each of the following events affect the outcome of mitosis or meiosis?

a. Mitotic cohesin fails to form early in mitosis.
b. Shugoshin is absent during meiosis.
c. Shugoshin does not break down after anaphase I of meiosis.
d. Separase is defective.

28. A cell in prophase II of meiosis contains 12 chromosomes. How many chromosomes would be present in a cell from the same organism if it were in prophase of mitosis? Prophase I of meiosis?

29. A cell has 8 chromosomes in G_1 of interphase. Draw a picture of this cell with its chromosomes at the following stages. Indicate how many DNA molecules are present at each stage.

a. Metaphase of mitosis   c. Anaphase II of meiosis
b. Anaphase of mitosis   d. Diplotene of meiosis I

*30. The fruit fly *Drosophila melanogaster* (left) has four pairs of chromosomes, whereas the house fly *Musca domestica* (right) has six pairs of chromosomes. Other things being equal, in which species would you expect to see more genetic variation among the progeny of a cross? Explain your answer.

[Herman Eisenbeiss/Photo Researchers; Materz Mali/iStockphoto.]

*31. A cell has two pairs of submetacentric chromosomes, which we will call chromosomes I_a, I_b, II_a, and II_b (chromosomes I_a and I_b are homologs, and chromosomes II_a and II_b are homologs). Allele M is located on the long arm of chromosome I_a, and allele m is located at the same position on chromosome I_b. Allele P is located on the short arm of chromosome I_a, and allele p is located at the same position on chromosome I_b. Allele R is located on chromosome II_a and allele r is located at the same position on chromosome II_b.

a. Draw these chromosomes, identifying genes M, m, P, p, R, and r, as they might appear in metaphase I of meiosis. Assume that there is no crossing over.

b. Taking into consideration the random separation of chromosomes in anaphase I, draw the chromosomes (with genes identified) present in all possible types of gametes that might result from this cell's undergoing meiosis. Assume that there is no crossing over.

32. A horse has 64 chromosomes and a donkey has 62 chromosomes. A cross between a female horse and a male donkey produces a mule, which is usually sterile. How many chromosomes does a mule have? Can you think of any reasons for the fact that most mules are sterile?

33. Normal somatic cells of horses have 64 chromosomes ($2n = 64$). How many chromosomes and DNA molecules will be present in the following types of horse cells?

[Tamara Didenko/iStockphoto.]

Cell type	Number of chromosomes	Number of DNA molecules
a. Spermatogonium	_____	_____
b. First polar body	_____	_____
c. Primary oocyte	_____	_____
d. Secondary spermatocyte	_____	_____

*34. A primary oocyte divides to give rise to a secondary oocyte and a first polar body. The secondary oocyte then divides to give rise to an ovum and a second polar body.

a. Is the genetic information found in the first polar body identical with that found in the secondary oocyte? Explain your answer.

b. Is the genetic information found in the second polar body identical with that in the ovum? Explain your answer.

CHALLENGE QUESTIONS

Section 2.3

35. From 80% to 90% of the most common human chromosome abnormalities arise because the chromosomes fail to divide properly in oogenesis. Can you think of a reason why failure of chromosome division might be more common in female gametogenesis than in male gametogenesis?

36. On average, what proportion of the genome in the following pairs of humans would be exactly the same if no crossing over took place? (For the purposes of this

question only, we will ignore the special case of the X and Y sex chromosomes and assume that all genes are located on nonsex chromosomes.)

a. Father and child

b. Mother and child

c. Two full siblings (offspring that have the same two biological parents)

d. Half siblings (offspring that have only one biological parent in common)

e. Uncle and niece

f. Grandparent and grandchild

*37. Female bees are diploid, and male bees are haploid. The haploid males produce sperm and can successfully mate with diploid females. Fertilized eggs develop into females and unfertilized eggs develop into males. How do you think the process of sperm production in male bees differs from sperm production in other animals?

3 Basic Principles of Heredity

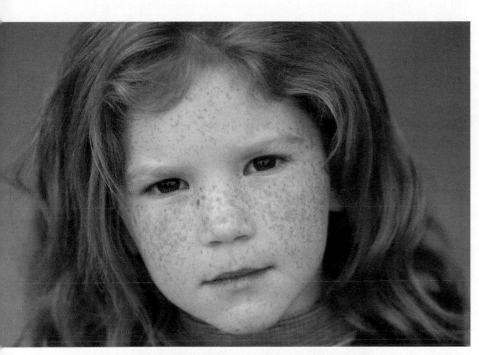

Red hair is caused by recessive mutations at the melanocortin 1 receptor gene. Reed Kaestner/Corbis.

THE GENETICS OF RED HAIR

Whether because of its exotic hue or its novelty, red hair has long been a subject of fascination for historians, poets, artists, and scientists. Historians made special note of the fact that Boudica, the Celtic queen who led a revolt against the Roman Empire, possessed a "great mass of red hair." Early Christian artists frequently portrayed Mary Magdalene as a striking red head (though there is no mention of her red hair in the Bible), and the famous artist Botticelli painted the goddess Venus as a red-haired beauty in his masterpiece *The Birth of Venus.* Queen Elizabeth I of England possessed curly red hair; during her reign, red hair was quite fashionable in London society.

The color of our hair is caused largely by a pigment called melanin that comes in two primary forms: eumelanin, which is black or brown, and pheomelanin, which is red or yellow. The color of a person's hair is determined by two factors: (1) the amount of melanin produced (more melanin causes darker hair; less melanin causes lighter hair) and (2) the relative amounts of eumelanin and pheomelanin (more eumelanin produces black or brown hair; more pheomelanin produces red or blond hair). The color of our hair is not just an academic curiosity; melanin protects against the harmful effects of sunlight, and people with red hair are usually fair skinned and particularly susceptible to skin cancer.

The inheritance of red hair has long been a subject of scientific debate. In 1909, Charles and Gertrude Davenport speculated on the inheritance of hair color in humans. Charles Davenport was an early enthusiast of genetics, particularly of inheritance in humans, and was the first director of the Biological Laboratory in Cold Spring Harbor, New York. He later became a leading proponent of eugenics, a movement—now discredited—that advocated improvement of the human race through genetics. The Davenports' study was based on family histories sent in by untrained amateurs and was methodologically flawed, but their results suggested that red hair is recessive to black and brown, meaning that a person must inherit two copies of a red-hair gene—one from each parent—to have red hair. Subsequent research contradicted this initial conclusion, suggesting that red hair is inherited instead as a dominant trait and that a person will have red hair even if possessing only a single red-hair gene. Controversy over whether red hair color is dominant or recessive or even dependent on combinations of several different genes continued for many years.

In 1993, scientists who were investigating a gene that affects the color of fur in mice discovered that the gene encodes the melanocortin-1 receptor. This receptor, when activated, increases the production of black eumelanin and decreases the production of red pheomelanin, resulting in black or brown fur. Shortly thereafter, the same melanocortin-1 receptor gene (*MC1R*) was located on human chromosome 16 and analyzed. When this gene is mutated in humans, red hair results. Most people with red hair carry two defective copies of the *MC1R* gene, which means that the trait is recessive (as originally proposed by the Davenports back in 1909). However, from 10% to 20% of red heads possess only a single mutant copy of *MC1R*, muddling the recessive interpretation of red hair (the people with a single mutant copy of the gene tend to have lighter red hair than those who harbor two mutant copies). The type and frequency of mutations at the *MC1R* gene vary widely among human populations, accounting for ethnic differences in the preponderance of red hair: among those of African and Asian descent, mutations for red hair are uncommon, whereas almost 40% of the people from the northern part of the United Kingdom carry at least one mutant copy of the gene for red hair.

Modern humans are not the only people with red hair. Analysis of DNA from ancient bones indicates that some Neanderthals also carried a mutation in the *MC1R* gene that almost certainly caused red hair, but the mutation is distinct from those seen in modern humans. This finding suggests that red hair arose independently in Neanderthals and modern humans.

This chapter is about the principles of heredity: how genes—such as the one for the melanocortin-1 receptor—are passed from generation to generation and how factors such as dominance influence that inheritance. The principles of heredity were first put forth by Gregor Mendel, and so we begin this chapter by examining Mendel's scientific achievements. We then turn to simple genetic crosses, those in which a single characteristic is examined. We will consider some techniques for predicting the outcome of genetic crosses and then turn to crosses in which two or more characteristics are examined. We will see how the principles applied to simple genetic crosses and the ratios of offspring that they produce serve as the key for understanding more-complicated crosses. The chapter ends with a discussion of statistical tests for analyzing crosses.

Throughout this chapter, a number of concepts are interwoven: Mendel's principles of segregation and independent assortment, probability, and the behavior of chromosomes. These concepts might at first appear to be unrelated, but they are actually different views of the same phenomenon, because the genes that undergo segregation and independent assortment are located on chromosomes. The principal aim of this chapter is to examine these different views and to clarify their relations.

3.1 Gregor Mendel Discovered the Basic Principles of Heredity

In 1909, when the Davenports speculated about the inheritance of red hair, the basic principles of heredity were just becoming widely known among biologists. Surprisingly, these principles had been discovered some 44 years earlier by Gregor Johann Mendel (1822–1884; **Figure 3.1**).

Mendel was born in what is now part of the Czech Republic. Although his parents were simple farmers with little money, he was able to achieve a sound education and was admitted to the Augustinian monastery in Brno in September 1843. After graduating from seminary, Mendel was ordained a priest and appointed to a teaching position in a local school. He excelled at teaching, and the abbot of the monastery recommended him for further study at

3.1 Gregor Johann Mendel, experimenting with peas, first discovered the principles of heredity. James King-Holmes/Photo Researchers.

the University of Vienna, which he attended from 1851 to 1853. There, Mendel enrolled in the newly opened Physics Institute and took courses in mathematics, chemistry, entomology, paleontology, botany, and plant physiology. It was probably there that Mendel acquired knowledge of the scientific method, which he later applied so successfully to his genetics experiments. After 2 years of study in Vienna, Mendel returned to Brno, where he taught school and began his experimental work with pea plants. He conducted breeding experiments from 1856 to 1863 and presented his results publicly at meetings of the Brno Natural Science Society in 1865. Mendel's paper from these lectures was published in 1866. In spite of widespread interest in heredity, the effect of his research on the scientific community was minimal. At the time, no one seemed to have noticed that Mendel had discovered the basic principles of inheritance.

In 1868, Mendel was elected abbot of his monastery, and increasing administrative duties brought an end to his teaching and eventually to his genetics experiments. He died at the age of 61 on January 6, 1884, unrecognized for his contribution to genetics.

The significance of Mendel's discovery was not recognized until 1900, when three botanists—Hugo de Vries, Erich von Tschermak, and Carl Correns—began independently conducting similar experiments with plants and arrived at conclusions similar to those of Mendel. Coming across Mendel's paper, they interpreted their results in accord with his principles and drew attention to his pioneering work.

Mendel's Success

Mendel's approach to the study of heredity was effective for several reasons. Foremost was his choice of experimental subject, the pea plant *Pisum sativum* (**Figure 3.2**), which offered clear advantages for genetic investigation. The plant is easy to cultivate, and Mendel had the monastery garden and greenhouse at his disposal. Compared with some other plants, peas grow relatively rapidly, completing an entire generation in a single growing season. By today's standards, one generation per year seems frightfully slow—fruit flies complete a generation in 2 weeks and bacteria in 20 minutes—but Mendel was under no pressure to publish quickly and was able to follow the inheritance of individual characteristics for several generations. Had he chosen to work on an organism with a longer generation time—horses, for example—he might never have discovered the basis of inheritance. Pea plants also produce many offspring—their seeds—which allowed Mendel to detect meaningful mathematical ratios in the traits that he observed in the progeny.

The large number of varieties of peas that were available to Mendel also was crucial, because these varieties differed in various traits and were genetically pure. Mendel was therefore able to begin with plants of variable, known genetic makeup.

Much of Mendel's success can be attributed to the seven characteristics that he chose for study (see Figure 3.2). He avoided characteristics that display a range of variation; instead, he focused his attention on those that exist in two easily differentiated forms, such as white versus gray seed coats, round versus wrinkled seeds, and inflated versus constricted pods.

Finally, Mendel was successful because he adopted an experimental approach and interpreted his results by using mathematics. Unlike many earlier investigators who just described the *results* of crosses, Mendel formulated *hypotheses* based on his initial observations and then conducted additional crosses to test his hypotheses. He kept careful records of the numbers of progeny possessing each type of trait and computed ratios of the different types. He was adept at seeing patterns in detail and was patient and thorough, conducting his experiments for 10 years before attempting to write up his results. TRY PROBLEM 13 →

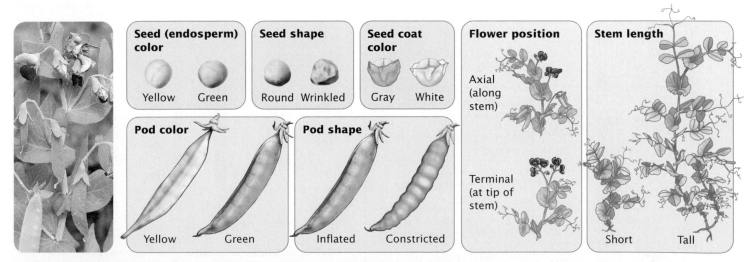

3.2 Mendel used the pea plant *Pisum sativum* in his studies of heredity. He examined seven characteristics that appeared in the seeds and in plants grown from the seeds. [Photograph by Wally Eberhart/Visuals Unlimited.]

CONCEPTS

Gregor Mendel put forth the basic principles of inheritance, publishing his findings in 1866. The significance of his work did not become widely known until 1900.

✔ CONCEPT CHECK 1

Which of the following factors did not contribute to Mendel's success in his study of heredity?

a. His use of the pea plant

b. His study of plant chromosomes

c. His adoption of an experimental approach

d. His use of mathematics

Genetic Terminology

Before we examine Mendel's crosses and the conclusions that he drew from them, a review of some terms commonly used in genetics will be helpful (**Table 3.1**). The term *gene* is a word that Mendel never knew. It was not coined until 1909, when Danish geneticist Wilhelm Johannsen first used it. The definition of a gene varies with the context of its use, and so its definition will change as we explore different aspects of heredity. For our present use in the context of genetic crosses, we will define a **gene** as an inherited factor that determines a characteristic.

Table 3.1	Summary of important genetic terms
Term	**Definition**
Gene	A genetic factor (region of DNA) that helps determine a characteristic
Allele	One of two or more alternative forms of a gene
Locus	Specific place on a chromosome occupied by an allele
Genotype	Set of alleles possessed by an individual organism
Heterozygote	An individual organism possessing two different alleles at a locus
Homozygote	An individual organism possessing two of the same alleles at a locus
Phenotype or trait	The appearance or manifestation of a character
Character or characteristic	An attribute or feature

Genes frequently come in different versions called **alleles** (**Figure 3.3**). In Mendel's crosses, seed shape was determined by a gene that exists as two different alleles: one allele encodes round seeds and the other encodes wrinkled seeds. All alleles for any particular gene will be found at a specific place on a chromosome called the **locus** for that gene. (The plural of locus is loci; it's bad form in genetics—and incorrect—to speak of locuses.) Thus, there is a specific place—a locus—on a chromosome in pea plants where the shape of seeds is determined. This locus might be occupied by an allele for round seeds or one for wrinkled seeds. We will use the term *allele* when referring to a specific version of a gene; we will use the term *gene* to refer more generally to any allele at a locus.

The **genotype** is the set of alleles that an individual organism possesses. A diploid organism with a genotype consisting of two identical alleles is **homozygous** for that locus. One that has a genotype consisting of two different alleles is **heterozygous** for the locus.

Another important term is **phenotype**, which is the manifestation or appearance of a characteristic. A phenotype can refer to any type of characteristic—physical, physiological, biochemical, or behavioral. Thus, the condition of having round seeds is a phenotype, a body weight of 50 kilograms (50 kg) is a phenotype, and having sickle-cell anemia is a phenotype. In this book, the term *characteristic* or *character* refers to a general feature such as eye color; the term *trait* or *phenotype* refers to specific manifestations of that feature, such as blue or brown eyes.

A given phenotype arises from a genotype that develops within a particular environment. The genotype determines the potential for development; it sets certain limits, or boundaries, on that development. How the phenotype develops within those limits is determined by the effects of other genes and of environmental factors, and the balance between these effects varies from character to character. For some characters, the differences between phenotypes are determined largely by differences in genotype. In Mendel's peas, for example, the genotype, not the environment, largely

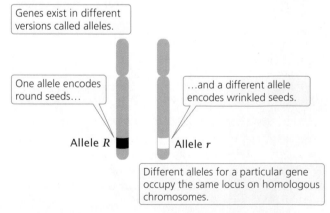

3.3 At each locus, a diploid organism possesses two alleles located on different homologous chromosomes. The alleles identified here refer to traits studied by Mendel.

determined the shape of the seeds. For other characters, environmental differences are more important. The height reached by an oak tree at maturity is a phenotype that is strongly influenced by environmental factors, such as the availability of water, sunlight, and nutrients. Nevertheless, the tree's genotype still imposes some limits on its height: an oak tree will never grow to be 300 meters (300 m) tall no matter how much sunlight, water, and fertilizer are provided. Thus, even the height of an oak tree is determined to some degree by genes. For many characteristics, both genes and environment are important in determining phenotypic differences.

An obvious but important concept is that only the alleles of the genotype are inherited. Although the phenotype is determined, at least to some extent, by genotype, organisms do not transmit their phenotypes to the next generation. The distinction between genotype and phenotype is one of the most important principles of modern genetics. The next section describes Mendel's careful observation of phenotypes through several generations of breeding experiments. These experiments allowed him to deduce not only the genotypes of the individual plants, but also the rules governing their inheritance.

CONCEPTS

Each phenotype results from a genotype developing within a specific environment. The alleles of the genotype, not the phenotype, are inherited.

✔ CONCEPT CHECK 2

Distinguish among the following terms: locus, allele, genotype, and phenotype.

3.2 Monohybrid Crosses Reveal the Principle of Segregation and the Concept of Dominance

Mendel started with 34 varieties of peas and spent 2 years selecting those varieties that he would use in his experiments. He verified that each variety was pure-breeding (homozygous for each of the traits that he chose to study) by growing the plants for two generations and confirming that all offspring were the same as their parents. He then carried out a number of crosses between the different varieties. Although peas are normally self-fertilizing (each plant crosses with itself), Mendel conducted crosses between different plants by opening the buds before the anthers (male sex organs) were fully developed, removing the anthers, and then dusting the stigma (female sex organs) with pollen from a different plant's anthers (**Figure 3.4**).

Mendel began by studying **monohybrid crosses**—those between parents that differed in a single characteristic. In one experiment, Mendel crossed a pure-breeding (homozygous) pea plant for round seeds with one that was pure-breeding for wrinkled seeds (see Figure 3.4). This first generation of a cross is the **P (parental) generation**.

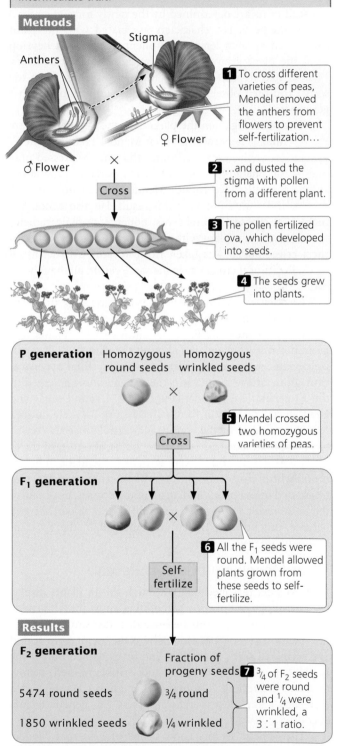

3.4 Mendel conducted monohybrid crosses.

After crossing the two varieties in the P generation, Mendel observed the offspring that resulted from the cross. In regard to seed characteristics, such as seed shape, the phenotype develops as soon as the seed matures, because the seed traits are determined by the newly formed embryo within the seed. For characters associated with the plant itself, such as stem length, the phenotype doesn't develop until the plant grows from the seed; for these characters, Mendel had to wait until the following spring, plant the seeds, and then observe the phenotypes on the plants that germinated.

The offspring from the parents in the P generation are the **F_1 (filial 1) generation**. When Mendel examined the F_1 generation of this cross, he found that they expressed only one of the phenotypes present in the parental generation: all the F_1 seeds were round. Mendel carried out 60 such crosses and always obtained this result. He also conducted **reciprocal crosses**: in one cross, pollen (the male gamete) was taken from a plant with round seeds and, in its reciprocal cross, pollen was taken from a plant with wrinkled seeds. Reciprocal crosses gave the same result: all the F_1 were round.

Mendel wasn't content with examining only the seeds arising from these monohybrid crosses. The following spring, he planted the F_1 seeds, cultivated the plants that germinated from them, and allowed the plants to self-fertilize, producing a second generation—the **F_2 (filial 2) generation**. Both of the traits from the P generation emerged in the F_2 generation; Mendel counted 5474 round seeds and 1850 wrinkled seeds in the F_2 (see Figure 3.4). He noticed that the number of the round and wrinkled seeds constituted approximately a 3 to 1 ratio; that is, about $^3/_4$ of the F_2 seeds were round and $^1/_4$ were wrinkled. Mendel conducted monohybrid crosses for all seven of the characteristics that he studied in pea plants and, in all of the crosses, he obtained the same result: all of the F_1 resembled only one of the two parents, but both parental traits emerged in the F_2 in an approximate ratio of 3 : 1.

What Monohybrid Crosses Reveal

First, Mendel reasoned that, although the F_1 plants display the phenotype of only one parent, they must inherit genetic factors from both parents because they transmit both phenotypes to the F_2 generation. The presence of both round and wrinkled seeds in the F_2 could be explained only if the F_1 plants possessed both round and wrinkled genetic factors that they had inherited from the P generation. He concluded that each plant must therefore possess two genetic factors encoding a character.

The genetic factors (now called alleles) that Mendel discovered are, by convention, designated with letters; the allele for round seeds is usually represented by R, and the allele for wrinkled seeds by r. The plants in the P generation of Mendel's cross possessed two identical alleles: RR in the round-seeded parent and rr in the wrinkled-seeded parent (**Figure 3.5a**).

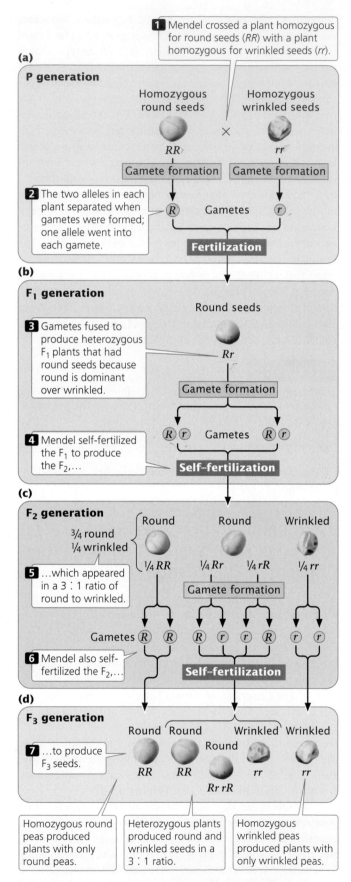

3.5 Mendel's monohybrid crosses revealed the principle of segregation and the concept of dominance.

The second conclusion that Mendel drew from his monohybrid crosses was that the two alleles in each plant separate when gametes are formed, and one allele goes into each gamete. When two gametes (one from each parent) fuse to produce a zygote, the allele from the male parent unites with the allele from the female parent to produce the genotype of the offspring. Thus, Mendel's F_1 plants inherited an R allele from the round-seeded plant and an r allele from the wrinkled-seeded plant (**Figure 3.5b**). However, only the trait encoded by the round allele (R) was *observed* in the F_1: all the F_1 progeny had round seeds. Those traits that appeared unchanged in the F_1 heterozygous offspring Mendel called **dominant**, and those traits that disappeared in the F_1 heterozygous offspring he called **recessive**. When dominant and recessive alleles are present together, the recessive allele is masked, or suppressed. The concept of dominance was the third important conclusion that Mendel derived from his monohybrid crosses.

Mendel's fourth conclusion was that the two alleles of an individual plant separate with equal probability into the gametes. When plants of the F_1 (with genotype Rr) produced gametes, half of the gametes received the R allele for round seeds and half received the r allele for wrinkled seeds. The gametes then paired randomly to produce the following genotypes in equal proportions among the F_2: RR, Rr, rR, rr (**Figure 3.5c**). Because round (R) is dominant over wrinkled (r), there were three round progeny in the F_2 (RR, Rr, rR) for every one wrinkled progeny (rr) in the F_2. This 3 : 1 ratio of round to wrinkled progeny that Mendel observed in the F_2 could be obtained only if the two alleles of a genotype separated into the gametes with equal probability.

The conclusions that Mendel developed about inheritance from his monohybrid crosses have been further developed and formalized into the principle of segregation and the concept of dominance. The **principle of segregation** (Mendel's first law, see **Table 3.2**) states that each individual diploid organism possesses two alleles for any particular characteristic. These two alleles segregate (separate) when gametes are formed, and one allele goes into each gamete. Furthermore, the two alleles segregate into gametes in equal proportions. The **concept of dominance** states that, when two different alleles are present in a genotype, only the trait encoded by one of them—the "dominant" allele—is observed in the phenotype.

Mendel confirmed these principles by allowing his F_2 plants to self-fertilize and produce an F_3 generation. He found that the plants grown from the wrinkled seeds—those displaying the recessive trait (rr)—produced an F_3 in which all plants produced wrinkled seeds. Because his wrinkled-seeded plants were homozygous for wrinkled alleles (rr), only wrinkled alleles could be passed on to their progeny (**Figure 3.5d**).

The plants grown from round seeds—the dominant trait—fell into two types (see Figure 3.5c). On self-fertilization, about $\frac{2}{3}$ of these plants produced both round and wrinkled seeds in the F_3 generation. These plants were heterozygous (Rr); so they produced $\frac{1}{4}$ RR (round), $\frac{1}{2}$ Rr

Table 3.2	Comparison of the principles of segregation and independent assortment		
Principle	Observation		Stage of Meiosis
Segregation (Mendel's first law)	1. Each individual organism possesses two alleles encoding a trait.		Before meiosis
	2. Alleles separate when gamets are formed.		Anaphase I
	3. Alleles separate in equal proportions.		Anaphase I
Independent assortment (Mendel's second law)	Alleles at different loci separate independently.		Anaphase I

(round), and $\frac{1}{4}$ rr (wrinkled) seeds, giving a 3 : 1 ratio of round to wrinkled in the F_3. About $\frac{1}{3}$ of the plants grown from round seeds were of the second type; they produced only the round-seeded trait in the F_3. These plants were homozygous for the round allele (RR) and could thus produce only round offspring in the F_3 generation. Mendel planted the seeds obtained in the F_3 and carried these plants through three more rounds of self-fertilization. In each generation, $\frac{2}{3}$ of the round-seeded plants produced round and wrinkled offspring, whereas $\frac{1}{3}$ produced only round offspring. These results are entirely consistent with the principle of segregation.

CONCEPTS

The principle of segregation states that each individual organism possesses two alleles that can encode a characteristic. These alleles segregate when gametes are formed, and one allele goes into each gamete. The concept of dominance states that, when the two alleles of a genotype are different, only the trait encoded by one of them—the "dominant" allele—is observed.

✔ CONCEPT CHECK 3

How did Mendel know that each of his pea plants carried two alleles encoding a characteristic?

CONNECTING CONCEPTS

Relating Genetic Crosses to Meiosis

We have now seen how the results of monohybrid crosses are explained by Mendel's principle of segregation. Many students find that they enjoy working genetic crosses but are frustrated by the abstract nature of the symbols. Perhaps you feel the same at this point. You may be asking, "What do these symbols really represent?

(a)

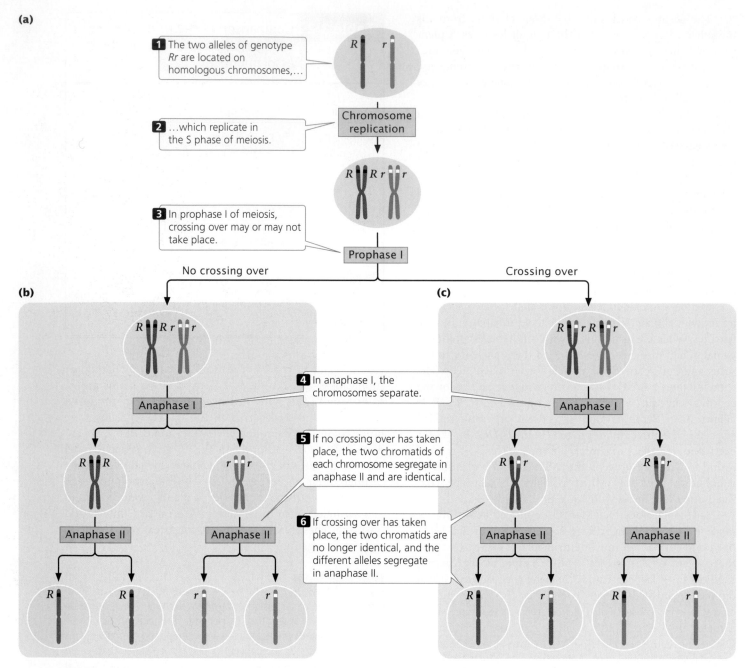

1 The two alleles of genotype *Rr* are located on homologous chromosomes,...

2 ...which replicate in the S phase of meiosis.

Chromosome replication

3 In prophase I of meiosis, crossing over may or may not take place.

Prophase I

No crossing over

Crossing over

(b)

(c)

4 In anaphase I, the chromosomes separate.

Anaphase I

Anaphase I

5 If no crossing over has taken place, the two chromatids of each chromosome segregate in anaphase II and are identical.

Anaphase II

Anaphase II

6 If crossing over has taken place, the two chromatids are no longer identical, and the different alleles segregate in anaphase II.

Anaphase II

Anaphase II

3.6 Segregation results from the separation of homologous chromosomes in meiosis.

What does the genotype *RR* mean in regard to the biology of the organism?" The answers to these questions lie in relating the abstract symbols of crosses to the structure and behavior of chromosomes, the repositories of genetic information (see Chapter 2).

In 1900, when Mendel's work was rediscovered and biologists began to apply his principles of heredity, the relation between genes and chromosomes was still unclear. The theory that genes are located on chromosomes (the **chromosome theory of heredity**) was developed in the early 1900s by Walter Sutton, then a graduate student at Columbia University. Through the careful study of meiosis in insects, Sutton documented the fact that each homologous pair of chromosomes consists of one maternal chromosome and one paternal chromosome. Showing that these pairs segregate

independently into gametes in meiosis, he concluded that this process is the biological basis for Mendel's principles of heredity. German cytologist and embryologist Theodor Boveri came to similar conclusions at about the same time.

The symbols used in genetic crosses, such as *R* and *r*, are just shorthand notations for particular sequences of DNA in the chromosomes that encode particular phenotypes. The two alleles of a genotype are found on different but homologous chromosomes. One chromosome of each homologous pair is inherited from the mother and the other is inherited from the father. In the S phase of meiotic interphase, each chromosome replicates, producing two copies of each allele, one on each chromatid (**Figure 3.6a**). The homologous chromosomes segregate in anaphase I, thereby separating the two different alleles (**Figure 3.6b** and **c**). This chromosome segregation is the basis of the principle of segregation. In anaphase II of meiosis, the two chromatids of each replicated chromosome separate; so each gamete resulting from meiosis carries only a single allele at each locus, as Mendel's principle of segregation predicts.

If crossing over has taken place in prophase I of meiosis, then the two chromatids of each replicated chromosome are no longer identical, and the segregation of different alleles takes place at anaphase I and anaphase II (see Figure 3.6c). However, Mendel didn't know anything about chromosomes; he formulated his principles of heredity entirely on the basis of the results of the crosses that he carried out. Nevertheless, we should not forget that these principles work because they are based on the behavior of actual chromosomes in meiosis. TRY PROBLEM 28 →

Predicting the Outcomes of Genetic Crosses

One of Mendel's goals in conducting his experiments on pea plants was to develop a way to predict the outcome of crosses between plants with different phenotypes. In this section, you will first learn a simple, shorthand method for predicting outcomes of genetic crosses (the Punnett square), and then you will learn how to use probability to predict the results of crosses.

The Punnett square   The Punnett square was developed by English geneticist Reginald C. Punnett in 1917. To illustrate the Punnett square, let's examine another cross carried out by Mendel. By crossing two varieties of peas that differed in height, Mendel established that tall (T) was dominant over short (t). He tested his theory concerning the inheritance of dominant traits by crossing an F_1 tall plant that was heterozygous (Tt) with the short homozygous parental variety (tt). This type of cross, between an F_1 genotype and either of the parental genotypes, is called a **backcross**.

To predict the types of offspring that result from this backcross, we first determine which gametes will be produced by each parent (**Figure 3.7a**). The principle of segregation tells us that the two alleles in each parent separate, and one allele passes to each gamete. All gametes from the homozygous tt short plant will receive a single short (t) allele. The tall plant in this cross is heterozygous (Tt); so 50% of its gametes will receive a tall allele (T) and the other 50% will receive a short allele (t).

A **Punnett square** is constructed by drawing a grid, putting the gametes produced by one parent along the upper edge and the gametes produced by the other parent down the left side (**Figure 3.7b**). Each cell (a block within the Punnett square) contains an allele from each of the corresponding gametes, generating the genotype of the progeny produced by fusion of those gametes. In the upper left-hand cell of the Punnett square in Figure 3.7b, a gamete containing T from the tall plant unites with a gamete containing t from the short plant, giving the genotype of the progeny (Tt). It is useful to write the phenotype expressed by each

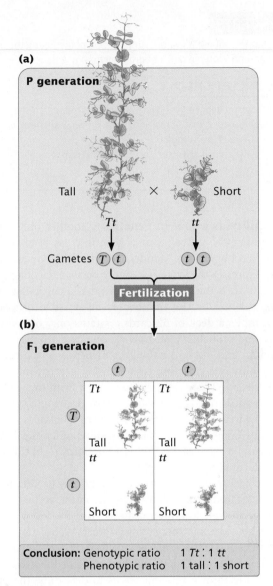

3.7 The Punnett square can be used to determine the results of a genetic cross.

genotype; here the progeny will be tall, because the tall allele is dominant over the short allele. This process is repeated for all the cells in the Punnett square.

By simply counting, we can determine the types of progeny produced and their ratios. In Figure 3.7b, two cells contain tall (Tt) progeny and two cells contain short (tt) progeny; so the genotypic ratio expected for this cross is 2 Tt to 2 tt (a 1 : 1 ratio). Another way to express this result is to say that we expect $\frac{1}{2}$ of the progeny to have genotype Tt (and phenotype tall) and $\frac{1}{2}$ of the progeny to have genotype tt (and phenotype short). In this cross, the genotypic ratio and the phenotypic ratio are the same, but this outcome need not be the case. Try completing a Punnett square for the cross in which the F_1 round-seeded plants in Figure 3.5 undergo self-fertilization (you should obtain a phenotypic ratio of 3 round to 1 wrinkled and a genotypic ratio of 1 RR to 2 Rr to 1 rr).

Probability as a tool in genetics   Another method for determining the outcome of a genetic cross is to use the rules of probability, as Mendel did with his crosses. **Probability** expresses the likelihood of the occurrence of a particular event. It is the number of times that a particular event occurs, divided by the number of all possible outcomes. For example, a deck of 52 cards contains only one king of hearts. The probability of drawing one card from the deck at random and obtaining the king of hearts is $^1/_{52}$, because there is only one card that is the king of hearts (one event) and there are 52 cards that can be drawn from the deck (52 possible outcomes). The probability of drawing a card and obtaining an ace is $^4/_{52}$, because there are four cards that are aces (four events) and 52 cards (possible outcomes). Probability can be expressed either as a fraction ($^4/_{52}$ in this case) or as a decimal number (0.077 in this case).

The probability of a particular event may be determined by knowing something about *how* the event occurs or *how often* it occurs. We know, for example, that the probability of rolling a six-sided die and getting a four is $^1/_6$, because the die has six sides and any one side is equally likely to end up on top. So, in this case, understanding the nature of the event—the shape of the thrown die—allows us to determine the probability. In other cases, we determine the probability of an event by making a large number of observations. When a weather forecaster says that there is a 40% chance of rain on a particular day, this probability was obtained by observing a large number of days with similar atmospheric conditions and finding that it rains on 40% of those days. In this case, the probability has been determined empirically (by observation).

The multiplication rule   Two rules of probability are useful for predicting the ratios of offspring produced in genetic crosses. The first is the **multiplication rule**, which states that the probability of two or more independent events occurring together is calculated by multiplying their independent probabilities.

To illustrate the use of the multiplication rule, let's again consider the roll of a die. The probability of rolling one die and obtaining a four is $^1/_6$. To calculate the probability of rolling a die twice and obtaining 2 fours, we can apply the multiplication rule. The probability of obtaining a four on the

(a) The multiplication rule

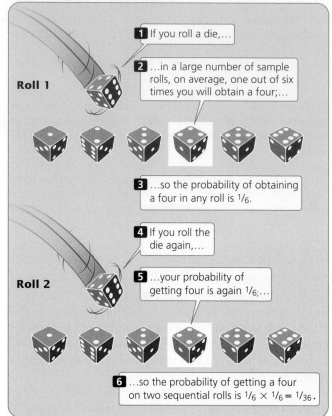

(b) The addition rule

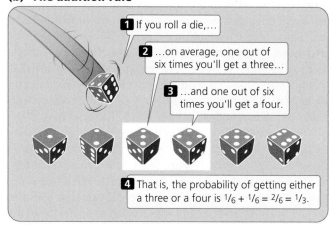

3.8 The multiplication and addition rules can be used to determine the probability of combinations of events.

first roll is $^1/_6$ and the probability of obtaining a four on the second roll is $^1/_6$; so the probability of rolling a four on both is $^1/_6 \times ^1/_6 = ^1/_{36}$ (**Figure 3.8a**). The key indicator for applying the multiplication rule is the word *and*; in the example just considered, we wanted to know the probability of obtaining a four on the first roll *and* a four on the second roll.

For the multiplication rule to be valid, the events whose joint probability is being calculated must be independent—the outcome of one event must not influence the outcome of the other. For example, the number that comes up on one roll of the die has no influence on the number that comes up on the other roll; so these events are independent. However, if we wanted to know the probability of being hit on the head with a hammer and going to the hospital on the same day, we could not simply multiply the probability of being hit on the head with a hammer by the probability of going to the hospital. The multiplication rule cannot be applied here, because the two events are not independent—being hit on the head with a hammer certainly influences the probability of going to the hospital.

The addition rule   The second rule of probability frequently used in genetics is the **addition rule**, which states that the probability of any one of two or more mutually exclusive events is calculated by adding the probabilities of these events. Let's look at this rule in concrete terms. To obtain the probability of throwing a die once and rolling *either* a three *or* a four, we would use the addition rule, adding the probability of obtaining a three ($^1/_6$) to the probability of obtaining a four (again, $^1/_6$), or $^1/_6 + ^1/_6 = ^2/_6 = ^1/_3$ (**Figure 3.8b**). The key indicators for applying the addition rule are the words *either* and *or*.

For the addition rule to be valid, the events whose probability is being calculated must be mutually exclusive, meaning that one event excludes the possibility of the occurrence of the other event. For example, you cannot throw a single die just once and obtain both a three and a four, because only one side of the die can be on top. These events are mutually exclusive.

CONCEPTS

The multiplication rule states that the probability of two or more independent events occurring together is calculated by multiplying their independent probabilities. The addition rule states that the probability that any one of two or more mutually exclusive events occurring is calculated by adding their probabilities.

✔ CONCEPT CHECK 5

If the probability of being blood-type A is $^1/_8$ and the probability of blood-type O is $^1/_2$, what is the probability of being either blood-type A or blood-type O?

a. $^5/_8$ c. $^1/_8$

b. $^1/_2$ d. $^1/_{16}$

The application of probability to genetic crosses
The multiplication and addition rules of probability can be used in place of the Punnett square to predict the ratios of progeny expected from a genetic cross. Let's first consider a cross between two pea plants heterozygous for the locus that determines height, $Tt \times Tt$. Half of the gametes produced by each plant have a T allele, and the other half have a t allele; so the probability for each type of gamete is $^1/_2$.

The gametes from the two parents can combine in four different ways to produce offspring. Using the multiplication rule, we can determine the probability of each possible type. To calculate the probability of obtaining TT progeny, for example, we multiply the probability of receiving a T allele from the first parent ($^1/_2$) times the probability of receiving a T allele from the second parent ($^1/_2$). The multiplication rule should be used here because we need the probability of receiving a T allele from the first parent *and* a T allele from the second parent—two independent events. The four types of progeny from this cross and their associated probabilities are:

TT	(T gamete and T gamete)	$^1/_2 \times ^1/_2 = ^1/_4$	tall
Tt	(T gamete and t gamete)	$^1/_2 \times ^1/_2 = ^1/_4$	tall
tT	(t gamete and T gamete)	$^1/_2 \times ^1/_2 = ^1/_4$	tall
tt	(t gamete and t gamete)	$^1/_2 \times ^1/_2 = ^1/_4$	short

Notice that there are two ways for heterozygous progeny to be produced: a heterozygote can either receive a T allele from the first parent and a t allele from the second or receive a t allele from the first parent and a T allele from the second.

After determining the probabilities of obtaining each type of progeny, we can use the addition rule to determine the overall phenotypic ratios. Because of dominance, a tall plant can have genotype TT, Tt, or tT; so, using the addition rule, we find the probability of tall progeny to be $^1/_4 + ^1/_4 + ^1/_4 = ^3/_4$. Because only one genotype encodes short (tt), the probability of short progeny is simply $^1/_4$.

Two methods have now been introduced to solve genetic crosses: the Punnett square and the probability method. At this point, you may be asking, "Why bother with probability rules and calculations? The Punnett square is easier to understand and just as quick." For simple monohybrid crosses, the Punnett square is simpler than the probability method and is just as easy to use. However, for tackling more-complex crosses concerning genes at two or more loci, the probability method is both clearer and quicker than the Punnett square.

The binomial expansion and probability   When probability is used, it is important to recognize that there may be several different ways in which a set of events can occur. Consider two parents who are both heterozygous for albinism, a recessive condition in humans that causes reduced pigmentation in the skin, hair, and eyes (**Figure 3.9**; see also the introduction to Chapter 1). When two parents heterozygous for albinism mate ($Aa \times Aa$), the probability of their having a child with albinism (aa) is $^1/_4$ and the probability of having a child with normal pigmentation (AA or Aa) is $^3/_4$. Suppose we want to know the probability of this couple having three children, all three with albinism. In this case, there is only one way in which they can have three children with albinism: their first child has albinism *and* their second child has albinism *and* their third child has albinism. Here, we simply apply the multiplication rule: $^1/_4 \times ^1/_4 \times ^1/_4 = ^1/_{64}$.

3.9 Albinism in human beings is usually inherited as a recessive trait. [Richard Dranitzke/SS/Photo Researchers.]

Suppose we now ask, What is the probability of this couple having three children, one with albinism and two with normal pigmentation? This situation is more complicated. The first child might have albinism, whereas the second and third are unaffected; the probability of this sequence of events is $\frac{1}{4} \times \frac{3}{4} \times \frac{3}{4} = \frac{9}{64}$. Alternatively, the first and third child might have normal pigmentation, whereas the second has albinism; the probability of this sequence is $\frac{3}{4} \times \frac{1}{4} \times \frac{3}{4} = \frac{9}{64}$. Finally, the first two children might have normal pigmentation and the third albinism; the probability of this sequence is $\frac{3}{4} \times \frac{3}{4} \times \frac{1}{4} = \frac{9}{64}$. Because *either* the first sequence *or* the second sequence *or* the third sequence produces one child with albinism and two with normal pigmentation, we apply the addition rule and add the probabilities: $\frac{9}{64} + \frac{9}{64} + \frac{9}{64} = \frac{27}{64}$.

If we want to know the probability of this couple having five children, two with albinism and three with normal pigmentation, figuring out *all* the different combinations of children and their probabilities becomes more difficult. This task is made easier if we apply the binomial expansion.

The binomial takes the form $(p + q)^n$, where p equals the probability of one event, q equals the probability of the alternative event, and n equals the number of times the event occurs. For figuring the probability of two out of five children with albinism:

p = the probability of a child having albinism ($\frac{1}{4}$)

q = the probability of a child having normal pigmentation ($\frac{3}{4}$)

The binomial for this situation is $(p + q)^5$ because there are five children in the family ($n = 5$). The expansion is:

$$(p + q)^5 = p^5 + 5p^4q + 10p^3q^2 + 10p^2q^3 + 5pq^4 + q^5$$

Each of the terms in the expansion provides the probability for one particular combination of traits in the children. The first term in the expansion (p^5) equals the probability of having five children all with albinism, because p is the probability of albinism. The second term ($5p^4q$) equals the probability of having four children with albinism and one with normal pigmentation, the third term ($10p^3q^2$) equals the probability of having three children with albinism and two with normal pigmentation, and so forth.

To obtain the probability of any combination of events, we insert the values of p and q; so the probability of having two out of five children with albinism is:

$$10p^2q^3 = 10(\tfrac{1}{4})^2(\tfrac{3}{4})^3 = \tfrac{270}{1024} = 0.26$$

We could easily figure out the probability of any desired combination of albinism and pigmentation among five children by using the other terms in the expansion.

How did we expand the binomial in this example? In general, the expansion of any binomial $(p + q)^n$ consists of a series of $n + 1$ terms. In the preceding example, $n = 5$; so there are $5 + 1 = 6$ terms: p^5, $5p^4q$, $10p^3q^2$, $10p^2q^3$, $5pq^4$, and q^5. To write out the terms, first figure out their exponents. The exponent of p in the first term always begins with the power to which the binomial is raised, or n. In our example, n equals 5, so our first term is p^5. The exponent of p decreases by one in each successive term; so the exponent of p is 4 in the second term (p^4), 3 in the third term (p^3), and so forth. The exponent of q is 0 (no q) in the first term and increases by 1 in each successive term, increasing from 0 to 5 in our example.

Next, determine the coefficient of each term. The coefficient of the first term is always 1; so, in our example, the first term is $1p^5$, or just p^5. The coefficient of the second term is always the same as the power to which the binomial is raised; in our example, this coefficient is 5 and the term is $5p^4q$. For the coefficient of the third term, look back at the preceding term; multiply the coefficient of the preceding term (5 in our example) by the exponent of p in that term (4) and then divide by the number of that term (second term, or 2). So the coefficient of the third term in our example is $(5 \times 4)/2 = 20/2 = 10$ and the term is $10p^3q^2$. Follow this procedure for each successive term.

Another way to determine the probability of any particular combination of events isn to use the following formula:

$$P = \frac{n!}{s!t!}p^sq^t$$

where P equals the overall probability of event X with probability p occurring s times and event Y with probability q occurring t times. For our albinism example, event X would be the occurrence of a child with albinism ($\frac{1}{4}$) and event Y would be the occurrence of a child with normal pigmentation ($\frac{3}{4}$); s would equal the number of children with albinism (2) and t would equal the number of children with normal pigmentation (3). The ! symbol stands for factorial, and it means the product of all the integers from n to 1. In this example, $n = 5$; so $n! = 5 \times 4 \times 3 \times 2 \times 1$. Applying this formula to obtain the probability of two out of five children having albinism, we obtain:

$$P = \frac{5!}{2!3!}(^1/_4)^2(^3/_4)^3$$

$$= \frac{5 \times 4 \times 3 \times 2 \times 1}{2 \times 1 \times 3 \times 2 \times 1}(^1/_4)^2(^3/_4)^3 = 0.26$$

This value is the same as that obtained with the binomial expansion. TRY PROBLEMS 23, 24, AND 25 →

The Testcross

A useful tool for analyzing genetic crosses is the **testcross**, in which one individual of unknown genotype is crossed with another individual with a homozygous recessive genotype for the trait in question. Figure 3.7 illustrates a testcross (in this case, it is also a backcross). A testcross tests, or reveals, the genotype of the first individual.

Suppose you were given a tall pea plant with no information about its parents. Because tallness is a dominant trait in peas, your plant could be either homozygous (TT) or heterozygous (Tt), but you would not know which. You could determine its genotype by performing a testcross. If the plant were homozygous (TT), a testcross would produce all tall progeny ($TT \times tt \rightarrow$ all Tt); if the plant were heterozygous (Tt), half of the progeny would be tall and half would be short ($Tt \times tt \rightarrow {}^1/_2\ Tt$ and ${}^1/_2\ tt$). When a testcross is performed, any recessive allele in the unknown genotype is expressed in the progeny, because it will be paired with a recessive allele from the homozygous recessive parent. TRY PROBLEMS 17 AND 19 →

CONCEPTS

The binomial expansion can be used to determine the probability of a particular set of events. A testcross is a cross between an individual with an unknown genotype and one with a homozygous recessive genotype. The outcome of the testcross can reveal the unknown genotype.

Genetic Symbols

As we have seen, genetic crosses are usually depicted with the use of symbols to designate the different alleles. Lowercase letters are traditionally used to designate recessive alleles, and uppercase letters are for dominant alleles. Two or three letters may be used for a single allele: the recessive allele for heart-shaped leaves in cucumbers is designated hl, and the recessive allele for abnormal sperm-head shape in mice is designated azh.

The common allele for a character—called the **wild type** because it is the allele usually found in the wild—is often symbolized by one or more letters and a plus sign ($+$). The letter or letters chosen are usually based on the mutant (unusual) phenotype. For example, the recessive allele for yellow eyes in the Oriental fruit fly is represented by ye, whereas the allele for wild-type eye color is represented by ye^+. At times, the letters for the wild-type allele are dropped and the allele is represented simply by a plus sign. In another way of distinguishing alleles, the first letter is lowercase if the mutant phenotype is recessive and it is uppercase if the mutant phenotype is dominant: for example, narrow leaflet (ln) in soybeans is recessive to broad leaflet (Ln). Superscripts and subscripts are sometimes added to distinguish between genes: Lfr_1 and Lfr_2 represent dominant mutant alleles at different loci that produce lacerate leaf margins in opium poppies; El^R represents an allele in goats that restricts the length of the ears.

A slash may be used to distinguish alleles present in an individual genotype. For example, the genotype of a goat that is heterozygous for restricted ears might be written El^+/El^R or simply $+/El^R$. If genotypes at more than one locus are presented together, a space separates the genotypes. For example, a goat heterozygous for a pair of alleles that produces restricted ears and heterozygous for another pair of alleles that produces goiter can be designated by $El^+/El^R\ G/g$.

CONNECTING CONCEPTS

Ratios in Simple Crosses

Now that we have had some experience with genetic crosses, let's review the ratios that appear in the progeny of simple crosses, in which a single locus is under consideration and one of the alleles is dominant over the other. Understanding these ratios and the parental genotypes that produce them will allow you to work simple genetic crosses quickly, without resorting to the Punnett square. Later, we will use these ratios to work more-complicated crosses entailing several loci.

There are only three phenotypic ratios to understand (**Table 3.3**). The 3 : 1 ratio arises in a simple genetic cross when both of the

Table 3.3	Phenotypic ratios for simple genetic crosses (crosses for a single locus) with dominance	
Phenotypic Ratio	Genotypes of Parents	Genotypes of Progeny
3 : 1	$Aa \times Aa$	$^3/_4\ A\_ : {}^1/_4\ aa$
1 : 1	$Aa \times aa$	$^1/_2\ Aa : {}^1/_2\ aa$
Uniform progeny	$AA \times AA$	All AA
	$aa \times aa$	All aa
	$AA \times aa$	All Aa
	$AA \times Aa$	All $A\_$

Note: A line in a genotype, such as $A\_$, indicates that any allele is possible.

Table 3.4	Genotypic ratios for simple genetic crosses (crosses for a single locus)	
Genotypic Ratio	Genotypes of Parents	Genotypes of Progeny
1 : 2 : 1	$Aa \times Aa$	$\frac{1}{4} AA : \frac{1}{2} Aa : \frac{1}{4} aa$
1 : 1	$Aa \times aa$	$\frac{1}{2} Aa : \frac{1}{2} aa$
	$Aa \times AA$	$\frac{1}{2} Aa : \frac{1}{2} AA$
Uniform progeny	$AA \times AA$	All AA
	$aa \times aa$	All aa
	$AA \times aa$	All Aa

parents are heterozygous for a dominant trait ($Aa \times Aa$). The second phenotypic ratio is the 1 : 1 ratio, which results from the mating of a homozygous parent and a heterozygous parent. The homozygous parent in this cross must carry two recessive alleles ($Aa \times aa$) to obtain a 1 : 1 ratio, because a cross between a homozygous dominant parent and a heterozygous parent ($AA \times Aa$) produces offspring displaying only the dominant trait.

The third phenotypic ratio is not really a ratio: all the offspring have the same phenotype (uniform progeny). Several combinations of parents can produce this outcome (see Table 3.3). A cross between any two homozygous parents—either between two of the same homozygotes ($AA \times AA$ and $aa \times aa$) or between two different homozygotes ($AA \times aa$)—produces progeny all having the same phenotype. Progeny of a single phenotype also can result from a cross between a homozygous dominant parent and a heterozygote ($AA \times Aa$).

If we are interested in the ratios of genotypes instead of phenotypes, there are only three outcomes to remember (**Table 3.4**): the 1 : 2 : 1 ratio, produced by a cross between two heterozygotes; the 1 : 1 ratio, produced by a cross between a heterozygote and a homozygote; and the uniform progeny produced by a cross between two homozygotes. These simple phenotypic and genotypic ratios and the parental genotypes that produce them provide the key to understanding crosses for a single locus and, as you will see in the next section, for multiple loci.

3.3 Dihybrid Crosses Reveal the Principle of Independent Assortment

We will now extend Mendel's principle of segregation to more-complex crosses entailing alleles at multiple loci. Understanding the nature of these crosses will require an additional principle, the principle of independent assortment.

Dihybrid Crosses

In addition to his work on monohybrid crosses, Mendel crossed varieties of peas that differed in *two* characteristics—a **dihybrid cross**. For example, he had one homozygous variety of pea with seeds that were round and yellow;

another homozygous variety with seeds that were wrinkled and green. When he crossed the two varieties, the seeds of all the F_1 progeny were round and yellow. He then self-fertilized the F_1 and obtained the following progeny in the F_2 : 315 round, yellow seeds; 101 wrinkled, yellow seeds; 108 round, green seeds; and 32 wrinkled, green seeds. Mendel recognized that these traits appeared approximately in a 9 : 3 : 3 : 1 ratio; that is, $\frac{9}{16}$ of the progeny were round and yellow, $\frac{3}{16}$ were wrinkled and yellow, $\frac{3}{16}$ were round and green, and $\frac{1}{16}$ were wrinkled and green.

The Principle of Independent Assortment

Mendel carried out a number of dihybrid crosses for pairs of characteristics and always obtained a 9 : 3 : 3 : 1 ratio in the F_2. This ratio makes perfect sense in regard to segregation and dominance if we add a third principle, which Mendel recognized in his dihybrid crosses: the **principle of independent assortment** (Mendel's second law). This principle states that alleles at different loci separate independently of one another (see Table 3.2).

A common mistake is to think that the principle of segregation and the principle of independent assortment refer to two different processes. The principle of independent assortment is really an extension of the principle of segregation. The principle of segregation states that the two alleles of a locus separate when gametes are formed; the principle of independent assortment states that, when these two alleles separate, their separation is independent of the separation of alleles at *other* loci.

Let's see how the principle of independent assortment explains the results that Mendel obtained in his dihybrid cross. Each plant possesses two alleles encoding each characteristic, and so the parental plants must have had genotypes $RR\ YY$ and $rr\ yy$ (**Figure 3.10a**). The principle of segregation indicates that the alleles for each locus separate, and one allele for each locus passes to each gamete. The gametes produced by the round, yellow parent therefore contain alleles RY, whereas the gametes produced by the wrinkled, green parent contain alleles ry. These two types of gametes unite to produce the F_1, all with genotype $Rr\ Yy$. Because round is dominant over wrinkled and yellow is dominant over green, the phenotype of the F_1 will be round and yellow.

When Mendel self-fertilized the F_1 plants to produce the F_2, the alleles for each locus separated, with one allele going into each gamete. This event is where the principle of independent assortment becomes important. Each pair of alleles can separate in two ways: (1) R separates with Y, and r separates with y, to produce gametes RY and ry or (2) R separates with y, and r separates with Y, to produce gametes Ry and rY. The principle of independent assortment tells us that the alleles at each locus separate independently; thus, both kinds of separation occur equally and all four type of gametes (RY, ry, Ry, and rY) are produced in equal proportions (**Figure 3.10b**). When these four types of gametes are combined to produce the F_2 generation, the progeny con-

Experiment

Question: Do alleles encoding different traits separate independently?

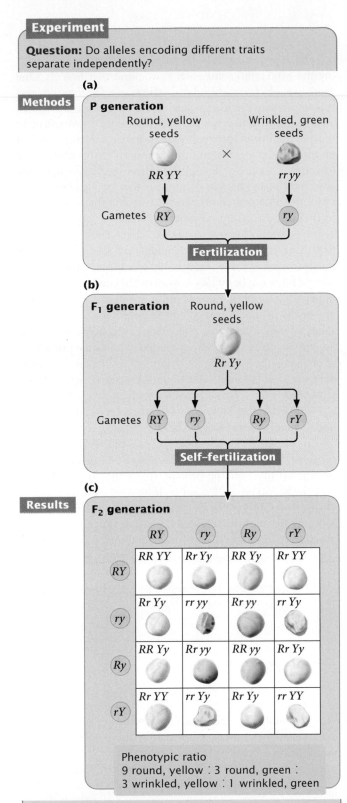

(a)

Methods

P generation

Round, yellow seeds × Wrinkled, green seeds

RR YY   *rr yy*

Gametes (*RY*) (*ry*)

Fertilization

(b)

F₁ generation   Round, yellow seeds

Rr Yy

Gametes (*RY*) (*ry*) (*Ry*) (*rY*)

Self–fertilization

(c)

Results

F₂ generation

	(*RY*)	(*ry*)	(*Ry*)	(*rY*)
(*RY*)	*RR YY*	*Rr Yy*	*RR Yy*	*Rr YY*
(*ry*)	*Rr Yy*	*rr yy*	*Rr yy*	*rr Yy*
(*Ry*)	*RR Yy*	*Rr yy*	*RR yy*	*Rr Yy*
(*rY*)	*Rr YY*	*rr Yy*	*Rr Yy*	*rr YY*

Phenotypic ratio
9 round, yellow : 3 round, green :
3 wrinkled, yellow : 1 wrinkled, green

Conclusion: The allele encoding color separated independently of the allele encoding seed shape, producing a 9 : 3 : 3 : 1 ratio in the F₂ progeny.

3.10 Mendel's dihybrid crosses revealed the principle of independent assortment.

www.whfreeman.com/pierce4e   Set up your own crosses and explore
Mendel's principles of heredity in Animation 3.1.

sist of $^9/_{16}$ round and yellow, $^3/_{16}$ wrinkled and yellow, $^3/_{16}$ round and green, and $^1/_{16}$ wrinkled and green, resulting in a 9 : 3 : 3 : 1 phenotypic ratio (**Figure 3.10c**).

Relating the Principle of Independent Assortment to Meiosis

An important qualification of the principle of independent assortment is that it applies to characters encoded by loci located on different chromosomes because, like the principle of segregation, it is based wholly on the behavior of chromosomes in meiosis. Each pair of homologous chromosomes separates independently of all other pairs in anaphase I of meiosis (see Figure 2.17); so genes located on different pairs of homologs will assort independently. Genes that happen to be located on the same chromosome will travel together during anaphase I of meiosis and will arrive at the same destination—within the same gamete (unless crossing over takes place). Genes located on the same chromosome therefore do not assort independently (unless they are located sufficiently far apart that crossing over takes place every meiotic division, as will be discussed fully in Chapter 7).

CONCEPTS

The principle of independent assortment states that genes encoding different characteristics separate independently of one another when gametes are formed, owing to the independent separation of homologous pairs of chromosomes in meiosis. Genes located close together on the same chromosome do not, however, assort independently.

✔ CONCEPT CHECK 6

How are the principles of segregation and independent assortment related and how are they different?

Applying Probability and the Branch Diagram to Dihybrid Crosses

When the genes at two loci separate independently, a dihybrid cross can be understood as two monohybrid crosses. Let's examine Mendel's dihybrid cross (*Rr Yy* × *Rr Yy*) by considering each characteristic separately (**Figure 3.11a**). If we consider only the shape of the seeds, the cross was *Rr* × *Rr*, which yields a 3 : 1 phenotypic ratio ($^3/_4$ round and $^1/_4$ wrinkled progeny, see Table 3.3). Next consider the other characteristic, the color of the seed. The cross was *Yy* × *Yy*, which produces a 3 : 1 phenotypic ratio ($^3/_4$ yellow and $^1/_4$ green progeny).

We can now combine these monohybrid ratios by using the multiplication rule to obtain the proportion of progeny with different combinations of seed shape and color. The proportion of progeny with round and yellow seeds is $^3/_4$ (the probability of round) × $^3/_4$ (the probability of yellow) = $^9/_{16}$. The proportion of progeny with round and

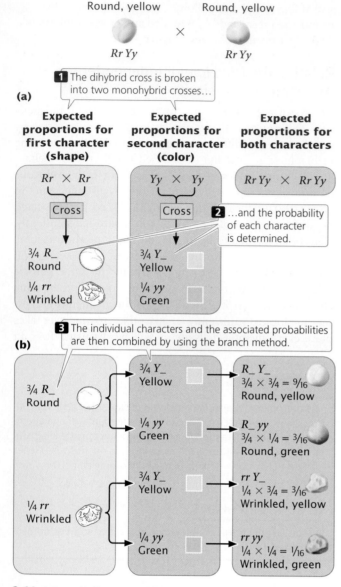

(a)

Round, yellow   Round, yellow

×

Rr Yy   *Rr Yy*

1 The dihybrid cross is broken into two monohybrid crosses…

Expected proportions for first character (shape)	Expected proportions for second character (color)	Expected proportions for both characters

Rr × *Rr*   *Yy* × *Yy*   *Rr Yy* × *Rr Yy*

Cross   Cross

2 …and the probability of each character is determined.

¾ *R_* Round   ¾ *Y_* Yellow

¼ *rr* Wrinkled   ¼ *yy* Green

(b)

3 The individual characters and the associated probabilities are then combined by using the branch method.

¾ *R_* Round

  ¾ *Y_* Yellow → *R_ Y_* ¾ × ¾ = ⁹⁄₁₆ Round, yellow

  ¼ *yy* Green → *R_ yy* ¾ × ¼ = ³⁄₁₆ Round, green

¼ *rr* Wrinkled

  ¾ *Y_* Yellow → *rr Y_* ¼ × ¾ = ³⁄₁₆ Wrinkled, yellow

  ¼ *yy* Green → *rr yy* ¼ × ¼ = ¹⁄₁₆ Wrinkled, green

3.11 A branch diagram can be used to determine the phenotypes and expected proportions of offspring from a dihybrid cross (*Rr Yy* × *Rr Yy*).

green seeds is $\frac{3}{4} \times \frac{1}{4} = \frac{3}{16}$; the proportion of progeny with wrinkled and yellow seeds is $\frac{1}{4} \times \frac{3}{4} = \frac{3}{16}$; and the proportion of progeny with wrinkled and green seeds is $\frac{1}{4} \times \frac{1}{4} = \frac{1}{16}$.

Branch diagrams are a convenient way of organizing all the combinations of characteristics (**Figure 3.11b**). In the first column, list the proportions of the phenotypes for one character (here, $\frac{3}{4}$ round and $\frac{1}{4}$ wrinkled). In the second column, list the proportions of the phenotypes for the second character ($\frac{3}{4}$ yellow and $\frac{1}{4}$ green) twice, next to each of the phenotypes in the first column: put $\frac{3}{4}$ yellow and $\frac{1}{4}$ green next to the round phenotype and again next to the wrinkled phenotype. Draw lines between the phenotypes in the first column and each of the phenotypes in the second

column. Now follow each branch of the diagram, multiplying the probabilities for each trait along that branch. One branch leads from round to yellow, yielding round and yellow progeny. Another branch leads from round to green, yielding round and green progeny, and so forth. We calculate the probability of progeny with a particular combination of traits by using the multiplication rule: the probability of round ($\frac{3}{4}$) and yellow ($\frac{3}{4}$) seeds is $\frac{3}{4} \times \frac{3}{4} = \frac{9}{16}$. The advantage of the branch diagram is that it helps keep track of all the potential combinations of traits that may appear in the progeny. It can be used to determine phenotypic or genotypic ratios for any number of characteristics.

Using probability is much faster than using the Punnett square for crosses that include multiple loci. Genotypic and phenotypic ratios can be quickly worked out by combining, with the multiplication rule, the simple ratios in Tables 3.3 and 3.4. The probability method is particularly efficient if we need the probability of only a *particular* phenotype or genotype among the progeny of a cross. Suppose we needed to know the probability of obtaining the genotype *Rr yy* in the F_2 of the dihybrid cross in Figure 3.10. The probability of obtaining the *Rr* genotype in a cross of *Rr* × *Rr* is $\frac{1}{2}$ and that of obtaining *yy* progeny in a cross of *Yy* × *Yy* is $\frac{1}{4}$ (see Table 3.4). Using the multiplication rule, we find the probability of *Rr yy* to be $\frac{1}{2} \times \frac{1}{4} = \frac{1}{8}$.

To illustrate the advantage of the probability method, consider the cross *Aa Bb cc Dd Ee* × *Aa Bb Cc dd Ee*. Suppose we wanted to know the probability of obtaining offspring with the genotype *aa bb cc dd ee*. If we used a Punnett square to determine this probability, we might be working on the solution for months. However, we can quickly figure the probability of obtaining this one genotype by breaking this cross into a series of single-locus crosses:

Progeny cross	Genotype	Probability
Aa × *Aa*	*aa*	$\frac{1}{4}$
Bb × *Bb*	*bb*	$\frac{1}{4}$
cc × *Cc*	*cc*	$\frac{1}{2}$
Dd × *dd*	*dd*	$\frac{1}{2}$
Ee × *Ee*	*ee*	$\frac{1}{4}$

The probability of an offspring from this cross having genotype *aa bb cc dd ee* is now easily obtained by using the multiplication rule: $\frac{1}{4} \times \frac{1}{4} \times \frac{1}{2} \times \frac{1}{2} \times \frac{1}{4} = \frac{1}{256}$. This calculation assumes that genes at these five loci all assort independently.

CONCEPTS

A cross including several characteristics can be worked by breaking the cross down into single-locus crosses and using the multiplication rule to determine the proportions of combinations of characteristics (provided that the genes assort independently).

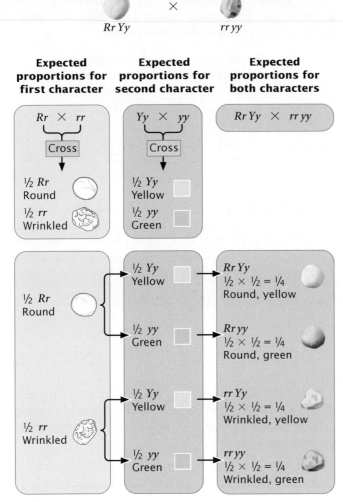

3.12 A branch diagram can be used to determine the phenotypes and expected proportions of offspring from a dihybrid testcross (*Rr Yy* × *rr yy*).

The Dihybrid Testcross

Let's practice using the branch diagram by determining the types and proportions of phenotypes in a dihybrid testcross between the round and yellow F$_1$ plants (*Rr Yy*) obtained by Mendel in his dihybrid cross and the wrinkled and green plants (*rr yy*), as depicted in **Figure 3.12**. Break the cross down into a series of single-locus crosses. The cross *Rr* × *rr* yields $^1/_2$ round (*Rr*) progeny and $^1/_2$ wrinkled (*rr*) progeny. The cross *Yy* × *yy* yields $^1/_2$ yellow (*Yy*) progeny and $^1/_2$ green (*yy*) progeny. Using the multiplication rule, we find the proportion of round and yellow progeny to be $^1/_2$ (the probability of round) × $^1/_2$ (the probability of yellow) = $^1/_4$. Four combinations of traits with the following proportions appear in the offspring: $^1/_4$ *Rr Yy*, round yellow; $^1/_4$ *Rr yy*, round green; $^1/_4$ *rr Yy*, wrinkled yellow; and $^1/_4$ *rr yy*, wrinkled green.

Worked Problem

Not only are the principles of segregation and independent assortment important because they explain how heredity works, but they also provide the means for predicting the outcome of genetic crosses. This predictive power has made genetics a powerful tool in agriculture and other fields, and the ability to apply the principles of heredity is an important skill for all students of genetics. Practice with genetic problems is essential for mastering the basic principles of heredity; no amount of reading and memorization can substitute for the experience gained by deriving solutions to specific problems in genetics.

Students may have difficulty with genetics problems when they are unsure of where to begin or how to organize the problem and plan a solution. In genetics, every problem is different, and so no common series of steps can be applied to all genetics problems. Logic and common sense must be used to analyze a problem and arrive at a solution. Nevertheless, certain steps can facilitate the process, and solving the following problem will serve to illustrate these steps.

In mice, black coat color (*B*) is dominant over brown (*b*), and a solid pattern (*S*) is dominant over white spotted (*s*). Color and spotting are controlled by genes that assort independently. A homozygous black, spotted mouse is crossed with a homozygous brown, solid mouse. All the F$_1$ mice are black and solid. A testcross is then carried out by mating the F$_1$ mice with brown, spotted mice.

a. Give the genotypes of the parents and the F$_1$ mice.

b. Give the genotypes and phenotypes, along with their expected ratios, of the progeny expected from the testcross.

• Solution

Step 1. Determine the questions to be answered. What question or questions is the problem asking? Is it asking for genotypes, genotypic ratios, or phenotypic ratios? This problem asks you to provide the *genotypes* of the parents and the F$_1$, the *expected genotypes* and *phenotypes* of the progeny of the testcross, and their *expected proportions*.

Step 2. Write down the basic information given in the problem. This problem provides important information about the dominance relations of the characters and about the mice being crossed. Black is dominant over brown, and solid is dominant over white spotted. Furthermore, the genes for the two characters assort independently. In this problem, symbols are provided for the different alleles (*B* for black, *b* for brown, *S* for solid, and *s* for spotted); had these symbols not been provided, you would need to choose symbols to represent these alleles.

It is useful to record these symbols at the beginning of the solution:

B—black	S—solid
b—brown	s—white spotted

Next, write out the crosses given in the problem.

P   Homozygous × Homozygous
         black, spotted   brown,solid

↓

F₁   Black, solid

Testcross   Black, solid × Brown, spotted

Step 3. Write down any genetic information that can be determined from the phenotypes alone. From the phenotypes and the statement that they are homozygous, you know that the P-generation mice must be $BB\ ss$ and $bb\ SS$. The F₁ mice are black and solid, both dominant traits, and so the F₁ mice must possess at least one black allele (B) and one solid allele (S). At this point, you cannot be certain about the other alleles; so represent the genotype of the F₁ as $B\_\ S\_$, where _ means that any allele is possible. The brown, spotted mice in the testcross must be $bb\ ss$, because both brown and spotted are recessive traits that will be expressed only if two recessive alleles are present. Record these genotypes on the crosses that you wrote out in step 2:

P   Homozygous × Homozygous
       black, spotted   brown, solid
          $BB\ ss$ $bb\ SS$

↓

F₁   Black, solid
             $B\_\ S\_$

Testcross   Black, solid × Brown, spotted
            $B\_\ S\_$ $bb\ ss$

Step 4. Break the problem down into smaller parts. First, determine the genotype of the F₁. After this genotype has been determined, you can predict the results of the testcross and determine the genotypes and phenotypes of the progeny from the testcross. Second, because this cross includes two independently assorting loci, it can be conveniently broken down into two single-locus crosses: one for coat color and the other for spotting. Third, use a branch diagram to determine the proportion of progeny of the testcross with different combinations of the two traits.

Step 5. Work the different parts of the problem. Start by determining the genotype of the F₁ progeny. Mendel's first law indicates that the two alleles at a locus separate, one

going into each gamete. Thus, the gametes produced by the black, spotted parent contain $B\ s$ and the gametes produced by the brown, solid parent contain $b\ S$, which combine to produce F₁ progeny with the genotype $Bb\ Ss$:

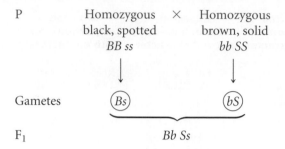

P   Homozygous × Homozygous
             black, spotted   brown, solid
              $BB\ ss$ $bb\ SS$
               ↓   ↓

Gametes   (Bs)   (bS)

F₁ $Bb\ Ss$

Use the F₁ genotype to work the testcross ($Bb\ Ss \times bb\ ss$), breaking it into two single-locus crosses. First, consider the cross for coat color: $Bb \times bb$. Any cross between a heterozygote and a homozygous recessive genotype produces a 1 : 1 phenotypic ratio of progeny (see Table 3.3):

$Bb \times bb$

↓

$\frac{1}{2}\ Bb$ black
$\frac{1}{2}\ bb$ brown

Next, do the cross for spotting: $Ss \times ss$. This cross also is between a heterozygote and a homozygous recessive genotype and will produce $\frac{1}{2}$ solid (Ss) and $\frac{1}{2}$ spotted (ss) progeny (see Table 3.3).

$Ss \times ss$

↓

$\frac{1}{2}\ Ss$ solid
$\frac{1}{2}\ ss$ spotted

Finally, determine the proportions of progeny with combinations of these characters by using the branch diagram.

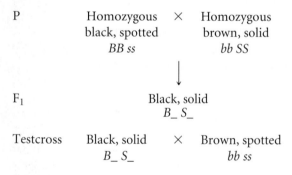

$\frac{1}{2}\ Bb$ black
    $\frac{1}{2}\ Ss$ solid ⟶ $Bb\ Ss$ black, solid
                   $\frac{1}{2} \times \frac{1}{2} = \frac{1}{4}$
    $\frac{1}{2}\ ss$ spotted ⟶ $Bb\ ss$ black, spotted
                   $\frac{1}{2} \times \frac{1}{2} = \frac{1}{4}$

$\frac{1}{2}\ bb$ brown
    $\frac{1}{2}\ Ss$ solid ⟶ $bb\ Ss$ brown, solid
                   $\frac{1}{2} \times \frac{1}{2} = \frac{1}{4}$
    $\frac{1}{2}\ ss$ spotted ⟶ $bb\ ss$ brown, spotted
                   $\frac{1}{2} \times \frac{1}{2} = \frac{1}{4}$

Step 6. Check all work. As a last step, reread the problem, checking to see if your answers are consistent with the information provided. You have used the genotypes $BB\ ss$ and *bb SS* in the P generation. Do these genotypes encode the phenotypes given in the problem? Are the F_1 progeny phenotypes consistent with the genotypes that you assigned? The answers are consistent with the information.

> → Now that we have stepped through a genetics problem together, try your hand at Problem 30 at the end of the chapter.

3.4 Observed Ratios of Progeny May Deviate from Expected Ratios by Chance

When two individual organisms of known genotype are crossed, we expect certain ratios of genotypes and phenotypes in the progeny; these expected ratios are based on the Mendelian principles of segregation, independent assortment, and dominance. The ratios of genotypes and phenotypes *actually* observed among the progeny, however, may deviate from these expectations.

For example, in German cockroaches, brown body color (*Y*) is dominant over yellow body color (*y*). If we cross a brown, heterozygous cockroach (*Yy*) with a yellow cockroach (*yy*), we expect a 1 : 1 ratio of brown (*Yy*) and yellow (*yy*) progeny. Among 40 progeny, we therefore expect to see 20 brown and 20 yellow offspring. However, the observed numbers might deviate from these expected values; we might in fact see 22 brown and 18 yellow progeny.

Chance plays a critical role in genetic crosses, just as it does in flipping a coin. When you flip a coin, you expect a 1 : 1 ratio—$\frac{1}{2}$ heads and $\frac{1}{2}$ tails. If you flip a coin 1000 times, the proportion of heads and tails obtained will probably be very close to that expected 1 : 1 ratio. However, if you flipped the coin 10 times, the ratio of heads to tails might be quite different from 1 : 1. You could easily get 6 heads and 4 tails, or 3 heads and 7 tails, just by chance. You might even get 10 heads and 0 tails. The same thing happens in genetic crosses. We may expect 20 brown and 20 yellow cockroaches, but 22 brown and 18 yellow progeny *could* arise as a result of chance.

The Goodness-of-Fit Chi-Square Test

If you expected a 1 : 1 ratio of brown and yellow cockroaches but the cross produced 22 brown and 18 yellow, you probably wouldn't be too surprised even though it wasn't a perfect 1 : 1 ratio. In this case, it seems reasonable to assume that chance produced the deviation between the expected and the observed results. But, if you observed 25 brown and 15 yellow, would the ratio still be 1 : 1? Something other than chance might have caused the deviation. Perhaps the inheritance of this character is more complicated than was assumed or perhaps some of the yellow progeny died before they were counted. Clearly, we need some means of evaluating how likely it is that chance is responsible for the deviation between the observed and the expected numbers.

To evaluate the role of chance in producing deviations between observed and expected values, a statistical test called the **goodness-of-fit chi-square test** is used. This test provides information about how well observed values fit expected values. Before we learn how to calculate the chi square, it is important to understand what this test does and does not indicate about a genetic cross.

The chi-square test cannot tell us whether a genetic cross has been correctly carried out, whether the results are correct, or whether we have chosen the correct genetic explanation for the results. What it does indicate is the *probability* that the difference between the observed and the expected values is due to chance. In other words, it indicates the likelihood that chance alone could produce the deviation between the expected and the observed values.

If we expected 20 brown and 20 yellow progeny from a genetic cross, the chi-square test gives the probability that we might observe 25 brown and 15 yellow progeny simply owing to chance deviations from the expected 20 : 20 ratio. This hypothesis, that chance alone is responsible for any deviations between observed and expected values, is sometimes called the null hypothesis. When the probability calculated from the chi-square test is high, we assume that chance alone produced the difference (the null hypothesis is true). When the probability is low, we assume that some factor other than chance—some significant factor—produced the deviation (the null hypothesis is false).

To use the goodness-of-fit chi-square test, we first determine the expected results. The chi-square test must always be applied to *numbers* of progeny, not to proportions or percentages. Let's consider a locus for coat color in domestic cats, for which black color (*B*) is dominant over gray (*b*). If we crossed two heterozygous black cats (*Bb* × *Bb*), we would expect a 3 : 1 ratio of black and gray kittens. A series of such crosses yields a total of 50 kittens—30 black and 20 gray. These numbers are our *observed* values. We can obtain the expected numbers by multiplying the *expected* proportions by the total number of observed progeny. In this case, the expected number of black kittens is $\frac{3}{4} \times 50 = 37.5$ and the expected number of gray kittens is $\frac{1}{4} \times 50 = 12.5$. The chi-square ($\chi^2$) value is calculated by using the following formula:

$$\chi^2 = \Sigma \frac{(\text{observed} - \text{expected})^2}{\text{expected}}$$

Table 3.5 Critical values of the χ^2 distribution

df	P 0.995	0.975	0.9	0.5	0.1	0.05*	0.025	0.01	0.005
1	0.000	0.000	0.016	0.455	2.706	3.841	5.024	6.635	7.879
2	0.010	0.051	0.211	1.386	4.605	5.991	7.378	9.210	10.597
3	0.072	0.216	0.584	2.366	6.251	7.815	9.348	11.345	12.838
4	0.207	0.484	1.064	3.357	7.779	9.488	11.143	13.277	14.860
5	0.412	0.831	1.610	4.351	9.236	11.070	12.832	15.086	16.750
6	0.676	1.237	2.204	5.348	10.645	12.592	14.449	16.812	18.548
7	0.989	1.690	2.833	6.346	12.017	14.067	16.013	18.475	20.278
8	1.344	2.180	3.490	7.344	13.362	15.507	17.535	20.090	21.955
9	1.735	2.700	4.168	8.343	14.684	16.919	19.023	21.666	23.589
10	2.156	3.247	4.865	9.342	15.987	18.307	20.483	23.209	25.188
11	2.603	3.816	5.578	10.341	17.275	19.675	21.920	24.725	26.757
12	3.074	4.404	6.304	11.340	18.549	21.026	23.337	26.217	28.300
13	3.565	5.009	7.042	12.340	19.812	22.362	24.736	27.688	29.819
14	4.075	5.629	7.790	13.339	21.064	23.685	26.119	29.141	31.319
15	4.601	6.262	8.547	14.339	22.307	24.996	27.488	30.578	32.801

P, probability; df, degrees of freedom.
*Most scientists assume that, when $P < 0.05$, a significant difference exists between the observed and the expected values in a chi-square test.

where Σ means the sum. We calculate the sum of all the squared differences between observed and expected and divide by the expected values. To calculate the chi-square value for our black and gray kittens, we first subtract the number of *expected* black kittens from the number of *observed* black kittens ($30 - 37.5 = -7.5$) and square this value: $-7.5^2 = 56.25$. We then divide this result by the expected number of black kittens, $56.25/37.5 = 1.5$. We repeat the calculations on the number of expected gray kittens: $(20 - 12.5)^2/12.5 = 4.5$. To obtain the overall chi-square value, we sum the (observed - expected)^2/expected values: $1.5 + 4.5 = 6.0$.

The next step is to determine the probability associated with this calculated chi-square value, which is the probability that the deviation between the observed and the expected results could be due to chance. This step requires us to compare the calculated chi-square value (6.0) with theoretical values that have the same degrees of freedom in a chi-square table. The degrees of freedom represent the number of ways in which the expected classes are free to vary. For a goodness-of-fit chi-square test, the degrees of freedom are equal to $n - 1$, where n is the number of different expected phenotypes. In our example, there are two expected phenotypes

(black and gray); so $n = 2$, and the degree of freedom equals $2 - 1 = 1$.

Now that we have our calculated chi-square value and have figured out the associated degrees of freedom, we are ready to obtain the probability from a chi-square table (**Table 3.5**). The degrees of freedom are given in the left-hand column of the table and the probabilities are given at the top; within the body of the table are chi-square values associated with these probabilities. First, find the row for the appropriate degrees of freedom; for our example with 1 degree of freedom, it is the first row of the table. Find where our calculated chi-square value (6.0) lies among the theoretical values in this row. The theoretical chi-square values increase from left to right and the probabilities decrease from left to right. Our chi-square value of 6.0 falls between the value of 5.024, associated with a probability of 0.025, and the value of 6.635, associated with a probability of 0.01.

Thus, the probability associated with our chi-square value is less than 0.025 and greater than 0.01. So there is less than a 2.5% probability that the deviation that we observed between the expected and the observed numbers of black and gray kittens could be due to chance.

Most scientists use the 0.05 probability level as their cutoff value: if the probability of chance being responsible for the deviation is greater than or equal to 0.05, they accept that chance may be responsible for the deviation between the observed and the expected values. When the probability is less than 0.05, scientists assume that chance is not responsible and a significant difference exists. The expression *significant difference* means that some factor other than chance is responsible for the observed values being different from the expected values. In regard to the kittens, perhaps one of the genotypes had a greater mortality rate before the progeny were counted or perhaps other genetic factors skewed the observed ratios.

In choosing 0.05 as the cutoff value, scientists have agreed to assume that chance is responsible for the deviations between observed and expected values unless there is strong evidence to the contrary. Bear in mind that, even if we obtain a probability of, say, 0.01, there is still a 1% probability that the deviation between the observed and the expected numbers is due to nothing more than chance. Calculation of the chi-square value is illustrated in **Figure 3.13**. TRY PROBLEM 35 →

CONCEPTS

Differences between observed and expected ratios can arise by chance. The goodness-of-fit chi-square test can be used to evaluate whether deviations between observed and expected numbers are likely to be due to chance or to some other significant factor.

✔ CONCEPT CHECK 7

A chi-square test comparing observed and expected progeny is carried out, and the probability associated with the calculated chi-square value is 0.72. What does this probability represent?

a. Probability that the correct results were obtained

b. Probability of obtaining the observed numbers

c. Probability that the difference between observed and expected numbers is significant

d. Probability that the difference between observed and expected numbers could be due to chance

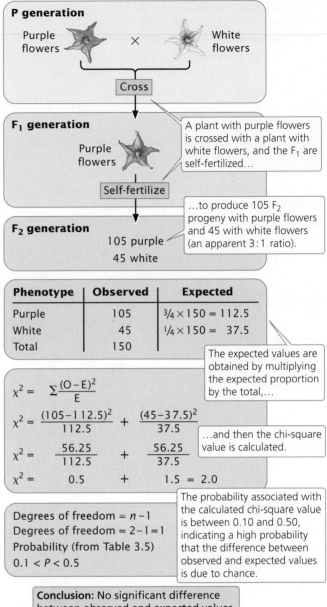

P generation

Purple flowers × White flowers

Cross

F₁ generation

Purple flowers

A plant with purple flowers is crossed with a plant with white flowers, and the F₁ are self-fertilized...

Self-fertilize

F₂ generation

105 purple

45 white

...to produce 105 F₂ progeny with purple flowers and 45 with white flowers (an apparent 3:1 ratio).

Phenotype	Observed	Expected
Purple	105	$\frac{3}{4} \times 150 = 112.5$
White	45	$\frac{1}{4} \times 150 = 37.5$
Total	150	

The expected values are obtained by multiplying the expected proportion by the total,...

$$\chi^2 = \sum \frac{(O-E)^2}{E}$$

$$\chi^2 = \frac{(105-112.5)^2}{112.5} + \frac{(45-37.5)^2}{37.5}$$

$$\chi^2 = \frac{56.25}{112.5} + \frac{56.25}{37.5}$$

$$\chi^2 = 0.5 + 1.5 = 2.0$$

...and then the chi-square value is calculated.

Degrees of freedom = $n - 1$
Degrees of freedom = $2 - 1 = 1$
Probability (from Table 3.5)
$0.1 < P < 0.5$

The probability associated with the calculated chi-square value is between 0.10 and 0.50, indicating a high probability that the difference between observed and expected values is due to chance.

Conclusion: No significant difference between observed and expected values.

3.13 A chi-square test is used to determine the probability that the difference between observed and expected values is due to chance.

CONCEPTS SUMMARY

- Gregor Mendel discovered the principles of heredity. His success can be attributed to his choice of the pea plant as an experimental organism, the use of characters with a few easily distinguishable phenotypes, his experimental approach, the use of mathematics to interpret his results, and careful attention to detail.

- Genes are inherited factors that determine a characteristic. Alternative forms of a gene are called alleles. The alleles are located at a specific place, a locus, on a chromosome, and the set of genes that an individual organism possesses is its genotype. Phenotype is the manifestation or appearance of a characteristic and may refer to a physical, biochemical, or behavioral characteristic. Only the genotype—not the phenotype—is inherited.

- The principle of segregation states that an individual organism possesses two alleles encoding a trait and that these two alleles separate in equal proportions when gametes are formed.

- The concept of dominance indicates that, when two different alleles are present in a heterozygote, only the trait of one of them, the dominant allele, is observed in the phenotype. The other allele is said to be recessive.

- The two alleles of a genotype are located on homologous chromosomes. The separation of homologous chromosomes in anaphase I of meiosis brings about the segregation of alleles.

- Probability is the likelihood that a particular event will occur. The multiplication rule of probability states that the probability of two or more independent events occurring together is calculated by multiplying the probabilities of the independent events. The addition rule of probability states that the probability of any of two or more mutually exclusive events occurring is calculated by adding the probabilities of the events.

- The binomial expansion can be used to determine the probability of a particular combination of events.

- A testcross reveals the genotype (homozygote or heterozygote) of an individual organism having a dominant trait and consists of crossing that individual with one having the homozygous recessive genotype.

- The principle of independent assortment states that genes encoding different characters assort independently when gametes are formed. Independent assortment is based on the random separation of homologous pairs of chromosomes in anaphase I of meiosis; it takes place when genes encoding two characters are located on different pairs of chromosomes.

- Observed ratios of progeny from a genetic cross may deviate from the expected ratios owing to chance. The goodness-of-fit chi-square test can be used to determine the probability that a difference between observed and expected numbers is due to chance.

IMPORTANT TERMS

gene (p. 46)
allele (p. 46)
locus (p. 46)
genotype (p. 46)
homozygous (p. 46)
heterozygous (p. 46)
phenotype (p. 46)
monohybrid cross (p. 47)
P (parental) generation (p. 47)
F_1 (filial 1) generation (p. 48)

reciprocal crosses (p. 48)
F_2 (filial 2) generation (p. 48)
dominant (p. 49)
recessive (p. 49)
principle of segregation
 (Mendel's first law) (p. 49)
concept of dominance (p. 49)
chromosome theory of heredity (p. 50)
backcross (p. 51)
Punnett square (p. 51)

probability (p. 52)
multiplication rule (p. 52)
addition rule (p. 53)
testcross (p. 55)
wild type (p. 55)
dihybrid cross (p. 56)
principle of independent assortment
 (Mendel's second law) (p. 56)
goodness-of-fit chi-square test (p. 61)

ANSWERS TO CONCEPT CHECKS

1. b

2. A locus is a place on a chromosome where genetic information encoding a character is located. An allele is a copy of a gene that encodes a specific trait. A genotype is the set of alleles possesed by an individual organism, and a phenotype is the manifestation or appearance of a character.

3. Because the traits for both alleles appeared in the F_2 progeny

4. d

5. a

6. The principle of segregation and the principle of independent assortment both refer to the separation of alleles in anaphase I of meiosis. The principle of segregation says that these alleles separate, and the principle of independent assortment says that they separate independently of alleles at other loci.

7. d

WORKED PROBLEMS

1. Short hair (*S*) in rabbits is dominant over long hair (*s*). The following crosses are carried out, producing the progeny shown. Give all possible genotypes of the parents in each cross.

Parents	Progeny
a. short × short	4 short and 2 long
b. short × short	8 short
c. short × long	12 short
d. short × long	3 short and 1 long
e. long × long	2 long

• Solution

For this problem, it is useful to first gather as much information about the genotypes of the parents as possible on the basis of their phenotypes. We can then look at the types of progeny produced to provide the missing information. Notice that the problem asks for *all* possible genotypes of the parents.

a. short × short   4 short and 2 long

Because short hair is dominant over long hair, a rabbit having short hair could be either *SS* or *Ss*. The 2 long-haired offspring must be homozygous (*ss*) because long hair is recessive and will appear in the phenotype only when both alleles for long hair are present. Because each parent contributes one of the two alleles found in the progeny, each parent must be carrying the *s* allele and must therefore be *Ss*.

b. short × short   8 short

The short-haired parents could be *SS* or *Ss*. All 8 of the offspring are short (*S_*), and so at least one of the parents is likely to be homozygous (*SS*); if both parents were heterozygous, we would expect $^{1}/_{4}$ of the progeny to be long haired (*ss*), but we do not observe any long-haired progeny. The other parent could be homozygous (*SS*) or heterozygous (*Ss*); as long as one parent is homozygous, all the offspring will be short haired. It is theoretically possible, although unlikely, that both parents are heterozygous (*Ss* × *Ss*). If both were heterozygous, we would expect 2 of the 8 progeny to be long haired. Although no long-haired progeny are observed, it is possible that just by chance no long-haired rabbits would be produced among the 8 progeny of the cross.

c. short × long   12 short

The short-haired parent could be *SS* or *Ss*. The long-haired parent must be *ss*. If the short-haired parent were heterozygous (*Ss*), half of the offspring would be expected to be long haired, but we don't see any long-haired progeny. Therefore, this parent is most likely homozygous (*SS*). It is theoretically possible, although unlikely, that the parent is heterozygous and just by chance no long-haired progeny were produced.

d. short × long   3 short and 1 long

On the basis of its phenotype, the short-haired parent could be homozygous (*SS*) or heterozygous (*Ss*), but the presence of one long-haired offspring tells us that the short-haired parent must be heterozygous (*Ss*). The long-haired parent must be homozygous (*ss*).

e. long × long   2 long

Because long hair is recessive, both parents must be homozygous for a long-hair allele (*ss*).

2. In cats, black coat color is dominant over gray. A female black cat whose mother is gray mates with a gray male. If this female has a litter of six kittens, what is the probability that three will be black and three will be gray?

• Solution

Because black (*G*) is dominant over gray (*g*), a black cat may be homozygous (*GG*) or heterozygous (*Gg*). The black female in this problem must be heterozygous (*Gg*) because her mother is gray (*gg*) and she must inherit one of her mother's alleles. The gray male is homozygous (*gg*) because gray is recessive. Thus the cross is:

$$\begin{array}{ccc} Gg & \times & gg \\ \text{Black female} & & \text{Gray male} \end{array}$$

$^{1}/_{2}$ *Gg* black
$^{1}/_{2}$ *gg* gray

We can use the binomial expansion to determine the probability of obtaining three black and three gray kittens in a litter of six. Let *p* equal the probability of a kitten being black and *q* equal the probability of a kitten being gray. The binomial is $(p + q)^6$, the expansion of which is:

$$(p + q)^6 = p^6 + 6p^5q + 15p^4q^2 + 20p^3q^3 + 15p^2q^4 + 6p^1q^5 + q^6$$

(See the section on The Binomial Expansion and Probability for an explanation of how to expand the binomial.) The probability of obtaining three black and three gray kittens in a litter of six is provided by the term $20p^3q^3$. The probabilities of *p* and *q* are both $^{1}/_{2}$; so the overall probability is $20(^{1}/_{2})^3(^{1}/_{2})^3 = {}^{20}/_{64} = {}^{5}/_{16}$.

3. The following genotypes are crossed:

$$Aa\ Bb\ Cc\ Dd \times Aa\ Bb\ Cc\ Dd$$

Give the proportion of the progeny of this cross having each of the following genotypes: (**a**) *Aa Bb Cc Dd*, (**b**) *aa bb cc dd*, (**c**) *Aa Bb cc Dd*.

• Solution

This problem is easily worked if the cross is broken down into simple crosses and the multiplication rule is used to find the different combinations of genotypes:

Locus 1	*Aa* × *Aa* = $^{1}/_{4}$ *AA*, $^{1}/_{2}$ *Aa*, $^{1}/_{4}$ *aa*
Locus 2	*Bb* × *Bb* = $^{1}/_{4}$ *BB*, $^{1}/_{2}$ *Bb*, $^{1}/_{4}$ *bb*
Locus 3	*Cc* × *Cc* = $^{1}/_{4}$ *CC*, $^{1}/_{2}$ *Cc*, $^{1}/_{4}$ *cc*
Locus 4	*Dd* × *Dd* = $^{1}/_{4}$ *DD*, $^{1}/_{2}$ *Dd*, $^{1}/_{4}$ *dd*

To find the probability of any combination of genotypes, simply multiply the probabilities of the different genotypes:

a. *Aa Bb Cc Dd* $^{1}/_{2}$ (*Aa*) × $^{1}/_{2}$ (*Bb*) × $^{1}/_{2}$ (*Cc*) × $^{1}/_{2}$ (*Dd*) = $^{1}/_{16}$

b. *aa bb cc dd* $^{1}/_{4}$ (*aa*) × $^{1}/_{4}$ (*bb*) × $^{1}/_{4}$ (*cc*) × $^{1}/_{4}$ (*dd*) = $^{1}/_{256}$

c. *Aa Bb cc Dd* $^{1}/_{2}$ (*Aa*) × $^{1}/_{2}$ (*Bb*) × $^{1}/_{4}$ (*cc*) × $^{1}/_{2}$ (*Dd*) = $^{1}/_{32}$

4. In corn, purple kernels are dominant over yellow kernels, and full kernels are dominant over shrunken kernels. A corn plant having purple and full kernels is crossed with a plant having yellow and shrunken kernels, and the following progeny are obtained:

purple, full	112
purple, shrunken	103
yellow, full	91
yellow, shrunken	94

What are the most likely genotypes of the parents and progeny? Test your genetic hypothesis with a chi-square test.

• Solution

The best way to begin this problem is by breaking the cross down into simple crosses for a single characteristic (seed color or seed shape):

P purple × yellow full × shrunken

F$_1$ 112 + 103 = 215 purple 112 + 91 = 203 full
   91 + 94 = 185 yellow 103 + 94 = 197 shrunken

Purple × yellow produces approximately $\frac{1}{2}$ purple and $\frac{1}{2}$ yellow. A 1 : 1 ratio is usually caused by a cross between a heterozygote and a homozygote. Because purple is dominant, the purple parent must be heterozygous (*Pp*) and the yellow parent must be homozygous (*pp*). The purple progeny produced by this cross will be heterozygous (*Pp*) and the yellow progeny must be homozygous (*pp*).

Now let's examine the other character. Full × shrunken produces $\frac{1}{2}$ full and $\frac{1}{2}$ shrunken, or a 1 : 1 ratio, and so these progeny phenotypes also are produced by a cross between a heterozygote (*Ff*) and a homozygote (*ff*); the full-kernel progeny will be heterozygous (*Ff*) and the shrunken-kernel progeny will be homozygous (*ff*).

Now combine the two crosses and use the multiplication rule to obtain the overall genotypes and the proportions of each genotype:

P Purple, full × Yellow, shrunken
   Pp Ff   *pp ff*

F$_1$ *Pp Ff* = $\frac{1}{2}$ purple × $\frac{1}{2}$ full = $\frac{1}{4}$ purple, full
   Pp ff = $\frac{1}{2}$ purple × $\frac{1}{2}$ shrunken = $\frac{1}{4}$ purple, shrunken
   pp Ff = $\frac{1}{2}$ yellow × $\frac{1}{2}$ full = $\frac{1}{4}$ yellow, full
   pp ff = $\frac{1}{2}$ yellow × $\frac{1}{2}$ shrunken = $\frac{1}{4}$ yellow, shrunken

Our genetic explanation predicts that, from this cross, we should see $\frac{1}{4}$ purple, full-kernel progeny; $\frac{1}{4}$ purple, shrunken-kernel progeny; $\frac{1}{4}$ yellow, full-kernel progeny; and $\frac{1}{4}$ yellow, shrunken-kernel progeny. A total of 400 progeny were produced; so $\frac{1}{4} \times 400 = 100$ of each phenotype are expected. These observed numbers do not fit the expected numbers exactly. Could the difference between what we observe and what we expect be due to chance? If the probability is high that chance alone is responsible for the difference between observed and expected,

we will assume that the progeny have been produced in the 1 : 1 : 1 : 1 ratio predicted by the cross. If the probability that the difference between observed and expected is due to chance is low, the progeny are not really in the predicted ratio and some other, *significant* factor must be responsible for the deviation.

The observed and expected numbers are:

Phenotype	Observed	Expected
purple, full	112	$\frac{1}{4} \times 400 = 100$
purple, shrunken	103	$\frac{1}{4} \times 400 = 100$
yellow, full	91	$\frac{1}{4} \times 400 = 100$
yellow, shrunken	94	$\frac{1}{4} \times 400 = 100$

To determine the probability that the difference between observed and expected is due to chance, we calculate a chi-square value with the formula $\chi^2 = \Sigma[(\text{observed} - \text{expected})^2/\text{expected}]$:

$$\chi^2 = \frac{(112-100)^2}{100} + \frac{(103-100)^2}{100} + \frac{(91-100)^2}{100} + \frac{(94-100)^2}{100}$$

$$= \frac{12^2}{100} + \frac{3^2}{100} + \frac{9^2}{100} + \frac{6^2}{100}$$

$$= \frac{144}{100} + \frac{9}{100} + \frac{81}{100} + \frac{36}{100}$$

$$= 1.44 + 0.09 + 0.81 + 0.36 = 2.70$$

Now that we have the chi-square value, we must determine the probability that this chi-square value is due to chance. To obtain this probability, we first calculate the degrees of freedom, which for a goodness-of-fit chi-square test are $n - 1$, where n equals the number of expected phenotypic classes. In this case, there are four expected phenotypic classes; so the degrees of freedom equal $4 - 1 = 3$. We must now look up the chi-square value in a chi-square table (see Table 3.5). We select the row corresponding to 3 degrees of freedom and look along this row to find our calculated chi-square value. The calculated chi-square value of 2.7 lies between 2.366 (a probability of 0.5) and 6.251 (a probability of 0.1). The probability (*P*) associated with the calculated chi-square value is therefore $0.5 < P < 0.1$. It is the probability that the difference between what we observed and what we expected is due to chance, which in this case is high, and so chance is likely responsible for the deviation. We can conclude that the progeny *do* appear in the 1 : 1 : 1 : 1 ratio predicted by our genetic explanation.

COMPREHENSION QUESTIONS

Section 3.1

*1. Why was Mendel's approach to the study of heredity so successful?

2. What is the difference between genotype and phenotype?

Section 3.2

*3. What is the principle of segregation? Why is it important?

4. How are Mendel's principles different from the concept of blending inheritance discussed in Chapter 1?

5. What is the concept of dominance?

6. What are the addition and multiplication rules of probability and when should they be used?

7. Give the genotypic ratios that may appear among the progeny of simple crosses and the genotypes of the parents that may give rise to each ratio.

*8. What is the chromosome theory of heredity? Why was it important?

Section 3.3

*9. What is the principle of independent assortment? How is it related to the principle of segregation?

10. In which phases of mitosis and meiosis are the principles of segregation and independent assortment at work?

Section 3.4

11. How is the goodness-of-fit chi-square test used to analyze genetic crosses? What does the probability associated with a chi-square value indicate about the results of a cross?

APPLICATION QUESTIONS AND PROBLEMS

Introduction

12. The inheritance of red hair was discussed in the introduction to this chapter. At times in the past, red hair in humans was thought to be a recessive trait and, at other times, it was thought to be a dominant trait. What features of heritance is red hair expected to exhibit as a recessive trait? What features would it exhibit if it were a dominant trait?

Section 3.1

13. What characteristics of an organism would make it suitable for studies of the principles of inheritance? Can you name several organisms that have these characteristics?

Section 3.2

*14. In cucumbers, orange fruit color (R) is dominant over cream fruit color (r). A cucumber plant homozygous for orange fruits is crossed with a plant homozygous for cream fruits. The F_1 are intercrossed to produce the F_2.

a. Give the genotypes and phenotypes of the parents, the F_1, and the F_2.

b. Give the genotypes and phenotypes of the offspring of a backcross between the F_1 and the orange parent.

c. Give the genotypes and phenotypes of a backcross between the F_1 and the cream parent.

15. J. W. McKay crossed a stock melon plant that produced tan seeds with a plant that produced red seeds and obtained the following results (J. W. McKay. 1936. *Journal of Heredity* 27:110–112).

Cross	F_1	F_2
tan ♀ × red ♂	13 tan seeds	93 tan, 24 red seeds

a. Explain the inheritance of tan and red seeds in this plant.

b. Assign symbols for the alleles in this cross and give genotypes for all the individual plants.

*16. White (w) coat color in guinea pigs is recessive to black (W). In 1909, W. E. Castle and J. C. Phillips transplanted an ovary from a black guinea pig into a white female whose ovaries had been removed. They then mated this white female with a white male. All the offspring from the mating were black in color (W. E. Castle and J. C. Phillips. 1909. *Science* 30:312–313).

[Wegner/ARCO/Nature Picture Library; Nigel Cattlin/Alamy.]

a. Explain results of this cross.

b. Give the genotype of the offspring of this cross.

c. What, if anything, does this experiment indicate about the validity of the pangenesis and the germ-plasm theories discussed in Chapter 1?

*17. In cats, blood-type A results from an allele (I^A) that is dominant over an allele (i^B) that produces blood-type B. There is no O blood type. The blood types of male and female cats that were mated and the blood types of their kittens follow. Give the most likely genotypes for the parents of each litter.

	Male parent	Female parent	Kittens
a.	A	B	4 with type A, 3 with type B
b.	B	B	6 with type B
c.	B	A	8 with type A
d.	A	A	7 with type A, 2 with type B
e.	A	A	10 with type A
f.	A	B	4 with type A, 1 with type B

18. Joe has a white cat named Sam. When Joe crosses Sam with a black cat, he obtains $1/2$ white kittens and $1/2$ black kittens. When the black kittens are interbred, all the kittens that they produce are black. On the basis of these results, would you conclude that white or black coat color in cats is a recessive trait? Explain your reasoning.

19. In sheep, lustrous fleece results from an allele (L) that is dominant over an allele (l) for normal fleece. A ewe (adult female) with lustrous fleece is mated with a ram (adult male) with normal fleece. The ewe then gives birth to a single lamb with normal fleece. From this single offspring, is it possible to determine the genotypes of the two parents? If so, what are their genotypes? If not, why not?

[Jeffrey van daele/FeaturePics.]

*20. Alkaptonuria is a metabolic disorder in which affected persons produce black urine. Alkaptonuria results from an allele (a) that is recessive to the allele for normal metabolism (A). Sally has normal metabolism, but her brother has alkaptonuria. Sally's father has alkaptonuria, and her mother has normal metabolism.

 a. Give the genotypes of Sally, her mother, her father, and her brother.

 b. If Sally's parents have another child, what is the probability that this child will have alkaptonuria?

 c. If Sally marries a man with alkaptonuria, what is the probability that their first child will have alkaptonuria?

21. Suppose that you are raising Mongolian gerbils. You notice that some of your gerbils have white spots, whereas others have solid coats. What type of crosses could you carry out to determine whether white spots are due to a recessive or a dominant allele?

*22. Hairlessness in American rat terriers is recessive to the presence of hair. Suppose that you have a rat terrier with hair. How can you determine whether this dog is homozygous or heterozygous for the hairy trait?

23. What is the probability of rolling one six-sided die and obtaining the following numbers?

 a. 2

 b. 1 or 2

 c. An even number

 d. Any number but a 6

*24. What is the probability of rolling two six-sided dice and obtaining the following numbers?

 a. 2 and 3

 b. 6 and 6

 c. At least one 6

 d. Two of the same number (two 1s, or two 2s, or two 3s, etc.)

 e. An even number on both dice

 f. An even number on at least one die

*25. In a family of seven children, what is the probability of obtaining the following numbers of boys and girls?

 a. All boys

 b. All children of the same sex

 c. Six girls and one boy

 d. Four boys and three girls

 e. Four girls and three boys

26. Phenylketonuria (PKU) is a disease that results from a recessive gene. Two normal parents produce a child with PKU.

 a. What is the probability that a sperm from the father will contain the PKU allele?

 b. What is the probability that an egg from the mother will contain the PKU allele?

 c. What is the probability that their next child will have PKU?

 d. What is the probability that their next child will be heterozygous for the PKU gene?

*27. In German cockroaches, curved wing (cv) is recessive to normal wing (cv^+). A homozygous cockroach having normal wings is crossed with a homozygous cockroach having curved wings. The F_1 are intercrossed to produce the F_2. Assume that the pair of chromosomes containing the locus for wing shape is metacentric. Draw this pair of chromosomes as it would appear in the parents, the F_1, and each class of F_2 progeny at metaphase I of meiosis. Assume that no crossing over takes place. At each stage, label a location for the alleles for wing shape (cv and cv^+) on the chromosomes.

*28. In guinea pigs, the allele for black fur (B) is dominant over the allele for brown (b) fur. A black guinea pig is crossed with a brown guinea pig, producing five F_1 black guinea pigs and six F_1 brown guinea pigs.

a. How many copies of the black allele (*B*) will be present in each cell of an F_1 black guinea pig at the following stages: G_1, G_2, metaphase of mitosis, metaphase I of meiosis, metaphase II of meiosis, and after the second cytokinesis following meiosis? Assume that no crossing over takes place.

b. How many copies of the brown allele (*b*) will be present in each cell of an F_1 brown guinea pig at the same stages as those listed in part *a*? Assume that no crossing over takes place.

Section 3.3

29. In watermelons, bitter fruit (*B*) is dominant over sweet fruit (*b*), and yellow spots (*S*) are dominant over no spots (*s*). The genes for these two characteristics assort independently. A homozygous plant that has bitter fruit and yellow spots is crossed with a homozygous plant that has sweet fruit and no spots. The F_1 are intercrossed to produce the F_2.

a. What are the phenotypic ratios in the F_2?

b. If an F_1 plant is backcrossed with the bitter, yellow-spotted parent, what phenotypes and proportions are expected in the offspring?

c. If an F_1 plant is backcrossed with the sweet, nonspotted parent, what phenotypes and proportions are expected in the offspring?

*30. In cats, curled ears result from an allele (*Cu*) that is dominant over an allele (*cu*) for normal ears. Black color results from an independently assorting allele (*G*) that is dominant over an allele for gray (*g*). A gray cat homozygous for curled ears is mated with a homozygous black cat with normal ears. All the F_1 cats are black and have curled ears.

a. If two of the F_1 cats mate, what phenotypes and proportions are expected in the F_2?

b. An F_1 cat mates with a stray cat that is gray and possesses normal ears. What phenotypes and proportions of progeny are expected from this cross?

[Biosphoto/J.-L. Klein & M.-L. Hubert/Peter Arnold.]

*31. The following two genotypes are crossed: *Aa Bb Cc dd Ee* × *Aa bb Cc Dd Ee*. What will the proportion of the following genotypes be among the progeny of this cross?

a. *Aa Bb Cc Dd Ee*

b. *Aa bb Cc dd ee*

c. *aa bb cc dd ee*

d. *AA BB CC DD EE*

32. In mice, an allele for apricot eyes (*a*) is recessive to an allele for brown eyes (*a*⁺). At an independently assorting locus, an allele for tan coat color (*t*) is recessive to an allele

for black coat color (*t*⁺). A mouse that is homozygous for brown eyes and black coat color is crossed with a mouse having apricot eyes and a tan coat. The resulting F_1 are intercrossed to produce the F_2. In a litter of eight F_2 mice, what is the probability that two will have apricot eyes and tan coats?

33. In cucumbers, dull fruit (*D*) is dominant over glossy fruit (*d*), orange fruit (*R*) is dominant over cream fruit (*r*), and bitter cotyledons (*B*) are dominant over nonbitter cotyledons (*b*). The three characters are encoded by genes located on different pairs of chromosomes. A plant homozygous for dull, orange fruit and bitter cotyledons is crossed with a plant that has glossy, cream fruit and nonbitter cotyledons. The F_1 are intercrossed to produce the F_2.

a. Give the phenotypes and their expected proportions in the F_2.

b. An F_1 plant is crossed with a plant that has glossy, cream fruit and nonbitter cotyledons. Give the phenotypes and expected proportions among the progeny of this cross.

*34. Alleles *A* and *a* are located on a pair of metacentric chromosomes. Alleles *B* and *b* are located on a pair of acrocentric chromosomes. A cross is made between individuals having the following genotypes: *Aa Bb* × *aa bb*.

a. Draw the chromosomes as they would appear in each type of gamete produced by the individuals of this cross.

b. For each type of progeny resulting from this cross, draw the chromosomes as they would appear in a cell at G_1, G_2, and metaphase of mitosis.

Section 3.4

*35. J. A. Moore investigated the inheritance of spotting patterns in leopard frogs (J. A. Moore. 1943. *Journal of Heredity* 34:3–7). The pipiens phenotype had the normal spots that give leopard frogs their name. In contrast, the burnsi phenotype lacked spots on its back. Moore carried out the following crosses, producing the progeny indicated.

Parent phenotypes	Progeny phenotypes
burnsi × burnsi	39 burnsi, 6 pipiens
burnsi × pipiens	23 burnsi, 3 pipiens
burnsi × pipiens	196 burnsi, 210 pipiens

a. On the basis of these results, what is the most likely mode of inheritance of the burnsi phenotype?

b. Give the most likely genotypes of the parent in each cross (use *B* for the burnsi allele and *B*⁺ for pipiens allele)

c. Use a chi-square test to evaluate the fit of the observed numbers of progeny to the number expected on the basis of your proposed genotypes.

36. In the 1800s, a man with dwarfism who lived in Utah produced a large number of descendants: 22 children, 49 grandchildren, and 250 great-grandchildren (see the illustration of a family pedigree on page 70), many of also were dwarfs (F. F. Stephens. 1943. *Journal of Heredity* 34:229–235). The type of dwarfism found in this family is called Schmid-type metaphyseal chondrodysplasia, although it was originally thought to be achondroplastic dwarfism. Among the families of this kindred, dwarfism appeared only in members who had one parent with dwarfism. When one parent was a dwarf, the following numbers of children were produced.

Family in which one parent had dwarfism	Children with normal stature	Children with dwarfism
A	15	7
B	4	6
C	1	6
D	6	2
E	2	2
F	8	4
G	4	4
H	2	1
I	0	1
J	3	1
K	2	3
L	2	1
M	2	0
N	1	0
O	0	2
Total	52	40

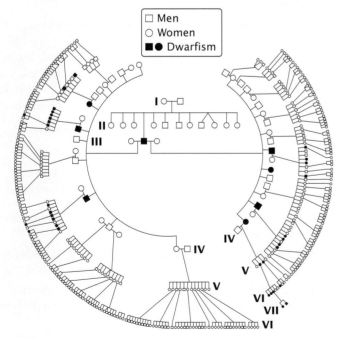

Legend:
□ Men
○ Women
■● Dwarfism

[Adapted from *The Journal of Heredity* 34:232.]

a. With the assumption that Schmid-type metaphyseal chondrodysplasia is rare, is this type of dwarfism inherited as a dominant or recessive trait? Explain your reasoning?

b. On the basis of your answer for part *a*, what is the expected ratio of dwarf and normal children in the families given in the table. Use a chi-square test to determine if the total number of children for these families (52 normal, 40 dwarfs) is significantly different from the number expected.

c. Use chi-square tests to determine if the number of children in family C (1 normal, 6 dwarf) and the number in family D (6 normal and 2 dwarf) are significantly different from the numbers expected on the basis of your proposed type of inheritance. How would you explain these deviations from the overall ratio expected?

37. Pink-eye and albino are two recessive traits found in the deer mouse *Peromyscus maniculatus*. In mice with pink-eye, the eye is devoid of color and appears pink from the blood vessels within it. Albino mice are completely lacking color both in their fur and in their eyes. F. H. Clark crossed pink-eyed mice with albino mice; the resulting F_1 had normal coloration in their fur and eyes. He then crossed these F_1 mice with mice that were pink eyed and albino and obtained the following mice. It is very hard to distinguish between mice that are albino and mice that are both pink-eye and albino, and so he combined these two phenotypes (F. H. Clark. 1936. *Journal of Heredity* 27: 259–260).

Phenotype	Number of progeny
wild-type fur, wild-type eye color	12
wild-type fur, pink-eye	62
albino pink-eye	78
Total	152

a. Give the expected numbers of progeny with each phenotype if the genes for pink-eye and albino assort independently.

b. Use a chi-square test to determine if the observed numbers of progeny fit the number expected with independent assortment.

*38. In the California poppy, an allele for yellow flowers (*C*) is dominant over an allele for white flowers (*c*). At an independently assorting locus, an allele for entire petals (*F*) is dominant over an allele for fringed petals (*f*). A plant that is homozygous for yellow and entire petals is crossed with a plant that is white and fringed. A resulting F_1 plant is then crossed with a plant that is white and fringed, and the following progeny are produced: 54 yellow and entire; 58 yellow and fringed, 53 white and entire, and 10 white and fringed.

a. Use a chi-square test to compare the observed numbers with those expected for the cross.

b. What conclusion can you make from the results of the chi-square test?

c. Suggest an explanation for the results.

CHALLENGE QUESTIONS

Section 3.2

39. Dwarfism is a recessive trait in Hereford cattle. A rancher in western Texas discovers that several of the calves in his herd are dwarfs, and he wants to eliminate this undesirable trait from the herd as rapidly as possible. Suppose that the rancher hires you as a genetic consultant to advise him on how to breed the dwarfism trait out of the herd. What crosses would you advise the rancher to conduct to ensure that the allele causing dwarfism is eliminated from the herd?

***40.** A geneticist discovers an obese mouse in his laboratory colony. He breeds this obese mouse with a normal mouse. All the F_1 mice from this cross are normal in size. When he interbreeds two F_1 mice, eight of the F_2 mice are normal in size and two are obese. The geneticist then intercrosses two of his obese mice, and he finds that all of the progeny from this cross are obese. These results lead the geneticist to conclude that obesity in mice results from a recessive allele.

A second geneticist at a different university also discovers an obese mouse in her laboratory colony. She carries out the same crosses as those done by the first geneticist and obtains the same results. She also concludes that obesity in mice results from a recessive allele. One day the two geneticists meet at a genetics conference, learn of each other's experiments, and decide to exchange mice. They both find that, when they cross two obese mice from the different laboratories, all the offspring are normal; however, when they cross two obese mice from the same laboratory, all the offspring are obese. Explain their results.

41. Albinism is a recessive trait in humans (see the introduction to Chapter 1). A geneticist studies a series of families in which both parents are normal and at least one child has albinism. The geneticist reasons that both parents in these families must be heterozygotes and that albinism should appear in $1/4$ of the children of these families. To his surprise, the geneticist finds that the frequency of albinism among the children of these families is considerably greater than $1/4$. Can you think of an explanation for the higher-than-expected frequency of albinism among these families?

42. Two distinct phenotypes are found in the salamander *Plethodon cinereus*: a red form and a black form. Some biologists have speculated that the red phenotype is due to an autosomal allele that is dominant over an allele for black. Unfortunately, these salamanders will not mate in captivity; so the hypothesis that red is dominant over black has never been tested.

[Phil Degginger/Alamy.]

One day a genetics student is hiking through the forest and finds 30 female salamanders, some red and some black, laying eggs. The student places each female and her eggs (from about 20 to 30 eggs per female) in separate plastic bags and takes them back to the lab. There, the student successfully raises the eggs until they hatch. After the eggs have hatched, the student records the phenotypes of the juvenile salamanders, along with the phenotypes of their mothers. Thus, the student has the phenotypes for 30 females and their progeny, but no information is available about the phenotypes of the fathers.

Explain how the student can determine whether red is dominant over black with this information on the phenotypes of the females and their offspring.

4 Sex Determination and Sex-Linked Characteristics

Sex in the platypus is determined by sex chromosomes. Females have 10 X chromosomes, whereas males have 5 X and 5 Y chromosomes. [Australian Geographic/Aflo Foto Agency/Photolibrary.]

THE STRANGE CASE OF PLATYPUS SEX

The platypus, *Ornithorhynchus anatinus,* is clearly one of life's strangest animals. It possesses a furry coat and, like a mammal, is warm blooded and produces milk to nourish its young, but it lacks teeth, has a bill, and lays eggs like a bird. The feet are webbed like those of a duck, and females have no nipples (offspring suck milk directly from the abdominal skin); males have spurs on their hind legs that deliver a deadly, snakelike venom. The platypus has such a hodgepodge of mammalian, avian, and reptilian traits that the first scientists to examine a platypus skin specimen thought it might be an elaborate hoax, produced by attaching parts culled from several different organisms. In spite of its strange appearance, the platypus is genetically a monotreme mammal, a side branch that diverged from the rest of mammals some 166 million years ago.

The platypus lives in eastern and southern Australia and on the island of Tasmania. An excellent swimmer, it spends most of its time in small rivers and streams foraging for worms, frogs, insect larvae, shrimp, and crayfish. Among other oddities, it locates its prey by using electroreception. The platypus genome was sequenced in 2008, providing a detailed view of the genetic makeup of this strange animal. It has a relatively small genome for a mammal, with 2.3 billion base pairs of DNA and about 18,500 protein-encoding genes. Almost 10% of its genes encode proteins that take part in odor and chemical reception. The platypus genome is a blend of mammalian, reptilian, and unique characteristics.

Platypus sex also is unusual. For most mammals, whether an individual organism is male or female is determined by sex chromosomes. Most females possess two X chromosomes, whereas males have a single X chromosome and a smaller sex chromosome called Y. This is the usual type of sex determination in mammals, but how sex is determined in the platypus remained a mystery for many years. The platypus possesses 52 chromosomes, and early geneticists observed a confusing mix of different chromosomes in male and female platypuses, including an unusual chainlike group of chromosomes in meiosis (**Figure 4.1**).

In 2004, Frank Grutzner and a group of other scientists created fluorescent paints to mark the platypus chromosomes so that they could follow the behavior of individual chromosomes in the course of meiosis. What they discovered was remarkable: platypuses

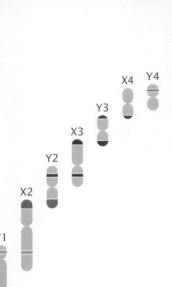

4.1 Sex chromosomes of the duckbill platypus. In meiosis, the sex chromosomes form chainlike structures. [Adapted from F. Veyrunes et al., *Genome Research* 18(6): 965–973, 2008. Copyright © 2008, Cold Spring Harbor Laboratory Press.]

possess ten sex chromosomes: female platypuses have ten X chromosomes, whereas male platypuses have five X chromosomes and five Y chromosomes. In meiosis, these sex chromosomes line up in a precise order, forming a long chain of sex chromosomes. In spite of what at first appears to be mass confusion, the platypus sex chromosomes pair and align with great precision so that each egg cell gets exactly five Xs; half the sperm get five Xs and the other half get five Ys. The mechanism that brings about this precise separation is not yet known. The complicated set of sex chromosomes in the platypus is just one example of the varied ways in which sex is determined and influences inheritance.

In Chapter 3, we studied Mendel's principles of segregation and independent assortment and saw how these principles explain much about the nature of inheritance. After Mendel's principles were rediscovered in 1900, biologists began to conduct genetic studies on a wide array of different organisms. As they applied Mendel's principles more widely, exceptions were observed, and it became necessary to devise extensions of his basic principles of heredity. In this chapter, we explore one of the major extensions of Mendel's principles: the inheritance of characteristics encoded by genes located on the sex chromosomes, which often differ in males and females (**Figure 4.2**). These characteristics and the genes that produce them are referred to as sex linked. To understand the inheritance of sex-linked characteristics, we must first know how sex is determined—why some members of a species are male and others are female. Sex determination is the focus of the first part of the chapter. The second part examines how characteristics encoded by genes on the sex chromosomes are inherited. In Chapter 5, we will explore some additional ways in which sex and inheritance interact.

As we consider sex determination and sex-linked characteristics, it will be helpful to think about two important principles. First, there are several different mechanisms of sex determination and, ultimately, the mechanism of sex determination controls the inheritance of sex-linked characteristics. Second, like other pairs of chromosomes, the X and Y sex chromosomes pair in the course of meiosis and segregate, but, throughout most of their length, they are not homologous (their gene sequences do not encode the same characteristics): most genes on the X chromosome are different from genes on the Y chromosome. Consequently, males and females do not possess the same number of alleles at sex-linked loci. This difference in the number of sex-linked alleles produces distinct patterns of inheritance in males and females. **TRY PROBLEM 14 →**

4.1 Sex Is Determined by a Number of Different Mechanisms

Sexual reproduction is the formation of offspring that are genetically distinct from their parents; most often, two parents contribute genes to their offspring and the genes are assorted into new combinations through meiosis. Among most eukaryotes, sexual reproduction consists of two processes that lead to an alternation of haploid and diploid cells: meiosis produces haploid gametes (spores in plants), and fertilization produces diploid zygotes (**Figure 4.3**).

The term **sex** refers to sexual phenotype. Most organisms have only two sexual phenotypes: male and female. The fundamental difference between males and females is gamete size: males produce small gametes; females produce relatively larger gametes (**Figure 4.4**).

The mechanism by which sex is established is termed **sex determination**. We define the sex of an individual organism in reference to its phenotype. Sometimes an individual organism has chromosomes or genes that are normally associated with one sex but a morphology corresponding to the opposite sex. For instance, the cells of female humans normally have two X chromosomes, and the cells

4.2 The male sex chromosome (Y, at the left) differs from the female sex chromosome (X, at the right) in size and shape.
[Biophoto Associates/Photo Researchers.]

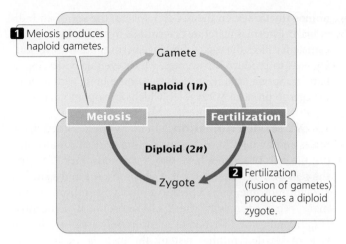

4.3 In most eukaryotic organisms, sexual reproduction consists of an alternation of haploid (1*n*) and diploid (2*n*) cells.

1 Meiosis produces haploid gametes.

Gamete

Haploid (1*n*)

Meiosis   Fertilization

Diploid (2*n*)

Zygote

2 Fertilization (fusion of gametes) produces a diploid zygote.

of males have one X chromosome and one Y chromosome. A few rare persons have male anatomy, although their cells each contain two X chromosomes. Even though these people are genetically female, we refer to them as male because their sexual phenotype is male. (As we will see later in the chapter, these XX males usually have a small piece of the Y chromosome, which is attached to another chromosome.)

CONCEPTS

In sexual reproduction, parents contribute genes to produce an offspring that is genetically distinct from both parents. In most eukaryotes, sexual reproduction consists of meiosis, which produces haploid gametes (or spores), and fertilization, which produces a diploid zygote.

✔ CONCEPT CHECK 1

What process causes the genetic variation seen in offspring produced by sexual reproduction?

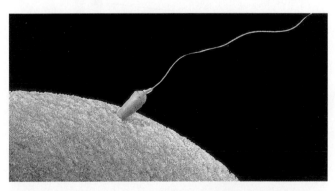

4.4 Male and female gametes (sperm and egg, respectively) differ in size. In this photograph, a human sperm (with flagellum) penetrates a human egg cell. [Francis Leroy, Biocosmos/Science Photo Library/Photo Researchers.]

There are many ways in which sex differences arise. In some species, both sexes are present in the same organism, a condition termed **hermaphroditism;** organisms that bear both male and female reproductive structures are said to be **monoecious** (meaning "one house"). Species in which the organism has either male or female reproductive structures are said to be **dioecious** (meaning "two houses"). Humans are dioecious. Among dioecious species, sex may be determined chromosomally, genetically, or environmentally.

Chromosomal Sex-Determining Systems

The chromosome theory of inheritance (discussed in Chapter 3) states that genes are located on chromosomes, which serve as vehicles for the segregation of genes in meiosis. Definitive proof of this theory was provided by the discovery that the sex of certain insects is determined by the presence or absence of particular chromosomes.

In 1891, Hermann Henking noticed a peculiar structure in the nuclei of cells from male insects. Understanding neither its function nor its relation to sex, he called this structure the X body. Later, Clarence E. McClung studied the X body in grasshoppers and recognized that it was a chromosome. McClung called it the accessory chromosome, but it eventually became known as the X chromosome, from Henking's original designation. McClung observed that the cells of female grasshoppers had one more chromosome than the number of chromosomes in the cells of male grasshoppers, and he concluded that accessory chromosomes played a role in sex determination. In 1905, Nettie Stevens and Edmund Wilson demonstrated that, in grasshoppers and other insects, the cells of females have two X chromosomes, whereas the cells of males have a single X. In some insects, they counted the same number of chromosomes in the cells of males and females but saw that one chromosome pair was different: two X chromosomes were found in female cells, whereas a single X chromosome plus a smaller chromosome, which they called Y, was found in male cells.

Stevens and Wilson also showed that the X and Y chromosomes separate into different cells in sperm formation; half of the sperm receive an X chromosome and the other half receive a Y. All egg cells produced by the female in meiosis receive one X chromosome. A sperm containing a Y chromosome unites with an X-bearing egg to produce an XY male, whereas a sperm containing an X chromosome unites with an X-bearing egg to produce an XX female. This distribution of X and Y chromosomes in sperm accounts for the 1 : 1 sex ratio observed in most dioecious organisms (**Figure 4.5**). Because sex is inherited like other genetically determined characteristics, Stevens and Wilson's discovery that sex is associated with the inheritance of a particular chromosome also demonstrated that genes are on chromosomes.

As Stevens and Wilson found for insects, sex in many organisms is determined by a pair of chromosomes, the **sex chromosomes**, which differ between males and females.

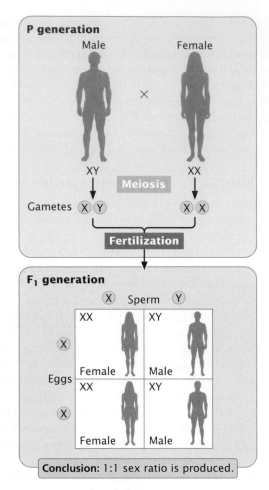

4.5 Inheritance of sex in organisms with X and Y chromosomes results in equal numbers of male and female offspring.

The nonsex chromosomes, which are the same for males and females, are called **autosomes**. We think of sex in these organisms as being determined by the presence of the sex chromosomes, but, in fact, the individual genes located on the sex chromosomes are usually responsible for the sexual phenotypes.

XX-XO sex determination   The mechanism of sex determination in the grasshoppers studied by McClung is one of the simplest mechanisms of chromosomal sex determination and is called the XX-XO system. In this system, females have two X chromosomes (XX), and males possess a single X chromosome (XO). There is no O chromosome; the letter O signifies the absence of a sex chromosome.

In meiosis in females, the two X chromosomes pair and then separate, with one X chromosome entering each haploid egg. In males, the single X chromosome segregates in meiosis to half the sperm cells; the other half receive no sex chromosome. Because males produce two different types of gametes with respect to the sex chromosomes, they are said to be the **heterogametic sex**. Females, which produce gametes that are all the same with respect to the sex chromosomes, are the

homogametic sex. In the XX-XO system, the sex of an individual organism is therefore determined by which type of male gamete fertilizes the egg. X-bearing sperm unite with X-bearing eggs to produce XX zygotes, which eventually develop as females. Sperm lacking an X chromosome unite with X-bearing eggs to produce XO zygotes, which develop into males.

XX-XY sex determination   In many species, the cells of males and females have the same number of chromosomes, but the cells of females have two X chromosomes (XX) and the cells of males have a single X chromosome and a smaller sex chromosome, the Y chromosome (XY). In humans and many other organisms, the Y chromosome is acrocentric (**Figure 4.6**), not Y shaped as is commonly assumed. In this type of sex-determining system, the male is the heterogametic sex—half of his gametes have an X chromosome and half have a Y chromosome. The female is the homogametic sex—all her egg cells contain a single X chromosome. Many organisms, including some plants, insects, and reptiles, and all mammals (including humans), have the XX-XY sex-determining system. Other organisms have variations of the XX-XY system of sex determination, including the duck-billed platypus (discussed in the introduction to this chapter) in which females have five pairs of X chromosomes and males have five pairs of X and Y chromosomes.

Although the X and Y chromosomes are not generally homologous, they do pair and segregate into different cells in meiosis. They can pair because these chromosomes are homologous in small regions called the **pseudoautosomal regions** (see Figure 4.6), in which they carry the same genes. In humans, there are pseudoautosomal regions at both tips of the X and Y chromosomes.

ZZ-ZW sex determination   In this system, the female is heterogametic and the male is homogametic. To prevent confusion with the XX-XY system, the sex chromosomes in this system are called Z and W, but the chromosomes do

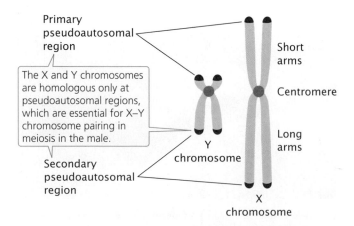

4.6 The X and Y chromosomes in humans differ in size and genetic content. They are homologous only in the pseudoautosomal regions.

not resemble Zs and Ws. Females in this system are ZW; after meiosis, half of the eggs have a Z chromosome and the other half have a W chromosome. Males are ZZ; all sperm contain a single Z chromosome. The ZZ-ZW system is found in birds, snakes, butterflies, some amphibians, and some fishes.

Genic Sex Determination

In some plants, fungi, and protozoans, sex is genetically determined, but there are no obvious differences in the chromosomes of males and females: there are no sex chromosomes. These organisms have **genic sex determination**; genotypes at one or more loci determine the sex of an individual plant, fungus, or protozoan.

It is important to understand that, even in chromosomal sex-determining systems, sex is actually determined by individual genes. For example, in mammals, a gene (*SRY*, discussed later in this chapter) located on the Y chromosome determines the male phenotype. In both genic sex determination and chromosomal sex determination, sex is controlled by individual genes; the difference is that, with chromosomal sex determination, the chromosomes also look different in males and females.

Environmental Sex Determination

Genes have had a role in all of the examples of sex determination discussed thus far. However, in a number of organisms, sex is determined fully or in part by environmental factors.

A fascinating example of environmental sex determination is seen in the marine mollusk *Crepidula fornicata*, also known as the common slipper limpet (**Figure 4.7**). Slipper limpets live in stacks, one on top of another. Each limpet begins life as a swimming larva. The first larva to settle on a solid, unoccupied substrate develops into a female limpet. It then produces chemicals that attract other larvae, which settle on top of it. These larvae develop into males, which then serve as mates for the limpet below. After a period of time, the males on top develop into females and, in turn, attract additional larvae that settle on top of the stack, develop into males, and serve as mates for the limpets under them. Limpets can form stacks of a dozen or more animals; the uppermost animals are always male. This type of sexual development is called **sequential hermaphroditism**; each individual animal can be both male and female, although not at the same time. In *Crepidula fornicata*, sex is determined environmentally by the limpet's position in the stack.

Environmental factors are also important in determining sex in many reptiles. Although most snakes and lizards have sex chromosomes, the sexual phenotype of many turtles, crocodiles, and alligators is affected by temperature during embryonic development. In turtles, for example,

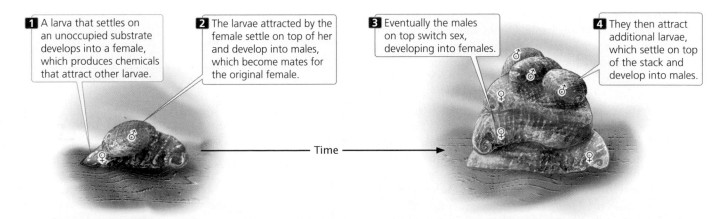

1 A larva that settles on an unoccupied substrate develops into a female, which produces chemicals that attract other larvae.

2 The larvae attracted by the female settle on top of her and develop into males, which become mates for the original female.

3 Eventually the males on top switch sex, developing into females.

4 They then attract additional larvae, which settle on top of the stack and develop into males.

Time →

4.7 In *Crepidula fornicata*, the common slipper limpet, sex is determined by an environmental factor—the limpet's position in a stack of limpets.

Table 4.1 Some sex-determining systems

System	Mechanism		Heterogametic Sex
XX-XO	Females XX	Males X	Male
XX-XY	Females XX	Males XY	Male
ZZ-ZW	Females ZW	Males ZZ	Female
Genic sex determination	No distinct sex chromosomes Sex determined by genes on undifferentiated chromosomes		Varies
Environmental sex determination	Sex determined by environmental factors		None

warm temperatures produce females during certain times of the year, whereas cool temperatures produce males. In alligators, the reverse is true. Some of the different types of sex determination are summarized in **Table 4.1**.

Now that we have surveyed some of the different ways in which sex can be determined, we will examine in detail one mechanism (the XX-XY system). Sex determination is XX-XY in both fruit flies and humans but, as we will see, the way in which the X and Y chromosomes determine sex in these two organisms is quite different. TRY PROBLEMS 4 AND 20 →

CONCEPTS

In genic sex determination, sex is determined by genes at one or more loci, but there are no obvious differences in the chromosomes of males and females. In environmental sex determination, sex is determined fully or in part by environmental factors.

✔ CONCEPT CHECK 3

How do chromosomal, genic, and environmental sex determination differ?

Sex Determination in *Drosophila melanogaster*

The fruit fly *Drosophila melanogaster* has eight chromosomes: three pairs of autosomes and one pair of sex chromosomes. Thus, it has inherited one haploid set of autosomes and one sex chromosome from each parent. Normally, females have two X chromosomes and males have an X chromosome and a Y chromosome. However, the presence of the Y chromosome does not determine maleness in *Drosophila;* instead, each fly's sex is determined by a balance between genes on the autosomes and genes on the X chromosome. This type of sex determination is called the **genic balance system**. In this system, a number of different genes influence sexual development. The X chromosome contains genes with female-producing effects, whereas the autosomes contain genes with male-producing effects. Consequently, a fly's sex is determined by the **X : A ratio**, the number of X chromosomes divided by the number of haploid sets of autosomal chromosomes.

An X : A ratio of 1.0 produces a female fly; an X : A ratio of 0.5 produces a male. If the X : A ratio is less than 0.5, a male phenotype is produced, but the fly is weak and sterile—such flies are sometimes called metamales. An X : A ratio between 1.0 and 0.5 produces an intersex fly, with a mixture of male and female characteristics. If the X : A ratio is greater than 1.0, a female phenotype is produced, but this fly (called a metafemale) has serious developmental problems and many never complete development. **Table 4.2** presents some different chromosome complements in *Drosophila* and their associated sexual phenotypes. Normal females have two X chromosomes and two sets of autosomes (XX, AA), and so their X : A ratio is 1.0. Males, on the other hand, normally have a single X and two sets of autosomes (XY, AA), and so their X : A ratio is 0.5. Flies with XXY sex chromosomes and two sets of autosomes (an X : A ratio of 1.0) develop as fully fertile females, in spite of the presence of a Y chromosome. Flies with only a single X and two sets of autosomes (XO, AA, for an X : A ratio of 0.5) develop as males, although they are sterile. These observations confirm that the Y chromosome does not determine sex in *Drosophila*. TRY PROBLEM 15 →

Table 4.2 Chromosome complements and sexual phenotypes in *Drosophila*

Sex-Chromosome Complement	Haploid Sets of Autosomes	X : A Ratio	Sexual Phenotype
XX	AA	1.0	Female
XY	AA	0.5	Male
XO	AA	0.5	Male
XXY	AA	1.0	Female
XXX	AA	1.5	Metafemale
XXXY	AA	1.5	Metafemale
XX	AAA	0.67	Intersex
XO	AAA	0.33	Metamale
XXXX	AAA	1.3	Metafemale

CONCEPTS

The sexual phenotype of a fruit fly is determined by the ratio of the number of X chromosomes to the number of haploid sets of autosomal chromosomes (the X : A ratio).

✔ CONCEPT CHECK 4

What is the sexual phenotype of a fruit fly that has XXYYYY sex chromosomes and two sets of autosomes?

a. Male   c. Intersex

b. Female   d. Metamale

Sex Determination in Humans

Humans, like *Drosophila,* have XX-XY sex determination, but, in humans, the presence of a gene (*SRY*) on the Y chromosome determines maleness. The phenotypes that result from abnormal numbers of sex chromosomes, which arise when the sex chromosomes do not segregate properly in meiosis or mitosis, illustrate the importance of the Y chromosome in human sex determination.

Turner syndrome   Persons who have **Turner syndrome** are female and often have underdeveloped secondary sex characteristics. This syndrome is seen in 1 of 3000 female births. Affected women are frequently short and have a low hairline, a relatively broad chest, and folds of skin on the neck. Their intelligence is usually normal. Most women who have Turner syndrome are sterile. In 1959, Charles Ford used new techniques to study human chromosomes and discovered that cells from a 14-year-old girl with Turner syndrome had only a single X chromosome (**Figure 4.8**); this chromosome complement is usually referred to as XO.

There are no known cases in which a person is missing both X chromosomes, an indication that at least one X chromosome is necessary for human development. Presumably,

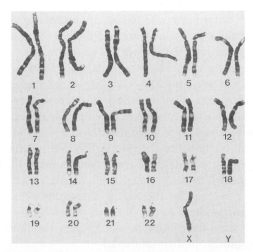

4.8 Persons with Turner syndrome have a single X chromosome in their cells. [Department of Clinical Cytogenetics, Addenbrookes Hospital/Science Photo Library/Photo Reseachers.]

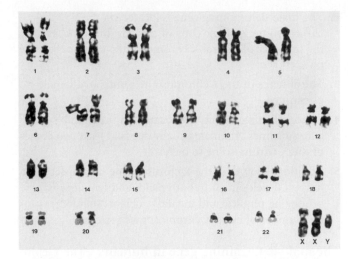

4.9 Persons with Klinefelter syndrome have a Y chromosome and two or more X chromosomes in their cells. [Biophoto Associates/Science Source/Photo Researchers.]

embryos missing both Xs spontaneously abort in the early stages of development.

Klinefelter syndrome   Persons who have **Klinefelter syndrome**, which occurs with a frequency of about 1 in 1000 male births, have cells with one or more Y chromosomes and multiple X chromosomes. The cells of most males having this condition are XXY (**Figure 4.9**), but the cells of a few Klinefelter males are XXXY, XXXXY, or XXYY. Men with this condition frequently have small testes and reduced facial and pubic hair. They are often taller than normal and sterile; most have normal intelligence.

Poly-X females   In about 1 in 1000 female births, the infant's cells possess three X chromosomes, a condition often referred to as **triplo-X syndrome**. These persons have no distinctive features other than a tendency to be tall and thin. Although a few are sterile, many menstruate regularly and are fertile. The incidence of mental retardation among triple-X females is slightly greater than that in the general population, but most XXX females have normal intelligence. Much rarer are females whose cells contain four or five X chromosomes. These females usually have normal female anatomy but are mentally retarded and have a number of physical problems. The severity of mental retardation increases as the number of X chromosomes increases beyond three.

The role of sex chromosomes   The phenotypes associated with sex-chromosome anomalies allow us to make several inferences about the role of sex chromosomes in human sex determination.

1. The X chromosome contains genetic information essential for both sexes; at least one copy of an X chromosome is required for human development.

2. The male-determining gene is located on the Y chromosome. A single copy of this chromosome, even in the presence of several X chromosomes, produces a male phenotype.

3. The absence of the Y chromosome results in a female phenotype.

4. Genes affecting fertility are located on the X and Y chromosomes. A female usually needs at least two copies of the X chromosome to be fertile.

5. Additional copies of the X chromosome may upset normal development in both males and females, producing physical and mental problems that increase as the number of extra X chromosomes increases.

The male-determining gene in humans   The Y chromosome in humans and all other mammals is of paramount importance in producing a male phenotype. However, scientists discovered a few rare XX males whose cells apparently lack a Y chromosome. For many years, these males presented an enigma: How could a male phenotype exist without a Y chromosome? Close examination eventually revealed a small part of the Y chromosome attached to another chromosome. This finding indicates that it is not the entire Y chromosome that determines maleness in humans; rather, it is a gene on the Y chromosome.

Early in development, all humans possess undifferentiated gonads and both male and female reproductive ducts. Then, about 6 weeks after fertilization, a gene on the Y chromosome becomes active. By an unknown mechanism, this gene causes the neutral gonads to develop into testes, which begin to secrete two hormones: testosterone and Mullerian-inhibiting substance. Testosterone induces the development of male characteristics, and Mullerian-inhibiting substance causes the degeneration of the female reproductive ducts. In the absence of this male-determining gene, the neutral gonads become ovaries, and female features develop.

The male-determining gene in humans, called the **sex-determining region Y** (*SRY*) **gene,** was discovered in 1990 (**Figure 4.10**). This gene is found in XX males and is missing from XY females; it is also found on the Y chromosome of other mammals. Definitive proof that *SRY* is the male-

determining gene came when scientists placed a copy of this gene into XX mice by means of genetic engineering. The XX mice that received this gene, although sterile, developed into anatomical males.

The *SRY* gene encodes a protein called a transcription factor (see Chapter 13) that binds to DNA and stimulates the transcription of other genes that promote the differentiation of the testes. Although *SRY* is the primary determinant of maleness in humans, other genes (some X linked, others Y linked, and still others autosomal) also have roles in fertility and the development of sex differences.

Androgen-insensitivity syndrome   Although the *SRY* gene is the primary determinant of sex in human embryos, several other genes influence sexual development, as illustrated by women with androgen-insensitivity syndrome. These persons have female external sexual characteristics and think of themselves as female. Indeed, most are unaware of their condition until they reach puberty and fail to menstruate. Examination by a gynecologist reveals that the vagina ends blindly and that the uterus, oviducts, and ovaries are absent. Inside the abdominal cavity, a pair of testes produce levels of testosterone normally seen in males. The cells of a woman with androgen-insensitivity syndrome contain an X and a Y chromosome.

How can a person be female in appearance when her cells contain a Y chromosome and she has testes that produce testosterone? The answer lies in the complex relation between genes and sex in humans. In a human embryo with a Y chromosome, the *SRY* gene causes the gonads to develop into testes, which produce testosterone. Testosterone stimulates embryonic tissues to develop male characteristics. But, for testosterone to have its effects, it must bind to an androgen receptor. This receptor is defective in females with androgen-insensitivity syndrome; consequently, their cells are insensitive to testosterone, and female characteristics develop. The gene for the androgen receptor is located on the X chromosome; so persons with this condition always inherit it from their mothers. (All XY persons inherit the X chromosome from their mothers.)

Androgen-insensitivity syndrome illustrates several important points about the influence of genes on a person's

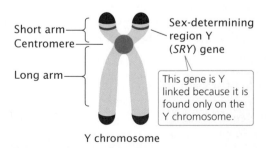

Short arm
Centromere

Long arm

Sex-determining region Y (*SRY*) gene

This gene is Y linked because it is found only on the Y chromosome.

Y chromosome

4.10 The *SRY* gene is on the Y chromosome and causes the development of male characteristics.

sex. First, this condition demonstrates that human sexual development is a complex process, influenced not only by the *SRY* gene on the Y chromosome, but also by other genes found elsewhere. Second, it shows that most people carry genes for both male and female characteristics, as illustrated by the fact that those with androgen-insensitivity syndrome have the capacity to produce female characteristics, even though they have male chromosomes. Indeed, the genes for most male and female secondary sex characteristics are present not on the sex chromosomes but on autosomes. The key to maleness and femaleness lies not in the genes but in the control of their expression. TRY PROBLEM 16 →

4.2 Sex-Linked Characteristics Are Determined by Genes on the Sex Chromosomes

In Chapter 3, we learned several basic principles of heredity that Mendel discovered from his crosses among pea plants. A major extension of these Mendelian principles is the pattern of inheritance exhibited by **sex-linked characteristics**, characteristics determined by genes located on the sex chromosomes. Genes on the X chromosome determine **X-linked characteristics**; those on the Y chromosome determine **Y-linked characteristics**. Because the Y chromosome of many organisms contains little genetic information, most sex-linked characteristics are X linked. Males and females differ in their sex chromosomes; so the pattern of inheritance for sex-linked characteristics differs from that exhibited by genes located on autosomal chromosomes.

X-Linked White Eyes in *Drosophila*

The first person to explain sex-linked inheritance was American biologist Thomas Hunt Morgan (**Figure 4.11**). Morgan began his career as an embryologist, but the discovery of Mendel's principles inspired him to begin conducting genetic experiments, initially on mice and rats. In 1909, Morgan switched to *Drosophila melanogaster;* a year later, he discovered among the flies of his laboratory colony a single male that possessed white eyes, in stark contrast with the red eyes of normal fruit flies. This fly had a tremendous effect on the future of genetics and on Morgan's career as a biologist.

To investigate the inheritance of the white-eyed characteristic in fruit flies, Morgan systematically carried out a series of genetic crosses. First, he crossed pure-breeding, red-eyed females with his white-eyed male, producing F_1 progeny of which all had red eyes (**Figure 4.12a**). (In fact, Morgan found 3 white-eyed males among the 1237 progeny, but he assumed that the white eyes were due to new mutations.) Morgan's results from this initial cross were consistent with Mendel's principles: a cross between a homozygous dominant individual and a homozygous recessive individual produces heterozygous offspring exhibiting the dominant trait. His results suggested that white eyes are a simple recessive trait. However, when Morgan crossed the F_1 flies with one another, he found that all the female F_2 flies possessed red eyes but that half the male F_2 flies had red eyes and the other half had white eyes. This finding was clearly not the expected result for a simple recessive trait, which should appear in $\frac{1}{4}$ of both male and female F_2 offspring.

To explain this unexpected result, Morgan proposed that the locus affecting eye color is on the X chromosome (i.e., eye color is X linked). He recognized that the eye-color alleles are present only on the X chromosome; no homologous allele is present on the Y chromosome. Because the cells of females possess two X chromosomes, females can be homozygous or heterozygous for the eye-color alleles. The cells of males, on the other hand, possess only a single X chromosome and can carry only a single eye-color allele. Males therefore cannot be either homozygous or heterozygous but are said to be **hemizygous** for X-linked loci.

(a)

(b)

4.11 Thomas Hunt Morgan's work with *Drosophila* **helped unravel many basic principles of genetics, including X-linked inheritance.** (a) Morgan. (b) The Fly Room, where Morgan and his students conducted genetic research. [Part a: AP/Wide World Photos. Part b: American Philosophical Society.]

Question: Are white eyes in fruit flies inherited as an autosomal recessive trait?

Methods Perform reciprocal crosses.

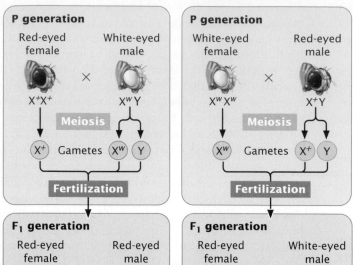

(a) Red-eyed female crossed with white-eyed male

(b) White-eyed female crossed with red-eyed male

Results

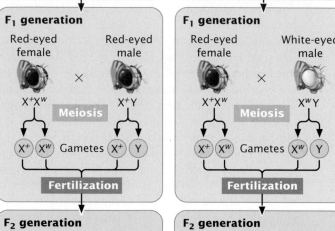

½ red-eyed females
¼ red-eyed males
¼ white-eyed males

¼ red-eyed females
¼ white-eyed females
¼ red-eyed males
¼ white-eyed males

Conclusion: No. The results of reciprocal crosses are consistent with X-linked inheritance.

4.12 Morgan's X-linked crosses for white eyes in fruit flies. (a) Original and F_1 crosses. (b) Reciprocal crosses.

▶ www.whfreeman.com/pierce4e   The results of Morgan's crosses are illustrated in Animation 4.1.

To verify his hypothesis that the white-eye trait is X linked, Morgan conducted additional crosses. He predicted that a cross between a white-eyed female and a red-eyed male would produce all red-eyed females and all white-eyed males (**Figure 4.12b**). When Morgan performed this cross, the results were exactly as predicted. Note that this cross is the reciprocal of the original cross and that the two reciprocal crosses produced different results in the F_1 and F_2 generations. Morgan also crossed the F_1 heterozygous females with their white-eyed father, the red-eyed F_2 females with white-eyed males, and white-eyed females with white-eyed males. In all of these crosses, the results were consistent with Morgan's conclusion that white eyes are an X-linked characteristic.

Nondisjunction and the Chromosome Theory of Inheritance

When Morgan crossed his original white-eyed male with homozygous red-eyed females, all 1237 of the progeny had red eyes, except for 3 white-eyed males. As already mentioned, Morgan attributed these white-eyed F_1 males to the occurrence of further random mutations. However, flies with these unexpected phenotypes continued to appear in his crosses. Although uncommon, they appeared far too often to be due to spontaneous mutation. Calvin Bridges, who was one of Morgan's students, set out to investigate the genetic basis of these exceptions.

Bridges found that the exceptions arose only in certain strains of white-eyed flies. When he crossed one of these exceptional white-eyed females with a red-eyed male, about 5% of the male offspring had red eyes and about 5% of the female offspring had white eyes. In this cross, the expected result is that every male fly should inherit its mother's X chromosome and should have the genotype $X^w Y$ and white eyes (see the F_1 progeny in Figure 4.12b). Every female fly should inherit a dominant red-eye allele on its father's X chromosome, along with a white-eye allele on its mother's X chromosome; thus, all the female progeny should be $X^+ X^w$ and have red eyes (see F_1 progeny in Figure 4.12b). The continual appearance of red-eyed males and white-eyed females in this cross was therefore unexpected.

Bridges's explanation To explain the appearance of red-eyed males and white-eyed females in his cross, Bridges hypothesized that the exceptional white-eyed females of this strain actually possessed two X chromosomes and a Y chromosome (X^wX^wY). Sex in *Drosophila* is determined by the X : A ratio (see Table 4.2); for XXY females, the X : A ratio is 1.0, and so these flies developed as females in spite of possessing a Y chromosome.

About 90% of the time, the two X chromosomes of the X^wX^wY females separate from each other in anaphase I of meiosis, with an X and a Y chromosome going into one gamete and a single X going into another other gamete (**Figure 4.13**). When these gametes are fertilized by sperm from a normal red-eyed male, white-eyed males and red-eyed females are produced. About 10% of the time, the two X chromosomes in the females fail to separate in anaphase I of meiosis, a phenomenon known as **nondisjunction.** When nondisjunction of the Xs occurs, half of the eggs receive two copies of the X chromosome and the other half receive only a Y chromosome (see Figure 4.13). When these eggs are fertilized by sperm from a normal red-eyed male, four combinations of sex chromosomes are produced. An egg with two X chromosomes that is fertilized by an X bearing sperm produces an $X^+X^wX^w$ zygote, which usually dies. When an egg carrying two X chromosomes is fertilized by a Y-bearing sperm, the resulting zygote is X^wX^wY, which develops into a white-eyed female. An egg with only a Y chromosome that is fertilized by an X-bearing sperm produces an X^+Y zygote, which develops into a normal red-eyed male. If the egg with only a Y chromosome is fertilized by a Y-bearing sperm, the resulting zygote has two Y chromosomes and no X chromosome and dies. Nondisjunction of the X chromosomes among X^wX^wY white-eyed females therefore produces a few white-eyed females and red-eyed males, which is exactly what Bridges found in his crosses.

Confirmation of Bridges's hypothesis   Bridges's hypothesis predicted that the white-eyed females from these crosses would possess two X chromosomes and one Y chromosome and that the red-eyed males would possess a single X chromosome. To verify his hypothesis, Bridges examined the chromosomes of his flies and found precisely what he had predicted. The significance of Bridges's study is not that it explained the appearance of an occasional odd fly in his culture but that he was able link the inheritance of a specific gene (*w*) to the presence of a specific chromosome (X). This association between genotype and chromosomes gave unequivocal evidence that sex-linked genes are located on the X chromosome and confirmed the chromosome theory of inheritance. **TRY PROBLEM 21 →**

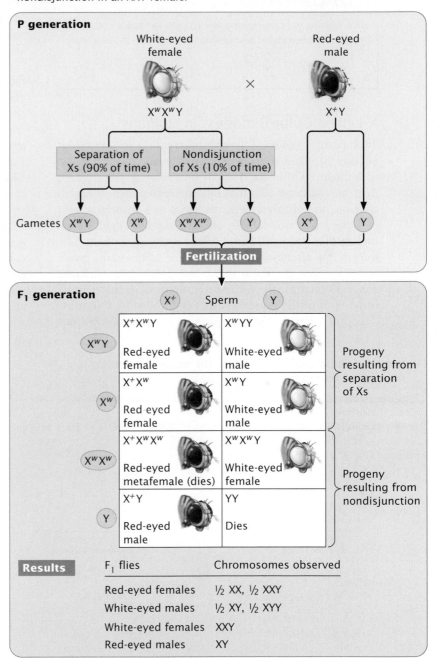

Experiment

Question: In a cross between a white-eyed female and a red-eyed male, why are a few white-eyed females and red-eyed males produced?

Methods

Hypothesis: White-eyed females and red-eyed males in F_1 result from nondisjunction in an XXY female.

Conclusion: The white-eyed females and red-eyed males in the F_1 result from nondisjunction of the X chromosomes in an XXY female.

4.13 Bridges conducted experiments that proved that the gene for white eyes is located on the X chromosome.

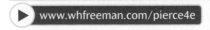

www.whfreeman.com/pierce4e   The results of Bridges's crosses are further explained in Animation 4.1.

CONCEPTS

By showing that the appearance of rare phenotypes is associated with the inheritance of particular chromosomes, Bridges proved that sex-linked genes are located on the X chromosome and that the chromosome theory of inheritance is correct.

✔ CONCEPT CHECK 6

What was the genotype of the few F_1 red-eyed males obtained by Bridges when he crossed a white-eyed female with a red-eyed male?

a. X^+ c. X^+Y

b. X^wX^+Y d. X^+X^+Y

X-Linked Color Blindness in Humans

To further examine X-linked inheritance, let's consider another X-linked characteristic: red–green color blindness in humans. Within the human eye, color is perceived in light-sensing cone cells that line the retina. Each cone cell contains one of three pigments capable of absorbing light of a particular wavelength; one absorbs blue light, a second absorbs red light, and a third absorbs green light. The human eye actually detects only three colors—red, green, and blue—but the brain mixes the signals from different cone cells to create the wide spectrum of colors that we perceive. Each of the three pigments is encoded by a separate locus; the locus for the blue pigment is found on chromosome 7, and those for the green and the red pigments lie close together on the X chromosome.

The most common types of human color blindness are caused by defects of the red and green pigments; we will refer to these conditions as red–green color blindness. Mutations that produce defective color vision are generally recessive and, because the genes encoding the red and the green pigments are located on the X chromosome, red–green color blindness is inherited as an X-linked recessive characteristic.

We will use the symbol X^c to represent an allele for red–green color blindness and the symbol X^+ to represent an allele for normal color vision. Females possess two X chromosomes; so there are three possible genotypes among females: X^+X^+ and X^+X^c, which produce normal vision, and X^cX^c, which produces color blindness. Males have only a single X chromosome and two possible genotypes: X^+Y, which produces normal vision, and X^cY which produces color blindness.

If a woman homozygous for normal color vision mates with a color-blind man (**Figure 4.14a**), all of the gametes produced by the woman will contain an allele for normal color vision. Half of the man's gametes will receive the X chromosome with the color-blind allele, and the other half will receive the Y chromosome, which carries no alleles affecting color vision. When an X^c-bearing sperm unites with the X^+-bearing egg, a heterozygous female with normal vision (X^+X^c) is produced. When a Y-bearing sperm unites with the X-bearing egg, a hemizygous male with normal vision (X^+Y) is produced.

In the reciprocal cross between a color-blind woman and a man with normal color vision (**Figure 4.14b**), the woman produces only X^c-bearing gametes. The man produces some gametes that contain the X chromosome and others that contain the Y chromosome. Males inherit the X chromosome from their mothers; because both of the mother's X chromosomes bear the X^c allele in this case, all the male offspring will be color blind. In contrast, females inherit an X chromosome from both parents; thus all the female offspring of this reciprocal cross will be heterozygous with normal vision. Females are color blind only when color-blind alleles have been inherited from both parents, whereas a color-blind male need inherit a color-blind allele from his mother only; for this reason, color blindness and most other rare X-linked recessive characteristics are more common in males.

In these crosses for color blindness, notice that an affected woman passes the X-linked recessive trait to her sons but not to her daughters, whereas an affected man passes the trait to his grandsons through his daughters but never to his sons. X-linked recessive characteristics may therefore appear to alternate

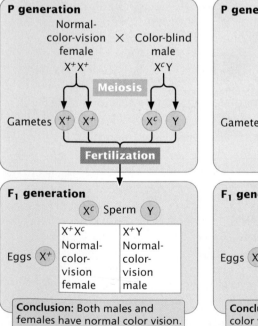
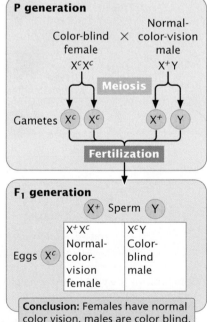

4.14 Red–green color blindness is inherited as an X-linked recessive trait in humans.

between the sexes, appearing in females one generation and in males the next generation.

Recall that the X and Y chromosomes pair in meiosis because they are homologous at the small pseudoautosomal regions. Genes in these regions of the X and Y chromosome are homologous, just like those on autosomes, and they exhibit autosomal patterns of inheritance rather than the sex-linked inheritance seen for most genes on the X and Y chromosomes.

Worked Problem

Now that we understand the pattern of X-linked inheritance, let's apply our knowledge to answer a specific question in regard to X-linked inheritance of color blindness in humans.

Betty has normal vision, but her mother is color blind. Bill is color blind. If Bill and Betty marry and have a child together, what is the probability that the child will be color blind?

• Solution

Because color blindness is an X-linked recessive characteristic, Betty's color-blind mother must be homozygous for the color-blind allele ($X^c X^c$). Females inherit one X chromosome from each of their parents; so Betty must have inherited a color-blind allele from her mother. Because Betty has normal color vision, she must have inherited an allele for normal vision (X^+) from her father; thus Betty is heterozygous ($X^+ X^c$). Bill is color blind. Because males are hemizygous for X-linked alleles, he must be ($X^c Y$). A mating between Betty and Bill is represented as:

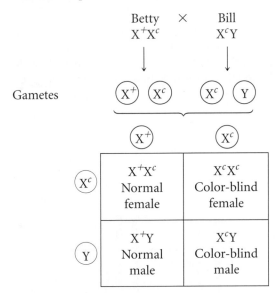

Thus, $\frac{1}{4}$ of the children are expected to be female with normal color vision, $\frac{1}{4}$ female with color blindness, $\frac{1}{4}$ male with normal color vision, and $\frac{1}{4}$ male with color blindness.

→ Get some additional practice with X-linked inheritance by working Problem 23 at the end of this chapter.

CONCEPTS

Characteristics determined by genes on the sex chromosomes are called sex-linked characteristics. Diploid females have two alleles at each X-linked locus, whereas diploid males possess a single allele at each X-linked locus. Females inherit X-linked alleles from both parents, but males inherit a single X-linked allele from their mothers.

✔ CONCEPT CHECK 7

Hemophilia (reduced blood clotting) is an X-linked recessive disease in humans. A woman with hemophilia mates with a man who exhibits normal blood clotting. What is the probability that their child will have hemophilia?

Symbols for X-Linked Genes

There are several different ways to record genotypes for X-linked traits. Sometimes the genotypes are recorded in the same fashion as they are for autosomal characteristics. In this case, the hemizygous males are simply given a single allele: the genotype of a female *Drosophila* with white eyes is *ww*, and the genotype of a white-eyed hemizygous male is *w*. Another method is to include the Y chromosome, designating it with a diagonal slash (/). With this method, the white-eyed female's genotype is still *ww* and the white-eyed male's genotype is *w/*. Perhaps the most useful method is to write the X and Y chromosomes in the genotype, designating the X-linked alleles with superscripts, as is done in this chapter. With this method, a white-eyed female is $X^w X^w$ and a white-eyed male is $X^w Y$. The use of Xs and Ys in the genotype has the advantage of reminding us that the genes are X linked and that the male must always have a single allele, inherited from the mother.

Z-Linked Characteristics

In organisms with ZZ-ZW sex determination, the males are the homogametic sex (ZZ) and carry two sex-linked (usually referred to as Z-linked) alleles; thus males may be homozygous or heterozygous. Females are the heterogametic sex (ZW) and possess only a single Z-linked allele. The inheritance of Z-linked characteristics is the same as that of X-linked characteristics, except that the pattern of inheritance in males and females is reversed.

An example of a Z-linked characteristic is the cameo plumage phenotype in Indian blue peafowl *(Pavo cristatus)*. In these birds, the wild-type plumage is a glossy metallic blue. The female peafowl is ZW and the male is ZZ. Cameo plumage, which produces brown feathers, results from a Z-linked allele (Z^{ca}) that is recessive to the wild-type blue allele (Z^{Ca+}). If a blue-colored female ($Z^{Ca+}W$) is crossed with a cameo male ($Z^{ca}Z^{ca}$), all the F_1 females are cameo ($Z^{ca}W$) and all the

F_1 males are blue ($Z^{Ca+}Z^{ca}$), as shown in **Figure 4.15**. When the F_1 are interbred, $1/4$ of the F_2 are blue males ($Z^{Ca+}Z^{ca}$), $1/4$ are blue females ($Z^{Ca+}W$), $1/4$ are cameo males ($Z^{ca}Z^{ca}$), and $1/4$ are cameo females ($Z^{ca}W$). The reciprocal cross of a cameo female with a homozygous blue male produces an F_1 generation in which all offspring are blue and an F_2 generation consisting of $1/2$ blue males ($Z^{Ca+}Z^{Ca+}$ and $Z^{Ca+}Z^{ca}$), $1/4$ blue females ($Z^{Ca+}W$), and $1/4$ cameo females ($Z^{ca}W$).

In organisms with ZZ-ZW sex determination, the female always inherits her W chromosome from her mother, and she inherits her Z chromosome, along with any Z-linked alleles, from her father. In this system, the male inherits Z chromosomes, along with any Z-linked alleles, from both his mother and his father. This pattern of inheritance is the reverse of that of X-linked alleles in organisms with XX-XY sex determination. ┃ TRY PROBLEM 32 →┃

Y-Linked Characteristics

Y-linked traits exhibit a distinct pattern of inheritance and are present only in males because only males possess a Y chromosome. All male offspring of a male with a Y-linked trait will display the trait, because every male inherits the Y chromosome from his father. ┃ TRY PROBLEM 42 →┃

Evolution of the Y chromosome Research on sex chromosomes has led to the conclusion the X and Y chromosomes in many organisms evolved from a pair of autosomes. The first step in this evolutionary process took place when one member of a pair of autosomes acquired a gene that determines maleness, such as the *SRY* gene found in humans today (**Figure 4.16**). This step took place in mammals about 250 millions years ago. Any individual organism with a copy of the chromosome containing this gene then became male. Additional mutations occurred on the proto-Y chromosome affecting traits that are beneficial only in males, such as the bright coloration used by male birds to attract females and the antlers used by a male elk in competition with other males. The genes that encode these types of traits are advantageous only if they are present in males. To prevent genes

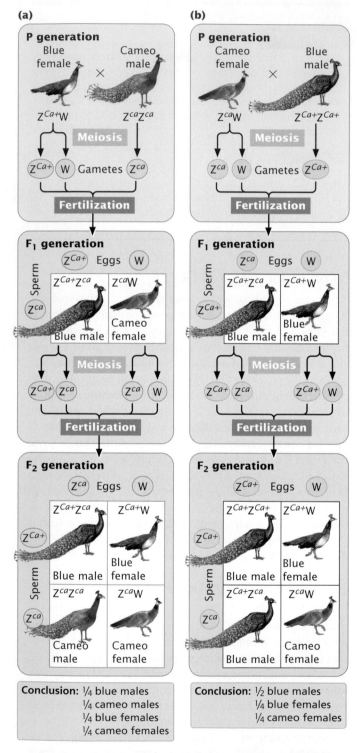

4.15 The cameo phenotype in Indian blue peafowl is inherited as a Z-linked recessive trait. (a) Blue female crossed with cameo male. (b) Reciprocal cross of cameo female crossed with homozygous blue male.

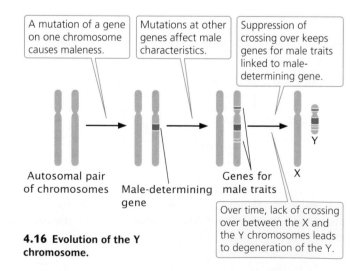

4.16 Evolution of the Y chromosome.

that encode male traits from appearing in females, crossing over was suppressed for most of the length of the X and Y chromosomes in meiosis. Crossing over can still take place between the two X chromosomes in females, but there is little crossing over between the X and the Y chromosomes, except for small pseudoautosomal regions in which the X and the Y chromosomes continue to pair and segregate in meiosis, as stated earlier.

For reasons that are beyond the discussion here, the lack of crossing over led to (and continues to lead to) an accumulation of mutations and the loss of genetic material from the Y chromosome (see Figure 4.16). Over millions of years, the Y chromosome slowly degenerated, losing DNA and genes until it became greatly reduced in size and contained little genetic information. This degeneration produced the Y chromosome found in males today. Indeed, the Y chromosomes of humans and many other organisms are usually small and contain little genetic information; therefore, few characteristics exhibit Y-linked inheritance. Some researchers have predicted that the human Y chromosome will continue to lose genetic information in the future and will completely disappear from the species in about 10 million years, a disheartening prospect for those of us with a Y chromosome. However, internal recombination within the Y chromosome (see next section) may prevent the ultimate demise of the human Y chromosome.

Characteristics of the human Y chromosome The genetic sequence of most of the human Y chromosome has now been determined as part of the Human Genome Project (see Chapter 19). This work reveals that about two-thirds of the Y chromosome consists of short DNA sequences that are repeated many times and contain no active genes. The other third consists of just a few genes. Only about 350 genes have been identified on the human Y chromosome, compared with thousands on most chromosomes, and only about half of those identified on the Y chromosome encode proteins. The function of most Y-linked genes is poorly understood; many appear to influence male sexual development and fertility. Some are expressed throughout the body, but many are expressed predominately or exclusively in the testes. Although the Y chromosome has relatively few genes, recent research in *Drosophila* suggests that the Y chromosome carries genetic elements that affect the expression of numerous genes on autosomal and X chromosomes.

A surprising feature revealed by sequencing is the presence of eight massive palindromic sequences on the Y chromosome. A palindrome is defined as a word, such as "rotator," or sentence that reads the same backward and forward. A palindromic sequence in DNA reads the same on both strands of the double helix, creating two nearly identical copies stretching out from a central point, such as:

$$\overbrace{}^{\text{Arm 1}}$$
$$5'\text{–TGGGAG}\ldots\text{CTCCCA–}3'$$
$$3'\text{–ACCCTC}\ldots\text{GAGGGT–}5'$$
$$\underbrace{}_{\text{Arm 2}}$$

Thus, a palindromic sequence in DNA appears twice, very much like the two copies of a DNA sequence that are found on two homologous chromosomes. Indeed, recombination takes place between the two palindromic sequences on the Y chromosome. As already mentioned, the X and the Y chromosomes are not homologous at almost all of their sequences, and most of the Y chromosome does not undergo crossing over with the X chromosome. This lack of interchromosomal recombination leads to an accumulation of deleterious mutations on the Y chromosome and the loss of genetic material. Evidence suggests that the two arms of the Y chromosome recombine with each other, which partly compensates for the absence of recombination between the X and the Y chromosomes. This internal recombination may help to maintain some sequences and functions of genes on the Y chromosome and prevent its total degeneration.

Although the palindromes afford opportunities for recombination, which helps prevent the decay of the Y chromosome over evolutionary time, they occasionally have harmful effects. Recent research has revealed that recombination between the palindromes can lead to rearrangements of the Y chromosome that cause anomalies of sexual development. In some cases, recombination between the palindromes leads to deletion of the *SRY* gene, producing an XY female. In other cases, recombination deletes other Y-chromosome genes that take part in sperm production. Sometimes, recombination produces a Y chromosome with two centromeres, which may break as the centromeres are pulled in opposite directions in mitosis. The broken Y chromosomes may be lost in mitosis, resulting in XO cells and Turner syndrome.

CONCEPTS

Y-linked characteristics exhibit a distinct pattern of inheritance: they are present only in males, and all male offspring of a male with a Y-linked trait inherit the trait. Palindromic sequences within the Y chromosome can undergo internal recombination, but such recombination may lead to chromosome anomalies.

✔ CONCEPT CHECK 8

What unusual feature of the Y chromosome allows some recombination among the genes found on it?

The use of Y-linked genetic markers DNA sequences in the Y chromosome undergo mutation with the passage of time and vary among individual males. These mutations create variations in DNA sequence—called genetic markers—that are passed from father to son like Y-linked traits and can be used to study male ancestry. Although the markers themselves do not encode any physical traits, they can be detected with the use of molecular methods. Much of the Y chromosome is nonfunctional; so mutations readily accumulate. Many of these mutations are unique; they arise only once and are passed down through the generations. Individual males possessing the same set of mutations are therefore related, and the distribution of these genetic markers on Y chromosomes provides clues about the genetic relationships of present-day people.

Y-linked markers have been used to study the offspring of Thomas Jefferson, principal author of the Declaration of Independence and third president of the United States. In 1802, Jefferson was accused by a political enemy of fathering a child by his slave Sally Hemings, but the evidence was circumstantial. Hemings, who worked in the Jefferson household and accompanied Jefferson on a trip to Paris, had five children. Jefferson was accused of fathering the first child, Tom, but rumors about the paternity of the other children circulated as well. Hemings's last child, Eston, bore a striking resemblance to Jefferson, and her fourth child, Madison, testified late in life that Jefferson was the father of all of Hemings's children. Ancestors of Hemings's children maintained that they were descendants of the Jefferson line, but some Jefferson descendants refused to recognize their claim.

To resolve this long-standing controversy, geneticists examined markers from the Y chromosomes of male-line descendants of Hemings's first son (Thomas Woodson), her last son (Eston Hemings), and a paternal uncle of Thomas Jefferson with whom Jefferson had Y chromosomes in common. (Descendants of Jefferson's uncle were used because Jefferson himself had no verified male descendants.) Geneticists determined that Jefferson possessed a rare and distinctive set of genetic markers on his Y chromosome. The same markers were also found on the Y chromosomes of the male-line descendants of Eston Hemings. The probability of such a match arising by chance is less than 1%. (The markers were not found on the Y chromosomes of the descendants of Thomas Woodson.) Together with the circumstantial historical evidence, these matching markers strongly suggest that Jefferson fathered Eston Hemings but not Thomas Woodson.

Y-chromosome sequences have also been used extensively to examine past patterns of human migration and the genetic relationships among different human populations.

CONNECTING CONCEPTS

Recognizing Sex-Linked Inheritance

What features should we look for to identify a trait as sex linked? A common misconception is that any genetic characteristic in which the phenotypes of males and females differ must be sex linked. In fact, the expression of many *autosomal* characteristics differs between males and females. The genes that encode these characteristics are the same in both sexes, but their expression is influenced by sex hormones. The different sex hormones of males and females cause the same genes to generate different phenotypes in males and females.

Another misconception is that any characteristic that is found more frequently in one sex is sex linked. A number of autosomal traits are expressed more commonly in one sex than in the other. These traits are said to be sex influenced. Some autosomal traits are expressed in only one sex; these traits are said to be sex limited. Both sex-influenced and sex-limited characteristics will be considered in more detail in Chapter 5.

Several features of sex-linked characteristics make them easy to recognize. Y-linked traits are found only in males, but this fact does not guarantee that a trait is Y linked, because some autosomal characteristics are expressed only in males. A Y-linked trait is unique, however, in that all the male offspring of an affected male will express the father's phenotype, and a Y-linked trait can be inherited only from the father's side of the family. Thus, a Y-linked trait can be inherited only from the paternal grandfather (the father's father), never from the maternal grandfather (the mother's father).

X-linked characteristics also exhibit a distinctive pattern of inheritance. X linkage is a possible explanation when reciprocal crosses give different results. If a characteristic is X-linked, a cross between an affected male and an unaffected female will not give the same results as a cross between an affected female and an unaffected male. For almost all autosomal characteristics, reciprocal crosses give the same result. We should not conclude, however, that, when the reciprocal crosses give different results, the characteristic is X linked. Other sex-associated forms of inheritance, considered in Chapter 5, also produce different results in reciprocal crosses. The key to recognizing X-linked inheritance is to remember that a male always inherits his X chromosome from his mother, not from his father. Thus, an X-linked characteristic is not passed directly from father to son; if a male clearly inherits a characteristic from his father—and the mother is not heterozygous—it cannot be X linked.

4.3 Dosage Compensation Equalizes the Amount of Protein Produced by X-Linked Genes in Males and Females

In species with XX-XY sex determination, the difference in the number of X chromosomes possessed by males and females presents a special problem in development. Because females have two copies of every X-linked gene and males have only one copy, the amount of gene product (protein) encoded by X-linked genes would differ in the two sexes: females would produce twice as much gene product as that produced by males. This difference could be highly detrimental because protein concentration plays a critical role in development. Animals overcome this potential problem through **dosage compensation,** which equalizes the amount of protein produced by X-linked genes in the two sexes. In fruit flies, dosage compensation is achieved by a doubling of the activity of the genes on the X chromosome of the male.

Table 4.3	Number of Barr bodies in human cells with different complements of sex chromosomes	
Sex Chromosomes	Syndrome	Number of Barr Bodies
XX	None	1
XY	None	0
XO	Turner	0
XXY	Klinefelter	1
XXYY	Klinefelter	1
XXXY	Klinefelter	2
XXXXY	Klinefelter	3
XXX	Triplo-X	2
XXXX	Poly-X female	3
XXXXX	Poly-X female	4

In the worm *Caenorhabditis elegans,* it is achieved by a halving of the activity of genes on both of the X chromosomes in the female. Placental mammals use yet another mechanism of dosage compensation: genes on one of the X chromosomes in the female are inactivated, as described later in this section.

Lyon Hypothesis

In 1949, Murray Barr observed condensed, darkly staining bodies in the nuclei of cells from female cats (**Figure 4.17**); this darkly staining structure became known as a **Barr body.** Mary Lyon proposed in 1961 that the Barr body was an inactive X chromosome; her hypothesis has become known as the **Lyon hypothesis.** She suggested that, within each female cell, one of the two X chromosomes becomes inactive; which

X chromosome is inactivated is random. If a cell contains more than two X chromosomes, all but one of them are inactivated. The number of Barr bodies present in human cells with different complements of sex chromosomes is shown in **Table 4.3**.

As a result of X inactivation, females are functionally hemizygous at the cellular level for X-linked genes. In females that are heterozygous at an X-linked locus, approximately 50% of the cells will express one allele and 50% will express the other allele; thus, in heterozygous females, proteins encoded by both alleles are produced, although not within the same cell. This functional hemizygosity means that cells in females are not identical with respect to the expression of the genes on the X chromosome; females are mosaics for the expression of X-linked genes.

Random X inactivation takes place early in development—in humans, within the first few weeks of development. After an X chromosome has become inactive in a cell, it remains inactive and is inactive in all somatic cells that descend from the cell. Thus, neighboring cells tend to have the same X chromosome inactivated, producing a patchy pattern (mosaic) for the expression of an X-linked characteristic in heterozygous females.

This patchy distribution can be seen in tortoiseshell and calico cats (**Figure 4.18**). Although many genes contribute to coat color and pattern in domestic cats, a single X-linked locus determines the presence of orange color. There are two possible alleles at this locus: X^+, which produces nonorange (usually black) fur, and X^o, which produces orange fur. Males are hemizygous and thus may be black (X^+Y) or orange (X^oY) but not black and orange. (Rare tortoiseshell males can arise from the presence of two X chromosomes, X^+X^oY.) Females may be black (X^+X^+), orange (X^oX^o), or tortoiseshell (X^+X^o), the tortoiseshell pattern arising from a patchy mixture of black and orange fur. Each orange patch is a clone of cells derived from an original cell in which the black allele is inactivated, and each black patch is a clone of cells derived from an original cell in which the orange allele is inactivated.

(a)

(b)

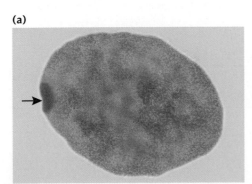 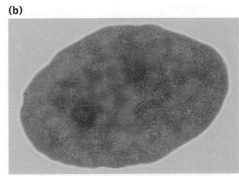

4.17 A Barr body is an inactivated X chromosome. (a) Female cell with a Barr body (indicated by arrow). (b) Male cell without a Barr body. [Chris Bjornberg/Photo Researchers.]

4.18 The patchy distribution of color on tortoiseshell cats results from the random inactivation of one X chromosome in females. [Chanan Photography.]

The Lyon hypothesis suggests that the presence of variable numbers of X chromosomes should not affect the phenotype in mammals, because any X chromosomes in excess of one X chromosome should be inactivated. However, persons with Turner syndrome (XO) differ from normal females, and those with Klinefelter syndrome (XXY) differ from normal males. How do these conditions arise in the face of dosage compensation?

These disorders probably arise because some X-linked genes escape inactivation. Indeed, the nature of X inactivation is more complex than originally envisioned. Studies of individual genes now reveal that only about 75% of X-linked human genes are permanently inactivated. About 15% completely escape X inactivation, meaning that these genes produce twice as much protein in females as they do in males. The remaining 10% are inactivated in some females but not in others. The reason for this variation among females in the activation of some X-linked genes is not known.

Mechanism of Random X Inactivation

Random inactivation of X chromosomes requires two steps. In the first step, the cell somehow assesses, or counts, how many X chromosomes are present. In the second step, one X chromosome is selected to become the active X chromosome and all others are silenced.

Although many details of X-chromosome inactivation remain unknown, several genes and sequences that participate in the process have been identified. Foremost among them is a gene called *Xist* (for X-inactivation-specific transcript). On the X chromosomes destined to become inactivated, the *Xist* gene is active, producing an RNA molecule that coats the X chromosome and inactivates the genes on it, probably by altering chromatin structure. On the X chromosome destined to become active, other genes repress the activity of *Xist* so that the *Xist* RNA never coats the X chromosome and genes on this chromosome remain active.

Dosage Imbalance Between X-Linked Genes and Autosomal Genes

X inactivation compensates for the difference in the number of X chromosomes in male and female mammals, but it creates another type of dosage problem. With X inactivation, males and females have a single active copy of each X-linked gene but possess two active copies of each autosomal gene, leading to different doses (amount) of proteins encoded by X-linked genes relative to proteins encoded by autosomal genes. Proteins often interact with one another and unbalanced dosage frequently leads to problems in development. Recent studies have shown that mammalian cells compensate for this imbalance by increasing the expression of (upregulating) genes on the single active X chromosome so that the total amount of protein produced by genes on the X chromosome is similar to that produced by genes on two copies of autosomal chromosomes. How genes on the single active X chromosome are upregulated is unknown at present. TRY PROBLEM 41 →

CONCEPTS

In mammals, dosage compensation ensures that the same amount of X-linked gene product will be produced in the cells of both males and females. All but one X chromosome are inactivated in each cell; which of the X chromosomes is inactivated is random and varies from cell to cell.

✔ CONCEPT CHECK 9

How many Barr bodies will a male with XXXYY chromosomes have in each of his cells? What are these Barr bodies?

CONCEPTS SUMMARY

- Sexual reproduction is the production of offspring that are genetically distinct from their parents. Most organisms have two sexual phenotypes—males and females. Males produce small gametes; females produce large gametes.

- The mechanism by which sex is specified is termed sex determination. Sex may be determined by differences in specific chromosomes, genotypes, or environment.

- The sex chromosomes of males and females differ in number and appearance. The homogametic sex produces gametes that are all identical with regard to sex chromosomes; the heterogametic sex produces gametes that differ in their sex-chromosome composition.

- In the XX-XO system of sex determination, females possess two X chromosomes, whereas males possess a single X chromosome. In the XX-XY system, females possess two X chromosomes, whereas males possess a single X chromosome and a single Y chromosome. In the ZZ-ZW system, males possess two Z chromosomes, whereas females possess a Z chromosome and a W chromosome.

- Some organisms have genic sex determination, in which genotypes at one or more loci determine the sex of an individual organism. Still others have environmental sex determination.

- In *Drosophila melanogaster*, sex is determined by a balance between genes on the X chromosomes and genes on the autosomes, the X : A ratio.

- In humans, sex is ultimately determined by the presence or absence of the *SRY* gene located on the Y chromosome.

- Sex-linked characteristics are determined by genes on the sex chromosomes; X-linked characteristics are encoded by genes on the X chromosome, and Y-linked characteristics are encoded by genes on the Y chromosome.

- A female inherits X-linked alleles from both parents; a male inherits X-linked alleles from his female parent only.

- The sex chromosomes evolved from autosomes. Crossing over between the X and the Y chromosomes has been suppressed, but palindromic sequences within the Y chromosome allow for internal recombination on the Y chromosome. This internal recombination sometimes leads to chromosome rearrangements that can adversely affect sexual development.

- Y-linked characteristics are found only in males and are passed from father to all sons.

- Dosage compensation equalizes the amount of protein produced by X-linked genes in males and females. In placental mammals, one of the two X chromosomes in females normally becomes inactivated. Which X chromosome is inactivated is random and varies from cell to cell. Some X-linked genes escape X inactivation, and other X-linked genes may be inactivated in some females but not in others. X inactivation is controlled by the *Xist* gene.

IMPORTANT TERMS

sex (p. 74)
sex determination (p. 74)
hermaphroditism (p. 75)
monoecious organism (p. 75)
dioecious organism (p. 75)
sex chromosome (p. 75)
autosome (p. 76)
heterogametic sex (p. 76)
homogametic sex (p. 76)
pseudoautosomal region (p. 76)

genic sex determination (p. 77)
sequential hermaphroditism (p. 77)
genic balance system (p. 78)
X : A ratio (p. 78)
Turner syndrome (p. 79)
Klinefelter syndrome (p. 79)
triplo-X syndrome (p. 79)
sex-determining region Y (*SRY*) gene (p. 80)
sex-linked characteristic (p. 81)

X-linked characteristic (p. 81)
Y-linked characteristic (p. 81)
hemizygosity (p. 81)
nondisjunction (p. 83)
dosage compensation (p. 88)
Barr body (p. 89)
Lyon hypothesis (p. 89)

ANSWERS TO CONCEPT CHECKS

1. Meiosis

2. b

3. In chromosomal sex determination, males and females have chromosomes that are distinguishable. In genic sex determination, sex is determined by genes, but the chromosomes of males and females are indistinguishable. In environmental sex determination, sex is determined fully or in part by environmental effects.

4. b

5. a

6. c

7. All male offspring will have hemophilia, and all female offspring will not have hemophilia; so the overall probability of hemophilia in the offspring is $^1/_2$.

8. Eight large palindromes that allow crossing over within the Y chromosome.

9. Two Barr bodies. A Barr body is an inactivated X chromosome.

WORKED PROBLEMS

1. A fruit fly has XXXYY sex chromosomes; all the autosomal chromosomes are normal. What sexual phenotype does this fly have?

• Solution

Sex in fruit flies is determined by the X : A ratio—the ratio of the number of X chromosomes to the number of haploid autosomal sets. An X : A ratio of 1.0 produces a female fly; an X : A ratio of 0.5 produces a male. If the X : A ratio is greater than 1.0, the fly is a metafemale; if it is less than 0.5, the fly is a metamale; if the X : A ratio is between 1.0 and 0.5, the fly is an intersex.

This fly has three X chromosomes and normal autosomes. Normal diploid flies have two autosomal sets of chromosomes; so the X : A ratio in this case is $^3/_2$, or 1.5. Thus, this fly is a metafemale.

2. Chickens, like all birds, have ZZ-ZW sex determination. The bar-feathered phenotype in chickens results from a Z-linked allele that is dominant over the allele for nonbar feathers. A barred female is crossed with a nonbarred male. The F_1 from this cross are intercrossed to produce the F_2. What are the phenotypes and their proportions in the F_1 and F_2 progeny?

• Solution

With the ZZ-ZW system of sex determination, females are the heterogametic sex, possessing a Z chromosome and a W chromosome; males are the homogametic sex, with two Z chromosomes. In this problem, the barred female is hemizygous for the bar phenotype (Z^BW). Because bar is dominant over nonbar, the nonbarred male must be homozygous for nonbar (Z^bZ^b). Crossing these two chickens, we obtain:

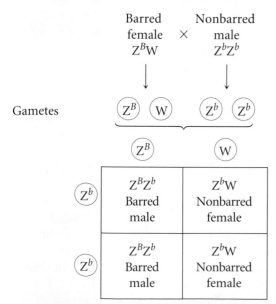

Thus, all the males in the F_1 are barred (Z^BZ^b), and all the females are nonbarred (Z^bW).

We now cross the F_1 to produce the F_2:

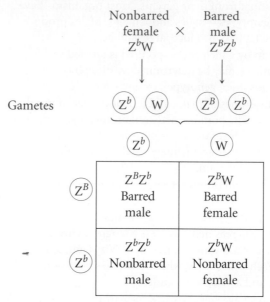

So, $^1/_4$ of the F_2 are barred males, $^1/_4$ are nonbarred males, $^1/_4$ are barred females, and $^1/_4$ are nonbarred females.

3. In *Drosophila melanogaster*, forked bristles are caused by an allele (X^f) that is X linked and recessive to an allele for normal bristles (X^+). Brown eyes are caused by an allele (b) that is autosomal and recessive to an allele for red eyes (b^+). A female fly that is homozygous for normal bristles and red eyes mates with a male fly that has forked bristles and brown eyes. The F_1 are intercrossed to produce the F_2. What will the phenotypes and proportions of the F_2 flies be from this cross?

• Solution

This problem is best worked by breaking the cross down into two separate crosses, one for the X-linked genes that determine the type of bristles and one for the autosomal genes that determine eye color.

Let's begin with the autosomal characteristics. A female fly that is homozygous for red eyes (b^+b^+) is crossed with a male with brown eyes. Because brown eyes are recessive, the male fly must be homozygous for the brown-eye allele (bb). All of the offspring of this cross will be heterozygous (b^+b) and will have red eyes:

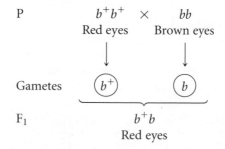

The F$_1$ are then intercrossed to produce the F$_2$. Whenever two individual organisms heterozygous for an autosomal recessive characteristic are crossed, $^3/_4$ of the offspring will have the dominant trait and $^1/_4$ will have the recessive trait; thus, $^3/_4$ of the F$_2$ flies will have red eyes and $^1/_4$ will have brown eyes:

F$_1$ b^+b × b^+b
        Red eyes   Red eyes

Gametes b^+ b b^+ b

F$_2$ $^1/_4\ b^+b^+$ red
        $^1/_2\ b^+b$ red
        $^1/_4\ bb$ brown
        $^3/_4$ red, $^1/_4$ brown

Next, we work out the results for the X-linked characteristic. A female that is homozygous for normal bristles (X$^+$X$^+$) is crossed with a male that has forked bristles (X^{f}Y). The female F$_1$ from this cross are heterozygous (X$^+$X^f), receiving an X chromosome with a normal-bristle allele from their mother (X$^+$) and an X chromosome with a forked-bristle allele (X^f) from their father. The male F$_1$ are hemizygous (X$^+$Y), receiving an X chromosome with a normal-bristle allele from their mother (X$^+$) and a Y chromosome from their father:

P   X$^+$X$^+$ ×   X^{f}Y
      Normal   Forked
      bristles   bristles

Gametes   X$^+$ X^f Y

F$_1$ $^1/_2$ X$^+$X^f normal bristle
        $^1/_2$ X$^+$Y   normal bristle

When these F$_1$ are intercrossed, $^1/_2$ of the F$_2$ will be normal-bristle females, $^1/_4$ will be normal-bristle males, and $^1/_4$ will be forked-bristle males:

F$_1$ X$^+$X^f ×   X$^+$Y

Gametes   X$^+$ X^f X$^+$ Y

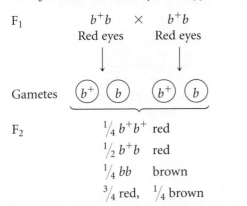

	X$^+$	X^f
X$^+$	X$^+$ X$^+$ Normal female	X$^+$X^f Normal female
Y	X$^+$Y Normal male	X^{f}Y Forked-bristle male

F$_2$

$^1/_2$ normal female, $^1/_4$ normal male, $^1/_4$ forked-bristle male

To obtain the phenotypic ratio in the F$_2$, we now combine these two crosses by using the multiplication rule of probability and the branch diagram:

Eye color	Bristle and sex	F$_2$ phenotype	Probability
red ($^3/_4$)	normal female ($^1/_2$)	red normal female	$^3/_4 × ^1/_2 = ^3/_8 = ^6/_{16}$
	normal male ($^1/_4$)	red normal male	$^3/_4 × ^1/_4 = ^3/_{16}$
	forked-bristle male ($^1/_4$)	red forked-bristle male	$^3/_4 × ^1/_4 = ^3/_{16}$
brown ($^1/_4$)	normal female ($^1/_2$)	brown normal female	$^1/_4 × ^1/_2 = ^1/_8 = ^2/_{16}$
	normal male ($^1/_4$)	brown normal male	$^1/_4 × ^1/_4 = ^1/_{16}$
	forked-bristle male ($^1/_4$)	brown forked-bristle male	$^1/_4 × ^1/_4 = ^1/_{16}$

COMPREHENSION QUESTIONS

Section 4.1

*1. What is the most defining difference between the males and the females of any organism?

2. How do monoecious organisms differ from dioecious organisms?

3. Describe the XX-XO system of sex determination. In this system, which is the heterogametic sex and which is the homogametic sex?

4. How does sex determination in the XX-XY system differ from sex determination in the ZZ-ZW system?

*5. What is the pseudoautosomal region? How does the inheritance of genes in this region differ from the inheritance of other Y-linked characteristics?

6. What is meant by genic sex determination?

7. How does sex determination in *Drosophila* differ from sex determination in humans?

8. Give the typical sex-chromosome complement found in the cells of people with Turner syndrome, with Klinefelter syndrome, and with androgen-insensitivity syndrome. What is the sex-chromosome complement of triplo-X females?

Section 4.2

*9. What characteristics are exhibited by an X-linked trait?

10. Explain how Bridges's study of nondisjunction in *Drosophila* helped prove the chromosome theory of inheritance.

*11. What characteristics are exhibited by a Y-linked trait?

APPLICATION QUESTIONS AND PROBLEMS

Introduction

14. As described in the introduction to this chapter, platypuses possess 10 sex chromosomes. Females have five pairs of X chromosomes ($X_1X_1X_2X_2X_3X_3X_4X_4X_5X_5$) and males have five pairs of X and Y chromosomes ($X_1Y_1X_2Y_2X_3Y_3X_4Y_4X_5Y_5$). Do each of the XY chromosome pairs in males assort independently in meiosis? Why or why not?

Section 4.1

*15. What is the sexual phenotype of fruit flies having the following chromosomes?

	Sex chromosomes	Autosomal chromosomes
a.	XX	all normal
b.	XY	all normal
c.	XO	all normal
d.	XXY	all normal
e.	XYY	all normal
f.	XXYY	all normal
g.	XXX	all normal
h.	XX	four haploid sets
i.	XXX	four haploid sets
j.	XXX	three haploid sets
k.	X	three haploid sets
l.	XY	three haploid sets
m.	XX	three haploid sets

16. For each of the following sex-chromosome complements, what is the phenotypic sex of a person who has

a. XY with the *SRY* gene deleted?

b. XY with the *SRY* gene located on an autosomal chromosome?

c. XX with a copy of the *SRY* gene on an autosomal chromosome?

d. XO with a copy of the *SRY* gene on an autosomal chromosome?

e. XXY with the *SRY* gene deleted?

f. XXYY with one copy of the *SRY* gene deleted?

17. A normal female *Drosophila* produces abnormal eggs that contain all (a complete diploid set) of her chromosomes. She mates with a normal male *Drosophila* that produces normal sperm. What will the sex of the progeny from this cross be?

18. In certain salamanders, the sex of a genetic female can be altered, changing her into a functional male; these salamanders are called sex-reversed males. When a sex-reversed male is mated with a normal female, approximately $2/3$ of the offspring are female and $1/3$ are male. How is sex determined in these salamanders? Explain the results of this cross.

19. In some mites, males pass genes to their grandsons, but they never pass genes to their sons. Explain.

*20. In organisms with the ZZ-ZW sex-determining system, from which of the following possibilities can a female inherit her Z chromosome?

	Yes	No
Her mother's mother	_____	_____
Her mother's father	_____	_____
Her father's mother	_____	_____
Her father's father	_____	_____

Section 4.2

21. When Bridges crossed white-eyed females with red-eyed males, he obtained a few red-eyed males and white-eyed females (see Figure 4.13). What types of offspring would be produced if these red-eyed males and white-eyed females were crossed with each other?

*22. Joe has classic hemophilia, an X-linked recessive disease. Could Joe have inherited the gene for this disease from the following persons?

	Yes	No
a. His mother's mother	_____	_____
b. His mother's father	_____	_____
c. His father's mother	_____	_____
d. His father's father	_____	_____

*23. In *Drosophila*, yellow body is due to an X-linked gene that is recessive to the gene for gray body.

a. A homozygous gray female is crossed with a yellow male. The F_1 are intercrossed to produce F_2. Give the genotypes and phenotypes, along with the expected proportions, of the F_1 and F_2 progeny.

b. A yellow female is crossed with a gray male. The F_1 are intercrossed to produce the F_2. Give the genotypes and phenotypes, along with the expected proportions, of the F_1 and F_2 progeny.

c. A yellow female is crossed with a gray male. The F_1 females are backcrossed with gray males. Give the genotypes and phenotypes, along with the expected proportions, of the F_2 progeny.

Section 4.3

12. Explain why tortoiseshell cats are almost always female and why they have a patchy distribution of orange and black fur.

13. What is a Barr body? How is it related to the Lyon hypothesis?

d. If the F_2 flies in part *b* mate randomly, what are the expected phenotypic proportions of flies in the F_3?

[Courtesy Masa-Toshi Yamamoto, Drosophila Genetic Resource Center, Kyoto Institute of Technology.]

24. Coat color in cats is determined by genes at several different loci. At one locus on the X chromosome, one allele (X^+) encodes black fur; another allele (X^o) encodes orange fur. Females can be black (X^+X^+), orange (X^oX^o), or a mixture of orange and black called tortoiseshell (X^+X^o). Males are either black (X^+Y) or orange (X^oY). Bill has a female tortoiseshell cat named Patches. One night Patches escapes from Bill's house, spends the night out, and mates with a stray male. Patches later gives birth to the following kittens: one orange male, one black male, two tortoiseshell females, and one orange female. Give the genotypes of Patches, her kittens, and the stray male with which Patches mated.

*25. Red–green color blindness in humans is due to an X-linked recessive gene. Both John and Cathy have normal color vision. After 10 years of marriage to John, Cathy gave birth to a color-blind daughter. John filed for divorce, claiming that he is not the father of the child. Is John justified in his claim of nonpaternity? Explain why. If Cathy had given birth to a color-blind son, would John be justified in claiming nonpaternity?

26. Red–green color blindness in humans is due to an X-linked recessive gene. A woman whose father is color blind possesses one eye with normal color vision and one eye with color blindness.

a. Propose an explanation for this woman's vision pattern. Assume that no new mutations have spontaneously arisen.

b. Would it be possible for a man to have one eye with normal color vision and one eye with color blindness?

*27. Bob has XXY chromosomes (Klinefelter syndrome) and is color blind. His mother and father have normal color vision, but his maternal grandfather is color blind. Assume that Bob's chromosome abnormality arose from nondisjunction in meiosis. In which parent and in which meiotic division did nondisjunction take place? Explain your answer.

28. Xg is an antigen found on red blood cells. This antigen is caused by an X-linked allele (X^a) that is dominant over an allele for the absence of the antigen (X^-). The inheritance of these X-linked alleles was studied in children with chromosome abnormalities to determine where nondisjunction of the sex chromosomes took place. For each type of mating in parts *a* through *d*, indicate whether nondisjunction took place in the mother or in the father and, if possible, whether it took place in meiosis I or meiosis II.

a. $X^aY \times X^-X^- \rightarrow X^a$ (Turner syndrome)

b. $X^aY \times X^aX^- \rightarrow X^-$ (Turner syndrome)

c. $X^aY \times X^-X^- \rightarrow X^aX^-Y$ (Klinefelter syndrome)

d. $X^aY \times X^aX^- \rightarrow X^-X^-Y$ (Klinefelter syndrome)

29. The Talmud, an ancient book of Jewish civil and religious laws, states that, if a woman bears two sons who die of bleeding after circumcision (removal of the foreskin from the penis), any additional sons that she has should not be circumcised. (The bleeding is most likely due to the X-linked disorder hemophilia.) Furthermore, the Talmud states that the sons of her sisters must not be circumcised, whereas the sons of her brothers should be. Is this religious law consistent with sound genetic principles? Explain your answer.

30. Craniofrontonasal syndrome (CFNS) is a birth defect in which premature fusion of the cranial sutures leads to abnormal head shape, widely spaced eyes, nasal clefts, and various other skeletal abnormalities. George Feldman and his colleagues, looked at several families in which offspring had CFNS and recorded the results shown in the following table (G. J. Feldman. 1997. *Human Molecular Genetics* 6:1937–1941).

Family number	Parents		Offspring			
			Normal		CFNS	
	Father	Mother	Male	Female	Male	Female
1	normal	CFNS	1	0	2	1
5	normal	CFNS	0	2	1	2
6	normal	CFNS	0	0	1	2
8	normal	CFNS	1	1	1	0
10a	CFNS	normal	3	0	0	2
10b	normal	CFNS	1	1	2	0
12	CFNS	normal	0	0	0	1
13a	normal	CFNS	0	1	2	1
13b	CFNS	normal	0	0	0	2
7b	CFNS	normal	0	0	0	2

a. On the basis of these results, what is the most likely mode of inheritance for CFNS?

b. Give the most likely genotypes of the parents in family 1 and in family 10a.

*31. Miniature wings (X^m) in *Drosophila* result from an X-linked allele that is recessive to the allele for long wings (X^+). Give the genotypes of the parents in each of the following crosses.

	Male parent	Female parent	Male offspring	Female offspring
a.	long	long	231 long, 250 miniature	560 long
b.	miniature	long	610 long	632 long
c.	miniature	long	410 long, 417 miniature	412 long, 415 miniature
d.	long	miniature	753 miniature	761 long
e.	long	long	625 long	630 long

*32. In chickens, congenital baldness is due to a Z-linked recessive gene. A bald rooster is mated with a normal hen. The F_1 from this cross are interbred to produce the F_2. Give the genotypes and phenotypes, along with their expected proportions, among the F_1 and F_2 progeny.

33. Red–green color blindness is an X-linked recessive trait in humans. Polydactyly (extra fingers and toes) is an autosomal dominant trait. Martha has normal fingers and toes and normal color vision. Her mother is normal in all respects, but her father is color blind and polydactylous. Bill is color blind and polydactylous. His mother has normal color vision and normal fingers and toes. If Bill and Martha marry, what types and proportions of children can they produce?

34. A *Drosophila* mutation called *singed* (s) causes the bristles to be bent and misshapen. A mutation called *purple* (p) causes the fly's eyes to be purple in color instead of the normal red. Flies homozygous for *singed* and *purple* were crossed with flies that were homozygous for normal bristles and red eyes. The F_1 were intercrossed to produce the F_2, and the following results were obtained.

Cross 1

P   male, singed bristles, purple eyes × female, normal bristles, red eyes

F_1 420 female, normal bristles, red eyes
        426 male, normal bristles, red eyes

F_2 337 female, normal bristles, red eyes
        113 female, normal bristles, purple eyes
        168 male, normal bristles, red eyes
        170 male, singed bristles, red eyes
        56 male, normal bristles, purple eyes
        58 male, singed bristles, purple eyes

Cross 2

P   female, singed bristles, purple eyes × male, normal bristles, red eyes

F_1 504 female, normal bristles, red eyes
        498 male, singed bristles, red eyes

F_2 227 female, normal bristles, red eyes
        223 female, singed bristles, red eyes
        225 male, normal bristles, red eyes
        225 male, singed bristles, red eyes
        78 female, normal bristles, purple eyes
        76 female, singed bristles, purple eyes
        74 male, normal bristles, purple eyes
        72 male, singed bristles, purple eyes

a. What are the modes of inheritance of *singed* and *purple*? Explain your reasoning.

b. Give genotypes for the parents and offspring in the P, F_1, and F_2 generations of Cross 1 and Cross 2.

35. The following two genotypes are crossed: *Aa Bb Cc* X^+X^r × *Aa BB cc* X^+Y, where *a, b,* and *c* represent alleles of autosomal genes and X^+ and X^r represent X-linked alleles in an organism with XX-XY sex determination. What is the probability of obtaining genotype *aa Bb Cc* X^+X^+ in the progeny?

*36. Miniature wings in *Drosophila melanogaster* are due to an X-linked gene (X^m) that is recessive to an allele for long wings (X^+). Sepia eyes are produced by an autosomal gene (s) that is recessive to an allele for red eyes (s^1).

a. A female fly that has miniature wings and sepia eyes is crossed with a male that has normal wings and is homozygous for red eyes. The F_1 flies are intercrossed to produce the F_2. Give the phenotypes, as well as their expected proportions, of the the F_1 and F_2 flies.

b. A female fly that is homozygous for normal wings and has sepia eyes is crossed with a male that has miniature wings and is homozygous for red eyes. The F_1 flies are intercrossed to produce the F_2. Give the phenotypes, as well as their expected proportions, of the F_2 flies.

37. Suppose that a recessive gene that produces a short tail in mice is located in the pseudoautosomal region. A short-tailed male mouse is mated with a female mouse that is homozygous for a normal tail. The F_1 mice from this cross are intercrossed to produce the F_2. Give the phenotypes, as well as their proportions, of the F_1 and F_2 mice?

*38. A color-blind woman and a man with normal vision have three sons and six daughters. All the sons are color blind. Five of the daughters have normal vision, but one of them is color blind. The color-blind daughter is 16 years old, is short for her age, and has not undergone puberty. Explain how this girl inherited her color blindness.

Section 4.3

*39. How many Barr bodies would you expect to see in a human cell containing the following chromosomes?

a. XX	**d.** XXY	**g.** XYY	
b. XY	**e.** XXYY	**h.** XXX	
c. XO	**f.** XXXY	**i.** XXXX	

40. A woman with normal chromosomes mates with a man who also has normal chromosomes.

a. Suppose that, in the course of oogenesis, the woman's sex chromosomes undergo nondisjunction in meiosis I; the man's chromosomes separate normally. Give all possible combinations of sex chromosomes that this couple's children might inherit and the number of Barr bodies that you would expect to see in each of the cells of each child.

b. What chromosome combinations and numbers of Barr bodies would you expect to see if the chromosomes separate normally in oogenesis, but nondisjunction of the sex chromosomes takes place in meiosis I of spermatogenesis?

41. Anhidrotic ectodermal dysplasia is an X-linked recessive disorder in humans characterized by small teeth, no sweat glands, and sparse body hair. The trait is usually seen in men, but women who are heterozygous carriers of the trait often have irregular patches of skin with few or no sweat glands (see the illustration on the right).

a. Explain why women who are heterozygous carriers of a recessive gene for anhidrotic ectodermal dysplasia have irregular patches of skin lacking sweat glands.

b. Why does the distribution of the patches of skin lacking sweat glands differ among the females depicted in the illustration, even between the identical twins?

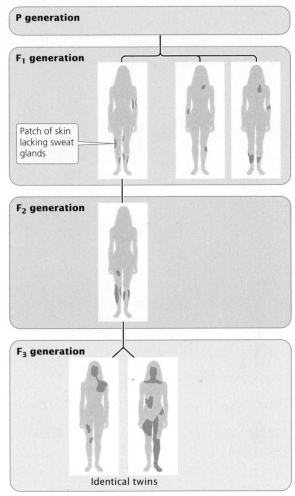

Patch of skin lacking sweat glands

Identical twins

[After A. P. Mance and J. Mance, *Genetics: Human Aspects* (Sinauer, 1990), p. 133.]

CHALLENGE QUESTIONS

Section 4.2

42. A geneticist discovers a male mouse with greatly enlarged testes in his laboratory colony. He suspects that this trait results from a new mutation that is either Y linked or autosomal dominant. How could he determine whether the trait is autosomal dominant or Y linked?

Section 4.3

43. Female humans who are heterozygous for X-linked recessive genes sometimes exhibit mild expression of the trait. However, such mild expression of X-linked traits in females who are heterozygous for X-linked alleles is not seen in *Drosophila*. What might cause this difference in the expression of X-linked genes in female humans and female *Drosophila*? (Hint: In *Drosophila*, dosage compensation is accomplished by doubling the activity of genes on the X chromosome of males.)

44. Identical twins (also called monozygotic twins) are derived from a single egg fertilized by a single sperm, creating a zygote that later divides into two (see Chapter 6). Because identical twins originate from a single zygote, they are genetically identical.

Caroline Loat and her colleagues examined 9 measures of social, behavioral, and cognitive ability in 1000 pairs of identical male twins and 1000 pairs of identical female twins (C. S. Loat et al. 2004. *Twin Research* 7:54–61). They found that, for three of the measures (prosocial behavior, peer problems, and verbal ability), the two male twins of a pair tended to be more alike in their scores than were two female twins of a pair. Propose a possible explanation for this observation. What might this observation indicate about the location of genes that influence prosocial behavior, peer problems, and verbal ability?

45. Occasionally, a mouse X chromosome is broken into two pieces and each piece becomes attached to a different autosomal chromosome. In this event, the genes on only one of the two pieces undergo X inactivation. What does this observation indicate about the mechanism of X-chromosome inactivation?

5

Extensions and Modifications of Basic Principles

Yellow coat color in mice is caused by a recessive lethal gene, producing distorted phenotypic ratios in the progeny of two yellow mice. William Castle and Clarence Little discovered the lethal nature of the yellow gene in 1910. [Reprinted with permission of Dr. Loan Phan and *In Vivo*, a publication of Columbia University Medical Center.]

CUÉNOT'S ODD YELLOW MICE

At the start of the twentieth century, Mendel's work on inheritance in pea plants became widely known (see Chapter 3), and a number of biologists set out to verify his conclusions by using other organisms to conduct crosses. One of these biologists was Lucien Cuénot, a French scientist working at the University of Nancy. Cuénot experimented with coat colors in mice and was among the first to show that Mendel's principles applied to animals.

Cuénot observed that the coat colors of his mice followed the same patterns of inheritance observed by Mendel in his pea plants. Cuénot found that, when he crossed pure-breeding gray mice with pure-breeding white mice, all of the F_1 progeny were gray, and interbreeding the F_1 produced a 3 : 1 ratio of gray and white mice in the F_2, as would be expected if gray were dominant over white. The results of Cuénot's breeding experiments perfectly fit Mendel's rules—with one exception. His crosses of yellow mice suggested that yellow coat color was dominant over gray, but he was never able to obtain true-breeding (homozygous) yellow mice. Whenever Cuénot crossed two yellow mice, he obtained yellow and gray mice in approximately a 3 : 1 ratio, suggesting that the yellow mice were heterozygous ($Yy \times Yy \rightarrow$ $^3/_4\ Y\_$ and $^1/_4\ yy$). If yellow were indeed dominant over gray, some of the yellow progeny from this cross should have been homozygous for yellow (YY) and crossing two of these mice should have yielded all yellow offspring ($YY \times YY \rightarrow YY$). However, he never obtained all yellow progeny in his crosses.

Cuénot was puzzled by these results, which failed to conform to Mendel's predictions. He speculated that yellow gametes were incompatible with each other and would not fuse to form a zygote. Other biologists thought that additional factors might affect the inheritance of the yellow coat color, but the genetics of the yellow mice remained a mystery.

In 1910, William Ernest Castle and his student Clarence Little solved the mystery of Cuénot's unusual results. They carried out a large series of crosses between two yellow mice and showed that the progeny appeared, not in the 3 : 1 ratio that Cuénot thought he had observed but actually in a 2 : 1 ratio of yellow and nonyellow. Castle and Little recognized that the allele for yellow was lethal when homozygous (**Figure 5.1**), and thus all the yellow mice were heterozygous (Yy). A cross between two yellow heterozygous mice produces an

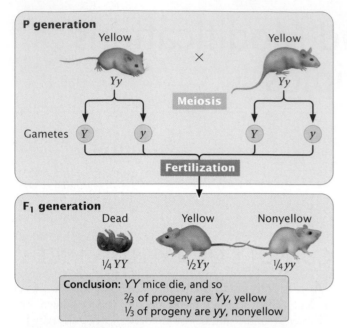

5.1 The 2 : 1 ratio produced by a cross between two yellow mice results from a lethal allele.

initial genotypic ratio of $\frac{1}{4}$ *YY*, $\frac{1}{2}$ *Yy*, and $\frac{1}{4}$ *yy*, but the homozygous *YY* mice die early in development and do not appear among the progeny, resulting in a 2 : 1 ratio of *Yy* (yellow) to *yy* (nonyellow) in offspring. Indeed, Castle and Little found that crosses of yellow × yellow mice resulted in smaller litters compared with litters of yellow × nonyellow mice. Because only mice homozygous for the *Y* allele die, the *yellow* allele is a recessive lethal. The *yellow* allele in mice is unusual in that it acts as a *recessive* allele in its effect on development but acts as a *dominant* allele in its effect on coat color.

Cuénot went on to make a number of other important contributions to genetics. He was the first to propose that more than two alleles could exist at a single locus, and he described how genes at different loci could interact in the determination of coat color in mice (aspects of inheritance that we will consider in this chapter). He observed that some types of cancer in mice display a hereditary predisposition; he also proposed, far ahead of his time, that genes might encode enzymes. Unfortunately, Cuénot's work brought him little recognition in his lifetime and was not well received by other French biologists, many of them openly hostile to the idea of Mendelian genetics. Cuénot's studies were interrupted by World War I, when foreign troops occupied his town and he was forced to abandon his laboratory at the university. He later returned to find his stocks of mice destroyed, and he never again took up genetic investigations.

Like a number of other genetic phenomena, the lethal *yellow* gene discovered by Cuénot does not produce the ratios predicted by Mendel's principles of heredity. This lack of adherence to Mendel's rules doesn't mean that Mendel was wrong; rather, it demonstrates that Mendel's principles are not, by themselves, sufficient to explain the inheritance of all genetic characteristics. Our modern understanding of genetics has been greatly enriched by the discovery of a number of modifications and extensions of Mendel's basic principles, which are the focus of this chapter. **TRY PROBLEM 12 →**

5.1 Additional Factors at a Single Locus Can Affect the Results of Genetic Crosses

In Chapter 3, we learned that the principle of segregation and the principle of independent assortment allow us to predict the outcomes of genetic crosses. Here, we examine several additional factors acting at individual loci that can alter the phenotypic ratios predicted by Mendel's principles.

Types of Dominance

One of Mendel's important contributions to the study of heredity is the concept of dominance—the idea that an individual organism possesses two different alleles for a characteristic but the trait encoded by only one of the alleles is observed in the phenotype. With dominance, the heterozygote possesses the same phenotype as one of the homozygotes.

Mendel observed dominance in all of the traits that he chose to study extensively, but he was aware that not all characteristics exhibit dominance. He conducted some crosses concerning the length of time that pea plants take to flower. For example, when he crossed two homozygous varieties that differed in their flowering time by an average of 20 days, the length of time taken by the F_1 plants to flower was intermediate between those of the two parents. When the heterozygote has a phenotype intermediate between the phenotypes of the two homozygotes, the trait is said to display *incomplete dominance.*

Complete and incomplete dominance Dominance can be understood in regard to how the phenotype of the heterozygote relates to the phenotypes of the homozygotes. In the example presented in the top panel of **Figure 5.2**, flower color potentially ranges from red to white. One homozygous genotype, A^1A^1, produces red pigment, resulting in red flowers; another, A^2A^2, produces no pigment, resulting in white flowers. Where the heterozygote falls in the range of phenotypes determines the type of dominance. If the heterozygote (A^1A^2) produces the same amount of pigment as the A^1A^1 homozygote, resulting in red, then the A^1 allele displays **complete dominance** over the A^2 allele; that is, red is dominant over white. If, on the other hand, the heterozygote produces no pigment resulting in flowers with the same color as the A^2A^2 homozygote (white), then the A^2 allele is completely dominant, and white is dominant over red.

When the heterozygote falls in between the phenotypes of the two homozygotes, dominance is incomplete. With

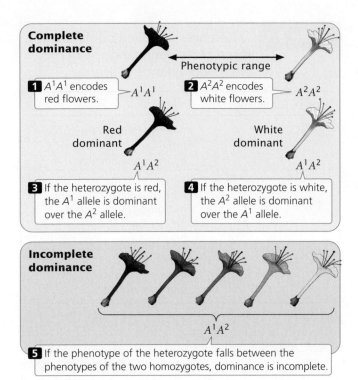

Complete dominance

1 A^1A^1 encodes red flowers. A^1A^1

$\longleftrightarrow$ Phenotypic range $\longrightarrow$

2 A^2A^2 encodes white flowers. A^2A^2

Red dominant A^1A^2

White dominant A^1A^2

3 If the heterozygote is red, the A^1 allele is dominant over the A^2 allele.

4 If the heterozygote is white, the A^2 allele is dominant over the A^1 allele.

Incomplete dominance

A^1A^2

5 If the phenotype of the heterozygote falls between the phenotypes of the two homozygotes, dominance is incomplete.

5.2 The type of dominance exhibited by a trait depends on how the phenotype of the heterozygote relates to the phenotypes of the homozygotes.

incomplete dominance, the heterozygote need not be exactly intermediate (see the bottom panel of Figure 5.2) between the two homozygotes; it might be a slightly lighter shade of red or a slightly pink shade of white. As long as the heterozygote's phenotype can be differentiated and falls within the range of the two homozygotes, dominance is incomplete.

Incomplete dominance is also exhibited in the fruit color of eggplant. When a homozygous plant that produces purple fruit (PP) is crossed with a homozygous plant that produces white fruit (pp), all the heterozygous F₁ (Pp) produce violet fruit (**Figure 5.3a**). When the F₁ are crossed with each other, $\frac{1}{4}$ of the F₂ are purple (PP), $\frac{1}{2}$ are violet (Pp), and $\frac{1}{4}$ are white (pp), as shown in **Figure 5.3b**. Note that this 1 : 2 : 1 ratio is different from the 3 : 1 ratio that we would observe if eggplant fruit color exhibited complete dominance. Another example of incomplete dominance is feather color in chickens. A cross between a homozygous black chicken and a homozygous white chicken produces F₁ chickens that are gray. If these gray F₁ are intercrossed, they produce F₂ birds in a ratio of 1 black : 2 gray : 1 white.

We should now add the 1 : 2 : 1 ratio to those phenotypic ratios for simple crosses presented in Chapter 3 (see Table 3.3). A 1 : 2 : 1 phenotypic ratio arises in the progeny of a cross between two parents heterozygous for a character that exhibits incomplete dominance ($Aa \times Aa$). The genotypic ratio among these progeny also is 1 : 2 : 1. When a trait displays incomplete dominance, the genotypic ratios and phenotypic ratios of the offspring are the *same*, because

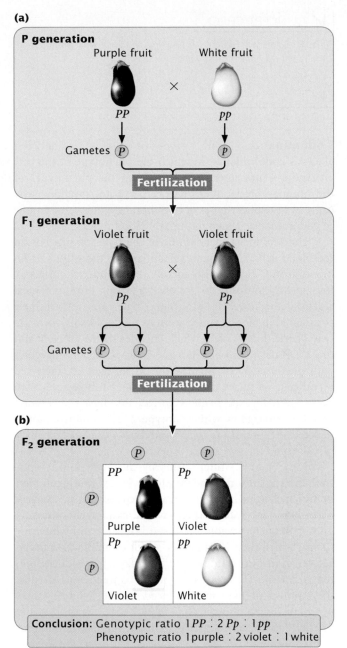

(a)

P generation

Purple fruit White fruit

PP pp

Gametes P P

Fertilization

F₁ generation

Violet fruit Violet fruit

Pp Pp

Gametes P P P P

Fertilization

(b)

F₂ generation

	P	P
P	PP Purple	Pp Violet
P	Pp Violet	pp White

Conclusion: Genotypic ratio 1 PP : 2 Pp : 1 pp
Phenotypic ratio 1 purple : 2 violet : 1 white

5.3 Fruit color in eggplant is inherited as an incompletely dominant trait.

each genotype has its own phenotype. The important thing to remember about dominance is that it affects the phenotype that genes produce but not the way in which genes are *inherited*.

CONCEPTS

Incomplete dominance is exhibited when the heterozygote has a phenotype intermediate between the phenotypes of the two homozygotes. When a trait exhibits incomplete dominance, a cross between two heterozygotes produces a 1 : 2 : 1 phenotypic ratio in the progeny.

✔ CONCEPT CHECK 1

If an F_1 eggplant in Figure 5.3 is used in a testcross, what proportion of the progeny from this cross will be white?

a. All the progeny
b. 1/2
c. 1/4
d. 0

Codominance Another type of interaction between alleles is **codominance**, in which the phenotype of the heterozygote is not intermediate between the phenotypes of the homozygotes; rather, the heterozygote simultaneously expresses the phenotypes of both homozygotes. An example of codominance is seen in the MN blood types.

The MN locus encodes one of the types of antigens on red blood cells. Unlike antigens foreign to the ABO and Rh blood groups (which also encode red-blood-cell antigens), foreign MN antigens do not elicit a strong immunological reaction; therefore, the MN blood types are not routinely considered in blood transfusions. At the MN locus, there are two alleles: the L^M allele, which encodes the M antigen; and the L^N allele, which encodes the N antigen. Homozygotes with genotype $L^M L^M$ express the M antigen on their red blood cells and have the M blood type. Homozygotes with genotype $L^N L^N$ express the N antigen and have the N blood type. Heterozygotes with genotype $L^M L^N$ exhibit codominance and express both the M and the N antigens; they have blood-type MN.

Some students might ask why the pink flowers illustrated in the bottom panel of Figure 5.2 exhibit incomplete dominance—that is, why is this outcome not an example of codominance? The flowers would exhibit codominance only if the heterozygote produced both red and white pigments, which then combined to produce a pink phenotype. However, in our example, the heterozygote produces only red pigment. The pink phenotype comes about because the amount of pigment produced by the heterozygote is less than the amount produced by the $A^1 A^1$ homozyogote. So, here, the alleles clearly exhibit incomplete dominance, not codominance. The differences between dominance, incomplete dominance, and codominance are summarized in **Table 5.1**. TRY PROBLEMS 14 AND 15 →

Table 5.1 Differences between dominance, incomplete dominance, and codominance

Type of Dominance	Definition
Dominance	Phenotype of the heterozygote is the same as the phenotype of one of the homozygotes.
Incomplete dominance	Phenotype of the heterozygote is intermediate (falls within the range) between the phenotypes of the two homozygotes.
Codominance	Phenotype of the heterozygote includes the phenotypes of both homozygotes.

Dependency of type of dominance on level of phenotype observed Phenotypes can frequently be observed at several different levels, including the anatomical level, the physiological level, and the molecular level. The type of dominance exhibited by a character depends on the level of the phenotype examined. This dependency is seen in cystic fibrosis, a common genetic disorder found in Caucasians and usually considered to be a recessive disease. People who have cystic fibrosis produce large quantities of thick, sticky mucus, which plugs up the airways of the lungs and clogs the ducts leading from the pancreas to the intestine, causing frequent respiratory infections and digestive problems. Even with medical treatment, patients with cystic fibrosis suffer chronic, life-threatening medical problems.

The gene responsible for cystic fibrosis resides on the long arm of chromosome 7. It encodes a protein termed *cystic fibrosis transmembrane conductance regulator*, abbreviated CFTR, which acts as a gate in the cell membrane and regulates the movement of chloride ions into and out of the cell. Patients with cystic fibrosis have a mutated, dysfunctional form of CFTR that causes the channel to stay closed, and so chloride ions build up in the cell. This buildup causes the formation of thick mucus and produces the symptoms of the disease.

Most people have two copies of the normal allele for CFTR and produce only functional CFTR protein. Those with cystic fibrosis possess two copies of the mutated CFTR allele and produce only the defective CFTR protein. Heterozygotes, having one normal and one defective CFTR allele, produce both functional and defective CFTR protein. Thus, at the molecular level, the alleles for normal and defective CFTR are codominant, because both alleles are expressed in the heterozygote. However, because one functional allele produces enough functional CFTR protein to allow normal chloride ion transport, the heterozygote exhibits no adverse effects, and the mutated CFTR allele appears to be recessive at the physiological level. The type of dominance expressed by an allele, as illustrated in this example, is a function of the phenotypic aspect of the allele that is observed.

Characteristics of dominance Several important characteristics of dominance should be emphasized. First, dominance is a result of interactions between genes at the same locus; in other words, dominance is *allelic* interaction. Second, dominance does not alter the way in which the genes are inherited; it only influences the way in which they are expressed as a phenotype. The allelic interaction that characterizes dominance is therefore interaction between the *products* of the genes. Finally, dominance is frequently "in the eye of the beholder," meaning that the classification of dominance depends on the level at which the phenotype is examined. As seen for cystic fibrosis, an allele may exhibit codominance at one level and be recessive at another level.

CONCEPTS

Dominance entails interactions between genes at the same locus (allelic genes) and is an aspect of the phenotype; dominance does not affect the way in which genes are inherited. The type of dominance exhibited by a characteristic frequently depends on the level of the phenotype examined.

✔ **CONCEPT CHECK 2**

How do complete dominance, incomplete dominance, and codominance differ?

Penetrance and Expressivity

In the genetic crosses presented thus far, we have considered only the interactions of alleles and have assumed that every individual organism having a particular genotype expresses the expected phenotype. We assumed, for example, that the genotype *Rr* always produces round seeds and that the genotype *rr* always produces wrinkled seeds. For some characters, however, such an assumption is incorrect: the genotype does not always produce the expected phenotype, a phenomenon termed **incomplete penetrance**.

Incomplete penetrance is seen in human polydactyly, the condition of having extra fingers and toes (**Figure 5.4**). There are several different forms of human polydactyly, but the trait is usually caused by a dominant allele. Occasionally, people possess the allele for polydactyly (as evidenced by the fact that their children inherit the polydactyly) but nevertheless have a normal number of fingers and toes. In these cases,

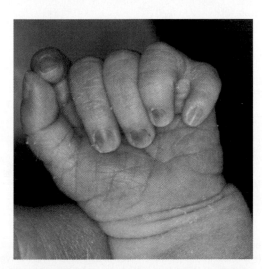

5.4 Human polydactyly (extra digits) exhibits incomplete penetrance and variable expressivity. [SPL/Photo Researchers.]

the gene for polydactyly is not fully penetrant. **Penetrance** is defined as the percentage of individual organisms having a particular genotype that express the expected phenotype. For example, if we examined 42 people having an allele for polydactyly and found that only 38 of them were polydactylous, the penetrance would be $^{38}/_{42} = 0.90$ (90%).

A related concept is that of **expressivity**, the degree to which a character is expressed. In addition to incomplete penetrance, polydactyly exhibits variable expressivity. Some polydactylous persons possess extra fingers and toes that are fully functional, whereas others possess only a small tag of extra skin.

Incomplete penetrance and variable expressivity are due to the effects of other genes and to environmental factors that can alter or completely suppress the effect of a particular gene. For example, a gene may encode an enzyme that produces a particular phenotype only within a limited temperature range. At higher or lower temperatures, the enzyme does not function and the phenotype is not expressed; the allele encoding such an enzyme is therefore penetrant only within a particular temperature range. Many characters exhibit incomplete penetrance and variable expressivity; thus the mere presence of a gene does not guarantee its expression. TRY PROBLEM 17 →

CONCEPTS

Penetrance is the percentage of individuals having a particular genotype that express the associated phenotype. Expressivity is the degree to which a trait is expressed. Incomplete penetrance and variable expressivity result from the influence of other genes and environmental factors on the phenotype.

✔ **CONCEPT CHECK 3**

How does incomplete dominance differ from incomplete penetrance?

a. Incomplete dominance refers to alleles at the same locus; incomplete penetrance refers to alleles at different loci.

b. Incomplete dominance ranges from 0% to 50%; incomplete penetance ranges from 51% to 99%.

c. In incomplete dominance, the heterozygote is intermediate to the homozygotes; in incomplete penetrance, heterozygotes express phenotypes of both homozygotes.

d. In incomplete dominance, the heterozygote is intermediate to the homozygotes; in incomplete penetrance, some individuals do not express the expected phenotype.

Lethal Alleles

As described in the introduction to this chapter, Lucien Cuénot reported the first case of a lethal allele, the allele for yellow coat color in mice (see Figure 5.1). A **lethal allele** causes death at an early stage of development—often before birth—and so some genotypes may not appear among the progeny.

Another example of a lethal allele, originally described by Erwin Baur in 1907, is found in snapdragons. The *aurea* strain in these plants has yellow leaves. When two plants with yellow leaves are crossed, $^2/_3$ of the progeny have yellow leaves and $^1/_3$ have green leaves. When green is crossed with green, all the progeny have green leaves; however, when yellow is crossed with green, $^1/_2$ of the progeny have green leaves and $^1/_2$ have yellow leaves, confirming that all yellow-leaved snapdragons are heterozygous. A 2 : 1 ratio is almost always produced by a recessive lethal allele; so observing this ratio among the progeny of a cross between individuals with the same phenotype is a strong clue that one of the alleles is lethal.

In this example, like that of yellow coat color in mice, the lethal allele is recessive because it causes death only in homozygotes. Unlike its effect on *survival*, the effect of the allele on *color* is dominant; in both mice and snapdragons, a single copy of the allele in the heterozygote produces a yellow color. This example illustrates the point made earlier that the type of dominance depends on the aspect of the phenotype examined.

Many lethal alleles in nature are recessive, but lethal alleles can also be dominant; in this case, homozygotes and heterozygotes for the allele die. Truly dominant lethal alleles cannot be transmitted unless they are expressed after the onset of reproduction, as in Huntington disease. TRY PROBLEMS 18 AND 40 →

CONCEPTS

A lethal allele causes death, frequently at an early developmental stage, and so one or more genotypes are missing from the progeny of a cross. Lethal alleles modify the ratio of progeny resulting from a cross.

✔ CONCEPT CHECK 4

A cross between two green corn plants yields $^2/_3$ progeny that are green and $^1/_3$ progeny that are white. What is the genotype of the green progeny?

a. *WW*   c. *ww*

b. *Ww*   d. *W_ (WW and Ww)*

Multiple Alleles

Most of the genetic systems that we have examined so far consist of two alleles. In Mendel's peas, for instance, one allele encoded round seeds and another encoded wrinkled seeds; in cats, one allele produced a black coat and another produced a gray coat. For some loci, more than two alleles are present within a group of organisms—the locus has **multiple alleles**. (Multiple alleles may also be referred to as an *allelic series*.) Although there may be more than two alleles present within a *group* of organisms, the genotype of each individual diploid organism still consists of only two alleles. The inheritance of characteristics encoded by multiple alleles is no different from the inheritance of characteristics encoded by two alleles, except that a greater variety of genotypes and phenotypes are possible.

Duck-feather patterns   An example of multiple alleles is at a locus that determines the feather pattern of mallard ducks. One allele, M, produces the wild-type *mallard* pattern. A second allele, M^R, produces a different pattern called *restricted*, and a third allele, m^d, produces a pattern termed *dusky*. In this allelic series, restricted is dominant over mallard and dusky, and mallard is dominant over dusky: $M^R > M > m^d$. The six genotypes possible with these three alleles and their resulting phenotypes are:

Genotype	Phenotype
$M^R M^R$	restricted
$M^R M$	restricted
$M^R m^d$	restricted
MM	mallard
Mm^d	mallard
$m^d m^d$	dusky

In general, the number of genotypes possible will be $[n(n + 1)]/2$, where n equals the number of different alleles at a locus. Working crosses with multiple alleles is no different from working crosses with two alleles; Mendel's principle of segregation still holds, as shown in the cross between a restricted duck and a mallard duck (**Figure 5.5**). TRY PROBLEM 19 →

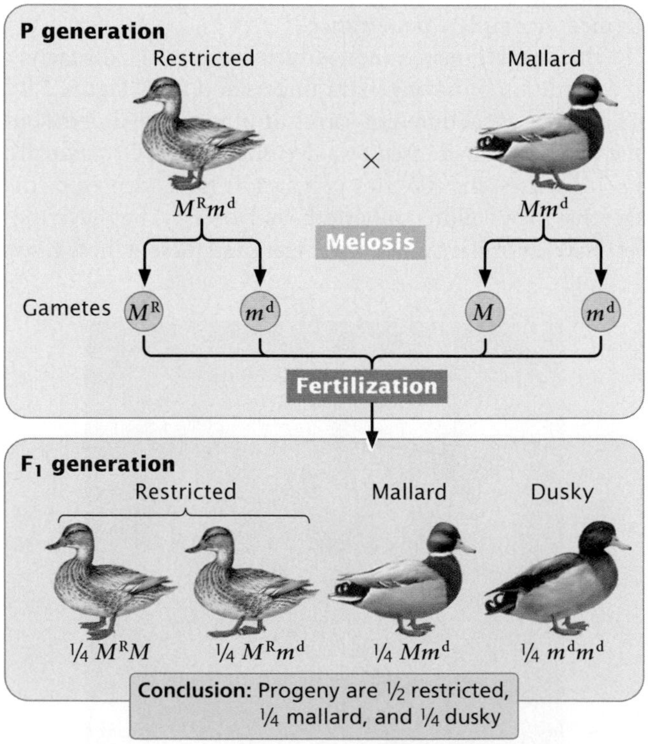

5.5 Mendel's principle of segregation applies to crosses with multiple alleles. In this example, three alleles determine the type of plumage in mallard ducks: M^R (restricted) $> M$ (mallard) $> m^d$ (dusky).

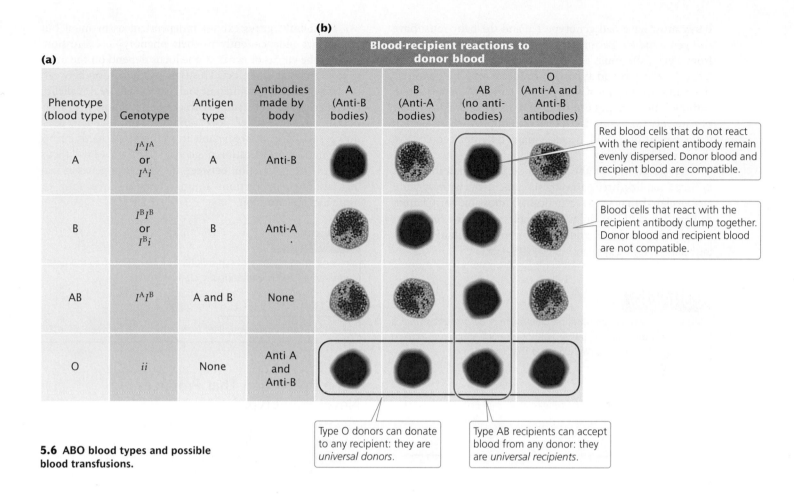

(b)

Phenotype (blood type)	Genotype	Antigen type	Antibodies made by body	Blood-recipient reactions to donor blood			
				A (Anti-B bodies)	B (Anti-A bodies)	AB (no anti-bodies)	O (Anti-A and Anti-B antibodies)
A	$I^A I^A$ or $I^A i$	A	Anti-B				
B	$I^B I^B$ or $I^B i$	B	Anti-A				
AB	$I^A I^B$	A and B	None				
O	ii	None	Anti A and Anti-B				

Red blood cells that do not react with the recipient antibody remain evenly dispersed. Donor blood and recipient blood are compatible.

Blood cells that react with the recipient antibody clump together. Donor blood and recipient blood are not compatible.

Type O donors can donate to any recipient: they are *universal donors*.

Type AB recipients can accept blood from any donor: they are *universal recipients*.

5.6 ABO blood types and possible blood transfusions.

The ABO blood group Another multiple-allele system is at the locus for the ABO blood group. This locus determines your ABO blood type and, like the MN locus, encodes antigens on red blood cells. The three common alleles for the ABO blood group locus are: I^A, which encodes the A antigen; I^B, which encodes the B antigen; and i, which encodes no antigen (O). We can represent the dominance relations among the ABO alleles as follows: $I^A > i$, $I^B > i$, $I^A = I^B$. Both the I^A and the I^B alleles are dominant over i and are codominant with each other; the AB phenotype is due to the presence of an I^A allele and an I^B allele, which results in the production of A and B antigens on red blood cells. A person with genotype ii produces neither antigen and has blood type O. The six common genotypes at this locus and their phenotypes are shown in **Figure 5.6a**.

Antibodies are produced against any foreign antigens (see Figure 5.6a). For instance, a person having blood-type A produces anti-B antibodies, because the B antigen is foreign. A person having blood-type B produces anti-A antibodies, and someone having blood-type AB produces neither anti-A nor anti-B antibodies, because neither A nor B antigen is foreign. A person having blood-type O possesses no A or B antigens; consequently, that person produces both anti-A

antibodies and anti-B antibodies. The presence of antibodies against foreign ABO antigens means that successful blood transfusions are possible only between persons with certain compatible blood types (**Figure 5.6b**).

The inheritance of alleles at the ABO locus is illustrated by a paternity suit against the famous movie actor Charlie Chaplin. In 1941, Chaplin met a young actress named Joan Barry, with whom he had an affair. The affair ended in February 1942 but, 20 months later, Barry gave birth to a baby girl and claimed that Chaplin was the father. Barry then sued for child support. At this time, blood typing had just come into widespread use, and Chaplin's attorneys had Chaplin, Barry, and the child blood typed. Barry had blood-type A, her child had blood-type B, and Chaplin had blood-type O. Could Chaplin have been the father of Barry's child?

Your answer should be no. Joan Barry had blood-type A, which can be produced by either genotype $I^A I^A$ or genotype $I^A i$. Her baby possessed blood-type B, which can be produced by either genotype $I^B I^B$ or genotype $I^B i$. The baby could not have inherited the I^B allele from Barry (Barry could not carry an I^B allele if she were blood-type A); therefore the baby must have inherited the i allele from her.

Barry must have had genotype $I^A i$, and the baby must have had genotype $I^B i$. Because the baby girl inherited her i allele from Barry, she must have inherited the I^B allele from her father. Having blood-type O, produced only by genotype ii, Chaplin could not have been the father of Barry's child. Although blood types can be used to exclude the possibility of paternity (as in this case), they cannot prove that a person is the parent of a child, because many different people have the same blood type.

In the course of the trial to settle the paternity suit against Chaplin, three pathologists came to the witness stand and declared that it was genetically impossible for Chaplin to have fathered the child. Nevertheless, the jury ruled that Chaplin was the father and ordered him to pay child support and Barry's legal expenses. **TRY PROBLEM 24** →

CONCEPTS

More than two alleles (multiple alleles) may be present within a group of individual organisms, although each individual diploid organism still has only two alleles at that locus.

✔ **CONCEPT CHECK 5**

How many genotypes are present at a locus with five alleles?

a. 30   c. 15

b. 27   d. 5

5.2 Gene Interaction Takes Place When Genes at Multiple Loci Determine a Single Phenotype

In the dihybrid crosses that we examined in Chapter 3, each locus had an independent effect on the phenotype. When Mendel crossed a homozygous round and yellow plant (*RR YY*) with a homozygous wrinkled and green plant (*rr yy*) and then self-fertilized the F_1, he obtained F_2 progeny in the following proportions:

$^9/_{16}$	$R\_\ Y\_$	round, yellow
$^3/_{16}$	$R\_\ yy$	round, green
$^3/_{16}$	$rr\ Y\_$	wrinkled, yellow
$^1/_{16}$	$rr\ yy$	wrinkled, green

In this example, the genes showed two kinds of independence. First, the genes at each locus are independent in their *assortment* in meiosis, which is what produces the 9 : 3 : 3 : 1 ratio of phenotypes in the progeny, in accord with Mendel's principle of independent assortment. Second, the genes are independent in their *phenotypic expression*, the R and r alleles affect only the shape of the seed and have no influence on the color of the seed; the Y and y alleles affect only color and have no influence on the shape of the seed.

Frequently, genes exhibit independent assortment but do not act independently in their phenotypic expression; instead, the effects of genes at one locus depend on the presence of genes at other loci. This type of interaction between the effects of genes at different loci (genes that are not allelic) is termed **gene interaction**. With gene interaction, the products of genes at different loci combine to produce new phenotypes that are not predictable from the single-locus effects alone. In our consideration of gene interaction, we will focus primarily on interaction between the effects of genes at two loci, although interactions among genes at three, four, or more loci are common.

CONCEPTS

In gene interaction, genes at different loci contribute to the determination of a single phenotypic characteristic.

✔ **CONCEPT CHECK 6**

How does gene interaction differ from dominance between alleles?

Gene Interaction That Produces Novel Phenotypes

Let's first examine gene interaction in which genes at two loci interact to produce a single characteristic. Fruit color in the pepper *Capsicum annuum* is determined in this way. Certain types of peppers produce fruits in one of four colors: red, peach, orange (sometimes called yellow), and cream (or white). If a homozygous plant with red peppers is crossed with a homozygous plant with cream peppers, all the F_1 plants have red peppers (**Figure 5.7a**). When the F_1 are crossed with one another, the F_2 are in a ratio of 9 red : 3 peach : 3 orange : 1 cream (**Figure 5.7b**). This dihybrid ratio (see Chapter 3) is produced by a cross between two plants that are both heterozygous for two loci ($Y^+y\ C^+c \times Y^+y\ C^+c$). In this example, the Y locus and the C locus interact to produce a single phenotype-the color of the pepper:

Genotype	Phenotype
$Y^+\_\ C^+\_$	red
$Y^+\_\ cc$	peach
$yy\ C^+\_$	orange
$yy\ cc$	cream

Color in peppers of *Capsicum annuum* results from the relative amounts of red and yellow carotenoids, compounds that are synthesized in a complex biochemical pathway. The Y locus encodes one enzyme (the first step in the pathway), and the C locus encodes a different enzyme (the last step in the pathway). When different loci influence different steps in a common biochemical pathway, gene interaction often arises because the product of one enzyme affects the substrate of another enzyme.

(a)

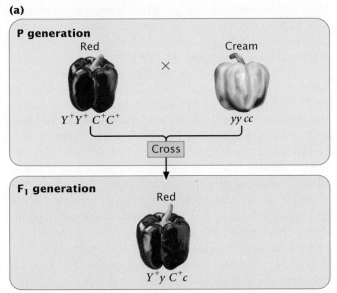

(b)

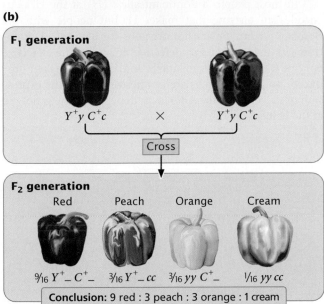

5.7 Gene interaction in which two loci determine a single characteristic, fruit color, in the pepper *Capsicum annuum*.

To illustrate how Mendel's rules of heredity can be used to understand the inheritance of characteristics determined by gene interaction, let's consider a testcross between an F_1 plant from the cross in Figure 5.7 ($Y^+y\ C^+c$) and a plant with cream peppers ($yy\ cc$). As outlined in Chapter 3 for independent loci, we can work this cross by breaking it down into two simple crosses. At the first locus, the heterozygote Y^+y is crossed with the homozygote yy; this cross produces $\frac{1}{2}\ Y^+y$ and $\frac{1}{2}\ yy$ progeny. Similarly, at the second locus, the heterozygous genotype C^+c is crossed with the homozygous genotype cc, producing $\frac{1}{2}\ C^+c$ and $\frac{1}{2}\ cc$ progeny. In accord with Mendel's principle of independent assortment, these single-locus ratios can be combined by using the multiplication rule: the probability of obtaining the genotype $Y^+y\ C^+c$

is the probability of Y^+y ($\frac{1}{2}$) multiplied by the probability of C^+c ($\frac{1}{2}$), or $\frac{1}{4}$. The probability of each progeny genotype resulting from the testcross is:

Progeny genotype	Probability at each locus		Overall probability	Phenotype
$Y^+y\ C^+c$	$\frac{1}{2} \times \frac{1}{2}$	$=$	$\frac{1}{4}$	red peppers
$Y^+y\ cc$	$\frac{1}{2} \times \frac{1}{2}$	$=$	$\frac{1}{4}$	peach peppers
$yy\ C^+c$	$\frac{1}{2} \times \frac{1}{2}$	$=$	$\frac{1}{4}$	orange peppers
$yy\ cc$	$\frac{1}{2} \times \frac{1}{2}$	$=$	$\frac{1}{4}$	cream peppers

When you work problems with gene interaction, it is especially important to determine the probabilities of single-locus genotypes and to multiply the probabilities of *genotypes*, not phenotypes, because the phenotypes cannot be determined without considering the effects of the genotypes at all the contributing loci. TRY PROBLEM 25 →

Gene Interaction with Epistasis

Sometimes the effect of gene interaction is that one gene masks (hides) the effect of another gene at a different locus, a phenomenon known as **epistasis**. In the examples of genic interaction that we have examined, alleles at different loci interact to determine a single phenotype. In those examples, one allele did not *mask* the effect of an allele at another locus, meaning that there was no epistasis. Epistasis is similar to dominance, except that dominance entails the masking of genes at the *same* locus (allelic genes). In epistasis, the gene that does the masking is called an **epistatic gene**; the gene whose effect is masked is a **hypostatic gene**. Epistatic genes may be recessive or dominant in their effects.

Recessive epistasis Recessive epistasis is seen in the genes that determine coat color in Labrador retrievers. These dogs may be black, brown, or yellow; their different coat colors are determined by interactions between genes at two loci (although a number of other loci also help to determine coat color; see pp. 113–115). One locus determines the type of pigment produced by the skin cells: a dominant allele *B* encodes black pigment, whereas a recessive allele *b* encodes brown pigment. Alleles at a second locus affect the *deposition* of the pigment in the shaft of the hair; dominant allele *E* allows dark pigment (black or brown) to be deposited, whereas recessive allele *e* prevents the deposition of dark pigment, causing the hair to be yellow. The presence of genotype *ee* at the second locus therefore masks the expression of the black and brown alleles at the first locus. The genotypes that determine coat color and their phenotypes are:

Genotype	Phenotype
*B_ E_*	black
*bb E_*	brown (frequently called chocolate)
B_ ee	yellow
bb ee	yellow

If we cross a black Labrador homozygous for the dominant alleles (*BB EE*) with a yellow Labrador homozygous for the recessive alleles (*bb ee*) and then intercross the F$_1$, we obtain progeny in the F$_2$ in a 9 : 3 : 4 ratio:

P *BB EE* × *bb ee*
   Black   Yellow

      ↓

F$_1$ *Bb Ee*
     Black

      ↓ Intercross

F$_2$ $^9/_{16}$ *B_ E_* black

   $^3/_{16}$ *bb E_* brown

   $^3/_{16}$ *B_ ee* yellow ⎱
   $^1/_{16}$ *bb ee* yellow ⎰ $^4/_{16}$ yellow

Notice that yellow dogs can carry alleles for either black or brown pigment, but these alleles are not expressed in their coat color.

In this example of gene interaction, allele *e* is epistatic to *B* and *b*, because *e* masks the expression of the alleles for black and brown pigments, and alleles *B* and *b* are hypostatic to *e*. In this case, *e* is a recessive epistatic allele, because two copies of *e* must be present to mask the expression of the black and brown pigments. **TRY PROBLEM 29** →

Another example of a recessive epistatic gene is the Bombay phenotype, which suppresses the expression of alleles at the ABO locus. As mentioned earlier in the chapter, the alleles at the ABO locus encode antigens on the red blood cells; the antigens consist of short chains of carbohydrates embedded in the membranes of red blood cells. The difference between the A and the B antigens is a function of chemical differences in the terminal sugar of the chain. The *I*^{A} and *I*^{B} alleles actually encode different enzymes, which add sugars designated A or B to the ends of the carbohydrate chains (**Figure 5.8**). The common substrate on which these enzymes act is a molecule called H. The enzyme encoded by the *i* allele apparently either adds no sugar to H or no functional enzyme is specified.

In most people, a dominant allele (*H*) at the H locus encodes an enzyme that makes H, but people with the Bombay phenotype are homozygous for a recessive mutation (*h*) that encodes a defective enzyme. The defective enzyme is incapable of making H and, because H is not produced, no ABO antigens are synthesized. Thus, the expression of the alleles at the ABO locus depends on the genotype at the H locus.

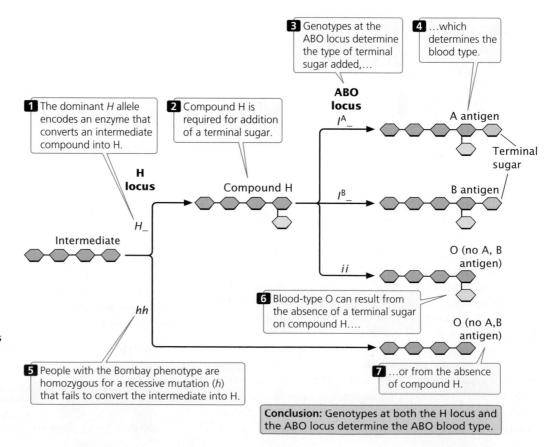

5.8 Expression of the ABO antigens depends on alleles at the H locus. The H locus encodes a precursor to the antigens called compound H. Alleles at the ABO locus determine which types of terminal sugars are added to compound H.

1 The dominant *H* allele encodes an enzyme that converts an intermediate compound into H.

2 Compound H is required for addition of a terminal sugar.

3 Genotypes at the ABO locus determine the type of terminal sugar added,...

4 ...which determines the blood type.

ABO locus

I^A_   A antigen

Terminal sugar

H locus

Compound H

I^B_   B antigen

*H*_

Intermediate

ii   O (no A, B antigen)

6 Blood-type O can result from the absence of a terminal sugar on compound H....

hh

5 People with the Bombay phenotype are homozygous for a recessive mutation (*h*) that fails to convert the intermediate into H.

O (no A,B antigen)

7 ...or from the absence of compound H.

Conclusion: Genotypes at both the H locus and the ABO locus determine the ABO blood type.

Genotype	H present	ABO phenotype
$H\_\ I^A I^A,\ H\_\ I^A i$	Yes	A
$H\_\ I^B I^B,\ H\_\ I^B i$	Yes	B
$H\_\ I^A I^B$	Yes	AB
$H\_\ ii$	Yes	O
$hh\ I^A I^A,\ hh\ I^A i,\ hh\ I^B I^B,\ hh\ I^B i,$ $\quad hh\ I^A I^B,$ and $hh\ ii$	No	O

In this example, the alleles at the ABO locus are hypostatic to the recessive h allele.

The Bombay phenotype provides us with a good opportunity for considering how epistasis often arises when genes affect a series of steps in a biochemical pathway. The ABO antigens are produced in a multistep biochemical pathway (see Figure 5.8), which depends on enzymes that make H and on other enzymes that convert H into the A or B antigen. Note that blood-type O may arise in one of two ways: (1) from failure to add a terminal sugar to compound H (genotype $H\_\ ii$) or (2) from failure to produce compound H (genotype $hh\_$). Many cases of epistasis arise in this way. A gene (such as h) that has an effect on an early step in a biochemical pathway will be epistatic to genes (such as I^A and I^B) that affect subsequent steps, because the effects of the genes in a later step depend on the product of the earlier reaction.

Dominant epistasis   In *recessive* epistasis, which we just considered, the presence of two recessive alleles (the homozygous genotype) inhibits the expression of an allele at a different locus. In *dominant* epistasis, only a single copy of an allele is required to inhibit the expression of the allele at a different locus.

Dominant epistasis is seen in the interaction of two loci that determine fruit color in summer squash, which is commonly found in one of three colors: yellow, white, or green. When a homozygous plant that produces white squash is crossed with a homozygous plant that produces green squash and the F_1 plants are crossed with each other, the following results are obtained:

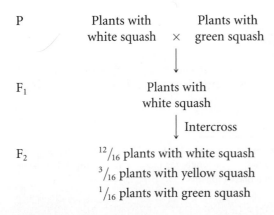

P   Plants with white squash   ×   Plants with green squash

F_1 Plants with white squash

Intercross

F_2 $^{12}/_{16}$ plants with white squash
$^{3}/_{16}$ plants with yellow squash
$^{1}/_{16}$ plants with green squash

How can gene interaction explain these results?

In the F_2, $^{12}/_{16}$, or $^{3}/_{4}$, of the plants produce white squash and $^{3}/_{16} + ^{1}/_{16} = ^{4}/_{16} = ^{1}/_{4}$ of the plants produce squash having color. This outcome is the familiar 3 : 1 ratio produced by a cross between two heterozygotes, which suggests that a dominant allele at one locus inhibits the production of pigment, resulting in white progeny. If we use the symbol W to represent the dominant allele that inhibits pigment production, then genotype $W\_$ inhibits pigment production and produces white squash, whereas ww allows pigment and results in colored squash.

Among those ww F_2 plants with pigmented fruit, we observe $^{3}/_{16}$ yellow and $^{1}/_{16}$ green (a 3 : 1 ratio). In this outcome, a second locus determines the type of pigment produced in the squash, with yellow ($Y\_$) dominant over green (yy). This locus is expressed only in ww plants, which lack the dominant inhibitory allele W. We can assign the genotype ww $Y\_$ to plants that produce yellow squash and the genotype $ww\ yy$ to plants that produce green squash. The genotypes and their associated phenotypes are:

$W\_\ Y\_$	white squash
$W\_\ yy$	white squash
$ww\ Y\_$	yellow squash
$ww\ yy$	green squash

Allele W is epistatic to Y and y: it suppresses the expression of these pigment-producing genes. Allele W is a dominant epistatic allele because, in contrast with e in Labrador retriever coat color and with h in the Bombay phenotype, a single copy of the allele is sufficient to inhibit pigment production.

Yellow pigment in the squash is most likely produced in a two-step biochemical pathway (**Figure 5.9**). A colorless (white) compound (designated A in Figure 5.9) is converted by enzyme I into green compound B, which is then converted into compound C by enzyme II. Compound C is the yellow pigment in the fruit. Plants with the genotype ww produce enzyme I and may be green or yellow, depending on whether enzyme II is present. When allele Y is present at a second locus, enzyme II is produced and compound B is converted into compound C, producing a yellow fruit. When two copies of allele y, which does not encode a functional form of enzyme II, are present, squash remain green. The presence of W at the first locus inhibits the conversion of compound A into compound B; plants with genotype $W\_$ do not make compound B and their fruit remains white, regardless of which alleles are present at the second locus.

Duplicate recessive epistasis   Finally, let's consider duplicate recessive epistasis, in which two recessive alleles at either of two loci are capable of suppressing a phenotype. This type of epistasis is illustrated by albinism in snails.

Albinism is the absence of pigment and is a common genetic trait in many plants and animals. Pigment is almost

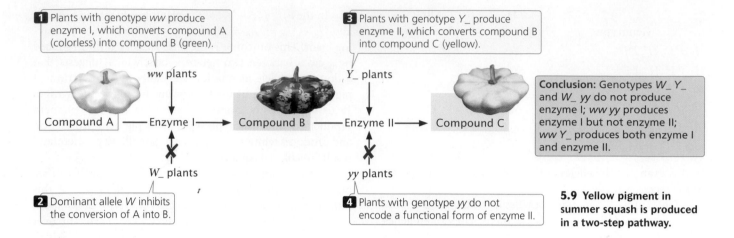

1 Plants with genotype *ww* produce enzyme I, which converts compound A (colorless) into compound B (green).

3 Plants with genotype *Y_* produce enzyme II, which converts compound B into compound C (yellow).

ww plants

*Y_* plants

Compound A —— Enzyme I ——→ Compound B —— Enzyme II——→ Compound C

*W_* plants

yy plants

2 Dominant allele *W* inhibits the conversion of A into B.

4 Plants with genotype *yy* do not encode a functional form of enzyme II.

Conclusion: Genotypes *W_ Y_* and *W_ yy* do not produce enzyme I; *ww yy* produces enzyme I but not enzyme II; *ww Y_* produces both enzyme I and enzyme II.

5.9 Yellow pigment in summer squash is produced in a two-step pathway.

always produced through a multistep biochemical pathway; thus, albinism may entail gene interaction. Robert T. Dillon and Amy R. Wethington found that albinism in the common freshwater snail *Physa heterostroha* can result from the presence of either of two recessive alleles at two different loci. Inseminated snails were collected from a natural population and placed in cups of water, where they laid eggs. Some of the eggs hatched into albino snails. When two albino snails were crossed, all of the F_1 were pigmented. When the F_1 were intercrossed, the F_2 consisted of $^9/_{16}$ pigmented snails and $^7/_{16}$ albino snails. How did this 9 : 7 ratio arise?

The 9 : 7 ratio seen in the F_2 snails can be understood as a modification of the 9 : 3 : 3 : 1 ratio obtained when two individuals heterozygous for two loci are crossed. The 9 : 7 ratio arises when dominant alleles at both loci (*A_ B_*) produce pigmented snails; any other genotype produces albino snails:

P   *aa BB*   ×   *AA bb*
           Albino   Albino

                      ↓

F_1 *Aa Bb*
                  Pigmented

                      ↓ Intercross

F_2 $^9/_{16}$ *A_ B_* pigmented

             $^3/_{16}$ *aa B_* albino

             $^3/_{16}$ *A_ bb* albino

             $^1/_{16}$ *aa bb* albino

The 9 : 7 ratio in these snails is probably produced by a two-step pathway of pigment production (**Figure 5.10**). Pigment (compound C) is produced only after compound

A has been converted into compound B by enzyme I and after compound B has been converted into compound C by enzyme II. At least one dominant allele *A* at the first locus is required to produce enzyme I; similarly, at least one dominant allele *B* at the second locus is required to produce enzyme II. Albinism arises from the absence of compound C, which may happen in one of three ways. First, two recessive alleles at the first locus (genotype *aa B_*) may prevent the production of enzyme I, and so compound B is never produced. Second, two recessive alleles at the second locus (genotype *A_ bb*) may prevent the production of enzyme II; in this case, compound B is never converted into compound C. Third, two recessive alleles may be present at both loci (*aa bb*), causing the absence of both enzyme I and enzyme II. In this example of gene interaction, *a* is epistatic to *B*, and *b* is epistatic to *A*; *both* are recessive epistatic alleles because the presence of two copies of either allele *a* or allele *b* is necessary to suppress pigment production. This example differs from the suppression of coat color in Labrador retrievers in that recessive alleles at either of two loci are capable of suppressing pigment production in the snails, whereas recessive alleles at a single locus suppress pigment expression in Labs.

CONCEPTS

Epistasis is the masking of the expression of one gene by another gene at a different locus. The epistatic gene does the masking; the hypostatic gene is masked. Epistatic genes can be dominant or recessive.

✔ CONCEPT CHECK 7

A number of all-white cats are crossed and they produce the following types of progeny: $^{12}/_{16}$ all-white, $^3/_{16}$ black, and $^1/_{16}$ gray. What is the genotype of the black progeny?

a. *Aa*   c. *A_ B_*

b. *Aa Bb*   d. *A_ bb*

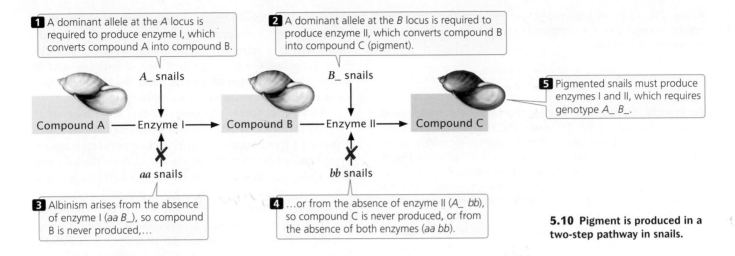

1 A dominant allele at the *A* locus is required to produce enzyme I, which converts compound A into compound B.

2 A dominant allele at the *B* locus is required to produce enzyme II, which converts compound B into compound C (pigment).

5 Pigmented snails must produce enzymes I and II, which requires genotype *A_ B_*.

*A_* snails

*B_* snails

Compound A — Enzyme I → Compound B — Enzyme II → Compound C

aa snails

bb snails

3 Albinism arises from the absence of enzyme I (*aa B_*), so compound B is never produced,…

4 …or from the absence of enzyme II (*A_ bb*), so compound C is never produced, or from the absence of both enzymes (*aa bb*).

5.10 Pigment is produced in a two-step pathway in snails.

CONNECTING CONCEPTS

Interpreting Ratios Produced by Gene Interaction

A number of modified ratios that result from gene interaction are shown in **Table 5.2**. Each of these examples represents a modification of the basic 9 : 3 : 3 : 1 dihybrid ratio. In interpreting the genetic basis of modified ratios, we should keep several points in mind. First, the inheritance of the genes producing these characteristics is no different from the inheritance of genes encoding simple genetic characters. Mendel's principles of segregation and independent assortment still apply; each individual organism possesses two alleles at each locus, which separate in meiosis, and genes at the different loci assort independently. The only difference is in how the *products* of the genotypes interact to produce the phenotype. Thus, we cannot consider the expression of genes at each locus separately; instead, we must take into consideration how the genes at different loci interact.

A second point is that, in the examples that we have considered, the phenotypic proportions were always in sixteenths because, in all the crosses, pairs of alleles segregated at two independently assorting loci. The probability of inheriting one of the two alleles at a locus is $1/2$. Because there are two loci, each with two alleles, the probability of inheriting any particular combination of genes is $(1/2)^4 = 1/16$. For a trihybrid cross, the progeny proportions should be in sixty-fourths, because $(1/2)^6 = 1/64$. In general, the progeny proportions should be in fractions of $(1/2)^{2n}$, where n equals the number of loci with two alleles segregating in the cross.

Crosses rarely produce exactly 16 progeny; therefore, modifications of a dihybrid ratio are not always obvious. Modified dihybrid ratios are more easily seen if the number of individuals of each phenotype is expressed in sixteenths:

$$\frac{x}{16} = \frac{\text{number of progeny with a phenotype}}{\text{total number of progeny}}$$

| | **Table 5.2** | Modified dihybrid phenotypic ratios due to gene interaction |

	Genotype					
Ratio*	*A_ B_*	*A_ bb*	*aa B_*	*aa bb*	Type of Interaction	Example
9 : 3 : 3 : 1	9	3	3	1	None	Seed shape and seed color in peas
9 : 3 : 4	9	3	4		Recessive epistasis	Coat color in Labrador retrievers
12 : 3 : 1	12		3	1	Dominant epistasis	Color in squash
9 : 7	9	7			Duplicate recessive epistasis	Albinism in snails
9 : 6 : 1	9	6		1	Duplicate interaction	—
15 : 1	15			1	Duplicate dominant epistasis	—
13 : 3	13		3		Dominant and recessive epistasis	—

*Each ratio is produced by a dihybrid cross (*Aa Bb* × *Aa Bb*). Shaded bars represent combinations of genotypes that give the same phenotype.

where $x/16$ equals the proportion of progeny with a particular phenotype. If we solve for x (the proportion of the particular phenotype in sixteenths), we have:

$$x = \frac{\text{number of progeny with a phenotype} \times 16}{\text{total number of progeny}}$$

For example, suppose that we cross two homozygotes, interbreed the F_1, and obtain 63 red, 21 brown, and 28 white F_2 individuals. Using the preceding formula, we find the phenotypic ratio in the F_2 to be: red = $(63 \times 16)/112 = 9$; brown = $(21 \times 16)/112 = 3$; and white = $(28 \times 16)/112 = 4$. The phenotypic ratio is 9 : 3 : 4.

A final point to consider is how to assign genotypes to the phenotypes in modified ratios that result from gene interaction. Don't try to *memorize* the genotypes associated with all the modified ratios in Table 5.2. Instead, practice relating modified ratios to known ratios, such as the 9 : 3 : 3 : 1 dihybrid ratio. Suppose that we obtain $^{15}/_{16}$ green progeny and $^{1}/_{16}$ white progeny in a cross between two plants. If we compare this 15 : 1 ratio with the standard 9 : 3 : 3 : 1 dihybrid ratio, we see that $^{9}/_{16} + ^{3}/_{16} + ^{3}/_{16}$ equals $^{15}/_{16}$. All the genotypes associated with these proportions in the dihybrid cross (A_ B_, A_ bb, and aa B_) must give the same phenotype, the green progeny. Genotype *aa BB* makes up $^{1}/_{16}$ of the progeny in a dihybrid cross, the white progeny in this cross.

In assigning genotypes to phenotypes in modified ratios, students sometimes become confused about which letters to assign to which phenotype. Suppose that we obtain the following phenotypic ratio: $^{9}/_{16}$ black : $^{3}/_{16}$ brown : $^{4}/_{16}$ white. Which genotype do we assign to the brown progeny, A_ bb or aa B_? Either answer is correct because the letters are just arbitrary symbols for the genetic information. The important thing to realize about this ratio is that the brown phenotype arises when two recessive alleles are present at one locus.

Worked Problem

A homozygous strain of yellow corn is crossed with a homozygous strain of purple corn. The F_1 are intercrossed, producing an ear of corn with 119 purple kernels and 89 yellow kernels (the progeny). What is the genotype of the yellow kernels?

• Solution

We should first consider whether the cross between yellow and purple strains might be a monohybrid cross for a simple dominant trait, which would produce a 3 : 1 ratio in the F_2 ($Aa \times Aa \rightarrow ^{3}/_{4} A\_$ and $^{1}/_{4} aa$). Under this hypothesis, we would expect 156 purple progeny and 52 yellow progeny:

Phenotype	Genotype	Observed number	Expected number
purple	A_	119	$^{3}/_{4} \times 208 = 156$
yellow	aa	89	$^{1}/_{4} \times 208 = 52$
Total		208	

We see that the expected numbers do not closely fit the observed numbers. If we performed a chi-square test (see

Chapter 3), we would obtain a calculated chi-square value of 35.08, which has a probability much less than 0.05, indicating that it is extremely unlikely that, when we expect a 3 : 1 ratio, we would obtain 119 purple progeny and 89 yellow progeny. Therefore, we can reject the hypothesis that these results were produced by a monohybrid cross.

Another possible hypothesis is that the observed F_2 progeny are in a 1 : 1 ratio. However, we learned in Chapter 3 that a 1 : 1 ratio is produced by a cross between a heterozygote and a homozygote ($Aa \times aa$) and, from the information given, the cross was not between a heterozygote and a homozygote, because both original parental strains were homozygous. Furthermore, a chi-square test comparing the observed numbers with an expected 1 : 1 ratio yields a calculated chi-square value of 4.32, which has a probability of less than 0.05.

Next, we should look to see if the results can be explained by a dihybrid cross ($Aa\, Bb \times Aa\, Bb$). A dihybrid cross results in phenotypic proportions that are in sixteenths. We can apply the formula given earlier in the chapter to determine the number of sixteenths for each phenotype:

$$x = \frac{\text{number of progeny with a phenotype} \times 16}{\text{total number of progeny}}$$

$$x_{(\text{purple})} = \frac{119 \times 16}{208} = 9.15$$

$$x_{(\text{yellow})} = \frac{89 \times 16}{208} = 6.85$$

Thus, purple and yellow appear in an approximate ratio of 9 : 7. We can test this hypothesis with a chi-square test:

Phenotype	Genotype	Observed number	Expected number
purple	?	119	$^{9}/_{16} \times 208 = 117$
yellow	?	89	$^{7}/_{16} \times 208 = 91$
Total		208	

$$x^2 = \sum \frac{(\text{observed} - \text{expected})^2}{\text{expected}}$$

$$= \frac{(119 - 117)^2}{117} + \frac{(89 - 91)^2}{91}$$

$$= 0.034 + 0.44 = 0.078$$

Degree of freedom $= n - 1 = 2 - 1 = 1$

$$P > 0.05$$

The probability associated with the chi-square value is greater than 0.05, indicating that there is a good fit between the observed results and a 9 : 7 ratio.

We now need to determine how a dihybrid cross can produce a 9 : 7 ratio and what genotypes correspond to the two phenotypes. A dihybrid cross without epistasis produces a 9 : 3 : 3 : 1 ratio:

$$Aa\ Bb\quad\times\quad Aa\ Bb$$

$$\downarrow$$

$$A\_\ B\_\ ^9/_{16}$$
$$A\_\ bb\ ^3/_{16}$$
$$aa\ B\_\ ^3/_{16}$$
$$aa\ bb\ ^1/_{16}$$

Because $^9/_{16}$ of the progeny from the corn cross are purple, purple must be produced by genotypes $A\_\ B\_$; in other words, individual kernels that have at least one dominant allele at the first locus and at least one dominant allele at the second locus are purple. The proportions of all the other genotypes ($A\_\ bb$, $aa\ B\_$, and $aa\ bb$) sum to $^7/_{16}$, which is the proportion of the progeny in the corn cross that are yellow, and so any individual kernel that does not have a dominant allele at both the first and the second locus is yellow.

> → Now test your understanding of epistasis by working Problem 26 at the end of the chapter.

Complementation: Determining Whether Mutations Are at the Same Locus or at Different Loci

How do we know whether different mutations that affect a characteristic occur at the same locus (are allelic) or at different loci? In fruit flies, for example, *white* is an X-linked recessive mutation that produces white eyes instead of the red eyes found in wild-type flies; *apricot* is an X-linked recessive mutation that produces light-orange-colored eyes. Do the *white* and *apricot* mutations occur at the same locus or at different loci? We can use the complementation test to answer this question.

To carry out a **complementation test** on recessive mutations, parents that are homozygous for different mutations are crossed, producing offspring that are heterozygous. If the mutations are allelic (occur at the same locus), then the heterozygous offspring have only mutant alleles ($a\ b$) and exhibit a mutant phenotype:

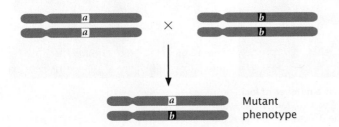

If, on the other hand, the mutations occur at different loci, each of the homozygous parents possesses wild-type genes at the other locus ($aa\ b^+b^+$ and $a^+a^+\ bb$); so the heterozygous offspring inherit a mutant allele and a wild-type allele at each locus. In this case, the mutations complement each other and the heterozygous offspring have the wild-type phenotype:

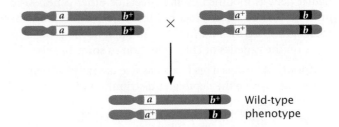

Complementation has occurred if an individual organism possessing two recessive mutations has a wild-type phenotype, indicating that the mutations are nonallelic genes. A lack of complementation occurs when two recessive mutations occur at the same locus, producing a mutant phenotype.

When the complementation test is applied to *white* and *apricot* mutations, all of the heterozygous offspring have light-colored eyes, demonstrating that white eyes and apricot eyes are produced by mutations that occur at the same locus and are allelic.

CONCEPTS

A complementation test is used to determine whether two mutations occur at the same locus (are allelic) or occur at different loci.

✔ CONCEPT CHECK 8

Brindle (tiger-striped appearance) is a recessive trait in bulldogs and in Chihuahuas. What types of crosses would you carry out to determine whether the *brindle* genes in bulldogs and in Chihuahuas are at the same locus?

The Complex Genetics of Coat Color in Dogs

The genetics of coat color in dogs is an excellent example of how complex interactions between genes may take part in the determination of a phenotype. Domestic dogs come in an amazing variety of shapes, sizes, and colors. For thousands of years, people have been breeding dogs for particular traits, producing the large number of types that we see today. Each breed of dog carries a selection of genes from the ancestral dog gene pool; these genes define the features of a particular breed. The genome of the domestic dog was completely sequenced in 2004, greatly facilitating the study of canine genetics.

An obvious difference between dogs is coat color. The genetics of coat color in dogs is quite complex; many genes participate, and there are numerous interactions between genes at different loci. We will consider four loci (in the list that follows) that are important in producing many of the noticeable differences in color and pattern among breeds of dogs. In interpreting the genetic basis of differences in the coat color of dogs, consider how the expression of a particular gene is modified by the effects of other genes. Keep in mind that additional loci not listed here can modify the colors produced by these four loci and that not all geneticists agree on the genetics of color variation in some breeds.

1. **Agouti (A) locus.** This locus has five common alleles that determine the depth and distribution of color in a dog's coat:

A^s — Solid black pigment.

a^w — Agouti, or wolflike gray. Hairs encoded by this allele have a "salt and pepper" appearance, produced by a band of yellow pigment on a black hair.

a^y — Yellow. The black pigment is markedly reduced; so the entire hair is yellow.

a^s — Saddle markings (dark color on the back, with extensive tan markings on the head and legs).

a^t — Bicolor (dark color over most of the body, with tan markings on the feet and eyebrows).

Alleles A^s and a^y are generally dominant over the other alleles, but the dominance relations are complex and not yet completely understood.

2. **Black (B) locus.** This locus determines whether black pigment can be formed. The actual color of a dog's coat depends on the effects of genes at other loci (such as the A and E loci). Two alleles are common:

B — Allows black pigment to be produced.

b — Black pigment cannot be produced; pigmented dogs can be chocolate, liver, tan, or red.

Allele B is dominant over allele b.

3. **Extension (E) locus.** Four alleles at this locus determine where the genotype at the A locus is expressed. For example, if a dog has the A^s allele (solid black) at the A locus, then black pigment will either be extended throughout the coat or be restricted to some areas, depending on the alleles present at the E locus. Areas where the A locus is not expressed may appear as yellow, red, or tan, depending on the presence of particular genes at other loci. When A^s is present at the A locus, the four alleles at the E locus have the following effects:

E^m — Black mask with a tan coat.

E — The A locus expressed throughout (solid black).

e^{br} — Brindle, in which black and yellow are in layers to give a tiger-striped appearance.

e — No black in the coat, but the nose and eyes may be black.

The dominance relations among these alleles are poorly known.

4. **Spotting (S) locus.** Alleles at this locus determine whether white spots will be present. There are four common alleles:

S — No spots.

s^i — Irish spotting; numerous white spots.

s^p — Piebald spotting; various amounts of white.

s^w — Extreme white piebald; almost all white.

Allele S is completely dominant over alleles s^i, s^p, and s^w; alleles s^i and s^p are dominant over allele s^w ($S > s^i, s^p > s^w$). The relation between s^i and s^p is poorly defined; indeed, they may not be separate alleles. Genes at other poorly known loci also modify spotting patterns.

To illustrate how genes at these loci interact in determining a dog's coat color, let's consider a few examples:

Labrador retriever Labrador retrievers (**Figure 5.11a**) may be black, brown, or yellow. Most are homozygous A^sA^s SS; thus, they vary only at the B and E loci. The A^s allele allows dark pigment to be expressed; whether a dog is black depends on which genes are present at the B and E loci. As discussed earlier in the chapter, all black Labradors must carry at least one B allele and one E allele ($B\_E\_$). Brown dogs are homozygous bb and have at least one E allele ($bb E\_$). Yellow dogs

(a)

(b)

(c)

5.11 Coat color in dogs is determined by interactions between genes at a number of loci. (a) Most Labrador retrievers are genotype A^sA^s SS, varying only at the B and E loci. (b) Most beagles are genotype a^sa^s BB s^ps^p. (c) Dalmatians are genotype A^sA^s EE s^ws^w, varying at the B locus, which makes the dogs black ($B\_$) or brown (bb). [Part a: Kent and Donna Dannen. Part b: Tara Darling. Part c: PhotoDisc.]

Table 5.3	Common genotypes in different breeds of dogs	
Breed	Usual Homozygous Genes*	Other Genes Present Within the Breed
Basset hound	$BB\ EE$	a^y, a^t S, s^p, s^i
Beagle	$a^s a^s\ BB\ s^p s^p$	E, e
English bulldog	BB	A^s, a^y, a^t E^m, E, e^{br} S, s^i, s^p, s^w
Chihuahua		A^s, a^y, a^s, a^t B, b E^m, E, e^{br}, e S, s^i, s^p, s^w
Collie	$BB\ EE$	a^y, a^t s^i, s^w
Dalmatian	$A^s A^s\ EE\ s^w s^w$	B, b
Doberman	$a^t a^t\ EE\ SS$	B, b
German shepherd	$BB\ SS$	a^y, a, a^s, a^t E^m, E, e
Golden retriever	$A^s A^s\ BB\ SS$	E, e
Greyhound	BB	A^s, a^y E, e^{br}, e S, s^p, s^w, s^i
Irish setter	$BB\ ee\ SS$	A, a^t
Labrador retriever	$A^s A^s\ SS$	B, b E, e
Poodle	SS	A^s, a^t B, b E, e
Rottweiler	$a^t a^t\ BB\ EE\ SS$	
St. Bernard	$a^y a^y\ BB$	E^m, E s^i, s^p, s^w

*Most dogs in the breed are homozygous for these genes; a few individual dogs may possess other alleles at these loci.

Source: Data from M. B. Willis, *Genetics of the Dog* (London: Witherby, 1989).

are a result of the presence of *ee* (*B_ ee* or *bb ee*). Labrador retrievers are homozygous for the *S* allele, which produces a solid color; the few white spots that appear in some dogs of this breed are due to other modifying genes.

Beagle   Most beagles (**Figure 5.11b**) are homozygous $a^s a^s$ $BB\ s^p s^p$, although other alleles at these loci are occasionally present. The a^s allele produces the saddle markings—dark back and sides, with tan head and legs—that are characteristic of the breed. Allele *B* allows black to be produced, but its distribution is limited by the a^s allele. Most beagles are *E_*, but the genotype *ee* does occasionally arise, leading to a few all-tan beagles. White spotting in beagles is due to the s^p allele.

Dalmatian   Dalmatians (**Figure 5.11c**) have an interesting genetic makeup. Most are homozygous $A^s A^s\ EE\ s^w s^w$; so they vary only at the *B* locus. Notice that these dogs possess genotype $A^s A^s\ EE$, which allows for a solid coat that would be black, if genotype *B_* were present, or brown (called liver), if genotype *bb* were present. However, the presence of the s^w allele produces a white coat, masking the expression of the solid color. The dog's color appears only in the pigmented spots, which are due to the presence of an allele at yet another locus that allows the color to penetrate in a limited number of spots.

Table 5.3 gives the common genotypes of other breeds of dogs. TRY PROBLEM 33 →

5.3 Sex Influences the Inheritance and Expression of Genes in a Variety of Ways

In Chapter 4, we considered characteristics encoded by genes located on the sex chromosomes (sex-linked traits) and how their inheritance differs from the inheritance of traits encoded by autosomal genes. X-linked traits, for example, are passed from father to daughter but never from father to son, and Y-linked traits are passed from father to all sons. Now, we will examine additional influences of sex, including the effect of the sex of an individual organism on the expression of genes on autosomal chromosomes, on characteristics determined by genes located in the cytoplasm, and on characteristics for which the genotype of only the maternal parent determines the phenotype of the offspring. Finally, we will look at situations in which the expression of genes on autosomal chromosomes is affected by the sex of the parent from whom the genes are inherited.

Sex-Influenced and Sex-Limited Characteristics

Sex-influenced characteristics are determined by autosomal genes and are inherited according to Mendel's principles, but they are expressed differently in males and females. In this case,

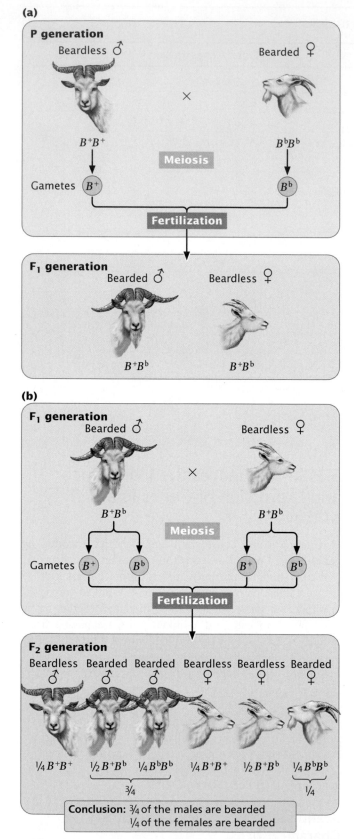

(a)

P generation

Beardless ♂   Bearded ♀

×

B^+B^+ B^bB^b

Meiosis

Gametes B^+ B^b

Fertilization

F₁ generation

Bearded ♂   Beardless ♀

B^+B^b B^+B^b

(b)

F₁ generation

Bearded ♂   Beardless ♀

×

B^+B^b B^+B^b

Gametes B^+ B^b B^+ B^b

Meiosis

Fertilization

F₂ generation

Beardless ♂ Bearded ♂ Bearded ♂ Beardless ♀ Beardless ♀ Bearded ♀

$\frac{1}{4}B^+B^+$ $\frac{1}{2}B^+B^b$ $\frac{1}{4}B^bB^b$ $\frac{1}{4}B^+B^+$ $\frac{1}{2}B^+B^b$ $\frac{1}{4}B^bB^b$

¾   ¼

Conclusion: ¾ of the males are bearded
¼ of the females are bearded

5.12 Genes that encode sex-influenced traits are inherited according to Mendel's principles but are expressed differently in males and females.

a particular trait is more readily expressed in one sex; in other words, the trait has higher penetrance in one of the sexes.

For example, the presence of a beard on some goats is determined by an autosomal gene (B^b) that is dominant in males and recessive in females. In males, a single allele is required for the expression of this trait: both the homozygote (B^bB^b) and the heterozygote (B^bB^+) have beards, whereas the B^+B^+ male is beardless.

Genotype	Males	Females
B^+B^+	beardless	beardless
B^+B^b	bearded	beardless
B^bB^b	bearded	bearded

In contrast, females require two alleles in order for this trait to be expressed: the homozygote B^bB^b has a beard, whereas the heterozygote (B^bB^+) and the other homozygote (B^+B^+) are beardless.

The key to understanding the expression of the bearded gene is to look at the heterozygote. In males (for which the presence of a beard is dominant), the heterozygous genotype produces a beard but, in females (for which the absence of a beard is dominant), the heterozygous genotype produces a goat without a beard.

Figure 5.12a illustrates a cross between a beardless male (B^+B^+) and a bearded female (B^bB^b). The alleles separate into gametes according to Mendel's principle of segregation, and all the F₁ are heterozygous (B^+B^b). Because the trait is dominant in males and recessive in females, all the F₁ males will be bearded and all the F₁ females will be beardless. When the F₁ are crossed with one another, ¼ of the F₂ progeny are B^bB^b, ½ are B^bB^+, and ¼ are B^+B^+ (**Figure 5.12b**). Because male heterozygotes are bearded, ¾ of the males in the F₂ possess beards; because female heterozygotes are beardless, only ¼ of the females in the F₂ are bearded.

An extreme form of sex-influenced inheritance, a **sex-limited characteristic** is encoded by autosomal genes that are expressed in only one sex; the trait has zero penetrance in the other sex. In domestic chickens, some males display a plumage pattern called cock feathering (**Figure 5.13a**). Other males and all females display a pattern called hen feathering (**Figure 5.13b** and **c**). Cock feathering is an autosomal recessive trait that is sex-limited to males. Because the trait is autosomal, the genotypes of males and females are the same, but the phenotypes produced by these genotypes differ in males and females:

Genotype	Male phenotype	Female phenotype
HH	hen feathering	hen feathering
Hh	hen feathering	hen feathering
hh	cock feathering	hen feathering

(a) 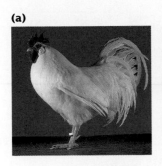   **(b)**    **(c)**

5.13 A sex-limited characteristic is encoded by autosomal genes that are expressed in only one sex. An example is cock feathering in chickens, an autosomal recessive trait that is limited to males. (a) Cock-feathered male. (b) Hen-feathered female. (c) Hen-feathered male. [Larry Lefever/Grant Heilman Photography.]

An example of a sex-limited characteristic in humans is male-limited precocious puberty. There are several types of precocious puberty in humans, most of which are not genetic. Male-limited precocious puberty, however, results from an autosomal dominant allele (P) that is expressed only in males; females with the gene are normal in phenotype. Males with precocious puberty undergo puberty at an early age, usually before the age of 4. At this time, the penis enlarges, the voice deepens, and pubic hair develops. There is no impairment of sexual function; affected males are fully fertile. Most are short as adults because the long bones stop growing after puberty.

Because the trait is rare, affected males are usually heterozygous (Pp). A male with precocious puberty who mates with a woman who has no family history of this condition will transmit the allele for precocious puberty to $^1/_2$ of their children (**Figure 5.14a**), but it will be expressed only in the sons. If one of the heterozygous daughters (Pp) mates with a male who has normal puberty (pp), $^1/_2$ of their sons will exhibit precocious puberty (**Figure 5.14b**). Thus a sex-limited characteristic can be inherited from either parent, although the trait appears in only one sex. TRY PROBLEM 35 →

CONCEPTS

Sex-influenced characteristics are encoded by autosomal genes that are more readily expressed in one sex. Sex-limited characteristics are encoded by autosomal genes whose expression is limited to one sex.

✔ CONCEPT CHECK 9

How do sex-influenced and sex-limited traits differ from sex-linked traits?

Cytoplasmic Inheritance

Mendel's principles of segregation and independent assortment are based on the assumption that genes are located on chromosomes in the nucleus of the cell. For most genetic characteristics, this assumption is valid, and Mendel's principles allow us to predict the types of offspring that will be produced in a genetic cross. However, not all the genetic

(a)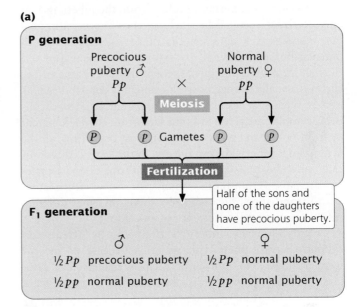

(b)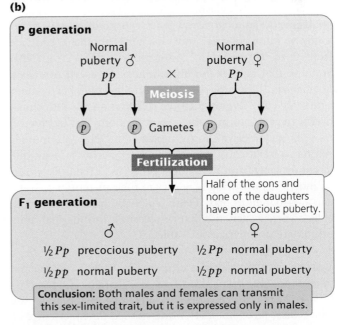

5.14 Sex-limited characteristics are inherited according to Mendel's principles. Precocious puberty is an autosomal dominant trait that is limited to males.

material of a cell is found in the nucleus; some characteristics are encoded by genes located in the cytoplasm. These characteristics exhibit **cytoplasmic inheritance**.

A few organelles, notably chloroplasts and mitochondria, contain DNA. The human mitochondrional genome contains about 15,000 nucleotides of DNA, encoding 37 genes. Compared with that of nuclear DNA, which contains some 3 billion nucleotides encoding some 20,000 to 25,000 genes, the size of the mitochondrial genome is very small; nevertheless, mitochondrial and chloroplast genes encode some important characteristics. The molecular details of this extranuclear DNA are discussed in Chapter 21; here, we will focus on *patterns* of cytoplasmic inheritance.

Cytoplasmic inheritance differs from the inheritance of characteristics encoded by nuclear genes in several important respects. A zygote inherits nuclear genes from both parents; but, typically, all its cytoplasmic organelles, and thus all its cytoplasmic genes, come from only one of the gametes, usually the egg. A sperm generally contributes only a set of nuclear genes from the male parent. In a few organisms, cytoplasmic genes are inherited from the male parent or from both parents; however, for most organisms, all the cytoplasm is inherited from the egg. In this case, cytoplasmically inherited traits are present in both males and females and are passed from mother to offspring, never from father to offspring. Reciprocal crosses, therefore, give different results when cytoplasmic genes encode a trait.

Cytoplasmically inherited characteristics frequently exhibit extensive phenotypic variation because no mechanism analogous to mitosis or meiosis ensures that cytoplasmic genes are evenly distributed in cell division. Thus, different cells and individual offspring will contain various proportions of cytoplasmic genes.

Consider mitochondrial genes. Most cells contain thousands of mitochondria, and each mitochondrion contains from 2 to 10 copies of mitochondrial DNA (mtDNA). Suppose that half of the mitochondria in a cell contain a normal wild-type copy of mtDNA and the other half contain a mutated copy (**Figure 5.15**). In cell division, the mitochondria segregate into progeny cells at random. Just by chance, one cell may receive mostly mutated mtDNA and another cell may receive mostly wild-type mtDNA. In this way, different progeny from the same mother and even cells within an individual offspring may vary in their phenotypes. Traits encoded by chloroplast DNA (cpDNA) are similarly variable. The characteristics that cytoplasmically inherited traits exhibit are summarized in **Table 5.4**.

Variegation in four-o'clocks In 1909, cytoplasmic inheritance was recognized by Carl Correns as an exception to Mendel's principles. Correns, one of the biologists who rediscovered Mendel's work, studied the inheritance of leaf variegation in the four-o'clock plant, *Mirabilis jalapa*. Correns found that the leaves and shoots of one variety of

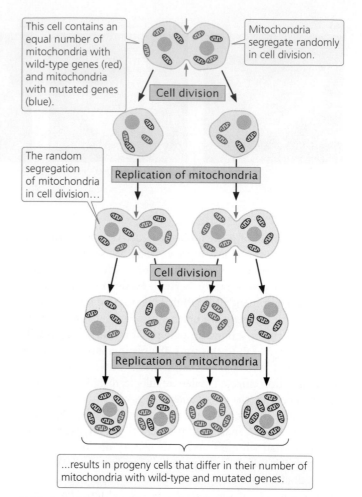

This cell contains an equal number of mitochondria with wild-type genes (red) and mitochondria with mutated genes (blue).

Mitochondria segregate randomly in cell division.

Cell division

Replication of mitochondria

The random segregation of mitochondria in cell division…

Cell division

Replication of mitochondria

…results in progeny cells that differ in their number of mitochondria with wild-type and mutated genes.

5.15 Cytoplasmically inherited characteristics frequently exhibit extensive phenotypic variation because cells and individual offspring contain various proportions of cytoplasmic genes. Mitochondria that have wild-type mtDNA are shown in red; those having mutant mtDNA are shown in blue.

four-o'clock were variegated, displaying a mixture of green and white splotches. He also noted that some branches of the variegated strain had all-green leaves; other branches had all-white leaves. Each branch produced flowers; so Correns was able to cross flowers from variegated, green, and

Table 5.4	Characteristics of cytoplasmically inherited traits.
1.	Present in males and females.
2.	Usually inherited from one parent, usually the maternal parent.
3.	Reciprocal crosses give different results.
4.	Exhibit extensive phenotypic variation, even within a single family.

white branches in all combinations (**Figure 5.16**). The seeds from green branches always gave rise to green progeny, no matter whether the pollen was from a green, white, or variegated branch. Similarly, flowers on white branches always produced white progeny. Flowers on the variegated branches gave rise to green, white, and variegated progeny, in no particular ratio.

Correns's crosses demonstrated cytoplasmic inheritance of variegation in the four-o'clocks. The phenotypes of the offspring were determined entirely by the maternal parent, never by the paternal parent (the source of the pollen). Furthermore, the production of all three phenotypes by flowers on variegated branches is consistent with cytoplasmic inheritance. Variegation in these plants is caused by a defective gene in the cpDNA, which results in a failure to produce the green pigment chlorophyll. Cells from green branches contain normal chloroplasts only, cells from white branches contain abnormal chloroplasts only, and cells from variegated branches contain a mixture of normal and abnormal chloroplasts. In the flowers from variegated branches, the random segregation of chloroplasts in the course of oogenesis produces some egg cells with normal cpDNA, which develop into green progeny; other egg cells with only abnormal cpDNA develop into white progeny; and, finally, still other egg cells with a mixture of normal and abnormal cpDNA develop into variegated progeny.

Mitochondrial diseases   A number of human diseases (mostly rare) that exhibit cytoplasmic inheritance have been identified. These disorders arise from mutations in mtDNA, most of which occur in genes encoding components of the electron-transport chain, which generates most of the ATP (adenosine triphosphate) in aerobic cellular respiration. One such disease is Leber hereditary optic neuropathy (LHON). Patients who have this disorder experience rapid loss of vision in both eyes, resulting from the death of cells in the optic nerve. This loss of vision typically occurs in early adulthood (usually between the ages of 20 and 24), but it can occur any time after adolescence. There is much clinical variability in the severity of the disease, even within the same family. Leber hereditary optic neuropathy exhibits cytoplasmic inheritance: the trait is passed from mother to all children, sons and daughters alike.

Genetic Maternal Effect

A genetic phenomenon that is sometimes confused with cytoplasmic inheritance is **genetic maternal effect**, in which the phenotype of the offspring is determined by the genotype of the mother. In cytoplasmic inheritance, the genes for a characteristic are inherited from only one parent, usually the mother. In genetic maternal effect, the genes are inherited from both parents, but the offspring's phenotype is determined not by its own genotype but by the genotype of its mother.

Genetic maternal effect frequently arises when substances present in the cytoplasm of an egg (encoded by the mother's nuclear genes) are pivotal in early development. An excellent example is the shell coiling of the snail *Limnaea*

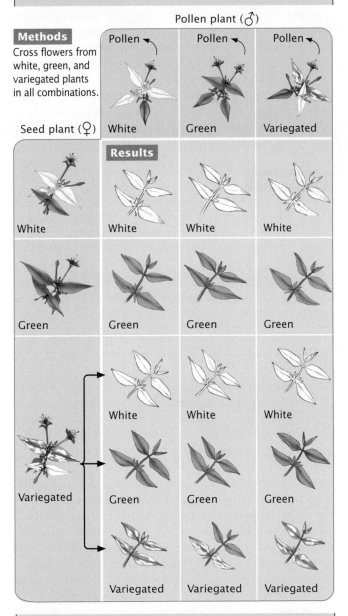

Experiment

Question: How is stem and leaf color inherited in the four-o'clock plant?

Methods
Cross flowers from white, green, and variegated plants in all combinations.

Pollen plant (♂)

Seed plant (♀)

Pollen / White   Pollen / Green   Pollen / Variegated

Results

White · White / White · White / White · White

Green · Green / Green · Green / Green · Green

Variegated · White / White / White

· Green / Green / Green

· Variegated / Variegated / Variegated

Conclusion: The phenotype of the progeny is determined by the phenotype of the branch from which the seed originated, not from the branch on which the pollen originated. Stem and leaf color exhibits cytoplasmic inheritance.

5.16 Crosses for leaf type in four-o'clocks illustrate cytoplasmic inheritance.

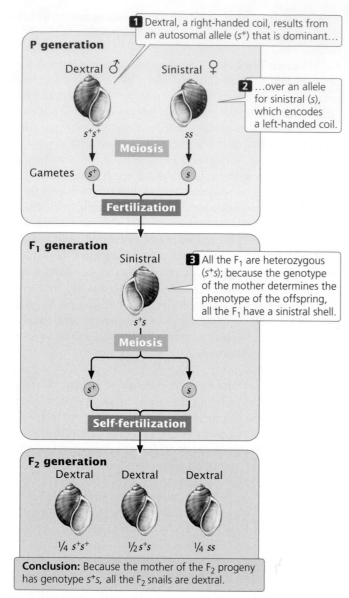

1 Dextral, a right-handed coil, results from an autosomal allele (s^+) that is dominant…

P generation

Dextral ♂   Sinistral ♀

2 …over an allele for sinistral (s), which encodes a left-handed coil.

s^+s^+ ss

Meiosis

Gametes s^+ s

Fertilization

F₁ generation

Sinistral

3 All the F₁ are heterozygous (s^+s); because the genotype of the mother determines the phenotype of the offspring, all the F₁ have a sinistral shell.

s^+s

Meiosis

s^+ s

Self-fertilization

F₂ generation

Dextral   Dextral   Dextral

¼ s^+s^+ ½ s^+s ¼ ss

Conclusion: Because the mother of the F₂ progeny has genotype s^+s, all the F₂ snails are dextral.

5.17 In genetic maternal effect, the genotype of the maternal parent determines the phenotype of the offspring. The shell coiling of a snail is a trait that exhibits genetic maternal effect.

peregra (**Figure 5.17**). In most snails of this species, the shell coils to the right, which is termed dextral coiling. However, some snails possess a left-coiling shell, exhibiting sinistral coiling. The direction of coiling is determined by a pair of alleles; the allele for dextral (s^+) is dominant over the allele for sinistral (s). However, the direction of coiling is determined not by that snail's own genotype, but by the genotype of its *mother*. The direction of coiling is affected by the way in which the cytoplasm divides soon after fertilization, which in turn is determined by a substance produced by the mother and passed to the offspring in the cytoplasm of the egg.

If a male homozygous for dextral alleles (s^+s^+) is crossed with a female homozygous for sinistral alleles (ss), all of the F₁ are heterozygous (s^+s) and have a sinistral shell because

the genotype of the mother (ss) encodes sinistral coiling (see Figure 5.17). If these F₁ snails are self-fertilized, the genotypic ratio of the F₂ is $1\ s^+s^+ : 2\ s^+s : 1\ ss$.

Notice that that the phenotype of all the F₂ snails is dextral coiled, regardless of their genotypes. The F₂ offspring are dextral coiled because the genotype of their mother (s^+s) encodes a right-coiling shell and determines their phenotype. With genetic maternal effect, the phenotype of the progeny is not necessarily the same as the phenotype of the mother, because the progeny's phenotype is determined by the mother's *genotype*, not her phenotype. Neither the male parent's nor the offspring's own genotype has any role in the offspring's phenotype. However, a male does influence the phenotype of the F₂ generation: by contributing to the genotypes of his daughters, he affects the phenotypes of their offspring. Genes that exhibit genetic maternal effect are therefore transmitted through males to future generations. In contrast, genes that exhibit cytoplasmic inheritance are always transmitted through only one of the sexes (usually the female). **TRY PROBLEM 38 →**

CONCEPTS

Characteristics exhibiting cytoplasmic inheritance are encoded by genes in the cytoplasm and are usually inherited from one parent, most commonly the mother. In genetic maternal effect, the genotype of the mother determines the phenotype of the offspring.

✔ **CONCEPT CHECK 10**

How might you determine whether a particular trait is due to cytoplasmic inheritance or to genetic maternal effect?

Genomic Imprinting

A basic tenet of Mendelian genetics is that the parental origin of a gene does not affect its expression and, therefore, reciprocal crosses give identical results. We have seen that there are some genetic characteristics—those encoded by X-linked genes and cytoplasmic genes—for which reciprocal crosses do not give the same results. In these cases, males and females do not contribute the same genetic material to the offspring. With regard to autosomal genes, males and females contribute the same number of genes, and paternal and maternal genes have long been assumed to have equal effects. However, the expression of some genes is significantly affected by their parental origin. This phenomenon, the differential expression of genetic material depending on whether it is inherited from the male or female parent, is called **genomic imprinting**.

A gene that exhibits genomic imprinting in both mice and humans is *Igf2*, which encodes a protein called insulin-like growth factor II (Igf2). Offspring inherit one *Igf2* allele from their mother and one from their father. The paternal copy of *Igf2* is actively expressed in the fetus and placenta, but the maternal copy is completely silent (**Figure 5.18**). Both male and female offspring possess *Igf2* genes; the key to whether the gene is expressed is the sex of the parent transmitting the gene. In the present example, the gene is expressed only when it is transmitted by a male parent. In other genomically imprinted traits, the trait is expressed only when the gene is transmitted by the female parent. In a way that is not completely understood, the paternal *Igf2* allele (but not the maternal allele) promotes placental and fetal growth; when the paternal copy of *Igf2* is deleted in mice, a small placenta and low-birth-weight offspring result.

Genomic imprinting has been implicated in several human disorders, including Prader–Willi and Angelman syndromes. Children with Prader–Willi syndrome have small hands and feet, short stature, poor sexual development, and mental retardation. These children are small at birth and suckle poorly; but, as toddlers, they develop voracious appetites and frequently become obese. Many persons with Prader–Willi syndrome are missing a small region on the long arm of chromosome 15. The deletion of this region is always inherited from the *father*. Thus, children with Prader–Willi syndrome lack a paternal copy of genes on the long arm of chromosome 15.

The deletion of this same region of chromosome 15 can also be inherited from the *mother*, but this inheritance results in a completely different set of symptoms, producing Angelman syndrome. Children with Angelman syndrome exhibit frequent laughter, uncontrolled muscle movement, a large mouth, and unusual seizures. They are missing a maternal copy of genes on the long arm of chromosome 15. For normal development to take place, copies of this region of chromosome 15 from both male and female parents are apparently required.

Many imprinted genes in mammals are associated with fetal growth. Imprinting has also been reported in plants, with differential expression of paternal and maternal genes in the endosperm, which, like the placenta in mammals, provides nutrients for the growth of the embryo. The mechanism of imprinting is still under investigation, but the methylation of DNA—the addition of methyl (CH_3) groups to DNA nucleotides (see Chapters 10 and 16)—is essential to the process. In mammals, methylation is erased in the germ cells each generation and then reestablished in the course of gamete formation, with sperm and eggs undergoing different levels of methylation, which then causes the differential expression of male and female alleles in the offspring.

Imprinting and genetic conflict Why does genomic imprinting occur? One possible answer is the **genetic-conflict hypothesis**, which suggests that there are different and conflicting evolutionary pressures acting on maternal and paternal alleles for genes (such as *Igf2*) that affect fetal growth. From an evolutionary standpoint, paternal alleles that maximize the size of the offspring are favored, because birth weight is strongly associated with infant mortality and adult health. Thus, it is to the advantage of the male parent to pass on alleles that promote maximum fetal growth of their offspring. In contrast, maternal alleles that cause more-limited fetal growth are favored because committing too many of the female parent's nutrients to any one fetus may limit her ability to reproduce in the future and because giving birth to very large babies is difficult and risky for the mother. This hypothesis predicts that genomic imprinting will evolve: paternal copies of genes that affect fetal growth should be maximally expressed, whereas maternal copies of the same genes should be less actively expressed or even silent. Indeed, *Igf2* follows this pattern: the paternal allele is active and promotes growth; the maternal allele is silent and does not contribute to growth. Recent findings demonstrate that the paternal copy of *Igf2* promotes fetal growth by

(a)

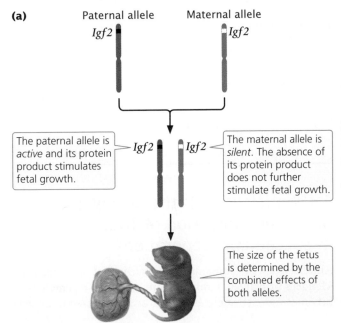

Paternal allele   Maternal allele
Igf2   *Igf2*

Igf2   *Igf2*

The paternal allele is *active* and its protein product stimulates fetal growth.

The maternal allele is *silent*. The absence of its protein product does not further stimulate fetal growth.

The size of the fetus is determined by the combined effects of both alleles.

(b)

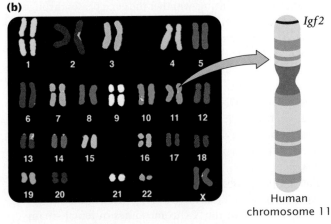

Igf2

Human chromosome 11

5.18 Genomic imprinting of the *Igf2* gene in mice and humans affects fetal growth. (a) The paternal *Igf2* allele is active in the fetus and placenta, whereas the maternal allele is silent. (b) The human *Igf2* locus is on the short arm of chromosome 11; the locus in mice is on chromosome 7 [Courtesy of Dr. Thomas Ried and Dr. Evelin Schrock.]

Table 5.5 Sex influences on heredity	
Genetic Phenomenon	**Phenotype determined by**
Sex-linked characteristic	Genes located on the sex chromosome
Sex-influenced characteristic	Genes on autosomal chromosomes that are more readily expressed in one sex
Sex-limited characteristic	Autosomal genes whose expression is limited to one sex
Genetic maternal effect	Nuclear genotype of the maternal parent
Cytoplasmic inheritance	Cytoplasmic genes, which are usually inherited entirely from only one parent
Genomic imprinting	Genes whose expression is affected by the sex of the transmitting parent

directing more maternal nutrients to the fetus through the placenta. Some of the different ways in which sex interacts with heredity are summarized in **Table 5.5**.

Epigenetics Genomic imprinting is just one form of a phenomenon known as **epigenetics**. Most traits are encoded by genetic information that resides in the sequence of nucleotide bases of DNA—the genetic code, which will be discussed in Chapter 15. However, some traits are caused by alterations to the DNA, such as DNA methylation, that affect the way in which the DNA sequences are expressed. These changes are often stable and heritable in the sense that they are passed from one cell to another.

In genomic imprinting, whether the gene passes through the egg or sperm determines how much methylation of the DNA takes place. The pattern of methylation on a gene is copied when the DNA is replicated and therefore remains on the gene as it is passed from cell to cell through mitosis. However, the pattern of methylation may be modified or removed when the DNA passes through a gamete, and so a gene methylated in sperm is unmethylated when it is eventually passed down to a daughter's egg. Ultimately, the amount of methylation determines whether the gene is expressed in the offspring.

These types of reversible changes to DNA that influence the expression of traits are termed epigenetic marks. The inactivation of one of the X chromosomes in female mammals (discussed in Chapter 4) is another type of epigenetic change.

A remarkable example of epigenetics is seen in honey bees. Queen bees and worker bees are both female, but there the resemblance ends. A queen is large and develops functional ovaries, whereas workers are small and sterile. The queen goes on a mating flight and spends her entire life reproducing, whereas workers spend all of their time collecting nectar and pollen, tending the queen, and raising her offspring. In spite of these significant differences in anatomy, physiology, and behavior, queens and workers are genetically the same; both develop from ordinary eggs. How they differ is in diet: worker bees produce and feed a few female larvae a special substance called royal jelly, which causes these larvae to develop as queens. Other larvae are fed ordinary bee food, and they develop as workers. This simple difference in diet greatly affects gene expression, causing different genes to be activated in queens and workers and resulting in a very different set of phenotypic traits. How royal jelly affects gene expression has long been a mystery, but recent research suggests that it changes an epigenetic mark.

Research conducted in 2008 by Ryszard Kucharski and his colleagues demonstrated that royal jelly silences the expression of a key gene called *Dnmt3*, whose product normally adds methyl groups to DNA. With *Dnmt3* shut down, bee DNA is less methylated and many genes that are normally silenced in workers are expressed, leading to the development of queen characteristics. Kucharski and his coworkers demonstrated the importance of DNA methylation in queen development by injecting into bee larvae small RNA molecules (small interfering RNAs, or siRNAs; see Chapter 13) that specifically inhibited the expression of *Dnmt3*. These larvae had lower levels of DNA methylation and many developed as queens with fully functional ovaries. This experiment demonstrates that royal jelley brings about epigenetic changes (less DNA methylation), which are transmitted through cell division and modify developmental pathways, eventually leading to a queen bee. We will consider epigenetic changes in more detail in Chapter 17.

CONCEPTS

In genomic imprinting, the expression of a gene is influenced by the sex of the parent transmitting the gene to the offspring. Epigenetic marks are reversible changes in DNA that do not alter the base sequence but may affect how a gene is expressed.

✔ **CONCEPT CHECK 11**

What type of epigenetic mark is responsible for genomic imprinting?

5.4 Anticipation Is the Stronger or Earlier Expression of Traits in Succeeding Generations

Another genetic phenomenon that is not explained by Mendel's principles is **anticipation**, in which a genetic trait becomes more strongly expressed or is expressed at an ear-

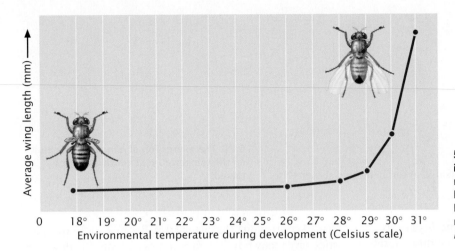

5.19 The expression of the *vestigial* mutation in *Drosophila* is temperature dependent. When reared at temperatures below 29°C, flies homozygous for *vestigial* have greatly reduced wings; but, at temperatures above 31°C, the flies develop normal wings. [Data from M. H. Harnly, *Journal of Experimental Zoology* 56:363–379, 1936.]

Average wing length (mm)

Environmental temperature during development (Celsius scale)

lier age as it is passed from generation to generation. In the early 1900s, several physicians observed that many patients with moderate to severe myotonic dystrophy—an autosomal dominant muscle disorder—had ancestors who were only mildly affected by the disease. These observations led to the concept of anticipation. However, the concept quickly fell out of favor with geneticists because there was no obvious mechanism to explain it; traditional genetics held that genes are passed unaltered from parents to offspring. Geneticists tended to attribute anticipation to observational bias.

Research has now reestablished anticipation as a legitimate genetic phenomenon. The mutation causing myotonic dystrophy consists of an unstable region of DNA that can increase in size as the gene is passed from generation to generation. The age of onset and the severity of the disease are correlated with the size of the unstable region; an increase in the size of the region through generations produces anticipation. The phenomenon has now been implicated in a number of genetic diseases. We will examine these interesting types of mutations in more detail in Chapter 18.

CONCEPTS

Anticipation is the stronger or earlier expression of a genetic trait in succeeding generations. It is caused by an unstable region of DNA that increases in size from generation to generation.

5.5 The Expression of a Genotype May Be Affected by Environmental Effects

In Chapter 3, we learned that each phenotype is the result of a genotype developing within a specific environment; the genotype sets the potential for development, but how the phenotype actually develops within the limits imposed by the genotype depends on environmental effects. Stated another way, each genotype may produce several different phenotypes, depending on the environmental conditions in which development takes place. For example, a fruit fly

homozygous for the vestigial mutation (*vg vg*) develops reduced wings when raised at a temperature below 29°C, but the same genotype develops much longer wings when raised at 31°C (**Figure 5.19**).

For most of the characteristics discussed so far, the effect of the environment on the phenotype has been slight. Mendel's peas with genotype *yy*, for example, developed green seeds regardless of the environment in which they were raised. Similarly, persons with genotype $I^A I^A$ have the A antigen on their red blood cells regardless of their diet, socioeconomic status, or family environment. For other phenotypes, however, environmental effects play a more important role.

Environmental Effects on the Phenotype

The phenotypic expression of some genotypes critically depends on the presence of a specific environment. For example, the *himalayan* allele in rabbits produces dark fur at the extremities of the body—on the nose, ears, and feet (**Figure 5.20**). The dark pigment develops, however, only when a rabbit is reared at a temperature of 25°C or lower; if a Himalayan rabbit is reared at 30°C, no dark patches develop. The expression of the *himalayan* allele is thus temperature dependent; an enzyme necessary for the production of dark pigment is inactivated at higher temperatures. The pigment is restricted to the nose, feet, and ears of a Himalayan rabbit because the animal's core body temperature is normally above 25°C and the enzyme is functional only in the cells of the relatively cool extremities. The *himalayan* allele is an example of a **temperature-sensitive allele**, an allele whose product is functional only at certain temperatures. Similarly, vestigial wings in *Drosophila melanogaster* is caused by a temperature-dependent mutation (see Figure 5.19).

Environmental factors also play an important role in the expression of a number of human genetic diseases. Phenylketonuria (PKU) is due to an autosomal recessive allele that causes mental retardation. The disorder arises from a defect in an enzyme that normally metabolizes the

Reared at 25°C or lower

Reared at temperatures above 30°C

5.20 The expression of *himalayan* depends on the temperature at which a rabbit is reared.

amino acid phenylalanine. When this enzyme is defective, phenylalanine is not metabolized, and its buildup causes brain damage in children. A simple environmental change, putting an affected child on a low-phenylalanine diet, prevents retardation.

These examples illustrate the point that genes and their products do not act in isolation; rather, they frequently interact with environmental factors. Occasionally, environmental factors alone can produce a phenotype that is the same as the phenotype produced by a genotype; this phenotype is called a **phenocopy**. In fruit flies, for example, the autosomal recessive mutation *eyeless* produces greatly reduced eyes. The eyeless phenotype can also be produced by exposing the larvae of normal flies to sodium metaborate.

> ### CONCEPTS
>
> The expression of many genes is modified by the environment. A phenocopy is a trait produced by environmental effects that mimics the phenotype produced by the genotype.
>
> ✔ **CONCEPT CHECK 12**
>
> How can you determine whether a phenotype such as reduced eyes in fruit flies is due to a recessive mutation or is a phenocopy?

The Inheritance of Continuous Characteristics

So far, we've dealt primarily with characteristics that have only a few distinct phenotypes. In Mendel's peas, for example, the seeds were either smooth or wrinkled, yellow or green; the coats of dogs were black, brown, or yellow; blood types were of four distinct types, A, B, AB, or O. Such characteristics, which have a few easily distinguished phenotypes, are called **discontinuous characteristics**.

Not all characteristics exhibit discontinuous phenotypes. Human height is an example of such a characteristic; people do not come in just a few distinct heights but, rather, display a continuum of heights. Indeed, there are so many possible phenotypes of human height that we must use a measurement to describe a person's height. Characteristics that exhibit a continuous distribution of phenotypes are termed **continuous characteristics**. Because such characteristics have many possible phenotypes and must be described in quantitative terms, continuous characteristics are also called **quantitative characteristics**.

Continuous characteristics frequently arise because genes at many loci interact to produce the phenotypes. When a single locus with two alleles encodes a characteristic, there are three genotypes possible: *AA*, *Aa*, and *aa*. With two loci, each with two alleles, there are $3^2 = 9$ genotypes possible. The number of genotypes encoding a characteristic is 3^n, where n equals the number of loci, each with two alleles, that influence the characteristic. For example, when a characteristic is determined by eight loci, each with two alleles, there are $3^8 = 6561$ different genotypes possible for this characteristic. If each genotype produces a different phenotype, many phenotypes will be possible. The slight differences between the phenotypes will be indistinguishable, and the characteristic will appear continuous. Characteristics encoded by genes at many loci are called **polygenic characteristics**.

The converse of polygeny is **pleiotropy**, in which one gene affects multiple characteristics. Many genes exhibit pleiotropy. Phenylketonuria, mentioned earlier, results from a recessive allele; persons homozygous for this allele, if untreated, exhibit mental retardation, blue eyes, and light skin color. The lethal allele that causes yellow coat color in mice (discussed in the introduction to the chapter) also is pleiotropic. In addition to its lethality and effect on hair color, the gene causes a diabetes-like condition, obesity, and increased propensity to develop tumors.

Frequently, the phenotypes of continuous characteristics are also influenced by environmental factors. Each genotype is capable of producing a range of phenotypes. In this situation, the particular phenotype that results depends on both the genotype and the environmental conditions in which the genotype develops. For example, only three genotypes may encode a characteristic, but, because each genotype produces a range of phenotypes associated with different environments, the phenotype of the characteristic exhibits a continuous distribution. Many continuous characteristics are both polygenic and influenced by environmental factors; such characteristics are called **multifactorial characteristics** because many factors help determine the phenotype.

The inheritance of continuous characteristics may appear to be complex, but the alleles at each locus follow Mendel's principles and are inherited in the same way as alleles encoding simple, discontinuous characteristics. However, because many genes participate, because environmental factors influence the phenotype, and because the phenotypes do not sort out into a few distinct types, we cannot observe the distinct ratios that have allowed us to interpret the genetic basis of discontinuous characteristics. To analyze continuous characteristics, we must employ special statistical tools, as will be discussed in Chapter 24. TRY PROBLEMS 42 AND 43 →

CONCEPTS

Discontinuous characteristics exhibit a few distinct phenotypes; continuous characteristics exhibit a range of phenotypes. A continuous characteristic is frequently produced when genes at many loci and environmental factors combine to determine a phenotype.

✔ CONCEPT CHECK 13

What is the difference between polygeny and pleiotropy?

CONCEPTS SUMMARY

- Dominance always refers to genes at the same locus (allelic genes) and can be understood in regard to how the phenotype of the heterozygote relates to the phenotypes of the homozygotes.

- Dominance is complete when a heterozygote has the same phenotype as a homozygote, is incomplete when the heterozygote has a phenotype intermediate between those of two parental homozygotes, and is codominant when the heterozygote exhibits traits of both parental homozygotes.

- The type of dominance does not affect the inheritance of an allele; it does affect the phenotypic expression of the allele. The classification of dominance depends on the level of the phenotype examined.

- Penetrance is the percentage of individuals having a particular genotype that exhibit the expected phenotype. Expressivity is the degree to which a character is expressed.

- Lethal alleles cause the death of an individual organism possessing them, usually at an early stage of development, and may alter phenotypic ratios.

- Multiple alleles refer to the presence of more than two alleles at a locus within a group. Their presence increases the number of genotypes and phenotypes possible.

- Gene interaction refers to the interaction between genes at different loci to produce a single phenotype. An epistatic gene at one locus suppresses, or masks, the expression of

hypostatic genes at other loci. Gene interaction frequently produces phenotypic ratios that are modifications of dihybrid ratios.

- Sex-influenced characteristics are encoded by autosomal genes that are expressed more readily in one sex. Sex-limited characteristics are encoded by autosomal genes expressed in only one sex.

- In cytoplasmic inheritance, the genes for the characteristic are found in the organelles and are usually inherited from a single (usually maternal) parent. Genetic maternal effect is present when an offspring inherits genes from both parents, but the nuclear genes of the mother determine the offspring's phenotype.

- Genomic imprinting refers to characteristics encoded by autosomal genes whose expression is affected by the sex of the parent transmitting the genes.

- Anticipation refers to a genetic trait that is more strongly expressed or is expressed at an earlier age in succeeding generations.

- Phenotypes are often modified by environmental effects. A phenocopy is a phenotype produced by an environmental effect that mimics a phenotype produced by a genotype.

- Continuous characteristics are those that exhibit a wide range of phenotypes; they are frequently produced by the combined effects of many genes and environmental effects.

IMPORTANT TERMS

complete dominance (p. 100)
incomplete dominance (p. 101)
codominance (p. 102)
incomplete penetrance (p. 103)
penetrance (p. 103)
expressivity (p. 103)
lethal allele (p. 103)
multiple alleles (p. 104)
gene interaction (p. 106)
epistasis (p. 107)

epistatic gene (p. 107)
hypostatic gene (p. 107)
complementation test (p. 113)
complementation (p. 113)
sex-influenced characteristic (p. 115)
sex-limited characteristic (p. 116)
cytoplasmic inheritance (p. 118)
genetic maternal effect (p. 119)
genomic imprinting (p. 120)
genetic-conflict hypothesis (p. 121)

epigenetics (p. 122)
anticipation (p. 122)
temperature-sensitive allele (p. 123)
phenocopy (p. 124)
discontinuous characteristic (p. 124)
continuous characteristic (p. 124)
quantitative characteristic (p. 124)
polygenic characteristic (p. 124)
pleiotropy (p. 124)
multifactorial characteristic (p. 124)

ANSWERS TO CONCEPT CHECKS

1. b

2. With complete dominance, the heterozygote expresses the same phenotype as that of one of the homozygotes. With incomplete dominance, the heterozygote has a phenotype that is intermediate between the two homozygotes. With codominance, the heterozygote has a phenotype that simultaneously expresses the phenotypes of both homozygotes.

3. d

4. b

5. c

6. Gene interaction is interaction between genes at different loci. Dominance is interaction between alleles at a single locus.

7. d

8. Cross a bulldog homozygous for *brindle* with a Chihuahua homozygous for *brindle*. If the two *brindle* genes are allelic, all the offspring will be brindle: $bb \times bb \rightarrow$ all bb (brindle). If, on the other hand, brindle in the two breeds is due to recessive genes at different loci, then none of the offspring will be brindle: $a^+a^+\ bb \times aa\ b^+b^+ \rightarrow a^+a\ b^+b$.

9. Both sex-influenced and sex-limited traits are encoded by autosomal genes whose expression is affected by the sex of the individual organism possessing the gene. Sex-linked traits are encoded by genes on the sex chromosomes.

10. Cytoplasmically inherited traits are encoded by genes in the cytoplasm, which is usually inherited only from the female parent. Therefore, a trait due to cytoplasmic inheritance will always be passed through females. Traits due to genetic maternal effect are encoded by autosomal genes and can therefore be passed through males, although any individual organism's trait is determined by the genotype of the maternal parent.

11. Methylation of DNA

12. Cross two eyeless flies, and cross an eyeless fly with a wild-type fly. Raise the offspring of both crosses in the same environment. If the trait is due to a recessive mutation, all the offspring of the two eyeless flies should be eyeless, whereas at least some of the offspring of the of the eyeless and wild-type flies should be wild type. If the trait is due to a phenocopy, there should be no differences in the progeny of the two crosses.

13. Polygeny refers to the influence of multiple genes on the expression of a single characteristic. Pleiotropy refers to the effect of a single gene on the expression of multiple characteristics.

WORKED PROBLEMS

1. The type of plumage found in mallard ducks is determined by three alleles at a single locus: M^R, which encodes restricted plumage; M, which encodes mallard plumage; and m^d, which encodes dusky plumage. The restricted phenotype is dominant over mallard and dusky; mallard is dominant over dusky ($M^R > M > m^d$). Give the expected phenotypes and proportions of offspring produced by the following crosses.

a. $M^R M \times m^d m^d$ **c.** $M^R m^d \times M^R M$

b. $M^R m^d \times Mm^d$ **d.** $M^R M \times Mm^d$

• Solution

We can determine the phenotypes and proportions of offspring by (1) determining the types of gametes produced by each parent and (2) combining the gametes of the two parents with the use of a Punnett square.

a.

Parents $M^R M \quad \times \quad m^d m^d$

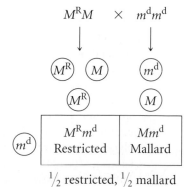

Gametes M^R M m^d

M^R M

	M^R	M
m^d	$M^R m^d$ Restricted	Mm^d Mallard

$\frac{1}{2}$ restricted, $\frac{1}{2}$ mallard

b.

Parents $M^R m^d \quad \times \quad Mm^d$

Gametes M^R m^d M m^d

M m^d

	M^R	m^d
M^R	$M^R M$ Restricted	$M^R m^d$ Restricted
m^d	Mm^d Mallard	$m^d m^d$ Dusky

$\frac{1}{2}$ restricted, $\frac{1}{4}$ mallard, $\frac{1}{4}$ dusky

c.

Parents $M^R m^d \quad \times \quad M^R M$

Gametes M^R m^d M^R M

M^R M

	M^R	M
M^R	$M^R M^R$ Restricted	$M^R M$ Restricted
m^d	$M^R m^d$ Restricted	Mm^d Mallard

$\frac{3}{4}$ restricted, $\frac{1}{4}$ mallard

d.

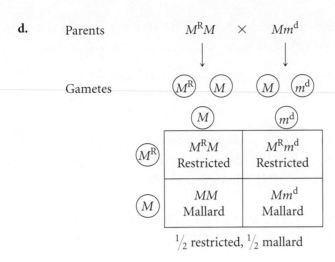

Parents $M^R M$ × $M m^d$

Gametes M^R M M m^d

	M	m^d
M^R	$M^R M$ Restricted	$M^R m^d$ Restricted
M	MM Mallard	$M m^d$ Mallard

$\frac{1}{2}$ restricted, $\frac{1}{2}$ mallard

2. A geneticist crosses two yellow mice with straight hair and obtains the following progeny:

$\frac{1}{2}$ yellow, straight

$\frac{1}{6}$ yellow, fuzzy

$\frac{1}{4}$ gray, straight

$\frac{1}{12}$ gray, fuzzy

a. Provide a genetic explanation for the results and assign genotypes to the parents and progeny of this cross.

b. What additional crosses might be carried out to determine if your explanation is correct?

• Solution

a. This cross concerns two separate characteristics—color and type of hair; so we should begin by examining the results for each characteristic separately. First, let's look at the inheritance of color. Two yellow mice are crossed, producing $\frac{1}{2} + \frac{1}{6} = \frac{3}{6} + \frac{1}{6} = \frac{4}{6} = \frac{2}{3}$ yellow mice and $\frac{1}{4} + \frac{1}{12} = \frac{3}{12} + \frac{1}{12} = \frac{4}{12} = \frac{1}{3}$ gray mice. We learned in this chapter that a 2 : 1 ratio is often produced when a recessive lethal gene is present:

$Yy \times Yy$

YY $\frac{1}{4}$ die

Yy $\frac{1}{2}$ yellow, becomes $\frac{2}{3}$

yy $\frac{1}{4}$ gray, becomes $\frac{1}{3}$

Now, let's examine the inheritance of the hair type. Two mice with straight hair are crossed, producing $\frac{1}{2} + \frac{1}{4} = \frac{2}{4} + \frac{1}{4} = \frac{3}{4}$ mice with straight hair and $\frac{1}{6} + \frac{1}{12} = \frac{2}{12} + \frac{1}{12} = \frac{3}{12} = \frac{1}{4}$ mice with fuzzy hair. We learned in Chapter 3 that a 3 : 1 ratio is usually produced by a cross between two individuals heterozygous for a simple dominant allele:

$Ss \times Ss$

SS $\frac{1}{4}$ straight

Ss $\frac{1}{2}$ straight $\Big\}$ $\frac{3}{4}$ straight

ss $\frac{1}{4}$ fuzzy

We can now combine both loci and assign genotypes to all the individual mice in the cross:

P $Yy\ Ss$ × $Yy\ Ss$
 Yellow, straight   Yellow, straight

Phenotype	Genotype	Probability at each locus	Combined probability
yellow, straight	$Yy\ S\_$	$\frac{2}{3} \times \frac{3}{4}$	$= \frac{6}{12} = \frac{1}{2}$
yellow, fuzzy	$Yy\ ss$	$\frac{2}{3} \times \frac{1}{4}$	$= \frac{2}{12} = \frac{1}{6}$
gray, straight	$yy\ S\_$	$\frac{1}{3} \times \frac{3}{4}$	$= \frac{3}{12} = \frac{1}{4}$
gray, fuzzy	$yy\ ss$	$\frac{1}{3} \times \frac{1}{4}$	$= \frac{1}{12}$

b. We could carry out a number of different crosses to test our hypothesis that yellow is a recessive lethal and straight is dominant over fuzzy. For example, a cross between any two yellow mice should always produce $\frac{2}{3}$ yellow and $\frac{1}{3}$ gray offspring, and a cross between two gray mice should produce gray offspring only. A cross between two fuzzy mice should produce fuzzy offspring only.

3. In some sheep, the presence of horns is produced by an autosomal allele that is dominant in males and recessive in females. A horned female is crossed with a hornless male. One of the resulting F_1 females is crossed with a hornless male. What proportion of the male and female progeny from this cross will have horns?

• Solution

The presence of horns in these sheep is an example of a sex-influenced characteristic. Because the phenotypes associated with the genotypes differ for the two sexes, let's begin this problem by writing out the genotypes and phenotypes for each sex. We will let H represent the allele that encodes horns and H^+ represent the allele that encodes hornless. In males, the allele for horns is dominant over the allele for hornless, which means that males homozygous (HH) and heterozygous (H^+H) for this gene are horned. Only males homozygous for the recessive hornless allele (H^+H^+) are hornless. In females, the allele for horns is recessive, which means that only females homozygous for this allele (HH) are horned; females heterozygous (H^+H) and homozygous (H^+H^+) for the hornless allele are hornless. The following table summarizes genotypes and associated phenotypes:

Genotype	Male phenotype	Female phenotype
HH	horned	horned
HH^+	horned	hornless
H^+H^+	hornless	hornless

In the problem, a horned female is crossed with a hornless male. From the preceding table, we see that a horned female must be homozygous for the allele for horns (HH) and a hornless male must be homozygous for the allele for hornless (H^+H^+); so all the F_1 will be heterozygous; the F_1 males will be horned and the F_1 females will be hornless, as shown in the following diagram:

$$P \qquad\qquad H^+H^+ \quad\times\quad HH$$

$$\downarrow$$

$$F_1 \qquad\qquad\qquad H^+H$$
Horned males and hornless females

A heterozygous hornless F_1 female (H^+H) is then crossed with a hornless male (H^+H^+):

$$\begin{array}{ccc} H^+H & \times & H^+H^+ \\ \text{Hornless female} & & \text{Hornless male} \end{array}$$

$$\downarrow$$

	Males	Females
$\frac{1}{2}\,H^+H^+$	hornless	hornless
$\frac{1}{2}\,H^+H$	horned	hornless

Therefore, $\frac{1}{2}$ of the male progeny will be horned, but none of the female progeny will be horned.

COMPREHENSION QUESTIONS

Section 5.1

*1. How do incomplete dominance and codominance differ?

*2. What is incomplete penetrance and what causes it?

Section 5.2

3. What is gene interaction? What is the difference between an epistatic gene and a hypostatic gene?

4. What is a recessive epistatic gene?

*5. What is a complementation test and what is it used for?

Section 5.3

*6. What characteristics are exhibited by a cytoplasmically inherited trait?

7. What is genomic imprinting?

8. What is the difference between genetic maternal effect and genomic imprinting?

9. What is the difference between a sex-influenced gene and a gene that exhibits genomic imprinting?

Section 5.4

10. What characteristics do you expect to see in a trait that exhibits anticipation?

Section 5.5

*11. What are continuous characteristics and how do they arise?

APPLICATION QUESTIONS AND PROBLEMS

Introduction

12. As discussed in the introduction to this chapter, Cuénot [DATA ANALYSIS] studied the genetic basis of yellow coat color in mice. He carried out a number of crosses between two yellow mice and obtained what he thought was a 3 : 1 ratio of yellow to gray mice in the progeny. The following table gives Cuénot's actual results, along with the results of a much larger series of crosses carried out by Castle and Little (W. E. Castle and C. C. Little. 1910. *Science* 32:868–870).

Progeny Resulting from Crosses of Yellow × Yellow Mice

Investigators	Yellow progeny	Non-yellow progeny	Total progeny
Cuénot	263	100	363
Castle and Little	800	435	1235
Both combined	1063	535	1598

a. Using a chi-square test, determine whether Cuénot's results are significantly different from the 3 : 1 ratio that he thought he observed. Are they different from a 2 : 1 ratio?

b. Determine whether Castle and Little's results are significantly different from a 3 : 1 ratio. Are they different from a 2 : 1 ratio?

c. Combine the results of Castle and Cuénot and determine whether they are significantly different from a 3 : 1 ratio and a 2 : 1 ratio.

d. Offer an explanation for the different ratios that Cuénot and Castle obtained.

Sections 5.1 through 5.4

13. Match each of the following terms with its correct definition (parts *a* through *i*).

____ phenocopy

____ pleiotrophy

____ polygenic trait

____ penetrance

____ sex-limited trait

____ genetic maternal effect

____ genomic imprinting

____ sex-influenced trait

____ anticipation

a. the percentage of individuals with a particular genotype that express the expected phenotype

b. a trait determined by an autosomal gene that is more easily expressed in one sex

c. a trait determined by an autosomal gene that is expressed in only one sex

d. a trait that is determined by an environmental effect and has the same phenotype as a genetically determined trait

e. a trait determined by genes at many loci

f. the expression of a trait is affected by the sex of the parent that transmits the gene to the offspring

g. the trait appears earlier or more severely in succeeding generations

h. a gene affects more than one phenotype

i. the genotype of the maternal parent influences the phenotype of the offspring

Section 5.1

*14. Palomino horses have a golden yellow coat, chestnut horses have a brown coat, and cremello horses have a coat that is almost white. A series of crosses between the three different types of horses produced the following offspring:

Cross	Offspring
palomino × palomino	13 palomino, 6 chestnut, 5 cremello
chestnut × chestnut	16 chestnut
cremello × cremello	13 cremello
palomino × chestnut	8 palomino, 9 chestnut
palomino × cremello	11 palomino, 11 cremello
chestnut × cremello	23 palomino

(a)

(b)

(c)

Coat color: (a) palomino; (b) chestnut; (c) cremello. [Part a: Keith J. Smith/Alamy; part b: jiri jura/iStock photo; part c: Caitlin Fagan.]

a. Explain the inheritance of the palomino, chestnut, and cremello phenotypes in horses.

b. Assign symbols for the alleles that determine these phenotypes, and list the genotypes of all parents and offspring given in the preceding table.

*15. The L^M and L^N alleles at the MN blood-group locus exhibit codominance. Give the expected genotypes and phenotypes and their ratios in progeny resulting from the following crosses.

a. $L^M L^M \times L^M L^N$

b. $L^N L^N \times L^N L^N$

c. $L^M L^N \times L^M L^N$

d. $L^M L^N \times L^N L^N$

e. $L^M L^M \times L^N L^N$

16. Assume that long ear lobes in humans are an autosomal dominant trait that exhibits 30% penetrance. A person who is heterozygous for long ear lobes mates with a person who is homozygous for normal ear lobes. What is the probability that their first child will have long ear lobes?

17. Club foot is one of the most common congenital skeletal abnormalities, with a worldwide incidence of about 1 in 1000 births. Both genetic and nongenetic factors are thought to be responsible for club foot. C. A. Gurnett et al. (2008. *American Journal of Human Genetics* 83:616–622) identified a family in which club foot was inherited as an autosomal dominant trait with reduced penetrance. They discovered a mutation in the *PITXI* gene that caused club foot in this family. Through DNA testing, they determined that 11 people in the family carried the *PITXI* mutation, but only 8 of these people had club foot. What is the penetrance of the *PITXI* mutation in this family?

*18. When a Chinese hamster with white spots is crossed with another hamster that has no spots, approximately $1/2$ of the offspring have white spots and $1/2$ have no spots. When two hamsters with white spots are crossed, $2/3$ of the offspring possess white spots and $1/3$ have no spots.

a. What is the genetic basis of white spotting in Chinese hamsters?

b. How might you go about producing Chinese hamsters that breed true for white spotting?

19. In the pearl-millet plant, color is determined by three alleles at a single locus: Rp^1 (red), Rp^2 (purple), and rp (green). Red is dominant over purple and green, and purple is dominant over green ($Rp^1 > Rp^2 > rp$). Give the expected phenotypes and ratios of offspring produced by the following crosses.

a. $Rp^1/Rp^2 \times Rp^1/rp$

b. $Rp^1/rp \times Rp^2/rp$

c. $Rp^1/Rp^2 \times Rp^1/Rp^2$

d. $Rp^2/rp \times rp/rp$

e. $rp/rp \times Rp^1/Rp^2$

20. If there are five alleles at a locus, how many genotypes may there be at this locus? How many different kinds of homozygotes will there be? How many genotypes and homozygotes may there be with eight alleles at a locus?

21. Turkeys have black, bronze, or black-bronze plumage. Examine the results of the following crosses:

Parents	Offspring
Cross 1: black and bronze	all black
Cross 2: black and black	$^3\!/_4$ black, $^1\!/_4$ bronze
Cross 3: black-bronze and black-bronze	all black-bronze
Cross 4: black and bronze	$^1\!/_2$ black, $^1\!/_4$ bronze, $^1\!/_4$ black-bronze
Cross 5: bronze and black-bronze	$^1\!/_2$ bronze, $^1\!/_2$ black-bronze
Cross 6: bronze and bronze	$^3\!/_4$ bronze, $^1\!/_4$ black-bronze

Do you think these differences in plumage arise from incomplete dominance between two alleles at a single locus? If yes, support your conclusion by assigning symbols to each allele and providing genotypes for all turkeys in the crosses. If your answer is no, provide an alternative explanation and assign genotypes to all turkeys in the crosses.

22. In rabbits, an allelic series helps to determine coat color: C (full color), c^{ch} (chinchilla, gray color), c^h (Himalayan, white with black extremities), and c (albino, all-white). The C allele is dominant over all others, c^{ch} is dominant over c^h and c, c^h is dominant over c, and c is recessive to all the other alleles. This dominance hierarchy can be summarized as $C > c^{ch} > c^h > c$. The rabbits in the following list are crossed and produce the progeny shown. Give the genotypes of the parents for each cross:

Phenotypes of parents	Phenotypes of offspring
a. full color × albino	$^1\!/_2$ full color, $^1\!/_2$ albino
b. Himalayan × albino	$^1\!/_2$ Himalayan, $^1\!/_2$ albino
c. full color × albino	$^1\!/_2$ full color, $^1\!/_2$ chinchilla
d. full color × Himalayan	$^1\!/_2$ full color, $^1\!/_4$ Himalayan, $^1\!/_4$ albino
e. full color × full color	$^3\!/_4$ full color, $^1\!/_4$ albino

23. In this chapter, we considered Joan Barry's paternity suit against Charlie Chaplin and how, on the basis of blood types, Chaplin could not have been the father of her child.

a. What blood types are possible for the father of Barry's child?

b. If Chaplin had possessed one of these blood types, would that prove that he fathered Barry's child?

*24. A woman has blood-type A M. She has a child with blood-type AB MN. Which of the following blood types could *not* be that of the child's father? Explain your reasoning.

George	O	N
Tom	AB	MN
Bill	B	MN
Claude	A	N
Henry	AB	M

Section 5.2

*25. In chickens, comb shape is determined by alleles at two loci (R, r and P, p). A walnut comb is produced when at least one dominant allele R is present at one locus and at least one dominant allele P is present at a second locus (genotype $R\_ P\_$). A rose comb is produced when at least one dominant allele is present at the first locus and two recessive alleles are present at the second locus (genotype $R\_ pp$). A pea comb is produced when two recessive alleles are present at the first locus and at least one dominant allele is present at the second (genotype $rr P\_$). If two recessive alleles are present at the first and at the second locus ($rr pp$), a single comb is produced. Progeny with what types of combs and in what proportions will result from the following crosses?

a. $RR\ PP \times rr\ pp$ **d.** $Rr\ pp \times Rr\ pp$

b. $Rr\ Pp \times rr\ pp$ **e.** $Rr\ pp \times rr\ Pp$

c. $Rr\ Pp \times Rr\ Pp$ **f.** $Rr\ pp \times rr\ pp$

(a)    **(b)**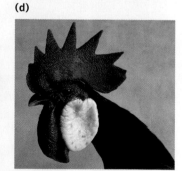

(c)    **(d)**

Comb shape: (a) walnut; (b) rose; (c) pea; (d) single. [Parts a and d: R. OSF Dowling/Animals Animals; part b: Robert Maier/Animals Animals; part c: George Godfrey/Animals Animals.]

*26. Tatuo Aida investigated the genetic basis of color variation in the Medaka *(Aplocheilus latipes)*, a small fish found naturally in Japan (T. Aida. 1921. *Genetics* 6:554–573). Aida found that genes at two loci (*B, b* and *R, r*) determine the color of the fish: fish with a dominant allele at both loci (*B_R_*) are brown, fish with a dominant allele at the *B* locus only (*B_ rr*) are blue, fish with a dominant allele at the *R* locus only (*bb R_*) are red, and fish with recessive alleles at both loci (*bb rr*) are white. Aida crossed a homozygous brown fish with a homozygous white fish. He then backcrossed the F_1 with the homozygous white parent and obtained 228 brown fish, 230 blue fish, 237 red fish, and 222 white fish.

a. Give the genotypes of the backcross progeny.

b. Use a chi-square test to compare the observed numbers of backcross progeny with the number expected. What conclusion can you make from your chi-square results?

c. What results would you expect for a cross between a homozygous red fish and a white fish?

d. What results would you expect if you crossed a homozygous red fish with a homozygous blue fish and then backcrossed the F_1 with a homozygous red parental fish?

27. A variety of opium poppy *(Papaver somniferum* L.) having lacerate leaves was crossed with a variety that has normal leaves. All the F_1 had lacerate leaves. Two F_1 plants were interbred to produce the F_2. Of the F_2, 249 had lacerate leaves and 16 had normal leaves. Give genotypes for all the plants in the P, F_1, and F_2 generations. Explain how lacerate leaves are determined in the opium poppy.

28. E. W. Lindstrom crossed two corn plants with green seedlings and obtained the following progeny: 3583 green seedlings, 853 virescent-white seedlings, and 260 yellow seedlings (E. W. Lindstrom. 1921. *Genetics* 6:91–110).

a. Give the genotypes for the green, virescent-white, and yellow progeny.

b. Explain how color is determined in these seedlings.

c. Is there epistasis among the genes that determine color in the corn seedlings? If so, which gene is epistatic and which is hypostatic.

*29. A dog breeder liked yellow and brown Labrador retrievers. In an attempt to produce yellow and brown puppies, he bought a yellow Labrador male and a brown Labrador female and mated them. Unfortunately, all the puppies produced in this cross were black. (See p. 107 for a discussion of the genetic basis of coat color in Labrador retrievers.)

a. Explain this result.

b. How might the breeder go about producing yellow and brown Labradors?

(a) 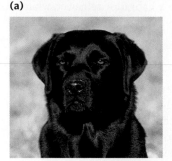 (b)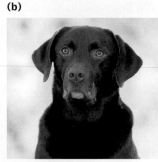

(c)

Coat color: (a) black; (b) brown; (c) yellow. [Part a: Cenorman/Dreamstime.com; parts b and c: Juniors Bildarchiv/Alamy.]

30. When a yellow female Labrador retriever was mated with a brown male, half of the puppies were brown and half were yellow. The same female, when mated with a different brown male, produced only brown offspring. Explain these results.

*31. A summer-squash plant that produces disc-shaped fruit is crossed with a summer-squash plant that produces long fruit. All the F_1 have disc-shaped fruit. When the F_1 are intercrossed, F_2 progeny are produced in the following ratio: $9/16$ disc-shaped fruit : $6/16$ spherical fruit : $1/16$ long fruit. Give the genotypes of the F_2 progeny.

32. Some sweet-pea plants have purple flowers and other plants have white flowers. A homozygous variety of pea that has purple flowers is crossed with a homozygous variety that has white flowers. All the F_1 have purple flowers. When these F_1 are self-fertilized, the F_2 appear in a ratio of $9/16$ purple to $7/16$ white.

a. Give genotypes for the purple and white flowers in these crosses.

b. Draw a hypothetical biochemical pathway to explain the production of purple and white flowers in sweet peas.

33. For the following questions, refer to pages 113–115 for a discussion of how coat color and pattern are determined in dogs.

a. Why are Irish setters reddish in color?

b. Can a poodle crossed with any other breed produce spotted puppies? Why or why not?

c. If a St. Bernard is crossed with a Doberman, what will be the coat color of the offspring: solid, yellow, saddle, or bicolor?

d. If a Rottweiler is crossed with a Labrador retriever, what will be the coat color of the offspring: solid, yellow, saddle, or bicolor?

Section 5.3

34. Male-limited precocious puberty results from a rare, sex-limited autosomal allele (P) that is dominant over the allele for normal puberty (p) and is expressed only in males. Bill undergoes precocious puberty, but his brother Jack and his sister Beth underwent puberty at the usual time, between the ages of 10 and 14. Although Bill's mother and father underwent normal puberty, two of his maternal uncles (his mother's brothers) underwent precocious puberty. All of Bill's grandparents underwent normal puberty. Give the most likely genotypes for all the relatives mentioned in this family.

35. In some goats, the presence of horns is produced by an autosomal gene that is dominant in males and recessive in females. A horned female is crossed with a hornless male. The F_1 offspring are intercrossed to produce the F_2. What proportion of the F_2 females will have horns?

36. In goats, a beard is produced by an autosomal allele that is dominant in males and recessive in females. We'll use the symbol B^b for the beard allele and B^+ for the beardless allele. Another independently assorting autosomal allele that produces a black coat (W) is dominant over the allele for white coat (w). Give the phenotypes and their expected proportions for the following crosses.

a. B^+B^b Ww male $\times$ B^+B^b Ww female

b. B^+B^b Ww male $\times$ B^+B^b ww female

c. B^+B^+ Ww male $\times$ B^bB^b Ww female

d. B^+B^b Ww male $\times$ B^bB^b ww female

37. J. K. Breitenbecher (1921. *Genetics* 6:65–86) investigated the genetic basis of color variation in the four-spotted cowpea weevil (*Bruchus quadrimaculatus*). The weevils were red, black, white, or tan. Breitenbecher found that four alleles (R, R^b, R^w, and r) at a single locus determine color. The alleles exhibit a dominance hierarchy, with red (R) dominant over all other alleles, black (R^b) dominant over white (R^w) and tan (r), white dominant over tan, and tan recessive to all others ($R > R^b > R^w > r$). The following genotypes encode each of the colors:

RR, RR^b, RR^w, Rr	red
R^bR^b, R^bR^w, R^br	black
R^wR^w, R^wr	white
rr	tan

Color variation in this species is sex-limited to females: males carry color genes but are always tan regardless of their genotype. For each of the following crosses carried out by Breitenbecher, give all possible genotypes of the parents.

Parents	Progeny
a. tan ♀ × tan ♂	78 reds ♀, 70 white ♀, 184 tan ♂
b. black ♀ × tan ♂	151 red ♀, 49 black ♀, 61 tan ♀, 249 tan ♂
c. white ♀ × tan ♂	32 red ♀, 31 tan ♂
d. black ♀ × tan ♂	3586 black ♀, 1282 tan ♀, 4791 tan ♂
e. white ♀ × tan ♂	594 white ♀, 189 tan ♀, 862 tan ♂
f. black ♀ × tan ♂	88 black ♀, 88 tan ♀, 186 tan ♂
g. tan ♀ × tan ♂	47 white ♀, 51 tan ♀, 100 tan ♂
h. red ♀ × tan ♂	1932 red ♀, 592 tan ♀, 2587 tan ♂
i. white ♀ × tan ♂	13 red ♀, 6 white ♀, 5 tan ♀, 19 tan ♂
j. red ♀ × tan ♂	190 red ♀, 196 black ♀, 311 tan ♂
k. black ♀ × tan ♂	1412 black ♀, 502 white ♀, 1766 tan ♂

38. Shell coiling of the snail *Limnaea peregra* results from a genetic maternal effect. An autosomal allele for a right-handed, or dextral, shell (s^+) is dominant over the allele for a left-handed, or sinistral, shell (s). A pet snail called Martha is sinistral and reproduces only as a female (the snails are hermaphroditic). Indicate which of the following statements are true and which are false. Explain your reasoning in each case.

a. Martha's genotype *must* be ss.

b. Martha's genotype *cannot* be s^+s^+.

c. All the offspring produced by Martha *must* be sinistral.

d. At least some of the offspring produced by Martha *must* be sinistral.

e. Martha's mother *must* have been sinistral.

f. All of Martha's brothers *must* be sinistral.

39. Hypospadias, a birth defect in male humans in which the urethra opens on the shaft instead of at the tip of the penis, results from an autosomal dominant gene in some families. Females who carry the gene show no effects. Is this birth defect an example of (a) an X-linked trait, (b) a Y-linked trait, (c) a sex-limited trait, (d) a sex-influenced trait, or (e) genetic maternal effect? Explain your answer.

40. In unicorns, two autosomal loci interact to determine the type of tail. One locus controls whether a tail is present at all; the allele for a tail (T) is dominant over the allele for tailless (t). If a unicorn has a tail, then alleles at a second locus determine whether the tail is curly or straight. Farmer Baldridge has two unicorns with curly tails. When he crosses these two unicorns, $^1/_2$ of the progeny have curly tails, $^1/_4$ have straight tails, and $^1/_4$ do not have a tail. Give the genotypes of the parents and progeny in Farmer Baldridge's cross. Explain how he obtained the 2 : 1 : 1 phenotypic ratio in his cross.

41. In 1983, a sheep farmer in Oklahoma noticed in his flock a ram that possessed increased muscle mass in his hindquarters. Many of the offspring of this ram possessed the same trait, which became known as the callipyge mutant (*callipyge* is Greek for "beautiful buttocks"). The mutation that caused the callipyge phenotype was eventually mapped to a position on the sheep chromosome 18.

    When the male callipyge offspring of the original mutant ram were crossed with normal females, they produced the following progeny: $\frac{1}{4}$ male callipyge, $\frac{1}{4}$ female callipyge, $\frac{1}{4}$ male normal, and $\frac{1}{4}$ female normal. When the female callipyge offspring of the original mutant ram were crossed with normal males, all of the offspring were normal. Analysis of the chromosomes of these offspring of callipyge females showed that half of them received a chromosome 18 with the allele encoding callipyge from their mother. Propose an explanation for the inheritance of the allele for callipyge. How might you test your explanation?

Section 5.4

42. Which of the following statements is an example of a phenocopy? Explain your reasoning.

    a. Phenylketonuria results from a recessive mutation that causes light skin as well as mental retardation.

    b. Human height is influenced by genes at many different loci.

    c. Dwarf plants and mottled leaves in tomatoes are caused by separate genes that are linked.

    d. Vestigial wings in *Drosophila* are produced by a recessive mutation. This trait is also produced by high temperature during development.

    e. Intelligence in humans is influenced by both genetic and environmental factors.

43. Long ears in some dogs are an autosomal dominant trait. Two dogs mate and produce a litter in which 75% of the puppies have long ears. Of the dogs with long ears in this litter, $\frac{1}{3}$ are known to be phenocopies. What are the most likely genotypes of the two parents of this litter?

CHALLENGE QUESTIONS

Section 5.1

44. Pigeons have long been the subject of genetic studies. Indeed, Charles Darwin bred pigeons in the hope of unraveling the principles of heredity but was unsuccessful. A series of genetic investigations in the early 1900s worked out the hereditary basis of color variation in these birds. W. R. Horlancher was interested in the genetic basis of kiteness, a color pattern that consists of a mixture of red and black stippling of the feathers. Horlancher knew from earlier experiments that black feather color was dominant over red. He carried out the following crosses to investigate the genetic relation of kiteness to black and red feather color (W. R. Horlancher. 1930. *Genetics* 15:312–346).

Cross	Offspring
kitey × kitey	16 kitey, 5 black, 3 red
kitey × black	6 kitey, 7 black
red × kitey	18 red, 9 kitey, 6 black

    a. On the basis of these results and the assumption that black is dominant over red, propose an hypothesis to explain the inheritance of kitey, black, and red feather color in pigeons.

    a. For each of the preceding crosses, test your hypothesis by using a chi-square test.

Section 5.3

45. Suppose that you are tending a mouse colony at a genetics research institute and one day you discover a mouse with twisted ears. You breed this mouse with twisted ears and find that the trait is inherited. Both male and female mice have twisted ears, but, when you cross a twisted-eared male with a normal-eared female, you obtain results that differ from those obtained when you cross a twisted-eared female with normal-eared male: the reciprocal crosses give different results. Describe how you would go about determining whether this trait results from a sex-linked gene, a sex-influenced gene, a genetic maternal effect, a cytoplasmically inherited gene, or genomic imprinting. What crosses would you conduct and what results would be expected with these different types of inheritance?

6 Pedigree Analysis, Applications, and Genetic Testing

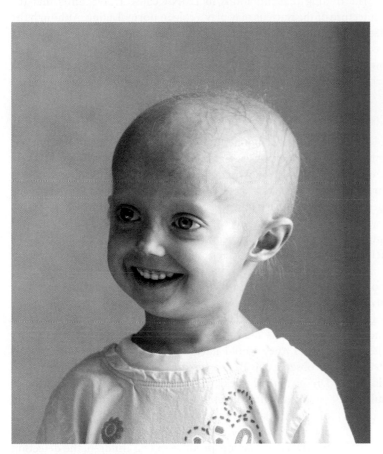

Megan, a 5-year-old with Hutchinson–Gilford progeria syndrome. Today, researchers seek to understand aging from the study of children with this rare autosomal dominant disorder that causes premature aging. [Courtesy of John Hurley for the Progeria Research Foundation (progeriaresearch.org).]

HUTCHINSON–GILFORD SYNDROME AND THE SECRET OF AGING

Spanish explorer Juan Ponce de León searched in vain for the fabled fountain of youth, the healing waters of which were reported to prevent old age. In the course of his explorations, Ponce de León discovered Florida, but he never located the source of perpetual youth. Like Ponce de León, scientists today seek to forestall aging, not in the waters of a mythical spring, but rather in understanding the molecular changes associated with the aging process. Curiously, their search has led to a small group of special children, who experience, not perpetual youth, but alarmingly premature old age.

Such children were first noticed by an English physician, Jonathan Hutchinson, who, in 1886, published a report of a young patient only 6 years old with the appearance of an 80-year-old man; the patient was bald with aged facial features and stiffness of joints. A few years later, another physician, Hastings Gilford, observed an additional case of the disorder, giving it the name progeria, derived from the Greek word for "prematurely old."

This exceedingly rare disease, known today as Hutchinson–Gilford progeria syndrome (HGPS), affects about 50 children worldwide and is characterized by the same features described by Hutchinson and Gilford more than 100 years ago. Children with HGPS appear healthy at birth but, by 2 years of age, usually begin to show features of accelerated aging, including cessation of growth, hair loss, aged skin, stiffness of joints, heart disease, and osteoporosis. By the age of 13, many have died, usually from coronary artery disease. HGPS is clearly genetic in origin and is usually inherited as an autosomal dominant trait.

In 1998, Leslie Gordon and Scott Berns's 2-year-old child Sam was diagnosed with HGPS. Gordon and Berns—both physicians—begin to seek information about the disease but, to their dismay, virtually none existed. Shocked, Gordon and Berns established the Progeria Research Foundation, and Gordon focused her career on directing the foundation and conducting research on HGPS. The efforts of the foundation stimulated a number of scientists to study HGPS and, in 2003, the gene that causes the disease was identified. Gordon was a part of the research team that made this important discovery.

6.1 Hutchinson–Gilford progeria syndrome is caused by mutations in the *lamin A* gene, which encodes the lamin A nuclear protein. A structural model of the lamin A protein.

HGPS results from mutations in a gene called *lamin A* (*LMNA*), which is located on human chromosome 1. The protein normally encoded by *LMNA* helps to form an internal scaffold that supports the nuclear membrane of the cell (**Figure 6.1**). The specific mutations that cause HGPS produce an abnormal, truncated form of the lamin A protein, which disrupts the nuclear membrane and, in some unknown way, results in premature aging.

The finding that mutations in *LMNA* cause HGPS hints that the architecture of the nucleus may play a role in normal aging. Although the relation of lamin A protein to normal aging in humans is still unclear, research on patients with HGPS may lead to a better understanding of how we all change with age. The fountain of youth may yet be found, perhaps in the cells of children with HGPS.

Hutchinson–Gilford progeria syndrome is just one of a large number of human diseases that are currently the focus of intensive genetic research. In this chapter, we consider human genetic characteristics and examine three important techniques used by geneticists to investigate these characteristics: pedigrees, twin studies, and adoption studies. At the end of the chapter, we will see how the information garnered with these techniques can be used in genetic counseling and prenatal diagnosis.

Keep in mind as you read this chapter that many important characteristics are influenced by both genes and environment, and separating these factors in humans is always difficult. Studies of twins and adopted persons are designed to distinguish the effects of genes and environment, but such studies are based on assumptions that may be difficult to meet for some human characteristics, particularly behavioral ones. Therefore, it's always prudent to interpret the results of such studies with caution. TRY PROBLEM 19 →

6.1 The Study of Genetics in Humans Is Constrained by Special Features of Human Biology and Culture

Humans are the best and the worst of all organisms for genetic study. On the one hand, we know more about human anatomy, physiology, and biochemistry than we know about most other organisms, and so many well-characterized traits are available for study. Families often keep detailed records about their members extending back many generations. Additionally, a number of important human diseases have a genetic component, and so the incentive for understanding human inheritance is tremendous. On the other hand, the study of human genetic characteristics presents some major obstacles.

First, controlled matings are not possible. With other organisms, geneticists carry out specific crosses to test their hypotheses about inheritance. We have seen, for example, how the testcross provides a convenient way to determine whether an individual organism having a dominant trait is homozygous or heterozygous. Unfortunately (for the geneticist at least), matings between humans are usually determined by romance, family expectations, or—occasionally—accident rather than by the requirements of a geneticist.

Another obstacle is that humans have a long generation time. Human reproductive age is not normally reached until 10 to 14 years after birth, and most people do not reproduce until they are 18 years of age or older; thus, generation time in humans is usually about 20 years. This long generation time means that, even if geneticists could control human crosses, they would have to wait on average 40 years just to observe the F_2 progeny. In contrast, generation time in *Drosophila* is 2 weeks; in bacteria, it's a mere 20 minutes.

Finally, human family size is generally small. Observation of even the simple genetic ratios that we learned in Chapter 3 would require a substantial number of progeny in each family. When parents produce only 2 children, the detection of a 3 : 1 ratio is impossible. Even an extremely large family of 10 to 15 children would not permit the recognition of a dihybrid 9 : 3 : 3 : 1 ratio.

Although these special constraints make genetic studies of humans more complex, understanding human heredity is tremendously important. So geneticists have been forced to develop techniques that are uniquely suited to human biology and culture. TRY PROBLEM 20 →

CONCEPTS

Although the principles of heredity are the same in humans and in other organisms, the study of human inheritance is constrained by the inability to control genetic crosses, the long generation time, and the small number of offspring.

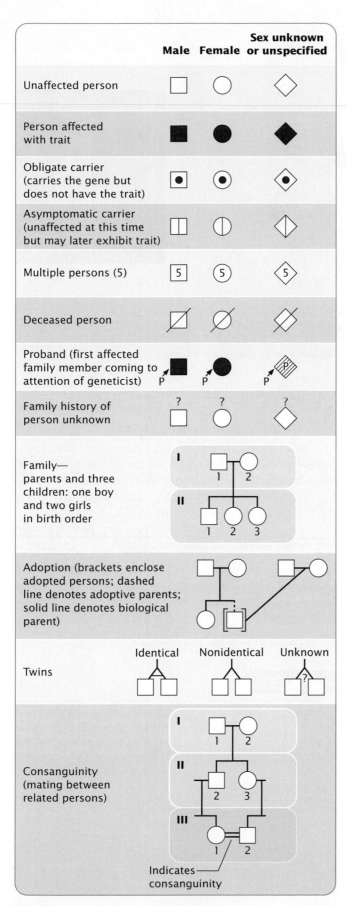

6.2 Standard symbols are used in pedigrees.

6.2 Geneticists Often Use Pedigrees to Study the Inheritance of Characteristics in Humans

An important technique used by geneticists to study human inheritance is the analysis of pedigrees. A **pedigree** is a pictorial representation of a family history, essentially a family tree that outlines the inheritance of one or more characteristics. When a particular characteristic or disease is observed in a person, a geneticist often studies the family of this affected person by drawing a pedigree.

Symbols Used in Pedigrees

The symbols commonly used in pedigrees are summarized in **Figure 6.2**. Males in a pedigree are represented by squares, females by circles. A horizontal line drawn between two symbols representing a man and a woman indicates a mating; children are connected to their parents by vertical lines extending downward from the parents. The pedigree shown in **Figure 6.3a** illustrates a family with Waardenburg syndrome, an autosomal dominant type of deafness that may be accompanied by fair skin, a white forelock, and visual problems (**Figure 6.3b**). Persons who exhibit the trait of interest are represented by filled circles and squares; in the pedigree of Figure 6.3a, the filled symbols represent members of the family who have Waardenburg syndrome. Unaffected members are represented by open circles and squares. The person from whom the pedigree is initiated is called the **proband** and is usually designated by an arrow (IV-2 in Figure 6.3a).

Let's look closely at Figure 6.3 and consider some additional features of a pedigree. Each generation in a pedigree is identified by a Roman numeral; within each generation, family members are assigned Arabic numerals, and children in each family are listed in birth order from left to right. Person II-4, a man with Waardenburg syndrome, mated with II-5, an unaffected woman, and they produced five children. The oldest of their children is III-8, a male with Waardenburg syndrome, and the youngest is III-14, an unaffected female. **TRY PROBLEM 21 →**

Analysis of Pedigrees

The limited number of offspring in most human families means that clear Mendelian ratios in a single pedigree are usually impossible to discern. Pedigree analysis requires a certain amount of genetic sleuthing, based on recognizing patterns associated with different modes of inheritance. For example, autosomal dominant traits should appear with equal frequency in both sexes and should not skip generations, provided that the trait is fully penetrant (see p. 103 in Chapter 5) and not sex influenced (see pp. 115–117 in Chapter 5).

Certain patterns may exclude the possibility of a particular mode of inheritance. For instance, a son inherits his X chromosome from his mother. If we observe that a trait

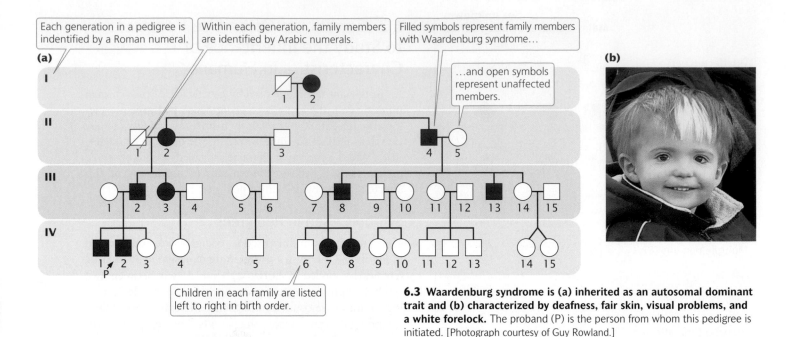

(a)

Each generation in a pedigree is indentified by a Roman numeral.

Within each generation, family members are identified by Arabic numerals.

Filled symbols represent family members with Waardenburg syndrome…

…and open symbols represent unaffected members.

(b)

Children in each family are listed left to right in birth order.

6.3 Waardenburg syndrome is (a) inherited as an autosomal dominant trait and (b) characterized by deafness, fair skin, visual problems, and a white forelock. The proband (P) is the person from whom this pedigree is initiated. [Photograph courtesy of Guy Rowland.]

is passed from father to son, we can exclude the possibility of X-linked inheritance. In the following sections, the traits discussed are assumed to be fully penetrant and rare.

Autosomal Recessive Traits

Autosomal recessive traits normally appear with equal frequency in both sexes (unless penetrance differs in males and females) and appear only when a person inherits two alleles for the trait, one from each parent. If the trait is uncommon, most parents of affected offspring are heterozygous and unaffected; consequently, the trait seems to skip generations (**Figure 6.4**). Frequently, a recessive allele may be passed for a number of generations without the trait appearing in a pedigree. Whenever both parents are heterozygous, approximately one-fourth of the offspring are expected to express

the trait, but this ratio will not be obvious unless the family is large. In the rare event that both parents are affected by an autosomal recessive trait, all the offspring will be affected.

When a recessive trait is rare, persons from outside the family are usually homozygous for the normal allele. Thus, when an affected person mates with someone outside the family ($aa \times AA$), usually none of the children will display the trait, although all will be carriers (i.e., heterozygous). A recessive trait is more likely to appear in a pedigree when two people within the same family mate, because there is a greater chance of both parents carrying the same recessive allele. Mating between closely related people is called **consanguinity**. In the pedigree shown in Figure 6.4, persons III-3 and III-4 are first cousins, and both are heterozygous for the recessive allele; when they mate, $^1/_4$ of their children are expected to have the recessive trait.

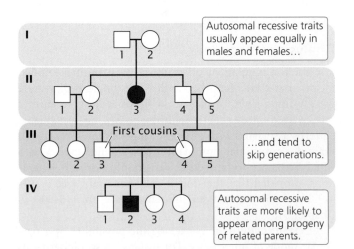

Autosomal recessive traits usually appear equally in males and females…

First cousins

…and tend to skip generations.

Autosomal recessive traits are more likely to appear among progeny of related parents.

6.4 Autosomal recessive traits normally appear with equal frequency in both sexes and seem to skip generations.

CONCEPTS

Autosomal recessive traits appear with equal frequency in males and females. Affected children are commonly born to unaffected parents who are carriers of the gene for the trait, and the trait tends to skip generations. Recessive traits appear more frequently among the offspring of consanguine matings.

✔ CONCEPT CHECK 1

Autosomal recessive traits often appear in pedigrees in which there have been consanguine matings, because these traits

a. tend to skip generations.

b. appear only when both parents carry a copy of the gene for the trait, which is more likely when the parents are related.

c. usually arise in children born to parents who are unaffected.

d. appear equally in males and females.

A number of human metabolic diseases are inherited as autosomal recessive traits. One of them is Tay–Sachs disease. Children with Tay–Sachs disease appear healthy at birth but become listless and weak at about 6 months of age. Gradually, their physical and neurological conditions worsen, leading to blindness, deafness, and, eventually, death at 2 to 3 years of age. The disease results from the accumulation of a lipid called G_{M2} ganglioside in the brain. A normal component of brain cells, G_{M2} ganglioside is usually broken down by an enzyme called hexosaminidase A, but children with Tay–Sachs disease lack this enzyme. Excessive G_{M2} ganglioside accumulates in the brain, causing swelling and, ultimately, neurological symptoms. Heterozygotes have only one normal copy of the allele encoding hexosaminidase A and produce only about half the normal amount of the enzyme. However, this amount is enough to ensure that G_{M2} ganglioside is broken down normally, and heterozygotes are usually healthy.

Autosomal Dominant Traits

Autosomal dominant traits appear in both sexes with equal frequency, and both sexes are capable of transmitting these traits to their offspring. Every person with a dominant trait must have inherited the allele from at least one parent; autosomal dominant traits therefore do not skip generations (**Figure 6.5**). Exceptions to this rule arise when people acquire the trait as a result of a new mutation or when the trait has reduced penetrance.

If an autosomal dominant allele is rare, most people displaying the trait are heterozygous. When one parent is affected and heterozygous and the other parent is unaffected, approximately $1/2$ of the offspring will be affected. If both parents have the trait and are heterozygous, approximately $3/4$ of the children will be affected. Unaffected people do not transmit the trait to their descendants, provided that the trait is fully penetrant. In Figure 6.5, we see that none of the descendants of II-4 (who is unaffected) have the trait.

CONCEPTS

Autosomal dominant traits appear in both sexes with equal frequency. An affected person has an affected parent (unless the person carries new mutations), and the trait does not skip generations. Unaffected persons do not transmit the trait.

✔ CONCEPT CHECK 2

When might you see an autosomal dominant trait skip generations?

A trait that is usually considered to be autosomal dominant is familial hypercholesterolemia, an inherited disease in which blood cholesterol is greatly elevated owing to a defect in cholesterol transport. Cholesterol is transported throughout the body in small soluble particles called lipoproteins (**Figure 6.6**). A principal lipoprotein in the transport of cholesterol is low-density lipoprotein (LDL). When an LDL molecule reaches a cell, it attaches to an LDL receptor, which

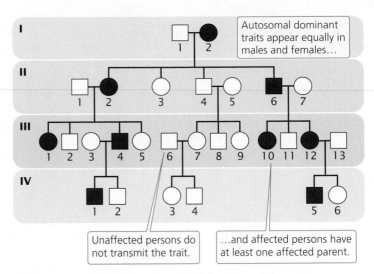

6.5 Autosomal dominant traits normally appear with equal frequency in both sexes and do not skip generations.

then moves the LDL through the cell membrane into the cytoplasm, where it is broken down and its cholesterol is released for use by the cell.

Familial hypercholesterolemia is due to a defect in the gene that normally encodes the LDL receptor. The disease is usually considered an autosomal dominant disorder because heterozygotes are deficient in LDL receptors and have elevated blood levels of cholesterol, leading to increased risk of coronary artery disease. Persons heterozygous for familial hypercholesterolemia have blood LDL levels that are twice normal and usually have heart attacks by the age of 35.

Very rarely, a person inherits two defective alleles for the LDL receptor. Such persons don't make any functional LDL receptors; their blood cholesterol levels are more than six times normal, and they may suffer a heart attack as early as age 2 and almost inevitably by age 20. Because homozygotes are more severely affected than heterozygotes, familial hypercholesterolemia is said to be incompletely dominant. However, homozygotes are rarely seen, and the common heterozygous form of the disease appears as a simple dominant trait in most pedigrees.

X-Linked Recessive Traits

X-linked recessive traits have a distinctive pattern of inheritance (**Figure 6.7**). First, these traits appear more frequently in males than in females because males need inherit only a single copy of the allele to display the trait, whereas females must inherit two copies of the allele, one from each parent, to be affected. Second, because a male inherits his X chromosome from his mother, affected males are usually born to unaffected mothers who carry an allele for the trait. Because the trait is passed from unaffected female to affected male to unaffected female, it tends to skip generations (see Figure 6.7). When a woman is heterozygous, approximately $1/2$ of her sons will be affected and $1/2$ of her

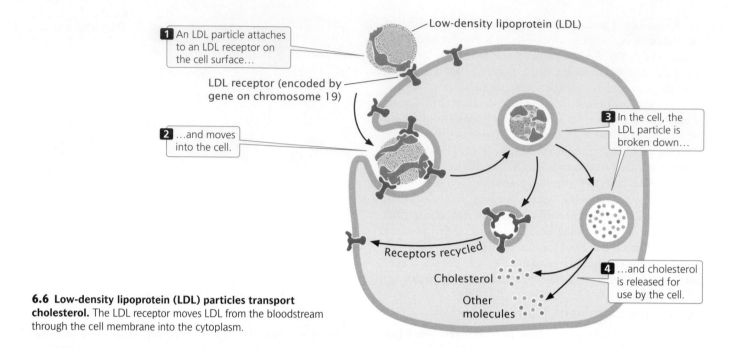

6.6 Low-density lipoprotein (LDL) particles transport cholesterol. The LDL receptor moves LDL from the bloodstream through the cell membrane into the cytoplasm.

daughters will be unaffected carriers. For example, we know that females I-2, II-2, and III-7 in Figure 6.7 are carriers because they transmit the trait to approximately half of their sons.

A third important characteristic of X-linked recessive traits is that they are not passed from father to son, because a son inherits his father's Y chromosome, not his X. In Figure 6.7, there is no case in which both a father and his son are affected. All daughters of an affected man, however, will be carriers (if their mother is homozygous for the normal allele). When a woman displays an X-linked recessive trait, she must be homozygous for the trait, and all of her sons also will display the trait.

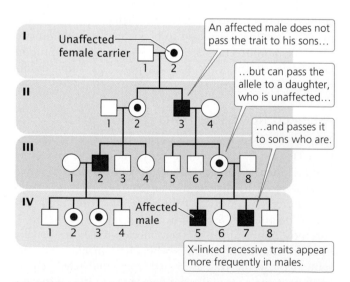

6.7 X-linked recessive traits appear more often in males than in females and are not passed from father to son.

CONCEPTS

Rare X-linked recessive traits appear more often in males than in females and are not passed from father to son. Affected sons are usually born to unaffected mothers who are carriers of the gene for the trait; thus X-linked recessive traits tend to skip generations.

✔ CONCEPT CHECK 3

How can you distinguish between an autosomal recessive trait with higher penetrance in males and an X-linked recessive trait?

An example of an X-linked recessive trait in humans is hemophilia A, also called classic hemophilia. Hemophila results from the absence of a protein necessary for blood to clot. The complex process of blood clotting consists of a cascade of reactions that includes more than 13 different factors. For this reason, there are several types of clotting disorders, each due to a glitch in a different step of the clotting pathway.

Hemophilia A results from abnormal or missing factor VIII, one of the proteins in the clotting cascade. The gene for factor VIII is located on the tip of the long arm of the X chromosome; so hemophilia A is an X-linked recessive disorder. People with hemophilia A bleed excessively; even small cuts and bruises can be life threatening. Spontaneous bleeding occurs in joints such as elbows, knees, and ankles, producing pain, swelling, and erosion of the bone. Fortunately, bleeding in people with hemophilia A can now be controlled by administering concentrated doses of factor VIII. The inheritance of hemophilia A is illustrated by the family of Queen Victoria of England (**Figure 6.8**).

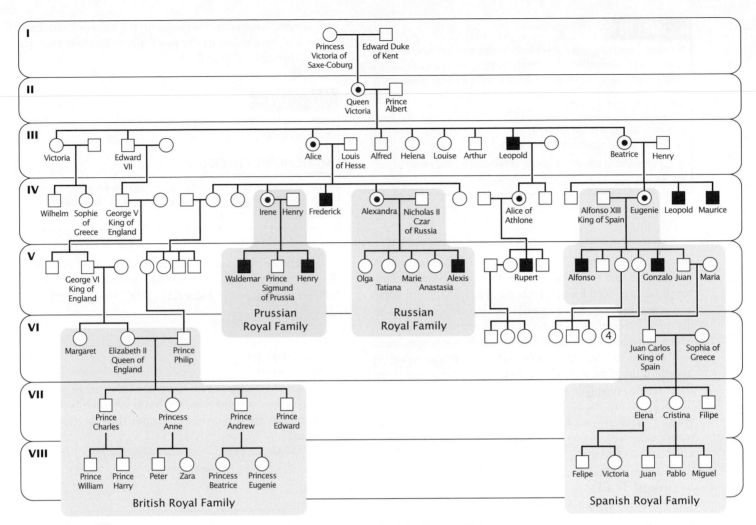

6.8 Classic hemophilia is inherited as an X-linked recessive trait. This pedigree is of hemophilia in the royal families of Europe.

X-Linked Dominant Traits

X-linked dominant traits appear in males and females, although they often affect more females than males. Each person with an X-linked dominant trait must have an affected parent (unless the person possesses a new mutation or the trait has reduced penetrance). X-linked dominant traits do not skip generations (**Figure 6.9**); affected men pass the trait on to all their daughters and none of their sons, as is seen in the children of I-1 in Figure 6.9. In contrast, affected women (if heterozygous) pass the trait on to about $1/2$ of their sons and about $1/2$ of their daughters, as seen in the children of III-6 in the pedigree. As with X-linked recessive traits, a male inherits an X-linked dominant trait only from his mother; the trait is not passed from father to son. This fact is what distinguishes X-linked dominant inheritance from autosomal dominant inheritance, in which a male can inherit the trait from his father. A female, on the other hand, inherits an X chromosome from both her mother and her father; so females can receive an X-linked trait from either parent.

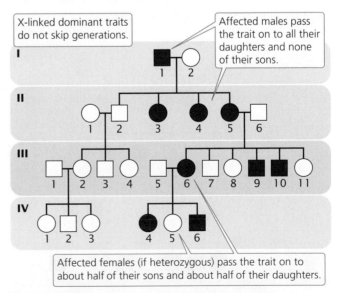

6.9 X-linked dominant traits affect both males and females. An affected male must have an affected mother.

An example of an X-linked dominant trait in humans is hypophosphatemia, or familial vitamin-D-resistant rickets. People with this trait have features that superficially resemble those produced by rickets: bone deformities, stiff spines and joints, bowed legs, and mild growth deficiencies. This disorder, however, is resistant to treatment with vitamin D, which normally cures rickets. X-linked hypophosphatemia results from the defective transport of phosphate, especially in cells of the kidneys. People with this disorder excrete large amounts of phosphate in their urine, resulting in low levels of phosphate in the blood and reduced deposition of minerals in the bone. As is common with X-linked dominant traits, males with hypophosphatemia are often more severely affected than females.

Y-Linked Traits

Y-linked traits exhibit a specific, easily recognized pattern of inheritance. Only males are affected, and the trait is passed from father to son. If a man is affected, all his male offspring also should be affected, as is the case for I-1, II-4, II-6, III-6, and III-10 of the pedigree in **Figure 6.10**. Y-linked traits do not skip generations. As mentioned in Chapter 4,

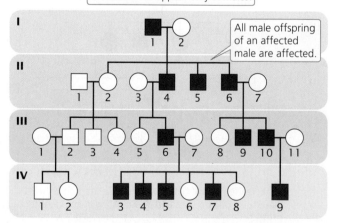
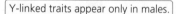

6.10 Y-linked traits appear only in males and are passed from a father to all his sons.

little genetic information is found on the human Y chromosome. Maleness is one of the few traits in humans that has been shown to be Y linked.

The major characteristics of autosomal recessive, autosomal dominant, X-linked recessive, X-linked dominant, and Y-linked traits are summarized in **Table 6.1**. **TRY PROBLEMS 23 AND 24 →**

Worked Problem

The following pedigree represents the inheritance of a rare disorder in an extended family. What is the most likely mode of inheritance for this disease? (Assume that the trait is fully penetrant.)

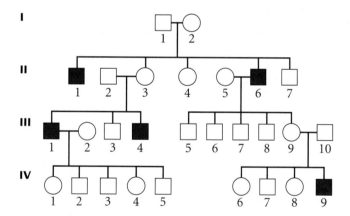

• Solution

To answer this question, we should consider each mode of inheritance and determine which, if any, we can eliminate. The trait appears only in males, and so autosomal dominant and autosomal recessive modes of inheritance are unlikely because traits with these modes appear equally in males and females. Additionally, autosomal dominance can be eliminated because some affected persons do not have an affected parent.

The trait is observed only among males in this pedigree, which might suggest Y-linked inheritance. However, for a Y-linked trait, affected men should pass the trait on to all their sons, which is not the case here; II-6 is an affected man

Table 6.1	Pedigree characteristics of autosomal recessive, autosomal dominant, X-linked recessive, X-linked dominant, and Y-linked traits

Autosomal recessive trait

1. Usually appears in both sexes with equal frequency.
2. Tends to skip generations.
3. Affected offspring are usually born to unaffected parents.
4. When both parents are heterozygous, approximately one-fourth of the offspring will be affected.
5. Appears more frequently among the children of consanguine marriages.

Autosomal dominant trait

1. Usually appears in both sexes with equal frequency.
2. Both sexes transmit the trait to their offspring.
3. Does not skip generations.
4. Affected offspring must have an affected parent unless they possess a new mutation.
5. When one parent is affected (heterozygous) and the other parent is unaffected, approximately half of the offspring will be affected.
6. Unaffected parents do not transmit the trait.

X-linked recessive trait

1. Usually more males than females are affected.
2. Affected sons are usually born to unaffected mothers; thus, the trait skips generations.
3. Approximately half of a carrier (heterozygous) mother's sons are affected.
4. Never passed from father to son.
5. All daughters of affected fathers are carriers.

X-linked dominant trait

1. Both males and females are usually affected; often more females than males are affected.
2. Does not skip generations. Affected sons must have an affected mother; affected daughters must have either an affected mother or an affected father.
3. Affected fathers will pass the trait on to all their daughters.
4. Affected mothers (if heterozygous) will pass the trait on to half of their sons and half of their daughters.

Y-linked trait

1. Only males are affected.
2. Passed from father to all sons.
3. Does not skip generations.

who has four unaffected male offspring. We can eliminate Y-linked inheritance.

X-linked dominance can be eliminated because affected men should pass an X-linked dominant trait on to all of their female offspring, and II-6 has an unaffected daughter (III-9).

X-linked recessive traits often appear more commonly in males, and affected males are usually born to unaffected female carriers; the pedigree shows this pattern of inheritance. For an X-linked trait, about half the sons of a heterozygous carrier mother should be affected. II-3 and III-9 are suspected carriers, and about half of their male children (three of five) are affected. Another important characteristic of an X-linked recessive trait is that it is not passed from father to son. We observe no father-to-son transmission in this pedigree. X-linked recessive is therefore the most likely mode of inheritance.

→ For additional practice, try to determine the mode of inheritance for the pedigrees in Problem 25 at the end of the chapter.

6.3 Studying Twins and Adoptions Can Help Assess the Importance of Genes and Environment

Twins and adoptions provide natural experiments for separating the effects of genes and environmental factors in determining differences in traits. These two techniques have been widely used in genetic studies.

Types of Twins

Twins are of two types: **dizygotic** (nonidentical) **twins** arise when two separate eggs are fertilized by two different sperm, producing genetically distinct zygotes; **monozygotic** (identical) **twins** result when a single egg, fertilized by a single sperm, splits early in development into two separate embryos.

Because monozygotic twins result from a single egg and sperm (a single, "mono," zygote), they're genetically identical (except for rare somatic mutations), having 100% of their genes in common (**Figure 6.11a**). Dizygotic twins (**Figure 6.11b**), on the other hand, have on average only 50% of their genes in common, which is the same percentage that any pair of siblings has in common. Like other siblings, dizygotic twins may be of the same sex or of different sexes. The only difference between dizygotic twins and other siblings is that dizygotic twins are the same age and shared the same uterine environment. Dizygotic twinning often runs in families and the tendancy to produce dizygotic twins is influenced by heredity, but there appears to be little genetic tendency for producing monozygotic twins.

(a)

(b)

6.11 Monozygotic twins (a) are identical; dizygotic twins (b) are nonidentical. [Part a: Joe Carini/Index Stock Imagery/PictureQuest. Part b: Courtesy of Randi Rossignol.]

Concordance in Twins

Comparisons of dizygotic and monozygotic twins can be used to assess the importance of genetic and environmental factors in producing differences in a characteristic. This assessment is often made by calculating the concordance for a trait. If both members of a twin pair have a trait, the twins are said to be *concordant*; if only one member of the pair has the trait, the twins are said to be *discordant*. **Concordance** is the percentage of twin pairs that are concordant for a trait. Because identical twins have 100% of their genes in common and dizygotic twins have on average only 50% in common, genetically influenced traits should exhibit higher concordance in monozygotic twins. For instance, when one member of a monozygotic twin pair has epilepsy (**Table 6.2**), the other twin of the pair has epilepsy about 59% of the time; so the monozygotic concordance for epilepsy is 59%. However, when a dizygotic twin has epilepsy, the other twin has epilepsy only 19% of the time (19% dizygotic concordance). The higher concordance in the monozygotic twins suggests that genes influence epilepsy, a finding supported by the results of other family studies of this disease. Concordance values for several additional human traits and diseases are listed in Table 6.2.

The hallmark of a genetic influence on a particular trait is higher concordance in monozygotic twins compared with concordance in dizygotic twins. High concordance in monozygotic twins by itself does not signal a genetic influence. Twins normally share the same environment—they are raised in the same home, have the same friends, attend the same school—and so high concordance may be due to common genes or to common environment. If the high concordance is due to environmental factors, then dizygotic twins, who also share the same environment, should have just as high a concordance as that of monozygotic twins. When genes influence the trait, however, monozygotic twin pairs should exhibit higher concordance than that of dizygotic twin pairs, because monozygotic twins have a greater percentage of genes in common. It is important to note that

Table 6.2	Concordance of monozygotic and dizygotic twins for several traits	
	Concordance (%)	
Trait	Monozygotic	Dizygotic
1. Heart attack (males)	39	26
2. Heart attack (females)	44	14
3. Bronchial asthma	47	24
4. Cancer (all sites)	12	15
5. Epilepsy	59	19
6. Rheumatoid arthritis	32	6
7. Multiple sclerosis	28	5

Sources: (1 and 2) B. Havald and M. Hauge, U.S. Public Health Service Publication 1103 (1963), pp. 61–67; (3, 4, and 5) B. Havald and M. Hauge, *Genetics and the Epidemiology of Chronic Diseases* (U.S. Department of Health, Education, and Welfare, 1965); (6) J. S. Lawrence, *Annals of Rheumatic Diseases* 26:357–379, 1970; (7) G. C. Ebers et al., *American Journal of Human Genetics* 36:495, 1984.

any discordance among monozygotic twins is usually due to environmental factors, because monozygotic twins are genetically identical.

The use of twins in genetic research rests on the important assumption that, when concordance for monozygotic twins is greater than that for dizygotic twins, it is because monozygotic twins are more similar in their genes and not because they have experienced a more similar environment. The degree of environmental similarity between monozygotic twins and dizygotic twins is assumed to be the same. This assumption may not always be correct, particularly for human behaviors. Because they look alike, identical twins may be treated more similarly by parents, teachers, and peers than are nonidentical twins. Evidence of this similar treatment is seen in the past tendency of parents to dress identical twins alike. In spite of this potential complication, twin studies have played a pivotal role in the study of human genetics. TRY PROBLEM 31 →

A Twin Study of Asthma

To illustrate the use of twins in genetic research, let's consider a study of asthma. Asthma is characterized by constriction of the airways and the secretion of mucus into the air passages, causing coughing, labored breathing, and wheezing (**Figure 6.12**). Severe cases can be life threatening. Asthma is a major health problem in industrialized countries and appears to be on the rise. The incidence of childhood asthma varies widely; the highest rates (from 17% to 30%) are in the United Kingdom, Australia, and New Zealand.

A number of environmental stimuli are known to precipitate asthma attacks, including dust, pollen, air pollu-

Table 6.3	Concordance values for asthma among monozygotic and dizygotic twins from the Twins Early Development Study	
Group	Number of Pairs	Concordance (%)
Monozygotic	1658	65
Dizygotic	3252	37

tion, respiratory infections, exercise, cold air, and emotional stress. Allergies frequently accompany asthma, suggesting that asthma is a disorder of the immune system, but the precise relation between immune function and asthma is poorly understood. Numerous studies have shown that genetic factors are important in asthma.

A genetic study of childhood asthma was conducted as a part of the Twins Early Development Study in England, an ongoing research project that studies more than 15,000 twins born in the United Kingdom between 1994 and 1996. These twins have been assessed for language, cognitive development, behavioral problems, and academic achievement at ages 7 and 9, and the genetic and environmental contributions to a number of their traits were examined. In the asthma study, researchers looked at a sample of 4910 twins at age 4. Parents of the twins were asked whether either of their twins had been prescribed medication to control asthma; those children receiving asthma medication were considered to have asthma.

Concordances of the monozygotic and dizygotic twins for asthma are shown in **Table 6.3**. The concordance value for the monozygotic twins (65%) was significantly higher than that for the dizygotic twins (37%), and the researchers concluded that, among the 4-year-olds included in the study, asthma was strongly influenced by genetic factors. The fact that even monozygotic twins were discordant 35% of the time indicates that environmental factors also play a role in asthma.

CONCEPTS

Higher concordance for monozygotic twins compared with that for dizygotic twins indicates that genetic factors play a role in determining differences in a trait. Low concordance for monozygotic twins indicates that environmental factors play a significant role.

✔ CONCEPT CHECK 7

A trait exhibits 100% concordance for both monozygotic and dizygotic twins. What conclusion can you draw about the role of genetic factors in determining differences in the trait?

a. Genetic factors are extremely important.

b. Genetic factors are somewhat important.

c. Genetic factors are unimportant.

d. Both genetic and environmental factors are important.

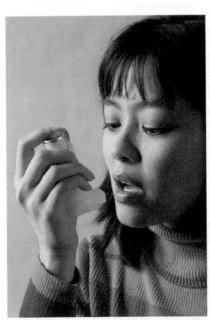

6.12 Twin studies show that asthma, characterized by constriction of the airways, is caused by a combination of genetic and environmental factors. [Custom Medical Stock/Photolibrary.]

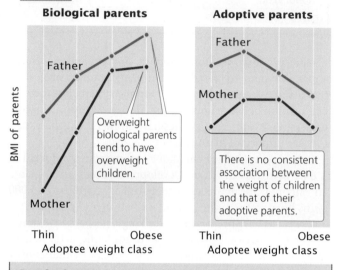

Experiment

Question: Is body-mass index (BMI) influenced by genetic factors?

Methods

Compare the body-mass index of adopted children with those of their adoptive and biological parents.

Results

Biological parents

Father

Mother

Overweight biological parents tend to have overweight children.

BMI of parents

Thin   Obese
Adoptee weight class

Adoptive parents

Father

Mother

There is no consistent association between the weight of children and that of their adoptive parents.

Thin   Obese
Adoptee weight class

Conclusion: Genetic factors influence body-mass index.

6.13 Adoption studies demonstrate that obesity has a genetic influence. [Redrawn with the permission of the *New England Journal of Medicine* 314:195, 1986.]

Adoption Studies

Another technique used by geneticists to analyze human inheritance is the study of adopted people. This approach is one of the most powerful for distinguishing the effects of genes and environment on characteristics.

For a variety of reasons, many children each year are separated from their biological parents soon after birth and adopted by adults with whom they have no genetic relationship. These adopted persons have no more genes in common with their adoptive parents than do two randomly chosen persons; however, they do share an environment with their adoptive parents. In contrast, the adopted persons have 50% of their genes in common with each of their biological parents but do not share the same environment with them. If adopted persons and their adoptive parents show similarities in a characteristic, these similarities can be attributed to environmental factors. If, on the other hand, adopted persons and their biological parents show similarities, these similarities are likely to be due to genetic factors. Comparisons of adopted persons with their adoptive parents and with their biological parents can therefore help to define the roles of genetic and environmental factors in the determination of human variation. For example, adoption stud-

ies were instrumental in showing that schizophrenia has a genetic basis. Adoption studies have also shown that obesity, as measured by body-mass index, is at least partly influenced by genetics (**Figure 6.13**).

Adoption studies assume that the environments of biological and adoptive families are independent (i.e., not more alike than would be expected by chance). This assumption may not always be correct, because adoption agencies carefully choose adoptive parents and may select a family that resembles the biological family. Thus, some of the similarity between adopted persons and their biological parents may be due to these similar environments and not due to common genetic factors. In addition, offspring and the biological mother share the same environment during prenatal development. **TRY PROBLEM 33 →**

CONCEPTS

Similarities between adopted persons and their genetically unrelated adoptive parents indicate that environmental factors affect a particular characteristic; similarities between adopted persons and their biological parents indicate that genetic factors influence the characteristic.

✔ CONCEPT CHECK 8

What assumptions underlie the use of adoption studies in genetics?

a. Adoptees have no contact with their biological parents after birth.

b. The foster parents and biological parents are not related.

c. The environments of biological and adopted parents are independent.

d. All of the above.

6.4 Genetic Counseling and Genetic Testing Provide Information to Those Concerned about Genetic Diseases and Traits

Our knowledge of human genetic diseases and disorders has expanded rapidly in recent years. The *Online Mendelian Inheritance in Man* now lists more than 19,000 human genetic diseases, disorders, genes, and traits that have a simple genetic basis. Research has provided a great deal of information about the inheritance, chromosomal location, biochemical basis, and symptoms of many of these genetic traits, diseases, and disorders. This information is often useful to people who have a genetic condition.

Genetic Counseling

Genetic counseling is a field that provides information to patients and others who are concerned about hereditary conditions. It is an educational process that helps patients

Table 6.4	Common reasons for seeking genetic counseling

1. A person knows of a genetic disease in the family.

2. A couple has given birth to a child with a genetic disease, birth defect, or chromosomal abnormality.

3. A couple has a child who is mentally retarded or has a close relative who is mentally retarded.

4. An older woman becomes pregnant or wants to become pregnant. There is disagreement about the age at which a prospective mother who has no other risk factor should seek genetic counseling; many experts suggest that any prospective mother age 35 or older should seek genetic counseling.

5. Husband and wife are closely related (e.g., first cousins).

6. A couple experiences difficulties achieving a successful pregnancy.

7. A pregnant woman is concerned about exposure to an environmental substance (drug, chemical, or virus) that causes birth defects.

8. A couple needs assistance in interpreting the results of a prenatal or other test.

9. Both prospective parents are known carriers for a recessive genetic disease.

and family members deal with many aspects of a genetic condition including a diagnosis, information about symptoms and treatment, and information about the mode of inheritance. Genetic counseling also helps a patient and the family cope with the psychological and physical stress that may be associated with the disorder. Clearly, all of these considerations cannot be handled by a single person; so most genetic counseling is done by a team that can include counselors, physicians, medical geneticists, and laboratory personnel. **Table 6.4** lists some common reasons for seeking genetic counseling.

Genetic counseling usually begins with a diagnosis of the condition. On the bases of a physical examination, biochemical tests, DNA testing, chromosome analysis, family history, and other information, a physician determines the cause of the condition. An accurate diagnosis is critical, because treatment and the probability of passing the condition on may vary, depending on the diagnosis. For example, there are a number of different types of dwarfism, which may be caused by chromosome abnormalities, single-gene mutations, hormonal imbalances, or environmental factors. People who have dwarfism resulting from an autosomal dominant gene have a 50% chance of passing the condition on to their children, whereas people who have dwarfism caused by a rare recessive gene have a low likelihood of passing the trait on to their children.

When the nature of the condition is known, a genetic counselor meets with the patient and members of the patient's family and explains the diagnosis. A family pedigree may be constructed with the use of a computer program or by hand, and the probability of transmitting the condition to future generations can be calculated for different family members. The counselor helps the family interpret the genetic risks and explains various available reproductive options, including prenatal diagnosis, artificial insemination, and in vitro fertilization. Often family members have questions about genetic testing that may be available to help determine whether they carry a genetic mutation. The counselor helps them decide whether genetic testing is appropriate and which tests to apply. After the test results are in, the genetic counselor usually helps family members interpret the results.

A family's decision about future pregnancies frequently depends on the magnitude of the genetic risk, the severity and effects of the condition, the importance of having children, and religious and cultural views. Traditionally, genetic counselors have been trained to apply *nondirected* counseling, which means that they provide information and facilitate discussion but do not bring their own opinions and values into the discussion. The goal of nondirected counseling is for the family to reach its own decision on the basis of the best available information.

Because of the growing number of genetic tests and the complexity of assessing genetic risk, there is now some movement away from completely nondirected counseling. The goal is still to provide the family with information about all options and to reach the best decision for the family, but that goal may sometimes require the counselor to recommend certain options, much as a physician recommends the most appropriate medical treatments for his or her patient.

Genetic conditions are often perceived differently from other diseases and medical problems, because genetic conditions are intrinsic to the individual person and can be passed on to children. Such perceptions may produce feelings of guilt about past reproductive choices and intense personal dilemmas about future choices. Genetic counselors are trained to help patients and their families recognize and cope with these feelings.

Who does genetic counseling? In the United States, about 4500 health professionals are currently certified in genetics by the American Board of Medical Genetics or the American Board of Genetic Counseling. About half of them are specifically trained in genetic counseling, usually by completing a special 2-year masters program that provides education in both genetics and counseling. Most of the remainder are physicians and scientists certified in medical or clinical genetics. Because of the shortage of genetic counselors and medical geneticists, information about genetic testing and genetic risk is often conveyed by primary care physicians, nurses, and social workers. TRY PROBLEMS 9 AND 10 →

CONCEPTS

Genetic counseling is an educational process that provides patients and their families with information about a genetic condition, its medical implications, the probabilities that others in the family may have the disease, and reproductive options. It also helps patients and their families cope with the psychological and physical stress associated with a genetic condition.

Genetic Testing

The ultimate goal of genetic testing is to recognize the potential for a genetic condition at an early stage. In some cases, genetic testing allows people to make informed choices about reproduction. In other cases, genetic testing allows early intervention that may lessen or even prevent the development of the condition. For those who know that they are at risk for a genetic condition, genetic testing may help alleviate anxiety associated with the uncertainty of their situation. Genetic testing includes prenatal testing and postnatal testing.

Prenatal genetic tests are those that are conducted before birth and now include procedures for diagnosing several hundred genetic diseases and disorders (**Table 6.5**). The major purpose of prenatal tests is to provide families with the information that they need to make choices during pregnancies and, in some cases, to prepare for the birth of a child with a genetic condition. The Human Genome Project (see Chapter 20) has accelerated the rate at which new genes are being isolated and new genetic tests are being developed. In spite of these advances, prenatal tests are still not available for many common genetic diseases, and no test can guarantee that a "perfect" child will be born. Several approaches to prenatal diagnosis are described in the following sections.

Ultrasonography Some genetic conditions can be detected through direct visualization of the fetus. Such visualization is most commonly done with the use of **ultrasonography**—usually referred to as ultrasound. In this technique, high-frequency sound is beamed into the uterus; when the sound waves encounter dense tissue, they bounce back and are transformed into a picture (**Figure 6.14**). The size of the fetus can be determined, as can genetic conditions such as neural-tube defects (defects in the development of the spinal column and the skull) and skeletal abnormalities.

Amniocentesis Most prenatal testing requires fetal tissue, which can be obtained in several ways. The most widely used method is **amniocentesis**, a procedure for obtaining a sample of amniotic fluid from a pregnant woman (**Figure 6.15**). Amniotic fluid—the substance that fills the amniotic sac and surrounds the developing fetus—contains fetal cells that can be used for genetic testing.

Amniocentesis is routinely performed as an outpatient procedure either with or without the use of a local anesthetic. First, ultrasonography is used to locate the position

Table 6.5 Examples of genetic diseases and disorders that can be detected prenatally and the techniques used in their detection

Disorder	Method of Detection
Chromosome abnormalities	Examination of a karyotype from cells obtained by amniocentesis or chorionic villus sampling
Cleft lip and palate	Ultrasound
Cystic fibrosis	DNA analysis of cells obtained by amniocentesis or chorionic villus sampling
Dwarfism	Ultrasound or X-ray; some forms can be detected by DNA analysis of cells obtained by amniocentesis or chorionic villus sampling
Hemophilia	Fetal blood sampling* or DNA analysis of cells obtained by amniocentesis or chorionic villus sampling
Lesch–Nyhan syndrome	Biochemical tests on cells obtained by amniocentesis or chorionic villus sampling
Neural-tube defects	Initial screening with maternal blood test, followed by biochemical tests on amniotic fluid obtained by amniocentesis or by the detection of birth defects with the use of ultrasound
Osteogenesis imperfecta (brittle bones)	Ultrasound or X-ray
Phenylketonuria	DNA analysis of cells obtained by amniocentesis or chorionic villus sampling
Sickle-cell anemia	Fetal blood sampling* or DNA analysis of cells obtained by amniocentesis or chorionic villus sampling
Tay–Sachs disease	Biochemical tests on cells obtained by amniocentesis or chorionic villus sampling

*A sample of fetal blood is obtained by inserting a needle into the umbilical cord.

these reasons, genetic information about the fetus may not be available until the 17th or 18th week of pregnancy. By this stage, abortion carries a risk of complications and is even more stressful for the parents. **Chorionic villus sampling** (CVS) can be performed earlier (between the 10th and 12th weeks of pregnancy) and collects a larger amount of fetal tissue, which eliminates the necessity of culturing the cells.

In CVS, a catheter—a soft plastic tube—is inserted into the vagina (**Figure 6.16**) and, with the use of ultrasonography for guidance, is pushed through the cervix into the uterus. The tip of the tube is placed into contact with the chorion, the outer layer of the placenta. Suction is then applied, and a small piece of the chorion is removed. Although the chorion is composed of fetal cells, it is a part of the placenta that is expelled from the uterus after birth; so the tissue that is removed is not actually from the fetus. This tissue contains millions of actively dividing cells that can be used directly in many genetic tests. Chorionic villus sampling has a somewhat higher risk of complication than that of amniocentesis; the results of several studies suggest that this procedure may increase the incidence of limb defects in the fetus when performed earlier than 10 weeks of gestation.

Fetal cells obtained by amniocentesis or by CVS can be used to prepare a **karyotype**, which is a picture of a complete set of metaphase chromosomes. Karyotypes can be studied for chromosome abnormalities (see Chapter 9). Biochemical analyses can be conducted on fetal cells to determine the presence of particular metabolic products of genes. For genetic diseases in which the DNA sequence of the causative gene has been determined, the DNA sequence (DNA testing; see Chapter 18) can be examined for defective alleles.

Maternal blood screening tests Increased risk of some genetic conditions can be detected by examining levels of certain substances in the blood of the mother (referred to as a **maternal blood screening test**). However, these tests do not determine the presence of a genetic problem; rather,

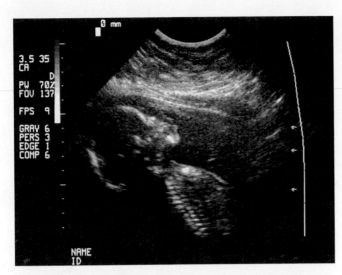

6.14 Ultrasonography can be used to detect some genetic disorders in a fetus and to locate the fetus during amniocentesis and chorionic villus sampling. [Getty Images.]

of the fetus in the uterus. Next, a long, sterile needle is inserted through the abdominal wall into the amniotic sac, and a small amount of amniotic fluid is withdrawn through the needle. Fetal cells are separated from the amniotic fluid and placed in a culture medium that stimulates them to grow and divide. Genetic tests are then performed on the cultured cells. Complications with amniocentesis (mostly miscarriage) are uncommon, arising in only about 1 in 400 procedures.

Chorionic villus sampling A major disadvantage of amniocentesis is that it is routinely performed at about the 15th to 18th week of a pregnancy (although some obstetricians successfully perform amniocentesis earlier). The cells obtained by amniocentesis must then be cultured before genetic tests can be performed, requiring yet more time. For

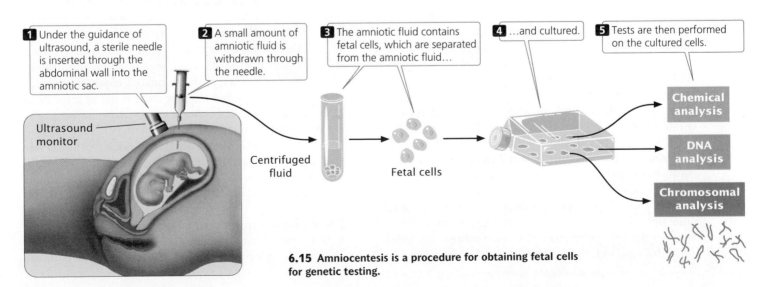

6.15 Amniocentesis is a procedure for obtaining fetal cells for genetic testing.

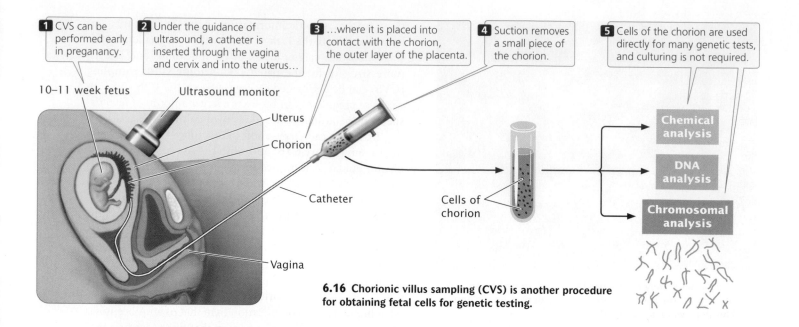

1 CVS can be performed early in pregnancy.

2 Under the guidance of ultrasound, a catheter is inserted through the vagina and cervix and into the uterus…

3 …where it is placed into contact with the chorion, the outer layer of the placenta.

4 Suction removes a small piece of the chorion.

5 Cells of the chorion are used directly for many genetic tests, and culturing is not required.

10–11 week fetus

Ultrasound monitor

Uterus

Chorion

Catheter

Vagina

Cells of chorion

Chemical analysis

DNA analysis

Chromosomal analysis

6.16 Chorionic villus sampling (CVS) is another procedure for obtaining fetal cells for genetic testing.

they simply indicate that the fetus is at increased risk and hence are referred to as *screening* tests. When increased risk is detected, follow-up tests (additional blood-screening tests, ultrasound, amniocentesis, or all three) are usually conducted.

One substance examined in maternal screening tests is α-fetoprotein, a protein that is normally produced by the fetus during development and is present in fetal blood, amniotic fluid, and the mother's blood during pregnancy. The level of α-fetoprotein is significantly higher than normal when the fetus has a neural-tube defect or one of several other disorders. Some chromosome abnormalities produce lower-than-normal levels of α-fetoprotein. Measuring the amount of α-fetoprotein in the mother's blood gives an indication of these conditions.

The American Collge of Obstetricians and Gynecologists recommends that physicians offer all pregnant women maternal blood screening tests. One typical test, carried out between 11 and 13 weeks of pregnancy, measures human chorionic gonadotropin (hCG, a pregnancy hormone) and a substance called pregnancy-associated plasma protein A (PAPP-A). When the fetus has certain chromosomal defects, the level of PAPP-A tends to be low and the level of hCG tends to be high. The risk of a chromosomal abnormality is calculated on the basis of the levels of hCG and PAPP-A in the mother's blood, along with the results of utrasound tests. Another test, referred to as the quad screen, is offered between 15 and 20 weeks of pregnancy; it measures the levels of four substances: α-fetoprotein, hCG, estriol, and inhibin. The risk of chromosomal abnormalities and certain other birth defects is calculated on the basis of the combined levels of the four substances plus the mother's age, weight, ethnic background, and number of fetuses. The quad screen

successfully detects Down syndrome (due to three copies of chromosome 21) 81% of the time.

Noninvasive fetal diagnosis Prenatal tests that utilize only maternal blood are highly desirable because they are noninvasive and pose no risk to the fetus. In addition to maternal blood screening tests, which measure chemical substances produced by the fetus or placenta, some newer procedures diagnose genetic conditions by examining fetal DNA obtained from fetal cells in maternal blood. During pregnancy, a few fetal cells are released into the mother's circulatory system, where they mix and circulate with her blood. Recent advances have made it possible to detect and separate fetal cells from maternal blood cells (a procedure called **fetal cell sorting**) with the use of lasers and automated cell-sorting machines. The fetal cells obtained can be cultured for chromosome analysis or used as a source of fetal DNA for molecular testing (see Chapter 18). Maternal blood also contains low levels of free-floating fetal DNA, which also can be tested for mutations. Although still experimental, noninvasive fetal diagnosis has now been used to detect Down syndrome and diseases such as cystic fibrosis and thalassemia (a blood disorder). A current limitation is that the procedure is not as accurate as amniocentesis or CVS, because enough fetal cells for genetic testing cannot always be obtained. Additionally, only mutated genes from the father can be detected, because there is no way to completely separate maternal DNA from the fetal cells. Thus, if the mother carries a copy of the mutation, determining whether the fetus also carries the gene is impossible.

Preimplantation genetic diagnosis Prenatal genetic tests provide today's prospective parents with increasing amounts of information about the health of their future children. New reproductive technologies provide couples with options for using this information. One of these technologies is in vitro fertilization. In this procedure, hormones are used to induce ovulation. The ovulated eggs are surgically removed from the surface of the ovary, placed in a laboratory dish, and fertilized with sperm. The resulting embryo is then implanted in the uterus. Thousands of babies resulting from in vitro fertilization have now been born.

Genetic testing can be combined with in vitro fertilization to allow the implantation of embryos that are free of a specific genetic defect. Called **preimplantation genetic diagnosis** (PGD), this technique enables people who carry a genetic defect to avoid producing a child with the disorder. For example, if a woman is a carrier of an X-linked recessive disease, approximately half of her sons are expected to have the disease. Through in vitro fertilization and preimplantation testing, an embryo without the disorder can be selected for implantation in her uterus.

The procedure begins with the production of several single-celled embryos through in vitro fertilization. The embryos are allowed to divide several times until they reach the 8- or 16-cell stage. At this point, one cell is removed from each embryo and tested for the genetic abnormality. Removing a single cell at this early stage does not harm the embryo. After determination of which embryos are free of the disorder, a healthy embryo is selected and implanted in the woman's uterus.

Preimplantation genetic diagnosis requires the ability to conduct a genetic test on a single cell. Such testing is possible with the use of the polymerase chain reaction through which minute quantities of DNA can be amplified (replicated) quickly (see Chapter 19). After amplification of the cell's DNA, the DNA sequence is examined. Preimplantation genetic diagnosis has resulted in the birth of thousands of healthy children. Its use raises a number of ethical concerns, because it can be used as a means of selecting for or against genetic traits that have nothing to do with medical concerns. For example, it can potentially be used to select for a child with genes for a certain eye color or genes for increased height.

Newborn screening Testing for genetic disorders in newborn infants is called **newborn screening**. All states in the United States and many other countries require by law that newborn infants be tested for some genetic diseases and conditions. In 2006, the American College of Medical Genetics recommended mandatory screening for 29 conditions (**Table 6.6**), and many states have now adopted this list for newborn testing These genetic conditions were chosen because early identification can lead to effective treatment. For example, as mentioned in Chapter 5, phenylketonuria is an autosomal recessive disease that, if not treated at an early age, can result in mental retardation. But early intervention, through the administration of a modified diet, prevents retardation.

Table 6.6	Genetic conditions recommended for mandatory screening by the American College of Medical Genetics
Medium-chain acyl-CoA dehydrogenase deficiency	
Congenital hypothyroidism	
Phenylketonuria	
Biotinidase deficiency	
Sickle-cell anemia (Hb SS disease)	
Congenital adrenal hyperplasia (21-hydroxylase deficiency)	
Isovaleric acidemia	
Very long chain acyl-CoA dehydrogenase deficiency	
Maple syrup (urine) disease	
Classical galactosemia	
Hb S β-thalassemia	
Hb S C disease	
Long-chain L-3-hydroxyacyl-CoA dehydrogenase deficiency	
Glutaric acidemia type I	
3-Hydroxy-3-methyl glutaric aciduria	
Trifunctional protein deficiency	
Multiple carboxylase deficiency	
Methylmalonic acidemia (mutase deficiency)	
Homocystinuria (due to cystathionine β-synthase deficiency)	
3-Methylcrotonyl-CoA carboxylase deficiency	
Hearing loss	
Methylmalonic acidemia	
Propionic acidemia	
Carnitine uptake defect	
β-Ketothiolase deficiency	
Citrullinemia	
Argininosuccinic acidemia	
Tyrosinemia type I	
Cystic fibrosis	

Most newborn testing is done by collecting a drop of an infant's blood on a card soon after birth. The card is then shipped to a laboratory for genetic analysis. These cards, which have been in use for more than 40 years, are often stored for many years—sometimes indefinitely—after their initial use in genetic testing. Recently, blood samples from archived cards have been used for research studies, such as examining genetic mutations that lead to childhood leukemia and birth defects and looking at the relation between genes and environmental toxins. These studies and other proposed studies have raised ethical concerns about the long-term storage of these blood samples and their use—in some cases, without consent—in some research studies. For example, in 2009, five families filed a federal lawsuit against the Texas Department of State Health Services, claiming that the agency violated privacy rights by collecting blood samples from Texas children and storing them for future research purposes, without the parents' knowledge or consent. This action prompted passage of a new Texas state law requiring that parents be informed at the time of a child's birth about potential uses of the samples; it also gives parents the right to withdraw their consent in the future and have the sample destroyed.

Presymptomatic testing In addition to testing for genetic diseases in fetuses and newborns, testing healthy adults for genes that might predispose them to a genetic condition in the future is now possible. This type of testing is known as **presymptomatic genetic testing**. For example, presymptomatic testing is available for members of families that have an autosomal dominant form of breast cancer. In this case, early identification of the disease-causing allele allows for closer surveillance and the early detection of tumors. Presymptomatic testing is also available for some genetic diseases for which no treatment is available, such as Huntington disease, an autosomal dominant disease that leads to slow physical and mental deterioration in middle age. Presymptomatic testing for untreatable conditions raises a number of social and ethical questions (see Chapter 18).

Heterozygote screening Another form of genetic testing in adults is **heterozygote screening**. In this type of screening, members of a population are tested to identify heterozygous carriers of recessive disease-causing alleles—people who are healthy but have the potential to produce children with a particular disease.

Testing for Tay–Sachs disease is a successful example of heterozygote screening. In the general population of North America, the frequency of Tay–Sachs disease is only about 1 person in 360,000. Among Ashkenazi Jews (descendants of Jewish people who settled in eastern and central Europe), the frequency is 100 times as great. A simple blood test is used to identify Ashkenazi Jews who carry the allele for Tay–Sachs disease. If a man and woman are both heterozygotes, approximately one in four of their children is expected to have Tay–Sachs disease. Couples identified as heterozygous carriers may use that information in deciding whether to have children. A prenatal test also is available for determining if the fetus of an at-risk couple will have Tay–Sachs disease. Screening programs have led to a significant decline

in the number of children of Ashkenazi ancestry born with Tay–Sachs disease (now fewer than 10 children per year in the United States). TRY PROBLEM 12 →

Interpreting Genetic Tests

Today, more than 1200 genetic tests are clinically available and several hundred more are available through research studies. Information from the Human Genome Project and research examining the association of genetic markers and disease will greatly increase the number and complexity of genetic tests available in the future. Many of these tests will be for complex multifactorial diseases that are influenced by both genetics and environment, such as coronary artery disease, diabetes, asthma, some types of cancer, and depression.

Interpreting the results of genetic tests is often complicated by several factors. First, some genetic diseases are caused by numerous different mutations. For example, more than 1000 different mutations at a single locus can cause cystic fibrosis, an autosomal recessive disease in which chloride ion transport is defective (see p. 102 in Chapter 5). Genetic tests typically screen for only the most common mutations; uncommon and rare mutations are not detected. Therefore, a negative result does not mean that a genetic defect is absent; it indicates only that the person does not have a common mutation. When family members have the disease, their DNA can be examined to determine the nature of the mutation and other family members can then be screened for the same mutation, but this option is not possible if affected family members are unavailable or unwilling to be tested.

A second problem lies in interpreting the results of genetic tests. For a classic genetic disease such as Tay–Sachs disease, the inheritance of two copies of the gene virtually ensures that a person will have the disease. However, this is not the case for many genetic diseases in which penetrance is incomplete and environmental factors play a role. For these conditions, carrying a disease-predisposing mutation only elevates a person's risk of acquiring the disease. The risk associated with a particular mutation is a statistical estimate, based on the average effect of the mutation on many people. In this case, the calculated risk may provide little useful information to a specific person.

> ### CONCEPTS
>
> Interpreting genetic tests is complicated by the presence of multiple causative mutations, incomplete penetrance, and the influence of environmental factors.

Direct-to-Consumer Genetic Testing

An increasing number of genetic tests are now being offered to anyone interested in investigating his or her own hereditary conditions, without requiring a health-care provider. These **direct-to-consumer genetic tests** are available for testing a large and growing array of genetic conditions in adults and children, everything from single-gene disorders such as cystic fibrosis to multifactorial conditions such as obesity, cardiovascular disease, athletic performance, and predisposition to nicotine addiction. Direct-to-consumer tests are also available for paternity testing and for determining ancestry.

Many direct-to-consumer genetic tests are advertised and ordered through the Internet. After a person orders a test, the company sends a kit for collecting a sample of DNA (usually cells from inside the cheek or a spot of blood). The person collects the sample and sends it back to the company, which performs the test and sends the results to the person. Geneticists, public health officials, and consumer advocates have raised a number of concerns about direct-to-consumer genetic testing, including concerns that some tests are offered without appropriate information and genetic counseling and that consumers are often not equipped to interpret the results. Other concerns focus on the accuracy of some tests, the confidentiality of the results, and whether indications of risk provided by the test are even useful.

Advocates of direct-to-consumer genetic tests contend that the tests provide greater access to testing and enhanced confidentiality. Many states do not regulate direct-to-consumer testing and, currently, there is little federal oversight. **TRY PROBLEM 15 →**

Genetic Discrimination and Privacy

With the development of many new genetic tests, concerns have been raised about privacy regarding genetic information and the potential for genetic discrimination. Research shows that many people at risk for genetic diseases avoid genetic testing because they fear that the results would make it difficult for them to obtain health insurance or that the information might adversely affect their employability. Some of those who do seek genetic testing pay for it themselves and use aliases to prevent the results from becoming part of their health records. Fears about genetic discrimination have been reinforced by past practices. In the 1970s, some African Americans were forced to undergo genetic testing for sickle-cell anemia (an autosomal recessive disorder) and those who were identified as carriers faced employment discrimination, in spite of the fact that carriers are healthy.

In response to these concerns, the U. S. Congress passed the **Genetic Information Nondiscrimination Act**, commonly called GINA, in 2008. This law prohibits health insurers from using genetic information to make decisions about health-insurance coverage and rates. It also prevents employers from using genetic information in employment decisions and prohibits health insurers and employers from asking or requiring a person to take a genetic test. Results of genetic testing receive some degree of protection by other federal regulations that cover the uses and disclosure of individual health information. However, GINA covers health insurance and employment only; it does not apply to life, disability, and long-term care insurance. **TRY PROBLEM 16 →**

> ### CONCEPTS
>
> The growing number of genetic tests and their complexity have raised several concerns, including those about direct-to-consumer tests, genetic discrimination, and privacy regarding test results.

6.5 Comparison of Human and Chimpanzee Genomes Is Helping to Reveal Genes That Make Humans Unique

Now that we've examined the inheritance of human traits and how some of these conditions can be detected with genetic tests, let's see what our genes can tell us about our place in evolution. Chimpanzees are our closest living relatives, yet we differ from chimps anatomically and in a huge number of behavioral, social, and intellectual skills. The perceived large degree of difference between chimpanzees and humans is manifested by the placement of these two species in entirely different primate families (humans in Hominidae and chimpanzees in Pongidae).

In spite of the large phenotypic gulf between humans and chimpanzees, findings in early DNA-hybridization studies suggested that they differ in only about 1% of their DNA sequences. Recent sequencing of human and chimpanzee genomes has provided more-precise estimates of the genetic differences that separate these species: about 1% of the two genomes differ in base sequences and about 3% differ in regard to deletions and insertions. Thus, more than 95% of the DNA of humans and chimpanzees is identical. But, clearly, humans are not chimpanzees. What, then, are the genetic differences that make us distinctly human? What are the genes that give us our human qualities?

Geneticists are now identifying genes that contribute to human uniqueness and potentially played an important role in the evolution of modern humans. In some cases, these genes have been identified through the study of mutations that cause abnormalities in our human traits, such as brain size or language ability. One set of genes that potentially contribute to human uniqueness regulates brain size. Mutations at six loci, known as *microcephalin 1* through *microcephalin 6* (*MCPH1–MCPH6*), cause microcephaly, a condition in which the brain is severely reduced in overall size but brain structure is not affected. The observation that mutations in *microcephalin* drastically affect brain size has led to the suggestion that, in the course of human evolution, selection for alleles encoding large brains at one or more of these genes might have led to enlarged brain size in humans. Geneticists have studied variation in the *microcephalin* genes of different primates and have come to the conclusion that there was strong selection in the recent past for the sequences of the *microcephalin* genes that are currently found in humans.

Another candidate gene for helping to make humans unique is called *FOXP2*, which is required for human speech. Mutations in the *FOXP2* gene often cause speech and language disabilities. Evolutionary studies of sequence variation in the *FOXP2* gene of mice, humans, and other primates reveal evidence of past selection and suggest that humans acquired their version of the gene (which enables human speech) no more than 200,000 years ago, about the time at which modern humans emerged. In 2007, scientists extracted DNA from Neanderthal fossils and found that these now-extinct humans possessed the same version of the *FOXP2* gene seen in modern humans, perhaps giving them the ability to talk.

Other genes that may contribute to human uniqueness have been identified on the basis of their rapid evolution in the human lineage. For example, the *HAR1* gene evolved slowly in other vertebrates but has undergone rapid evolution since humans diverged from chimpanzees. This gene is expressed in brain cells and some evidence suggests that it may be important in the development of the cerebral cortex, an area of the brain that is greatly enlarged in humans.

CONCEPTS SUMMARY

- Constraints on the genetic study of human traits include the inability to conduct controlled crosses, long generation time, small family size, and the difficulty of separating genetic and environmental influences. Pedigrees are often used to study the inheritance of traits in humans.

- Autosomal recessive traits typically appear with equal frequency in both sexes and tend to skip generations. When both parents are heterozygous for a particular autosomal recessive trait, approximately one-fourth of their offspring will have the trait. Recessive traits are more likely to appear in families with consanguinity (mating between closely related persons).

- Autosomal dominant traits usually appear equally in both sexes and do not skip generations. When one parent is affected and heterozygous for an autosomal dominant trait, approximately half of the offspring will have the trait. Unaffected people do not normally transmit an autosomal dominant trait to their offspring.

- X-linked recessive traits appear more frequently in males than in females. When a woman is a heterozygous carrier for an X-linked recessive trait and a man is unaffected, approximately half of their sons will have the trait and half of their daughters will be unaffected carriers. X-linked traits are not passed on from father to son.

- X-linked dominant traits appear in males and females but more frequently in females. They do not skip generations. Affected men pass an X-linked dominant trait to all of their daughters but none of their sons. Heterozygous women pass the trait to half of their sons and half of their daughters.

- Y-linked traits appear only in males and are passed from father to all sons.

- A trait's higher concordance in monozygotic than in dizygotic twins indicates a genetic influence on the trait; less than 100% concordance in monozygotic twins indicates environmental influences on the trait.

- Similarities between adopted children and their biological parents indicate the importance of genetic factors in the expression of a trait; similarities between adopted children and their genetically unrelated adoptive parents indicate the influence of environmental factors.

- Genetic counseling provides information and support to people concerned about hereditary conditions in their families.

- Genetic testing includes prenatal diagnosis, screening for disease-causing alleles in newborns, the detection of people heterozygous for recessive alleles, and presymptomatic testing for the presence of a disease-causing allele in at-risk people.

- The interpretation of genetic tests may be complicated by the presence of numerous causitive mutations, incomplete penetrance, and the influence of environmental factors.

- The recent availability of direct-to-consumer genetic tests has raised concerns about the adequacey of information provided, the absence of genetic counseling, acccuracy, privacy, and the practical uses of some tests.

- Genetic testing has raised concerns about genetic descrimination and privacy regarding test results. The Genetic Information Nondiscrimination Act prohibits the use of genetic information in deciding health insurability and employment.

- Genetic research has identified a number of genes that may contribute to human uniqueness, including genes that influence brain size and speech.

IMPORTANT TERMS

pedigree (p. 137)
proband (p. 137)
consanguinity (p. 138)
dizygotic twins (p. 143)
monozygotic twins (p. 143)
concordance (p. 144)
genetic counseling (p. 146)
ultrasonography (p. 148)

amniocentesis (p. 148)
chorionic villus sampling (CVS) (p. 149)
karyotype (p. 149)
maternal blood screening test (p. 149)
fetal cell sorting (p. 150)
preimplantation genetic diagnosis (PGD)
  (p. 151)
newborn screening (p. 151)

presymptomatic genetic testing (p. 152)
heterozygote screening (p. 152)
direct-to-consumer genetic test (p. 153)
Genetic Information Nondiscrimination
  Act (GINA) (p. 153)

ANSWERS TO CONCEPT CHECKS

1. b

2. It might skip generations when a new mutation arises or the trait has reduced penetrance.

3. If X-linked recessive, the trait will not be passed from father to son.

4. c

5. If the trait were Y linked, an affected male would pass it on to all his sons, whereas, if the trait were autosomal and sex-limited, affected heterozygous males would pass it on to only half of their sons on average.

6. d

7. c

8. d

9. Preimplantation genetic diagnosis determines the presence of disease-causing genes in an embryo at an early stage, before it is implanted in the uterus and initiates pregnancy. Prenatal genetic diagnosis determines the presence of disease-causing genes or chromosomes in a developing fetus.

WORKED PROBLEMS

1. Joanna has "short fingers" (brachydactyly). She has two older brothers who are identical twins; both have short fingers. Joanna's two younger sisters have normal fingers. Joanna's mother has normal fingers, and her father has short fingers. Joanna's paternal grandmother (her father's mother) has short fingers; her paternal grandfather (her father's father), who is now deceased, had normal fingers. Both of Joanna's maternal grandparents (her mother's parents) have normal fingers. Joanna marries Tom, who has normal fingers; they adopt a son named Bill who has normal fingers. Bill's biological parents both have normal fingers. After adopting Bill, Joanna and Tom produce two children: an older daughter with short fingers and a younger son with normal fingers.

a. Using standard symbols and labels, draw a pedigree illustrating the inheritance of short fingers in Joanna's family.

b. What is the most likely mode of inheritance for short fingers in this family?

c. If Joanna and Tom have another biological child, what is the probability (based on your answer to part *b*) that this child will have short fingers?

• Solution

a. In the pedigree for the family, identify persons with the trait (short fingers) by filled circles (females) and filled squares (males). Connect Joanna's identical twin brothers to the line above by drawing diagonal lines that have a horizontal line between them. Enclose the adopted child of Joanna and

Tom in brackets; connect him to his biological parents by drawing a diagonal line and to his adopted parents by a dashed line.

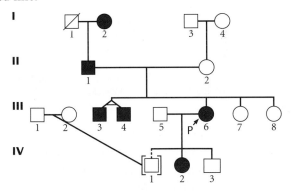

b. The most likely mode of inheritance for short fingers in this family is autosomal dominant. The trait appears equally in males and females and does not skip generations. When one parent has the trait, it appears in approximately half of that parent's sons and daughters, although the number of children in the families is small. We can eliminate Y-linked inheritance because the trait is found in females as well as males. If short fingers were X-linked recessive, females with the trait would be expected to pass the trait to all their sons, but Joanna (III-6), who has short fingers, produced a son with normal fingers. For X-linked dominant traits, affected men should pass the trait to all their daughters; because male II-1 has short fingers and produced two daughters without short fingers (III-7 and III-8), we know that the trait cannot be

X-linked dominant. The trait is unlikely to be autosomal recessive, because it does not skip generations and approximately half the children of affected parents have the trait.

c.   If having short fingers is autosomal dominant, Tom must be homozygous (*bb*) because he has normal fingers. Joanna must be heterozygous (*Bb*) because she and Tom have produced both short- and normal-fingered offspring. In a cross between a heterozygote and homozygote, half the progeny are expected to be heterozygous and the other half homozygous (*Bb* × *bb* → $\frac{1}{2}$ *Bb*, $\frac{1}{2}$ *bb*); so the probability that Joanna's and Tom's next biological child will have short fingers is $\frac{1}{2}$.

2.   Concordance values for a series of traits were measured in monozygotic twins and dizygotic twins; the results are shown in the following table. For each trait, indicate whether the rates of concordance suggest genetic influences, environmental influences, or both. Explain your reasoning.

Characteristic	Concordance (%)	
	Monozygotic	Dizygotic
a. ABO blood type	100	65
b. Diabetes	85	36
c. Coffee drinking	80	80
d. Smoking	75	42
e. Schizophrenia	53	16

• Solution

a.   The concordance for ABO blood type in the monozygotic twins is 100%. This high concordance in monozygotic twins does not, by itself, indicate a genetic basis for the trait. An important indicator of a genetic influence on the trait is lower concordance in dizygotic twins. Because concordance for ABO blood type is substantially lower in the dizygotic twins, we would be safe in concluding that genes play a role in determining differences in ABO blood types.

b.   The concordance for diabetes is substantially higher in monozygotic twins than in dizygotic twins; therefore, we can conclude that genetic factors play some role in susceptibility to diabetes. The fact that monozygotic twins show a concordance less than 100% suggests that environmental factors also play a role.

c.   Both monozygotic and dizygotic twins exhibit the same high concordance for coffee drinking; so we can conclude that there is little genetic influence on coffee drinking. The fact that monozygotic twins show a concordance less than 100% suggests that environmental factors play a role.

d.   The concordance for smoking is lower in dizygotic twins than in monozygotic twins; so genetic factors appear to influence the tendency to smoke. The fact that monozygotic twins show a concordance less than 100% suggests that environmental factors also play a role.

e.   Monozygotic twins exhibit substantially higher concordance for schizophrenia than do dizygotic twins; so we can conclude that genetic factors influence this psychiatric disorder. Because the concordance of monozygotic twins is substantially less than 100%, we can also conclude that environmental factors play a role in the disorder as well.

COMPREHENSION QUESTIONS

Section 6.1

*1. What three factors complicate the task of studying the inheritance of human characteristics?

Section 6.2

2. Who is the proband in a pedigree? Is the proband always found in the last generation of the pedigree? Why or why not?

*3. For each of the following modes of inheritance, describe the features that will be exhibited in a pedigree in which the trait is present: autosomal recessive, autosomal dominant, X-linked recessive, X-linked dominant, and Y-linked inheritance.

4. How does the pedigree of an autosomal recessive trait differ from the pedigree of an X-linked recessive trait?

5. Other than the fact that a Y-linked trait appears only in males, how does the pedigree of a Y-linked trait differ from the pedigree of an autosomal dominant trait?

Section 6.3

*6. What are the two types of twins and how do they arise?

7. Explain how a comparison of concordance in monozygotic and dizygotic twins can be used to determine the extent to which the expression of a trait is influenced by genes or by environmental factors.

8. How are adoption studies used to separate the effects of genes and environment in the study of human characteristics?

Section 6.4

*9. What is genetic counseling?

10. Give at least four different reasons for seeking genetic counseling.

11. Briefly define newborn screening, heterozygote screening, presymptomatic testing, and prenatal diagnosis.

12. Compare the advantages and disadvantages of aminocentesis versus chorionic villus sampling for prenatal diagnosis.

13. What is preimplantation genetic diagnosis?

14. How does heterozygote screening differ from presymptomatic genetic testing?

15. What are direct-to-consumer genetic tests? What are some of the concerns about these tests?

16. What activities does the Genetic Information Nondiscrimination Act prohibit?

17. How might genetic testing lead to genetic discrimination?

APPLICATION QUESTIONS AND PROBLEMS

Introduction

19. Hutchinson–Gilford progeria syndrome (HGPS) is caused by an autosomal dominant gene. Sam, the child discussed in the introduction to this chapter, has HGPS, but both of his parents are unaffected. Later in the chapter, we learned that autosomal dominant traits do not normally skip generations and that affected children have at least one affected parent. Why does HGPS not display these characteristics of an autosomal dominant trait?

Section 6.1

20. If humans have characteristics that make them unsuitable for genetic analysis, such as long generation time, small family size, and uncontrolled crosses, why do geneticists study humans? Give several reasons why humans have been the focus of so much genetic study.

Section 6.2

*21. Joe is color blind. Both his mother and his father have normal vision, but his mother's father (Joe's maternal grandfather) is color blind. All Joe's other grandparents have normal color vision. Joe has three sisters—Patty, Betsy, and Lora—all with normal color vision. Joe's oldest sister, Patty, is married to a man with normal color vision; they have two children, a 9-year-old color-blind boy and a 4-year-old girl with normal color vision.

a. Using standard symbols and labels, draw a pedigree of Joe's family.

b. What is the most likely mode of inheritance for color blindness in Joe's family?

c. If Joe marries a woman who has no family history of color blindness, what is the probability that their first child will be a color-blind boy?

d. If Joe marries a woman who is a carrier of the color-blind allele, what is the probability that their first child will be a color-blind boy?

e. If Patty and her husband have another child, what is the probability that the child will be a color-blind boy?

22. Many studies have suggested a strong genetic predisposition to migraine headaches, but the mode of inheritance is not clear. L. Russo and colleagues examined migraine headaches in several families, two of which are shown below (L. Russo et al. 2005. *American Journal of Human Genetics* 76:327–333). What is the most likely mode of inheritance for migraine headaches in these families? Explain your reasoning.

Section 6.5

18. Briefly describe some of the recently discovered genes that contribute to human uniqueness and their possible importance in human evolution.

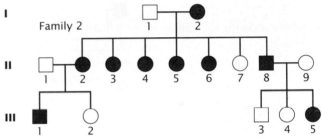

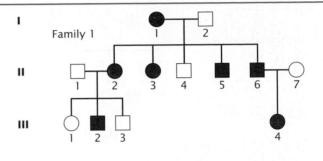

23. Dent disease is a rare disorder of the kidney, in which reabsorption of filtered solutes is impaired and there is progressive renal failure. R. R. Hoopes and colleagues studied mutations associated with Dent disease in the following family (R. R. Hoopes et al. 2005. *American Journal of Human Genetics* 76:260–267).

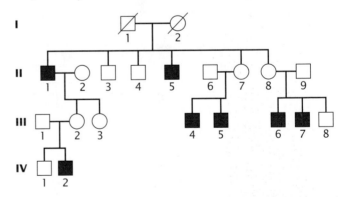

a. On the basis of this pedigree, what is the most likely mode of inheritance for the disease? Explain your reasoning.

b. From your answer to part *a*, give the most likely genotypes for all persons in the pedigree.

24. A man with a specific unusual genetic trait marries an unaffected woman and they have four children. Pedigrees of this family are shown in parts *a* through *e*, but the presence or absence of the trait in the children is not indicated. For each type of inheritance, indicate how many children of each sex are expected to express the trait by filling in the appropriate circles and squares. Assume that the trait is rare and fully penetrant.

a. Autosomal recessive trait

b. Autosomal dominant trait

c. X-linked recessive trait

d. X-linked dominant trait

e. Y-linked trait

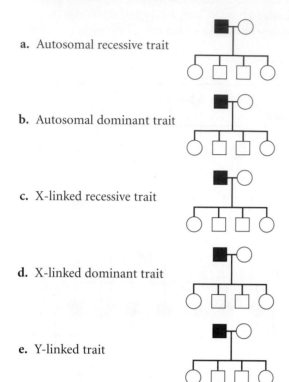

*25. For each of the following pedigrees, give the most likely mode of inheritance, assuming that the trait is rare. Carefully explain your reasoning.

a.

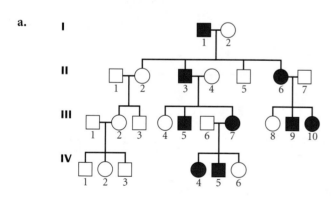

b.

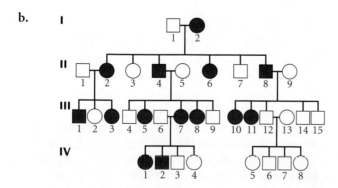

c.

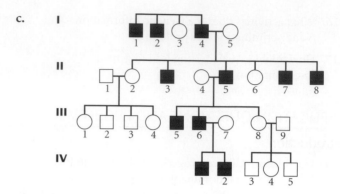

d.

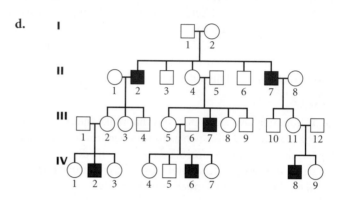

e.

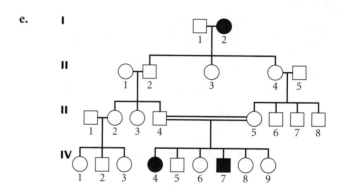

26. The trait represented in the following pedigree is expressed only in the males of the family. Is the trait Y linked? Why or why not? If you believe that the trait is not Y linked, propose an alternate explanation for its inheritance.

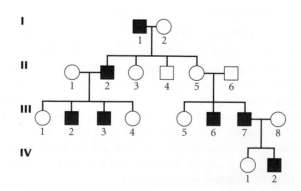

*27. The following pedigree illustrates the inheritance of Nance–Horan syndrome, a rare genetic condition in which affected persons have cataracts and abnormally shaped teeth.

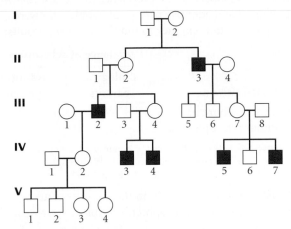

[Pedigree adapted from D. Stambolian, R. A. Lewis, K. Buetow, A. Bond, and R. Nussbaum. 1990. *American Journal of Human Genetics* 47:15.]

a. On the basis of this pedigree, what do you think is the most likely mode of inheritance for Nance–Horan syndrome?

b. If couple III-7 and III-8 have another child, what is the probability that the child will have Nance–Horan syndrome?

c. If III-2 and III-7 were to mate, what is the probability that one of their children would have Nance–Horan syndrome?

28. The following pedigree illustrates the inheritance of ringed hair, a condition in which each hair is differentiated into light and dark zones. What mode or modes of inheritance are possible for the ringed-hair trait in this family?

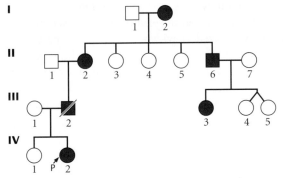

[Pedigree adapted from L. M. Ashley and R. S. Jacques. 1950. *Journal of Heredity* 41:83.]

29. Ectrodactyly is a rare condition in which the fingers are absent and the hand is split. This condition is usually inherited as an autosomal dominant trait. Ademar Freire-Maia reported the appearance of ectrodactyly in a family in São Paulo, Brazil, whose pedigree is shown here. Is this pedigree consistent with autosomal dominant inheritance? If not, what mode of inheritance is most likely? Explain your reasoning.

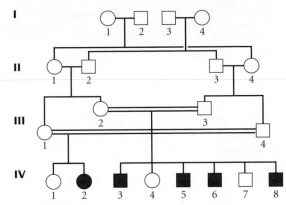

[Pedigree adapted from A. Freire-Maia. 1971. *Journal of Heredity* 62:53.]

30. The complete absence of one or more teeth (tooth agenesis) is a common trait in humans—indeed, more than 20% of humans lack one or more of their third molars. However, more-severe absence of teeth, defined as missing six or more teeth, is less common and frequently an inherited condition. L. Lammi and colleagues examined tooth agenesis in the Finnish family shown in the pedigree on the next page (L. Lammi. 2004. *American Journal of Human Genetics* 74:1043–1050).

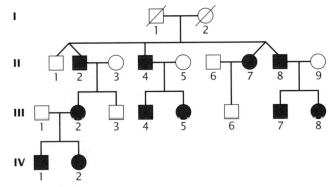

[Pedigree adapted from L. Lammi. 2004. *American Journal of Human Genetics* 74:1043–1050.]

a. What is the most likely mode of inheritance for tooth agenesis in this family? Explain your reasoning.

b. Are the two sets of twins in this family monozygotic or dizygotic twins? What is the basis of your answer?

c. If IV-2 married a man who has a full set of teeth, what is the probability that their child would have tooth agenesis?

d. If III-2 and III-7 married and had a child, what is the probability that their child would have tooth agenesis?

Section 6.4

*31. A geneticist studies a series of characteristics in monozygotic twins and dizygotic twins, obtaining the concordances listed on page 160. For each characteristic, indicate whether the rates of concordance suggest genetic influences, environmental influences, or both. Explain your reasoning.

	Concordance (%)	
Characteristic	**Monozygotic**	**Dizygotic**
Migraine headaches	60	30
Eye color	100	40
Measles	90	90
Clubfoot	30	10
High blood pressure	70	40
Handedness	70	70
Tuberculosis	5	5

32. M. T. Tsuang and colleagues studied drug dependence in male twin pairs (M. T. Tsuang et al. 1996. *American Journal of Medical Genetics* 67:473–477). They found that 4 out of 30 monozygotic twins were concordant for dependence on opioid drugs, whereas 1 out of 34 dizygotic twins were concordant for the same trait. Calculate the concordance rates for opioid dependence in these monozygotic and dizygotic twins. On the basis of these data, what conclusion can you make concerning the roles of genetic and environmental factors in opioid dependence?

CHALLENGE QUESTIONS

Section 6.1

35. Many genetic studies, particularly those of recessive traits, have focused on small isolated human populations, such as those on islands. Suggest one or more advantages that isolated populations might have for the study of recessive traits.

Section 6.2

36. Draw a pedigree that represents an autosomal dominant trait, sex-limited to males, and that excludes the possibility that the trait is Y linked.

37. A. C. Stevenson and E. A. Cheeseman studied deafness in a family in Northern Ireland and recorded the following pedigree (A. C. Stevenson and E. A. Cheeseman. 1956. *Annals of Human Genetics* 20:177–231).

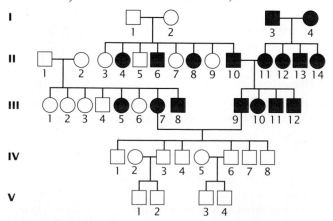

[Pedigree adapted from A. C. Stevenson and E. A. Cheeseman. 1956. *Annals of Human Genetics* 20:177–231.]

Section 6.5

33. In a study of schizophrenia (a mental disorder including disorganization of thought and withdrawal from reality), researchers looked at the prevalence of the disorder in the biological and adoptive parents of people who were adopted as children; they found the following results:

	Prevalence of schizophrenia	
Adopted Persons	**Biological parents**	**Adoptive parents**
With schizophrenia	12	2
Without schizophrenia	6	4

[Source: S. S. Kety et al. 1978. in *The Nature of Schizophrenia: New Approaches to Research and Treatment*, L. C. Wynne, R. L. Cromwell, and S. Matthysse, Eds. New York: Wiley, 1978, pp. 25–37.]

What can you conclude from these results concerning the role of genetics in schizophrenia? Explain your reasoning.

Section 6.3

34. What, if any, ethical issues might arise from the widespread use of noninvasive fetal diagnosis, which can be carried out much earlier than amniocentesis or chorionic villus sampling?

a. If you consider only generations I through III, what is the most likely mode of inheritance for this type of deafness?

b. Provide a possible explanation for the cross between III-7 and III-9 and the results for generations IV through V.

Section 6.3

38. Dizygotic twinning often runs in families and its frequency varies among ethnic groups, whereas monozygotic twinning rarely runs in families and its frequency is quite constant among ethnic groups. These observations have been interpreted as evidence for a genetic basis for variation in dizygotic twinning but for little genetic basis for variation in monozygotic twinning. Can you suggest a possible reason for these differences in the genetic tendencies of dizygotic and monozygotic twinning?

7 Linkage, Recombination, and Eukaryotic Gene Mapping

(a)

(b)

(c)

Pattern baldness is a hereditary trait.
Recent research demonstrated that a gene for pattern baldness is linked to genetic markers located on the X chromosome, leading to the discovery that pattern baldness is influenced by variation in the androgen-receptor gene. The trait is seen in three generations of the Adams family: (a) John Adams (1735–1826), the second president of the United States, was father to (b) John Quincy Adams (1767–1848), sixth U.S. president, who was father to (c) Charles Francis Adams (1807–1886). [Part a: National Museum of American Art, Washington, D.C./Art Resource, New York. Part b: National Portrait Gallery, Washington, D.C./Art Resource, New York. Part c: Bettmann/Corbis.]

LINKED GENES AND BALD HEADS

For many, baldness is the curse of manhood. Twenty-five percent of men begin balding by age 30 and almost half are bald to some degree by age 50. In the United States, baldness affects more than 40 million men, who spend hundreds of millions of dollars each year on hair-loss treatment. Baldness is not just a matter of vanity: bald males are more likely to suffer from heart disease, high blood pressure, and prostate cancer.

Baldness can arise for a number of different reasons, including illness, injury, drugs, and heredity. The most-common type of baldness seen in men is pattern baldness—technically known as androgenic alopecia—in which hair is lost prematurely from the front and top of the head. More than 95% of hair loss in men is pattern baldness. Although pattern baldness is also seen in women, it is usually expressed weakly as mild thinning of the hair. The trait is clearly stimulated by male sex hormones (androgens), as evidenced by the observation that males castrated at an early age rarely become bald (but this is not recommended as a preventive treatment for baldness).

A strong hereditary influence on pattern baldness has long been recognized, but the exact mode of inheritance has been controversial. An early study suggested that it was autosomal dominant in males and recessive in females, an example of a sex influenced trait (see Chapter 4). Other evidence and common folklore suggested that a man inherits baldness from his mother's side of the family, exhibiting X-linked inheritance.

In 2005, geneticist Axel Hillmer and his colleagues set out to locate the gene that causes pattern baldness. They suspected that the gene for pattern baldness might be located on the X chromosome, but they had no idea where on the X chromosome the gene might reside. To identify the location of the gene, they conducted a linkage analysis study, in which they looked for an association between the inheritance of pattern baldness and the inheritance of genetic variants known to be located on the X chromosome. The genetic variants used in the study were single-nucleotide polymorphisms (SNPs, pronounced "snips"), which are positions in the genome where individuals vary in a single nucleotide. The geneticists studied the inheritance of pattern baldness and SNPs in 95 families in which at least two brothers developed pattern baldness at an early age.

What Hillmer and his colleagues found was that pattern baldness and SNPS from the X chromosome were not inherited independently, as predicted by Mendel's principle of independent assortment. Instead, they tended to be inherited together, which occurs when genes are physically linked on the same chromosome and segregate together in meiosis.

As we will learn in this chapter, linkage between genes is broken down by recombination (crossing over), and the amount of recombination between genes is directly related to the distance between them. In 1911, Thomas Hunt Morgan and his student Alfred Sturtevant demonstrated in fruit flies that genes can be mapped by determining the rates of recombination between the genes. Using this method for the families with pattern baldness, Hillmer and his colleagues demonstrated that the gene for pattern baldness is closely linked to SNPs located at position p12–22, a region that includes the androgen-receptor gene, on the X chromosome. The androgen receptor gene encodes a protein that binds male sex hormones. Given the clear involvement of male hormones in the development of pattern baldness, the androgen-receptor gene seemed a likely candidate for causing pattern baldness. Further analysis revealed that certain alleles of the androgen-receptor gene were closely associated with the inheritance of pattern baldness, and the androgen-receptor gene is almost certainly responsible for much of the differences in pattern baldness seen in the families examined. Additional studies conducted in 2008 found that genes on chromosomes 3 and 20 also appear to contribute to the expression of pattern baldness. TRY PROBLEM 13 →

This chapter explores the inheritance of genes located on the same chromosome. These linked genes do not strictly obey Mendel's principle of independent assortment; rather, they tend to be inherited together. This tendency requires a new approach to understanding their inheritance and predicting the types of offspring produced. A critical piece of information necessary for predicting the results of these crosses is the arrangement of the genes on the chromosomes; thus, it will be necessary to think about the relation between genes and chromosomes. A key to understanding the inheritance of linked genes is to make the conceptual connection between the genotypes in a cross and the behavior of chromosomes in meiosis.

We will begin our exploration of linkage by first comparing the inheritance of two linked genes with the inheritance of two genes that assort independently. We will then examine how crossing over, or recombination, breaks up linked genes. This knowledge of linkage and recombination will be used for predicting the results of genetic crosses in which genes are linked and for mapping genes. Later in the chapter, we will focus on physical methods of determining the chromosomal locations of genes. The final section examines variation in rates of recombination.

7.1 Linked Genes Do Not Assort Independently

Chapter 3 introduced Mendel's principles of segregation and independent assortment. Let's take a moment to review these two important concepts. The principle of segregation states that each individual diploid organism possesses two alleles at a locus that separate in meiosis, with one allele going into each gamete. The principle of independent assortment provides additional information about the process of segrega-

tion: it tells us that, in the process of separation, the two alleles at a locus act independently of alleles at other loci.

The independent separation of alleles results in *recombination*, the sorting of alleles into new combinations. Consider a cross between individuals homozygous for two different pairs of alleles: $AA\ BB \times aa\ bb$. The first parent, $AA\ BB$, produces gametes with the alleles $A\ B$, and the second parent, $aa\ bb$, produces gametes with the alleles $a\ b$, resulting in F_1 progeny with genotype $Aa\ Bb$ (**Figure 7.1**).

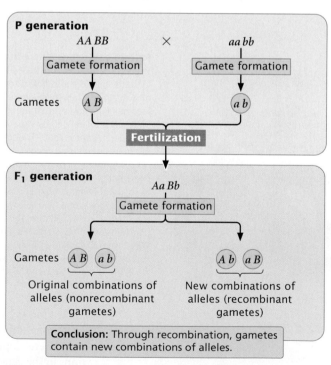

P generation

$AA\ BB$ × $aa\ bb$

Gamete formation   Gamete formation

Gametes $A\ B$ $a\ b$

Fertilization

F_1 generation

$Aa\ Bb$

Gamete formation

Gametes $A\ B$ $a\ b$ $A\ b$ $a\ B$

Original combinations of alleles (nonrecombinant gametes)   New combinations of alleles (recombinant gametes)

Conclusion: Through recombination, gametes contain new combinations of alleles.

7.1 Recombination is the sorting of alleles into new combinations.

Recombination means that, when one of the F_1 progeny reproduces, the combination of alleles in its gametes may differ from the combinations in the gametes from its parents. In other words, the F_1 may produce gametes with alleles A b or a B in addition to gametes with A B or a b.

Mendel derived his principles of segregation and independent assortment by observing the progeny of genetic crosses, but he had no idea of what biological processes produced these phenomena. In 1903, Walter Sutton proposed a biological basis for Mendel's principles, called the chromosome theory of heredity (see Chapter 3). This theory holds that genes are found on chromosomes. Let's restate Mendel's two principles in relation to the chromosome theory of heredity. The principle of segregation states that a diploid organism possesses two alleles for a trait, each of which is located at the same position, or locus, on each of the two homologous chromosomes. These chromosomes segregate in meiosis, with each gamete receiving one homolog. The principle of independent assortment states that, in meiosis, each pair of homologous chromosomes assorts independently of other homologous pairs. With this new perspective, it is easy to see that the number of chromosomes in most organisms is limited and that there are certain to be more genes than chromosomes; so some genes must be present on the same chromosome and should not assort independently. Genes located close together on the same chromosome are called **linked genes** and belong to the same **linkage group**. Linked genes travel together in meiosis, eventually arriving at the same destination (the same gamete), and are not expected to assort independently.

All of the characteristics examined by Mendel in peas did display independent assortment and, after the rediscovery of Mendel's work, the first genetic characteristics studied in other organisms also seemed to assort independently. How could genes be carried on a limited number of chromosomes and yet assort independently?

The apparent inconsistency between the principle of independent assortment and the chromosome theory of heredity soon disappeared as biologists began finding genetic characteristics that did not assort independently. One of the first cases was reported in sweet peas by William Bateson, Edith Rebecca Saunders, and Reginald C. Punnett in 1905. They crossed a homozygous strain of peas having purple flowers and long pollen grains with a homozygous strain having red flowers and round pollen grains. All the F_1 had purple flowers and long pollen grains, indicating that purple was dominant over red and long was dominant over round. When they intercrossed the F_1, the resulting F_2 progeny did not appear in the 9 : 3 : 3 : 1 ratio expected with independent assortment (**Figure 7.2**). An excess of F_2 plants had purple flowers and long pollen or red flowers and round pollen (the parental phenotypes). Although Bateson, Saunders, and Punnett were unable to explain these results, we now know that the two loci that they examined lie close together on the same chromosome and therefore do not assort independently.

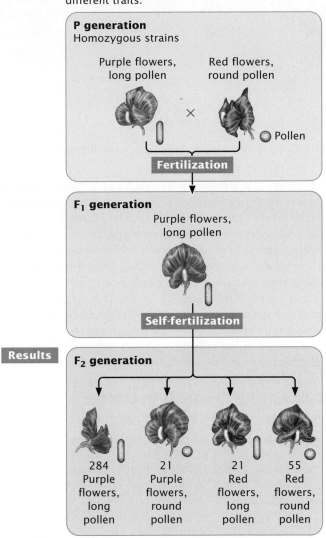

Experiment

Question: Do the genes for flower color and pollen shape in sweet peas assort independently?

Methods Cross two strains homozygous for two different traits.

P generation
Homozygous strains

Purple flowers, long pollen × Red flowers, round pollen

Pollen

Fertilization

F_1 generation
Purple flowers, long pollen

Self-fertilization

Results

F_2 generation

| 284 Purple flowers, long pollen | 21 Purple flowers, round pollen | 21 Red flowers, long pollen | 55 Red flowers, round pollen |

Conclusion: F_2 progeny do not appear in the 9 : 3 : 3 : 1 ratio expected with independent assortment.

7.2 Nonindependent assortment of flower color and pollen shape in sweet peas.

7.2 Linked Genes Segregate Together and Crossing Over Produces Recombination Between Them

Genes that are close together on the same chromosome usually segregate as a unit and are therefore inherited together. However, genes occasionally switch from one homologous chromosome to the other through the process of crossing over (see Chapter 2), as illustrated

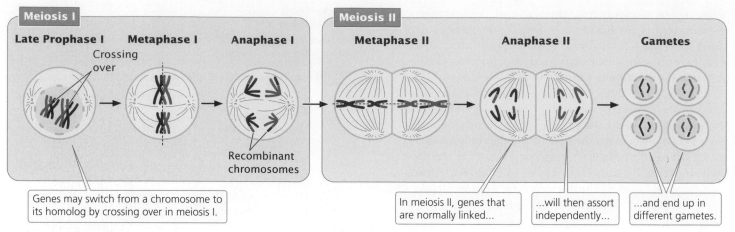

7.3 Crossing over takes place in meiosis and is responsible for recombination.

in **Figure 7.3**. Crossing over results in recombination; it breaks up the associations of genes that are close together on the same chromosome. Linkage and crossing over can be seen as processes that have opposite effects: linkage keeps particular genes together, and crossing over mixes them up. In Chapter 5, we considered a number of exceptions and extensions to Mendel's principles of heredity. The concept of linked genes adds a further complication to interpretations of the results of genetic crosses. However, with an understanding of how linkage affects heredity, we can analyze crosses for linked genes and successfully predict the types of progeny that will be produced.

Notation for Crosses with Linkage

In analyzing crosses with linked genes, we must know not only the genotypes of the individuals crossed, but also the arrangement of the genes on the chromosomes. To keep track of this arrangement, we introduce a new system of notation for presenting crosses with linked genes. Consider a cross between an individual homozygous for dominant alleles at two linked loci and another individual homozygous for recessive alleles at those loci ($AA\ BB \times aa\ bb$). For linked genes, it's necessary to write out the specific alleles as they are arranged on each of the homologous chromosomes:

$$\frac{A \qquad B}{A \qquad B} \times \frac{a \qquad b}{a \qquad b}$$

In this notation, each line represents one of the two homologous chromosomes. Inheriting one chromosome from each parent, the F_1 progeny will have the following genotype:

$$\frac{A \qquad B}{a \qquad b}$$

Here, the importance of designating the alleles on each chromosome is clear. One chromosome has the two dominant alleles A and B, whereas the homologous chromosome has the two recessive alleles a and b. The notation can be simplified by drawing only a single line, with the understanding that genes located on the same side of the line lie on the same chromosome:

$$\frac{A \qquad B}{a \qquad b}$$

This notation can be simplified further by separating the alleles on each chromosome with a slash: AB/ab.

Remember that the two alleles at a locus are always located on different homologous chromosomes and therefore must lie on opposite sides of the line. Consequently, we would *never* write the genotypes as

$$\frac{A \qquad a}{B \qquad b}$$

because the alleles A and a can *never* be on the same chromosome. It is also important to always keep the same order of the genes on both sides of the line; thus, we should *never* write

$$\frac{A \qquad B}{b \qquad a}$$

because it would imply that alleles A and b are allelic (at the same locus).

Complete Linkage Compared with Independent Assortment

We will first consider what happens to genes that exhibit complete linkage, meaning that they are located very close together on the same chromosome and do not exhibit crossing over. Genes are rarely completely linked but, by assuming that no crossing over occurs, we can see the effect of linkage more clearly. We will then consider what happens when genes assort independently. Finally, we will consider the results obtained if the genes are linked but exhibit some crossing over.

A testcross reveals the effects of linkage. For example, if a heterozygous individual is test-crossed with a homozygous recessive individual ($Aa\ Bb \times aa\ bb$), the alleles that are present in the gametes contributed by the heterozygous parent will be expressed in the phenotype of the offspring,

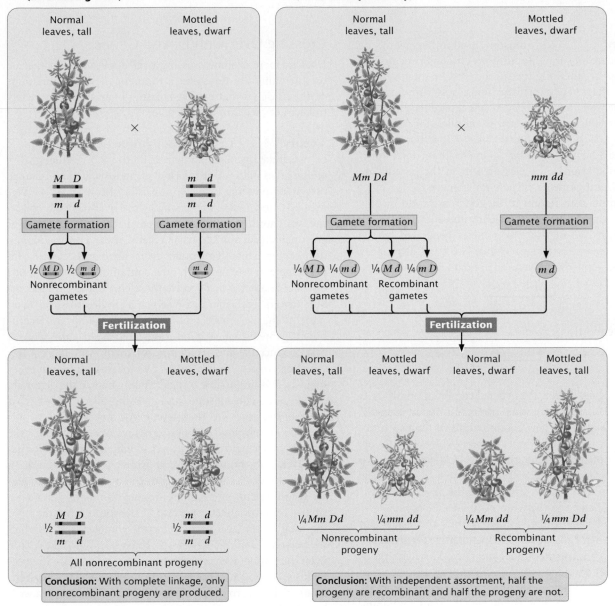

(a) If genes are completely linked (no crossing over)

Normal leaves, tall   ×   Mottled leaves, dwarf

M D / m d   m d / m d

Gamete formation   Gamete formation

½ MD ½ md   md
Nonrecombinant gametes

Fertilization

Normal leaves, tall   Mottled leaves, dwarf

½ M D / m d   ½ m d / m d

All nonrecombinant progeny

Conclusion: With complete linkage, only nonrecombinant progeny are produced.

(b) If genes are unlinked (assort independently)

Normal leaves, tall   ×   Mottled leaves, dwarf

Mm Dd   mm dd

Gamete formation   Gamete formation

¼ MD ¼ md   ¼ Md ¼ mD   md
Nonrecombinant   Recombinant
gametes   gametes

Fertilization

Normal leaves, tall   Mottled leaves, dwarf   Normal leaves, dwarf   Mottled leaves, tall

¼ Mm Dd   ¼ mm dd   ¼ Mm dd   ¼ mm Dd
Nonrecombinant progeny   Recombinant progeny

Conclusion: With independent assortment, half the progeny are recombinant and half the progeny are not.

7.4 A testcross reveals the effects of linkage. Results of a testcross for two loci in tomatoes that determine leaf type and plant height.

because the homozygous parent could not contribute dominant alleles that might mask them. Consequently, traits that appear in the progeny reveal which alleles were transmitted by the heterozygous parent.

Consider a pair of linked genes in tomato plants. One of the genes affects the type of leaf: an allele for mottled leaves (m) is recessive to an allele that produces normal leaves (M). Nearby on the same chromosome the other gene determines the height of the plant: an allele for dwarf (d) is recessive to an allele for tall (D).

Testing for linkage can be done with a testcross, which requires a plant heterozygous for both characteristics. A geneticist might produce this heterozygous plant by crossing a variety of tomato that is homozygous for normal leaves and tall height with a variety that is homozygous for mottled leaves and dwarf height:

$$P \qquad \frac{M \quad D}{M \quad D} \times \frac{m \quad d}{m \quad d}$$

$$\downarrow$$

$$F_1 \qquad \frac{M \quad D}{m \quad d}$$

The geneticist would then use these F_1 heterozygotes in a testcross, crossing them with plants homozygous for mottled leaves and dwarf height:

$$\frac{M \quad D}{m \quad d} \times \frac{m \quad d}{m \quad d}$$

The results of this testcross are diagrammed in **Figure 7.4a**. The heterozygote produces two types of gametes: some

165

with the $\underline{M \qquad D}$ chromosome and others with the $\underline{m \qquad d}$ chromosome. Because no crossing over occurs, these gametes are the only types produced by the heterozygote. Notice that these gametes contain only combinations of alleles that were present in the original parents: either the allele for normal leaves together with the allele for tall height (M and D) or the allele for mottled leaves together with the allele for dwarf height (m and d). Gametes that contain only original combinations of alleles present in the parents are **nonrecombinant gametes**, or *parental* gametes.

The homozygous parent in the testcross produces only one type of gamete; it contains chromosome $\underline{m \qquad d}$ and pairs with one of the two gametes generated by the heterozygous parent (see Figure 7.4a). Two types of progeny result: half have normal leaves and are tall:

$$\frac{M \qquad D}{m \qquad d}$$

and half have mottled leaves and are dwarf:

$$\frac{m \qquad d}{m \qquad d}$$

These progeny display the original combinations of traits present in the P generation and are **nonrecombinant progeny**, or *parental* progeny. No new combinations of the two traits, such as normal leaves with dwarf or mottled leaves with tall, appear in the offspring, because the genes affecting the two traits are completely linked and are inherited together. New combinations of traits could arise only if the physical connection between M and D or between m and d were broken.

These results are distinctly different from the results that are expected when genes assort independently (**Figure 7.4b**). If the M and D loci assorted independently, the heterozygous plant ($Mm\ Dd$) would produce four types of gametes: two nonrecombinant gametes containing the original combinations of alleles ($M\ D$ and $m\ d$) and two gametes containing new combinations of alleles ($M\ d$ and $m\ D$). Gametes with new combinations of alleles are called **recombinant gametes**. With independent assortment, nonrecombinant and recombinant gametes are produced in equal proportions. These four types of gametes join with the single type of gamete produced by the homozygous parent of the testcross to produce four kinds of progeny in equal proportions (see Figure 7.4b). The progeny with new combinations of traits formed from recombinant gametes are termed **recombinant progeny**.

CONCEPTS

A testcross in which one of the plants is heterozygous for two completely linked genes yields two types of progeny, each type displaying one of the original combinations of traits present in the P generation. Independent assortment, in contrast, produces four types of progeny in a 1 : 1 : 1 : 1 ratio—two types of recombinant progeny and two types of nonrecombinant progeny in equal proportions.

Crossing Over with Linked Genes

Usually, there is some crossing over between genes that lie on the same chromosome, producing new combinations of traits. Genes that exhibit crossing over are incompletely linked. Let's see how it takes place.

Theory   The effect of crossing over on the inheritance of two linked genes is shown in **Figure 7.5**. Crossing over, which takes place in prophase I of meiosis, is the exchange of genetic material between nonsister chromatids (see Figures 2.16 and 2.18). After a single crossover has taken place, the two chromatids that did not participate in crossing over are unchanged; gametes that receive these chromatids are nonrecombinants. The other two chromatids, which did participate in crossing over, now contain new combinations of alleles; gametes that receive these chromatids are recombinants. For each meiosis in which a single crossover takes place, then, two nonrecombinant gametes and two recombinant gametes will be produced. This result is the same as that produced by independent assortment (see Figure 7.4b); so, when crossing over between two loci takes place in every meiosis, it is impossible to determine whether the genes are on the same chromosome and crossing over took place or whether the genes are on different chromosomes.

For closely linked genes, crossing over does not take place in every meiosis. In meioses in which there is no crossing over, only nonrecombinant gametes are produced. In meioses in which there is a single crossover, half the gametes are recombinants and half are nonrecombinants (because a single crossover affects only two of the four chromatids); so the total percentage of recombinant gametes is always half the percentage of meioses in which crossing over takes place. Even if crossing over between two genes takes place in every meiosis, only 50% of the resulting gametes will be recombinants. Thus, the frequency of recombinant gametes is always half the frequency of crossing over, and the maximum proportion of recombinant gametes is 50%.

CONCEPTS

Linkage between genes causes them to be inherited together and reduces recombination; crossing over breaks up the associations of such genes. In a testcross for two linked genes, each crossover produces two recombinant gametes and two nonrecombinants. The frequency of recombinant gametes is half the frequency of crossing over, and the maximum frequency of recombinant gametes is 50%.

✔ CONCEPT CHECK 1

For single crossovers, the frequency of recombinant gametes is half the frequency of crossing over because

a. a test cross between a homozygote and heterozygote produces $^1/_2$ heterozygous and $^1/_2$ homozygous progeny.

(a) No crossing over

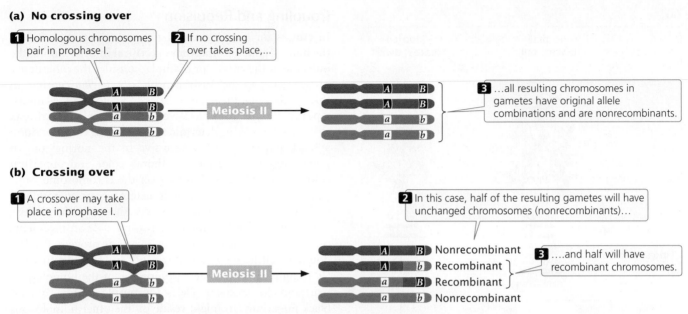

(b) Crossing over

7.5 A single crossover produces half nonrecombinant gametes and half recombinant gametes.

b. the frequency of recombination is always 50%.

c. each crossover takes place between only two of the four chromatids of a homologous pair.

d. crossovers occur in about 50% of meioses.

Application   Let's apply what we have learned about linkage and recombination to a cross between tomato plants that differ in the genes that encode leaf type and plant height. Assume now that these genes are linked and that some crossing over takes place between them. Suppose a geneticist carried out the testcross described earlier:

$$\frac{M \quad D}{m \quad d} \times \frac{m \quad d}{m \quad d}$$

When crossing over takes place in the genes for leaf type and height, two of the four gametes produced are recombinants. When there is no crossing over, all four resulting gametes are nonrecombinants. Thus, over all meioses, the majority of gametes will be nonrecombinants (**Figure 7.6a**). These gametes then unite with gametes produced by the homozygous recessive parent, which contain only the recessive alleles, resulting in mostly nonrecombinant progeny and a few recombinant progeny (**Figure 7.6b**). In this cross, we see that 55 of the testcross progeny have normal leaves and are tall and 53 have mottled leaves and are dwarf. These plants are the nonrecombinant progeny, containing the original combinations of traits that were present in the parents. Of the 123 progeny, 15 have new combinations of traits that were not seen in the parents: 8 are normal leaved

and dwarf, and 7 are mottle leaved and tall. These plants are the recombinant progeny.

The results of a cross such as the one illustrated in Figure 7.6 reveal several things. A testcross for two independently assorting genes is expected to produce a 1 : 1 : 1 : 1 phenotypic ratio in the progeny. The progeny of this cross clearly do not exhibit such a ratio; so we might suspect that the genes are not assorting independently. When linked genes undergo some crossing over, the result is mostly nonrecombinant progeny and fewer recombinant progeny. This result is what we observe among the progeny of the testcross illustrated in Figure 7.6; so we conclude that the two genes show evidence of linkage with some crossing over.

Calculating Recombination Frequency

The percentage of recombinant progeny produced in a cross is called the **recombination frequency**, which is calculated as follows:

$$\frac{\text{recombinant}}{\text{frequency}} = \frac{\text{number of recombinant progeny}}{\text{total number of progeny}} \times 100\%$$

In the testcross shown in Figure 7.6, 15 progeny exhibit new combinations of traits; so the recombination frequency is:

$$\frac{8+7}{55+53+8+7} \times 100\% = \frac{15}{123} \times 100\% = 12.2\%$$

Thus, 12.2% of the progeny exhibit new combinations of traits resulting from crossing over. The recombination frequency can also be expressed as a decimal fraction (0.122).

TRY PROBLEM 15 →

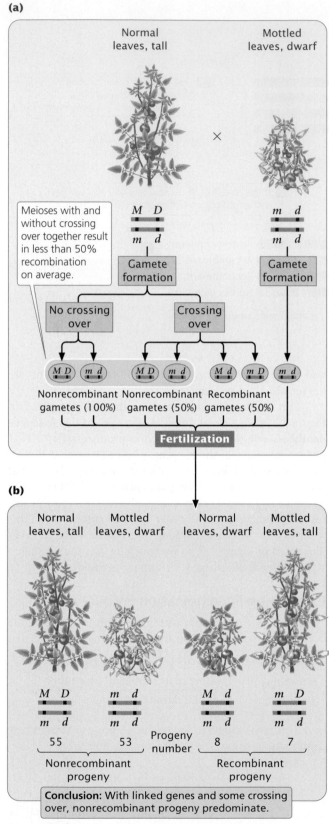

(a)

Normal leaves, tall × Mottled leaves, dwarf

Meioses with and without crossing over together result in less than 50% recombination on average.

M D / m d

m d / m d

Gamete formation

Gamete formation

No crossing over

Crossing over

M D | m d   M D | m d   M d | m D | m d

Nonrecombinant gametes (100%)

Nonrecombinant gametes (50%)

Recombinant gametes (50%)

Fertilization

(b)

| Normal leaves, tall | Mottled leaves, dwarf | Normal leaves, dwarf | Mottled leaves, tall |

M D / m d   m d / m d   M d / m d   m D / m d

55   53   Progeny number   8   7

Nonrecombinant progeny

Recombinant progeny

Conclusion: With linked genes and some crossing over, nonrecombinant progeny predominate.

7.6 Crossing over between linked genes produces nonrecombinant and recombinant offspring. In this testcross, genes are linked and there is some crossing over.

Coupling and Repulsion

In crosses for linked genes, the arrangement of alleles on the homologous chromosomes is critical in determining the outcome of the cross. For example, consider the inheritance of two genes in the Australian blowfly, *Lucilia cuprina*. In this species, one locus determines the color of the thorax: a purple thorax (p) is recessive to the normal green thorax (p^+). A second locus determines the color of the puparium: a black puparium (b) is recessive to the normal brown puparium (b^+). The loci for thorax color and puparium color are located close together on the chromosome. Suppose we test-cross a fly that is heterozygous at both loci with a fly that is homozygous recessive at both. Because these genes are linked, there are two possible arrangements on the chromosomes of the heterozygous progeny fly. The dominant alleles for green thorax (p^+) and brown puparium (b^+) might reside on one chromosome of the homologous pair, and the recessive alleles for purple thorax (p) and black puparium (b) might reside on the other homologous chromosome:

$$\frac{p^+ \qquad b^+}{p \qquad b}$$

This arrangement, in which wild-type alleles are found on one chromosome and mutant alleles are found on the other chromosome, is referred to as the **coupling**, or **cis**, **configuration**. Alternatively, one chromosome might bear the alleles for green thorax (p^+) and black puparium (b), and the other chromosome carries the alleles for purple thorax (p) and brown puparium (b^+):

$$\frac{p^+ \qquad b}{p \qquad b^+}$$

This arrangement, in which each chromosome contains one wild-type and one mutant allele, is called the **repulsion**, or **trans**, **configuration**. Whether the alleles in the heterozygous parent are in coupling or repulsion determines which phenotypes will be most common among the progeny of a testcross.

When the alleles are in the coupling configuration, the most numerous progeny types are those with a green thorax and brown puparium and those with a purple thorax and black puparium (**Figure 7.7a**); but, when the alleles of the heterozygous parent are in repulsion, the most numerous progeny types are those with a green thorax and black puparium and those with a purple thorax and brown puparium (**Figure 7.7b**). Notice that the genotypes of the parents in Figure 7.7a and b are the same ($p^+p\ b^+b \times pp\ bb$) and that the dramatic difference in the phenotypic ratios of the progeny in the two crosses results entirely from the configuration—coupling or repulsion—of the chromosomes. Knowledge of the arrangement of the alleles on the chromosomes is essential to accurately predict the outcome of crosses in which genes are linked.

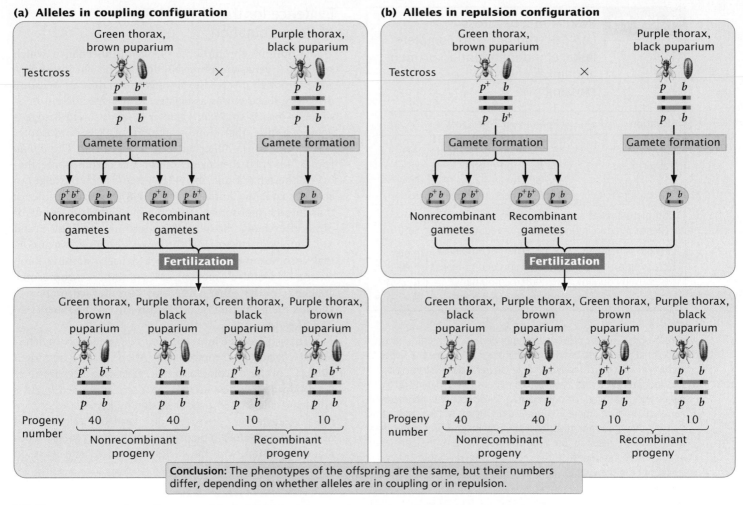

(a) Alleles in coupling configuration

(b) Alleles in repulsion configuration

Conclusion: The phenotypes of the offspring are the same, but their numbers differ, depending on whether alleles are in coupling or in repulsion.

7.7 The arrangement (coupling or repulsion) of linked genes on a chromosome affects the results of a testcross.
Linked loci in the Australian blowfly, *Luciliá cuprina,* determine the color of the thorax and that of the puparium.

CONCEPTS

In a cross, the arrangement of linked alleles on the chromosomes is critical for determining the outcome. When two wild-type alleles are on one homologous chromosome and two mutant alleles are on the other, they are in the coupling configuration; when each chromosome contains one wild-type allele and one mutant allele, the alleles are in repulsion.

✔ CONCEPT CHECK 2

The following testcross produces the progeny shown: *Aa Bb* × *aa bb* → 10 *Aa Bb*, 40 *Aa bb*, 40 *aa Bb*, 10 *aa bb*. Were the genes in the *Aa Bb* parent in coupling or in repulsion?

CONNECTING CONCEPTS

Relating Independent Assortment, Linkage, and Crossing Over

We have now considered three situations concerning genes at different loci. First, the genes may be located on different chromo-

somes; in this case, they exhibit independent assortment and combine randomly when gametes are formed. An individual heterozygous at two loci (*Aa Bb*) produces four types of gametes (*A B, a b, A b,* and *a B*) in equal proportions: two types of nonrecombinants and two types of recombinants, In a testcross, these gametes will result in four types of progeny in equal proportions (**Table 7.1**).

Second, the genes may be completely linked—meaning that they are on the same chromosome and lie so close together that crossing over between them is rare. In this case, the genes do not recombine. An individual heterozygous for two closely linked genes in the coupling configuration

$$\frac{A \qquad B}{a \qquad b}$$

produces only the nonrecombinant gametes containing alleles *A B* or *a b*; the alleles do not assort into new combinations such as *A b* or *a B*. In a testcross, completely linked genes will produce only two types of progeny, both nonrecombinants, in equal proportions (see Table 7.1).

The third situation, incomplete linkage, is intermediate between the two extremes of independent assortment and complete linkage. Here, the genes are physically linked on the same

Table 7.1	Results of a testcross ($Aa\ Bb \times aa\ bb$) with complete linkage, independent assortment, and linkage with some crossing over		
Situation	**Progeny of Test Cross**		
Independent assortment	$Aa\ Bb$ (nonrecombinant)	25%	
	$aa\ bb$ (nonrecombinant)	25%	
	$Aa\ bb$ (recombinant)	25%	
	$aa\ Bb$ (recombinant)	25%	
Complete linkage (genes in coupling)	$Aa\ Bb$ (nonrecombinant)	50%	
	$aa\ bb$ (nonrecombinant)	50%	
Linkage with some crossing over (genes in coupling)	$Aa\ Bb$ (nonrecombinant)	} more than 50%	
	$aa\ bb$ (nonrecombinant)		
	$Aa\ bb$ (recombinant)	} less than 50%	
	$aa\ Bb$ (recombinant)		

chromosome, which prevents independent assortment. However, occasional crossovers break up the linkage and allow the genes to recombine. With incomplete linkage, an individual heterozygous at two loci produces four types of gametes—two types of recombinants and two types of nonrecombinants—but the nonrecombinants are produced more frequently than the recombinants because crossing over does not take place in every meiosis. In the testcross, these gametes result in four types of progeny, with the nonrecombinants more frequent than the recombinant (see Table 7.1).

Earlier in the chapter, the term recombination was defined as the sorting of alleles into new combinations. We've now considered two types of recombination that differ in the mechanism that generates these new combinations of alleles. **Interchromosomal recombination** takes place between genes located on *different* chromosomes. It arises from independent assortment—the random segregation of chromosomes in anaphase I of meiosis—and is the kind of recombination that Mendel discovered while studying dihybrid crosses. A second type of recombination, **intrachromosomal recombination**, takes place between genes located on the *same* chromosome. This recombination arises from crossing over—the exchange of genetic material in prophase I of meiosis. Both types of recombination produce new allele combinations in the gametes; so they cannot be distinguished by examining the types of gametes produced. Nevertheless, they can often be distinguished by the *frequencies* of types of gametes: interchromosomal recombination produces 50% nonrecombinant gametes and 50% recombinant gametes, whereas intrachromosomal recombination frequently produces less than 50% recombinant gametes. However, when the genes are very far apart on the same chromosome, they assort independently, as if they were on different chromosomes. In this case, intrachromosomal recombination also produces 50% recombinant gametes. Intrachromosomal recombination of genes that lie far apart on the same chromosome and interchromosomal recombination are genetically indistinguishable.

Evidence for the Physical Basis of Recombination

Walter Sutton's chromosome theory of inheritance, which stated that genes are physically located on chromosomes, was supported by Nettie Stevens and Edmund Wilson's discovery that sex was associated with a specific chromosome in insects (p. 75 in Chapter 4) and Calvin Bridges's demonstration that nondisjunction of X chromosomes was related to the inheritance of eye color in *Drosophila* (pp. 82–83). Further evidence for the chromosome theory of heredity came in 1931, when Harriet Creighton and Barbara McClintock (**Figure 7.8**) obtained evidence that intrachromosomal recombination was the result of physical exchange between chromosomes. Creighton and McClintock discovered a strain of corn that had an abnormal chromosome 9, containing a densely staining knob at one end and a small piece of another chromosome attached to the other end. This aberrant chromosome allowed them to visually distinguish the two members of a homologous pair.

They studied the inheritance of two traits in corn determined by genes on chromosome 9. At one locus, a dominant allele (C) produced colored kernels, whereas a recessive allele (c) produced colorless kernels. At a second, linked locus, a dominant allele (Wx) produced starchy kernels, whereas a recessive allele (wx) produced waxy kernels. Creighton and McClintock obtained a plant that was heterozygous at both loci in repulsion, with the alleles for colored and waxy on

7.8 Barbara McClintock (left) and Harriet Creighton (right) provided evidence that genes are located on chromosomes.
[Karl Maramorosch/Courtesy of Cold Spring Harbor Laboratory Archives.]

the aberrant chromosome and the alleles for colorless and starchy on the normal chromosome:

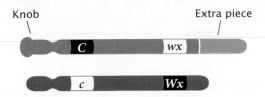

They crossed this heterozygous plant with a plant that was homozygous for colorless and heterozygous for waxy (with both chromosomes normal):

$$\frac{C \quad wx}{c \quad Wx} \times \frac{c \quad Wx}{c \quad wx}$$

This cross will produce different combinations of traits in the progeny, but the only way that colorless and waxy progeny can arise is through crossing over in the doubly heterozygous parent:

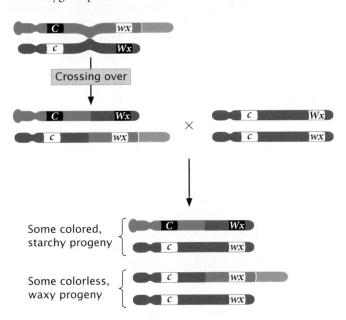

Note: Not all progeny genotypes are shown.

Notice that, if crossing over entails physical exchange between the chromosomes, then the colorless, waxy progeny resulting from recombination should have a chromosome with an extra piece but not a knob. Furthermore, some of the colored, starchy progeny should possess a knob but not the extra piece. This outcome is precisely what Creighton and McClintock observed, confirming the chromosomal theory of inheritance. Curt Stern provided a similar demonstration by using chromosomal markers in *Drosophila* at about the same time. We will examine the molecular basis of recombination in more detail in Chapter 12.

Predicting the Outcomes of Crosses with Linked Genes

Knowing the arrangement of alleles on a chromosome allows us to predict the types of progeny that will result from a cross entailing linked genes and to determine which of these types will be the most numerous. Determining the *proportions* of the types of offspring requires an additional piece of information—the recombination frequency. The recombination frequency provides us with information about how often the alleles in the gametes appear in new combinations and therefore allows us to predict the proportions of offspring phenotypes that will result from a specific cross with linked genes.

In cucumbers, smooth fruit (*t*) is recessive to warty fruit (*T*) and glossy fruit (*d*) is recessive to dull fruit (*D*). Geneticists have determined that these two genes exhibit a recombination frequency of 16%. Suppose we cross a plant homozygous for warty and dull fruit with a plant homozygous for smooth and glossy fruit and then carry out a testcross by using the F_1:

$$\frac{T \quad D}{t \quad d} \times \frac{t \quad d}{t \quad d}$$

What types and proportions of progeny will result from this testcross?

Four types of gametes will be produced by the heterozygous parent, as shown in **Figure 7.9**: two types of nonrecombinant gametes ($T \quad D$ and $t \quad d$) and two types of recombinant gametes ($T \quad d$ and $t \quad D$). The recombination frequency tells us that 16% of the gametes produced by the heterozygous parent will be recombinants. Because there are two types of recombinant gametes, each should arise with a frequency of 16%/2 = 8%. This frequency can also be represented as a probability of 0.08. All the other gametes will be nonrecombinants; so they should arise with a frequency of 100% − 16% = 84%. Because there are two types of nonrecombinant gametes, each should arise with a frequency of 84%/2 = 42% (or 0.42). The other parent in the testcross is homozygous and therefore produces only a single type of gamete ($t \quad d$) with a frequency of 100% (or 1.00).

Four types of progeny result from the testcross (see Figure 7.9). The expected proportion of each type can be determined by using the multiplication rule, multiplying together the probability of each gamete. Testcross progeny with warty and dull fruit

$$\frac{T \quad D}{t \quad d}$$

appear with a frequency of 0.42 (the probability of inheriting a gamete with chromosome $T \quad D$ from the heterozygous parent) × 1.00 (the probability of inheriting a gamete

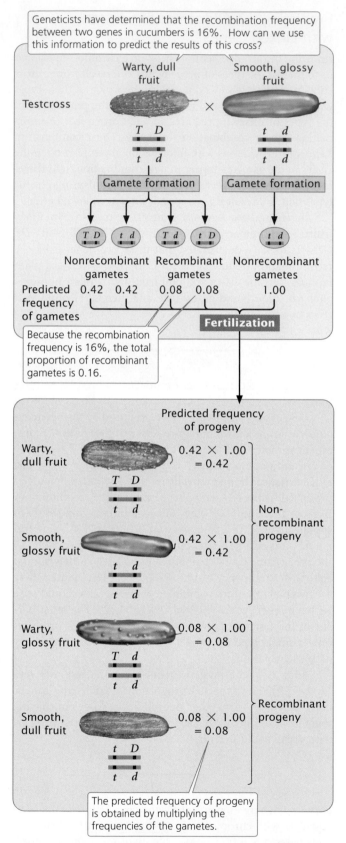

Geneticists have determined that the recombination frequency between two genes in cucumbers is 16%. How can we use this information to predict the results of this cross?

Testcross

Warty, dull fruit   ×   Smooth, glossy fruit

T D
t d

t d
t d

Gamete formation   Gamete formation

T D   t d   T d   t D   t d

Nonrecombinant gametes   Recombinant gametes   Nonrecombinant gametes

Predicted frequency of gametes   0.42   0.42   0.08   0.08   1.00

Because the recombination frequency is 16%, the total proportion of recombinant gametes is 0.16.

Fertilization

Predicted frequency of progeny

Warty, dull fruit   0.42 × 1.00 = 0.42
T D
t d

Smooth, glossy fruit   0.42 × 1.00 = 0.42
t d
t d

Non-recombinant progeny

Warty, glossy fruit   0.08 × 1.00 = 0.08
T d
t d

Smooth, dull fruit   0.08 × 1.00 = 0.08
t D
t d

Recombinant progeny

The predicted frequency of progeny is obtained by multiplying the frequencies of the gametes.

7.9 The recombination frequency allows a prediction of the proportions of offspring expected for a cross entailing linked genes.

with chromosome <u> t   d </u> from the recessive parent) = 0.42. The proportions of the other types of F_2 progeny can be calculated in a similar manner (see Figure 7.9). This method can be used for predicting the outcome of any cross with linked genes for which the recombination frequency is known. **TRY PROBLEM 17** →

Testing for Independent Assortment

In some crosses, the genes are obviously linked because there are clearly more nonrecombinant progeny than recombinant progeny. In other crosses, the difference between independent assortment and linkage is not so obvious. For example, suppose we did a testcross for two pairs of genes, such as $Aa\ Bb \times aa\ bb$, and observed the following numbers of progeny: 54 $Aa\ Bb$, 56 $aa\ bb$, 42 $Aa\ bb$, and 48 $aa\ Bb$. Is this outcome the $1 : 1 : 1 : 1$ ratio we would expect if A and B assorted independently? Not exactly, but it's pretty close. Perhaps these genes assorted independently and chance produced the slight deviations between the observed numbers and the expected $1 : 1 : 1 : 1$ ratio. Alternatively, the genes might be linked, with considerable crossing over taking place between them, and so the number of nonrecombinants is only slightly greater than the number of recombinants. How do we distinguish between the role of chance and the role of linkage in producing deviations from the results expected with independent assortment?

We encountered a similar problem in crosses in which genes were unlinked—the problem of distinguishing between deviations due to chance and those due to other factors. We addressed this problem (in Chapter 3) with the goodness-of-fit chi-square test, which helps us evaluate the likelihood that chance alone is responsible for deviations between the numbers of progeny that we observed and the numbers that we expected by applying the principles of inheritance. Here, we are interested in a different question: Is the inheritance of alleles at one locus independent of the inheritance of alleles at a second locus? If the answer to this question is yes, then the genes are assorting independently; if the answer is no, then the genes are probably linked.

A possible way to test for independent assortment is to calculate the expected probability of each progeny type, assuming independent assortment, and then use the goodness-of-fit chi-square test to evaluate whether the observed numbers deviate significantly from the expected numbers. With independent assortment, we expect $^1/_4$ of each phenotype: $^1/_4\ Aa\ Bb$, $^1/_4\ aa\ bb$, $^1/_4\ Aa\ bb$, and $^1/_4\ aa\ Bb$. This expected probability of each genotype is based on the multiplication rule of probability, which we considered in Chapter 3. For example, if the probability of Aa is $^1/_2$ and the probability of Bb is $^1/_2$, then the probability of $Aa\ Bb$ is $^1/_2 \times ^1/_2 = ^1/_4$. In this calculation, we are making two assumptions: (1) the probability of each single-locus genotype is $^1/_2$, and (2) genotypes at the two loci are inherited independently $(^1/_2 \times ^1/_2 = ^1/_4)$.

One problem with this approach is that a significant chi-square value can result from a violation of either assumption. If the genes are linked, then the inheritance of genotypes at the two loci are not independent (assumption 2), and we will get a significant deviation between observed and expected numbers. But we can also get a significant deviation if the probability of each single-locus genotype is not $\frac{1}{2}$ (assumption 1), even when the genotypes are assorting independently. We may obtain a significant deviation, for example, if individuals with one genotype have a lower probability of surviving or the penetrance of a genotype is not 100%. We could test both assumptions by conducting a series of chi-square tests, first testing the inheritance of genotypes at each locus separately (assumption 1) and then testing for independent assortment (assumption 2). However, a faster method is to test for independence in genotypes with a *chi-square test of independence*.

The chi-square test of independence The chi-square test of independence allows us to evaluate whether the segregation of alleles at one locus is independent of the segregation of alleles at another locus, without making any assumption about the probability of single-locus genotypes. To illustrate this analysis, we will examine results from a cross between German cockroaches, in which yellow body (y) is recessive to brown body (y^+) and curved wings (cv) are recessive to straight wings (cv^+). A testcross ($y^+y\ cv^+cv \times yy\ cvcv$) produced the progeny shown in **Figure 7.10a**.

To carry out the chi-square test of independence, we first construct a table of the observed numbers of progeny, somewhat like a Punnett square, except, here, we put the genotypes that result from the segregation of alleles at one locus along the top and the genotypes that result from the segregation of alleles at the other locus along the side (**Figure 7.10b**). Next, we compute the total for each row, the total for each column, and the grand total (the sum of all row totals or the sum of all column totals, which should be the same). These totals will be used to compute the expected values for the chi-square test of independence.

Now, we compute the values expected if the segregation of alleles at the y locus is independent of the segregation of alleles at the cv locus. If the segregation of alleles at each locus is independent, then the proportion of progeny with y^+y and yy genotypes should be the same for cockroaches with genotype cv^+cv and for cockroaches with genotype $cvcv$. The converse is also true; the proportions of progeny with cv^+cv and

7.10 A chi-square test of independence can be used to determine if genes at two loci are assorting independently.

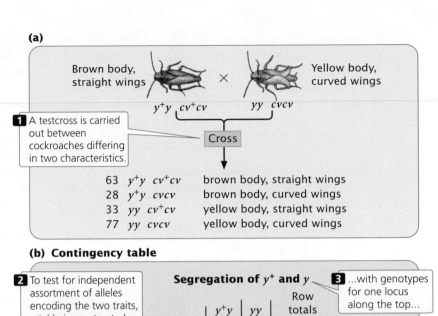

(a)

Brown body, straight wings × Yellow body, curved wings

$y^+y\ cv^+cv$ $yy\ cvcv$

1 A testcross is carried out between cockroaches differing in two characteristics.

Cross

63	$y^+y\ cv^+cv$	brown body, straight wings
28	$y^+y\ cvcv$	brown body, curved wings
33	$yy\ cv^+cv$	yellow body, straight wings
77	$yy\ cvcv$	yellow body, curved wings

(b) Contingency table

2 To test for independent assortment of alleles encoding the two traits, a table is constructed...

3 ...with genotypes for one locus along the top...

4 ...and genotypes for the other locus along the left side.

5 Numbers of each genotype are placed in the table cells, and the row totals, column totals, and grand total are computed.

Segregation of y^+ and y

Segregation of cv^+ and cv		y^+y	yy	Row totals
	cv^+cv	63	33	96
	$cv\ cv$	28	77	105
	Column totals	91	110	201
				Grand total

(c)

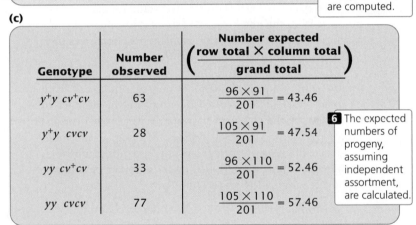

Genotype	Number observed	Number expected $\left(\dfrac{\text{row total} \times \text{column total}}{\text{grand total}}\right)$
$y^+y\ cv^+cv$	63	$\dfrac{96 \times 91}{201} = 43.46$
$y^+y\ cvcv$	28	$\dfrac{105 \times 91}{201} = 47.54$
$yy\ cv^+cv$	33	$\dfrac{96 \times 110}{201} = 52.46$
$yy\ cvcv$	77	$\dfrac{105 \times 110}{201} = 57.46$

6 The expected numbers of progeny, assuming independent assortment, are calculated.

(d)

7 A chi-square value is calculated.

$$\chi^2 = \Sigma\ \frac{(\text{observed} - \text{expected})^2}{\text{expected}}$$

$$= \frac{(63 - 43.46)^2}{43.46} + \frac{(28 - 47.54)^2}{47.54} + \frac{(33 - 52.54)^2}{52.54} + \frac{(77 - 57.46)^2}{57.46}$$

$$= 8.79 + 8.03 + 7.27 + 6.64$$

$$= 30.73$$

(e)

8 The probability is less than 0.005, indicating that the *difference* between numbers of observed and expected progeny is probably not due to chance.

$df = (\text{number of rows} - 1) \times (\text{number of columns} - 1)$

$df = (2 - 1) \times (2 - 1) = 1 \times 1 = 1$

$P < 0.005$

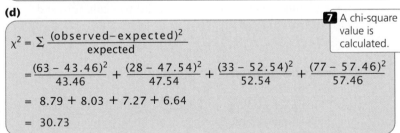

Conclusion: The genes for body color and type of wing are not assorting independently and must be linked.

cvcv genotypes should be the same for cockroaches with genotype y^+y and for cockroaches with genotype *yy*. With the assumption that the alleles at the two loci segregate independently, the expected number for each cell of the table can be computed by using the following formula:

$$\text{expected number} = \frac{\text{row total} \times \text{column total}}{\text{grand total}}$$

For the cell of the table corresponding to genotype $y^+y\ cv^+cv$ (the upper-left-hand cell of the table in Figure 7.10b) the expected number is:

$$\frac{96\ (\text{row total}) \times 91\ (\text{column total})}{201\ (\text{grand total})} = \frac{8736}{201} = 43.46$$

With the use of this method, the expected numbers for each cell are given in **Figure 7.10c**.

We now calculate a chi-square value by using the same formula that we used for the goodness-of-fit chi-square test in Chapter 3:

$$x^2 = \sum \frac{(\text{observed} - \text{expected})^2}{\text{expected}}$$

Recall that $\sum$ means "sum," and that we are adding together the $(\text{observed} - \text{expected})^2$/expected value for each type of progeny. With the observed and expected numbers of cockroaches from the testcross, the calculated chi-square value is 30.73 (**Figure 7.10d**).

To determine the probability associated with this chi-square value, we need the degrees of freedom. Recall from Chapter 3 that the degrees of freedom are the number of ways in which the observed classes are free to vary from the expected values. In general, for the chi-square test of independence, the degrees of freedom equal the number of rows in the table minus 1 multiplied by the number of columns in the table minus 1 (**Figure 7.10e**), or

$$df = (\text{number of rows} - 1) \times (\text{number of columns} - 1)$$

In our example, there were two rows and two columns, and so the degrees of freedom are:

$$df = (2 - 1) \times (2 - 1) = 1 \times 1 = 1$$

Therefore, our calculated chi-square value is 30.73, with 1 degree of freedom. We can use Table 3.5 to find the associated probability. Looking at Table 3.5, we find that our calculated chi-square value is larger than the largest chi-square value given for 1 degree of freedom, which has a probability of 0.005. Thus, our calculated chi-square value has a probability less than 0.005. This very small probability indicates that the genotypes are not in the proportions that we would expect if independent assortment were taking place. Our conclusion, then, is that these genes are not assorting independently and must be linked. As is the case for the goodness-of-fit chi-square test, geneticists generally consider that any chi-square value for the test of independence with a probability less than 0.05 is significantly different from the expected values and is therefore evidence that the genes are not assorting independently. **TRY PROBLEM 16** →

Gene Mapping with Recombination Frequencies

Thomas Hunt Morgan and his students developed the idea that physical distances between genes on a chromosome are related to the rates of recombination. They hypothesized that crossover events occur more or less at random up and down the chromosome and that two genes that lie far apart are more likely to undergo a crossover than are two genes that lie close together. They proposed that recombination frequencies could provide a convenient way to determine the order of genes along a chromosome and would give estimates of the relative distances between the genes. Chromosome maps calculated by using the genetic phenomenon of recombination are called **genetic maps**. In contrast, chromosome maps calculated by using physical distances along the chromosome (often expressed as numbers of base pairs) are called **physical maps**.

Distances on genetic maps are measured in **map units** (abbreviated m.u.); one map unit equals 1% recombination. Map units are also called **centiMorgans** (cM), in honor of Thomas Hunt Morgan; 100 centiMorgans equals 1 **Morgan**. Genetic distances measured with recombination rates are approximately additive: if the distance from gene *A* to gene *B* is 5 m.u., the distance from gene *B* to gene *C* is 10 m.u., and the distance from gene *A* to gene *C* is 15 m.u., then gene *B* must be located between genes *A* and *C*. On the basis of the map distances just given, we can draw a simple genetic map for genes *A*, *B*, and *C*, as shown here:

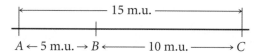

We could just as plausibly draw this map with *C* on the left and *A* on the right:

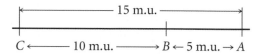

Both maps are correct and equivalent because, with information about the relative positions of only three genes, the most that we can determine is which gene lies in the middle. If we obtained distances to an additional gene, then we could position *A* and *C* relative to that gene. An additional gene *D*, examined through genetic crosses, might yield the following recombination frequencies:

Gene pair	Recombination frequency (%)
A and *D*	8
B and *D*	13
C and *D*	23

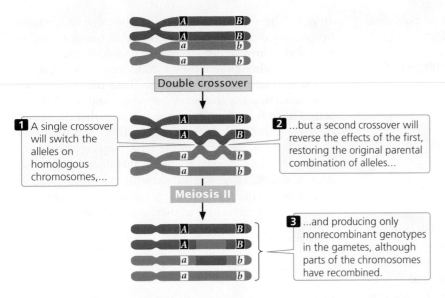

1 A single crossover will switch the alleles on homologous chromosomes,...

2 ...but a second crossover will reverse the effects of the first, restoring the original parental combination of alleles...

3 ...and producing only nonrecombinant genotypes in the gametes, although parts of the chromosomes have recombined.

7.11 A two-strand double crossover between two linked genes produces only nonrecombinant gametes.

Notice that C and D exhibit the greatest amount of recombination; therefore, C and D must be farthest apart, with genes A and B between them. Using the recombination frequencies and remembering that 1 m.u. = 1% recombination, we can now add D to our map:

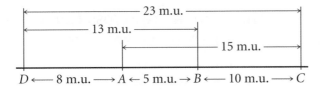

By doing a series of crosses between pairs of genes, we can construct genetic maps showing the linkage arrangements of a number of genes.

Two points should be emphasized about constructing chromosome maps from recombination frequencies. First, recall that we cannot distinguish between genes on different chromosomes and genes located far apart on the same chromosome. If genes exhibit 50% recombination, the most that can be said about them is that they belong to different groups of linked genes (different linkage groups), either on different chromosomes or far apart on the same chromosome.

The second point is that a testcross for two genes that are far apart on the same chromosome tends to underestimate the true physical distance, because the cross does not reveal double crossovers that might take place between the two genes (**Figure 7.11**). A double crossover arises when two separate crossover events take place between two loci. (For now, we will consider only double crossovers that occur between two of the four chromatids of a homologus pair—a two-strand double crossover. Double crossovers that take place among three and four chromatids will be considered later.) Whereas a single crossover produces combinations of alleles that were not present on the original parental chromosomes, a second crossover

between the same two genes reverses the effects of the first, thus restoring the original parental combination of alleles (see Figure 7.11). We therefore cannot distinguish between the progeny produced by two-strand double crossovers and the progeny produced when there is no crossing over at all. As we shall see in the next section, we can detect double crossovers if we examine a third gene that lies between the two crossovers. Because double crossovers between two genes go undetected, map distances will be underestimated whenever double crossovers take place. Double crossovers are more frequent between genes that are far apart; therefore genetic maps based on short distances are usually more accurate than those based on longer distances.

CONCEPTS

A genetic map provides the order of the genes on a chromosome and the approximate distances from one gene to another based on recombination frequencies. In genetic maps, 1% recombination equals 1 map unit, or 1 centiMorgan. Double crossovers between two genes go undetected; so map distances between distant genes tend to underestimate genetic distances.

✔ CONCEPT CHECK 3

How does a genetic map differ from a physical map?

Constructing a Genetic Map with the Use of Two-Point Testcrosses

Genetic maps can be constructed by conducting a series of testcrosses. In each testcross, one of the parents is heterozygous for a different pair of genes, and recombination frequencies are calculated between pairs of genes. A testcross between two genes is called a **two-point testcross** or a two-point cross for short. Suppose that we carried out a series of two-point crosses for four genes, *a*, *b*, *c*, and *d*, and obtained the following recombination frequencies:

Gene loci in testcross	Recombination frequency (%)
a and *b*	50
a and *c*	50
a and *d*	50
b and *c*	20
b and *d*	10
c and *d*	28

We can begin constructing a genetic map for these genes by considering the recombination frequencies for each pair of genes. The recombination frequency between *a* and *b* is 50%, which is the recombination frequency expected with independent assortment. Therefore, genes *a* and *b* may either be on different chromosomes or be very far apart on the same chromosome; so we will place them in different linkage groups with the understanding that they may or may not be on the same chromosome:

Linkage group 1

a

Linkage group 2

b

The recombination frequency between *a* and *c* is 50%, indicating that they, too, are in different linkage groups. The recombination frequency between *b* and *c* is 20%; so these genes are linked and separated by 20 map units:

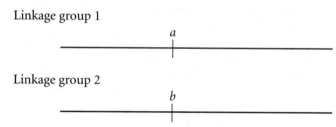

Linkage group 1

a

Linkage group 2

b   c
|←——— 20 m.u. ———→|

The recombination frequency between *a* and *d* is 50%, indicating that these genes belong to different linkage groups, whereas genes *b* and *d* are linked, with a recombination frequency of 10%. To decide whether gene *d* is 10 m.u. to the left or to the right of gene *b*, we must consult the *c*-to-*d* distance. If gene *d* is 10 m.u. to the left of gene *b*, then the distance between *d* and *c* should be approximately the sum of the distance between b and c and between c and d: 20 m.u. + 10 m.u. = 30 m.u. If, on the other hand, gene *d* lies to the right of gene *b*, then the distance between gene *d* and gene *c* will be much shorter, approximately 20 m.u. − 10 m.u. =

10 m.u. The summed distances will be only approximate because any double crossovers between the two genes will be missed and the map distance will be underestimated.

By examining the recombination frequency between *c* and *d*, we can distinguish between these two possibilities. The recombination frequency between *c* and *d* is 28%; so gene *d* must lie to the left of gene *b*. Notice that the sum of the recombination between *d* and *b* (10%) and between *b* and *c* (20%) is greater than the recombination between *d* and *c* (28%). As already discussed, this discrepancy arises because double crossovers between the two outer genes go undetected, causing an underestimation of the true map distance. The genetic map of these genes is now complete:

Linkage group 1

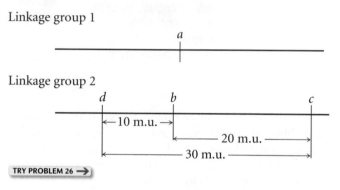

a

Linkage group 2

d   b   c
|←—10 m.u.—→|
|←——————— 20 m.u. ———————→|
|←——————————— 30 m.u. ———————————→|

TRY PROBLEM 26 →

7.3 A Three-Point Testcross Can Be Used to Map Three Linked Genes

Genetic maps can be constructed from a series of testcrosses for pairs of genes, but this approach is not particularly efficient, because numerous two-point crosses must be carried out to establish the order of the genes and because double crossovers are missed. A more efficient mapping technique is a testcross for three genes—a **three-point testcross**, or three-point cross. With a three-point cross, the order of the three genes can be established in a single set of progeny and some double crossovers can usually be detected, providing more-accurate map distances.

Consider what happens when crossing over takes place among three hypothetical linked genes. **Figure 7.12** illustrates a pair of homologous chromosomes of an individual that is heterozygous at three loci (*Aa Bb Cc*). Notice that the genes are in the coupling configuration; that is, all the dominant alleles are on one chromosome (*A   B   C*) and all the recessive alleles are on the other chromosome (*a   b   c*). Three types of crossover events can take place between these three genes: two types of single crossovers (see Figure 7.12a and b) and a double crossover (see Figure 7.12c). In each type of crossover, two of the resulting chromosomes are recombinants and two are non-recombinants.

Notice that, in the recombinant chromosomes resulting from the double crossover, the outer two alleles are the

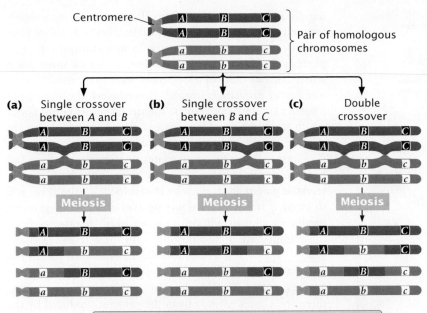

Conclusion: Recombinant chromosomes resulting from the double crossover have only the middle gene altered.

7.12 Three types of crossovers can take place among three linked loci.

same as in the nonrecombinants, but the middle allele is different. This result provides us with an important clue about the order of the genes. In progeny that result from a double crossover, only the middle allele should differ from the alleles present in the nonrecombinant progeny.

Constructing a Genetic Map with the Three-Point Testcross

To examine gene mapping with a three-point testcross, we will consider three recessive mutations in the fruit fly *Drosophila melanogaster*. In this species, scarlet eyes (st) are recessive to wild-type red eyes (st^+), ebony body color (e) is recessive to wild-type gray body color (e^+), and spineless (ss)—that is, the presence of small bristles—is recessive to wild-type normal bristles (ss^+). The loci encoding these three characteristics are linked and located on chromosome 3.

We will refer to these three loci as st, e, and ss, but keep in mind that either the recessive alleles (st, e, and ss) or the dominant alleles (st^+, e^+, and ss^+) may be present at each locus. So, when we say that there are 10 m.u. between st and ss, we mean that there are 10 m.u. between the loci at which mutations st and ss occur; we could just as easily say that there are 10 m.u. between st^+ and ss^+.

To map these genes, we need to determine their order on the chromosome and the genetic distances between them. First, we must set up a three-point testcross, a cross between a fly heterozygous at all three loci and a fly homozygous for recessive alleles at all three loci. To produce flies heterozygous for all three loci, we might cross a stock of flies that are homozygous for wild-type alleles at all three

loci with flies that are homozygous for recessive alleles at all three loci:

$$P \qquad \frac{st^+ \quad e^+ \quad ss^+}{st^+ \quad e^+ \quad ss^+} \times \frac{st \quad e \quad ss}{st \quad e \quad ss}$$

$$F_1 \qquad \frac{st^+ \quad e^+ \quad ss^+}{st \quad e \quad ss}$$

The order of the genes has been arbitrarily assigned because, at this point, we do not know which is the middle gene. Additionally, the alleles in these heterozygotes are in coupling configuration (because all the wild-type dominant alleles were inherited from one parent and all the recessive mutations from the other parent), although the testcross can also be done with alleles in repulsion.

In the three-point testcross, we cross the F_1 heterozygotes with flies that are homozygous for all three recessive mutations. In many organisms, it makes no difference whether the heterozygous parent in the testcross is male or female (provided that the genes are autosomal) but, in *Drosophila*, no crossing over takes place in males. Because crossing over in the heterozygous parent is essential for determining recombination frequencies, the heterozygous flies in our testcross must be female. So we mate female F_1 flies that are heterozygous for all three traits with male flies that are homozygous for all the recessive traits:

$$\frac{st^+ \quad e^+ \quad ss^+}{st \quad e \quad ss} \text{ Female} \times \frac{st \quad e \quad ss}{st \quad e \quad ss} \text{ Male}$$

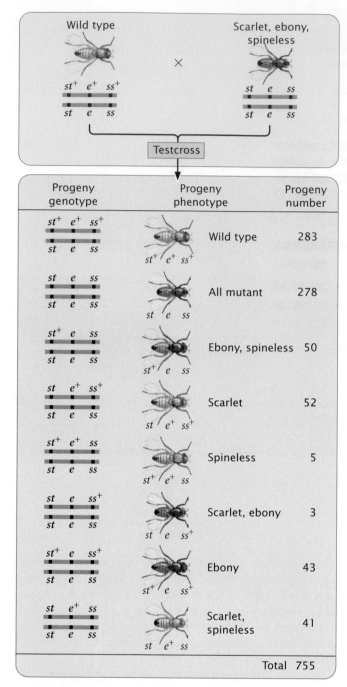

7.13 The results of a three-point testcross can be used to map linked genes. In this three-point testcross of *Drosophila melanogaster*, the recessive mutations scarlet eyes (*st*), ebony body color (*e*), and spineless bristles (*ss*) are at three linked loci. The order of the loci has been designated arbitrarily, as has the sex of the progeny flies.

 www.whfreeman.com/pierce4e

Animation 7.1 illustrates how to determine the order of the three linked genes.

The progeny produced from this cross are listed in **Figure 7.13**. For each locus, two classes of progeny are produced: progeny that are heterozygous, displaying the dominant trait, and progeny that are homozygous, display-

ing the recessive trait. With two classes of progeny possible for each of the three loci, there will be $2^3 = 8$ classes of phenotypes possible in the progeny. In this example, all eight phenotypic classes are present but, in some three-point crosses, one or more of the phenotypes may be missing if the number of progeny is limited. Nevertheless, the absence of a particular class can provide important information about which combination of traits is least frequent and, ultimately, about the order of the genes, as we will see.

To map the genes, we need information about where and how often crossing over has taken place. In the homozygous recessive parent, the two alleles at each locus are the same, and so crossing over will have no effect on the types of gametes produced; with or without crossing over, all gametes from this parent have a chromosome with three recessive alleles ($\underline{st \quad e \quad ss}$). In contrast, the heterozygous parent has different alleles on its two chromosomes, and so crossing over can be detected. The information that we need for mapping, therefore, comes entirely from the gametes produced by the heterozygous parent. Because chromosomes contributed by the homozygous parent carry only recessive alleles, whatever alleles are present on the chromosome contributed by the heterozygous parent will be expressed in the progeny.

As a shortcut, we often do not write out the complete genotypes of the testcross progeny, listing instead only the alleles expressed in the phenotype, which are the alleles inherited from the heterozygous parent. This convention is used in the discussion that follows.

CONCEPTS

To map genes, information about the location and number of crossovers in the gametes that produced the progeny of a cross is needed. An efficient way to obtain this information is to use a three-point testcross, in which an individual heterozygous at three linked loci is crossed with an individual that is homozygous recessive at the three loci.

✔ CONCEPT CHECK 4

Write the genotypes of all recombinant and nonrecombinant progeny expected from the following three-point cross:

$$\frac{m^+ \quad p^+ \quad s^+}{m \quad p \quad s} \times \frac{m \quad p \quad s}{m \quad p \quad s}$$

Determining the gene order The first task in mapping the genes is to determine their order on the chromosome. In Figure 7.13, we arbitrarily listed the loci in the order *st*, *e*, *ss*, but we had no way of knowing which of the three loci was between the other two. We can now identify the middle locus by examining the double-crossover progeny.

First, determine which progeny are the nonrecombinants; they will be the two most-numerous classes of progeny. (Even if crossing over takes place in every meiosis, the nonrecombinants will constitute at least 50% of the progeny.) Among the progeny of the testcross in Figure 7.13, the most numerous are those with all three dominant traits (st^+ e^+ ss^+) and those with all three recessive traits (st e ss).

Next, identify the double-crossover progeny. These progeny should always have the two least-numerous phenotypes, because the probability of a double crossover is always less than the probability of a single crossover. The least-common progeny among those listed in Figure 7.13 are progeny with spineless bristles (st^+ e^+ ss) and progeny with scarlet eyes and ebony body (st e ss^+); so they are the double-crossover progeny.

Three orders of genes on the chromosome are possible: the eye-color locus could be in the middle (e st ss), the body-color locus could be in the middle (st e ss), or the bristle locus could be in the middle (st ss e). To determine which gene is in the middle, we can draw the chromosomes of the heterozygous parent with all three possible gene orders and then see if a double crossover produces the combination of genes observed in the double-crossover progeny. The three possible gene orders and the types of progeny produced by their double crossovers are:

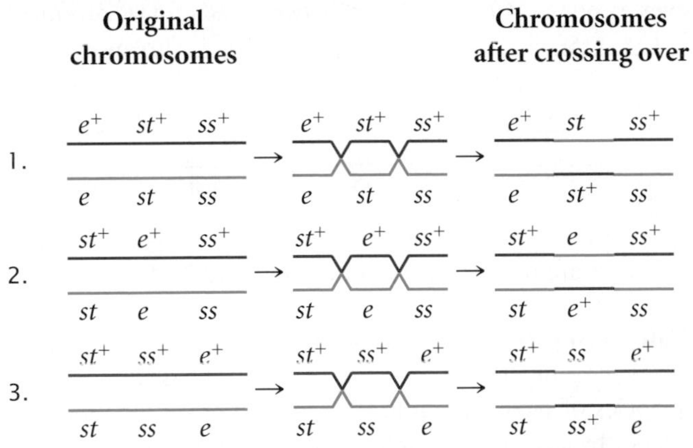

Original chromosomes		Chromosomes after crossing over

The only gene order that produces chromosomes with the set of alleles observed in the least-numerous progeny or double crossovers (st^+ e^+ ss and st e ss^+ in Figure 7.13) is the one in which the ss locus for bristles lies in the middle (gene-order 3). Therefore, this order (st ss e) must be the correct sequence of genes on the chromosome.

With a little practice, we can quickly determine which locus is in the middle without writing out all the gene orders. The phenotypes of the progeny are expressions of the alleles inherited from the heterozygous parent. Recall that, when we looked at the results of double crossovers (see Figure 7.12), only the alleles at the middle locus differed from the nonrecombinants. If we compare the nonrecombinant progeny with double-crossover progeny, they should differ only in alleles of the middle locus (**Table 7.2**).

Let's compare the alleles in the double-crossover progeny st^+ e^+ ss with those in the nonrecombinant progeny st^+ e^+ ss^+. We see that both have an allele

for red eyes (st^+) and both have an allele for gray body (e^+), but the nonrecombinants have an allele for normal bristles (ss^+), whereas the double crossovers have an allele for spineless bristles (ss). Because the bristle locus is the only one that differs, it must lie in the middle. We would obtain the same results if we compared the other class of double-crossover progeny (st e ss^+) with other nonrecombinant progeny (st e ss). Again, the only locus that differs is the one for bristles. Don't forget that the nonrecombinants and the double crossovers should differ at only one locus; if they differ at two loci, the wrong classes of progeny are being compared.

Table 7.2 Steps in determining gene order in a three-point cross

1. Identify the nonrecombinant progeny (two most-numerous phenotypes).

2. Indentify the double-crossover progeny (two least-numerous phenotypes).

3. Compare the phenotype of double-crossover progeny with the phenotype of nonrecombinant progeny. They should be alike in two characteristics and differ in one.

4. The characteristic that differs between the double crossover and the nonrecombinant progeny is encoded by the middle gene.

CONCEPTS

To determine the middle locus in a three-point cross, compare the double-crossover progeny with the nonrecombinant progeny. The double crossovers will be the two least-common classes of phenotypes; the nonrecombinants will be the two most-common classes of phenotypes. The double-crossover progeny should have the same alleles as the nonrecombinant types at two loci and different alleles at the locus in the middle.

✔ CONCEPT CHECK 5

A three-point test cross is carried out between three linked genes. The resulting nonrecombinant progeny are s^+ r^+ c^+ and s r c and the double-crossover progeny are s r c^+ and s^+ r^+ c. Which is the middle locus?

Determining the locations of crossovers When we know the correct order of the loci on the chromosome, we should rewrite the phenotypes of the testcross progeny in Figure 7.13 with the alleles in the correct order so that we can determine where crossovers have taken place (**Figure 7.14**).

Among the eight classes of progeny, we have already identified two classes as nonrecombinants (st^+ ss^+ e^+ and st ss e) and two classes as double crossovers (st^+ ss e^+ and st ss^+ e). The other

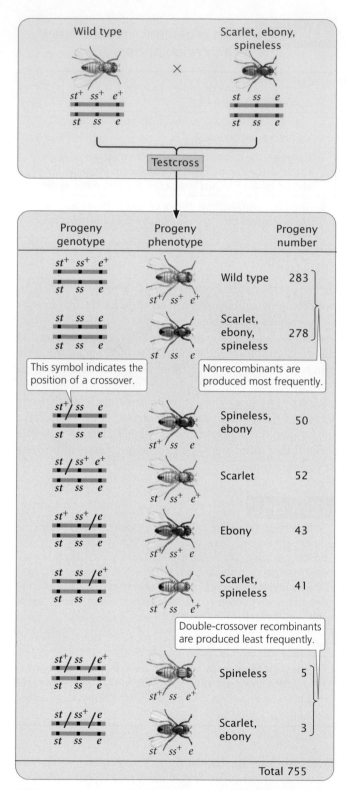

7.14 Writing the results of a three-point testcross with the loci in the correct order allows the locations of crossovers to be determined. These results are from the testcross illustrated in Figure 7.13, with the loci shown in the correct order. The location of a crossover is indicated by a slash (/). The sex of the progeny flies has been designated arbitrarily.

four classes include progeny that resulted from a chromosome that underwent a single crossover: two underwent single crossovers between *st* and *ss*, and two underwent single crossovers between *ss* and *e*.

To determine where the crossovers took place in these progeny, compare the alleles found in the single-crossover progeny with those found in the nonrecombinants, just as we did for the double crossovers. For example, consider progeny with chromosome $\underline{st^+ \quad ss \quad e}$. The first allele ($st^+$) came from the nonrecombinant chromosome $\underline{st^+ \quad ss^+ \quad e^+}$ and the other two alleles (ss and e) must have come from the other nonrecombinant chromosome $\underline{st \quad ss \quad e}$ through crossing over:

$$\frac{st^+ \quad ss^+ \quad e^+}{st \quad ss \quad e} \rightarrow \frac{st^+ \quad ss^+ \quad e^+}{st \quad ss \quad e} \rightarrow \frac{st^+ \quad ss \quad e}{st \quad ss^+ \quad e^+}$$

This same crossover also produces the $\underline{st \quad ss^+ \quad e^+}$ progeny.

This method can also be used to determine the location of crossing over in the other two types of single-crossover progeny. Crossing over between *ss* and *e* produces $\underline{st^+ \quad ss^+ \quad e}$ and $\underline{st \quad ss \quad e^+}$ chromosomes:

$$\frac{st^+ \quad ss^+ \quad e^+}{st \quad ss \quad e} \rightarrow \frac{st^+ \quad ss^+ \quad e^+}{st \quad ss \quad e} \rightarrow \frac{st^+ \quad ss^+ \quad e}{st \quad ss \quad e^+}$$

We now know the locations of all the crossovers; their locations are marked with a slash in Figure 7.14.

Calculating the recombination frequencies Next, we can determine the map distances, which are based on the frequencies of recombination. We calculate recombination frequency by adding up all of the recombinant progeny, dividing this number by the total number of progeny from the cross, and multiplying the number obtained by 100%. To determine the map distances accurately, we must include all crossovers (both single and double) that take place between two genes.

Recombinant progeny that possess a chromosome that underwent crossing over between the eye-color locus (*st*) and the bristle locus (*ss*) include the single crossovers ($\underline{st^+ \ / \ ss \quad e}$ and $\underline{st \ / \ ss^+ \quad e^+}$) and the two double crossovers ($\underline{st^+ \ / \ ss \ / \ e^+}$ and $\underline{st \ / \ ss^+ \ / \ e}$); see Figure 7.14. There are a total of 755 progeny; so the recombination frequency between *ss* and *st* is:

st–ss recombination frequency =

$$\frac{(50 + 52 + 5 + 3)}{755} \times 100\% = 14.6\%$$

The distance between the *st* and *ss* loci can be expressed as 14.6 m.u.

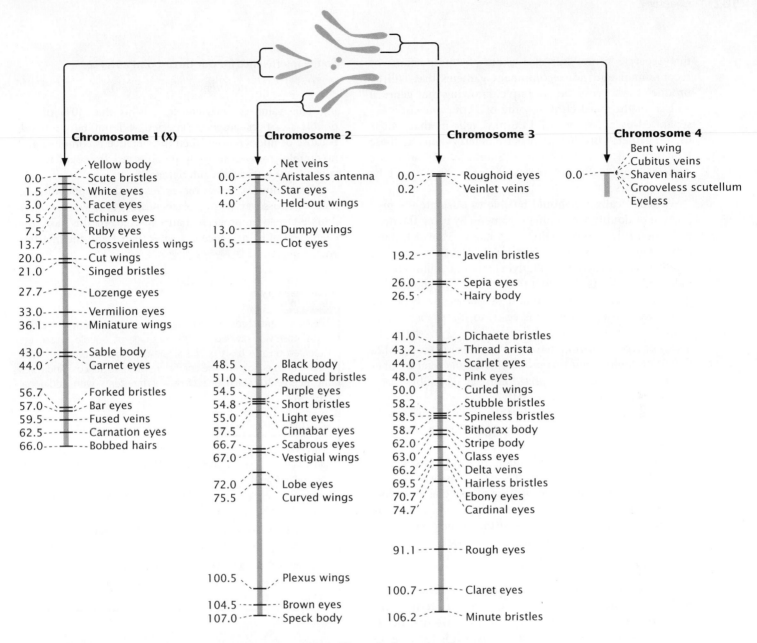

7.15 Drosophila melanogaster has four linkage groups corresponding to its four pairs of chromosomes. Distances between genes within a linkage group are in map units.

The map distance between the bristle locus (*ss*) and the body locus (*e*) is determined in the same manner. The recombinant progeny that possess a crossover between *ss* and *e* are the single crossovers $\underline{st^+ \quad ss^+ \: / \: e}$ and $\underline{st \quad ss \: / \: e^+}$ and the double crossovers $\underline{st \: / \: ss \: / \: e^+}$ and $\underline{st \: / \: ss^+ \: / \: e}$. The recombination frequency is:

ss−e recombination frequency =

$$\frac{(43 + 41 + 5 + 3)}{755} \times 100\% = 12.2\%$$

Thus, the map distance between *ss* and *e* is 12.2 m.u.

Finally, calculate the map distance between the outer two loci, *st* and *e*. This map distance can be obtained by summing the map distances between *st* and *ss* and between *ss* and *e* (14.6 m.u. + 12.2 m.u. = 26.8 m.u.). We can now use the map distances to draw a map of the three genes on the chromosome:

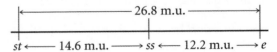

A genetic map of *D. melanogaster* is illustrated in **Figure 7.15**.

Interference and the coefficient of coincidence Map distances give us information not only about the distances that separate genes, but also about the proportions of recombinant and nonrecombinant gametes that will be produced in a cross. For example, knowing that genes *st* and *ss* on the third chromosome of *D. melanogaster* are separated by a distance of 14.6 m.u. tells us that 14.6% of the gametes produced by a fly heterozygous at these two loci will be recombinants. Similarly, 12.2% of the gametes from a fly heterozygous for *ss* and *e* will be recombinants.

Theoretically, we should be able to calculate the proportion of double-recombinant gametes by using the multiplication rule of probability (see Chapter 3), which states that the probability of two independent events occurring together is calculated by multiplying the probabilities of the independent events. Applying this principle, we should find that the proportion (probability) of gametes with double crossovers between *st* and *e* is equal to the probability of recombination between *st* and *ss* multiplied by the probability of recombination between *ss* and *e*, or $0.146 \times 0.122 = 0.0178$. Multiplying this probability by the total number of progeny gives us the *expected* number of double-crossover progeny from the cross: $0.0178 \times 755 = 13.4$. Only 8 double crossovers—considerably fewer than the 13 expected—were observed in the progeny of the cross (see Figure 7.14).

This phenomenon is common in eukaryotic organisms. The calculation assumes that each crossover event is independent and that the occurrence of one crossover does not influence the occurrence of another. But crossovers are frequently *not* independent events: the occurrence of one crossover tends to inhibit additional crossovers in the same region of the chromosome, and so double crossovers are less frequent than expected.

The degree to which one crossover interferes with additional crossovers in the same region is termed the **interference**. To calculate the interference, we first determine the **coefficient of coincidence**, which is the ratio of observed double crossovers to expected double crossovers:

coefficient of coincidence =

$$\frac{\text{number of observed double crossovers}}{\text{number of expected double crossovers}}$$

For the loci that we mapped on the third chromosome of *D. melanogaster* (see Figure 7.14), we find that the

coefficient of coincidence =

$$\frac{5+3}{0.146 \times 0.122 \times 755} = \frac{8}{13.4} = 0.6$$

which indicates that we are actually observing only 60% of the double crossovers that we expected on the basis of the single-crossover frequencies. The interference is calculated as

$$\text{interference} = 1 - \text{coefficient of coincidence}$$

So the interference for our three-point cross is:

$$\text{interference} = 1 - 0.6 = 0.4$$

This value of interference tells us that 40% of the double-crossover progeny expected will not be observed, because of interference. When interference is complete and no double-crossover progeny are observed, the coefficient of coincidence is 0 and the interference is 1.

Sometimes a crossover *increases* the probability of another crossover taking place nearby and we see *more* double-crossover progeny than expected. In this case, the coefficient of coincidence is greater than 1 and the interference is negative. **TRY PROBLEM 28 →**

CONCEPTS

The coefficient of coincidence equals the number of double crossovers observed divided by the number of double crossovers expected on the basis of the single-crossover frequencies. The interference equals 1 − the coefficient of coincidence; it indicates the degree to which one crossover interferes with additional crossovers.

✔ CONCEPT CHECK 6

In analyzing the results of a three-point testcross, a student determines that the interference is −0.23. What does this negative interference value indicate?

a. Fewer double crossovers took place than expected on the basis of single-crossover frequencies.

b. More double crossovers took place than expected on the basis of single-crossover frequencies.

c. Fewer single crossovers took place than expected.

d. A crossover in one region interferes with additional crossovers in the same region.

CONNECTING CONCEPTS

Stepping Through the Three-Point Cross

We have now examined the three-point cross in considerable detail and have seen how the information derived from the cross can be used to map a series of three linked genes. Let's briefly review the steps required to map genes from a three-point cross.

1. **Write out the phenotypes and numbers of progeny produced in the three-point cross.** The progeny phenotypes will be easier to interpret if you use allelic symbols for the traits (such as st^+ e^+ ss).

2. **Write out the genotypes of the original parents used to produce the triply heterozygous individual** in the testcross and, if known, the arrangement (coupling or repulsion) of the alleles on their chromosomes.

3. **Determine which phenotypic classes among the progeny are the nonrecombinants and which are the double crossovers.** The nonrecombinants will be the two most-

common phenotypes; double crossovers will be the two least-common phenotypes.

4. **Determine which locus lies in the middle.** Compare the alleles present in the double crossovers with those present in the nonrecombinants; each class of double crossovers should be like one of the nonrecombinants for two loci and should differ for one locus. The locus that differs is the middle one.

5. **Rewrite the phenotypes with the genes in correct order.**

6. **Determine where crossovers must have taken place to give rise to the progeny phenotypes.** To do so, compare each phenotype with the phenotype of the nonrecombinant progeny.

7. **Determine the recombination frequencies.** Add the numbers of the progeny that possess a chromosome with a crossover between a pair of loci. Add the double crossovers to this number. Divide this sum by the total number of progeny from the cross, and multiply by 100%; the result is the recombination frequency between the loci, which is the same as the map distance.

8. **Draw a map of the three loci.** Indicate which locus lies in the middle, and indicate the distances between them.

9. **Determine the coefficient of coincidence and the interference.** The coefficient of coincidence is the number of observed double-crossover progeny divided by the number of expected double-crossover progeny. The expected number can be obtained by multiplying the product of the two single-recombination probabilities by the total number of progeny in the cross. TRY PROBLEM 31 ➡

Worked Problem

In *D. melanogaster*, cherub wings (*ch*), black body (*b*), and cinnabar eyes (*cn*) result from recessive alleles that are all located on chromosome 2. A homozygous wild-type fly was mated with a cherub, black, and cinnabar fly, and the resulting F_1 females were test-crossed with cherub, black, and cinnabar males. The following progeny were produced from the testcross:

ch	b^+	*cn*	105
ch^+	b^+	cn^+	750
ch^+	*b*	*cn*	40
ch^+	b^+	*cn*	4
ch	*b*	*cn*	753
ch	b^+	cn^+	41
ch^+	*b*	cn^+	102
ch	*b*	cn^+	5
Total			1800

a. Determine the linear order of the genes on the chromosome (which gene is in the middle?).

b. Calculate the recombinant distances between the three loci.

c. Determine the coefficient of coincidence and the interference for these three loci.

• Solution

a. We can represent the crosses in this problem as follows:

P $\quad \dfrac{ch^+ \quad b^+ \quad cn^+}{ch^+ \quad b^+ \quad cn^+} \times \dfrac{ch \quad b \quad cn}{ch \quad b \quad cn}$

$\downarrow$

$F_1 \quad \dfrac{ch^+ \quad b^+ \quad cn^+}{ch \quad b \quad cn}$

Testcross $\quad \dfrac{ch^+ \quad b^+ \quad cn^+}{ch \quad b \quad cn} \times \dfrac{ch \quad b \quad cn}{ch \quad b \quad cn}$

Note that we do not know, at this point, the order of the genes; we have arbitrarily put *b* in the middle.

The next step is to determine which of the testcross progeny are nonrecombinants and which are double crossovers. The nonrecombinants should be the most-frequent phenotype; so they must be the progeny with phenotypes encoded by $\underline{ch^+ \; b^+ \; cn^+}$ and $\underline{ch \; b \; cn}$. These genotypes are consistent with the genotypes of the parents, given earlier. The double crossovers are the least-frequent phenotypes and are encoded by $\underline{ch^+ \; b^+ \; cn}$ and $\underline{ch \; b \; cn^+}$.

We can determine the gene order by comparing the alleles present in the double crossovers with those present in the nonrecombinants. The double-crossover progeny should be like one of the nonrecombinants at two loci and unlike it at one locus; the allele that differs should be in the middle. Compare the double-crossover progeny $\underline{ch \; b \; cn^+}$ with the nonrecombinant $\underline{ch \; b \; cn}$. Both have cherub wings (*ch*) and black body (*b*), but the double-crossover progeny have wild-type eyes (cn^+), whereas the nonrecombinants have cinnabar eyes (*cn*). The locus that determines cinnabar eyes must be in the middle.

b. To calculate the recombination frequencies among the genes, we first write the phenotypes of the progeny with the genes encoding them in the correct order. We have already identified the nonrecombinant and double-crossover progeny; so the other four progeny types must have resulted from single crossovers. To determine *where* single crossovers took place, we compare the alleles found in the single-crossover progeny with those in the nonrecombinants. Crossing over must have taken place where the alleles switch from those found in one nonrecombinant to those found in the other nonrecombinant. The locations of the crossovers are indicated with a slash:

ch		*cn*	/ b^+	105	single crossover
ch^+		cn^+	b^+	750	nonrecombinant
ch^+ /	*cn*		*b*	40	single crossover
ch^+ /	*cn*	/ b^+		4	double crossover
ch		*cn*	*b*	753	nonrecombinant
ch /	cn^+		b^+	41	single crossover
ch^+	cn^+ /	*b*		102	single crossover
ch /	cn^+ /	*b*		5	double crossover
Total				1800	

Next, we determine the recombination frequencies and draw a genetic map:

ch-cn recombination frequency =

$$\frac{40+4+41+5}{1800} \times 100\% = 5\%$$

cn-b recombination frequency =

$$\frac{105+4+102+5}{1800} \times 100\% = 12\%$$

ch-b map distance = 5% + 12% = 17%

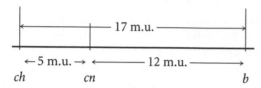

c. The coefficient of coincidence is the number of observed double crossovers divided by the number of expected double crossovers. The number of expected double crossovers is obtained by multiplying the probability of a crossover between *ch* and *cn* (0.05) × the probability of a crossover between *cn* and *b* (0.12) × the total number of progeny in the cross (1800):

$$\text{coefficient of coincidence} = \frac{4+5}{0.05 \times 0.12 \times 1800} = 0.83$$

Finally, the interference is equal to 1 − the coefficient of coincidence:

$$\text{interference} = 1 - 0.83 = 0.17$$

→ To increase your skill with three-point crosses, try working Problem 29 at the end of this chapter.

Effect of Multiple Crossovers

So far, we have examined the effects of double crossovers taking place between only two of the four chromatids (strands) of a homologous pair. These crossovers are called two-strand crossovers. Double crossovers including three and even four of the chromatids of a homologous pair also may take place (**Figure 7.16**). If we examine only the alleles at loci on either side of both crossover events, two-strand double crossovers result in no new combinations of alleles, and no recombinant gametes are produced (see Figure 7.16). Three-strand double crossovers result in two of the four gametes being recombinant, and four-strand double crossovers result in all four gametes being recombinant. Thus, two-strand double crossovers produce 0% recombination, three-strand double crossovers produce 50% recombination, and four-strand double crossovers produce 100% recombination. The overall result is that all types of double crossovers, taken together, produce an average of 50% recombinant progeny.

As we have seen, two-strand double crossover cause alleles on either side of the crossovers to remain the same and produce no recombinant progeny. Three-strand and four-strand crossovers produce recombinant progeny, but these progeny are the same types produced by single crossovers. Consequently, some multiple crossovers go unde-

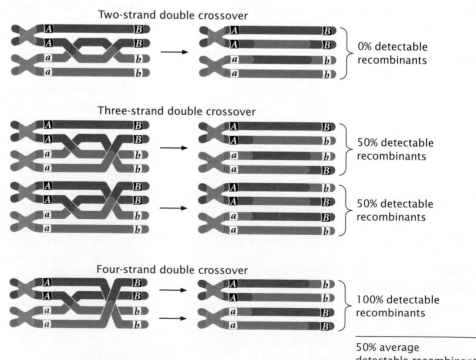

7.16 Results of two-, three-, and four-strand double crossovers on recombination between two genes.

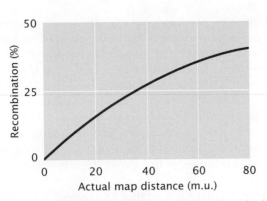

7.17 Percent recombination underestimates the true physical distance between genes at higher map distances.

tected when the progeny of a genetic cross are observed. Therefore, map distances based on recombination rates will underestimate the true physical distances between genes, because some multiple crossovers are not detected among the progeny of a cross. When genes are very close together, multiple crossovers are unlikely, and the distances based on recombination rates accurately correspond to the physical distances on the chromosome. But, as the distance between genes increases, more multiple crossovers are likely, and the discrepancy between genetic distances (based on recombination rates) and physical distances increases. To correct for this discrepancy, geneticists have developed mathematical **mapping functions**, which relate recombination frequencies to actual physical distances between genes (**Figure 7.17**). Most of these functions are based on the Poisson distribution, which predicts the probability of multiple rare events. With the use of such mapping functions, map distances based on recombination rates can be more accurately estimated.

Mapping Human Genes

Efforts in mapping human genes are hampered by the inability to perform desired crosses and the small number of progeny in most human families. Geneticists are restricted to analyses of pedigrees, which are often incomplete and provide limited information. Nevertheless, a large number of human traits have been successfully mapped with the use of pedigree data to analyze linkage. Because the number of progeny from any one mating is usually small, data from several families and pedigrees are usually combined to test for independent assortment. The methods used in these types of analysis are beyond the scope of this book, but an example will illustrate how linkage can be detected from pedigree data.

One of the first documented demonstrations of linkage in humans was between the locus for nail–patella syndrome and the locus that determines the ABO blood types. Nail–patella syndrome is an autosomal dominant disorder characterized by abnormal fingernails and absent or rudimentary kneecaps. The ABO blood types are determined by an autosomal locus with multiple alleles (see Chapter 5). Linkage between the genes encoding these traits was established in families in which both traits segregate. Part of one such family is illustrated in **Figure 7.18**.

Nail–patella syndrome is rare, and so we can assume that people having this trait are heterozygous (Nn); unaffected people are homozygous (nn). The ABO genotypes can be inferred from the phenotypes and the types of offspring produced. Person I-2 in Figure 7.18, for example, has blood-type B, which has two possible genotypes: $I^B I^B$ or $I^B i$ (see Figure 5.6). Because some of her offspring are blood-type O (genotype ii) and must have therefore inherited an i allele from each parent, female I-2 must have genotype $I^B i$. Similarly, the presence of blood-type O offspring

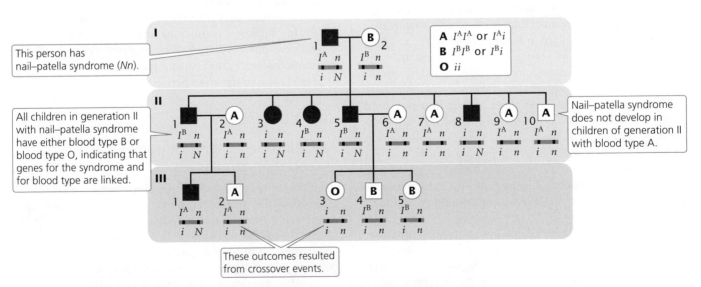

7.18 Linkage between ABO blood types and nail–patella syndrome was established by examining families in whom both traits segregate. The pedigree shown here is for one such family. The ABO blood type is indicated in each circle or square. The genotype, inferred from phenotype, is given below each circle or square.

in generation II indicates that male I-1, with blood-type A, also must carry an i allele and therefore has genotype $I^A i$. The parents of this family are:

$$I^A i \, Nn \times I^B i \, nn$$

From generation II, we can see that the genes for nail–patella syndrome and the blood types do not appear to assort independently. All children in generation II with nail–patella syndrome have either blood-type B or blood-type O; all those with blood-type A have normal nails and kneecaps. If the genes encoding nail–patella syndrome and the ABO blood types assorted independently, we would expect that some children in generation II would have blood-type A and nail–patella syndrome, inheriting both the I^A and N alleles from their father. This outcome indicates that the arrangements of the alleles on the chromosomes of the crossed parents are:

$$\frac{I^A \qquad n}{i \qquad N} \times \frac{I^B \qquad n}{i \qquad n}$$

The pedigree indicates that there is no recombination among the offspring (generation II) of these parents, but there are two instances of recombination among the persons in generation III. Persons II-1 and II-2 have the following genotypes:

$$\frac{i^B \qquad n}{i \qquad N} \times \frac{I^A \qquad n}{i \qquad n}$$

Their child III-2 has blood-type A and does not have nail–patella syndrome; so he must have genotype

$$\frac{I^A \qquad n}{i \qquad n}$$

and must have inherited both the i and the n alleles from his father. These alleles are on different chromosomes in the father; so crossing over must have taken place. Crossing over also must have taken place to produce child III-3.

In the pedigree of Figure 7.18, 13 children are from matings in which the genes encoding nail–patella syndrome and ABO blood types segregate; 2 of them are recombinants. On this basis, we might assume that the loci for nail–patella syndrome and ABO blood types are linked, with a recombination frequency of $^2/_{13} = 0.154$. However, it is possible that the genes *are* assorting independently and that the small number of children just makes it seem as though the genes are linked. To determine the probability that genes are actually linked, geneticists often calculate **lod** (logarithm of odds) scores.

To obtain a lod score, we calculate both the probability of obtaining the observed results with the assumption that the genes are linked with a specified degree of recombination and the probability of obtaining the observed results with the assumption of independent assortment. We then determine the ratio of these two probabilities, and the logarithm of this ratio is the lod score. Suppose that the

probability of obtaining a particular set of observations with the assumption of linkage and a certain recombination frequency is 0.1 and that the probability of obtaining the same observations with the assumption of independent assortment is 0.0001. The ratio of these two probabilities is $^{0.1}/_{0.0001} = 1000$, the logarithm of which (the lod score) is 3. Thus, linkage with the specified recombination is 1000 times as likely as independent assortment to produce what was observed. A lod score of 3 or higher is usually considered convincing evidence for linkage. TRY PROBLEM 33 →

Mapping with Molecular Markers

For many years, gene mapping was limited in most organisms by the availability of **genetic markers**—that is, variable genes with easily observable phenotypes for which inheritance could be studied. Traditional genetic markers include genes that encode easily observable characteristics such as flower color, seed shape, blood types, and biochemical differences. The paucity of these types of characteristics in many organisms limited mapping efforts.

In the 1980s, new molecular techniques made it possible to examine variations in DNA itself, providing an almost unlimited number of genetic markers that can be used for creating genetic maps and studying linkage relations. The earliest of these molecular markers consisted of restriction fragment length polymorphisms (RFLPs), which are variations in DNA sequence detected by cutting the DNA with restriction enzymes (see Chapter 19). Later, methods were developed for detecting variable numbers of short DNA sequences repeated in tandem, called microsatellites. More recently, DNA sequencing allows the direct detection of individual variations in the DNA nucleotides. All of these methods have expanded the availability of genetic markers and greatly facilitated the creation of genetic maps.

Gene mapping with molecular markers is done essentially in the same manner as mapping performed with traditional phenotypic markers: the cosegregation of two or more markers is studied, and map distances are based on the rates of recombination between markers. These methods and their use in mapping are presented in more detail in Chapters 19 and 20.

Locating Genes with Genomewide Association Studies

The traditional approach to mapping genes, which we have learned in this chapter, is to examine progeny phenotypes in genetic crosses or among individuals in a pedigree, looking for associations between the inheritance of particular phenotype and the inheritance of alleles at other loci. This type of gene mapping is called **linkage analysis**, because it is based on the detection of physical linkage between genes, as measured by the rate of recombination, in progeny from a cross. Linkage analysis has been a powerful tool in the genetic analysis of many different types of organisms.

Another alternative approach to mapping genes is to conduct **genomewide association studies**, looking for nonrandom associations between the presence of a trait and alleles at many different loci scattered across the genome. Unlike linkage analysis, this approach does not trace the inheritance of genetic markers and a trait in a genetic cross or family. Rather, it looks for associations between traits and particular suites of alleles in a *population*.

Imagine that we are interested in finding genes that contribute to bipolar disease, a psychiatric illness characterized by severe depression and mania. When a mutation that predisposes a person to bipolar disease first arises in a population, it will occur on a particular chromosome and be associated with a specific set of alleles on that chromosome (**Figure 7.19**). A specific set of linked alleles, such as this, is called a **haplotype**, and the nonrandom association between alleles in a haplotype is called **linkage disequilibrium**. Because of the physical linkage between the bipolar mutation and the other alleles of the haplotype, bipolar illness and the haplotype will tend to be inherited together. Crossing over, however, breaks up the association between the alleles of the haplotype (see Figure 7.19), reducing the linkage disequilibrium between them. How long the linkage disequilibrium persists depends on the amount of recombination between alleles at different loci. When the loci are far apart, linkage disequilibrium breaks down quickly; when the loci are close together, crossing over is less common and linkage disequilibrium will persist longer. The important point is that linkage disequilibrium—the nonrandom association between alleles—provides information about the distance between genes. A strong association between a trait such as bipolar illness and a set of linked genetic markers indicates that one or more genes contributing to bipolar illness are likely to be near the genetic markers.

In recent years, geneticists have mapped millions of genetic variants called **single-nucleotide polymorphisms** (SNPs), which are positions in the genome where people vary in a single nucleotide base (see Chapter 20). It is now possible to quickly and inexpensively genotype people for hundreds of thousands or millions of SNPs. This genotyping has provided genetic markers needed for conducting genomewide association studies, in which SNP haplotypes of people with a particular disease, such as bipolar illness, are compared with the haplotypes of healthy persons. Nonrandom associations between SNPs and the disease suggest that one or more genes that contribute to the disease are closely linked to the SNPs. Genomewide association studies do not usually locate specific genes, but rather associate the inheritance of a trait or disease with a specific chromosomal region. After such an association has been established, geneticists can examine the chromosomal region for genes that might be responsible for the trait. Genomewide association studies have been instrumental in the discovery of genes or chromosomal regions that affect a number of genetic diseases and important human traits,

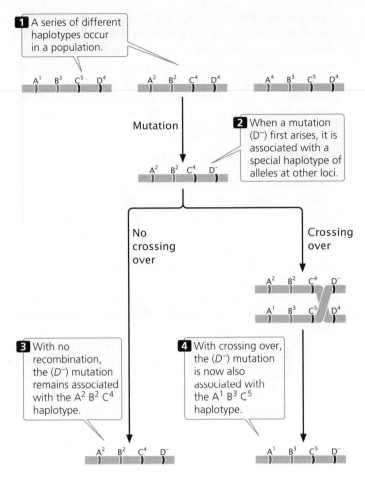

7.19 Genomewide association studies are based on the nonrandom association of a mutation (D^-) that produces a trait and closely linked genes that constitute a haplotype.

including bipolar disease, height, skin pigmentation, eye color, body weight, coronary artery disease, blood-lipid concentrations, diabetes, heart attacks, bone density, and glaucoma, among others.

CONCEPTS

The development of molecular techniques for examining variation in DNA sequences has provided a large number of genetic markers that can be used to create genetic maps and study linkage relations. Genomewide association studies examine the nonrandom association of genetic markers and phenotypes to locate genes that contribute to the expression of traits.

7.4 Physical-Mapping Methods Are Used to Determine the Physical Positions of Genes on Particular Chromosomes

Genetic maps reveal the relative positions of genes on a chromosome on the basis of frequencies of recombination, but they do not provide information that allow us to place groups of linked genes on particular chromosomes. Furthermore, the units of a genetic map do not always precisely correspond to physical distances on the chromosome, because a number of factors other than physical distances between genes (such as the type and sex of the organism) can influence recombination. Because of these limitations, physical-mapping methods that do not rely on recombination frequencies have been developed.

Deletion Mapping

One method for determining the chromosomal location of a gene is **deletion mapping**. Special staining methods have been developed that reveal characteristic banding patterns on the chromosomes (see Chapter 9). The absence of one or more of the bands that are normally on a chromosome reveals the presence of a chromosome deletion, a mutation in which a part of a chromosome is missing. Genes can be assigned to regions of chromosomes by studying the association between a gene's phenotype or product and particular chromosome deletions.

In deletion mapping, an individual that is homozygous for a recessive mutation in the gene of interest is crossed with an individual that is heterozygous for a deletion (**Figure 7.20**). If the gene of interest is in the region of the chromosome represented by the deletion (the red part of the chromosomes in Figure 7.20), approximately half of the progeny will display the mutant phenotype (see Figure

7.20a). If the gene is not within the deleted region, all of the progeny will be wild type (see Figure 7.20b).

Deletion mapping has been used to reveal the chromosomal locations of a number of human genes. For example, Duchenne muscular dystrophy is a disease that causes progressive weakening and degeneration of the muscles. From its X-linked pattern of inheritance, the mutated allele causing this disorder was known to be on the X chromosome, but its precise location was uncertain. Examination of a number of patients having Duchenne muscular dystrophy, who also possessed small deletions, allowed researchers to position the gene on a small segment of the short arm of the X chromosome. **TRY PROBLEM 34** →

Somatic-Cell Hybridization

Another method used for positioning genes on chromosomes is **somatic-cell hybridization**, which requires the fusion of different types of cells. Most mature somatic (nonsex) cells can undergo only a limited number of divisions and therefore cannot be grown continuously. However, cells that have been altered by viruses or derived from tumors that have lost the normal constraints on cell division will divide indefinitely; this type of cell can be cultured in the laboratory to produce a **cell line**.

Cells from two different cell lines can be fused by treating them with polyethylene glycol or other agents that alter their plasma membranes. After fusion, the cell possesses

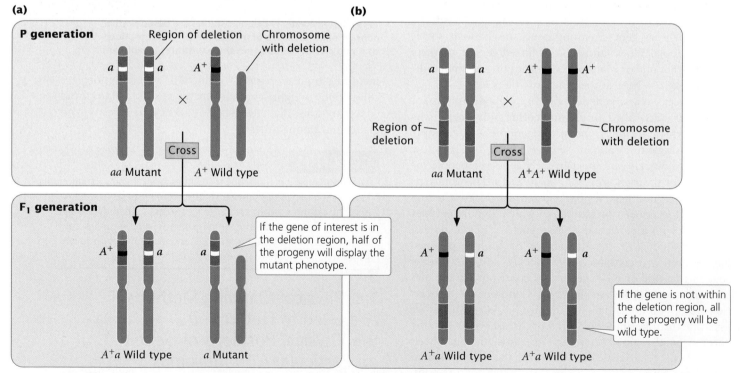

7.20 Deletion mapping can be used to determine the chromosomal location of a gene. An individual homozygous for a recessive mutation in the gene of interest (*aa*) is crossed with an individual heterozygous for a deletion.

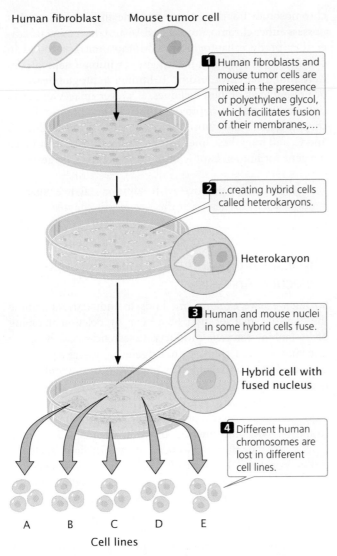

1 Human fibroblasts and mouse tumor cells are mixed in the presence of polyethylene glycol, which facilitates fusion of their membranes,...

2 ...creating hybrid cells called heterokaryons.

Heterokaryon

3 Human and mouse nuclei in some hybrid cells fuse.

Hybrid cell with fused nucleus

4 Different human chromosomes are lost in different cell lines.

A   B   C   D   E

Cell lines

7.21 Somatic-cell hybridization can be used to determine which chromosome contains a gene of interest.

two nuclei and is called a **heterokaryon**. The two nuclei of a heterokaryon eventually also fuse, generating a hybrid cell that contains chromosomes from both cell lines. If human and mouse cells are mixed in the presence of polyethylene glycol, fusion results in human–mouse somatic-cell hybrids (**Figure 7.21**). The hybrid cells tend to lose chromosomes as they divide and, for reasons that are not understood, chromosomes from one of the species are lost preferentially. In human–mouse somatic-cell hybrids, the human chromosomes tend to be lost, whereas the mouse chromosomes are retained. Eventually, the chromosome number stabilizes when all but a few of the human chromosomes have been lost. Chromosome loss is random and differs among cell lines. The presence of these "extra" human chromosomes in the mouse genome makes it possible to assign human genes to specific chromosomes.

To map genes by using somatic-cell hybridization requires a panel of different hybrid cell lines. Each cell line is examined microscopically and the human chromosomes that it contains are identified. The cell lines of the panel are chosen so that they differ in the human chromosomes that they have retained. For example, one cell line might possess human chromosomes 2, 4, 7, and 8, whereas another might possess chromosomes 4, 19, and 20. Each cell line in the panel is examined for evidence of a particular human gene. The human gene can be detected by looking either for the for the gene itself (discussed in Chapter 19) or for the protein that it produces. Correlation of the presence of the gene with the presence of specific human chromosomes often allows the gene to be assigned to the correct chromosome. For example, if a gene is detected in both of the aforementioned cell lines, the gene must be on chromosome 4, because it is the only human chromosome common to both cell lines (**Figure 7.22**).

Sometimes somatic-cell hybridization can be used to position a gene on a specific part of a chromosome. Some

Human chromosomes present

Cell line	Gene product present	1	2	3	4	5	6	7	8	9	10	11	12	13	14	15	16	17	18	19	20	21	22	X
A	+		+		+			+	+															
B	+	+	+		+				+	+	+	+	+	+										
C	−														+		+		+			+		
D	+		+		+		+	+	+															
E	−												+							+				
F	+				+																+	+		

7.22 Somatic-cell hybridization is used to assign a gene to a particular human chromosome. A panel of six cell lines, each line containing a different subset of human chromosomes, is examined for the presence of the gene product (such as an enzyme). Four of the cell lines (A, B, D, and F) have the gene product. The only chromosome common to all four of these cell lines is chromosome 4, indicating that the gene is located on this chromosome.

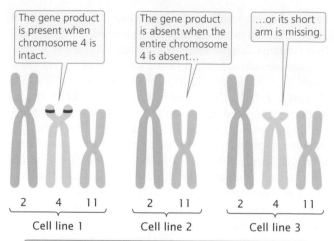

The gene product is present when chromosome 4 is intact.

The gene product is absent when the entire chromosome 4 is absent…

…or its short arm is missing.

| 2 | 4 | 11 |
Cell line 1

| 2 | 11 |
Cell line 2

| 2 | 4 | 11 |
Cell line 3

Conclusion: If the gene product is present in a cell line with an intact chromosome but missing from a line with a chromosome deletion, the gene for that product must be located in the deleted region.

7.23 Genes can be localized to a specific part of a chromosome by using somatic-cell hybridization.

hybrid cell lines carry a human chromosome with a chromosome mutation such as a deletion or a translocation. If the gene is present in a cell line with the intact chromosome but missing from a line with a chromosome deletion, the gene must be located in the deleted region (**Figure 7.23**). Similarly, if a gene is usually absent from a chromosome but consistently appears whenever a translocation (a piece of another chromosome that has broken off and attached itself to the chromosome in question) is present, it must be present on the translocated part of the chromosome.

Worked Problem

A panel of cell lines was created from human–mouse somatic-cell fusions. Each line was examined for the presence of human chromosomes and for the production of human haptoglobin (a protein). The following results were obtained:

Cell line	Human haptoglobin	Human chromosomes						
		1	2	3	14	15	16	21
A	−	+	−	+	−	+	−	−
B	+	+	−	+	−	−	+	−
C	+	+	−	−	−	+	+	−
D	−	+	+	−	−	+	−	−

On the basis of these results, which human chromosome carries the gene for haptoglobin?

• Solution

First, identify the cell lines that are positive for the protein (human haptoglobin) and determine the chromosomes that they have in common. Lines B and C produce human haptoglobin; the only chromosomes that they have in common are

chromosomes 1 and 16. Next, examine all the cell lines that possess either chromosomes 1 and 16 and determine whether they produce haptoglobin. Chromosome 1 is found in cell lines A, B, C, and D. If the gene for human haptoglobin were found on chromosome 1, human haptoglobin would be present in all of these cell lines. However, lines A and D do not produce human haptoglobin; so the gene cannot be on chromosome 1. Chromosome 16 is found only in cell lines B and C, and only these lines produce human haptoglobin; so the gene for human haptoglobin lies on chromosome 16.

→ For more practice with somatic-cell hybridizations, work Problem 35 at the end of this chapter.

Physical Chromosome Mapping Through Molecular Analysis

So far, we have explored methods to indirectly determine the chromosomal location of a gene by deletion mapping or by looking for gene products. Researchers now have the information and technology to actually see where a gene lies. Described in more detail in Chapter 19, in situ hybridization is a method for determining the chromosomal location of a particular gene through molecular analysis. This method requires the creation of a probe for the gene, which is a single-stranded DNA complement to the gene of interest. The probe is radioactive or fluoresces under ultraviolet light so that it can be visualized. The probe binds to the DNA sequence of the gene on the chromosome. The presence of radioactivity or fluorescence from the bound probe reveals the location of the gene on a particular chromosome (**Figure 7.24**).

In addition to allowing us to see where a gene is located on a chromosome, modern laboratory techniques now allow researchers to identify the precise location of a gene at the nucleotide level. For example, with DNA sequencing (described fully in Chapter 19), physical distances between genes can be determined in numbers of base pairs.

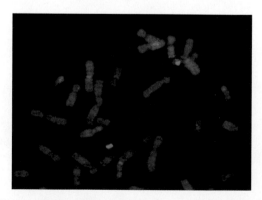

7.24 In situ hybridization is another technique for determining the chromosomal location of a gene. The red fluorescence is produced by a probe for sequences on chromosome 9; the green fluorescence is produced by a probe for sequences on chromosome 22. [Courtesy of Applied Imaging Corp.]

CONCEPTS

Physical-mapping methods determine the physical locations of genes on chromosomes and include deletion mapping, somatic-cell hybridization, in situ hybridization, and direct DNA sequencing.

7.5 Recombination Rates Exhibit Extensive Variation

In recent years, geneticists have studied variation in rates of recombination and found that levels of recombination vary widely—among species, among and along chromosomes of a single species, and even between males and females of the same species. For example, about twice as much recombination takes place in humans as in the mouse and the rat. Within the human genome, recombination varies among chromosomes, with chromosomes 21 and 22 having the highest rates and chromosomes 2 and 4 having the lowest rates. Researchers have also detected differences between male and female humans in rates of recombination: the autosomal chromosomes of females undergo about 50% more recombination than do the autosomal chromosomes of males.

Geneticists have found numerous recombination *hotspots*, where recombination is at least 10 times as high as the average elsewhere in the genome. The human genome may contain an estimated 25,000 to 50,000 such recombination hotspots. Approximately 60% of all crossovers take place in hotspots. For humans, recombination hotspots tend to be found near but not within active genes. Recombination hotspots have been detected in the genomes of other organisms as well. In comparing recombination hotspots in the genomes of humans and chimpanzees (our closest living relative), geneticists have determined that, although the DNA sequences of humans and chimpanzees are extremely similar, their recombination hotspots are at entirely different places.

CONCEPTS

Rates of recombination vary among species, among and along chromosomes, and even between males and females.

CONCEPTS SUMMARY

- Linked genes do not assort independently. In a testcross for two completely linked genes (no crossing over), only nonrecombinant progeny are produced. When two genes assort independently, recombinant progeny and nonrecombinant progeny are produced in equal proportions. When two genes are linked with some crossing over between them, more nonrecombinant progeny than recombinant progeny are produced.

- Recombination frequency is calculated by summing the number of recombinant progeny, dividing by the total number of progeny produced in the cross, and multiplying by 100%. The recombination frequency is half the frequency of crossing over, and the maximum frequency of recombinant gametes is 50%.

- Coupling and repulsion refer to the arrangement of alleles on a chromosome. Whether genes are in coupling configuration or in repulsion determines which combination of phenotypes will be most frequent in the progeny of a testcross.

- Interchromosomal recombination takes place among genes located on different chromosomes through the random segregation of chromosomes in meiosis. Intrachromosomal recombination takes place among genes located on the same chromosome through crossing over.

- A chi-square test of independence can be used to determine if genes are linked.

- Recombination rates can be used to determine the relative order of genes and distances between them on a chromosome. One percent recombination equals one map unit. Maps based on recombination rates are called genetic maps; maps based on physical distances are called physical maps.

- Genetic maps can be constructed by examining recombination rates from a series of two-point crosses or by examining the progeny of a three-point testcross.

- Some multiple crossovers go undetected; thus, genetic maps based on recombination rates underestimate the true physical distances between genes.

- Human genes can be mapped by examining the cosegregation of traits in pedigrees.

- A lod score is the logarithm of the ratio of the probability of obtaining the observed progeny with the assumption of linkage to the probability of obtaining the observed progeny with the assumption of independent assortment. A lod score of 3 or higher is usually considered evidence for linkage.

- Molecular techniques that allow the detection of variable differences in DNA sequence have greatly facilitated gene mapping.

- Genomewide association studies locate genes that affect particular traits by examining the nonrandom association of a trait with genetic markers from across the genome.

- Nucleotide sequencing is another method of physically mapping genes.

- Rates of recombination vary widely, differing among species, among and along chromosomes within a single species, and even between males and females of the same species.

IMPORTANT TERMS

linked genes (p. 163)
linkage group (p. 163)
nonrecombinant (parental) gamete
(p. 166)
nonrecombinant (parental) progeny
(p. 166)
recombinant gamete (p. 166)
recombinant progeny (p. 166)
recombination frequency (p. 167)
coupling (cis) configuration (p. 168)
repulsion (trans) configuration (p. 168)
interchromosomal recombination
(p. 170)

intrachromosomal recombination
(p. 170)
genetic map (p. 174)
physical map (p. 174)
map unit (m.u.) (p. 174)
centiMorgan (p. 174)
Morgan (p. 174)
two-point testcross (p. 175)
three-point testcross (p. 176)
interference (p. 182)
coefficient of coincidence (p. 182)
mapping function (p. 185)
lod (logarithm of odds) score (p. 186)

genetic marker (p. 186)
linkage analysis (p. 186)
genomewide association studies (p. 187)
haplotype (p. 187)
linkage disequilibrium (p. 187)
single-nucleotide polymorphism (SNP)
(p. 187)
deletion mapping (p. 188)
somatic-cell hybridization (p. 188)
cell line (p. 188)
heterokaryon (p. 189)

ANSWERS TO CONCEPT CHECKS

1. c

2. Repulsion

3. Genetic maps are based on rates of recombination; physical maps are based on physical distances.

4. $\dfrac{m^+\ p^+\ s^+}{m\ p\ s}\ \dfrac{m^+\ p\ s}{m\ p\ s}\ \dfrac{m\ p^+\ s^+}{m\ p\ s}\ \dfrac{m^+\ p^+s}{m\ p\ s}\ \dfrac{m\ p\ s^+}{m\ p\ s}\ \dfrac{m^+\ p\ s^+}{m\ p\ s}\ \dfrac{m\ p^+\ s}{m\ p\ s}\ \dfrac{m\ p\ s}{m\ p\ s}$

5. The c locus

6. b

WORKED PROBLEMS

1. In guinea pigs, white coat (w) is recessive to black coat (W) and wavy hair (v) is recessive to straight hair (V). A breeder crosses a guinea pig that is homozygous for white coat and wavy hair with a guinea pig that is black with straight hair. The F$_1$ are then crossed with guinea pigs having white coats and wavy hair in a series of testcrosses. The following progeny are produced from these testcrosses:

black, straight	30
black, wavy	10
white, straight	12
white, wavy	31
Total	83

a. Are the genes that determine coat color and hair type assorting independently? Carry out chi-square tests to test your hypothesis.

b. If the genes are not assorting independently, what is the recombination frequency between them?

• **Solution**

a. Assuming independent assortment, outline the crosses conducted by the breeder:

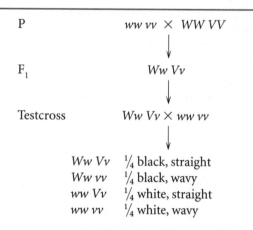

Because a total of 83 progeny were produced in the testcrosses, we expect $\frac{1}{4} \times 83 = 20.75$ of each. The observed numbers of progeny from the testcross (30, 10, 12, 31) do not appear to fit the expected numbers (20.75, 20.75, 20.75, 20.75) well; so independent assortment may not have taken place.

To test the hypothesis, carry out a chi-square test of independence. Construct a table, with the genotypes of the first locus along the top and the genotypes of the second locus along the side. Compute the totals for the rows and columns and the grand total.

	Ww	ww	Row totals
Vv	30	12	42
vv	10	31	41
Column totals	40	43	83 ← Grand total

The expected value for each cell of the table is calculated with the following formula:

$$\text{expected number} = \frac{\text{row total} \times \text{column total}}{\text{grand total}}$$

Using this formula, we find the expected values (given in parentheses) to be:

	Ww	ww	Row totals
Vv	30 (20.24)	12 (21.76)	42
vv	10 (19.76)	31 (21.24)	41
Column totals	40	43	83 ← Grand total

Using these observed and expected numbers, we find the calculated chi-square value to be:

$$\chi^2 = \sum \frac{(\text{observed} - \text{expected})^2}{\text{expected}}$$

$$= \frac{(30 - 20.24)^2}{20.24} + \frac{(10 - 19.76)^2}{19.76} + \frac{(12 - 21.76)^2}{21.76} + \frac{(31 - 21.24)^2}{21.24}$$

$$= 4.71 + 4.82 + 4.38 + 4.48 = 18.39$$

The degrees of freedom for the chi-square test of independence are $df = (\text{number of rows} - 1) \times (\text{number of columns} - 1)$. There are two rows and two columns, and so the degrees of freedom are:

$$df = (2 - 1) \times (2 - 1) = 1 \times 1 = 1$$

In Table 3.5, the probability associated with a chi-square value of 18.39 and 1 degree of freedom is less than 0.005, indicating that chance is very unlikely to be responsible for the differences between the observed numbers and the numbers expected with independent assortment. The genes for coat color and hair type have therefore not assorted independently.

b. To determine the recombination frequencies, identify the recombinant progeny. Using the notation for linked genes, write the crosses:

P $\dfrac{w \quad v}{w \quad v}$ × $\dfrac{W \quad V}{W \quad V}$

↓

F_1 $\dfrac{W \quad V}{w \quad v}$

Testcross $\dfrac{W \quad V}{w \quad v}$ × $\dfrac{w \quad v}{w \quad v}$

↓

$\dfrac{W \quad V}{w \quad v}$ 30 black, straight (nonrecombinant progeny)

$\dfrac{w \quad v}{w \quad v}$ 31 white, wavy (nonrecombinant progeny)

$\dfrac{W \quad v}{w \quad v}$ 10 black, wavy (recombinant progeny)

$\dfrac{w \quad V}{w \quad v}$ 12 white, straight (recombinant progeny)

The recombination frequency is:

$$\frac{\text{number of recombinant progeny}}{\text{total number progeny}} \times 100\%$$

or

recombinant frequency =

$$\frac{10 + 12}{30 + 31 + 10 + 12} \times 100\% = \frac{22}{83} \times 100\% = 26.5$$

2. A series of two-point crosses were carried out among seven loci (*a*, *b*, *c*, *d*, *e*, *f*, and *g*), producing the following recombination frequencies. Using these recombination frequencies, map the seven loci, showing their linkage groups, the order of the loci in each linkage group, and the distances between the loci of each group:

Loci	Recombination frequency (%)	Loci	Recombination frequency (%)
a and *b*	10	*c* and *d*	50
a and *c*	50	*c* and *e*	8
a and *d*	14	*c* and *f*	50
a and *e*	50	*c* and *g*	12
a and *f*	50	*d* and *e*	50
a and *g*	50	*d* and *f*	50
b and *c*	50	*d* and *g*	50
b and *d*	4	*e* and *f*	50
b and *e*	50	*e* and *g*	18
b and *f*	50	*f* and *g*	50
b and *g*	50		

• Solution

To work this problem, remember that 1% recombination equals 1 map unit and a recombination frequency of 50% means that genes at the two loci are assorting independently (located in different linkage groups).

The recombination frequency between *a* and *b* is 10%; so these two loci are in the same linkage group, approximately 10 m.u. apart.

The recombination frequency between *a* and *c* is 50%; so *c* must lie in a second linkage group.

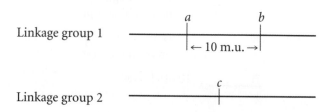

The recombination frequency between *a* and *d* is 14%; so *d* is located in linkage group 1. Is locus *d* 14 m.u. to the right or to the left of gene *a*? If *d* is 14 m.u. to the left of *a*, then the *b*-to-*d* distance should be 10 m.u. + 14 m.u. = 24 m.u. On the other hand, if *d* is to the right of *a*, then the distance between *b* and *d* should be 14 m.u. − 10 m.u. = 4 m.u. The *b*–*d* recombination frequency is 4%; so *d* is 14 m.u. to the right of *a*. The updated map is:

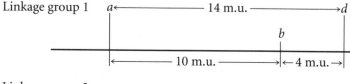

The recombination frequencies between each of loci *a*, *b*, and *d*, and locus *e* are all 50%; so *e* is not in linkage group 1 with *a*, *b*, and *d*. The recombination frequency between *e* and *c* is 8 m.u.; so *e* is in linkage group 2:

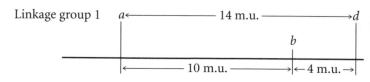

There is 50% recombination between *f* and all the other genes; so *f* must belong to a third linkage group:

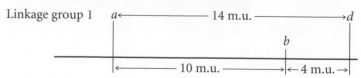

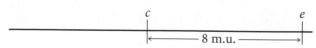

Finally, position locus *g* with respect to the other genes. The recombination frequencies between *g* and loci *a*, *b*, and *d* are all 50%; so *g* is not in linkage group 1. The recombination frequency between *g* and *c* is 12 m.u.; so *g* is a part of linkage group 2. To determine whether *g* is 12 m.u. to the right or left of *c*, consult the *g*–*e* recombination frequency. Because this recombination frequency is 18%, *g* must lie to the left of *c*:

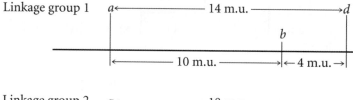

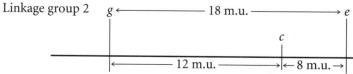

Note that the *g*-to-*e* distance (18 m.u.) is shorter than the sum of the *g*-to-*c* (12 m.u.) and *c*-to-*e* distances (8 m.u.) because of undetectable double crossovers between *g* and *e*.

3. Ebony body color (*e*), rough eyes (*ro*), and brevis bristles (*bv*) are three recessive mutations that occur in fruit flies. The loci for these mutations have been mapped and are separated by the following map distances:

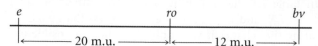

The interference between these genes is 0.4.

A fly with ebony body, rough eyes, and brevis bristles is crossed with a fly that is homozygous for the wild-type traits. The resulting F₁ females are test-crossed with males that have ebony body, rough eyes, and brevis bristles; 1800 progeny are produced. Give the expected numbers of phenotypes in the progeny of the testcross.

• Solution

The crosses are:

P $\dfrac{e^+ \quad ro^+ \quad bv^+}{e^+ \quad ro^+ \quad bv^+}$ $\times$ $\dfrac{e \quad ro \quad bv}{e \quad ro \quad bv}$

$\downarrow$

F_1 $\dfrac{e^+ \quad ro^+ \quad bv^+}{e \quad ro \quad bv}$

$\downarrow$

Testcross $\dfrac{e^+ \quad ro^+ \quad bv^+}{e \quad ro \quad bv}$ $\times$ $\dfrac{e \quad ro \quad bv}{e \quad ro \quad bv}$

In this case, we know that *ro* is the middle locus because the genes have been mapped. Eight classes of progeny will be produced from this cross:

e^+	ro^+	bv^+	nonrecombinant
e	ro	bv	nonrecombinant
e^+ / ro	bv		single crossover between *e* and *ro*
e / ro^+	bv^+		single crossover between *e* and *ro*
e^+	ro^+ / bv		single crossover between *ro* and *bv*
e	ro / bv^+		single crossover between *ro* and *bv*
e^+ / ro / bv^+			double crossover
e / ro^+ / bv			double crossover

To determine the numbers of each type, use the map distances, starting with the double crossovers. The expected number of double crossovers is equal to the product of the single-crossover probabilities:

expected number of double crossovers $= 0.20 \times 0.12 \times 1800$

$= 43.2$

However, there is some interference; so the observed number of double crossovers will be less than the expected. The interference is $1 -$ coefficient of coincidence; so the coefficient of coincidence is:

coefficient of coincidence $= 1 -$ interference

The interference is given as 0.4; so the coefficient of coincidence equals $1 - 0.4 = 0.6$. Recall that the coefficient of coincidence is:

coefficient of coincidence $=$

$$\dfrac{\text{number of observed double crossovers}}{\text{number of expected double crossovers}}$$

Rearranging this equation, we obtain:

number of observed double crossovers $=$

coefficient of coincidence $\times$ number of expected double crossovers

number of observed double crossovers $= 0.6 \times 43.2 = 26$

A total of 26 double crossovers should be observed. Because there are two classes of double crossovers (e^+ / ro / bv^+ and e / ro^+ / bv), we expect to observe 13 of each class.

Next, we determine the number of single-crossover progeny. The genetic map indicates that the distance between *e* and *ro* is 20 m.u.; so 360 progeny (20% of 1800) are expected to have resulted from recombination between these two loci. Some of them will be single-crossover progeny and some will be double-crossover progeny. We have already determined that the number of double-crossover progeny is 26; so the number of progeny resulting from a single crossover between *e* and *ro* is $360 - 26 = 334$, which will be divided equally between the two single-crossover phenotypes (e / ro^+ / bv^+ and e^+ / ro bv).

The distance between *ro* and *bv* is 12 m.u.; so the number of progeny resulting from recombination between these two genes is $0.12 \times 1800 = 216$. Again, some of these recombinants will be single-crossover progeny and some will be double-crossover progeny. To determine the number of progeny resulting from a single crossover, subtract the double crossovers: $216 - 26 = 190$. These single-crossover progeny will be divided between the two single-crossover phenotypes (e^+ ro^+ / bv and e ro / bv^+); so there will be $^{190}/_2 = 95$ of each of these phenotypes. The remaining progeny will be nonrecombinants, and they can be obtained by subtraction: $1800 - 26 - 334 - 190 = 1250$; there are two nonrecombinants (e^+ ro^+ bv^+ and e ro bv); so there will be $^{1250}/_2 = 625$ of each. The numbers of the various phenotypes are listed here:

e^+	ro^+	bv^+	625	nonrecombinant
e	ro	bv	625	nonrecombinant
e^+ / ro	bv		167	single crossover between *e* and *ro*
e / ro^+	bv^+		167	single crossover between *e* and *ro*
e^+	ro^+ / bv		95	single crossover between *ro* and *bv*
e	ro / bv^+		95	single crossover between *ro* and *bv*
e^+ / ro / bv^+			13	double crossover
e / ro^+ / bv			13	double crossover
Total			1800	

4. The locations of six deletions have been mapped to the *Drosophila* chromosome as shown in the following diagram.

Chromosome

Deletion 1 ————————
　 Deletion 2 ————————
　　　 Deletion 3 ————————
　　　 Deletion 4 ————————————
　　　　　　　 Deletion 5 ————————
　　 Deletion 6 ————————————

Recessive mutations *a*, *b*, *c*, *d*, *e*, *f*, and *g* are known to be located in the same regions as the deletions, but the order of the mutations on the chromosome is not known. When flies homozygous for the recessive mutations are crossed with flies homozygous for the deletions, the following results are obtained, where the letter "m" represents a mutant phenotype and a plus sign ($+$) represents the wild type. On the basis of these data, determine the relative order of the seven mutant genes on the chromosome:

Deletion	Mutations						
	a	*b*	*c*	*d*	*e*	*f*	*g*
1	+	m	m	m	+	+	+
2	+	+	m	m	+	+	+
3	+	+	+	m	m	+	+
4	m	+	+	m	m	+	+
5	m	+	+	+	+	m	m
6	m	+	+	m	m	m	+

• Solution

The offspring of the cross will be heterozygous, possessing one chromosome with the deletion and wild-type alleles and its homolog without the deletion and recessive mutant alleles. For loci within the deleted region, only the recessive mutations will be present in the offspring, which will exhibit the mutant phenotype. The presence of a mutant trait in the offspring therefore indicates that the locus for that trait is within the region covered by the deletion. We can map the genes by examining the expression of the recessive mutations in the flies with different deletions.

Mutation *a* is expressed in flies with deletions 4, 5, and 6 but not in flies with other deletions; so *a* must be in the area that is unique to deletions 4, 5, and 6:

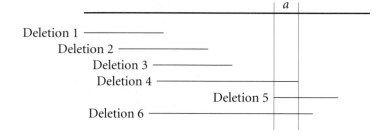

Mutation *b* is expressed only when deletion 1 is present; so *b* must be located in the region of the chromosome covered by deletion 1 and none of the other deletions:

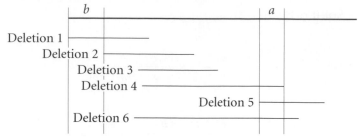

Using this procedure, we can map the remaining mutations. For each mutation, we look for the areas of overlap among deletions that express the mutations and exclude any areas of overlap that are covered by other deletions that do not express the mutation:

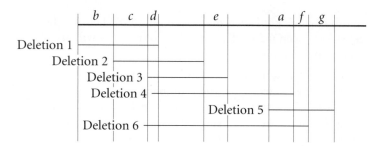

COMPREHENSION QUESTIONS

Section 7.1

*1. What does the term recombination mean? What are two causes of recombination?

Section 7.2

*2. In a testcross for two genes, what types of gametes are produced with (a) complete linkage, (b) independent assortment, and (c) incomplete linkage?

3. What effect does crossing over have on linkage?

4. Why is the frequency of recombinant gametes always half the frequency of crossing over?

*5. What is the difference between genes in coupling configuration and genes in repulsion? What effect does the arrangement of linked genes (whether they are in coupling configuration or in repulsion) have on the results of a cross?

6. How would you test to see if two genes are linked?

7. What is the difference between a genetic map and a physical map?

*8. Why do calculated recombination frequencies between pairs of loci that are located far apart underestimate the true genetic distances between loci?

Section 7.3

9. Explain how to determine, using the numbers of progeny from a three-point cross, which of three linked loci is the middle locus.

*10. What does the interference tell us about the effect of one crossover on another?

Section 7.4

11. What is a lod score and how is it calculated?

12. List some of the methods for physically mapping genes and explain how they are used to position genes on chromosomes.

APPLICATION QUESTIONS AND PROBLEMS

Introduction

13. The introduction to this chapter described the search for genes that determine pattern baldness in humans. In 1916, Dorothy Osborn suggested that pattern baldness is a sex-influenced trait (see Chapter 5) that is dominant in males and recessive in females. More research suggested that pattern baldness is an X-linked recessive trait. Would you expect to see independent assortment between genetic markers on the X chromosome and pattern baldness if (a) pattern baldness is sex-influenced and (b) if pattern baldness is X-linked recessive? Explain your answer.

Section 7.2

14. In the snail *Cepaea nemoralis,* an autosomal allele causing a banded shell (B^B) is recessive to the allele for an unbanded shell (B^O). Genes at a different locus determine the background color of the shell; here, yellow (C^Y) is recessive to brown (C^{Bw}). A banded, yellow snail is crossed with a homozygous brown, unbanded snail. The F_1 are then crossed with banded, yellow snails (a testcross).

   a. What will the results of the testcross be if the loci that control banding and color are linked with no crossing over?

   b. What will the results of the testcross be if the loci assort independently?

   c. What will the results of the testcross be if the loci are linked and 20 m.u. apart?

[Otto Hahn/Photolibrary.]

15. In silkmoths (*Bombyx mori*), red eyes (*re*) and white-banded wing (*wb*) are encoded by two mutant alleles that are recessive to those that produce wild-type traits (*re*⁺ and *wb*⁺); these two genes are on the same chromosome. A moth homozygous for red eyes and white-banded wings is crossed with a moth homozygous for the wild-type traits. The F_1 have normal eyes and normal wings. The F_1 are crossed with moths that have red eyes and white-banded wings in a testcross. The progeny of this testcross are:

wild-type eyes, wild-type wings	418
red eyes, wild-type wings	19
wild-type eyes, white-banded wings	16
red eyes, white-banded wings	426

   a. What phenotypic proportions would be expected if the genes for red eyes and for white-banded wings were located on different chromosomes?

   b. What is the percent recombination between the genes for red eyes and those for white-banded wings?

*16. A geneticist discovers a new mutation in *Drosophila melanogaster* that causes the flies to shake and quiver. She calls this mutation spastic (*sps*) and determines that spastic is due to an autosomal recessive gene. She wants to determine if the gene encoding spastic is linked to the recessive gene for vestigial wings (*vg*). She crosses a fly homozygous for spastic and vestigial traits with a fly homozygous for the wild-type traits and then uses the resulting F_1 females in a testcross. She obtains the following flies from this testcross.

vg^+	sps^+	230
vg	sps	224
vg	sps^+	97
vg^+	sps	99
Total		650

   Are the genes that cause vestigial wings and the spastic mutation linked? Do a chi-square test of independence to determine whether the genes have assorted independently.

17. In cucumbers, heart-shaped leaves (*hl*) are recessive to normal leaves (*Hl*) and having numerous fruit spines (*ns*) is recessive to having few fruit spines (*Ns*). The genes for leaf shape and for number of spines are located on the same chromosome; findings from mapping experiments indicate that they are 32.6 m.u. apart. A cucumber plant having heart-shaped leaves and numerous spines is

crossed with a plant that is homozygous for normal leaves and few spines. The F$_1$ are crossed with plants that have heart-shaped leaves and numerous spines. What phenotypes and proportions are expected in the progeny of this cross?

*18. In tomatoes, tall (D) is dominant over dwarf (d) and smooth fruit (P) is dominant over pubescent fruit (p), which is covered with fine hairs. A farmer has two tall and smooth tomato plants, which we will call plant A and plant B. The farmer crosses plants A and B with the same dwarf and pubescent plant and obtains the following numbers of progeny:

	Progeny of	
	Plant A	**Plant B**
Dd Pp	122	2
Dd pp	6	82
dd Pp	4	82
dd pp	124	4

a. What are the genotypes of plant A and plant B?

b. Are the loci that determine height of the plant and pubescence linked? If so, what is the percent recombination between them?

c. Explain why different proportions of progeny are produced when plant A and plant B are crossed with the same dwarf pubescent plant.

19. Alleles *A* and *a* are at a locus on the same chromosome as is a locus with alleles *B* and *b*. *Aa Bb* is crossed with *aa bb* and the following progeny are produced:

Aa Bb	5
Aa bb	45
aa Bb	45
aa bb	5

What conclusion can be made about the arrangement of the genes on the chromosome in the *Aa Bb* parent?

20. Daniel McDonald and Nancy Peer determined that eyespot (a clear spot in the center of the eye) in flour beetles is caused by an X-linked gene (*es*) that is recessive to the allele for the absence of eyespot (*es$^+$*). They conducted a series of crosses to determine the distance between the gene for eyespot and a dominant X-linked gene for striped (*St*), which causes white stripes on females and acts as a recessive lethal (is lethal when homozygous in females or hemizygous in males). The following cross was carried out (D. J. McDonald and N. J. Peer. 1961. *Journal of Heredity* 52:261–264).

$$\female \; \frac{es^+ \quad St}{es \quad St^+} \times \frac{es \quad St^+}{Y} \; \male$$

$$\downarrow$$

$\dfrac{es^+ \quad St}{es \quad St^+}$	1630
$\dfrac{es \quad St^+}{es \quad St^+}$	1665
$\dfrac{es \quad St}{es \quad St^+}$	935
$\dfrac{es^+ \quad St^+}{es \quad St^+}$	1005
$\dfrac{es \quad St^+}{Y}$	1661
$\dfrac{es^+ \quad St^+}{Y}$	1024

a. Which progeny are the recombinants and which progeny are the nonrecombinants?

b. Calculate the recombination frequency between *es* and *St*.

c. Are some potential genotypes missing among the progeny of the cross? If so, which ones and why?

*21. Recombination rates between three loci in corn are shown here.

Loci	**Recombination rate**
R and W_2	17%
R and L_2	35%
W_2 and L_2	18%

What is the order of the genes on the chromosome?

22. In tomatoes, dwarf (d) is recessive to tall (D) and opaque (light-green) leaves (op) are recessive to green leaves (Op). The loci that determine height and leaf color are linked and separated by a distance of 7 m.u. For each of the following crosses, determine the phenotypes and proportions of progeny produced.

a. $\dfrac{D \quad Op}{d \quad op} \times \dfrac{d \quad op}{d \quad op}$

b. $\dfrac{D \quad op}{d \quad Op} \times \dfrac{d \quad op}{d \quad op}$

c. $\dfrac{D \quad Op}{d \quad op} \times \dfrac{D \quad Op}{d \quad op}$

d. $\dfrac{D \quad op}{d \quad Op} \times \dfrac{D \quad op}{d \quad Op}$

23. In German cockroaches, bulging eyes (*bu*) are recessive to normal eyes (*bu*⁺) and curved wings (*cv*) are recessive to straight wings (*cv*⁺). Both traits are encoded by autosomal genes that are linked. A cockroach has genotype $bu^+bu\ cv^+cv$, and the genes are in repulsion. Which of the following sets of genes will be found in the most-common gametes produced by this cockroach?

[W. Willner/Photolibrary.]

a. $bu^+\ cv^+$

b. $bu\ cv$

c. $bu^+\ bu$

d. $cv^+\ cv$

e. $bu\ cv^+$

Explain your answer.

*24. In *Drosophila melanogaster*, ebony body (*e*) and rough eyes (*ro*) are encoded by autosomal recessive genes found on chromosome 3; they are separated by 20 m.u. The gene that encodes forked bristles (*f*) is X-linked recessive and assorts independently of *e* and *ro*. Give the phenotypes of progeny and their expected proportions when a female of each of the following genotypes is test-crossed with a male.

a. $\dfrac{e^+\quad ro^+}{e\quad ro}\ \dfrac{f^+}{f}$

b. $\dfrac{e^+\quad ro}{e\quad ro^+}\ \dfrac{f^+}{f}$

25. Honeybees have haplodiploid sex determination: females are diploid, developing from fertilized eggs, whereas males are haploid, developing from unfertilized eggs. Otto Mackensen studied linkage relations among eight mutations in honeybees (O. Mackensen. 1958. *Journal of Heredity* 49:99–102). The following table gives the results of two of MacKensen's crosses including three recessive mutations: *cd* (cordovan body color), *h* (hairless), and *ch* (chartreuse eye color).

Queen genotype	Phenotypes of drone (male) progeny
$\dfrac{cd\quad h^+}{cd^+\quad h}$	294 cordovan, 236 hairless, 262 cordovan and hairless, 289 wild .type
$\dfrac{h\quad ch^+}{h^+\quad ch}$	3131 hairless, 3064 chartreuse, 96 chartreuse and hairless, 132 wild type

a. Only the genotype of the queen is given. Why is the genotype of the male parent not needed for mapping these genes? Would the genotype of the male parent be required if we examined female progeny instead of male progeny?

b. Determine the nonrecombinant and recombinant progeny for each cross and calculate the map distances between *cd*, *h*, and *ch*. Draw a linkage map illustrating the linkage arrangements among these three genes.

*26. A series of two-point crosses were carried out among seven loci (*a*, *b*, *c*, *d*, *e*, *f*, and *g*), producing the following recombination frequencies. Map the seven loci, showing their linkage groups, the order of the loci in each linkage group, and the distances between the loci of each group.

Loci	Percent recombination	Loci	Percent recombination
a and *b*	50	*c* and *d*	50
a and *c*	50	*c* and *e*	26
a and *d*	12	*c* and *f*	50
a and *e*	50	*c* and *g*	50
a and *f*	50	*d* and *e*	50
a and *g*	4	*d* and *f*	50
b and *c*	10	*d* and *g*	8
b and *d*	50	*e* and *f*	50
b and *e*	18	*e* and *g*	50
b and *f*	50	*f* and *g*	50
b and *g*	50		

27. R. W. Allard and W. M. Clement determined recombination rates for a series of genes in lima beans (R. W. Allard and W. M. Clement. 1959. *Journal of Heredity* 50:63–67). The following table lists paired recombination rates for eight of the loci (*D*, *Wl*, *R*, *S*, L_1, *Ms*, *C*, and *G*) that they mapped. On the basis of these data, draw a series of genetic maps for the different linkage groups of the genes, indicating the distances between the genes. Keep in mind that these rates are estimates of the true recombination rates and that some error is associated with each estimate. An asterisk beside a recombination frequency indicates that the recombination frequency is significantly different from 50%.

Recombination Rates (%) among Seven Loci in Lima Beans

	Wl	R	S	L_1	Ms	C	G
D	2.1*	39.3*	52.4	48.1	53.1	51.4	49.8
Wl		38.0*	47.3	47.7	48.8	50.3	50.4
R			51.9	52.7	54.6	49.3	52.6
S				26.9*	54.9	52.0	48.0
L_1					48.2	45.3	50.4
Ms						14.7*	43.1
C							52.0

*Significantly different from 50%.

Section 7.3

28. Raymond Popp studied linkage among genes for pink eye (*p*), shaker-1 (*sh-1*, which causes circling behavior, head tossing, and deafness), and hemoglobin (*Hb*) in mice (R. A. Popp. 1962. *Journal of Heredity* 53:73–80). He performed a series of testcrosses, in which mice heterozygous for pink eye, shaker-1, and hemoglobin 1 and 2 were crossed with mice that were homozygous for pink eye, shaker-1, and hemoglobin 2.

$$\frac{P\ Sh\text{-}1\ Hb^1}{p\ sh\text{-}1\ Hb^2} \times \frac{p\ sh\text{-}1\ Hb^2}{p\ sh\text{-}1\ Hb^2}$$

The following progeny were produced.

Progeny genotype	Number
$\dfrac{p\ sh\text{-}1\ Hb^2}{p\ sh\text{-}1\ Hb^2}$	274
$\dfrac{P\ Sh\text{-}1\ Hb^1}{p\ sh\text{-}1\ Hb^2}$	320
$\dfrac{P\ sh\text{-}1\ Hb^2}{p\ sh\text{-}1\ Hb^2}$	57
$\dfrac{p\ Sh\text{-}1\ Hb^1}{p\ sh\text{-}1\ Hb^2}$	45
$\dfrac{p\ Sh\text{-}1\ Hb^2}{p\ sh\text{-}1\ Hb^2}$	6
$\dfrac{p\ sh\text{-}1\ Hb^1}{p\ sh\text{-}1\ Hb^2}$	5
$\dfrac{p\ Sh\text{-}1\ Hb^2}{p\ sh\text{-}1\ Hb^2}$	0
$\dfrac{P\ sh\text{-}1\ Hb^1}{p\ sh\text{-}1\ Hb^2}$	1
Total	708

a. Determine the order of these genes on the chromosome.

b. Calculate the map distances between the genes.

c. Determine the coefficient of coincidence and the interference among these genes.

*29. Waxy endosperm (wx), shrunken endosperm (sh), and yellow seedling (v) are encoded by three recessive genes in corn that are linked on chromosome 5. A corn plant homozygous for all three recessive alleles is crossed with a plant homozygous for all the dominant alleles. The resulting F_1 are then crossed with a plant homozygous for the recessive alleles in a three-point testcross. The progeny of the testcross are:

wx	sh	V	87
Wx	Sh	v	94
Wx	Sh	V	3,479
wx	sh	v	3,478
Wx	sh	V	1,515
wx	Sh	v	1,531
wx	Sh	V	292
Wx	sh	v	280
Total			10,756

a. Determine the order of these genes on the chromosome.

b. Calculate the map distances between the genes.

c. Determine the coefficient of coincidence and the interference among these genes.

30. Priscilla Lane and Margaret Green studied the linkage relations of three genes affecting coat color in mice: mahogany (mg), agouti (a), and ragged (Rg). They carried out a series of three-point crosses, mating mice that were heterozygous at all three loci with mice that were homozygous for the recessive alleles at these loci (P. W. Lane and M. C. Green. 1960. *Journal of Heredity* 51:228–230). The following table lists the results of the testcrosses.

Phenotype			Number
a	Rg	$+$	1
$+$	$+$	mg	1
a	$+$	$+$	15
$+$	Rg	mg	9
$+$	$+$	$+$	16
a	Rg	mg	36
a	$+$	mg	76
$+$	Rg	$+$	69
Total			213

Note: $+$ represents a wild-type allele.

a. Determine the order of the loci that encode mahogany, agouti, and ragged on the chromosome, the map distances between them, and the interference and coefficient of coincidence for these genes.

b. Draw a picture of the two chromosomes in the triply heterozygous mice used in the testcrosses, indicating which of the alleles are present on each chromosome.

31. Fine spines (s), smooth fruit (tu), and uniform fruit color (u) are three recessive traits in cucumbers, the genes of which are linked on the same chromosome. A cucumber plant heterozygous for all three traits is used in a testcross, and the following progeny are produced from this testcross:

S	U	Tu	2
s	u	Tu	70
S	u	Tu	21
s	u	tu	4
S	U	tu	82
s	U	tu	21
s	U	Tu	13
S	u	tu	17
Total			230

a. Determine the order of these genes on the chromosome.

b. Calculate the map distances between the genes.

c. Determine the coefficient of coincidence and the interference among these genes.

d. List the genes found on each chromosome in the parents used in the testcross.

*32. In *Drosophila melanogaster*, black body (*b*) is recessive to gray body (*b⁺*), purple eyes (*pr*) are recessive to red eyes (*pr⁺*), and vestigial wings (*vg*) are recessive to normal wings (*vg⁺*). The loci encoding these traits are linked, with the following map distances:

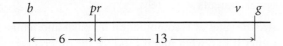

The interference among these genes is 0.5. A fly with a black body, purple eyes, and vestigial wings is crossed with a fly homozygous for a gray body, red eyes, and normal wings. The female progeny are then crossed with males that have a black body, purple eyes, and vestigial wings. If 1000 progeny are produced from this testcross, what will be the phenotypes and proportions of the progeny?

33. A group of geneticists are interested in identifying genes that may play a role in susceptibility to asthma. They study the inheritance of genetic markers in a series of families that have two or more asthmatic children. They find an association between the presence or absence of asthma and a genetic marker on the short arm of chromosome 20 and calculate a lod score of 2 for this association. What does this lod score indicate about genes that may influence asthma?

Section 7.4

*34. The locations of six deletions have been mapped to the *Drosophila* chromosome, as shown in the following deletion map. Recessive mutations *a*, *b*, *c*, *d*, *e*, and *f* are known to be located in the same region as the deletions, but the order of the mutations on the chromosome is not known.

```
                Chromosome
            _____

Deletion 1  ———————————
      Deletion 2  ——
            Deletion 3  ——————————————
            Deletion 4  ————————
                Deletion 5  ————
                    Deletion 6  ——————————
```

When flies homozygous for the recessive mutations are crossed with flies homozygous for the deletions, the

following results are obtained, in which "m" represents a mutant phenotype and a plus sign (+) represents the wild type. On the basis of these data, determine the relative order of the seven mutant genes on the chromosome:

			Mutations			
Deletion	*a*	*b*	*c*	*d*	*e*	*f*
1	m	+	m	+	+	m
2	m	+	+	+	+	+
3	+	m	m	m	m	+
4	+	+	m	m	m	+
5	+	+	+	m	m	+
6	+	m	+	m	+	+

35. A panel of cell lines was created from human–mouse somatic-cell fusions. Each line was examined for the presence of human chromosomes and for the production of an enzyme. The following results were obtained:

Cell line	Enzyme	1	2	3	4	5	6	7	8	9	10	17	22
A	–	+	–	–	–	+	–	–	–	–	–	+	–
B	+	+	+	–	–	–	–	–	+	–	–	+	+
C	–	+	–	–	–	+	–	–	–	–	–	–	+
D	–	–	–	–	+	–	–	–	–	–	–	–	–
E	+	+	–	–	–	–	–	–	+	–	+	+	–

On the basis of these results, which chromosome has the gene that encodes the enzyme?

*36. A panel of cell lines was created from human–mouse somatic-cell fusions. Each line was examined for the presence of human chromosomes and for the production of three enzymes. The following results were obtained.

Cell line	Enzyme 1	Enzyme 2	Enzyme 3	4	8	9	12	15	16	17	22	X
A	+	–	+	–	–	+	–	+	+	–	–	+
B	+	–	–	–	–	+	–	–	+	+	–	–
C	–	+	+	+	–	–	–	–	–	+	–	+
D	–	+	+	+	+	–	–	–	+	–	–	+

On the basis of these results, give the chromosome location of enzyme 1, enzyme 2, and enzyme 3.

CHALLENGE QUESTION

Section 7.5

37. Transferrin is a blood protein that is encoded by the transferrin locus (*Trf*). In house mice, the two alleles at this locus (*Trf ᵃ* and *Trf ᵇ*) are codominant and encode three types of transferrin:

Genotype	Phenotype
Trf ᵃ/Trf ᵃ	Trf-a
Trf ᵃ/Trf ᵇ	Trf-ab
Trf ᵇ/Trf ᵇ	Trf-b

The dilution locus, found on the same chromosome, determines whether the color of a mouse is diluted or full; an allele for dilution (*d*) is recessive to an allele for full color (*d⁺*):

Genotype	Phenotype
d^+d^+	d^+ (full color)
d^+d	d^+ (full color)
dd	d (dilution)

Donald Shreffler conducted a series of crosses to determine the map distance between the tranferrin locus and the dilution locus (D. C. Shreffler. 1963 *Journal of Heredity* 54:127–129). The following table presents a series of crosses carried out by Shreffler and the progeny resulting from these crosses.

			Progeny phenotypes				
Cross	♂	♀	d^+ Trf-ab	d^+ Trf-b	d Trf-ab	d Trf-b	Total
1	$\dfrac{d^+\ Trf^a}{d\ Trf^b}$	$\times\ \dfrac{d\ Trf^b}{d\ Trf^b}$	32	3	6	21	62
2	$\dfrac{d\ Trf^b}{d\ Trf^b}$	$\times\ \dfrac{d^+\ Trf^a}{d\ Trf^b}$	16	0	2	20	38
3	$\dfrac{d^+\ Trf^a}{d\ Trf^b}$	$\times\ \dfrac{d\ Trf^b}{d\ Trf^b}$	35	9	4	30	78
4	$\dfrac{d\ Trf^b}{d\ Trf^b}$	$\times\ \dfrac{d^+\ Trf^a}{d\ Trf^b}$	21	3	2	19	45
5	$\dfrac{d^+\ Trf^b}{d\ Trf^a}$	$\times\ \dfrac{d\ Trf^b}{d\ Trf^b}$	8	29	22	5	64
6	$\dfrac{d\ Trf^b}{d\ Trf^b}$	$\times\ \dfrac{d^+\ Trf^b}{d\ Trf^a}$	4	14	11	0	29

a. Calculate the recombinant frequency between the *Trf* and the *d* loci by using the pooled data from all the crosses.

b. Which crosses represent recombination in male gamete formation and which crosses represent recombination in female gamete formation?

c. On the basis of your answer to part *b*, calculate the frequency of recombination among male parents and female parents separately.

d. Are the rates of recombination in males and females the same? If not, what might produce the difference?

A mouse with the dilution trait. [L. Montoliu et al., *Color Genes*. (European Society for Pigment Cell Research, 2010).]

8 Bacterial and Viral Genetic Systems

Bacteria account for most of life's diversity and exist in almost every conceivable environment, including inhospitable habitats such as the highly saline Dead Sea. [PhotoStock-Israel/Alamy.]

LIFE IN A BACTERIAL WORLD

Humans like to think that that they rule the world but, compared with bacteria, we are clearly in a minor position. Bacteria first evolved some 2.5 billion years ago, a billion years before the first eukaryotes appeared. Today, bacteria are found in every conceivable environment, including boiling springs, highly saline lakes, and beneath more than 2 miles of ice in Antarctica. They are found at the top of Mt. Everest and at the bottoms of the deepest oceans. They are also present on and in *us*—in alarming numbers! Within the average human gut, there are about 10 trillion bacteria or more, ten times the total number of cells in the entire human body. No one knows how many bacteria populate the world, but an analysis conducted by scientists in 1998 estimated that the total number of living bacteria on Earth cxcccdcd 5 million trillion (5×10^{30}).

Not only are bacteria numerically vast, they also constitute the majority of life's diversity. The total number of described species of bacteria is only about 6800, compared with about 1.4 million plants, animals, fungi, and single-celled eukaryotes. But the number of described species falls far short of the true microbial diversity. Species of bacteria are typically described only after they are cultivated and studied in the laboratory, and microbiologists now recognize that only a small fraction of all bacterial species are amenable to laboratory culture. For many years, it was impossible to identify and study most bacteria because they could not be grown in the laboratory. Then, in the 1970s, molecular techniques for analyzing DNA became available and opened up a whole new vista on microbial diversity. These techniques revealed several important facts about bacteria. First, many of the relations among bacteria that microbiologists had worked out on the basis of physical and biochemical traits turned out to be incorrect. Bacteria once thought to be related were in fact genetically quite different. Second, molecular analysis showed that one whole group of microbes—now called the archaea—were as different from other bacteria (the eubacteria) as they are from eukaryotes. Third, molecular analysis revealed that the number of different types of bacteria is astounding.

In 2007, Luiz Roesch and his colleagues set out to determine exactly how many types of bacteria exist in a gram of soil. They obtained soil samples from four locations: Brazil, Florida, Illinois, and Canada. From the soil samples, they extracted and purified bacterial DNA. From this DNA, they determined the sequences of a gene present in all bacteria, the 16S rRNA gene. Each different species of bacteria has a unique 16S rRNA gene sequence, and so they could determine how many species of bacteria there were in each soil sample by counting the number of different DNA sequences.

Roesch's results were amazing. The number of different eubacterial species in each gram of soil ranged from 26,140 for samples from Brazil to 53,533 for Canadian samples. Many unusual bacteria were detected that appeared dissimilar to all previously described groups of bacteria. Another interesting finding was that soil from agricultural fields harbored considerably fewer species than did soil from forests.

This study and others demonstrate that bacterial diversity far exceeds that of multicellular organisms and, undoubtedly, numerous groups of bacteria have yet to be discovered. Like it or not, we truly live in a bacterial world.

In this chapter, we examine some of the genetic properties of bacteria and viruses and the mechanisms by which they exchange and recombine their genes. Since the 1940s, the genetic systems of bacteria and viruses have contributed to the discovery of many important concepts in genetics. The study of molecular genetics initially focused almost entirely on their genes; today, bacteria and viruses are still essential tools for probing the nature of genes in more-complex organisms, in part because they possess a number of characteristics that make them suitable for genetic studies (**Table 8.1**).

The genetic systems of bacteria and viruses are also studied because these organisms play important roles in human society. Bacteria are found naturally in the mouth, gut, and on the skin, where they play essential roles in human ecology. They have been harnessed to produce a number of economically important substances, and they are of immense medical significance, causing many human diseases. In this chapter, we focus on several unique aspects of bacterial and viral genetic systems. Important processes of gene transfer and recombination will be described, and we will see how these processes can be used to map bacterial and viral genes. TRY PROBLEM 16 →

Table 8.1	Advantages of using bacteria and viruses for genetic studies
1.	Reproduction is rapid.
2.	Many progeny are produced.
3.	The haploid genome allows all mutations to be expressed directly.
4.	Asexual reproduction simplifies the isolation of genetically pure strains.
5.	Growth in the laboratory is easy and requires little space.
6.	Genomes are small.
7.	Techniques are available for isolating and manipulating their genes.
8.	They have medical importance.
9.	They can be genetically engineered to produce substances of commercial value.

8.1 Genetic Analysis of Bacteria Requires Special Methods

Heredity in bacteria is fundamentally similar to heredity in more-complex organisms, but the bacterial haploid genome and the small size of bacteria (which makes observation of their phenotypes difficult) require different approaches and methods.

Bacterial Diversity

Prokaryotes are unicellular organisms that lack nuclear membranes and membrane-bounded cell organelles. For many years, biologists considered all prokaryotes to be related, but DNA sequence information now provides convincing evidence that prokaryotes are divided into at least two distinct groups: the archaea and the eubacteria. The archaea are a group of diverse prokaryotes that are frequently found in extreme environments, such as hot springs and at the bottoms of oceans. The eubacteria are the remaining prokaryotes, including most of the familiar bacteria. Though superficially similar in their cell structure, eubacteria and archaea are distinct in their genetic makeup, and are as different from one another as each is to eukaryotic organisms. In fact, the archaea are *more* similar to eukaryotes than to eubacteria in a number of molecular features and genetic processes. In this book, the term bacteria is generally used to refer to eubacteria.

Bacteria are extremely diverse and come in a variety of shapes and sizes. Some are rod shaped, whereas others are spherical. Most are much smaller than eukaryotic cells, but at least one species isolated from the gut of fish is almost 1 mm long and can be seen with the naked eye. Some bacteria are photosynthetic. Others produce stalks and spores, superficially resembling fungi.

Bacteria have long been considered simple organisms that lack much of the cellular complexity of eukaryotes. However, recent evidence points to a number of similarities and parallels in bacterial and eukaryotic structure. For example, a bacterial protein termed FtsZ, which plays an integral part in cell division, is structurally similar to eukaryotic tubulin proteins, which are subunits of microtubules and help segregate chromosomes in mitosis and meiosis, as described in Chapter 2. Like eukaryotes, bacteria have

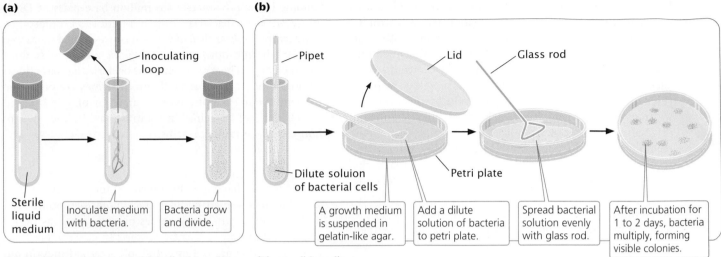

(a)

Inoculating loop

Sterile liquid medium

| Inoculate medium with bacteria. | Bacteria grow and divide. |

(b)

Pipet

Lid

Glass rod

Dilute soluion of bacterial cells

Petri plate

| A growth medium is suspended in gelatin-like agar. | Add a dilute solution of bacteria to petri plate. | Spread bacterial solution evenly with glass rod. | After incubation for 1 to 2 days, bacteria multiply, forming visible colonies. |

8.1 Bacteria can be grown (a) in liquid medium or (b) on solid medium.

proteins that help condense DNA. Other bacterial proteins function much as cytoskeletal proteins do in eukaryotes, helping to give bacterial cells shape and structure. And, although bacteria don't go through mitosis and meiosis, replication of the bacterial chromosome precedes binary fission, and there are bacterial processes that ensure that one copy of the chromosome is allocated to each cell.

Techniques for the Study of Bacteria

The culture and study of bacteria require special techniques. Microbiologists have defined the nutritional needs of a number of bacteria and developed culture media for growing them in the laboratory. Culture media typically contain a carbon source, essential elements such as nitrogen and phosphorus, certain vitamins, and other required ions and nutrients. Wild-type, or **prototrophic**, bacteria can use these simple ingredients to synthesize all the compounds that they need for growth and reproduction. A medium that contains only the nutrients required by prototrophic bacteria is termed **minimal medium**. Mutant strains called **auxotrophs** lack one or more enzymes necessary for synthesizing essential molecules and will grow only on medium supplemented with these essential molecules. For example, auxotrophic strains that are unable to synthesize the amino acid leucine will not grow on minimal medium but *will* grow on medium to which leucine has been added. **Complete medium** contains all the substances, such as the amino acid leucine, required by bacteria for growth and reproduction.

Cultures of bacteria are often grown in test tubes that contain sterile liquid medium (**Figure 8.1a**). A few bacteria are added to a tube, and they grow and divide until all the nutrients are used up or—more commonly—until the concentration of their waste products becomes toxic. Bacteria are also grown on petri plates (**Figure 8.1b**). Growth medium suspended in agar is poured into the bottom half of the petri plate, providing a solid, gel-like base for bacterial growth. In a process called plating, a dilute solution of bacteria is spread over the surface of an agar-filled petri plate. As each bacterium grows and divides, it gives rise to a visible clump of genetically identical cells (a **colony**). Genetically pure strains of the bacteria can be isolated by collecting bacteria from a single colony and transferring them to a new test tube or petri plate. The chief advantage of this method is that it allows one to isolate and count bacteria, which individually are too small to see without a microscope.

Because individual bacteria are too small to be seen directly, it is often easier to study phenotypes that affect the appearance of the colony (**Figure 8.2**) or can be detected

(a)

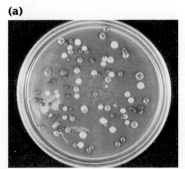

(b)

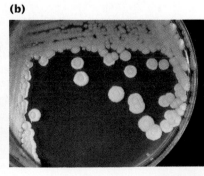

8.2 Bacteria have a variety of phenotypes. (a) *Serratia marcescens* with color variation. (b) *Bacillus cereus.* [Part a: Dr. E. Bottone/Peter Arnold. Part b: Biophoto Associates/Photo Researchers.]

by simple chemical tests. Auxotrophs are commonly studied phenotypes. Suppose we want to detect auxotrophs that cannot synthesize leucine (*leu*⁻ mutants). We first spread the bacteria on a petri plate containing medium that includes leucine; both prototrophs that have the *leu*⁺ allele and auxotrophs that have *leu*⁻ allele will grow on it (**Figure 8.3**). Next, using a technique called replica plating, we transfer a few cells from each of the colonies on the original plate to two new replica plates: one plate contains medium to which leucine has been added; the other plate contains a medium lacking leucine. A medium that lacks an essential nutrient, such as the medium lacking leucine, is called a selective medium. The *leu*⁺ bacteria will grow on both media, but the *leu*⁻ mutants will grow only on the selective medium supplemented by leucine, because they cannot synthesize their own leucine. Any colony that grows on medium that contains leucine but not on medium that lacks leucine consists of *leu*⁻ bacteria. The auxotrophs that grow on the supplemented medium can then be cultured for further study.

The Bacterial Genome

Most bacterial genomes consist of a circular chromosome that contains a single DNA molecule several million base pairs (bp) in length (**Figure 8.4**). For example, the *E. coli* genome has approximately 4.6 million base pairs of DNA. However, some bacteria contain multiple chromosomes. For example, *Vibrio cholerae*, which causes cholera, has two circular chromosomes, and *Rhizobium meliloti* has three chromosomes. There are even a few bacteria that have linear chromosomes. Many bacterial chromosomes are organized efficiently. For example, more than 90% of the DNA in *E. coli* encodes proteins. In contrast, only about 1% of human DNA encodes proteins.

Plasmids

In addition to having a chromosome, many bacteria possess **plasmids**—small, circular DNA molecules. Some plasmids are present in many copies per cell, whereas others are present in only one or two copies. In general, plasmids carry genes that are not essential to bacterial function but that may play an important role in the life cycle and growth of their bacterial hosts. There are many different types of plasmids; *E. coli* alone is estimated to have more than 270 different naturally occurring plasmids. Some plasmids promote mating between bacteria; others contain genes that kill other bacteria. Of great importance, plasmids are used extensively in genetic engineering (see Chapter 19), and some of them play a role in the spread of antibiotic resistance among bacteria.

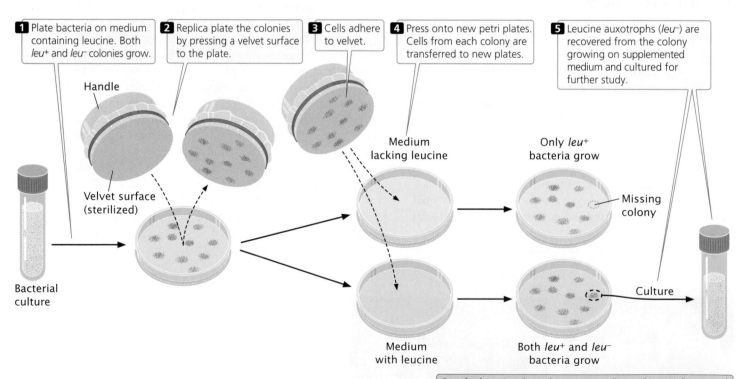

1. Plate bacteria on medium containing leucine. Both *leu*⁺ and *leu*⁻ colonies grow.

2. Replica plate the colonies by pressing a velvet surface to the plate.

3. Cells adhere to velvet.

4. Press onto new petri plates. Cells from each colony are transferred to new plates.

5. Leucine auxotrophs (*leu*⁻) are recovered from the colony growing on supplemented medium and cultured for further study.

Handle

Velvet surface (sterilized)

Bacterial culture

Medium lacking leucine

Only *leu*⁺ bacteria grow

Missing colony

Medium with leucine

Both *leu*⁺ and *leu*⁻ bacteria grow

Culture

8.3 Mutant bacterial strains can be isolated on the basis of their nutritional requirements.

Conclusion: A colony that grows only on the supplemented medium has a mutation in a gene that encodes the synthesis of an essential nutrient.

Most plasmids are circular and several thousand base pairs in length, although plasmids consisting of several hundred thousand base pairs also have been found. Each plasmid possesses an origin of replication, a specific DNA sequence where DNA replication is initiated (see Chapter 2). The origin allows a plasmid to replicate independently of the bacterial chromosome (**Figure 8.5**). **Episomes** are plasmids that are capable of freely replicating and able to integrate into the bacterial chromosomes. The **F** (fertility) **factor** of *E. coli* (**Figure 8.6**) is an episome that controls mating and gene exchange between *E. coli* cells, as will be discussed in the next section.

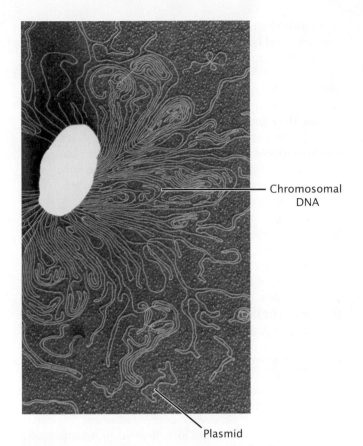

— Chromosomal DNA

— Plasmid

8.4 Most bacterial cells possess a single, circular chromosome, shown here emerging from a ruptured bacterial cell. Many bacteria contain plasmids—small, circular molecules of DNA. [Drs. H. Potter and D. Dressler/Visuals Unlimited.]

CONCEPTS

Bacteria can be studied in the laboratory by growing them on defined liquid or solid medium. A typical bacterial genome consists of a single circular chromosome that contains several million base pairs. Some bacterial genes may be present on plasmids, which are small, circular DNA molecules that replicate independently of the bacterial chromosome.

✔ CONCEPT CHECK 1

Which is true of plasmids?

a. They are composed of RNA.

b. They normally exist outside of bacterial cells.

c. They possess only a single strand of DNA.

d. They contain an origin of replication.

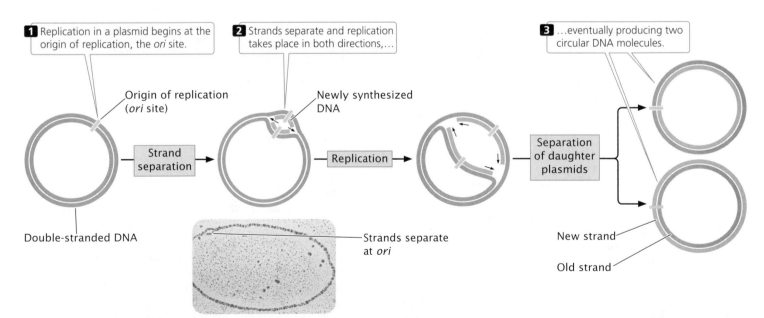

1 Replication in a plasmid begins at the origin of replication, the *ori* site.

2 Strands separate and replication takes place in both directions,…

3 …eventually producing two circular DNA molecules.

Origin of replication (*ori* site)

Newly synthesized DNA

Strand separation

Replication

Separation of daughter plasmids

Double-stranded DNA

Strands separate at *ori*

New strand

Old strand

8.5 A plasmid replicates independently of its bacterial chromosome. Replication begins at the origin of replication (*ori*) and continues around the circle. In this diagram, replication is taking place in both directions; in some plasmids, replication is in one direction only.

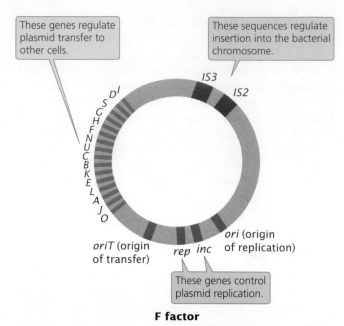

These genes regulate plasmid transfer to other cells.

These sequences regulate insertion into the bacterial chromosome.

oriT (origin of transfer)

ori (origin of replication)

These genes control plasmid replication.

F factor

8.6 The F factor, a circular episome of _E. coli_, contains a number of genes that regulate transfer into the bacterial cell, replication, and insertion into the bacterial chromosome. Replication is initiated at _ori_. Insertion sequences (see Chapter 11) _IS3_ and _IS2_ control insertion into the bacterial chromosome and excision from it.

8.2 Bacteria Exchange Genes Through Conjugation, Transformation, and Transduction

Bacteria exchange genetic material by three different mechanisms, all entailing some type of DNA transfer and recombination between the transferred DNA and the bacterial chromosome.

1. **Conjugation** takes place when genetic material passes directly from one bacterium to another (**Figure 8.7a**). In conjugation, two bacteria lie close together and a connection forms between them. A plasmid or a part of the bacterial chromosome passes from one cell (the donor) to the other (the recipient). Subsequent to conjugation, crossing over may take place between homologous sequences in the transferred DNA and the chromosome of the recipient cell. In conjugation, DNA is transferred only from donor to recipient, with no reciprocal exchange of genetic material.

2. **Transformation** takes place when a bacterium takes up DNA from the medium in which it is growing (**Figure 8.7b**). After transformation, recombination may take place between the introduced genes and those of the bacterial chromosome.

3. **Transduction** takes place when bacterial viruses (bacteriophages) carry DNA from one bacterium to

another (**Figure 8.7c**). Inside the bacterium, the newly introduced DNA may undergo recombination with the bacterial chromosome.

Not all bacterial species exhibit all three types of genetic transfer. Conjugation takes place more frequently in some species than in others. Transformation takes place to a limited extent in many species of bacteria, but laboratory techniques increase the rate of DNA uptake. Most bacteriophages have a limited host range; so transduction is normally between bacteria of the same or closely related species only.

These processes of genetic exchange in bacteria differ from diploid eukaryotic sexual reproduction in two important ways. First, DNA exchange and reproduction are not coupled in bacteria. Second, donated genetic material that is not recombined into the host DNA is usually degraded, and so the recipient cell remains haploid. Each type of genetic transfer can be used to map genes, as will be discussed in the following sections.

CONCEPTS

DNA may be transferred between bacterial cells through conjugation, transformation, or transduction. Each type of genetic transfer consists of a one-way movement of genetic information to the recipient cell, sometimes followed by recombination. These processes are not connected to cellular reproduction in bacteria.

✔ CONCEPT CHECK 2

Which process of DNA transfer in bacteria requires a virus?

a. Conjugation

b. Transduction

c. Transformation

d. All of the above

Conjugation

In 1946, Joshua Lederberg and Edward Tatum demonstrated that bacteria can transfer and recombine genetic information, paving the way for the use of bacteria in genetic studies. In the course of their research, Lederberg and Tatum studied auxotrophic strains of _E. coli_. The Y10 strain required the amino acids threonine (and was genotypically thr^-) and leucine (leu^-) and the vitamin thiamine (thi^-) for growth but did not require the vitamin biotin (bio^+) or the amino acids phenylalanine (phe^+) and cysteine (cys^+); the genotype of this strain can be written as thr^- leu^- thi^- bio^+ phe^+ cys^+. The Y24 strain had the opposite set of alleles: it required biotin, phenylalanine, and cysteine in its medium, but it did not require threonine, leucine, or thiamine; its genotype was thr^+ leu^+ thi^+ bio^- phe^- cys^-. In one experiment, Lederberg and Tatum mixed Y10 and Y24

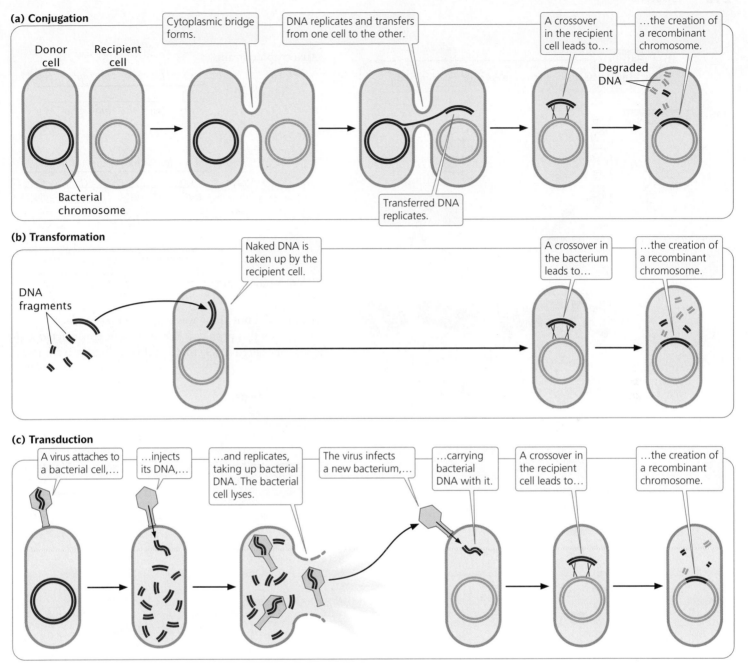

(a) Conjugation

Cytoplasmic bridge forms.

DNA replicates and transfers from one cell to the other.

A crossover in the recipient cell leads to...

...the creation of a recombinant chromosome.

Donor cell

Recipient cell

Bacterial chromosome

Transferred DNA replicates.

Degraded DNA

(b) Transformation

Naked DNA is taken up by the recipient cell.

A crossover in the bacterium leads to...

...the creation of a recombinant chromosome.

DNA fragments

(c) Transduction

A virus attaches to a bacterial cell,...

...injects its DNA,...

...and replicates, taking up bacterial DNA. The bacterial cell lyses.

The virus infects a new bacterium,...

...carrying bacterial DNA with it.

A crossover in the recipient cell leads to...

...the creation of a recombinant chromosome.

8.7 Conjugation, transformation, and transduction are three processes of gene transfer in bacteria. For the transferred DNA to be stably inherited, all three processes require the transferred DNA to undergo recombination with the bacterial chromosome.

bacteria together and plated them on minimal medium (**Figure 8.8**). Each strain was also plated separately on minimal medium.

Alone, neither Y10 nor Y24 grew on minimal medium. Strain Y10 was unable to grow, because it required threonine, leucine, and thiamine, which were absent in the minimal medium; strain Y24 was unable to grow, because it required biotin, phenylalanine, and cysteine, which also were absent from the minimal medium. When Lederberg and Tatum mixed the two strains, however, a few colonies did grow on the minimal medium. These prototrophic bacteria

must have had genotype $thr^+ leu^+ thi^+ bio^+ phe^+ cys^+$. Where had they come from?

If mutations were responsible for the prototrophic colonies, then some colonies should also have grown on the plates containing Y10 or Y24 alone, but no bacteria grew on these plates. Multiple simultaneous mutations ($thr^- \rightarrow thr^+$, $leu^- \rightarrow leu^+$, and $thi^- \rightarrow thi^+$ in strain Y10 or $bio^- \rightarrow bio^+$, $phe^- \rightarrow phe^+$, and $cys^- \rightarrow cys^+$ in strain Y24) would have been required for either strain to become prototrophic by mutation, which was very improbable. Lederberg and Tatum concluded that some type of genetic transfer and recombination had taken place:

Experiment

Question: Do bacteria exchange genetic information?

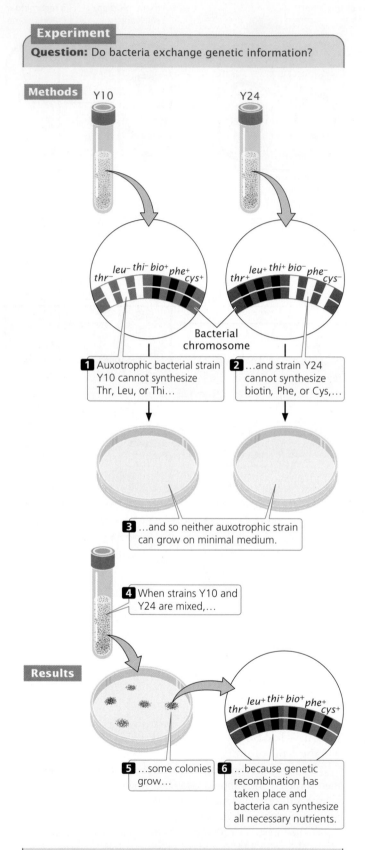

Methods

1 Auxotrophic bacterial strain Y10 cannot synthesize Thr, Leu, or Thi...

2 ...and strain Y24 cannot synthesize biotin, Phe, or Cys,...

3 ...and so neither auxotrophic strain can grow on minimal medium.

Bacterial chromosome

4 When strains Y10 and Y24 are mixed,...

Results

5 ...some colonies grow...

6 ...because genetic recombination has taken place and bacteria can synthesize all necessary nutrients.

Conclusion: Yes, genetic exchange and recombination took place between the two mutant strains.

8.8 Lederberg and Tatum's experiment demonstrated that bacteria undergo genetic exchange.

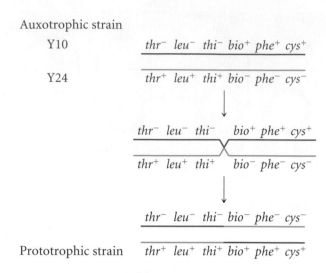

Auxotrophic strain

Y10 $\quad$ thr^- leu^- thi^- bio^+ phe^+ cys^+

Y24 $\quad$ thr^+ leu^+ thi^+ bio^- phe^- cys^-

thr^- leu^- thi^- $\quad$ bio^+ phe^+ cys^+

thr^+ leu^+ thi^+ $\quad$ bio^- phe^- cys^-

thr^- leu^- thi^- bio^- phe^- cys^-

Prototrophic strain $\quad$ thr^+ leu^+ thi^+ bio^+ phe^+ cys^+

What they did not know was *how* it had taken place.

To study this problem, Bernard Davis constructed a U-shaped tube (**Figure 8.9**) that was divided into two compart-

Experiment

Question: How did the genetic exchange seen in Lederberg and Tatum's experiment take place?

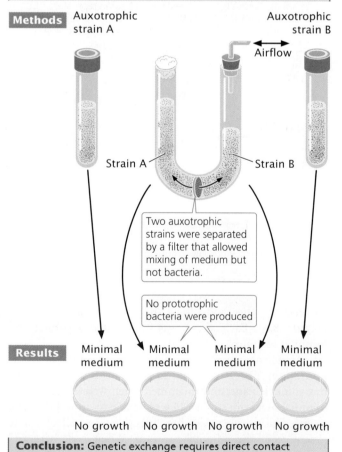

Methods

Auxotrophic strain A

Auxotrophic strain B

Airflow

Strain A

Strain B

Two auxotrophic strains were separated by a filter that allowed mixing of medium but not bacteria.

No prototrophic bacteria were produced

Results

Minimal medium — No growth

Minimal medium — No growth

Minimal medium — No growth

Minimal medium — No growth

Conclusion: Genetic exchange requires direct contact between bacterial cells.

8.9 Davis's U-tube experiment.

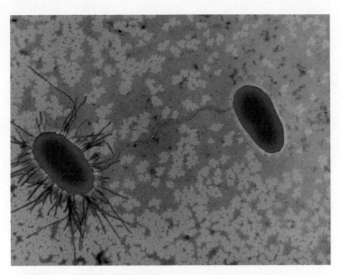

8.10 A sex pilus connects F⁺ and F⁻ cells during bacterial conjugation. [Dr. Dennis Kunkel/Phototake.]

ments by a filter having fine pores. This filter allowed liquid medium to pass from one side of the tube to the other, but the pores of the filter were too small to allow the passage of bacteria. Two auxotrophic strains of bacteria were placed on opposite sides of the filter, and suction was applied alternately to the ends of the U-tube, causing the medium to flow back and forth between the two compartments. Despite hours of incubation in the U-tube, bacteria plated out on minimal medium did not grow; there had been no genetic exchange between the strains. The exchange of bacterial genes clearly required direct contact, or conjugation, between the bacterial cells.

F⁺ and F⁻ cells In most bacteria, conjugation depends on a fertility (F) factor that is present in the donor cell and

absent in the recipient cell. Cells that contain F are referred to as F⁺, and cells lacking F are F⁻.

The F factor contains an origin of replication and a number of genes required for conjugation (see Figure 8.6). For example, some of these genes encode sex **pili** (singular, pilus), slender extensions of the cell membrane. A cell containing F produces the sex pili, one of which makes contact with a receptor on an F⁻ cell (**Figure 8.10**) and pulls the two cells together. DNA is then transferred from the F⁺ cell to the F⁻ cell. Conjugation can take place only between a cell that possesses F and a cell that lacks F.

In most cases, the only genes transferred during conjugation between an F⁺ and F⁻ cell are those on the F factor (**Figure 8.11a** and **b**). Transfer is initiated when one of the DNA strands on the F factor is nicked at an origin (*oriT*). One end of the nicked DNA separates from the circle and passes into the recipient cell (**Figure 8.11c**). Replication takes place on the nicked strand, proceeding around the circular plasmid in the F⁺ cell and replacing the transferred strand (**Figure 8.11d**). Because the plasmid in the F⁺ cell is always nicked at the *oriT* site, this site always enters the recipient cell first, followed by the rest of the plasmid. Thus, the transfer of genetic material has a defined direction. Inside the recipient cell, the single strand replicates, producing a circular, double-stranded copy of the F plasmid (**Figure 8.11e**). If the entire F factor is transferred to the recipient F⁻ cell, that cell becomes an F⁺ cell.

Hfr cells Conjugation transfers genetic material in the F plasmid from F⁺ to F⁻ cells but does not account for the transfer of chromosomal genes observed by Lederberg and Tatum. In Hfr (high-frequency) strains, the F factor is integrated into the bacterial chromosome (**Figure 8.12**). Hfr cells behave as F⁺ cells, forming sex pili and undergoing conjugation with F⁻ cells.

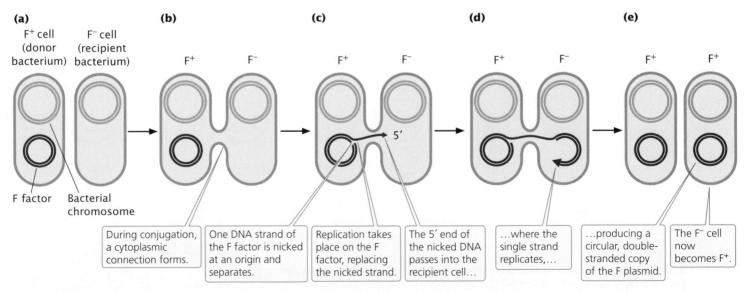

8.11 The F factor is transferred during conjugation between an F⁺ and F⁻ cell.

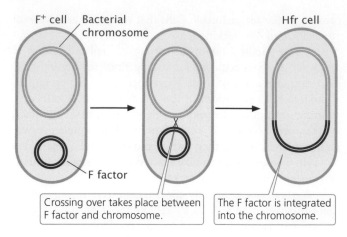

F+ cell   Bacterial chromosome   Hfr cell

F factor

Crossing over takes place between F factor and chromosome.

The F factor is integrated into the chromosome.

8.12 The F factor is integrated into the bacterial chromosome in an Hfr cell.

In conjugation between Hfr and F⁻ cells (**Figure 8.13a**), the integrated F factor is nicked, and the end of the nicked strand moves into the F⁻ cell (**Figure 8.13b**), just as it does in conjugation between F⁺ and F⁻ cells. Because, in an Hfr cell, the F factor has been integrated into the bacterial chromosome, the chromosome follows it into the recipient cell. How much of the bacterial chromosome is transferred depends on the length of time that the two cells remain in conjugation.

Inside the recipient cell, the donor DNA strand replicates (**Figure 8.13c**), and crossing over between it and the original chromosome of the F⁻ cell (**Figure 8.13d**) may take place. This gene transfer between Hfr and F⁻ cells is how the recombinant prototrophic cells observed by Lederberg and

Tatum were produced. After crossing over has taken place in the recipient cell, the donated chromosome is degraded and the recombinant recipient chromosome remains (**Figure 8.13e**), to be replicated and passed on to later generations by binary fission.

In a mating of Hfr × F⁻, the F⁻ cell almost never becomes F⁺ or Hfr, because the F factor is nicked in the middle in the initiation of strand transfer, placing part of F at the beginning and part at the end of the strand to be transferred. To become F⁺ or Hfr, the recipient cell must receive the entire F factor, requiring the entire bacterial chromosome to be transferred. This event happens rarely, because most conjugating cells break apart before the entire chromosome has been transferred.

The F plasmid in F⁺ cells integrates into the bacterial chromosome, causing an F⁺ cell to become Hfr, at a frequency of only about $\frac{1}{10,000}$. This low frequency accounts for the low rate of recombination observed by Lederberg and Tatum in their F⁺ cells. The F factor is excised from the bacterial chromosome at a similarly low rate, causing a few Hfr cells to become F⁺.

F′ cells   When an F factor does excise from the bacterial chromosome, a small amount of the bacterial chromosome may be removed with it, and these chromosomal genes will then be carried with the F plasmid (**Figure 8.14**). Cells containing an F plasmid with some bacterial genes are called F prime (F′). For example, if an F factor integrates into a chromosome adjacent to the *lac* genes (genes that enable a cell to metabolize the sugar lactose), the F factor may pick up *lac* genes when it excises, becoming F′*lac*. F′ cells can conjugate with F⁻ cells because F′ cells possess the F plasmid with all the genetic information necessary for conjugation

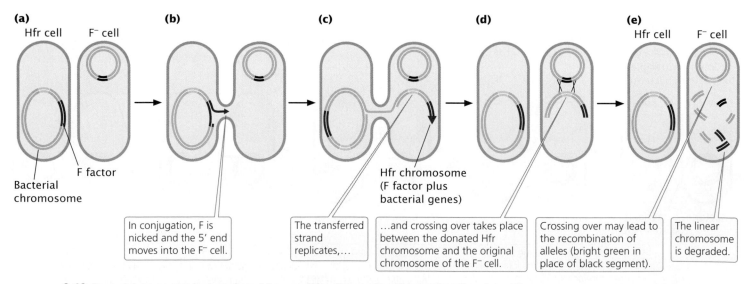

(a)   Hfr cell   F⁻ cell
(b)
(c)
(d)
(e)   Hfr cell   F⁻ cell

Bacterial chromosome   F factor

In conjugation, F is nicked and the 5′ end moves into the F⁻ cell.

The transferred strand replicates,...

Hfr chromosome (F factor plus bacterial genes)

...and crossing over takes place between the donated Hfr chromosome and the original chromosome of the F⁻ cell.

Crossing over may lead to the recombination of alleles (bright green in place of black segment).

The linear chromosome is degraded.

8.13 Bacterial genes may be transferred from an Hfr cell to an F⁻ cell in conjugation. In an Hfr cell, the F factor has been integrated into the bacterial chromosome.

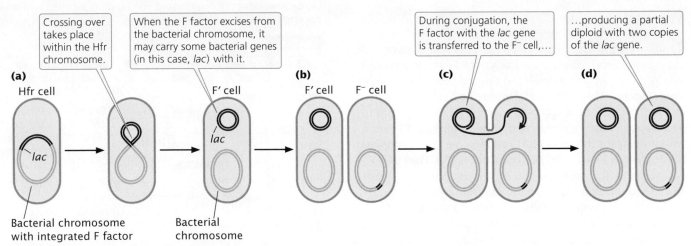

8.14 An Hfr cell may be converted into an F′ cell when the F factor excises from the bacterial chromosome and carries bacterial genes with it. Conjugation produces a partial diploid.

and gene transfer. Characteristics of different mating types of *E. coli* (cells with different types of F) are summarized in **Table 8.2**.

During conjugation between an F′*lac* cell and an F⁻ cell, the F plasmid is transferred to the F⁻ cell, which means that any genes on the F plasmid, including those from the bacterial chromosome, may be transferred to F⁻ recipient cells. This process is called sexduction. It produces partial diploids, or merozygotes, which are cells with two copies of some genes, one on the bacterial chromosome and one on the newly introduced F plasmid. The outcomes of conjugation between different mating types of *E. coli* are summarized in **Table 8.3**.

Table 8.2	Characteristics of *E. coli* cells with different types of F factor	
Type	**F Factor Characteristics**	**Role in Conjugation**
F⁺	Present as separate circular DNA	Donor
F⁻	Absent	Recipient
Hfr	Present, integrated into bacterial chromosome	High-frequency donor
F′	Present as separate circular DNA, carrying some bacterial genes	Donor

Table 8.3	Results of conjugation between cells with different F factors
Conjugating	**Cell Types Present after Conjugation**
F⁺ × F⁻	Two F⁺ cells (F⁻ cell becomes F⁺)
Hfr × F⁻	One Hfr cell and one F⁻ (no change)*
F′ × F⁻	Two F′ cells (F⁻ cell becomes F′)

*Rarely, the F⁻ cell becomes F⁺ in an Hfr × F⁻ conjugation if the entire chromosome is transferred during conjugation.

CONCEPTS

Conjugation in *E. coli* is controlled by an episome called the F factor. Cells containing F (F⁺ cells) are donors during gene transfer; cells lacking F (F⁻ cells) are recipients. Hfr cells possess F integrated into the bacterial chromosome; they donate DNA to F⁻ cells at a high frequency. F′ cells contain a copy of F with some bacterial genes.

✔ CONCEPT CHECK 3

Conjugation between an F⁺ and an F⁻ cell usually results in

a. two F⁺ cells.   b. two F⁻ cells.

c. an F⁺ and an F⁻ cell.   d. an Hfr cell and an F⁺ cell.

Mapping bacterial genes with interrupted conjugation

The transfer of DNA that takes place during conjugation between Hfr and F⁻ cells allows bacterial genes to be mapped. In conjugation, the chromosome of the Hfr cell is transferred to the F⁻ cell. Transfer of the entire *E. coli* chromosome requires about 100 minutes; if conjugation is interrupted before 100 minutes have elapsed, only part of the chromosome will have passed into the F⁻ cell and have had an opportunity to recombine with the recipient chromosome.

Chromosome transfer always begins within the integrated F factor and proceeds in a continuous direction; so genes are transferred according to their sequence on the chromosome. The time required for individual genes to be transferred indicates their relative positions on the chromosome. In most genetic maps, distances are expressed as percent recombination; but, in bacterial maps constructed with interrupted conjugation, the basic unit of distance is a minute.

Worked Problem

To illustrate the method of mapping genes with interrupted conjugation, let's look at a cross analyzed by François Jacob and Elie Wollman, who first developed this method of gene mapping (**Figure 8.15a**). They used donor Hfr cells that were sensitive to the antibiotic streptomycin (genotype str^s), resistant to sodium azide (azi^r) and infection by bacteriophage T1 (ton^r), prototrophic for threonine (thr^+) and leucine (leu^+), and able to break down lactose (lac^+) and galactose (gal^+). They used F⁻ recipient cells that were resistant to streptomycin (str^r), sensitive to sodium azide (azi^s) and to infection by bacteriophage T1 (ton^s), auxotrophic for threonine (thr^-) and leucine (leu^-), and unable to break down lactose (lac^-) and galactose (gal^-). Thus, the genotypes of the donor and recipient cells were:

Donor Hfr cells: $str^s\ leu^+\ thr^+\ azi^r\ ton^r\ lac^+\ gal^+$

Recipient F⁻ cells: $str^r\ leu^-\ thr^-\ azi^s\ ton^s\ lac^-\ gal^-$

The two strains were mixed in nutrient medium and allowed to conjugate. After a few minutes, the medium was diluted to prevent any new pairings. At regular intervals, a sample of cells was removed and agitated vigorously in a kitchen blender to halt all conjugation and DNA transfer. The cells from each sample were plated on a selective medium that contained streptomycin and lacked leucine and threonine. The original donor cells were streptomycin sensitive (str^s) and would not grow on this medium. The F⁻ recipient cells were auxotrophic for leucine and threonine, and they also failed to grow on this medium. Only cells that underwent conjugation and received at least the leu^+ and thr^+ genes from the Hfr donors could grow on this medium. All $str^r\ leu^+\ thr^+$ cells were then tested for the presence of other genes that might have been transferred from the donor Hfr strain.

All of the cells that grew on this selective medium were $str^r\ leu^+\ thr^+$; so we know that these genes were transferred. The percentage of $str^r\ leu^+\ thr^+$ cells receiving specific alleles (azi^r, ton^r, leu^+, and gal^+) from the Hfr strain are plotted against the duration of conjugation (**Figure 8.15b**). What are the order in which the genes are transferred and the distances among them?

• Solution

The first donor gene to appear in all of the recipient cells (at about 9 minutes) was azi^r. Gene ton^r appeared next (after about 10 minutes), followed by lac^+ (at about 18 minutes) and by gal^+ (after 25 minutes). These transfer times indicate the order of gene transfer and the relative distances among the genes (see Figure 8.15b).

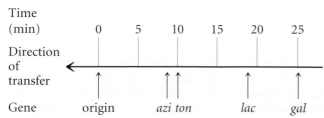

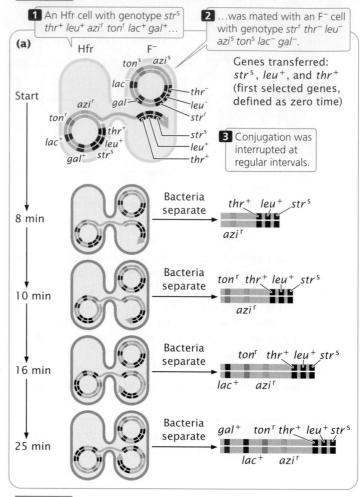

Results

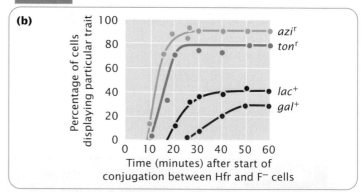

Conclusion: The transfer times indicate the order and relative distances between genes and can be used to construct a genetic map.

8.15 Jacob and Wollman used interrupted conjugation to map bacterial genes.

www.whfreeman.com/pierce4e   Animation 8.1 demonstrates the process of mapping genes through interrupted conjugation.

Notice that the frequency of gene transfer from donor to recipient cells decreased for the more distant genes. For example, about 90% of the recipients received the *azi^t* allele, but only about 30% received the *gal^+* allele. The lower percentage for *gal^+* is due to the fact that some conjugating cells spontaneously broke apart before they were disrupted by the blender. The probability of spontaneous disruption increases with time; so fewer cells had an opportunity to receive genes that were transferred later.

→ For additional practice mapping bacterial genes with interrupted conjugation, try Problem 21 at the end of the chapter.

Directional transfer and mapping Different Hfr strains of a given species of bacteria have the F factor integrated into the bacterial chromosome at different sites and in different orientations. Gene transfer always begins within F, and the orientation and position of F determine the direction and starting point of gene transfer. **Figure 8.16a** shows that, in strain Hfr1, F is integrated between *leu* and *azi*; the orientation of F at this site dictates that gene transfer will proceed in a counterclockwise direction around the circular chromosome. Genes from this strain will be transferred in the order of:

$$\longleftarrow leu-thr-thi-his-gal-lac-pro-azi$$

In strain Hfr5, F is integrated between the *thi* and the *his* genes (**Figure 8.16b**) and in the opposite orientation. Here gene transfer will proceed in a clockwise direction:

$$\longleftarrow thi-thr-leu-azi-pro-lac-gal-his$$

Although the starting point and direction of transfer may differ between two strains, the relative distance in time between any two pairs of genes is constant.

Notice that the order of gene transfer is not the same for different Hfr strains (**Figure 8.17a**). For example, *azi* is transferred just after *leu* in strain HfrH but long after *leu* in strain Hfr1. Aligning the sequences (**Figure 8.17b**) shows that the two genes on either side of *azi* are always the same: *leu* and *pro*. That they are the same makes sense when we recognize that the bacterial chromosome is circular and the starting point of transfer varies from strain to strain. These data provided the first evidence that the bacterial chromosome is circular (**Figure 8.17c**). TRY PROBLEM 20 →

CONCEPTS

Conjugation can be used to map bacterial genes by mixing Hfr and F⁻ cells of different genotypes and interrupting conjugation at regular intervals. The amount of time required for individual genes to be transferred from the Hfr to the F⁻ cells indicates the relative positions of the genes on the bacterial chromosome.

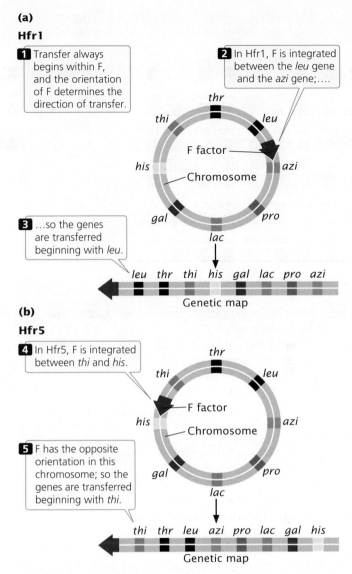

(a)

Hfr1

1 Transfer always begins within F, and the orientation of F determines the direction of transfer.

2 In Hfr1, F is integrated between the *leu* gene and the *azi* gene;....

3 ...so the genes are transferred beginning with *leu*.

Genes: thr, thi, leu, his, F factor, Chromosome, azi, gal, lac, pro

Genetic map: leu thr thi his gal lac pro azi

(b)

Hfr5

4 In Hfr5, F is integrated between *thi* and *his*.

5 F has the opposite orientation in this chromosome; so the genes are transferred beginning with *thi*.

Genes: thr, thi, leu, his, F factor, Chromosome, azi, gal, pro, lac

Genetic map: thi thr leu azi pro lac gal his

8.16 The orientation of the F factor in an Hfr strain determines the direction of gene transfer. Arrowheads indicate the origin and direction of transfer.

✔ CONCEPT CHECK 4

Interrupted conjugation was used to map three genes in *E. coli*. The donor genes first appeared in the recipient cells at the following times: *gal*, 10 minutes; *his*, 8 minutes; *pro*, 15 minutes. Which gene is in the middle?

Natural Gene Transfer and Antibiotic Resistance

Antibiotics are substances that kill bacteria. Their development and widespread use has greatly reduced the threat of infectious disease and saved countless lives. But many pathogenic bacteria have developed resistance to antibiotics,

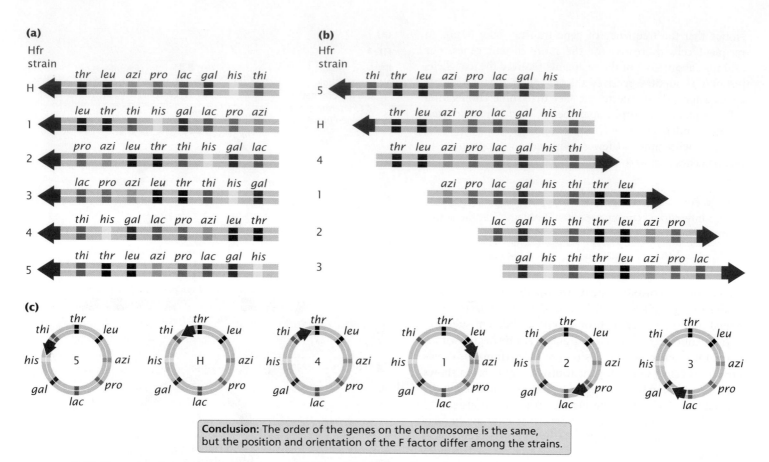

(a)

Hfr strain

H   thr leu azi pro lac gal his thi

1   leu thr thi his gal lac pro azi

2   pro azi leu thr thi his gal lac

3   lac pro azi leu thr thi his gal

4   thi his gal lac pro azi leu thr

5   thi thr leu azi pro lac gal his

(b)

Hfr strain

5   thi thr leu azi pro lac gal his

H   thr leu azi pro lac gal his thi

4   thr leu azi pro lac gal his thi

1   azi pro lac gal his thi thr leu

2   lac gal his thi thr leu azi pro

3   gal his thi thr leu azi pro lac

(c)

Conclusion: The order of the genes on the chromosome is the same, but the position and orientation of the F factor differ among the strains.

8.17 The order of gene transfer in a series of different Hfr strains indicates that the *E. coli* chromosome is circular.

particularly in environments where antibiotics are routinely used, such as hospitals, livestock operations, and fish farms. In these enviornments where antibiotics are continually present, the only bacteria to survive are those that possess resistance to antibiotics. No longer in competition with other bacteria, resistant bacteria multiply quickly and spread. In this way, the presence of antibiotics selects for resistant bacteria and reduces the effectiveness of antibiotic treatment for medically important infections.

Antibiotic resistance in bacteria frequently results from the action of genes located on *R plasmids*, small circular plasmids that can be transferred by conjugation. Drug-resistant R plasmids have evolved in the past 60 years (since the beginning of widespread use of antibiotics), and some of them convey resistance to several antibiotics simultaneously. Ironic but plausible sources of some of the resistance genes found in R plasmids are the microbes that produce antibiotics in the first place. R plasmids can spread easily throughout the environment, passing between related and unrelated bacteria in a variety of common situations. For example, research shows that plasmids carrying genes for resistance to multiple antibiotics were transferred from a cow udder infected with *E. coli* to a human strain of *E. coli* on a hand towel: a farmer wiping his hands after milking an infected cow might unwittingly transfer antibiotic resistance from bovine- to human-inhabiting microbes. Conjugation taking place in minced meat on a cutting board allowed R plasmids to be passed from porcine (pig) to human *E. coli*. The transfer of R plasmids also takes place in sewage, soil, lake water, and marine sediments. The fact that R plasmids can easily spread throughout the environment and pass between related and unrelated bacteria underscores both the importance of limiting antibiotic use to the treatment of medically important infections and the importance of hygiene in everyday life.

Transformation in Bacteria

A second way in which DNA can be transferred between bacteria is through transformation (see Figure 8.7b). Transformation played an important role in the initial identification of DNA as the genetic material, which will be discussed in Chapter 10.

Transformation requires both the uptake of DNA from the surrounding medium and its incorporation into a bacterial chromosome or a plasmid. It may occur naturally when dead bacteria break up and release DNA fragments into the environment. In soil and marine environments, this means may be an important route of genetic exchange for some bacteria.

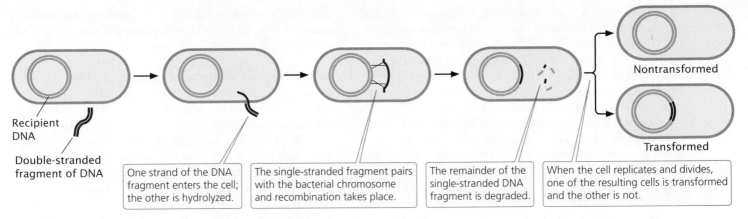

One strand of the DNA fragment enters the cell; the other is hydrolyzed.

The single-stranded fragment pairs with the bacterial chromosome and recombination takes place.

The remainder of the single-stranded DNA fragment is degraded.

When the cell replicates and divides, one of the resulting cells is transformed and the other is not.

8.18 Genes can be transferred between bacteria through transformation.

Mechanism of transformation   Cells that take up DNA through their membranes are said to be **competent**. Some species of bacteria take up DNA more easily than others: competence is influenced by growth stage, the concentration of available DNA in the environment, and other environmental factors. The DNA that a competent cell takes up need not be bacterial: virtually any type of DNA (bacterial or otherwise) can be taken up by competent cells under the appropriate conditions.

As a DNA fragment enters the cell in the course of transformation (**Figure 8.18**), one of the strands is broken up, whereas the other strand moves across the membrane and may pair with a homologous region and become integrated into the bacterial chromosome. This integration requires two crossover events, after which the remaining single-stranded DNA is degraded by bacterial enzymes.

Bacterial geneticists have developed techniques for increasing the frequency of transformation in the labora-tory in order to introduce particular DNA fragments into cells. They have developed strains of bacteria that are more competent than wild-type cells. Treatment with calcium chloride, heat shock, or an electrical field makes bacterial membranes more porous and permeable to DNA, and the efficiency of transformation can also be increased by using high concentrations of DNA. These techniques enable researchers to transform bacteria such as *E. coli*, which are not naturally competent.

Gene mapping with transformation   Transformation, like conjugation, is used to map bacterial genes, especially in those species that do not undergo conjugation or trans-duction (see Figure 8.7a and c). Transformation mapping requires two strains of bacteria that differ in several genetic traits; for example, the recipient strain might be $a^-\ b^-\ c^-$ (auxotrophic for three nutrients), and the donor cell pro-totrophic with alleles $a^+\ b^+\ c^+$ (**Figure 8.19**). DNA from

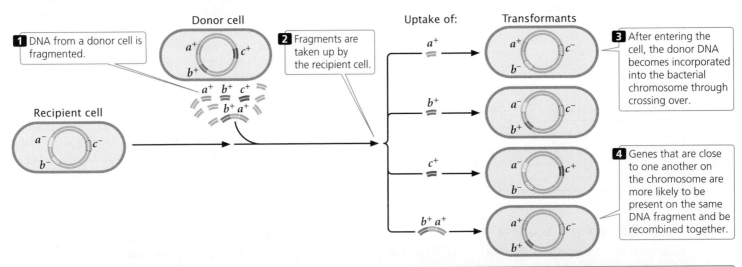

1. DNA from a donor cell is fragmented.

2. Fragments are taken up by the recipient cell.

3. After entering the cell, the donor DNA becomes incorporated into the bacterial chromosome through crossing over.

4. Genes that are close to one another on the chromosome are more likely to be present on the same DNA fragment and be recombined together.

Conclusion: The rate of cotransformation is inversely proportional to the distances between genes.

8.19 Transformation can be used to map bacterial genes.

the donor strain is isolated, purified, and fragmented. The recipient strain is treated to increase competency, and DNA from the donor strain is added to the medium. Fragments of the donor DNA enter the recipient cells and undergo recombination with homologous DNA sequences on the bacterial chromosome. Cells that receive genetic material through transformation are called **transformants**.

Genes can be mapped by observing the rate at which two or more genes are transferred to the host chromosome, or **cotransformed**, in transformation. When the donor DNA is fragmented before transformation, genes that are physically close on the chromosome are more likely to be present on the same DNA fragment and transferred together, as shown for genes a^+ and b^+ in Figure 8.19. Genes that are far apart are unlikely to be present on the same DNA fragment and will rarely be transferred together. Inside the cell, DNA becomes incorporated into the bacterial chromosome through recombination. If two genes are close together on the same fragment, any two crossovers are likely to take place on either side of the two genes, allowing both to become part of the recipient chromosome. If the two genes are far apart, there may be one crossover between them, allowing one gene but not the other to recombine with the bacterial chromosome. Thus, two genes are more likely to be transferred together when they are close together on the chromosome, and genes located far apart are rarely cotransformed. Therefore, the frequency of cotransformation can be used to map bacterial genes. If genes a and b are frequently cotransformed, and genes b and c are frequently cotransformed, but genes a and c are rarely cotransformed, then gene b must be between a and c—the gene order is $a\ b\ c$. **TRY PROBLEMS 22 AND 23 →**

CONCEPTS

Genes can be mapped in bacteria by taking advantage of transformation—the ability of cells to take up DNA from the environment and incorporate it into their chromosomes through crossing over. The relative rate at which pairs of genes are cotransformed indicates the distance between them: the higher the rate of cotransformation, the closer the genes are on the bacterial chromosome.

✔ CONCEPT CHECK 5

DNA from a bacterial strain with genotype $his^-\ leu^-\ thr^-$ is transformed with DNA from a strain that is $his^+\ leu^+\ thr^+$. A few $leu^+\ thr^+$ cells and a few $his^+\ thr^+$ cells are found, but no $his^+\ leu^+$ cells are observed. Which genes are farthest apart?

Bacterial Genome Sequences

Genetic maps serve as the foundation for more-detailed information provided by DNA sequencing, such as gene content and organization (see Chapter 19 for a discussion of gene sequencing). Geneticists have now determined the complete nucleotide sequence of hundreds of bacterial genomes (see Table 20.2), and many additional microbial sequencing projects are underway.

Characteristics of bacterial genomes Most bacterial genomes contain from 1 million to 4 million base pairs of DNA, but a few are much smaller (e.g., 580,000 bp in *Mycoplasma genitalium*) and some are considerably larger (e.g., more than 7 million base pairs in *Mesorhizobium loti*). The small size of bacterial genomes relative to those found in multicellular eukaryotes, which often have billions of base pairs of DNA, is thought to be an adaptation for rapid cell division, because the rate of cell division is limited by the time required to replicate the DNA. On the other hand, a lack of mobility in most bacteria requires metabolic and environmental flexibility, and so genome size and content are likely to be a balance between the opposing evolutionary forces of gene loss to maintain rapid reproduction and gene acquisition to ensure flexibility.

The function of a substantial proportion of genes in all bacteria has not been determined. Certain genes, particularly those with related functions, tend to reside next to one another, but these clusters are in very different locations in different species, suggesting that bacterial genomes are constantly being reshuffled. Comparisons of the gene sequences of disease-causing and benign bacteria are helping to identify genes implicated in disease and may suggest new targets for antibiotics and other antimicrobial agents.

Horizontal Gene Transfer

The availability of genome sequences has provided evidence that many bacteria have acquired genetic information from other species of bacteria—and sometimes even from eukaryotic organisms—in a process called **horizontal gene transfer**. In most eukaryotes, genes are passed only among members of the same species through the process of reproduction (called vertical transmission); that is, genes are passed down from one generation to the next. In horizontal transfer, genes can be passed between individual members of different species by nonreproductive mechanisms, such as conjugation, transformation, and transduction. Evidence suggests that horizontal gene transfer has taken place repeatedly among bacteria. For example, as much as 17% of *E. coli*'s genome has been acquired from other bacteria through horizontal gene transfer. Of medical significance, some pathogenic bacteria have acquired, through horizontal gene transfer, the genes necessary for infection, whereas others have acquired genes that confer resistance to antibiotics.

Because of the widespread occurence of horizontal gene transfer, many bacterial chromosomes are a mixture of genes inherited through vertical transmission and genes acquired through horizontal transfer. This situation has caused some biologists to question whether the species concept is even appropriate for bacteria. A species is often defined as a group of organisms that are reproductively isolated from other

(a)    **(b)**

8.20 Viruses come in different structures and sizes.
(a) T4 bacteriophage (bright orange).
(b) Influenza A virus (green structures).
[(a) Biozentrum, University of Basel/Photo Researchers; (b) Eye of Science/Photo Researchers.]

groups, have a set of genes in common, and evolve together (see Chapter 26). Because of horizontal gene flow, the genes of one bacterial species are not isolated from the genes of other species, making the traditional species concept difficult to apply to bacteria. Horizontal gene flow also muddies the determination of the ancestral relationships among bacteria. The reconstruction of ancestrial relationships is usually based on genetic similarities and differences: organisms that are genetically similar are assumed to have descended from a recent common ancestor, whereas organisms that are genetically distinct are assumed to be more distantly related. Through horizontal gene flow, however, even distantly related bacteria may have genes in common and thus appear to have descended from a recent common ancestor. The nature of species and how to classify bacteria are currently controversial topics within the field of microbiology.

8.3 Viruses Are Simple Replicating Systems Amenable to Genetic Analysis

All organisms—plants, animals, fungi, and bacteria—are infected by viruses. A **virus** is a simple replicating structure made up of nucleic acid surrounded by a protein coat (see Figure 2.4). Viruses come in a great variety of shapes and sizes (**Figure 8.20**). Some have DNA as their genetic material, whereas others have RNA; the nucleic acid may be double stranded or single stranded, linear or circular.

Viruses that infect bacteria (bacteriophages, or phages for short) have played a central role in genetic research since the late 1940s. They are ideal for many types of genetic research because they have small and easily manageable genomes, reproduce rapidly, and produce large numbers of progeny. Bacteriophages have two alternative life cycles: the *lytic* and the *lysogenic* cycles. In the lytic cycle, a phage attaches to a receptor on the bacterial cell wall and injects its DNA into the cell (**Figure 8.21**). Inside the host cell, the phage DNA is replicated, transcribed, and translated, producing more phage DNA and phage proteins. New phage particles are assembled from these components. The phages then produce an enzyme that breaks open the host cell, releasing the new phages. **Virulent phages** reproduce strictly through the lytic cycle and always kill their host cells.

Temperate phages can undergo either the lytic or the lysogenic cycle. The lysogenic cycle begins like the lytic cycle (see Figure 8.21) but, inside the cell, the phage DNA integrates into the bacterial chromosome, where it remains as an inactive **prophage**. The prophage is replicated along with the bacterial DNA and is passed on when the bacterium divides. Certain stimuli can cause the prophage to dissociate from the bacterial chromosome and enter into the lytic cycle, producing new phage particles and lysing the cell.

Techniques for the Study of Bacteriophages

Viruses reproduce only within host cells; so bacteriophages must be cultured in bacterial cells. To do so, phages and bacteria are mixed together and plated on solid medium on a petri plate. A high concentration of bacteria is used so that the colonies grow into one another and produce a continuous layer of bacteria, or "lawn," on the agar. An individual phage infects a single bacterial cell and goes through its lytic cycle. Many new phages are released from the lysed cell and infect additional cells; the cycle is then repeated. Because the bacteria grow on solid medium, the diffusion of the phages is restricted and only nearby cells are infected. After several rounds of phage reproduction, a clear patch of lysed cells, or **plaque**, appears on the plate (**Figure 8.22**). Each plaque represents a single phage that multiplied and lysed many cells. Plating a known volume of a dilute solution of phages on a bacterial lawn and counting the number of plaques that appear can be used to determine the original concentration of phage in the solution.

CONCEPTS

Viral genomes may be DNA or RNA, circular or linear, and double or single stranded. Bacteriophages are used in many types of genetic research.

✔ CONCEPT CHECK 6

In which bacteriophage life cycle does the phage DNA become incorporated into the bacterial chromosome?

a. Lytic   c. Both lytic and lysogenic

b. Lysogenic   d. Neither lytic or lysogenic

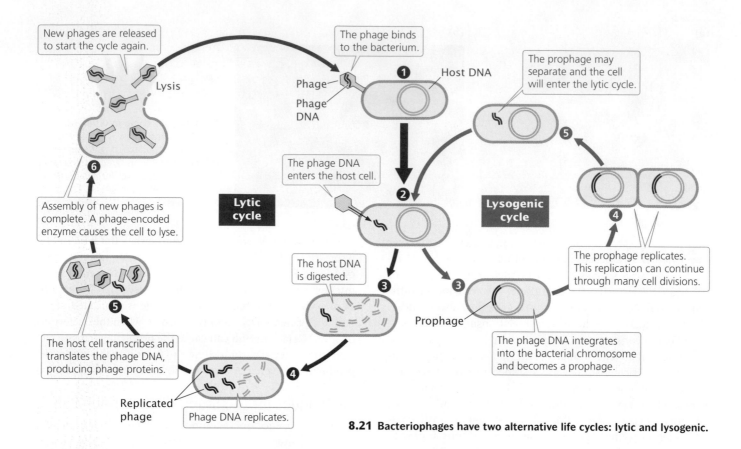

8.21 Bacteriophages have two alternative life cycles: lytic and lysogenic.

Transduction: Using Phages to Map Bacterial Genes

In the discussion of bacterial genetics, three mechanisms of gene transfer were identified: conjugation, transformation, and transduction (see Figure 8.7). Let's take a closer look at transduction,

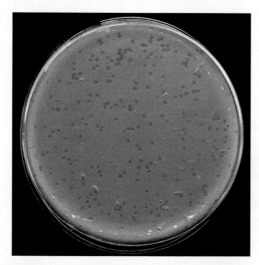

8.22 Plaques are clear patches of lysed cells on a lawn of bacteria.
[Internationally Genetically Engineered Machine Competition, 2007.]

in which genes are transferred between bacteria by viruses. In **generalized transduction**, any gene may be transferred. In **specialized transduction**, only a few genes are transferred.

Generalized transduction   Joshua Lederberg and Norton Zinder discovered generalized transduction in 1952. They were trying to produce recombination in the bacterium *Salmonella typhimurium* by conjugation. They mixed a strain of *S. typhimurium* that was $phe^+ trp^+ tyr^+ met^- his^-$ with a strain that was $phe^- trp^- tyr^- met^+ his^+$ (**Figure 8.23**) and plated them on minimal medium. A few prototrophic recombinants ($phe^+ trp^+ tyr^+ met^+ his^+$) appeared, suggesting that conjugation had taken place. However, when they tested the two strains in a U-shaped tube similar to the one used by Davis, some $phe^+ trp^+ tyr^+ met^+ his^+$ prototrophs were obtained on one side of the tube (compare Figure 8.23 with Figure 8.9). This apparatus separated the two strains by a filter with pores too small for the passage of bacteria; so how were genes being transferred between bacteria in the absence of conjugation? The results of subsequent studies revealed that the agent of transfer was a bacteriophage.

In the lytic cycle of phage reproduction, the bacterial chromosome is broken into random fragments (**Figure 8.24**). For some types of bacteriophage, a piece of the bacterial chromosome instead of phage DNA occasionally gets

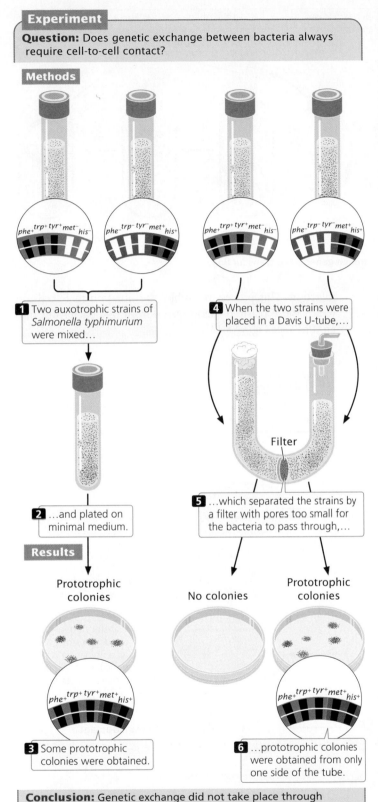

Question: Does genetic exchange between bacteria always require cell-to-cell contact?

Methods

$phe^+ trp^+ tyr^+ met^- his^-$

$phe^- trp^- tyr^- met^+ his^+$

$phe^+ trp^+ tyr^+ met^- his^-$

$phe^- trp^- tyr^- met^+ his^+$

1 Two auxotrophic strains of *Salmonella typhimurium* were mixed…

4 When the two strains were placed in a Davis U-tube,…

Filter

2 …and plated on minimal medium.

5 …which separated the strains by a filter with pores too small for the bacteria to pass through,…

Results

Prototrophic colonies

No colonies

Prototrophic colonies

$phe^+ trp^+ tyr^+ met^+ his^+$

$phe^+ trp^+ tyr^+ met^+ his^+$

3 Some prototrophic colonies were obtained.

6 …prototrophic colonies were obtained from only one side of the tube.

Conclusion: Genetic exchange did not take place through conjugation. A phage was later shown to be the agent of transfer.

8.23 The Lederberg and Zinder experiment.

packaged into a phage coat; these phage particles are called **transducing phages.** The transducing phage infects a new cell, releasing the bacterial DNA, and the introduced genes may then become integrated into the bacterial chromosome by a double crossover. Bacterial genes can, by this process, be moved from one bacterial strain to another, producing recombinant bacteria called **transductants.**

Not all phages are capable of transduction, a rare event that requires (1) that the phage degrade the bacterial chromosome, (2) that the process of packaging DNA into the phage protein not be specific for phage DNA, and (3) that the bacterial genes transferred by the virus recombine with the chromosome in the recipient cell. The overall rate of transduction ranges from only about 1 in 100,000 to 1 in 1,000,000.

Because of the limited size of a phage particle, only about 1% of the bacterial chromosome can be transduced. Only genes located close together on the bacterial chromosome will be transferred together, or **cotransduced.** Because the chance of a cell being transduced by two separate phages is exceedingly small, any cotransduced genes are usually located close together on the bacterial chromosome. Thus, rates of cotransduction, like rates of cotransformation, give an indication of the physical distances between genes on a bacterial chromosome.

To map genes by using transduction, two bacterial strains with different alleles at several loci are used. The donor strain is infected with phages (**Figure 8.25**), which reproduce within the cell. When the phages have lysed the donor cells, a suspension of the progeny phages is mixed with a recipient strain of bacteria, which is then plated on several different kinds of media to determine the phenotypes of the transducing progeny phages.

TRY PROBLEM 29 ➡

CONCEPTS

In transduction, bacterial genes become packaged into a viral coat, are transferred to another bacterium by the virus, and become incorporated into the bacterial chromosome by crossing over. Bacterial genes can be mapped with the use of generalized transduction.

✔ CONCEPT CHECK 7

In gene mapping experiments using generalized transduction, bacterial genes that are cotransduced are

a. far apart on the bacterial chromosome.

b. on different bacterial chromosomes.

c. close together on the bacterial chromosome.

d. on a plasmid.

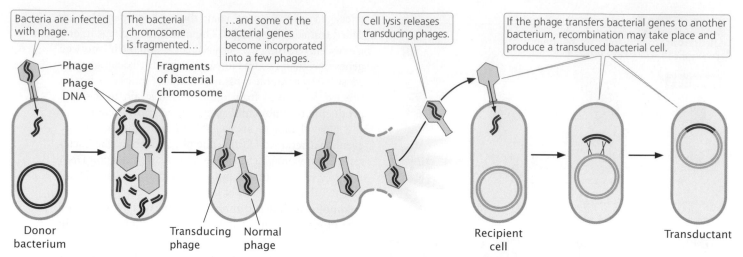

8.24 Genes can be transferred from one bacterium to another through generalized transduction.

Specialized transduction   Like generalized transduction, specialized transduction requires gene transfer from one bacterium to another through phages, but, here, only genes near particular sites on the bacterial chromosome are transferred. This process requires lysogenic bacteriophages. The prophage may imperfectly excise from the bacterial chromosome, carrying with it a small part of the bacterial DNA adjacent to the site of prophage integration. A phage carrying this DNA will then inject it into another bacterial cell in the next round of infection. This process resembles the situation in F′ cells, in which the F plasmid carries genes from one bacterium into another (see Figure 8.14).

CONCEPTS

Specialized transduction transfers only those bacterial genes located near the site of prophage insertion.

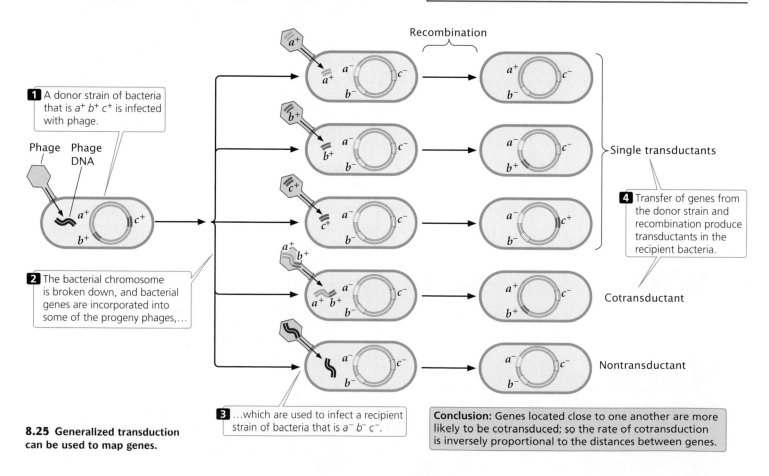

8.25 Generalized transduction can be used to map genes.

Conclusion: Genes located close to one another are more likely to be cotransduced; so the rate of cotransduction is inversely proportional to the distances between genes.

CONNECTING CONCEPTS

Three Methods for Mapping Bacterial Genes

Three methods of mapping bacterial genes have now been outlined: (1) interrupted conjugation; (2) transformation; and (3) transduction. These methods have important similarities and differences.

Mapping with interrupted conjugation is based on the time required for genes to be transferred from one bacterium to another by means of cell-to-cell contact. The key to this technique is that the bacterial chromosome itself is transferred, and the order of genes and the time required for their transfer provide information about the positions of the genes on the chromosome. In contrast with other mapping methods, the distance between genes is measured not in recombination frequencies but in units of time required for genes to be transferred. Here, the basic unit of conjugation mapping is a minute.

In gene mapping with transformation, DNA from the donor strain is isolated, broken up, and mixed with the recipient strain. Some fragments pass into the recipient cells, where the transformed DNA may recombine with the bacterial chromosome. The unit of transfer here is a random fragment of the chromosome. Loci that are close together on the donor chromosome tend to be on the same DNA fragment; so the rates of cotransformation provide information about the relative positions of genes on the chromosome.

Transduction mapping also relies on the transfer of genes between bacteria that differ in two or more traits, but, here, the vehicle of gene transfer is a bacteriophage. In a number of respects, transduction mapping is similar to transformation mapping. Small fragments of DNA are carried by the phage from donor to recipient bacteria, and the rates of cotransduction, like the rates of cotransformation, provide information about the relative distances between the genes.

All of the methods use a common strategy for mapping bacterial genes. The movement of genes from donor to recipient is detected by using strains that differ in two or more traits, and the transfer of one gene relative to the transfer of others is examined. Additionally, all three methods rely on recombination between the transferred DNA and the bacterial chromosome. In mapping with interrupted conjugation, the relative order and timing of gene transfer provide the information necessary to map the genes; in transformation and transduction, the rate of cotransfer provides this information.

In conclusion, the same basic strategies are used for mapping with interrupted conjugation, transformation, and transduction. The methods differ principally in their mechanisms of transfer: in conjugation mapping, DNA is transferred though contact between bacteria; in transformation, DNA is transferred as small naked fragments; and, in transduction, DNA is transferred by bacteriophages.

Gene Mapping in Phages

Mapping genes in the bacteriophages themselves depends on homolgous recombination between phage chromosomes and therefore requires the application of the same principles as those applied to mapping genes in eukaryotic organisms (see Chapter 7). Crosses are made between viruses that differ in two or more genes, and recombinant progeny phages are identified and counted. The proportion of recombinant progeny is then used to estimate the distances between genes and their linear order on the chromosome.

In 1949, Alfred Hershey and Raquel Rotman examined rates of recombination in the T2 bacteriophage, which has single-stranded DNA. They studied recombination between genes in two strains that differed in plaque appearance and host range (the bacterial strains that the phages could infect). One strain was able to infect and lyse type B *E. coli* cells but not B/2 cells (making this strain of bacteria wild type with normal host range, or h^+) and produced an abnormal plaque that was large with distinct borders (r^-). The other strain was able to infect and lyse *both* B *and* B/2 cells (mutant host range, h^-) and produced wild-type plaques that were small with fuzzy borders (r^+).

Hershey and Rotman crossed the $h^+\ r^-$ and $h^-\ r^+$ strains of T2 by infecting type B *E. coli* cells with a mixture of the two strains. They used a high concentration of phages so that most cells could be simultaneously infected by both strains (**Figure 8.26**). Within the bacteria cells, homologous recombination occasionally took place between the chromosomes of the different bacteriophage strains, producing $h^+\ r^+$ and $h^-\ r^-$ chromosomes, which were then packaged into new phage particles. When the cells lysed, the recombinant phages were released, along with the nonrecombinant $h^+\ r^-$ phages and $h^-\ r^+$ phages.

Hershey and Rotman diluted and plated the progeny phages on a bacterial lawn that consisted of a *mixture* of B and B/2 cells. Phages carrying the h^+ allele (which conferred the ability to infect only B cells) produced a cloudy plaque because the B/2 cells did not lyse. Phages carrying the h^- allele produced a clear plaque because all the cells within the plaque were lysed. The r^+ phages produced small plaques, whereas the r^- phages produced large plaques. The genotypes of these progeny phages could therefore be determined by the appearance of the plaque (see Figure 8.26 and **Table 8.4**).

In this type of phage cross, the recombination frequency (*RF*) between the two genes can be calculated by using the following formula:

$$RF = \frac{\text{recombinant plaques}}{\text{total plaques}}$$

In Hershey and Rotman's cross, the recombinant plaques were $h^+\ r^+$ and $h^-\ r^-$; so the recombination frequency was

$$RF = \frac{(h^+\ r^+) + (h^-\ r^-)}{\text{total plaques}}$$

Recombination frequencies can be used to determine the distances between genes and their order on the phage chromosome, just as recombination frequencies are used to map genes in eukaryotes. TRY PROBLEMS 28 AND 32 →

CONCEPTS

To map phage genes, bacterial cells are infected with viruses that differ in two or more genes. Recombinant plaques are counted, and rates of recombination are used to determine the linear order of the genes on the chromosome and the distances between them.

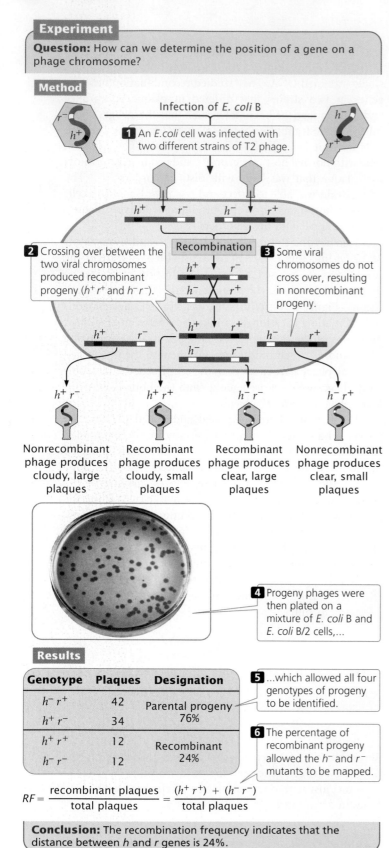

Question: How can we determine the position of a gene on a phage chromosome?

Method

Infection of *E. coli* B

1 An *E.coli* cell was infected with two different strains of T2 phage.

2 Crossing over between the two viral chromosomes produced recombinant progeny ($h^+ r^+$ and $h^- r^-$).

Recombination

3 Some viral chromosomes do not cross over, resulting in nonrecombinant progeny.

$h^+ r^-$

Nonrecombinant phage produces cloudy, large plaques

$h^+ r^+$

Recombinant phage produces cloudy, small plaques

$h^- r^-$

Recombinant phage produces clear, large plaques

$h^- r^+$

Nonrecombinant phage produces clear, small plaques

4 Progeny phages were then plated on a mixture of *E. coli* B and *E. coli* B/2 cells,...

Results

Genotype	Plaques	Designation
$h^- r^+$	42	Parental progeny
$h^+ r^-$	34	76%
$h^+ r^+$	12	Recombinant
$h^- r^-$	12	24%

5 ...which allowed all four genotypes of progeny to be identified.

6 The percentage of recombinant progeny allowed the h^- and r^- mutants to be mapped.

$$RF = \frac{\text{recombinant plaques}}{\text{total plaques}} = \frac{(h^+ r^+) + (h^- r^-)}{\text{total plaques}}$$

Conclusion: The recombination frequency indicates that the distance between *h* and *r* genes is 24%.

8.26 Hershey and Rotman developed a technique for mapping viral genes. [Photograph from G. S. Stent, *Molecular Biology of Bacterial Viruses.* © 1963 by W. H. Freeman and Company.]

Table 8.4	Progeny phages produced from $h^- r^+ \times h^+ r^-$

Phenotype	Genotype
Clear and small	$h^- r^+$
Cloudy and large	$h^+ r^-$
Cloudy and small	$h^+ r^+$
Clear and large	$h^- r^-$

Fine-Structure Analysis of Bacteriophage Genes

In the 1950s and 1960s, Seymour Benzer conducted a series of experiments to examine the structure of a gene. Because no molecular techniques were available at the time for directly examining nucleotide sequences, Benzer was forced to infer gene structure from analyses of mutations and their effects. The results of his studies showed that different mutational sites *within* a single gene could be mapped (referred to as **intragenic mapping**) by using techniques similar to those described for mapping bacterial genes by transduction. Different sites within a single gene are very close together; so recombination between them takes place at a very low frequency. Because large numbers of progeny are required to detect these recombination events, Benzer used the bacteriophage T4, which reproduces rapidly and produces large numbers of progeny.

Benzer's mapping techniques Wild-type T4 phages normally produce small plaques with rough edges when grown on a lawn of *E. coli*. Certain mutants, called *r* for rapid lysis, produce larger plaques with sharply defined edges. Benzer isolated phages with a number of different *r* mutations, concentrating on one particular subgroup called *rII* mutants.

Wild-type T4 phages produce typical plaques (**Figure 8.27**) on *E. coli* strains B and K. In contrast, the *rII* mutants produce *r* plaques on strain B and do not form plaques at all on strain K. Benzer recognized the *r* mutants by their

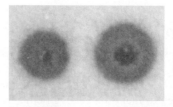

8.27 T4 phage *rII* mutants produce distinct plaques when grown on *E. coli* B cells. (Left) Plaque produced by wild-type phage. (Right) Plaque produced by *rII* mutant. [Dr. D. P. Snustad, College of Biological Sciences, University of Minnesota.]

Experiment

Question: How can *rII* phage mutants be mapped and what can they reveal about the structure of the gene?

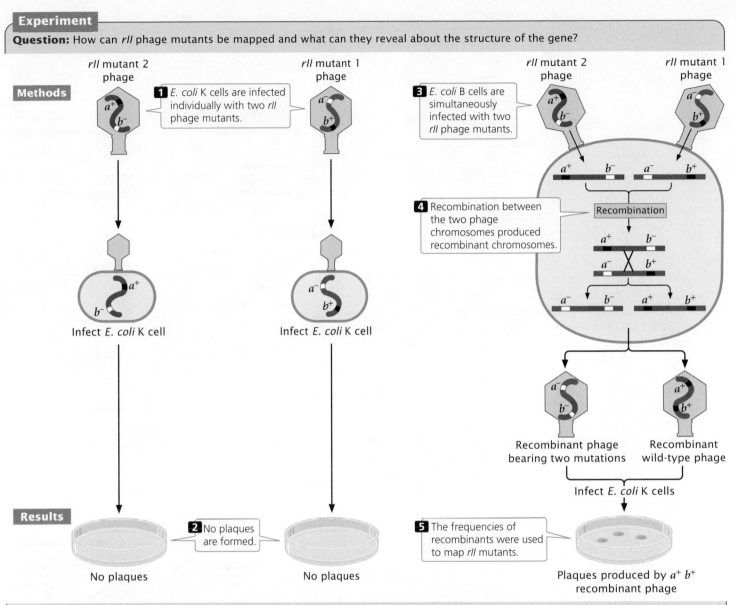

Conclusion: Mapping more than 2400 *rII* mutants provided information about the internal structure of a gene at the base-pair level—the first view of the molecular structure of a gene.

8.28 Benzer developed a procedure for mapping *rII* mutants. Two different *rII* mutants ($a^- b^+$ and $a^+ b^-$) are isolated on *E. coli* B cells. Only the $a^+ b^+$ recombinant can grow on *E. coli* K, allowing these recombinants to be identified.

distinctive plaques when grown on *E. coli* B. He then collected lysate from these plaques and used it to infect *E. coli* K. Phages that did not produce plaques on *E. coli* K were defined as the *rII* type.

Benzer collected thousands of *rII* mutations. He simultaneously infected bacterial cells with two different mutants and looked for recombinant progeny (**Figure 8.28**).

Consider two *rII* mutations, a^- and b^- (their wild-type alleles are a^+ and b^+). Benzer infected *E. coli* B cells with two different strains of phages, one $a^- b^+$ and the other $a^+ b^-$ (see Figure 8.28, step 3). Neither of these mutations is able to grow on *E. coli* K cells. While reproducing within the B cells, a few phages of the two strains recombined (see Figure 8.28, step 4). A single crossover produces two recombinant chromosomes; one with genotype $a^+ b^+$ and the other with genotype $a^- b^-$:

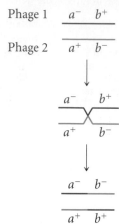

Phage 1 $\quad a^- \quad b^+$

Phage 2 $\quad a^+ \quad b^-$

The resulting recombinant chromosomes, along with the nonrecombinant (parental) chromosomes, were incorporated into progeny phages, which were then used to infect *E. coli* K cells. The resulting plaques were examined to determine the genotype of the infecting phage and map the *rII* mutants (see Figure 8.28, step 5).

Neither of the *rII* mutants grew on *E. coli* K (see Figure 8.28, step 2), but wild-type phages grew; so progeny phages that produced plaques on *E. coli* K must have the recombinant genotype $a^+ b^+$. Each recombination event produces equal numbers of double mutants ($a^- b^-$) and wild-type chromosomes ($a^+ b^+$); so the number of recombinant progeny should be twice the number of wild-type plaques that appeared on *E. coli* K. The recombination frequency between the two *rII* mutants would be:

$$RF = \frac{2 \times \text{number of plaques on } E.\,coli\,\text{K}}{\text{total number of plaques on } E.\,coli\,\text{B}}$$

Because phages produce large numbers of progeny, Benzer was able to detect a single recombinant among billions of progeny phages. Recombination frequencies are proportional to physical distances along the chromosome (see p. 174 in Chapter 7), revealing the positions of the different mutations within the *rII* region of the phage chromosome. In this way, Benzer eventually mapped more than 2400 *rII* mutations, many corresponding to single base pairs in the viral DNA. His work provided the first molecular view of a gene.

Complementation experiments Benzer's mapping experiments demonstrated that some *rII* mutations were very closely linked. This finding raised the question whether they were at the same locus or at different loci. To determine whether different *rII* mutations belonged to different functional loci, Benzer used the complementation (cis–trans) test (see p. 113 in Chapter 5).

To carry out the complementation test in bacteriophage, Benzer infected cells of *E. coli* K with large numbers of two mutant strains of phage (**Figure 8.29**, step 1) so that cells would become doubly infected with both strains. Consider

two *rII* mutations: r_{101}^- and r_{104}^-. Cells infected with both mutants

$$\frac{r_{101}^- \qquad r_{104}^+}{r_{101}^+ \qquad r_{104}^-}$$

were effectively heterozygous for the phage genes, with the mutations in the trans configuration (see Figure 8.29, step 2). In the complementation testing, the phenotypes of progeny

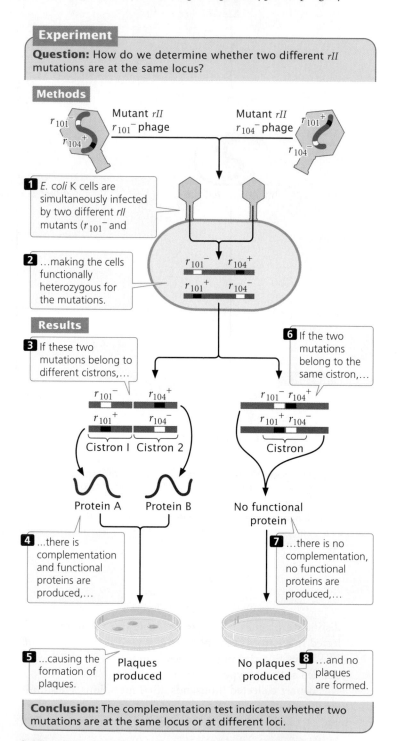

Experiment

Question: How do we determine whether two different *rII* mutations are at the same locus?

Methods

Results

Conclusion: The complementation test indicates whether two mutations are at the same locus or at different loci.

8.29 Complementation tests are used to determine whether different mutations are at the same functional gene.

phages were examined on the K strain, rather than the B strain as illustrated in Figure 8.28.

If the r_{101}^- and r_{104}^- mutations are at different functional loci that encode different proteins, then, in bacterial cells infected by both mutants, the wild-type sequences on the chromosome opposite each mutation will overcome the effects of the recessive mutations; the phages will produce normal plaques on *E. coli* K cells (Figure 8.29, steps 3, 4, and 5). If, on the other hand, the mutations are at the same locus, no functional protein is produced by either chromosome, and no plaques develop in the *E. coli* K cells (Figure 8.29, steps 6, 7, and 8). Thus, the absence of plaques indicates that the two mutations are at the same locus. Benzer coined the term *cistron* to designate a functional gene defined by the complementation test.

In the complementation test, the cis heterozygote is used as a control. Benzer simultaneously infected bacteria with wild-type phages ($r_{101}^+ \quad r_{104}^+$) and with phages carrying both mutations ($r_{101}^- \quad r_{104}^-$). This test produced cells that were heterozygous and in cis configuration for the phage genes:

$$\frac{r_{101}^+ \qquad r_{104}^+}{r_{101}^- \qquad r_{104}^-}$$

Regardless of whether the r_{101}^- and r_{104}^- mutations are in the same functional unit, these cells contain a copy of the wild-type phage chromosome ($\overline{r_{101}^+ \quad r_{104}^+}$) and will produce normal plaques in *E. coli* K.

Benzer carried out complementation testing on many pairs of *rII* mutants. He found that the *rII* region consists of two loci, designated *rIIA* and *rIIB*. Mutations belonging to the *rIIA* and *rIIB* groups complemented each other, but mutations in the *rIIA* group did not complement others in *rIIA*; nor did mutations in the *rIIB* group complement others in *rIIB*.

CONCEPTS

In a series of experiments with the bacteriophage T4, Seymour Benzer showed that recombination could take place within a single gene and created the first molecular map of a gene. Benzer used the complementation test to distinguish between functional genes (loci).

✔ CONCEPT CHECK 8

In complementation tests, Benzer simultaneously infected *E. coli* cells with two phages, each of which carried a different mutation. What conclusion did he make when the progeny phages produced normal plaques?

a. The mutations were at the same locus.

b. The mutations were at different loci.

c. The mutations were close together on the chromosome.

d. The genes were in the cis configuration.

At the time of Benzer's research, the relation between genes and DNA structure was unknown. A gene was defined as a functional unit of heredity that encoded a phenotype.

Many geneticists believed that genes were indivisible and that recombination could not take place within them. Benzer demonstrated that intragenic recombination did indeed take place (although at a very low rate) and gave geneticists their first glimpse at the structure of an individual gene.

TRY PROBLEMS 36 AND 37 →

RNA Viruses

Viral genomes may be encoded in either DNA or RNA, as stated earlier. RNA is the genetic material of some medically important human viruses, including those that cause influenza, common colds, polio, and AIDS. Almost all viruses that infect plants have RNA genomes. The medical and economic importance of RNA viruses has encouraged their study.

RNA viruses capable of integrating into the genomes of their hosts, much as temperate phages insert themselves into bacterial chromosomes, are called **retroviruses** (**Figure 8.30a**). Because the retroviral genome is RNA, whereas that of the host is DNA, a retrovirus must produce **reverse transcriptase**, an enzyme that synthesizes complementary DNA (cDNA) from either an RNA or a DNA template. A retrovirus uses reverse transcriptase to copy its RNA genome into a single-stranded DNA molecule, which it then copies to make double-stranded DNA. The DNA copy of the viral genome then integrates into the host chromosome. A viral genome incorporated into the host chromosome is called a **provirus**. The provirus is replicated by host enzymes when the host chromosome is duplicated (**Figure 8.30b**).

When conditions are appropriate, the provirus undergoes transcription to produce numerous copies of the original viral RNA genome. This RNA encodes viral proteins and serves as genomic RNA for new viral particles. As these viruses escape the cell, they collect patches of the cell membrane to use as their envelopes.

All known retroviral genomes have in common three genes: *gag*, *pol*, and *env*, each encoding a precursor protein that is cleaved into two or more functional proteins. The *gag* gene encodes proteins that make up the viral protein coat. The *pol* gene encodes reverse transcriptase and an enzyme called **integrase** that inserts the viral DNA into the host chromosome. The *env* gene encodes the glycoproteins that appear on the surface of the virus.

Some retroviruses contain **oncogenes** (see Chapter 23) that may stimulate cell division and cause the formation of tumors. The first retrovirus to be isolated, the Rous sarcoma virus, was originally recognized by its ability to produce connective-tissue tumors (sarcomas) in chickens.

Human Immunodeficiency Virus and AIDS

The human immunodeficiency virus (HIV) causes acquired immune deficiency syndrome (AIDS), a disease that killed 2 million people worldwide in 2008 alone. AIDS was first recognized in 1982, when a number of homosexual males in the United States began to exhibit symptoms of a new immune-system-deficiency disease. In that year, Robert

(a)

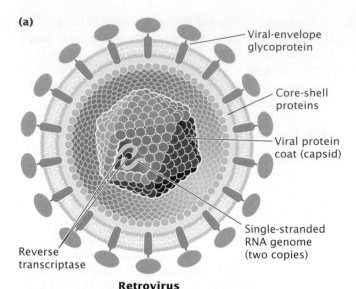

- Viral-envelope glycoprotein
- Core-shell proteins
- Viral protein coat (capsid)
- Single-stranded RNA genome (two copies)
- Reverse transcriptase

Retrovirus

8.30 A retrovirus uses reverse transcription to incorporate its RNA into the host DNA. (a) Structure of a typical retrovirus. Two copies of the single-stranded RNA genome and the reverse transcriptase enzyme are shown enclosed within a protein capsid. The capsid is surrounded by a viral envelope that is studded with viral glycoproteins. (b) The retrovirus life cycle.

Gallo proposed that AIDS was caused by a retrovirus. Between 1983 and 1984, as the AIDS epidemic became widespread, the HIV retrovirus was isolated from AIDS patients. AIDS is now known to be caused by two different immunodeficiency viruses, HIV-1 and HIV-2, which together have infected more than 65 million people worldwide. Of those infected, 25 million have died. Most cases of AIDS are caused by HIV-1, which now has a global distribution; HIV-2 is found primarily in western Africa.

HIV illustrates the importance of genetic recombination in viral evolution. Studies of the DNA sequences of HIV and other retroviruses reveal that HIV-1 is closely related to the simian immunodeficiency virus found in chimpanzees (SIV$_{cpz}$). Many wild chimpanzees in Africa are infected with SIV$_{cpz}$, although it doesn't cause AIDS-like symptoms in chimps. SIV$_{cpz}$ is itself a hybrid that resulted from recombination between a retrovirus found in the red-capped mangabey (a monkey) and a retrovirus found in the greater spot-nosed monkey (**Figure 8.31**). Apparently, one or more chimpanzees became infected with both viruses; recombination between the viruses produced SIV$_{cpz}$, which was then transmitted to humans through contact with infected chimpanzees. In humans, SIV$_{cpz}$ underwent significant evolution to become HIV-1, which then spread throughout the world to produce the AIDS epidemic. Several independent transfers of SIV$_{cpz}$ to humans gave rise to different strains of HIV-1. HIV-2 evolved from a different retrovirus, SIV$_{sm}$, found in sooty mangabeys.

HIV is transmitted by sexual contact between humans and through any type of blood-to-blood contact, such as that caused by the sharing of dirty needles by drug addicts.

(b)

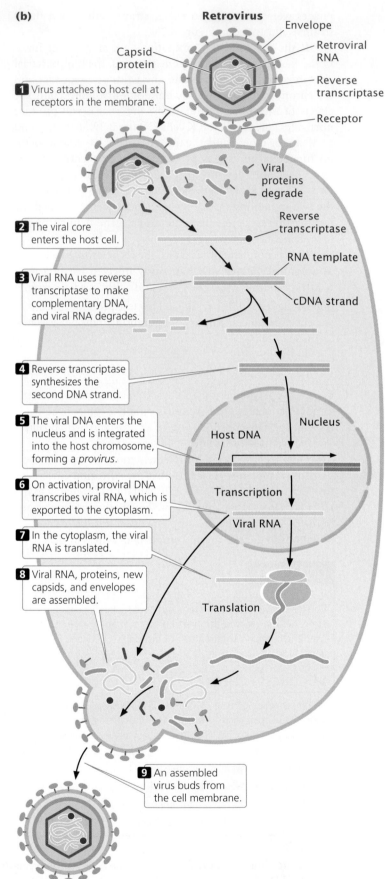

Retrovirus

- Capsid protein
- Envelope
- Retroviral RNA
- Reverse transcriptase
- Receptor

1 Virus attaches to host cell at receptors in the membrane.

2 The viral core enters the host cell.

Viral proteins degrade

Reverse transcriptase

RNA template

3 Viral RNA uses reverse transcriptase to make complementary DNA, and viral RNA degrades.

cDNA strand

4 Reverse transcriptase synthesizes the second DNA strand.

5 The viral DNA enters the nucleus and is integrated into the host chromosome, forming a *provirus*.

Host DNA

Nucleus

6 On activation, proviral DNA transcribes viral RNA, which is exported to the cytoplasm.

Transcription

Viral RNA

7 In the cytoplasm, the viral RNA is translated.

8 Viral RNA, proteins, new capsids, and envelopes are assembled.

Translation

9 An assembled virus buds from the cell membrane.

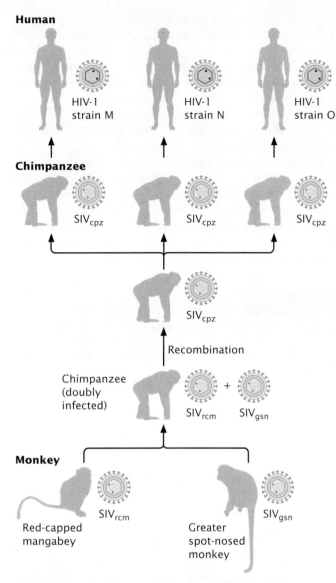

8.31 HIV-1 evolved from a similar virus (SIV_cpz) found in chimpanzees and was transmitted to humans. SIV_cpz arose from recombination taking place between retroviruses in red-capped mangabeys and greater spot-nosed monkeys.

Until screening tests could identify HIV-infected blood, transfusions and clotting factors used by hemophiliacs also were sources of infection.

HIV principally attacks a class of blood cells called helper T lymphocytes or, simply, helper T cells (**Figure 8.32**). HIV enters a helper T cell, undergoes reverse transcription, and integrates into the chromosome. The virus reproduces rapidly, destroying the T cell as new virus particles escape from the cell. Because helper T cells are central to immune function and are destroyed in the infection, AIDS patients have a diminished immune response; most AIDS patients die of secondary infections that develop because they have lost the ability to fight off pathogens.

The HIV genome is 9749 nucleotides long and carries *gag, pol, env*, and six other genes that regulate the life cycle of

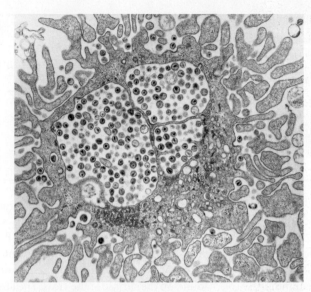

8.32 HIV principally attacks T lymphocytes. Electron micrograph showing a T cell infected with HIV, visible as small circles with dark centers. [Courtesy of Dr. Hans Gelderblom.]

the virus. HIV's reverse transcriptase is very error prone, giving the virus a high mutation rate and allowing it to evolve rapidly, even within a single host. This rapid evolution makes the development of an effective vaccine against HIV particularly difficult. Genetic variation within the human population also affects the virus. To date, more than 10 loci in humans that affect HIV infection and the progression of AIDS have been identified.

CONCEPTS

A retrovirus is an RNA virus that integrates into its host's chromosome by making a DNA copy of its RNA genome through the process of reverse transcription. Human immunodeficiency virus, the causative agent of AIDS, is a retrovirus. It evolved from related retroviruses found in other primates.

✔ **CONCEPT CHECK 9**

What enzyme is used by a retrovirus to make a DNA copy of its genome?

Influenza Virus

Influenza demonstrates how rapid changes in a pathogen can arise through recombination of its genetic material. Influenza, commonly called flu, is a respiratory disease caused by influenza viruses. In the United States, from 5% to 20% of the entire population is infected with influenza annually and, though most cases are mild, 36,000 people die from influenza-related causes each year. At certain times, particularly when new strains of influenza virus enter the human population, worldwide epidemics (called pandemics) take place; for example, the Spanish flu virus in 1918 killed an estimated 20 million to 100 million people worldwide.

Table 8.5	Strains of influenza virus responsible for major flu pandemics	
Year	Influenza Pandemic	Strain
1918	Spanish flu	H1N1
1957	Asian flu	H2N2
1968	Hong Kong flu	H3N2
2009	Swine flu	H1N1

Influenza viruses are RNA viruses that infect birds and mammals. They come in three main types: influenza A, influenza B, and influenza C. Most cases of the common flu are caused by influenza A and B. Influenza A is divided into subtypes on the basis of two proteins, hemagglutinin (HA) and neuraminidase (NA), found on the surface of the virus. The HA and NA proteins affect the ability of the virus to enter host cells and the host organism's immune response to infection. There are 16 types of HA and 9 types of NA, which can exist in a virus in different combinations. For example, common strains of influenza A circulating in humans today are H1N1 and H3N2 (**Table 8.5**), along with several strains of influenza B. Most of the different subtypes of influenza A are found in birds.

Although influenza is an RNA virus, it is not a retrovirus: its genome is not copied into DNA and incorporated into the host chromosome as is that of a retrovirus. The influenza viral genome consists of 7 or 8 pieces of RNA that are enclosed in a viral envelope. Each piece of RNA encodes one or two of the virus's proteins. The virus enters a host cell by attaching to specific receptors on the cell membrane. After the viral particle has entered the cell, the viral RNA is released, copied, and translated into viral proteins. Viral RNA molecules and viral proteins are then assembled into new viral particles, which exit the cell and infect additional cells.

One of the dangers of the influenza virus is that it evolves every rapid, with new strains appearing frequently. Influenza evolves in two ways. First, each strain continually changes through mutations arising in the viral RNA. The enzyme that copies the RNA is especially prone to making mistakes, and so new mutations are continually introduced into the viral genome. This type of continual change is call **antigenic drift**. Occasionally, major changes in the viral genome take place through **antigenic shift**, in which genetic material from different strains is combined in a process called reassortment. Reassortment takes place when a host is simultaneously infected with two different strains. The RNAs of both strains are replicated within the cell, and RNA segments from two different strains are incorporated into the same viral particle, creating a new strain. For example,

in 2002, reassortment occurred between the H1N1 and H3N2 subtypes, creating a new H1N2 strain that contained the hemagluttinin from H1N1 and the neuraminidase from H3N2. New strains produced by antigenic shift are responsible for most pandemics, because no one has immunity to the radically different virus that is produced.

Most different strains of influenza A infect birds, but humans are not easily infected with bird influenza. The appearance of new strains in humans is thought to often arise from viruses that reassort in pigs, which can be infected by viruses from both humans and birds. In 2009, a new strain of H1N1 influenza (called swine flu) emerged in Mexico and quickly spread throughout the world. This virus arose from a series of reassortment events that combined gene sequences from human, bird, and pig influenza viruses to produce the new H1N1 virus (**Figure 8.33**).

CONCEPTS

Influenza is caused by RNA influenza viruses. New strains of influenza appear through antigenic shift, in which new viral genomes are created through the reassortment of RNA molecules of different strains.

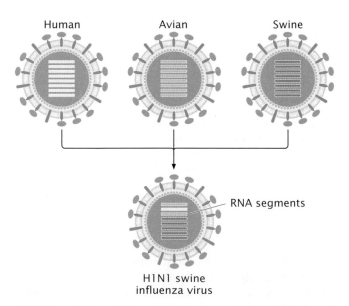

8.33 New strains of influenza virus are created by reassortment of genetic material from different strains. A new H1N1 virus (swine flu virus) that appeared in 2009 contained genetic material from avian, swine, and human viruses.

SUMMARY

- Bacteria and viruses are well suited to genetic studies: they are small, have a small haploid genome, undergo rapid reproduction, and produce large numbers of progeny through asexual reproduction.

- The bacterial genome normally consists of a single, circular molecule of double-stranded DNA. Plasmids are small pieces of bacterial DNA that can replicate independently of the chromosome.

- DNA can be transferred between bacteria by conjugation, transformation, or transduction.

- Conjugation is the union of two bacterial cells and the transfer of genetic material between them. It is controlled by an episome called F. The rate at which individual genes are transferred during conjugation provides information about the order of the genes and the distances between them on the bacterial chromosome.

- Bacteria take up DNA from the environment through the process of transformation. Frequencies of the cotransformation of genes provide information about the physical distances between chromosomal genes.

- Complete DNA sequences of a number of bacterial species have been determined. This sequence information indicates that horizontal gene transfer—the movement of DNA between species—is common in bacteria.

- Viruses are replicating structures with DNA or RNA genomes that may be double stranded or single stranded and linear or circular.

- Bacterial genes become incorporated into phage coats and are transferred to other bacteria by phages through the process of transduction. Rates of cotransduction can be used to map bacterial genes.

- Phage genes can be mapped by infecting bacterial cells with two different phage strains and counting the number of recombinant plaques produced by the progeny phages.

- Benzer mapped a large number of mutations that occurred within the *rII* region of phage T4. The results of his complementation studies demonstrated that the *rII* region consists of two functional units that he called cistrons. He showed that intragenic recombination takes place.

- A number of viruses have RNA genomes. Retroviruses encode reverse transcriptase, an enzyme used to make a DNA copy of the viral genome, which then integrates into the host genome as a provirus. HIV is a retrovirus that is the causative agent for AIDS.

- Influenza is caused by RNA influenza viruses that evolve through small changes taking place through mutation (antigenic drift) and major changes taking place through reassortment of the genetic material from different strains.

IMPORTANT TERMS

prototrophic bacteria (p. 205)
minimal medium (p. 205)
auxotrophic bacteria (p. 205)
complete medium (p. 205)
colony (p. 205)
plasmid (p. 206)
episome (p. 207)
F factor (p. 207)
conjugation (p. 208)
transformation (p. 208)
transduction (p. 208)
pili (singular, pilus) (p. 211)

competent cell (p. 217)
transformant (p. 218)
cotransformation (p. 218)
horizontal gene transfer (p. 218)
virus (p. 219)
virulent phage (p. 219)
temperate phage (p. 219)
prophage (p. 219)
plaque (p. 219)
generalized transduction (p. 220)
specialized transduction (p. 222)
transducing phage (p. 221)

transductant (p. 221)
cotransduction (p. 221)
intragenic mapping (p. 224)
retrovirus (p. 227)
reverse transcriptase (p. 227)
provirus (p. 227)
integrase (p. 227)
oncogene (p. 227)
antigenic drift (p. 230)
antigenic shift (p. 230)

ANSWERS TO CONCEPT CHECKS

1. d
2. b
3. a
4. *gal*
5. *his* and *leu*

6. b
7. c
8. b
9. Reverse transcriptase

WORKED PROBLEMS

1. DNA from a strain of bacteria with genotype $a^+ b^+ c^+ d^+ e^+$ was isolated and used to transform a strain of bacteria that was $a^- b^- c^- d^- e^-$. The transformed cells were tested for the presence of donated genes. The following genes were cotransformed:

$$a^+ \text{ and } d^+ \qquad c^+ \text{ and } d^+$$
$$b^+ \text{ and } e^+ \qquad c^+ \text{ and } e^+$$

What is the order of genes a, b, c, d, and e on the bacterial chromosome?

• Solution

The rate at which genes are cotransformed is inversely proportional to the distance between them: genes that are close together are frequently cotransformed, whereas genes that are far apart are rarely cotransformed. In this transformation experiment, gene c^+ is cotransformed with both genes e^+ and d^+, but genes e^+ and d^+ are not cotransformed; therefore the c locus must be between the d and e loci:

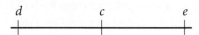

Gene e^+ is also cotransformed with gene b^+; so the e and b loci must be located close together. Locus b could be on either side of locus e. To determine whether locus b is on the same side of e as locus c, we look to see whether genes b^+ and c^+ are cotransformed. They are not; so locus b must be on the opposite side of e from c:

Gene a^+ is cotransformed with gene d^+; so they must be located close together. If locus a were located on the same side of d as locus c, then genes a^+ and c^+ would be cotransformed. Because these genes display no cotransformation, locus a must be on the opposite side of locus d:

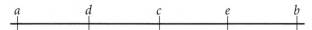

2. Consider three genes in *E. coli*: thr^+ (the ability to synthesize threonine), ara^+ (the ability to metabolize arabinose), and leu^+ (the ability to synthesize leucine). All three of these genes are close together on the *E. coli* chromosome. Phages are grown in a $thr^+ ara^+ leu^+$ strain of bacteria (the donor strain). The phage lysate is collected and used to infect a strain of bacteria that is $thr^- ara^- leu^-$. The recipient bacteria are then tested on medium lacking leucine. Bacteria that grow and form colonies on this

medium (leu^+ transductants) are then replica plated onto medium lacking threonine and onto medium lacking arabinose to see which are thr^+ and which are ara^+.

Another group of recipient bacteria are tested on medium lacking threonine. Bacteria that grow and form colonies on this medium (thr^+ transductants) are then replica plated onto medium lacking leucine and onto medium lacking arabinose to see which are ara^+ and which are leu^+. Results from these experiments are as follows:

Selected marker	Cells with cotransduced genes (%)
leu^+	3 thr^+
	76 ara^+
thr^+	3 leu^+
	0 ara^+

How are the loci arranged on the chromosome?

• Solution

Notice that, when we select for leu^+ (the top half of the table), most of the selected cells are also ara^+. This finding indicates that the leu and ara genes are located close together, because they are usually cotransduced. In contrast, thr^+ is only rarely cotransduced with leu^+, indicating that leu and thr are much farther apart. On the basis of these observations, we know that leu and ara are closer together than are leu and thr, but we don't yet know the order of three genes—whether thr is on the same side of ara as leu or on the opposite side, as shown here:

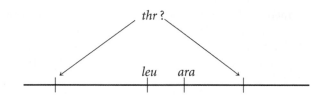

We can determine the position of thr with respect to the other two genes by looking at the cotransduction frequencies when thr^+ is selected (the bottom half of the preceding table). Notice that, although the cotransduction frequency for thr and leu also is 3%, no $thr^+ ara^+$ cotransductants are observed. This finding indicates that thr is closer to leu than to ara, and therefore thr must be to the left of leu, as shown here:

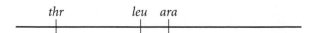

COMPREHENSION QUESTIONS

Section 8.1

1. Explain how auxotrophic bacteria are isolated.

Section 8.2

2. Briefly explain the differences between F$^+$, F$^-$, Hfr, and F$'$ cells.

*3. What types of matings are possible between F⁺, F⁻, Hfr, and F′ cells? What outcomes do these matings produce? What is the role of F factor in conjugation?

*4. Explain how interrupted conjugation, transformation, and transduction can be used to map bacterial genes. How are these methods similar and how are they different?

5. What is horizontal gene transfer and how might it take place?

Section 8.2

*6. List some of the characteristics that make bacteria and viruses ideal organisms for many types of genetic studies.

7. What types of genomes do viruses have?

8. Briefly describe the differences between the lytic cycle of virulent phages and the lysogenic cycle of temperate phages.

9. Briefly explain how genes in phages are mapped.

*10. How does specialized transduction differ from generalized transduction?

*11. Briefly explain the method used by Benzer to determine whether two different mutations occurred at the same locus.

*12. Explain how a retrovirus, which has an RNA genome, is able to integrate its genetic material into that of a host having a DNA genome.

13. Briefly describe the genetic structure of a typical retrovirus.

14. What are the evolutionary origins of HIV-1 and HIV-2?

15. Most humans are not easily infected by avian influenza. How then do DNA sequences from avian influenza become incorporated into human influenza?

APPLICATION QUESTIONS AND PROBLEMS

Introduction

16. Suppose you want to compare the species of bacteria that exist in a polluted stream with the species that exist in an unpolluted stream. Traditionally, bacteria have been identified by growing them in the laboratory and comparing their physical and biochemical properties. You recognize that you will be unable to culture most of the bacteria that reside in the streams. How might you go about identifying the species in the two streams without culturing them in the lab?

Section 8.2

*17. John Smith is a pig farmer. For the past 5 years, Smith has been adding vitamins and low doses of antibiotics to his pig food; he says that these supplements enhance the growth of the pigs. Within the past year, however, several of his pigs died from infections of common bacteria, which failed to respond to large doses of antibiotics. Can you offer an explanation for the increased rate of mortality due to infection in Smith's pigs? What advice might you offer Smith to prevent this problem in the future?

18. Rarely, the conjugation of Hfr and F⁻ cells produces two Hfr cells. Explain how this event takes place.

19. Austin Taylor and Edward Adelberg isolated some new strains of Hfr cells that they then used to map several genes in *E. coli* by using interrupted conjugation (A. L. Taylor and E. A. Adelberg. 1960. *Genetics* 45:1233–1243). In one experiment, they mixed cells of Hfr strain AB-312, which were *xyl⁺ mtl⁺ mal⁺ met⁺* and sensitive to phage T6, with F⁻ strain AB-531, which was *xyl⁻ mtl⁻ mal⁻ met⁻* and resistant to phage T6. The cells were allowed to undergo conjugation. At regular intervals, the researchers removed a sample of cells and interrupted conjugation by killing the Hfr cells with phage T6. The F⁻ cells, which were resistant to phage T6, survived and were then tested

for the presence of genes transferred from the Hfr strain. The results of this experiment are shown in the accompanying graph. On the basis of these data, give the order of the *xyl, mtl, mal*, and *met* genes on the bacterial chromosome and indicate the minimum distances between them.

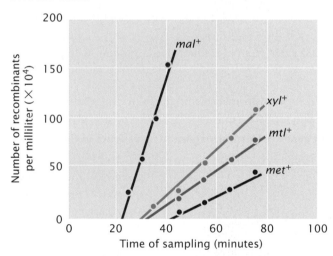

*20. A series of Hfr strains that have genotype *m⁺ n⁺ o⁺ p⁺ q⁺ r⁺* are mixed with an F⁻ strain that has genotype *m⁻ n⁻ o⁻ p⁻ q⁻ r⁻*. Conjugation is interrupted at regular intervals and the order of the appearance of genes from the Hfr strain is determined in the recipient cells. The order of gene transfer for each Hfr strain is:

Hfr5	*m⁺ q⁺ p⁺ n⁺ r⁺ o⁺*
Hfr4	*n⁺ r⁺ o⁺ m⁺ q⁺ p⁺*
Hfr1	*o⁺ m⁺ q⁺ p⁺ n⁺ r⁺*
Hfr9	*q⁺ m⁺ o⁺ r⁺ n⁺ p⁺*

What is the order of genes on the circular bacterial chromosome? For each Hfr strain, give the location of the F factor in the chromosome and its polarity.

*21. Crosses of three different Hfr strains with separate samples of an F⁻ strain are carried out, and the following mapping data are provided from studies of interrupted conjugation:

Appearance of genes in F⁻ cells

Hfr1:	Genes	b^+	d^+	c^+	f^+	g^+
	Time	3	5	16	27	59
Hfr2:	Genes	e^+	f^+	c^+	d^+	b^+
	Time	6	24	35	46	48
Hfr3:	Genes	d^+	c^+	f^+	e^+	g^+
	Time	4	15	26	44	58

Construct a genetic map for these genes, indicating their order on the bacterial chromosome and the distances between them.

22. DNA from a strain of *Bacillus subtilis* with the genotype trp^+ tyr^+ was used to transform a recipient strain with the genotype trp^- tyr^-. The following numbers of transformed cells were recovered:

Genotype	Number of transformed cells
trp^+ tyr^-	154
trp^- tyr^+	312
trp^+ tyr^+	354

What do these results suggest about the linkage of the *trp* and *tyr* genes?

23. DNA from a strain of *Bacillus subtilis* with genotype a^+ b^+ c^+ d^+ e^+ is used to transform a strain with genotype a^- b^- c^- d^- e^-. Pairs of genes are checked for cotransformation and the following results are obtained:

Pair of genes	Cotransformation	Pair of genes	Cotransformation
a^+ and b^+	no	b^+ and d^+	no
a^+ and c^+	no	b^+ and e^+	yes
a^+ and d^+	yes	c^+ and d^+	no
a^+ and e^+	yes	c^+ and e^+	yes
b^+ and c^+	yes	d^+ and e^+	no

On the basis of these results, what is the order of the genes on the bacterial chromosome?

24. DNA from a bacterial strain that is his^+ leu^+ lac^+ is used to transform a strain that is his^- leu^- lac^-. The following percentages of cells were transformed:

Donor strain	Recipient strain	Genotype of transformed cells	Percentage
his^+ leu^+ lac^+	his^- leu^- lac^-	his^+ leu^+ lac^+	0.02
		his^+ leu^+ lac^-	0.00
		his^+ leu^- lac^+	2.00
		his^+ leu^- lac^-	4.00
		his^- leu^+ lac^+	0.10
		his^- leu^- lac^+	3.00
		his^- leu^+ lac^-	1.50

a. What conclusions can you make about the order of these three genes on the chromosome?

b. Which two genes are closest?

25. Rollin Hotchkiss and Julius Marmur studied transformation in the bacterium *Streptococcus pneumoniae* (R. D. Hotchkiss and J. Marmur. 1954. *Proceedings of the National Academy of Sciences* 40:55–60). They examined four mutations in this bacterium: penicillin resistance (*P*), streptomycin resistance (*S*), sulfanilamide resistance (*F*), and the ability to utilize mannitol (*M*). They extracted DNA from strains of bacteria with different combinations of different mutations and used this DNA to transform wild-type bacterial cells (P^+ S^+ F^+ M^+). The results from one of their transformation experiments are shown here.

Donor DNA	Recipient DNA	Transformants	Percentage of all cells
M S F	M^+ S^+ F^+	M^+ S F^+	4.0
		M^+ S^+ F	4.0
		M S^+ F^+	2.6
		M S F^+	0.41
		M^+ S F	0.22
		M S^+ F	0.0058
		M S F	0.0071

a. Hotchkiss and Marmur noted that the percentage of cotransformation was higher than would be expected on a random basis. For example, the results show that the 2.6% of the cells were transformed into *M* and 4% were transformed into *S*. If the *M* and *S* traits were inherited independently, the expected probability of cotransformation of *M* and *S* (*M S*) would be 0.026 × 0.04 = 0.001, or 0.1%. However, they observed 0.41% *M S* cotransformants, four times more than they expected. What accounts for the relatively high frequency of cotransformation of the traits they observed?

b. On the basis of the results, what conclusion can you make about the order of the *M*, *S*, and *F* genes on the bacterial chromosome?

c. Why is the rate of cotransformation for all three genes (*M S F*) almost the same as the rate of cotransformation for *M F* alone?

26. In the course of a study on the effects of the mechanical shearing of DNA, Eugene Nester, A. T. Ganesan, and Joshua Lederberg studied the transfer, by transformation, of sheared DNA from a wild-type strain of *Bacillus subtilis* (his_2^+ aro_3^+ try_2^+ aro_1^+ tyr_1^+ aro_2^+) to strains of bacteria carrying a series of mutations (E. W. Nester, A. T. Ganesan, and J. Lederberg. 1963. *Proceedings of the National Academy of Sciences* 49:61–68). They reported the following rates of cotransformation between $his_2^|$ and the other genes (expressed as cotransfer rate).

Genes	Rate of cotransfer
his_2^+ and aro_3^+	0.015
his_2^+ and try_2^+	0.10
his_2^+ and aro_1^+	0.12
his_2^+ and tyr_1^+	0.23
his_2^+ and aro_2^+	0.05

On the basis of these data, which gene is farthest from his_2^+? Which gene is closest?

27. C. Anagnostopoulos and I. P. Crawford isolated and studied a series of mutations that affected several steps in the biochemical pathway leading to tryptophan in the bacterium *Bacillus subtilis* (C. Anagnostopoulos and I. P. Crawford. 1961. *Proceedings of the National Academy of Sciences* 47:378–390). Seven of the strains that they used in their study are listed here, along with the mutation found in that strain.

Strain	Mutation
T3	T^-
168	I^-
168PT	I^-
TI	I^-
TII	I^-
T8	A^-
H25	H^-

To map the genes for tryptophan synthesis, they carried out a series of transformation experiments on strains having different mutations and determined the percentage of recombinants among the transformed bacteria. Their results were as follows:

Recipient	Donor	Percent recombinants
T3	168PT	12.7
T3	T11	11.8
T3	T8	43.5
T3	H25	28.6
168	H25	44.9
TII	H25	41.4
TI	H25	31.3
T8	H25	67.4
H25	T3	19.0
H25	TII	26.3
H25	TI	13.4
H25	T8	45.0

On the basis of these two-point recombination frequencies, determine the order of the genes and the distances between them. Where more than one cross was completed for a pair of genes, average the recombination rates from the different crosses. Draw a map of the genes on the chromosome.

Section 8.3

28. Two mutations that affect plaque morphology in phages (a^- and b^-) have been isolated. Phages carrying both mutations ($a^- b^-$) are mixed with wild-type phages ($a^+ b^+$) and added to a culture of bacterial cells. Subsequent to infection and lysis, samples of the phage lysate are collected and cultured on bacterial cells. The following numbers of plaques are observed:

Plaque phenotype	Number
$a^+ b^+$	2043
$a^+ b^-$	320
$a^- b^+$	357
$a^- b^-$	2134

What is the frequency of recombination between the *a* and *b* genes?

29. T. Miyake and M. Demerec examined proline-requiring mutations in the bacterium *Salmonella typhimurium* (T. Miyake and M. Demerec. 1960. *Genetics* 45:755–762). On the basis of complementation studies, they found four proline auxotrophs: *proA*, *proB*, *proC*, and *proD*. To determine whether *proA*, *proB*, *proC*, and *proD* loci were located close together on the bacterial chromosome, they conducted a transduction experiment. Bacterial strains that were $proC^+$ and had mutations at *proA*, *proB*, or *proD*, were used as donors. The donors were infected with bacteriophages, and progeny phages were allowed to infect recipient

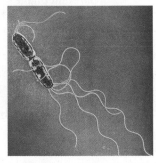

S. typhimurium. [Kwangshin Kim/ Photo Researchers.]

bacteria with genotype $proC^- proA^+ proB^+ proD^+$. The bacteria were then plated on a selective medium that allowed only $proC^+$ bacteria to grow. The following results were obtained:

Donor genotype	Transductant genotype	Number
$proC^+ proA^- proB^+ proD^+$	$proC^+ proA^+ proB^+ proD^+$	2765
	$proC^+ proA^- proB^+ proD^+$	3
$proC^+ proA^+ proB^- proD^+$	$proC^+ proA^+ proB^+ proD^+$	1838
	$proC^+ proA^+ proB^- proD^+$	2
$proC^+ proA^+ proB^+ proD^-$	$proC^+ proA^+ proB^+ proD^+$	1166
	$proC^+ proA^+ proB^+ proD^-$	0

a. Why are there no *proC*⁻ genotypes among the transductants?

b. Which genotypes represent single transductants and which represent cotransductants?

c. Is there evidence that *proA*, *proB*, and *proD* are located close to *proC*? Explain your answer.

*30. A geneticist isolates two mutations in a bacteriophage. One mutation causes clear plaques (*c*), and the other produces minute plaques (*m*). Previous mapping experiments have established that the genes responsible for these two mutations are 8 m.u. apart. The geneticist mixes phages with genotype $c^+ m^+$ and genotype $c^- m^-$ and uses the mixture to infect bacterial cells. She collects the progeny phages and cultures a sample of them on plated bacteria. A total of 1000 plaques are observed. What numbers of the different types of plaques ($c^+ m^+, c^- m^-, c^+ m^-, c^- m^+$) should she expect to see?

31. The geneticist carries out the same experiment described in Problem 30, but this time she mixes phages with genotypes $c^+ m^-$ and $c^- m^+$. What results are expected from *this* cross?

*32. A geneticist isolates two bacteriophage *r* mutants (r_{13} and r_2) that cause rapid lysis. He carries out the following crosses and counts the number of plaques listed here:

Genotype of parental phage	Progeny	Number of plaques
$h^+ r_{13}^- \times h^- r_{13}^+$	$h^+ r_{13}^+$	1
	$h^- r_{13}^+$	104
	$h^+ r_{13}^-$	110
	$h^- r_{13}^-$	2
Total		216
$h^+ r_2^- \times h^- r_2^+$	$h^+ r_2^+$	6
	$h^- r_2^+$	86
	$h^+ r_2^-$	81
	$h^- r_2^-$	7
Total		180

a. Calculate the recombination frequencies between r_2 and *h* and between r_{13} and *h*.

b. Draw all possible linkage maps for these three genes.

*33. *E. coli* cells are simultaneously infected with two strains of phage λ. One strain has a mutant host range, is temperature sensitive, and produces clear plaques (genotype is *h st c*); another strain carries the wild-type alleles (genotype is $h^+ st^+ c^+$). Progeny phages are collected from the lysed cells and are plated on bacteria. The genotypes of the progeny phages are given here:

Progeny phage genotype	Number of plaques
$h^+ c^+ st^+$	321
$h\ c\ st$	338
$h^+ c\ st$	26
$h\ c^+ st^+$	30
$h^+ c\ st^+$	106
$h\ c^+ st$	110
$h^+ c^+ st$	5
$h\ c\ st^+$	6

a. Determine the order of the three genes on the phage chromosome.

b. Determine the map distances between the genes.

c. Determine the coefficient of coincidence and the interference (see p. 182 in Chapter 7).

34. A donor strain of bacteria with genes $a^+ b^+ c^+$ is infected with phages to map the donor chromosome with generalized transduction. The phage lysate from the bacterial cells is collected and used to infect a second strain of bacteria that are $a^- b^- c^-$. Bacteria with the a^+ gene are selected, and the percentage of cells with cotransduced b^+ and c^+ genes are recorded.

Donor	Recipient	Selected gene	Cells with cotransduced gene (%)
$a^+ b^+ c^+$	$a^- b^- c^-$	a^+	25 b^+
		a^+	3 c^+

Is gene *b* or gene *c* closer to gene *a*? Explain your reasoning.

35. A donor strain of bacteria with genotype $leu^+ gal^- pro^+$ is infected with phages. The phage lysate from the bacterial cells is collected and used to infect a second strain of bacteria that are $leu^- gal^+ pro^-$. The second strain is selected for leu^+, and the following cotransduction data are obtained:

Donor	Recipient	Selected gene	Cells with cotransduced gene (%)
$leu^+ gal^- pro^+$	$leu^- gal^+ pro^-$	leu^+	47 pro^+
		leu^+	26 gal^-

Which genes are closest, *leu* and *gal* or *leu* and *pro*?

36. A geneticist isolates two new mutations, called rII_x and rII_y, from the *rII* region of bacteriophage T4. *E. coli* B cells are simultaneously infected with phages carrying the rII_x mutation and with phages carrying the rII_y mutation. After the cells have lysed, samples of the phage lysate are collected. One sample is grown on *E. coli* K cells; a

second sample is grown on *E. coli* B cells. There are 8322 plaques on *E. coli* B and 3 plaques on *E. coli* K. What is the recombination frequency between these two mutations?

37. A geneticist is working with a new bacteriophage called phage Y3 that infects *E. coli*. He has isolated eight mutant phages that fail to produce plaques when grown on *E. coli* strain K. To determine whether these mutations occur at the same functional gene, he simultaneously infects *E. coli* K cells with paired combinations of the mutants and looks to see whether plaques are formed. He obtains the following results. (A plus sign means that plaques were formed on *E. coli* K; a minus sign means that no plaques were formed on *E. coli* K.)

Mutant	1	2	3	4	5	6	7	8
1								
2	+							
3	+	+						
4	+	−	+					
5	−	+	+	+				
6	−	+	+	+	−			
7	+	−	+	−	+	+		
8	−	+	+	+	−	−	+	

a. To how many functional genes (cistrons) do these mutations belong?

b. Which mutations belong to the same functional gene?

CHALLENGE QUESTIONS

Section 8.2

38. As a summer project, a microbiology student independently isolates two mutations in *E. coli* that are auxotrophic for glycine (*gly⁻*). The student wants to know whether these two mutants are at the same functional unit. Outline a procedure that the student could use to determine whether these two *gly⁻* mutations occur within the same functional unit.

39. A group of genetics students mix two auxotrophic strains of bacteria: one is *leu⁺ trp⁺ his⁻ met⁻* and the other is *leu⁻ trp⁻ his⁺ met⁺*. After mixing the two strains, they plate the bacteria on minimal medium and observe a few prototrophic colonies (*leu⁺ trp⁺ his⁺ met⁺*). They assume that some gene transfer has taken place between the two strains. How can they determine whether the transfer of genes is due to conjugation, transduction, or transformation?

9 Chromosome Variation

Down syndrome is caused by the presence of three copies of one or more genes located on chromosome 21. [Stockbyte.]

TRISOMY 21 AND THE DOWN-SYNDROME CRITICAL REGION

In 1866, John Langdon Down, physician and medical superintendent of the Earlswood Asylum in Surrey, England, noticed a remarkable resemblance among a number of his mentally retarded patients: all of them possessed a broad, flat face, a thick tongue, a small nose, and oval-shaped eyes. Their features were so similar, in fact, that he felt that they might easily be mistaken as children of the same family. Down did not understand the cause of their retardation, but his original description faithfully records the physical characteristics of this most common genetic form of mental retardation. In his honor, the disorder is today known as Down syndrome.

As early as the 1930s, geneticists suggested that Down syndrome might be due to a chromosome abnormality, but not until 1959 did researchers firmly establish the cause of Down syndrome: most people with the disorder have three copies of chromosome 21, a condition known as trisomy 21. In a few rare cases, people having the disorder are trisomic for just specific parts of chromosome 21. By comparing the parts of chromosome 21 that these people have in common, geneticists have established that a specific segment—called the Down-syndrome critical region or DSCR—probably contains one or more genes responsible for the features of Down syndrome. Sequencing of the human genome has established that the DSCR consists of a little more than 5 million base pairs and only 33 genes.

Research by several groups in 2006 was a source of insight into the roles of two genes in the DSCR. Mice that have deficiencies in the regulatory pathway controlled by two DSCR genes, called *DSCR1* and *DYRK1A*, exhibit many of the features seen in Down syndrome. Mice genetically engineered to express *DYRK1A* and *DSCR1* at high levels have abnormal heart development, similar to congenital heart problems seen in many people with Down syndrome.

In spite of this exciting finding, the genetics of Down syndrome appears to be more complex than formerly thought. Geneticist Lisa Olson and her colleagues had conducted an earlier study on mice to determine whether genes within the DSCR are solely responsible for Down syndrome. Mouse breeders have developed several strains of mice that are trisomic for most of the genes found on human chromosome 21 (the equivalent mouse genes are found on mouse chromosome 16). These mice display many of the same anatomical features found in people with Down syndrome, as well as altered behavior, and they are considered an animal model for Down syndrome. Olson and her colleagues carefully created mice that were trisomic for genes in the DSCR but possessed the normal two copies of other genes found on human chromosome 21. If one or more of the genes in the DSCR are indeed responsible for Down syndrome, these mice should exhibit the features of Down syndrome. Surprisingly, the engineered mice exhibited none of the anatomical features of Down syndrome, demonstrating that three copies of genes in the DSCR are not the sole cause of the features of Down syndrome, at least in mice. Another study examined a human gene called *APP*, which lies outside of the DSCR. This gene appears to be responsible for at least some of the Alzheimer-like features observed in older Down-syndrome people.

Taken together, findings from these studies suggest that Down syndrome is not due to a single gene but is instead caused by complex interactions among multiple genes that are affected when an extra copy of chromosome 21 is present. Research on Down syndrome illustrates the principle that chromosome abnormalities often affect many genes that interact in complex ways.

Most species have a characteristic number of chromosomes, each with a distinct size and structure, and all the tissues of an organism (except for gametes) generally have the same set of chromosomes. Nevertheless, variations in chromosome number—such as the extra chromosome 21 that leads to Down syndrome—do periodically arise. Variations may also arise in chromosome structure: individual chromosomes may lose or gain parts and the order of genes within a chromosome may become altered. These variations in the number and structure of chromosomes are termed **chromosome mutations**, and they frequently play an important role in evolution.

We begin this chapter by briefly reviewing some basic concepts of chromosome structure, which we learned in Chapter 2. We then consider the different types of chromosome mutations, their definitions, features, phenotypic effects, and influence on evolution.

9.1 Chromosome Mutations Include Rearrangements, Aneuploids, and Polyploids

Before we consider the different types of chromosome mutations, their effects, and how they arise, we will review the basics of chromosome structure.

Chromosome Morphology

Each functional chromosome has a centromere, to which spindle fibers attach, and two telomeres, which stabilize the chromosome (see Figure 2.7). Chromosomes are classified into four basic types:

1. **Metacentric.** The centromere is located approximately in the middle, and so the chromosome has two arms of equal length.

2. **Submetacentric.** The centromere is displaced toward one end, creating a long arm and a short arm. (On human chromosomes, the short arm is designated by the letter p and the long arm by the letter q.)

3. **Acrocentric.** The centromere is near one end, producing a long arm and a knob, or satellite, at the other end.

4. **Telocentric.** The centromere is at or very near the end of the chromosome (see Figure 2.8).

The complete set of chromosomes possessed by an organism is called its *karyotype* and is usually presented as a picture of metaphase chromosomes lined up in descending order of their size (**Figure 9.1**). Karyotypes are prepared from actively dividing cells, such as white blood cells, bone-marrow cells, or cells from meristematic tissues of plants. After treatment with a chemical (such as colchicine) that prevents them from entering anaphase, the cells are chemically preserved, spread on a microscope slide, stained, and photographed. The photograph is then enlarged, and the individual chromosomes are cut out and arranged in a karyotype. For human chromosomes, karyotypes are often routinely prepared by automated machines, which scan a slide with a video camera attached to a microscope, looking for chromosome spreads. When a spread has been located, the camera

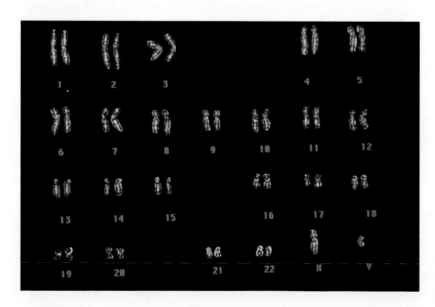

9.1 A human karyotype consists of 46 chromosomes. A karyotype for a male is shown here; a karyotype for a female would have two X chromosomes. [CNRI/Photo Researchers.]

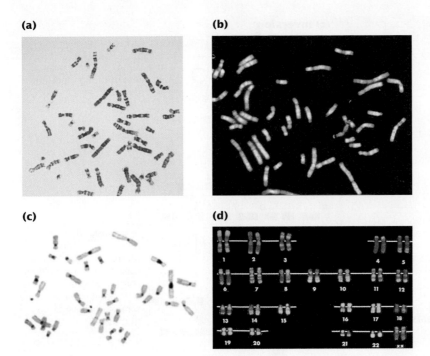

(a)

(b)

(c)

(d)

9.2 Chromosome banding is revealed by special staining techniques. (a) G banding. (b) Q banding. (c) C banding. (d) R banding. [Part a: Leonard Lessin/Peter Arnold. Parts b and c: University of Washington Pathology Department. Part d: Dr. Ram Verma/Phototake.]

mosomes under ultraviolet light; variation in the brightness of Q bands results from differences in the relative amounts of cytosine–guanine (C–G) and adenine–thymine base pairs. Other techniques reveal C bands (**Figure 9.2c**), which are regions of DNA occupied by centromeric heterochromatin, and R bands (**Figure 9.2d**), which are rich in cytosine–guanine base pairs.

Types of Chromosome Mutations

Chromosome mutations can be grouped into three basic categories: chromosome rearrangements, aneuploids, and polyploids (**Figure 9.3**). Chromosome rearrangements alter the *structure* of chromosomes; for example, a piece of a chromosome might be duplicated, deleted, or inverted. In aneuploidy, the *number* of chromosomes is altered: one or more individual chromosomes are added or deleted. In polyploidy, one or more complete *sets* of chromosomes are added. Some organisms (such as yeast) possess a single chromosome set ($1n$) for most of their life cycles and are referred to as haploid, whereas others possess two chromosome sets and are referred to as diploid ($2n$). A polyploid is any organism that has more than two sets of chromosomes ($3n$, $4n$, $5n$, or more).

takes a picture of the chromosomes, the image is digitized, and the chromosomes are sorted and arranged electronically by a computer.

Preparation and staining techniques help to distinguish among chromosomes of similar size and shape. For instance, the staining of chromosomes with a special dye called Giemsa reveals G bands, which distinguish areas of DNA that are rich in adenine–thymine (A–T) base pairs (**Figure 9.2a**; see Chapter 10). Q bands (**Figure 9.2b**) are revealed by staining chromosomes with quinacrine mustard and viewing the chro-

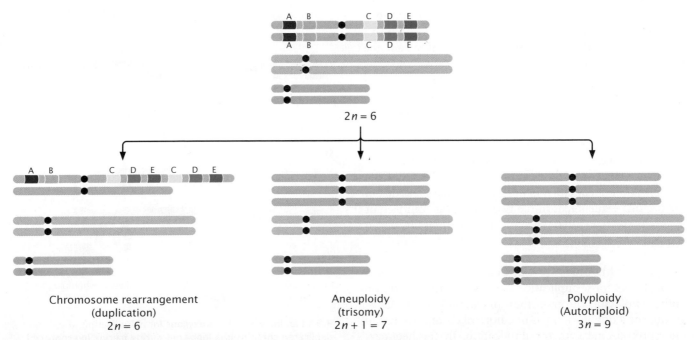

$2n = 6$

Chromosome rearrangement
(duplication)
$2n = 6$

Aneuploidy
(trisomy)
$2n + 1 = 7$

Polyploidy
(Autotriploid)
$3n = 9$

9.3 Chromosome mutations consist of chromosome rearrangements, aneuploids, and polyploids.
Duplications, trisomy, and autotriploids are examples of each category of mutation.

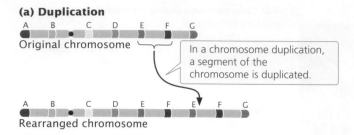

(a) Duplication

Original chromosome

In a chromosome duplication, a segment of the chromosome is duplicated.

Rearranged chromosome

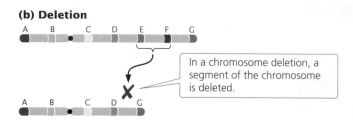

(b) Deletion

In a chromosome deletion, a segment of the chromosome is deleted.

9.4 The four basic types of chromosome rearrangements are duplication, deletion, inversion, and translocation.

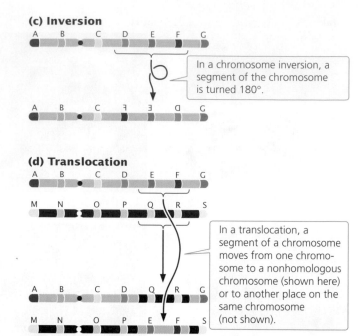

(c) Inversion

In a chromosome inversion, a segment of the chromosome is turned 180°.

(d) Translocation

In a translocation, a segment of a chromosome moves from one chromosome to a nonhomologous chromosome (shown here) or to another place on the same chromosome (not shown).

9.2 Chromosome Rearrangements Alter Chromosome Structure

Chromosome rearrangements are mutations that change the structures of individual chromosomes. The four basic types of rearrangements are duplications, deletions, inversions, and translocations (**Figure 9.4**).

Duplications

A **chromosome duplication** is a mutation in which part of the chromosome has been doubled (see Figure 9.4a). Consider a chromosome with segments AB•CDEFG, in which • represents the centromere. A duplication might include the EF segments, giving rise to a chromosome with segments AB•CDEFEFG. This type of duplication, in which the duplicated region is immediately adjacent to the original segment, is called a **tandem duplication**. If the duplicated segment is located some distance from the original segment, either on the same chromosome or on a different one, the chromosome rearrangement is called a **displaced duplication**. An example of a displaced duplication would be AB•CDEFGEF. A duplication can be either in the same orientation as that of the original sequence, as in the two preceding examples, or inverted: AB•CDEFFEG. When the duplication is inverted, it is called a **reverse duplication**.

Effects of chromosome duplications An individual homozygous for a duplication carries the duplication on both homologous chromosomes, and an individual heterozygous for a duplication has one normal chromosome and one chromosome with the duplication. In the heterozy-

gotes (**Figure 9.5a**), problems arise in chromosome pairing at prophase I of meiosis, because the two chromosomes are not homologous throughout their length. The pairing and synapsis of homologous regions require that one or

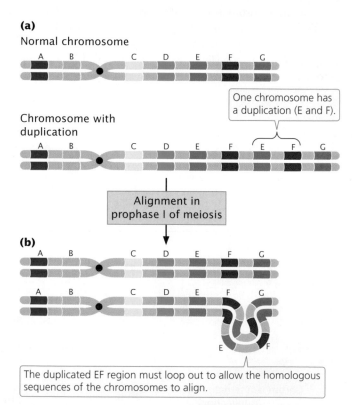

(a)
Normal chromosome

Chromosome with duplication

One chromosome has a duplication (E and F).

Alignment in prophase I of meiosis

(b)

The duplicated EF region must loop out to allow the homologous sequences of the chromosomes to align.

9.5 In an individual heterozygous for a duplication, the duplicated chromosome loops out during pairing in prophase I.

both chromosomes loop and twist so that these regions are able to line up (**Figure 9.5b**). The appearance of this characteristic loop structure in meiosis is one way to detect duplications.

Duplications may have major effects on the phenotype. Among fruit flies (*Drosophila melanogaster*), for example, a fly having a *Bar* mutation has a reduced number of facets in the eye, making the eye smaller and bar shaped instead of oval (**Figure 9.6**). The *Bar* mutation results from a small duplication on the X chromosome that is inherited as an incompletely dominant, X-linked trait: heterozygous female flies have somewhat smaller eyes (the number of facets is reduced; see Figure 9.6b), whereas, in homozygous female and hemizygous male flies, the number of facets is greatly reduced (see Figure 9.6c). Occasionally, a fly carries three copies of the *Bar* duplication on its X chromosome; for flies carrying such mutations, which are termed *double Bar*, the number of facets is extremely reduced (see Figure 9.6d). The *Bar* mutation arises from unequal crossing over, a duplication-generating process (**Figure 9.7**; see also Figure 18.14).

Unbalanced gene dosage How does a chromosome duplication alter the phenotype? After all, gene sequences are not altered by duplications, and no genetic information is missing; the only change is the presence of additional copies of normal sequences. The answer to this question is not well understood, but the effects are most likely due to imbalances in the amounts of gene products (abnormal gene dosage). The amount of a particular protein synthesized by a cell is often directly related to the number of copies of its corresponding gene: an individual organism with three functional copies of a gene often produces 1.5 times as much of the protein encoded by that gene as that produced by an individual with two copies. Because developmental processes require the interaction of many proteins, they often depend critically on proper gene dosage. If the amount of one protein increases while the amounts

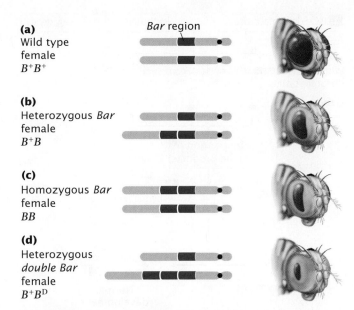

9.6 The Bar phenotype in *Drosophila melanogaster* results from an X-linked duplication. (a) Wild-type fruit flies have normal-size eyes. (b) Flies heterozygous and (c) homozygous for the *Bar* mutation have smaller, bar-shaped eyes. (d) Flies with *double Bar* have three copies of the duplication and much smaller bar-shaped eyes.

of others remain constant, problems can result (**Figure 9.8**). Although duplications can have severe consequences when the precise balance of a gene product is critical to cell function, duplications have arisen frequently throughout the evolution of many eukaryotic organisms and are a source of new genes that may provide novel functions. For example, humans have a series of genes that encode different globin chains, some of which function as an oxygen carrier during adult stages and others that function during embryonic and fetal development. All of these globin genes arose from an original ancestral gene that underwent a

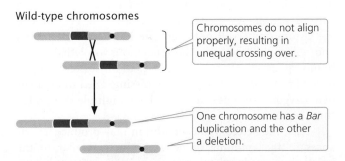

9.7 Unequal crossing over produces *Bar* and *double-Bar* mutations.

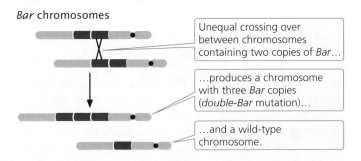

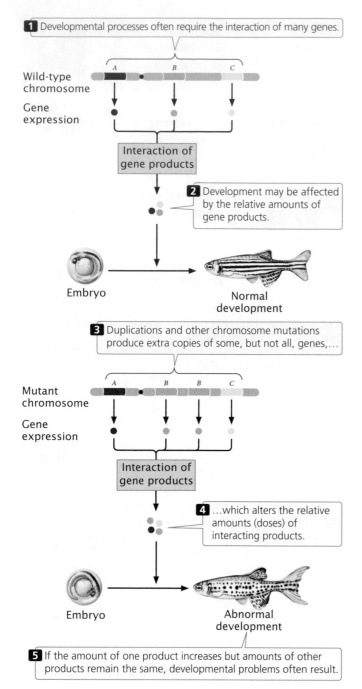

1 Developmental processes often require the interaction of many genes.

Wild-type chromosome

Gene expression

Interaction of gene products

2 Development may be affected by the relative amounts of gene products.

Embryo

Normal development

3 Duplications and other chromosome mutations produce extra copies of some, but not all, genes,…

Mutant chromosome

Gene expression

Interaction of gene products

4 …which alters the relative amounts (doses) of interacting products.

Embryo

Abnormal development

5 If the amount of one product increases but amounts of other products remain the same, developmental problems often result.

9.8 Unbalanced gene dosage leads to developmental abnormalities.

series of duplications. Human phenotypes associated with some duplications are summarized in **Table 9.1**.

Segmental duplications The human genome consists of numerous duplicated sequences called **segmental duplications**, which are defined as duplications greater than 1000 bp in length. In most segmental duplications, the two copies are found on the same chromosome (an intrachromosomal duplication) but, in others, the two copies are found on different chromosomes (an interchromosomal duplication). Many segmental duplications have been detected with molecular

techniques that examine DNA sequences on a chromosome. These techniques reveal that about 4% of the human genome consists of segmental duplications. In the human genome, the average size of segmental duplications is 15,000 bp.

Deletions

A second type of chromosome rearrangement is a **chromosome deletion**, the loss of a chromosome segment (see Figure 9.4b). A chromosome with segments AB•CDEFG that undergoes a deletion of segment EF would generate the mutated chromosome AB•CDG.

A large deletion can be easily detected because the chromosome is noticeably shortened. In individuals heterozygous for deletions, the normal chromosome must loop during the pairing of homologs in prophase I of meiosis (**Figure 9.9**) to allow the homologous regions of the two chromosomes to align and undergo synapsis. This looping out generates a structure that looks very much like that seen for individuals heterozygous for duplications.

Effects of deletions The phenotypic consequences of a deletion depend on which genes are located in the deleted region. If the deletion includes the centromere, the chromosome will not segregate in meiosis or mitosis and will usually be lost. Many deletions are lethal in the homozygous state because all copies of any essential genes located in the deleted region are missing. Even individuals heterozygous for a deletion may have multiple defects for three reasons.

First, the heterozygous condition may produce imbalances in the amounts of gene products, similar to the imbalances produced by extra gene copies. Second, normally recessive mutations on the homologous chromosome lacking the deletion may be expressed when the wild-type

Table 9.1	Effects of some human chromosome rearrangements		
Type of Rearrangement	Chromosome	Disorder	Symptoms
Duplication	4, short arm	—	Small head, short neck, low hairline, growth and mental retardation
Duplication	4, long arm	—	Small head, sloping forehead, hand abnormalities
Duplication	7, long arm	—	Delayed development, asymmetry of the head, fuzzy scalp, small nose, low-set ears
Duplication	9, short arm	—	Characteristic face, variable mental retardation, high and broad forehead, hand abnormalities
Deletion	5, short arm	*Cri-du-chat* syndrome	Small head, distinctive cry, widely spaced eyes, round face, mental retardation
Deletion	4, short arm	Wolf–Hirschhorn syndrome	Small head with high forehead, wide nose, cleft lip and palate, severe mental retardation
Deletion	4, long arm	—	Small head, from mild to moderate mental retardation, cleft lip and palate, hand and foot abnormalities
Deletion	7, long arm	Williams–Beuren syndrome	Facial features, heart defects, mental impairment
Deletion	15, long arm	Prader–Willi syndrome	Feeding difficulty at early age, but becoming obese after 1 year of age, from mild to moderate mental retardation
Deletion	18, short arm	—	Round face, large low-set ears, from mild to moderate mental retardation
Deletion	18, long arm	—	Distinctive mouth shape, small hands, small head, mental retardation

allele has been deleted (and is no longer present to mask the recessive allele's expression). The expression of a normally recessive mutation is referred to as **pseudodominance**, and it is an indication that one of the homologous chromosomes has a deletion. Third, some genes must be present in two copies for normal function. When a single copy of a gene is not sufficient to produce a wild-type phenotype, it is said to be a **haploinsufficient gene**. *Notch* is a series of X-linked wing mutations in *Drosophila* that often result from chromosome deletions. *Notch* deletions behave as

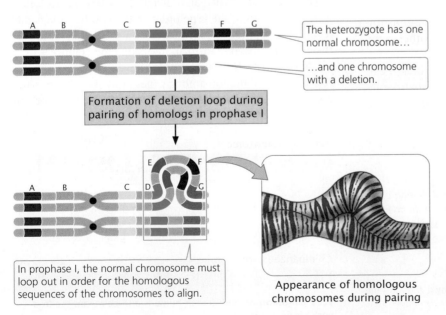

The heterozygote has one normal chromosome…

…and one chromosome with a deletion.

Formation of deletion loop during pairing of homologs in prophase I

In prophase I, the normal chromosome must loop out in order for the homologous sequences of the chromosomes to align.

Appearance of homologous chromosomes during pairing

9.9 In an individual heterozygous for a deletion, the normal chromosome loops out during chromosome pairing in prophase I.

9.10 The Notch phenotype is produced by a chromosome deletion that includes the *Notch* gene. (Left) Normal wing veination. (Right) Wing veination produced by a *Notch* mutation. [Spyros Artavanis-Tsakonas, Kenji Matsuno, and Mark E. Fortini.]

dominant mutations: when heterozygous for a *Notch* deletion, a fly has wings that are notched at the tips and along the edges (**Figure 9.10**). The *Notch* mutation is therefore haploinsufficient. Females that are homozygous for a *Notch* deletion (or males that are hemizygous) die early in embryonic development. The *Notch* locus, which is deleted in *Notch* mutations, encodes a receptor that normally transmits signals received from outside the cell to the cell's interior and is important in fly development. The deletion acts as a recessive lethal because loss of all copies of the *Notch* gene prevents normal development.

Chromosome deletions in humans In humans, a deletion on the short arm of chromosome 5 is responsible for *cri-du-chat* syndrome. The name (French for "cry of the cat") derives from the peculiar, catlike cry of infants with this syndrome. A child who is heterozygous for this deletion has a small head, widely spaced eyes, and a round face and is mentally retarded. Deletion of part of the short arm of chromosome 4 results in another human disorder—Wolf–Hirschhorn syndrome, which is characterized by seizures and severe mental and growth retardation. A deletion of a tiny segment of chromosome 7 causes haploinsufficiency of the gene encoding elastin and a few other genes and leads to a condition known as Williams–Beuren syndrome, characterized by distinctive facial features, heart defects, high blood pressure, and cognitive impairments. Effects of deletions in human chromosomes are summarized in Table 9.1.

CONCEPTS

A chromosomal deletion is a mutation in which a part of a chromosome is lost. In individuals heterozygous for a deletion, the normal chromosome loops out during prophase I of meiosis. Deletions cause recessive genes on the homologous chromosome to be expressed and may cause imbalances in gene products.

✔ CONCEPT CHECK 2

What is pseudodominance and how is it produced by a chromosome deletion?

Inversions

A third type of chromosome rearrangement is a **chromosome inversion**, in which a chromosome segment is inverted—turned 180 degrees (see Figure 9.4c). If a chromosome originally had segments AB•CDEFG, then chromosome AB•CFEDG represents an inversion that includes segments DEF. For an inversion to take place, the chromosome must break in two places. Inversions that do not include the centromere, such as AB•CFEDG, are termed **paracentric inversions** (*para* meaning "next to"), whereas inversions that include the centromere, such as ADC•BEFG, are termed **pericentric inversions** (*peri* meaning "around").

Inversion heterozygotes are common in many organisms, including a number of plants, some species of *Drosophila*, mosquitoes, and grasshoppers. Inversions may have played an important role in human evolution: G-banding patterns reveal that several human chromosomes differ from those of chimpanzees by only a pericentric inversion (**Figure 9.11**).

Effects of inversions Individual organisms with inversions have neither lost nor gained any genetic material; just the DNA sequence has been altered. Nevertheless, these mutations often have pronounced phenotypic effects. An inversion may break a gene into two parts, with one part moving to a new location and destroying the function of that gene. Even when the chromosome breaks are between genes, phenotypic effects may arise from the inverted gene order in an inversion. Many genes are regulated in a position-dependent manner; if their positions are altered by an inversion, they may be expressed at inappropriate times or in inappropriate tissues, an outcome referred to as a **position effect**.

Inversions in meiosis When an individual is homozygous for a particular inversion, no special problems arise in meiosis, and the two homologous chromosomes can pair and separate normally. When an individual is heterozygous for an inversion, however, the gene order of the two homologs differs, and the homologous sequences can align and pair only if the two chromosomes form an inversion loop (**Figure 9.12**).

Individuals heterozygous for inversions also exhibit reduced recombination among genes located in the inverted region. The frequency of crossing over within the inversion

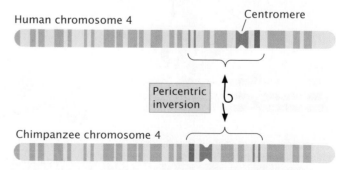

9.11 Chromosome 4 differs in humans and chimpanzees in a pericentric inversion.

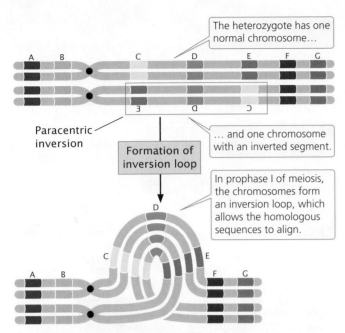

9.12 In an individual heterozygous for a paracentric inversion, the chromosomes form an inversion loop during pairing in prophase I.

The heterozygote has one normal chromosome…

Paracentric inversion

… and one chromosome with an inverted segment.

Formation of inversion loop

In prophase I of meiosis, the chromosomes form an inversion loop, which allows the homologous sequences to align.

is not actually diminished but, when crossing over does take place, the result is abnormal gametes that result in nonviable offspring, and thus no recombinant progeny are observed. Let's see why this result occurs.

Figure 9.13 illustrates the results of crossing over within a paracentric inversion. The individual is heterozygous for an inversion (see Figure 9.13a), with one wild-type, unmutated chromosome (AB•CDEFG) and one inverted chromosome (AB•EDCFG). In prophase I of meiosis, an inversion loop forms, allowing the homologous sequences to pair up (see Figure 9.13b). If a single crossover takes place in the inverted region (between segments C and D in Figure 9.13), an unusual structure results (see Figure 9.13c). The two outer chromatids, which did not participate in crossing over, contain original, nonrecombinant gene sequences. The two inner chromatids, which did cross over, are highly abnormal: each has two copies of some genes and no copies of others. Furthermore, one of the four chromatids now has two centromeres and is said to be a **dicentric chromatid**; the other lacks a centromere and is an **acentric chromatid**.

In anaphase I of meiosis, the centromeres are pulled toward opposite poles and the two homologous chromosomes separate. This action stretches the dicentric chromatid across the center of the nucleus, forming a structure called a **dicentric bridge** (see Figure 9.13d). Eventually, the dicentric bridge breaks, as the two centromeres are pulled farther apart. Spindle fibers do not attach to the acentric fragment, and so this fragment does not segregate into a nucleus in meiosis and is usually lost.

In the second division of meiosis, the sister chromatids separate and four gametes are produced (see Figure 9.13e). Two of the gametes contain the original, nonrecombinant chromosomes (AB•CDEFG and AB•EDCFG). The other

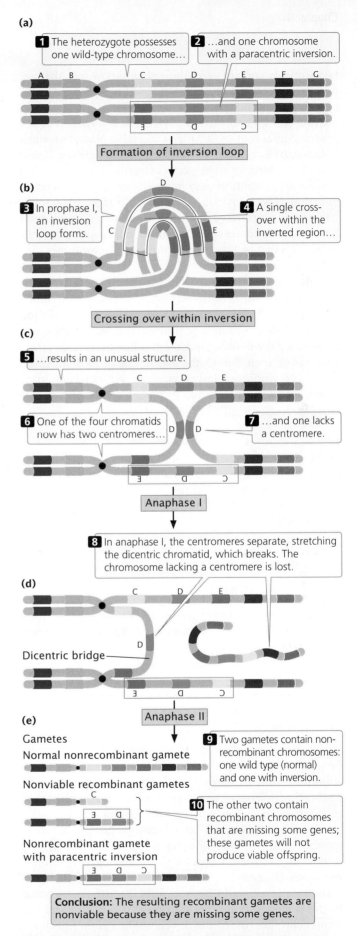

(a)

1 The heterozygote possesses one wild-type chromosome…

2 …and one chromosome with a paracentric inversion.

Formation of inversion loop

(b)

3 In prophase I, an inversion loop forms.

4 A single crossover within the inverted region…

Crossing over within inversion

(c)

5 …results in an unusual structure.

6 One of the four chromatids now has two centromeres…

7 …and one lacks a centromere.

Anaphase I

8 In anaphase I, the centromeres separate, stretching the dicentric chromatid, which breaks. The chromosome lacking a centromere is lost.

(d)

Dicentric bridge

Anaphase II

(e)

Gametes

Normal nonrecombinant gamete

9 Two gametes contain nonrecombinant chromosomes: one wild type (normal) and one with inversion.

Nonviable recombinant gametes

10 The other two contain recombinant chromosomes that are missing some genes; these gametes will not produce viable offspring.

Nonrecombinant gamete with paracentric inversion

Conclusion: The resulting recombinant gametes are nonviable because they are missing some genes.

9.13 In a heterozygous individual, a single crossover within a paracentric inversion leads to abnormal gametes.

two gametes contain recombinant chromosomes that are missing some genes; these gametes will not produce viable offspring. Thus, no recombinant progeny result when crossing over takes place within a paracentric inversion. The key is to recognize that crossing over still takes place, but, when it does so, the resulting recombinant gametes are not viable; so no recombinant progeny are observed.

Recombination is also reduced within a pericentric inversion (**Figure 9.14**). No dicentric bridges or acentric fragments are produced, but the recombinant chromosomes have too many copies of some genes and no copies of others; so gametes that receive the recombinant chromosomes cannot produce viable progeny.

Figures 9.13 and 9.14 illustrate the results of single crossovers within inversions. Double crossovers in which both crossovers are on the same two strands (two-strand double crossovers) result in functional recombinant chromosomes. (Try drawing out the results of a double crossover.) Thus, even though the overall rate of recombination is reduced within an inversion, some viable recombinant progeny may still be produced through two-strand double crossovers. TRY PROBLEM 26 →

CONCEPTS

In an inversion, a segment of a chromosome is turned 180 degrees. Inversions cause breaks in some genes and may move others to new locations. In individuals heterozygous for a chromosome inversion, the homologous chromosomes form a loop in prophase I of meiosis. When crossing over takes place within the inverted region, nonviable gametes are usually produced, resulting in a depression in observed recombination frequencies.

✔ CONCEPT CHECK 3

A dicentric chromosome is produced when crossing over takes place in an individual heterozygous for which type of chromosome rearrangement?

a. Duplication   c. Paracentric inversion

b. Deletion   d. Pericentric inversion

Translocations

A **translocation** entails the movement of genetic material between nonhomologous chromosomes (see Figure 9.4d) or within the same chromosome. Translocation should not be confused with crossing over, in which there is an exchange of genetic material between *homologous* chromosomes.

In a **nonreciprocal translocation**, genetic material moves from one chromosome to another without any reciprocal exchange. Consider the following two nonhomologous chromosomes: AB•CDEFG and MN•OPQRS. If chromosome segment EF moves from the first chromosome to the second without any transfer of segments from the second chromosome to the first, a nonreciprocal translocation has taken place, producing chromosomes AB•CDG and MN•OPEFQRS. More commonly, there is a two-way

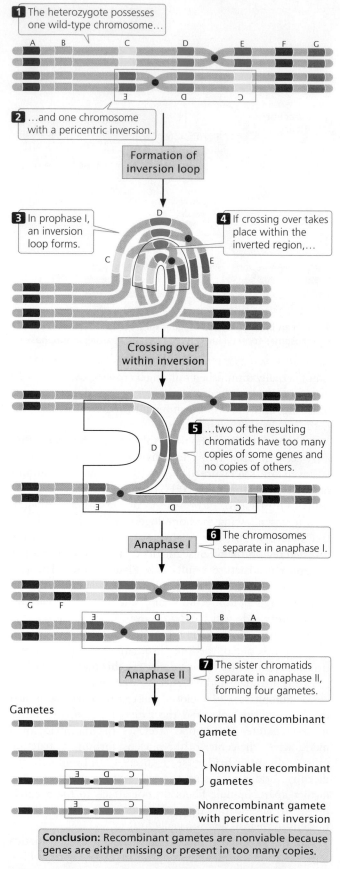

1 The heterozygote possesses one wild-type chromosome…

2 …and one chromosome with a pericentric inversion.

Formation of inversion loop

3 In prophase I, an inversion loop forms.

4 If crossing over takes place within the inverted region,…

Crossing over within inversion

5 …two of the resulting chromatids have too many copies of some genes and no copies of others.

Anaphase I

6 The chromosomes separate in anaphase I.

Anaphase II

7 The sister chromatids separate in anaphase II, forming four gametes.

Gametes

Normal nonrecombinant gamete

Nonviable recombinant gametes

Nonrecombinant gamete with pericentric inversion

Conclusion: Recombinant gametes are nonviable because genes are either missing or present in too many copies.

9.14 In a heterozygous individual, a single crossover within a pericentric inversion leads to abnormal gametes.

exchange of segments between the chromosomes, resulting in a **reciprocal translocation**. A reciprocal translocation between chromosomes AB•CDEFG and MN•OPQRS might give rise to chromosomes AB•CDQRG and MN•OPEFS.

Effects of translocations Translocations can affect a phenotype in several ways. First, they can physically link genes that were formerly located on different chromosomes. These new linkage relations may affect gene expression (a position effect): genes translocated to new locations may come under the control of different regulatory sequences or other genes that affect their expression.

Second, the chromosomal breaks that bring about translocations may take place within a gene and disrupt its function. Molecular geneticists have used these types of effects to map human genes. Neurofibromatosis is a genetic disease characterized by numerous fibrous tumors of the skin and nervous tissue; it results from an autosomal dominant mutation. Linkage studies first placed the locus that, when mutated, causes neurofibromatosis on chromosome 17, but the precise location of the locus was unknown. Geneticists later narrowed down the location when they identified two patients with neurofibromatosis who possessed a translocation affecting chromosome 17. These patients were assumed to have developed neurofibromatosis because one of the chromosome breaks that occurred in the translocation disrupted a particular gene, resulting in neurofibromatosis. DNA from the regions around the breaks was sequenced, eventually leading to the identification of the gene responsible for neurofibromatosis.

Deletions frequently accompany translocations. In a **Robertsonian translocation**, for example, the long arms of two acrocentric chromosomes become joined to a common centromere through a translocation, generating a metacentric chromosome with two long arms and another chromosome with two very short arms (**Figure 9.15**). The

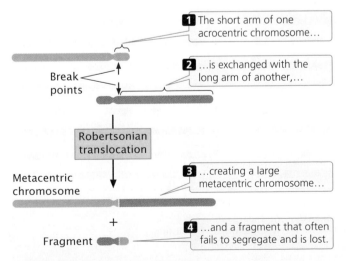

9.15 In a Robertsonian translocation, the short arm of one acrocentric chromosome is exchanged with the long arm of another.

smaller chromosome often fails to segregate, leading to an overall reduction in chromosome number. As we will see, Robertsonian translocations are the cause of some cases of Down syndrome, a chromosome disorder discussed in the introduction to this chapter.

Translocations in meiosis The effects of a translocation on chromosome segregation in meiosis depend on the nature of the translocation. Let's consider what happens in an individual heterozygous for a reciprocal translocation. Suppose that the original chromosomes were AB•CDEFG and MN•OPQRS (designated N_1 and N_2 respectively, for normal chromosomes 1 and 2) and that a reciprocal translocation takes place, producing chromosomes AB•CDQRS and MN•OPEFG (designated T_1 and T_2, respectively, for translocated chromosomes 1 and 2). An individual heterozygous for this translocation would possess one normal copy of each chromosome and one translocated copy (**Figure 9.16a**). Each of these chromosomes contains segments that are homologous to two other chromosomes. When the homologous sequences pair in prophase I of meiosis, crosslike configurations consisting of all four chromosomes (**Figure 9.16b**) form.

Notice that N_1 and T_1 have homologous centromeres (in both chromosomes, the centromere is between segments B and C); similarly, N_2 and T_2 have homologous centromeres (between segments N and O). Normally, homologous centromeres separate and move toward opposite poles in anaphase I of meiosis. With a reciprocal translocation, the chromosomes may segregate in three different ways. In **alternate segregation** (**Figure 9.16c**), N_1 and N_2 move toward one pole and T_1 and T_2 move toward the opposite pole. In **adjacent-1 segregation**, N_1 and T_2 move toward one pole and T_1 and N_2 move toward the other pole. In both alternate and adjacent-1 segregation, homologous centromeres segregate toward opposite poles. **Adjacent-2 segregation**, in which N_1 and T_1 move toward one pole and T_2 and N_2 move toward the other, is rare because the two homologous chromosomes usually separate in meiosis.

The products of the three segregation patterns are illustrated in **Figure 9.16d**. As you can see, the gametes produced by alternate segregation possess one complete set of the chromosome segments. These gametes are therefore functional and can produce viable progeny. In contrast, gametes produced by adjacent-1 and adjacent-2 segregation are not viable, because some chromosome segments are present in two copies, whereas others are missing. Because adjacent-2 segregation is rare, most gametes are produced by alternate or adjacent-1 segregation. Therefore, approximately half of the gametes from an individual heterozygous for a reciprocal translocation are expected to be functional.

The role of translocations in evolution Translocations frequently play an important role in the evolution of karyotypes. Chimpanzees, gorillas, and orangutans all

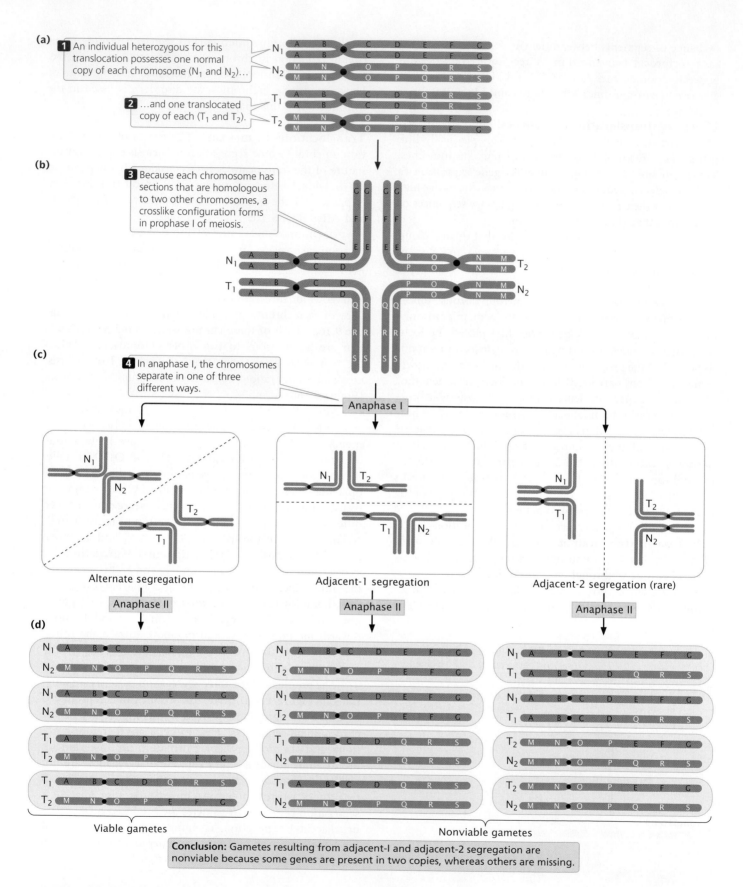

(a)

1 An individual heterozygous for this translocation possesses one normal copy of each chromosome (N_1 and N_2)...

2 ...and one translocated copy of each (T_1 and T_2).

(b)

3 Because each chromosome has sections that are homologous to two other chromosomes, a crosslike configuration forms in prophase I of meiosis.

(c)

4 In anaphase I, the chromosomes separate in one of three different ways.

Anaphase I

Alternate segregation

Adjacent-1 segregation

Adjacent-2 segregation (rare)

Anaphase II   Anaphase II   Anaphase II

(d)

Viable gametes

Nonviable gametes

Conclusion: Gametes resulting from adjacent-1 and adjacent-2 segregation are nonviable because some genes are present in two copies, whereas others are missing.

9.16 In an individual heterozygous for a reciprocal translocation, crosslike structures form in homologous pairing.

Human chromosome 2

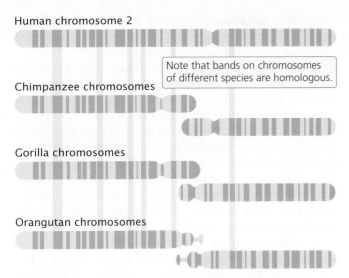

Note that bands on chromosomes of different species are homologous.

Chimpanzee chromosomes

Gorilla chromosomes

Orangutan chromosomes

9.17 Human chromosome 2 contains a Robertsonian translocation that is not present in chimpanzees, gorillas, or orangutans. G-banding reveals that a Robertsonian translocation in a human ancestor switched the long and short arms of the two acrocentric chromosomes that are still found in the other three primates. This translocation created the large metacentric human chromosome 2.

have 48 chromosomes, whereas humans have 46. Human chromosome 2 is a large, metacentric chromosome with G-banding patterns that match those found on two different acrocentric chromosomes of the apes (**Figure 9.17**). Apparently, a Robertsonian translocation took place in a human ancestor, creating a large metacentric chromosome from the two long arms of the ancestral acrocentric chromosomes and a small chromosome consisting of the two short arms. The small chromosome was subsequently lost, leading to the reduced chromosome number in humans relative to that of the other apes. TRY PROBLEM 27 →

CONCEPTS

In translocations, parts of chromosomes move to other, nonhomologous chromosomes or to other regions of the same chromosome. Translocations can affect the phenotype by causing genes to move to new locations, where they come under the influence of new regulatory sequences, or by breaking genes and disrupting their function.

✔ CONCEPT CHECK 4

What is the outcome of a Robertsonian translocation?

a. Two acrocentric chromosomes

b. One metacentric chromosome and one chromosome with two very short arms

c. One metacentric and one acrocentric chromosome

d. Two metacentric chromosomes

Fragile Sites

Chromosomes of cells grown in culture sometimes develop constrictions or gaps at particular locations called **fragile sites** (**Figure 9.18**) because they are prone to breakage under certain conditions. More than 100 fragile sites have been identified on human chromosomes.

Fragile sites fall into two groups. *Common fragile sites* are present in all humans and are a normal feature of chromosomes. Common fragile sites are often the location of chromosome breakage and rearrangements in cancer cells, leading to chromosome deletions, translocations, and other chromosome rearrangements. *Rare fragile sites* are found in few people and are inherited as a Mendelian trait. Rare fragile sites are often associated with genetic disorders, such as mental retardation. Most of them consist of expanding nucleotide repeats, in which the number of repeats of a set of nucleotides is increased (see Chapter 18).

One of the most intensively studied rare fragile sites is located on the human X chromosome. This site is associated with mental retardation and is known as the **fragile-X syndrome**. Exhibiting X-linked inheritance and arising with a frequency of about 1 in 1250 male births, fragile-X syndrome has been shown to result from an increase in the number of repeats of a CGG trinucleotide.

Molecular studies of fragile sites have shown that many of these sites are more than 100,000 bp in length and include one or more genes. Fragile sites are often late in being replicated. At these places, the enzymes that replicate DNA may stall while unwinding of the DNA continues (see Chapter 12), leading to long stretches of DNA that are unwound and vulnerable to breakage. In spite of recent advances in our understanding of fragile sites, much about their nature remains poorly understood.

9.18 Fragile sites are chromosomal regions susceptible to breakage under certain conditions. Shown here is a fragile site on human chromosome X. [Courtesy of Dr. Christine Harrison.]

Copy-Number Variations

Chromosome rearrangements have traditionally been detected by examination of the chromosomes with a microscope. Visual examination identifies chromosome rearrangements on the basis of changes in the overall size of a chromosome, alteration of banding patterns revealed by chromosome staining, or the behavior of chromosomes in meiosis. Microscopy, however, can detect only large chromosome rearrangements, typically those that are at least 5 million bp in length.

With the completion of the Human Genome Project (see Chapter 20), detailed information about DNA sequences found on individual chromosomes has become available. Using this information, geneticists can now examine the number of copies of specific DNA sequences present in a cell and detect duplications, deletions, and other chromosome rearrangements that cannot be observed with microscopy alone. The work has been greatly facilitated by the availability of microarrays (see Chapter 20), which allow the simultaneous detection of hundreds of thousands of DNA sequences across the genome. Because these methods measure the number of copies of particular DNA sequences, the variations that they detect are called **copy-number variations** (CNVs). Copy-number variations include duplications and deletions that range in size from thousands of base pairs to several million base pairs. Many of these variants encompass at least one gene and may encompass several genes.

Recent studies of copy-number variation have revealed that submicroscopic chromosome duplications and deletions are quite common: research suggests that each person may possess as many as 1000 copy-number variations. Many of them probably have no observable phenotypic effects, but some copy-number variations have now been implicated in causing a number of diseases and disorders. For example, Janine Wagenstaller and her colleagues studied copy-number variation in 67 children with unexplained mental retardation and found that 11 (16%) of them had duplications and deletions. Copy-number variations have also been associated with osteoporosis, autism, schizophrenia, and a number of other diseases and disorders. TRY PROBLEM 19 ➡️

> ### CONCEPTS
>
> Variations in the number of copies of particular DNA sequences (copy-number variations) are surprisingly common in the human genome.

9.3 Aneuploidy Is an Increase or Decrease in the Number of Individual Chromosomes

In addition to chromosome rearrangements, chromosome mutations include changes in the number of chromosomes. Variations in chromosome number can be classified into two basic types: **aneuploidy**, which is a change in the number of individual chromosomes, and **polyploidy**, which is a change in the number of chromosome sets.

Aneuploidy can arise in several ways. First, a chromosome may be lost in the course of mitosis or meiosis if, for example, its centromere is deleted. Loss of the centromere prevents the spindle fibers from attaching; so the chromosome fails to move to the spindle pole and does not become incorporated into a nucleus after cell division. Second, the small chromosome generated by a Robertsonian translocation may be lost in mitosis or meiosis. Third, aneuploids may arise through nondisjunction, the failure of homologous chromosomes or sister chromatids to separate in meiosis or mitosis (see p. 83 in Chapter 4). Nondisjunction leads to some gametes or cells that contain an extra chromosome and others that are missing a chromosome (**Figure 9.19**). TRY PROBLEM 28 ➡️

Types of Aneuploidy

We will consider four types of common aneuploid conditions in diploid individuals: nullisomy, monosomy, trisomy, and tetrasomy.

1. **Nullisomy** is the loss of both members of a homologous pair of chromosomes. It is represented as $2n - 2$, where n refers to the haploid number of chromosomes. Thus, among humans, who normally possess $2n = 46$ chromosomes, a nullisomic zygote has 44 chromosomes.

2. **Monosomy** is the loss of a single chromosome, represented as $2n - 1$. A human monosomic zygote has 45 chromosomes.

3. **Trisomy** is the gain of a single chromosome, represented as $2n + 1$. A human trisomic zygote has 47 chromosomes. The gain of a chromosome means that there are three homologous copies of one chromosome. Most cases of Down syndrome, discussed in the introduction to the chapter, result from trisomy of chromosome 21.

4. **Tetrasomy** is the gain of two homologous chromosomes, represented as $2n + 2$. A human tetrasomic zygote has 48 chromosomes. Tetrasomy is not the gain of *any* two extra chromosomes, but rather the gain of two homologous chromosomes; so there will be four homologous copies of a particular chromosome.

More than one aneuploid mutation may occur in the same individual organism. An individual that has an extra copy of two different (nonhomologous) chromosomes is referred to as being double trisomic and represented as $2n + 1 + 1$. Similarly, a double monosomic has two fewer nonhomologous chromosomes ($2n - 1 - 1$), and a double tetrasomic has two extra pairs of homologous chromosomes ($2n + 2 + 2$).

Effects of Aneuploidy

One of the first aneuploids to be recognized was a fruit fly with a single X chromosome and no Y chromosome discov-

(a) Nondisjunction in meiosis I

Gametes

Zygotes

(c) Nondisjunction in mitosis

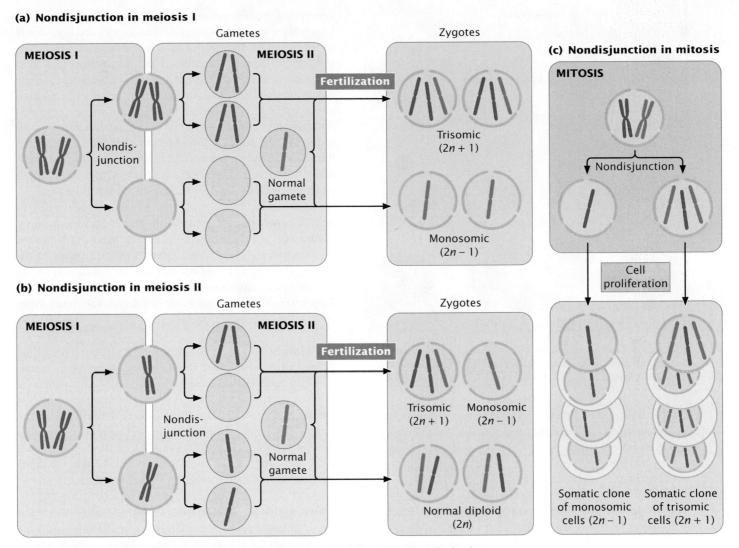

MEIOSIS I

MEIOSIS II

Fertilization

MITOSIS

Nondis-
junction

Normal
gamete

Trisomic
(2n + 1)

Monosomic
(2n − 1)

Nondisjunction

Cell
proliferation

(b) Nondisjunction in meiosis II

Gametes

Zygotes

MEIOSIS I

MEIOSIS II

Fertilization

Nondis-
junction

Normal
gamete

Trisomic
(2n + 1)

Monosomic
(2n − 1)

Normal diploid
(2n)

Somatic clone
of monosomic
cells (2n − 1)

Somatic clone
of trisomic
cells (2n + 1)

9.19 Aneuploids can be produced through nondisjunction in meiosis I, meiosis II, and mitosis.
The gametes that result from meioses with nondisjunction combine with a gamete (with blue
chromosome) that results from normal meiosis to produce the zygotes.

ered by Calvin Bridges in 1913 (see pp. 82–83 in Chapter 4). Another early study of aneuploidy focused on mutants in the Jimson weed, *Datura stramonium*. A. Francis Blakeslee began breeding this plant in 1913, and he observed that crosses with several Jimson mutants produced unusual ratios of progeny. For example, the *globe* mutant (producing a seedcase globular in shape) was dominant but was inherited primarily from the female parent. When plants having the *globe* mutation were self-fertilized, only 25% of the progeny had the globe phenotype. If the *globe* mutant were strictly dominant, Blakeslee should have seen 75% of the progeny with the trait (see Chapter 3), and so the 25% that he observed was unusal. Blakeslee isolated 12 different mutants (**Figure 9.20**) that exhibited peculiar patterns of inheritance. Eventually, John Belling demonstrated that these 12 mutants are in fact trisomics. *Datura stramonium* has 12 pairs of chromosomes (2n = 24), and each of the

12 mutants is trisomic for a different chromosome pair. The aneuploid nature of the mutants explained the unusual ratios that Blakeslee had observed in the progeny. Many of the extra chromosomes in the trisomics were lost in meiosis; so fewer than 50% of the gametes carried the extra chromosome, and the proportion of trisomics in the progeny was low. Furthermore, the pollen containing an extra chromosome was not as successful in fertilization, and trisomic zygotes were less viable.

Aneuploidy usually alters the phenotype drastically. In most animals and many plants, aneuploid mutations are lethal. Because aneuploidy affects the number of gene copies but not their nucleotide sequences, the effects of aneuploidy are most likely due to abnormal gene dosage. Aneuploidy alters the dosage for some, but not all, genes, disrupting the relative concentrations of gene products and often interfering with normal development.

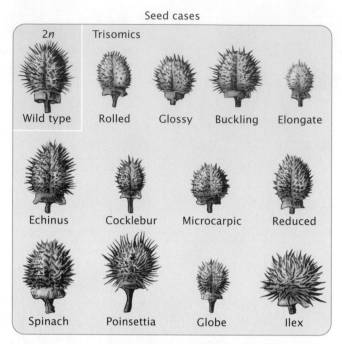

Seed cases

2n | Trisomics

Wild type   Rolled   Glossy   Buckling   Elongate

Echinus   Cocklebur   Microcarpic   Reduced

Spinach   Poinsettia   Globe   Ilex

9.20 Mutant capsules in Jimson weed (*Datura stramonium*) result from different trisomies. Each type of capsule is a phenotype that is trisomic for a different chromosome.

A major exception to the relation between gene number and protein dosage pertains to genes on the mammalian X chromosome. In mammals, X-chromosome inactivation ensures that males (who have a single X chromosome) and females (who have two X chromosomes) receive the same functional dosage for X-linked genes (see pp. 88–90 in Chapter 4 for further discussion of X-chromosome inactivation). Extra X chromosomes in mammals are inactivated; so we might expect that aneuploidy of the sex chromosomes would be less detrimental in these animals. Indeed, it is the case for mice and humans, for whom aneuploids of the sex chromosomes are the most common form of aneuploidy seen in living organisms. Y-chromosome aneuploids are probably common because there is so little information in the Y-chromosome.

CONCEPTS

Aneuploidy, the loss or gain of one or more individual chromosomes, may arise from the loss of a chromosome subsequent to translocation or from nondisjunction in meiosis or mitosis. It disrupts gene dosage and often has severe phenotypic effects.

✔ CONCEPT CHECK 5

A diploid organism has $2n = 36$ chromosomes. How many chromosomes will be found in a trisomic member of this species?

Aneuploidy in Humans

For unknown reasons, an incredibly high percentage of all human embryos that are conceived possess chromosome abnormalities. Findings from studies of women who are attempting pregnancy suggest that more than 30% of all conceptions spontaneously abort (miscarry), usually so early in development that the mother is not even aware of her pregnancy. Chromosome defects are present in at least 50% of spontaneously aborted human fetuses, with aneuploidy accounting for most of them. This rate of chromosome abnormality in humans is higher than in other organisms that have been studied; in mice, for example, aneuploidy is found in no more than 2% of fertilized eggs. Aneuploidy in humans usually produces such serious developmental problems that spontaneous abortion results. Only about 2% of all fetuses with a chromosome defect survive to birth.

Sex-chromosome aneuploids The most common aneuploidy seen in living humans has to do with the sex chromosomes. As is true of all mammals, aneuploidy of the human sex chromosomes is better tolerated than aneuploidy of autosomal chromosomes. Both Turner syndrome and Klinefelter syndrome (see Figures 4.8 and 4.9) result from aneuploidy of the sex chromosomes.

Autosomal aneuploids Autosomal aneuploids resulting in live births are less common than sex-chromosome aneuploids in humans, probably because there is no mechanism of dosage compensation for autosomal chromosomes. Most autosomal aneuploids are spontaneously aborted, with the exception of aneuploids of some of the small autosomes such as chromosome 21. Because these chromosomes are small and carry fewer genes, the presence of extra copies is less detrimental than it is for larger chromosomes.

Down syndrome The most common autosomal aneuploidy in humans is **trisomy 21**, also called **Down syndrome** (discussed in the introduction to the chapter). The incidence of Down syndrome in the United States is similar to that of the world, about 1 in 700 human births, although the incidence increases among children born to older mothers. Approximately 92% of those who have Down syndrome have three full copies of chromosome 21 (and therefore a total of 47 chromosomes), a condition termed **primary Down syndrome** (**Figure 9.21**). Primary Down syndrome usually arises from spontaneous nondisjunction in egg formation: about 75% of the nondisjunction events that cause Down syndrome are maternal in origin, most arising in meiosis I. Most children with Down syndrome are born to normal parents, and the failure of the chromosomes to divide has little hereditary tendency. A couple who has conceived one child with primary Down syndrome has only a slightly higher risk of conceiving a second child with Down syndrome (compared with other couples of similar age who have not had any Down-syndrome children). Similarly, the couple's relatives are not more likely to have a child with primary Down syndrome.

About 4% of people with Down syndrome are not trisomic for a complete chromosome 21. Instead, they have 46 chromosomes, but an extra copy of part of chromosome 21 is attached to another chromosome through a translocation.

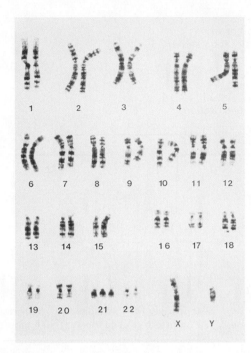

9.21 Primary Down syndrome is caused by the presence of three copies of chromosome 21. [L. Willatt, East Anglian Regional Genetics Service/Science Photo Library/Photo Researchers.]

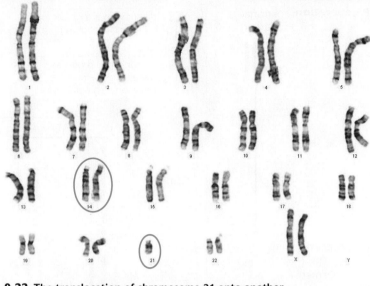

9.22 The translocation of chromosome 21 onto another chromosome results in familial Down syndrome. Here, the long arm of chromosome 21 is attached to chromosome 14. This karyotype is from a translocation carrier, who is phenotypically normal but is at increased risk for producing children with Down syndrome. [Copyright Centre for Genetics Education for and on behalf of the Crown in right of the State of New South Wales.]

This condition is termed **familial Down syndrome** because it has a tendency to run in families. The phenotypic characteristics of familial Down syndrome are the same as those of primary Down syndrome.

Familial Down syndrome arises in offspring whose parents are carriers of chromosomes that have undergone a Robertsonian translocation, most commonly between chromosome 21 and chromosome 14: the long arm of 21 and the short arm of 14 exchange places (**Figure 9.22**). This exchange produces a chromosome that includes the long arms of chromosomes 14 and 21, and a very small chromosome that consists of the short arms of chromosomes 21 and 14. The small chromosome is generally lost after several cell divisions. Although exchange between chromosomes 21 and 14 is the most common cause of familial Down syndrome, the condition can also be caused by translocations between 21 and other chromosomes such as 15.

Persons with the translocation, called **translocation carriers**, do not have Down syndrome. Although they possess only 45 chromosomes, their phenotypes are normal because they have two copies of the long arms of chromosomes 14 and 21, and apparently the short arms of these chromosomes (which are lost) carry no essential genetic information. Although translocation carriers are completely healthy, they have an increased chance of producing children with Down syndrome.

When a translocation carrier produces gametes, the translocation chromosome may segregate in three different ways. First, it may separate from the normal chromosomes 14 and 21 in anaphase I of meiosis (**Figure 9.23a**). In this type of segregation, half of the gametes will have the trans-

location chromosome and no other copies of chromosomes 21 and 14; the fusion of such a gamete with a normal gamete will give rise to a translocation carrier. The other half of the gametes produced by this first type of segregation will be normal, each with a single copy of chromosomes 21 and 14, and will result in normal offspring.

Alternatively, the translocation chromosome may separate from chromosome 14 and pass into the same cell with the normal chromosome 21 (**Figure 9.23b**). This type of segregation produces abnormal gametes only; half will have two functional copies of chromosome 21 (one normal and one attached to chromosome 14) and the other half will lack chromosome 21. If a gamete with the two functional copies of chromosome 21 fuses with a normal gamete carrying a single copy of chromosome 21, the resulting zygote will have familial Down syndrome. If a gamete lacking chromosome 21 fuses with a normal gamete, the resulting zygote will have monosomy 21 and will be spontaneously aborted.

In the third type of segregation, the translocation chromosome and the normal copy of chromosome 14 segregate together (**Figure 9.23c**). This pattern is presumably rare, because the two centromeres are both derived from chromosome 14 and usually separate from each other. All the gametes produced by this process are abnormal: half result in monosomy 14 and the other half result in trisomy 14. All are spontaneously aborted.

Thus, only three of the six types of gametes that can be produced by a translocation carrier will result in the birth of a baby and, theoretically, these gametes should arise with equal frequency. One-third of the offspring of a translocation carrier should be translocation carriers like their

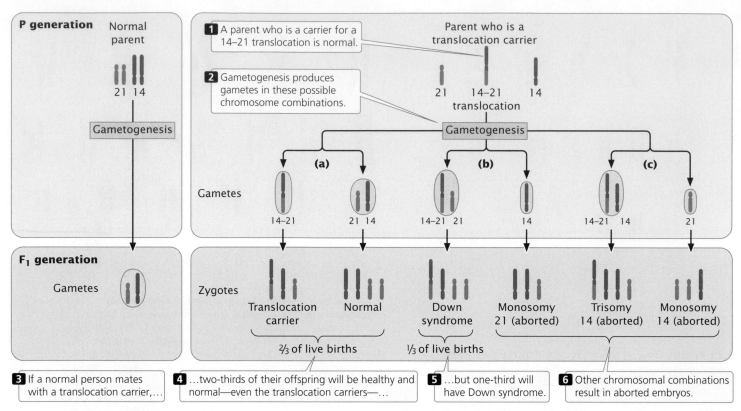

1 A parent who is a carrier for a 14–21 translocation is normal.

2 Gametogenesis produces gametes in these possible chromosome combinations.

P generation Normal parent

21 14

Gametogenesis

Parent who is a translocation carrier

21 14–21 translocation 14

Gametogenesis

Gametes

(a) 14–21   21 14

(b) 14–21 21   14

(c) 14–21 14   21

F₁ generation

Gametes

Zygotes

Translocation carrier   Normal   Down syndrome   Monosomy 21 (aborted)   Trisomy 14 (aborted)   Monosomy 14 (aborted)

⅔ of live births   ⅓ of live births

3 If a normal person mates with a translocation carrier,…

4 …two-thirds of their offspring will be healthy and normal—even the translocation carriers—…

5 …but one-third will have Down syndrome.

6 Other chromosomal combinations result in aborted embryos.

9.23 Translocation carriers are at increased risk for producing children with Down syndrome.

parent, one-third should have familial Down syndrome, and one-third should be normal. In reality, however, fewer than one-third of the children born to translocation carriers have Down syndrome, which suggests that some of the embryos with Down syndrome are spontaneously aborted.

TRY PROBLEM 31 →

Other human trisomies Few autosomal aneuploids besides trisomy 21 result in human live births. **Trisomy 18**, also known as **Edward syndrome**, arises with a frequency of approximately 1 in 8000 live births. Babies with Edward syndrome are severely retarded and have low-set ears, a short neck, deformed feet, clenched fingers, heart problems, and other disabilities. Few live for more than a year after birth. **Trisomy 13** has a frequency of about 1 in 15,000 live births and produces features that are collectively known as **Patau syndrome**. Characteristics of this condition include severe mental retardation, a small head, sloping forehead, small eyes, cleft lip and palate, extra fingers and toes, and numerous other problems. About half of children with trisomy 13 die within the first month of life, and 95% die by the age of 3. Rarer still is **trisomy 8**, which arises with a frequency ranging from about 1 in 25,000 to 1 in 50,000 live births. This aneuploid is characterized by mental retardation, contracted fingers and toes, low-set malformed ears, and a prominent forehead. Many who have this condition have normal life expectancy.

Aneuploidy and maternal age Most cases of Down syndrome and other types of aneuploidy in humans arise from maternal nondisjunction, and the frequency of aneuploidy increases with maternal age (**Figure 9.24**). Why maternal age is associated with nondisjunction is not known for certain. Female mammals are born with primary oocytes suspended in the diplotene substage of prophase I of meiosis. Just before ovulation, meiosis resumes and the first division is completed, producing a secondary oocyte. At this point, meiosis is suspended again and remains so until the secondary oocyte is penetrated by a sperm. The second meiotic division takes place immediately before the nuclei of egg and sperm unite to form a zygote.

Primary oocytes may remain suspended in diplotene for many years before ovulation takes place and meiosis recommences. Components of the spindle and other structures required for chromosome segregation may break down in the long arrest of meiosis, leading to more aneuploidy in children born to older mothers. According to this theory, no age effect is seen in males, because sperm are produced continuously after puberty with no long suspension of the meiotic divisions.

Aneuploidy and cancer Many tumor cells have extra or missing chromosomes or both; some types of tumors are consistently associated with specific chromosome mutations, including aneuploidy and chromosome rearrangements. The role of chromosome mutations in cancer will be explored in Chapter 23.

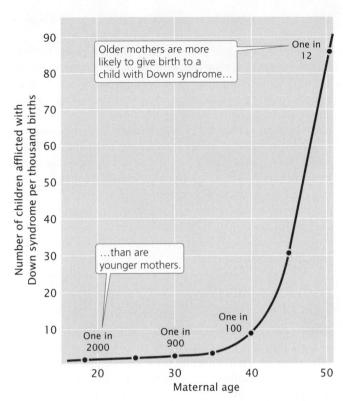

9.24 The incidence of primary Down syndrome and other aneuploids increases with maternal age.

CONCEPTS

In humans, sex-chromosome aneuploids are more common than autosomal aneuploids. X-chromosome inactivation prevents problems of gene dosage for X-linked genes. Down syndrome results from three functional copies of chromosome 21, either through trisomy (primary Down syndrome) or a Robertsonian translocation (familial Down syndrome).

✔ CONCEPT CHECK 6

Briefly explain why, in humans and mammals, sex-chromosome aneuploids are more common than autosomal aneuploids?

Uniparental Disomy

Normally, the two chromosomes of a homologous pair are inherited from different parents—one from the father and one from the mother. The development of molecular techniques that facilitate the identification of specific DNA sequences (see Chapter 19) has made the determination of the parental origins of chromosomes possible. Surprisingly, sometimes both chromosomes are inherited from the same parent, a condition termed **uniparental disomy**.

Many cases of uniparental disomy probably originate as a trisomy. Although most autosomal trisomies are lethal, a trisomic embryo can survive if one of the three chromosomes is lost early in development. If, just by chance, the two remaining chromosomes are both from the same parent, uniparental disomy results.

Uniparental disomy violates the rule that children affected with a recessive disorder appear only in families where both parents are carriers. For example, cystic fibrosis is an autosomal recessive disease; typically, both parents of an affected child are heterozygous for the cystic fibrosis mutation on chromosome 7. However, for a small proportion of people with cystic fibrosis, only one of the parents is heterozygous for the cystic fibrosis mutation. In these cases, cystic fibrosis is due to uniparental disomy: the person who has cystic fibrosis inherited from the heterozygous parent two copies of the chromosome 7 that carries the defective cystic fibrosis allele and no copy of the normal allele from the other parent.

Uniparental disomy has also been observed in Prader–Willi syndrome, a rare condition that arises when a paternal copy of a gene on chromosome 15 is missing. Although most cases of Prader–Willi syndrome result from a chromosome deletion that removes the paternal copy of the gene (see p. 121 in Chapter 5), from 20% to 30% arise when both copies of chromosome 15 are inherited from the mother and no copy is inherited from the father.

Mosaicism

Nondisjunction in a mitotic division may generate patches of cells in which every cell has a chromosome abnormality and other patches in which every cell has a normal karyotype. This type of nondisjunction leads to regions of tissue with different chromosome constitutions, a condition known as **mosaicism**. Growing evidence suggests that mosaicism is common.

Only about 50% of those diagnosed with Turner syndrome have the 45,X karyotype (presence of a single X chromosome) in all their cells; most others are mosaics, possessing some 45,X cells and some normal 46,XX cells. A few may even be mosaics for two or more types of abnormal karyotypes. The 45,X/46,XX mosaic usually arises when an X chromosome is lost soon after fertilization in an XX embryo.

Fruit flies that are XX/XO mosaics (O designates the absence of a homologous chromosome; XO means that the cell has a single X chromosome and no Y chromosome) develop a mixture of male and female traits, because the presence of two X chromosomes in fruit flies produces female traits and the presence of a single X chromosome produces male traits (**Figure 9.25**). In fruit flies, sex is determined independently in each cell in the course of development. Those cells that are XX express female traits; those that are XO express male traits. Such sexual mosaics are called **gynandromorphs**. Normally, X-linked recessive genes are masked in heterozygous females but, in XX/XO mosaics, any X-linked recessive genes present in the cells with a single X chromosome will be expressed.

CONCEPTS

In uniparental disomy, an individual organism has two copies of a chromosome from one parent and no copy from the other. Uniparental disomy may arise when a trisomic embryo loses one of the triplicate chromosomes early in development. In mosaicism, different cells within the same individual organism have different chromosome constitutions.

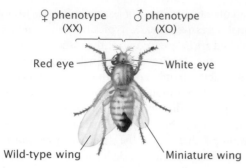

♀ phenotype (XX)   ♂ phenotype (XO)

Red eye — White eye

Wild-type wing — Miniature wing

9.25 Mosaicism for the sex chromosomes produces a gynandromorph. This XX/XO gynandromorph fruit fly carries one wild-type X chromosome and one X chromosome with recessive alleles for white eyes and miniature wings. The left side of the fly has a normal female phenotype, because the cells are XX and the recessive alleles on one X chromosome are masked by the presence of wild-type alleles on the other. The right side of the fly has a male phenotype with white eyes and miniature wing, because the cells are missing the wild-type X chromosome (are XO), allowing the white and miniature alleles to be expressed.

9.4 Polyploidy Is the Presence of More Than Two Sets of Chromosomes

Most eukaryotic organisms are diploid ($2n$) for most of their life cycles, possessing two sets of chromosomes. Occasionally, whole sets of chromosomes fail to separate in meiosis or mitosis, leading to polyploidy, the presence of more than two genomic sets of chromosomes. Polyploids include *trip-*

loids ($3n$), *tetraploids* ($4n$), *pentaploids* ($5n$), and even higher numbers of chromosome sets.

Polyploidy is common in plants and is a major mechanism by which new plant species have evolved. Approximately 40% of all flowering-plant species and from 70% to 80% of grasses are polyploids. They include a number of agriculturally important plants, such as wheat, oats, cotton, potatoes, and sugar cane. Polyploidy is less common in animals but is found in some invertebrates, fishes, salamanders, frogs, and lizards. No naturally occurring, viable polyploids are known in birds, but at least one polyploid mammal—a rat in Argentina—has been reported.

We will consider two major types of polyploidy: **autopolyploidy**, in which all chromosome sets are from a single species; and **allopolyploidy**, in which chromosome sets are from two or more species.

Autopolyploidy

Autopolyploidy is due to accidents of meiosis or mitosis that produce extra sets of chromosomes, all derived from a single species. Nondisjunction of all chromosomes in mitosis in an early $2n$ embryo, for example, doubles the chromosome number and produces an autotetraploid ($4n$), as depicted in **Figure 9.26a**. An autotriploid ($3n$) may arise when nondisjunction in meiosis produces a diploid gamete that then fuses with a normal haploid gamete to produce a triploid zygote (**Figure 9.26b**). Alternatively, triploids may arise from a cross between an autotetraploid that produces $2n$ gametes and a diploid that produces $1n$ gametes.

(a) Autopolyploidy through mitosis

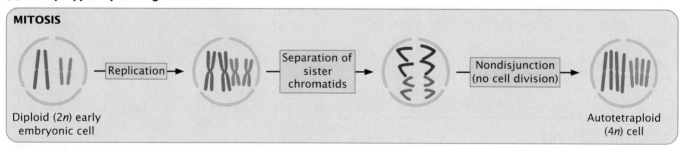

(b) Autopolyploidy through meiosis

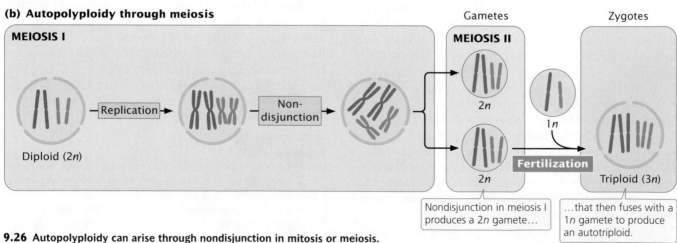

9.26 Autopolyploidy can arise through nondisjunction in mitosis or meiosis.

Because all the chromosome sets in autopolyploids are from the same species, they are homologous and attempt to align in prophase I of meiosis, which usually results in sterility. Consider meiosis in an autotriploid (**Figure 9.27**). In meiosis in a diploid cell, two chromosome homologs pair and align, but, in autotriploids, three homologs are present. One of the three homologs may fail to align with the other two, and this unaligned chromosome will segregate randomly (see Figure 9.27a). Which gamete gets the extra chromosome will be determined by chance and will differ for each homologous group of chromosomes. The resulting gametes will have two copies of some chromosomes and one copy of others. Even if all three chromosomes do align, two chromosomes must segregate to one gamete and one chromosome to the other (see Figure 9.27b). Occasionally, the presence of a third chromosome interferes with normal alignment, and all three chromosomes move to the same gamete (see Figure 9.27c).

No matter how the three homologous chromosomes align, their random segregation will create **unbalanced gametes**, with various numbers of chromosomes. A gamete produced by meiosis in such an autotriploid might receive, say, two copies of chromosome 1, one copy of chromosome 2, three copies of chromosome 3, and no copies of chromosome 4. When the unbalanced gamete fuses with a normal gamete (or with another unbalanced gamete), the resulting zygote has different numbers of the four types of chromosomes. This difference in number creates unbalanced gene dosage in the zygote, which is often lethal. For this reason, triploids do not usually produce viable offspring.

In even-numbered autopolyploids, such as autotetraploids, the homologous chromosomes can theoretically form pairs and divide equally. However, this event rarely happens; so these types of autotetraploids also produce unbalanced gametes.

The sterility that usually accompanies autopolyploidy has been exploited in agriculture. Wild diploid bananas ($2n = 22$), for example, produce seeds that are hard and inedible, but triploid bananas ($3n = 33$) are sterile and produce no seeds; they are the bananas sold commercially. Similarly, seedless triploid watermelons have been created and are now widely sold.

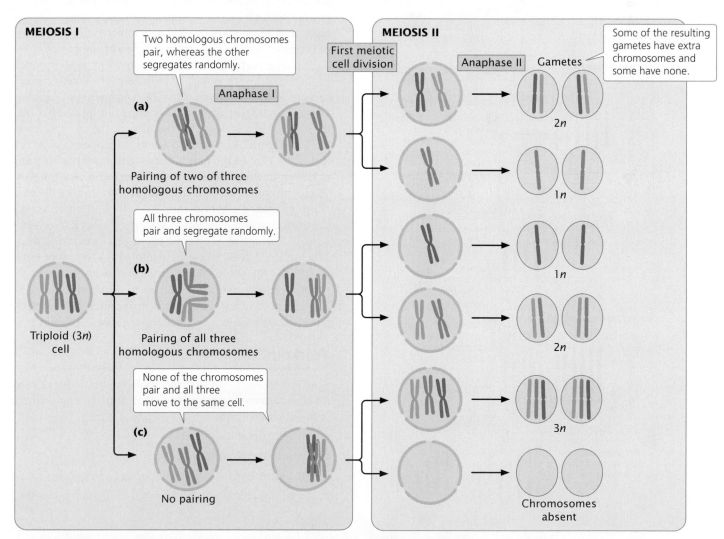

9.27 In meiosis of an autotriploid, homologous chromosomes can pair or not pair in three ways.
This example illustrates the pairing and segregation of a single homologous set of chromosomes.

Allopolyploidy

Allopolyploidy arises from hybridization between two species; the resulting polyploid carries chromosome sets derived from two or more species. **Figure 9.28** shows how allopolyploidy can arise from two species that are sufficiently related

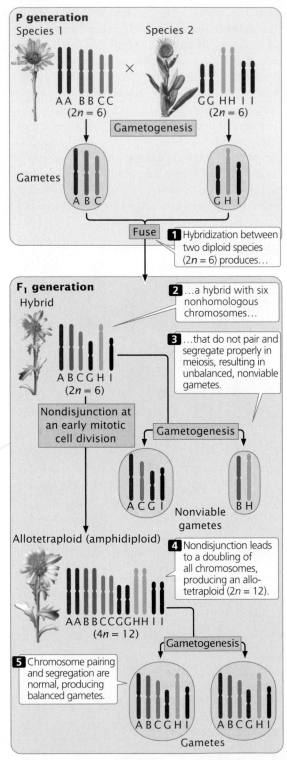

P generation

Species 1 × Species 2

A A B B C C
($2n = 6$)

Gametogenesis

GG HH I I
($2n = 6$)

Gametes

A B C

G H I

Fuse

1 Hybridization between two diploid species ($2n = 6$) produces...

F_1 generation

Hybrid

2 ...a hybrid with six nonhomologous chromosomes...

3 ...that do not pair and segregate properly in meiosis, resulting in unbalanced, nonviable gametes.

A B C G H I
($2n = 6$)

Nondisjunction at an early mitotic cell division

Gametogenesis

A C G I   B H

Nonviable gametes

Allotetraploid (amphidiploid)

4 Nondisjunction leads to a doubling of all chromosomes, producing an allotetraploid ($2n = 12$).

A A B B C C G G H H I I
($4n = 12$)

Gametogenesis

5 Chromosome pairing and segregation are normal, producing balanced gametes.

A B C G H I   A B C G H I

Gametes

9.28 Most allopolyploids arise from hybridization between two species followed by chromosome doubling.

that hybridization takes place between them. Species 1 (AABBCC, $2n = 6$) produces haploid gametes with chromosomes ABC, and species 2 (GGHHII, $2n = 6$) produces gametes with chromosomes GHI. If gametes from species 1 and 2 fuse, a hybrid with six chromosomes (ABCGHI) is created. The hybrid has the same chromosome number as that of both diploid species; so the hybrid is considered diploid. However, because the hybrid chromosomes are not homologous, they will not pair and segregate properly in meiosis; so this hybrid is functionally haploid and sterile.

The sterile hybrid is unable to produce viable gametes through meiosis, but it may be able to perpetuate itself through mitosis (asexual reproduction). On rare occasions, nondisjunction takes place in a mitotic division, which leads to a doubling of chromosome number and an allotetraploid with chromosomes AABBCCGGHHII. This type of allopolyploid, consisting of two combined diploid genomes, is sometimes called an **amphidiploid**. Although the chromosome number has doubled compared with what was present in each of the parental species, the amphidiploid is functionally diploid: every chromosome has one and only one homologous partner, which is exactly what meiosis requires for proper segregation. The amphidiploid can now undergo normal meiosis to produce balanced gametes having six chromosomes.

George Karpechenko created polyploids experimentally in the 1920s. Today, as well as in the early twentieth century, cabbage (*Brassica oleracea*, $2n = 18$) and radishes (*Raphanus sativa*, $2n = 18$) are agriculturally important plants, but only the leaves of the cabbage and the roots of the radish are normally consumed. Karpechenko wanted to produce a plant that had cabbage leaves and radish roots so that no part of the plant would go to waste. Because both cabbage and radish possess 18 chromosomes, Karpechenko was able to successfully cross them, producing a hybrid with $2n = 18$, but, unfortunately, the hybrid was sterile. After several crosses, Karpechenko noticed that one of his hybrid plants produced a few seeds. When planted, these seeds grew into plants that were viable and fertile. Analysis of their chromosomes revealed that the plants were allotetraploids, with $2n = 36$ chromosomes. To Karpechencko's great disappointment, however, the new plants possessed the roots of a cabbage and the leaves of a radish.

Worked Problem

Species I has $2n = 14$ and species II has $2n = 20$. Give all possible chromosome numbers that may be found in the following individuals.

a. An autotriploid of species I

b. An autotetraploid of species II

c. An allotriploid formed from species I and species II

d. An allotetraploid formed from species I and species II

• Solution

The haploid number of chromosomes (n) for species I is 7 and for species II is 10.

a. A triploid individual is 3*n*. A common mistake is to assume that 3*n* means three times as many chromosomes as in a normal individual, but remember that normal individuals are 2*n*. Because *n* for species I is 7 and all genomes of an autopolyploid are from the same species, $3n = 3 \times 7 = 21$.

b. An autotetraploid is 4*n* with all genomes from the same species. The *n* for species II is 10, so $4n = 4 \times 10 = 40$.

c. A triploid is 3*n*. By definition, an allopolyploid must have genomes from two different species. An allotriploid could have 1*n* from species I and 2*n* from species II or $(1 \times 7) + (2 \times 10) = 27$. Alternatively, it might have 2*n* from species I and 1*n* from species II, or $(2 \times 7) + (1 \times 10) = 24$. Thus, the number of chromosomes in an allotriploid could be 24 or 27.

d. A tetraploid is 4*n*. By definition, an allotetraploid must have genomes from at least two different species. An allotetraploid could have 3*n* from species I and 1*n* from species II or $(3 \times 7) + (1 \times 10) = 31$; or 2*n* from species I and 2*n* from species II or $(2 \times 7) + (2 \times 10) = 34$; or 1*n* from species I and 3*n* from species II or $(1 \times 7) + (3 \times 10) = 37$. Thus, the number of chromosomes could be 31, 34, or 37.

→ For additional practice, try Problem 37 at the end of this chapter.

The Significance of Polyploidy

In many organisms, cell volume is correlated with nuclear volume, which, in turn, is determined by genome size. Thus, the increase in chromosome number in polyploidy is often associated with an increase in cell size, and many polyploids are physically larger than diploids. Breeders have used this effect to produce plants with larger leaves, flowers, fruits, and seeds. The hexaploid (6*n* = 42) genome of wheat probably contains chromosomes derived from three different wild species (**Figure 9.29**). As a result, the seeds of modern wheat are larger than those of its ancestors. Many other cultivated plants also are polyploid (**Table 9.2**).

Polyploidy is less common in animals than in plants for several reasons. As discussed, allopolyploids require hybridization between different species, which happens less frequently in animals than in plants. Animal behavior often prevents interbreeding among species, and the complexity of animal development causes most interspecific hybrids to be nonviable. Many of the polyploid animals that do arise are in groups that reproduce through parthenogenesis (a type of reproduction in which the animal develops from an unfertilized egg). Thus, asexual reproduction may facilitate the development of polyploids, perhaps because the perpetuation of hybrids through asexual reproduction provides greater opportunities for nondisjunction than does sexual reproduction. Only a few human polyploid babies have been reported, and most died within a few days of birth. Polyploidy—usually triploidy—is seen in about 10% of all spontaneously aborted human fetuses.

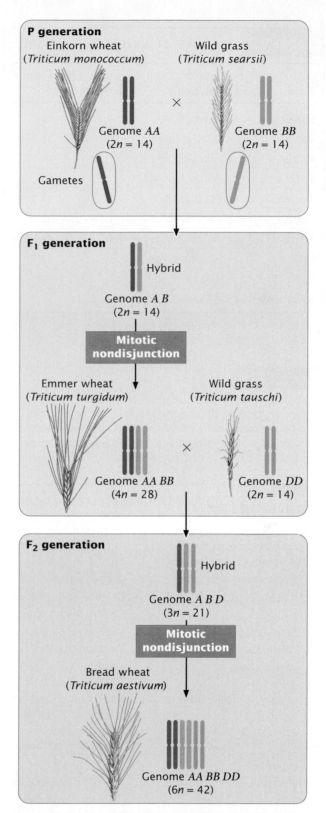

9.29 Modern bread wheat, *Triticum aestivum*, is a hexaploid with genes derived from three different species. Two diploid species, *T. monococcum* (*n* = 14) and probably *T. searsii* (*n* = 14), originally crossed to produce a diploid hybrid (2*n* = 14) that underwent chromosome doubling to create *T. turgidum* (4*n* = 28). A cross between *T. turgidum* and *T. tauschi* (2*n* = 14) produced a triploid hybrid (3*n* = 21) that then underwent chromosome doubling to produce *T. aestivum*, which is a hexaploid (6*n* = 42).

Table 9.2	Examples of polyploid crop plants		
Plant	**Type of Polyploidy**	**Ploidy**	**Chromosome Number**
Potato	Autopolyploid	$4n$	48
Banana	Autopolyploid	$3n$	33
Peanut	Autopolyploid	$4n$	40
Sweet potato	Autopolyploid	$6n$	90
Tobacco	Allopolyploid	$4n$	48
Cotton	Allopolyploid	$4n$	52
Wheat	Allopolyploid	$6n$	42
Oats	Allopolyploid	$6n$	42
Sugar cane	Allopolyploid	$8n$	80
Strawberry	Allopolyploid	$8n$	56

Source: After F. C. Elliot, *Plant Breeding and Cytogenetics* (New York: McGraw-Hill, 1958).

CONCEPTS

Polyploidy is the presence of extra chromosome sets: autopolyploids possess extra chromosome sets from the same species; allopolyploids possess extra chromosome sets from two or more species. Problems in chromosome pairing and segregation often lead to sterility in autopolyploids, but many allopolyploids are fertile.

✔ CONCEPT CHECK 7

Species A has $2n = 16$ chromosomes and species B has $2n = 14$. How many chromosomes would be found in an allotriploid of these two species?

a. 21 or 24   c. 22 or 23

b. 42 or 48   d. 45

9.5 Chromosome Variation Plays an Important Role in Evolution

Chromosome variations are potentially important in evolution and, within a number of different groups of organisms, have clearly played a significant role in past evolution. Chromosome duplications provide one way in which new genes evolve. In many cases, existing copies of a gene are not free to vary, because they encode a product that is essential to development or function. However, after a chromosome undergoes duplication, extra copies of genes within the duplicated region are present. The original copy can provide the essential function, whereas an extra copy from the duplication is free to undergo mutation and change. Over evolutionary time, the extra copy may acquire enough mutations to assume a new function that benefits the organism.

Inversions also can play important evolutionary roles by suppressing recombination among a set of genes. As we have seen, crossing over within an inversion in an individual

Table 9.3	Different types of chromosome mutations
Chromosome Mutation	**Definition**
Chromosome rearrangement	Change in chromosome structure
Chromosome duplication	Duplication of a chromosome segment
Chromosome deletion	Deletion of a chromosome segment
Inversion	Chromosome segment inverted 180 degrees
Paracentric inversion	Inversion that does not include the centromere in the inverted region
Pericentric inversion	Inversion that includes the centromere in the inverted region
Translocation	Movement of a chromosome segment to a nonhomologous chromosome or to another region of the same chromosome
Nonreciprocal translocation	Movement of a chromosome segment to a nonhomologous chromosome or to another region of the same chromosome without reciprocal exchange
Reciprocal translocation	Exchange between segments of nonhomologous chromosomes or between regions of the same chromosome
Aneuploidy	Change in number of individual chromosomes
Nullisomy	Loss of both members of a homologous pair
Monosomy	Loss of one member of a homologous pair
Trisomy	Gain of one chromosome, resulting in three homologous chromosomes
Tetrasomy	Gain of two homologous chromosomes, resulting in four homologous chromosomes
Polyploidy	Addition of entire chromosome sets
Autopolyploidy	Polyploidy in which extra chromosome sets are derived from the same species
Allopolyploidy	Polyploidy in which extra chromosome sets are derived from two or more species

heterozygous for a pericentric or paracentric inversion leads to unbalanced gametes and no recombinant progeny. This suppression of recombination allows particular sets of co-adapted alleles that function well together to remain intact, unshuffled by recombination.

Polyploidy, particularly allopolyploidy, often gives rise to new species and has been particularly important in the evolution of flowering plants. Occasional genome doubling through polyploidy has been a major contributor to evolutionary suc-cess in animal groups. For example, *Saccharomyces cerevisiae* (yeast) is a tetraploid, having undergone whole-genome dupli-cation about 100 million years ago. The vertebrate genome has duplicated twice, once in the common ancestor to jawed vertebrates and again in the ancestor of fishes. Certain groups of vertebrates, such as some frogs and some fishes, have under-gone additional polyploidy. Cereal plants have undergone several genome-duplication events. Different types of chromo-some mutations are summarized in **Table 9.3**.

CONCEPTS SUMMARY

- Three basic types of chromosome mutations are: (1) chromosome rearrangements, which are changes in the structures of chromosomes; (2) aneuploidy, which is an increase or decrease in chromosome number; and (3) polyploidy, which is the presence of extra chromosome sets.

- Chromosome rearrangements include duplications, deletions, inversions, and translocations.

- In individuals heterozygous for a duplication, the duplicated region will form a loop when homologous chromosomes pair in meiosis. Duplications often have pronounced effects on the phenotype owing to unbalanced gene dosage.

- Segmental duplications are common in the human genome.

- In individuals heterozygous for a deletion, one of the chromosomes will loop out during pairing in meiosis. Deletions may cause recessive alleles to be expressed.

- Pericentric inversions include the centromere; paracentric inversions do not. In individuals heterozygous for an inversion, the homologous chromosomes form inversion loops in meiosis, with reduced recombination taking place within the inverted region.

- In translocation heterozygotes, the chromosomes form crosslike structures in meiosis, and the segregation of chromosomes produces unbalanced gametes.

- Fragile sites are constrictions or gaps that appear at particular regions on the chromosomes of cells grown in culture and are prone to breakage under certain conditions.

- Copy-number variations (CNVs) are differences in the number of copies of DNA sequences and include duplications and deletions. These variants are common in the human genome; some are associated with diseases and disorders.

- Nullisomy is the loss of two homologous chromosomes; monosomy is the loss of one homologous chromosome; trisomy is the addition of one homologous chromosome; tetrasomy is the addition of two homologous chromosomes.

- Aneuploidy usually causes drastic phenotypic effects because it leads to unbalanced gene dosage.

- Primary Down syndrome is caused by the presence of three full copies of chromosome 21, whereas familial Down syndrome is caused by the presence of two normal copies of chromosome 21 and a third copy that is attached to another chromosome through a translocation.

- Uniparental disomy is the presence of two copies of a chromosome from one parent and no copy from the other. Mosaicism is caused by nondisjunction in an early mitotic division that leads to different chromosome constitutions in different cells of a single individual.

- All the chromosomes in an autopolyploid derive from one species; chromosomes in an allopolyploid come from two or more species.

- Chromosome variations have played an important role in the evolution of many groups of organisms.

IMPORTANT TERMS

chromosome mutation (p. 240)
metacentric chromosome (p. 240)
submetacentric chromosome (p. 240)
acrocentric chromosome (p. 240)
telocentric chromosome (p. 240)
chromosome rearrangement (p. 242)
chromosome duplication (p. 242)
tandem duplication (p. 242)
displaced duplication (p. 242)
reverse duplication (p. 242)
segmental duplication (p. 244)

chromosome deletion (p. 244)
pseudodominance (p. 245)
haploinsufficient gene (p. 245)
chromosome inversion (p. 246)
paracentric inversion (p. 246)
pericentric inversion (p. 246)
position effect (p. 246)
dicentric chromatid (p. 247)
acentric chromatid (p. 247)
dicentric bridge (p. 247)
translocation (p. 248)

nonreciprocal translocation (p. 248)
reciprocal translocation (p. 249)
Robertsonian translocation (p. 249)
alternate segregation (p. 249)
adjacent-1 segregation (p. 249)
adjacent-2 segregation (p. 249)
fragile site (p. 251)
fragile-X syndrome (p. 251)
copy-number variation (CNV) (p. 252)
aneuploidy (p. 252)
polyploidy (p. 252)

ANSWERS TO CONCEPT CHECKS

1. a

2. Pseudodominance is the expression of a recessive mutation. It is produced when the dominant wild-type allele in a heterozygous individual is absent due to a deletion on one chromosome.

3. c

4. b

5. 37

6. Dosage compensation prevents the expression of additional copies of X-linked genes in mammals, and there is little information in the Y chromosome; so extra copies of the X and Y chromosomes do not have major effects on development. In contrast, there is no mechanism of dosage compensation for autosomes, and so extra copies of autosomal genes are expressed, upsetting development and causing the spontaneous abortion of aneuploid embryos.

7. c

WORKED PROBLEMS

1. A chromosome has the following segments, where • represents the centromere.

$$A\ B\ C\ D\ E\ \bullet\ F\ G$$

What types of chromosome mutations are required to change this chromosome into each of the following chromosomes? (In some cases, more than one chromosome mutation may be required.)

a. $\underline{A\ B\ E\ \bullet\ F\ G}$

b. $\underline{A\ E\ D\ C\ B\ \bullet\ F\ G}$

c. $\underline{A\ B\ A\ B\ C\ D\ E\ \bullet\ F\ G}$

d. $\underline{A\ F\ \bullet\ E\ D\ C\ B\ G}$

e. $\underline{A\ B\ C\ D\ E\ E\ D\ C\ \bullet\ F\ G}$

• **Solution**

The types of chromosome mutations are identified by comparing the mutated chromosome with the original, wild-type chromosome.

a. The mutated chromosome ($\underline{A\ B\ E\ \bullet\ F\ G}$) is missing segment $\underline{C\ D}$; so this mutation is a deletion.

b. The mutated chromosome ($\underline{A\ E\ D\ C\ B\ \bullet\ F\ G}$) has one and only one copy of all the gene segments, but segment $\underline{B\ C\ D\ E}$ has been inverted 180 degrees. Because the centromere has not changed location and is not in the inverted region, this chromosome mutation is a paracentric inversion.

c. The mutated chromosome ($\underline{A\ B\ A\ B\ C\ D\ E\ \bullet\ F\ G}$) is longer than normal, and we see that segment $\underline{A\ B}$ has been duplicated. This mutation is a tandem duplication.

d. The mutated chromosome ($\underline{A\ F\ \bullet\ E\ D\ C\ B\ G}$) is normal length, but the gene order and the location of the centromere

have changed; this mutation is therefore a pericentric inversion of region ($\underline{B\ C\ D\ E\ \bullet\ F}$).

e. The mutated chromosome ($\underline{A\ B\ C\ D\ E\ E\ D\ C\ \bullet\ F\ G}$) contains a duplication ($\underline{C\ D\ E}$) that is also inverted; so this chromosome has undergone a duplication and a paracentric inversion.

2. Species I is diploid ($2n = 4$) with chromosomes AABB; related species II is diploid ($2n = 6$) with chromosomes MMNNOO. Give the chromosomes that would be found in individuals with the following chromosome mutations.

a. Autotriploidy in species I

b. Allotetraploidy including species I and II

c. Monosomy in species I

d. Trisomy in species II for chromosome M

e. Tetrasomy in species I for chromosome A

f. Allotriploidy including species I and II

g. Nullisomy in species II for chromosome N

• **Solution**

To work this problem, we should first determine the haploid genome complement for each species. For species I, $n = 2$ with chromosomes AB and, for species II, $n = 3$ with chromosomes MNO.

a. An autotriploid is $3n$, with all the chromosomes coming from a single species; so an autotriploid of species I would have chromosomes AAABBB ($3n = 6$).

b. An allotetraploid is $4n$, with the chromosomes coming from more than one species. An allotetraploid could consist of $2n$ from species I and $2n$ from species II, giving the allotetraploid

$(4n = 2 + 2 + 3 + 3 = 10)$ chromosomes AABBMMNNOO. An allotetraploid could also possess $3n$ from species I and $1n$ from species II $(4n = 2 + 2 + 2 + 3 = 9;$ AAABBBMNO) or $1n$ from species I and $3n$ from species II $(4n = 2 + 3 + 3 + 3 = 11;$ ABMMMNNNOOO).

c. A monosomic is missing a single chromosome; so a monosomic for species I would be $2n - 1 = 4 - 1 = 3$. The monosomy might include either of the two chromosome pairs, with chromosomes ABB or AAB.

d. Trisomy requires an extra chromosome; so a trisomic of species II for chromosome M would be $2n + 1 = 6 + 1 = 7$ (MMMNNOO).

e. A tetrasomic has two extra homologous chromosomes; so a tetrasomic of species I for chromosome A would be $2n + 2 = 4 + 2 = 6$ (AAAABB).

f. An allotriploid is $3n$ with the chromosomes coming from two different species; so an allotriploid could be $3n = 2 + 2 + 3 = 7$ (AABBMNO) or $3n = 2 + 3 + 3 = 8$ (ABMMNOO).

g. A nullisomic is missing both chromosomes of a homologous pair; so a nullisomic of species II for chromosome N would be $2n - 2 = 6 - 2 = 4$ (MMOO).

COMPREHENSION QUESTIONS

Section 9.1

*1. List the different types of chromosome mutations and define each one.

Section 9.2

*2. Why do extra copies of genes sometimes cause drastic phenotypic effects?

3. Draw a pair of chromosomes as they would appear during synapsis in prophase I of meiosis in an individual heterozygous for a chromosome duplication.

4. What is haploinsufficiency?

*5. What is the difference between a paracentric and a pericentric inversion?

6. How do inversions cause phenotypic effects?

7. Explain, with the aid of a drawing, how a dicentric bridge is produced when crossing over takes place in an individual heterozygous for a paracentric inversion.

8. Explain why recombination is suppressed in individuals heterozygous for paracentric and pericentric inversions.

*9. How do translocations produce phenotypic effects?

10. Sketch the chromosome pairing and the different segregation patterns that can arise in an individual heterozygous for a reciprocal translocation.

11. What is a Robertsonian translocation?

Section 9.3

12. List four major types of aneuploidy.

*13. What is the difference between primary Down syndrome and familial Down syndrome? How does each type arise?

*14. What is uniparental disomy and how does it arise?

15. What is mosaicism and how does it arise?

Section 9.4

*16. What is the difference between autopolyploidy and allopolyploidy? How does each arise?

17. Explain why autopolyploids are usually sterile, whereas allopolyploids are often fertile.

APPLICATION QUESTIONS AND PROBLEMS

Section 9.1

*18. Which types of chromosome mutations

a. increase the amount of genetic material in a particular chromosome?

b. increase the amount of genetic material in all chromosomes?

c. decrease the amount of genetic material in a particular chromosome?

d. change the position of DNA sequences in a single chromosome without changing the amount of genetic material?

e. move DNA from one chromosome to a nonhomologous chromosome?

Section 9.2

*19. A chromosome has the following segments, where • represents the centromere:

$$A\ B\ \bullet\ C\ D\ E\ F\ G$$

What types of chromosome mutations are required to change this chromosome into each of the following chromosomes? (In some cases, more than one chromosome mutation may be required.)

a. A B A B • C D E F G

b. A B • C D E A B F G

c. A B • C F E D G

d. A • C D E F G

e. A B · C D E

f. A B · E D C F G

g. C · B A D E F G

h. A B · C F E D F E D G

i. A B · C D E F C D F E G

20. A chromosome initially has the following segments:

A B · C D E F G

Draw the chromosome, identifying its segments, that would result from each of the following mutations.

a. Tandem duplication of DEF

b. Displaced duplication of DEF

c. Deletion of FG

d. Paracentric inversion that includes DEFG

e. Pericentric inversion of BCDE

21. The following diagram represents two nonhomologous chromosomes:

A B · C D E F G

R S · T U V W X

What type of chromosome mutation would produce each of the following chromosomes?

a. A B · C D
   R S · T U V W X E F G

b. A U V B · C D E F G
   R S · T W X

c. A B · T U V F G
   R S · C D E W X

d. A B · C W G
   R S · T U V D E F X

*22. The *Notch* mutation is a deletion on the X chromosome of *Drosophila melanogaster*. Female flies heterozygous for *Notch* have an indentation on the margins of their wings; *Notch* is lethal in the homozygous and hemizygous conditions. The *Notch* deletion covers the region of the X chromosome that contains the locus for white eyes, an X-linked recessive trait. Give the phenotypes and proportions of progeny produced in the following crosses.

a. A red-eyed, Notch female is mated with a white-eyed male.

b. A white-eyed, Notch female is mated with a red-eyed male.

c. A white-eyed, Notch female is mated with a white-eyed male.

23. The green-nose fly normally has six chromosomes, two metacentric and four acrocentric. A geneticist examines the chromosomes of an odd-looking green-nose fly and discovers that it has only five chromosomes; three of them are metacentric and two are acrocentric. Explain how this change in chromosome number might have taken place.

*24. A wild-type chromosome has the following segments:

A B C · D E F G H I

An individual is heterozygous for the following chromosome mutations. For each mutation, sketch how the wild-type and mutated chromosomes would pair in prophase I of meiosis, showing all chromosome strands.

a. A B C · D E F D E F G H I

b. A B C · D H I

c. A B C · D G F E H I

d. A B E D · C F G H I

25. Draw the chromatids that would result from a two-strand double crossover between E and F in Problem 24.

26. As discussed in this chapter, crossing over within a pericentric inversion produces chromosomes that have extra copies of some genes and no copies of other genes. The fertilization of gametes containing such duplication or deficient chromosomes often results in children with syndromes characterized by developmental delay, mental retardation, abnormal development of organ systems, and early death. Maarit Jaarola and colleagues examined individual sperm cells of a male who was heterozygous for a pericentric inversion on chromosome 8 and determined that crossing over took place within the pericentric inversion in 26% of the meiotic divisions (M. Jaarola, R. H. Martin, and T. Ashley. 1998. *American Journal of Human Genetics* 63:218–224).

   Assume that you are a genetic counselor and that a couple seeks counseling from you. Both the man and the woman are phenotypically normal, but the woman is heterozygous for a pericentric inversion on chromosome 8. The man is karyotypically normal. What is the probability that this couple will produce a child with a debilitating syndrome as the result of crossing over within the pericentric inversion?

*27. An individual heterozygous for a reciprocal translocation possesses the following chromosomes:

A B · C D E F G

A B · C D V W X

R S · T U E F G

R S · T U V W X

a. Draw the pairing arrangement of these chromosomes in prophase I of meiosis.

b. Diagram the alternate, adjacent-1, and adjacent-2 segregation patterns in anaphase I of meiosis.

c. Give the products that result from alternate, adjacent-1, and adjacent-2 segregation.

Section 9.3

28. Red–green color blindness is a human X-linked recessive disorder. A young man with a 47,XXY karyotype

(Klinefelter syndrome) is color blind. His 46,XY brother also is color blind. Both parents have normal color vision. Where did the nondisjunction that gave rise to the young man with Klinefelter syndrome take place?

29. Junctional epidermolysis bullosa (JEB) is a severe skin disorder that results in blisters over the entire body. The disorder is caused by autosomal recessive mutations at any one of three loci that help to encode laminin 5, a major component in the dermal–epidermal basement membrane. Leena Pulkkinen and colleagues described a male newborn who was born with JEB and died at 2 months of age (L. Pulkkinen et al. 1997. *American Journal of Human Genetics* 61:611–619); the child had healthy unrelated parents. Chromosome analysis revealed that the infant had 46 normal-appearing chromosomes. Analysis of DNA showed that his mother was heterozygous for a JEB-causing allele at the *LAMB3* locus, which is on chromosome 1. The father had two normal alleles at this locus. DNA fingerprinting demonstrated that the male assumed to be the father had, in fact, conceived the child.

a. Assuming that no new mutations occurred in this family, explain the presence of an autosomal recessive disease in the child when the mother is heterozygous and the father is homozygous normal.

b. How might you go about proving your explanation? Assume that a number of genetic markers are available for each chromosome.

30. Some people with Turner syndrome are 45,X/46,XY mosaics. Explain how this mosaicism could arise.

***31.** Bill and Betty have had two children with Down syndrome. Bill's brother has Down syndrome and his sister has two children with Down syndrome. On the basis of these observations, indicate which of the following statements are most likely correct and which are most likely incorrect. Explain your reasoning.

a. Bill has 47 chromosomes.

b. Betty has 47 chromosomes.

c. Bill and Betty's children each have 47 chromosomes.

d. Bill's sister has 45 chromosomes.

e. Bill has 46 chromosomes.

f. Betty has 45 chromosomes.

g. Bill's brother has 45 chromosomes.

***32.** In mammals, sex-chromosome aneuploids are more common than autosomal aneuploids but, in fish, sex-chromosome aneuploids and autosomal aneuploids are found with equal frequency. Explain these differences in mammals and fish.

***33.** A young couple is planning to have children. Knowing that there have been a substantial number of stillbirths, miscarriages, and fertility problems on the husband's side of the family, they see a genetic counselor. A chromosome analysis reveals that, whereas the woman has a normal karyotype, the man possesses only 45 chromosomes and is a carrier of a Robertsonian translocation between chromosomes 22 and 13.

a. List all the different types of gametes that might be produced by the man.

b. What types of zygotes will develop when each of gametes produced by the man fuses with a normal gamete produced by the woman?

c. If trisomies and monosomies entailing chromosomes 13 and 22 are lethal, what proportion of the surviving offspring will be carriers of the translocation?

34. Using breeding techniques, Andrei Dyban and V. S. Baranov (*Cytogenetics of Mammalian Embryonic Development.* Oxford: Oxford University Press, Clarendon Press; New York: Oxford University Press, 1987) created mice that were trisomic for each of the different mouse chromosomes. They found that only mice with trisomy 19 developed. Mice trisomic for all other chromosomes died in the course of development. For some of these trisomics, they compared the length of development (number of days after conception before the embryo died) as a function of the size of the mouse chromosome that was present in three copies (see the adjoining graph). Summarize their findings as presented in this graph and provide a possible explanation for the results.

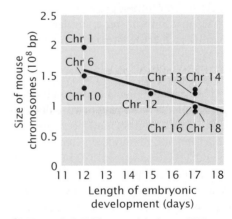

[E. Torres, B. R. Williams, and A. Amon. 2008. *Genetics* 179:737–746, Fig. 2B.]

Section 9.4

***35.** Species I has $2n = 16$ chromosomes. How many chromosomes will be found per cell in each of the following mutants in this species?

a. Monosomic   **e.** Double monosomic

b. Autotriploid   **f.** Nullisomic

c. Autotetraploid   **g.** Autopentaploid

d. Trisomic   **h.** Tetrasomic

36. Species I is diploid ($2n = 8$) with chromosomes AABBCCDD; related species II is diploid ($2n = 8$) with chromosomes MMNNOOPP. What types of chromosome mutations do individual organisms with the following sets of chromosomes have?

a. AAABBCCDD

e. AAABBCCDDD

b. MMNNOOOOPP

f. AABBDD

c. AABBCDD

g. AABBCCDDMMNNOOPP

d. AAABBBCCCDDD

h. AABBCCDDMNOP

37. Species I has $2n = 8$ chromosomes and species II has $2n = 14$ chromosomes. What would the expected chromosome numbers be in individual organisms with the following chromosome mutations? Give all possible answers.

a. Allotriploidy including species I and II

b. Autotetraploidy in species II

c. Trisomy in species I

d. Monosomy in species II

e. Tetrasomy in species I

f. Allotetraploidy including species I and II

38. Consider a diploid cell that has $2n = 4$ chromosomes—one pair of metacentric chromosomes and one pair of acrocentric chromosomes. Suppose this cell undergoes nondisjunction giving rise to an autotriploid cell ($3n$). The triploid cell then undergoes meiosis. Draw the different types of gametes that may result from meiosis in the triploid cell, showing the chromosomes present in each type. To distinguish between the different metacentric and acrocentric chromosomes, use a different color to draw each metacentric chromosome; similarly, use a different color to draw each acrocentric chromosome. (Hint: See Figure 9.27).

39. *Nicotiana glutinosa* ($2n = 24$) and *N. tabacum* ($2n = 48$) are two closely related plants that can be intercrossed, but the F_1 hybrid plants that result are usually sterile. In 1925, Roy Clausen and Thomas Goodspeed crossed *N. glutinosa* and *N. tabacum* and obtained one fertile F_1 plant (R. E. Clausen and T. H. Goodspeed. 1925 *Genetics* 10:278–284). They were able to self-pollinate the flowers of this plant to produce an F_2 generation. Surprisingly, the F_2 plants were fully fertile and produced viable seed. When Clausen and Goodspeed examined the chromosomes of the F_2 plants, they observed 36 pairs of chromosomes in metaphase I and 36 individual chromosomes in metaphase II. Explain the origin of the F_2 plants obtained by Clausen and Goodspeed and the numbers of chromosomes observed.

40. What would be the chromosome number of progeny resulting from the following crosses in wheat (see Figure 9.29)? What type of polyploid (allotriploid, allotetraploid, etc.) would result from each cross?

a. Einkorn wheat and emmer wheat

b. Bread wheat and emmer wheat

c. Einkorn wheat and bread wheat

41. Karl and Hally Sax crossed *Aegilops cylindrica* ($2n = 28$), a wild grass found in the Mediterranean region, with *Triticum vulgare* ($2n = 42$), a type of wheat (K. Sax and H. J. Sax. 1924. *Genetics* 9:454–464). The resulting F_1 plants from this cross had 35 chromosomes. Examination of metaphase I in the F_1 plants revealed the presence of 7 pairs of chromosomes (bivalents) and 21 unpaired chromosomes (univalents).

a. If the unpaired chromosomes segregate randomly, what possible chromosome numbers will appear in the gametes of the F_1 plants?

b. What does the appearance of the bivalents in the F_1 hybrids suggest about the origin of *Triticum vulgare* wheat?

Aegilops cylindrica, jointed goatgrass. [Sam Brinker, OMNR-NHIC, 2008. Canadian Food Inspection Agency.]

Triticum vulgare, wheat. [Michael Hieber/123RE.com.]

CHALLENGE QUESTIONS

Section 9.3

42. Red–green color blindness is a human X-linked recessive disorder. Jill has normal color vision, but her father is color blind. Jill marries Tom, who also has normal color vision. Jill and Tom have a daughter who has Turner syndrome and is color blind.

a. How did the daughter inherit color blindness?

b. Did the daughter inherit her X chromosome from Jill or from Tom?

43. Progeny of triploid tomato plants often contain parts of an extra chromosome, in addition to the normal complement of 24 chromosomes (J. W. Lesley and M. M. Lesley. 1929. *Genetics* 14:321–336). Mutants with a part of an extra chromosome are referred to as secondaries. James and Margaret Lesley observed that secondaries arise from triploid ($3n$), trisomic ($3n + 1$), and double trisomic ($3n + 1 + 1$) parents, but never from diploids ($2n$). Give one or more possible reasons that secondaries arise from parents that have unpaired chromosomes but not from parents that are normal diploids.

44. Mules result from a cross between a horse ($2n = 64$) and a donkey ($2n = 62$), have 63 chromosomes, and are almost always sterile. However, in the summer of 1985, a female mule named Krause who was pastured with a male donkey gave birth to a newborn foal (O. A. Ryder et al. 1985. *Journal of Heredity* 76:379–381). Blood tests established that the male foal, appropriately named Blue Moon, was the offspring of Krause and that Krause was indeed a mule. Both Blue Moon and Krause were fathered by the same donkey (see the illustration). The foal, like his mother, had 63 chromosomes—half of them horse chromosomes and the other half donkey chromosomes. Analyses of genetic markers showed that, remarkably, Blue Moon seemed to have inherited a complete set of horse chromosomes from his mother, instead of the random mixture of horse and donkey chromosomes that would be expected with normal meiosis. Thus, Blue Moon and Krause were not only mother and son, but also brother and sister.

a. With the use of a diagram, show how, if Blue Moon inherited only horse chromosomes from his mother, Blue Moon and Krause are both mother and son as well as brother and sister.

b. Although rare, additional cases of fertile mules giving births to offspring have been reported. In these cases, when a female mule mates with a male horse, the offspring is horselike in appearance but, when a female mule mates with a male donkey, the offspring is mulelike in appearance. Is this observation consistent with the idea that the offspring of fertile female mules inherit only a set of horse chromosomes from their mule mothers? Explain your reasoning.

c. Can you suggest a possible mechanism for how the offspring of fertile female mules might pass on a complete set of horse chromosomes to their offspring?

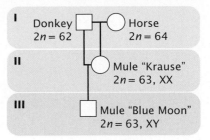

Section 9.5

45. Humans and many other complex organisms are diploid, possessing two sets of genes, one inherited from the mother and one from the father. However, a number of eukaryotic organisms spend most of their life cycles in a haploid state. Many of these eukaryotes, such as *Neurospora* and yeast, still undergo meiosis and sexual reproduction, but most of the cells that make up the organism are haploid.

Considering that haploid organisms are fully capable of sexual reproduction and generating genetic variation, why are most complex eukaryotes diploid? In other words, what might be the evolutionary advantage of existing in a diploid state instead of a haploid state? And why might a few organisms, such as *Neurospora* and yeast, exist as haploids?

10 DNA: The Chemical Nature of the Gene

The remarkable stability of DNA makes the extraction and analysis of DNA from ancient remains possible, including Neanderthal bones that are more than 30,000 years old. [John Reader/Photo Researchers.]

NEANDERTHAL'S DNA

DNA, with its double-stranded spiral, is among the most elegant of all biological molecules. But the double helix is not just a beautiful structure; it also gives DNA incredible stability and permanence, providing geneticists with a unique window to the past.

In 1856, a group of men working a limestone quarry in the Neander Valley of Germany discovered a small cave containing a number of bones. The workers assumed that the bones were those of a cave bear, but a local schoolteacher recognized them as human, although they were clearly unlike any human bones the teacher had ever seen. The bones appeared to be those of a large person with great muscular strength, a low forehead, a large nose with broad nostrils, and massive protruding brows. Experts confirmed that the bones belonged to an extinct human, who became known as Neanderthal. In the next 100 years, similar fossils were discovered in Spain, Belgium, France, Croatia, and the Middle East.

Research has now revealed that Neanderthals roamed Europe and western Asia for at least 200,000 years, disappearing abruptly 30,000 to 40,000 years ago. During the last years of this period, Neanderthals coexisted with the direct ancestors of modern humans, the Cro-Magnons. The fate of the Neanderthals—why they disappeared—has captured the imagination of scientists and laypersons alike. Did Cro-Magnons, migrating out of Africa with a superior technology, cause the demise of the Neanderthal people, either through competition or perhaps through deliberate extermination? Or did the Neanderthals interbreed with Cro-Magnons, their genes becoming assimilated into the larger gene pool of modern humans? Support for the latter hypothesis came from the discovery of fossils that appeared to be transitional between Neanderthals and Cro-Magnons. Unfortunately, the meager fossil record of Neanderthals and Cro-Magnons did not allow a definitive resolution of these questions.

Whether Neanderthals interbred with the ancestors of modern humans has now become a testable question. In 1997, scientists extracted DNA from a bone of the original Neanderthal specimen collected from the Neander Valley more than 140 years earlier. This Neanderthal bone was at least 30,000 years old. Using a technique called the polymerase chain reaction (see Chapter 19), the scientists amplified 378 nucleotides of the Neanderthal's mitochondrial DNA a millionfold. They then determined the base sequence of this amplified DNA, and compared it with mitochondrial sequences from living humans. Later, the squences of the entire mitochondrial DNA were determined for five Neanderthal specimens from Croatia, Germany, Spain, and Russia. And, in 2009, German scientists announced that they had completed a rough draft of the entire *nuclear* Neanderthal genome (about 3 billion base pairs of DNA) isolated from bones of two 38,000-year-old Neanderthal females.

Analyses of mitochondrial DNA and nuclear DNA have demonstrated that Neanderthals were a unique species, similar to but genetically distinct from modern humans. Comparison of the genomes of modern humans and Neanderthals reveal evidence of some interbreeding between the two groups. The genomes of non-African populations of modern humans contain from 1% to 4% of DNA sequences inherited from Neanderthals.

That Neanderthal DNA persists and reveals its genetic affinities, even 40,000 years later, is testimony to the remarkable stability of the double helix. This chapter focuses on how DNA was identified as the source of genetic information and how this elegant molecule encodes the genetic instructions. We begin by considering the basic requirements of the genetic material and the history of our understanding of DNA—how its relation to genes was uncovered and its structure determined. The history of DNA illustrates several important points about the nature of scientific research. As with so many important scientific advances, the structure of DNA and its role as the genetic material were not discovered by any single person but were gradually revealed over a period of almost 100 years, thanks to the work of many investigators. Our understanding of the relation between DNA and genes was enormously enhanced in 1953, when James Watson and Francis Crick proposed a three-dimensional structure for DNA that brilliantly illuminated its role in genetics. As illustrated by Watson and Crick's discovery, major scientific advances are often achieved, not through the collection of new data, but through the interpretation of old data in new ways.

After reviewing the discoveries that led to our current understanding of DNA, we will examine DNA structure. The structure of DNA is important in its own right, but the key genetic concept is the relation between the structure and the function of DNA—how its structure allows it to serve as the genetic material.

10.1 Genetic Material Possesses Several Key Characteristics

Life is characterized by tremendous diversity, but the coding instructions of all living organisms are written in the same genetic language—that of nucleic acids. Surprisingly, the idea that genes are made of nucleic acids was not widely accepted until after 1950. This skepticism was due in part to a lack of knowledge about the structure of deoxyribonucleic acid (DNA). Until the structure of DNA was understood, how DNA could store and transmit genetic information was unclear. Even before nucleic acids were identified as the genetic material, biologists recognized that, whatever the nature of the genetic material, it must possess three important characteristics.

1. **Genetic material must contain complex information.**
   First and foremost, the genetic material must be capable of storing large amounts of information—instructions for all

the traits and functions of an organism. This information must have the capacity to vary, because different species and even individual members of a species differ in their genetic makeup. At the same time, the genetic material must be stable, because alterations to the genetic instructions (mutations) are frequently detrimental.

2. **Genetic material must replicate faithfully.** A second necessary feature is that genetic material must have the capacity to be copied accurately. Every organism begins life as a single cell, which must undergo billions of cell divisions to produce a complex, multicellular creature like yourself. At each cell division, the genetic instructions must be transmitted to descendant cells with great accuracy. When organisms reproduce and pass genes on to their progeny, the coding instructions must be copied with fidelity.

3. **Genetic material must encode the phenotype.** The genetic material (the genotype) must have the capacity to "code for" (determine) traits (the phenotype). The product of a gene is often a protein; so there must be a mechanism for genetic instructions to be translated into the amino acid sequence of a protein.

CONCEPTS

The genetic material must be capable of carrying large amounts of information, replicating faithfully, and translating its coding instructions into phenotypes.

✔ CONCEPT CHECK 1

Why was the discovery of the structure of DNA so important for understanding genetics?

10.2 All Genetic Information Is Encoded in the Structure of DNA or RNA

Although our understanding of how DNA encodes genetic information is relatively recent, the study of DNA structure stretches back more than 100 years (**Figure 10.1**).

Early Studies of DNA

In 1868, Johann Friedrich Miescher graduated from medical school in Switzerland. Influenced by an uncle who believed that the key to understanding disease lay in

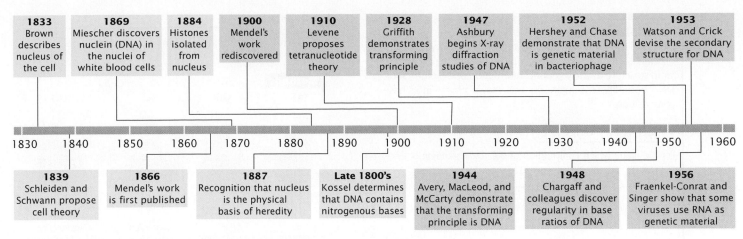

10.1 Many people have contributed to our understanding of the structure of DNA.

the chemistry of tissues, Miescher traveled to Tubingen, Germany, to study under Ernst Felix Hoppe-Seyler, an early leader in the emerging field of biochemistry. Under Hoppe-Seyler's direction, Miescher turned his attention to the chemistry of pus, a substance of clear medical importance. Pus contains white blood cells with large nuclei; Miescher developed a method for isolating these nuclei. The minute amounts of nuclear material that he obtained were insufficient for a thorough chemical analysis, but he did establish that the nuclear material contained a novel substance that was slightly acidic and high in phosphorus. This material, which consisted of DNA and protein, Miescher called *nuclein*. The substance was later renamed *nucleic acid* by one of his students.

By 1887, researchers had concluded that the physical basis of heredity lies in the nucleus. Chromatin was shown to consist of nucleic acid and proteins, but which of these substances is actually the genetic information was not clear. In the late 1800s, further work on the chemistry of DNA was carried out by Albrecht Kossel, who determined that DNA contains four nitrogenous bases: adenine, cytosine, guanine, and thymine (abbreviated A, C, G, and T).

In the early twentieth century, the Rockefeller Institute in New York City became a center for nucleic acid research. Phoebus Aaron Levene joined the Institute in 1905 and spent the next 40 years studying the chemistry of DNA. He discovered that DNA consists of a large number of linked, repeating units, called **nucleotides**; each nucleotide contains a sugar, a phosphate, and a base.

Phosphate ◯ — ⬠ Sugar — ⬛ Base

Nucleotide

Levene incorrectly proposed that DNA consists of a series of four-nucleotide units, each unit containing all four bases—adenine, guanine, cytosine, and thymine—in a fixed

sequence. This concept, known as the tetranucleotide theory, implied that the structure of DNA is not variable enough to be the genetic material. The tetranucleotide theory contributed to the idea that protein is the genetic material because, with its 20 different amino acids, protein structure could be highly variable.

As additional studies of the chemistry of DNA were completed in the 1940s and 1950s, this notion of DNA as a simple, invariant molecule began to change. Erwin Chargaff and his colleagues carefully measured the amounts of the four bases in DNA from a variety of organisms and found that DNA from different organisms varies greatly in base composition. This finding disproved the tetranucleotide theory. They discovered that, within each species, there is some regularity in the ratios of the bases: the amount of adenine is always equal to the amount of thymine (A = T), and the amount of guanine is always equal to the amount of cytosine (G = C; **Table 10.1**). These findings became known as **Chargaff's rules**.

Table 10.1	Base composition of DNA from different sources and ratios of bases						
Source of DNA	A	T	G	C	A/T	G/C	(A + G)/(T + C)
E. coli	26.0	23.9	24.9	25.2	1.09	0.99	1.04
Yeast	31.3	32.9	18.7	17.1	0.95	1.09	1.00
Sea urchin	32.8	32.1	17.7	18.4	1.02	0.96	1.00
Rat	28.6	28.4	21.4	21.5	1.01	1.00	1.00
Human	30.3	30.3	19.5	19.9	1.00	0.98	0.99

DNA As the Source of Genetic Information

While chemists were working out the structure of DNA, biologists were attempting to identify the source of genetic information. Mendel identified the basic rules of heredity in 1866, but he had no idea about the physical nature of hereditary information. By the early 1900s, biologists had concluded that genes resided on chromosomes, which were known to contain both DNA and protein. Two sets of experiments, one conducted on bacteria and the other on viruses, provided pivotal evidence that DNA, rather than protein, was the genetic material.

The discovery of the transforming principle The first clue that DNA was the carrier of hereditary information came with the demonstration that DNA was responsible for a phenomenon called *transformation*. The phenomenon was first observed in 1928 by Fred Griffith, an English physician whose special interest was the bacterium that causes pneumonia, *Streptococcus pneumoniae*. Griffith had succeeded in isolating several different strains of *S. pneumoniae* (type I, II, III, and so forth). In the virulent (disease-causing) forms of a strain, each bacterium is surrounded by a polysaccharide coat, which makes the bacterial colony appear smooth when grown on an agar plate; these forms are referred to as S, for smooth. Griffith found that these virulent forms occasionally mutated to nonvirulent forms, which lack a polysaccharide coat and produce a rough-appearing colony; these forms are referred to as R, for rough.

Griffith observed that small amounts of living type IIIS bacteria injected into mice caused the mice to develop pneumonia and die; on autopsy, he found large amounts of type IIIS bacteria in the blood of the mice (**Figure 10.2a**). When Griffith injected type IIR bacteria into mice, the mice lived, and no bacteria were recovered from their blood (**Figure 10.2b**). Griffith knew that boiling killed all the bacteria and destroyed their virulence; when he injected large amounts of heat-killed type IIIS bacteria into mice, the mice lived and no type IIIS bacteria were recovered from their blood (**Figure 10.2c**).

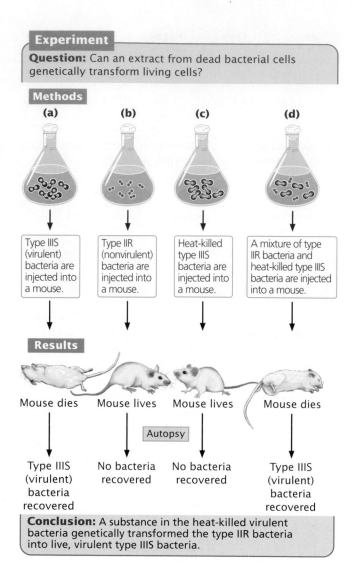

Experiment

Question: Can an extract from dead bacterial cells genetically transform living cells?

Methods

(a) Type IIIS (virulent) bacteria are injected into a mouse.

(b) Type IIR (nonvirulent) bacteria are injected into a mouse.

(c) Heat-killed type IIIS bacteria are injected into a mouse.

(d) A mixture of type IIR bacteria and heat-killed type IIIS bacteria are injected into a mouse.

Results

(a) Mouse dies — Type IIIS (virulent) bacteria recovered

(b) Mouse lives — Autopsy — No bacteria recovered

(c) Mouse lives — No bacteria recovered

(d) Mouse dies — Type IIIS (virulent) bacteria recovered

Conclusion: A substance in the heat-killed virulent bacteria genetically transformed the type IIR bacteria into live, virulent type IIIS bacteria.

10.2 Griffith's experiments demonstrated transformation in bacteria.

The results of these experiments were not unusual. However, Griffith got a surprise when he infected his mice with a small amount of living type IIR bacteria along with a large amount of heat-killed type IIIS bacteria. Because both the type IIR bacteria and the heat-killed type IIIS bacteria were nonvirulent, he expected these mice to live. Surprisingly, 5 days after the injections, the mice became infected with pneumonia and died (**Figure 10.2d**). When Griffith examined blood from the hearts of these mice, he observed live type IIIS bacteria. Furthermore, these bacteria retained their type IIIS characteristics through several generations; so the infectivity was heritable.

Griffith's results had several possible interpretations, all of which he considered. First, it could have been the case that he had not sufficiently sterilized the type IIIS bacteria and thus a few live bacteria remained in the culture. Any live bacteria injected into the mice would have multiplied and caused pneumonia. Griffith knew that this possibility was unlikely, because he had used only heat-killed type IIIS bacteria in the control experiment, and they never produced pneumonia in the mice.

A second interpretation was that the live, type IIR bacteria had mutated to the virulent S form. Such a mutation would cause pneumonia in the mice, but it would produce

type IIS bacteria, not the type IIIS that Griffith found in the dead mice. Many mutations would be required for type II bacteria to mutate to type III bacteria, and the chance of all the mutations occurring simultaneously was impossibly low.

Griffith finally concluded that the type IIR bacteria had somehow been transformed, acquiring the genetic virulence of the dead type IIIS bacteria. This transformation had produced a permanent, genetic change in the bacteria. Although Griffith didn't understand the nature of transformation, he theorized that some substance in the polysaccharide coat of the dead bacteria might be responsible. He called this substance the **transforming principle**. TRY PROBLEM 19 →

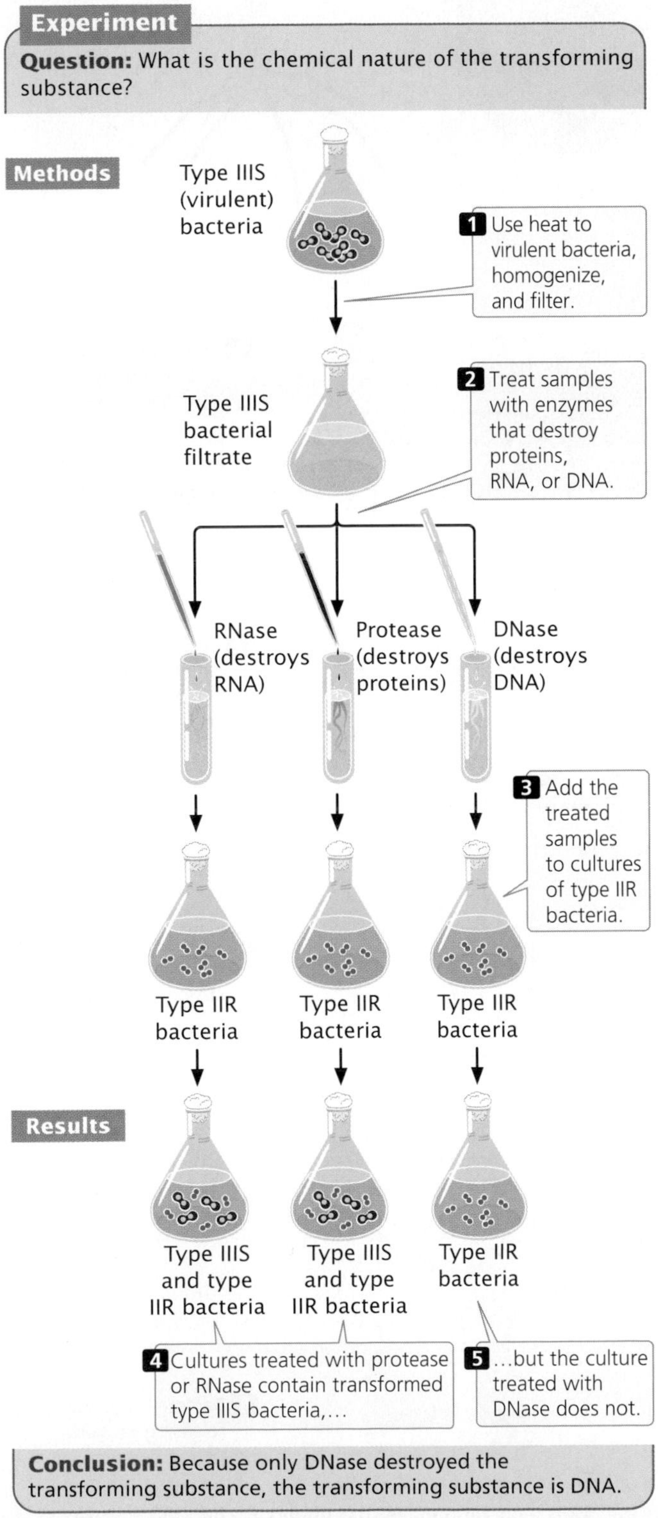

Identification of the transforming principle

At the time of Griffith's report, Oswald Avery (see Figure 10.1) was a microbiologist at the Rockefeller Institute. At first Avery was skeptical but, after other microbiologists successfully repeated Griffith's experiments with other bacteria, Avery set out to identify the nature of the transforming substance.

After 10 years of research, Avery, Colin MacLeod, and Maclyn McCarty succeeded in isolating and purifying the transforming substance. They showed that it had a chemical composition closely matching that of DNA and quite different from that of proteins. Enzymes such as trypsin and chymotrypsin, known to break down proteins, had no effect on the transforming substance. Ribonuclease, an enzyme that destroys RNA, also had no effect. Enzymes capable of destroying DNA, however, eliminated the biological activity of the transforming substance (**Figure 10.3**). Avery, MacLeod, and McCarty showed that purified transforming substance precipitated at about the same rate as purified DNA and that it absorbed ultraviolet light at the same wavelengths as DNA. These results, published in 1944, provided compelling evidence that the transforming principle—and therefore genetic information—resides in DNA. However, new theories in science are rarely accepted on the basis of a single experiment, and many biologists continued to prefer the hypothesis that the genetic material is protein.

CONCEPTS

The process of transformation indicates that some substance—the transforming principle—is capable of genetically altering bacteria. Avery, MacLeod, and McCarty demonstrated that the transforming principle is DNA, providing the first evidence that DNA is the genetic material.

✔ CONCEPT CHECK 3

If Avery, MacLeod, and McCarty had found that samples of heat-killed bacteria treated with RNase and DNase transformed bacteria, but samples treated with protease did not, what conclusion would they have made?

a. Protease carries out transformation.

b. RNA and DNA are the genetic materials.

c. Protein is the genetic material.

d. RNase and DNase are necessary for transformation.

10.3 Avery, MacLeod, and McCarty's experiment revealed the nature of the transforming principle.

The Hershey–Chase experiment A second piece of evidence implicating DNA as the genetic material resulted from a study of the T2 virus conducted by Alfred Hershey and Martha Chase. The T2 virus is a *bacteriophage* (phage) that infects the bacterium *Escherichia coli* (**Figure 10.4a**). As stated in Chapter 8, a phage reproduces by attaching to the outer wall of a bacterial cell and injecting its DNA into the cell, where it replicates and directs the cell to synthesize phage protein. The phage DNA becomes encapsulated within the proteins, producing progeny phages that lyse (break open) the cell and escape (**Figure 10.4b**).

At the time of the Hershey–Chase study (their paper was published in 1952), biologists did not understand exactly how phages reproduce. What they did know was that the T2 phage is approximately 50% protein and 50% nucleic acid, that a phage infects a cell by first attaching to the cell wall, and that progeny phages are ultimately produced within the cell. Because the progeny carry the same traits as the infecting phage, genetic material from the infecting phage must be transmitted to the progeny, but how this genetic transmission takes place was unknown.

Hershey and Chase designed a series of experiments to determine whether the phage *protein* or the phage *DNA* is transmitted in phage reproduction. To follow the fate of protein and DNA, they used radioactive forms, or **isotopes**, of phosphorus and sulfur. A radioactive isotope can be used as a tracer to identify the location of a specific molecule because any molecule containing the isotope will be radioactive and therefore easily detected. DNA contains phosphorus but not sulfur; so Hershey and Chase used ^{32}P to follow phage DNA during reproduction. Protein contains sulfur but not phosphorus; so they used ^{35}S to follow the protein.

Hershey and Chase grew one batch of *E. coli* in a medium containing ^{32}P and infected the bacteria with T2 phage so that all the new phages would have DNA labeled with ^{32}P (**Figure 10.5**). They grew a second batch of *E. coli* in a medium containing ^{35}S and infected these bacteria with T2 phage so that all these new phages would have protein labeled with ^{35}S. Hershey and Chase then infected separate batches of unlabeled *E. coli* with the ^{35}S- and ^{32}P-labeled phages. After allowing time for the phages to infect the cells, they placed the *E. coli* cells in a blender and sheared off the then-empty protein coats (ghosts) from the cell walls. They separated out the protein coats and cultured the infected bacterial cells.

When phages labeled with ^{35}S infected the bacteria, most of the radioactivity was detected in the protein ghosts and little was detected in the cells. Furthermore, when new phages emerged from the cell, they contained almost no ^{35}S (see Figure 10.5). This result indicated that, although the protein component of a phage is necessary for infection, it does not enter the cell and is not transmitted to progeny phages.

In contrast, when Hershey and Chase infected bacteria with ^{32}P-labeled phages and removed the protein ghosts, the bacteria were still radioactive. Most significantly, after the cells lysed and new progeny phages emerged, many of these

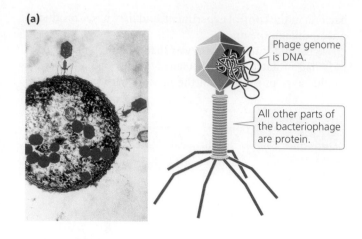

(a)

Phage genome is DNA.

All other parts of the bacteriophage are protein.

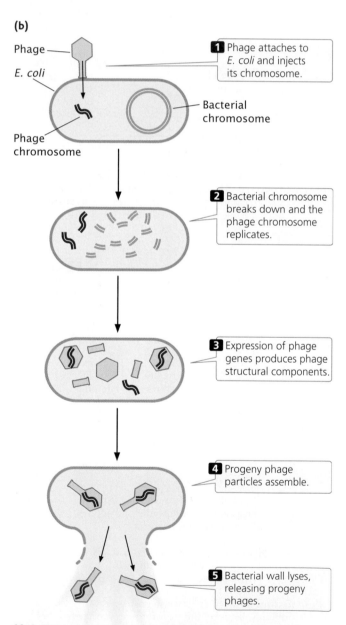

(b)

Phage

E. coli

Phage chromosome

1 Phage attaches to *E. coli* and injects its chromosome.

Bacterial chromosome

2 Bacterial chromosome breaks down and the phage chromosome replicates.

3 Expression of phage genes produces phage structural components.

4 Progeny phage particles assemble.

5 Bacterial wall lyses, releasing progeny phages.

10.4 T2 is a bacteriophage that infects *E. coli*. (a) T2 phage. (b) Its life cycle. [Micrograph: © Lee D. Simon/Photo Researchers.]

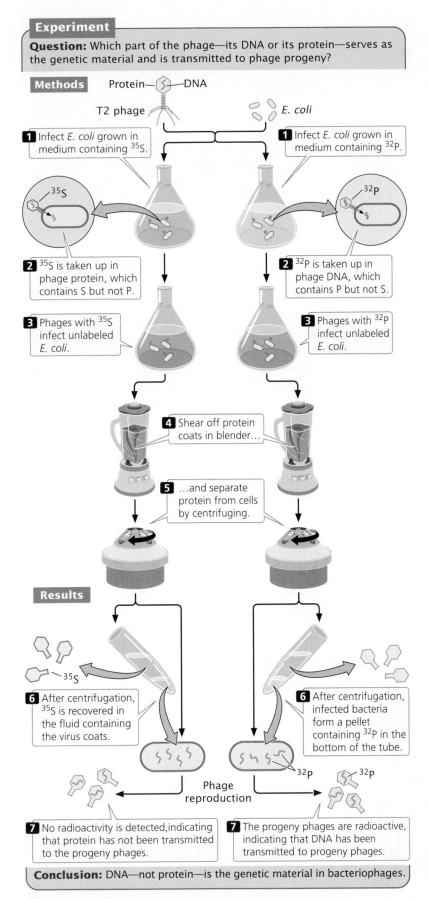

Experiment

Question: Which part of the phage—its DNA or its protein—serves as the genetic material and is transmitted to phage progeny?

Methods   Protein—⬡—DNA

T2 phage   *E. coli*

1 Infect *E. coli* grown in medium containing ^{35}S.

1 Infect *E. coli* grown in medium containing ^{32}P.

^{35}S

^{32}P

2 ^{35}S is taken up in phage protein, which contains S but not P.

2 ^{32}P is taken up in phage DNA, which contains P but not S.

3 Phages with ^{35}S infect unlabeled *E. coli.*

3 Phages with ^{32}P infect unlabeled *E. coli.*

4 Shear off protein coats in blender…

5 …and separate protein from cells by centrifuging.

Results

^{35}S

6 After centrifugation, ^{35}S is recovered in the fluid containing the virus coats.

6 After centrifugation, infected bacteria form a pellet containing ^{32}P in the bottom of the tube.

^{32}P

Phage reproduction

^{32}P

7 No radioactivity is detected, indicating that protein has not been transmitted to the progeny phages.

7 The progeny phages are radioactive, indicating that DNA has been transmitted to progeny phages.

Conclusion: DNA—not protein—is the genetic material in bacteriophages.

10.5 Hershey and Chase demonstrated that DNA carries the genetic information in bacteriophages.

phages emitted radioactivity from ^{32}P, demonstrating that DNA from the infecting phages had been passed on to the progeny (see Figure 10.5). These results confirmed that DNA, not protein, is the genetic material of phages. **TRY PROBLEM 22** →

CONCEPTS

Using radioactive isotopes, Hershey and Chase traced the movement of DNA and protein during phage infection. They demonstrated that DNA, not protein, enters the bacterial cell during phage reproduction and that only DNA is passed on to progeny phages.

✔ CONCEPT CHECK 4

Could Hershey and Chase have used a radioactive isotope of carbon instead of ^{32}P? Why or why not?

Watson and Crick's Discovery of the Three-Dimensional Structure of DNA

The experiments on the nature of the genetic material set the stage for one of the most important advances in the history of biology—the discovery of the three-dimensional structure of DNA by James Watson and Francis Crick in 1953.

Before Watson and Crick's breakthrough, much of the basic chemistry of DNA had already been determined by Miescher, Kossel, Levene, Chargaff, and others, who had established that DNA consists of nucleotides and that each nucleotide contains a sugar, a base, and a phosphate group. However, how the nucleotides fit together in the three-dimensional structure of the molecule was not at all clear.

In 1947, William Ashbury began studying the three-dimensional structure of DNA by using a technique called **X-ray diffraction** (**Figure 10.6**), in which X-rays beamed at a molecule are reflected in specific patterns that reveal aspects of the structure of the molecule. But his diffraction pictures did not provide enough resolution to reveal the structure. A research group at King's College in London, led by Maurice Wilkins and Rosalind Franklin, also used X-ray diffraction to study DNA and obtained strikingly better pictures of the molecule. Wilkins and Franklin, however, were unable to develop a complete structure of the molecule; their progress was impeded by the personal discord that existed between them.

Watson and Crick investigated the structure of DNA, not by collecting new data but by using all available information about the chemistry of

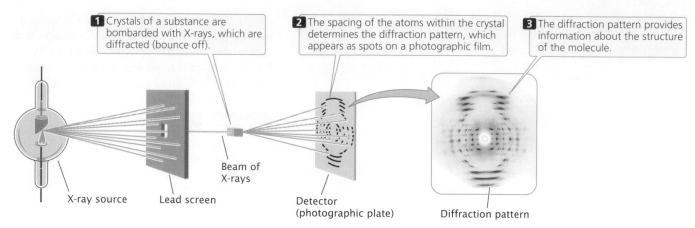

1 Crystals of a substance are bombarded with X-rays, which are diffracted (bounce off).

2 The spacing of the atoms within the crystal determines the diffraction pattern, which appears as spots on a photographic film.

3 The diffraction pattern provides information about the structure of the molecule.

X-ray source   Lead screen

Beam of X-rays

Detector (photographic plate)

Diffraction pattern

10.6 X-ray diffraction provides information about the structures of molecules. [Photograph from M. H. F. Wilkins, Department of Biophysics, King's College, University of London.]

DNA to construct molecular models (**Figure 10.7**). By applying the laws of structural chemistry, they were able to limit the number of possible structures that DNA could assume. They tested various structures by building models made of wire and metal plates. With their models, they were able to see whether a structure was compatible with chemical principles and with the X-ray images.

The key to solving the structure came when Watson recognized that an adenine base could bond with a thymine base and that a guanine base could bond with a cytosine base; these pairings accounted for the base ratios that Chargaff had discovered earlier. The model developed by Watson and Crick showed that DNA consists of two strands of nucleotides wound around each other to form a

10.7 Watson (left) and Crick (right) provided a three-dimensional model of the structure of DNA. [A. Barrington Brown/Science Photo Library/Photo Researchers.]

right-handed helix, with the sugars and phosphates on the outside and the bases in the interior. They recognized that the double-stranded structure of DNA with its specific base pairing provided an elegant means by which the DNA can be replicated. Watson and Crick published an electrifying description of their model in *Nature* in 1953. At the same time, Wilkins and Franklin published their X-ray diffraction data, which demonstrated experimentally the theory that DNA was helical in structure.

Many have called the solving of DNA's structure the most important biological discovery of the twentieth century. For their discovery, Watson and Crick, along with Maurice Wilkins, were awarded a Nobel Prize in 1962. Rosalind Franklin had died of cancer in 1957 and thus could not be considered a candidate for the shared prize.

After the discovery of DNA's structure, much research focused on how genetic information is encoded within the base sequence and how this information is copied and expressed. Even today, the details of DNA structure and function continue to be the subject of active research.

CONCEPTS

By collecting existing information about the chemistry of DNA and building molecular models, Watson and Crick were able to discover the three-dimensional structure of the DNA molecule.

✔ CONCEPT CHECK 5

What did Watson and Crick use to help solve the structure of DNA?

a. X-ray diffraction.   c. Models of DNA

b. Laws of structural chemistry   d. All the above

RNA As Genetic Material

In most organisms, DNA carries the genetic information. However, a few viruses use RNA, not DNA, as their genetic material. This fact was demonstrated in 1956 by Heinz Fraenkel-Conrat and Bea Singer, who worked with tobacco

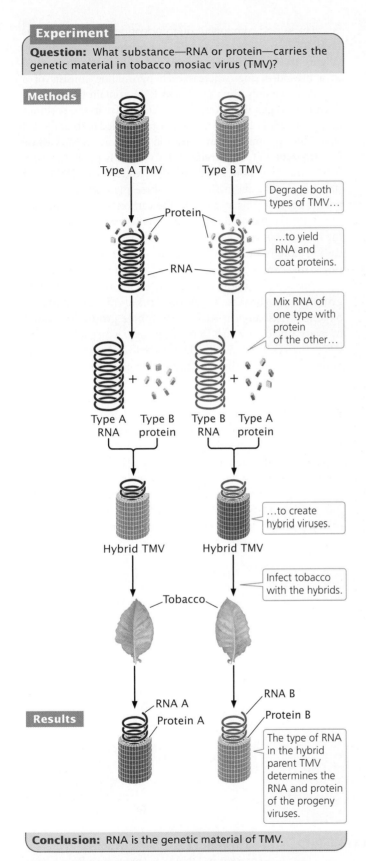

Experiment

Question: What substance—RNA or protein—carries the genetic material in tobacco mosaic virus (TMV)?

Methods

Type A TMV   Type B TMV

Degrade both types of TMV...

Protein

...to yield RNA and coat proteins.

RNA

Mix RNA of one type with protein of the other...

Type A RNA   Type B protein   Type B RNA   Type A protein

...to create hybrid viruses.

Hybrid TMV   Hybrid TMV

Infect tobacco with the hybrids.

Tobacco

Results

RNA A
Protein A

RNA B
Protein B

The type of RNA in the hybrid parent TMV determines the RNA and protein of the progeny viruses.

Conclusion: RNA is the genetic material of TMV.

10.8 Fraenkel-Conrat and Singer's experiment demonstrated that RNA in the tobacco mosaic virus carries the genetic information.

mosaic virus (TMV), a virus that infects and causes disease in tobacco plants (**Figure 10.8**). The tobacco mosaic virus possesses a single molecule of RNA surrounded by a helically arranged cylinder of protein molecules. Fraenkel-Conrat found that, after separating the RNA and protein of TMV, he could remix the RNA and protein of different strains of TMV and obtain intact, infectious viral particles.

With Singer, Fraenkel-Conrat then created hybrid viruses by mixing RNA and protein from different strains of TMV. When these hybrid viruses infected tobacco leaves, new viral particles were produced. The new viral progeny were identical with the strain from which the RNA had been isolated and did not exhibit the characteristics of the strain that donated the protein. These results showed that RNA carries the genetic information in TMV.

Also in 1956, Alfred Gierer and Gerhard Schramm demonstrated that just RNA isolated from TMV is sufficient to infect tobacco plants and direct the production of new TMV particles. This finding confirmed that RNA carries genetic instructions. **TRY PROBLEM 18** ➡️

CONCEPTS

RNA serves as the genetic material in some viruses.

10.3 DNA Consists of Two Complementary and Antiparallel Nucleotide Strands That Form a Double Helix

DNA, though relatively simple in structure, has an elegance and beauty unsurpassed by other large molecules. It is useful to consider the structure of DNA at three levels of increasing complexity, known as the primary, secondary, and tertiary structures of DNA. The primary structure of DNA refers to its nucleotide structure and how the nucleotides are joined together. The secondary structure refers to DNA's stable three-dimensional configuration, the helical structure worked out by Watson and Crick. In Chapter 11, we will consider DNA's tertiary structures, which are the complex packing arrangements of double-stranded DNA in chromosomes.

The Primary Structure of DNA

The primary structure of DNA consists of a string of nucleotides joined together by phosphodiester linkages.

Nucleotides DNA is typically a very long molecule and is therefore termed a macromolecule. For example, within each human chromosome is a single DNA molecule that, if stretched out straight, would be several centimeters in length, thousands of times longer than the cell itself. In spite of its large size, DNA has a quite simple structure: it is a polymer—that is, a chain made up of many repeating units

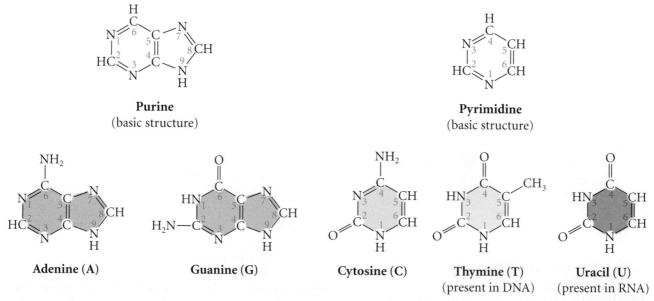

10.9 A nucleotide contains either a ribose sugar (in RNA) or a deoxyribose sugar (in DNA). The carbon atoms are assigned primed numbers.

linked together. The repeating units of DNA are *nucleotides*, each comprising three parts: (1) a sugar, (2) a phosphate, and (3) a nitrogen-containing base.

The sugars of nucleic acids—called pentose sugars—have five carbon atoms, numbered $1'$, $2'$, $3'$, and so forth (**Figure 10.9**). The sugars of DNA and RNA are slightly different in structure. RNA's sugar, called **ribose**, has a hydroxyl group (–OH) attached to the $2'$-carbon atom, whereas DNA's sugar, or **deoxyribose**, has a hydrogen atom (–H) at this position and therefore contains one oxygen atom fewer overall. This difference gives rise to the names ribonucleic acid (RNA) and *deoxy*ribonucleic acid (DNA). This minor chemical difference is recognized by all the cellular enzymes that interact with DNA or RNA, thus yielding specific functions for each nucleic acid. Furthermore, the additional oxygen atom in the RNA nucleotide makes it more reactive and less chemically stable than DNA. For this reason, DNA is better suited to serve as the long-term repository of genetic information.

The second component of a nucleotide is its **nitrogenous base**, which may be of two types—a **purine** or a **pyrimidine** (**Figure 10.10**). Each purine consists of a six-sided ring attached to a five-sided ring, whereas each pyrimidine consists of a six-sided ring only. Both DNA and RNA contain two purines, **adenine** and **guanine** (A and G), which differ in the positions of their double bonds and in the groups attached to the six-sided ring. Three pyrimidines are common in nucleic acids: **cytosine** (C), **thymine** (T), and **uracil** (U). Cytosine is present in both DNA and RNA; however, thymine is restricted to DNA, and uracil is found only in RNA. The three pyrimidines differ in the groups or atoms attached to the carbon atoms of the ring and in the number of double bonds in the ring. In a nucleotide, the nitrogenous base always forms a covalent bond with the $1'$-carbon atom of the sugar (see Figure 10.9). A deoxyribose or a ribose sugar and a base together are referred to as a **nucleoside**.

The third component of a nucleotide is the **phosphate group**, which consists of a phosphorus atom bonded to four oxygen atoms (**Figure 10.11**). Phosphate groups are found in every nucleotide and frequently carry a negative charge, which makes DNA acidic. The phosphate group is always bonded to the $5'$-carbon atom of the sugar (see Figure 10.9) in a nucleotide.

The DNA nucleotides are properly known as **deoxyribonucleotides** or deoxyribonucleoside $5'$-monophosphates. Because there are four types of bases, there are four different kinds of DNA nucleotides (**Figure 10.12**). The equivalent RNA nucleotides are termed **ribonucleotides** or ribonucleoside $5'$-monophosphates. RNA molecules sometimes contain additional rare bases, which are modified forms of the four common bases. These modified bases will be discussed in more detail when we examine the function of RNA molecules in Chapter 14. **TRY PROBLEM 23** →

Purine
(basic structure)

Pyrimidine
(basic structure)

Adenine (A)

Guanine (G)

Cytosine (C)

Thymine (T)
(present in DNA)

Uracil (U)
(present in RNA)

10.10 A nucleotide contains either a purine or a pyrimidine base. The atoms of the rings in the bases are assigned unprimed numbers.

Phosphate

10.11 A nucleotide contains a phosphate group.

CONCEPTS

The primary structure of DNA consists of a string of nucleotides. Each nucleotide consists of a five-carbon sugar, a phosphate, and a base. There are two types of DNA bases: purines (adenine and guanine) and pyrimidines (thymine and cytosine).

✔ CONCEPT CHECK 6

How do the sugars of RNA and DNA differ?

a. RNA has a six-carbon sugar; DNA has a five-carbon sugar.

b. The sugar of RNA has a hydroxyl group that is not found in the sugar of DNA.

c. RNA contains uracil; DNA contains thymine.

d. DNA's sugar has a phosphorus atom; RNA's sugar does not.

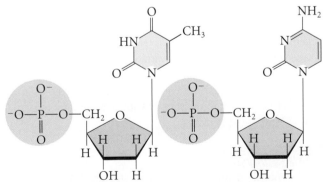

Deoxyadenosine 5′-monophosphate (dAMP)　　**Deoxyguanosine 5′-monophosphate (dGMP)**

Deoxythymidine 5′-monophosphate (dTMP)　　**Deoxycytidine 5′-monophosphate (dCMP)**

10.12 There are four types of DNA nucleotides.

Polynucleotide strands DNA is made up of many nucleotides connected by covalent bonds, which join the 5′-phosphate group of one nucleotide to the 3′-carbon atom of the next nucleotide (**Figure 10.13**; note that the structures shown in Figure 10.13 are flattened into two dimensions but the molecule itself is three dimensional, as shown in Figure 10.14). These bonds, called **phosphodiester linkages**, are strong covalent bonds; a series of nucleotides linked in this way constitutes a **polynucleotide strand**. The backbone of the polynucleotide strand is composed of alternating sugars and phosphates; the bases project away from the long axis of the strand. The negative charges of the phosphate groups are frequently neutralized by the association of positive charges on proteins, metals, or other molecules.

An important characteristic of the polynucleotide strand is its direction, or polarity. At one end of the strand, a free (meaning that it's unattached on one side) phosphate group is attached to the 5′-carbon atom of the sugar in the nucleotide. This end of the strand is therefore referred to as the **5′ end**. The other end of the strand, referred to as the **3′ end**, has a free OH group attached to the 3′-carbon atom of the sugar.

RNA nucleotides also are connected by phosphodiester linkages to form similar polynucleotide strands (see Figure 10.13).

CONCEPTS

The nucleotides of DNA are joined in polynucleotide strands by phosphodiester bonds that connect the 3′-carbon atom of one nucleotide to the 5′-phosphate group of the next. Each polynucleotide strand has polarity, with a 5′ direction and a 3′ direction.

Secondary Structures of DNA

The secondary structure of DNA refers to its three-dimensional configuration—its fundamental helical structure. DNA's secondary structure can assume a variety of configurations, depending on its base sequence and the conditions in which it is placed.

The double helix A fundamental characteristic of DNA's secondary structure is that it consists of two polynucleotide strands wound around each other—it's a *double* helix. The sugar–phosphate linkages are on the outside of the helix, and the bases are stacked in the interior of the molecule (see Figure 10.13). The two polynucleotide strands run in opposite directions—they are **antiparallel**, which means that the 5′ end of one strand is opposite the 3′ end of the other strand.

The strands are held together by two types of molecular forces. Hydrogen bonds link the bases on opposite strands (see Figure 10.13). These bonds are relatively weak compared with the covalent phosphodiester bonds that connect the sugar and phosphate groups of adjoining nucleotides on

DNA polynucleotide strand

RNA polynucleotide strand

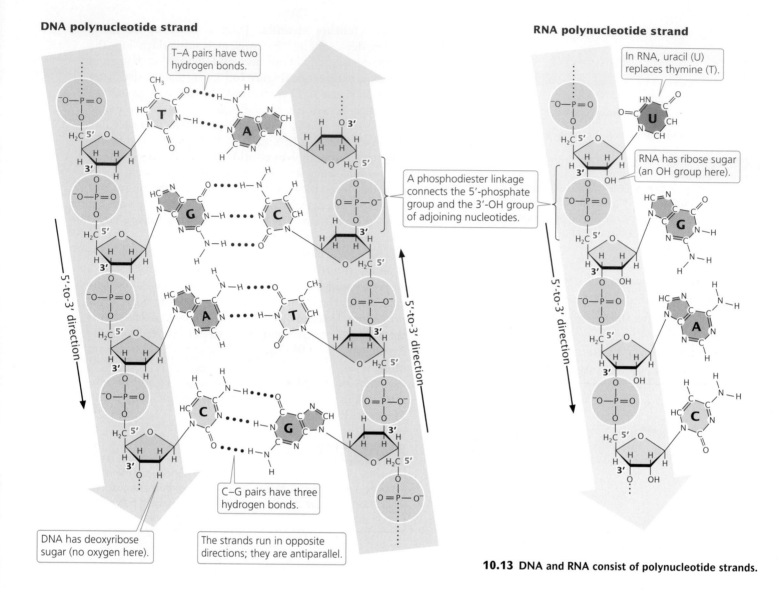

T–A pairs have two hydrogen bonds.

In RNA, uracil (U) replaces thymine (T).

A phosphodiester linkage connects the 5'-phosphate group and the 3'-OH group of adjoining nucleotides.

RNA has ribose sugar (an OH group here).

5'-to-3' direction

5'-to-3' direction

5'-to-3' direction

C–G pairs have three hydrogen bonds.

DNA has deoxyribose sugar (no oxygen here).

The strands run in opposite directions; they are antiparallel.

10.13 DNA and RNA consist of polynucleotide strands.

the same strand. As we will see, several important functions of DNA require the separation of its two nucleotide strands, and this separation can be readily accomplished because of the relative ease of breaking and reestablishing the hydrogen bonds.

The nature of the hydrogen bond imposes a limitation on the types of bases that can pair. Adenine normally pairs only with thymine through two hydrogen bonds, and cytosine normally pairs only with guanine through three hydrogen bonds (see Figure 10.13). Because three hydrogen bonds form between C and G and only two hydrogen bonds form between A and T, C–G pairing is stronger than A–T pairing. The specificity of the base pairing means that, wherever there is an A on one strand, there must be a T in the corresponding position on the other strand, and, wherever there is a G on one strand, a C must be on the other. The two polynucleotide strands of a DNA molecule are therefore not identical but are **complementary DNA strands**. The

complementary nature of the two nucleotide strands provides for efficient and accurate DNA replication, as will be discussed in Chapter 12.

The second force that holds the two DNA strands together is the interaction between the stacked base pairs. These stacking interactions contribute to the stability of the DNA molecule but do not require that any particular base follow another. Thus, the base sequence of the DNA molecule is free to vary, allowing DNA to carry genetic information. **TRY PROBLEMS 28 AND 33** →

CONCEPTS

DNA consists of two polynucleotide strands. The sugar–phosphate groups of each polynucleotide strand are on the outside of the molecule, and the bases are in the interior. Hydrogen bonding joins the bases of the two strands: guanine pairs with cytosine, and adenine pairs with thymine. The two polynucleotide strands of a DNA molecule are complementary and antiparallel.

✔ CONCEPT CHECK 7

The antiparallel nature of DNA refers to

a. its charged phosphate groups.

b. the pairing of bases on one strand with bases on the other strand.

c. the formation of hydrogen bonds between bases from opposite strands.

d. the opposite direction of the two strands of nucleotides.

Different secondary structures   As we have seen, DNA normally consists of two polynucleotide strands that are antiparallel and complementary (exceptions are single-stranded DNA molecules in a few viruses). The precise three-dimensional shape of the molecule can vary, however, depending on the conditions in which the DNA is placed and, in some cases, on the base sequence itself.

The three-dimensional structure of DNA described by Watson and Crick is termed the **B-DNA** structure (**Figure 10.14**). This structure exists when plenty of water surrounds the

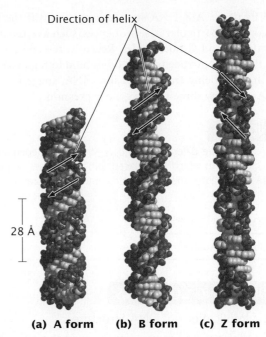

Direction of helix

(a) A form   (b) B form   (c) Z form

10.15 DNA can assume several different secondary structures.
[After J. M. Berg, J. L. Tymoczko, and L. Stryer, *Biochemistry*, 6th ed. (New York: W. H. Freeman and Company, 2002), pp. 785 and 787.]

molecule and there is no unusual base sequence in the DNA—conditions that are likely to be present in cells. The B-DNA structure is the most stable configuration for a random sequence of nucleotides under physiological conditions, and most evidence suggests that it is the predominate structure in the cell.

B-DNA is an alpha helix, meaning that it has a right-handed, or clockwise, spiral. There are approximately 10 base pairs (bp) per 360-degree rotation of the helix; so each base pair is twisted 36 degrees relative to the adjacent bases (see Figure 10.14b). The base pairs are 0.34 nanometer (nm) apart; so each complete rotation of the molecule encompasses 3.4 nm. The diameter of the helix is 2 nm, and the bases are perpendicular to the long axis of the DNA molecule. A space-filling model shows that B-DNA has a slim and elongated structure (see Figure 10.14a). The spiraling of the nucleotide strands creates major and minor grooves in the helix, features that are important for the binding of some proteins that regulate the expression of genetic information (see Chapter 16).

Another secondary structure that DNA can assume is the **A-DNA** structure, which exists if less water is present. Like B-DNA, A-DNA is an alpha (right-handed) helix (**Figure 10.15a**), but it is shorter and wider than B-DNA (**Figure 10.15b**) and its bases are tilted away from the main axis of the molecule. There is little evidence that A-DNA exists under physiological conditions.

A radically different secondary structure, called **Z-DNA** (**Figure 10.15c**), forms a left-handed helix. In this form, the sugar–phosphate backbone zigzags back and forth, giving

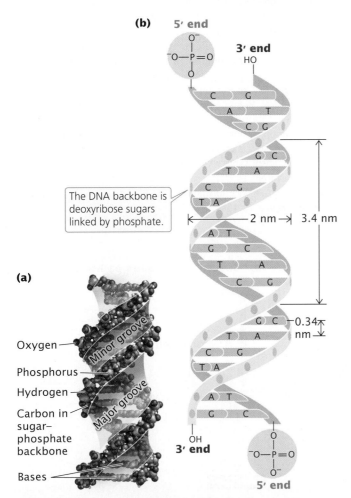

(b)

5′ end

3′ end

HO

The DNA backbone is deoxyribose sugars linked by phosphate.

2 nm   3.4 nm

0.34 nm

3′ end

5′ end

(a)

Oxygen

Phosphorus

Hydrogen

Carbon in sugar–phosphate backbone

Bases

Minor groove

Major groove

10.14 B-DNA consists of an alpha helix with approximately 10 bases per turn. (a) Space-filling model of B-DNA showing major and minor grooves. (b) Diagrammatic representation.

28 Å

rise to its name. A Z-DNA structure can result if the molecule contains particular base sequences, such as stretches of alternating C and G nucleotides. Recently, researchers have found that Z-DNA-specific antibodies bind to regions of the DNA that are being transcribed into RNA, suggesting that Z-DNA may play some role in gene expression.

CONCEPTS

DNA can assume different secondary structures, depending on the conditions in which it is placed and on its base sequence. B-DNA is thought to be the most common configuration in the cell.

✔ CONCEPT CHECK 8

How does Z-DNA differ from B-DNA?

CONNECTING CONCEPTS

Genetic Implications of DNA Structure

Watson and Crick's great contribution was their elucidation of the genotype's chemical structure, making it possible for geneticists to begin to examine genes directly, instead of looking only at the phenotypic consequences of gene action. The determination of the structure of DNA led to the birth of molecular genetics—the study of the chemical and molecular nature of genetic information.

Watson and Crick's structure did more than just create the potential for molecular genetic studies; it was an immediate source of insight into key genetic processes. At the beginning of this chapter, three fundamental properties of the genetic material were identified. First, it must be capable of carrying large amounts of information; so it must vary in structure. Watson and Crick's model suggested that genetic instructions are encoded in the base sequence, the only variable part of the molecule.

The second necessary property of genetic material is its ability to replicate faithfully. The complementary polynucleotide strands of DNA make this replication possible. Watson and Crick proposed that, in replication, the two polynucleotide strands unzip, breaking the weak hydrogen bonds between the two strands, and each strand serves as a template on which a new strand is synthesized. The specificity of the base pairing means that only one possible sequence of bases—the complementary sequence—can be synthesized from each template. Newly replicated double-stranded DNA molecules will therefore be identical with the original double-stranded DNA molecule (see Chapter 12 on DNA replication).

The third essential property of genetic material is the ability to translate its instructions into the phenotype. DNA expresses its genetic instructions by first transferring its information to an RNA molecule, in a process termed **transcription** (see Chapter 13). The term *transcription* is appropriate because, although the information is transferred from DNA to RNA, the information remains in the language of nucleic acids. The RNA molecule then transfers the genetic information to a protein by specifying its amino acid sequence. This process is termed **translation** (see Chapter 15) because the information must be *translated* from the language of nucleotides into the language of amino acids.

We can now identify three major pathways of information flow in the cell (**Figure 10.16a**): in **replication**, information passes from one DNA molecule to other DNA molecules; in transcription, information passes from DNA to RNA; and, in translation, information passes from RNA to protein. This concept of information flow was formalized by Francis Crick in a concept that he called the **central dogma** of molecular biology. The central dogma states that genetic information passes from DNA to protein in a one-way information pathway. We now realize, however, that the central dogma is an oversimplification. In addition to the three general information pathways of replication, transcription, and translation, other transfers may take place in certain organisms or under special circumstances. Retroviruses (see Chapter 8) and some transposable elements (see Chapter 11) transfer information from RNA to DNA (in **reverse transcription**), and some RNA viruses transfer information from RNA to RNA (in **RNA replication; Figure 10.16b**).

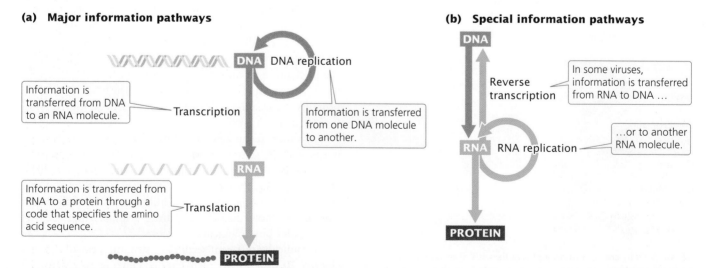

(a) Major information pathways

DNA — DNA replication

Information is transferred from DNA to an RNA molecule. — Transcription

Information is transferred from one DNA molecule to another.

RNA

Information is transferred from RNA to a protein through a code that specifies the amino acid sequence. — Translation

PROTEIN

(b) Special information pathways

DNA

Reverse transcription — In some viruses, information is transferred from RNA to DNA …

RNA RNA replication — …or to another RNA molecule.

PROTEIN

10.16 Pathways of information transfer within the cell.

10.4 Special Structures Can Form in DNA and RNA

Sequences *within* a single strand of nucleotides may be complementary to each other and can pair by forming hydrogen bonds, producing double-stranded regions (**Figure 10.17**). This internal base pairing imparts a secondary structure to a single-stranded molecule. One common type of secondary structure found in single strands of nucleotides is a **hairpin**, which forms when sequences of nucleotides on the same strand are inverted complements (see Figure 10.17a). A hairpin consists of a region of paired bases (the stem) and sometimes includes intervening unpaired bases (the loop). When the complementary sequences are contiguous, the hairpin has a stem but no loop (see Figure 10.17b). RNA molecules may contain numerous hairpins, allowing them to fold up into complex structures (see Figure 10.17c). Secondary structures play important roles in the functions of many RNA molecules, as we will see in Chapters 14 and 15. **TRY PROBLEM 36 ➡**

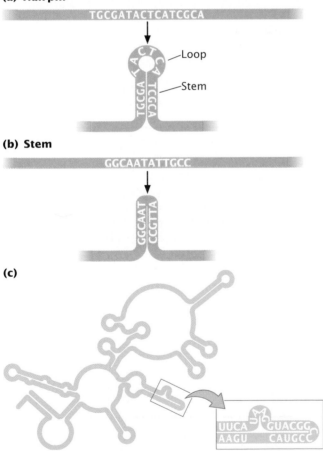

(a) Hairpin

(b) Stem

(c)

10.17 Both DNA and RNA can form special secondary structures. (a) A hairpin, consisting of a region of paired bases (which form the stem) and a region of unpaired bases between the complementary sequences (which form a loop at the end of the stem). (b) A stem with no loop. (c) Secondary structure of RNA component of RNase P of *E. coli*. RNA molecules often have complex secondary structures.

CONCEPTS

In DNA and RNA, base pairing between nucleotides on the same strand produces special secondary structures such as hairpins.

✔ **CONCEPT CHECK 9**

Hairpins are formed in DNA as a result of

a. sequences on the same strand that are inverted and complementary.

b. sequences on the opposite strand that are complements.

c. sequences on the same strand that are identical.

d. sequences on the opposite strand that are identical.

The primary structure of DNA can be modified in various ways. One such modification is **DNA methylation**, a process in which methyl groups (–CH₃) are added (by specific enzymes) to certain positions on the nucleotide bases. Bacterial DNA is frequently methylated to distinguish it from foreign, unmethylated DNA that may be introduced by viruses; bacteria use proteins called restriction enzymes to cut up any unmethylated viral DNA (see Chapter 19). In eukaryotic cells, methylation is often related to gene expression. Sequences that are methylated typically show low levels of transcription while sequences lacking methylation are actively being transcribed (see Chapter 17). Methylation can also affect the three-dimensional structure of the DNA molecule.

Adenine and cytosine are commonly methylated in bacteria. In eukaryotic DNA, cytosine bases are often methylated to form **5-methylcytosine** (**Figure 10.18**). The extent of cytosine methylation varies among eukaryotic organisms; in most animal cells, about 5% of the cytosine bases are methylated, but there is no methylation of cytosine in yeast and more than 50% of the cytosine bases in some plants are methylated. Why eukaryotic organisms differ so widely in their degree of methylation is not clear.

CONCEPTS

Methyl groups may be added to certain bases in DNA, depending on the positions of the bases in the molecule. Both prokaryotic and eukaryotic DNA can be methylated. In eukaryotes, cytosine bases are most often methylated to form 5-methylcytosine and methylation is often related to gene expression.

5-Methylcytosine

10.18 In eukaryotic DNA, cytosine bases are often methylated to form 5-methylcytosine.

CONCEPTS SUMMARY

- Genetic material must contain complex information, be replicated accurately, and have the capacity to be translated into the phenotype.

- Evidence that DNA is the source of genetic information came from the finding by Avery, MacLeod, and McCarty that transformation depended on DNA and from the demonstration by Hershey and Chase that viral DNA is passed on to progeny phages. The results of experiments with tobacco mosaic virus showed that RNA carries genetic information in some viruses.

- James Watson and Francis Crick proposed a new model for the three-dimensional structure of DNA in 1953.

- A DNA nucleotide consists of a deoxyribose sugar, a phosphate group, and a nitrogenous base. RNA consists of a ribose sugar, a phosphate group, and a nitrogenous base.

- The bases of a DNA nucleotide are of two types: purines (adenine and guanine) and pyrimidines (cytosine and thymine). RNA contains the pyrimidine uracil instead of thymine.

- Nucleotides are joined together by phosphodiester linkages in a polynucleotide strand. Each polynucleotide strand has a free phosphate group at its 5' end and a free hydroxyl group at its 3' end.

- DNA consists of two nucleotide strands that wind around each other to form a double helix. The sugars and phosphates lie on the outside of the helix, and the bases are stacked in the interior. The two strands are joined together by hydrogen bonding between bases in each strand. The two strands are antiparallel and complementary.

- DNA molecules can form a number of different secondary structures, depending on the conditions in which the DNA is placed and on its base sequence.

- The structure of DNA has several important genetic implications. Genetic information resides in the base sequence of DNA, which ultimately specifies the amino acid sequence of proteins. Complementarity of the bases on DNA's two strands allows genetic information to be replicated.

- The central dogma of molecular biology proposes that information flows in a one-way direction, from DNA to RNA to protein. Exceptions to the central dogma are now known.

- Pairing between bases on the same nucleotide strand can lead to hairpins and other secondary structures.

- DNA may be modified by the addition of methyl groups to the nucleotide bases.

IMPORTANT TERMS

nucleotide (p. 273)
Chargaff's rules (p. 273)
transforming principle (p. 275)
isotope (p. 276)
X-ray diffraction (p. 277)
ribose (p. 280)
deoxyribose (p. 280)
nitrogenous base (p. 280)
purine (p. 280)
pyrimidine (p. 280)
adenine (A) (p. 280)
guanine (G) (p. 280)
cytosine (C) (p. 280)

thymine (T) (p. 280)
uracil (U) (p. 280)
nucleoside (p. 280)
phosphate group (p. 280)
deoxyribonucleotide (p. 280)
ribonucleotide (p. 280)
phosphodiester linkage (p. 281)
polynucleotide strand (p. 281)
5' end (p. 281)
3' end (p. 281)
antiparallel (p. 281)
complementary DNA strands (p. 282)
B-DNA (p. 283)

A-DNA (p. 283)
Z-DNA (p. 283)
transcription (p. 284)
translation (p. 284)
replication (p. 284)
central dogma (p. 284)
reverse transcription (p. 284)
RNA replication (p. 284)
hairpin (p. 285)
DNA methylation (p. 285)
5-methylcytosine (p. 285)

ANSWERS TO CONCEPT CHECKS

1. Without knowledge of the structure of DNA, an understanding of how genetic information was encoded or expressed was impossible.

2. c

3. c

4. No, because carbon is found in both protein and nucleic acid.

5. d

6. b

7. d

8. Z-DNA has a left-handed helix; B-DNA has a right-handed helix. The sugar–phosphate backbone of Z-DNA zigzags back and forth, whereas the sugar–phosphate backbone of B-DNA forms a smooth continuous ribbon.

9. a

WORKED PROBLEMS

1. The percentage of cytosine in a double-stranded DNA molecule is 40%. What is the percentage of thymine?

• Solution

In double-stranded DNA, A pairs with T, whereas G pairs with C; so the percentage of A equals the percentage of T, and the percentage of G equals the percentage of C. If C = 40%, then G also must be 40%. The total percentage of C + G is therefore 40% + 40% = 80%. All the remaining bases must be either A or T; so the total percentage of A + T = 100% − 80% = 20%; because the percentage of A equals the percentage of T, the percentage of T is 20%/2 = 10%.

2. Which of the following relations will be true for the percentage of bases in double-stranded DNA?

a. $C + T = A + G$ **b.** $\dfrac{C}{A} = \dfrac{T}{G}$

• Solution

An easy way to determine whether the relations are true is to arbitrarily assign percentages to the bases, remembering that, in double-stranded DNA, A = T and G = C. For example, if the percentages of A and T are each 30%, then the percentages of G and C are each 20%. We can substitute these values into the equations to see if the relations are true.

a. $20 + 30 = 30 + 20$ This relation is true.

b. $^{20}/_{30} \neq \, ^{30}/_{20}$. This relation is not true.

COMPREHENSION QUESTIONS

Section 10.1

*1. What three general characteristics must the genetic material possess?

Section 10.2

2. Briefly outline the history of our knowledge of the structure of DNA until the time of Watson and Crick. Which do you think were the principle contributions and developments?

*3. What experiments demonstrated that DNA is the genetic material?

4. What is transformation? How did Avery and his colleagues demonstrate that the transforming principle is DNA?

*5. How did Hershey and Chase show that DNA is passed to new phages in phage reproduction?

6. Why was Watson and Crick's discovery so important?

Section 10.3

*7. Draw and identify the three parts of a DNA nucleotide.

8. How does an RNA nucleotide differ from a DNA nucleotide?

9. How does a purine differ from a pyrimidine? What purines and pyrimidines are found in DNA and RNA?

*10. Draw a short segment of a single polynucleotide strand, including at least three nucleotides. Indicate the polarity of the strand by identifying the 5′ end and the 3′ end.

11. Which bases are capable of forming hydrogen bonds with each other?

12. What different types of chemical bonds are found in DNA and where are they found?

13. What are some of the important genetic implications of the DNA structure?

*14. What are the major transfers of genetic information?

Section 10.4

15. What are hairpins and how do they form?

16. What is DNA methylation?

APPLICATION QUESTIONS AND PROBLEMS

Introduction

17. The introduction to this chapter about Neanderthal DNA emphasizes the extreme stability of DNA, which is able to persist for tens of thousands of years. What aspects of DNA's structure contribute to the stability of the molecule? Why is RNA less stable than DNA?

Section 10.2

18. Match the researchers (a–j) with the discoveries listed.

a. Kossel   **f.** Hershey and Chase

b. Fraenkel-Conrat   **g.** Avery, MacLeod, and McCarty

c. Watson and Crick   **h.** Griffith

d. Levene   **i.** Wilkins and Franklin

e. Miescher   **j.** Chargaff

____ Took X-ray diffraction pictures used in constructing the structure of DNA.

____ Determined that DNA contains nitrogenous bases.

____ Identified DNA as the genetic material in bacteriophage.

_____ Discovered regularity in the ratios of different bases in DNA.

_____ Determined that DNA is responsible for transformation in bacteria.

_____ Worked out the helical structure of DNA by building models.

_____ Discovered that DNA consists of repeating nucleotides.

_____ Determined that DNA is acidic and high in phosphorous.

_____ Conducted experiments showing that RNA can serve as the genetic material in some viruses.

_____ Demonstrated that heat-killed material from bacteria can genetically transform live bacteria.

19. A student mixes some heat-killed type IIS *Streptococcus pneumoniae* bacteria with live type IIR bacteria and injects the mixture into a mouse. The mouse develops pneumonia and dies. The student recovers some type IIS bacteria from the dead mouse. It is the only experiment conducted by the student. Has the student demonstrated that transformation has taken place? What other explanations might explain the presence of the type IIS bacteria in the dead mouse?

*20. Explain how heat-killed type IIIS bacteria in Griffith's experiment genetically altered the live type IIR bacteria. (Hint: See the discussion of transformation in Chapter 8.)

21. What results would you expect if the Hershey and Chase experiment were conducted on tobacco mosaic virus?

22. Imagine that you are a student in Alfred Hershey and Martha Chase's lab in the late 1940s. You are given five test tubes containing *E. coli* bacteria that were infected with T2 bacteriophage that have been labeled with either ^{32}P or ^{35}S. Unfortunately, you forgot to mark the tubes and are now uncertain about which were labeled with ^{32}P and which with ^{35}S. You place the contents of each tube in a blender and turn it on for a few seconds to shear off the protein coats. You then centrifuge the contents to separate the protein coats and the cells. You check for the presence of radioactivity and obtain the following results. Which tubes contained *E. coli* infected with ^{32}P-labeled phage? Explain your answer.

Tube number	Presence of radioactivity in
1	cells
2	protein coats
3	protein coats
4	cells
5	cells

Section 10.3

23. DNA molecules of different sizes are often separated with the use of a technique called electrophoresis (see Chapter 19). With this technique, DNA molecules are placed in a gel, an electrical current is applied to the gel, and the DNA molecules migrate toward the positive (+) pole of the current. What aspect of its structure causes a DNA molecule to migrate toward the positive pole?

*24. Each nucleotide pair of a DNA double helix weighs about 1×10^{-21} g. The human body contains approximately 0.5 g of DNA. How many nucleotide pairs of DNA are in the human body? If you assume that all the DNA in human cells is in the B-DNA form, how far would the DNA reach if stretched end to end?

*25. Erwin Chargaff collected data on the proportions of nucleotide bases from the DNA of a variety of different organisms and tissues (E. Chargaff, in *The Nucleic Acids: Chemistry and Biology*, vol. 1, E. Chargaff and J. N. Davidson, Eds. New York: Academic Press, 1955). The following data are from the DNA of several organisms analyzed by Chargaff.

Erwin Chargaff. [Courtesy Columbia University Medical Center.]

	Percent			
Organism and tissue	**A**	**G**	**C**	**T**
Sheep thymus	29.3	21.4	21.0	28.3
Pig liver	29.4	20.5	20.5	29.7
Human thymus	30.9	19.9	19.8	29.4
Rat bone marrow	28.6	21.4	20.4	28.4
Hen erythrocytes	28.8	20.5	21.5	29.2
Yeast	31.7	18.3	17.4	32.6
E. coli	26.0	24.9	25.2	23.9
Human sperm	30.9	19.1	18.4	31.6
Salmon sperm	29.7	20.8	20.4	29.1
Herring sperm	27.8	22.1	20.7	27.5

a. For each organism, compute the ratio of $(A + G)/(T + C)$ and the ratio of $(A + T)/(C + G)$.

b. Are these ratios constant or do they vary among the organisms? Explain why.

c. Is the $(A + G)/(T + C)$ ratio different for the sperm samples? Would you expect it to be? Why or why not?

26. Boris Magasanik collected data on the amounts of the bases of RNA isolated from a number of sources, expressed relative to a value of 10 for adenine (B. Magasanik, in *The Nucleic Acids: Chemistry and Biology*, vol. 1, E. Chargaff and J. N. Davidson, Eds. New York: Academic Press, 1955).

	Percent			
Organism and tissue	**A**	**G**	**C**	**U**
Rat liver nuclei	10	14.8	14.3	12.9
Rabbit liver nuclei	10	13.6	13.1	14.0
Cat brain	10	14.7	12.0	9.5
Carp muscle	10	21.0	19.0	11.0
Yeast	10	12.0	8.0	9.8

a. For each organism, compute the ratio of $(A + G)/(U + C)$.

b. How do these ratios compare with the $(A + G)/(T + C)$ ratio found in DNA (see Problem 25)? Explain.

*27. Which of the following relations will be found in the percentages of bases of a double-stranded DNA molecule?

a. $A + T = G + C$ **e.** $\dfrac{A + G}{C + T} = 1.0$

b. $A + T = T + C$ **f.** $\dfrac{A}{C} = \dfrac{G}{T}$

c. $A + C = G + T$ **g.** $\dfrac{A}{G} = \dfrac{T}{C}$

d. $\dfrac{A + T}{C + G}$ **h.** $\dfrac{A}{T} = \dfrac{G}{C}$

*28. If a double-stranded DNA molecule is 15% thymine, what are the percentages of all the other bases?

29. Suppose that each of the bases in DNA were capable of pairing with any other base. What effect would this capability have on DNA's capacity to serve as the source of genetic information?

30. Heinz Shuster collected the following data on the base composition of ribgrass virus (H. Shuster, in *The Nucleic Acids: Chemistry and Biology*, vol. 3, E. Chargaff and J. N. Davidson, Eds. New York: Academic Press, 1955). On the basis of this information, is the hereditary information of the ribgrass virus RNA or DNA? Is it likely to be single stranded or double stranded?

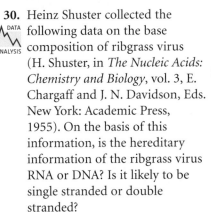

Ribgrass mosaic virus. [Jena Library of Biological Macromolecules.]

	Percent				
	A	**G**	**C**	**T**	**U**
Ribgrass virus	29.3	25.8	18.0	0.0	27.0

*31. The relative amounts of each nucleotide base are tabulated here for four different viruses. For each virus listed in the following table, indicate whether its genetic material is DNA or RNA and whether it is single stranded or double stranded. Explain your reasoning.

Virus	T	C	U	G	A
I	0	12	9	12	9
II	23	16	0	16	23
III	34	42	0	18	39
IV	0	24	35	27	17

*32. A B-DNA molecule has 1 million nucleotide pairs. How many complete turns are there in this molecule?

33. For entertainment on a Friday night, a genetics professor proposed that his children diagram a polynucleotide strand of DNA. Having learned about DNA in preschool, his 5-year-old daughter was able to draw a polynucleotide strand, but she made a few mistakes. The daughter's diagram (represented here) contained at least 10 mistakes.

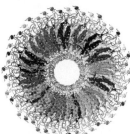

a. Make a list of all the mistakes in the structure of this DNA polynucleotide strand.

b. Draw the correct structure for the polynucleotide strand.

*34. Chapter 1 considered the theory of the inheritance of acquired characteristics and noted that this theory is no longer accepted. Is the central dogma consistent with the theory of the inheritance of acquired characteristics? Why or why not?

Section 10.4

35. Write a sequence of bases in an RNA molecule that will produce a hairpin structure.

36. Write a sequence of nucleotides on a strand of DNA that will form a hairpin structure.

CHALLENGE QUESTIONS

Introduction

37. The sequencing of the Neanderthal genome has prompted speculation that it might be possible in the future to genetically engineer a living Neanderthal. What ethical issues might be raised by creating a living Neanderthal?

Section 10.1

*38. Suppose that an automated, unmanned probe is sent into deep space to search for extraterrestrial life. After wandering for many light-years among the far reaches of the universe, this probe arrives on a distant planet and detects life. The chemical composition of life on this planet is completely different from that of life on Earth, and its genetic material is not composed of nucleic acids. What predictions can you make about the chemical properties of the genetic material on this planet?

Section 10.2

39. How might ^{32}P and ^{35}S be used to demonstrate that the transforming principle is DNA? Briefly outline an experiment that would show that DNA rather than protein is the transforming principle.

Secti on 10.3

*40. Researchers have proposed that early life on Earth used RNA as its source of genetic information and that DNA eventually replaced RNA as the source of genetic information. What aspects of DNA structure might make it better suited than RNA to be the genetic material?

41. Scientists have reportedly isolated short fragments of DNA from fossilized dinosaur bones hundreds of millions of years old. The technique used to isolate this DNA is the polymerase chain reaction, which is capable of amplifying very small amounts of DNA a millionfold (see Chapter 19). Critics have claimed that the DNA isolated from dinosaur bones is not purely of ancient origin but instead has been contaminated by DNA from present-day organisms such as bacteria, mold, or humans. What precautions, analyses, and control experiments could be carried out to ensure that DNA recovered from fossils is truly of ancient origin?

42. Using methods of organic chemistry, scientists have created DNA-like molecules with different chemical structures, called DNA analogs. For example, chemists have created DNA analogs that lack negative charges on the phosphate groups. (Neutral DNA molecules are desirable because the large negative charge on DNA impedes its passage through the cell membrane.) However, when DNA analogs lack a negative charge on the phosphates, they bend, fold up, and aggregate. Explain why the repeating negative charge on the phosphates of DNA are necessary to maintain the stable structure of the molecule.

11 Chromosome Structure and Transposable Elements

Tomatoes come in a great variety of shapes, sizes, and colors. Elongated tomoatoes in some varieties are produced by a duplication that arose as a result of the presence of the transposable element *Rider*. [Photolibrary.]

JUMPING GENES IN ELONGATED TOMATOES

Juicy and colorful, tomatoes are a delight to the eye and the mouth. Modern cultivated tomatoes (*Solanum lycopersicum*) originated in the highlands of South America, where a number of native species still grow today. Early cultivated varieties of tomatoes are believed to have been transported by Native Americans to Mexico and cultivated there by Aztecs. Early Spanish explorers found them delicious and carried tomato plants to Europe and throughout the world. Today, more than 100 million tons of tomatoes are grown annually.

As is known to anyone who visits the produce section of a grocery store, tomatoes come in a variety shapes, sizes, and colors. Some varieties are tiny, barely 5 mm in diameter. Others are huge, more than 10 centimeters across. They come in yellow, orange, and red colors. Some are round; others are elongate. The genetic basis of these different types of tomatoes has obvious economic importance and has long been of interest to plant biologists and geneticists.

Recent research indicates that the shape of the tomato fruit in some varieties has a peculiar origin. The Sun1642 tomato variety produces an elongated fruit, whereas the common and ancestral shape of tomatoes is round. Molecular analysis of Sun1642 showed that the elongated shape arises from the presence of an extra 24,700-bp segment of DNA, which is missing from other varieties with round fruits. The extra segment in Sun1642 is a chromosome duplication: a copy of this segment is found on chromosome 10 of all tomato varieties, but Sun1642 has an extra copy of the segment on its chromosome 7. How did Sun1642 acquire an extra copy of the segment?

Sequencing DNA from the extra segment revealed that it contains a transposable element called *Rider*. Transposable elements are sequences that can move about in the genome. These mobile DNA sequences have been given a variety of names, including transposons, transposable genetic elements, movable genes, controlling elements, and jumping genes. We will refer to them as **transposable elements** and, by this term, include any DNA sequence capable of moving from one place to another within the genome. Transposable elements have played an important role in shaping the structure of chromosomes and genomes, as will be discussed in this chapter. As they move about the genome, transposable elements often cause chromosome duplications, deletions, and other types of rearrangements. In regard to Sun1642, the presence of *Rider* caused a piece of DNA from chromosome 10 to duplicate and

E. coli bacterium

Bacterial chromosome

11.1 The DNA in *E. coli* is about 1000 times as long as the cell itself.

move to chromosome 7. One of the genes within this duplicated region is *IQD12*. The movement of *IQD12* to a new location led to it being overexpressed, resulting in an elongated tomato. The pear-shaped, elongated fruit of Sun1642 tomatoes illustrates the importance of transposable elements in shaping the structure of chromosomes.

In this chapter, we examine the molecular structure of chromosomes, including transposable elements. The first part of the chapter focuses on a storage problem: how to cram tremendous amounts of DNA into the limited confines of a cell. Even in those organisms having the smallest amounts of DNA, the length of genetic material far exceeds the length of the cell. Thus, cellular DNA must be highly folded and tightly packed, but this packing creates problems: it renders the DNA inaccessible, unable to be copied or read. Functional DNA must be capable of partly unfolding and expanding so that individual genes can undergo replication and transcription. The flexible, dynamic nature of DNA packing is a major theme of this chapter. We first consider supercoiling, an important tertiary structure of DNA found in both prokaryotic and eukaryotic cells. After a brief look at the bacterial chromosome, we examine the structure of eukaryotic chromosomes. After considering chromosome structure, we pay special attention to the working parts of a chromosome—specifically, centromeres and telomeres. We also consider the types of DNA sequences present in many eukaryotic chromosomes.

The second part of this chapter focuses on transposable elements. We begin by considering some of the general features of transposable elements and the processes by which they move from place to place. We then examine several different types of transposable elements found in prokaryotic and eukaryotic genomes. Finally, we consider the evolutionary significance of transposable elements.

11.1 Large Amounts of DNA Are Packed into a Cell

The packaging of tremendous amounts of genetic information into the small space within a cell has been called the ultimate storage problem. Consider the chromosome of the bacterium *E. coli,* a single molecule of DNA with approximately 4.6 million base pairs. Stretched out straight, this DNA would be about 1000 times as long as the cell within which it resides (**Figure 11.1**). Human cells contain more than 6 billion base pairs of DNA, which would measure over 2 meters (over 6 feet) stretched end to end. Even DNA in the smallest human chromosome would stretch 14,000 times the length of the nucleus. Clearly, DNA molecules must be tightly packed to fit into such small spaces.

The structure of DNA can be considered at three hierarchical levels: the primary structure of DNA is its nucleotide sequence; the secondary structure is the double-stranded helix; and the tertiary structure refers to higher-order folding that allows DNA to be packed into the confined space of a cell.

CONCEPTS

Chromosomal DNA exists in the form of very long molecules, which must be tightly packed to fit into the small confines of a cell.

Supercoiling

One type of DNA tertiary structure is **supercoiling**, which takes place when the DNA helix is subjected to strain by being overwound or underwound. The lowest energy state for B-DNA is when it has approximately 10 bp per turn of its helix. In this **relaxed state**, a stretch of 100 bp of DNA would assume about 10 complete turns (**Figure 11.2a**). If energy is used to add or remove any turns, strain is placed on the molecule, causing the helix to supercoil, or twist, on itself (**Figure 11.2b** and **c**). Molecules that are overrotated exhibit **positive supercoiling** (see Figure 11.2b). Underrotated molecules exhibit **negative supercoiling** (see Figure 11.2c). Supercoiling is a partial solution to the cell's DNA packing problem because supercoiled DNA occupies less space than relaxed DNA.

Supercoiling takes place when the strain of overrotating or underrotating cannot be compensated by the turning of the ends of the double helix, which is the case if the DNA is circular—that is, there are no free ends. If the chains *can* turn freely, their ends will simply turn as extra rotations are added or removed, and the molecule will spontaneously revert to the relaxed state. Both bacterial and eukaryotic DNA usually fold into loops stabilized by proteins (which prevent free rotation of the ends, see Figure 11.3), and supercoiling takes place within the loops.

Supercoiling relies on **topoisomerases,** enzymes that add or remove rotations from the DNA helix by temporarily breaking the nucleotide strands, rotating the ends around each other, and then rejoining the broken ends. Thus topoisomerases can both induce and relieve supercoiling.

Most DNA found in cells is negatively supercoiled, which has two advantages over nonsupercoiled DNA. First, supercoiling makes the separation of the two strands of DNA easier during replication and transcription. Negatively supercoiled DNA is underrotated; so separation of the two strands during replication and transcription is more rapid and requires less energy. Second, supercoiled DNA can be packed into a smaller space than can relaxed DNA.

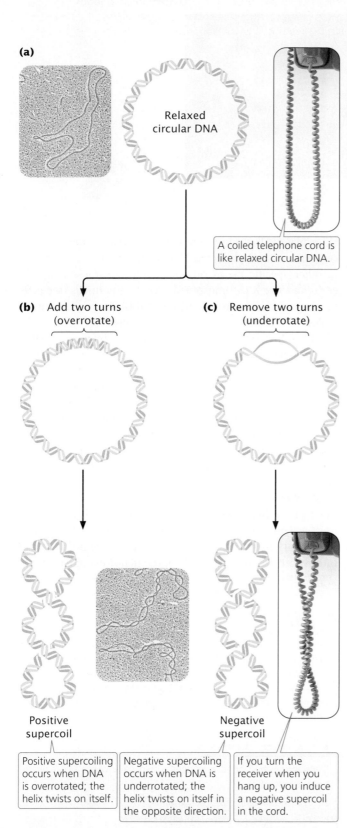

(a)

Relaxed circular DNA

A coiled telephone cord is like relaxed circular DNA.

(b) Add two turns (overrotate)

(c) Remove two turns (underrotate)

Positive supercoil

Negative supercoil

Positive supercoiling occurs when DNA is overrotated; the helix twists on itself.

Negative supercoiling occurs when DNA is underrotated; the helix twists on itself in the opposite direction.

If you turn the receiver when you hang up, you induce a negative supercoil in the cord.

11.2 Supercoiled DNA is overwound or underwound, causing it to twist on itself. Electron micrographs are of relaxed DNA (top) and supercoiled DNA (bottom). [Dr. Gopal Murti/Phototake.]

CONCEPTS

Overrotation or underrotation of a DNA double helix places strain on the molecule, causing it to supercoil. Supercoiling is controlled by topoisomerase enzymes. Most cellular DNA is negatively supercoiled, which eases the separation of nucleotide strands during replication and transcription and allows DNA to be packed into small spaces.

✔ CONCEPT CHECK 1

A DNA molecule 300 bp long has 20 complete rotations. This DNA molecule is

a. positively supercoiled.   c. relaxed.

b. negatively supercoiled.

The Bacterial Chromosome

Most bacterial genomes consist of a single circular DNA molecule, although linear DNA molecules have been found in a few species. In circular bacterial chromosomes, the DNA does not exist in an open, relaxed circle; the 3 million to 4 million base pairs of DNA found in a typical bacterial genome would be much too large to fit into a bacterial cell (see Figure 11.1). Bacterial DNA is not attached to histone proteins as is eukaryotic DNA (discussed later in the chapter), but bacterial DNA is complexed to a number of proteins that help to compact it.

When a bacterial cell is viewed with the electron microscope, its DNA frequently appears as a distinct clump, the **nucleoid**, which is confined to a definite region of the cytoplasm. If a bacterial cell is broken open gently, its DNA spills out in a series of twisted loops (**Figure 11.3a**). The ends of the loops are most likely held in place by proteins (**Figure 11.3b**). Many bacteria contain additional DNA in the form of small circular molecules called plasmids, which replicate independently of the chromosome (see Chapter 8).

CONCEPTS

A typical bacterial chromosome consists of a large, circular molecule of DNA that is a series of twisted loops. Bacterial DNA appears as a distinct clump, the nucleoid, within the bacterial cell.

✔ CONCEPT CHECK 2

How does bacterial DNA differ from eukaryotic DNA?

Eukaryotic Chromosomes

Individual eukaryotic chromosomes contain enormous amounts of DNA. Like the bacterial chromosome, each eukaryotic chromosome consists of a single, extremely long molecule of DNA. For all of this DNA to fit into the nucleus, tremendous packing and folding are required, the extent of which must change in the course of the cell cycle. The

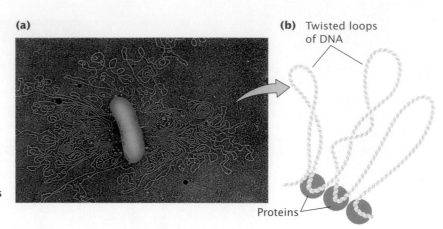

(a)

(b) Twisted loops of DNA

Proteins

11.3 Bacterial DNA is highly folded into a series of twisted loops. [Part a: Dr. Gopal Murti/Photo Researchers.]

chromosomes are in an elongated, relatively uncondensed state during interphase of the cell cycle (see pp. 21–22 in Chapter 2), but the term *relatively* is an important qualification here. Although the DNA of interphase chromosomes is less tightly packed than the DNA of mitotic chromosomes, it is still highly condensed; it's just *less* condensed. In the course of the cell cycle, the level of DNA packing changes: chromosomes progress from a highly packed state to a state of extreme condensation. DNA packing also changes locally in replication and transcription, when the two nucleotide strands must unwind so that particular base sequences are exposed. Thus, the packing of eukaryotic DNA (its tertiary chromosomal structure) is not static but changes regularly in response to cellular processes.

Chromatin   Eukaryotic DNA in the cell is closely associated with proteins. This combination of DNA and protein is called chromatin. The two basic types of chromatin are **euchromatin**, which undergoes the normal process of condensation and decondensation in the cell cycle, and **heterochromatin**, which remains in a highly condensed state throughout the cell cycle, even during interphase. Euchromatin constitutes the majority of the chromosomal material and is where most transcription takes place. All chromosomes have heterochromatin at the centromeres and telomeres. Heterochromatin is also present at other specific places on some chromosomes, along the entire inactive X chromosome in female mammals (see p. 90 in Chapter 4) and throughout most of the Y chromosome in males. In addition to remaining condensed throughout the cell cycle, heterochromatin is characterized by a general lack of transcription, the absence of crossing over, and replication late in the S stage.

The most abundant proteins in chromatin are the *histones*, which are small, positively charged proteins of five major types: H1, H2A, H2B, H3, and H4 (**Table 11.1**). All histones have a high percentage of arginine and lysine, positively charged amino acids that give the histones a net positive charge. The positive charges attract the negative charges on the phosphates of DNA; this attraction holds the DNA in contact with the histones. A heterogeneous assortment of **nonhistone chromosomal proteins** also are found in eukaryotic chromosomes. At times, variant histones, with somewhat different amino acid sequences, are incorporated into chromatin in place of one of the major histone proteins.

TRY PROBLEM 24 →

Table 11.1	Characteristics of histone proteins	
Histone Protein	Molecular Weight	Number of Amino Acids
H1	21,130	223
H2A	13,960	129
H2B	13,774	125
H3	15,273	135
H4	11,236	102

Note: The sizes of H1, H2A, and H2B histones vary somewhat from species to species. The values given are for bovine histones.
Source: Data are from D. L. Nelson and M. M. Cox, *Lehninger Principles of Biochemistry*, 5th ed. (New York: W. H. Freeman and Company, 2009), p. 963.

CONCEPTS

Chromatin, which consists of DNA complexed to proteins, is the material that makes up eukaryotic chromosomes. The most abundant of these proteins are the five types of positively charged histone proteins: H1, H2A, H2B, H3, and H4. Variant histones may at times be incorporated into chromatin in place of the normal histones.

✔ CONCEPT CHECK 3

Neutralizing their positive charges would have which effect on the histone proteins?

a. They would bind the DNA tighter.

b. They would separate from the DNA.

c. They would no longer be attracted to each other.

d. They would cause supercoiling of the DNA.

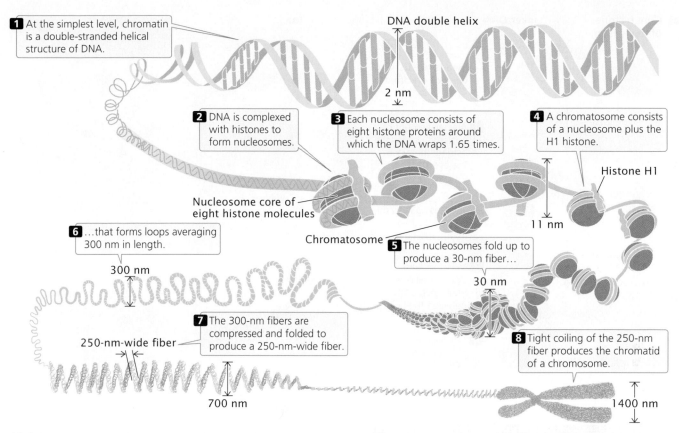

1 At the simplest level, chromatin is a double-stranded helical structure of DNA.

DNA double helix

2 nm

2 DNA is complexed with histones to form nucleosomes.

3 Each nucleosome consists of eight histone proteins around which the DNA wraps 1.65 times.

4 A chromatosome consists of a nucleosome plus the H1 histone.

Histone H1

Nucleosome core of eight histone molecules

Chromatosome

11 nm

6 ...that forms loops averaging 300 nm in length.

300 nm

5 The nucleosomes fold up to produce a 30-nm fiber...

30 nm

7 The 300-nm fibers are compressed and folded to produce a 250-nm-wide fiber.

250-nm-wide fiber

8 Tight coiling of the 250-nm fiber produces the chromatid of a chromosome.

700 nm

1400 nm

11.4 Chromatin has a highly complex structure with several levels of organization.

The nucleosome   Chromatin has a highly complex structure with several levels of organization (**Figure 11.4**). The simplest level is the double-helical structure of DNA discussed in Chapter 10. At a more complex level, the DNA molecule is associated with proteins and is highly folded to produce a chromosome.

When chromatin is isolated from the nucleus of a cell and viewed with an electron microscope, it frequently looks like beads on a string (**Figure 11.5a**). If a small amount of nuclease is added to this structure, the enzyme cleaves the "string" between the "beads," leaving individual beads attached to about 200 bp of DNA (**Figure 11.5b**). If more nuclease is added, the enzyme chews up all of the DNA between the beads and leaves a core of proteins attached to a fragment of DNA (**Figure 11.5c**). Such experiments demonstrated that chromatin is not a random association of proteins and DNA but has a fundamental repeating structure.

The repeating core of protein and DNA produced by digestion with nuclease enzymes is the simplest level of chromatin structure, the **nucleosome** (see Figure 11.4). The nucleosome is a core particle consisting of DNA wrapped about two times around an octamer of eight histone proteins (two copies each of H2A, H2B, H3, and H4), much like thread wound around a spool (**Figure 11.5d**). The DNA in direct contact with the histone octamer is between 145 and 147 bp in length.

Each of the histone proteins that make up the nucleosome core particle has a flexible "tail," containing from 11 to 37 amino acids, that extends out from the nucleosome. Positively charged amino acids in the tails of the histones interact with the negative charges of the phosphates on the DNA, keeping the DNA and histones tightly associated. The tails of one nucleosome may also interact with neighboring nucleosomes, which facilitates compaction of the nucleosomes themselves. Chemical modifications of the histone tails bring about changes in chromatin structure (discussed in the next section) that are necessary for gene expression.

The fifth type of histone, H1, is not a part of the core particle but plays an important role in nucleosome structure. H1 binds to 20 to 22 bp of DNA where the DNA joins and leaves the octamer (see Figure 11.4) and helps to lock the DNA into place, acting as a clamp around the nucleosome octamer.

Together, the core particle and its associated H1 histone are called the **chromatosome** (see Figure 11.4), the next level of chromatin organization. Each chromatosome encompasses about 167 bp of DNA. Chromatosomes are located at regular intervals along the DNA molecule and are separated from one another by **linker DNA**, which varies in size among cell types; in most cells, linker DNA comprises from

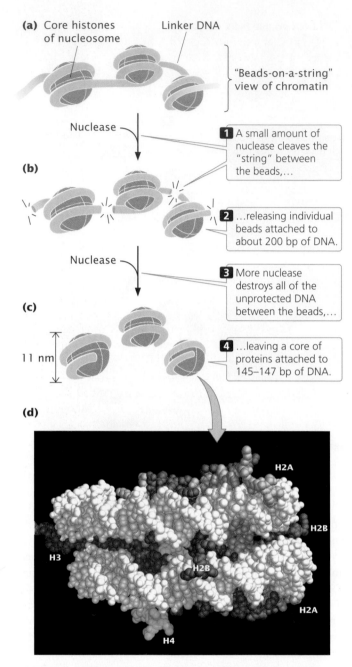

(a) Core histones of nucleosome   Linker DNA

"Beads-on-a-string" view of chromatin

Nuclease

1 A small amount of nuclease cleaves the "string" between the beads,...

(b)

2 ...releasing individual beads attached to about 200 bp of DNA.

Nuclease

3 More nuclease destroys all of the unprotected DNA between the beads,...

(c)

11 nm

4 ...leaving a core of proteins attached to 145–147 bp of DNA.

(d)

H2A

H2B

H3

H2B

H2A

H4

11.5 The nucleosome is the fundamental repeating unit of chromatin. The space-filling model shows that the nucleosome core particle consists of two copies each of H2A, H2B, H3, and H4, around which DNA (white) coils. [Part d: From K. Luger et al., *Nature* 389:251, 1997; courtesy of T. H. Richmond.]

about 30 to 40 bp. Nonhistone chromosomal proteins may be associated with this linker DNA, and a few also appear to bind directly to the core particle. **TRY PROBLEMS 25 AND 27 →**

Higher-order chromatin structure When chromatin is in a condensed form, adjacent nucleosomes are not separated by space equal to the length of the linker DNA; rather, nucleosomes fold on themselves to form a dense, tightly

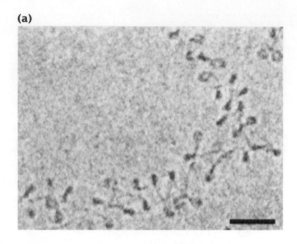

(a)

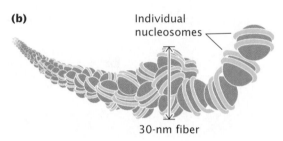

(b)   Individual nucleosomes

30-nm fiber

11.6 Adjacent nucleosomes pack together to form a 30-nm fiber. (a) Electron micrograph of nucleosomes. (b) One model of how nucleosomes associate to form the helical fiber. [Part a: Jan Bednar et al., *PNAS* 1998; 95:14173–14178. Copyright 2004 National Academy of Sciences, U. S. A.]

packed structure (see Figure 11.4) that makes up a fiber with a diameter of about 30 nm (**Figure 11.6a**). Two different models have been proposed for the 30-nm fiber: a solenoid model, in which a linear array of nucleosomes are coiled, and a helix model, in which nucleosomes are arranged in a zigzag ribbon that twists or supercoils. Recent evidence supports the helix model (**Figure 11.6b**).

The next-higher level of chromatin structure is a series of loops of 30-nm fibers, each anchored at its base by proteins in the nuclear scaffold (see Figure 11.4). On average, each loop encompasses some 20,000 to 100,000 bp of DNA and is about 300 nm in length, but the individual loops vary considerably. The 300-nm loops are packed and folded to produce a 250-nm-wide fiber. Tight helical coiling of the 250-nm fiber, in turn, produces the structure that appears in metaphase—individual chromatids approximately 700 nm in width.

CONCEPTS

The nucleosome consists of a core particle of eight histone proteins and DNA that wraps around the core. Chromatosomes, which are nucleosomes bound to an H1 histone, are separated by linker DNA. Nucleosomes fold to form a 30-nm chromatin fiber, which appears as a series of loops that pack to create a 250-nm-wide fiber. Helical coiling of the 250-nm fiber produces a chromatid.

Changes in Chromatin Structure

Although eukaryotic DNA must be tightly packed to fit into the cell nucleus, it must also periodically unwind to undergo transcription and replication.

Polytene chromosomes   Giant chromosomes found in certain tissues of *Drosophila* and some other organisms (**Figure 11.7**), **polytene chromosomes** have provided researchers with evidence of the changing nature of chromatin structure. These large, unusual chromosomes arise when repeated rounds of DNA replication take place without accompanying cell divisions, producing thousands of copies of DNA that lie side by side. When polytene chromosomes are stained with dyes, numerous bands are revealed. Under certain conditions, the bands may exhibit **chromosomal puffs**—localized swellings of the chromosome. Each puff is a region of the chromatin having a relaxed structure and, consequently, a more open state. If radioactively labeled uridine (a precursor to RNA) is briefly added to a *Drosophila* larva, radioactivity accumulates in chromosomal puffs, indicating that they are regions of active transcription. Additionally, the appearance of puffs at particular locations on the chromosome can be stimulated by exposure to hormones and other compounds that are known to induce the transcription of genes at those locations. This correlation between the occurrence of transcription and the relaxation of chromatin at a puff site indicates that chromatin structure undergoes dynamic change associated with gene activity.   TRY PROBLEM 26 →

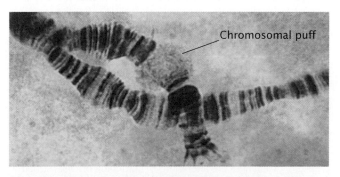

11.7 Chromosomal puffs are regions of relaxed chromatin where active transcription is taking place. Pictured here are chromosomal puffs on giant polytene chromosomes isolated from the salivary glands of larval *Drosophila*. [Robert Calentine/Visuals Unlimited.]

DNase I sensitivity   A second piece of evidence indicating that chromatin structure changes with gene activity is sensitivity to DNase I, an enzyme that digests DNA. The ability of this enzyme to digest DNA depends on chromatin structure: when DNA is tightly bound to histone proteins, it is less sensitive to DNase I, whereas unbound DNA is more sensitive to digestion by DNase I. The results of experiments that examine the effect of DNase I on specific genes show that DNase sensitivity is correlated with gene activity. Studies of chicken globin genes give evidence for this correlation. Globin genes encode hemoglobin in the erythroblasts (precursors of red blood cells) of chickens (**Figure 11.8**). No hemoglobin is synthesized in chick embryos in the first 24 hours after fertilization. If DNase I is applied to chromatin from chick erythroblasts in this first 24-hour period, all the globin genes are insensitive to digestion. From day 2 to day 6 after fertilization, after hemoglobin synthesis has begun, the globin genes become sensitive to DNase I, and the genes that encode embryonic hemoglobin are the most sensitive. After 14 days of development, embryonic hemoglobin is replaced by the adult forms of hemoglobin. The most-sensitive regions now lie near the genes that produce the adult hemoglobins. DNA from brain cells, which produce no hemoglobin, remains insensitive to DNase digestion throughout development. In summary, when genes become transcriptionally active, they also become sensitive to DNase I, indicating that the chromatin structure is more exposed during transcription.

What is the nature of the change in chromatin structure that produces chromosome puffs and DNase I sensitivity? In both cases, the chromatin relaxes; presumably, the histones loosen their grip on the DNA. One process that alters chromatin structure is acetylation. Enzymes called acetyltransferases attach acetyl groups to lysine amino acids on the histone tails. This modification reduces the positive charges that normally exist on lysine and destabilizes the nucleosome structure, and so the histones hold the DNA less tightly. Other chemical modifications of the histone proteins, such as methylation and phosphorylation, also alter chromatin structure, as do special chromatin-remodeling proteins that bind to the DNA.

Epigenetic changes associated with chromatin modifications   We have now seen how chromatin structure can be altered by chemical modification of the histone proteins. A number of other changes also can affect chromatin structure, including the methylation of DNA (see Chapter 10), the use of variant histone proteins in the nucleosome, and the binding of proteins to DNA and chromatin. Although these changes do not alter the DNA sequence, they often have major effects on the expression of genes, which will be discussed in more detail in Chapter 17.

Some changes to chromatin structure are retained through cell division, and so they are passed on future generations of cells and even occasionally to future generations of organisms. Stable alterations of chromatin structure that may

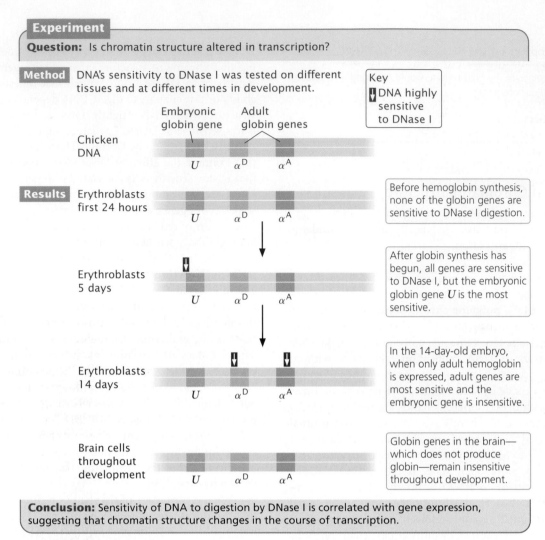

Experiment

Question: Is chromatin structure altered in transcription?

Method DNA's sensitivity to DNase I was tested on different tissues and at different times in development.

Key
⬇ DNA highly sensitive to DNase I

Embryonic globin gene   Adult globin genes

Chicken DNA

U α^D α^A

Results Erythroblasts first 24 hours

U α^D α^A

Before hemoglobin synthesis, none of the globin genes are sensitive to DNase I digestion.

Erythroblasts 5 days

U α^D α^A

After globin synthesis has begun, all genes are sensitive to DNase I, but the embryonic globin gene U is the most sensitive.

Erythroblasts 14 days

U α^D α^A

In the 14-day-old embryo, when only adult hemoglobin is expressed, adult genes are most sensitive and the embryonic gene is insensitive.

Brain cells throughout development

U α^D α^A

Globin genes in the brain—which does not produce globin—remain insensitive throughout development.

Conclusion: Sensitivity of DNA to digestion by DNase I is correlated with gene expression, suggesting that chromatin structure changes in the course of transcription.

11.8 DNase I sensitivity is correlated with the transcription of globin genes in erythroblasts of chick embryos. The U gene encodes embryonic hemoglobin; the α^D and α^A genes encode adult hemoglobin.

be passed on to cells or individual organisms are frequently referred to as **epigenetic changes** or simply as epigenetics (see Chapter 5). For example, the *agouti* locus helps determine coat color in mice. Parents that have identical DNA sequences but have different degrees of methylation on their DNA may give rise to offspring with different coat colors. Such epigenetic changes have been observed in a number of organisms and are responsible for a variety of phenotypic effects. Unlike mutations, epigenetic changes do not alter the DNA sequence, are capable of being reversed, and are often influenced by environmental factors.

One type of epigenetic change is genomic imprinting, in which an allele is differentially expressed, depending on whether it is inherited from the maternal or paternal parent (see Chapter 5). More than 100 mammalian genes are imprinted, including many that play important roles in growth and development and in genetic diseases. Recall from Chapter 5, for example, that genomic imprinting is thought to affect fetal growth and birth weight. Genomic imprint-

ing is caused by differences in DNA methylation between oocytes and sperm. The DNA of sperm and oocytes is differentially methylated at various imprinting control regions in gamete formation, and the different methylation patterns of paternal and maternal alleles are then maintained and passed on to all resulting cells in the zygote. Recent research indicates that genomic imprinting depends not only on the methylation of DNA but also on methylation patterns of some histone proteins. Interestingly, demethylation of some histones is required before methylation of imprinting control regions in the DNA can take place.

CONCEPTS

Epigenetic changes are alterations of chromatin or DNA structure that do not include changes in the base sequence but are stable and passed on to cells or organisms. Some epigenetic changes result from alterations of histone proteins.

11.2 Eukaryotic Chromosomes Possess Centromeres and Telomeres

Chromosomes segregate in mitosis and meosis and remain stable over many cell divisions. These properties of chromosomes arise, in part, from special structural features of chromosomes, including centromeres and telomeres.

Centromere Structure

The centromere is a constricted region of the chromosome to which spindle fibers attach and is essential for proper chromosome movement in mitosis and meiosis (see Chapter 2). The essential role of the centromere in chromosome movement was recognized by early geneticists, who observed what happens when a chromosome breaks in two. A chromosome break produces two fragments, one with a centromere and one without (**Figure 11.9**). In mitosis, the chromosome fragment containing the centromere attaches to spindle fibers and moves to the spindle pole, whereas the fragment lacking a centromere fails to connect to a spindle fiber and is usually lost because it fails to move into the nucleus of a daughter cell (see Figure 11.9).

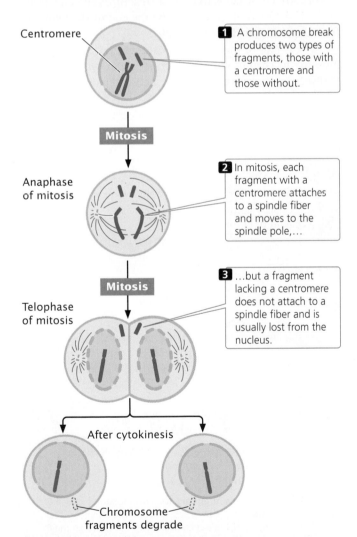

11.9 Chromosome fragments that lack centromeres are lost in mitosis.

The first centromeres to be isolated and studied at the molecular level came from yeast, which has small linear chromosomes. When molecular biologists attached DNA sequences from yeast centromeres to plasmids, the plasmids behaved in mitosis as if they were eukaryotic chromosomes, segregating and dividing as the yeast cell replicated.

The centromeres of different organisms exhibit considerable variation in centromeric sequences. These **centromeric sequences** are the binding sites for the kinetochore, to which spindle fibers attach. Some organisms have chromosomes with diffuse centromeres, and spindle fibers attach along the entire length of each chromosome. Most have chromosomes with localized centromeres; in these organisms, spindle fibers attach at a specific place on the chromosome, but there can also be secondary constrictions at places that do not have centromeric functions.

In *Drosophila*, *Arabidopsis*, and humans, centromeres span hundreds of thousands of base pairs. Most of the centromere is made up of heterochromatin, consisting of short sequences of DNA that are repeated thousands of times in tandem. Surprisingly, there are no specific sequences that are found in all centromeres, which raises the question of what exactly determines where the centromere is. Today, experts agree that most centromeres are not defined by DNA sequence but rather by epigenetic changes in chromatin structure. Nucleosomes in the centromeres of most eukaryotes have a variant histone protein called CenH3, which takes the place of the usual H3 histone. The CenH3 histone brings about a change in the nucleosome and chromatin structure, which is believed to promote the formation of the kinetochore and the attachment of spindle fibers to the chromosome.

In addition to their roles in the attachment of the spindle fibers and the movement of chromosomes, centromeres help control the cell cycle (see p. 21 in Chapter 2). In mitosis, spindle fibers attach to the kinetochore of the centromere and orient each chromosome on the metaphase plate. If anaphase is initiated before each chromosome is attached to the spindle fibers, chromosomes will not move toward the spindle pole and will be lost. Research findings indicate that the commencement of anaphase is inhibited by a signal from the centromere. This inhibitory signal disappears only after the centromere of each chromosome is attached to spindle fibers from opposite poles.

CONCEPTS

The centromere is a region of the chromosome to which spindle fibers attach. Centromeres display considerable variation in structure and are distinguished by epigenetic alterations to chromatin structure, including the use of a variant H3 histone in the nucleosome. In addition to their role in chromosome movement, centromeres help control the cell cycle by inhibiting anaphase until chromosomes are attached to spindle fibers from both poles.

✔ CONCEPT CHECK 5

What happens to a chromosome that loses its centromere?

Telomere Structure

Telomeres are the natural ends of a chromosome (see p. 20 in Chapter 2). Pioneering work by Hermann Muller (on fruit flies) and Barbara McClintock (on corn) showed that chromosome breaks produce unstable ends that have a tendency to stick together and enable the chromosome to be degraded. Because attachment and degradation do not happen to the ends of a chromosome that has telomeres, each telomere must serve as a cap that stabilizes the chromosome, much as the plastic tips on the ends of a shoelace prevent the lace from unraveling. Telomeres also provide a means of replicating the ends of the chromosome, which will be discussed in Chapter 12. In 2009, Elizabeth Blackburn, Carol Greider, and Jack Szostak were awarded the Nobel Prize in physiology and medicine for discovering the structure of telomeres and how they are replicated (discussed in Chapter 12).

Telomeres have now been isolated from protozoans, plants, humans, and other organisms; most are similar in structure (**Table 11.2**). These **telomeric sequences** usually consist of repeated units of a series of adenine or thymine nucleotides followed by several guanine nucleotides, taking the form $5'-(A \text{ or } T)_m G_n-3'$, where m is from 1 to 4 and n is 2 or more. For example, the repeating unit in human telomeres is 5'-TTAGGG-3', which may be repeated from

hundreds to thousands of times. The sequence is always oriented with the string of Gs and Cs toward the end of the chromosome, as shown here:

$$
\begin{array}{ccc}
\text{toward} & 5'\text{–TTAGGG–}3' & \text{end of} \\
\text{centromere} \to & 3'\text{–AATCCC–}5' & \to \text{chromosome}
\end{array}
$$

The G-rich strand often protrudes beyond the complementary C-rich strand at the end of the chromosome (**Figure 11.10a**) and is called the 3' overhang. The 3' overhang in the telomeres of mammals is from 50 to 500 nucleotides long. Special proteins bind to the G-rich single-stranded sequence, protecting the telomere from degradation and preventing the ends of chromosomes from sticking together. A multiprotein complex called **shelterin** binds to mammalian telomeres and protects the ends of the DNA from being inadvertently repaired as a double-stranded break in the DNA. In mammalian cells, the single-stranded overhang may fold over and pair with a short stretch of DNA to form a structure called a t-loop, which also functions in protecting the end of the telomere from degradation (**Figure 11.10b**).

The length of the telomeric sequence varies from chromosome to chromosome and from cell to cell, suggesting that each telomere is a dynamic structure that actively grows

Table 11.2	DNA sequences typically found in telomeres of various organisms
Organism	**Sequence**
Tetrahymena (protozoan)	5'–TTGGGG–3' 3'–AACCCC–5'
Oxytricha (protozoan)	5'–TTTTGGGG–3' 3'–AAAACCCC–5'
Trypanosoma (protozoan)	5'–TTAGGG–3' 3'–AATCCC–5'
Saccharomyces (yeast)	$5'-T_{1-6}GTG_{2-3}-3'$ $3'-A_{1-6}CTG_{2-3}-5'$
Neurospora (fungus)	5'–TTAGGG–3' 3'–AATCCC–5'
Caenorhabditis (nematode)	5'–TTAGGC–3' 3'–AATCCG–5'
Bombyx (insect)	5'–TTAGG–3' 3'–AAGCC–5'
Vertebrate	5'–TTAGGG–3' 3'–AATCCC–5'
Arabidopsis (plant)	5'–TTTAGGG–3' 3'–AAATCCC–5'

Source: V. A. Zakian, *Science* 270:1602, 1995.

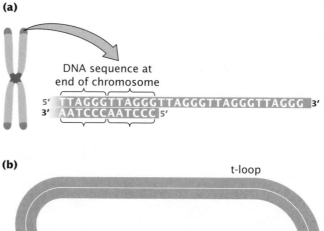

(a)

DNA sequence at end of chromosome

5' TTAGGGTTAGGGTTAGGGTTAGGGTTAGGG 3'
3' AATCCCAATCCC 5'

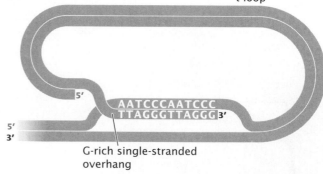

(b)

t-loop

5'

AATCCCAATCCC
TTAGGGTTAGGG 3'

5'
3'

G-rich single-stranded overhang

11.10 DNA at the ends of eukaryotic chromosomes consists of telomeric sequences. (a) The G-rich strand at the telomere is longer than the C-rich strand. (b) In mammalian cells, the G-rich strand folds over and pairs with a short stretch of DNA to form a t-loop.

and shrinks. The telomeres of *Drosophila* chromosomes are different from those of most other organisms. Telomeres in *Drosophila* consist of multiple copies of two different transposable elements, *Het-A* and *Tart*, arranged in tandem repeats. Apparently, in *Drosophila*, the loss of telomeric sequences in the course of replication is balanced by the insertion of additional copies of the *Het-A* and *Tart* elements into the telomere.

Farther away from the ends of chromosomes are **telomere-associated sequences**, comprising from several thousand to hundreds of thousands of base pairs. They, too, contain repeated sequences, but the repeats are longer, more varied, and more complex than those found in telomeric sequences.

CONCEPTS

A telomere is the stabilizing end of a chromosome. At the end of each telomere are many short telomeric sequences. Longer, more complex telomere-associated sequences are found adjacent to the telomeric sequences.

✔ **CONCEPT CHECK 6**

Which is a characteristic of DNA sequences at the telomeres?

a. They consist of guanine and adenine (or thymine) nucleotides.

b. They consist of repeated sequences.

c. One strand protrudes beyond the other, creating some single-standed DNA at the end.

d. All of the above.

Artificial Chromosomes

In 1983, geneticists constructed the first artificial chromosomes from parts culled from yeast and protozoans. In 1987, David Burke and Maynard Olson (at Washington University, St. Louis) used yeast to create much larger artificial chromosomes called yeast artificial chromosomes or YACs. Artificial chromosomes have also been made from chromosomal components of bacteria (BACs) and mammals (MACs). Each eukaryotic artificial chromosome includes the three essential elements of a chromosome: a centromere, a pair of telomeres, and an origin of replication. These elements ensure that artificial chromosomes will segregate in mitosis and meiosis, will not be degraded, and will replicate successfully. Large chunks of extra DNA (as many as a million base pairs) from any source can be added, and the new artificial chromosome can be inserted into a cell. BACs, YACs, and MACs are now routinely used in genetic engineering to clone large fragments of DNA, and they played an important role in the sequencing of the human genome (see Chapters 19 and 20).

11.3 Eukaryotic DNA Contains Several Classes of Sequence Variation

Eukaryotic organisms differ dramatically in the amount of DNA per cell, a quantity termed an organism's **C value** (**Table 11.3**). Each cell of a fruit fly, for example, contains 35 times the amount of DNA found in a cell of the bacterium *E. coli*. In general, eukaryotic cells contain more DNA than prokaryotic cells do, but variability in the C values of different eukaryotes is huge. Human cells contain more than 10 times the amount of DNA found in *Drosophila* cells, whereas some salamander cells contain 20 times as much DNA as that in human cells. Clearly, these differences in C value cannot be explained simply by differences in organismal complexity. So, what is all this extra DNA in eukaryotic cells doing? We do not yet have a complete answer to this question, but eukaryotic DNA sequences reveal a complexity that is absent from prokaryotic DNA.

The Denaturation and Renaturation of DNA

The first clue that eukaryotic DNA contains several types of sequences not present in prokaryotic DNA came from studies in which double-stranded DNA was separated and then allowed to reassociate. When double-stranded DNA in solution is heated, the hydrogen bonds that hold the two strands together are weakened and, with enough heat, the two nucleotide strands separate completely, a process called **denaturation** or melting. The temperature at which DNA denatures, called the **melting temperature** (T_m), depends on the base sequence of the particular sample of DNA: G–C base pairs have three hydrogen bonds, whereas A–T base pairs only have two; so the separation of G–C pairs requires more heat (energy) than does the separation of A–T pairs.

The denaturation of DNA by heating is reversible; if single-stranded DNA is slowly cooled, single strands will collide and hydrogen bonds will again form between

Table 11.3	Genome sizes of various organisms
Organism	Approximate Genome Size (bp)
λ (bacteriophage)	50,000
Escherichia coli (bacterium)	4,640,000
Saccharomyces cerevisiae (yeast)	12,000,000
Arabidopsis thaliana (plant)	125,000,000
Drosophila melanogaster (insect)	170,000,000
Homo sapiens (human)	3,200,000,000
Zea mays (corn)	4,500,000,000
Amphiuma (salamander)	765,000,000,000

complementary base pairs, producing double-stranded DNA. This reaction is called **renaturation** or reannealing.

Two single-stranded molecules of DNA from different sources, such as different organisms, will anneal if they are complementary, a process termed **hybridization**. For hybridization to take place, the two strands do not have to be complementary at all their bases—just at enough bases to hold the two strands together. The extent of hybridization can be used to measure the similarity of nucleic acids from two different sources and is a common tool for assessing evolutionary relationships. The rate at which hybridization takes place also provides information about the sequence complexity of DNA. TRY PROBLEM 32 →

Types of DNA Sequences in Eukaryotes

Eukaryotic DNA consists of at least three types of sequences: unique-sequence DNA, moderately repetitive DNA, and highly repetitive DNA. **Unique-sequence DNA** consists of sequences that are present only once or, at most, a few times in the genome. This DNA includes sequences that encode proteins, as well as a great deal of DNA whose function is unknown. Genes that are present in a single copy constitute from roughly 25% to 50% of the protein-encoding genes in most multicellular eukaryotes. Other genes within unique-sequence DNA are present in several similar, but not identical, copies and together are referred to as a **gene family**. Most gene families arose through duplication of an existing gene and include just a few member genes, but some, such as those that encode immunoglobulin proteins in vertebrates, contain hundreds of members. The genes that encode β-like globins are another example of a gene family. In humans, there are seven β-globin genes, clustered together on chromosome 11. The polypeptides encoded by these genes join with α-globin polypeptides to form hemoglobin molecules, which transport oxygen in the blood.

Other sequences exist in many copies and are called **repetitive DNA**. Some eukaryotic organisms have large amounts of repetitive DNA; for example, almost half of the human genome consists of repetitive DNA. A major class of repetitive DNA is called **moderately repetitive DNA**, which typically consists of sequences from 150 to 300 bp in length (although they may be longer) that are repeated many thousands of times. Some of these sequences perform important functions for the cell; for example, the genes for ribosomal RNAs (rRNAs) and transfer RNAs (tRNAs) make up a part of the moderately repetitive DNA. However, much of the moderately repetitive DNA has no known function in the cell. Moderately repetitive DNA itself is of two types of repeats. **Tandem repeat sequences** appear one after another and tend to be clustered at particular locations on the chromosomes. **Interspersed repeat sequences** are scattered throughout the genome. An example of an interspersed repeat is the *Alu* sequence, a 200-bp sequence that is present more than a million times

and comprises 11% of the human genome, although it has no obvious ceullar function. Short repeats, such as the *Alu* sequences, are called **SINEs (short interspersed elements)**. Longer interspersed repeats consisting of several thousand base pairs are called **LINEs (long interspersed elements)**. One class of LINE, called LINE1, comprises about 17% of the human genome. Most interspersed repeats are transposable elements, sequences that can multiply and move (see next section).

The other major class of repetitive DNA is **highly repetitive DNA**. These short sequences, often less than 10 bp in length, are present in hundreds of thousands to millions of copies that are repeated in tandem and clustered in certain regions of the chromosome, especially at centromeres and telomeres. Highly repetitive DNA is sometimes called satellite DNA, because its percentages of the four bases differ from those of other DNA sequences and, therefore, it separates as a satellite fraction when centrifuged at high speeds. Highly repetitive DNA is rarely transcribed into RNA. Although these sequences may contribute to centromere and telomere function, most highly repetitive DNA has no known function.

DNA renaturation reactions and, more recently, direct sequencing of eukaryotic genomes also tell us a lot about how genetic information is organized within chromosomes. We now know that the density of genes varies greatly among and within chromosomes. For example, human chromosome 19 has a high density of genes, with about 26 genes per million base pairs. Chromosome 13, on the other hand, has only about 6.5 genes per million base pairs. Gene density can also vary within different regions of the same chromosome: some parts of the long arm of chromosome 13 have only 3 genes per million base pairs, whereas other parts have almost 30 genes per million base pairs. And the short arm of chromosome 13 contains almost no genes, consisting entirely of heterochromatin.

CONCEPTS

Eukaryotic DNA comprises three major classes: unique-sequence DNA, moderately repetitive DNA, and highly repetitive DNA. Unique-sequence DNA consists of sequences that exist in one or only a few copies; moderately repetitive DNA consists of sequences that may be several hundred base pairs in length and is present in thousands to hundreds of thousands of copies. Highly repetitive DNA consists of very short sequences repeated in tandem and is present in hundreds of thousands to millions of copies. The density of genes varies greatly among and even within chromosomes.

✔ CONCEPT CHECK 7

Most of the genes that encode proteins are found in

a. unique-sequence DNA.

b. moderately repetitive DNA.

c. highly repetitive DNA.

d. all of the above.

11.4 Transposable Elements Are DNA Sequences Capable of Moving

Transposable elements are mobile DNA sequences found in the genomes of all organisms. In many genomes, they are quite abundant: for example, they make up at least 45% of human DNA. Most transposable elements are able to insert at many different locations, relying on mechanisms that are distinct from homologous recombination. They often cause mutations, either by inserting into another gene and disrupting it or by promoting DNA rearrangements such as chromosome deletions, duplications, and inversions (see Chapter 9).

General Characteristics of Transposable Elements

There are many different types of transposable elements: some have simple structures, encompassing only those sequences necessary for their own transposition (movement), whereas others have complex structures and encode a number of functions not directly related to transposition. Despite this variation, many transposable elements have certain features in common.

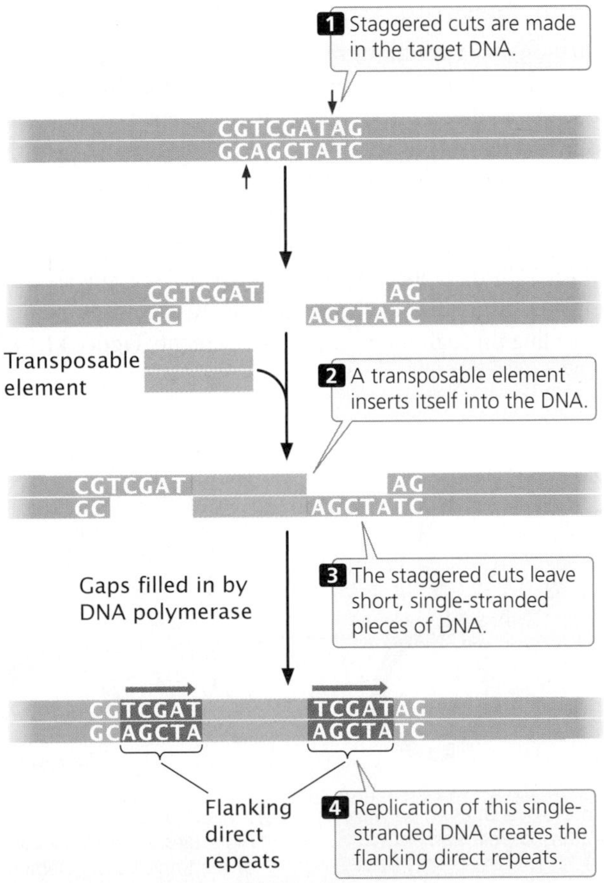

1 Staggered cuts are made in the target DNA.
CGTCGATAG
GCAGCTATC
CGTCGAT
GC
AG
AGCTATC
Transposable element
2 A transposable element inserts itself into the DNA.
CGTCGAT
GC
AG
AGCTATC
Gaps filled in by DNA polymerase
3 The staggered cuts leave short, single-stranded pieces of DNA.
CGTCGAT
GCAGCTA
TCGATAG
AGCTATC
Flanking direct repeats
4 Replication of this single-stranded DNA creates the flanking direct repeats.

11.11 Flanking direct repeats are generated when a transposable element inserts into DNA.

Short **flanking direct repeats** from 3 to 12 bp long are present on both sides of most transposable elements. The sequences of these repeats vary, but the length is constant for each type of transposable element. They are not a part of a transposable element and do not travel with it. Rather, they are generated in the process of transposition, at the point of insertion. In the course of transposition, flanking repeats are created when staggered cuts are made in the target DNA, as shown in **Figure 11.11**. The staggered cuts leave short, single-stranded pieces of DNA on either side of the transposable element. Replication of the single-stranded DNA then creates the flanking direct repeats.

At the ends of many, but not all, transposable elements are **terminal inverted repeats**—sequences from 9 to 40 bp in length that are inverted complements of one another. For example, the following sequences are inverted repeats:

5′–ACAGTTCAG . . . CTGAACTGT–3′
3′–TGTCAAGTC . . . GACTTGACA–5′

On the same strand, the two sequences are not simple inversions, as their name might imply; rather, they are both inverted and complementary. (Notice that the sequence from left to right in the top strand is the same as the sequence from right to left in the bottom strand.) Terminal inverted repeats are recognized by enzymes that catalyze transposition and are required for transposition to take place. **Figure 11.12** summarizes the general characteristics of transposable elements. TRY PROBLEM 33 →

CONCEPTS

Transposable elements are mobile DNA sequences that often cause mutations. There are many different types of transposable elements; most generate short flanking direct repeats at the target sites as they insert. Many transposable elements also possess short terminal inverted repeats.

✔ CONCEPT CHECK 8

How are flanking direct repeats created in transposition?

Transposition

Transposition is the movement of a transposable element from one location to another. Several different mechanisms are used for transposition in both prokaryotic and eukaryotic cells. Nevertheless, all types of transposition have several features in common: (1) staggered breaks are made in the target DNA (see Figure 11.11); (2) the transposable element is joined to single-stranded ends of the target DNA; and (3) DNA is replicated at the single-strand gaps.

Some transposable elements transpose as DNA and are referred to as **DNA transposons** (also called Class II transposable elements). Other transposable elements transpose through an RNA intermediate. In this case, RNA

(a)

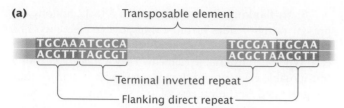

Transposable element

```
TGCAAATCGCA          TGCGATTGCAA
ACGTTTAGCGT          ACGCTAACGTT
```

Terminal inverted repeat

Flanking direct repeat

11.12 Many transposable elements have common characteristics. (a) Most transposable elements generate flanking direct repeats on each side of the point of insertion into target DNA.

(b)

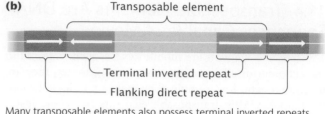

Transposable element

Terminal inverted repeat

Flanking direct repeat

Many transposable elements also possess terminal inverted repeats. (b) These representations of direct and indirect repeats are used in illustrations throughout this chapter.

is transcribed from the transposable element (DNA) and is then copied back into DNA by a special enzyme called reverse transcriptase. Elements that transpose through an RNA intermediate are called **retrotransposons** (also called Class I transposons). Most transposable elements found in bacteria are DNA transposons. Both DNA transposons and retrotransposons are found in eukaryotes, although retrotransposons are more common.

Among DNA transposons, transposition may be replicative or nonreplicative. In **replicative transposition** (also called copy-and-paste transpostion), a new copy of the transposable element is introduced at a new site while the old copy remains behind at the original site, and so the number of copies of the transposable element increases as a result of transposition. In **nonreplicative transposition** (also called cut-and-paste transposition), the transposable element excises from the old site and inserts at a new site without any increase in the number of its copies. Nonreplicative transposition requires the replication of only the few nucleotides that constitute the direct repeats. Retrotransposons use replicative transposition only.

Replicative transposition Replicative transposition can be either between two different DNA molecules or between two parts of the same DNA molecule. **Figure 11.13** summa-

rizes the steps of transposition between two circular DNA molecules. Before transposition (see Figure 11.13a), a single copy of the transposable element is on one molecule. In the first step, the two DNA molecules are joined, and the transposable element is replicated, producing the **cointegrate structure** that consists of molecules A + B fused together with two copies of the transposable element (see Figure 11.13b). We'll soon see how the copy is produced, but let's first look at the second step of the replicative-transposition process. After the cointegrate has formed, crossing over at regions within the copies of the transposable element produces two molecules, each with a single copy of the transposable element (see Figure 11.13c). This second step is known as resolution of the cointegrate.

How are the steps of replicative transposition (cointegrate formation and resolution) brought about? Cointegrate formation requires four events. First, a **transposase** enzyme (often encoded by the transposable element) makes single-strand breaks at each end of the transposable element and on either side of the target sequence where the element inserts (**Figure 11.14a** and **b**). Second, the free ends of the transposable element attach to the free ends of the target sequence (**Figure 11.14c**). Third, replication takes place on the single-strand templates, beginning at the 3′ ends of the single strands and proceeding through the transposable element (**Figure 11.14d**).

(a)

(b) A + B cointegrate

(c) Resolution

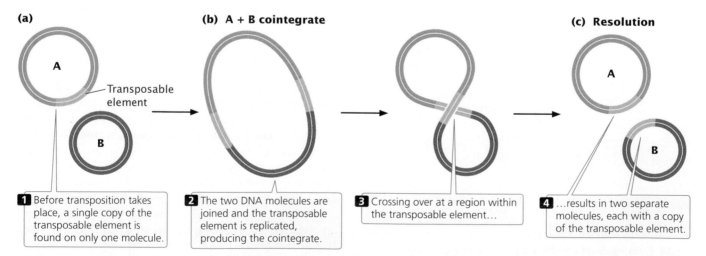

Transposable element

1 Before transposition takes place, a single copy of the transposable element is found on only one molecule.

2 The two DNA molecules are joined and the transposable element is replicated, producing the cointegrate.

3 Crossing over at a region within the transposable element…

4 …results in two separate molecules, each with a copy of the transposable element.

11.13 Replicative transposition increases the number of copies of the transposable element.

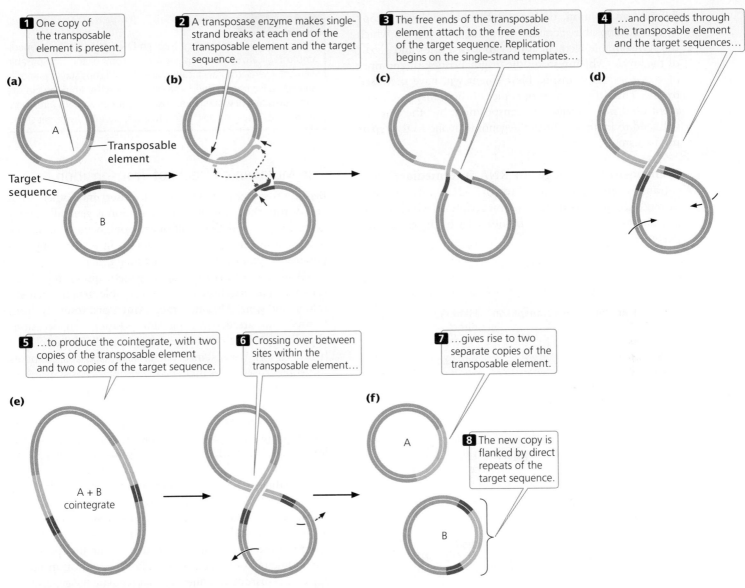

1 One copy of the transposable element is present.

(a)

A

Transposable element

Target sequence

B

2 A transposase enzyme makes single-strand breaks at each end of the transposable element and the target sequence.

(b)

3 The free ends of the transposable element attach to the free ends of the target sequence. Replication begins on the single-strand templates…

(c)

4 …and proceeds through the transposable element and the target sequences…

(d)

5 …to produce the cointegrate, with two copies of the transposable element and two copies of the target sequence.

(e)

A + B cointegrate

6 Crossing over between sites within the transposable element…

7 …gives rise to two separate copies of the transposable element.

(f)

A

8 The new copy is flanked by direct repeats of the target sequence.

B

11.14 Replicative transposition requires single-strand breaks, replication, and resolution.

This replication creates the cointegrate, with its two copies of both the transposable element and the sequence at the target site, which is now on one side of each copy (**Figure 11.14e**). The enzymes that perform the replication and ligation functions are cellular enzymes that function in replication and DNA repair. After the cointegrate has formed, it undergoes resolution, which requires crossing over between sites located within the transposon. Resolution gives rise to two copies of the transposable element (**Figure 11.14f**). The resolution step is brought about by **resolvase** enzymes (encoded in some cases by the transposable element and in other cases by a cellular gene) that function in homologous recombination.

Nonreplicative transposition   In nonreplicative transposition, the transposable element moves from one site to another. There is no replication of the transposable element, although short sequences in the target DNA are replicated, generating flanking direct repeats. Nonreplicative transposition requires only that the transposable element and the target DNA be cleaved and joined together. Cleavage requires a transposase enzyme produced by the transposable element. The joining of the transposable element and the target DNA is probably carried out by normal replication and repair enzymes. If a transposable element moves by nonreplicative transposition, how does it increase in number in the genome? The answer comes from examining the fate of the original site of the element. After excision of the transposable element, a break will be left at the original insertion site. Such breaks are harmful to the cell, and so they are repaired efficiently (see Chapter 18). A common method of repair is to remove and then replicate the broken segment of DNA by using the homologous template on the sister chromatid.

Before transposition, both sister chromatids have a copy of the transposable element. After transposition (in which the transposable element moves to a new site) and repair of the break (which restores the original copy), the number of copies of the transposable element will have increased by one. Thus, the number of copies of the transposable element does not increase by transposition, but the number will tend to increase within the genome owing to the repair mechanism.

Transposition through an RNA intermediate Retrotransposons transpose through RNA intermediates through the process of reverse transcription, which generates flanking direct repeats on both sides of the retrotransposon (**Figure 11.15**).

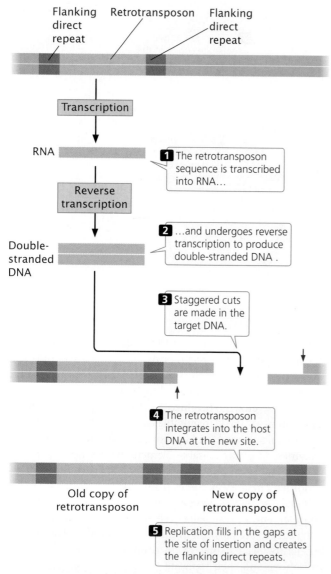

11.15 Retrotransposons transpose through RNA intermediates.

Old copy of retrotransposon   New copy of retrotransposon

CONCEPTS

Transposition may take place through DNA or an RNA intermediate. In replicative transposition, a new copy of the transposable element inserts in a new location and the old copy stays behind; in nonreplicative transposition, the old copy excises from the old site and moves to a new site. Transposition through an RNA intermediate requires reverse transcription to integrate into the target site.

The Mutagenic Effects of Transposition

Because transposable elements can insert into other genes and disrupt their function, transposition is generally mutagenic. In fact, more than half of all spontaneously occurring mutations in *Drosophila* result from the insertion of a transposable element in or near a functional gene.

A number of cases of human genetic disease have been traced to the insertion of a transposable genetic element into a vital gene. Although most mutations resulting from transposition are detrimental, transposition may occasionally activate a gene or change the phenotype of the cell in a beneficial way. For instance, bacterial transposable elements sometimes carry genes that encode antibiotic resistance, and several transposable elements have created mutations that confer insecticide resistance in insects.

A dramatic example of the mutagenic effect of transposable elements is seen in the color of grapes, which come in black, red, and white varieties (**Figure 11.16**). Black and red grapes result from the production of red pigments—called anthocyanins—in the skin, which are lacking in white grapes. White grapes resulted from a mutation in black grapes that turned off the production of anthocyanin pigments. This mutation consisted of the insertion of a 10,422-bp retrotransposon called *Gret1* near a gene that promotes the production of anthocyanins. The *Gret1* retrotransposon apparently disrupted sequences that regulate the gene, effectively shutting down pigment production and producing a white grape with no anthocyanins. Interestingly, red grapes resulted from a second mutation taking place in the white grapes (see Figure 11.16). This mutation (probably resulting from faulty recombination) removed most but not all of the retrotransposon, switching pigment production back on, but not as intensely as in the original black grapes.

Because transposition entails the exchange of DNA sequences and recombination, it often leads to DNA rearrangements. Homologous recombination between multiple copies of transposons can lead to duplications, deletions, and inversions, as shown in **Figure 11.17**. The *Bar* mutation in *Drosophila* (see Figures 9.6 and 9.7) is a tandem duplication thought to have arisen through homologous recombination between two copies of a transposable element present in different locations on the X chromosome. Similarly, recombination between copies of the transposable element *Rider* caused a duplication that results in elongated fruit in tomatoes (see the introduction to this chapter).

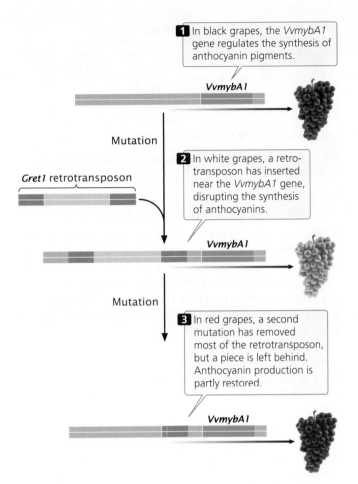

1 In black grapes, the *VvmybA1* gene regulates the synthesis of anthocyanin pigments.

VvmybA1

Mutation

Gret1 retrotransposon

2 In white grapes, a retrotransposon has inserted near the *VvmybA1* gene, disrupting the synthesis of anthocyanins.

VvmybA1

Mutation

3 In red grapes, a second mutation has removed most of the retrotransposon, but a piece is left behind. Anthocyanin production is partly restored.

VvmybA1

11.16 Red and white color in grapes resulted from insertion and deletion of a retrotransposon.

DNA rearrangements can also be caused by the excision of transposable elements in a cut-and-paste transposition. If the broken DNA is not repaired properly, a chromosome rearrangement can be generated. This type of chromosome breakage led to the discovery of transposable elements by Barbara McClintock (described later in this chapter). She named the gene that appeared at these sites *Dissociation*, because of the tendency for it to cause chromosome breakage and the loss of a fragment.

Because most transposable elements insert randomly into DNA sequences, they provide researchers with a powerful tool for inducing mutations throughout the genome, allowing them to determine the functions of genes, study genetic phenomena, and map genes. Furthermore, because the transposable element being used by a researcher has a known sequence, it can serve as a "tag" for locating the gene in which the mutation has occurred. For example, researchers engineered a transposable element named *Sleeping Beauty* to induce mutations in mice and used it to search for genes that cause cancer. *Sleeping Beauty* was introduced into a strain of mice that produce the transposase needed for transposition, and the transposable element

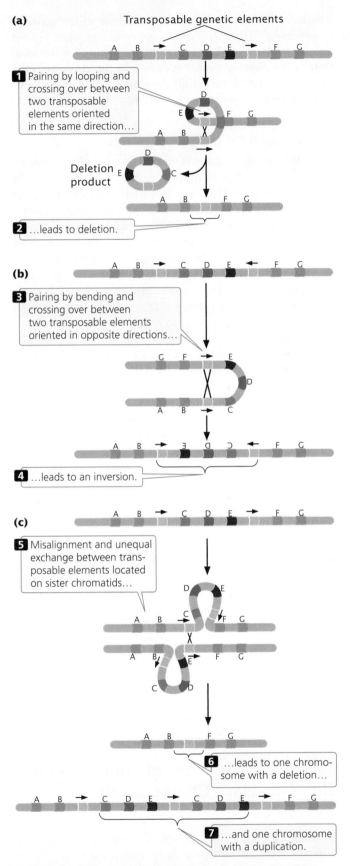

(a)   Transposable genetic elements

1 Pairing by looping and crossing over between two transposable elements oriented in the same direction...

Deletion product

2 ...leads to deletion.

(b)

3 Pairing by bending and crossing over between two transposable elements oriented in opposite directions...

4 ...leads to an inversion.

(c)

5 Misalignment and unequal exchange between transposable elements located on sister chromatids...

6 ...leads to one chromosome with a deletion...

7 ...and one chromosome with a duplication.

11.17 Many chromosomal rearrangements are generated by transposition.

inserted randomly into different locations in the genome. Occasionally, it inserted into a gene that protects against cancer and destroyed its function. By looking for the location of the *Sleeping Beauty* sequence in the DNA from the tumor cells that subsequently developed, geneticists identified a number of genes that protect against cancer.

The Regulation of Transposition

Many transposable elements move through replicative transposition and increase in number with each transposition. In the absence of mechanisms to restrict transposition, the number of copies of transposable elements would increase continuously, and the host DNA would be harmed by the resulting high rate of mutation (caused by the frequent insertion of transposable elements). Furthermore, large amounts of energy and resources would be required to replicate the "extra" DNA in the proliferating transposable elements. For these reasons, it isn't surprising that cells have evolved defense mechanisms to regulate transposition.

Many organims limit transposition by methylating the DNA in regions where transposons are common. DNA methylation usually suppresses transcription (see Chapter 17), preventing the production of the transposase enzyme necessary for transposition. Alterations of chromatin structure also are used to prevent the transcription of transposons. In other cases, translation of the transposase mRNA is controlled (see Chapter 16). For example, fruit flies use small RNA molecules and a process called RNA interference (see Chapter 13) to silence transpose genes and prevent transposition. Other regulatory mechanisms do not affect the level of transposase; rather, they directly inhibit the transposition event.

CONCEPTS

Transposable elements frequently cause mutations and DNA rearrangements. Many cells regulate transposition by altering DNA or chromatin structure, by controlling the amount of transposase produced, or by direct inhibition of the transposition event.

✔ CONCEPT CHECK 9

Briefly explain how transposition causes mutations and chromosome rearrangements.

11.5 Different Types of Transposable Elements Have Characteristic Structures

Bacteria and eukaryotic organisms possess a number of different types of transposable elements, the structures of which vary extensively. In this section, we consider the structures of representative types of transposable elements.

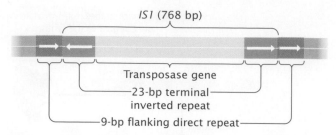

11.18 Insertion sequences are simple transposable elements found in bacteria.

Transposable Elements in Bacteria

The DNA transposons found in bacteria (there are no retrotransposons in bacteria) constitute two major groups: (1) simple transposable elements, called insertion sequences, that carry only the information required for movement and (2) more-complex transposable elements, called composite transposons, that contain DNA sequences not directly related to transposition.

Insertion sequences   The simplest type of transposable element in bacterial chromosomes and plasmids is an **insertion sequence** (IS). This type of element carries only the genetic information necessary for its movement. Insertion sequences are common constituents of bacteria; they can also infect plasmids and viruses and, in this way, can be passed from one cell to another. Geneticists designate each type of insertion sequence with *IS* followed by an identifying number. For example, *IS1* is a common insertion sequence found in *E. coli.*

A number of different insertion sequences have been found in bacteria. They are typically from 800 to 2000 bp in length and possess the two hallmarks of transposable elements: terminal inverted repeats and the generation of flanking direct repeats at the site of insertion. Most insertion sequences contain one or two genes that encode transposase. *IS1*, a typical insertion sequence, is shown in **Figure 11.18**. **Table 11.4** summarizes these features for several bacterial insertion sequences. **TRY PROBLEM 36** →

Table 11.4	Structures of some common insertion sequences		
		Length of	
Insertion Sequence	Total Length (bp)	Inverted Repeats (bp)	Flanking Direct Repeats (bp)
IS1	768	23	9
IS2	1327	41	5
IS4	1428	18	11 or 12
IS5	1195	16	4

Source: B. Lewin, *Genes*, 3d ed. (New York: Wiley, 1987), p. 591.

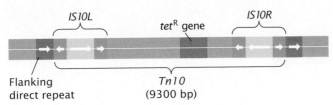

11.19 *Tn10* **is a composite transposon in bacteria.**

Composite transposons   Any segment of DNA that becomes flanked by two copies of an insertion sequence may itself transpose and is called a **composite transposon**. Each composite transposon is designated by the abbreviation *Tn*, followed by a number. The composite transposon *Tn10*, for example, is consists of about 9300 bp that carries a gene for tetracycline resistance between two *IS10* insertion sequences (**Figure 11.19**). The insertion sequences have terminal inverted repeats; so the composite transposon also ends in inverted repeats. Composite transposons also generate flanking direct repeats at their sites of insertion (see Figure 11.19).The insertion sequences at the ends of a composite transposon may be in the same orientation or they may be inverted relative to one other (as in *Tn10*).

The insertion sequences at the ends of a composite transposon are responsible for transposition. The DNA between the insertion sequences is not required for movement and may carry additional information (such as antibiotic resistance). Presumably, composite transposons evolve when one insertion sequence transposes to a location close to another of the same type. The transposase produced by one of the insertion sequences catalyzes the transposition of both insertions sequences, allowing them to move together and carry along the DNA that lies between them. In some composite transposons (such as *Tn10*), one of the insertion sequences may be defective; so its movement depends on the transposase produced by the other. Characteristics of several composite transposons are listed in **Table 11.5**.

Noncomposite transposons   Some transposable elements in bacteria lack insertion sequences and are referred to as noncomposite transposons. Noncomposite transposons possess a gene for transposase and have inverted repeats at their ends. For instance, the noncomposite transposon *Tn3* carries genes for transposase and resolvase (mentioned

Table 11.5	Characteristics of several composite transposons		
Composite Transposon	Total Length (bp)	Associated IS Elements	Other Genes Within the Transposon
Tn9	2500	*IS1*	Chloramphenicol resistance
Tn10	9300	*IS10*	Tetracycline resistance
Tn5	5700	*IS50*	Kanamycin resistance
Tn903	3100	*IS903*	Kanamycin resistance

earlier in this chapter), plus a gene that encodes the enzyme β-lactamase, which provides resistance to the antibiotic ampicillin.

A few bacteriophage genomes reproduce by transposition and use transposition to insert themselves into a bacterial chromosome in their lysogenic cycle; the best studied of these transposing bacteriophages is Mu (**Figure 11.20**). Although Mu does not possess terminal inverted repeats, it does generate short (5-bp) flanking direct repeats when it inserts randomly into DNA. Mu replicates through transposition and causes mutations at the site of insertion, properties characteristic of transposable elements.

CONCEPTS

Insertion sequences are prokaryotic transposable elements that carry only the information needed for transposition. A composite transposon consists of two insertion sequences plus intervening DNA. Noncomposite transposons in bacteria lack insertion sequences but have terminal inverted repeats and carry information not related to transposition. All of these transposable elements generate flanking direct repeats at their points of insertion.

✔ CONCEPT CHECK 10

Which type of transposable element possesses terminal inverted repeats?

a. Insertion sequence   c. Noncomposite transposon *Tn3*

b. Composite transposons   d. All the above

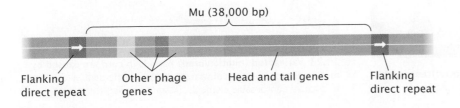

11.20 Mu is a transposing bacteriophage.

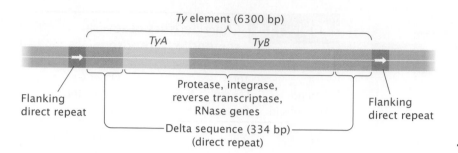

Ty element (6300 bp)

TyA   *TyB*

Flanking
direct repeat

Protease, integrase,
reverse transcriptase,
RNase genes

Flanking
direct repeat

Delta sequence (334 bp)
(direct repeat)

11.21 *Ty* **is a transposable element in yeast.**

Transposable Elements in Eukaryotes

Eukaryotic transposable elements can be divided into two groups. One group is structurally similar to transposable elements found in bacteria, typically ending in short inverted repeats and transposing as DNA. Examples of this type include the *P* elements in *Drosophila* and the *Ac* and *Ds* elements in maize (corn). The other group comprises retrotransposons (see Figure 11.15); they use RNA intermediates, and many are similar in structure and movement to retroviruses (see pp. 227–229 in Chapter 8). On the basis of their structure, function, and genomic sequences, some retrotransposons are clearly evolutionarily related to retroviruses. Although their mechanism of movement is fundamentally different from that of other transposable elements, retrotransposons also generate direct repeats at the point of insertion. Retrotransposons include the *Ty* elements in yeast, the *copia* elements in *Drosophila*, and the *Alu* sequences in humans.

Ty elements in yeast *Ty* (for transposon yeast) elements are a family of common retrotransposons found in yeast; many yeast cells have 30 copies of *Ty* elements (**Figure 11.21**).

At each end of a *Ty* element are direct repeats called **delta sequences**, which are 334 bp long. The delta sequences are analogous to the long terminal repeats found in retroviruses and contain several genes that are related to the *gag* and *pol* genes present in retroviruses (see p. 227 in Chapter 8). The delta sequences also contain promoters required for the transcription of *Ty* genes, and the promoters can also stimulate the transcription of genes that lie downstream of the *Ty* element.

Ac and Ds elements in maize Transposable elements were first identified in maize more than 50 years ago by Barbara McClintock (**Figure 11.22**). McClintock spent much of her long career studying their properties, and her work stands among the landmark discoveries of genetics. Her results, however, were misunderstood and ignored for many years. Not until molecular techniques were developed in the late 1960s and 1970s did the importance of transposable elements become widely accepted. The significance of McClintock's early discoveries was finally recognized in 1983, when she was awarded the Nobel Prize in physiology or medicine.

McClintock's discovery of transposable elements had its genesis in the early work of Rollins A. Emerson on the maize genes that caused variegated (multicolored) kernels. Most corn kernels are either wholly pigmented or colorless (yellow), but Emerson noted that some yellow kernels had spots or streaks of color (**Figure 11.23**). He proposed that these kernels resulted from an unstable mutation: a mutation in

11.22 Barbara McClintock was the first to discover transposable elements. [Courtesy of Cold Spring Harbor Laboratory Archives.]

11.23 Variegated (multicolored) kernels in corn are caused by mobile genes. The study of variegated corn led Barbara McClintock to discover transposable elements. [Matt Meadows/Peter Arnold.]

(a) *Ac* **element**

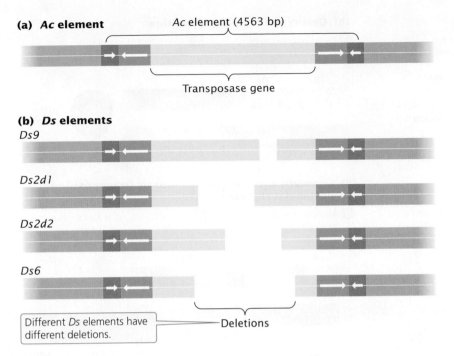

11.24 *Ac* **and** *Ds* **are transposable elements in maize.**

the wild-type gene for pigment produced a colorless kernel but, in some cells, the mutation reverted back to the wild type, causing a spot of pigment. However, Emerson didn't know why these mutations were unstable.

McClintock discovered that the cause of the unstable mutation was a gene that moved. She noticed that chromosome breakage in maize often occurred at a gene that she called *Dissociation* (*Ds*) but only if another gene, the *Activator* (*Ac*), also was present. *Ds* and *Ac* exhibited unusual patterns of inheritance; occasionally, the genes moved together. McClintock called these moving genes controlling elements, because they controlled the expression of other genes.

Since the significance of McClintock's work was recognized, *Ac* and *Ds* elements in maize have been examined in detail. They are DNA transposons that possess terminal inverted repeats and generate flanking direct repeats at the points of insertion (**Figure 11.24a**). Each *Ac* element contains a single gene that encodes a transposase enzyme. Thus *Ac* elements are *autonomous*—that is, able to transpose. *Ds* elements are *Ac* elements with one or more deletions that have inactivated the transposase gene (**Figure 11.24b**). Unable to transpose on their own (*nonautonomous*), *Ds* elements can transpose in the presence of *Ac* elements because they still possess terminal inverted repeats recognized by *Ac* transposase.

Each kernel in an ear of corn is an individual, originating as an ovule fertilized by a pollen grain. A kernel's pigment pattern is determined by several loci. A pigment-encoding allele at one of these loci can be designated *C*, and an allele at the same locus that does not confer pigment can be designated *c*. A kernel with genotype *cc* will be color-

less—that is, yellow or white (**Figure 11.25a**); a kernel with genotype *CC* or *Cc* will produce pigment and be purple (**Figure 11.25b**).

A *Ds* element, transposing under the influence of a nearby *Ac* element, may insert into the *C* allele, destroying its ability to produce pigment (**Figure 11.25c**). An allele inactivated by a transposable element is designated by a subscript "t"; so, in this case, the allele would be C_t. After the transposition of *Ds* into the *C* allele, the kernel cell has genotype $C_t c$. This kernel will be colorless (white or yellow), because neither the C_t allele nor the *c* allele confers pigment.

As the original one-celled maize embryo develops and divides by mitosis, additional transpositions may take place in some cells. In any cell in which the transposable element excises from the C_t allele and moves to a new location, the *C* allele is rendered functional again: all cells derived from those in which this event has taken place will have the genotype *Cc* and be purple. The presence of these pigmented cells, surrounded by the colorless ($C_t c$) cells, produces a purple spot or streak (called a sector) in the otherwise yellow kernel (**Figure 11.25d**). The size of the sector varies, depending on when the excision of the transposable element from the C_t allele takes place. If excision is early in development, then many cells will contain the functional *C* allele and the pigmented sector will be large; if excision is late in development, few cells will have the functional *C* allele and the pigmented sector will be small. TRY PROBLEM 42 →

Transposable elements in *Drosophila* A number of different transposable elements are found in *Drosophila*. One family of *Drosophila* transposable elements comprises the *P* elements. Most functional *P* elements

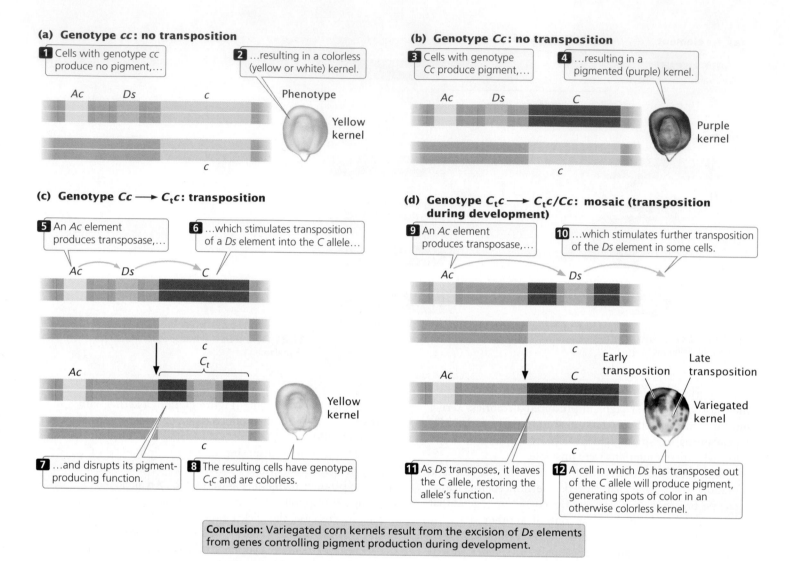

(a) Genotype _cc_: no transposition

1 Cells with genotype _cc_ produce no pigment,…

2 …resulting in a colorless (yellow or white) kernel.

_Ac_   _Ds_   _c_

Phenotype

Yellow kernel

_c_

(b) Genotype _Cc_: no transposition

3 Cells with genotype _Cc_ produce pigment,…

4 …resulting in a pigmented (purple) kernel.

_Ac_   _Ds_   _C_

Purple kernel

_c_

(c) Genotype _Cc_ ⟶ _C_t_c_: transposition

5 An _Ac_ element produces transposase,…

6 …which stimulates transposition of a _Ds_ element into the _C_ allele…

_Ac_   _Ds_   _C_

_c_

_C_t

_Ac_

Yellow kernel

_c_

7 …and disrupts its pigment-producing function.

8 The resulting cells have genotype _C_t_c_ and are colorless.

(d) Genotype _C_t_c_ ⟶ _C_t_c_/_Cc_: mosaic (transposition during development)

9 An _Ac_ element produces transposase,…

10 …which stimulates further transposition of the _Ds_ element in some cells.

_Ac_   _Ds_

_c_

Early transposition

Late transposition

_Ac_   _C_

Variegated kernel

_c_

11 As _Ds_ transposes, it leaves the _C_ allele, restoring the allele's function.

12 A cell in which _Ds_ has transposed out of the _C_ allele will produce pigment, generating spots of color in an otherwise colorless kernel.

Conclusion: Variegated corn kernels result from the excision of _Ds_ elements from genes controlling pigment production during development.

11.25 Transposition results in variegated maize kernels.

are about 2900 bp long, although shorter _P_ elements containing deletions also exist. Each _P_ element possesses terminal inverted repeats and generates flanking direct repeats at the site of insertion. Like transposable elements in bacteria, _P_ elements are DNA transposons, transposing as DNA. Each element encodes both a transposase and a repressor of transposition.

The role of this repressor in controlling transposition is demonstrated dramatically in **hybrid dysgenesis**, which is the sudden appearance of numerous mutations, chromosome aberrations, and sterility in the offspring of a cross between a P^+ male fly (_with P elements_) and a P^- female fly (_without_ them). The reciprocal cross between a P^+ female and a P^- male produces normal offspring.

Hybrid dysgenesis arises from a burst of transposition when _P_ elements are introduced into a cell that does not possess them. In a cell that contains _P_ elements, a repressor in the cytoplasm inhibits transposition. When a P^+ female

produces eggs, the repressor protein is incorporated into the egg cytoplasm, which prevents further transposition in the embryo and thus prevents mutations from arising. The resulting offspring are fertile as adults (**Figure 11.26a**). However, a P^- female does not produce the repressor protein; so none is stored in the cytoplasm of her eggs. Sperm contain little or no cytoplasm, so a P^+ male does not contribute the repressor protein to his offspring. When eggs from a P^- female are fertilized by sperm from a P^+ male, the absence of repression allows the _P_ elements contributed by the sperm to undergo rapid transposition in the embryo, causing hybrid dysgenesis (**Figure 11.26b**). **TRY PROBLEM 37 →**

Transposable elements in humans About 45% of the human genome consists of sequences derived from transposable elements, although most of these elements are now inactive and no longer capable of transposing. A comparison of human and chimpanzee genomes suggests

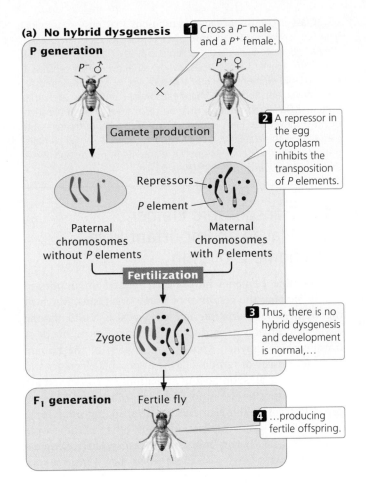

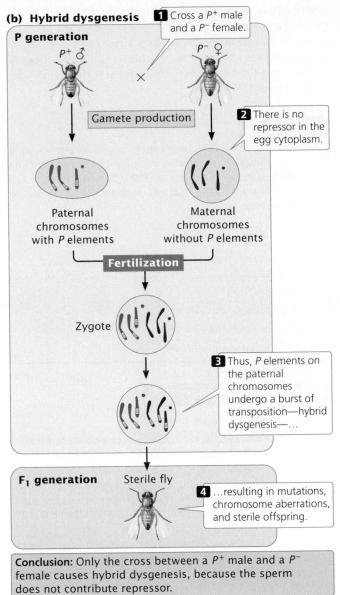

(a) No hybrid dysgenesis

P generation

P^- ♂   × P^+ ♀

1 Cross a P^- male and a P^+ female.

Gamete production

2 A repressor in the egg cytoplasm inhibits the transposition of *P* elements.

Repressors

P element

Paternal chromosomes without *P* elements

Maternal chromosomes with *P* elements

Fertilization

Zygote

3 Thus, there is no hybrid dysgenesis and development is normal,…

F₁ generation   Fertile fly

4 …producing fertile offspring.

(b) Hybrid dysgenesis

P generation

P^+ ♂   × P^- ♀

1 Cross a P^+ male and a P^- female.

Gamete production

2 There is no repressor in the egg cytoplasm.

Paternal chromosomes with *P* elements

Maternal chromosomes without *P* elements

Fertilization

Zygote

3 Thus, *P* elements on the paternal chromosomes undergo a burst of transposition—hybrid dysgenesis—…

F₁ generation   Sterile fly

4 …resulting in mutations, chromosome aberrations, and sterile offspring.

11.26 Hybrid dysgenesis in *Drosophila* is caused by the transposition of *P* elements. [After W. Y. Chooi, *Genetics* 68:213–230, 1971.]

Conclusion: Only the cross between a P^+ male and a P^- female causes hybrid dysgenesis, because the sperm does not contribute repressor.

that almost 11,000 transposition events have taken place since these two species diverged approximately 5 million to 6 million years ago.

One of the most common transposable elements in the human genome is *Alu*. Every human cell contains more than 1 million related, but not identical, copies of *Alu* in its chromosomes. *Alu* sequences are similar to the gene that encodes the 7S RNA molecule, which transports newly synthesized proteins across the endoplasmic reticulum. *Alu* sequences create short flanking direct repeats when they insert into DNA and have characteristics that suggest that they have transposed through an RNA intermediate.

Alu belongs to a class of repetitive sequences found frequently in mammalian and some other genomes. These sequences are collectively referred to as SINEs and constitute about 11% of the human genome. Most SINEs are copies

of transposable elements that have been shortened at the 5′ end, probably because the reverse-transcription process used in their transposition terminated before the entire sequence was copied. SINEs have been identified as the cause of mutations in more than 20 cases of human genetic disease.

The human genome also has many transposons classified as LINEs, which are somewhat more similar in structure to retroviruses. Like SINEs, most LINEs in the human genome have been shortened at the 5′ end. The longest LINEs are usually about 6000 bp but, because most copies are shortened, the average LINE is only about 900 bp. Human DNA contains three major families of LINEs that differ in sequences. There are approximately 900,000 copies of LINEs in the human genome, collectively constituting 21% of the total human DNA. One of every 600 mutations that cause significant disease in humans results from the transposition of a LINE or a SINE element. The human genome contains evidence for several classes of transposable elements that transpose as DNA by a cut-and-paste mechanism. However, all appear to have been inactive for about 50 million years; the nonfunctional sequences that remain are referred to as DNA fossils.

CONCEPTS

A great variety of transposable elements exist in eukaryotes. Some resemble transposable elements in prokaryotes, having terminal inverted repeats, and transpose as DNA. Others are retrotransposons with long direct repeats at their ends and transpose through an RNA intermediate.

✔ CONCEPT CHECK 11

Hybrid dysgenesis results when

a. a male fly with P elements (P^+) mates with a female fly that lacks P elements (P^-).

b. a P^- male mates with a P^+ female.

c. a P^+ male mates with a P^+ female.

d. a P^- male mates with a P^- female.

CONNECTING CONCEPTS

Classes of Transposable Elements

Now that we have looked at some examples of transposable elements, let's take a moment to review their major classes (**Table 11.6**).

Transposable elements can be divided into two major classes on the basis of structure and movement. Class I comprises the retrotransposons, which possess terminal direct repeats and transpose through RNA intermediates. They generate flanking direct repeats at their points of insertion when they transpose into DNA. Retrotransposons do not encode transposase, but some types are similar in structure to retroviruses and carry sequences that produce reverse transcriptase. Transposition takes place when transcription produces an RNA intermediate, which is then transcribed into DNA by reverse transcriptase and inserted into the target site. Examples of retrotransposons include *Ty* elements in yeast and *Alu* sequences in humans. Retrotransposons are not found in prokaryotes.

Class II consists of DNA transposons that possess terminal inverted repeats and transpose as DNA. Like Class I transposons, they all generate flanking direct repeats at their points of insertion into DNA. Unlike Class I transposons, all active forms of Class II transposable elements encode transposase, which is required for their movement. Some also encode resolvase, repressors, and other proteins. Their transposition may be replicative or nonreplicative, but they never use RNA intermediates. Examples of transposable elements in this class include insertion sequences and all complex transposons in bacteria, *Ac* and *Ds* elements in maize, and *P* elements in *Drosophila*. TRY PROBLEM 41 →

11.6 Transposable Elements Have Played an Important Role in Genome Evolution

Transposable elements have clearly played an important role in shaping the genomes of many organisms. Much of the tremendous variation in genome size found among eukaryotic organisms is due to differences in numbers of transposable elements. Approximately 45% of the human genome consists of remnants of transposable elements and about 50% of all spontaneous mutations in *Drosophila* are due to transposition. Homologous recombination between copies of transposable elements has been an important force in producing gene duplications and other chromosome rearrangements. Furthermore, some transposable elements may carry extra DNA with them when they transpose to a new site, providing the potential to move DNA sequences that regulate genes to new sites, where they may alter the expression of genes.

The Evolution of Transposable Elements

Transposable elements exist in all organisms, often in large numbers. Why are they so common? Several different ideas have been proposed to explain their widespread presence.

Transposable elements as genomic parasites   As we have seen, many transposble elements leave a copy behind when they transpose to a new location (copy-and-paste

Table 11.6	Characteristics of two major classes of transposable genetic elements			
	Structure	Genes Encoded	Transposition	Examples
Class I (retrotransposon)	Long terminal direct repeats; short flanking direct repeats at target site	Reverse-transcriptase gene (and sometimes others)	By RNA intermediate	*Ty* (yeast) *copia* (*Drosophila*) *Alu* (human)
Class II	Short terminal inverted repeats; short flanking direct repeats at target site	Transposase gene (and sometimes others)	Through DNA (replicative or nonreplicative)	*IS1* (*E. coli*) *Tn3* (*E. coli*) *Ac, Ds* (maize) *P* elements (*Drosophila*)

transposition) and therefore increase in number within a genome with the passage of time. This ability to replicate and spread means that many transposable elements may serve no purpose for the cell; they exist simply because they are capable of replicating and spreading. The presence of transposable elements in the genome may actually be detrimental. The insertion of transposable elements into a gene will often destroy its function, with harmful consequences for the cell. Furthermore, the time and energy required to replicate large numbers of transposable elements are likely to place a metabolic burden on the cell. Thus, transposable elements can be thought of as genomic parasites that provide no benefit to the cell and may even be harmful. Their capacity to reproduce and spread is what makes them common, not their necessity to the cell.

Transposable elements and genetic variation The process of evolution requires the presence of genetic variation within a population (see Chapter 26). Because transposable elements induce mutations and chromosome rearrangements, some scientists have argued that they exist because they generate genetic variation, which facilitates evolutionary adaptation. This idea suggests that a certain amount of genetic variation is useful because it allows a species to adapt to environmental change. However, much of what we know about transposable elements and evolution works against this hypothesis. Some mutations caused by transposable elements may allow species to evolve beneficial traits, but most mutations generated by random transposition have deleterious immediate effects and will likely be eliminated by natural selection. Furthermore, the fact that many organisms have evolved mechanisms to regulate transposition suggests that there is selective pressure to limit the extent of transposition. Thus, although transposable elments do generate genetic variation that is occasionally beneficial to the species, it is unlikely to be the only reason for their existence.

Domestication of Transposable Elements

Regardless of the evolutionary reasons for their existence, some transposable elements have clearly evolved to serve useful purposes for their host cells. These transposons are sometimes refered to as domesticated, implying that their parasitic tendencies have been replaced by properties useful to the cell.

In the plant *Arabidopsis thaliana*, a transposase protein encoded by a transposable element regulates plant genes and has become essential for normal plant growth. Another example is the mechanism that generates antibody diversity in the immune systems of vertebrates (see Chapter 22). Immune cells called lymphocytes have the ability to unite several DNA segments that encode antigen-recognition proteins. This mechanism may have arisen from a transposable element that inserted into the germ line of a vertebrate ancestor some 450 million years ago.

Another cellular function that may have originated as the result of a transposable element is the process that maintains the ends of chromosomes in eukaryotic organisms. As mentioned earlier in this chapter, DNA polymerases are unable to replicate the ends of chromosomes. In germ cells and single-celled eukaryotic organisms, chromosome length is maintained by the enzyme telomerase. The mechanism used by telomerase is similar to the reverse-transcription process used in retrotransposition, and telomerase is evolutionarily related to the reverse transcriptases encoded by certain retrotransposons. These findings suggest that an invading retrotransposon in an ancestral eukaryotic cell may have provided the ability to copy the ends of chromosomes and eventually evolved into the gene that encodes the modern telomerase enzyme. *Drosophila* lacks the telomerase enzyme; retrotransposons appear to have resumed the role of telomere maintenance in this case.

The evolution of new genes through tranposons

Transposable elements that employ the cut-and-paste mechanism of transposition often imprecisely excise from the DNA. Imprecise excision may result in part of the transposon sequence being left behind when the transposon moves, leaving a genetic footprint of the transposable element at the site of exision. Alternatively, transposable elements may pick up parts of cellular genes when they excise, which then become part of the transposable element and are moved to new locations in the genome.

Mutator-like transposable elements (MULEs) are found in plants and use cut-and-paste transposition. MULEs often carry gene fragments when they excise; gene-carrying MULEs are referred to as Pack-MULEs. Recent sequencing of the rice genome has revealed the presence of more than 3000 Pack-MULEs. These Pack-MULEs often contain multiple fragments of genes from different chromosomal locations, which are shuffled into different combinations and fused together. This ability to pick up pieces of genes, shuffle them into new combinations, and move them to new locations provides a mechanism by which new genes may evolve. The extent to which new genes actually evolve in this way is unknown.

CONCEPTS

Many transposable elements appear to be genomic parasites, existing in large numbers because of their ability to efficiently increase in copy number. Increases in copy number of transposable elements have contributed to the large size of may eukaryotic genomes. In several cases, transposable elements and their ability to transpose have been adopted for specific cellular functions.

✔ **CONCEPT CHECK 12**

What evidence suggests that the ability to replicate telomeres may have evolved from a retrotransposon?

CONCEPTS SUMMARY

- Chromosomes contain very long DNA molecules that are tightly packed.

- Supercoiling results from strain produced when rotations are added to a relaxed DNA molecule or removed from it. Overrotation produces positive supercoiling; underrotation produces negative supercoiling. Supercoiling is controlled by topoisomerase enzymes.

- A bacterial chromosome consists of a single, circular DNA molecule that is bound to proteins and exists as a series of large loops. It usually appears in the cell as a distinct clump known as the nucleoid.

- Each eukaryotic chromosome contains a single, long linear DNA molecule that is bound to histone and nonhistone chromosomal proteins. Euchromatin undergoes the normal cycle of decondensation and condensation in the cell cycle. Heterochromatin remains highly condensed throughout the cell cycle.

- The nucleosome is a core of eight histone proteins and the DNA that wraps around the core.

- Nucleosomes are folded into a 30-nm fiber that forms a series of 300-nm-long loops; these loops are anchored at their bases by proteins associated with the nuclear scaffold. The 300-nm loops are condensed to form a fiber that is itself tightly coiled to produce a chromatid.

- Chromosome regions that are undergoing active transcription are sensitive to digestion by DNase I, indicating that DNA unfolds during transcription.

- Epigenetic changes are stable alterations of gene expression that do not require changes in DNA sequences. Epigenetic changes can take place through alterations of DNA or chromatin structure.

- Centromeres are chromosomal regions where spindle fibers attach; chromosomes without centromeres are usually lost in the course of cell division. Most centromeres are defined by epigenetic changes to chromatin structure. Telomeres stabilize the ends of chromosomes.

- Eukaryotic DNA exhibits three classes of sequences. Unique-sequence DNA exists in very few copies. Moderately repetitive DNA consists of moderately long sequences that are repeated from hundreds to thousands of times. Highly repetitive DNA consists of very short sequences that are repeated in tandem from many thousands to millions of times.

- Transposable elements are mobile DNA sequences that insert into many locations within a genome and often cause mutations and DNA rearrangements.

- Most transposable elements have two common characteristics: terminal inverted repeats and the generation of short direct repeats in DNA at the point of insertion.

- Transposition may take place through a DNA molecule or through the production of an RNA molecule that is then reverse transcribed into DNA. Transposition may be replicative, in which the transposable element is copied and the copy moves to a new site, or nonreplicative, in which the transposable element excises from the old site and moves to a new site.

- Retrotransposons transpose through RNA molecules that undergo reverse transcription to produce DNA.

- Insertion sequences are small bacterial transposable elements that carry only the information needed for their own movement. Composite transposons in bacteria are more complex elements that consist of DNA between two insertion sequences. Some complex transposable elements in bacteria do not contain insertion sequences.

- DNA transposons in eukaryotic cells are similar to those found in bacteria, ending in short inverted repeats and producing flanking direct repeats at the point of insertion. Others are retrotransposons, similar in structure to retroviruses and transposing through RNA intermediates.

- Transposons have played an important role in genome evolution.

IMPORTANT TERMS

transposable element (p. 291)
supercoiling (p. 292)
relaxed state of DNA (p. 292)
positive supercoiling (p. 292)
negative supercoiling (p. 292)
topoisomerase (p. 292)
nucleoid (p. 293)
euchromatin (p. 294)
heterochromatin (p. 294)
nonhistone chromosomal protein (p. 294)
nucleosome (p. 295)

chromatosome (p. 295)
linker DNA (p. 295)
polytene chromosome (p. 297)
chromosomal puff (p. 297)
epigenetic change (p. 298)
centromeric sequence (p. 299)
telomeric sequence (p. 300)
shelterin (p. 300)
telomere-associated sequence (p. 301)
C value (p. 301)
denaturation (melting) (p. 301)
melting temperature (T_m) (p. 301)

renaturation (reannealing) (p. 302)
hybridization (p. 302)
unique-sequence DNA (p. 302)
gene family (p. 302)
repetitive DNA (p. 302)
moderately repetitive DNA (p. 302)
tandem repeat sequence (p. 302)
interspersed repeat sequence (p. 302)
short interspersed element (SINE) (p. 302)
long interspersed element (LINE) (p. 302)
highly repetitive DNA (p. 302)

flanking direct repeat (p. 303)
terminal inverted repeat (p. 303)
transposition (p. 303)
DNA transposon (p. 303)
retrotransposon (p. 304)

replicative transposition (p. 304)
nonreplicative transposition (p. 304)
cointegrate structure (p. 304)
transposase (p. 304)
resolvase (p. 305)

insertion sequence (IS) (p. 308)
composite transposon (p. 309)
delta sequence (p. 310)
hybrid dysgenesis (p. 312)

ANSWERS TO CONCEPT CHECKS

1. b

2. Bacterial DNA is not complexed to histone proteins and is circular.

3. b

4. d

5. A chromosome that loses its centromere will not segregate into the nucleus in mitosis and is usually lost.

6. d

7. a

8. In transposition, staggered cuts are made in DNA and the transposable element inserts into the cut. Later, replication of the single-stranded pieces of DNA creates short flanking direct repeats on either side of the inserted transposable element.

9. Transposition often results in mutations because the transposable element inserts into a gene, destroying its function. Chromosome rearrangements arise because transposition includes the breaking and exchange of DNA sequences. Additionally, multiple copies of a transposable element may undergo homologous recombination, producing chromosome rearrangements.

10. d

11. a

12. The mechanism used by the telomerase enzyme is similar to the reverse-transcription process used in retrotransposition; additionally, telomerase is evolutionarily related to the reverse transcriptases encoded by certain retrotransposons.

WORKED PROBLEMS

1. A diploid plant cell contains 2 billion base pairs of DNA.

a. How many nucleosomes are present in the cell?

b. Give the numbers of molecules of each type of histone protein associated with the genomic DNA.

• Solution

Each nucleosome encompasses about 200 bp of DNA: from 145 to 147 bp of DNA wrapped around the histone core, from 20 to 22 bp of DNA associated with the H1 protein, and another 30 to 40 bp of linker DNA.

a. To determine how many nucleosomes are present in the cell, we simply divide the total number of base pairs of DNA (2×10^9 bp) by the number of base pairs per nucleosome:

$$\frac{2 \times 10^9 \text{ nucleotides}}{2 \times 10^2 \text{ nucleotides per nucleosome}} = 1 \times 10^7 \text{ nucleosomes}$$

Thus, there are approximately 10 million nucleosomes in the cell.

b. Each nucleosome includes two molecules each of H2A, H2B, H3, and H4 histones. Therefore, there are 2×10^7 molecules each of H2A, H2B, H3, and H4 histones. Each nucleosome has associated with it one copy of the H1 histone; so there are 1×10^7 molecules of H1.

2. Certain repeated sequences in eukaryotes are flanked by short direct repeats, suggesting that they originated as transposable elements. These same sequences lack introns and possess a string of thymine nucleotides at their 3′ ends. Have these elements transposed through DNA or RNA sequences? Explain your reasoning.

• Solution

The absence of introns and the string of thymine nucleotides (which would be complementary to adenine nucleotides in RNA) at the 3′ end are characteristics of processed RNA. These similarities to RNA suggest that the element was originally transcribed into mRNA, processed to remove the introns and to add a poly(A) tail, and then reverse transcribed into a complementary DNA that was inserted into the chromosome.

3. Which of the following pairs of sequences might be found at the ends of an insertion sequence?

a.	5′–TAAGGCCG–3′	and	5′–TAAGGCCG–3′
b.	5′–AAAGGGCTA–3′	and	5′–ATCGGGAAA–3′
c.	5′–GATCCCAGTT–3′	and	5′–CTAGGGTCAA–3′
d.	5′–GATCCAGGT–3′	and	5′–ACCTGGATC–3′
e.	5′–AAAATTTT–3′	and	5′–TTTTAAAA–3′
f.	5′–AAAATTTT–3′	and	5′–AAAATTTT–3′

• Solution

The correct answer is parts *d* and *f*. The ends of all insertion sequences have inverted repeats, which are sequences on the same strand that are inverted and complementary. The sequences in part *a* are direct repeats, which are generated on the outside of an insertion sequence but are not part of the transposable element itself. The sequences in part *b* are inverted but not complementary. The sequences in part *c* are complementary but not inverted. The sequences in part *d* are both inverted and complementary. The sequences in part *e* are complementary but not inverted. Interestingly, the sequences in part *f* are both inverted complements and direct repeats.

COMPREHENSION QUESTIONS

Section 11.1

*1. How does supercoiling arise? What is the difference between positive and negative supercoiling?

2. What functions does supercoiling serve for the cell?

*3. Describe the composition and structure of the nucleosome. How do core particles differ from chromatosomes?

4. Describe in steps how the double helix of DNA, which is 2 nm in width, gives rise to a chromosome that is 700 nm in width.

5. What are polytene chromosomes and chromosomal puffs?

6. What are epigenetic changes and how are they brought about?

Section 11.2

*7. Describe the function of the centromere. How are centromeres different from other regions of the chromosome?

*8. Describe the function and molecular structure of a telomere.

9. What is the difference between euchromatin and heterochromatin?

Section 11.3

10. What is the C value of an organism?

*11. Describe the different types of DNA sequences that exist in eukaryotes.

Section 11.6

*12. What general characteristics are found in many transposable elements? Describe the differences between replicative and nonreplicative transposition.

*13. What is a retrotransposon and how does it move?

*14. Describe the process of replicative transposition through DNA. What enzymes are required?

Section 11.5

*15. Draw the structure of a typical insertion sequence and identify its parts.

16. Draw the structure of a typical composite transposon in bacteria and identify its parts.

17. How are composite transposons and retrotransposons alike and how are they different?

18. Explain how *Ac* and *Ds* elements produce variegated corn kernels.

19. Briefly explain hybrid dysgenesis and how *P* elements lead to hybrid dysgenesis.

20. What are some differences between LINEs and SINEs?

21. What are some differences between class I and class II transposable elements?

Section 11.6

*22. Why are transposable elements often called genomic parasites?

APPLICATION QUESTIONS AND PROBLEMS

Introduction

23. The introduction to this chapter described how *Rider* transposons caused a segment of chromosome 10 in tomatoes to be duplicated and transported to chromosome 7, which resulted in the overexpression of *IQD12* and the production of elongated fruits. How might movement of a gene such as *IQD12* to a new location cause it to be overexpressed?

Section 11.2

*24. Compare and contrast prokaryotic and eukaryotic chromosomes. How are they alike and how do they differ?

25. (a) In a typical eukaryotic cell, would you expect to find more molecules of the H1 histone or more molecules of the H2A histone? Explain your reasoning. (b) Would you expect to find more molecules of H2A or more molecules of H3? Explain your reasoning.

26. Suppose you examined polytene chromosomes from the salivary glands of fruit fly larvae and counted the number of chromosomal puffs observed in different regions of DNA.

   a. Would you expect to observe more puffs from euchromatin or from heterochromatin? Explain your answer.

   b. Would you expect to observe more puffs in unique-sequence DNA, in moderately repetitive DNA, or in repetitive DNA? Why?

*27. A diploid human cell contains approximately 6.4 billion base pairs of DNA.

   a. How many nucleosomes are present in such a cell? (Assume that the linker DNA encompasses 40 bp.)

   b. How many histone proteins are complexed to this DNA?

*28. Would you expect to see more or less acetylation in regions of DNA that are sensitive to digestion by DNase I? Why?

29. Gunter Korge examined several proteins that are secreted from the salivary glands of *Drosophila melanogaster* during larval development (G. Korge. 1975. *Proceedings of the National Academy of Sciences of the United States of America* 72:4550–4554). One protein, called protein fraction 4, was encoded by a gene found by deletion mapping to be located on the X chromosome at position 3C. Korge observed that, about 5 hours before the first synthesis of protein fraction 4, an expanded and puffed-out region formed on the X chromosome at position 3C. This chromosome puff disappeared before the end of the third larval instar stage, when the synthesis of protein fraction 4 ceased. He observed that there was no puff at position 3C in a special strain of flies that lacked secretion of protein fraction 4. Explain these results. What is the chromosome puff at region 3 and why does its appearance and disappearance roughly coincide with the secretion of protein fraction 4?

30. Suppose a chemist develops a new drug that neutralizes the positive charges on the tails of histone proteins. What would be the most likely effect of this new drug on chromatin structure? Would this drug have any effect on gene expression? Explain your answers.

Section 11.3

*31. Which of the following two molecules of DNA has the lower melting temperature? Why?

AGTTACTAAAGCAATACATC
TCAATGATTTCGTTATGTAG

AGGCGGGTAGGCACCCTTA
TCCGCCCATCCGTGGGAAT

32. In a DNA hybridization study, DNA was isolated from a particular species, labeled with ^{32}P, and sheared into small fragments (S. K. Dutta et al. 1967. *Genetics* 57:719–727). Hybridizations between these labeled fragments and denatured DNA from different species were then compared. The following table gives the percentages of labeled wheat DNA that hybridized to DNA molecules of wheat, corn, radish, and cabbage.

Species	Percentage of bound wheat DNA hybridized relative to wheat
Wheat	100
Cabbage	23
Corn	63
Radish	30

What do these results indicate about the evolutionary differences among these organisms?

Section 11.5

*33. A particular transposable element generates flanking direct repeats that are 4 bp long. Give the sequence that will be found on both sides of the transposable element if this transposable element inserts at the position indicated on each of the following sequences.

a.

Transposable element

5′–ATTCGAACTGACCGATCA–3′

b.

Transposable element

5′–ATTCGAACTGACCGATCA–3′

*34. White eyes in *Drosophila melanogaster* result from an X-linked recessive mutation. Occasionally, white-eyed mutants give rise to offspring that possess white eyes with small red spots. The number, distribution, and size of the red spots are variable. Explain how a transposable element could be responsible for this spotting phenomenon.

*35. What factor might potentially determine the length of the flanking direct repeats that are produced in transposition?

Section 11.5

36. Which of the following pairs of sequences might be found at the ends of an insertion sequence?

   a. 5′–GGGCCAATT–3′   and   5′–CCCGGTTAA–3′
   b. 5′–AAACCCTTT–3′   and   5′–AAAGGGTTT–3′
   c. 5′–TTTCGAC–3′   and   5′–CAGCTTT–3′
   d. 5′–ACGTACG–3′   and   5′–CGTACGT–3′
   e. 5′–GCCCCAT–3′   and   5′–GCCCAT–3′

37. Two different strains of *Drosophila melanogaster* are mated in reciprocal crosses. When strain A males are crossed with strain B females, the progeny are normal. However, when strain A females are crossed with strain B males, there are many mutations and chromosome rearrangements in the gametes of the F_1 progeny and the F_1 generation is effectively sterile. Explain these results.

*38. An insertion sequence contains a large deletion in its transposase gene. Under what circumstances would this insertion sequence be able to transpose?

39. A transposable element is found to encode a transposase enzyme. On the basis of this information, what conclusions can you make about the likely structure and method of transposition of this element?

40. Zidovudine (AZT) is a drug used to treat patients with AIDS. AZT works by blocking the reverse transcriptase enzyme used by the human immunodeficiency virus (HIV), the causative agent of AIDS. Do you expect that AZT would have any effect on transposable elements? If so, what type of transposable elements would be affected and what would be the most likely effect?

41. A transposable element is found to encode a reverse transcriptase enzyme. On the basis of this information, what conclusions can you make about the likely structure and method of transposition of this element?

42. A geneticist examines an ear of corn in which most kernels are yellow, but he finds a few kernels with purple spots, as shown here. Give a possible explanation for the appearance of the purple spots in these otherwise yellow kernels, accounting for their different sizes. (Hint: See the section on *Ac* and *Ds* elements in maize on pp. 310–311.)

43. A geneticist studying the DNA of the Japanese bottle fly finds many copies of a particular sequence that appears similar to the *copia* transposable element in *Drosophila* (see Table 11.6). Using recombinant DNA techniques, the geneticist places an intron into a copy of this DNA sequence and inserts it into the genome of a Japanese bottle fly. If the sequence is a transposable element similar to *copia*, what prediction would you make concerning the fate of the introduced sequence in the genomes of offspring of the fly receiving it?

CHALLENGE QUESTIONS

Section 11.1

44. An explorer discovers a strange new species of plant and sends some of the plant tissue to a geneticist to study. The geneticist isolates chromatin from the plant and examines it with an electron microscope. She observes what appear to be beads on a string. She then adds a small amount of nuclease, which cleaves the string into individual beads that each contain 280 bp of DNA. After digestion with more nuclease, a 120-bp fragment of DNA remains attached to a core of histone proteins. Analysis of the histone core reveals histones in the following proportions:

H1	12.5%
H2A	25%
H2B	25%
H3	0%
H4	25%
H7 (a new histone)	12.5%

On the basis of these observations, what conclusions could the geneticist make about the probable structure of the nucleosome in the chromatin of this plant?

Section 11.3

45. Although highly repetitive DNA is common in eukaryotic chromosomes, it does not encode proteins; in fact, it is probably never transcribed into RNA. If highly repetitive DNA does not encode RNA or proteins, why is it present in eukaryotic genomes? Suggest some possible reasons for the widespread presence of highly repetitive DNA.

46. In DNA-hybridization experiments on six species of plants in the genus *Vicia*, DNA was isolated from each of the six species, denatured by heating, and sheared into small fragments (W. Y. Chooi. 1971. *Genetics* 68:213–230). In one experiment, DNA from each species and from *E. coli* was allowed to renature. The adjoining graph shows the results of this renaturation experiment.

a. Can you explain why the *E. coli* DNA renatures at a much faster rate than does DNA from all of the *Vicia* species?

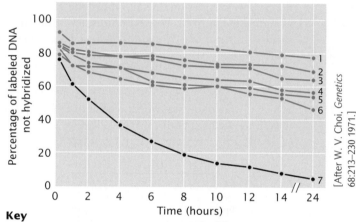

[After W. V. Choi, *Genetics* 68:213–230 1971.]

Key
1 = *V. melanops*, 2 = *V. sativa*, 3 = *V. benghalensis*, 4 = *V. atropurpurea*, 5 = *V. faba*, 6 = *V. narbonensis*, 7 = *E. coli*

b. Notice that, for the *Vicia* species, the rate of renaturation is much faster in the first hour and then slows down. What might cause this initial rapid renaturation and the subsequent slowdown?

Section 11.5

47. Marilyn Houck and Margaret Kidwell proposed that *P* elements were carried from *Drosophila willistoni* to *Drosophila melanogaster* by mites that fed on fruit flies (M. A. Houck et al. 1991. *Science* 253:1125–1129). What evidence do you think would be required to demonstrate that *D. melanogaster* acquired *P* elements in this way? Propose a series of experiments to provide such evidence.

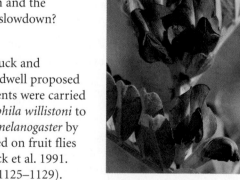

Vodder vetch (*Vicia sativa*).
[Bob Gibbons/Alamy.]

12 DNA Replication and Recombination

The happy tree, *Camptotheca acuminata*, contains camptothecin, a substanced used to treat cancer. Camptothecin inhibits cancer by blocking an important component of the replication machinery. [Gary Retherford/Photo Researchers.]

TOPOISOMERASE, REPLICATION, AND CANCER

In 1966, Monroe Wall and Mansukh Wani found a potential cure for cancer in the bark of the happy tree (*Camptotheca acuminata*), a rare plant native to China. Wall and Wani were in the process of screening a large number of natural substances for anticancer activity, hoping to find chemicals that might prove effective in the treatment of cancer. They discovered that an extract from the happy tree was effective in treating leukemia in mice. Through chemical analysis, they were able to isolate the active compound, which was dubbed camptothecin.

In the 1970s, physicians administered camptothecin to patients with incurable cancers. Although the drug showed some anticancer activity, it had toxic side effects. Eventually, chemists synthesized several analogs of camptothecin that were less toxic and more effective in cancer treatment. Two of these analogs, topotecan and irinotecan, are used today for the treatment of ovarian cancer, small-cell lung cancer, and colon cancer.

For many years, the mechanism by which camptothecin compounds inhibited cancer was unknown. In 1985, almost 20 years after its discovery, scientists at Johns Hopkins University and Smith Kline and French Laboratories showed that camptothecin worked by inhibiting an important component of the DNA-synthesizing machinery in humans, an enzyme called topoisomerase I.

Cancer chemotherapy is a delicate task, because the target cells are the patient's own and the drugs must kill the cancer cells without killing the patient. One of the hallmarks of cancer is proliferation: cancer-cell division is unregulated and many cancer cells divide at a rapid rate, giving rise to tumors with the ability to grow and spread. As we learned in Chapter 2, before a cell can divide it must successfully replicate its DNA so that each daughter cell receives an exact copy of the genetic material. Checkpoints in the cell cycle ensure that cell division does not proceed if DNA replication is inhibited or faulty, and many cancer treatments focus on interfering with the process of DNA replication.

DNA replication is a complex process that requires a large number of components, the actions of which must be exquisitely coordinated to ensure that DNA is faithfully and accurately copied. An essential component of replication is topoisomerase. As the DNA unwinds in the course of replication, strain builds up ahead of the separation and the two strands writhe around each other, much as a rope knots up as you pull apart two of its strands. This writhing of the DNA is called supercoiling (see Chapter 11). If the supercoils are not removed, they eventually stop strand separation and replication comes to a halt. Topoisomerase enzymes remove the supercoils by clamping tightly to the DNA and breaking one or both of its strands. The strands then revolve around one another, removing the supercoiling and strain. When the DNA is relaxed, the topoisomerase that removed the supercoils reseals the the broken ends of the DNA.

Camptothecin works by interfering with topoisomerase I. The drug inserts itself into the gap created by the break in the DNA strand, blocking the topoisomerase from resealing the broken ends. Researchers originally assumed that camptothecin trapped the topoisomerase on DNA and blocked the action of other enzymes needed for synthesizing DNA. However, recent research indicates that camptothecin poisons the topoisomerase so that it is unable to remove supercoils ahead of replication. Accumulating supercoils halt the replication machinery and prevent the proliferation of cancer cells. Like many other cancer drugs, camptothecin also inhibits the replication of normal, noncancerous cells, which is why chemotherapy makes many patients sick.

This chapter focuses on DNA replication, the process by which a cell doubles its DNA before division. We begin with the basic mechanism of replication that emerged from the Watson-and-Crick structure of DNA. We then examine several different modes of replication, the requirements of replication, and the universal direction of DNA synthesis. We also examine the enzymes and proteins that participate in the process. Finally, we consider the molecular details of recombination, which is closely related to replication and is essential for the segregation of homologous chromosomes, for the production of genetic variation, and for DNA repair.

12.1 Genetic Information Must Be Accurately Copied Every Time a Cell Divides

In a schoolyard game, a verbal message, such as "John's brown dog ran away from home," is whispered to a child, who runs to a second child and repeats the message. The message is relayed from child to child around the schoolyard until it returns to the original sender. Inevitably, the last child returns with an amazingly transformed message, such as "Joe Brown has a pig living under his porch." The larger the number of children playing the game, the more garbled the message becomes. This game illustrates an important principle: errors arise whenever information is copied; the more times it is copied, the greater the potential number of errors.

A complex, multicellular organism faces a problem analogous to that of the children in the schoolyard game: how to faithfully transmit genetic instructions each time its cells divide. The solution to this problem is central to replication. A huge amount of genetic information and an enormous number of cell divisions are required to produce a multicellular adult organism; even a low rate of error during copying would be catastrophic. A single-celled human zygote contains 6.4 billion base pairs of DNA. If a copying error were made only once per million base pairs, 6400 mistakes would be made every time a cell divided—errors that

would be compounded at each of the millions of cell divisions that take place in human development.

Not only must the copying of DNA be astoundingly accurate, it must also take place at breakneck speed. The single, circular chromosome of *E. coli* contains about 4.6 million base pairs. At a rate of more than 1000 nucleotides per minute, replication of the entire chromosome would require almost 3 days. Yet, as already stated, these bacteria are capable of dividing every 20 minutes. *E. coli* actually replicates its DNA at a rate of 1000 nucleotides per second, with less than one error in a billion nucleotides. How is this extraordinarily accurate and rapid process accomplished?

12.2 All DNA Replication Takes Place in a Semiconservative Manner

From the three-dimensional structure of DNA proposed by Watson and Crick in 1953 (see Figure 10.7), several important genetic implications were immediately apparent. The complementary nature of the two nucleotide strands in a DNA molecule suggested that, during replication, each strand can serve as a template for the synthesis of a new strand. The specificity of base pairing (adenine with thymine; guanine with cytosine) implied that only one sequence of bases can be specified by each template, and so the two DNA molecules built on the pair of templates will be identical with the original. This process is called **semiconservative replication** because each of the original nucleotide strands remains intact (conserved), despite no longer being combined in the same molecule; the original DNA molecule is half (semi) conserved during replication.

Initially, three models were proposed for DNA replication. In conservative replication (**Figure 12.1a**), the entire double-stranded DNA molecule serves as a template for a whole new molecule of DNA, and the original DNA molecule is *fully* conserved during replication. In dispersive replication (**Figure 12.1b**), both nucleotide strands break down (disperse) into fragments, which serve as templates for the synthesis of new DNA fragments, and then somehow reassemble into two complete DNA molecules. In this model, each resulting DNA

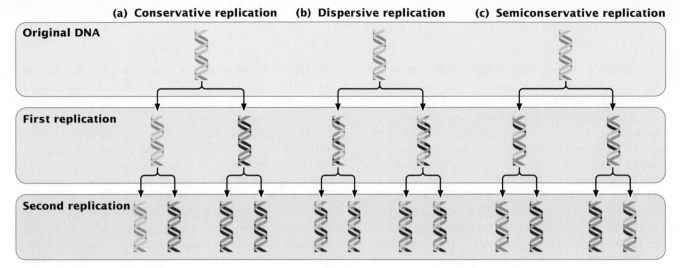

(a) Conservative replication   **(b) Dispersive replication**   **(c) Semiconservative replication**

Original DNA

First replication

Second replication

12.1 Three proposed models of replication are conservative replication, dispersive replication, and semiconservative replication.

molecule is interspersed with fragments of old and new DNA; none of the original molecule is conserved. Semiconservative replication (**Figure 12.1c**) is intermediate between these two models; the two nucleotide strands unwind and each serves as a template for a new DNA molecule.

These three models allow different predictions to be made about the distribution of original DNA and newly synthesized DNA after replication. With conservative replication, after one round of replication, 50% of the molecules would consist entirely of the original DNA and 50% would consist entirely of new DNA. After a second round of replication, 25% of the molecules would consist entirely of the original DNA and 75% would consist entirely of new DNA. With each additional round of replication, the proportion of molecules with new DNA would increase, although the number of molecules with the original DNA would remain constant. Dispersive replication would always produce hybrid molecules, containing some original and some new DNA, but the proportion of new DNA within the molecules would increase with each replication event. In contrast, with semiconservative replication, one round of replication would produce two hybrid molecules, each consisting of half original DNA and half new DNA. After a second round of replication, half the molecules would be hybrid, and the other half would consist of new DNA only. Additional rounds of replication would produce more and more molecules consisting entirely of new DNA, and a few hybrid molecules would persist.

Meselson and Stahl's Experiment

To determine which of the three models of replication applied to *E. coli* cells, Matthew Meselson and Franklin Stahl needed a way to distinguish old and new DNA. They did so by using two isotopes of nitrogen, ^{14}N (the common form) and ^{15}N (a rare, heavy form). Meselson and Stahl grew a culture of *E. coli* in a medium that contained ^{15}N as the sole nitrogen source; after many generations, all the *E. coli* cells had ^{15}N incorporated into all of the purine and pyrimidine bases of their DNA (see Figure 10.10). Meselson and Stahl took a sample of these bacteria, switched the rest of the bacteria to a medium that contained only ^{14}N, and then

took additional samples of bacteria over the next few cellular generations. In each sample, the bacterial DNA that was synthesized before the change in medium contained ^{15}N and was relatively heavy, whereas any DNA synthesized after the switch contained ^{14}N and was relatively light.

Meselson and Stahl distinguished between the heavy ^{15}N-laden DNA and the light ^{14}N-containing DNA with the use of **equilibrium density gradient centrifugation (Figure 12.2)**. In

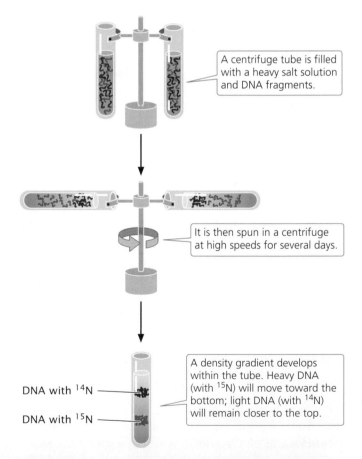

A centrifuge tube is filled with a heavy salt solution and DNA fragments.

It is then spun in a centrifuge at high speeds for several days.

DNA with ^{14}N

DNA with ^{15}N

A density gradient develops within the tube. Heavy DNA (with ^{15}N) will move toward the bottom; light DNA (with ^{14}N) will remain closer to the top.

12.2 Meselson and Stahl used equilibrium density gradient centrifugation to distinguish between heavy, ^{15}N-laden DNA and lighter, ^{14}N-laden DNA.

323

this technique, a centrifuge tube is filled with a heavy salt solution and a substance of unkown density—in this case, DNA fragments. The tube is then spun in a centrifuge at high speeds. After several days of spinning, a gradient of density develops within the tube, with high density at the bottom and low density at the top. The density of the DNA fragments matches that of the salt: light molecules rise and heavy molecules sink.

Meselson and Stahl found that DNA from bacteria grown only on medium containing ^{15}N produced a single band at the position expected of DNA containing only ^{15}N (**Figure 12.3a**). DNA from bacteria transferred to the medium with ^{14}N and allowed one round of replication also produced a single band but at a position intermediate between that expected of DNA containing only ^{15}N and that expected of DNA containing only ^{14}N (**Figure 12.3b**). This result is inconsistent with the conservative replication model, which

predicts one heavy band (the original DNA molecules) and one light band (the new DNA molecules). A single band of intermediate density is predicted by both the semiconservative and the dispersive models.

To distinguish between these two models, Meselson and Stahl grew the bacteria in medium containing ^{14}N for a second generation. After a second round of replication in medium with ^{14}N, two bands of equal intensity appeared, one in the intermediate position and the other at the position expected of DNA containing only ^{14}N (**Figure 12.3c**). All samples taken after additional rounds of replication produced the same two bands, and the band representing light DNA became progressively stronger (**Figure 12.3d**). Meselson and Stahl's results were exactly as expected for semiconservative replication and are incompatible with those predicated for both conservative and dispersive replication. TRY PROBLEM 20 →

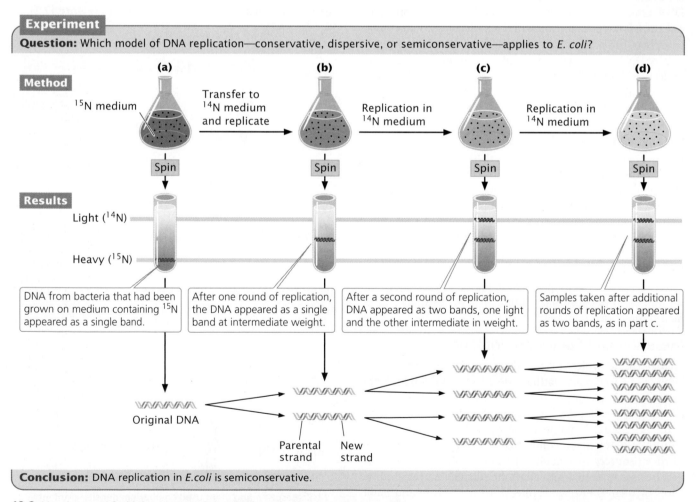

Experiment

Question: Which model of DNA replication—conservative, dispersive, or semiconservative—applies to *E. coli*?

Method

(a)   (b)   (c)   (d)

^{15}N medium

Transfer to ^{14}N medium and replicate

Replication in ^{14}N medium

Replication in ^{14}N medium

Spin   Spin   Spin   Spin

Results

Light (^{14}N)

Heavy (^{15}N)

DNA from bacteria that had been grown on medium containing ^{15}N appeared as a single band.

After one round of replication, the DNA appeared as a single band at intermediate weight.

After a second round of replication, DNA appeared as two bands, one light and the other intermediate in weight.

Samples taken after additional rounds of replication appeared as two bands, as in part c.

Original DNA

Parental strand   New strand

Conclusion: DNA replication in *E.coli* is semiconservative.

12.3 Meselson and Stahl demonstrated that DNA replication is semiconservative.

Replication is semiconservative: each DNA strand serves as a template for the synthesis of a new DNA molecule. Meselson and Stahl convincingly demonstrated that replication in *E. coli* is semiconservative.

✔ CONCEPT CHECK 1

How many bands of DNA would be expected in Meselson and Stahl's experiment after two rounds of *conservative* replication?

Modes of Replication

After Meselson and Stahl's work, investigators confirmed that other organisms also use semiconservative replication. No evidence was found for conservative or dispersive replication. There are, however, several different ways in which semiconservative replication can take place, differing principally in the nature of the template DNA—that is, whether it is linear or circular.

Individual units of replication are called **replicons**, each of which contains a **replication origin**. Replication starts at the origin and continues until the entire replicon has been replicated. Bacterial chromosomes have a single replication origin, whereas eukaryotic chromosomes contain many.

Theta replication   A common type of replication that takes place in circular DNA, such as that found in *E. coli* and other bacteria, is called **theta replication** (**Figure 12.4a**) because it generates a structure that resembles the Greek letter theta (θ). In theta replication, double-stranded DNA begins to unwind at the replication origin, producing single-stranded nucleotide strands that then serve as templates on which new DNA can be synthesized. The unwinding of the double helix generates a loop, termed a **replication bubble**. Unwinding may be at one or both ends of the bubble, making it progressively larger. DNA replication on both of the template strands is simultaneous with unwinding. The point of unwinding, where the two single nucleotide strands separate from the double-stranded DNA helix, is called a **replication fork**.

If there are two replication forks, one at each end of the replication bubble, the forks proceed outward in both directions in a process called **bidirectional replication**,

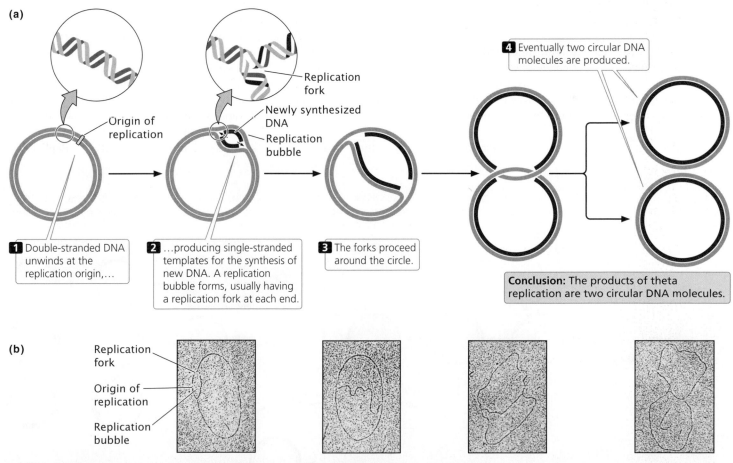

(a)

Replication fork

Origin of replication

Newly synthesized DNA

Replication bubble

4 Eventually two circular DNA molecules are produced.

1 Double-stranded DNA unwinds at the replication origin,…

2 …producing single-stranded templates for the synthesis of new DNA. A replication bubble forms, usually having a replication fork at each end.

3 The forks proceed around the circle.

Conclusion: The products of theta replication are two circular DNA molecules.

(b)

Replication fork

Origin of replication

Replication bubble

12.4 Theta replication is a type of replication common in *E. coli* and other organisms possessing circular DNA. [Electron micrographs from Bernard Hirt, L'Institut Suisse de Recherche Expérimentale sur le Cancer.]

simultaneously unwinding and replicating the DNA until they eventually meet. If a single replication fork is present, it proceeds around the entire circle to produce two complete circular DNA molecules, each consisting of one old and one new nucleotide strand.

John Cairns provided the first visible evidence of theta replication in 1963 by growing bacteria in the presence of radioactive nucleotides. After replication, each DNA molecule consisted of one "hot" (radioactive) strand and one "cold" (nonradioactive) strand. Cairns isolated DNA from the bacteria after replication and placed it on an electron-microscope grid, which was then covered with a photographic emulsion. Radioactivity present in the sample exposes the emulsion and produces a picture of the molecule (called an autoradiograph), similar to the way in which light exposes a photographic film. Because the newly synthesized DNA contained radioactive nucleotides, Cairns was able to produce an electron micrograph of the replication process, similar to those shown in **Figure 12.4b**.

Rolling-circle replication   Another form of replication, called **rolling-circle replication** (**Figure 12.5**), takes place in some viruses and in the F factor (a small circle of extrachromosomal DNA that controls mating, discussed in Chapter 8) of *E. coli*. This form of replication is initiated by a break in one of the nucleotide strands that creates a 3'-OH group and a 5'-phosphate group. New nucleotides are added to the 3' end of the broken strand, with the inner (unbroken) strand used as a template. As new nucleotides are added to the 3' end, the 5' end of the broken strand is displaced from the template, rolling out like thread being pulled off a spool. The 3' end grows around the circle, giving rise to the name rolling-circle model.

The replication fork may continue around the circle a number of times, producing several linked copies of the same sequence. With each revolution around the circle, the growing 3' end displaces the nucleotide strand synthesized in the preceding revolution. Eventually, the linear DNA molecule is cleaved from the circle, resulting in a double-stranded circular DNA molecule and a single-stranded linear DNA molecule. The linear molecule circularizes either before or after serving as a template for the synthesis of a complementary strand.

Linear eukaryotic replication   Circular DNA molecules that undergo theta or rolling-circle replication have a single origin of replication. Because of the limited size of these DNA molecules, replication starting from one origin can traverse the entire chromosome in a reasonable amount of time. The large linear chromosomes in eukaryotic cells, however, contain far too much DNA to be replicated speedily from a single origin. Eukaryotic replication proceeds at a rate ranging from 500 to 5000 nucleotides per minute at each replication fork (considerably slower than bacterial replication). Even at 5000 nucleotides per minute at each fork, DNA synthesis starting from a single origin would require 7 days to replicate a typical human chromosome consisting of 100 million base pairs of DNA. The replication of eukaryotic chromosomes actually takes place in a matter of minutes or hours, not days. This rate is possible because replication initiates at thousands of origins.

Typical eukaryotic replicons are from 20,000 to 300,000 base pairs in length (**Table 12.1**). At each replication origin, the DNA unwinds and produces a replication bubble. Replication takes place on both strands at each end of the bubble, with the two replication forks spreading outward. Eventually, the replication forks of adjacent replicons run into each other, and the replicons fuse to form long stretches of newly synthesized DNA (**Figure 12.6**). Replication

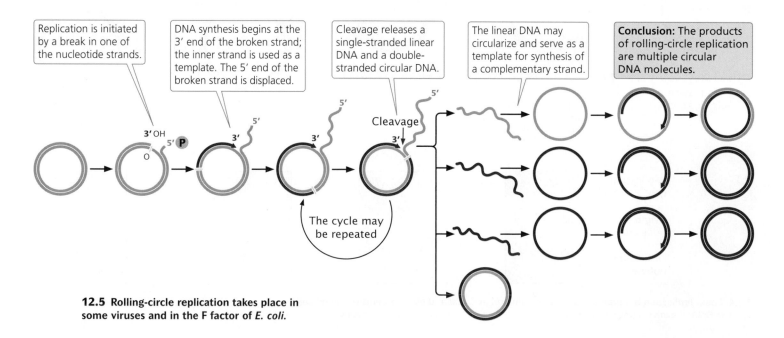

| Replication is initiated by a break in one of the nucleotide strands. | DNA synthesis begins at the 3' end of the broken strand; the inner strand is used as a template. The 5' end of the broken strand is displaced. | Cleavage releases a single-stranded linear DNA and a double-stranded circular DNA. | The linear DNA may circularize and serve as a template for synthesis of a complementary strand. | **Conclusion:** The products of rolling-circle replication are multiple circular DNA molecules. |

12.5 Rolling-circle replication takes place in some viruses and in the F factor of *E. coli*.

and fusion of all the replicons leads to two identical DNA molecules. Important features of theta replication, rolling-circle replication, and linear eukaryotic replication are summarized in **Table 12.2**. TRY PROBLEM 21 →

CONCEPTS

Theta replication, rolling-circle replication, and linear replication differ with respect to the initiation and progression of replication, but all produce new DNA molecules by semiconservative replication.

✔ CONCEPT CHECK 2

Which type of replication requires a break in the nucleotide strand to get started?

a. Theta replication   c. Linear eukaryotic replication

b. Rolling-circle replication   d. All of the above

Table 12.1 Number and length of replicons

Organism	Number of Replication Origins	Average Length of Replicon (bp)
Escherichia coli (bacterium)	1	4,600,000
Saccharomyces cerevisiae (yeast)	500	40,000
Drosophila melanogaster (fruit fly)	3,500	40,000
Xenopus laevis (frog)	15,000	200,000
Mus musculus (mouse)	25,000	150,000

Source: Data from B. L. Lewin, *Genes V* (Oxford: Oxford University Press, 1994), p. 536.

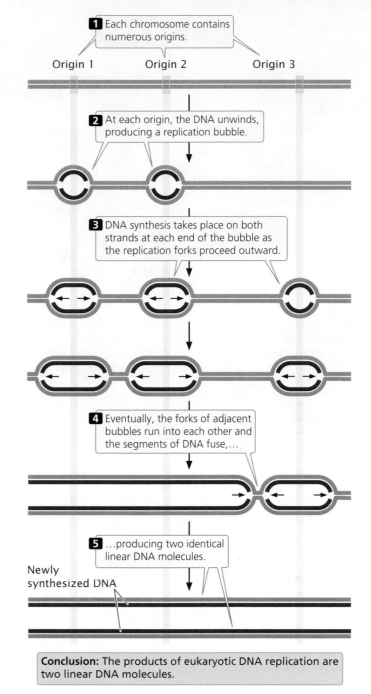

1 Each chromosome contains numerous origins.

Origin 1   Origin 2   Origin 3

2 At each origin, the DNA unwinds, producing a replication bubble.

3 DNA synthesis takes place on both strands at each end of the bubble as the replication forks proceed outward.

4 Eventually, the forks of adjacent bubbles run into each other and the segments of DNA fuse,…

5 …producing two identical linear DNA molecules.

Newly synthesized DNA

Conclusion: The products of eukaryotic DNA replication are two linear DNA molecules.

12.6 Linear DNA replication takes place in eukaryotic chromosomes.

Table 12.2 Characteristics of theta, rolling-circle, and linear eukaryotic replication

Replication Model	DNA Template	Breakage of Nucleotide Strand	Number of Replicons	Unidirectional or Bidirectional	Products
Theta	Circular	No	1	Unidirectional or bidirectional	Two circular molecules
Rolling circle	Circular	Yes	1	Unidirectional	One circular molecule and one linear molecule that may circularize
Linear eukaryotic	Linear	No	Many	Bidirectional	Two linear molecules

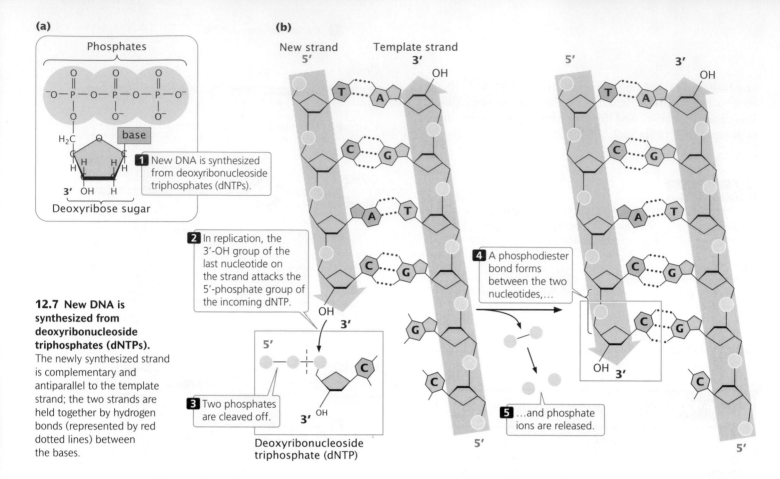

(a)

Phosphates

base

Deoxyribose sugar

1 New DNA is synthesized from deoxyribonucleoside triphosphates (dNTPs).

2 In replication, the 3′-OH group of the last nucleotide on the strand attacks the 5′-phosphate group of the incoming dNTP.

3 Two phosphates are cleaved off.

Deoxyribonucleoside triphosphate (dNTP)

12.7 New DNA is synthesized from deoxyribonucleoside triphosphates (dNTPs). The newly synthesized strand is complementary and antiparallel to the template strand; the two strands are held together by hydrogen bonds (represented by red dotted lines) between the bases.

(b)

New strand 5′   Template strand 3′

4 A phosphodiester bond forms between the two nucleotides,...

5 ...and phosphate ions are released.

Requirements of Replication

Although the process of replication includes many components, they can be combined into three major groups:

1. a template consisting of single-stranded DNA,
2. raw materials (substrates) to be assembled into a new nucleotide strand, and
3. enzymes and other proteins that "read" the template and assemble the substrates into a DNA molecule.

Because of the semiconservative nature of DNA replication, a double-stranded DNA molecule must unwind to expose the bases that act as a template for the assembly of new polynucleotide strands, which will be complementary and antiparallel to the template strands. The raw materials from which new DNA molecules are synthesized are deoxy-ribonucleoside triphosphates (dNTPs), each consisting of a deoxyribose sugar and a base (a nucleoside) attached to three phosphate groups (**Figure 12.7a**). In DNA synthesis, nucleotides are added to the 3′-OH group of the growing nucleotide strand (**Figure 12.7b**). The 3′-OH group of the last nucleotide on the strand attacks the 5′-phosphate group of the incoming dNTP. Two phosphate groups are cleaved from the incoming dNTP, and a phosphodiester bond is created between the two nucleotides.

DNA synthesis does not happen spontaneously. Rather, it requires a host of enzymes and proteins that function in a coordinated manner. We will examine this complex array of proteins and enzymes as we consider the replication process in more detail.

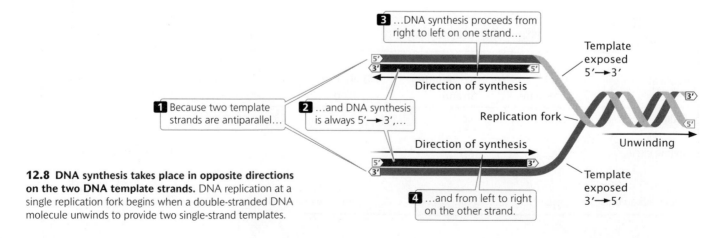

3 ...DNA synthesis proceeds from right to left on one strand...

Direction of synthesis

Template exposed 5′→3′

1 Because two template strands are antiparallel...

2 ...and DNA synthesis is always 5′→3′,...

Replication fork

Direction of synthesis

Unwinding

4 ...and from left to right on the other strand.

Template exposed 3′→5′

12.8 DNA synthesis takes place in opposite directions on the two DNA template strands. DNA replication at a single replication fork begins when a double-stranded DNA molecule unwinds to provide two single-strand templates.

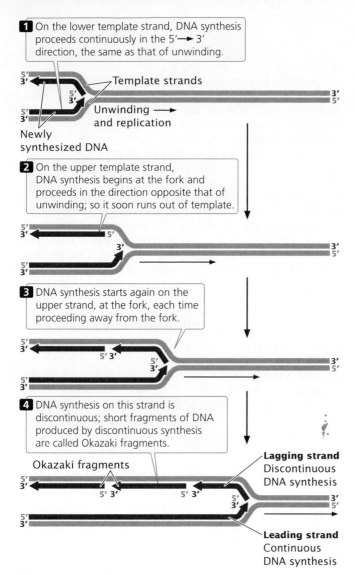

1 On the lower template strand, DNA synthesis proceeds continuously in the 5′⟶ 3′ direction, the same as that of unwinding.

Template strands

Unwinding ⟶ and replication

Newly synthesized DNA

2 On the upper template strand, DNA synthesis begins at the fork and proceeds in the direction opposite that of unwinding; so it soon runs out of template.

3 DNA synthesis starts again on the upper strand, at the fork, each time proceeding away from the fork.

4 DNA synthesis on this strand is discontinuous; short fragments of DNA produced by discontinuous synthesis are called Okazaki fragments.

Okazaki fragments

Lagging strand Discontinuous DNA synthesis

Leading strand Continuous DNA synthesis

12.9 DNA synthesis is continuous on one template strand of DNA and discontinuous on the other.

CONCEPTS

DNA synthesis requires a single-stranded DNA template, deoxyribonucleoside triphosphates, a growing nucleotide strand, and a group of enzymes and proteins.

Direction of Replication

In DNA synthesis, new nucleotides are joined one at a time to the 3′ end of the newly synthesized strand. **DNA polymerases**, the enzymes that synthesize DNA, can add nucleotides only to the 3′ end of the growing strand (not the 5′ end), and so new DNA strands always elongate in the same 5′-to-3′ direction (5′⟶3′). Because the two single-stranded DNA templates are antiparallel and strand elongation is always 5′⟶3′, if synthesis on one template proceeds from, say, right to left, then synthesis on the other template must proceed in the opposite direction, from left to right (**Figure 12.8**). As DNA unwinds during replication, the antiparallel nature of the two DNA strands means that one template is exposed in the 5′⟶3′ direction and the other template is exposed in the

3′⟶5′ direction (see Figure 12.8); so how can synthesis take place simultaneously on both strands at the fork?

Continuous and discontinuous replication As the DNA unwinds, the template strand that is exposed in the 3′⟶5′ direction (the lower strand in Figures 12.8 and 12.9) allows the new strand to be synthesized continuously, in the 5′⟶3′ direction. This new strand, which undergoes **continuous replication**, is called the **leading strand**.

The other template strand is exposed in the 5′⟶3′ direction (the upper strand in Figures 12.8 and 12.9). After a short length of the DNA has been unwound, synthesis must proceed 5′⟶3′; that is, in the direction *opposite* that of unwinding (**Figure 12.9**). Because only a short length of DNA needs to be unwound before synthesis on this strand gets started, the replication machinery soon runs out of template. By that time, more DNA has unwound, providing new template at the 5′ end of the new strand. DNA synthesis must start anew at the replication fork and proceed in the direction opposite that of the movement of the fork until it runs into the previously replicated segment of DNA. This process is repeated again and again, and so synthesis of this strand is in short, discontinuous bursts. The newly made strand that undergoes **discontinuous replication** is called the **lagging strand**.

Okazaki fragments The short lengths of DNA produced by discontinuous replication of the lagging strand are called **Okazaki fragments**, after Reiji Okazaki, who discovered them. In bacterial cells, each Okazaki fragment ranges in length from about 1000 to 2000 nucleotides; in eukaryotic cells, they are about 100 to 200 nucleotides long. Okazaki fragments on the lagging strand are linked together to create a continuous new DNA molecule.

CONCEPTS

All DNA synthesis is 5′⟶3′, meaning that new nucleotides are always added to the 3′ end of the growing nucleotide strand. At each replication fork, synthesis of the leading strand proceeds continuously and that of the lagging strand proceeds discontinuously.

✔ **CONCEPT CHECK 3**

Discontinuous replication is a result of which property of DNA?

a. Complementary bases   c. Antiparallel nucleotide strands

b. Charged phosphate group   d. Five-carbon sugar

CONNECTING CONCEPTS

The Direction of Replication in Different Models of Replication

Let's relate the direction of DNA synthesis to the modes of replication examined earlier. In the theta model (**Figure 12.10a**), the DNA unwinds at one particular location, the origin, and a replication bubble is formed. If the bubble has two forks, one at each end, synthesis takes place simultaneously at both forks (bidirectional replication). At each fork, synthesis on one of the template strands proceeds in the same direction as that of unwinding; this newly replicated strand is the leading strand with continuous replication. On the other template strand, synthesis proceeds in the direction opposite that of unwinding; this newly synthesized strand is the lagging strand with discontinuous replication. Focus on just one of the template strands within the bubble. Notice that synthesis on this template strand is continuous at one fork but discontinuous at the other. This difference arises because DNA synthesis is always in the same direction (5′⟶ 3′), but the two forks are moving in opposite directions.

(a) Theta model

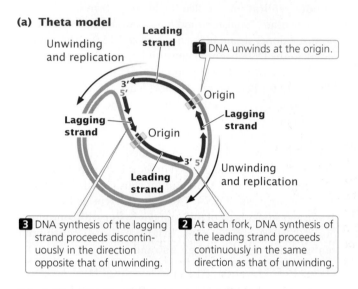

1 DNA unwinds at the origin.

3 DNA synthesis of the lagging strand proceeds discontinuously in the direction opposite that of unwinding.

2 At each fork, DNA synthesis of the leading strand proceeds continuously in the same direction as that of unwinding.

(b) Rolling-circle model

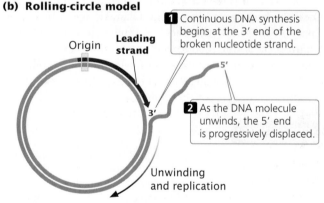

1 Continuous DNA synthesis begins at the 3′ end of the broken nucleotide strand.

2 As the DNA molecule unwinds, the 5′ end is progressively displaced.

(c) Linear eukaryotic replication

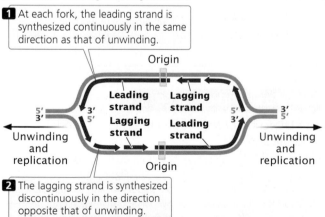

1 At each fork, the leading strand is synthesized continuously in the same direction as that of unwinding.

2 The lagging strand is synthesized discontinuously in the direction opposite that of unwinding.

12.10 The process of replication differs in theta replication, rolling-circle replication, and linear replication.

Replication in the rolling-circle model (**Figure 12.10b**) is somewhat different, because there is no replication bubble. Replication begins at the 3′ end of the broken nucleotide strand. Continuous replication takes place on the circular template as new nucleotides are added to this 3′ end.

The replication of linear molecules of DNA, such as those found in eukaryotic cells, produces a series of replication bubbles (**Figure 12.10c**). DNA synthesis in these bubbles is the same as that in the single replication bubble of the theta model; it begins at the center of each replication bubble and proceeds at two forks, one at each end of the bubble. At both forks, synthesis of the leading strand proceeds in the same direction as that of unwinding, whereas synthesis of the lagging strand proceeds in the direction opposite that of unwinding. TRY PROBLEM 23 →

12.3 Bacterial Replication Requires a Large Number of Enzymes and Proteins

Replication takes place in four stages: initiation, unwinding, elongation, and termination. The following discussion of the process of replication will focus on bacterial systems, where replication has been most thoroughly studied and is best understood. Although many aspects of replication in eukaryotic cells are similar to those in prokaryotic cells, there are some important differences. We will compare bacterial and eukaryotic replication later in the chapter.

Initiation

The circular chromosome of *E. coli* has a single replication origin (*oriC*). The minimal sequence required for *oriC* to function consists of 245 bp that contain several critical sites. An **initiator protein** (known as DnaA in *E. coli*) binds to *oriC* and causes a short section of DNA to unwind. This unwinding allows helicase and other single-strand-binding proteins to attach to the polynucleotide strand (**Figure 12.11**).

Unwinding

Because DNA synthesis requires a single-stranded template and because double-stranded DNA must be unwound before DNA synthesis can take place, the cell relies on several proteins and enzymes to accomplish the unwinding.

DNA helicase A **DNA helicase** breaks the hydrogen bonds that exist between the bases of the two nucleotide strands of a DNA molecule. Helicase cannot *initiate* the unwinding of double-stranded DNA; the initiator protein first separates DNA strands at the origin, providing a short stretch of single-stranded DNA to which a helicase binds. Helicase binds to the lagging-strand template at each replication fork and moves in the 5′⟶3′ direction along this strand, thus also moving the replication fork (**Figure 12.12**).

Single-strand-binding proteins After DNA has been unwound by helicase, **single-strand-binding proteins** (SSBs) attach tightly to the exposed single-stranded DNA (see Figure 12.12). These proteins protect the single-stranded nucleotide

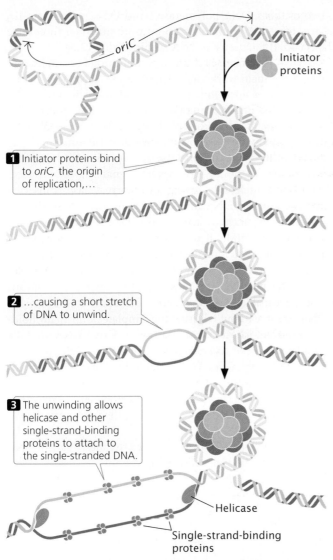

1 Initiator proteins bind to *oriC,* the origin of replication,…

Initiator proteins

2 …causing a short stretch of DNA to unwind.

3 The unwinding allows helicase and other single-strand-binding proteins to attach to the single-stranded DNA.

Helicase

Single-strand-binding proteins

12.11 *E. coli* **DNA replication begins when initiator proteins bind to *oriC,* the origin of replication.**

chains and prevent the formation of secondary structures such as hairpins (see Figure 10.17) that interfere with replication. Unlike many DNA-binding proteins, SSBs are indifferent to base sequence: they will bind to any single-stranded DNA. Single-strand-binding proteins form tetramers (groups of four); each tetramer covers from 35 to 65 nucleotides.

DNA gyrase Another protein essential for the unwinding process is the enzyme **DNA gyrase**, a topoisomerase. As discussed in Chapter 11, topoisomerases control the supercoiling of DNA. In replication, DNA gyrase reduces the torsional strain (torque) that builds up ahead of the replication fork as a result of unwinding (see Figure 12.12). It reduces torque by making a double-stranded break in one segment of the DNA helix, passing another segment of the helix through the break, and then resealing the broken ends of the DNA. This action removes a twist in the DNA and reduces the supercoiling.

A group of antibiotics called 4-quinolones kill bacteria by binding to DNA gyrase and inhibiting its action. The inhibition of DNA gyrase results in the cessation of DNA synthesis and bacterial growth. Many bacteria have acquired resistance to quinolones through mutations in the gene for DNA gyrase.

CONCEPTS

Replication is initiated at a replication origin, where an initiator protein binds and causes a short stretch of DNA to unwind. DNA helicase breaks hydrogen bonds at a replication fork, and single-strand-binding proteins stabilize the separated strands. DNA gyrase reduces the torsional strain that develops as the two strands of double-helical DNA unwind.

✔ CONCEPT CHECK 4

Place the following components in the order in which they are first used in the course of replication: helicase, single-strand-binding protein, DNA gyrase, initiator protein.

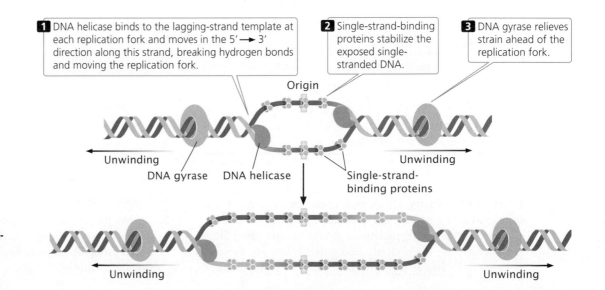

1 DNA helicase binds to the lagging-strand template at each replication fork and moves in the 5′ → 3′ direction along this strand, breaking hydrogen bonds and moving the replication fork.

2 Single-strand-binding proteins stabilize the exposed single-stranded DNA.

3 DNA gyrase relieves strain ahead of the replication fork.

Origin

Unwinding

DNA gyrase   DNA helicase

Single-strand-binding proteins

Unwinding

Unwinding

Unwinding

12.12 DNA helicase unwinds DNA by binding to the lagging-strand template at each replication fork and moving in the 5′ → 3′ direction.

Elongation

During the elongation phase of replication, single-stranded DNA is used as a template for the synthesis of DNA. This process requires a series of enzymes.

Synthesis of primers All DNA polymerases require a nucleotide with a 3′-OH group to which a new nucleotide can be added. Because of this requirement, DNA polymerases cannot initiate DNA synthesis on a bare template; rather, they require a primer—an existing 3′-OH group—to get started. How, then, does DNA synthesis begin?

An enzyme called **primase** synthesizes short stretches of nucleotides, or **primers**, to get DNA replication started. Primase synthesizes a short stretch of RNA nucleotides (about 10–12 nucleotides long), which provides a 3′-OH group to which DNA polymerase can attach DNA nucleotides. (Because primase is an RNA polymerase, it does not require a 3′-OH group to which nucleotides can be added.) All DNA molecules initially have short RNA primers embedded within them; these primers are later removed and replaced by DNA nucleotides.

On the leading strand, where DNA synthesis is continuous, a primer is required only at the 5′ end of the newly synthesized strand. On the lagging strand, where replication is discontinuous, a new primer must be generated at the beginning of each Okazaki fragment (**Figure 12.13**). Primase forms a complex with helicase at the replication fork and moves along the template of the lagging strand. The single primer on the leading strand is probably synthesized by the primase–helicase complex on the template of the lagging strand of the *other* replication fork, at the opposite end of the replication bubble.

TRY PROBLEM 26 →

CONCEPTS

Primase synthesizes a short stretch of RNA nucleotides (primers), which provides a 3′-OH group for the attachment of DNA nucleotides to start DNA synthesis.

✔ CONCEPT CHECK 5

Primers are synthesized where on the lagging strand?

a. Only at the 5′ end of the newly synthesized strand

b. Only at the 3′ end of the newly synthesized strand

c. At the beginning of every Okazaki fragment

d. At multiple places within an Okazaki fragment

DNA synthesis by DNA polymerases After DNA is unwound and a primer has been added, DNA polymerases elongate the new polynucleotide strand by catalyzing DNA polymerization. The best-studied polymerases are those of *E. coli*, which has at least five different DNA polymerases.

Two of them, DNA polymerase I and DNA polymerase III, carry out DNA synthesis in replication; the other three have specialized functions in DNA repair (**Table 12.3**).

DNA polymerase III is a large multiprotein complex that acts as the main workhorse of replication. DNA polymerase III synthesizes nucleotide strands by adding new nucleotides to the 3′ end of a growing DNA molecule. This enzyme has two enzymatic activities (see Table 12.3). Its 5′→3′ polymerase activity allows it to add new nucleotides in the 5′→3′ direction. Its 3′→5′ exonuclease activity allows it to remove nucleotides in the 3′→5′ direction, enabling it to correct errors. If a nucleotide having an incorrect base is inserted into the growing DNA molecule, DNA polymerase III uses its 3′→5′ exonuclease activity to back up and remove the incorrect nucleotide. It then resumes its 5′→3′ polymerase activity. These two functions together allow DNA polymerase III to efficiently and accurately synthesize new DNA molecules. DNA polymerase III has high processivity, which means that it is capable of adding many nucleotides to the growing DNA strand without releasing the template: it normally holds on to the template and continues synthesizing DNA until the template has been completely replicated. The high processivity of DNA polymerase III is ensured by one of the polypeptides that constitutes the enzyme. This polypeptide, termed the β subunit, serves as a clamp for the polymerase enzyme: it encircles the DNA and keeps the DNA polymerase attached to the template strand during replication. DNA polymerase III adds DNA nucleotides to the primer, synthesizing the DNA of both the leading and the lagging strands.

The first *E. coli* polymerase to be discovered, **DNA polymerase I**, also has 5′→3′ polymerase and 3′→5′ exonuclease activities (see Table 12.3), permitting the enzyme to synthesize DNA and to correct errors. Unlike DNA polymerase III, however, DNA polymerase I also possesses 5′→3′ exonuclease activity, which is used to remove the primers laid down by primase and to replace them with DNA nucleotides by synthesizing in a 5′→3′ direction. DNA polymerase I has lower processivity than DNA polymerase III. The removal and replacement of primers appear to constitute the main function of DNA polymerase I. After DNA polymerase III has initiated synthesis at the primer and moved downstream, DNA polymerase I removes the RNA nucleotides of the primer, replacing them with DNA nucleotides. DNA polymerases II, IV, and V function in DNA repair.

Despite their differences, all of *E. coli*'s DNA polymerases

1. synthesize any sequence specified by the template strand;

2. synthesize in the 5′→3′ direction by adding nucleotides to a 3′-OH group;

3. use dNTPs to synthesize new DNA;

4. require a primer to initiate synthesis;

5. catalyze the formation of a phosphodiester bond by joining the 5′-phosphate group of the incoming nucleotide to the 3′-OH group of the preceding nucleotide on the growing strand, cleaving off two phosphates in the process;

6. produce newly synthesized strands that are complementary and antiparallel to the template strands; and

7. are associated with a number of other proteins.
   TRY PROBLEM 24 →

CONCEPTS

DNA polymerases synthesize DNA in the 5′→3′ direction by adding new nucleotides to the 3′ end of a growing nucleotide strand.

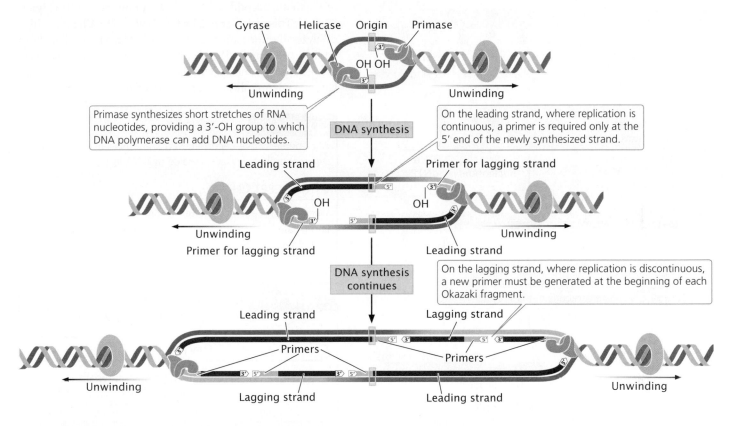

12.13 Primase synthesizes short stretches of RNA nucleotides, providing a 3′-OH group to which DNA polymerase can add DNA nucleotides.

 www.whfreeman.com/pierce4e   Animation 12.2 visualizes DNA synthesis on the leading and lagging strands.

Table 12.3 Characteristics of DNA Polymerases in *E. coli*

DNA Polymerase	5′→3′ Polymerization	3′→5′ Exonuclease	5′→3′ Exonuclease	Function
I	Yes	Yes	Yes	Removes and replaces primers
II	Yes	Yes	No	DNA repair; restarts replication after damaged DNA halts synthesis
III	Yes	Yes	No	Elongates DNA
IV	Yes	No	No	DNA repair
V	Yes	No	No	DNA repair; translesion DNA synthesis

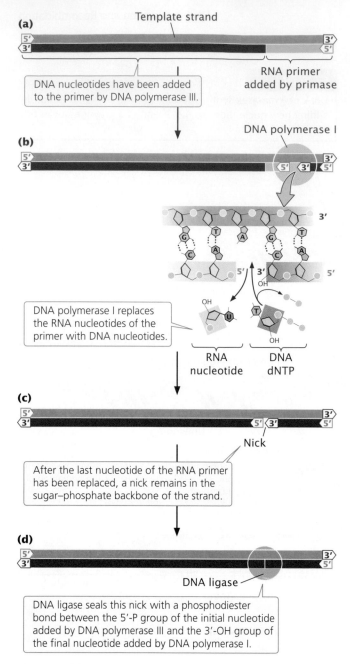

(a)

Template strand

DNA nucleotides have been added to the primer by DNA polymerase III.

RNA primer added by primase

(b)

DNA polymerase I

DNA polymerase I replaces the RNA nucleotides of the primer with DNA nucleotides.

RNA nucleotide | DNA dNTP

(c)

Nick

After the last nucleotide of the RNA primer has been replaced, a nick remains in the sugar–phosphate backbone of the strand.

(d)

DNA ligase

DNA ligase seals this nick with a phosphodiester bond between the 5′-P group of the initial nucleotide added by DNA polymerase III and the 3′-OH group of the final nucleotide added by DNA polymerase I.

12.14 DNA ligase seals the nick left by DNA polymerase I in the sugar–phosphate backbone.

▶ www.whfreeman.com/pierce4e To view the entire process of replication in action, take a look at Animations 12.1 and 12.4.

DNA ligase After DNA polymerase III attaches a DNA nucleotide to the 3′-OH group on the last nucleotide of the RNA primer, each new DNA nucleotide then provides the 3′-OH group needed for the next DNA nucleotide to be added. This process continues as long as template is available (**Figure 12.14a**). DNA polymerase I follows DNA polymerase III and, using its 5′→3′ exonuclease activity, removes the RNA primer. It then uses its 5′→3′ polymerase activity to replace the RNA nucleotides with DNA nucleotides. DNA polymerase I attaches the first nucleotide to the OH group at the 3′ end of the preced-

ing Okazaki fragment and then continues, in the 5′→3′ direction along the nucleotide strand, removing and replacing, one at a time, the RNA nucleotides of the primer (**Figure 12.14b**).

After polymerase I has replaced the last nucleotide of the RNA primer with a DNA nucleotide, a nick remains in the sugar–phosphate backbone of the new DNA strand. The 3′-OH group of the last nucleotide to have been added by DNA polymerase I is not attached to the 5′-phosphate group of the first nucleotide added by DNA polymerase III (**Figure 12.14c**). This nick is sealed by the enzyme **DNA ligase**, which catalyzes the formation of a phosphodiester bond without adding another nucleotide to the strand (**Figure 12.14d**). Some of the major enzymes and proteins required for prokaryotic DNA replication are summarized in **Table 12.4**.

CONCEPTS

After primers have been removed and replaced, the nick in the sugar–phosphate linkage is sealed by DNA ligase.

✔ CONCEPT CHECK 6

Which bacterial enzyme removes the primers?

a. Primase

b. DNA polymerase I

c. DNA polymerase II

d. Ligase

Table 12.4 Components required for replication in bacterial cells

Component	Function
Initiator protein	Binds to origin and separates strands of DNA to initiate replication
DNA helicase	Unwinds DNA at replication fork
Single-strand-binding proteins	Attach to single-stranded DNA and prevent secondary structures from forming
DNA gyrase	Moves ahead of the replication fork, making and resealing breaks in the double-helical DNA to release the torque that builds up as a result of unwinding at the replication fork
DNA primase	Synthesizes a short RNA primer to provide a 3′-OH group for the attachment of DNA nucleotides
DNA polymerase III	Elongates a new nucleotide strand from the 3′-OH group provided by the primer
DNA polymerase I	Removes RNA primers and replaces them with DNA
DNA ligase	Joins Okazaki fragments by sealing nicks in the sugar–phosphate backbone of newly synthesized DNA

Elongation at the replication fork Now that the major enzymatic components of elongation—DNA polymerases, helicase, primase, and ligase—have been introduced, let's consider how these components interact at the replication fork. Because the synthesis of both strands takes place simultaneously, two units of DNA polymerase III must be present at the replication fork, one for each strand. In one model of the replication process, the two units of DNA polymerase III are connected (**Figure 12.15**); the lagging-strand template loops around so that it is in position for $5' \rightarrow 3'$ replication. In this way, the DNA polymerase III complex is able to carry out $5' \rightarrow 3'$ replication simultaneously on both templates, even though they run in opposite directions. After about 1000 bp of new DNA has been synthesized, DNA polymerase III releases the lagging-strand template, and a new loop forms (see Figure 12.15). Primase synthesizes a new primer on the lagging strand and DNA polymerase III then synthesizes a new Okazaki fragment.

In summary, each active replication fork requires five basic components

1. helicase to unwind the DNA,

2. single-strand-binding proteins to protect the single nucleotide strands and prevent secondary structures,

3. the topoisomerase gyrase to remove strain ahead of the replication fork,

4. primase to synthesize primers with a $3'$-OH group at the beginning of each DNA fragment, and

5. DNA polymerase to synthesize the leading and lagging nucleotide strands.

Termination

In some DNA molecules, replication is terminated whenever two replication forks meet. In others, specific termination sequences block further replication. A termination protein, called Tus in *E. coli*, binds to these sequences. Tus blocks the movement of helicase, thus stalling the replication fork and preventing further DNA replication.

The Fidelity of DNA Replication

Overall, the error rate in replication is less than one mistake per billion nucleotides. How is this incredible accuracy achieved?

DNA polymerases are very particular in pairing nucleotides with their complements on the template strand. Errors in nucleotide selection by DNA polymerase arise only about once per 100,000 nucleotides. Most of the errors that do arise in nucleotide selection are corrected in a second process called **proofreading**. When a DNA polymerase inserts an incorrect nucleotide into the growing strand, the $3'$-OH group of the mispaired nucleotide is not correctly positioned in the

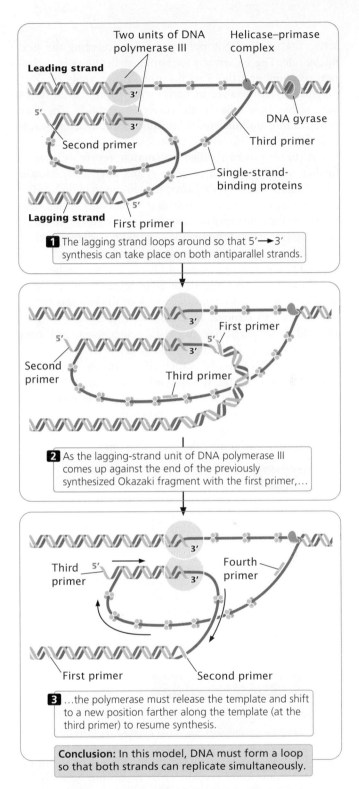

1 The lagging strand loops around so that $5' \rightarrow 3'$ synthesis can take place on both antiparallel strands.

2 As the lagging-strand unit of DNA polymerase III comes up against the end of the previously synthesized Okazaki fragment with the first primer,…

3 …the polymerase must release the template and shift to a new position farther along the template (at the third primer) to resume synthesis.

Conclusion: In this model, DNA must form a loop so that both strands can replicate simultaneously.

12.15 In one model of DNA replication in *E. coli*, the two units of DNA polymerase III are connected. The lagging-strand template forms a loop so that replication can take place on the two antiparallel DNA strands. Components of the replication machinery at the replication fork are shown at the top.

 www.whfreeman.com/pierce4e To better understand the coordination of leading- and lagging-strand synthesis, see Animation 12.3.

active site of the DNA polymerase for accepting the next nucleotide. The incorrect positioning stalls the polymerization reaction, and the $3' \rightarrow 5'$ exonuclease activity of DNA polymerase removes the incorrectly paired nucleotide. DNA polymerase then inserts the correct nucleotide. Together, proofreading and nucleotide selection result in an error rate of only one in 10 million nucleotides.

A third process, called **mismatch repair** (discussed further in Chapter 18), corrects errors after replication is complete. Any incorrectly paired nucleotides remaining after replication produce a deformity in the secondary structure of the DNA; the deformity is recognized by enzymes that excise an incorrectly paired nucleotide and use the original nucleotide strand as a template to replace the incorrect nucleotide. Mismatch repair requires the ability to distinguish between the old and the new strands of DNA, because the enzymes need some way of determining which of the two incorrectly paired bases to remove. In *E. coli*, methyl groups ($-CH_3$) are added to particular nucleotide sequences, but only *after* replication. Thus, immediately after DNA synthesis, only the old DNA strand is methylated. It can therefore be distinguished from the newly synthesized strand, and mismatch repair takes place preferentially on the unmethylated nucleotide strand. No single process could produce this level of accuracy; a series of processes are required, each process catching errors missed by the preceding ones.

CONCEPTS

Replication is extremely accurate, with less than one error per billion nucleotides. This accuracy is due to the processes of nucleotide selection, proofreading, and mismatch repair.

✔ **CONCEPT CHECK 7**

Which mechanism requires the ability to distinguish between newly synthesized and template strands of DNA?

a. Nucleotide selection   c. Mismatch repair

b. DNA proofreading   d. All of the above

CONNECTING CONCEPTS

The Basic Rules of Replication

Bacterial replication requires a number of enzymes (see Table 12.4), proteins, and DNA sequences that function together to synthesize a new DNA molecule. These components are important, but we must not become so immersed in the details of the process that we lose sight of the general principles of replication.

1. Replication is always semiconservative.

2. Replication begins at sequences called origins.

3. DNA synthesis is initiated by short segments of RNA called primers.

4. The elongation of DNA strands is always in the $5' \rightarrow 3'$ direction.

5. New DNA is synthesized from dNTPs; in the polymerization of DNA, two phosphate groups are cleaved from a dNTP and the

resulting nucleotide is added to the 3'-OH group of the growing nucleotide strand.

6. Replication is continuous on the leading strand and discontinuous on the lagging strand.

7. New nucleotide strands are complementary and antiparallel to their template strands.

8. Replication takes place at very high rates and is astonishingly accurate, thanks to precise nucleotide selection, proofreading, and repair mechanisms.

12.4 Eukaryotic DNA Replication Is Similar to Bacterial Replication but Differs in Several Aspects

Although eukaryotic replication resembles bacterial replication in many respects, replication in eukaryotic cells presents several additional challenges. First, the much greater size of eukaryotic genomes requires that replication be initiated at multiple origins. Second, eukaryotic chromosomes are linear, whereas prokaryotic chromosomes are circular. Third, the DNA template is associated with histone proteins in the form of nucleosomes, and nucleosome assembly must immediately follow DNA replication.

Eukaryotic Origins

Researchers first isolated eukaryotic origins of replication from yeast cells by demonstrating that certain DNA sequences confer the ability to replicate when transferred from a yeast chromosome to small circular pieces of DNA (plasmids). These **autonomously replicating sequences** (ARSs) enabled any DNA to which they were attached to replicate. They were subsequently shown to be the origins of replication in yeast chromosomes. The origins of replication of different organisms vary greatly in sequence, although they usually contain numerous A–T base pairs. In yeast, origins consist of 100 to 120 bp of DNA. A multiprotein complex, the origin-recognition complex (ORC), binds to origins and unwinds the DNA in this region. Interestingly, ORCs also function in regulating transcription.

CONCEPTS

Eukaryotic DNA contains many origins of replication. At each origin, a multiprotein origin-recognition complex binds to initiate the unwinding of the DNA.

✔ **CONCEPT CHECK 8**

In comparison with prokaryotes, what are some differences in the genome structure of eukaryotic cells that affect how replication takes place?

The Licensing of DNA Replication

Eukaryotic cells utilize thousands of origins, and so the entire genome can be replicated in a timely manner. The use of multiple origins, however, creates a special problem in the timing of replication: the entire genome must be precisely replicated once and only once in each cell cycle so that no genes are left unreplicated and no genes are replicated more than once. How does a cell ensure that replication is initiated at thousands of origins only once per cell cycle?

The precise replication of DNA is accomplished by the separation of the initiation of replication into two distinct steps. In the first step, the origins are licensed, meaning that they are approved for replication. This step is early in the cell cycle when a **replication licensing factor** attaches to an origin. In the second step, the replication machinery initiates replication at each *licensed* origin. The key is that the replication machinery functions only at licensed origins. As the replication forks move away from the origin, the licensing factor is removed, leaving the origin in an unlicensed state, where replication cannot be initiated again until the license is renewed. To ensure that replication takes place only once per cell cycle, the licensing factor is active only after the cell has completed mitosis and before the replication is initiated.

The eukaryotic licensing factor is a complex called MCM (for minichromosome maintenance), which contains a DNA helicase that unwinds a short stretch of DNA in the initiation of replication. MCM must bind to the DNA for replication to initiate at an origin. After replication has begun at an origin, a protein called Geminin prevents MCM from binding to DNA and reinitiating replication at that origin. At the end of mitosis, Geminin is degraded, allowing MCM to bind once again to DNA and relicense the origin.

Unwinding

Several different helicases that separate double-stranded DNA have been isolated from eukaryotic cells, as have single-strand-binding proteins and topoisomerases (which have a function equivalent to the DNA gyrase in bacterial cells). These enzymes and proteins are assumed to function in unwinding eukaryotic DNA in much the same way as their bacterial counterparts do.

Eukaryotic DNA Polymerases

Some significant differences in the processes of bacterial and eukaryotic replication are in the number and functions of DNA polymerases. Eukaryotic cells contain a number of different DNA polymerases that function in replication, recombination, and DNA repair (**Table 12.5**).

Three DNA polymerases carry out most of nuclear DNA synthesis during replication: DNA polymerase α,

Table 12.5   DNA polymerases in eukaryotic cells

DNA Polymerase	5′⟶3′ Polymerase Activity	3′⟶5′ Exonuclease Activity	Cellular Function
α (alpha)	Yes	No	Initiation of nuclear DNA synthesis and DNA repair; has primase activity
β (beta)	Yes	No	DNA repair and recombination of nuclear DNA
γ (gamma)	Yes	Yes	Replication and repair of mitochondrial DNA
δ (delta)	Yes	Yes	Lagging-strand synthesis of nuclear DNA, DNA repair, and translesion DNA synthesis
ε (epsilon)	Yes	Yes	Leading-strand synthesis
ξ (zeta)	Yes	No	Translesion DNA synthesis
η (eta)	Yes	No	Translesion DNA synthesis
θ (theta)	Yes	No	DNA repair
ι (iota)	Yes	No	Translesion DNA synthesis
κ (kappa)	Yes	No	Translesion DNA synthesis
λ (lambda)	Yes	No	DNA repair
μ (mu)	Yes	No	DNA repair
σ (sigma)	Yes	No	Nuclear DNA replication (possibly), DNA repair, and sister-chromatid cohesion
ϕ (phi)	Yes	No	Translesion DNA synthesis
Rev1	Yes	No	DNA repair

DNA polymerase δ, and DNA polymerase ε. **DNA polymerase α** contains primase activity and initiates nuclear DNA synthesis by synthesizing an RNA primer, followed by a short string of DNA nucleotides. After DNA polymerase α has laid down from 30 to 40 nucleotides, **DNA polymerase δ** completes replication on the lagging strand. Similar in structure and function to DNA polymerase δ, **DNA polymerase ε** also takes part in the replication of nuclear DNA. For many years, its precise role was not clear. Recent research shows that DNA polymerase ε replicates the leading strand. The other DNA polymerases take part in repair and recombination or catalyze the replication of organelle DNA (see Table 12.5).

Some DNA polymerases, such as DNA polymerase δ and DNA polymerase ε, are capable of replicating DNA at high speed and with high fidelity (few mistakes). These polymerases operate with high fidelity because they have active sites that snugly and exclusively accommodate the four normal DNA nucleotides, adenosine, guanosine, cytidine, and thymidine monophosphates. As a result of this specificity, distorted DNA templates and abnormal bases are not readily accommodated within the active site of the enzyme. When these errors are encountered in the DNA template, the high-fidelity DNA polymerases stall and are unable to bypass the lesion.

Other DNA polymerases have lower fidelity but are able to bypass distortions in the DNA template. These specialized translesion DNA polymerases generally have a more open active site and are able to accommodate and copy templates with abnormal bases, distorted structures, and bulky lesions. Thus, these specialized enzymes can bypass such errors but, because their active sites are more open and accommodating, they tend to make more errors. In replication, high-speed, high-fidelity enzymes are generally used until they encounter a replication block. At that point, one or more of the translesion polymerases takes over, bypasses the lesion, and continues replicating a short section of DNA. Then, the translesion polymerases detach from the replication fork and high-fidelity enzymes resume replication with high speed and accuracy. DNA-repair enzymes often repair errors produced by the translesion polymerases, although some of these errors may escape detection and lead to mutations.

CONCEPTS

There are a large number of different DNA polymerases in eukaryotic cells. DNA polymerases α, δ, and ε carry out replication on the leading and lagging strands. Other DNA polymerases carry out DNA repair. Specialized translesion polymerases are used to bypass distortions of the DNA template that normally stall the main DNA polymerases.

✔ **CONCEPT CHECK 9**

Some of the eukaryotic DNA polymerases have a tendency to make errors in replication. Why would a cell use an error-prone DNA polymerase instead of one that is more accurate?

Nucleosome Assembly

Eukaryotic DNA is complexed to histone proteins in nucleosome structures that contribute to the stability and packing of the DNA molecule (see Figure 11.4). In replication, chromatin structure is disrupted by the replication fork, but nucleosomes are quickly reassembled on the two new DNA molecules. Electron micrographs of eukaryotic DNA, such as that in **Figure 12.16**, show recently replicated DNA already covered with nucleosomes, indicating that nucleosomes are reassembled quickly.

The creation of new nucleosomes requires three steps: (1) the disruption of the original nucleosomes on the parental DNA molecule ahead of the replication fork; (2) the redistribution of preexisting histones on the new DNA molecules; and (3) the addition of newly synthesized histones to complete the formation of new nucleosomes. Before replication, a single DNA molecule is associated with histone proteins. After replication and nucleosome assembly, two DNA molecules are associated with histone proteins. Do the original histones of a nucleosome remain together, attached to one of the new DNA molecules, or do they disassemble and mix with new histones on both DNA molecules?

Techniques similar to those employed by Meselson and Stahl to determine the mode of DNA replication were used to address this question. Cells were cultivated for several generations in a medium containing amino acids labeled with a heavy isotope. The histone proteins incorporated

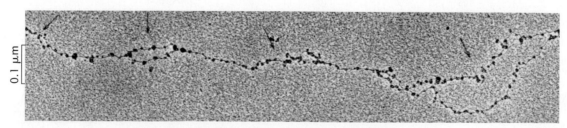

12.16 Nucleosomes are quickly assembled onto newly synthesized DNA. This electron micrograph of eukaryotic DNA in the process of replication clearly shows that newly replicated DNA is already covered with nucleosomes (dark circles). [Victoria Foe.]

these heavy amino acids and were dense (**Figure 12.17**). The cells were then transferred to a culture medium that contained amino acids labeled with a light isotope. Histones assembled after the transfer possessed the new, light amino acids and were less dense.

After replication, the histone octamers were isolated and centrifuged in a density gradient. Results showed that, after replication, the octamers were in a continuous band between high density (representing old octamers) and low density (representing new octamers). This finding indicates that newly assembled octamers consist of a mixture of old and new histones. Further evidence indicates that reconstituted nucleosomes appear on the new DNA molecules quickly after the new DNA emerges from the replication machinery.

The reassembly of nucleosomes during replication is facilitated by proteins called histone chaperones, which are associated with the helicase enzyme that unwinds the DNA. The histone chaperones accept old histones from the original DNA molecule and deposit them, along with newly synthesized histones, on the two new DNA molecules. Current evidence suggests that the original nucleosome is broken down into two H2A-H2B dimers (each dimer consisting of one H2A and one H2B) and a single H3-H4 tetramer (each tetramer consisting of two H3 histones and two H4 histones). The old H3-H4 tetramer is then transferred randomly to one of the new DNA molecules and serves as a foundation onto which either new or old copies of H2A-H2B dimers are added. Newly synthesized H3-H4 tetramers and H2A-H2b dimers also are added to each new DNA molecule to complete the formation of new nucleosomes. The assembly of the new nucleosomes is facilitated by a protein called chromatin-assembly factor 1 (CAF-1). **TRY PROBLEM 29 →**

In Chapter 11, we considered epigenetic changes—alterations to chromatin structure that are stably passed on to future generations of cells and sometimes organisms. The altered chromatin structure—the epigenetic marks—include DNA and histone methylation, acetylation, phosphorylation, and other chemical modifications of histone proteins; the use of variant histone proteins; and changes in the positions of nucleosomes. How are these epigenetic marks maintained during DNA replication and histone synthesis? The details of how epigenetic marks are maintained during replication are poorly understood. One possibility is that, after replication, the epigenetic marks remain on the original histones and the template strand of DNA. These marks then recruit enzymes that make similar changes in the new histones and newly replicated DNA strand.

CONCEPTS

After DNA replication, new nucleosomes quickly reassemble on the molecules of DNA. Nucleosomes break down in the course of replication and reassemble from a mixture of old and new histones. The reassembly of nucleosomes during replication is facilitated by histone chaperones and chromatin-assembly factors. Epigenetic changes in chromatin structure are often maintained during replication.

The Location of Replication Within the Nucleus

The DNA polymerases that carry out replication are frequently depicted as moving down the DNA template, much as a locomotive travels along a train track. Recent evidence suggests that this view is incorrect. A more accurate view is that the polymerase is fixed in location and template DNA is threaded through it, with newly synthesized DNA molecules emerging from the other end.

Techniques of fluorescence microscopy, which are able to reveal active sites of DNA synthesis, show that most replication in the nucleus of a eukaryotic cell takes place at a limited number of fixed sites, often referred to as replication factories. Time-lapse micrographs reveal that newly duplicated DNA is extruded from these particular sites. Similar results have been obtained for bacterial cells.

DNA Synthesis and the Cell Cycle

In rapidly dividing bacteria, DNA replication is continuous. In eukaryotic cells, however, replication is coordinated with the cell cycle. Passage through the cell cycle, including the onset of replication, is controlled by cell-cycle checkpoints.

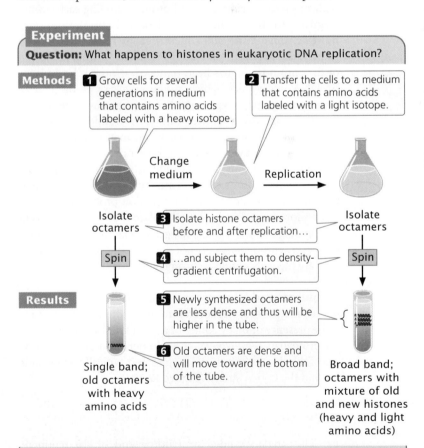

Experiment

Question: What happens to histones in eukaryotic DNA replication?

Methods
1. Grow cells for several generations in medium that contains amino acids labeled with a heavy isotope.
2. Transfer the cells to a medium that contains amino acids labeled with a light isotope.

Change medium → Replication →

Isolate octamers

3. Isolate histone octamers before and after replication…

Spin

4. …and subject them to density-gradient centrifugation.

Results

5. Newly synthesized octamers are less dense and thus will be higher in the tube.

6. Old octamers are dense and will move toward the bottom of the tube.

Single band; old octamers with heavy amino acids

Isolate octamers

Spin

Broad band; octamers with mixture of old and new histones (heavy and light amino acids)

Conclusion: After DNA replication, the new reassembled octamers are a random mixture of old and new histones.

12.17 Experimental procedure for studying how nucleosomes dissociate and reassociate in the course of replication.

The important G_1/S checkpoint (see Chapter 2) holds the cell cycle in G_1 until the DNA is ready to be replicated. After the G_1/S checkpoint is passed, the cell enters S phase and the DNA is replicated. The replication licensing system then ensures that the DNA is not replicated again until after the cell has passed through mitosis.

Replication at the Ends of Chromosomes

A fundamental difference between eukaryotic and bacterial replication arises because eukaryotic chromosomes are linear and thus have ends. As already stated, the 3′-OH group needed for replication by DNA polymerases is provided at the initiation of replication by RNA primers that are synthesized by primase. This solution is temporary because, eventually, the primers must be removed and replaced by DNA nucleotides. In a circular DNA molecule, elongation around the circle eventually provides a 3′-OH group immediately in front of the primer (**Figure 12.18a**). After the primer has been removed, the replacement DNA nucleotides can be added to this 3′-OH group.

The end-replication problem In linear chromosomes with multiple origins, the elongation of DNA in adjacent replicons also provides a 3′-OH group preceding each primer (**Figure 12.18b**). At the very end of a linear chromosome, however, there is no adjacent stretch of replicated DNA to provide this crucial 3′-OH group. When the primer at the end of the chromosome has been removed, it cannot be replaced by DNA nucleotides, which produces a gap at the end of the chromosome, suggesting that the chromosome should become progressively shorter with each round of replication. Chromosome shortening would mean that, when an organism reproduced, it would pass on shorter chromosomes than it had inherited. Chromosomes would become shorter with each new generation and would eventually destabilize. This situation has been termed the end-replication problem. Chromosome shortening does in fact take place in many somatic cells but, in single-celled organisms, germ cells, and early embryonic cells, chromosomes do not shorten and self-destruct. So how are the ends of linear chromosomes replicated?

Telomeres and telomerase The ends of chromosomes—the telomeres—possess several unique features, one of which is the presence of many copies of a short repeated sequence. In the protozoan *Tetrahymena*, this telomeric repeat is TTGGGG (see Table 11.2), with this G-rich strand typically protruding beyond the C-rich strand (**Figure 12.19a**):

$$\text{toward centromere} \xleftarrow{\hspace{1cm}} \begin{array}{l} 5'\text{–TTGGGGTTGGGG–}3' \\ 3'\text{–AACCCC–}5' \end{array} \xrightarrow{\hspace{1cm}} \text{end of chromosome}$$

The single-stranded protruding end of the telomere, known as the **G overhang** can be extended by **telomerase**, an enzyme with both a protein and an RNA component (also known as a ribonucleoprotein). The RNA part of the enzyme contains from 15 to 22 nucleotides that are complementary to the sequence on the G-rich strand. This

sequence pairs with the overhanging 3′ end of the DNA (**Figure 12.19b**) and provides a template for the synthesis of additional DNA copies of the repeats. DNA nucleotides are added to the 3′ end of the strand one at a time (**Figure 12.19c**) and, after several nucleotides have been added, the RNA template moves down the DNA and more nucleotides are added to the 3′ end (**Figure 12.19d**). Usually, from 14 to 16 nucleotides are added to the 3′ end of the G-rich strand.

In this way, the telomerase can extend the 3′ end of the chromosome without the use of a complementary DNA template (**Figure 12.19e**). How the complementary C-rich

(a) Circular DNA

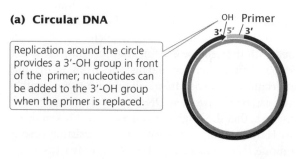

Replication around the circle provides a 3′-OH group in front of the primer; nucleotides can be added to the 3′-OH group when the primer is replaced.

(b) Linear DNA

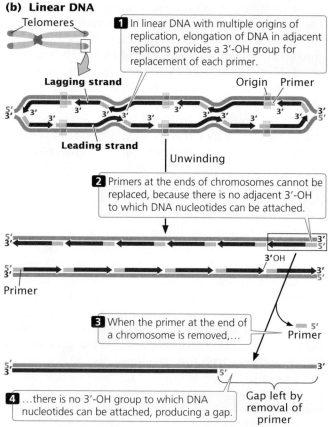

1 In linear DNA with multiple origins of replication, elongation of DNA in adjacent replicons provides a 3′-OH group for replacement of each primer.

2 Primers at the ends of chromosomes cannot be replaced, because there is no adjacent 3′-OH to which DNA nucleotides can be attached.

3 When the primer at the end of a chromosome is removed,…

4 …there is no 3′-OH group to which DNA nucleotides can be attached, producing a gap.

Gap left by removal of primer

Conclusion: In the absence of special mechanisms, DNA replication would leave gaps due to the removal of primers at the ends of chromosomes.

12.18 DNA synthesis at the ends of circular and linear chromosomes must differ.

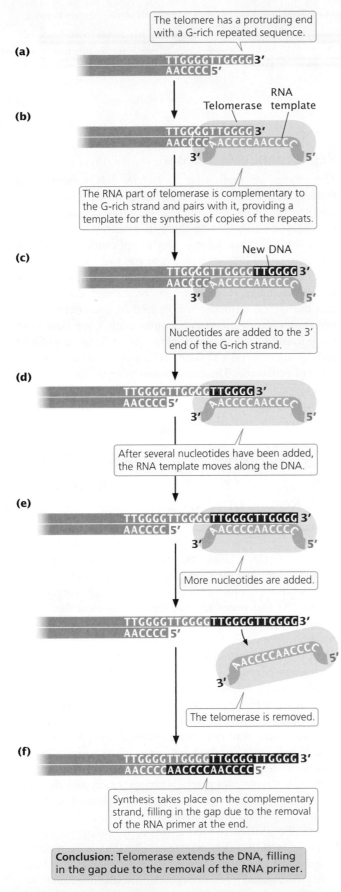

The telomere has a protruding end with a G-rich repeated sequence.

(a)

```
TTGGGGTTGGGG 3'
AACCCC 5'
```

(b)

Telomerase   RNA template

```
TTGGGGTTGGGG 3'
AACCCC ACCCCAACCC
    3'              5'
```

The RNA part of telomerase is complementary to the G-rich strand and pairs with it, providing a template for the synthesis of copies of the repeats.

(c)

New DNA

```
TTGGGGTTGGGG TTGGGG 3'
AACCCC ACCCCAACCC
    3'              5'
```

Nucleotides are added to the 3' end of the G-rich strand.

(d)

```
TTGGGGTTGGGG TTGGGG 3'
AACCCC 5'        ACCCCAACCC
              3'              5'
```

After several nucleotides have been added, the RNA template moves along the DNA.

(e)

```
TTGGGGTTGGGG TTGGGGTTGGGG 3'
AACCCC 5'        ACCCCAACCC
              3'              5'
```

More nucleotides are added.

```
TTGGGGTTGGGG TTGGGGTTGGGG 3'
AACCCC 5'
                ACCCCAACCC
                          5'
              3'
```

The telomerase is removed.

(f)

```
TTGGGGTTGGGG TTGGGGTTGGGG 3'
AACCCCAACCCCAACCCC 5'
```

Synthesis takes place on the complementary strand, filling in the gap due to the removal of the RNA primer at the end.

Conclusion: Telomerase extends the DNA, filling in the gap due to the removal of the RNA primer.

12.19 **The enzyme telomerase is responsible for the replication of chromosome ends.**

strand is synthesized (**Figure 12.19f**) is not yet clear. It may be synthesized by conventional replication, with DNA polymerase α synthesizing an RNA primer on the 5' end of the extended (G-rich) template. The removal of this primer once again leaves a gap at the 5' end of the chromosome, but this gap does not matter, because the end of the chromosome is extended at each replication by telomerase; so, the chromosome does not become shorter overall.

Telomerase is present in single-celled organisms, germ cells, early embryonic cells, and certain proliferative somatic cells (such as bone-marrow cells and cells lining the intestine), all of which must undergo continuous cell division. Most somatic cells have little or no telomerase activity, and chromosomes in these cells progressively shorten with each cell division. These cells are capable of only a limited number of divisions; when the telomeres have shortened beyond a critical point, a chromosome becomes unstable, has a tendency to undergo rearrangements, and is degraded. These events lead to cell death.

CONCEPTS

The ends of eukaryotic chromosomes are replicated by an RNA–protein enzyme called telomerase. This enzyme adds extra nucleotides to the G-rich DNA strand of the telomere.

✔ CONCEPT CHECK 10

What would be the result if an organism's telomerase were mutated and nonfunctional?

a. No DNA replication would take place.

b. The DNA polymerase enzyme would stall at the telomere.

c. Chromosomes would shorten with each new generation.

d. RNA primers could not be removed.

Telomerase, aging, and disease The shortening of telomeres may contribute to the process of aging. The telomeres of genetically engineered mice that lack a functional telomerase gene (and therefore do not express telomerase in somatic or germ cells) undergo progressive shortening in successive generations. After several generations, these mice show some signs of premature aging, such as graying, hair loss, and delayed wound healing. Through genetic engineering, it is also possible to create somatic cells that express telomerase. In these cells, telomeres do not shorten, cell aging is inhibited, and the cells will divide indefinitely. Although these observations suggest that telomere length is associated with aging, the precise role of telomeres in human aging is uncertain and controversial.

Some diseases are associated with abnormalities of telomere replication. People who have Werner syndrome, an autosomal recessive disease, show signs of premature aging that begins in adolescence or early adulthood, including wrinkled skin, graying of the hair, baldness, cataracts, and

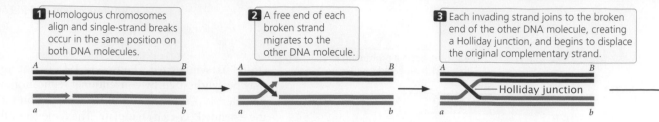

1 Homologous chromosomes align and single-strand breaks occur in the same position on both DNA molecules.

2 A free end of each broken strand migrates to the other DNA molecule.

3 Each invading strand joins to the broken end of the other DNA molecule, creating a Holliday junction, and begins to displace the original complementary strand.

Holliday junction

muscle atrophy. They often develop cancer, osteoporosis, heart and artery disease, and other ailments typically associated with aging. The causative gene, *WRN*, has been mapped to human chromosome 8 and normally encodes a RecQ helicase enzyme. This enzyme is necessary for the efficient replication of telomeres. In people who have Werner syndrome, this helicase is defective and, consequently, the telomeres shorten prematurely.

Another disease associated with abnormal maintenance of telomeres is dyskeratosis congenita, which leads to progressive bone-marrow failure, in which the bone marrow fails to produce enough new blood cells. People with an X-linked form of the disease have a mutation in a gene that encodes dyskerin. This protein normally helps process the RNA component of telomerase. People who have the disease typically inherit short telomeres from a parent who carries the mutation and who is unable to maintain telomere length in his or her germ cells owing to defective dyskerin. In families that carry this mutation, telomere length typically shortens with each successive generation, leading to anticipation, a progressive increase in the severity of the disease over generations (see Chapter 5).

Telomerase also appears to play a role in cancer. Cancer tumor cells have the capacity to divide indefinitely, and the telomerase enzyme is expressed in 90% of all cancers. Some recent evidence indicates that telomerase may stimulate cell proliferation independently of its effect on telomere length, and so the mechanism by which telomerase contributes to cancer is not clear. As will be discussed in Chapter 23, cancer is a complex, multistep process that usually requires mutations in at least several genes. Telomerase activation alone does not lead to cancerous growth in most cells, but it does appear to be required, along with other mutations, for cancer to develop. Some experimental cancer drugs work by inhibiting the action of telomerase.

One of the difficulties in studying the effect of telomere shortening on the aging process is that the expression of telomerase in somatic cells also promotes cancer, which may shorten a person's life span. To circumvent this problem, Antonia Tomas-Loba and colleagues created genetically engineered mice that expresssed telomerase and carried genes that made them resistant to cancer. These mice had longer telomeres, lived longer, and exhibited fewer age-related changes, such as skin alterations, a decrease in neuromuscular coordination, and degenerative diseases. These results support the idea that telomere shortening does contribute to aging. **TRY PROBLEM 31** →

Replication in Archaea

The process of replication in archaea has a number of features in common with replication in eukaryotic cells; many of the proteins taking part are more similar to those in eukaryotic cells than to those in eubacteria. Like eubacteria, some archaea have a single replication origin, but the archaean *Sulfolobus solfataricus* has two origins of replication, similar to the multiple origins seen in eukaryotic genomes. The replication origins of archaea do not contain the typical sequences recognized by bacterial initiator proteins; instead, they have sequences that are similar to those found in eukaryotic origins. The initiator proteins of archaea also are more similar to those of eukaryotes than to those of eubacteria. These similarities in replication between archaeal and eukaryotic cells reinforce the conclusion that the archaea are more closely related to eukaryotic cells than to the prokaryotic eubacteria.

12.5 Recombination Takes Place Through the Breakage, Alignment, and Repair of DNA Strands

Recombination is the exchange of genetic information between DNA molecules; when the exchange is between homologous DNA molecules, it is called **homologous recombination**. This process takes place in crossing over, in which homologous regions of chromosomes are exchanged (see Figure 7.5) and genes are shuffled into new combinations. Recombination is an extremely important genetic process because it increases genetic variation. Rates of recombination provide important information about linkage relations among genes, which is used to create genetic maps (see Figures 7.13 and 7.14). Recombination is also essential for some types of DNA repair (as will be discussed in Chapter 18).

Homologous recombination is a remarkable process: a nucleotide strand of one chromosome aligns precisely with a nucleotide strand of the homologous chromosome, breaks arise in corresponding regions of different DNA molecules, parts of the molecules precisely change place, and then the pieces are correctly joined. In this complicated series of events, no genetic information is lost or gained. Although the precise molecular mechanism of homologous recombination is still poorly known, the exchange is probably accomplished through the pairing of complementary bases. A single-stranded DNA molecule of one chromosome pairs with a single-stranded DNA molecule of another, forming **heteroduplex DNA**.

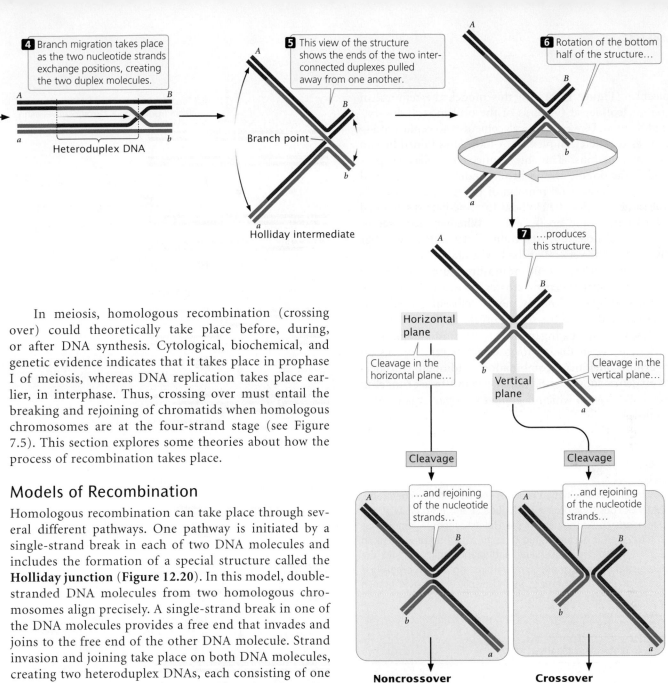

4 Branch migration takes place as the two nucleotide strands exchange positions, creating the two duplex molecules.

Heteroduplex DNA

5 This view of the structure shows the ends of the two interconnected duplexes pulled away from one another.

Branch point

Holliday intermediate

6 Rotation of the bottom half of the structure...

7 ...produces this structure.

Horizontal plane

Cleavage in the horizontal plane...

Vertical plane

Cleavage in the vertical plane...

Cleavage

Cleavage

...and rejoining of the nucleotide strands...

...and rejoining of the nucleotide strands...

Noncrossover recombinants

Crossover recombinants

...produces noncrossover recombinants consisting of two heteroduplex molecules.

...produces crossover recombinants consisting of two heteroduplex molecules.

Conclusion: The Holliday model predicts noncrossover or crossover recombinant DNA, depending on whether cleavage is in the horizontal or the vertical plane.

12.20 The Holliday model of homologous recombination. In this model, recombination takes place through a single-strand break in each DNA duplex, strand displacement, branch migration, and resolution of a single Holliday junction.

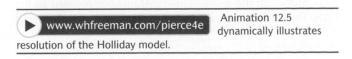

www.whfreeman.com/pierce4e

Animation 12.5 dynamically illustrates resolution of the Holliday model.

In meiosis, homologous recombination (crossing over) could theoretically take place before, during, or after DNA synthesis. Cytological, biochemical, and genetic evidence indicates that it takes place in prophase I of meiosis, whereas DNA replication takes place earlier, in interphase. Thus, crossing over must entail the breaking and rejoining of chromatids when homologous chromosomes are at the four-strand stage (see Figure 7.5). This section explores some theories about how the process of recombination takes place.

Models of Recombination

Homologous recombination can take place through several different pathways. One pathway is initiated by a single-strand break in each of two DNA molecules and includes the formation of a special structure called the **Holliday junction** (**Figure 12.20**). In this model, double-stranded DNA molecules from two homologous chromosomes align precisely. A single-strand break in one of the DNA molecules provides a free end that invades and joins to the free end of the other DNA molecule. Strand invasion and joining take place on both DNA molecules, creating two heteroduplex DNAs, each consisting of one original strand plus one new strand from the other DNA molecule. The point at which nucleotide strands pass from one DNA molecule to the other is the Holliday junction (see Figure 12.20). The junction moves along the molecules in a process called branch migration. The exchange of nucleotide strands and branch migration produce a structure termed the Holliday intermediate, which can be cleaved in one of two ways. Cleavage may be in the horizontal plane, followed by rejoining of the strands, producing noncrossover recombinants, in which the genes on either end of the molecules are identical with those originally present (gene A with gene B, and gene a with gene b). Cleavage in the vertical plane, followed by rejoining, produces crossover recombinants, in which the genes on either end of the molecules are different from those originally present (gene A with gene b, and gene a with gene B).

Another pathway for recombination is initiated by double-strand breaks in one of the two aligned DNA molecules (**Figure 12.21**). In this model, the removal of some nucleotides at the ends of the broken strands—followed by strand invasion, displacement, and replication—produces two heteroduplex DNA molecules joined by two Holliday junctions. The interconnected molecules produced in the double-strand-break model can be separated by further cleavage and reunion of the nucleotide strands in the same way that the Holliday intermediate is separated in the single-strand-break model. Whether crossover or noncrossover molecules are produced depends on whether cleavage is in the vertical or the horizontal plane.

Evidence for the double-strand-break model originally came from results of genetic crosses in yeast that could not be explained by the Holliday model. Subsequent observations showed that double-strand breaks appear in yeast in prophase I, when crossing over takes place, and that mutant strains that are unable to form double-strand breaks do not exhibit meiotic recombination. Although considerable evidence supports the double-strand-break model in yeast, the extent to which it applies to other organisms is not known.

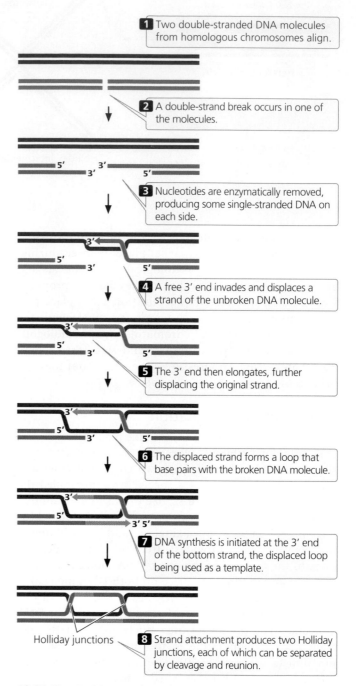

1 Two double-stranded DNA molecules from homologous chromosomes align.

2 A double-strand break occurs in one of the molecules.

3 Nucleotides are enzymatically removed, producing some single-stranded DNA on each side.

4 A free 3′ end invades and displaces a strand of the unbroken DNA molecule.

5 The 3′ end then elongates, further displacing the original strand.

6 The displaced strand forms a loop that base pairs with the broken DNA molecule.

7 DNA synthesis is initiated at the 3′ end of the bottom strand, the displaced loop being used as a template.

Holliday junctions

8 Strand attachment produces two Holliday junctions, each of which can be separated by cleavage and reunion.

12.21 The double-strand-break model of recombination. In this model, recombination takes place through a double-strand break in one DNA duplex, strand displacement, DNA synthesis, and the resolution of two Holliday junctions.

CONCEPTS

Homologous recombination requires the formation of heteroduplex DNA consisting of one nucleotide strand from each of two homologous chromosomes. In the Holliday model, homologous recombination is accomplished through a single-strand break in the DNA, strand displacement, and branch migration. In the double-strand-break model, recombination is accomplished through double-strand breaks, strand displacement, and branch migration.

✔ **CONCEPT CHECK 11**

Why is recombination important?

Enzymes Required for Recombination

Recombination between DNA molecules requires the unwinding of DNA helices, the cleavage of nucleotide strands, strand invasion, and branch migration, followed by further strand cleavage and union to remove Holliday junctions. Much of what we know about these processes arises from studies of gene exchange in *E. coli*. Although bacteria do not undergo meiosis, they do have a type of sexual reproduction (conjugation), in which one bacterium donates its chromosome to another (discussed more fully in Chapter 8). Subsequent to conjugation, the recipient bacterium has two chromosomes, which may undergo homologous recombination. Geneticists have isolated mutant strains of *E. coli* that are deficient in recombination; the study of these strains has resulted in the identification of genes and proteins that take part in bacterial recombination, revealing several different pathways by which it can take place.

Three genes that play pivotal roles in *E. coli* recombination are *recB*, *recC*, and *recD*, which encode three polypeptides that together form the RecBCD protein. This protein unwinds double-stranded DNA and is capable of cleaving nucleotide strands. The *recA* gene encodes the RecA protein; this protein allows a single strand to invade a DNA helix and the subsequent displacement of one of the original strands. In eukaryotes, the formation and branch migration of Holliday structures is facilitated by the enzyme Rad51.

In *E. coli*, *ruvA* and *ruvB* genes encode proteins that catalyze branch migration, and the *ruvC* gene produces a protein, called resolvase, that cleaves Holliday structures. Cleavage and resolution of Holliday structures in eukaryotes is carried out by an anlagous enzyme called GEN1. Single-strand-binding proteins, DNA ligase, DNA polymerases, and DNA gyrase also play roles in various types of recombination, in addition to their functions in DNA replication.

CONCEPTS

A number of proteins have roles in recombination, including RecA, RecBCD, RuvA, RuvB, resolvase, single-strand-binding proteins, ligase, DNA polymerases, and gyrase.

✔ CONCEPT CHECK 12

What is the function of resolvase in recombination?

a. Unwinds double-stranded DNA.

b. Allows a single DNA strand to invade a DNA helix.

c. Displaces one of the original DNA strands during branch migration.

d. Cleaves the Holliday structure.

Gene Conversion

As we have seen, homologous recombination is the mechanism that produces crossing over. It is also responsible for a related phenomenon known as **gene conversion**, a process of nonreciprocal genetic exchange that can produce abnormal ratios of gametes following meiosis. For example, an individual organism with genotype *Aa* is expected to produce $\frac{1}{2}$ *A* gametes and $\frac{1}{2}$ *a* gametes. Sometimes, however, meiosis in an *Aa* individual produces $\frac{3}{4}$ *A* and $\frac{1}{4}$ *a* or $\frac{1}{4}$ *A* and $\frac{3}{4}$ *a*. Gene conversion arises from heteroduplex formation that takes place in recombination. During heteroduplex formation, a single-stranded DNA molecule of one chromosome pairs with a single-stranded DNA molecule of another chromosome. If the two strands in a heteroduplex come from chromosomes with different alleles, there will be a mismatch of bases in the heteroduplex DNA (**Figure 12.22**). Such mismatches are often repaired by the cell. Repair mechanisms frequently excise nucleotides on one of the strands and replace them with new DNA by using the complementary strand as

a template. One copy of an allele may be converted into the other allele, leading to a gene-conversion event (see Figure 12.22), depending on which strand serves as a template.

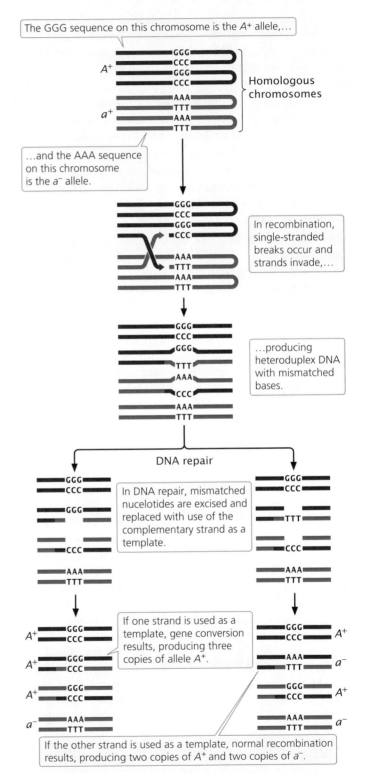

12.22 Gene conversion takes place through the repair of mismatched bases in heteroduplex DNA.

CONCEPTS SUMMARY

- Replication is semiconservative: DNA's two nucleotide strands separate, and each serves as a template on which a new strand is synthesized.

- In theta replication of DNA, the two nucleotide strands of a circular DNA molecule unwind, creating a replication bubble; within each replication bubble, DNA is normally synthesized on both strands and at both replication forks, producing two circular DNA molecules.

- Rolling-circle replication is initiated by a nick in one strand of circular DNA, which produces a 3′-OH group to which new nucleotides are added while the 5′ end of the broken strand is displaced from the circle.

- Linear eukaryotic DNA contains many origins of replication. Unwinding and replication take place on both templates at both ends of the replication bubble until adjacent replicons meet, resulting in two linear DNA molecules.

- All DNA synthesis is in the 5′→3′ direction. Because the two nucleotide strands of DNA are antiparallel, replication takes place continuously on one strand (the leading strand) and discontinuously on the other (the lagging strand).

- Replication begins when an initiator protein binds to a replication origin and unwinds a short stretch of DNA to which DNA helicase attaches. DNA helicase unwinds the DNA at the replication fork, single-strand-binding proteins bind to single nucleotide strands to prevent secondary structures, and DNA gyrase (a topoisomerase) removes the strain ahead of the replication fork that is generated by unwinding.

- During replication, primase synthesizes short primers of RNA nucleotides, providing a 3′-OH group to which DNA polymerase can add DNA nucleotides.

- DNA polymerase adds new nucleotides to the 3′ end of a growing polynucleotide strand. Bacteria have two DNA polymerases that have primary roles in replication: DNA polymerase III, which synthesizes new DNA on the leading and lagging strands, and DNA polymerase I, which removes and replaces primers.

- DNA ligase seals the nicks that remain in the sugar–phosphate backbones when the RNA primers are replaced by DNA nucleotides.

- Several mechanisms ensure the high rate of accuracy in replication, including precise nucleotide selection, proofreading, and mismatch repair.

- Precise replication at multiple origins in eukaryotes is ensured by a licensing factor that must attach to an origin before replication can begin.

- Eukaryotic nucleosomes are quickly assembled on new molecules of DNA; newly assembled nucleosomes consist of a random mixture of old and new histone proteins.

- The ends of linear eukaryotic DNA molecules are replicated by the enzyme telomerase.

- Homologous recombination takes place through breaks in nucleotide strands, alignment of homologous DNA segments, and rejoining of the strands. Homologous recombination requires a number of enzymes and proteins.

- Gene conversion is nonreciprocal genetic exchange and produces abnormal ratios of gametes.

IMPORTANT TERMS

semiconservative replication (p. 322)
equilibrium density gradient centrifugation (p. 323)
replicon (p. 325)
replication origin (p. 325)
theta replication (p. 325)
replication bubble (p. 325)
replication fork (p. 325)
bidirectional replication (p. 325)
rolling-circle replication (p. 326)
DNA polymerase (p. 329)
continuous replication (p. 329)
leading strand (p. 329)
discontinuous replication (p. 329)

lagging strand (p. 329)
Okazaki fragment (p. 329)
initiator protein (p. 330)
DNA helicase (p. 330)
single-strand-binding protein (SSB) (p. 330)
DNA gyrase (p. 331)
primase (p. 332)
primer (p. 332)
DNA polymerase III (p. 332)
DNA polymerase I (p. 332)
DNA ligase (p. 334)
proofreading (p. 335)
mismatch repair (p. 336)

autonomously replicating sequence (ARS) (p. 336)
replication licensing factor (p. 337)
DNA polymerase α (p. 338)
DNA polymerase δ (p. 338)
DNA polymerase ε (p. 338)
G overhang (p. 340)
telomerase (p. 340)
homologous recombination (p. 342)
heteroduplex DNA (p. 342)
Holliday junction (p. 343)
gene conversion (p. 345)

ANSWERS TO CONCEPT CHECKS

1. Two bands
2. b
3. c
4. Initiator protein, helicase, single-strand-binding protein, DNA gyrase

5. c
6. b
7. c
8. The size of eukaryotic genomes, the linear structure of eukaryotic chromosomes, and the association of DNA with histone proteins

9. Because error-prone DNA polymerases can bypass bulky lesions in the DNA helix that stall accurate, high-speed DNA polymerases

10. c

11. Recombination is important for genetic variation and for some types of DNA repair.

12. d

WORKED PROBLEMS

1. The following diagram represents the template strands of a replication bubble in a DNA molecule. Draw in the newly synthesized strands and identify the leading and lagging strands.

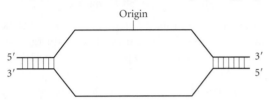

• Solution

To determine the leading and lagging strands, first note which end of each template strand is 5′ and which end is 3′. With a pencil, draw in the strands being synthesized on these templates, and identify their 5′ and 3′ ends, recalling that the newly synthesized strands must be antiparallel to the templates.

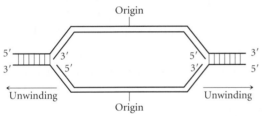

Next, determine the direction of replication for each new strand, which must be 5′→3′. You might draw arrows on the new strands to indicate the direction of replication. After you have established the direction of replication for each strand, look at each fork and determine whether the direction of replication for a strand is the same as the direction of unwinding. The strand on which replication is in the same direction as unwinding is the leading strand. The strand on which replication is in the direction opposite that of unwinding is the lagging strand. Make sure that you have one leading strand and one lagging strand for each fork.

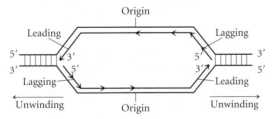

2. Consider the experiment conducted by Meselson and Stahl in which they used ^{14}N and ^{15}N in cultures of *E. coli* and equilibrium density gradient centrifugation. Draw pictures to represent the bands produced by bacterial DNA in the density-gradient tube before the switch to medium containing ^{14}N and after one, two, and three rounds of replication after the switch to the medium containing ^{14}N. Use a separate set of drawings to show the bands that would appear if replication were (a) semiconservative; (b) conservative; (c) dispersive.

• Solution

DNA labeled with ^{15}N will be denser than DNA labeled with ^{14}N; therefore ^{15}N-labeled DNA will sink lower in the density-gradient tube. Before the switch to medium containing ^{14}N, all DNA in the bacteria will contain ^{15}N and will produce a single band in the lower end of the tube.

a. With semiconservative replication, the two strands separate, and each serves as a template on which a new strand is synthesized. After one round of replication, the original template strand of each molecule will contain ^{15}N and the new strand of each molecule will contain ^{14}N; so a single band will appear in the density gradient halfway between the positions expected of DNA containing only ^{15}N and of DNA containing only ^{14}N. In the next round of replication, the two strands again separate and serve as templates for new strands. Each of the new strands contains only ^{14}N, thus some DNA molecules will contain one strand with the original ^{15}N and one strand with new ^{14}N, whereas the other molecules will contain two strands with ^{14}N. This labeling will produce two bands, one at the intermediate position and one at a higher position in the tube. Additional rounds of replication should produce increasing amounts of DNA that contains only ^{14}N; so the higher band will get darker.

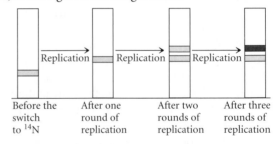

b. With conservative replication, the entire molecule serves as a template. After one round of replication, some molecules will consist entirely of ^{15}N, and others will consist entirely of ^{14}N; so two bands should be present. Subsequent rounds of replication will increase the fraction of DNA consisting entirely of new ^{14}N; thus the upper band will get darker. However, the original DNA with ^{15}N will remain, and so two bands will be present.

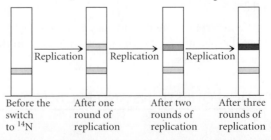

c. In dispersive replication, both nucleotide strands break down into fragments that serve as templates for the synthesis of new DNA. The fragments then reassemble into DNA molecules. After one round of replication, all DNA should contain approximately half ^{15}N and half ^{14}N, producing a single band that is halfway between the positions expected of DNA labeled with ^{15}N and of DNA labeled with ^{14}N. With further rounds of replication, the proportion of ^{14}N in each molecule increases; so a single hybrid band remains, but its position in the density gradient will move upward. The band is also expected to get darker as the total amount of DNA increases.

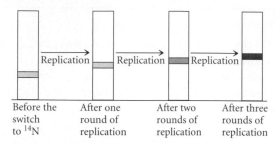

| Before the switch to ^{14}N | After one round of replication | After two rounds of replication | After three rounds of replication |

3.   The *E. coli* chromosome contains 4.6 million base pairs of DNA. If synthesis at each replication fork takes place at a rate of 1000 nucleotides per second, how long will it take to completely replicate the *E. coli* chromosome with theta replication?

• Solution

Bacterial chromosomes contain a single origin of replication, and theta replication usually employs two replication forks, which proceed around the chromosome in opposite directions. Thus, the overall rate of replication for the whole chromosome is 2000 nucleotides per second. With a total of 4.6 million base pairs of DNA, the entire chromosome will be replicated in:

$$4{,}600{,}000 \text{ bp} \times \frac{1 \text{ second}}{2000 \text{ bp}} = 2300 \text{ seconds} \times \frac{1 \text{ minute}}{60 \text{ seconds}}$$

$$= 38.33 \text{ minutes}$$

At the beginning of this chapter, *E. coli* was said to be capable of dividing every 20 minutes. How is this rate possible if it takes almost twice as long to replicate its genome? The answer is that a second round of replication begins before the first round has finished. Thus, when an *E. coli* cell divides, the chromosomes that are passed on to the daughter cells are already partly replicated. In contrast, a eukaryotic cell replicates its entire genome once, and only once, in each cell cycle.

COMPREHENSION QUESTIONS

Section 12.2

1.   What is semiconservative replication?

*2.   How did Meselson and Stahl demonstrate that replication in *E. coli* takes place in a semiconservative manner?

*3.   Draw a molecule of DNA undergoing theta replication. On your drawing, identify (a) origin, (b) polarity (5′ and 3′ ends) of all template strands and newly synthesized strands, (c) leading and lagging strands, (d) Okazaki fragments, and (e) location of primers.

4.   Draw a molecule of DNA undergoing rolling-circle replication. On your drawing, identify (a) origin, (b) polarity (5′ and 3′ ends) of all template and newly synthesized strands, (c) leading and lagging strands, (d) Okazaki fragments, and (e) location of primers.

5.   Draw a molecule of DNA undergoing eukaryotic linear replication. On your drawing, identify (a) origin, (b) polarity (5′ and 3′ ends) of all template and newly synthesized strands, (c) leading and lagging strands, (d) Okazaki fragments, and (e) location of primers.

6.   What are three major requirements of replication?

*7.   What substrates are used in the DNA-synthesis reaction?

Section 12.3

8.   List the different proteins and enzymes taking part in bacterial replication. Give the function of each in the replication process.

9.   What similarities and differences exist in the enzymatic activities of DNA polymerases I, II, and III? What is the function of each type of DNA polymerase in bacterial cells?

*10.   Why is primase required for replication?

11.   What three mechanisms ensure the accuracy of replication in bacteria?

Section 12.4

12.   How does replication licensing ensure that DNA is replicated only once at each origin per cell cycle?

*13.   In what ways is eukaryotic replication similar to bacterial replication, and in what ways is it different?

14.   What is the end-of-chromosome problem for replication? Why, in the absence of telomerase, do the ends of chromosomes get progressively shorter each time the DNA is replicated?

15.   Outline in words and pictures how telomeres at the ends of eukaryotic chromosomes are replicated.

Section 12.5

*16.   What are some of the enzymes taking part in recombination in *E. coli* and what roles do they play?

17.   What is gene conversion? How does it arise?

APPLICATION QUESTIONS AND PROBLEMS

Section 12.2

*18.   Suppose a future scientist explores a distant planet and discovers a novel form of double-stranded nucleic acid. When this nucleic acid is exposed to DNA polymerases from *E. coli*, replication takes place continuously on both strands. What conclusion can you make about the structure of this novel nucleic acid?

*19. Phosphorus is required to synthesize the deoxyribonucleoside triphosphates used in DNA replication. A geneticist grows some *E. coli* in a medium containing nonradioactive phosphorous for many generations. A sample of the bacteria is then transferred to a medium that contains a radioactive isotope of phosphorus (^{32}P). Samples of the bacteria are removed immediately after the transfer and after one and two rounds of replication. What will be the distribution of radioactivity in the DNA of the bacteria in each sample? Will radioactivity be detected in neither, one, or both strands of the DNA?

20. A line of mouse cells is grown for many generations in a medium with ^{15}N. Cells in G_1 are then switched to a new medium that contains ^{14}N. Draw a pair of homologous chromosomes from these cells at the following stages, showing the two strands of DNA molecules found in the chromosomes. Use different colors to represent strands with ^{14}N and ^{15}N.

  a. Cells in G_1, before switching to medium with ^{14}N

  b. Cells in G_2, after switching to medium with ^{14}N

  c. Cells in anaphase of mitosis, after switching to medium with ^{14}N

  d. Cells in metaphase I of meiosis, after switching to medium with ^{14}N

  e. Cells in anaphase II of meiosis, after switching to medium with ^{14}N

*21. A circular molecule of DNA contains 1 million base pairs. If the rate of DNA synthesis at a replication fork is 100,000 nucleotides per minute, how much time will theta replication require to completely replicate the molecule, assuming that theta replication is bidirectional? How long will replication of this circular chromosome take by rolling-circle replication? Ignore replication of the displaced strand in rolling-circle replication.

22. A bacterium synthesizes DNA at each replication fork at a rate of 1000 nucleotides per second. If this bacterium completely replicates its circular chromosome by theta replication in 30 minutes, how many base pairs of DNA will its chromosome contain?

Section 12.3

*23. The following diagram represents a DNA molecule that is undergoing replication. Draw in the strands of newly synthesized DNA and identify (a) the polarity of newly synthesized strands, (b) the leading and lagging strands, (c) Okazaki fragments, and (d) RNA primers.

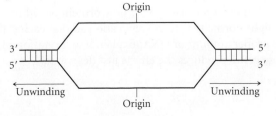

*24. What would be the effect on DNA replication of mutations that destroyed each of the following activities in DNA polymerase I?

  a. $3' \rightarrow 5'$ exonuclease activity

  b. $5' \rightarrow 3'$ exonuclease activity

  c. $5' \rightarrow 3'$ polymerase activity

25. How would DNA replication be affected in a cell that is lacking topoisomerase?

26. If the gene for primase were mutated so that no functional primase was produced, what would be the effect on theta replication? On rolling-circle replication?

27. DNA polymerases are not able to prime replication, yet primase and other RNA polymerases can. Some geneticists have speculated that the inability of DNA polymerase to prime replication is due to its proofreading function. This hypothesis argues that proofreading is essential for the faithful transmission of genetic information and that, because DNA polymerases have evolved the ability to proofread, they cannot prime DNA synthesis. Explain why proofreading and priming functions in the same enzyme might be incompatible?

Section 12.4

*28. Marina Melixetian and her colleagues suppressed the expression of Geminin protein in human cells by treating the cells with small interfering RNAs (siRNAs) complementary to Geminin messenger RNA (M. Melixetian et al. 2004. *Journal of Cell Biology* 165:473–482). (Small interfering RNAs cleave mRNAs with which they are complementary, and so there is no translation of the mRNA; pp. 394–396 in Chapter 14). Forty-eight hours after treatment with siRNA, the Geminin-depleted cells were enlarged and contained a single giant nucleus. Analysis of DNA content showed that many of these Geminin-depleted cells were 4*n* or greater. Explain these results.

29. What results would be expected in the experiment outlined in Figure 12.17 if, during replication, all the original histone proteins remained on one strand of the DNA and new histones attached to the other strand?

30. A number of scientists who study ways to treat cancer have become interested in telomerase. Why would they be interested in telomerase? How might cancer-drug therapies that target telomerase work?

31. The enzyme telomerase is part protein and part RNA. What would be the most likely effect of a large deletion in the gene that encodes the RNA part of telomerase? How would the function of telomerase be affected?

32. Dyskeratosis congenita (DKC) is a rare genetic disorder characterized by abnormal fingernails and skin pigmentation, the formation of white patches on the tongue and cheek, and progressive failure of the bone marrow. An autosomal dominant form of DKC results from mutations in the gene that encodes the

RNA component of telomerase. Tom Vulliamy and his colleagues examined 15 families with autosomal dominant DKC (T. Vulliamy et al. 2004. *Nature Genetics* 36:447–449). They observed that the median age of onset of DKC in parents was 37 years, whereas the median age of onset in the children of affected parents was 14.5 years. Thus, DKC in these families arose at progressively younger ages in successive generations, a phenomenon known as anticipation (see p. 122–123 in Chapter 5). The researchers measured telomere length of members of these families; the measurements are given in the adjoining table. Telomere length normally shortens with age, and so telomere length was adjusted for age. Note that the age-adjusted telomere length of all members of these families is negative, indicating that their telomeres are shorter than normal. For age-adjusted telomere length, the more negative the number, the shorter the telomere.

a. How do the telomere lengths of parents compare with the telomere lengths of their children?

Parent telomere length	Child telomere length
−4.7	−6.1
	−6.6
	−6.0
−3.9	−0.6
−1.4	−2.2
−5.2	−5.4
−2.2	−3.6
−4.4	−2.0
−4.3	−6.8
−5.0	−3.8
−5.3	−6.4
−0.6	−2.5
−1.3	−5.1
	−3.9
−4.2	−5.9

b. Explain why the telomeres of people with DKC are shorter than normal and why DKC arises at an earlier age in subsequent generations.

CHALLENGE QUESTIONS

Section 12.3

33. A conditional mutation expresses its mutant phenotype only under certain conditions (the restrictive conditions) and expresses the normal phenotype under other conditions (the permissive conditions). One type of conditional mutation is a temperature-sensitive mutation, which expresses the mutant phenotype only at certain temperatures.

Strains of *E. coli* have been isolated that contain temperature-sensitive mutations in the genes encoding different components of the replication machinery. In each of these strains, the protein produced by the mutated gene is nonfunctional under the restrictive conditions. These strains are grown under permissive conditions and then abruptly switched to the restrictive condition. After one round of replication under the restrictive condition, the DNA from each strain is isolated and analyzed. What characteristics would you expect to see in the DNA isolated from each strain with a temperature-sensitive mutation in its gene that encodes in the following?

a. DNA ligase   **d.** Primase

b. DNA polymerase I   **e.** Initiator protein

c. DNA polymerase III

Section 12.4

34. DNA topoisomerases play important roles in DNA replication and supercoiling (see Chapter 11). These enzymes are also the targets for certain anticancer drugs. Eric Nelson and his colleagues studied m-AMSA, one of the

DATA ANALYSIS

anticancer compounds that acts on topisomerase enzymes (E. M. Nelson, K. M. Tewey, and L. F. Liu. 1984. *Proceedings of the National Academy of Sciences* 81:1361–1365). They found that m-AMSA stabilizes an intermediate produced in the course of the topoisomerase's action. The intermediate consisted of the topoisomerase bound to the broken ends of the DNA. Breaks in DNA that are produced by anticancer compounds such as m-AMSA inhibit the replication of the cellular DNA and thus stop cancer cells from proliferating. Propose a mechanism for how m-AMSA and other anticancer agents that target topoisomerase enzymes taking part in replication might lead to DNA breaks and chromosome rearrangments.

***35.** The regulation of replication is essential to genomic stability, and, normally, the DNA is replicated just once every cell cycle (in the S phase). Normal cells produce protein A, which increases in concentration in the S phase. In cells that have a mutated copy of the gene for protein A, the protein is not functional and replication takes place continuously throughout the cell cycle, with the result that cells may have 50 times the normal amount of DNA. Protein B is normally present in G_1 but disappears from the cell nucleus in the S phase. In cells with a mutated copy of the gene for protein A, the levels of protein B fail to disappear in the S phase and, instead, remain high throughout the cell cycle. When the gene for protein B is mutated, no replication takes place.

Propose a mechanism for how protein A and protein B might normally regulate replication so that each cell gets the proper amount of DNA. Explain how mutation of these genes produces the effects just described.

13 Transcription

The death cap mushroom, _Amanita phalloides_, causes death by inhibiting the process of transcription. [Ron Wolf/Tom Stack & Associates.]

DEATH CAP POISONING

On November 8, 2009, 31-year-old Tomasa was hiking the Lodi Lake nature trail east of San Francisco with her husband and cousin when they came across some large white mushrooms that looked very much like the edible mushrooms that they enjoyed in their native Mexico. They picked the mushrooms and took them home, cooking and consuming them for dinner. Within hours, Tomasa and her family were sick and went to the hospital. They were later transferred to the critical care unit at California Pacific Medical Center in San Francisco, where Tomasa died of liver failure 3 weeks later. Her husband eventually recovered after a lengthy hospitalization; her cousin required a liver transplant to survive.

The mushrooms consumed by Tomasa and her family were _Amanita phalloides_, commonly known as the death cap. A single death cap mushroom contains enough toxin to kill an adult human. The death rate among those who consume death caps is 22%; among children under the age of 10, it's more than 50%. Death cap mushrooms appear to be spreading in California, leading to a recent surge in the number of mushroom poisonings.

Death cap poisoning is insidious. Gastrointestinal symptoms—abdominal pain, cramping, vomiting, diarrhea—begin within 6 to 12 hours of consuming the mushrooms, but these symptoms usually subside within a few hours and the patient seems to recover. Because of this initial remission, the poisoning is often not taken seriously until it's too late to pump the stomach and remove the toxin from the body. After a day or two, serious symptoms begin. Cells in the liver die, often causing permanent liver damage and death within a few days. There is no effective treatment, other than a liver transplant to replace the damaged organ.

How do death caps kill? Their deadly toxin, contained within the fruiting bodies that produce reproductive spores, is the protein α-amanitin, which consists of a short peptide of eight amino acids that forms a circular loop. α-Amanitin is a potent inhibitor of RNA polymerase II, the enzyme that transcribes protein-encoding genes in eukaryotes. RNA polymerase II binds to genes and synthesizes RNA molecules that are complementary to the DNA template. In the process of transcription, the RNA polymerase moves down the DNA template, adding one nucleotide at a time to the growing RNA chain. α-Amanitin binds to RNA polymerase and jams the moving parts of the enzyme, interfering with its ability to move along the DNA template. In the presence of α-amanitin, RNA synthesis slows from its normal rate of several thousand nucleotides per minute to just a few nucleotides per minute. The results are catastrophic. Without transcription, protein synthesis—required for cellular function—ceases and cells die. The liver, where the toxin accumulates, is irreparably damaged and stops functioning. In severe cases, the patient dies.

Death cap poisoning illustrates the extreme importance of transcription and the central role that RNA polymerase plays in the process. This chaper is about the process of transcription, the first step in the central dogma, the pathway of information transfer from DNA (genotype) to protein (phenoypte). Transcription is a complex process that requires precursors to RNA nucleotides, a DNA template, and a number of protein components. As we examine the stages of transcription, try to keep all the details in perspective; focus on understanding how the details relate to the overall purpose of transcription—the selective synthesis of an RNA molecule.

This chapter begins with a brief review of RNA structure and a discussion of the different classes of RNA. We then consider the major components required for transcription. Finally, we explore the process of transcription in eubacteria, eukaryotic cells, and archaea. At several points in the text, we'll pause to absorb some general principles that emerge.

13.1 RNA, Consisting of a Single Strand of Ribonucleotides, Participates in a Variety of Cellular Functions

Before we begin our study of transcription, we will consider the past and present importance of RNA, review the structure of RNA, and examine some of the different types of RNA molecules.

An Early RNA World

Life requires two basic functions. First, living organisms must be able to store and faithfully transmit genetic information during reproduction. Second, they must have the ability to catalyze the chemical transformations that drive life processes. A long-held belief was that the functions of information storage and chemical transformation are handled by two entirely different types of molecules: genetic information is stored in nucleic acids, whereas chemical transformations are catalyzed by protein enzymes. This biochemical dichotomy—nucleic acid for information, proteins for catalysts—created a dilemma. Which came first: proteins or nucleic acids? If nucleic acids carry the coding instructions for proteins, how could proteins be generated without them? Because nucleic acids are unable to copy themselves, how could they be generated without proteins? If DNA and proteins each require the other, how could life begin?

This apparent paradox was given a possible answer in 1981 when Thomas Cech and his colleagues discovered that RNA can serve as a biological catalyst. They found that RNA from the protozoan *Tetrahymena thermophila* can excise 400 nucleotides from its RNA in the absence of any protein. Other examples of catalytic RNAs have now been discovered in different types of cells. Called **ribozymes**, these catalytic RNA molecules can cut out parts of their own sequences, connect some RNA molecules together, replicate others, and even catalyze the formation of peptide bonds between amino acids. The discovery of ribozymes complements other evidence suggesting that the original genetic material was RNA.

Self-replicating ribozymes probably first arose between 3.5 billion and 4 billion years ago and may have begun the evolution of life on Earth. Early life was probably an RNA world, with RNA molecules serving both as carriers of genetic information and as catalysts that drove the chemical reactions needed to sustain and perpetuate life. These catalytic RNAs may have acquired the ability to synthesize protein-based enzymes, which are more efficient catalysts. With enzymes taking over more and more of the catalytic functions, RNA probably became relegated to the role of information storage and transfer. DNA, with its chemical stability and faithful replication, eventually replaced RNA as the primary carrier of genetic information. Nevertheless, RNA still plays a vital role in many biological processes, including replication, RNA processing, and translation. And research in the past 10 years has determined that newly discovered small RNA molecules play a fundamental role in many basic biological processes, demonstrating that life today is still very much an RNA world. These small RNA molecules will be discussed in more detail in Chapter 14.

CONCEPTS

Early life was probably centered on RNA, which served as the original genetic material and as biological catalysts.

The Structure of RNA

RNA, like DNA, is a polymer consisting of nucleotides joined together by phosphodiester bonds (see Chapter 10 for a discussion of RNA structure). However, there are several important differences in the structures of DNA and RNA. Whereas DNA nucleotides contain deoxyribose sugars, RNA nucleotides have ribose sugars (**Figure 13.1a**). With a free hydroxyl group on the $2'$-carbon atom of the ribose sugar, RNA is degraded rapidly under alkaline conditions. The deoxyribose sugar of DNA lacks this free hydroxyl group; so

(a)

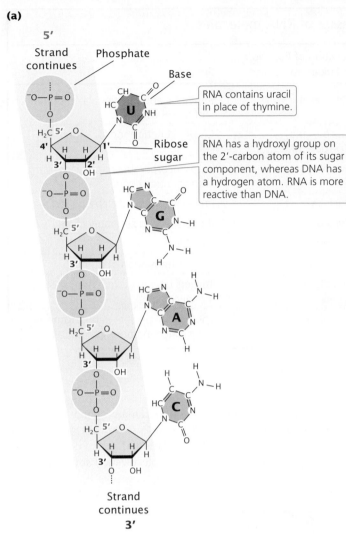

RNA contains uracil in place of thymine.

RNA has a hydroxyl group on the 2′-carbon atom of its sugar component, whereas DNA has a hydrogen atom. RNA is more reactive than DNA.

(b) Primary structure

5′ AUGCGGCUACGUAACGAGCUUAGCGCGUAUACCGAAAGGGUAGAAC 3′

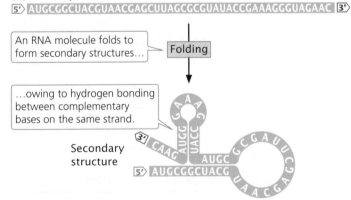

An RNA molecule folds to form secondary structures…

Folding

…owing to hydrogen bonding between complementary bases on the same strand.

Secondary structure

13.1 RNA has a primary and a secondary structure.

DNA is a more stable molecule. Another important difference is that thymine, one of the two pyrimidines found in DNA, is replaced by uracil in RNA.

A final difference in the structures of DNA and RNA is that RNA is usually single stranded, consisting of a single polynucleotide strand (**Figure 13.1b**), whereas DNA

normally consists of two polynucleotide strands joined by hydrogen bonding between complementary bases. Some viruses contain double-stranded RNA genomes, as discussed in Chapter 8. Although RNA is usually single stranded, short complementary regions within a nucleotide strand can pair and form secondary structures (see Figure 13.1b). These RNA secondary structures are often called hairpin-loop or stem-loop structures. When two regions within a single RNA molecule pair up, the strands in those regions must be antiparallel, with pairing between cytosine and guanine and between adenine and uracil (although, occasionally, guanine pairs with uracil).

The formation of secondary structures plays an important role in RNA function. Secondary structure is determined by the base sequence of the nucleotide strand; so different RNA molecules can assume different structures. Because their structure determines their function, RNA molecules have the potential for tremendous variation in function. With its two complementary strands forming a helix, DNA is much more restricted in the range of secondary structures that it can assume and so has fewer functional roles in the cell. Similarities and differences in DNA and RNA structures are summarized in **Table 13.1**. TRY PROBLEM 15 →

Classes of RNA

RNA molecules perform a variety of functions in the cell. **Ribosomal RNA** (rRNA) and ribosomal protein subunits make up the ribosome, the site of protein assembly. We'll take a more detailed look at the ribosome in Chapter 14. **Messenger RNA** (mRNA) carries the coding instructions for polypeptide chains from DNA to a ribosome. After attaching to the ribosome, an mRNA molecule specifies the sequence of the amino acids in a polypeptide chain and provides a template for joining amino acids. Large precursor molecules, which are termed **pre-messenger RNAs** (pre-mRNAs), are the immediate products of transcription

Table 13.1	The structures of DNA and RNA compared	
Characteristic	DNA	RNA
Composed of nucleotides	Yes	Yes
Type of sugar	Deoxyribose	Ribose
Presence of 2′-OH group	No	Yes
Bases	A, G, C, T	A, G, C, U
Nucleotides joined by phosphodiester bonds	Yes	Yes
Double or single stranded	Usually double	Usually single
Secondary structure	Double helix	Many types
Stability	Stable	Easily degraded

Table 13.2 Location and functions of different classes of RNA molecules

Class of RNA	Cell Type	Location of Function in Eukaryotic Cells*	Function
Ribosomal RNA (rRNA)	Bacterial and eukaryotic	Cytoplasm	Structural and functional components of the ribosome
Messenger RNA (mRNA)	Bacterial and eukaryotic	Nucleus and cytoplasm	Carries genetic code for proteins
Transfer RNA (tRNA)	Bacterial and eukaryotic	Cytoplasm	Helps incorporate amino acids into polypeptide chain
Small nuclear RNA (snRNA)	Eukaryotic	Nucleus	Processing of pre-mRNA
Small nucleolar RNA (snoRNA)	Eukaryotic	Nucleus	Processing and assembly of rRNA
Small cytoplasmic RNA (scRNA)	Eukaryotic	Cytoplasm	Variable
MicroRNA (miRNA)	Eukaryotic	Cytoplasm	Inhibits translation of mRNA
Small interfering RNA (siRNA)	Eukaryotic	Cytoplasm	Triggers degradation of other RNA molecules
Piwi-interacting RNA (piRNA)	Eukaryotic	Cytoplasm	Thought to regulate gametogenesis, but function poorly defined

*All eukaryotic RNAs are transcribed in the nucleus.

in eukaryotic cells. Pre-mRNAs are modified extensively before becoming mRNA and exiting the nucleus for translation into protein. Bacterial cells do not possess pre-mRNA; in these cells, transcription takes place concurrently with translation.

Transfer RNA (tRNA) serves as the link between the coding sequence of nucleotides in the mRNA and the amino acid sequence of a polypeptide chain. Each tRNA attaches to one particular type of amino acid and helps to incorporate that amino acid into a polypeptide chain (discussed in Chapter 15).

Additional classes of RNA molecules are found in the nuclei of eukaryotic cells. **Small nuclear RNAs** (snRNAs) combine with small protein subunits to form **small nuclear ribonucleoproteins** (snRNPs, affectionately known as "snurps"). Some snRNAs participate in the processing of RNA, converting pre-mRNA into mRNA. **Small nucleolar RNAs** (snoRNAs) take part in the processing of rRNA. Small RNA molecules are also found in the cytoplasm of eukaryotic cells; these molecules, called **small cytoplasmic RNAs** (scRNAs), have varied and often unknown function.

A class of very small and abundant RNA molecules, termed **microRNAs** (miRNAs) and **small interfering RNAs** (siRNAs), are found in bacteria and eukaryotic cells and carry out RNA interference (RNAi), a process in which these small RNA molecules help trigger the degradation of mRNA or inhibit its translation into protein. More will be said about RNA interference in Chapter 14. Recent research has uncovered another class of small RNA molecules called **Piwi-interacting RNAs** (piRNAs; named after Piwi proteins,

with which they interact). Found in mammalian testes, these RNA molecules are similar to miRNAs and siRNAs; they are thought have a role in the regulation of sperm development. The different classes of RNA molecules are summarized in **Table 13.2**.

CONCEPTS

RNA differs from DNA in that RNA possesses a hydroxyl group on the 2′-carbon atom of its sugar, contains uracil instead of thymine, and is normally single stranded. Several classes of RNA exist within bacterial and eukaryotic cells.

✔ **CONCEPT CHECK 1**

Which class of RNA is correctly paired with its function?

a. Small nuclear RNA (snRNA): processes rRNA

b. Transfer RNA (tRNA): attaches to an amino acid

c. MicroRNA (miRNA): carries information for the amino acid sequence of a protein

d. Ribosomal RNA (rRNA): carries out RNA interference

13.2 Transcription Is the Synthesis of an RNA Molecule from a DNA Template

All cellular RNAs are synthesized from DNA templates through the process of transcription (**Figure 13.2**). Transcription is in many ways similar to the process of replica-

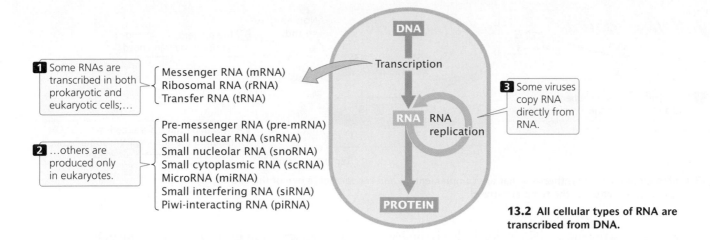

1 Some RNAs are transcribed in both prokaryotic and eukaryotic cells;...

Messenger RNA (mRNA)
Ribosomal RNA (rRNA)
Transfer RNA (tRNA)

2 ...others are produced only in eukaryotes.

Pre-messenger RNA (pre-mRNA)
Small nuclear RNA (snRNA)
Small nucleolar RNA (snoRNA)
Small cytoplasmic RNA (scRNA)
MicroRNA (miRNA)
Small interfering RNA (siRNA)
Piwi-interacting RNA (piRNA)

Transcription

DNA

RNA   RNA replication

3 Some viruses copy RNA directly from RNA.

PROTEIN

13.2 All cellular types of RNA are transcribed from DNA.

tion, but a fundamental difference relates to the length of the template used. In replication, all the nucleotides in the DNA template are copied, but, in transcription, only small parts of the DNA molecule—usually a single gene or, at most, a few genes—are transcribed into RNA. Because not all gene products are needed at the same time or in the same cell, the constant transcription of all of a cell's genes would be highly inefficient. Furthermore, much of the DNA does not encode a functional product, and transcription of such sequences would be pointless. Transcription is, in fact, a highly selective process: individual genes are transcribed only as their products are needed. But this selectivity imposes a fundamental problem on the cell—the problem of how to recognize individual genes and transcribe them at the proper time and place.

Like replication, transcription requires three major components:

1. a DNA template;

2. the raw materials (substrates) needed to build a new RNA molecule; and

3. the transcription apparatus, consisting of the proteins necessary to catalyze the synthesis of RNA.

The Template

In 1970, Oscar Miller, Jr., Barbara Hamkalo, and Charles Thomas used electron microscopy to examine cellular contents and demonstrate that RNA is transcribed from a DNA template. They saw within the cell Christmas-tree-like structures: thin central fibers (the trunk of the tree), to which were attached strings (the branches) with granules (**Figure 13.3**). The addition of deoxyribonuclease (an enzyme that degrades DNA) caused the central fibers to disappear, indicating that the "tree trunks" were DNA molecules. Ribonuclease (an enzyme that degrades RNA) removed the granular strings, indicating that the branches were RNA. Their conclusion was that each "Christmas tree" represented a gene undergoing transcription. The transcription of each

gene begins at the top of the tree; there, little of the DNA has been transcribed and the RNA branches are short. As the transcription apparatus moves down the tree, transcribing more of the template, the RNA molecules lengthen, producing the long branches at the bottom.

The transcribed strand The template for RNA synthesis, as for DNA synthesis, is a single strand of the DNA double helix. Unlike replication, however, the transcription of a gene takes place on only one of the two nucleotide strands of DNA (**Figure 13.4**). The nucleotide strand used for transcription is termed the **template strand**. The other strand, called the **nontemplate strand**, is not ordinarily transcribed. Thus, within a gene, only one of the nucleotide strands is normally transcribed into RNA (there are some exceptions to this rule).

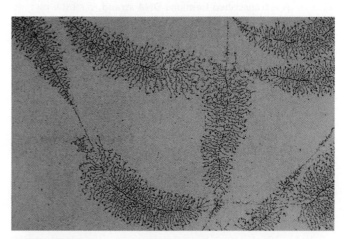

13.3 Under the electron microscope, DNA molecules undergoing transcription exhibit Christmas-tree-like structures. The trunk of each "Christmas tree" (a transcription unit) represents a DNA molecule; the tree branches (granular strings attached to the DNA) are RNA molecules that have been transcribed from the DNA. As the transcription apparatus moves down the DNA, transcribing more of the template, the RNA molecules become longer and longer. [Dr. Thomas Broker/Phototake.]

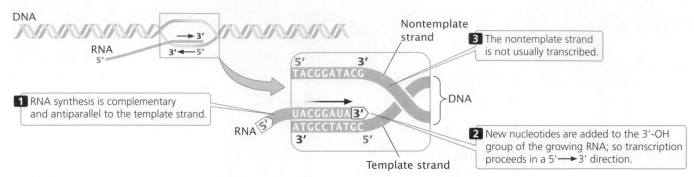

13.4 RNA molecules are synthesized that are complementary and antiparallel to one of the two nucleotide strands of DNA, the template strand.

During transcription, an RNA molecule that is complementary and antiparallel to the DNA template strand is synthesized (see Figure 13.4). The RNA transcript has the same polarity and base sequence as that of the nontemplate strand, with the exception that RNA contains U rather than T. In most organisms, each gene is transcribed from a single strand, but different genes may be transcribed from different strands, as shown in **Figure 13.5**. TRY PROBLEM 16 →

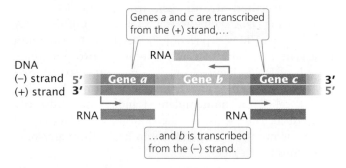

13.5 RNA is transcribed from one DNA strand. In most organisms, each gene is transcribed from a single DNA strand, but different genes may be transcribed from one or the other of the two DNA strands.

CONCEPTS

Within a single gene, only one of the two DNA strands, the template strand, is usually transcribed into RNA.

✔ CONCEPT CHECK 2

What is the difference between the template strand and the nontemplate strand?

The transcription unit   A **transcription unit** is a stretch of DNA that encodes an RNA molecule and the sequences necessary for its transcription. How does the complex of enzymes and proteins that performs transcription—the transcription apparatus—recognize a transcription unit? How does it know which DNA strand to read and where to start and stop? This information is encoded by the DNA sequence.

Included within a transcription unit are three critical regions: a promoter, an RNA-coding sequence, and a terminator (**Figure 13.6**). The **promoter** is a DNA sequence that the transcription apparatus recognizes and binds. It indicates which of the two DNA strands is to be read as the template and the direction of transcription. The promoter also determines the transcription start site, the first nucleotide that will be transcribed into RNA. In most transcription units, the promoter is located next to the transcription start site but is not, itself, transcribed.

The second critical region of the transcription unit is the **RNA-coding region**, a sequence of DNA nucleotides that is copied into an RNA molecule. The third component of the transcription unit is the **terminator**, a sequence of nucleotides that signals where transcription is to end. Terminators are usually part of the RNA-coding sequence; that is, transcription stops only after the terminator has been copied into RNA.

Molecular biologists often use the terms *upstream* and *downstream* to refer to the direction of transcription and the location of nucleotide sequences surrounding the RNA-coding sequence. The transcription apparatus is said to move downstream during transcription: it binds to the

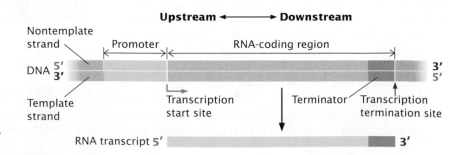

13.6 A transcription unit includes a promoter, an RNA-coding region, and a terminator.

promoter (which is usually upstream of the start site) and moves toward the terminator (which is downstream of the start site).

When DNA sequences are written out, often the sequence of only one of the two strands is listed. Molecular biologists typically write the sequence of the nontemplate strand, because it will be the same as the sequence of the RNA transcribed from the template (with the exception that U in RNA replaces T in DNA). By convention, the sequence on the nontemplate strand is written with the 5′ end on the left and the 3′ end on the right. The first nucleotide transcribed (the transcription start site) is numbered +1; nucleotides downstream of the start site are assigned positive numbers, and nucleotides upstream of the start site are assigned negative numbers. So, nucleotide +34 would be 34 nucleotides downstream of the start site, whereas nucleotide −75 would be 75 nucleotides upstream of the start site. There is no nucleotide numbered 0.

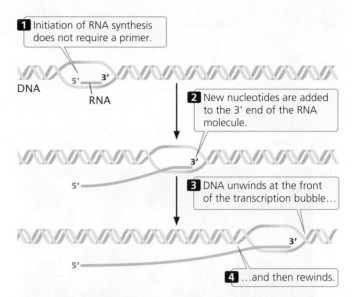

13.8 In transcription, nucleotides are always added to the 3′ end of the RNA molecule.

CONCEPTS

A transcription unit is a piece of DNA that encodes an RNA molecule and the sequences necessary for its proper transcription. Each transcription unit includes a promoter, an RNA-coding region, and a terminator.

✔ CONCEPT CHECK 3

Which of the following phrases does *not* describe a function of the promoter?

a. Serves as sequence to which transcription apparatus binds

b. Determines the first nucleotide that is transcribed into RNA

c. Determines which DNA strand is template

d. Signals where transcription ends

The Substrate for Transcription

RNA is synthesized from **ribonucleoside triphosphates** (rNTPs; **Figure 13.7**). In synthesis, nucleotides are added one at a time to the 3′-OH group of the growing RNA molecule. Two phosphate groups are cleaved from the incoming ribonucleoside triphosphate; the remaining phosphate group participates in a phosphodiester bond that connects

Triphosphate

13.7 Ribonucleoside triphosphates are substrates used in RNA synthesis.

the nucleotide to the growing RNA molecule. The overall chemical reaction for the addition of each nucleotide is:

$$RNA_n + rNTP \rightarrow RNA_{n+1} + PP_i$$

where PP_i represents pyrophosphate. Nucleotides are always added to the 3′ end of the RNA molecule, and the direction of transcription is therefore 5′→3′ (**Figure 13.8**), the same as the direction of DNA synthesis during replication. The synthesis of RNA is complementary and antiparallel to one of the DNA strands (the template strand). Unlike DNA synthesis, RNA synthesis does not require a primer.

CONCEPTS

RNA is synthesized from ribonucleoside triphosphates. Transcription is 5′→3′: each new nucleotide is joined to the 3′-OH group of the last nucleotide added to the growing RNA molecule.

The Transcription Apparatus

Recall that DNA replication requires a number of different enzymes and proteins. Although transcription might initially appear to be quite different because a single enzyme—**RNA polymerase**—carries out all the required steps of transcription, on closer inspection, the processes are actually similar. The action of RNA polymerase is enhanced by a number of accessory proteins that join and leave the polymerase at different stages of the process. Each accessory protein is responsible for providing or regulating a special function. Thus, transcription, like replication, requires an array of proteins.

Bacterial RNA polymerase Bacterial cells typically possess only one type of RNA polymerase, which catalyzes the synthesis of all classes of bacterial RNA: mRNA, tRNA, and rRNA.

(a)

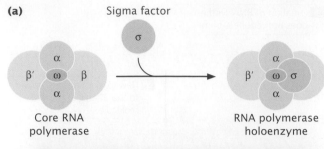

(b)

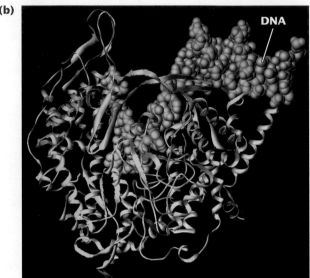

13.9 In bacterial RNA polymerase, the core enzyme consists of five subunits: two copies of alpha (α), a single copy of beta (β), a single copy of beta prime (β′), and a single copy of omega (ω). The core enzyme catalyzes the elongation of the RNA molecule by the addition of RNA nucleotides. (a) The sigma factor (σ) joins the core to form the holoenzyme, which is capable of binding to a promoter and initiating transcription. (b) The molecular model shows RNA polymerase (in yellow) binding DNA.

Bacterial RNA polymerase is a large, multimeric enzyme (meaning that it consists of several polypeptide chains).

At the heart of most bacterial RNA polymerases are five subunits (individual polypeptide chains) that make up the **core enzyme**: two copies of a subunit called alpha (α) and single copies of subunits beta (β), beta prime (β′), and omega (ω) (**Figure 13.9**). The ω subunit is not essential for transcription, but it helps stabilize the enzyme. The core enzyme catalyzes the elongation of the RNA molecule by the addition of RNA nucleotides. Other functional subunits join and leave the core enzyme at particular stages of the transcription process. The **sigma (σ) factor** controls the binding of RNA polymerase to the promoter. Without sigma, RNA polymerase will initiate transcription at a random point along the DNA. After sigma has associated with the core enzyme (forming a **holoenzyme**), RNA polymerase binds stably only to the promoter region and initiates transcription at the proper start site. Sigma is required only for promoter binding and initiation; when

a few RNA nucleotides have been joined together, sigma usually detaches from the core enzyme. Many bacteria have multiple types of sigma factors; each type of sigma initiates the binding of RNA polymerase to a particular set of promoters.

Rifamycins are a group of antibiotics that kill bacterial cells by inhibiting RNA polymerase. These antibiotics are widely used to treat tuberculosis, a disease that kills almost 2 million people worldwide each year. The structures of bacterial and eukaryotic RNA polymerases are sufficiently different that rifamycins inhibit bacterial RNA polymerases without interfering with eukaryotic RNA polymerases. Recent research has demonstrated that several rifamycins inibit RNA polymerase by binding to the part of the RNA polymerase that clamps on to DNA and jamming it, thus preventing the RNA polymerase from interacting with the promoter on the DNA.

Eukaryotic RNA polymerases   Most eukaryotic cells possess three distinct types of RNA polymerase, each of which is responsible for transcribing a different class of RNA: **RNA polymerase I** transcribes rRNA; **RNA polymerase II** transcribes pre-mRNAs, snoRNAs, some miRNAs, and some snRNAs; and **RNA polymerase III** transcribes other small RNA molecules—specifically tRNAs, small rRNA, some miRNAs, and some snRNAs (**Table 13.3**). RNA polymerases I, II, and III are found in all eukaryotes. Two additional RNA polymerases, **RNA polymerase IV** and **RNA polymerase V**, have been found in plants. RNA polymerases IV and V transcribe RNAs that play a role in DNA methylation and chromatin structure.

All eukaryotic polymerases are large, multimeric enzymes, typically consisting of more than a dozen subunits. Some sub-

Table 13.3	Eukaryotic RNA polymerases	
Type	Present in	Transcribes
RNA polymerase I	All eukaryotes	Large rRNAs
RNA polymerase II	All eukaryotes	Pre-mRNA, some snRNAs, snoRNAs, some miRNAs
RNA polymerase III	All eukaryotes	tRNAs, small rRNAs, some snRNAs, some miRNAs
RNA polymerase IV	Plants	Some siRNAs
RNA polymerase V	Plants	RNA molecules taking part in heterochromatin formation

units are common to all RNA polymerases, whereas others are limited to one of the polymerases. As in bacterial cells, a number of accessory proteins bind to the core enzyme and affect its function.

CONCEPTS

Bacterial cells possess a single type of RNA polymerase, consisting of a core enzyme and other subunits that participate in various stages of transcription. Eukaryotic cells possess several distinct types of RNA polymerase that transcribe different kinds of RNA molecules.

✔ CONCEPT CHECK 4

What is the function of the sigma factor?

13.3 The Process of Bacterial Transcription Consists of Initiation, Elongation, and Termination

Now that we've considered some of the major components of transcription, we're ready to take a detailed look at the process. Transcription can be conveniently divided into three stages:

1. initiation, in which the transcription apparatus assembles on the promoter and begins the synthesis of RNA;

2. elongation, in which DNA is threaded through RNA polymerase, the polymerase unwinding the DNA and adding new nucleotides, one at a time, to the 3' end of the growing RNA strand; and

3. termination, the recognition of the end of the transcription unit and the separation of the RNA molecule from the DNA template.

We will first examine each of these steps in bacterial cells, where the process is best understood; then we will consider eukaryotic and archaeal transcription.

Initiation

Initiation comprises all the steps necessary to begin RNA synthesis, including (1) promoter recognition, (2) formation of the transcription bubble, (3) creation of the first bonds between rNTPs, and (4) escape of the transcription apparatus from the promoter.

Transcription initiation requires that the transcription apparatus recognize and bind to the promoter. At this step, the selectivity of transcription is enforced; the binding of RNA polymerase to the promoter determines which parts of the DNA template are to be transcribed and how often. Different genes are transcribed with different frequencies, and promoter binding is primarily responsible for determining the frequency of transcription for a particular gene.

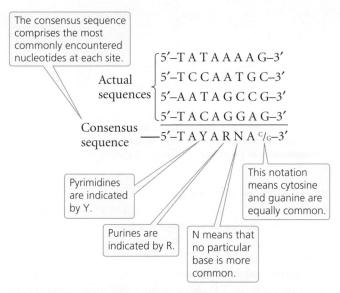

13.10 A consensus sequence consists of the most commonly encountered bases at each position in a group of related sequences.

Promoters also have different affinities for RNA polymerase. Even within a single promoter, the affinity can vary with the passage of time, depending on the promoter's interaction with RNA polymerase and a number of other factors.

Bacterial promoters Essential information for the transcription unit—where it will start transcribing, which strand is to be read, and in what direction the RNA polymerase will move—is imbedded in the nucleotide sequence of the promoter. Promoters are DNA sequences that are recognized by the transcription apparatus and are required for transcription to take place. In bacterial cells, promoters are usually adjacent to an RNA-coding sequence.

An examination of many promoters in *E. coli* and other bacteria reveals a general feature: although most of the nucleotides within the promoters vary in sequence, short stretches of nucleotides are common to many. Furthermore, the spacing and location of these nucleotides relative to the transcription start site are similar in most promoters. These short stretches of common nucleotides are called **consensus sequences**; "consensus sequence" refers to sequences that possess considerable similarity, or consensus (**Figure 13.10**). The presence of consensus in a set of nucleotides usually implies that the sequence is associated with an important function. TRY PROBLEM 19 →

The most commonly encountered consensus sequence, found in almost all bacterial promoters, is centered about 10 bp upsteam of the start site. Called the **−10 consensus sequence** or, sometimes, the Pribnow box, its consensus sequence is

5′–T A T A A T–3′
3′–A T A T T A–5′

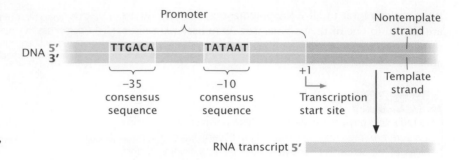

13.11 In bacterial promoters, consensus sequences are found upstream of the start site, approximately at positions −10 and −35.

and is often written simply as TATAAT (**Figure 13.11**). Remember that TATAAT is just the *consensus* sequence—representing the most commonly encountered nucleotides at each of these positions. In most prokaryotic promoters, the actual sequence is not TATAAT.

Another consensus sequence common to most bacterial promoters is TTGACA, which lies approximately 35 nucleotides upstream of the start site and is termed the **−35 consensus sequence** (see Figure 13.11). The nucleotides on either side of the −10 and −35 consensus sequences and those between them vary greatly from promoter to promoter, suggesting that these nucleotides are not very important in promoter recognition.

The function of these consensus sequences in bacterial promoters has been studied by inducing mutations at various positions within the consensus sequences and observing the effect of the changes on transcription. The results of these studies reveal that most base substitutions within the −10 and −35 consensus sequences reduce the rate of transcription; these substitutions are termed *down mutations* because they slow down the rate of transcription. Occasionally, a particular change in a consensus sequence increases the rate of transcription; such a change is called an *up mutation.*

As mentioned earlier, the sigma factor associates with the core enzyme (**Figure 13.12a**) to form a holoenzyme, which binds to the −35 and −10 consensus sequences in the DNA promoter (**Figure 13.12b**). Although it binds only the nucleotides of consensus sequences, the enzyme extends from −50 to +20 when bound to the promoter. The holoenzyme initially binds weakly to the promoter but then undergoes a change in structure that allows it to bind more tightly and unwind the double-stranded DNA (**Figure 13.12c**). Unwinding begins within the −10 consensus sequence and extends downstream for about 14 nucleotides, including the start site (from nucleotides −12 to +2).

Some bacterial promoters contain a third consensus sequence that also takes part in the initiation of transcription. Called the **upstream element**, this sequence contains a number of A–T pairs and is found at about −40 to −60. A number of proteins may bind to sequences in and near the promoter; some stimulate the rate of transcription and others repress it. We will consider these proteins, which regulate gene expression, in Chapter 16. **TRY PROBLEM 22 →**

CONCEPTS

A promoter is a DNA sequence that is adjacent to a gene and required for transcription. Promoters contain short consensus sequences that are important in the initiation of transcription.

✔ CONCEPT CHECK 5

What binds to the −10 consensus sequence found in most bacterial promoters?

a. The holoenzyme (core enzyme + sigma)

b. The sigma factor alone

c. The core enzyme alone

d. mRNA

Initial RNA synthesis   After the holoenzyme has attached to the promoter, RNA polymerase is positioned over the start site for transcription (at position +1) and has unwound the DNA to produce a single-stranded template. The orientation and spacing of consensus sequences on a DNA strand determine which strand will be the template for transcription and thereby determine the direction of transcription.

The position of the start site is determined not by the sequences located there but by the location of the consensus sequences, which positions RNA polymerase so that the enzyme's active site is aligned for the initiation of transcription at +1. If the consensus sequences are artificially moved upstream or downstream, the location of the starting point of transcription correspondingly changes.

To begin the synthesis of an RNA molecule, RNA polymerase pairs the base on a ribonucleoside triphosphate with its complementary base at the start site on the DNA template strand (**Figure 13.12d**). No primer is required to initiate the synthesis of the 5′ end of the RNA molecule. Two of the three phosphate groups are cleaved from the ribonucleoside triphosphate as the nucleotide is added to the 3′ end of the growing RNA molecule. However, because the 5′ end of the first ribonucleoside triphosphate does not take part in the formation of a phosphodiester bond, all three of its phosphate groups remain. An RNA molecule therefore possesses, at least initially, three phosphate groups at its 5′ end (**Figure 13.12e**).

Often, in the course of initiation, RNA polymerase repeatedly generates and releases short transcripts, from 2 to 6 nucleotides in length, while still bound to the promoter. After

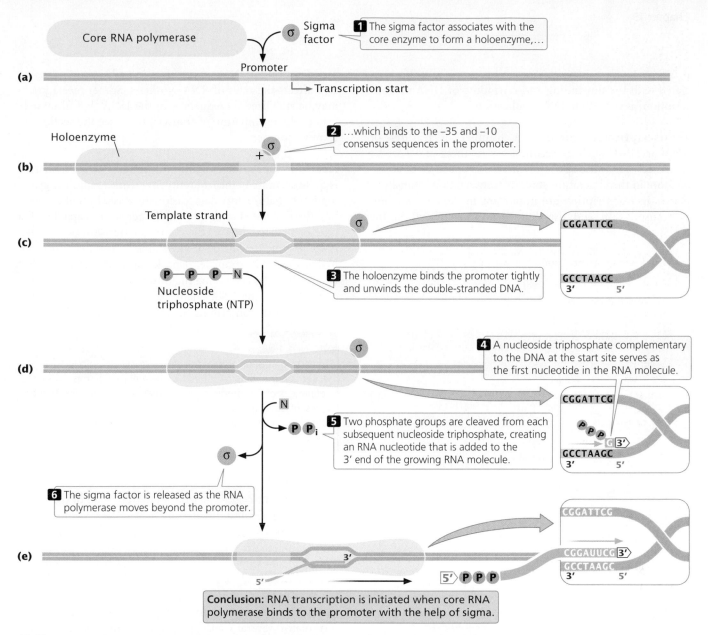

13.12 Transcription in bacteria is carried out by RNA polymerase, which must bind to the sigma factor to initiate transcription.

several abortive attempts, the polymerase synthesizes an RNA molecule from 9 to 12 nucleotides in length, which allows the RNA polymerase to transition to the elongation stage.

Elongation

At the end of initiation, RNA polymerase undergoes a change in conformation (shape) and thereafter is no longer able to bind to the consensus sequences in the promoter. This change allows the polymerase to escape from the promoter and begin transcribing downstream. The sigma subunit is usually released after initiation, although some populations of RNA polymerase may retain sigma throughout elongation.

As it moves downstream along the template, RNA polymerase progressively unwinds the DNA at the leading (downstream) edge of the transcription bubble, joining nucleotides to the RNA molecule according to the sequence on the template, and rewinds the DNA at the trailing (upstream) edge of the bubble. In bacterial cells at 37°C, about 40 nucleotides are added per second. This rate of RNA synthesis is much lower than that of DNA synthesis, which is 1000 to 2000 nucleotides per second in bacterial cells.

The transcription bubble Transcription takes place within a short stretch of about 18 nucleotides of unwound DNA—the transcription bubble. Within this region, RNA is continuously synthesized, with single-stranded DNA used as a template. About 8 nucleotides of newly synthesized RNA are paired with the DNA-template nucleotides at any one time. As the transcription apparatus moves down the DNA template, it generates positive supercoiling ahead of the transcription bubble and negative supercoiling behind it. Topoisomerase enzymes probably relieve the stress associated with the unwinding and rewinding of DNA in transcription, as they do in DNA replication.

Transcriptional pausing   A number of features of RNA or DNA, such as secondary structures and specific sequences, can cause RNA polymerase to pause for a second or more in the elongation stage of transcription. Transitory pauses in transcription are important in the coordination of transcription and translation in bacteria and in the

1. A rho-independent terminator contains an inverted repeat followed by a string of approximately six adenine nucleotides.

Inverted repeats

DNA

AGCCCGCC   GGCGGGCT
TCGGGCGG   CCGCCCGAAAAAAAA

GGCGGGCT

AGCCCGCC
3′ TCGGGCGG

GGCGGGCUUUUUUUU 3′
CCGCCCGAAAAAAAA
5′

2. The inverted repeats are transcribed into RNA.

5′ AGCCCGCC

RNA transcript

3. The string of Us causes the RNA polymerase to pause…

AGCCCGCC   GGCGGGCT
TCGGGCGG   CCGCCCGA

UUUUUUU 3′
AAAAAAA

5′

4. …and the inverted repeats in RNA fold into a hairpin loop,…

5. …which destabilizes the DNA-RNA pairing.

AGCCCGCC   GGCGGGCT
TCGGGCGG   CCGCCCGAAAAAAAA

6. The RNA transcript separates from the template, terminating transcription.

5′ CC   UUUUUUU 3′

Hairpin

Conclusion: Transcription terminates when inverted repeats form a hairpin followed by a string of uracils.

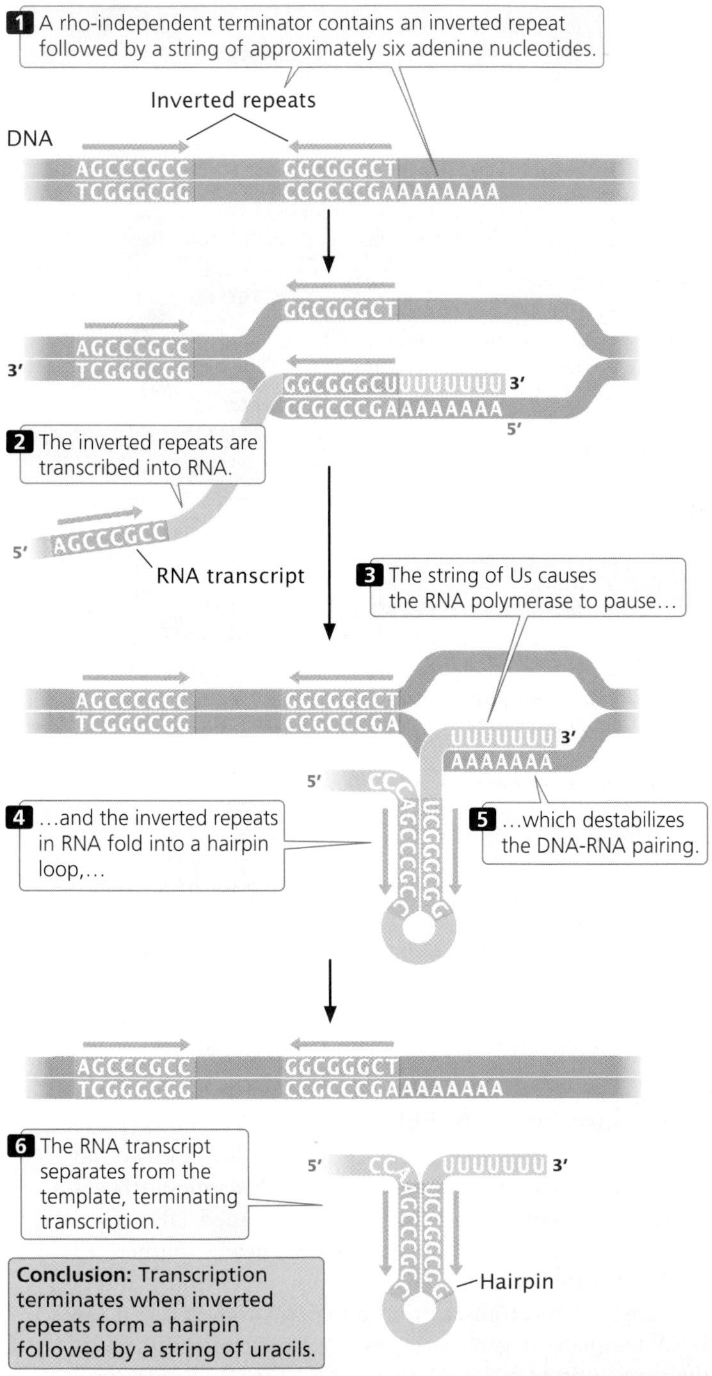

13.13 Rho-independent termination in bacteria is a multistep process.

coordination of RNA processing in eukaryotes. Pausing also affects the rates of RNA synthesis. Sometimes a pause may be stabilized by sequences in the DNA that ultimately lead to the termination of transcription (see the section on termination next).

Accuracy of transcription   Although RNA polymerase is quite accurate in incorporating nucleotides into the growing RNA chain, errors do occasionally arise. Recent research has demonstrated that RNA polymerase is capable of a type of proofreading in the course of transcription. When RNA polymerase incorporates a nucleotide that does not match the DNA template, it backs up and cleaves the last two nucleotides (including the misincorporated nucleotide) from the growing RNA chain. RNA polymerase then proceeds forward, transcribing the DNA template again.

CONCEPTS

Transcription is initiated at the start site, which, in bacterial cells, is set by the binding of RNA polymerase to the consensus sequences of the promoter. No primer is required. Transcription takes place within the transcription bubble. DNA is unwound ahead of the bubble and rewound behind it. There are frequent pauses in the process of transcription.

Termination

RNA polymerase adds nucleotides to the 3′ end of the growing RNA molecule until it transcribes a terminator. Most terminators are found upstream of the site at which termination actually takes place. Transcription therefore does not suddenly stop when polymerase reaches a terminator, as does a car stopping at a stop sign. Rather, transcription stops after the terminator has been transcribed, like a car that stops only after running over a speed bump. At the terminator, several overlapping events are needed to bring an end to transcription: RNA polymerase must stop synthesizing RNA, the RNA molecule must be released from RNA polymerase, the newly made RNA molecule must dissociate fully from the DNA, and RNA polymerase must detach from the DNA template.

Bacterial cells possess two major types of terminators. **Rho-dependent terminators** are able to cause the termination of transcription only in the presence of an ancillary protein called the **rho factor**. **Rho-independent terminators** (also known as intrinsic terminators) are able to cause the end of transcription in the absence of rho.

Rho-independent terminators   Rho-independent terminators, which make up about 50 percent of all terminators in prokaryotes, have two common features. First, they contain inverted repeats (sequences of nucleotides on one strand that are inverted and complementary). When inverted repeats have been transcribed into RNA, a hairpin secondary structure forms (**Figure 13.13**). Second, in rho-independent terminators, a string of seven to nine adenine nucleotides follows the second inverted repeat in the template DNA. Their transcription produces a string of uracil nucleotides after the hairpin in the transcribed RNA.

The string of uracils in the RNA molecule causes the RNA polymerase to pause, allowing time for the hairpin structure to form. Evidence suggests that the formation of the hairpin destablizes the DNA–RNA pairing, causing the RNA molecule to separate from its DNA template. Separation may be facilitated by the adenine–uracil base pairings, which are relatively weak compared with other types of base pairings. When the RNA transcript has separated from the template, RNA synthesis can no longer continue (see Figure 13.13).

Rho-dependent terminators   Rho-dependent termina-tors have two features: (1) DNA sequences that produce a pause in transcription and (2) a DNA sequence that encodes a stretch of RNA upstream of the terminator that is devoid of any secondary structures. This unstructured RNA serves as a binding site for the rho protein, which binds the RNA and moves toward its 3′ end, following the RNA polymerase (**Figure 13.14**). When RNA polymerase encounters the ter-minator, it pauses, allowing rho to catch up. The rho protein has helicase activity, which it uses to unwind the RNA–DNA hybrid in the transcription bubble, bringing transcription to an end. **TRY PROBLEM 26** →

Polycistronic mRNA   In bacteria, a group of genes is often transcribed into a single RNA molecule, which is termed a **polycistronic RNA**. Thus, polycistronic RNA is produced when a single terminator is present at the end of a group of several genes that are transcribed together, instead of each gene having its own terminator. Typically, each eukaryotic gene is transcribed and terminated separately, and so polycistronic mRNA is uncommon in eukaryotes.

 www.whfreeman.com/pierce4e   The process of transcriptional initiation, elongation, and termination are illustrated in Animation 13.1, which shows how the different parts of the transcriptional unit interact to bring about the complete synthesis of an RNA molecule.

CONCEPTS

Transcription ends after RNA polymerase transcribes a terminator. Bacterial cells possess two types of terminator: a rho-independent terminator, which RNA polymerase can recognize by itself; and a rho-dependent terminator, which RNA polymerase can recognize only with the help of the rho protein.

✔ CONCEPT CHECK 6

What characteristics are most commonly found in rho-independent terminators?

CONNECTING CONCEPTS

The Basic Rules of Transcription

Before we examine the process of eukaryotic transcription, let's pause to summarize some of the general principles of bacterial transcription.

1. Transcription is a selective process; only certain parts of the DNA are transcribed at any one time.

2. RNA is transcribed from single-stranded DNA. Within a gene, only one of the two DNA strands—the template strand—is normally copied into RNA.

3. Ribonucleoside triphosphates are used as the substrates in RNA synthesis. Two phosphate groups are cleaved from a ribonucleoside triphosphate, and the resulting nucleotide is joined to the 3′-OH group of the growing RNA strand.

4. RNA molecules are antiparallel and complementary to the DNA template strand. Transcription is always in the 5′→3′ direction, meaning that the RNA molecule grows at the 3′ end.

5. Transcription depends on RNA polymerase—a complex, multimeric enzyme. RNA polymerase consists of a core enzyme, which is capable of synthesizing RNA, and other subunits that may join transiently to perform additional functions.

6. A sigma factor enables the core enzyme of RNA polymerase to bind to a promoter and initiate transcription.

7. Promoters contain short sequences crucial in the binding of RNA polymerase to DNA; these consensus sequences are interspersed with nucleotides that play no known role in transcription.

8. RNA polymerase binds to DNA at a promoter, begins transcribing at the start site of the gene, and ends transcription after a terminator has been transcribed.

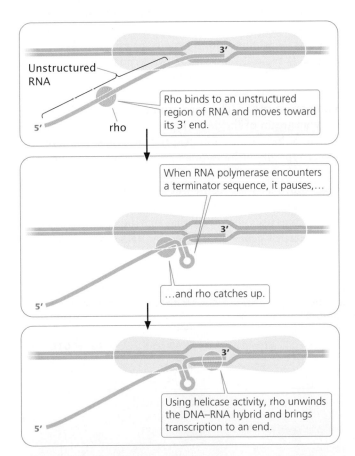

Unstructured RNA

5′   rho

Rho binds to an unstructured region of RNA and moves toward its 3′ end.

When RNA polymerase encounters a terminator sequence, it pauses,…

…and rho catches up.

Using helicase activity, rho unwinds the DNA–RNA hybrid and brings transcription to an end.

13.14 The termination of transcription in some bacterial genes requires the presence of the rho protein.

13.4 Eukaryotic Transcription Is Similar to Bacterial Transcription but Has Some Important Differences

Transcription in eukaryotes is similar to bacterial transcription in that it includes initiation, elongation, and termination, and the basic principles of transcription already outlined apply to eukaryotic transcription. However, there are some important differences. Eukaryotic cells possess three different RNA polymerases, each of which transcribes a different class of RNA and recognizes a different type of promoter. Thus, a generic promoter cannot be described for eukaryotic cells, as was done for bacterial cells; rather, a promoter's description depends on whether the promoter is recognized by RNA polymerase I, II, or III. Another difference is in the nature of promoter recognition and initiation. Many proteins take part in the binding of eukaryotic RNA polymerases to DNA templates, and the different types of promoters require different proteins.

Transcription and Nucleosome Structure

Transcription requires sequences on DNA to be accessible to RNA polymerase and other proteins. However, in eukaryotic cells, DNA is complexed with histone proteins in highly compressed chromatin (see Figure 11.4). How can the proteins necessary for transcription gain access to eukaryotic DNA when it is complexed with histones?

The answer to this question is that chromatin structure is modified before transcription so that the DNA is in a more open configuration and is more accessible to the transcription machinery. Several types of proteins have roles in chromatin modification. Acetyltransferases add acetyl groups to amino acids at the ends of the histone proteins, which destabilizes nucleosome structure and makes the DNA more accessible. Other types of histone modification also can affect chromatin packing. In addition, proteins called chromatin-remodeling proteins may bind to the chromatin and displace nucleosomes from promoters and other regions important for transcription. We will take a closer look at the role of changes in chromatin structure associated with gene expression in Chapter 17.

CONCEPTS

The initiation of transcription requires modification of chromatin structure so that DNA is accessible to the transcriptional machinery.

Promoters

A significant difference between bacterial and eukaryotic transcription is the existence of three different eukaryotic RNA polymerases, which recognize different types of promoters. In bacterial cells, the holoenzyme (RNA polymerase plus the sigma factor) recognizes and binds directly to sequences in the promoter. In eukaryotic cells, promoter recognition is carried out by accessory proteins that bind to the promoter and then recruit a specific RNA polymerase (I, II, or III) to the promoter.

One class of accessory proteins comprises **general transcription factors**, which, along with RNA polymerase, form the **basal transcription apparatus**—a group of proteins that assemble near the start site and are sufficient to initiate minimal levels of transcription. Another class of accessory proteins consists of **transcriptional activator proteins**, which bind to specific DNA sequences and bring about higher levels of transcription by stimulating the assembly of the basal transcription apparatus at the start site.

We will focus our attention on promoters recognized by RNA polymerase II, which transcribes the genes that encode proteins. A promoter for a gene transcribed by RNA polymerase II typically consists of two primary parts: the core promoter and the regulatory promoter.

Core promoter   The **core promoter** is located immediately upstream of the gene (**Figure 13.15**) and is the site to which the basal transcription apparatus binds. The core promoter typically includes one or more consensus sequences. One of the most common of these sequences is the **TATA box**, which has the consensus sequence TATAAA and is located from −25 to −30 bp upstream of the start site. Additional consensus sequences that may be found in the core promoters of genes transcribed by RNA polymerase II are

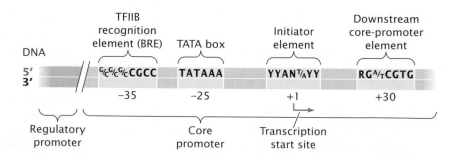

13.15 The promoters of genes transcribed by RNA polymerase II consist of a core promoter and a regulatory promoter that contain consensus sequences. Not all the consensus sequences shown are found in all promoters.

shown in Figure 13.15. These consensus sequences are recognized by transcription factors that bind to them and serve as a platform for the assembly of the basal transcription apparatus.

Regulatory promoter The **regulatory promoter** is located immediately upstream of the core promoter. A variety of different consensus sequences can be found in the regulatory promoters, and they can be mixed and matched in different combinations (**Figure 13.16**). Transcriptional activator proteins bind to these sequences and either directly or indirectly make contact with the basal transcription apparatus and affect the rate at which transcription is initiated. Transcriptional activator proteins also regulate transcription by binding to more-distant sequences called **enhancers**. The DNA between an enhancer and the promoter loops out, and so transcriptional activator proteins bound to the enhancer can interact with the basal transcription machinery at the core promoter. Enhancers will be discussed in more detail in Chapter 17.

Polymerase I and III promoters RNA polymerase I and RNA polymerase III each recognize promoters that are distinct from those recognized by RNA polymerase II. For example, promoters for small rRNA and tRNA genes, transcribed by RNA polymerase III, contain **internal promoters** that are downstream of the start site and are transcribed into the RNA.

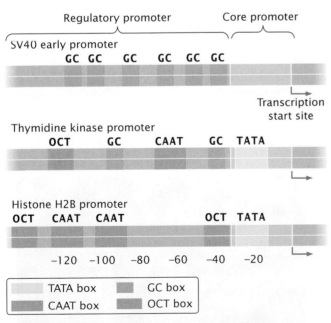

13.16 Consensus sequences in the promoters of three eukaryotic genes. These promoters illustrate the principle that consensus sequences can be mixed and matched in different combinations to yield a functional eukaryotic promoter.

✔ **CONCEPT CHECK 7**

What is the difference between the core promoter and the regulatory promoter?

a. Only the core promoter has consensus sequences.

b. The regulatory promoter is farther upstream of the gene.

c. Transcription factors bind to the core promoter; transcriptional activator proteins bind to the regulatory promoter.

d. Both b and c.

Initiation

Transcription in eukaryotes is initiated through the assembly of the transcriptional machinery on the promoter. This machinery consists of RNA polymerase II and a series of transcription factors that form a giant complex consisting of 50 or more polypeptides. Assembly of the transcription machinery begins when regulatory proteins bind DNA near the promoter and modify the chromatin structure so that transcription can take place. These proteins and other regulatory proteins then recruit the basal transcriptional apparatus to the core promoter.

The basal transcription apparatus consists of RNA polymerase, a series of general transcription factors, and a complex of proteins known as the mediator (**Figure 13.17**). The general transcription factors include TFIIA, TFIIB, TFIID, TFIIE, TFIIF, and TFIIH, in which TFII stands for transcription factor for RNA polymerase II and the final letter designates the individual factor.

RNA polymerase II and the general transcription factors assemble at the core promoter, forming a preinitiation complex that is analogous to the closed complex seen in bacterial initiation. Recall that, in bacteria, the sigma factor recognizes and binds to the promoter sequence. In eukaryotes, the function of sigma is replaced by that of the general transcription factors. A first step in initiation is the binding of TFIID to the TATA box on the DNA template. TFIID consists of at least nine polypeptides. One of them is the **TATA-binding protein** (TBP), which recognizes and binds to the TATA consensus sequence. The TATA-binding protein binds to the minor groove and straddles the DNA as a

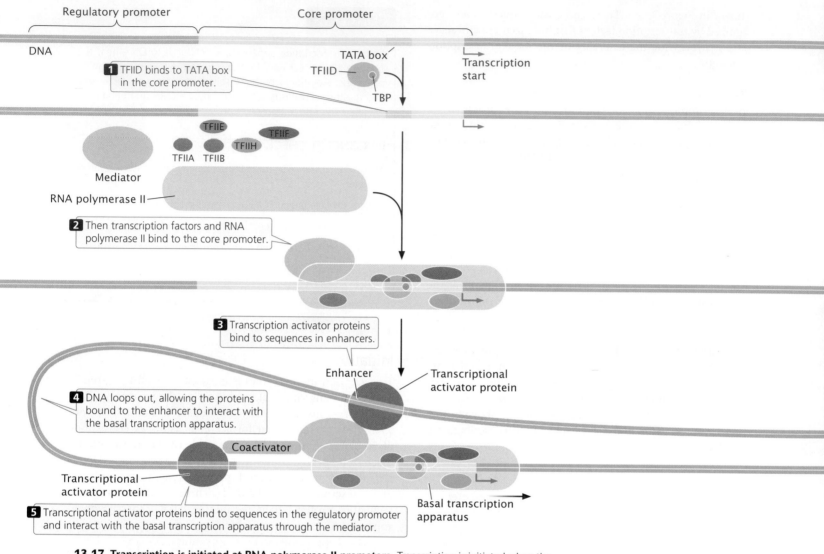

13.17 Transcription is initiated at RNA polymerase II promoters. Transcription is initiated when the TFIID transcription factor binds to the TATA box, followed by the binding of a preassembled holoenzyme containing general transcription factors, RNA polymerase II, and the mediator. TBP stands for TATA-binding protein.

molecular saddle (**Figure 13.18**), bending the DNA and partly unwinding it. Other transcription factors bind to additional consensus sequences in the core promoter and to RNA polymerase and position the polymerase over the transcription start site.

After the RNA polymerase and transcription factors have assembled on the core promoter, conformational changes take place in both the DNA and the polymerase. These changes cause from 11 to 15 bp of DNA surrounding the transcription start site to separate, producing the single-stranded DNA that will serve as a template for transcription.

The single-stranded DNA template is positioned within the active site of RNA polymerase, creating a structure called the open complex. After the open complex has formed, the synthesis of RNA begins as phosphate groups are cleaved off nucleoside triphosphates and nucleotides are joined

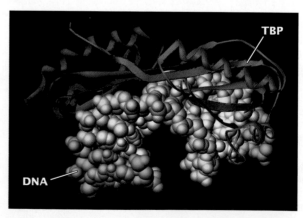

13.18 The TATA-binding protein (TBP) binds to the minor groove of DNA, straddling the double helix of DNA like a saddle.

together to form an RNA molecule. As in bacterial transcription, RNA polymerase may generate and release several short RNA molecules in abortive transcription before the polymerase initiates the synthesis of a full-length RNA molecule.

TRY PROBLEM 32 →

CONCEPTS

Transcription is initiated when the basal transcription apparatus, consisting of RNA polymerase and transcription factors, assembles on the core promoter and becomes an open complex.

✔ **CONCEPT CHECK 8**

What is the role of TFIID in transcription initiation?

Elongation

After about 30 bp of RNA have been synthesized, the RNA polymerase leaves the promoter and enters the elongation stage of transcription. Many of the transcription factors are left behind at the promoter and can serve to quickly reinitiate transcription with another RNA polymerase enzyme.

The molecular structure of eukaryotic RNA polymerase II and how it functions during elongation have been revealed through the work of Roger Kornberg and his colleagues, for which Kornberg was awarded a Nobel Prize in chemistry in 2006. The RNA polymerase maintains a transcription bubble during elongation, in which about eight nucleotides of RNA remain base paired with the DNA template strand. The DNA double helix enters a cleft in the polymerase and is gripped by jawlike extensions of the enzyme (**Figure 13.19**). The two strands of the DNA are unwound and RNA nucleotides that are complementary to the template strand are added to the growing 3′ end of the RNA molecule. As it funnels through the polymerase, the DNA–RNA hybrid hits a wall of amino acids and bends at almost a right angle; this bend positions the end of the DNA–RNA hybrid at the active site of the polymerase, and new nucleotides are added to the 3′ end of the growing RNA molecule. The newly synthesized RNA is separated from the DNA and runs through another groove before exiting from the polymerase.

Termination

The three eukaryotic RNA polymerases use different mechanisms for termination. RNA polymerase I requires a termination factor like the rho factor utilized in the termination of some bacterial genes. Unlike rho, which binds to the newly transcribed RNA molecule, the termination factor for RNA polymerase I binds to a DNA sequence downstream of the termination site.

RNA polymerase III ends transcription after transcribing a terminator sequence that produces a string of uracil

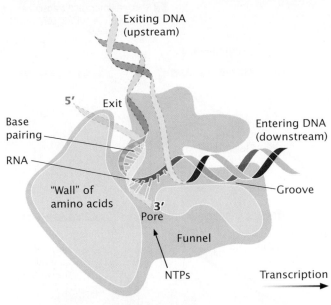

RNA polymerase II

13.19 The structure of RNA polymerase II is a source of insight into its function. The DNA double helix enters the polymerase through a groove and unwinds. The DNA–RNA duplex is bent at a right angle, which positions the 3′ end of the RNA at the active site of the enzyme. New nucleotides are added to the 3′ end of the RNA.

nucleotides in the RNA molecule, like that produced by the rho-independent terminators of bacteria. Unlike rho-independent terminators in bacterial cells, however, RNA polymerase III does not require that a hairpin structure precede the string of Us.

The termination of transcription by RNA polymerase II is not at specific sequences. Instead, RNA polymerase II often continues to synthesize RNA hundreds or even thousands of nucleotides past the coding sequence necessary to produce the mRNA. As we will see in Chapter 14, the end of pre-mRNA is cleaved at a specific site, designated by a consensus sequence, while transcription is still taking place at the 3′ end of the molecule. Cleavage cuts the pre-mRNA into two pieces: the mRNA that will eventually encode the protein and another piece of RNA that has its 5′ end trailing out of the RNA polymerase (**Figure 13.20**). An enzyme called Rat1 attaches to the 5′ end of this RNA and moves toward the 3′ end where RNA polymerase continues the transcription of RNA. Rat1 is a 5′→3′ exonuclease—an enzyme capable of degrading RNA in the 5′→3′ direction. Like a guided torpedo, Rat1 homes in on the polymerase, chewing up the RNA as it moves. When Rat1 reaches the transcriptional machinery, transcription terminates. Note that this mechanism is similar to that of rho-dependent termination in bacteria (see Figure 13.14), except that rho does not degrade the RNA molecule.

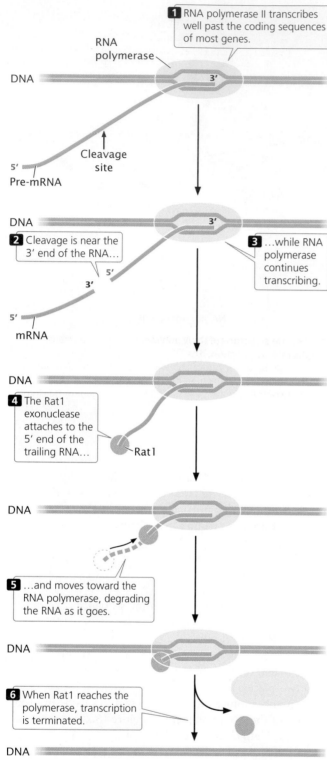

1 RNA polymerase II transcribes well past the coding sequences of most genes.

RNA polymerase

DNA

3'

Cleavage site

5'

Pre-mRNA

DNA

3'

2 Cleavage is near the 3' end of the RNA…

3'

5'

3'

5'

mRNA

3 …while RNA polymerase continues transcribing.

DNA

4 The Rat1 exonuclease attaches to the 5' end of the trailing RNA…

Rat1

DNA

5 …and moves toward the RNA polymerase, degrading the RNA as it goes.

DNA

6 When Rat1 reaches the polymerase, transcription is terminated.

DNA

DNA

13.20 Termination of transcription by RNA polymerase II requires the Rat1 exonuclease. Cleavage of the pre-mRNA produces a 5' end to which Rat1 attaches. Rat1 degrades the RNA molecule in the 5→3' direction. When Rat1 reaches the polymerase, transcription is halted.

CONCEPTS

The different eukaryotic RNA polymerases use different mechanisms of termination. Transcription at genes transcribed by RNA polymerase II is terminated when an exonuclease enzyme attaches to the cleaved 5' end of the RNA, moves down the RNA, and reaches the polymerase enzyme.

✔ CONCEPT CHECK 9

How are the processes of RNA polymerase II termination and rho-dependent termination in bacteria similar and how are they different?

13.5 Transcription in Archaea Is More Similar to Transcription in Eukaryotes Than to Transcription in Eubacteria

Some 2 billion to 3 billion years ago, life diverged into three lines of evolutionary descent: the eubacteria, the archaea, and the eukaryotes (see Chapter 2). Although eubacteria and archaea are superficially similar—both are unicellular and lack a nucleus—the results of studies of their DNA sequences and other biochemical properties indicate that they are as distantly related to each other as they are to eukaryotes. The evolutionary distinction between archaea, eubacteria, and eukaryotes is clear. However, did eukaryotes first diverge from an ancestral prokaryote, with the later separation of prokaryotes into eubacteria and archaea, or did the archaea and the eubacteria split first, with the eukaryotes later evolving from one of these groups?

Studies of transcription in eubacteria, archaea, and eukaryotes have yielded important findings about the evolutionary relationships of these organisms. Archaea, like eubacteria, have a single RNA polymerase, but this enzyme is most similar to the RNA polymerases of eukaryotes. Archaea possess a TATA-binding protein, a critical transcription factor in all three of the eukaryotic polymerases but not present in eubacteria. TBP binds the TATA box in archaea with the help of another transcription factor, TFIIB, which also is found in eukaryotes but not in eubacteria.

Transcription, one of the most basic of life processes, has strong similarities in eukaryotes and archaea, suggesting that these two groups are more closely related to each other than either is to the eubacteria. This conclusion is supported by other data, including those obtained from a comparison of gene sequences.

CONCEPTS

The process of transcription in archaea has many similarities to transcription in eukaryotes.

CONCEPTS SUMMARY

- RNA is a polymer, consisting of nucleotides joined together by phosphodiester bonds. Each RNA nucleotide consists of a ribose sugar, a phosphate, and a base. RNA contains the base uracil and is usually single stranded, which allows it to form secondary structures.

- Cells possess a number of different classes of RNA. Ribosomal RNA is a component of the ribosome, messenger RNA carries coding instructions for proteins, and transfer RNA helps incorporate the amino acids into a polypeptide chain.

- Early life used RNA as both the carrier of genetic information and as biological catalysts.

- The template for RNA synthesis is single-stranded DNA. In transcription, RNA synthesis is complementary and antiparallel to the DNA template strand. A transcription unit consists of a promoter, an RNA-coding region, and a terminator.

- The substrates for RNA synthesis are ribonucleoside triphosphates.

- RNA polymerase in bacterial cells consists of a core enzyme, which catalyzes the addition of nucleotides to an RNA molecule, and other subunits. The sigma factor controls the binding of the core enzyme to the promoter.

- Eukaryotic cells contain several different RNA polymerases.

- The process of transcription consists of three stages: initiation, elongation, and termination.

- Transcription begins at the start site, which is determined by consensus sequences. A short stretch of DNA is unwound near the start site, RNA is synthesized from a single strand of DNA as a template, and the DNA is rewound at the lagging end of the transcription bubble.

- RNA synthesis ceases after a terminator sequence has been transcribed. Bacterial cells have two types of terminators: rho-independent terminators and rho-dependent terminators.

- RNA polymerases are capable of proofreading.

- The initiation of transcription in eukaryotes requires the modification of chromatin structure. Different types of RNA polymerases in eukaryotes recognize different types of promoters.

- For genes transcribed by RNA polymerase II, general transcription factors bind to the core promoter and are part of the basal transcription apparatus. Transcriptional activator proteins bind to sequences in regulatory promoters and enhancers and interact with the basal transcription apparatus at the core promoter.

- The three RNA polymerases found in all eukaryotic cells use different mechanisms of termination.

- Transcription in archaea has many similarities to transcription in eukaryotes.

IMPORTANT TERMS

ribozyme (p. 352)
ribosomal RNA (rRNA) (p. 353)
messenger RNA (mRNA) (p. 353)
pre-messenger RNA (pre-mRNA) (p. 353)
transfer RNA (tRNA) (p. 354)
small nuclear RNA (snRNA) (p. 354)
small nuclear ribonucleoprotein (snRNP) (p. 354)
small nucleolar RNA (snoRNA) (p. 354)
small cytoplasmic RNA (scRNA) (p. 354)
microRNA (miRNA) (p. 354)
small interfering RNA (siRNA) (p. 354)
Piwi-interacting RNA (piRNA) (p. 354)
template strand (p. 355)
nontemplate strand (p. 355)
transcription unit (p. 356)

promoter (p. 356)
RNA-coding region (p. 356)
terminator (p. 356)
ribonucleoside triphosphate (rNTP) (p. 357)
RNA polymerase (p. 357)
core enzyme (p. 358)
sigma factor (p. 358)
holoenzyme (p. 358)
RNA polymerase I (p. 358)
RNA polymerase II (p. 358)
RNA polymerase III (p. 358)
RNA polymerase IV (p. 358)
RNA polymerase V (p. 358)
consensus sequence (p. 359)
−10 consensus sequence (Pribnow box) (p. 359)

−35 consensus sequence (p. 360)
upstream element (p. 360)
rho-dependent terminator (p. 362)
rho factor (p. 362)
rho-independent terminator (p. 362)
polycistronic mRNA (p. 363)
general transcription factor (p. 364)
basal transcription apparatus (p. 364)
transcriptional activator protein (p. 364)
core promoter (p. 364)
TATA box (p. 364)
regulatory promoter (p. 365)
enhancer (p. 365)
internal promoter (p. 365)
TATA-binding protein (TBP) (p. 365)

ANSWERS TO CONCEPT CHECKS

1. b

2. The template strand is the DNA strand that is copied into an RNA molecule, whereas the nontemplate strand is not copied.

3. d

4. The sigma factor recognizes the promoter and controls the binding of RNA polymerase to the promoter.

5. a

6. Inverted repeats followed by a string of adenine nucleotides

7. d

8. TFIID binds to the TATA box and helps to center the RNA polymerase over the start site of transcription.

9. Both processes use a protein that binds to the RNA molecule and moves down the RNA toward the RNA polymerase. They differ in that rho does not degrade the RNA, whereas Rat1 does so.

WORKED PROBLEMS

1. The diagram at right represents a sequence of nucleotides surrounding an RNA-coding sequence.

```
5′–CATGTT…TTGATGT– | RNA-     | –GACGA…TTTATA…GGCGCGC–3′
3′–GTACAA…AACTACA– | coding   | –CTGCT…AAATAT…CCGCGCG–5′
                   | sequence |
```

a. Is the RNA-coding sequence likely to be from a bacterial cell or from a eukaryotic cell? How can you tell?

b. Which DNA strand will serve as the template strand during the transcription of the RNA-coding sequence?

• Solution

a. Bacterial and eukaryotic cells use the same DNA bases (A, T, G, and C); so the bases themselves provide no clue to the origin of the sequence. The RNA-coding sequence must have a promoter, and bacterial and eukaryotic cells do differ in the consensus sequences found in their promoters; so we should examine the sequences for the presence of familiar consensus sequences. On the bottom strand to the right of the RNA-coding sequence, we find AAATAT, which, written in the conventional manner (5′ on the left), is 5′–TATAAA–3′. This sequence is the TATA box found in most eukaryotic promoters. However, the sequence is also quite similar to the −10 consensus sequence (5′–TATAAT–3′) found in bacterial promoters.

Farther to the right on the bottom strand, we also see 5′–GCGCGCC–3′, which is the TFIIB recognition element (BRE, see Figure 13.15) in eukaryotic RNA polymerase II promoters. No similar consensus sequence is found in bacterial promoters; so we can be fairly certain that this sequence is a eukaryotic promoter and an RNA-coding sequence.

b. The TATA box and BRE of RNA polymerase II promoters are upstream of the RNA-coding sequences; so RNA polymerase must bind to these sequences and then proceed downstream, transcribing the RNA-coding sequence. Thus RNA polymerase must proceed from right (upstream) to left (downstream). The RNA molecule is always synthesized in the 5′→3′ direction and is antiparallel to the DNA template strand; so the template strand must be read 3′→5′. If the enzyme proceeds from right to left and reads the template in the 3′→5′ direction, the upper strand must be the template, as shown in the diagram below.

2. Suppose that a consensus sequence in the regulatory promoter of a gene that encodes enzyme A were deleted. Which of the following effects would result from this deletion?

a. Enzyme A would have a different amino acid sequence.

b. The mRNA for enzyme A would be abnormally short.

c. Enzyme A would be missing some amino acids.

d. The mRNA for enzyme A would be transcribed but not translated.

e. The amount of mRNA transcribed would be affected. Explain your reasoning.

• Solution

The correct answer is part *e*. The regulatory protein contains binding sites for transcriptional activator proteins. These sequences are not part of the RNA-coding sequence for enzyme A; so the mutation would have no effect on the length or the amino acid sequence of the enzyme, eliminating answers *a*, *b*, and *c*. The TATA box is the binding site for the basal transcription apparatus. Transcriptional activator proteins bind to the regulatory promoter and affect the amount of transcription that takes place through interactions with the basal transcription apparatus at the core promoter.

For Worked Problem 1 (Solution b)

```
        Template strand | RNA-     | RNA polymerase
5′–CATGTT…TTGATGT–      | coding   | –GACGA…TTTATA…GGCGCGC–3′
3′–GTACAA…AACTACA–      | sequence | –CTGCT…AAATAT…CCGCGCG–5′
```

←——— **Direction of transcription**

COMPREHENSION QUESTIONS

Section 13.1

*1. Draw an RNA nucleotide and a DNA nucleotide, highlighting the differences. How is the structure of RNA similar to that of DNA? How is it different?

2. What are the major classes of cellular RNA? Where would you expect to find each class of RNA within eukaryotic cells?

3. Why is DNA more stable than RNA?

Section 13.2

*4. What parts of DNA make up a transcription unit? Draw a typical bacterial transcription unit and identify its parts.

5. What is the substrate for RNA synthesis? How is this substrate modified and joined together to produce an RNA molecule?

6. Describe the structure of bacterial RNA polymerase.

*7. Give the names of the RNA polymerases found in eukaryotic cells and the types of RNA that they transcribe.

Section 13.3

8. What are the three basic stages of transcription? Describe what happens at each stage.

*9. Draw a typical bacterial promoter and identify any common consensus sequences.

10. What are the two basic types of terminators found in bacterial cells? Describe the structure of each type.

Section 13.4

11. How does the process of transcription in eukaryotic cells differ from that in bacterial cells?

12. Compare the roles of general transcription factors and transcriptional activator proteins.

*13. Compare and contrast transcription and replication. How are these processes similar and how are they different?

14. How are the processes of transcription in bacteria and eukaryotes different? How are they similar?

APPLICATION QUESTIONS AND PROBLEMS

Section 13.1

15. An RNA molecule has the following percentages of bases: A = 23%, U = 42%, C = 21%, and G = 14%.

   a. Is this RNA single stranded or double stranded? How can you tell?

   b. What would be the percentages of bases in the template strand of the DNA that contains the gene for this RNA?

Section 13.2

*16. The following diagram represents DNA that is part of the RNA-coding sequence of a transcription unit. The bottom strand is the template strand. Give the sequence found on the RNA molecule transcribed from this DNA and identify the 5′ and 3′ ends of the RNA.

   5′–ATAGGCGATGCCA–3′
   3′–TATCCGCTACGGT–5′ ← Template strand

17. The following sequence of nucleotides is found in a single-stranded DNA template:

   ATTGCCAGATCATCCCAATAGAT

   Assume that RNA polymerase proceeds along this template from left to right.

   a. Which end of the DNA template is 5′ and which end is 3′?

   b. Give the sequence and identify the 5′ and 3′ ends of the RNA copied from this template.

18. RNA polymerases carry out transcription at a much lower rate than that at which DNA polymerases carry out replication. Why is speed more important in replication than in transcription?

Section 13.3

19. Write the consensus sequence for the following set of nucleotide sequences.

   AGGAGTT
   AGCTATT
   TGCAATA
   ACGAAAA
   TCCTAAT
   TGCAATT

*20. List at least five properties that DNA polymerases and RNA polymerases have in common. List at least three differences.

21. Most RNA molecules have *three* phosphate groups at the 5′ end, but DNA molecules never do. Explain this difference.

22. Write a hypothetical sequence of bases that might be found in the first 20 nucleotides of a promoter of a bacterial gene. Include both strands of DNA and identify the 5′ and 3′ ends of both strands. Be sure to include the start site for transcription and any consensus sequences found in the promoter.

*23. What would be the most likely effect of a mutation at the following locations in an *E. coli* gene?

   a. −8   c. −20

   b. −35   d. Start site

24. A strain of bacteria possesses a temperature-sensitive mutation in the gene that encodes the sigma factor. The mutant bacteria produce a sigma factor that is unable to bind to RNA polymerase at elevated temperatures. What effect will this mutation have on the process of transcription when the bacteria are raised at elevated temperatures?

*25. The following diagram represents a transcription unit on a DNA molecule.

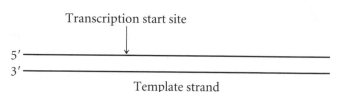

Transcription start site

Template strand

a. Assume that this DNA molecule is from a bacterial cell. Draw the approximate location of the promoter and terminator for this transcription unit.

b. Assume that this DNA molecule is from a eukaryotic cell. Draw the approximate location of an RNA polymerase II promoter.

26. The following DNA nucleotides are found near the end of a bacterial transcription unit. Find the terminator in this sequence.

3′–AGCATACAGCAGACCGTTGGTCTGAAAAAAGCATACA–5′

a. Mark the point at which transcription will terminate.

b. Is this terminator rho independent or rho dependent?

c. Draw a diagram of the RNA that will be transcribed from this DNA, including its nucleotide sequence and any secondary structures that form.

*27. A strain of bacteria possesses a temperature-sensitive mutation in the gene that encodes the rho subunit of RNA polymerase. At high temperatures, rho is not functional. When these bacteria are raised at elevated temperatures, which of the following effects would you expect to see?

a. Transcription does not take place.

b. All RNA molecules are shorter than normal.

c. All RNA molecules are longer than normal.

d. Some RNA molecules are longer than normal.

e. RNA is copied from both DNA strands.

Explain your reasoning for accepting or rejecting each of these five options.

28. The following diagram represents the Christmas-tree-like structure of active transcription observed by Miller, Hamkalo, and Thomas (see Figure 13.3).

On the diagram, identify parts a through i:

a. DNA molecule

b. 5′ and 3′ ends of the template strand of DNA

c. At least one RNA molecule

d. 5′ and 3′ ends of at least one RNA molecule

e. Direction of movement of the transcription apparatus on the DNA molecule

f. Approximate location of the promoter

g. Possible location of a terminator

h. Upstream and downstream directions

i. Molecules of RNA polymerase (use dots to represent these molecules)

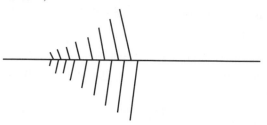

29. Suppose that the string of A nucleotides following the inverted repeat in a rho-independent terminator was deleted but that the inverted repeat was left intact. How will this deletion affect termination? What will happen when RNA polymerase reached this region?

Section 13.4

30. The following diagram represents a transcription unit in a hypothetical DNA molecule.

5′...TTGACA...TATAAT...3′
3′...AACTGT...ATATTA...5′

a. On the basis of the information given, is this DNA from a bacterium or from a eukaryotic organism?

b. If this DNA molecule is transcribed, which strand will be the template strand and which will be the nontemplate strand?

c. Where, approximately, will the start site of transcription be?

31. Computer programmers, working with molecular geneticists, have developed computer programs that can identify genes within long stretches of DNA sequences. Imagine that you are working with a computer programmer on such a project. On the basis of what you know about the process of transcription, what sequences should be used to identify the beginning and end of a gene with the use of this computer program?

*32. Through genetic engineering, a geneticist mutates the gene that encodes TBP in cultured human cells. This mutation destroys the ability of TBP to bind to the TATA box. Predict the effect of this mutation on cells that possess it.

33. Elaborate repair mechanisms are associated with replication to prevent permanent mutations in DNA, yet no similar repair is associated with transcription. Can you think of a reason for this difference in replication and transcription? (Hint: Think about the relative effects of a permanent mutation in a DNA molecule compared with one in an RNA molecule.)

CHALLENGE QUESTIONS

Section 13.3

34. Many genes in both bacteria and eukaryotes contain numerous sequences that potentially cause pauses or premature terminations of transcription. Nevertheless, the transcription of these genes within a cell normally produces multiple RNA molecules thousands of nucleotides long without pausing or terminating prematurely. However, when a single round of transcription takes place on such templates in a test tube, RNA synthesis is frequently interrupted by pauses and premature terminations, which reduce the rate at which transcription takes place and frequently shorten the length of the mRNA molecules produced. Most pauses and premature terminations occur when RNA polymerase temporarily backtracks (i.e., backs up) for one or two nucleotides along the DNA. Experimental findings have demonstrated that most transcriptional delays and premature terminations disappear if several RNA polymerases are simultaneously transcribing the DNA molecule. Propose an explanation for faster transcription and longer mRNA when the template DNA is being transcribed by multiple RNA polymerases.

Section 13.4

35. Enhancers are sequences that affect the initiation of the transcription of genes that are hundreds or thousands of nucleotides away. Transcriptional activator proteins that bind to enhancers usually interact directly with transcription factors at promoters by causing the intervening DNA to loop out. An enhancer of bacteriophage T4 does not function by looping of the DNA (D. R. Herendeen et al. 1992. *Science* 256:1298–1303). Propose some additional mechanisms (other than DNA looping) by which this enhancer might affect transcription at a gene thousands of nucleotides away.

*36. The locations of the TATA box in two species of yeast, *Saccharomyces pombe* and *Saccharomyces cerevisiae*, differ dramatically. The TATA box of *S. pombe* is about 30 nucleotides upstream of the start site, similar to the location for most other eukaryotic cells. However, the TATA box of *S. cerevisiae* is 40 to 120 nucleotides upstream of the start site.

To better understand what sets the start site in these organisms, researchers at Stanford University conducted a series of experiments to determine which components of the transcription apparatus of these two species could be interchanged (Y. Li et al. 1994. *Science* 263:805–807). In these experiments, different transcription factors and RNA polymerases were switched in *S. pombe* and *S. cerevisiae*, and the effects of the switch on the level of RNA synthesis and on the start point of transcription were observed. The results from one set of experiments are shown in the table below. Components cTFIIB, cTFIIE, cTFIIF, cTFIIH are transcription factors from *S. cerevisiae*. Components pTFIIB, pTFIIE, pTFIIF, pTFIIH are transcription factors from *S. pombe*. Components cPol II and pPol II are RNA polymerase II from *S. cerevisiae* and *S. pombe*, respectively. The table indicates whether the component was present (+) or missing (−) in the experiment. In the accompanying gel, the presence of a band indicates that RNA was produced and the position of the band indicates whether it was the length predicated when transcription begins 30 bp downstream from the TATA box or from 40 to 120 bp downstream from the TATA box.

	Experiment						
Components	1	2	3	4	5	6	7
cTFIIE	+	−	+	+	+	+	−
cTFIIH + cTFIIF	+	−	+	+	+	−	+
cTFIIB + cPol II	+	−	−	−	−	−	−
pPol II	−	+	+	+	−	+	+
pTFIIB	−	+	+	−	+	+	+
pTFIIF + pTFIIH + pTFIIF	−	+	−	−	−	−	−

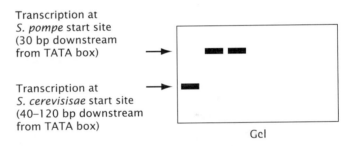

Transcription at *S. pompe* start site (30 bp downstream from TATA box) →

Transcription at *S. cerevisisae* start site (40–120 bp downstream from TATA box) →

Gel

a. What conclusion can you draw from these data about what components determine the start site for transcription?

b. What conclusions can you draw about the interactions of the different components of the transcription apparatus?

c. Propose a mechanism for why the start site for transcription in *S. pombe* is about 30 bp downstream from the TATA box, whereas the start site for transcription in *S. cerevisiae* is 40 to 120 bp downstream from the TATA box.

37. Glenn Croston and his colleagues studied the

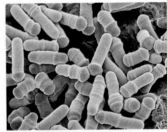

S. pombe. [Steve Gschmeissner/ Photo Researchers.]

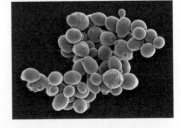

S. cerevisiae. [Science Photo Library/PhotoLibrary.]

relation between chromatin structure and transcription activity. In one set of experiments, they measured the level of in vitro transcription of a *Drosophila* gene by RNA polymerase II with the use of DNA and various combinations of histone proteins (G. E. Croston et al. 1991. *Science* 251:643–649).

First, they measured the level of transcription for naked DNA, with no associated histone proteins. Then, they measured the level of transcription after nucleosome octamers (without H1) were added to the DNA. The addition of the octamers caused the level of transcription to drop by 50%. When both the nucleosome octamers and the H1 proteins were added to the DNA, transcription was greatly repressed, dropping to less than 1% of that obtained with naked DNA, as shown in the adjoining table.

GAL4-VP16 is a protein that binds to the DNA of certain eukaryotic genes. When GAL4-VP16 is added to DNA, the level of RNA polymerase II transcription is greatly elevated.

Treatment	Relative amount of transcription
Naked DNA	100
DNA + octamers	50
DNA + octamers + H1	<1
DNA + GAL4-VP16	1000
DNA + octamers + GAL4-VP16	1000
DNA + octamers + H1 + GAL4-VP16	1000

Even in the presence of the H1 protein, GAL4-VP16 stimulates high levels of transcription.

Propose a mechanism for how the H1 protein represses transcription and how GAL4-VP16 overcomes this repression. Explain how your proposed mechanism would produce the results obtained in these experiments.

14 RNA Molecules and RNA Processing

In *Drosophila melanogaster*, RNA transcribed from the *fru* gene is spliced differently in male and female flies, producing courtship behaviors in males and their absence in females. [Dr. Dennis Kunkel/Visuals Unlimited.]

SEX THROUGH SPLICING

A male fruit fly has a single thought—to win the heart of a female fly. On spotting a female, he performs a carefully choreographed dance, orienting toward the female, following her, and tapping on her with his legs. He sings a courtship song by vibrating one wing and licks her genitalia. If his execution is proper and the female is aroused, she accepts his advances and a successful copulation ensues.

Geneticists have long been intrigued by the genetic basis of this complex and yet essential behavior in male fruit flies. In the 1960s, researchers discovered a gene, called *fruitless* (*fru*), that is necessary for male sexual behavior. Male fruit flies with a mutated copy of *fru* seem unable to discriminate between males and females; they court males and females with equal intensity and even copulate with other males. This failure to discriminate indicates that *fru* is critical for male courtship behavior. However, both males and females possess copies of the *fru* gene; so why do males sing, dance, and lick, whereas females don't?

Further research findings have shown that the *fru* gene of males is transcribed and translated into a protein, which turns on other genes participating in male courtship behavior. In females, the *fru* gene is transcribed, but a shortened form of the protein or no protein at all is produced. That the same gene produces different proteins in males and females is quite remarkable, given that males and females have the exact same *fru* gene. How does the same sequence of nucleotides in the *fru* gene specify different proteins in males and females?

The answer lies, not in the sequences of the gene, but in what happens to the RNA transcribed from those sequences. In eukaryotic cells, RNA molecules are often extensively modified after transcription. For genes that encode proteins, a special nucleotide called the cap is added to the 5′ end, a tail of adenine nucleotides is added to the 3′ end, and long stretches of nucleotides called introns are cut out of the middle. The nature of this processing often varies, and so different proteins are produced from the same DNA sequence. These types of alternative pathways for RNA processing are surprisingly common: more than 90% of protein-encoding genes in humans are capable of being processed in different ways.

The RNA transcribed from the fruit-fly *fru* gene undergoes alternative splicing. In females, some extra nucleotides are included in the resulting mRNA. These extra nucleotides include a stop codon that prematurely halts translation, resulting in the production of a truncated protein in females that is nonfunctional and fails to produce male courtship behavior. The tremendous role that this small splicing difference has on fruit-fly behavior was dramatically illustrated in 2005, when scientists modified several bases in the *fru* genes of female flies, causing their cells to splice the *fru* RNA in the typical male fashion. These engineered flies were anatomically female in all respects, but they courted other females with the same behaviors used by males. Male flies engineered to splice their *fru* RNA in the female way lost their interest in females and exhibited little male courtship behavior.

The experiments just described illustrate the important influence of RNA modification on expression of the genotype. The *fru* RNA, like many eukaryotic RNAs, undergoes extensive processing after transcription, and this modification is essential for normal gene function. In Chapter 13, we focused on transcription—the process by which RNA molecules are synthesized. In this chapter, we examine the function and processing of RNA.

We begin by taking a careful look at the nature of the gene. We then examine messenger RNA (mRNA), its structure, and how it is modified in eukaryotes after transcription. Then, we turn to transfer RNA (tRNA), the adapter molecule that forms the interface between amino acids and mRNA in protein synthesis. We examine ribosomal RNA (rRNA), the structure and organization of rRNA genes, and how rRNAs are processed. Finally, we consider a newly discovered group of very small RNAs that play important roles in numerous biological functions.

As we explore the world of RNA and its role in gene function, we will see evidence of two important characteristics of this nucleic acid. First, RNA is extremely versatile, both structurally and biochemically. It can assume a number of different secondary structures, which provide the basis for its functional diversity. Second, RNA processing and function frequently include interactions between two or more RNA molecules.

14.1 Many Genes Have Complex Structures

What is a gene? As noted in Chapter 3, the definition of a gene often changes as we explore different aspects of heredity. A gene was defined in Chapter 3 as an inherited factor that determines a characteristic. This definition may have seemed vague because it says only what a gene does but nothing about what a gene is. Nevertheless, this definition was appropriate for our purposes at the time because our focus was on how genes influence the inheritance of traits. We did not have to consider the physical nature of the gene in learning the rules of inheritance.

Knowing something about the chemical structure of DNA and the process of transcription now enables us to be more precise about what a gene is. Chapter 10 described how genetic information is encoded in the base sequence of DNA; so a gene consists of a set of DNA nucleotides. But how many nucleotides constitute a gene, and how is the information in these nucleotides organized? In 1902, Archibald Garrod suggested, correctly, that genes encode proteins. Proteins are made of amino acids; so a gene contains the nucleotides that specify the amino acids of a protein. We could, then, define a gene as a set of nucleotides that specifies the amino acid sequence of a protein, which indeed was, for many years, the working definition of a gene. As geneticists learned more about the structure of genes, however, it became clear that this concept of a gene was an oversimplification.

Gene Organization

Early work on gene structure was carried out largely through the examination of mutations in bacteria and viruses. This research led Francis Crick in 1958 to propose that genes and proteins are **colinear**—that there is a direct correspondence between the nucleotide sequence of DNA and the amino acid sequence of a protein (**Figure 14.1**). The concept of

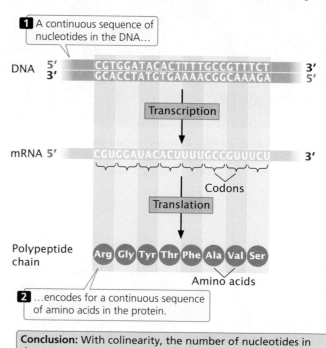

1 A continuous sequence of nucleotides in the DNA…

DNA 5′ CGTGGATACACTTTTGCCGTTTCT 3′
3′ GCACCTATGTGAAAACGGCAAAGA 5′

Transcription

mRNA 5′ CGUGGAUACACUUUUGCCGUUUCU 3′

Codons

Translation

Polypeptide chain Arg Gly Tyr Thr Phe Ala Val Ser

Amino acids

2 …encodes for a continuous sequence of amino acids in the protein.

Conclusion: With colinearity, the number of nucleotides in the gene is proportional to the number of amino acids in the protein.

14.1 The concept of colinearity suggests that a continuous sequence of nucleotides in DNA encodes a continuous sequence of amino acids in a protein.

colinearity suggests that the number of nucleotides in a gene should be proportional to the number of amino acids in the protein encoded by that gene. In a general sense, this concept is true for genes found in bacterial cells and many viruses, although these genes are slightly longer than would be expected if colinearity were strictly applied (the mRNAs encoded by the genes contain sequences at their ends that do not specify amino acids). At first, eukaryotic genes and proteins also were generally assumed to be colinear, but there were hints that eukaryotic gene structure is fundamentally different. Eukaryotic cells contain far more DNA than is required to encode proteins (see Chapter 11). Furthermore, many large RNA molecules observed in the nucleus were absent from the cytoplasm, suggesting that nuclear RNAs undergo some type of change before they are exported to the cytoplasm.

Most geneticists were nevertheless surprised by the announcement in the 1970s that not all genes are continuous. Researchers observed four coding sequences in a gene from a eukaryotic virus that were interrupted by nucleotides that did not specify amino acids. This discovery was made when the viral DNA was hybridized with the mRNA transcribed from it and the hybridized structure was examined with the use of an electron microscope (**Figure 14.2**). The DNA was clearly much longer than the mRNA because regions of DNA looped out from the hybridized molecules. These regions contained nucleotides in the DNA that were absent from the coding nucleotides in the mRNA. Many other examples of interrupted genes were subsequently discovered; it quickly became apparent that most eukaryotic genes consist of stretches of coding and noncoding nucleotides.

CONCEPTS

When a continuous sequence of nucleotides in DNA encodes a continuous sequence of amino acids in a protein, the two are said to be colinear. In eukaryotes, not all genes are colinear with the proteins that they encode.

✔ **CONCEPT CHECK 1**

What evidence indicated that eukaryotic genes are not colinear with their proteins?

Introns

Many eukaryotic genes contain coding regions called **exons** and noncoding regions called intervening sequences or **introns**. For example, the gene encoding the protein ovalbumin has eight exons and seven introns; the gene for cytochrome *b* has five exons and four introns (**Figure 14.3**). The average human gene contains from eight to nine introns. All the introns and the exons are initially transcribed into RNA but, after transcription, the introns are removed by splicing and the exons are joined to yield the mature RNA.

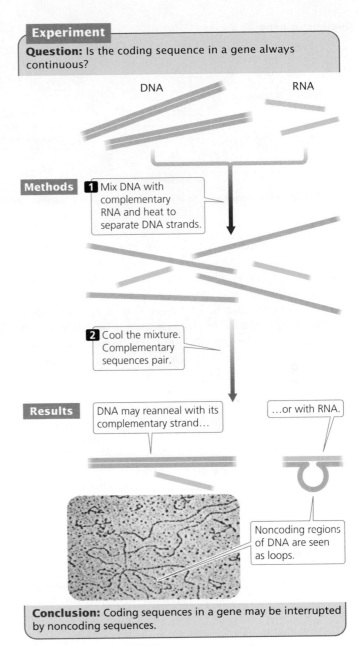

Experiment

Question: Is the coding sequence in a gene always continuous?

DNA | RNA

Methods **1** Mix DNA with complementary RNA and heat to separate DNA strands.

2 Cool the mixture. Complementary sequences pair.

Results DNA may reanneal with its complementary strand…

…or with RNA.

Noncoding regions of DNA are seen as loops.

Conclusion: Coding sequences in a gene may be interrupted by noncoding sequences.

14.2 The noncolinearity of eukaryotic genes was discovered by hybridizing DNA and mRNA. [Electromicrograph from O. L. Miller, B. R. Beatty, D. W. Fawcett/Visuals Unlimited.]

Introns are common in eukaryotic genes but are rare in bacterial genes. For a number of years after their discovery, introns were thought to be entirely absent from prokaryotic genomes, but they have now been observed in archaea, bacteriophages, and even some eubacteria. Introns are present in mitochondrial and chloroplast genes as well as the nuclear genes of eukaryotes. In eukaryotic genomes, the size and number of introns appear to be directly related to increasing organismal complexity. Yeast genes contain only a few short introns; *Drosophila* introns are longer and more numerous; and most vertebrate genes are interrupted by long introns. All classes of eukaryotic genes—those that encode rRNA,

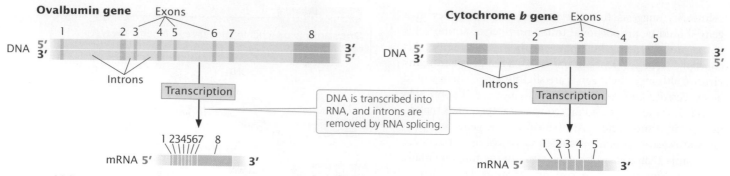

14.3 The coding sequences of many eukaryotic genes are disrupted by noncoding introns.

tRNA, and proteins—may contain introns. The number and size of introns vary widely: some eukaryotic genes have no introns, whereas others may have more than 60; intron length varies from fewer than 200 nucleotides to more than 50,000. Introns tend to be longer than exons, and most eukaryotic genes contain more noncoding nucleotides than coding nucleotides. Finally, most introns do not encode proteins (an intron of one gene is not usually an exon for another), although geneticists are finding a growing number of exceptions.

There are four major types of introns (**Table 14.1**). **Group I introns**, found in some rRNA genes, are self-splicing: they can catalyze their own removal. **Group II introns** are present in some protein-encoding genes of mitochondria, chloroplasts, and a few eubacteria; they also are self-splicing, but their mechanism of splicing differs from that of the group I introns. **Nuclear pre-mRNA introns** are the best studied; they include introns located in the protein-encoding genes of the eukaryotic nucleus. The splicing mechanism by which these introns are removed is similar to that of the group II introns, but nuclear introns are not self-splicing; their removal requires snRNAs (discussed later in the chapter) and a number of proteins. **Transfer RNA introns**, found in tRNA genes, utilize yet another splicing mechanism that relies on enzymes to cut and reseal the RNA. In addition to these major groups, there are several other types of introns.

Table 14.1	Major types of introns	
Type of Intron	Location	Splicing
Group I	Some rRNA genes	Self-splicing
Group II	Protein-encoding genes in mitochondria and chloroplasts	Self-splicing
Nuclear pre-mRNA	Protein-encoding genes in the nucleus	Spliceosomal
tRNA	tRNA genes	Enzymatic

Note: There are also several types of minor introns, including group III introns, twintrons, and archaeal introns.

We'll take a detailed look at the chemistry and mechanics of RNA splicing later in the chapter. For now, we should keep in mind two general characteristics of the splicing process: (1) the splicing of all pre-mRNA introns takes place in the nucleus; and (2) the order of exons in DNA is usually maintained in the spliced RNA: the coding sequences of a gene may be split up, but they are not usually jumbled up. **TRY PROBLEM 19** →

CONCEPTS

Many eukaryotic genes contain exons and introns, both of which are transcribed into RNA, but introns are later removed by RNA processing. The number and size of introns vary from gene to gene; they are common in many eukaryotic genes but uncommon in bacterial genes.

✔ **CONCEPT CHECK 2**

What are the four major types of introns?

The Concept of the Gene Revisited

How does the presence of introns affect our concept of a gene? To define a gene as a sequence of nucleotides that encodes amino acids in a protein no longer seems appropriate, because this definition excludes introns, which do not specify amino acids. This definition also excludes nucleotides that encode the 5′ and 3′ ends of an mRNA molecule, which are required for translation but do not encode amino acids. And defining a gene in these terms also excludes sequences that encode rRNA, tRNA, and other RNAs that do not encode proteins. In view of our current understanding of DNA structure and function, we need a more precise definition of gene.

Many geneticists have broadened the concept of a gene to include all sequences in the DNA that are transcribed into a single RNA molecule. Defined in this way, a gene includes all exons, introns, and those sequences at the beginning and end of the RNA that are not translated into a protein. This definition also includes DNA sequences that encode rRNAs, tRNAs, and other types of nonmessenger RNA. Some geneticists have expanded the definition of a gene even further, to include the entire transcription unit—the promoter, the RNA coding sequence, and the terminator. However, new evidence now calls into question even this definition. Recent research suggests that much of the genome is transcribed into RNA, although it is unclear what, if anything, much of this RNA does (many of these RNA molecules are possibly artifacts of faulty transcription). What is certain is that the process of transcription is more complex than formerly thought and that defining a gene as a sequence that is transcribed into an RNA molecule is not as straightforward as formerly thought. The more we learn about the nature of genetic information, the more elusive the definition of a gene seems to become.

CONCEPTS

The discovery of introns forced a reevaluation of the definition of the gene. Today, a gene is often defined as a DNA sequence that encodes an RNA molecule or the entire DNA sequence required to transcribe and encode an RNA molecule.

14.2 Messenger RNAs, Which Encode the Amino Acid Sequences of Proteins, Are Modified after Transcription in Eukaryotes

As soon as DNA was identified as the source of genetic information, it became clear that DNA cannot directly encode proteins. In eukaryotic cells, DNA resides in the nucleus, yet protein synthesis takes place in the cytoplasm. Geneticists recognized that an additional molecule must take part in the transfer of genetic information.

The results of studies of bacteriophage infection conducted in the late 1950s and early 1960s pointed to RNA as a likely candidate for this transport function. Bacteriophages inject their DNA into bacterial cells, where the DNA is replicated, and large amounts of phage protein are produced on the bacterial ribosomes. As early as 1953, Alfred Hershey discovered a type of RNA that was synthesized rapidly after bacteriophage infection. Findings from later studies showed that this short-lived RNA had a nucleotide composition similar to that of the phage DNA but quite different from that of the bacterial RNA. These observations were consistent with the idea that RNA was copied from DNA and that this RNA then directed the synthesis of proteins.

At the time, ribosomes were known to be *somehow* implicated in protein synthesis, and much of the RNA in a cell was known to be in the form of ribosomes. Ribosomes were believed to be the agents by which genetic information was moved to the cytoplasm for the production of protein. Using equilibrium density gradient centrifugation (see Figure 12.2), Sydney Brenner, François Jacob, and Matthew Meselson demonstrated in 1961 that it is not the case. They showed that new ribosomes are *not* produced during the burst of protein synthesis that accompanies phage infection (**Figure 14.4**). The genetic information needed to produce new phage proteins was not carried by the ribosomes.

In a related experiment, François Gros and his colleagues infected *E. coli* cells with bacteriophages while adding radio-

Experiment

Question: Do ribosomes carry genetic information?

Methods

1 *E. coli* were grown in medium containing heavy isotopes through several generations so that the heavy isotopes would become incorporated into all *E. coli* ribosomes.

Medium with ^{15}N and ^{13}C
E. coli culture

2 The cells were moved into medium containing light isotopes (^{14}N and ^{12}C)...

3 ... and infected with bacteriophage.

Medium with ^{14}N and ^{12}C
E. coli culture

4 If new ribosomes were produced after phage infection, they would contain ^{14}N and ^{12}C and would be relatively light.

5 After phage proteins were produced, ribosomes were separated by equilibrium density gradient centrifugation.

Spin

Results

Increasing density

6 Only old ribosomes containing heavy isotopes (^{15}N and ^{13}C) were found.

Conclusion: Ribosomes are not produced in phage reproduction and therefore do not carry genetic information.

14.4 Brenner, Jacob, and Meselson demonstrated that ribosomes do not carry genetic information.

actively labeled ("hot") uracil, which would become incorporated into newly produced phage RNA, to the medium. After a few minutes, they transferred the cells to a medium that contained unlabeled ("cold") uracil. This type of experiment is called a pulse–chase experiment: the cells are exposed to a brief pulse of label, which is then "chased" by cold, unlabeled precursor. Pulse–chase experiments allow researchers to follow, by tracking the presence of the radioactivity, products of short-term biochemical events, such as RNA synthesis immediately following phage infection.

Gros and his coworkers found that the newly produced phage RNA was short-lived, lasting only a few minutes, and was associated with ribosomes but was distinct from them. They concluded that newly synthesized, short-lived RNA carries the genetic information for protein structure to the ribosome. The term *messenger RNA* was coined for this carrier.

The Structure of Messenger RNA

Messenger RNA functions as the template for protein synthesis; it carries genetic information from DNA to a ribosome and helps to assemble amino acids in their correct order. In bacteria, mRNA is transcribed directly from DNA but, in eukaryotes, a pre-mRNA (also called the primary transcript) is first transcribed from DNA and then processed to yield the mature mRNA. We will reserve the term mRNA for RNA molecules that have been completely processed and are ready to undergo translation.

In the mature mRNA, each amino acid in a protein is specified by a set of three nucleotides called a **codon**. Both prokaryotic and eukaryotic mRNAs contain three primary regions (**Figure 14.5**). The **5′ untranslated region** (5′ UTR; sometimes called the leader) is a sequence of nucleotides at the 5′ end of the mRNA that does not encode any of the amino acids of a protein. In bacterial mRNA, this region contains a consensus sequence called the **Shine–Dalgarno** sequence, which serves as the ribosome-binding site during translation; it is found approximately seven nucleotides upstream of the first codon translated into an amino acid (called the start codon). Eukaryotic mRNA has no equiva-

lent consensus sequence in its 5′ untranslated region. In eukaryotic cells, ribosomes bind to a modified 5′ end of mRNA, as discussed later in the chapter.

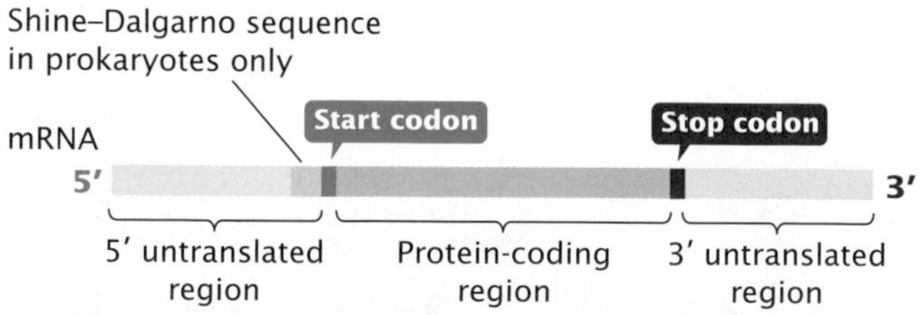

The next section of mRNA is the **protein-coding region**, which comprises the codons that specify the amino acid sequence of the protein. The protein-coding region begins with a start codon and ends with a stop codon. The last region of mRNA is the **3′ untranslated region** (3′ UTR; sometimes called a trailer), a sequence of nucleotides at the 3′ end of the mRNA and not translated into protein. The 3′ UTR affects the stability of mRNA and the translation of the mRNA protein-coding sequence.

CONCEPTS

Messenger RNA molecules contain three main regions: a 5′ untranslated region, a protein-coding region, and a 3′ untranslated region. The 5′ and 3′ untranslated regions do not encode any amino acids of a protein.

Pre-mRNA Processing

In bacterial cells, transcription and translation take place simultaneously; while the 3′ end of an mRNA is undergoing transcription, ribosomes attach to the Shine–Dalgarno sequence near the 5′ end and begin translation. Because transcription and translation are coupled, bacterial mRNA has little opportunity to be modified before protein synthesis. In contrast, transcription and translation are separated in both time and space in eukaryotic cells. Transcription takes place in the nucleus, whereas translation takes place in the cytoplasm; this separation provides an opportunity for eukaryotic RNA to be modified before it is translated. Indeed, eukaryotic mRNA is extensively altered after transcription. Changes are made to the 5′ end, the 3′ end, and the protein-coding section of the RNA molecule (**Table 14.2**).

Shine–Dalgarno sequence in prokaryotes only

14.5 Three primary regions of mature mRNA are the 5′ untranslated region, the protein-coding region, and the 3′ untranslated region.

www.whfreeman.com/pierce4e   Animation 14.2 shows how mutations in these different regions affect the flow of information from genotype to phenotype.

Table 14.2	Posttranscriptional modifications to eukaryotic pre-mRNA
Modification	Function
Addition of 5′ cap	Facilitates binding of ribosome to 5′ end of mRNA, increases mRNA stability, enhances RNA splicing
3′ cleavage and addition of poly(A) tail	Increases stability of mRNA, facilitates binding of ribosome to mRNA
RNA splicing	Removes noncoding introns from pre-mRNA, facilitates export of mRNA to cytoplasm, allows for multiple proteins to be produced through alternative splicing
RNA editing	Alters nucleotide sequence of mRNA

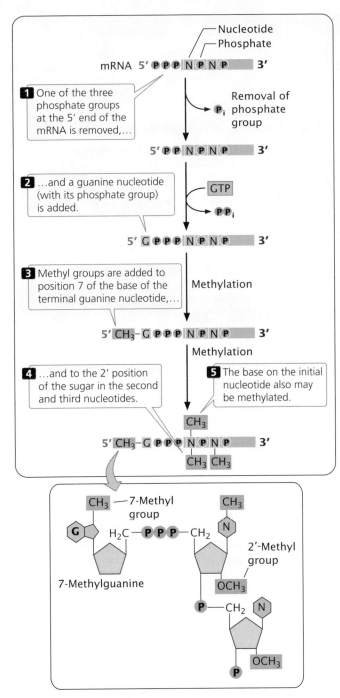

14.6 Most eukaryotic mRNAs have a 5′ cap. The cap consists of a nucleotide with 7-methylguanine attached to the pre-mRNA by a unique 5′–5′ bond (shown in detail in the bottom box).

The Addition of the 5′ Cap

One type of modification of eukaryotic pre-mRNA is the addition at its 5′ end of a structure called a **5′ cap**. The cap consists of an extra nucleotide at the 5′ end of the mRNA and methyl groups (CH_3) on the base in the newly added nucleotide and on the 2′-OH group of the sugar of one or more nucleotides at the 5′ end (**Figure 14.6**). The

addition of the cap takes place rapidly after the initiation of transcription and, as will be discussed in more depth in Chapter 15, the 5′ cap functions in the initiation of translation. Cap-binding proteins recognize the cap and attach to it; a ribosome then binds to these proteins and moves downstream along the mRNA until the start codon is reached and translation begins. The presence of a 5′ cap also increases the stability of mRNA and influences the removal of introns.

As noted in the discussion of transcription in Chapter 13, three phosphate groups are present at the 5′ end of all RNA molecules because phosphate groups are not cleaved from the first ribonucleoside triphosphate in the transcription reaction. The 5′ end of pre-mRNA can be represented as 5′–pppNpNpN . . . , in which the letter "N" represents a ribonucleotide and "p" represents a phosphate. Shortly after the initiation of transcription, one of these phosphate groups is removed and a guanine nucleotide is added (see Figure 14.6). This guanine nucleotide is attached to the pre-mRNA by a unique 5′–5′ bond, which is quite different from the usual 5′–3′ phosphodiester bond that joins all the other nucleotides in RNA. One or more methyl groups are then added to the 5′ end; the first of these methyl groups is added to position 7 of the base of the terminal guanine nucleotide, making the base 7-methylguanine. Next, a methyl group may be added to the 2′ position of the sugar in the second and third nucleotides, as shown in Figure 14.6. Rarely, additional methyl groups may be attached to the bases of the second and third nucleotides of the pre-mRNA.

Several different enzymes take part in the addition of the 5′ cap. The initial step is carried out by an enzyme that associates with RNA polymerase II. Because neither RNA polymerase I nor RNA polymerase III have this associated enzyme, RNA molecules transcribed by these polymerases (rRNAs, tRNAs, and some snRNAs) are not capped.

The Addition of the Poly(A) Tail

A second type of modification to eukaryotic mRNA is the addition of 50 to 250 or more adenine nucleotides at the 3′ end, forming a **poly(A) tail**. These nucleotides are not encoded in the DNA but are added after transcription (**Figure 14.7**) in a process termed polyadenylation. Many eukaryotic genes transcribed by RNA polymerase II are transcribed well beyond the end of the coding sequence (see Chapter 13); most of the extra material at the 3′ end is then cleaved and the poly(A) tail is added. For some pre-mRNA molecules, more than 1000 nucleotides may be removed from the 3′ end before polyadenylation.

Processing of the 3′ end of pre-mRNA requires sequences both upstream and downstream of the cleavage site. The consensus sequence AAUAAA is usually from 11 to 30 nucleotides upstream of the cleavage site (see Figure 14.7) and determines the point at which cleavage will take place. A sequence rich in uracil nucleotides (or in guanine and uracil nucleotides) is typically downstream of the cleavage site.

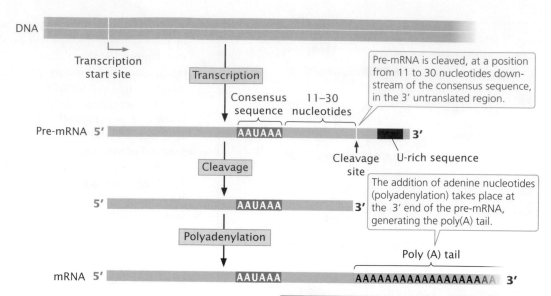

14.7 Most eukaryotic mRNAs have a 3′ poly(A) tail.

A large number of proteins take part in finding the cleavage site and removing the 3′ end. After cleavage has been completed, adenine nucleotides are added to the new 3′ end, creating the poly(A) tail.

The poly(A) tail confers stability on many mRNAs, increasing the time during which the mRNA remains intact and available for translation before it is degraded by cellular enzymes. The stability conferred by the poly(A) tail depends on the proteins that attach to the tail and on its length. The poly(A) tail also facilitates attachment of the ribosome to the mRNA.

Poly(U) tails are added to the 3′ ends of some mRNAs, microRNAs, and small nuclear RNAs. Although the function of poly(U) tails is still under investigation, evidence suggests that poly(U) tails on some mRNAs may facilitate their degradation. TRY PROBLEM 25 →

CONCEPTS

Eukaryotic pre-mRNAs are processed at their 5′ and 3′ ends. A cap, consisting of a modified nucleotide and several methyl groups, is added to the 5′ end. The cap facilitates the binding of a ribosome, increases the stability of the mRNA, and may affect the removal of introns. Processing at the 3′ end includes cleavage

downstream of an AAUAAA consensus sequence and the addition of a poly(A) tail.

✔ CONCEPT CHECK 3

Why are pre-mRNAs capped, but tRNAs and rRNAs aren't?

RNA Splicing

The other major type of modification of eukaryotic pre-mRNA is the removal of introns by **RNA splicing**. This modification takes place in the nucleus, before the RNA moves to the cytoplasm.

Consensus sequences and the spliceosome Splicing requires the presence of three sequences in the intron. One end of the intron is referred to as the **5′ splice site**, and the other end is the **3′ splice site** (**Figure 14.8**); these splice sites possess short consensus sequences. Most introns in pre-mRNAs begin with GU and end with AG, indicating that these sequences play a crucial role in splicing. Indeed, changing a single nucleotide at either of these sites prevents splicing.

The third sequence important for splicing is at the **branch point**, which is an adenine nucleotide that lies from 18 to 40 nucleotides upstream of the 3′ splice site

14.8 Splicing of pre-mRNA requires consensus sequences. Critical consensus sequences are present at the 5′ splice site and the 3′ splice site. A weak consensus sequence (not shown) exists at the branch point.

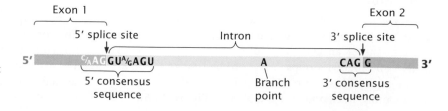

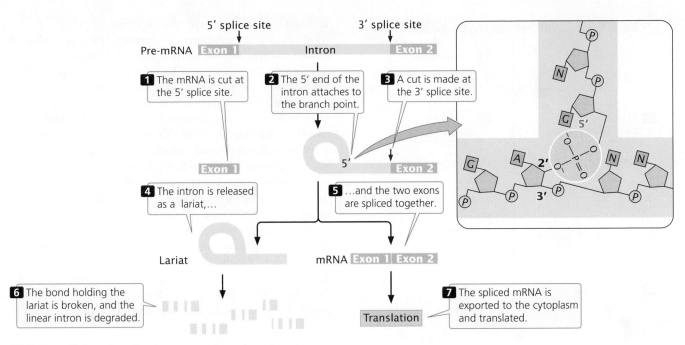

14.9 The splicing of nuclear introns requires a two-step process.

(see Figure 14.8). The sequence surrounding the branch point does not have a strong consensus. The deletion or mutation of the adenine nucleotide at the branch point prevents splicing.

Splicing takes place within a large structure called the **spliceosome**, which is one of the largest and most complex of all molecular complexes. The spliceosome consists of five RNA molecules and almost 300 proteins. The RNA components are small nuclear RNAs (snRNAs, see Chapter 13) ranging in length from 107 to 210 nucleotides; these snRNAs associate with proteins to form small ribonucleoprotein particles (snRNPs). Each snRNP contains a single snRNA molecule and multiple proteins. The spliceosome is composed of five snRNPs (U1, U2, U4, U5, and U6), and some proteins not associated with an snRNA.

CONCEPTS

Introns in nuclear genes contain three consensus sequences critical to splicing: a 5′ splice site, a 3′ splice site, and a branch point. The splicing of pre-mRNA takes place within a large complex called the spliceosome, which consists of snRNAs and proteins.

✔ CONCEPT CHECK 4

If a splice site were mutated so that splicing did not take place, what would be the effect on the mRNA?

a. It would be shorter than normal.

b. It would be longer than normal.

c. It would be the same length but would encode a different protein.

The process of splicing   Before splicing takes place, an intron lies between an upstream exon (exon 1) and a downstream exon (exon 2), as shown in **Figure 14.9**. Pre-mRNA is spliced in two distinct steps. In the first step of splicing, the pre-mRNA is cut at the 5′ splice site. This cut frees exon 1 from the intron, and the 5′ end of the intron attaches to the branch point; that is, the intron folds back on itself, forming a structure called a **lariat**. In this reaction, the guanine nucleotide in the consensus sequence at the 5′ splice site bonds with the adenine nucleotide at the branch point through a transesterification reaction. The result is that the 5′ phosphate group of the guanine nucleotide is now attached to the 2′-OH group of the adenine nucleotide at the branch point (see Figure 14.9).

In the second step of RNA splicing, a cut is made at the 3′ splice site and, simultaneously, the 3′ end of exon 1 becomes covalently attached (spliced) to the 5′ end of exon 2. The intron is released as a lariat. The intron becomes linear when the bond breaks at the branch point and is then rapidly degraded by nuclear enzymes. The mature mRNA consisting of the exons spliced together is exported to the cytoplasm, where it is translated.

These splicing reactions take place within the spliceosome, which assembles on the pre-mRNA in a step-by-step fashion and carries out the splicing reactions (**Figure 14.10**). A key feature of the process is a series of interactions between the mRNA and the snRNAs and between different snRNAs. These interactions depend on complementary base pairing between the different RNA molecules and bring the essential components of the pre-mRNA transcript and the spliceosome close together, which make splicing possible.

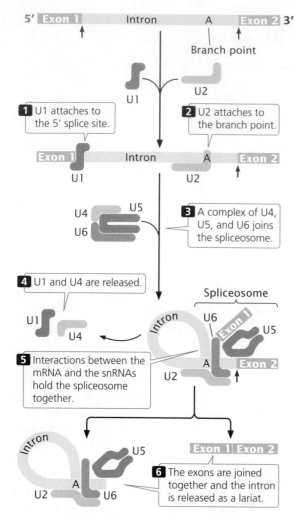

1 U1 attaches to the 5′ splice site.

2 U2 attaches to the branch point.

3 A complex of U4, U5, and U6 joins the spliceosome.

4 U1 and U4 are released.

5 Interactions between the mRNA and the snRNAs hold the spliceosome together.

6 The exons are joined together and the intron is released as a lariat.

14.10 RNA splicing takes place within the spliceosome. The spliceosome assembles sequentially.

Key catalytic steps in the splicing process are carried out by the snRNAs that constitute the spliceosome.

Most mRNAs are produced from a single pre-mRNA molecule from which the exons are spliced together. However, in a few organisms (principally nematodes and trypanosomes), mRNAs may be produced by splicing together sequences from two or more different RNA molecules; this process is called **trans-splicing**.

Many human genetic diseases arise from mutations that affect pre-mRNA splicing; indeed, about 15% of single-base substitutions that result in human genetic diseases alter pre-mRNA splicing. Some of these mutations interfere with recognition of the normal 5′ and 3′ splice sites. Others create new splice sites. TRY PROBLEM 21 →

Nuclear organization RNA splicing, which takes place in the nucleus, must be done before the RNA can move into the cytoplasm. Incompletely spliced RNAs remain in the nucleus until splicing is complete or until the pre-mRNA is degraded. Immediately after splicing, a group of proteins called the exon-junction complex (EJC) is deposited approximately 20 nucleotides upstream of each exon–exon junction on the mRNA. The EJC promotes the export of the mRNA from the nucleus into the cytoplasm.

For many years, the nucleus was viewed as a biochemical soup, in which components such as the spliceosome diffused and reacted randomly. Now, the nucleus is believed to have a highly ordered internal structure, with transcription and RNA processing taking place at particular locations within it. By attaching fluorescent tags to pre-mRNA and using special imaging techniques, researchers have been able to observe the location of pre-mRNA as it is transcribed and processed. The results of these studies reveal that intron removal and other processing reactions take place at the same sites as those of transcription (**Figure 14.11**), suggesting that these processes may be physically coupled. This suggestion is supported by the observation that part of RNA polymerase II is also required for the splicing and 3′ processing of pre-mRNA.

CONCEPTS

Intron splicing of nuclear genes is a two-step process: (1) the 5′ end of the intron is cleaved and attached to the branch point to form a lariat and (2) the 3′ end of the intron is cleaved and the two ends of the exon are spliced together. These reactions take place within the spliceosome.

Minor splicing A few introns in the pre-mRNAs of multicellular eukaryotes utilize a different process of intron removal known as minor splicing. Introns that undergo

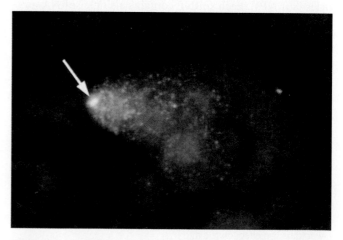

14.11 Intron removal, processing, and transcription take place at the same site. RNA tracks can be seen in the nucleus of a eukaryotic cell. Fluorescent tags were attached to DNA (red) and RNA (green). Transcribed RNA does not disperse; rather, it accumulates near the site of synthesis and follows a defined track during processing. [From R. W. Dirks, K. C. Daniël, and A. K. Raap.]

minor splicing have different consensus sequences at the 5′ splice site and branch point and use a minor spliceosome, which contains a somewhat different set of snRNAs.

Self-splicing introns Some introns are self-splicing, meaning that they possess the ability to remove themselves from an RNA molecule. These self-splicing introns fall into two major categories. Group I introns are found in a variety of genes, including some rRNA genes in protists, some mitochondrial genes in fungi, and even some bacteriophage genes. Although the lengths of group I introns vary, all of them fold into a common secondary structure

with nine looped stems (**Figure 14.12a**), which are necessary for splicing.

Group II introns, present in some mitochondrial genes, also have the ability to self-splice. All group II introns also fold into secondary structures (**Figure 14.12b**). The splicing of group II introns is accomplished by a mechanism that has some similarities to the spliceosomal-mediated splicing of nuclear genes, and splicing generates a lariat structure. Because of these similarities, group II introns and nuclear pre-mRNA introns have been suggested to be evolutionarily related; perhaps the nuclear introns evolved from self-splicing group II introns and later adopted the proteins and snRNAs of the spliceosome to carry out the splicing reaction.

CONCEPTS

A few introns are removed by the minor splicing system. Other introns are self-splicing and consist of two types: group I introns and group II introns. These introns have complex secondary structures that enable them to catalyze their excision from RNA molecules without the aid of enzymes or other proteins.

Alternative Processing Pathways

A finding that complicates the view of a gene as a sequence of nucleotides that specifies the amino acid sequence of a protein (see section on The Concept of the Gene Revisited) is the existence of **alternative processing pathways**. In these pathways, a single pre-mRNA is processed in different ways to produce alternative types of mRNA, resulting in the production of different proteins from the same DNA sequence.

One type of alternative processing is **alternative splicing**, in which the same pre-mRNA can be spliced in more than one way to yield multiple mRNAs that are translated into different amino acid sequences and thus different proteins (**Figure 14.13a**). Another type of alternative processing requires the use of **multiple 3′ cleavage sites** (**Figure 14.13b**), where two or more potential sites for cleavage and polyadenylation are present in the pre-mRNA. In the example in Figure 14.13b, cleavage at the first site produces a relatively short mRNA compared with the mRNA produced through cleavage at the second site. The use of an alternative cleavage site may or may not produce a different protein, depending on whether the position of the site is before or after the termination codon.

Both alternative splicing and multiple 3′ cleavage sites can exist in the same pre-mRNA transcript. An example is seen in the mammalian gene that encodes calcitonin; this gene contains six exons and five introns (**Figure 14.14a**). The entire gene is transcribed into pre-mRNA (**Figure 14.14b**). There are two possible 3′ cleavage sites. In cells of the thyroid gland, 3′ cleavage and polyadenylation take place after the fourth exon to produce a mature mRNA consisting of exons 1, 2, 3, and 4 (**Figure 14.14c**). This mRNA is translated into

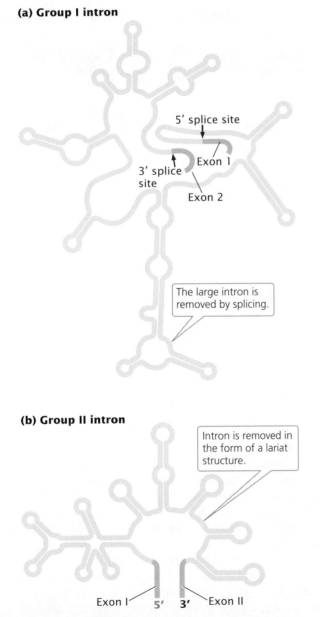

(a) Group I intron

5′ splice site

Exon 1

3′ splice site

Exon 2

The large intron is removed by splicing.

(b) Group II intron

Intron is removed in the form of a lariat structure.

Exon I   5′   3′   Exon II

14.12 Group I and group II introns fold into characteristic secondary structures.

(a) Alternative splicing

(b) Multiple 3′ cleavage sites

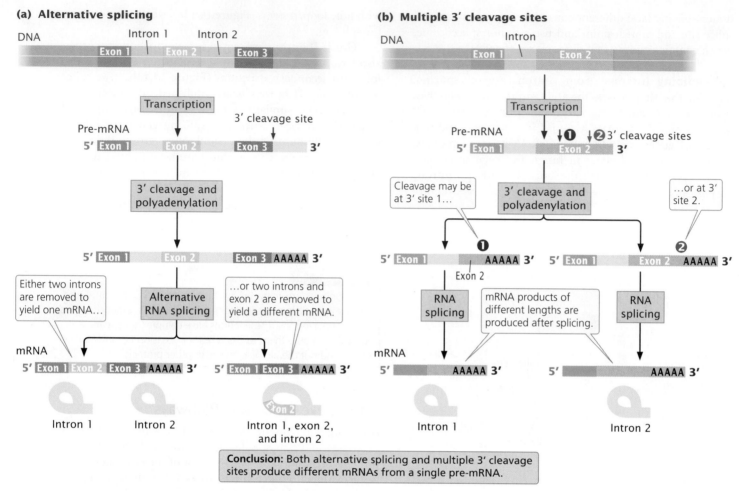

14.13 Eukaryotic cells have alternative pathways for processing pre-mRNA. (a) With alternative splicing, pre-mRNA can be spliced in different ways to produce different mRNAs. (b) With multiple 3′ cleavage sites, there are two or more potential sites for cleavage and polyadenylation; use of the different sites produces mRNAs of different lengths.

the hormone calcitonin. In brain cells, the *identical* pre-mRNA is transcribed from DNA, but cleavage and polyadenylation take place after the sixth exon, yielding an initial transcript that includes all six exons. During splicing, exon 4 is removed, and so only exons 1, 2, 3, 5, and 6 are present in the mature mRNA (**Figure 14.14d**). When translated, this mRNA produces a protein called calcitonin-gene-related peptide (CGRP), which has an amino acid sequence quite different from that of calcitonin. Alternative splicing may produce different combinations of exons in the mRNA, but the order of the exons is not usually changed.

Alternative processing of pre-mRNAs is common in multicellular eukaryotes. For example, researchers estimate that more than 90% of all human genes undergo alternative splicing. Often the form of splicing differs between human tissues; human brain and liver tissues have more alternatively spliced RNA compared with other tissues. Sometimes splicing even varies from one person to another. Different

processing pathways contribute to gene regulation, as will be discussed in Chapter 17.

Alternative splicing may play a role in organism complexity. The complete sequencing of the genomes of numerous organisms (see Chapter 20) has led to the conclusion that the number of genes possessed by an organism is not correlated with the complexity of the organism. For example, fruit flies have only about 14,000 genes, whereas anatomically simpler nematode worms have 19,000 genes. The plant *Arabidopsis thaliana* has about 20,000 genes, almost as many as humans have. If anatomically simple organisms have as many genes as complex organisms have, how is developmental complexity encoded in the genome? A possible answer is alternative processing, which can produce multiple proteins from a single gene. In fact, alternative processing is an important source of protein diversity in vertebrates; an estimated 90% of all human genes are alternatively processed.

(a)

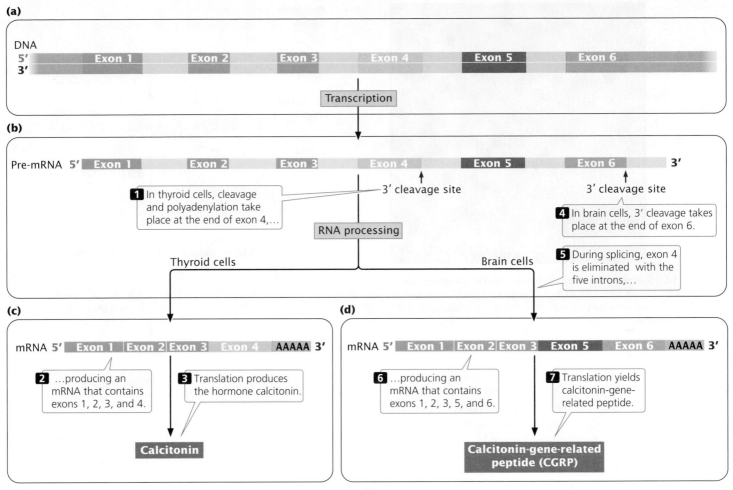

14.14 Pre-mRNA encoded by the gene for calcitonin undergoes alternative processing.

CONCEPTS

Alternative splicing enables exons to be spliced together in different combinations to yield mRNAs that encode different proteins. Alternative 3' cleavage sites allow pre-mRNA to be cleaved at different sites.

✔ CONCEPT CHECK 5

Alternative 3' cleavage sites result in

a. multiple genes of different lengths.

b. multiple pre-mRNAs of different lengths.

c. multiple mRNAs of different lengths.

d. all of the above.

RNA Editing

The assumption that all information about the amino acid sequence of a protein resides in DNA is violated by a process called **RNA editing**. In RNA editing, the coding sequence of an mRNA molecule is altered after transcription, and so the protein has an amino acid sequence that differs from that encoded by the gene.

RNA editing was first detected in 1986 when the coding sequences of mRNAs were compared with the coding sequences of the DNAs from which they had been transcribed. In some nuclear genes in mammalian cells and in some mitochondrial genes in plant cells, there had been substitutions in some of the nucleotides of the mRNA. More-extensive RNA editing has been found in the mRNA for some mitochondrial genes in trypanosome parasites (which cause African sleeping sickness; **Figure 14.15**). In some mRNAs of these organisms, more than 60% of the sequence is determined by RNA editing. Different types of RNA editing have now been observed in mRNAs, tRNAs, and rRNAs from a wide range of organisms; they include the insertion and the deletion of nucleotides and the conversion of one base into another.

If the modified sequence in an edited RNA molecule doesn't come from a DNA template, then how is it specified? A variety of mechanisms can bring about changes in RNA sequences. In some cases, molecules called

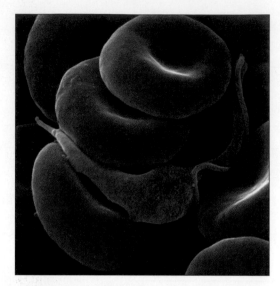

14.15 *Trypanosoma brucei* causes African sleeping sickness. Messenger RNA produced from mitochondrial genes of this parasite (in purple) undergoes extensive RNA editing. [Davic Spears/Last Refuge/Phototake.]

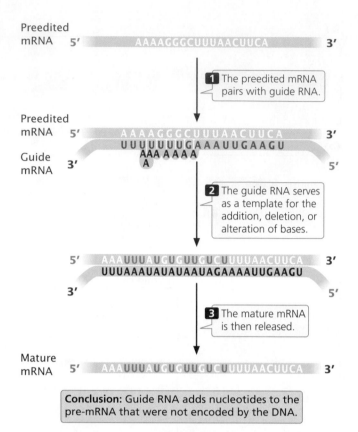

14.16 RNA editing is carried out by guide RNAs. The guide mRNA has sequences that are partly complementary to those of the preedited mRNA and pairs with it. After pairing, the mRNA undergoes cleavage and new nucleotides are added, with sequences in the gRNA serving as a template. The ends of the mRNA are then joined together.

guide RNAs (gRNAs) play a crucial role. A gRNA contains sequences that are partly complementary to segments of the preedited RNA, and the two molecules undergo base pairing in these sequences (**Figure 14.16**). After the mRNA is anchored to the gRNA, the mRNA undergoes cleavage and nucleotides are added, deleted, or altered according to the template provided by gRNA. In other cases, enzymes bring about base conversion. In humans, for example, a gene is transcribed into mRNA that encodes a lipid-transporting polypeptide called apolipoprotein-B100, which has 4563 amino acids and is synthesized in liver cells. A truncated form of the protein called apolipoprotein-B48—with only 2153 amino acids—is synthesized in intestinal cells through editing of the apolipoprotein-B100 mRNA. In this editing, an enzyme deaminates a cytosine base, converting it into uracil. This conversion changes a codon that specifies the amino acid glutamine into a stop codon that prematurely terminates translation, resulting in the shortened protein. TRY PROBLEM 29 →

CONCEPTS

Individual nucleotides in the interior of pre-mRNA may be changed, added, or deleted by RNA editing. The amino acid sequence produced by the edited mRNA is not the same as that encoded by DNA.

✔ CONCEPT CHECK 6

What specifies the modified sequence of nucleotides found in an edited RNA molecule?

CONNECTING CONCEPTS

Eukaryotic Gene Structure and Pre-mRNA Processing

Chapters 13 and 14 introduced a number of different components of genes and RNA molecules, including promoters, 5′ untranslated regions, coding sequences, introns, 3′ untranslated regions, poly(A) tails, and caps. Let's see how some of these components are combined to create a typical eukaryotic gene and how a mature mRNA is produced from them.

The promoter, which typically lies upstream of the transcription start site, is necessary for transcription to take place but is itself not usually transcribed when protein-encoding genes are transcribed by RNA polymerase II (**Figure 14.17a**). Farther upstream or downstream of the start site, there may be enhancers that also regulate transcription.

In transcription, all the nucleotides between the transcription start site and the termination site are transcribed into pre-mRNA, including exons, introns, and a long 3′ end that is later cleaved from the transcript (**Figure 14.17b**). Notice that the 5′ end of the first exon contains the sequence that encodes the 5′ untranslated region and that the 3′ end of the last exon contains the sequence that encodes the 3′ untranslated region.

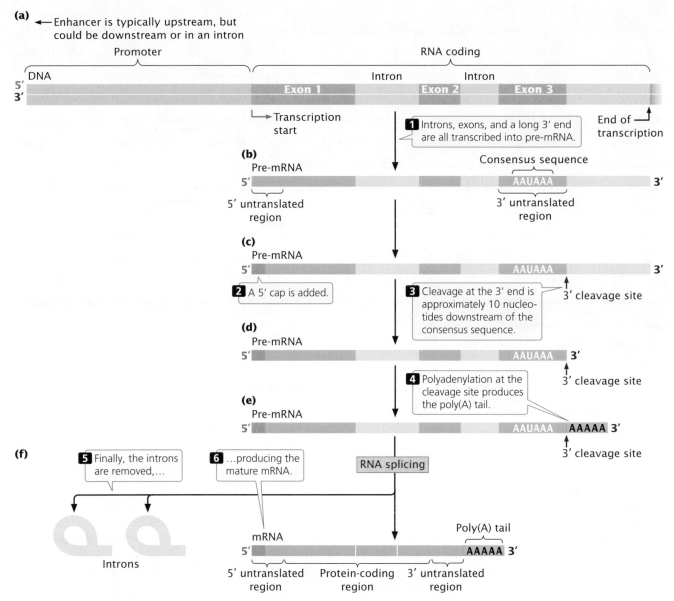

(a) ← Enhancer is typically upstream, but could be downstream or in an intron

14.17 Mature eukaryotic mRNA is produced when pre-mRNA is transcribed and undergoes several types of processing.

▶ www.whfreeman.com/pierce4e   Discover the consequences of failed RNA processing by viewing and interacting with Animation 14.1.

The pre-mRNA is then processed to yield a mature mRNA. The first step in this processing is the addition of a cap to the 5′ end of the pre-mRNA (**Figure 14.17c**). Next, the 3′ end is cleaved at a site downstream of the AAUAAA consensus sequence in the last exon (**Figure 14.17d**). Immediately after cleavage, a poly(A) tail is added to the 3′ end (**Figure 14.17e**). Finally, the introns are removed to yield the mature mRNA (**Figure 14.17f**). The mRNA now contains 5′ and 3′ untranslated regions, which are not translated into amino acids, and the nucleotides that carry the protein-coding sequences. The nucleotide sequence of a small gene (the human interleukin 2 gene), with these components identified, is presented in **Figure 14.18**.

14.3 Transfer RNAs, Which Attach to Amino Acids, Are Modified after Transcription in Bacterial and Eukaryotic Cells

In 1956, Francis Crick proposed the idea of a molecule that transports amino acids to the ribosome and interacts with codons in mRNA, placing amino acids in their proper order in protein synthesis. By 1963, the existence of such an adapter molecule, called transfer RNA, had been confirmed. Transfer RNA serves as a link between the genetic code in

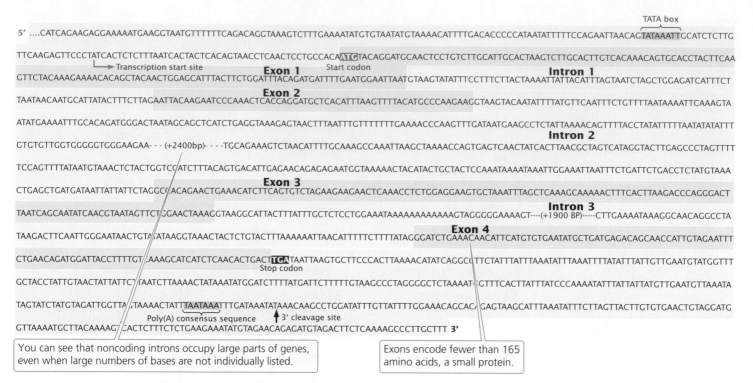

You can see that noncoding introns occupy large parts of genes, even when large numbers of bases are not individually listed.

Exons encode fewer than 165 amino acids, a small protein.

14.18 This representation of the nucleotide sequence of the gene for human interleukin 2 includes the TATA box, transcription start site, start and stop codons, introns, exons, poly(A) consensus sequence, and 3′ cleavage site.

mRNA and the amino acids that make up a protein. Each tRNA attaches to a particular amino acid and carries it to the ribosome, where the tRNA adds its amino acid to the growing polypeptide chain at the position specified by the genetic instructions in the mRNA. We'll take a closer look at the mechanism of this process in Chapter 15.

Each tRNA is capable of attaching to only one type of amino acid. The complex of tRNA plus its amino acid can be written in abbreviated form by adding a three-letter superscript representing the amino acid to the term tRNA. For example, a tRNA that attaches to the amino acid alanine is written as tRNA[Ala]. Because 20 different amino acids are found in proteins, there must be a minimum of 20 different types of tRNA. In fact, most organisms possess from at least 30 to 40 different types of tRNA, each encoded by a different gene (or, in some cases, multiple copies of a gene) in DNA.

The Structure of Transfer RNA

A unique feature of tRNA is the occurrence of rare **modified bases**. All RNAs have the four standard bases (adenine, cytosine, guanine, and uracil) specified by DNA, but tRNAs have additional bases, including ribothymine, pseudouridine (which is also occasionally present in snRNAs and rRNA), and dozens of others. The structures of two of these modified bases are shown in **Figure 14.19**.

If there are only four bases in DNA and if all RNA molecules are transcribed from DNA, how do tRNAs acquire these additional bases? Modified bases arise from chemical changes made to the four standard bases after transcription. These changes are carried out by special **tRNA-modifying enzymes**. For example, the addition of a methyl group to uracil creates the modified base ribothymidine.

14.19 Two of the modified bases found in tRNAs. All the modified bases in tRNAs are produced by the chemical alteration of the four standard RNA bases.

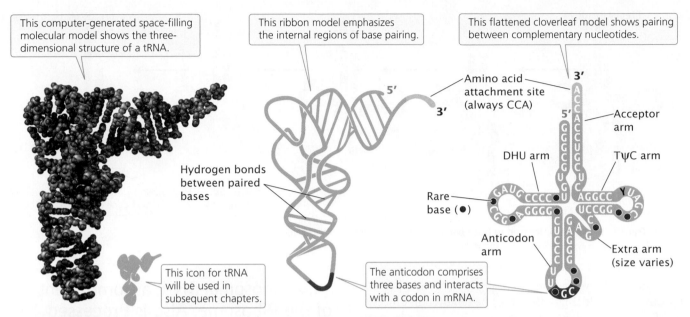

This computer-generated space-filling molecular model shows the three-dimensional structure of a tRNA.

This ribbon model emphasizes the internal regions of base pairing.

This flattened cloverleaf model shows pairing between complementary nucleotides.

Hydrogen bonds between paired bases

This icon for tRNA will be used in subsequent chapters.

Amino acid attachment site (always CCA)

Acceptor arm

DHU arm

TΨC arm

Rare base (●)

Anticodon arm

Extra arm (size varies)

The anticodon comprises three bases and interacts with a codon in mRNA.

14.20 All tRNAs possess a common secondary structure, the cloverleaf structure. The base sequence in the flattened model is for tRNA^{Ala}.

The structures of all tRNAs are similar, a feature critical to tRNA function. Most tRNAs contain between 74 and 95 nucleotides, some of which are complementary to each other and form intramolecular hydrogen bonds. As a result, each tRNA has a **cloverleaf structure** (**Figure 14.20**). The cloverleaf has four major arms. If we start at the top and proceed clockwise around the tRNA shown at the right in Figure 14.20, the four major arms are the acceptor arm, the TΨC arm, the anticodon arm, and the DHU arm. Three of the arms (the TΨC, anticodon, and DHU arms) consist of a stem and a loop. The stem is formed by the pairing of complementary nucleotides, and the loop lies at the terminus of the stem, where there is no nucleotide pairing.

Instead of having a loop, the acceptor arm includes the 5′ and 3′ ends of the tRNA molecule. All tRNAs have the same sequence (CCA) at the 3′ end, where the amino acid attaches to the tRNA; so, clearly, this sequence is not responsible for specifying which amino acid will attach to the tRNA.

The TΨC arm is named for the bases of three nucleotides in the loop of this arm: thymine (T), pseudouridine (Ψ), and cytosine (C). The anticodon arm lies at the bottom of the tRNA. Three nucleotides at the end of this arm make up the **anticodon**, which pairs with the corresponding codon on mRNA to ensure that the amino acids link in the correct order. The DHU arm is so named because it often contains the modified base dihydrouridine.

Although each tRNA molecule folds into a cloverleaf owing to the complementary pairing of bases, the cloverleaf is not the three-dimensional (tertiary) structure of tRNAs found in the cell. The results of X-ray crystallographic studies have shown that the cloverleaf folds on itself to form an L-shaped structure, as illustrated by the space-filling and ribbon models in Figure 14.20. Notice that the acceptor stem is at one end of the tertiary structure and the anticodon is at the other end.

Transfer RNA Gene Structure and Processing

The genes that produce tRNAs may be in clusters or scattered about the genome. In *E. coli*, the genes for some tRNAs are present in a single copy, whereas the genes for other tRNAs are present in several copies; eukaryotic cells usually have many copies of each tRNA gene. All tRNA molecules in both bacterial and eukaryotic cells undergo processing after transcription.

In *E. coli*, several tRNAs are usually transcribed together as one large precursor tRNA, which is then cut up into pieces, each containing a single tRNA. Additional nucleotides may then be removed one at a time from the 5′ and 3′ ends of the tRNA in a process known as trimming. Base-modifying enzymes may then change some of the standard bases into modified bases (**Figure 14.21**). In some prokaryotes, the CCA sequences found at the 3′ ends of tRNAs are encoded in the tRNA gene and are transcribed into the tRNA; in other prokaryotes and in eukaryotes, these sequence are added by a special enzyme that adds the nucleotides without the use of any template.

There is no generic processing pathway for all tRNAs: different tRNAs are processed in different ways. Eukaryotic tRNAs are processed in a manner similar to that for bacterial tRNAs: most are transcribed as larger precursors that are then cleaved, trimmed, and modified to produce mature tRNAs.

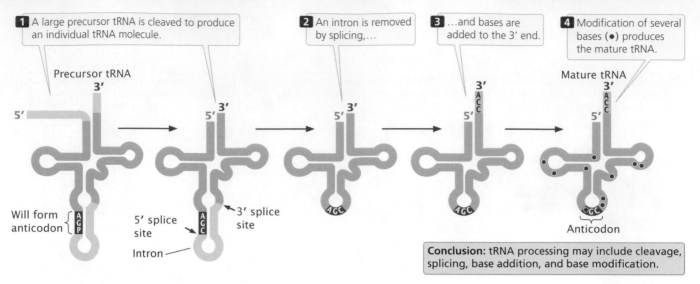

1 A large precursor tRNA is cleaved to produce an individual tRNA molecule.

2 An intron is removed by splicing,…

3 …and bases are added to the 3′ end.

4 Modification of several bases (•) produces the mature tRNA.

Precursor tRNA

Mature tRNA

Will form anticodon

5′ splice site

3′ splice site

Intron

Anticodon

Conclusion: tRNA processing may include cleavage, splicing, base addition, and base modification.

14.21 Transfer RNAs are processed in both bacterial and eukaryotic cells. Different tRNAs are modified in different ways. One example is shown here.

Some eukaryotic and archeal tRNA genes possess introns of variable length that must be removed in processing. For example, about 40 of the 400 tRNA genes in yeast contain a single intron that is always found adjacent to the 3′ side of the anticodon. The tRNA introns are shorter than those found in pre-mRNA and do not have the consensus sequences found at the intron–exon junctions of pre-mRNAs. The splicing process for tRNA genes is quite different from the spliceosome-mediated reactions that remove introns from protein-encoding genes.

CONCEPTS

All tRNAs are similar in size and have a common secondary structure known as the cloverleaf. Transfer RNAs contain modified bases and are extensively processed after transcription in both bacterial and eukaryotic cells.

✔ **CONCEPT CHECK 7**

How are rare bases incorporated into tRNAs?

a. Encoded by guide RNAs

b. By chemical changes to one of the standard bases

c. Encoded by rare bases in DNA

d. Encoded by sequences in introns

14.4 Ribosomal RNA, a Component of the Ribosome, Also Is Processed after Transcription

Within ribosomes, the genetic instructions contained in mRNA are translated into the amino acid sequences of polypeptides. Thus, ribosomes play an integral part in the transfer of genetic information from genotype to phenotype. We will examine the role of ribosomes in the process of translation in Chapter 15. Here, we consider ribosome structure and examine how ribosomes are processed before becoming functional.

The Structure of the Ribosome

The ribosome is one of the most abundant molecular complexes in the cell: a single bacterial cell may contain as many as 20,000 ribosomes, and eukaryotic cells possess even more. Ribosomes typically contain about 80% of the total cellular RNA. They are complex structures, each consisting of more than 50 different proteins and RNA molecules (**Table 14.3**). A functional ribosome consists of two subunits, a **large ribosomal subunit** and a **small ribosomal subunit**, each of which consists of one or more pieces of RNA and a number of proteins. The sizes of the

Table 14.3	Composition of ribosomes in bacterial and eukaryotic cells			
Cell Type	Ribosome Size	Subunit	rRNA Component	Proteins
Bacterial	70S	Large (50S)	23S (2900 nucleotides), 5S (120 nucleotides)	31
		Small (30S)	16S (1500 nucleotides)	21
Eukaryotic	80S	Large (60S)	28S (4700 nucleotides), 5.8S (160 nucleotides), 5S (120 nucleotides)	49
		Small (40S)	18S (1900 nucleotides)	33

Note: The letter "S" stands for Svedberg unit.

ribosomes and their RNA components are given in Svedberg (S) units (a measure of how rapidly an object sediments in a centrifugal field). It is important to note that Svedberg units are not additive; in other words, combining a 10S structure and a 20S structure does not necessarily produce a 30S structure, because the sedimentation rate is affected by the three-dimensional structure as well as the mass. The three-dimensional structure of the bacterial ribosome has been elucidated in great detail through the use of X-ray cystallography. More will be said about the ribosome's structure in Chapter 15.

Ribosomal RNA Gene Structure and Processing

The genes for rRNA, like those for tRNA, can be present in multiple copies, and the numbers vary among species (**Table 14.4**); all copies of the rRNA gene in a species are identical or nearly identical. In bacteria, rRNA genes are dispersed; but, in eukaryotic cells, they are clustered, with the genes arrayed in tandem, one after another.

Eukaryotic cells possess two types of rRNA genes: a large gene that encodes 18S rRNA, 28S rRNA, and 5.8S rRNA, and a small gene that encodes the 5S rRNA. All three bacterial rRNAs (23S rRNA, 16S rRNA, and 5S rRNA) are encoded by a single type of gene.

Ribosomal RNA is processed in both bacterial and eukaryotic cells. In *E. coli*, the immediate product of tran-

Table 14.4	Number of rRNA genes in different organisms
Species	**Copies of rRNA Genes per Genome**
Escherichia coli	1
Yeast	100–200
Human	280
Frog	450

scription is a 30S rRNA precursor (**Figure 14.22a**). This 30S precursor is methylated in several places, and then cleaved and trimmed to produce 16S rRNA, 23S rRNA, and 5S rRNA, along with one or more tRNAs.

Eukaryotic rRNAs undergo similar processing (**Figure 14.22b**). Small nucleolar RNAs (snoRNAs) help to cleave and modify eukaryotic rRNAs and assemble the processed rRNAs into mature ribosomes. Like the snRNAs taking part in pre-mRNA splicing, snoRNAs associate with proteins to form ribonucleoprotein particles (snoRNPs). The snoRNAs have extensive complementarity to the rRNA sequences in which modification takes place. Interestingly, some snoRNAs are encoded by sequences in the introns of other protein-encoding genes. The processing of rRNA and ribosome assembly in eukaryotes take place in the nucleolus.

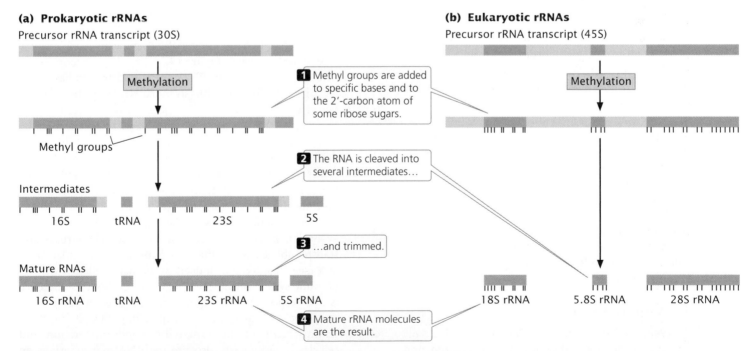

(a) Prokaryotic rRNAs
Precursor rRNA transcript (30S)

Methylation

1 Methyl groups are added to specific bases and to the 2'-carbon atom of some ribose sugars.

Methyl groups

2 The RNA is cleaved into several intermediates…

Intermediates

16S   tRNA   23S   5S

3 …and trimmed.

Mature RNAs

16S rRNA   tRNA   23S rRNA   5S rRNA

4 Mature rRNA molecules are the result.

(b) Eukaryotic rRNAs
Precursor rRNA transcript (45S)

Methylation

18S rRNA   5.8S rRNA   28S rRNA

14.22 Ribosomal RNA is processed after transcription. Prokaryotic rRNA (a) and eukaryotic rRNA (b) are produced from precursor RNA transcripts that are methylated, cleaved, and processed to produce mature rRNAs. Eukaryotic 5S rRNA is transcribed separately from a different gene.

14.5 Small RNA Molecules Participate in a Variety of Functions

Much evidence suggests that the first genetic material was RNA and early life was dominated by RNA molecules (see p. 352 in Chapter 13). This time period, when RNA dominated life's essential processes, has been termed the "early RNA world." The common perception is that this RNA world died out billions of years ago, when many of RNA's functions were replaced by more-stable DNA molecules and more-efficient protein catalysts. However, within the past 10 years, a group of small RNA molecules (most of them 20–30 nucleotides long) have been discovered that greatly influence many basic biological processes, including the formation of chromatin structure, transcription, and translation. Found in all plants and animals, these small RNA molecules play important roles in gene expression, development, and cancer. They are also being harnessed by researchers to study gene function and treat genetic diseases. The discovery of small RNA molecules has greatly influenced our understanding of how genes are regulated and the importance of DNA sequences that do not encode proteins. These new findings demonstrate that we still live very much in an RNA world.

In this last section of the chapter, we will examine RNA interference (which led to the discovery of small RNAs), different types of small RNAs, and how miRNAs are processed. In Chapter 17, we will look further at the role of small RNA molecules in controlling gene expression; in Chapter 19, we will see how small RNAs are being used as important tools in biotechnology.

RNA Interference

In 1998, Andrew Fire, Craig Mello, and their colleagues observed what appeared to be a strange phenomenon. They were inhibiting the expression of genes in the nematode *Caenorhabditis elegans* by putting into the animals single-stranded RNA molecules that were complementary to a gene's DNA sequence. Called antisense RNA, such molecules are known to inhibit gene expression by binding to the mRNA sequences and inhibiting translation. Fire, Mello, and colleagues found that even more potent gene silencing was triggered when double-stranded RNA was injected into the animals. This finding was puzzling, because no mechanism by which double-stranded RNA could inhibit translation was known. Several other, previously described types of gene silencing also were found to be triggered by double-stranded RNA. These initial studies led to the discovery of small RNA molecules that are important in gene silencing.

Subsequent research revealed an astonishing array of small RNA molecules with important cellular functions in eukaryotes, which now include at least three major classes: small interfering RNAs (siRNAs), microRNAs (miRNAs), and Piwi-interacting RNAs (piRNAs), depending on their origin and mode of function. These small RNAs are found in many eukaryotes and are responsible for a variety of different functions, including the regulation of gene expression, defense against viruses, suppression of transposons, and modification of chromatin structure. Additional types of small RNAs have been discovered in eukaryotes with their own unique functions. An analogous group of small RNAs with silencing functions—called prokaryotic silencing RNAs (psiRNAs)—have recently been detected in prokaryotes. For their discovery of RNA interference, Andrew Fire and Craig Mello were awarded the Nobel Prize in physiology or medicine in 2006.

RNA interference (RNAi) is a powerful and precise mechanism used by eukaryotic cells to limit the invasion of foreign genes (from viruses and transposons) and to censor the expression of their own genes. RNA interference is triggered by double-stranded RNA molecules, which may arise in several ways (**Figure 14.23**): by the transcription of inverted repeats into an RNA molecule that then base pairs with itself to form double-stranded RNA; by the simultaneous transcription of two different RNA molecules that are complementary to one another and that pair, forming double-stranded RNA; or by infection by viruses that make double-stranded RNA. These double-stranded RNA molecules are chopped up by an enzyme appropriately called Dicer, resulting in tiny RNA molecules that are unwound to produce siRNAs and miRNAs (see Figure 14.23).

Some geneticists speculate that RNA interference evolved as a defense mechanism against RNA viruses and transposable elements that move through RNA intermediates (see Chapter 11); indeed, some have called RNAi the immune system of the genome. However, RNA interference is also responsible for regulating a number of key genetic and developmental processes, including changes in chromatin structure, translation, cell fate and proliferation, and cell death. Geneticists also use the RNAi machinery as an effective tool for blocking the expression of specific genes (see Chapter 19).

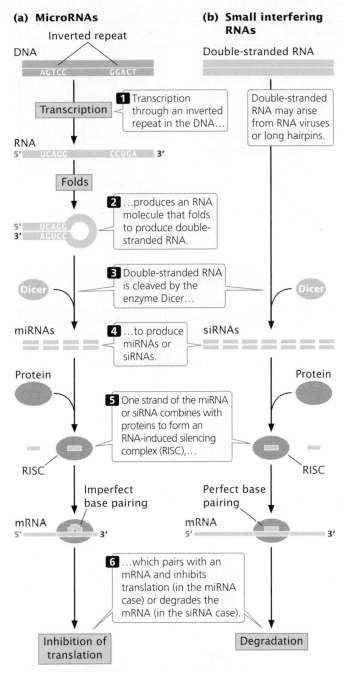

(a) MicroRNAs

Inverted repeat

DNA

AGTCC   GGACT

1 Transcription through an inverted repeat in the DNA...

Transcription

RNA

5' UCAGG   CCUGA 3'

Folds

5' UCAGG
3' AGUCC

2 ...produces an RNA molecule that folds to produce double-stranded RNA.

Dicer

3 Double-stranded RNA is cleaved by the enzyme Dicer...

miRNAs

4 ...to produce miRNAs or siRNAs.

Protein

5 One strand of the miRNA or siRNA combines with proteins to form an RNA-induced silencing complex (RISC),...

RISC

Imperfect base pairing

mRNA
5'   3'

6 ...which pairs with an mRNA and inhibits translation (in the miRNA case) or degrades the mRNA (in the siRNA case).

Inhibition of translation

(b) Small interfering RNAs

Double-stranded RNA

Double-stranded RNA may arise from RNA viruses or long hairpins.

Dicer

siRNAs

Protein

RISC

Perfect base pairing

mRNA
5'   3'

Degradation

14.23 Small interfering RNAs and microRNAs are produced from double-stranded RNAs.

▶ www.whfreeman.com/pierce4e

To see how small interfering RNAs and microRNAs affect gene expression, see Animation 14.3.

Types of Small RNAs

Two abundant classes of RNA molecules that function in RNA interference in eukaryotes are small interfering RNAs and microRNAs. Although these two types of RNA differ in how they originate (**Table 14.5**; see Figure 14.23), they

have a number of features in common and their functions overlap considerably. Both are about 22 nucleotides long. Small interfering RNAs arise from the cleavage of mRNAs, RNA transposons, and RNA viruses. Some miRNAs are cleaved from RNA molecules transcribed from sequences that encode miRNA only, but others are encoded in the introns and exons of mRNAs. Each miRNA is cleaved from a single-stranded RNA precursor that forms small hairpins, whereas multiple siRNAs are produced from the cleavage of an RNA duplex consisting of two different RNA molecules.

Usually, siRNAs have exact complementarity with their target mRNA or DNA sequences and suppress gene expression by degrading mRNA or inhibiting transcription, whereas miRNAs often have limited complementarity with their target mRNAs and often suppress gene expression by inhibiting translation. Finally, miRNAs usually silence genes that are distinct from those from which the miRNAs were transcribed, whereas siRNAs typically silence the genes from which the siRNAs were transcribed. Note, however, that these differences between siRNAs and miRNAs are not hard and fast, and scientists are increasingly finding small RNAs that exhibit characteristics of both. For example, some miRNAs have exact complementarity with mRNA sequences and cleave these sequences, characteristics that are usually associated with siRNAs.

Both siRNA and miRNA molecules combine with proteins to form an **RNA-induced silencing complex** (RISC; see Figure 14.23). Key to the functioning of the RISCs is a protein called Argonaute. The RISC pairs with an mRNA molecule that possesses a sequence complementary to its siRNA or miRNA component and either cleaves the mRNA, leading to degradation of the mRNA, or represses translation of the mRNA. Some siRNAs also serve as guides for the methylation of complementary sequences in DNA and others alter chromatin structure, both of which affect transcription.

Less is known about Piwi-interacting RNAs (piRNAs). They are somewhat longer than siRNAs and miRNAs and are derived from long, single-stranded RNA transcripts, in contrast with siRNAs and miRNAs, which are processed from double-stranded RNA. Piwi-interacting RNAs combine with Piwi proteins, which are related to Argonaute, and inhibit the expression of transposons, primarily in the germ cells of animals. Although the mechanism of transposon silencing by piRNAs is not well understood, it appears to entail degradation of mRNA transcribed from transposons and changes in chromatin structure that inhibit the transcription of transposons.

Processing and Function of MicroRNAs

MicroRNAs have been found in all eukaryotic organisms examined to date, as well as viruses; they control the expression of genes taking part in many biological

Table 14.5 Differences between siRNAs, miRNAs, and piRNAs

Feature	siRNA	miRNA	piRNA
Origin	mRNA, transposon, or virus	RNA transcribed from distinct gene	Transposons
Cleavage of	RNA duplex or single-stranded RNA that forms long hairpins	Single-stranded RNA that forms short hairpins of double-stranded RNA	Single-stranded RNA from transposons
Size	21–25 nucleotides	21–25 nucleotides	24–31 nucleotides
Action	Degradation of mRNA, inhibition of transcription, chromatin modification	Degradation of mRNA, inhibition of translation, chromatin modification	Degradation of RNA, chromatin modification
Target	Genes from which they were transcribed	Genes other than those from which they were transcribed	Transposons

processes, including growth, development, and metabolism. Humans have more than 450 distinct miRNAs; scientists estimate that more than one-third of all human genes are regulated by miRNAs. Most miRNA genes are found in regions of noncoding DNA or within the introns of other genes.

The genes that encode miRNAs are transcribed into longer precursors, called primary miRNA (pri-miRNA), that range from several hundred to several thousand nucleotides in length (**Figure 14.24**). The pri-miRNA is then cleaved into one or more smaller RNA molecules with a hairpin. Dicer binds to this hairpin structure and removes the terminal loop. One of the miRNA strands is incorporated into the RISC; the other strand is released and degraded.

The RISC attaches to a complementary sequence on the mRNA, usually in the 3′ untranslated region of the mRNA. The region of close complementarity, called the seed region, is quite short, usually only about seven nucleotides long. Because the seed sequence is so short, each miRNA can potentially pair with sequences on hundreds of different mRNAs. Furthermore, a single mRNA molecule may possess multiple miRNA-binding sites. The inhibition of translation may require binding by several RISC complexes to the same mRNA molecule.

CONCEPTS

Small interfering RNAs and microRNAs are tiny RNAs produced when larger double-stranded RNA molecules are cleaved by the enzyme Dicer. Small interfering RNAs and microRNAs participate in a variety of processes, including mRNA degradation, the inhibition of translation, the methylation of DNA, and chromatin remodeling. Piwi-interacting RNAs are found in the germ cells of animals and inhibit transposons.

✔ CONCEPT CHECK 9

How do siRNAs and miRNAs target specific mRNAs for degradation or for the repression of translation?

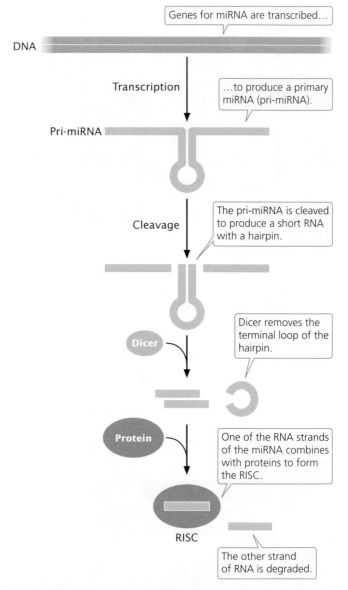

Genes for miRNA are transcribed…

DNA

Transcription

…to produce a primary miRNA (pri-miRNA).

Pri-miRNA

Cleavage

The pri-miRNA is cleaved to produce a short RNA with a hairpin.

Dicer

Dicer removes the terminal loop of the hairpin.

Protein

One of the RNA strands of the miRNA combines with proteins to form the RISC.

RISC

The other strand of RNA is degraded.

14.24 MicroRNAs are cleaved from larger precursors (pri-miRNAs).

 www.whfreeman.com/pierce4e

Animation 14.2 illustrates the process by which microRNAs are produced.

CONCEPTS SUMMARY

- A gene is often defined as a sequence of DNA nucleotides that is transcribed into a single RNA molecule.

- Introns—noncoding sequences that interrupt the coding sequences (exons) of genes—are common in eukaryotic cells but rare in bacterial cells.

- An mRNA molecule has three primary parts: a 5' untranslated region, a protein-coding sequence, and a 3' untranslated region.

- Bacterial mRNA is translated immediately after transcription and undergoes little processing. The pre-mRNA of a eukaryotic protein-encoding gene is extensively processed: a modified nucleotide and methyl groups, collectively termed the cap, are added to the 5' end of pre-mRNA; the 3' end is cleaved and a poly(A) tail is added; and introns are removed. Introns are removed within a structure called the spliceosome, which is composed of several small nuclear RNAs and proteins.

- Some introns found in rRNA genes and mitochondrial genes are self-splicing.

- Some pre-mRNAs undergo alternative splicing, in which different combinations of exons are spliced together or different 3' cleavage sites are used.

- Messenger RNAs may be altered by the addition, deletion, or modification of nucleotides in the coding sequence, a process called RNA editing.

- Transfer RNAs, which attach to amino acids, are short molecules that assume a common secondary structure and contain modified bases.

- Ribosomes, the sites of protein synthesis, are composed of several ribosomal RNA molecules and numerous proteins.

- Small interfering RNAs, microRNAs, and Piwi-interacting RNAs play important roles in gene silencing and in a number of other biological processes.

IMPORTANT TERMS

colinearity (p. 376)
exon (p. 377)
intron (p. 377)
group I intron (p. 378)
group II intron (p. 378)
nuclear pre-mRNA intron (p. 378)
transfer RNA intron (p. 378)
codon (p. 380)
5' untranslated region (5' UTR) (p. 380)
Shine–Dalgarno sequence (p. 380)
protein-coding region (p. 380)
3' untranslated region (3' UTR) (p. 380)

5' cap (p. 381)
poly(A) tail (p. 381)
RNA splicing (p. 382)
5' splice site (p. 382)
3' splice site (p. 382)
branch point (p. 382)
spliceosome (p. 383)
lariat (p. 383)
trans-splicing (p. 384)
alternative processing pathway (p. 385)
alternative splicing (p. 385)
multiple 3' cleavage sites (p. 385)

RNA editing (p. 387)
guide RNA (p. 388)
modified base (p. 390)
tRNA-modifying enzyme (p. 390)
cloverleaf structure (p. 391)
anticodon (p. 391)
large ribosomal subunit (p. 392)
small ribosomal subunit (p. 392)
RNA interference (RNAi) (p. 394)
RNA-induced silencing complex (RISC) (p. 395)

ANSWERS TO CONCEPT CHECKS

1. When DNA was hybridized to the mRNA transcribed from it, regions of DNA that did not correspond to RNA looped out.

2. Group I introns, group II introns, nuclear pre-mRNA introns, and transfer RNA introns

3. A protein that adds the 5' cap is associated with RNA polymerase II, which transcribes pre-mRNAs but is absent from RNA polymerase I and III, which transcribe rRNA and tRNAs.

4. b

5. c

6. Guide RNA

7. b

8. d

9. An siRNA or miRNA combines with proteins to form RISC, which then pairs with mRNA through complementary pairing between bases on the siRNA or miRNA and bases on the mRNA.

WORKED PROBLEMS

1. DNA from a eukaryotic gene was isolated, denatured, and hybridized to the mRNA transcribed from the gene; the hybridized structure was then observed with the use of an electron microscope. The following structure was observed.

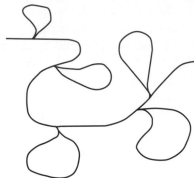

a. How many introns and exons are there in this gene? Explain your answer.

b. Identify the exons and introns in this hybridized structure.

• Solution

a. Each of the loops represents a region in which sequences in the DNA do not have corresponding sequences in the RNA; these regions are introns. There are five loops in the hybridized structure; so there must be five introns in the DNA and six exons.

b.

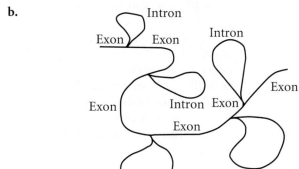

2. Draw a typical bacterial mRNA and the gene from which it was transcribed. Identify the 5′ and 3′ ends of the RNA and DNA molecules, as well as the following regions or sequences:

a. Promoter
b. 5′ untranslated region
c. 3′ untranslated region
d. Protein-coding sequence
e. Transcription start site
f. Terminator
g. Shine–Dalgarno sequence
h. Start and stop codons

• Solution

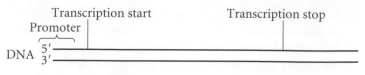

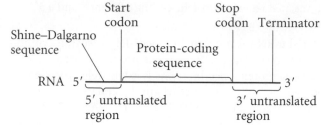

3. A test-tube splicing system contains all the components (snRNAs, proteins, splicing factors) necessary for the splicing of nuclear genes. When a piece of RNA containing an intron and two exons is added to the system, the intron is removed as a lariat and the exons are spliced together. If the RNA molecule added to the system has the following mutations, what intermediate products of the splicing reactions will accumulate? Explain your answer.

a. GU at the 5′ splice site is deleted.

b. A at the branch point is deleted.

c. AG at the 3′ splice site is deleted.

• Solution

a. The GU sequence at the 5′ splice site is required for the attachment of the spliceosome and the first cleavage reaction. If this sequence is mutated, cleavage will not take place. Thus, the original pre-mRNA with the intron will accumulate.

b. After cleavage at the 5′ splice site, the 5′ end of the intron attaches to the A at the branch point in a transesterification reaction. If the A at the branch point is deleted, no lariat structure will form. The separated first exon and the intron attached to the second exon will accumulate as intermediate products.

c. The AG sequence at the 3′ splice site is required for cleavage at the 3′ splice site. If this sequence is mutated, accumulated intermediate products will be: (1) the separated first exon and (2) the intron attached to the second exon, with the 5′ end of the intron attached to the branch point to form a lariat structure.

COMPREHENSION QUESTIONS

Section 14.1

*1. What is the concept of colinearity? In what way is this concept fulfilled in bacterial and eukaryotic cells?

2. What are some characteristics of introns?

*3. What are the four basic types of introns? In which genes are they found?

Section 14.2

*4. What are the three principal elements in mRNA sequences in bacterial cells?

5. What is the function of the Shine–Dalgarno consensus sequence?

*6. (a) What is the 5′ cap? (b) How is the 5′ cap added to eukaryotic pre-mRNA? (c) What is the function of the 5′ cap?

7. How is the poly(A) tail added to pre-mRNA? What is the purpose of the poly(A) tail?

*8. What makes up the spliceosome? What is the function of the spliceosome?

9. Explain the process of pre-mRNA splicing in nuclear genes.

10. Describe two types of alternative processing pathways. How do they lead to the production of multiple proteins from a single gene?

*11. What is RNA editing? Explain the role of guide RNAs in RNA editing.

*12. Summarize the different types of processing that can take place in pre-mRNA.

Section 14.3

*13. What are some of the modifications in tRNA that take place through processing?

Section 14.4

14. Describe the basic structure of ribosomes in bacterial and eukaryotic cells.

*15. Explain how rRNA is processed.

Section 14.5

16. What is the origin of small interfering RNAs, microRNAs, and Piwi-interacting RNAs? What do these RNA molecules do in the cell?

17. What are some similarities and differences between siRNAs and miRNAs?

18. How are miRNAs processed?

APPLICATION QUESTIONS AND PROBLEMS

Section 14.1

*19. Duchenne muscular dystrophy is caused by a mutation in a gene that comprises 2.5 million nucleotides and specifies a protein called dystrophin. However, less than 1% of the gene actually encodes the amino acids in the dystrophin protein. On the basis of what you now know about gene structure and RNA processing in eukaryotic cells, provide a possible explanation for the large size of the dystrophin gene.

Section 14.2

20. How do the mRNAs of bacterial cells and the pre-mRNAs of eukaryotic cells differ? How do the mature mRNAs of bacterial and eukaryotic cells differ?

*21. Draw a typical eukaryotic gene and the pre-mRNA and mRNA derived from it. Assume that the gene contains three exons. Identify the following items and, for each item, give a brief description of its function:

- **a.** 5′ untranslated region
- **b.** Promoter
- **c.** AAUAAA consensus sequence
- **d.** Transcription start site
- **e.** 3′ untranslated region
- **f.** Introns
- **g.** Exons
- **h.** Poly(A) tail
- **i.** 5′ cap

22. How would the deletion of the Shine–Dalgarno sequence affect a bacterial mRNA?

*23. How would the deletion of the following sequences or features most likely affect a eukaryotic pre-mRNA?

- **a.** AAUAAA consensus sequence
- **b.** 5′ cap
- **c.** Poly(A) tail

24. Suppose that a mutation occurs in an intron of a gene encoding a protein. What will the most likely effect of the mutation be on the amino acid sequence of that protein? Explain your answer.

25. A geneticist induces a mutation in a line of cells growing in the laboratory. The mutation occurs in one of the genes that encodes proteins that participate in the cleavage and polyadenylation of eukaryotic mRNA. What will the immediate effect of this mutation be on RNA molecules in the cultured cells?

*26. A geneticist mutates the gene for proteins that bind to the poly(A) tail in a line of cells growing in the laboratory. What will the immediate effect of this mutation be in the cultured cells?

27. A geneticist isolates a gene that contains five exons. He then isolates the mature mRNA produced by this gene. After making the DNA single stranded, he mixes the single-stranded DNA and RNA. Some of the single-stranded DNA hybridizes (pairs) with the complementary mRNA. Draw a picture of what the DNA–RNA hybrids will look like under the electron microscope.

28. A geneticist discovers that two different proteins are encoded by the same gene. One protein has 56 amino acids, and the other has 82 amino acids. Provide a possible explanation for how the same gene can encode both of these proteins.

29. Explain how each of the following processes complicates the concept of colinearity.

   a. Trans-splicing

   b. Alternative splicing

   c. RNA editing

Section 14.5

30. In the early 1990s, Carolyn Napoli and her colleagues were working on petunias, attempting to genetically engineer a variety with dark purple petals by introducing numerous copies of a gene that codes for purple petals (C. Napoli, C. Lemieux, and R. Jorgensen. 1990. *Plant Cell* 2:279–289). Their thinking was that extra copies of the gene would cause more purple pigment to be produced and would result in a petunia with an even darker hue of purple. However, much to their surprise, many of the plants carrying extra copies of the purple gene were completely white or had only patches of color. Molecular analysis revealed that the level of the mRNA produced by the purple gene was reduced 50-fold in the engineered plants compared with levels of mRNA in wild-type plants. Somehow, the introduction of extra copies of the purple gene silenced both the introduced copies and the plant's own purple genes. Provide a possible explanation for how the introduction of numerous copies of the purple gene silenced all copies of the purple gene.

Petunia. [Mattiola C/Albany.].

CHALLENGE QUESTIONS

Section 14.2

31. Alternative splicing takes place in more than 90% of the human genes that encode proteins. Researchers have found that how a pre-mRNA is spliced is affected by the pre-mRNA's promoter sequence (D. Auboeuf et al. 2002. *Science* 298:416–419). In addition, factors that affect the rate of elongation of the RNA polymerase during transcription affect the type of splicing that takes place. These findings suggest that the process of transcription affects splicing. Propose one or more mechanisms that would explain how transcription might affect alternative splicing.

32. Duchenne muscular dystrophy (DMD) is an X-linked recessive genetic disease caused by mutations in the gene that encodes dystrophin, a large protein that plays an important role in the development of normal muscle fibers. The dystrophin gene is immense, spanning 2.5 million base pairs, and includes 79 exons and 78 introns. Many of the mutations that cause DMD produce premature stop codons, which bring protein synthesis to a halt, resulting in a greatly shortened and nonfunctional form of dystrophin. Some geneticists have proposed treating DMD patients by introducing small RNA molecules that cause the spliceosome to skip the exon containing the stop codon. The introduction of the small RNAs will produce a protein that is somewhat shortened (because an exon is skipped and some amino acids are missing) but may still result in a protein that has some function (A. Goyenvalle et al. 2004. *Science* 306:1796–1799). The small RNAs used for exon skipping are complementary to bases in the pre-mRNA. If you were designing small RNAs to bring about exon skipping for the treatment of DMD, what sequences should the small RNAs contain?

33. In eukaryotic cells, a poly(A) tail is normally added to pre-mRNA molecules but not to rRNA or tRNA. With the use of recombinant DNA techniques, a protein-encoding gene (which is normally transcribed by RNA polymerase II) can be connected to a promoter for RNA polymerase I. This hybrid gene is subsequently transcribed by RNA polymerase I and the appropriate pre-mRNA is produced, but this pre-mRNA is not cleaved at the 3′ end and a poly(A) tail is not added.

   Propose a mechanism to explain how the type of promoter found at the 5′ end of a gene can affect whether a poly(A) tail is added to the 3′ end.

34. SR proteins are essential to proper spliceosome assembly and are known to take part in the regulation of alternative splicing. Surprisingly, the role of SR proteins in splice-site selection and alternative splicing is affected by the promoter used for the transcription of the pre-mRNA. For example, through genetic engineering, RNA polymerase II promoters that have somewhat different sequences can be created. When pre-mRNAs with exactly the same sequences are transcribed from two different RNA polymerase II promoters that differ slightly in sequence, which promoter is used can affect how the pre-mRNA is spliced.

   Propose a mechanism for how the DNA sequence of an RNA polymerase II promoter could affect alternative splicing of the pre-mRNA.

15 The Genetic Code and Translation

The Hutterites are a religious branch of Anabaptists who live on communal farms in the prairie states and provinces of North America. A small number of founders, coupled with a tendency to intermarry, has resulted in a high frequency of the mutation for Bowen–Conradi syndrome among Hutterites. Bowen–Conradi syndrome results from defective ribosome biosynthesis, affecting the process of translation. [Kevin Fleming/Corbis.]

HUTTERITES, RIBOSOMES, AND BOWEN–CONRADI SYNDROME

The essential nature of the ribosome—the cell's protein factory—is poignantly illustrated by children with Bowen–Conradi syndrome. Born with a prominent nose, small head, and an unusual curvature of the small finger, these children fail to thrive and gain weight, usually dying within the first year of life.

Almost all children with Bowen–Conradi syndrome are Hutterites, a branch of Anabaptists who originated in the 1500s in the Tyrolean Alps of Austria. After years of persecution, the Hutterites immigrated to South Dakota in the 1870s and subsequently spread to neighboring prairie states and Canadian provinces. Today, the Hutterites in North America number about 40,000 persons. They live on communal farms, are strict pacifists, and rarely marry outside of the Hutterite community.

Bowen–Conradi syndrome is inherited as an autosomal recessive disorder, and the association of Bowen–Conradi syndrome with the Hutterite community is a function of the group's unique genetic history. The gene pool of present-day Hutterites in North America can be traced to fewer than 100 persons who immigrated to South Dakota in the late 1800s. The increased incidence of Bowen–Conradi syndrome in Hutterites today is due to the founder effect—the presence of the gene in one or more of the original founders—and its spread as Hutterites intermarried within their community. Because of the founder effect and inbreeding, many Hutterites today are as closely related as first cousins. This close genetic relationship among the Hutterites increases the probability that a child will inherit two copies of the recessive gene and have Bowen–Conradi syndrome; indeed, almost 1 in 10 Hutterites is a heterozygous carrier of the gene that causes the disease.

Although Bowen–Conradi syndrome was described in 1976, the genetic and biochemical basis of the disease long remained a mystery. Following a 7-year quest to find the causative gene, researchers at the University of Manitoba determined in 2009 that Bowen–Conradi syndrome results from the mutation of a single base pair in the *EMG1* gene, located on chromosome 12.

The discovery of the gene for Bowen–Conradi syndrome was a source of immediate insight into the biochemical nature of the disease. Although little is know about the function of the *EMG1* gene in humans, earlier studies in yeast revealed that it encodes a protein that helps to assemble the ribosome. As discussed in Chapter 14, the ribosome comprises small and large subunits. The small subunit in humans consists of a single piece of ribosomal RNA, known as 18S rRNA, and a large number of proteins. The protein encoded by the *EMG1* gene plays an essential role in processing 18S rRNA and helps to assemble it into the small subunit of the ribosome. Because of a mutation in the *EMG1* gene, babies with Bowen–Conradi syndrome produce ribosomes that function poorly and the process of protein synthesis is universally affected.

Bowen–Conradi syndrome illustrates the extreme importance of translation, the process of protein synthesis, which is the focus of this chapter. We begin by examining the molecular relation between genotype and phenotype. Next, we study the genetic code—the instructions that specify the amino acid sequence of a protein—and then examine the mechanism of protein synthesis. Our primary focus will be on protein synthesis in bacterial cells, but we will examine some of the differences in eukaryotic cells. At the end of the chapter, we look at some additional aspects of protein synthesis.

15.1 Many Genes Encode Proteins

The first person to suggest the existence of a relation between genotype and proteins was English physician Archibald Garrod. In 1908, Garrod correctly proposed that genes encode enzymes, but, unfortunately, his theory made little impression on his contemporaries. Not until the 1940s, when George Beadle and Edward Tatum examined the genetic basis of biochemical pathways in *Neurospora*, did the relation between genes and proteins become widely accepted.

The One Gene, One Enzyme Hypothesis

Beadle and Tatum used the bread mold *Neurospora* to study the biochemical results of mutations. *Neurospora* is easy to cultivate in the laboratory. The main vegetative part of the fungus is haploid, which allows the effects of recessive mutations to be easily observed (**Figure 15.1**).

Wild-type *Neurospora* grows on minimal medium, which contains only inorganic salts, nitrogen, a carbon source such as sucrose, and the vitamin biotin. The fungus can synthesize all the biological molecules that it needs from these basic compounds. However, mutations may arise that disrupt fungal growth by destroying the fungus's ability to synthesize one or more essential biological molecules. These nutritionally deficient mutants, termed auxotrophs, will not grow on minimal medium, but they can grow on medium that contains the substance that they are no longer able to synthesize.

Beadle and Tatum first irradiated spores of *Neurospora* to induce mutations (**Figure 15.2**). After irradiation, they placed individual spores into different culture tubes containing complete medium (medium having all the biological substances needed for growth). Next, they transferred spores from each culture to tubes containing minimal medium. Fungi containing auxotrophic mutations grew on complete medium but would not grow on minimal medium, which allowed Beadle and Tatum to identify cultures that contained mutations.

After they had determined that a particular culture had an auxotrophic mutation, Beadle and Tatum set out to determine the specific *effect* of the mutation. They transferred spores of each mutant strain from complete medium to a series of tubes (see Figure 15.2), each of which possessed minimal medium plus one of a variety of essential biological molecules, such as an amino acid. If the spores in a tube grew, Beadle and Tatum were able to identify the added substance as the biological molecule whose synthesis had been affected by the mutation. For example, an auxotrophic mutant that would grow only on minimal medium to which arginine had been added must have possessed a mutation that disrupts the synthesis of arginine.

Adrian Srb and Norman H. Horowitz patiently applied this procedure to genetically dissect the multistep biochemical pathway of arginine synthesis (**Figure 15.3**). They first isolated a series of auxotrophic mutants whose growth required arginine. They then tested these mutants for their ability to grow on minimal medium supplemented with

15.1 Beadle and Tatum used the fungus *Neurospora*, which has a complex life cycle, to work out the relation of genes to proteins. [Namboori B. Raju.]

15.2 Beadle and Tatum developed a method for isolating auxotrophic mutants in *Neurospora*.

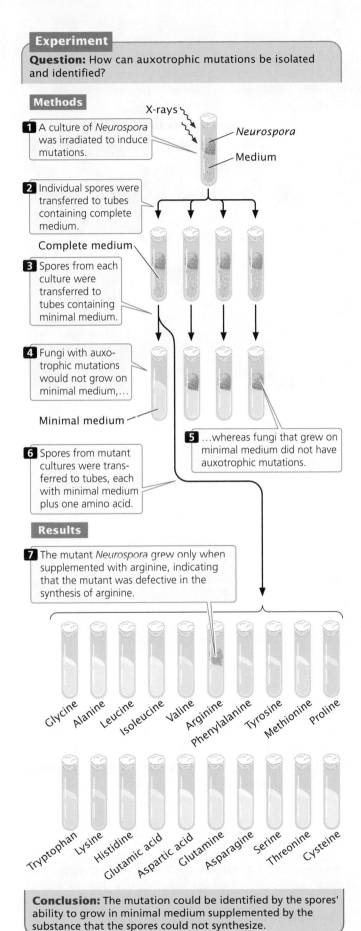

Experiment

Question: How can auxotrophic mutations be isolated and identified?

Methods

X-rays

1 A culture of *Neurospora* was irradiated to induce mutations.

Neurospora

Medium

2 Individual spores were transferred to tubes containing complete medium.

Complete medium

3 Spores from each culture were transferred to tubes containing minimal medium.

4 Fungi with auxotrophic mutations would not grow on minimal medium,…

Minimal medium

5 …whereas fungi that grew on minimal medium did not have auxotrophic mutations.

6 Spores from mutant cultures were transferred to tubes, each with minimal medium plus one amino acid.

Results

7 The mutant *Neurospora* grew only when supplemented with arginine, indicating that the mutant was defective in the synthesis of arginine.

Glycine · Alanine · Leucine · Isoleucine · Valine · Arginine · Phenylalanine · Tyrosine · Methionine · Proline

Tryptophan · Lysine · Histidine · Glutamic acid · Aspartic acid · Glutamine · Asparagine · Serine · Threonine · Cysteine

Conclusion: The mutation could be identified by the spores' ability to grow in minimal medium supplemented by the substance that the spores could not synthesize.

three compounds: ornithine, citrulline, and arginine. From the results, they were able to place the mutants into three groups (**Table 15.1**) on the basis of which of the substances allowed growth. Group I mutants grew on minimal medium supplemented with ornithine, citrulline, or arginine. Group II mutants grew on minimal medium supplemented with either arginine or citrulline but did not grow on medium supplemented only with ornithine. Finally, group III mutants grew only on medium supplemented with arginine.

Srb and Horowitz therefore proposed that the biochemical pathway leading to the amino acid arginine has at least three steps:

$$\text{precursor} \xrightarrow[\text{1}]{\text{Step}} \text{ornithine} \xrightarrow[\text{2}]{\text{Step}} \text{citrulline} \xrightarrow[\text{3}]{\text{Step}} \text{arginine}$$

They concluded that the mutations in group I affect step 1 of this pathway, mutations in group II affect step 2, and mutations in group III affect step 3. But how did they know that the order of the compounds in the biochemical pathway was correct?

Notice that, if step 1 is blocked by a mutation, then the addition of either ornithine or citrulline allows growth, because these compounds can still be converted into arginine (see Figure 15.3). Similarly, if step 2 is blocked, the addition of citrulline allows growth, but the addition of ornithine has no effect. If step 3 is blocked, the spores will grow only if arginine is added to the medium. The underlying principle is that an auxotrophic mutant cannot synthesize any compound that comes after the step blocked by a mutation.

Using this reasoning with the information in Table 15.1, we can see that the addition of arginine to the medium allows all three groups of mutants to grow. Therefore, biochemical steps affected by all the mutants precede the step that results

Table 15.1	Growth of arginine auxotrophic mutants on minimal medium with various supplements		
Mutant Strain Number	Ornithine	Citrulline	Arginine
Group I	+	+	+
Group II	−	+	+
Group III	−	−	+

Note: A plus sign (+) indicates growth; a minus sign (−) indicates no growth.

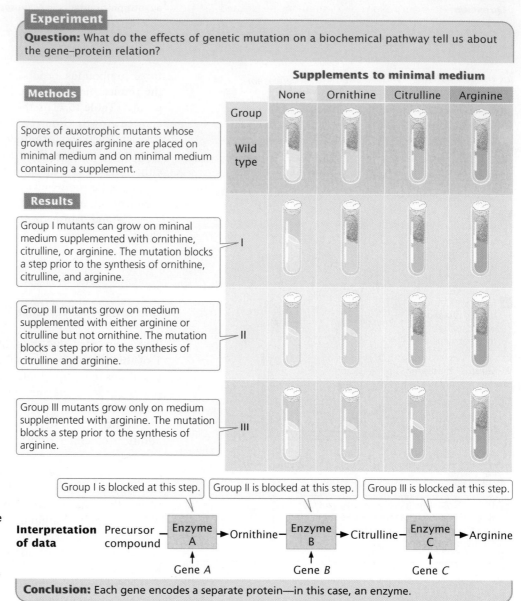

15.3 Method used to determine the relation between genes and enzymes in _Neurospora_. This biochemical pathway leads to the synthesis of arginine in _Neurospora_. Steps in the pathway are catalyzed by enzymes affected by mutations.

in arginine. The addition of citrulline allows group I and group II mutants to grow but not group III mutants; therefore, group III mutations must affect a biochemical step that takes place after the production of citrulline but before the production of arginine.

$$\text{citrulline} \xrightarrow[\text{mutations}]{\overset{\text{Group III}}{}} \text{arginine}$$

The addition of ornithine allows the growth of group I mutants but not group II or group III mutants; thus, mutations in groups II and III affect steps that come after the production of ornithine. We've already established that group II mutations affect a step before the production of citrulline; so group II mutations must block the conversion of ornithine into citrulline.

$$\text{ornithine} \xrightarrow[\text{mutations}]{\overset{\text{Group II}}{}} \text{citrulline} \xrightarrow[\text{mutations}]{\overset{\text{Group III}}{}} \text{arginine}$$

Because group I mutations affect some step before the production of ornithine, we can conclude that they must affect the conversion of some precursor into ornithine. We can now outline the biochemical pathway yielding ornithine, citrulline, and arginine.

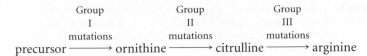

$$\text{precursor} \xrightarrow[\text{mutations}]{\substack{\text{Group} \\ \text{I}}} \text{ornithine} \xrightarrow[\text{mutations}]{\substack{\text{Group} \\ \text{II}}} \text{citrulline} \xrightarrow[\text{mutations}]{\substack{\text{Group} \\ \text{III}}} \text{arginine}$$

Importantly, this procedure does not necessarily detect all steps in a pathway; rather, it detects only the steps producing the compounds tested.

Using mutations and this type of reasoning, Beadle, Tatum, and others were able to identify genes that control several biosynthetic pathways in *Neurospora*. They established that each step in a pathway is controlled by a different enzyme, as shown in Figure 15.3 for the arginine pathway. The results of genetic crosses and mapping studies demonstrated that mutations affecting any one step in a pathway always map to the same chromosomal location. Beadle and Tatum reasoned that mutations affecting a particular biochemical step occurred at a single locus that encoded a particular enzyme. This idea became known as the **one gene, one enzyme hypothesis**: genes function by encoding enzymes, and each gene encodes a separate enzyme. When research findings showed that some proteins are composed of more than one polypeptide chain and that different polypeptide chains are encoded by separate genes, this model was modified to become the **one gene, one polypeptide hypothesis**. TRY PROBLEM 16 →

CONCEPTS

Beadle and Tatum's studies of biochemical pathways in the fungus *Neurospora* helped define the relation between genotype and phenotype by establishing the one gene, one enzyme hypothesis, the idea that each gene encodes a separate enzyme. This hypothesis was later modified to the one gene, one polypeptide hypothesis.

✔ CONCEPT CHECK 1

Auxotrophic mutation 103 grows on minimal medium supplemented with A, B, or C; mutation 106 grows on medium supplemented with A and C but not B; and mutation 102 grows only on medium supplemented with C. What is the order of A, B, and C in a biochemical pathway?

The Structure and Function of Proteins

Proteins are central to all living processes (**Figure 15.4**). Many proteins are enzymes, the biological catalysts that drive the chemical reactions of the cell; others are structural components, providing scaffolding and support for membranes, filaments, bone, and hair. Some proteins help transport substances; others have a regulatory, communication, or defense function.

Amino acids All proteins are composed of **amino acids**, linked end to end. Twenty common amino acids are found in proteins; these amino acids are shown in **Figure 15.5** with both their three-letter and their one-letter abbreviations. (Other amino acids sometimes found in proteins are modified forms of the common amino acids.) The 20 common amino acids are similar in structure: each consists of a central carbon atom bonded to an amino group, a hydrogen atom, a carboxyl group, and an R (radical) group that differs for each amino acid. The amino acids in proteins are joined together by **peptide bonds** (**Figure 15.6**) to form **polypeptide** chains, and a protein consists of one or more polypeptide chains. Like nucleic acids, polypeptides have polarity, with one end having a free amino group (NH_3^+) and the other end possessing a free carboxyl group (COO^-). Some proteins consist of only a few amino acids, whereas others may have thousands.

Protein structure Like that of nucleic acids, the molecular structure of proteins has several levels of organization. The *primary structure* of a protein is its sequence of amino acids (**Figure 15.7a**). Through interactions between neighboring amino acids, a polypeptide chain folds and twists into a *secondary structure* (**Figure 15.7b**); two common secondary structures found in proteins are the beta (β) pleated sheet and the alpha (α) helix. Secondary structures interact and fold further to form a *tertiary structure* (**Figure 15.7c**), which is the overall, three-dimensional shape of the protein. The secondary and tertiary structures of a protein

(a)

(b)

(c)

15.4 Proteins serve a number of biological functions. (a) The light produced by fireflies is the result of a light-producing reaction between luciferin and ATP catalyzed by the enzyme luciferase. (b) The protein fibroin is the major structural component of spider webs. (c) Castor beans contain a highly toxic protein called ricin. [Part a: Phil Degginger/Alamy. Part b: Rosemary Calvert/Imagestate. Part c: Gerald & Buff Corsi/Visuals Unlimited.]

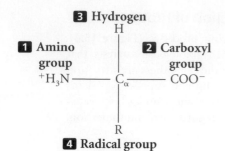

3 Hydrogen

1 Amino group

2 Carboxyl group

4 Radical group (side chain)

15.5 The common amino acids have similar structures. Each amino acid consists of a central carbon atom (C_α) attached to: (1) an amino group (NH_3^+); (2) a carboxyl group (COO^-); (3) a hydrogen atom (H); and (4) a radical group, designated R. In the structures of the 20 common amino acids, the parts in black are common to all amino acids and the parts in red are the R groups.

Nonpolar, aliphatic R groups

Glycine (Gly, G)

Alanine (Ala, A)

Valine (Val, V)

Leucine (Leu, L)

Isoleucine (Ile, I)

Methionine (Met, M)

Polar, uncharged R groups

Serine (Ser, S)

Threonine (Thr, T)

Cysteine (Cys, C)

Proline (Pro, P)

Asparagine (Asn, N)

Glutamine (Gln, Q)

Aromatic R groups

Phenylalanine (Phe, F)

Tyrosine (Try, Y)

Tryptophan (Trp, W)

Positively charged R groups

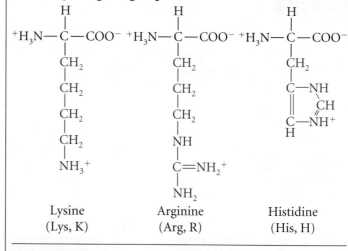

Lysine (Lys, K)

Arginine (Arg, R)

Histidine (His, H)

Negatively charged R groups

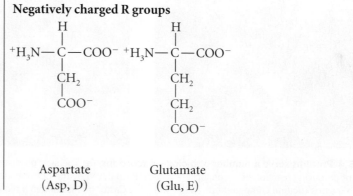

Aspartate (Asp, D)

Glutamate (Glu, E)

15.6 Amino acids are joined together by peptide bonds. In a peptide bond (red), the carboxyl group of one amino acid is covalently attached to the amino group of another amino acid.

are largely determined by the primary structure—the amino acid sequence—of the protein. Finally, some proteins consist of two or more polypeptide chains that associate to produce a *quaternary structure* (**Figure 15.7d**).

CONCEPTS

The products of many genes are proteins whose actions produce the traits encoded by these genes. Proteins are polymers consisting of amino acids linked by peptide bonds. The amino acid sequence of a protein is its primary structure. This structure folds to create the secondary and tertiary structures; two or more polypeptide chains may associate to create a quaternary structure.

✔ **CONCEPT CHECK 2**

What primarily determines the secondary and tertiary structures of a protein?

15.2 The Genetic Code Determines How the Nucleotide Sequence Specifies the Amino Acid Sequence of a Protein

In 1953, James Watson and Francis Crick solved the structure of DNA and identified the base sequence as the carrier of genetic information. However, the way in which the base sequence of DNA specifies the amino acid sequences of proteins (the genetic code) was not immediately obvious and remained elusive for another 10 years.

One of the first questions about the genetic code to be addressed was: How many nucleotides are necessary to specify a single amino acid? This basic unit of the genetic code—the set of bases that encode a single amino acid—is a *codon* (see p. 380 in Chapter 14). Many early investigators recognized that codons must contain a minimum of three nucleotides. Each nucleotide position in mRNA can be occupied by one of four bases: A, G, C, or U. If a codon consisted of a single nucleotide, only four different codons (A, G, C, and U) would be possible, which is not enough to encode the 20 different amino acids commonly found in proteins. If codons were made up of two nucleotides each (i.e., GU, AC, etc.), there would be $4 \times 4 = 16$ possible codons—still not enough to encode all 20 amino acids. With three nucleotides per codon, there are $4 \times 4 \times 4 = 64$ possible codons, which is more than enough to specify 20 different amino acids. Therefore, a *triplet code* requiring three nucleotides per codon is the most efficient way to encode all 20 amino acids. Using mutations in bacteriophage, Francis Crick and his colleagues confirmed in 1961 that the genetic code is indeed a triplet code. **TRY PROBLEMS 18 AND 19** ➡

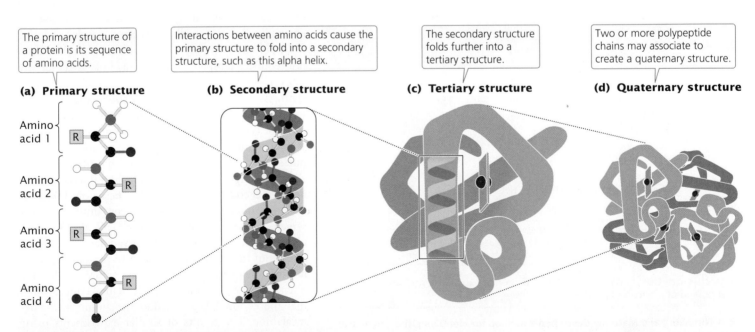

| The primary structure of a protein is its sequence of amino acids. | Interactions between amino acids cause the primary structure to fold into a secondary structure, such as this alpha helix. | The secondary structure folds further into a tertiary structure. | Two or more polypeptide chains may associate to create a quaternary structure. |

(a) Primary structure   **(b) Secondary structure**   **(c) Tertiary structure**   **(d) Quaternary structure**

Amino acid 1
Amino acid 2
Amino acid 3
Amino acid 4

15.7 Proteins have several levels of structural organization.

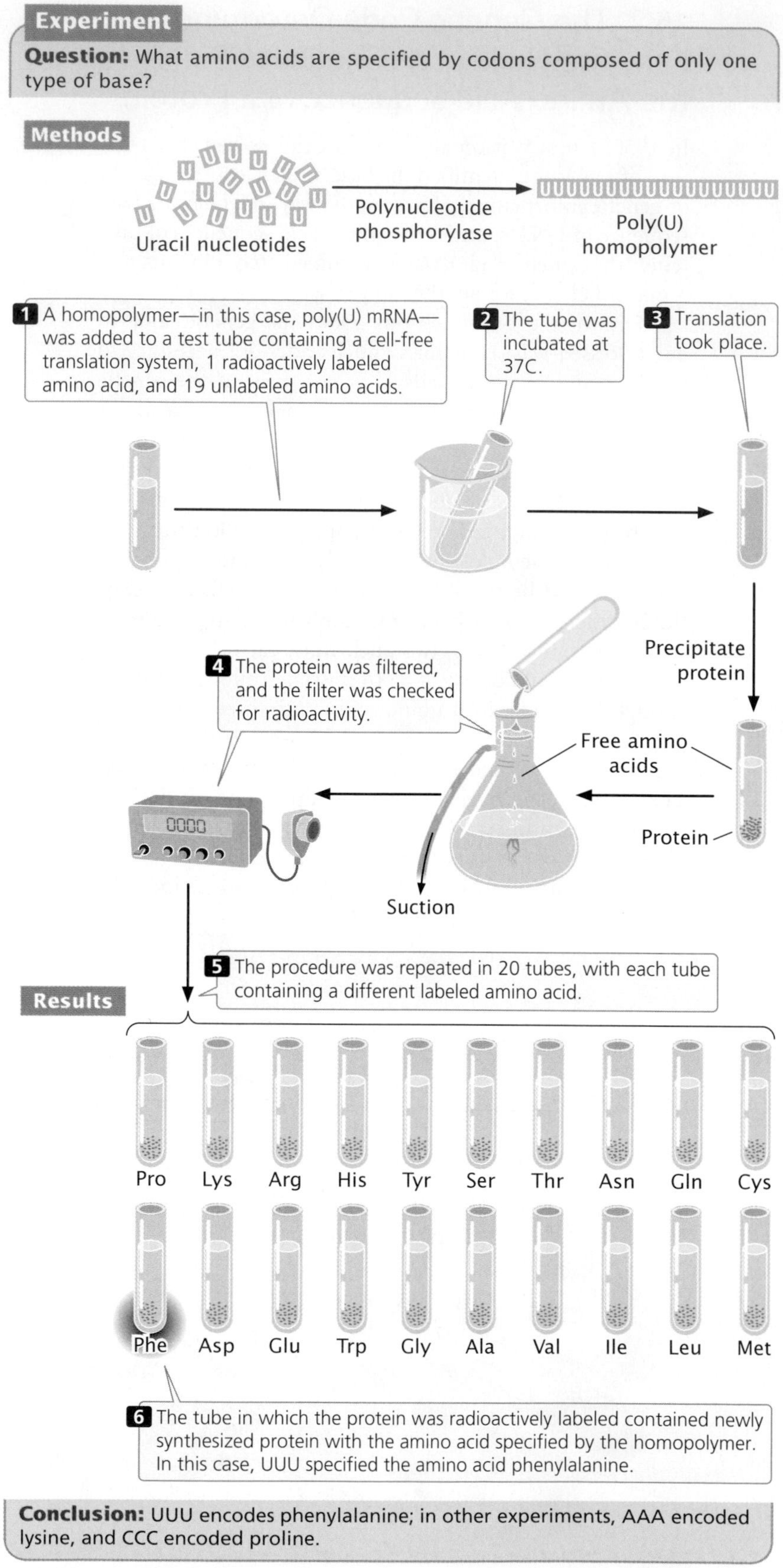

Experiment

Question: What amino acids are specified by codons composed of only one type of base?

Methods

Uracil nucleotides → Polynucleotide phosphorylase → Poly(U) homopolymer

1 A homopolymer—in this case, poly(U) mRNA—was added to a test tube containing a cell-free translation system, 1 radioactively labeled amino acid, and 19 unlabeled amino acids.

2 The tube was incubated at 37C.

3 Translation took place.

4 The protein was filtered, and the filter was checked for radioactivity.

Precipitate protein

Free amino acids

Protein

Suction

5 The procedure was repeated in 20 tubes, with each tube containing a different labeled amino acid.

Results

Pro Lys Arg His Tyr Ser Thr Asn Gln Cys

Phe Asp Glu Trp Gly Ala Val Ile Leu Met

6 The tube in which the protein was radioactively labeled contained newly synthesized protein with the amino acid specified by the homopolymer. In this case, UUU specified the amino acid phenylalanine.

Conclusion: UUU encodes phenylalanine; in other experiments, AAA encoded lysine, and CCC encoded proline.

15.8 Nirenberg and Matthaei developed a method for identifying the amino acid specified by a homopolymer.

CONCEPTS

The genetic code is a triplet code, in which three nucleotides encode each amino acid in a protein.

✔ **CONCEPT CHECK 3**

A codon is

a. one of three nucleotides that encode an amino acid.

b. three nucleotides that encode an amino acid.

c. three amino acids that encode a nucleotide.

d. one of four bases in DNA.

Breaking the Genetic Code

When it had been firmly established that the genetic code consists of codons that are three nucleotides in length, the next step was to determine which groups of three nucleotides specify which amino acids. Logically, the easiest way to break the code would have been to determine the base sequence of a piece of RNA, add it to a test tube containing all the components necessary for translation, and allow it to direct the synthesis of a protein. The amino acid sequence of the newly synthesized protein could then be determined, and its sequence could be compared with that of the RNA. Unfortunately, there was no way at that time to determine the nucleotide sequence of a piece of RNA; so indirect methods were necessary to break the code.

The use of homopolymers The first clues to the genetic code came in 1961, from the work of Marshall Nirenberg and Johann Heinrich Matthaei. These investigators created synthetic RNAs by using an enzyme called polynucleotide phosphorylase. Unlike RNA polymerase, polynucleotide phosphorylase does not require a template; it randomly links together any RNA nucleotides that happen to be available. The first synthetic mRNAs used by Nirenberg and Matthaei were homopolymers, RNA molecules consisting of a single type of nucleotide. For example, by adding polynucleotide phosphorylase to a solution of uracil nucleotides, they generated RNA molecules that consisted entirely of uracil nucleotides and thus contained only UUU codons (**Figure 15.8**). These poly(U) RNAs were then added to 20 tubes, each containing the components necessary for translation and all 20 amino acids. A different amino acid was radioactively labeled in each of the 20 tubes. Radioactive protein appeared in only one of the tubes—the one containing labeled phenylalanine (see Figure 15.8). This result showed that the codon UUU specifies the amino acid phenylalanine. The results of similar experiments using poly(C) and poly(A) RNA demonstrated that CCC encodes proline and AAA encodes lysine; for technical reasons, the results from poly(G) were uninterpretable.

The use of random copolymers To gain information about additional codons, Nirenberg and his colleagues created synthetic RNAs containing two or three different bases. Because polynucleotide phosphorylase incorporates nucleotides randomly, these RNAs contained random mixtures of the bases and are thus called random copolymers. For example, when adenine and cytosine nucleotides are mixed with polynucleotide phosphorylase, the RNA molecules produced have eight different codons: AAA, AAC, ACC, ACA, CAA, CCA, CAC, and CCC. These poly(AC) RNAs produced proteins containing six different amino acids: asparagine, glutamine, histidine, lysine, proline, and threonine.

The proportions of the different amino acids in the proteins depended on the ratio of the two nucleotides used in creating the synthetic mRNA, and the theoretical probability of finding a particular codon could be calculated from the ratios of the bases. If a 4 : 1 ratio of C to A were used in making the RNA, then the probability of C being at any given position in a codon is $^4/_5$ and the probability of A being in it is $^1/_5$. With random incorporation of bases, the probability of any one of the codons with two Cs and one A (CCA, CAC, or ACC) should be $^4/_5 \times ^4/_5 \times ^1/_5 = {}^{16}/_{125} = 0.13$, or 13%, and the probability of any codon with two As and one C (AAC, ACA, or CAA) should be $^1/_5 \times ^1/_5 \times ^4/_5 = {}^4/_{125} = 0.032$, or about 3%. Therefore, an amino acid encoded by two Cs and one A should be more common than an amino acid encoded by two As and one C. By comparing the percentages of amino acids in proteins produced by random copolymers with the theoretical frequencies expected for the codons, Nirenberg and his colleagues could derive information about the *base composition* of the codons. These experiments revealed nothing, however, about the codon base *sequence*; histidine was clearly encoded by a codon with two Cs and one A, but whether that codon was ACC, CAC, or CCA was unknown. There were other problems with this method: the theoretical calculations depended on the random incorporation of bases, which did not always occur, and, because the genetic code is redundant, sometimes several different codons specify the same amino acid.

The use of ribosome-bound tRNAs To overcome the limitations of random copolymers, Nirenberg and Philip Leder developed another technique in 1964 that used ribosome-bound tRNAs. They found that a very short sequence of mRNA—even one consisting of a single codon—would bind to a ribosome. The codon on the short mRNA would then base pair with the matching anticodon on a transfer RNA that carried the amino acid specified by the codon (**Figure 15.9**). Short mRNAs that were bound to ribosomes were mixed with tRNAs and amino acids, and this mixture was passed through a nitrocellulose filter. The tRNAs that were paired with the ribosome-bound mRNA stuck to the filter,

whereas unbound tRNAs passed through it. The advantage of this system was that it could be used with very short synthetic mRNA molecules that could be synthesized with a known sequence. Nirenberg and Leder synthesized more than 50 short mRNAs with known codons and added them individually to a mixture of ribosomes and tRNAs. They then isolated the tRNAs that were bound to the mRNA and ribosomes and

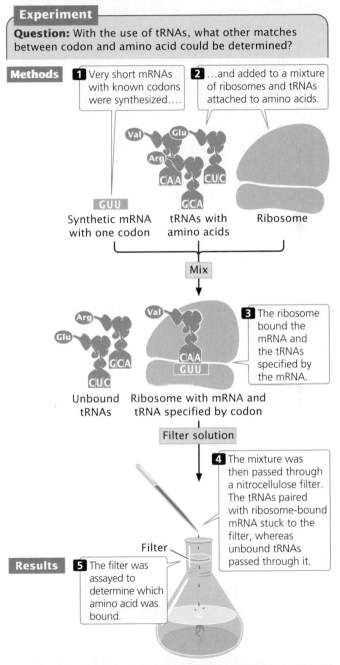

Experiment

Question: With the use of tRNAs, what other matches between codon and amino acid could be determined?

Methods

1 Very short mRNAs with known codons were synthesized….

2 …and added to a mixture of ribosomes and tRNAs attached to amino acids.

Synthetic mRNA with one codon

tRNAs with amino acids

Ribosome

Mix

3 The ribosome bound the mRNA and the tRNAs specified by the mRNA.

Unbound tRNAs

Ribosome with mRNA and tRNA specified by codon

Filter solution

4 The mixture was then passed through a nitrocellulose filter. The tRNAs paired with ribosome-bound mRNA stuck to the filter, whereas unbound tRNAs passed through it.

Filter

Results

5 The filter was assayed to determine which amino acid was bound.

Conclusion: When an mRNA with GUU was added, the tRNAs on the filter were bound to valine; therefore the codon GUU specifies valine. Many other codons were determined by using this method.

15.9 Nirenberg and Leder used ribosome-bound tRNAs to provide additional information about the genetic code.

determined which amino acids were present on the bound tRNAs. For example, synthetic RNA with the codon GUU retained a tRNA to which valine was attached, whereas RNAs with the codons UGU and UUG did not. Using this method, Nirenberg and his colleagues were able to determine the amino acids encoded by more than 50 codons.

Other experiments provided additional information about the genetic code, and the code was fully deciphered by 1968. In the next section, we will examine some of the features of the code, which is so important to modern biology that Francis Crick compared its place to that of the periodic table of the elements in chemistry.

The Degeneracy of the Code

One amino acid is encoded by three consecutive nucleotides in mRNA, and each nucleotide can have one of four possible bases (A, G, C, and U) at each nucleotide position, thus permitting $4^3 = 64$ possible codons (**Figure 15.10**). Three of these codons are stop codons, specifying the end of translation. Thus, 61 codons, called **sense codons**, encode amino acids. Because there are 61 sense codons and only 20 different amino acids commonly found in proteins, the code contains more information than is needed to specify the amino acids and is said to be a **degenerate code**. This expression does not mean that the genetic code is depraved; *degenerate* is a term that Francis Crick borrowed from quantum physics, where it describes multiple physical states that have equivalent meaning. The degeneracy of the

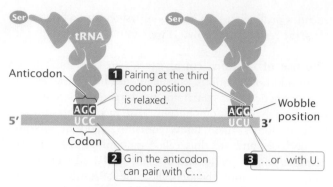

15.11 Wobble may exist in the pairing of a codon and anticodon. The mRNA and tRNA pair in an antiparallel fashion. Pairing at the first and second codon positions is in accord with the Watson-and-Crick pairing rules (A with U, G with C); however, pairing rules are relaxed at the third position of the codon, and G on the anticodon can pair with either U or C on the codon in this example.

genetic code means that amino acids may be specified by more than one codon. Only tryptophan and methionine are encoded by a single codon (see Figure 15.10). Other amino acids are specified by two codons, and some, such as leucine, are specified by six different codons. Codons that specify the same amino acid are said to be **synonymous**, just as synonymous words are different words that have the same meaning.

As we learned in Chapter 14, tRNAs serve as adapter molecules, binding particular amino acids and delivering them to a ribosome, where the amino acids are then assembled into polypeptide chains. Each type of tRNA attaches to a single type of amino acid. The cells of most organisms possess from about 30 to 50 different tRNAs, and yet there are only 20 different amino acids in proteins. Thus, some amino acids are carried by more than one tRNA. Different tRNAs that accept the same amino acid but have different anticodons are called **isoaccepting tRNAs**. Some synonymous codons specify different isoacceptors.

Even though some amino acids have multiple (isoaccepting) tRNAs, there are still more codons than anticodons, because different codons can sometimes pair with the same anticodon through flexibility in base pairing at the third position of the codon. Examination of Figure 15.10 reveals that many synonymous codons differ only in the third position. For example, alanine is encoded by the codons GCU, GCC, GCA, and GCG, all of which begin with GC. When the codon on the mRNA and the anticodon of the tRNA join (**Figure 15.11**), the first (5′) base of the codon pairs with the third (3′) base of the anticodon, strictly according to Watson-and-Crick rules: A with U; C with G. Next, the middle bases of codon and anticodon pair, also strictly following the Watson-and-Crick rules. After these pairs have hydrogen bonded, the third bases pair weakly: there may be flexibility, or **wobble**, in their pairing.

In 1966, Francis Crick developed the wobble hypothesis, which proposed that some nonstandard pairings of bases could take place at the third position of a codon. For example, a G in the anticodon may pair with either a C or a U in

Second base

		U	C	A	G	
U		UUU UUC Phe	UCU UCC Ser	UAU UAC Tyr	UGU UGC Cys	U C
		UUA UUG Leu	UCA UCG	UAA Stop UAG Stop	UGA Stop UGG Trp	A G
C		CUU CUC CUA CUG Leu	CCU CCC CCA CCG Pro	CAU CAC His CAA CAG Gln	CGU CGC CGA CGG Arg	U C A G
A		AUU AUC Ile AUA AUG Met	ACU ACC ACA ACG Thr	AAU AAC Asn AAA AAG Lys	AGU AGC Ser AGA AGG Arg	U C A G
G		GUU GUC Val GUA GUG	GCU GCC GCA GCG Ala	GAU GAC Asp GAA GAG Glu	GGU GGC GGA GGG Gly	U C A G

15.10 The genetic code consists of 64 codons. The amino acids specified by each codon are given in their three-letter abbreviation. The codons are written 5′→3′, as they appear in the mRNA. AUG is an initiation codon; UAA, UAG, and UGA are termination (stop) codons.

Table 15.2	The wobble rules, indicating which bases in the third position (3′ end) of the mRNA codon can pair with bases at the first position (5′ end) of the anticodon of the tRNA	
First Position of Anticodon	Third Position of Codon	Pairing
C	G	Anticodon 3′–X—Y—C–5′ | | | 5′–Y—X—G–3′ Codon
G	U or C	Anticodon 3′–X—Y—G–5′ | | | 5′–Y—X—U–3′ C Codon
A	U	Anticodon 3′–X—Y—A–5′ | | | 5′–Y—X—U–3′ Codon
U	A or G	Anticodon 3′–X—Y—U–5′ | | | 5′–Y—X—A–3′ G Codon

I (inosine)	A, U, or C	Anticodon
		3′–X—Y—I–5′
		| | |
		5′–Y—X—A–3′
		          U
		          C
		Codon

the third position of the codon (**Table 15.2**). The important thing to remember about wobble is that it allows some tRNAs to pair with more than one codon on an mRNA; thus from 30 to 50 tRNAs can pair with 61 sense codons. Some codons are synonymous through wobble. **TRY PROBLEM 24** ➡

CONCEPTS

The genetic code consists of 61 sense codons that specify the 20 common amino acids; the code is degenerate, meaning that some amino acids are encoded by more than one codon. Isoaccepting tRNAs are different tRNAs with different anticodons that specify the same amino acid. Wobble exists when more than one codon can pair with the same anticodon.

✔ CONCEPT CHECK 4

Through wobble, a single _____ can pair with more than one _____.

a. codon, anticodon

b. group of three nucleotides in DNA, codon in mRNA

c. tRNA, amino acid

d. anticodon, codon

The Reading Frame and Initiation Codons

Findings from early studies of the genetic code indicated that the code is generally **nonoverlapping**. An overlapping code is one in which a single nucleotide may be included in more than one codon, as follows:

Nucleotide sequence   A U A C G A G U C

Nonoverlapping code   A U A C G A G U C
                             Ile   Arg   Val

Overlapping code   A U A C G A G U
                            Ile
                           U A C
                           Tyr
                             A C G
                              Thr

Usually, however, each nucleotide is part of a single codon. A few overlapping genes are found in viruses, but codons within the same gene do not overlap, and the genetic code is generally considered to be nonoverlapping.

For any sequence of nucleotides, there are three potential sets of codons—three ways in which the sequence can be read in groups of three. Each different way of reading the sequence is called a **reading frame**, and any sequence of nucleotides has three potential reading frames. The three reading frames have completely different sets of codons and will therefore specify proteins with entirely different amino acid sequences. Thus, it is essential for the translational machinery to use the correct reading frame. How is the correct reading frame established? The reading frame is set by the **initiation codon**, which is the first codon of the mRNA to specify an amino acid. After the initiation codon, the other codons are read as successive groups of three nucleotides. No bases are skipped between the codons; so there are no punctuation marks to separate the codons.

The initiation codon is usually AUG, although GUG and UUG are used on rare occasions. The initiation codon is not just a sequence that marks the beginning of translation; it specifies an amino acid. In bacterial cells, AUG encodes a modified type of methionine, *N*-formylmethionine; all proteins in bacteria initially begin with this amino acid, but the formyl group (or, in some cases, the entire amino acid) may be removed after the protein has been synthesized. When the codon AUG is at an internal position in a gene, it encodes unformylated methionine. In archaeal and eukaryotic cells, AUG specifies unformylated methionine both at the initiation position and at internal positions.

Termination Codons

Three codons—UAA, UAG, and UGA—do not encode amino acids. These codons signal the end of the protein in both bacterial and eukaryotic cells and are called **stop codons, termination codons,** or **nonsense codons.** No tRNA molecules have anticodons that pair with termination codons.

The Universality of the Code

For many years the genetic code was assumed to be **universal,** meaning that each codon specifies the same amino acid in all organisms. We now know that the genetic code is almost, but not completely, universal; a few exceptions have been found. Most of these exceptions are termination codons, but there are a few cases in which one sense codon substitutes for another. Most exceptions are found in mitochondrial genes; a few nonuniversal codons have also been detected in the nuclear genes of protozoans and in bacterial DNA (**Table 15.3**). TRY PROBLEM 20 →

CONCEPTS

Each sequence of nucleotides possesses three potential reading frames. The correct reading frame is set by the initiation codon. The end of a protein-encoding sequence is marked by a termination codon. With a few exceptions, all organisms use the same genetic code.

✔ CONCEPT CHECK 5

Do the initiation and termination codons specify an amino acid? If so, which ones?

Table 15.3 Some exceptions to the universal genetic code

Genome	Codon	Universal Code	Altered Code
Bacterial DNA			
Mycoplasma capricolum	UGA	Stop	Trp
Mitochondrial DNA			
Human	UGA	Stop	Trp
Human	AUA	Ile	Met
Human	AGA, AGG	Arg	Stop
Yeast	UGA	Stop	Trp
Trypanosomes	UGA	Stop	Trp
Plants	CGG	Arg	Trp
Nuclear DNA			
Tetrahymena	UAA	Stop	Gln
Paramecium	UAG	Stop	Gln

CONNECTING CONCEPTS

Characteristics of the Genetic Code

We have now considered a number of characteristics of the genetic code. Let's pause for a moment and review these characteristics.

1. The genetic code consists of a sequence of nucleotides in DNA or RNA. There are four letters in the code, corresponding to the four bases—A, G, C, and U (T in DNA).

2. The genetic code is a triplet code. Each amino acid is encoded by a sequence of three consecutive nucleotides, called a codon.

3. The genetic code is degenerate; that is, of 64 codons, 61 codons encode only 20 amino acids in proteins (3 codons are termination codons). Some codons are synonymous, specifying the same amino acid.

4. Isoaccepting tRNAs are tRNAs with different anticodons that accept the same amino acid; wobble allows the anticodon on one type of tRNA to pair with more than one type of codon on mRNA.

5. The code is generally nonoverlapping; each nucleotide in an mRNA sequence belongs to a single reading frame.

6. The reading frame is set by an initiation codon, which is usually AUG.

7. When a reading frame has been set, codons are read as successive groups of three nucleotides.

8. Any one of three termination codons (UAA, UAG, and UGA) can signal the end of a protein; no amino acids are encoded by the termination codons.

9. The code is almost universal.

15.3 Amino Acids Are Assembled into a Protein Through the Mechanism of Translation

Now that we are familiar with the genetic code, we can begin to study the mechanism by which amino acids are assembled into proteins. Because more is known about translation in bacteria, we will focus primarily on bacterial translation. In most respects, eukaryotic translation is similar, although some significant differences will be noted as we proceed through the stages of translation.

Translation takes place on ribosomes; indeed, ribosomes can be thought of as moving protein-synthesizing machines. Through a variety of techniques, a detailed view of the structure of the ribosome has been produced in recent years, which has greatly improved our understanding of the translational process. A ribosome attaches near the 5′ end of an mRNA strand and moves toward the 3′ end, translating the codons as it goes (**Figure 15.12**). Synthesis begins at the amino end of the protein, and the protein is elongated by the addition of new amino acids to the carboxyl end. Protein synthesis includes a series of RNA–RNA interactions: interactions between the mRNA and the rRNA that hold the mRNA in the ribosome, interactions between the codon on the mRNA and the anticodon on the tRNA, and interactions between the tRNA and the rRNAs of the ribosome.

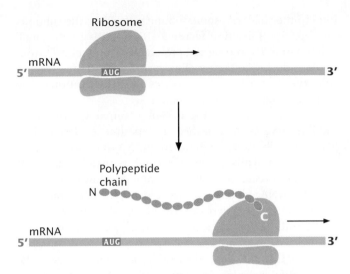

15.12 The translation of an mRNA molecule takes place on a ribosome. The letter N represents the amino end of the protein; C represents the carboxyl end.

Protein synthesis can be conveniently divided into four stages: (1) tRNA charging, in which tRNAs bind to amino acids; (2) initiation, in which the components necessary for translation are assembled at the ribosome; (3) elongation, in which amino acids are joined, one at a time, to the growing polypeptide chain; and (4) termination, in which protein synthesis halts at the termination codon and the translation components are released from the ribosome.

The Binding of Amino Acids to Transfer RNAs

The first stage of translation is the binding of tRNA molecules to their appropriate amino acids, called tRNA charging. Each tRNA is specific for a particular amino acid. All tRNAs have the sequence CCA at the 3′ end, and the carboxyl group (COO⁻) of the amino acid is attached to the adenine nucleotide at the 3′ end of the tRNA (**Figure 15.13**). If each tRNA is specific for a particular amino acid but all amino acids are attached to the same nucleotide (A) at the 3′ end of a tRNA, how does a tRNA link up with its appropriate amino acid?

The key to specificity between an amino acid and its tRNA is a set of enzymes called **aminoacyl-tRNA synthetases**. A cell has 20 different aminoacyl-tRNA synthetases, one for each of the 20 amino acids. Each synthetase recognizes a particular amino acid, as well as all the tRNAs that accept that amino acid. Recognition of the appropriate amino acid by a synthetase is based on the different sizes, charges, and R groups of the amino acids. The recognition of tRNAs by a synthetase depends on the differing nucleotide sequences of the tRNAs. Researchers have identified which nucleotides are important in recognition by altering different nucleotides in a particular tRNA and determining whether the altered tRNA is still recognized by its synthetase (**Figure 15.14**).

The attachment of a tRNA to its appropriate amino acid, termed **tRNA charging**, requires energy, which is supplied by adenosine triphosphate (ATP):

$$\text{amino acid} + \text{tRNA} + \text{ATP} \longrightarrow \text{aminoacyl-tRNA} + \text{AMP} + \text{PP}_i$$

This reaction takes place in two steps (**Figure 15.15**). To identify the resulting aminoacylated tRNA, we write the three-letter abbreviation for the amino acid in front of the tRNA; for example, the amino acid alanine (Ala) attaches to its tRNA (tRNA^{Ala}), giving rise to its aminoacyl-tRNA (Ala-tRNA^{Ala}).

Errors in tRNA charging are rare; they occur in only about 1 in 10,000 to 1 in 100,000 reactions. This fidelity is due in part to the presence of editing (proofreading) activity in many of the synthetases. Editing activity detects and removes incorrectly paired amino acids from the tRNAs. Some antifungal chemical agents work by trapping tRNAs in the editing site of the enzyme, preventing their release and thus inhibiting the process of translation in the fungi.

CONCEPTS

Amino acids are attached to specific tRNAs by aminoacyl-tRNA synthetases in a two-step reaction that requires ATP.

✔ CONCEPT CHECK 6

Amino acids bind to which part of the tRNA?

a. anticodon   c. 3′ end

b. DHU arm   d. 5′ end

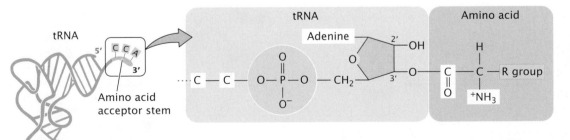

15.13 An amino acid attaches to the 3′ end of a tRNA. The carboxyl group (COO⁻) of the amino acid attaches to the hydroxyl group of the 2′- or 3′-carbon atom of the final nucleotide at the 3′ end of the tRNA, in which the base is always adenine.

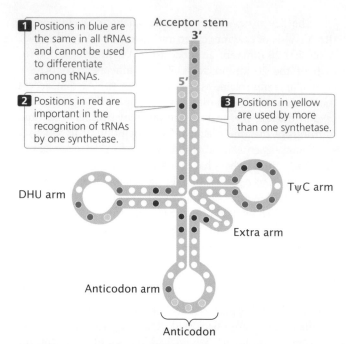

1 Positions in blue are the same in all tRNAs and cannot be used to differentiate among tRNAs.

2 Positions in red are important in the recognition of tRNAs by one synthetase.

3 Positions in yellow are used by more than one synthetase.

Acceptor stem

3'

5'

DHU arm

TψC arm

Extra arm

Anticodon arm

Anticodon

15.14 Certain positions on tRNA molecules are recognized by the appropriate aminoacyl-tRNA synthetase.

The Initiation of Translation

The second stage in the process of protein synthesis is initiation. At this stage, all the components necessary for protein synthesis assemble: (1) mRNA; (2) the small and large subunits of the ribosome; (3) a set of three proteins called initiation factors; (4) initiator tRNA with *N*-formylmethionine attached (fMet-tRNA^fMet); and (5) guanosine triphosphate (GTP). Initiation comprises three major steps. First, mRNA binds to the small subunit of the ribosome. Second, initiator tRNA binds to the mRNA through base pairing between the codon and the anticodon. Third, the large ribosome joins the initiation complex. Let's look at each of these steps more closely.

Initiation in bacteria   The functional ribosome of bacteria exists as two subunits, the small 30S subunit and the large 50S subunit (**Figure 15.16a**). An mRNA molecule can bind to the small ribosome subunit only when the subunits are separate. **Initiation factor 3** (IF-3) binds to the small subunit of the ribosome and prevents the large subunit from binding during initiation (**Figure 15.16b**). Another factor, **initiation factor 1** (IF-1), enhances the disassociation of the large and small ribosomal subunits.

Key sequences on the mRNA required for ribosome binding have been identified in experiments designed to allow the ribosome to bind to mRNA but not proceed with protein synthesis; the ribosome is thereby stalled at the initiation site. After the ribosome is allowed to attach to the mRNA, ribonuclease is added, which degrades all the mRNA except the region covered by the ribosome. The intact mRNA can be separated from the ribosome and studied. The sequence covered by the ribosome during initiation is from 30 to 40 nucleotides long and includes the AUG initiation codon. Within the ribosome-binding site is the Shine–Dalgarno consensus sequence (**Figure 15.17**; see also Chapter 14), which is complementary to a sequence of nucleotides at the 3' end of 16S rRNA (part of the small subunit of the ribosome). During initiation, the nucleotides in the Shine–Dalgarno sequence pair with their complementary nucleotides in the 16S rRNA, allowing the small subunit of the ribosome to attach to the mRNA and positioning the ribosome directly over the initiation codon.

Next, the initiator tRNA, fMet-tRNA^fMet, attaches to the initiation codon (see Figure 15.16c). This step requires **initiation factor 2** (IF-2), which forms a complex with GTP.

At this point, the initiation complex consists of (1) the small subunit of the ribosome; (2) the mRNA; (3) the initiator tRNA with its amino acid (fMet-tRNA^fMet); (4) one molecule of GTP; and (5) several initiation factors. These components are collectively known as the **30S initiation complex** (see Figure 15.16c). In the final step of initiation, IF-3 dissociates from the small subunit, allowing the large subunit of the ribosome to join the initiation complex. The molecule of GTP (provided by IF-2) is hydrolyzed to guanosine diphosphate (GDP), and the initiation factors dissociate (see Figure 15.16d). When the large subunit has joined the initiation complex, the complex is called the **70S initiation complex**.

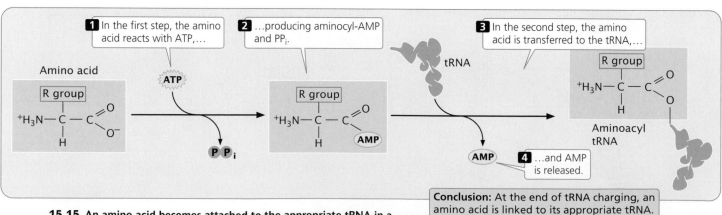

1 In the first step, the amino acid reacts with ATP,...

2 ...producing aminocyl-AMP and PPi.

3 In the second step, the amino acid is transferred to the tRNA,...

4 ...and AMP is released.

Amino acid

R group

ATP

PPi

R group

AMP

tRNA

R group

Aminoacyl tRNA

AMP

Conclusion: At the end of tRNA charging, an amino acid is linked to its appropriate tRNA.

15.15 An amino acid becomes attached to the appropriate tRNA in a two-step reaction.

Shine–Dalgarno sequence

mRNA 5′ AUGUAC **UAAGGAGGU** UGUA **AUG** GAACAAGACG 3′

AUUCCUCCA

Initiation codon

16S rRNA 3′ 5′

15.17 The Shine–Dalgarno consensus sequence in mRNA is required for the attachment of the small subunit of the ribosome.

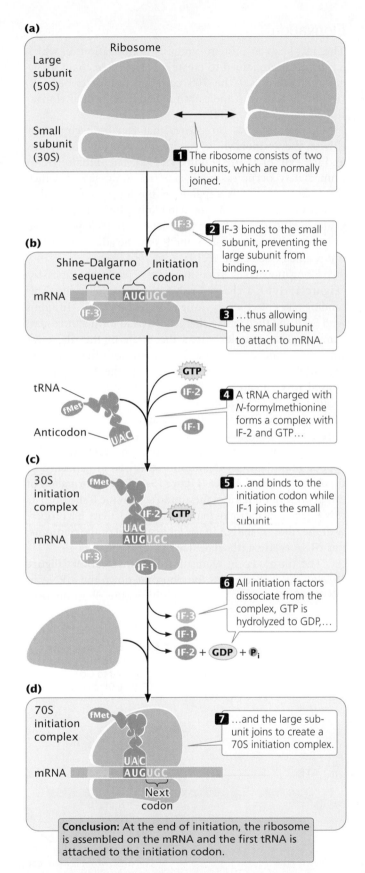

(a)

Ribosome

Large subunit (50S)

Small subunit (30S)

1 The ribosome consists of two subunits, which are normally joined.

(b)

2 IF-3 binds to the small subunit, preventing the large subunit from binding,…

Shine–Dalgarno sequence   Initiation codon

mRNA AUGUGC

IF-3

3 …thus allowing the small subunit to attach to mRNA.

GTP

tRNA

fMet

Anticodon UAC

IF-2

IF-1

4 A tRNA charged with *N*-formylmethionine forms a complex with IF-2 and GTP…

(c)

30S initiation complex

fMet

IF-2 GTP

UAC

mRNA AUGUGC

IF-3

IF-1

5 …and binds to the initiation codon while IF-1 joins the small subunit.

6 All initiation factors dissociate from the complex, GTP is hydrolyzed to GDP,…

IF-3

IF-1

IF-2 + GDP + Pi

(d)

70S initiation complex

fMet

UAC

mRNA AUGUGC

Next codon

7 …and the large subunit joins to create a 70S initiation complex.

Conclusion: At the end of initiation, the ribosome is assembled on the mRNA and the first tRNA is attached to the initiation codon.

15.16 The initiation of translation requires several initiation factors and GTP.

Initiation in eukaryotes Similar events take place in the initiation of translation in eukaryotic cells, but there are some important differences. In bacterial cells, sequences in 16S rRNA of the small subunit of the ribosome bind to the Shine–Dalgarno sequence in mRNA. No analogous consensus sequence exists in eukaryotic mRNA. Instead, the cap at the 5′ end of eukaryotic mRNA plays a critical role in the initiation of translation. In a series of steps, the small subunit of the eukaryotic ribosome, initiation factors, and the initiator tRNA with its amino acid (Met-tRNA$_i$^{Met}) form an initiation complex that recognizes the cap and binds there. The initiation complex then moves along (scans) the mRNA until it locates the first AUG codon. The identification of the start codon is facilitated by the presence of a consensus sequence (called the Kozak sequence) that surrounds the start codon:

Kozak sequence

5′–ACCAUGG–3′

Start codon

Another important difference is that eukaryotic initiation requires at least seven initiation factors. Some factors keep the ribosomal subunits separated, just as IF-3 does in bacterial cells. Others recognize the 5′ cap on mRNA and allow the small subunit of the ribosome to bind there. Still others possess RNA helicase activity, which is used to unwind secondary structures that may exist in the 5′ untranslated region of mRNA, allowing the small subunit to move down the mRNA until the initiation codon is reached. Other initiation factors help bring Met-tRNA$_i$^{Met} to the initiation complex.

The poly(A) tail at the 3′ end of eukaryotic mRNA also plays a role in the initiation of translation. During initiation, proteins that attach to the poly(A) tail interact with proteins that bind to the 5′ cap, enhancing the binding of the small subunit of the ribosome to the 5′ end of the mRNA. This interaction indicates that the 3′ end of mRNA bends over and associates with the 5′ cap during the initiation of translation, forming a circular structure known as the closed loop (**Figure 15.18**). A few eukaryotic mRNAs contain internal ribosome entry sites, where ribosomes can bind directly without first attaching to the 5′ cap.

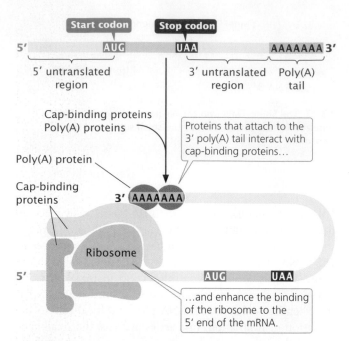

15.18 The poly(A) tail of eukaryotic mRNA plays a role in the initiation of translation.

Start codon / Stop codon

5′ untranslated region

3′ untranslated region

Poly(A) tail

Cap-binding proteins
Poly(A) proteins

Proteins that attach to the 3′ poly(A) tail interact with cap-binding proteins…

Poly(A) protein

Cap-binding proteins

3′ AAAAAAA

Ribosome

5′

AUG   UAA

…and enhance the binding of the ribosome to the 5′ end of the mRNA.

CONCEPTS

In the initiation of translation in bacterial cells, the small ribosomal subunit attaches to mRNA, and initiator tRNA attaches to the initiation codon. This process requires several initiation factors (IF-1, IF-2, and IF-3) and GTP. In the final step, the large ribosomal subunit joins the initiation complex.

✔ CONCEPT CHECK 7

During the initiation of translation, the small ribosome binds to which consensus sequence in bacteria?

Elongation

The next stage in protein synthesis is elongation, in which amino acids are joined to create a polypeptide chain. Elongation requires (1) the 70S complex just described; (2) tRNAs charged with their amino acids; (3) several elongation factors; and (4) GTP.

A ribosome has three sites that can be occupied by tRNAs; the **aminoacyl**, or **A**, **site**, the **peptidyl**, or **P**, **site**, and the **exit**, or **E**, **site** (**Figure 15.19a**). The initiator tRNA immediately occupies the P site (the only site to which the fMet-tRNA^fMet is capable of binding), but all other tRNAs first enter the A site. After initiation, the ribosome is attached to the mRNA, and fMet-tRNA^fMet is positioned over the AUG start codon in the P site; the adjacent A site is unoccupied (see Figure 15.19a).

Elongation takes place in three steps. In the first step (**Figure 15.19b**), a charged tRNA binds to the A site. This binding takes place when **elongation factor Tu** (EF-Tu) joins with GTP and then with a charged tRNA to form a three-part complex. This complex enters the A site of the ribosome, where the anticodon on the tRNA pairs with the codon on the mRNA. After the charged tRNA is in the A site, GTP is cleaved to GDP, and the EF-Tu–GDP complex is released (**Figure 15.19c**). **Elongation factor Ts** (EF-Ts) regenerates EF-Tu–GDP to EF-Tu–GTP. In eukaryotic cells, a similar set of reactions delivers the charged tRNA to the A site.

The second step of elongation is the formation of a peptide bond between the amino acids that are attached to tRNAs in the P and A sites (**Figure 15.19d**). The formation of this peptide bond releases the amino acid in the P site from its tRNA. Evidence indicates that the catalytic activity is a property of the ribosomal RNA in the large subunit of the ribosome; this rRNA acts as a ribozyme (see p. 352 in Chapter 13).

The third step in elongation is **translocation** (**Figure 15.19e**), the movement of the ribosome down the mRNA in the 5′→3′ direction. This step positions the ribosome over

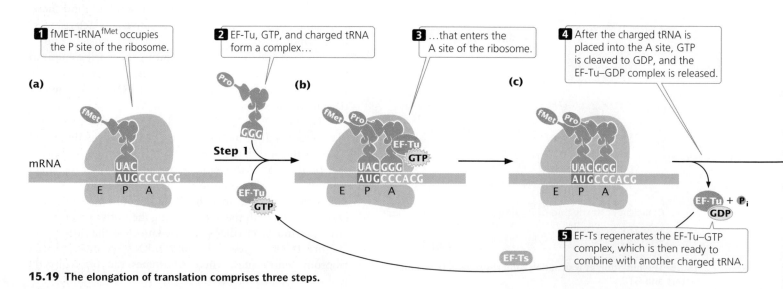

1 fMET-tRNA^fMet occupies the P site of the ribosome.

2 EF-Tu, GTP, and charged tRNA form a complex…

3 …that enters the A site of the ribosome.

4 After the charged tRNA is placed into the A site, GTP is cleaved to GDP, and the EF-Tu–GDP complex is released.

(a)   (b)   (c)

Step 1

mRNA

UAC
AUGCCCACG
E P A

UACGGG
AUGCCCACG
E P A

UACGGG
AUGCCCACG
E P A

EF-Tu + Pᵢ
GDP

5 EF-Ts regenerates the EF-Tu–GTP complex, which is then ready to combine with another charged tRNA.

15.19 The elongation of translation comprises three steps.

the next codon and requires **elongation factor G** (EF-G) and the hydrolysis of GTP to GDP. Because the tRNAs in the P and A sites are still attached to the mRNA through codon–anticodon pairing, they do not move with the ribosome as it translocates. Consequently, the ribosome shifts so that the tRNA that previously occupied the P site now occupies the E site, from which it moves into the cytoplasm where it can be recharged with another amino acid. Translocation also causes the tRNA that occupied the A site (which is attached to the growing polypeptide chain) to be in the P site, leaving the A site open. Thus, the progress of each tRNA through the ribosome in the course of elongation can be summarized as follows: cytoplasm → A site → P site → E site → cytoplasm. As discussed earlier, the initiator tRNA is an exception: it attaches directly to the P site and never occupies the A site.

After translocation, the A site of the ribosome is empty and ready to receive the tRNA specified by the next codon. The elongation cycle (see Figure 15.19b through e) repeats itself: a charged tRNA and its amino acid occupy the A site, a peptide bond is formed between the amino acids in the A and P sites, and the ribosome translocates to the next codon. Throughout the cycle, the polypeptide chain remains attached to the tRNA in the P site. Messenger RNAs, although single stranded, often contain secondary structures formed by pairing of complementary bases on different parts of the mRNA (see Figure 13.1b). As the ribosome moves along the mRNA, these secondary structures are unwound by helicase activity located in the small subunit of the ribosome.

Recently, researchers have developed methods for following a single ribosome as it translates individual codons of an mRNA molecule. These studies revealed that translation does not take place in a smooth continuous fashion. Each translocation step typically requires less than a tenth of a second, but sometimes there are distinct pauses, often lasting a few seconds, between each translocation event when the ribosome moves from one codon to another. Thus, transla-

tion takes place in a series of quick translocations interrupted by brief pauses. In addition to the short pauses between translocation events, translation may be interrupted by longer pauses—lasting from 1 to 2 minutes—that may play a role in regulating the process of translation.

Elongation in eukaryotic cells takes place in a similar manner. Eukaryotes possess at least three elongation factors, one of which also acts in initiation and termination. Another of the elongation factors used in eukaryotes, called elongation factor 2 (EF-2) is the target of a toxin produced by bacteria that causes diphtheria, a disease that, until recently, was a leading killer of children. The diphtheria toxin inhibits EF-2, preventing translocation of the ribosome along the mRNA, and protein synthesis ceases.

CONCEPTS

Elongation consists of three steps: (1) a charged tRNA enters the A site, (2) a peptide bond is created between amino acids in the A and P sites, and (3) the ribosome translocates to the next codon. Elongation requires several elongation factors and GTP.

✔ CONCEPT CHECK 8

In elongation, the creation of peptide bonds between amino acids is catalyzed by

a. rRNA.   c. protein in the large subunit.

b. protein in the small subunit.   d. tRNA.

Termination

Protein synthesis terminates when the ribosome translocates to a termination codon. Because there are no tRNAs with anticodons complementary to the termination codons, no tRNA enters the A site of the ribosome when a termination codon is encountered (**Figure 15.20a**). Instead, proteins called **release factors** bind to the ribosome (**Figure 15.20b**). *E. coli* has three release factors—RF_1, RF_2, and RF_3. Release factor

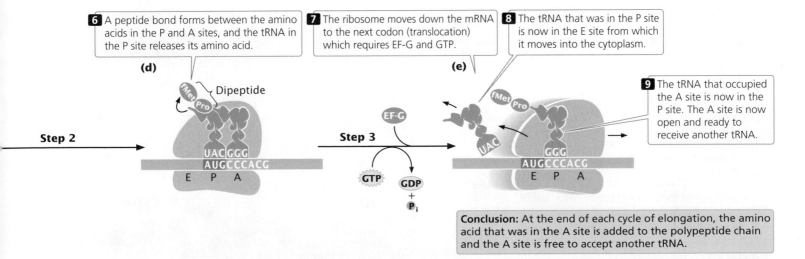

6 A peptide bond forms between the amino acids in the P and A sites, and the tRNA in the P site releases its amino acid.

7 The ribosome moves down the mRNA to the next codon (translocation) which requires EF-G and GTP.

8 The tRNA that was in the P site is now in the E site from which it moves into the cytoplasm.

9 The tRNA that occupied the A site is now in the P site. The A site is now open and ready to receive another tRNA.

(d)

Dipeptide

Step 2

Step 3

EF-G

GTP   GDP + P_i

Conclusion: At the end of each cycle of elongation, the amino acid that was in the A site is added to the polypeptide chain and the A site is free to accept another tRNA.

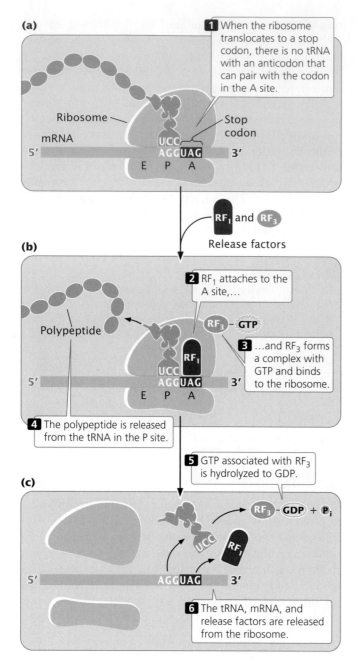

(a)

Ribosome

mRNA

5′ 3′

E P A

1 When the ribosome translocates to a stop codon, there is no tRNA with an anticodon that can pair with the codon in the A site.

Stop codon

RF₁ and RF₃

Release factors

(b)

Polypeptide

5′ 3′

E P A

RF₃ – GTP

RF₁

2 RF₁ attaches to the A site,...

3 ...and RF₃ forms a complex with GTP and binds to the ribosome.

4 The polypeptide is released from the tRNA in the P site.

5 GTP associated with RF₃ is hydrolyzed to GDP.

(c)

5′ 3′

AGGUAG

RF₃ – GDP + Pᵢ

RF₁

6 The tRNA, mRNA, and release factors are released from the ribosome.

15.20 Translation ends when a stop codon is encountered. Because UAG is the termination codon in this illustration, the release factor is RF₁.

15.21 Translation consists of tRNA charging, initiation, elongation, and termination. In this process, amino acids are linked together in the order specified by mRNA to create a polypeptide chain. A number of initiation, elongation, and release factors take part in the process, and energy is supplied by ATP and GTP.

▶ www.whfreeman.com/pierce4e Explore the process of bacterial translation by examining the consequences of various mutations in the coding region of a gene in Animation 15.1.

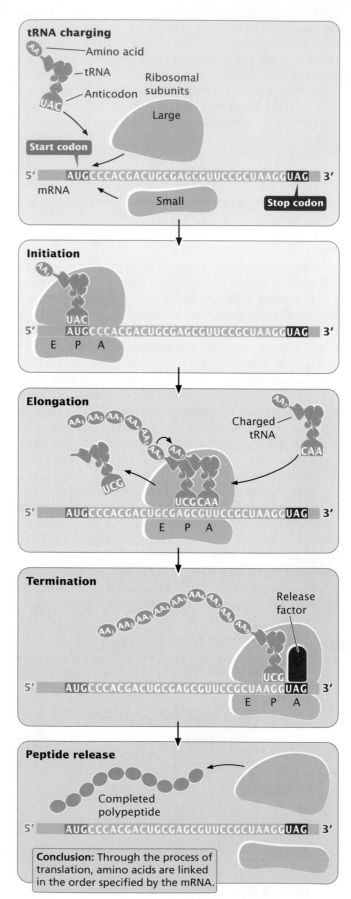

tRNA charging

Amino acid

tRNA

Anticodon

Ribosomal subunits

Large

Start codon

5′ AUGCCCACGACUGCGAGCGUUCCGCUAAGGUAG 3′

mRNA

Small

Stop codon

Initiation

AA₁

5′ AUGCCCACGACUGCGAGCGUUCCGCUAAGGUAG 3′

E P A

Elongation

AA₁ AA₂ AA₃ AA₄ AA₅ AA₆ AA₇

Charged tRNA

CAA

5′ AUGCCCACGACUGCGAGCGUUCCGCUAAGGUAG 3′

E P A

Termination

AA₁ AA₂ AA₃ AA₄ AA₅ AA₆ AA₇ AA₈ AA₉

Release factor

5′ AUGCCCACGACUGCGAGCGUUCCGCUAAGGUAG 3′

E P A

Peptide release

Completed polypeptide

5′ AUGCCCACGACUGCGAGCGUUCCGCUAAGGUAG 3′

Conclusion: Through the process of translation, amino acids are linked in the order specified by the mRNA.

1 binds to the termination codons UAA and UAG, and RF$_2$ binds to UGA and UAA. The binding of release factor RF$_1$ or RF$_2$ to the A site of the ribosome promotes the cleavage of the tRNA in the P site from the polypeptide chain and the release of the polypeptide. Release factor 3 binds to the ribosome and forms a complex with GTP. This binding brings about a conformational change in the ribosome, releasing RF$_1$ or RF$_2$ from the A site and causing the tRNA in the P site to move to the E site; in the process, GTP is hydrolyzed to GDP. Additional factors help bring about the release of the tRNA from the P site, the release of the mRNA from the ribosome, and the dissociation of the ribosome (**Figure 15.20c**).

Translation in eukaryotic cells terminates in a similar way, except that there are two release factors: eRF$_1$, which recognizes all three termination codons, and eRF$_2$, which binds GTP and stimulates the release of the polypeptide from the ribosome. TRY PROBLEMS 27 AND 30 →

CONCEPTS

Termination takes place when the ribosome reaches a termination codon. Release factors bind to the termination codon, causing the release of the polypeptide from the last tRNA, of this tRNA from the ribosome, and of the mRNA from the ribosome.

The overall process of protein synthesis, including tRNA charging, initiation, elongation, and termination, is summarized in **Figure 15.21**. The components taking part in this process are listed in **Table 15.4**.

CONNECTING CONCEPTS

A Comparison of Bacterial and Eukaryotic Translation

We have now considered the process of translation in bacterial cells and noted some distinctive differences that exist in eukaryotic cells. Let's take a few minutes to reflect on some of the important similarities and differences of protein synthesis in bacterial and eukaryotic cells.

First, we should emphasize that the genetic code of bacterial and eukaryotic cells is virtually identical; the only difference is in the amino acid specified by the initiation codon. In bacterial cells, the initiation AUG encodes a modified type of methionine, *N*-formylmethionine, whereas, in eukaryotic cells, initiation AUG encodes unformylated methionine. One consequence of the fact that bacteria and eukaryotes use the same code is that eukaryotic genes can be translated in bacterial systems, and vice versa; this feature makes genetic engineering possible, as we will see in Chapter 19.

Another difference is that transcription and translation take place simultaneously in bacterial cells, but the nuclear envelope separates these processes in eukaryotic cells. The physical separation

Table 15.4 Components required for protein synthesis in bacterial cells

Stage	Component	Function
Binding of amino acid to tRNA	Amino acids	Building blocks of proteins
	tRNAs	Deliver amino acids to ribosomes
	Aminoacyl-tRNA synthetase	Attaches amino acids to tRNAs
	ATP	Provides energy for binding amino acid to tRNA
Initiation	mRNA	Carries coding instructions
	fMet-tRNA^{fMet}	Provides first amino acid in peptide
	30S ribosomal subunit	Attaches to mRNA
	50S ribosomal subunit	Stabilizes tRNAs and amino acids
	Initiation factor 1	Enhances dissociation of large and small subunits of ribosome
	Initiation factor 2	Binds GTP; delivers fMet tRNA^{fMet} to initiation codon
	Initiation factor 3	Binds to 30S subunit and prevents association with 50S subunit
Elongation	70S initiation complex	Functional ribosome with A, P, and E sites where protein synthesis takes place
	Charged tRNAs	Bring amino acids to ribosome and help assemble them in order specified by mRNA
	Elongation factor Tu	Binds GTP and charged tRNA; delivers charged tRNA to A site
	Elongation factor Ts	Generates active elongation factor Tu
	Elongation factor G	Stimulates movement of ribosome to next codon
	GTP	Provides energy
	Peptidyl transferase center	Creates peptide bond between amino acids in A site and P site
Termination	Release factors 1, 2, and 3	Bind to ribosome when stop codon is reached and terminate translation

of transcription and translation has important implications for the control of gene expression, which we will consider in Chapter 17, and it allows for extensive modification of eukaryotic mRNAs, as discussed in Chapter 14.

Yet another difference is that mRNA in bacterial cells is short-lived, typically lasting only a few minutes, but the longevity of mRNA in eukaryotic cells is highly variable and is frequently hours or days.

In both bacterial and eukaryotic cells, aminoacyl-tRNA synthetases attach amino acids to their appropriate tRNAs and the chemical process is the same. There are significant differences in the sizes and compositions of bacterial and eukaryotic ribosomal subunits. For example, the large subunit of the eukaryotic ribosome contains three rRNAs, whereas the bacterial ribosome contains only two. These differences allow antibiotics and other substances to inhibit bacterial translation while having no effect on the translation of eukaryotic nuclear genes, as will be discussed later in this chapter.

Other fundamental differences lie in the process of initiation. In bacterial cells, the small subunit of the ribosome attaches directly to the region surrounding the start codon through hydrogen bonding between the Shine–Dalgarno consensus sequence in the 5′ untranslated region of the mRNA and a sequence at the 3′ end of the 16S rRNA. In contrast, the small subunit of a eukaryotic ribosome first binds to proteins attached to the 5′ cap on mRNA and then migrates down the mRNA, scanning the sequence until it encounters the first AUG initiation codon. Additionally, a larger number of initiation factors take part in eukaryotic initiation than in bacterial initiation.

Elongation and termination are similar in bacterial and eukaryotic cells, although different elongation and termination factors are used. In both types of organisms, mRNAs are translated multiple times and are simultaneously attached to several ribosomes, forming polyribosomes, as discussed in Section 15.4.

Much less is known about the process of translation in archaea, but they appear to possess a mixture of eubacterial and eukaryotic features. Because archaea lack nuclear membranes, transcription and translation take place simultaneously, just as they do in eubacterial cells. Archaea utilize unformylated methionine as the initiator amino acid, a characteristic of eukaryotic translation. Some of the initiation and release factors in archaea are similar to those found in eubacteria, whereas others are similar to those found in eukaryotes. Finally, some of the antibiotics that inhibit translation in eubacteria have no effect on translation in archaea, providing further evidence of the fundamental differences between eubacteria and archaea.

15.4 Additional Properties of RNA and Ribosomes Affect Protein Synthesis

Now that we have considered in some detail the process of translation, we will examine some additional aspects of protein synthesis and the protein-synthesis machinery.

The Three-Dimensional Structure of the Ribosome

The central role of the ribosome in protein synthesis was recognized in the 1950s, and many aspects of its structure have been studied since then. Nevertheless, many details of ribosome structure and function remained a mystery until detailed, three-dimensional reconstructions were completed recently. In 2009, the Nobel Prize in chemistry was awarded to Venkatraman Ramakrishnan, Thomas Steitz, and Ada Yonath for their research on the molecular structure of the ribosome.

Figure 15.22a shows a model of the *E. coli* ribosome at low resolution. A high-resolution image of the bacterial ribosome as determined by X-ray crystallography is represented in the model depicted in **Figure 15.22b**. The mRNA is bound to the small subunit of the ribosome, and the tRNAs are located in the A, P, and E sites (**Figure 15.22c**) that bridge the small and large subunits (see Figure 15.22a). Initiation factors 1 and 3 bind to sites on the outside of the small subunit of the ribosome. EF-Tu, EF-G, and other factors complexed with GTP interact with a *factor-binding center*. High-resolution crystallographic images provide information indicating that a *decoding center* resides in the small subunit of the ribosome (the decoding center cannot be seen in Figure 15.22b). This center senses the fit between the codon on the mRNA and the anticodon on the incoming charged tRNA. Only tRNAs with the correct anticodon are bound tightly by the ribosome. The structural analyses also indicate that the large subunit of the ribosome contains the *peptidyl transferase center*, where peptide-bond formation takes place. There are no ribosomal proteins in the peptidyl

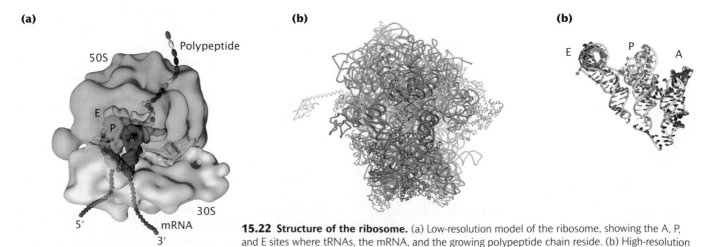

15.22 Structure of the ribosome. (a) Low-resolution model of the ribosome, showing the A, P, and E sites where tRNAs, the mRNA, and the growing polypeptide chain reside. (b) High-resolution model of the ribosome. (c) Positions of tRNAs in E, P, and A sites of the ribosome shown in part *b*.

transferase center; peptide-bond formation is carried out by the RNA molecule. These analyses also reveal that a tunnel connects the site of peptide-bond formation with the back of the ribosome; the growing polypeptide chain passes through this tunnel to the outside of the ribosome. The tunnel can accommodate about 35 amino acids of the growing polypeptide chain.

Polyribosomes

In both prokaryotic and eukaryotic cells, mRNA molecules are translated simultaneously by multiple ribosomes (**Figure 15.23**). The resulting structure—an mRNA with several ribosomes attached—is called a **polyribosome**. Each ribosome successively attaches to the ribosome-binding site at the 5′ end of the mRNA and moves toward the 3′ end; the polypeptide associated with each ribosome becomes progressively longer as the ribosome moves along the mRNA.

In prokaryotic cells, transcription and translation are simultaneous; so multiple ribosomes may be attached to the 5′ end of the mRNA while transcription is still taking place at the 3′ end, as shown in Figure 15.23.

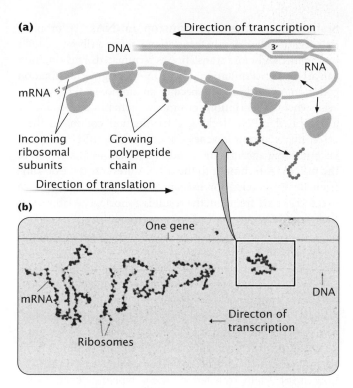

15.23 An mRNA molecule may be transcribed simultaneously by several ribosomes. (a) Four ribosomes are translating an mRNA molecule; the ribosomes move from the 5′ end to the 3′ end of the mRNA. (b) In this electron micrograph of a polyribosome, the dark-staining spheres are ribosomes, and the long, thin filament connecting the ribosomes is mRNA. The 5′ end of the mRNA is toward the left-hand side of the micrograph. [Part b: O. L. Miller, Jr., and Barbara A. Hamaklo.]

CONCEPTS

In both prokaryotic and eukaryotic cells, multiple ribosomes may be attached to a single mRNA, generating a structure called a polyribosome.

✔ **CONCEPT CHECK 9**

In a polyribosome, the polypeptides associated with which ribosomes will be the longest?

a. Those at the 5′ end of mRNA

b. Those at the 3′ end of mRNA

c. Those in the middle of mRNA

d. All polypeptides will be the same length.

Messenger RNA Surveillance

The accurate transfer of genetic information from one generation to the next and from genotype to phenotype is critical for the proper development and functioning of an organism. Consequently, cells have evolved a number of quality-control mechanisms to ensure the accuracy of information transfer. Protein synthesis is no exception: several mechanisms, collectively termed **mRNA surveillance**, exist to detect and deal with errors in mRNAs that may create problems in the course of translation.

Nonsense-mediated mRNA decay A common mutation is one in which a codon that specifies an amino acid is altered to become a termination codon (called a nonsense mutation, see Chapter 18). A nonsense mutation does not affect transcription, but translation ends prematurely when the termination codon is encountered. The resulting protein is truncated and often nonfunctional. A common way that nonsense mutations arise in eukaryotic cells is when one or more of the exons is skipped or improperly spliced. Improper splicing leads to the deletion or addition of nucleotides in the mRNA, which alters the reading frame and often introduces premature termination codons.

To prevent the synthesis of aberrant proteins resulting from nonsense mutations, eukaryotic cells have evolved a mechanism called **nonsense-mediated mRNA decay** (NMD), which results in the rapid elimination of mRNA containing premature termination codons. The mechanism responsible for nonsense-mediated mRNA decay is still unclear. In mammals, it appears to entail proteins that bind to the exon–exon junctions. These exon-junction proteins may interact with enzymes that degrade the mRNA. One possibility is that the first ribosome to translate the mRNA removes the exon-junction proteins, thus protecting the mRNA from degradation. However, when the ribosome encounters a premature termination codon, the ribosome does not traverse the entire mRNA, and some of the exon-junction proteins are not removed, resulting in nonsense-mediated mRNA decay. **TRY PROBLEM 33 ➡**

Stalled ribosomes and nonstop mRNAs A problem that occasionally arises in translation is when a ribosome stalls on the mRNA before translation is terminated. This situation can arise when a mutation in the DNA changes a termination codon into a codon that specifies an amino acid. It can also arise when transcription terminates prematurely, producing a truncated mRNA lacking a termination codon. In these cases, the ribosome reaches the end of the mRNA without encountering a termination codon and stalls, still attached to the mRNA. Attachment to the mRNA prevents the ribosome from being recycled for use on other mRNAs and, if such occurrences are frequent, the result is a shortage of ribosomes that diminishes overall levels of protein synthesis.

To deal with stalled ribosomes, bacteria have evolved a kind of molecular tow truck called **transfer–messenger RNA** (tmRNA). This RNA molecule has properties of both tRNA and mRNA; its tRNA component is normally charged with the amino acid alanine. When a ribosome becomes stalled on an mRNA, EF-Tu delivers the tmRNA to the ribosome's A site, where tmRNA acts as surrogate tRNA (**Figure 15.24**). A peptide bond is created between the amino acid in the P site of the ribosome and alanine (attached to tmRNA) now in the A site, transferring the polypeptide chain to the tmRNA and releasing the tRNA in the P site.

The ribosome then resumes translation, switching from the original, aberrant mRNA to the mRNA part of tmRNA. Translation adds 10 amino acids encoded by the tmRNA, and then a termination codon is reached at the 3′ end of the tmRNA, which terminates translation and releases the ribosome. The added amino acids are a special tag that targets the incomplete polypeptide chain for degradation. Some evidence suggests that the tmRNA also targets the aberrant mRNA for degradation. How stalled ribosomes are recognized by the tmRNA is not clear, but this method is efficient at recycling stalled ribosomes and eliminating abnormal proteins that result from truncated transcription.

Eukaryotes have evolved a different mechanism to deal with mRNAs that are missing termination codons. Instead of restarting the stalled ribosome and degrading the abnormal protein that results, eukaryotic cells use a mechanism called **nonstop mRNA decay**, which results in the rapid degradation of abnormal mRNA. In this mechanism, the codon-free A site of the stalled ribosome is recognized by a special protein that binds to the A site and recruits other proteins, which then degrade the mRNA from its 3′ end.

CONCEPTS

Cells possess mRNA surveillance mechanisms to detect and eliminate mRNA molecules containing errors that create problems in the course of translation. These mechanisms include tmRNA, which rescues ribosomes stalled on the mRNA, nonstop mRNA decay, which degrades mRNAs lacking a termination codon, and nonsense-mediated mRNA decay, which eliminates mRNAs that possess a premature stop codon.

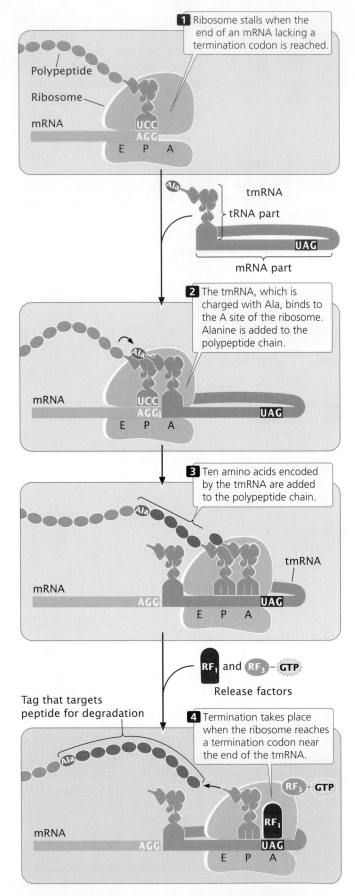

15.24 The tmRNA in bacteria allows stalled ribosomes to resume translation.

The Posttranslational Modifications of Proteins

After translation, proteins in both prokaryotic and eukaryotic cells may undergo alterations termed posttranslational modifications. A number of different types of modifications are possible. As mentioned earlier, the formyl group or the entire methionine residue may be removed from the amino end of a protein. Some proteins are synthesized as larger precursor proteins and must be cleaved and trimmed by enzymes before the proteins can become functional. For others, the attachment of carbohydrates may be required for activation. Amino acids within a protein may be modified: phosphates, carboxyl groups, and methyl groups are added to some amino acids. In eukaryotic cells, the amino end of a protein is often acetylated after translation.

The functions of many proteins critically depend on the proper folding of the polypeptide chain; some proteins spontaneously fold into their correct shapes, but, for others, correct folding may initially require the participation of other molecules called **molecular chaperones**. Some molecular chaperones are associated with the ribosome and fold newly synthesized polypeptide chains as they emerge from the ribosome tunnel, in which case protein folding takes place during ongoing translation.

A common posttranslational modification in eukaryotes is the attachment of a protein called ubiquitin, which targets the protein for degradation. Another modification of some proteins is the removal of 15 to 30 amino acids, called the **signal sequence**, at the amino end of the protein. The signal sequence helps direct a protein to a specific location within the cell, after which the sequence is removed by special enzymes.

> **CONCEPTS**
>
> Many proteins undergo posttranslational modifications after their synthesis.

Translation and Antibiotics

Antibiotics are drugs that kill microorganisms. To make an effective antibiotic, not just any poison will do: the trick is to kill the microbe without harming the patient. Antibiotics must be carefully chosen so that they destroy bacterial cells but not the eukaryotic cells of their host.

Translation is frequently the target of antibiotics because translation is essential to all living organisms and differs significantly between bacterial and eukaryotic cells. For example, as already mentioned, bacterial and eukaryotic ribosomes differ in size and composition. A number of antibiotics bind selectively to bacterial ribosomes and inhibit various steps in translation, but they do not affect eukaryotic ribosomes. Tetracyclines, for instance, are a class of antibiotics that bind to the A site of a bacterial ribosome and block the entry of charged tRNAs, yet they have no effect on eukaryotic ribosomes. Neomycin binds to the ribosome near the A site and induces translational errors, probably by causing mistakes in the binding of charged tRNAs to the A site. Chloramphenicol binds to the large subunit of the ribosome and blocks peptide-bond formation. Streptomycin binds to the small subunit of the ribosome and inhibits initiation, and erythromycin blocks translocation. Although chloramphenicol and streptomycin are potent inhibitors of translation in bacteria, they do not inhibit translation in archaea.

The three-dimensional structure of puromycin resembles the 3' end of a charged tRNA, permitting puromycin to enter the A site of a ribosome efficiently and inhibit the entry of tRNAs. A peptide bond can form between the puromycin molecule in the A site and an amino acid on the tRNA in the P site of the ribosome, but puromycin cannot bind to the P site and translocation does not take place, blocking further elongation of the protein. Because tRNA structure is similar in all organisms, puromycin inhibits translation in both bacterial and eukaryotic cells; consequently, puromycin kills eukaryotic cells along with bacteria and is sometimes used in cancer therapy to destroy tumor cells.

Many antibiotics act by blocking specific steps in translation, and different antibiotics affect different steps in protein synthesis. Because of this specificity, antibiotics are frequently used to study the process of protein synthesis.

Nonstandard Protein Synthesis

The process of translation that we have considered thus far is part of the common genetic machinery used by all organisms. Many bacteria and fungi also possess an alternative protein-synthesis pathway, which they use to synthesize a few short, specialized proteins. Remarkably, this system does not rely on the ribosome; instead, it uses enzymes—some of which are the largest known—to assemble amino acids. Because this pathway does not rely on the ribosome, it is able to produce some proteins with unusual structures. For example, proteins produced by nonribosomal protein synthesis may contain some unusual molecular forms of amino acids and even some molecular relatives of amino acids. Because these proteins have unusual structures, they often resist degradation and may go undetected by pathogens.

CONCEPTS SUMMARY

- George Beadle and Edward Tatum developed the one gene, one enzyme hypothesis, which proposed that each gene specifies one enzyme; this hypothesis was later modified to become the one gene, one polypeptide hypothesis.

- Twenty different amino acids are used in the composition of proteins. The amino acids in a protein are linked together by peptide bonds. Chains of amino acids fold and associate to produce the secondary, tertiary, and quaternary structures of proteins.

- Solving the genetic code required several different approaches including the use of synthetic mRNAs with random sequences and short mRNAs that bind charged tRNAs.

- The genetic code is a triplet code: three nucleotides specify a single amino acid. It is also degenerate (meaning that more than one codon may specify an amino acid), nonoverlapping, and universal (almost).

- Different tRNAs (isoaccepting tRNAs) may accept the same amino acid. Different codons may pair with the same anticodon through wobble, which can exist at the third position of the codon and allows some nonstandard pairing of bases in this position.

- The reading frame is set by the initiation codon. The end of the protein-coding section of an mRNA is marked by one of three termination codons.

- Protein synthesis comprises four steps: (1) the binding of amino acids to the appropriate tRNAs, (2) initiation, (3) elongation, and (4) termination.

- The binding of an amino acid to a tRNA requires the presence of a specific aminoacyl-tRNA synthetase and ATP. The amino acid is attached by its carboxyl end to the 3′ end of the tRNA.

- In bacterial translation initiation, the small subunit of the ribosome attaches to the mRNA and is positioned over the initiation codon. It is joined by the first tRNA and its associated amino acid (*N*-formylmethionine in bacterial cells) and, later, by the large subunit of the ribosome. Initiation requires several initiation factors and GTP.

- In elongation, a charged tRNA enters the A site of a ribosome, a peptide bond is formed between amino acids in the A and P sites, and the ribosome moves (translocates) along the mRNA to the next codon. Elongation requires several elongation factors and GTP.

- Translation is terminated when the ribosome encounters one of the three termination codons. Release factors and GTP are required to bring about termination.

- Each mRNA may be simultaneously translated by several ribosomes, producing a structure called a polyribosome.

- Cells possess RNA surveillance mechanisms that eliminate mRNAs with errors that may create problems in translation.

- Antibiotics frequently work by interfering with translation, because many aspects of translation differ in bacteria and eukaryotes.

- Many proteins undergo posttranslational modification. A few peptides are synthesized by nonribosomal pathways that utilize large enzymes.

IMPORTANT TERMS

one gene, one enzyme hypothesis (p. 405)
one gene, one polypeptide hypothesis (p. 405)
amino acid (p. 405)
peptide bond (p. 405)
polypeptide (p. 405)
sense codon (p. 410)
degenerate genetic code (p. 410)
synonymous codons (p. 410)
isoaccepting tRNAs (p. 410)
wobble (p. 410)
nonoverlapping genetic code (p. 411)
reading frame (p. 411)
initiation codon (p. 411)

stop (termination or nonsense) codon (p. 412)
universal genetic code (p. 412)
aminoacyl-tRNA synthetase (p. 413)
tRNA charging (p. 413)
initiation factor (IF-1, IF-2, IF-3) (pp. 414)
30S initiation complex (p. 414)
70S initiation complex (p. 414)
aminoacyl (A) site (p. 416)
peptidyl (P) site (p. 416)
exit (E) site (p. 416)
elongation factor Tu (EF-Tu) (p. 416)
elongation factor Ts (EF-Ts) (p. 416)

translocation (p. 416)
elongation factor G (EF-G) (p. 417)
release factor (RF$_1$, RF$_2$, RF$_3$) (p. 417)
polyribosome (p. 421)
mRNA surveillance (p. 421)
nonsense-mediated mRNA decay (NMD) (p. 421)
transfer–messenger RNA (tmRNA) (p. 422)
nonstop mRNA decay (p. 422)
molecular chaperone (p. 423)
signal sequence (p. 423)

ANSWERS TO CONCEPT CHECKS

1. B → A → C

2. The amino acid sequence (primary structure) of the protein

3. b

4. d

5. The initiation codon in bacteria encodes *N*-formylmethionine; in eukaryotes, it encodes methionine. Termination codons do not specify amino acids.

6. c

7. The Shine–Dalgarno sequence

8. a

9. b

WORKED PROBLEMS

1. A series of auxotrophic mutants were isolated in *Neurospora*. Examination of fungi containing these mutations revealed that they grew on minimal medium to which various compounds (A, B, C, D) were added; growth responses to each of the four compounds are presented in the following table. Give the order of compounds A, B, C, and D in a biochemical pathway. Outline a biochemical pathway that includes these four compounds and indicate which step in the pathway is affected by each of the mutations.

	Compound			
Mutation number	A	B	C	D
134	+	+	−	+
276	+	+	+	+
987	−	−	−	+
773	+	+	+	+
772	−	−	−	+
146	+	+	−	+
333	+	+	−	+
123	−	+	−	+

• Solution

To solve this problem, we should first group the mutations by which compounds allow growth, as follows:

Mutation		Compound			
Group	Number	A	B	C	D
I	276	+	+	+	+
	773	+	+	+	+
II	134	+	+	−	+
	146	+	+	−	+
	333	+	+	−	+
III	123	−	+	−	+
IV	987	−	−	−	+
	772	−	−	−	+

The underlying principle used to determine the order of the compounds in the pathway is as follows: if a compound is added after the block, it will allow the mutant to grow, whereas, if a compound is added before the block, it will have no effect. Applying this principle to the data in the table, we see that mutants in group I will grow if compound A, B, C, or D is added

to the medium; so these mutations must affect a step before the production of all four compounds:

$$\xrightarrow{\text{Group I mutations}} \text{compounds A, B, C, D}$$

Group II mutants will grow if compound A, B, or D is added but not if compound C is added. Thus, compound C comes before A, B, and D; and group II mutations affect the conversion of compound C into one of the other compounds:

$$\xrightarrow{\text{Group I mutations}} \text{compound C} \xrightarrow{\text{Group II mutations}} \text{compounds A, B, D}$$

Group III mutants allow growth if compound B or D is added but not if compound A or C is added. Thus, group III mutations affect steps that follow the production of A and C; we have already determined that compound C precedes A in the pathway, and so A must be the next compound in the pathway:

$$\xrightarrow{\text{Group I mutations}} \text{compound C} \xrightarrow{\text{Group II mutations}}$$

$$\text{compound A} \xrightarrow{\text{Group III mutations}} \text{compounds B, D}$$

Finally, mutants in group IV will grow if compound D is added but not if compound A, B, or C is added. Thus, compound D is the fourth compound in the pathway, and mutations in group IV block the conversion of B into D:

$$\xrightarrow{\text{Group I mutations}} \text{compound C} \xrightarrow{\text{Group II mutations}} \text{compound A} \xrightarrow{\text{Group III mutations}}$$

$$\text{compound B} \xrightarrow{\text{Group IV mutations}} \text{compound D}$$

2. If there were five different types of bases in mRNA instead of four, what would be the minimum codon size (number of nucleotides) required to specify the following numbers of different amino acid types: (a) 4, (b) 20, (c) 30?

• Solution

To answer this question, we must determine the number of combinations (codons) possible when there are different numbers of bases and different codon lengths. In general, the number of different codons possible will be equal to:

$$b^{lg} = \text{number of codons}$$

where b equals the number of different types of bases and lg equals the number of nucleotides in each codon (codon length). If there are five different types of bases, then:

$$5^1 = \quad 5 \text{ possible codons}$$
$$5^2 = \quad 25 \text{ possible codons}$$
$$5^3 = 125 \text{ possible codons}$$

The number of possible codons must be greater than or equal to the number of amino acids specified. Therefore, a codon length of one nucleotide could specify 4 different amino acids, a codon length of two nucleotides could specify 20 different amino acids, and a codon length of three nucleotides could specify 30 different amino acids: (a) one, (b) two, (c) three.

3. A template strand in bacterial DNA has the following base sequence:

$$5'\text{–AGGTTTAACGTGCAT–}3'$$

What amino acids are encoded by this sequence?

• Solution

To answer this question, we must first work out the mRNA sequence that will be transcribed from this DNA sequence. The mRNA must be antiparallel and complementary to the DNA template strand:

DNA template strand: $5'$–AGGTTTA ACGTGC AT–$3'$
mRNA copied from DNA: $3'$–UCCAAAUUGCACGUA–$5'$

An mRNA is translated $5' \rightarrow 3'$; so it will be helpful if we turn the RNA molecule around with the $5'$ end on the left:

mRNA copied from DNA: $5'$–AUGCACGUUAAACCU–$3'$

The codons consist of groups of three nucleotides that are read successively after the first AUG codon; using Figure 15.10, we can determine that the amino acids are

$$5'\text{–AUG—CAC—GUU—AAA—CCU–}3'$$

$$\downarrow \quad\quad \downarrow \quad\quad \downarrow \quad\quad \downarrow \quad\quad \downarrow$$

$$\text{fMet——His——Val——Lys——Pro}$$

4. The following triplets constitute anticodons found on a series of tRNAs. Name the amino acid carried by each of these tRNAs.

a. $5'$–UUU–$3'$

b. $5'$–GAC–$3'$

c. $5'$–UUG–$3'$

d. $5'$–CAG–$3'$

• Solution

To solve this problem, we first determine the codons with which these anticodons pair and then look up the amino acid specified by the codon in Figure 15.10. The codons are antiparallel and complementary to the anticodons. For part a, the anticodon is $5'$–UUU–$3'$. According to the wobble rules in Table 15.2, U in the first position of the anticodon can pair with either A or G in the third position of the codon, so there are two codons that can pair with this anticodon:

Anticodon: $5'$–UUU–$3'$
Codon: $3'$–AAA–$5'$
Codon: $3'$–GAA–$5'$

Listing these codons in the conventional manner, with the $5'$ end on the right, we have:

Codon: $5'$–AAA–$3'$
Codon: $5'$–AAG–$3'$

According to Figure 15.10, both codons specify the amino acid lysine (Lys). Recall that the wobble in the third position allows more than one codon to specify the same amino acid; so any wobble that exists should produce the same amino acid as the standard base pairings would, and we do not need to figure the wobble to answer this question. The answers for parts b, c, and d are:

b. Anticodon: $5'$–GAC–$3'$
     Codon: $3'$–CUG–$5'$
                  $5'$–GUC–$3'$ encodes Val

c. Anticodon: $5'$–UUG–$3'$
     Codon: $3'$–AAC–$5'$
                  $5'$–CAA–$3'$ encodes Gln

d. Anticodon: $5'$–CAG–$3'$
     Codon: $3'$–GUC–$5'$
                  $5'$–CUG–$3'$ encodes Leu

COMPREHENSION QUESTIONS

Section 15.1

1. What is the one gene, one enzyme hypothesis? Why was this hypothesis an important advance in our understanding of genetics?

Section 15.2

*2. What different methods were used to help break the genetic code? What did each method reveal and what were the advantages and disadvantages of each one?

3. What are isoaccepting tRNAs?

*4. What is the significance of the fact that many synonymous codons differ only in the third nucleotide position?

*5. Define the following terms as they apply to the genetic code:

a. Reading frame
b. Overlapping code
c. Nonoverlapping code
d. Initiation codon
e. Termination codon

f. Sense codon
g. Nonsense codon
h. Universal code
i. Nonuniversal codons

6. How is the reading frame of a nucleotide sequence set?

Section 15.3

*7. How are tRNAs linked to their corresponding amino acids?

8. What role do the initiation factors play in protein synthesis?

9. How does the process of initiation differ in bacterial and eukaryotic cells?

*10. Give the elongation factors used in bacterial translation and explain the role played by each factor in translation.

11. What events bring about the termination of translation?

12. Compare and contrast the process of protein synthesis in bacterial and eukaryotic cells, giving similarities and differences in the process of translation in these two types of cells.

Section 15.4

13. How do prokaryotic cells overcome the problem of a stalled ribosome on an mRNA that has no termination codon? How do eukaryotic cells solve this problem?

14. What are some types of posttranslational modification of proteins?

*15. Explain how some antibiotics work by affecting the process of protein synthesis.

APPLICATION QUESTIONS AND PROBLEMS

Section 15.1

*16. Sydney Brenner isolated *Salmonella typhimurium* mutants that were implicated in the biosynthesis of tryptophan and would not grow on minimal medium. When these mutants were tested on minimal medium to which one of four compounds (indole glycerol phosphate, indole, anthranilic acid, and tryptophan) had been added, the growth responses shown in the following table were obtained.

Mutant	Minimal medium	Anthranilic acid	Indole glycerol phosphate	Indole	Tryptophan
trp-1	−	−	−	−	+
trp-2	−	−	+	+	+
trp-3	−	−	−	+	+
trp-4	−	−	+	+	+
trp-6	−	−	−	−	+
trp-7	−	−	−	−	+
trp-8	−	+	+	+	+
trp-9	−	−	−	−	+
trp-10	−	−	−	−	+
trp-11	−	−	−	−	+

Give the order of indole glycerol phosphate, indole, anthranilic acid, and tryptophan in a biochemical pathway leading to the synthesis of tryptophan. Indicate which step in the pathway is affected by each of the mutations.

17. Compounds I, II, and III are in the following biochemical pathway:

$$\text{precursor} \xrightarrow[\substack{\text{enzyme} \\ A}]{} \text{compound I} \xrightarrow[\substack{\text{enzyme} \\ B}]{}$$

$$\text{compound II} \xrightarrow[\substack{\text{enzyme} \\ C}]{} \text{compound III}$$

Mutation *a* inactivates enzyme A, mutation *b* inactivates enzyme B, and mutation *c* inactivates enzyme C. Mutants, each having one of these defects, were tested on minimal medium to which compound I, II, or III was added. Fill in the results expected of these tests by placing a plus sign (+) for growth or a minus sign (−) for no growth in the table on page 428:

	Minimal medium to which is added		
Strain with mutation	Compound I	Compound II	Compound III
a			
b			
c			

Section 15.2

*18. Assume that the number of different types of bases in RNA is four. What would be the minimum codon size (number of nucleotides) required if the number of different types of amino acids in proteins were: (a) 2, (b) 8, (c) 17, (d) 45, (e) 75?

19. How many codons would be possible in a triplet code if only three bases (A, C, and U) were used?

*20. Using the genetic code presented in Figure 15.10, give the amino acids specified by the following bacterial mRNA sequences, and indicate the amino and carboxyl ends of the polypeptide produced.

   a. 5′–AUGUUUAAAUUUAAAUUUUGA–3′

   b. 5′–AUGUAUAUAUAUAUAUGA–3′

   c. 5′–AUGGAUGAAAGAUUUCUCGCUUGA–3′

   d. 5′–AUGGGUUAGGGGACAUCAUUUUGA–3′

21. A nontemplate strand on bacterial DNA has the following base sequence. What amino acid sequence will be encoded by this sequence?

   5′–ATGATACTAAGGCCC–3′

*22. The following amino acid sequence is found in a tripeptide: Met-Trp-His. Give all possible nucleotide sequences on the mRNA, on the template strand of DNA, and on the nontemplate strand of DNA that can encode this tripeptide.

23. How many different mRNA sequences can encode a polypeptide chain with the amino acid sequence Met-Leu-Arg? (Be sure to include the stop codon.)

*24. A series of tRNAs have the following anticodons. Consider the wobble rules listed in Table 15.2 and give all possible codons with which each tRNA can pair.

   a. 5′–GGC–3′

   b. 5′–AAG–3′

   c. 5′–IAA–3′

   d. 5′–UGG–3′

   e. 5′–CAG–3′

25. An anticodon on a tRNA has the sequence 5′–GCA–3′.

   a. What amino acid is carried by this tRNA?

   b. What would be the effect if the G in the anticodon were mutated to a U?

26. Which of the following amino acid changes could result from a mutation that changed a single base? For each change that could result from the alteration of a single base, determine which position of the codon (first, second, or third nucleotide) in the mRNA must be altered for the change to result.

   a. Leu → Gln

   b. Phe → Ser

   c. Phe → Ile

   d. Pro → Ala

   e. Asn → Lys

   f. Ile → Asn

Section 15.3

27. Arrange the following components of translation in the approximate order in which they would appear or be used in protein synthesis:

   70S initiation complex
   30S initiation complex
   release factor 1
   elongation factor G
   initiation factor 3
   elongation factor Tu
   fMet-tRNA^{fMet}

28. The following diagram illustrates a step in the process of translation. Sketch the diagram and identify the following elements on it.

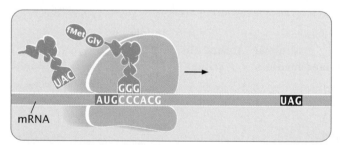

   a. 5′ and 3′ ends of the mRNA

   b. A, P, and E sites

   c. Start codon

   d. Stop codon

e. Amino and carboxyl ends of the newly synthesized polypeptide chain

f. Approximate location of the next peptide bond that will be formed

g. Place on the ribosome where release factor 1 will bind

29. Refer to the diagram in Problem 28 to answer the following questions.

a. What will be the anticodon of the next tRNA added to the A site of the ribosome?

b. What will be the next amino acid added to the growing polypeptide chain?

*30. A synthetic mRNA added to a cell-free protein-synthesizing system produces a peptide with the following amino acid sequence: Met-Pro-Ile-Ser-Ala. What would be the effect on translation if the following components were omitted from the cell-free protein-synthesizing system? What, if any, type of protein would be produced? Explain your reasoning.

a. Initiation factor 1

b. Initiation factor 2

c. Elongation factor Tu

d. Elongation factor G

e. Release factors RF_1, RF_2, and RF_3

f. ATP

g. GTP

31. For each of the following sequences, place a check mark in the appropriate space to indicate the process *most immediately* affected by deleting the sequence. Choose only one process for each sequence (i.e., one check mark per sequence).

Process most immediately affected by deletion

Sequence deleted	Replication	Transcription	RNA processing	Translation
a. *ori* site	_____	_____	_____	_____
b. 3′ splice-site consensus	_____	_____	_____	_____
c. poly(A) tail	_____	_____	_____	_____
d. terminator	_____	_____	_____	_____
e. start codon	_____	_____	_____	_____
f. −10 consensus	_____	_____	_____	_____
g. Shine–Dalgarno	_____	_____	_____	_____

32. MicroRNAs are small RNA molecules that bind to the 3′ end of mRNAs and suppress translation (see Chapter 14). How miRNAs suppress translation is still being investigated. Some eukaryotic mRNAs have internal ribosome-binding sites downstream of the 5′ cap, where ribosomes normally bind. In one investigation, miRNAs did not suppress the translation of ribosomes that attach to internal ribosome-binding sites (R. S. Pillai et al. 2005. *Science* 309:1573–1576). What does this finding suggest about how miRNAs suppress translation?

Section 15.4

33. Mutations that introduce stop codons cause a number of genetic diseases. For example, from 2% to 5% of the people who have cystic fibrosis possess a mutation that causes a premature stop codon in the gene encoding the cystic fibrosis transmembrane conductance regulator (CFTR). This premature stop codon produces a truncated form of CFTR that is nonfunctional and results in the symptoms of cystic fibrosis. One possible way to treat people with genetic diseases caused by these types of mutations is to trick the ribosome into reading through the stop codon, inserting an amino acid into its place. Although the protein produced may have one altered amino acid, it is more likely to be at least partly functional than is the truncated protein produced when the ribosome stalls at the stop codon. Indeed, geneticists have conducted clinical trials on people with cystic fibrosis with the use of a drug called PTC124, which interferes with the ribosome's ability to correctly read stop codons (C. Ainsworth. 2005. *Nature* 438:726–728). On the basis of what you know about the mechanism of nonsense-mediated mRNA decay (NMD), would you expect NMD to be a problem with this type of treatment? Why or why not?

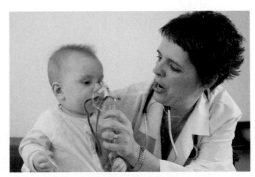

Child with cystic fibrosis. [Lisa Eastman/Alamy.]

CHALLENGE QUESTIONS

Section 15.2

34. The redundancy of the genetic code means that some amino acids are specified by more than one codon. For example, the amino acid leucine is encoded by six different codons. Within a genome, synonymous codons are not present in equal numbers; some synonymous codons appear much more frequently than others, and the preferred codons differ among different species. For example, in one species, the codon UUA might be used most often to encode leucine, whereas, in another species, the codon CUU might be used most often. Speculate on a reason for this bias in codon usage and why the preferred codons are not the same in all organisms.

Section 15.3

35. In what ways are spliceosomes and ribosomes similar? In what ways are they different? Can you suggest some possible reasons for their similarities.

*36. Several experiments were conducted to obtain information about how the eukaryotic ribosome recognizes the AUG start codon. In one experiment, the gene that encodes methionine initiator tRNA (tRNA$_i^{Met}$) was located and changed. The nucleotides that specify the anticodon on tRNA$_i^{Met}$ were mutated so that the anticodon in the tRNA was 5′-CCA-3′ instead of 5′-CAU-3′. When this mutated gene was placed in a eukaryotic cell, protein synthesis took place but the proteins produced were abnormal. Some of the proteins produced contained extra amino acids, and others contained fewer amino acids than normal.

a. What do these results indicate about how the ribosome recognizes the starting point for translation in eukaryotic cells? Explain your reasoning.

b. If the same experiment had been conducted on bacterial cells, what results would you expect?

16 Control of Gene Expression in Prokaryotes

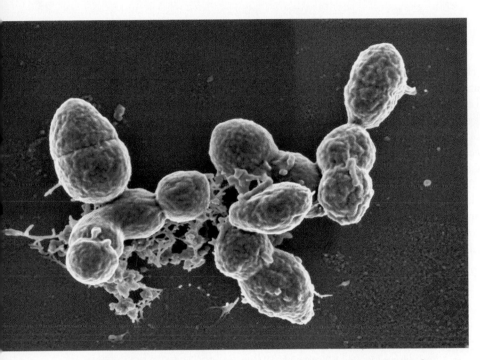

The bacterium *Streptococcus pneumoniae* is responsible each year for millions of cases of pneumonia, inner-ear infection, sinus infection, and meningitis. Gene regulation is required for transformation, a process by which many strains of *S. pneumoniae* have become resistant to antibiotics. [Dr. Gary Gaugler/Photo Researchers.]

STRESS, SEX, AND GENE REGULATION IN BACTERIA

Each year, *Streptococcus pneumoniae*, commonly known as pneumococcus, is responsible for millions of cases of pneumonia, inner-ear infection, sinus infection, and meningitis. Before the widespread use of antibiotics, pneumonia resulting from *S. pneumoniae* infection was a leading cause of death, particularly among the young and the elderly. The development of penicillin and other antibiotics, beginning in the 1940s, provided an effective tool against *Streptococcus* infections, and deaths from pneumonia and other life-threatening bacterial infections plummeted. In recent years, however, many strains of *S. pneumoniae* and other pathogenic bacteria have developed resistance to penicillin and other antibiotics. Some bacterial strains have rapidly acquired resistance to multiple antibiotics, making treatment of *S. pneumoniae* difficult. How has antibiotic resistance spread so rapidly among bacteria?

Although bacteria are simple organisms, they engage in complex and ingenious mechanisms to exchange genetic information and rapidly adapt to changing conditions. One such mechanism is transformation, first demonstrated by Fred Griffith in 1928, in which bacteria take up new DNA released by dead cells and integrate the foreign DNA into their own genomes (see pp. 274–275 in Chapter 10). Transformation is considered a type of sexuality in bacteria because it results in the exchange of genetic material between individual cells. It is widespread among bacteria and has been a major means by which antibiotic resistance has spread.

In *S. pneumoniae*, the ability to carry out transformation—called competence—requires from 105 to 124 genes, collectively termed the *com* (for competence) regulon. Genes within the *com* regulon facilitate the passage of DNA through the cell membrane and the DNA's recombination with homologous genes in the bacterial chromosome. For competence to take place, genes in the *com* regulon must be precisely turned on and off in a carefully regulated sequence.

The *com* regulon is activated in response to a protein called competence-stimulating peptide (CSP), which is produced by bacteria and is exported into the surrounding medium. When enough CSP accumulates, it attaches to a receptor on the bacterial cell membrane, which then activates a regulator protein that stimulates the transcription of genes within the *com* regulon and sets in motion a complex cascade of reactions that ultimately results in the development of competence.

Early steps in the development of competence are poorly understood, but most experts formerly assumed that CSP is continually produced and slowly accumulates in the medium; the assumption was that, when CSP reaches a critical threshold, transformation is induced. However, recent research has altered the view that the induction of transformation is a passive process. This work has demonstrated that the production of CSP is itself regulated. When *S. pneumoniae* is subjected to stress, it produces more CSP, which causes transformation and allows the bacteria to genetically adapt to the stress by acquiring new genetic material from the environment. In fact, the presence of antibiotics is one type of stress that stimulates the production of CSP and induces transformation. This reaction to stress is probably a primary reason that bacteria such as *S. pneumoniae* have so quickly evolved resistance to antibiotics. TRY PROBLEM 10 →

The regulation of gene expression not only is important for the development of transformation, but is critical for the control of numerous life processes in all organisms. This chapter is the first of two chapters about **gene regulation**, the mechanisms and systems that control the expression of genes. In this chapter, we consider systems of gene regulation in bacteria. We begin by considering the necessity for gene regulation, the levels at which gene expression is controlled, and the difference between genes and regulatory elements. We then examine the structure and function of operons, groups of genes that are regulated together in bacteria, and consider gene regulation in some specific examples of operons. Finally, we study several types of bacterial gene regulation that are facilitated by RNA molecules.

16.1 The Regulation of Gene Expression Is Critical for All Organisms

A major theme of molecular genetics is the central dogma, which states that genetic information flows from DNA to RNA to proteins (see Figure 10.16). Although the central dogma provided a molecular basis for the connection between genotype and phenotype, it failed to address a critical question: How is the flow of information along the molecular pathway *regulated*?

Consider *Escherichia coli*, a bacterium that resides in your large intestine. Your eating habits completely determine the nutrients available to this bacterium: it can neither seek out nourishment when nutrients are scarce nor move away when confronted with an unfavorable environment. *E. coli* makes up for its inability to alter the external environment by being internally flexible. For example, if glucose is present, *E. coli* uses it to generate ATP; if there's no glucose, it utilizes lactose, arabinose, maltose, xylose, or any of a number of other sugars. When amino acids are available, *E. coli* uses them to synthesize proteins; if a particular amino acid is absent, *E. coli* produces the enzymes needed to synthesize that amino acid. Thus, *E. coli* responds to environmental changes by rapidly altering its biochemistry. This biochemical flexibility, however, has a high price. Producing all the enzymes necessary for every environmental condition would be energetically expensive. So how does *E. coli* maintain biochemical flexibility while optimizing energy efficiency?

The answer is through gene regulation. Bacteria carry the genetic information for synthesizing many proteins, but only a subset of this genetic information is expressed at any time. When the environment changes, new genes are expressed, and proteins appropriate for the new environment are synthesized. For example, if a carbon source appears in the environment, genes encoding enzymes that take up and metabolize this carbon source are quickly transcribed and translated. When this carbon source disappears, the genes that encode these enzymes are shut off.

Multicellular eukaryotic organisms face a different dilemma. Individual cells in a multicellular organism are specialized for particular tasks. The proteins produced by a nerve cell, for example, are quite different from those produced by a white blood cell. The problem that a eukaryotic cell faces is how to specialize. Although they are quite different in shape and function, a nerve cell and a blood cell still carry the same genetic instructions.

A multicellular organism's challenge is to bring about the specialization of cells that have a common set of genetic instructions (the process of development). This challenge is met through gene regulation: all of an organism's cells carry the same genetic information, but only a subset of genes are expressed in each cell type. Genes needed for other cell types are not expressed. Gene regulation is therefore the key to both unicellular flexibility and multicellular specialization, and it is critical to the success of all living organisms.

CONCEPTS

In bacteria, gene regulation maintains internal flexibility, turning genes on and off in response to environmental changes. In multicellular eukaryotic organisms, gene regulation brings about cellular differentiation.

The mechanisms of gene regulation were first investigated in bacterial cells, in which the availability of mutants and the ease of laboratory manipulation made it possible to unravel the mechanisms. When the study of these mechanisms in eukaryotic cells began, bacterial gene regulation clearly seemed to differ from eukaryotic gene regulation. As more and more information has accumulated about gene regulation, however, a number of common themes have emerged, and, today, many aspects of gene regulation in bacterial and eukaryotic cells are recognized to be similar. Before examining specific elements of bacterial gene regulation (this chapter) and eukaryotic gene regulation (see Chapter 17), we will briefly consider some themes of gene regulation common to all organisms.

Genes and Regulatory Elements

In considering gene regulation in both bacteria and eukaryotes, we must distinguish between the DNA sequences that are transcribed and the DNA sequences that regulate the expression of other sequences. As discussed in Chapter 14, one definition of a *gene* is any DNA sequence that is transcribed into an RNA molecule. Genes include DNA sequences that encode proteins, as well as sequences that encode rRNA, tRNA, snRNA, and other types of RNA. **Structural genes** encode proteins that are used in metabolism or biosynthesis or that play a structural role in the cell. **Regulatory genes** are genes whose products, either RNA or proteins, interact with other DNA sequences and affect the transcription or translation of those sequences. In many cases, the products of regulatory genes are DNA-binding proteins. Bacteria and eukaryotes use regulatory genes to control the expression of many of their structural genes. However, a few structural genes, particularly those that encode essential cellular functions, are expressed continually and are said to be **constitutive**. Constitutive genes are therefore not regulated.

We will also encounter DNA sequences that are not transcribed at all but still play a role in regulating genes and other DNA sequences. These **regulatory elements** affect the expression of sequences to which they are physically linked. Regulatory elements are common in both bacterial and eukaryotic cells, and much of gene regulation in both types of organisms takes place through the action of proteins produced by regulatory genes that recognize and bind to regulatory elements.

The regulation of gene expression can be through processes that stimulate gene expression, termed *positive control*, or through processes that inhibit gene expression, termed *negative control*. Bacteria and eukaryotes use both positive and negative control mechanisms to regulate their genes. However, negative control is more important in bacteria, whereas eukaryotes are more likely to use positive control mechanisms.

CONCEPTS

Genes are DNA sequences that are transcribed into RNA. Regulatory elements are DNA sequences that are not transcribed but affect the expression of genes. Positive control includes mechanisms that stimulate gene expression, whereas negative control inhibits gene expression.

✔ **CONCEPT CHECK 1**

What is a constitutive gene?

Levels of Gene Regulation

In both bacteria and eukaryotes, genes can be regulated at a number of points along the pathway of information flow from genotype to phenotype (**Figure 16.1**). First, regulation

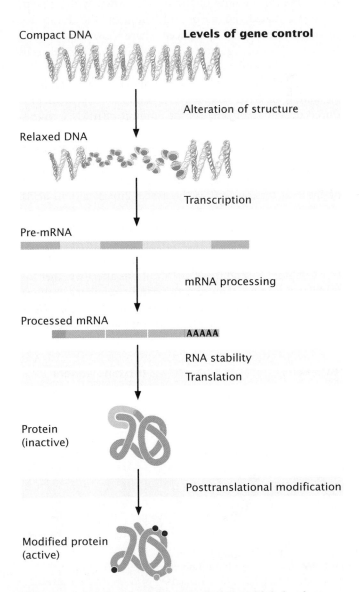

16.1 Gene expression can be controlled at multiple levels.

can be through the alteration of DNA or chromatin structure; this type of gene regulation takes place primarily in eukaryotes. Modifications to DNA or its packaging can help to determine which sequences are available for transcription or the rate at which sequences are transcribed. DNA methylation and changes in chromatin are two processes that play a pivotal role in gene regulation.

A second point at which a gene can be regulated is at the level of transcription. For the sake of cellular economy, limiting the production of a protein early in the process makes sense, and transcription is an important point of gene regulation in both bacterial and eukaryotic cells. A third potential point of gene regulation is mRNA processing. Eukaryotic mRNA is extensively modified before it is translated: a 5′ cap is added, the 3′ end is cleaved and polyadenylated, and introns are removed (see Chapter 14). These modifications determine the stability of the mRNA, whether the mRNA can be translated, the rate of translation, and the amino acid sequence of the protein produced. There is growing evidence that a number of regulatory mechanisms in eukaryotic cells operate at the level of mRNA processing.

A fourth point for the control of gene expression is the regulation of RNA stability. The amount of protein produced depends not only on the amount of mRNA synthesized, but also on the rate at which the mRNA is degraded. A fifth point of gene regulation is at the level of translation, a complex process requiring a large number of enzymes, protein factors, and RNA molecules (see Chapter 15). All of these factors, as well as the availability of amino acids, affect the rate at which proteins are produced and therefore provide points at which gene expression can be controlled. Translation can also be affected by sequences in mRNA.

Finally, many proteins are modified after translation (see Chapter 15), and these modifications affect whether the proteins become active; so genes can be regulated through processes that affect posttranslational modification. Gene expression can be affected by regulatory activities at any or all of these points.

CONCEPTS

Gene expression can be controlled at any of a number of points along the molecular pathway from DNA to protein, including DNA or chromatin structure, transcription, mRNA processing, RNA stability, translation, and posttranslational modification.

✔ **CONCEPT CHECK 2**

Why is transcription a particularly important level of gene regulation in both bacteria and eukaryotes?

DNA-Binding Proteins

Much of gene regulation in bacteria and eukaryotes is accomplished by proteins that bind to DNA sequences and affect their expression. These regulatory proteins generally have discrete functional parts—called **domains**, typically consisting of 60 to 90 amino acids—that are responsible for binding to DNA. Within a domain, only a few amino acids actually make contact with the DNA. These amino acids (most commonly asparagine, glutamine, glycine, lysine, and arginine) often form hydrogen bonds with the bases or interact with the sugar–phosphate backbone of the DNA. Many regulatory proteins have additional domains that can bind other molecules such as other regulatory proteins. By physically attaching to DNA, these proteins can affect the expression of a gene. Most DNA-binding proteins bind dynamically, which means that they are transiently binding and unbinding DNA and other regulatory proteins. Thus, although they may spend most of their time bound to DNA, they are never permanently attached. This dynamic nature means that other molecules can compete with DNA-binding proteins for regulatory sites on the DNA.

DNA-binding proteins can be grouped into several distinct types on the basis of a characteristic structure, called a motif, found within the binding domain. Motifs are simple structures, such as alpha helices, that can fit into the major groove of the DNA. For example, the helix-turn-helix motif (**Figure 16.2a**), consisting of two alpha helices connected

(a) Helix-turn-helix

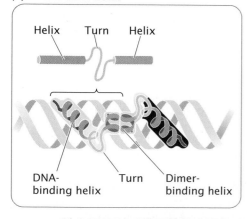

Helix   Turn   Helix

DNA-binding helix   Turn   Dimer-binding helix

(b) Zinc fingers

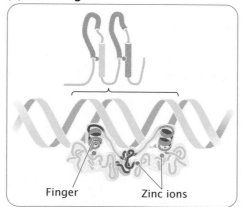

Finger   Zinc ions

(c) Leucine zipper

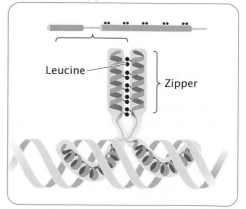

Leucine   Zipper

16.2 DNA-binding proteins can be grouped into several types on the basis of their structures, or motifs.

Table 16.1	Common DNA-binding motifs		
Motif	Location	Characteristics	Binding Site in DNA
Helix-turn-helix	Bacterial regulatory proteins; related motifs in eukaryotic proteins	Two alpha helices	Major groove
Zinc-finger	Eukaryotic regulatory and other proteins	Loop of amino acids with zinc at base	Major groove
Steroid receptor	Eukaryotic proteins	Two perpendicular alpha helices with zinc surrounded by four cysteine residues	Major groove and DNA backbone
Leucine-zipper	Eukaryotic transcription factors	Helix of leucine residues and a basic arm; two leucine residues interdigitate	Two adjacent major grooves
Helix-loop-helix	Eukaryotic proteins	Two alpha helices separated by a loop of amino acids	Major groove
Homeodomain	Eukaryotic regulatory proteins	Three alpha helices	Major groove

by a turn, is common in bacterial regulatory proteins. The zinc-finger motif (**Figure 16.2b**), common to many eukaryotic regulatory proteins, consists of a loop of amino acids containing a zinc ion. The leucine zipper (**Figure 16.2c**) is another motif found in a variety of eukaryotic binding proteins. These common DNA-binding motifs and others are summarized in **Table 16.1**.

CONCEPTS

Regulatory proteins that bind DNA have common motifs that interact with sequences in the DNA.

✔ **CONCEPT CHECK 3**

How do amino acids in DNA-binding proteins interact with DNA?

a. By forming covalent bonds with DNA bases

b. By forming hydrogen bonds with DNA bases

c. By forming covalent bonds with DNA sugars

16.2 Operons Control Transcription in Bacterial Cells

A significant difference between bacterial and eukaryotic gene control lies in the organization of functionally related genes. Many bacterial genes that have related functions are clustered and under the control of a single promoter. These genes are often transcribed together into a single mRNA. A group of bacterial structural genes that are transcribed together (along with their promoter and additional sequences that control transcription) is called an **operon**. The operon regulates the expression of the structural genes by controlling transcription, which, in bacteria, is usually the most important level of gene regulation.

Operon Structure

The organization of a typical operon is illustrated in **Figure 16.3**. At one end of the operon is a set of structural genes, shown in Figure 16.3 as gene *a*, gene *b*, and gene *c*. These structural genes are transcribed into a single mRNA, which is translated to produce enzymes A, B, and C. These enzymes carry out a series of biochemical reactions that convert precursor molecule X into product Y. The transcription of structural genes *a*, *b*, and *c* is under the control of a promoter, which lies upstream of the first structural gene. RNA polymerase binds to the promoter and then moves downstream, transcribing the structural genes.

A **regulator gene** helps to control the transcription of the structural genes of the operon. Although it affects operon function, the regulator gene is not considered part of the operon. The regulator gene has its own promoter and is transcribed into a short mRNA, which is translated into a small protein. This **regulator protein** can bind to a region of the operon called the **operator** and affect whether transcription can take place. The operator usually overlaps the 3' end of the promoter and sometimes the 5' end of the first structural gene (see Figure 16.3).

CONCEPTS

Functionally related genes in bacterial cells are frequently clustered together as a single transcriptional unit termed an operon. A typical operon includes several structural genes, a promoter for the structural genes, and an operator site to which the product of a regulator gene binds.

✔ **CONCEPT CHECK 4**

What is the difference between a structural gene and a regulator gene?

a. Structural genes are transcribed into mRNA, but regulator genes aren't.

b. Structural genes have complex structures; regulator genes have simple structures.

c. Structural genes encode proteins that function in the structure of the cell; regulator genes carry out metabolic reactions.

d. Structural genes encode proteins; regulator genes control the transcription of structural genes.

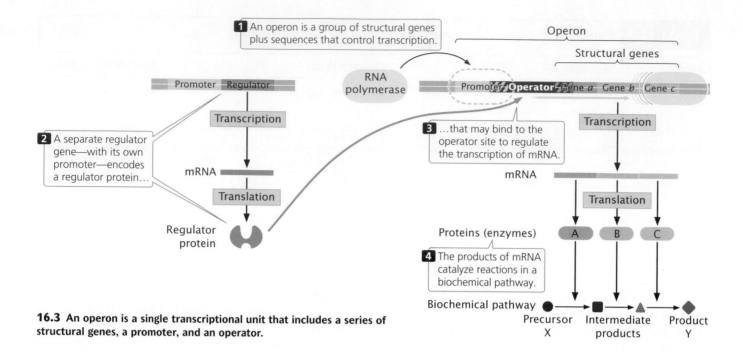

16.3 An operon is a single transcriptional unit that includes a series of structural genes, a promoter, and an operator.

Negative and Positive Control: Inducible and Repressible Operons

There are two types of transcriptional control: **negative control**, in which a regulatory protein is a repressor, binding to DNA and inhibiting transcription; and **positive control**, in which a regulatory protein is an activator, stimulating transcription. Operons can also be either inducible or repressible. **Inducible operons** are those in which transcription is normally off (not taking place); something must happen to induce transcription, or turn it on. **Repressible operons** are those in which transcription is normally on (taking place); something must happen to repress transcription, or turn it off. In the next sections, we will consider several varieties of these basic control mechanisms.

Negative inducible operons In a negative inducible operon, the regulator gene encodes an active *repressor* that readily binds to the operator (**Figure 16.4a**). Because the operator site overlaps the promoter site, the binding of this protein to the operator physically blocks the binding of RNA polymerase to the promoter and prevents transcription. For transcription to take place, something must happen to prevent the binding of the repressor at the operator site. This type of system is said to be *inducible*, because transcription is normally off (inhibited) and must be turned on (induced).

Transcription is turned on when a small molecule, an **inducer**, binds to the repressor (**Figure 16.4b**). Regulatory proteins frequently have two binding sites: one that binds to DNA and another that binds to a small molecule such as an inducer. The binding of the inducer (precursor V in Figure

16.4b) alters the shape of the repressor, preventing it from binding to DNA. Proteins of this type, which change shape on binding to another molecule, are called **allosteric proteins**.

When the inducer is absent, the repressor binds to the operator, the structural genes are not transcribed, and enzymes D, E, and F (which metabolize precursor V) are not synthesized (see Figure 16.4a). This mechanism is an adaptive one: because no precursor V is available, synthesis of the enzymes would be wasteful when they have no substrate to metabolize. As soon as precursor V becomes available, some of it binds to the repressor, rendering the repressor inactive and unable to bind to the operator site. RNA polymerase can now bind to the promoter and transcribe the structural genes. The resulting mRNA is then translated into enzymes D, E, and F, which convert substrate V into product W (see Figure 16.4b). So, an operon with negative inducible control regulates the synthesis of the enzymes economically: the enzymes are synthesized only when their substrate (V) is available.

Inducible operons usually control proteins that carry out degradative processes—proteins that break down molecules. For these types of proteins, inducible control makes sense because the proteins are not needed unless the substrate (which is broken down by the proteins) is present.

Negative repressible operons Some operons with negative control are *repressible*, meaning that transcription *normally* takes place and must be turned off, or repressed. The regulator protein in this type of operon also is a repressor but is synthesized in an *inactive* form that cannot by itself bind to the operator. Because no repressor is bound to the operator,

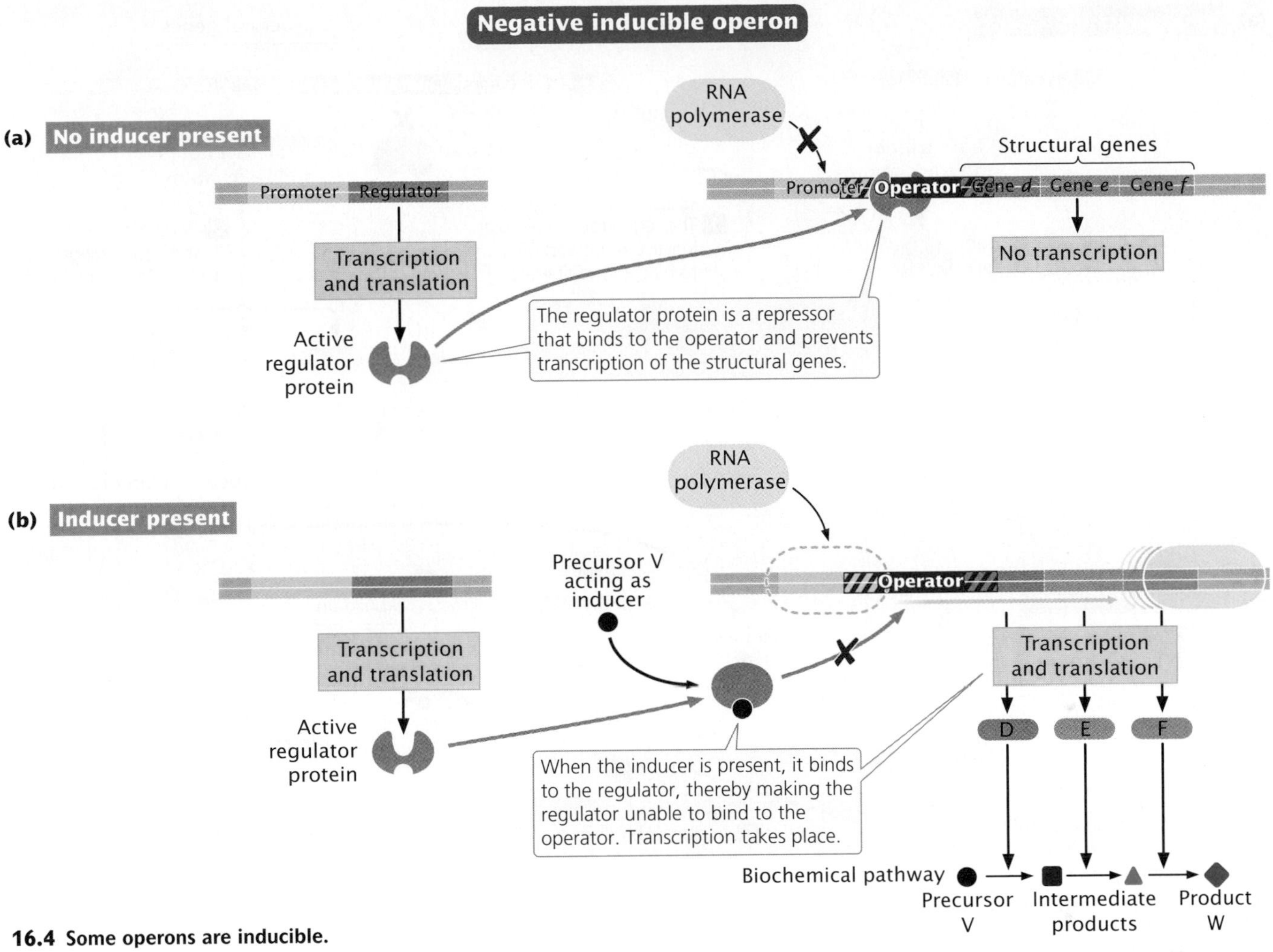

Negative inducible operon

(a) No inducer present

RNA polymerase

Promoter   Regulator

Structural genes

Promoter   **Operator**   Gene *d*   Gene *e*   Gene *f*

Transcription and translation

No transcription

Active regulator protein

The regulator protein is a repressor that binds to the operator and prevents transcription of the structural genes.

(b) Inducer present

RNA polymerase

Precursor V acting as inducer

Operator

Transcription and translation

Transcription and translation

Active regulator protein

D   E   F

When the inducer is present, it binds to the regulator, thereby making the regulator unable to bind to the operator. Transcription takes place.

Biochemical pathway

Precursor V → Intermediate products → Product W

16.4 Some operons are inducible.

RNA polymerase readily binds to the promoter and transcription of the structural genes takes place (**Figure 16.5a**).

To turn transcription off, something must happen to make the repressor active. A small molecule called a **corepressor** binds to the repressor and makes it capable of binding to the operator. In the example illustrated (see Figure 16.5a), the product (U) of the metabolic reaction is the corepressor. As long as the level of product U is high, it is available to bind to the repressor and activate it, preventing transcription (**Figure 16.5b**). With the operon repressed, enzymes G, H, and I are not synthesized, and no more U is produced from precursor T. However, when all of product U is used up, the repressor is no longer activated by U and cannot bind to the operator. The inactivation of the repressor allows the transcription of the structural genes and the synthesis of enzymes G, H, and I, resulting in the conversion of precursor T into product U. Like inducible operons, repressible operons are economical: the enzymes are synthesized only as needed.

Repressible operons usually control proteins that carry out the biosynthesis of molecules needed in the cell, such as

amino acids. For these types of operons, repressible control makes sense because the product produced by the proteins is always needed by the cell. Thus, these operons are normally on and are turned off when there are adequate amounts of the product already present.

Note that both the inducible and the repressible systems that we have considered are forms of negative control, in which the regulatory protein is a repressor. We will now consider positive control, in which a regulator protein stimulates transcription.

Positive control With positive control, a regulatory protein is an activator: it binds to DNA (usually at a site other than the operator) and stimulates transcription. Positive control can be inducible or repressible.

In a positive *inducible* operon, transcription is normally turned off because the regulator protein (an activator) is produced in an inactive form. Transcription takes place when an inducer has become attached to the regulatory protein, rendering the regulator active. Logically, the inducer should be the precursor of the reaction controlled by the

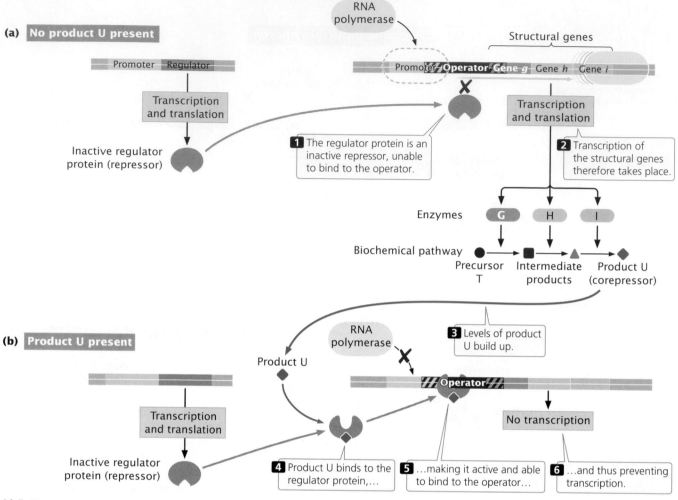

16.5 Some operons are repressible.

operon so that the necessary enzymes would be synthesized only when the substrate for their reaction was present.

A positive operon can also be repressible; the regulatory protein is produced in a form that readily binds to DNA, meaning that transcription normally takes place and has to be repressed. Transcription is inhibited when a substance becomes attached to the activator and renders it unable to bind to the DNA so that transcription is no longer stimulated. Here, the product (P) of the reaction controlled by the operon would logically be the repressing substance, because it would be economical for the cell to prevent the transcription of genes that allow the synthesis of P when plenty of P was already available. The characteristics of positive and negative control in inducible and repressible operons are summarized in **Figure 16.6.** TRY PROBLEM 11 ➡️

CONCEPTS

There are two basic types of transcriptional control: negative and positive. In negative control, when a regulatory protein (repressor) binds to DNA, transcription is inhibited; in positive control, when a regulatory protein (activator) binds to DNA, transcription is stimulated. Some operons are inducible; transcription is normally off and must be turned on. Other operons are repressible; transcription is normally on and must be turned off.

✔ CONCEPT CHECK 5

In a negative repressible operon, the regulator protein is synthesized as

a. an active activator.

b. an inactive activator.

c. an active repressor.

d. an inactive repressor.

The *lac* Operon of *E. coli*

In 1961, François Jacob and Jacques Monod described the "operon model" for the genetic control of lactose metabolism in *E. coli*. This work and subsequent research on the genetics of lactose metabolism established the operon as the basic unit of transcriptional control in bacteria. Despite the fact

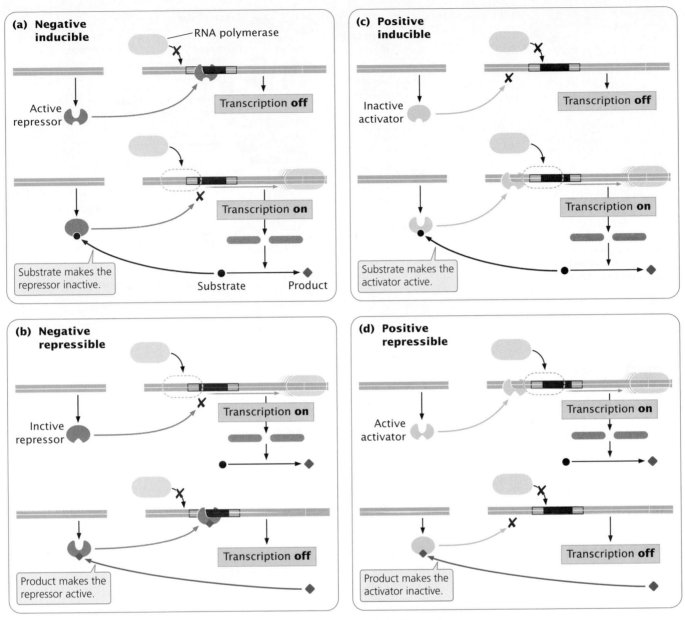

16.6 A summary of the characteristics of positive and negative control in inducible and repressible operons.

that, at the time, no methods were available for determining nucleotide sequences, Jacob and Monod deduced the structure of the operon *genetically* by analyzing the interactions of mutations that interfered with the normal regulation of lactose metabolism. We will examine the effects of some of these mutations after seeing how the *lac* operon regulates lactose metabolism.

Lactose metabolism   Lactose is a major carbohydrate found in milk; it can be metabolized by *E. coli* bacteria that reside in the mammalian gut. Lactose does not easily diffuse across the *E. coli* cell membrane and must be actively transported into the cell by the protein permease

(**Figure 16.7**). To utilize lactose as an energy source, *E. coli* must first break it into glucose and galactose, a reaction catalyzed by the enzyme β-galactosidase. This enzyme can also convert lactose into allolactose, a compound that plays an important role in regulating lactose metabolism. A third enzyme, thiogalactoside transacetylase, also is produced by the *lac* operon, but its function in lactose metabolism is not yet known.

Regulation of the lac operon   The *lac* operon is an example of a negative inducible operon. The enzymes β-galactosidase, permease, and transacetylase are encoded by adjacent structural genes in the *lac* operon of *E. coli*

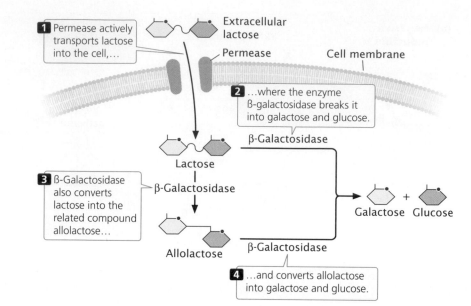

16.7 Lactose, a major carbohydrate found in milk, consists of 2 six-carbon sugars linked together.

The *lac* operon

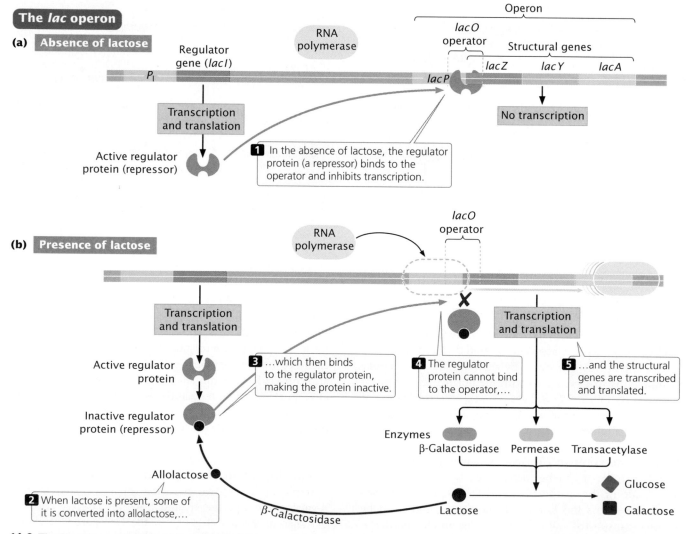

16.8 The *lac* operon regulates lactose metabolism.

(**Figure 16.8a**) and have a common promoter (*lacP* in Figure 16.8a). β-Galactosidase is encoded by the *lacZ* gene, permease by the *lacY* gene, and transacetylase by the *lacA* gene. When lactose is absent from the medium in which *E. coli* grows, few molecules of each protein are produced. If lactose is added to the medium and glucose is absent, the rate of synthesis of all three proteins simultaneously increases about a thousandfold within 2 to 3 minutes. This boost in protein synthesis results from the transcription of *lacZ*, *lacY*, and *lacA* and exemplifies **coordinate induction**, the simultaneous synthesis of several proteins, stimulated by a specific molecule, the inducer (**Figure 16.8b**).

Although lactose appears to be the inducer here, allolactose is actually responsible for induction. Upstream of *lacP* is a regulator gene, *lacI*, which has its own promoter (P_I). The *lacI* gene is transcribed into a short mRNA that is translated into a repressor. The repressor consists of four identical polypeptides and has two types of binding sites; one type of site binds to allolactose and the other binds to DNA. In the absence of lactose (and, therefore, allolactose), the repressor binds to the *lac* operator site *lacO* (see Figure 16.8a). Jacob and Monod mapped the operator to a position adjacent to the *lacZ* gene; more-recent nucleotide sequencing has demonstrated that the operator actually overlaps the 3′ end of the promoter and the 5′ end of *lacZ* (**Figure 16.9**).

RNA polymerase binds to the promoter and moves down the DNA molecule, transcribing the structural genes. When the repressor is bound to the operator, the binding of RNA polymerase is blocked, and transcription is prevented. When lactose is present, some of it is converted into allolactose, which binds to the repressor and causes the repressor to be released from the DNA. In the presence of lactose, then, the repressor is inactivated, the binding of RNA polymerase is no longer blocked, the transcription of *lacZ*, *lacY*, and *lacA* takes place, and the *lac* proteins are produced.

Have you spotted the flaw in the explanation just given for the induction of the *lac* proteins? You might recall that permease is required to transport lactose into the cell. If the *lac* operon is repressed and no permease is being produced, how does lactose get into the cell to inactivate the repressor and turn on transcription? Furthermore, the inducer is actually allolactose, which must be produced from lactose by β-galactosidase. If β-galactosidase production is repressed, how can lactose metabolism be induced?

The answer is that repression never *completely* shuts down transcription of the *lac* operon. Even with active repressor bound to the operator, there is a low level of transcription and a few molecules of β-galactosidase, permease, and transacetylase are synthesized. When lactose appears in the medium, the permease that is present transports a small amount of lactose into the cell. There, the few molecules of β-galactosidase that are present convert some of the lactose into allolactose, which then induces transcription.

Several compounds related to allolactose also can bind to the *lac* repressor and induce transcription of the *lac* operon. One such inducer is isopropylthiogalactoside (IPTG). Although IPTG inactivates the repressor and allows the transcription of *lacZ*, *lacY*, and *lacA*, this inducer is not metabolized by β-galactosidase; for this reason, IPTG is often used in research to examine the effects of induction, independent of metabolism.

CONCEPTS

The *lac* operon of *E. coli* controls the transcription of three genes needed in lactose metabolism: the *lacZ* gene, which encodes β-galactosidase; the *lacY* gene, which encodes permease; and the *lacA* gene, which encodes thiogalactoside transacetylase. The *lac* operon is negative inducible: a regulator gene produces a repressor that binds to the operator site and prevents the transcription of the structural genes. The presence of allolactose inactivates the repressor and allows the transcription of the *lac* operon.

✔ **CONCEPT CHECK 6**

In the presence of allolactose, the *lac* repressor

a. binds to the operator. c. cannot bind to the operator.

b. binds to the promoter. d. binds to the regulator gene.

lac Mutations

Jacob and Monod worked out the structure and function of the *lac* operon by analyzing mutations that affected lactose metabolism. To help define the roles of the different components of the operon, they used **partial diploid** strains of *E. coli*. The cells of these strains possessed two different DNA molecules: the full bacterial chromosome and an extra piece

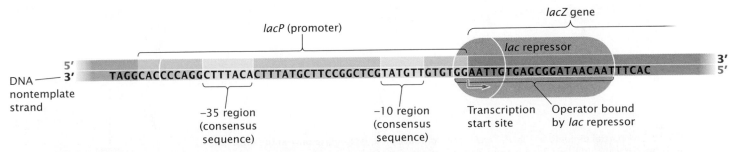

16.9 In the *lac* operon, the operator overlaps the promoter and the 5′ end of the first structural gene.

of DNA. Jacob and Monod created these strains by allowing conjugation to take place between two bacteria (see Chapter 8). In conjugation, a small circular piece of DNA (the F plasmid, see Chapter 8) is transferred from one bacterium to another. The F plasmid used by Jacob and Monod contained the *lac* operon; so the recipient bacterium became partly diploid, possessing two copies of the *lac* operon. By using different combinations of mutations on the bacterial and plasmid DNA, Jacob and Monod determined that some parts of the *lac* operon are cis acting (able to control the expression of genes only when on the same piece of DNA), whereas other parts are trans acting (able to control the expression of genes on other DNA molecules).

Structural-gene mutations   Jacob and Monod first discovered some mutant strains that had lost the ability to synthesize either β-galactosidase or permease. (They did not study in detail the effects of mutations on the transacetylase enzyme, and so transacetylase will not be considered here.) The mutations in the mutant strains mapped to the *lacZ* or *lacY* structural genes and altered the amino acid sequences of the proteins encoded by the genes. These mutations clearly affected the *structure* of the proteins and not the regulation of their synthesis.

Through the use of partial diploids, Jacob and Monod were able to establish that mutations at the *lacZ* and *lacY* genes were independent and usually affected only the product of the gene in which they occurred. Partial diploids with *lacZ⁺ lacY⁻* on the bacterial chromosome and *lacZ⁻ lacY⁺* on the plasmid functioned normally, producing β-galactosidase and permease in the presence of lactose. (The genotype of a partial diploid is written by separating the genes on each DNA molecule with a slash: *lacZ⁺ lacY⁻/lacZ⁻ lacY⁺*.) In this partial diploid, a single functional β-galactosidase gene (*lacZ⁺*) is sufficient to produce β-galactosidase; whether the functional β-galactosidase gene is coupled to a functional (*lacY⁺*) or a defective (*lacY⁻*) permease gene makes no difference. The same is true of the *lacY⁺* gene.

Regulator-gene mutations   Jacob and Monod also isolated mutations that affected the *regulation* of protein production. Mutations in the *lacI* gene affect the production of both β-galactosidase and permease because genes for both proteins are in the same operon and are regulated coordinately.

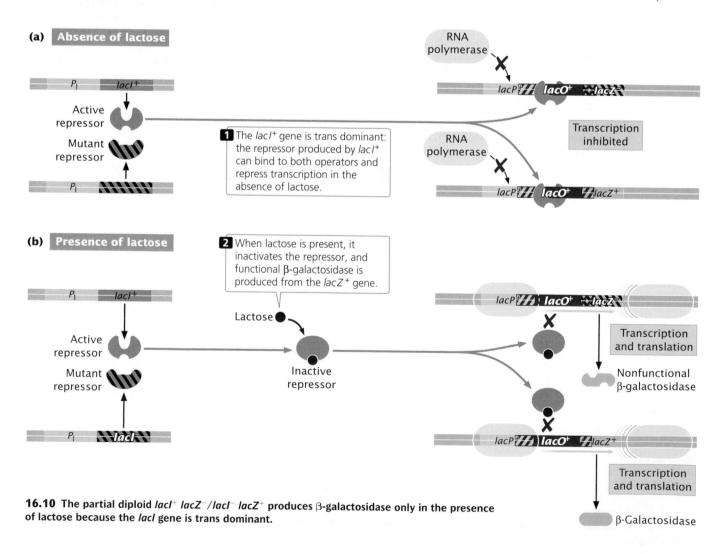

16.10 The partial diploid *lacI⁺ lacZ⁻/lacI⁻ lacZ⁺* produces β-galactosidase only in the presence of lactose because the *lacI* gene is trans dominant.

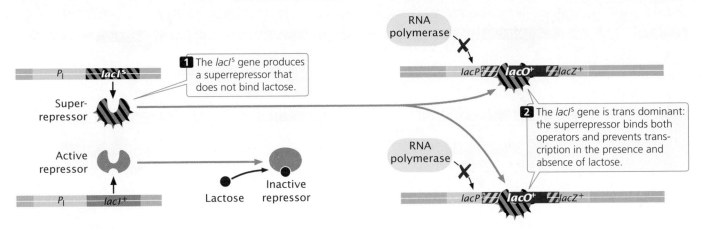

16.11 The partial diploid *lacI^s lacZ^+/lacI^+ lacZ^+* fails to produce β-galactosidase in the presence and absence of lactose because the *lacI^s* gene encodes a superrepressor.

Some of these mutations were constitutive, causing the *lac* proteins to be produced all the time, whether lactose was present or not. Such mutations in the regulator gene were designated *lacI^-*. The construction of partial diploids demonstrated that a *lacI^+* gene is dominant over a *lacI^-* gene; a single copy of *lacI^+* (genotype *lacI^+/lacI^-*) was sufficient to bring about normal regulation of protein production. Furthermore, *lacI^+* restored normal control to an operon even if the operon was located on a different DNA molecule, showing that *lacI^+* can be trans acting. A partial diploid with genotype *lacI^+ lacZ^-/lacI^- lacZ^+* functioned normally, synthesizing β-galactosidase only when lactose was present (**Figure 16.10**). In this strain, the *lacI^+* gene on the bacterial chromosome was functional, but the *lacZ^-* gene was defective; on the plasmid, the *lacI^-* gene was defective, but the *lacZ^+* gene was functional. The fact that a *lacI^+* gene could regulate a *lacZ^+* gene located on a different DNA molecule indicated to Jacob and Monod that the *lacI^+* gene product was able to operate on either the plasmid or the chromosome.

Some *lacI* mutations isolated by Jacob and Monod prevented transcription from taking place even in the presence of lactose and other inducers such as IPTG. These mutations were referred to as superrepressors (*lacI^s*), because they produced defective repressors that could not be inactivated by an inducer. The *lacI^s* mutations produced a repressor with an altered inducer-binding site, which made the inducer unable to bind to the repressor; consequently, the repressor was always able to attach to the operator site and prevent transcription of the *lac* genes. Superrepressor mutations were dominant over *lacI^+*; partial diploids with genotype *lacI^s lacZ^+/lacI^+ lacZ^+* were unable to synthesize either β-galactosidase or permease, whether or not lactose was present (**Figure 16.11**).

Operator mutations Jacob and Monod mapped another class of constitutive mutants to a site adjacent to *lacZ*. These mutations occurred at the operator site and were referred to as *lacO^c* (*O* stands for operator and "c" for constitutive). The *lacO^c* mutations altered the sequence of DNA at the operator so that the repressor protein was no longer able to bind. A partial diploid with genotype *lacI^+ lacO^c lacZ^+/lacI^+ lacO^+ lacZ^+* exhibited constitutive synthesis of β-galactosidase, indicating that *lacO^c* is dominant over *lacO^+*.

Analysis of other partial diploids showed that the *lacO* gene is cis acting, affecting only genes on the same DNA molecule. For example, a partial diploid with genotype *lacI^+ lacO^+ lacZ^-/lacI^+ lacO^c lacZ^+* was constitutive, producing β-galactosidase in the presence or absence of lactose (**Figure 16.12a**), but a partial diploid with genotype *lacI^+ lacO^+ lacZ^+/lacI^+ lacO^c lacZ^-* produced β-galactosidase only in the presence of lactose (**Figure 16.12b**). In the constitutive partial diploid (*lacI^+ lacO^+ lacZ^-/lacI^+ lacO^c lacZ^+*; see Figure 16.12a), the *lacO^c* mutation and the functional *lacZ^+* gene are present on the same DNA molecule; but, in *lacI^+ lacO^+ lacZ^+/lacI^+ lacO^c lacZ^-* (see Figure 16.12b), the *lacO^c* mutation and the functional *lacZ^+* gene are on different molecules. The *lacO* mutation affects only genes to which it is physically connected, as is true of all operator mutations. They prevent the binding of a repressor protein to the operator and thereby allow RNA polymerase to transcribe genes on the same DNA molecule. However, they cannot prevent a repressor from binding to normal operators on other DNA molecules. **TRY PROBLEM 20 →**

Promoter mutations Mutations affecting lactose metabolism have also been isolated at the promoter site; these mutations are designated *lacP^-*, and they interfere with the binding of RNA polymerase to the promoter. Because this binding is essential for the transcription of the structural genes, *E. coli* strains with *lacP^-* mutations don't produce *lac* proteins either in the presence or in the absence of lactose. Like operator mutations, *lacP^-* mutations are cis acting and thus affect only genes on the same DNA molecule. The partial diploid *lacI^+ lacP^+ lacZ^+/lacI^+ lacP^- lacZ^+* exhibits

(a) Partial diploid _lacI⁺ lacO⁺ lacZ⁻/lacI⁺ lacOᶜ lacZ⁺_

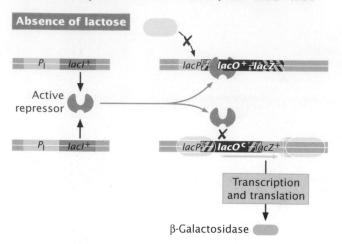

(b) Partial diploid _lacI⁺ lacO⁺ lacZ⁺/lacI⁺ lacOᶜ lacZ⁻_

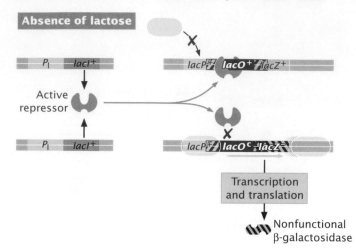

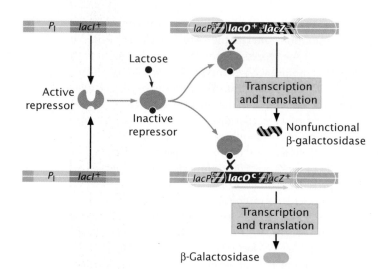

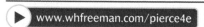

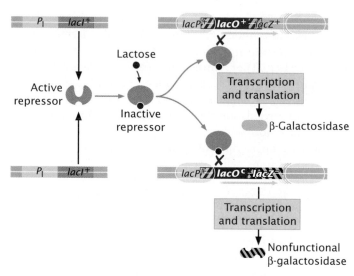

16.12 Mutations in _lacO_ are constitutive and cis acting. (a) The partial diploid _lacI⁺ lacO⁺ lacZ⁻ / lacI⁺ lacOᶜ lacZ⁺_ is constitutive, producing β-galactosidase in the presence and absence of lactose. (b) The partial diploid _lacI⁺ lacO⁺ lacZ⁺/lacI⁺ lacOᶜ lacZ⁻_ is inducible (produces β-galactosidase only when lactose is present), demonstrating that the _lacO_ gene is cis acting.

> ▶ www.whfreeman.com/pierce4e   See Animation 16.1 to explore the effects of different combinations of _lacI_ and _lacO_ mutations on the expression of the _lac_ operon.

normal synthesis of β-galactosidase, whereas _lacI⁺ lacP⁻ lacZ⁺ / lacI⁺ lacP⁺ lacZ⁻_ fails to produce β-galactosidase whether or not lactose is present.

Worked Problem

For _E. coli_ strains with the following _lac_ genotypes, use a plus sign (+) to indicate the synthesis of β-galactosidase and per-

mease and a minus sign (−) to indicate no synthesis of the proteins when lactose is absent and when it is present.

Genotype of strain

a. _lacI⁺ lacP⁺ lacO⁺ lacZ⁺ lacY⁺_

b. _lacI⁺ lacP⁺ lacOᶜ lacZ⁻ lacY⁺_

c. _lacI⁺ lacP⁻ lacO⁺ lacZ⁺ lacY⁻_

d. _lacI⁺ lacP⁺ lacO⁺ lacZ⁻ lacY⁻/lacI⁻ lacP⁺ lacO⁺ lacZ⁺ lacY⁺_

• Solution

Genotype of strain	Lactose absent		Lactose present	
	β-Galactosidase	Permease	β-Galactosidase	Permease
a. $lacI^+$ $lacP^+$ $lacO^+$ $lacZ^+$ $lacY^+$	−	−	+	+
b. $lacI^+$ $lacP^+$ $lacO^c$ $lacZ^-$ $lacY^+$	−	+	−	+
c. $lacI^+$ $lacP^-$ $lacO^+$ $lacZ^+$ $lacY^-$	−	−	−	−
d. $lacI^+$ $lacP^+$ $lacO^+$ $lacZ^-$ $lacY^-$/$lacI^-$ $lacP^+$ $lacO^+$ $lacZ^+$ $lacY^+$	−	−	+	+

a. All the genes possess normal sequences, and so the *lac* operon functions normally: when lactose is absent, the regulator protein binds to the operator and inhibits the transcription of the structural genes, and so β-galactosidase and permease are not produced. When lactose is present, some of it is converted into allolactose, which binds to the repressor and makes it inactive; the repressor does not bind to the operator, and so the structural genes are transcribed and β-galactosidase and permease are produced.

b. The structural *lacZ* gene is mutated; so β-galactosidase will not be produced under any conditions. The *lacO* gene has a constitutive mutation, which means that the repressor is unable to bind to *lacO*, and so transcription takes place at all times. Therefore, permease will be produced in both the presence and the absence of lactose.

c. In this strain, the promoter is mutated, and so RNA polymerase is unable to bind and transcription does not take place. Therefore, β-galactosidase and permease are not produced under any conditions.

d. This strain is a partial diploid, which consists of two copies of the *lac* operon—one on the bacterial chromosome and the other on a plasmid. The *lac* operon represented in the upper part of the genotype has mutations in both the *lacZ* and the *lacY* genes, and so it is not capable of encoding β-galactosidase or permease under any conditions. The *lac* operon in the lower part of the genotype has a defective regulator gene, but the normal regulator gene in the upper operon produces a diffusible repressor (trans acting) that binds to the lower operon in the absence of lactose and inhibits transcription. Therefore, no β-galactosidase or permease is produced when lactose is absent. In the presence of lactose, the repressor cannot bind to the operator, and so the lower operon is transcribed and β-galactosidase and permease are produced.

> → Now try your own hand at predicting the outcome of different *lac* mutations by working Problem 18 at the end of the chapter.

Positive Control and Catabolite Repression

E. coli and many other bacteria metabolize glucose preferentially in the presence of lactose and other sugars. They do so because glucose enters glycolysis without further modification and therefore requires less energy to metabolize than do other sugars. When glucose is available, genes that participate in the metabolism of other sugars are repressed, in a phenomenon known as **catabolite repression**. For example, the efficient transcription of the *lac* operon takes place only if lactose is present and glucose is absent. But how is the expression of the *lac* operon influenced by glucose? What brings about catabolite repression?

Catabolite repression results from positive control in response to glucose. (This regulation is in addition to the negative control brought about by the repressor binding at the operator site of the *lac* operon when lactose is absent.) Positive control is accomplished through the binding of a dimeric protein called the **catabolite activator protein** (CAP) to a site that is about 22 nucleotides long and is located within or slightly upstream of the promoter of the *lac* genes (**Figure 16.13**). RNA polymerase does not bind efficiently to many promoters unless CAP is first bound to the DNA. Before CAP can bind to DNA, it must form a complex with a modified nucleotide called **adenosine-3′,5′-cyclic monophosphate** (cyclic AMP, or cAMP), which is important in cellular signaling processes in both bacterial and eukaryotic cells. In *E. coli*, the concentration of cAMP is regulated so that its concentration is inversely proportional to the level of available glucose. A high concentration of glucose within the cell lowers the amount of cAMP, and so little cAMP–CAP complex is available to bind to the DNA. Subsequently, RNA polymerase has poor affinity for the *lac* promoter, and little transcription of the *lac* operon takes place. Low concentrations of glucose stimulate high levels of cAMP, resulting in increased cAMP–CAP binding to DNA. This increase enhances the binding of RNA polymerase to the promoter and increases transcription of the *lac* genes by some 50-fold.

The catabolite activator protein exerts positive control in more than 20 operons of *E. coli*. The response to CAP varies among these promoters; some operons are activated by low levels of CAP, whereas others require high levels.

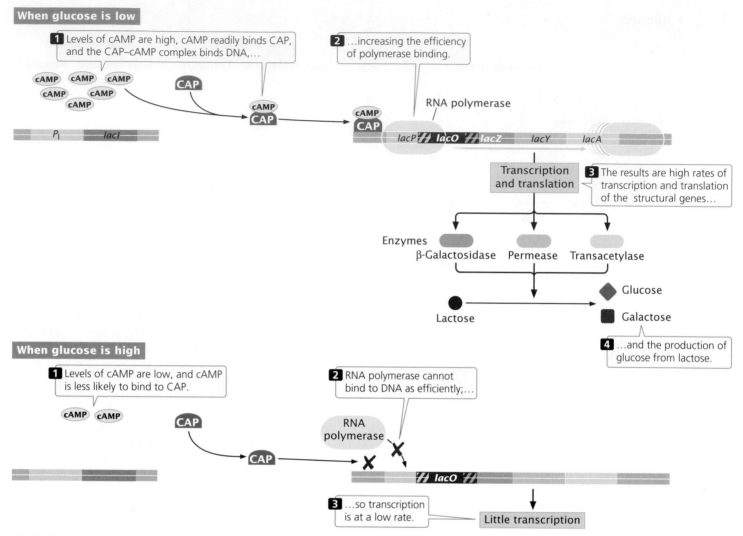

16.13 The catabolite activator protein (CAP) binds to the promoter of the *lac* operon and stimulates transcription. CAP must complex with adenosine-3′,5′-cyclic monophosphate (cAMP) before binding to the promoter of the *lac* operon. The binding of cAMP–CAP to the promoter activates transcription by facilitating the binding of RNA polymerase. Levels of cAMP are inversely related to glucose: low glucose stimulates high cAMP; high glucose stimulates low cAMP.

CAP contains a helix-turn-helix DNA-binding motif and, when it binds at the CAP site on DNA, it causes the DNA helix to bend (**Figure 16.14**). The bent helix enables CAP to interact directly with the RNA polymerase enzyme bound to the promoter and facilitate the initiation of transcription.

TRY PROBLEM 14 ➡

CONCEPTS

In spite of its name, catabolite repression is a type of positive control in the *lac* operon. The catabolite activator protein (CAP), complexed with cAMP, binds to a site near the promoter and stimulates the binding of RNA polymerase. Cellular levels of cAMP are controlled by glucose; a low glucose level increases the abundance of cAMP and enhances the transcription of the *lac* structural genes.

✔ CONCEPT CHECK 7

What is the effect of high levels of glucose on the *lac* operon?

a. Transcription is stimulated.

b. Little transcription takes place.

c. Transcription is not affected.

d. Transcription may be stimulated or inhibited, depending on the levels of lactose.

The *trp* Operon of *E. coli*

The *lac* operon just discussed is an inducible operon, one in which transcription does not normally take place and must be turned on. Other operons are repressible; transcription in

(a)

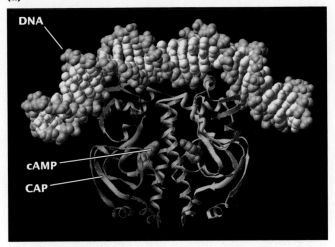

(b)

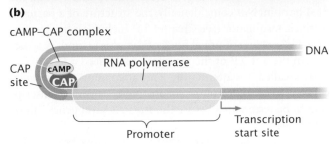

16.14 The binding of the cAMP–CAP complex to DNA produces a sharp bend in DNA that activates transcription.

The *trp* operon contains five structural genes (*trpE*, *trpD*, *trpC*, *trpB*, and *trpA*) that produce the components of three enzymes (two of the enzymes consist of two polypeptide chains). These enzymes convert chorismate into tryptophan (**Figure 16.15**). The first structural gene, *trpE*, contains a long 5′ untranslated region (5′ UTR) that is transcribed but does not encode any of these enzymes. Instead, this 5′ UTR plays an important role in another regulatory mechanism, discussed in the next section. Upstream of the 5′ UTR is the *trp* promoter. When tryptophan levels are low, RNA polymerase binds to the promoter and transcribes the five structural genes into a single mRNA, which is then translated into enzymes that convert chorismate into tryptophan.

Some distance from the *trp* operon is a regulator gene, *trpR*, which encodes a repressor that alone cannot bind DNA (see Figure 16.15). Like the *lac* repressor, the tryptophan repressor has two binding sites, one that binds to DNA at the operator site and another that binds to tryptophan (the activator). Binding with tryptophan causes a conformational change in the repressor that makes it capable of binding to DNA at the operator site, which overlaps the promoter. When the operator is occupied by the tryptophan repressor, RNA polymerase cannot bind to the promoter and the structural genes cannot be transcribed. Thus, when cellular levels of tryptophan are low, transcription of the *trp* operon takes place and more tryptophan is synthesized; when cellular levels of tryptophan are high, transcription of the *trp* operon is inhibited and the synthesis of more tryptophan does not take place.

these operons is normally turned on and must be repressed. The tryptophan (*trp*) operon in *E. coli*, which controls the biosynthesis of the amino acid tryptophan, is an example of a negative repressible operon.

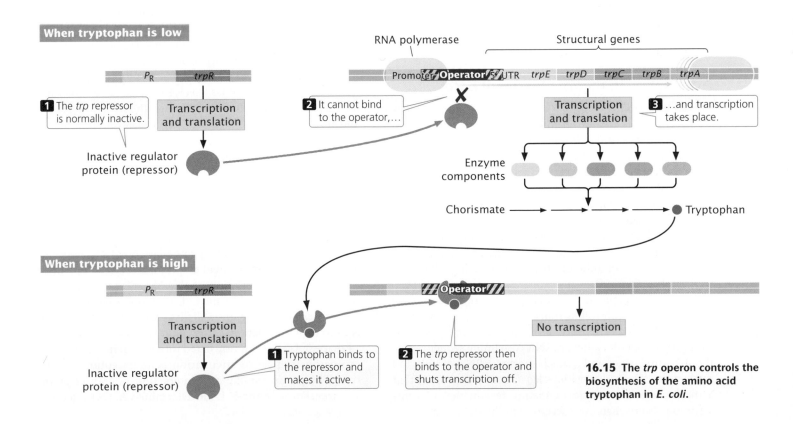

16.15 The *trp* operon controls the biosynthesis of the amino acid tryptophan in *E. coli*.

CONCEPTS

The *trp* operon is a negative repressible operon that controls the biosynthesis of tryptophan. In a repressible operon, transcription is normally turned on and must be repressed. Repression is accomplished through the binding of tryptophan to the repressor, which renders the repressor active. The active repressor binds to the operator and prevents RNA polymerase from transcribing the structural genes.

✔ CONCEPT CHECK 8

In the *trp* operon, what happens to the *trp* repressor in the absence of tryptophan?

a. It binds to the operator and represses transcription.

b. It cannot bind to the operator and transcription takes place.

c. It binds to the regulator gene and represses transcription.

d. It cannot bind to the regulator gene and transcription takes place.

16.3 Some Operons Regulate Transcription Through Attenuation, the Premature Termination of Transcription

We've now seen several different ways in which a cell regulates the initiation of transcription in an operon. Some operons have an additional level of control that affects the *continuation* of transcription rather than its initiation. In **attenuation**, transcription begins at the start site, but termination takes place prematurely, before the RNA polymerase even reaches the structural genes. Attenuation takes place in a number of operons that encode enzymes participating in the biosynthesis of amino acids.

Attenuation in the *trp* Operon of *E. coli*

We can understand the process of **attenuation** most easily by looking at one of the best-studied examples, which is found in the *trp* operon of *E. coli*. The *trp* operon is unusual in that it is regulated both by repression and by attenuation. Most operons are regulated by one of these mechanisms but not by both of them.

Attenuation first came to light when Charles Yanofsky and his colleagues made several observations in the early 1970s that indicated that repression at the operator site is not the only method of regulation in the *trp* operon. They isolated a series of mutants that exhibited high levels of transcription, yet control at the operator site was unaffected, suggesting that some mechanism other than repression at the operator site was controlling transcription. Furthermore, they observed that two mRNAs of different sizes were transcribed from the *trp* operon: a long mRNA containing sequences for the structural genes and a much shorter mRNA of only 140 nucleotides. These observations led Yanofsky to propose that a mechanism that caused premature termination of transcription also regulates transcription in the *trp* operon.

Close examination of the *trp* operon reveals a region of 162 nucleotides that corresponds to the long 5′ UTR of the mRNA (mentioned earlier) transcribed from the *trp* operon (**Figure 16.16a**). The 5′ UTR (also called a leader) contains four regions: region 1 is complementary to region 2, region 2 is complementary to region 3, and region 3 is complementary to region 4. These complementarities allow the 5′ UTR to fold into two different secondary structures (**Figure 16.16b**). Only one of these secondary structures causes attenuation.

One of the secondary structures contains one hairpin produced by the base pairing of regions 1 and 2 and another hairpin produced by the base pairing of regions 3 and 4. Notice that a string of uracil nucleotides follows the 3+4 hairpin. Not coincidentally, the structure of a bacterial intrinsic terminator (see Chapter 13) includes a hairpin followed by a string of uracil nucleotides; this secondary structure in the 5′ UTR of the *trp* operon is indeed a terminator and is called an **attenuator**. The attenuator forms when cellular levels of tryptophan are high, causing transcription to be terminated before the *trp* structural genes can be transcribed.

When cellular levels of tryptophan are low, however, the alternative secondary structure of the 5′ UTR is produced by the base pairing of regions 2 and 3 (see Figure 16.16b). This base pairing also produces a hairpin, but this hairpin is not followed by a string of uracil nucleotides; so this structure does *not* function as a terminator. RNA polymerase continues past the 5′ UTR into the coding section of the structural genes, and the enzymes that synthesize tryptophan are produced. Because it prevents the termination of transcription, the 2+3 structure is called an **antiterminator**.

To summarize, the 5′ UTR of the *trp* operon can fold into one of two structures. When the tryptophan level is high, the 3+4 structure forms, transcription is terminated within the 5′ UTR, and no additional tryptophan is synthesized. When the tryptophan level is low, the 2+3 structure forms, transcription continues through the structural genes, and tryptophan is synthesized. The critical question, then, is: Why does the 3+4 structure arise when the level of tryptophan in the cell is high, whereas the 2+3 structure arises when the level is low?

To answer this question, we must take a closer look at the nucleotide sequence of the 5′ UTR. At the 5′ end, upstream of region 1, is a ribosome-binding site (see Figure 16.16a). Region 1 encodes a small protein. Within the coding sequence for this protein are two UGG codons, which specify the amino acid tryptophan; so tryptophan is required for the translation of this 5′ UTR sequence. The small protein encoded by the 5′ UTR has not been isolated and is presumed to be unstable; its only apparent function is to control attenuation. Although it was stated in Chapter 14 that a 5′ UTR is not translated into a protein, the 5′ UTR of operons subject to attenuation is an exception to this rule. The precise timing and interaction of transcription and translation in the 5′ UTR determine whether attenuation takes place.

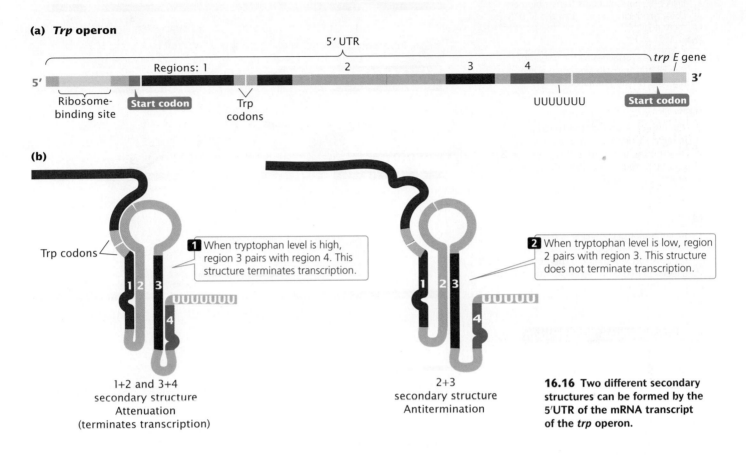

16.16 Two different secondary structures can be formed by the 5′UTR of the mRNA transcript of the *trp* operon.

Transcription when tryptophan levels are low Let's first consider what happens when intracellular levels of tryptophan are low. Recall that, in prokaryotic cells, transcription and translation are coupled: while transcription is taking place at the 3′ end of the mRNA, translation is initiated at the 5′ end. RNA polymerase begins transcribing the DNA, producing region 1 of the 5′ UTR (**Figure 16.17a**). Closely following RNA polymerase, a ribosome binds to the 5′ UTR and begins to translate the coding region. Meanwhile, RNA polymerase is transcribing region 2 (**Figure 16.17b**). Region 2 is complementary to region 1 but, because the ribosome is translating region 1, the nucleotides in regions 1 and 2 cannot base pair.

RNA polymerase begins to transcribe region 3, and the ribosome reaches the UGG tryptophan codons in region 1. When it reaches the tryptophan codons, the ribosome stalls (**Figure 16.17c**) because the level of tryptophan is low and tRNAs charged with tryptophan are scarce or even unavailable. The ribosome sits at the tryptophan codons, awaiting the arrival of a tRNA charged with tryptophan. Stalling of the ribosome does not, however, hinder transcription; RNA polymerase continues to move along the DNA, and transcription gets ahead of translation.

Because the ribosome is stalled at the tryptophan codons in region 1, region 2 is free to base pair with region 3, forming the 2+3 hairpin (**Figure 16.17d**). This hairpin does not cause termination, and so transcription continues. Because region 3 is already paired with region 2, the 3+4 hairpin (the attenuator) never forms, and so attenuation

does not take place and transcription continues. RNA polymerase continues along the DNA, past the 5′ UTR, transcribing all the structural genes into mRNA, which is translated into the enzymes encoded by the *trp* operon. These enzymes then synthesize more tryptophan.

Transcription when tryptophan levels are high Now, let's see what happens when intracellular levels of tryptophan are high. Once again, RNA polymerase begins transcribing the DNA, producing region 1 of the 5′ UTR (**Figure 16.17e**). Closely following RNA polymerase, a ribosome binds to the 5′ UTR and begins to translate the coding region (**Figure 16.17f**). When the ribosome reaches the two UGG tryptophan codons, it doesn't slow or stall, because tryptophan is abundant and tRNAs charged with tryptophan are readily available (**Figure 16.17g**). This point is critical to note: because tryptophan is abundant, translation can keep up with transcription.

As it moves past region 1, the ribosome partly covers region 2 (**Figure 16.17h**); meanwhile, RNA polymerase completes the transcription of region 3. Although regions 2 and 3 are complementary, the ribosome physically blocks their pairing.

RNA polymerase continues to move along the DNA, eventually transcribing region 4 of the 5′ UTR. Region 4 is complementary to region 3, and, because region 3 cannot base pair with region 2, it pairs with region 4. The pairing of regions 3 and 4 (see Figure 16.17h) produces the

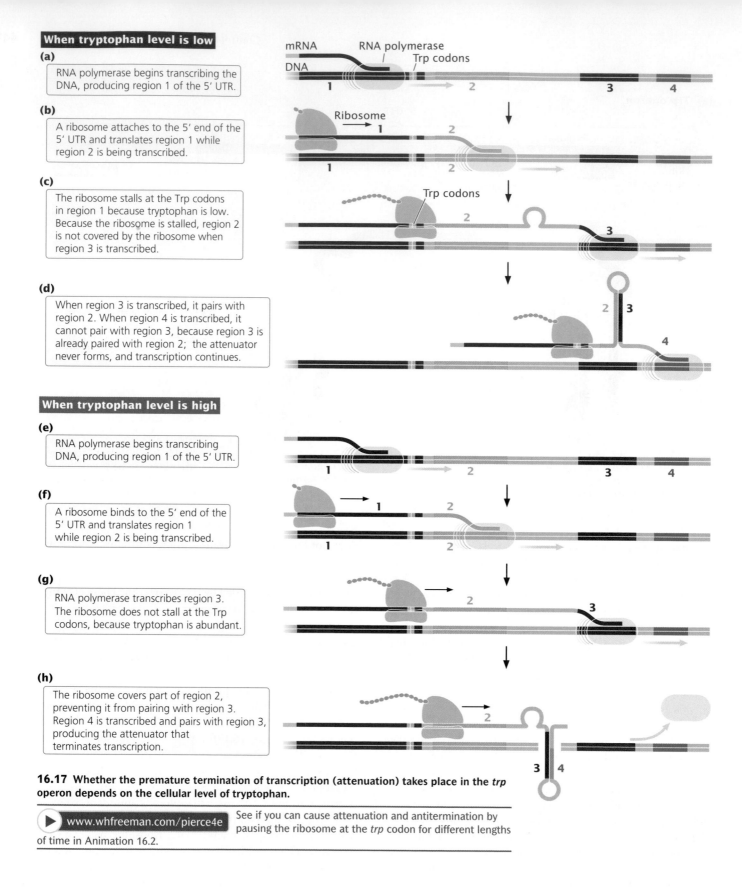

When tryptophan level is low

(a)

RNA polymerase begins transcribing the DNA, producing region 1 of the 5′ UTR.

(b)

A ribosome attaches to the 5′ end of the 5′ UTR and translates region 1 while region 2 is being transcribed.

(c)

The ribosome stalls at the Trp codons in region 1 because tryptophan is low. Because the ribosome is stalled, region 2 is not covered by the ribosome when region 3 is transcribed.

(d)

When region 3 is transcribed, it pairs with region 2. When region 4 is transcribed, it cannot pair with region 3, because region 3 is already paired with region 2; the attenuator never forms, and transcription continues.

When tryptophan level is high

(e)

RNA polymerase begins transcribing DNA, producing region 1 of the 5′ UTR.

(f)

A ribosome binds to the 5′ end of the 5′ UTR and translates region 1 while region 2 is being transcribed.

(g)

RNA polymerase transcribes region 3. The ribosome does not stall at the Trp codons, because tryptophan is abundant.

(h)

The ribosome covers part of region 2, preventing it from pairing with region 3. Region 4 is transcribed and pairs with region 3, producing the attenuator that terminates transcription.

16.17 Whether the premature termination of transcription (attenuation) takes place in the *trp* operon depends on the cellular level of tryptophan.

www.whfreeman.com/pierce4e — See if you can cause attenuation and antitermination by pausing the ribosome at the *trp* codon for different lengths of time in Animation 16.2.

attenuator and transcription terminates just beyond region 4. The structural genes are not transcribed, no tryptophan-producing enzymes are translated, and no additional tryptophan is synthesized. Important events in the process of attenuation are summarized in **Table 16.2**.

A key factor controlling attenuation is the number of tRNA molecules charged with tryptophan, because their availability is what determines whether the ribosome stalls at the tryptophan codons. A second factor concerns the synchronization of transcription and translation, which is

Table 16.2 Events in the process of attenuation

Intracellular Level of Tryptophan	Ribosome Stalls at Trp Codons	Position of Ribosome When Region 3 Is Transcribed	Secondary Structure of 5′ UTR	Termination of Transcription of *trp* Operon
High	No	Covers region 2	3+4 hairpin	Yes
Low	Yes	Covers region 1	2+3 hairpin	No

critical to attenuation. Synchronization is achieved through a pause site located in region 1 of the 5′ UTR. After initiating transcription, RNA polymerase stops temporarily at this site, which allows time for a ribosome to bind to the 5′ end of the mRNA so that translation can closely follow transcription. It is important to point out that ribosomes do not traverse the convoluted hairpins of the 5′ UTR to translate the structural genes. Ribosomes that attach to the 5′ end of region 1 of the mRNA encounter a stop codon at the end of region 1. New ribosomes translating the structural genes attach to a different ribosome-binding site located near the beginning of the *trpE* gene. **TRY PROBLEM 24** →

Why Does Attenuation Take Place in the *trp* Operon?

Why do bacteria need attenuation in the *trp* operon? Shouldn't repression at the operator site prevent transcription from taking place when tryptophan levels in the cell are high? Why does the cell have two types of control? Part of the answer is that repression is never complete; some transcription is initiated even when the *trp* repressor is active; repression reduces transcription only as much as 70-fold. Attenuation can further reduce transcription another 8- to 10-fold; so together the two processes are capable of reducing transcription of the *trp* operon more than 600-fold. Both mechanisms provide *E. coli* with a much finer degree of control over tryptophan synthesis than either could achieve alone.

Another reason for the dual control is that attenuation and repression respond to different signals: repression responds to the cellular levels of tryptophan, whereas attenuation responds to the number of tRNAs charged with tryptophan. There may be times when a cell's ability to respond to these different signals is advantageous. Finally, the *trp* repressor affects several operons other than the *trp* operon. At an earlier stage in the evolution of *E. coli*, the *trp* operon may have been controlled only by attenuation. The *trp* repressor may have evolved primarily to control the other operons and only incidentally affects the *trp* operon.

Attenuation is a difficult process to grasp because you must simultaneously visualize how two dynamic processes—transcription and translation—interact, and it's easy to confuse the two processes. Remember that attenuation refers to the early termination of *transcription*, not translation (although events in translation bring about the termination of transcription). Attenuation often causes confusion because we know that transcription must precede translation. We're comfortable with the idea that transcription might affect translation, but it's harder to imagine that the effects of translation could influence transcription, as they do in attenuation. The reality is that transcription and translation are closely coupled in prokaryotic cells, and events in one process can easily affect the other.

CONCEPTS

In attenuation, transcription is initiated but terminates prematurely. When tryptophan levels are low, the ribosome stalls at the tryptophan codons and transcription continues. When tryptophan levels are high, the ribosome does not stall at the tryptophan codons, and the 5′ UTR adopts a secondary structure that terminates transcription before the structural genes can be copied into RNA (attenuation).

✔ CONCEPT CHECK 9

Attenuation results when which regions of the 5′ UTR pair?

a. 1 and 3   c. 2 and 4

b. 2 and 3   d. 3 and 4

16.4 RNA Molecules Control the Expression of Some Bacterial Genes

All the regulators of gene expression that we have considered so far have been proteins. Several examples of RNA regulators also have been discovered.

Antisense RNA

Some small RNA molecules are complementary to particular sequences on mRNAs and are called **antisense RNA**. They control gene expression by binding to sequences on mRNA and inhibiting translation. Translational control by antisense RNA is seen in the regulation of the *ompF* gene of *E. coli* (**Figure 16.18a**). This gene encodes an outer-membrane protein that functions as a channel for the passive diffusion of small polar molecules, such as water and ions, across the cell membrane. Under most conditions, the *ompF* gene is transcribed and translated and the OmpF protein is synthesized. However, when the osmolarity of the medium increases, the cell depresses the production of OmpF protein to help maintain cellular osmolarity. A regulator gene named *micF*—for

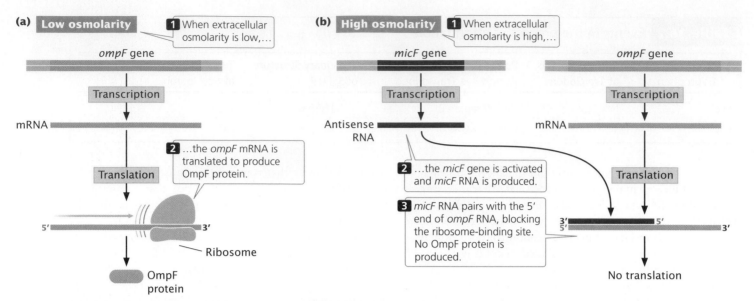

16.18 Antisense RNA can regulate translation.

mRNA-interfering complementary RNA—is activated and *micF* RNA is produced (**Figure 16.18b**). The *micF* RNA, an antisense RNA, binds to a complementary sequence in the 5′ UTR of the *ompF* mRNA and inhibits the binding of the ribosome. This inhibition reduces the amount of translation (see Figure 16.18b), which results in fewer OmpF proteins in the outer membrane and thus reduces the detrimental movement of substances across the membrane owing to the changes in osmolarity. A number of examples of antisense RNA controlling gene expression have now been identified in bacteria and bacteriophages. **TRY PROBLEM 27** →

Riboswitches

We have seen that operons of bacteria contain DNA sequences (promoters and operator sites) where the binding of small molecules induces or represses transcription.

Some mRNA molecules contain regulatory sequences called **riboswitches**, where molecules can bind and affect gene expression by influencing the formation of secondary structures in the mRNA (**Figure 16.19**). Riboswitches were first discovered in 2002 and now appear to be common in bacteria, regulating about 4% of all bacterial genes. They are also present in archaea, fungi, and plants.

Riboswitches are typically found in the 5′ UTR of the mRNA and fold into compact RNA secondary structures with a base stem and several branching hairpins. In some cases, a small regulatory molecule binds to the riboswitch and stabilizes a terminator, which causes premature termination of transcription. In other cases, the binding of a regulatory molecule stabilizes a secondary structure that masks the ribosome-binding site, preventing the initiation of translation. When not bound by the regulatory molecule, the riboswitch assumes an alternative structure that elimi-

(a) Regulatory protein present

Regulatory protein

A regulatory protein binds to the riboswitch and stabilizes a secondary structure…

Riboswitch

mRNA

Ribosome-binding site

…that masks a ribosome-binding site.

5′   3′
Gene

No translation takes place.

No translation

(b) Regulatory protein absent

Without the regulatory protein, the riboswitch assumes an alternative secondary structure…

…that makes the ribosome-binding site available.

5′

Ribosome-binding site

Gene   3′

Translation takes place.

Translation

16.19 Riboswitches are RNA sequences in mRNA that affect gene expression.

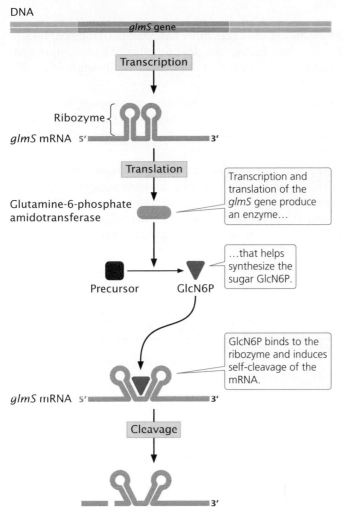

DNA

glmS gene

Transcription

Ribozyme

glmS mRNA 5'   3'

Translation

Glutamine-6-phosphate
amidotransferase

Transcription and
translation of the
glmS gene produce
an enzyme…

Precursor   GlcN6P

…that helps
synthesize the
sugar GlcN6P.

glmS mRNA 5'   3'

GlcN6P binds to the
ribozyme and induces
self-cleavage of the
mRNA.

Cleavage

3'

16.20 Ribozymes, when bound with small regulatory molecules, can induce the cleavage and degradation of mRNA.

nates the premature terminator or makes the ribosome-binding site available.

A riboswitch and the synthesis of vitamin B12   An example of a riboswitch is seen in bacterial genes that encode enzymes having roles in the synthesis of vitamin B12. The genes for these enzymes are transcribed into an mRNA molecule that has a riboswitch. When the activated form of

vitamin B12—called coenzyme B12—is present, it binds to the riboswitch, the mRNA folds into a secondary structure that obstructs the ribosome-binding site, and no translation of the mRNA takes place. In the absence of coenzyme B12, the mRNA assumes a different secondary structure. This secondary structure does not obstruct the ribosome-binding site, and so translation is initiated, the enzymes are synthesized, and more vitamin B12 is produced. For some riboswitches, the regulatory molecule acts as a repressor (as just described) by inhibiting transcription or translation; for others, the regulatory molecule acts as an inducer by causing the formation of a secondary structure that allows transcription or translation to take place.

Riboswitches That Function As Ribozymes

Another type of gene control is carried out by mRNA molecules called ribozymes, which possess catalytic activity (see Chapter 14). Termed RNA-mediated repression, this type of control has been demonstrated in the *glmS* gene of the bacterium *Bacillus subtilis*. Transcription of this gene produces an mRNA molecule that encodes the enzyme glutamine-fructose-6-phosphate amidotransferase (**Figure 16.20**), which helps synthesize a small sugar called glucosamine-6-phosphate (GlcN6P). Within the 5′ UTR of the *glmS* mRNA are about 75 nucleotides that act as a ribozyme. When GlcN6P is absent, the *glmS* gene is transcribed and translated to produce the enzyme, which synthesizes more GlcN6P. However, when sufficient GlcN6P is present, it binds to the ribozyme part of the mRNA, which then induces self-cleavage of the mRNA and prevents its translation.

CONCEPTS

Antisense RNA is complementary to other RNA or DNA sequences. In bacterial cells, it can inhibit translation by binding to sequences in the 5′ UTR of mRNA and preventing the attachment of the ribosome. Riboswitches are sequences in mRNA molecules that bind regulatory molecules and induce changes in the secondary structure of the mRNA that affect gene expression. In RNA-mediated repression, a ribozyme sequence on the mRNA induces self-cleavage and degradation of the mRNA when bound by a regulatory molecule.

CONCEPTS SUMMARY

- Gene expression can be controlled at different levels, including the alteration of DNA or chromatin structure, transcription, mRNA processing, RNA stability, translation, and posttranslational modification. Much of gene regulation is through the action of regulatory proteins binding to specific sequences in DNA.

- Genes in bacterial cells are typically clustered into operons—groups of functionally related structural genes and the

sequences that control their transcription. Structural genes in an operon are transcribed together as a single mRNA molecule.

- In negative control, a repressor protein binds to DNA and inhibits transcription. In positive control, an activator protein binds to DNA and stimulates transcription. In inducible operons, transcription is normally off and must be turned on; in repressible operons, transcription is normally on and must be turned off.

- The *lac* operon of *E. coli* is a negative inducible operon. In the absence of lactose, a repressor binds to the operator and prevents the transcription of genes that encode β-galactosidase, permease, and transacetylase. When lactose is present, some of it is converted into allolactose, which binds to the repressor and makes it inactive, allowing the structural genes to be transcribed.

- Positive control in the *lac* and other operons is through catabolite repression. When complexed with cyclic AMP, the catabolite activator protein binds to a site in or near the promoter and stimulates the transcription of the structural genes. Levels of cAMP are inversely correlated with glucose; so low levels of glucose stimulate transcription and high levels inhibit transcription.

- The *trp* operon of *E. coli* is a negative repressible operon that controls the biosynthesis of tryptophan.

- Attenuation causes premature termination of transcription. It takes place through the close coupling of transcription and translation and depends on the secondary structure of the 5′ UTR sequence.

- Antisense RNAs are complementary to sequences in mRNA and may inhibit translation by binding to these sequences, thereby preventing the attachment or progress of the ribosome.

- When bound by a regulatory molecule, riboswitches in mRNA molecles induce changes in the secondary structure of the mRNA, which affects gene expression. Some mRNAs possess ribozyme sequences that induce self-cleavage and degradation when bound by a regulatory molecule.

IMPORTANT TERMS

gene regulation (p. 432)
structural gene (p. 433)
regulatory gene (p. 433)
constitutive gene (p. 433)
regulatory element (p. 433)
domain (p. 434)
operon (p. 435)
regulator gene (p. 435)
regulator protein (p. 435)
operator (p. 435)

negative control (p. 436)
positive control (p. 436)
inducible operon (p. 436)
repressible operon (p. 436)
inducer (p. 436)
allosteric protein (p. 436)
corepressor (p. 437)
coordinate induction (p. 441)
partial diploid (p. 441)
catabolite repression (p. 445)

catabolite activator protein (CAP) (p. 445)
adenosine-3′,5′-cyclic monophosphate (cAMP) (p. 445)
attenuation (p. 448)
attenuator (p. 448)
antiterminator (p. 448)
antisense RNA (p. 451)
riboswitch (p. 452)

ANSWERS TO CONCEPT CHECKS

1. A constitutive gene is not regulated and is expressed continually.

2. Because it is the first step in the process of information transfer from DNA to protein. For cellular efficiency, gene expression is often regulated early in the process of protein production.

3. b

4. d

5. d

6. c

7. b

8. b

9. d

WORKED PROBLEMS

1. A regulator gene produces a repressor in an inducible operon. A geneticist isolates several constitutive mutations affecting this operon. Where might these constitutive mutations occur? How would the mutations cause the operon to be constitutive?

• Solution

An inducible operon is normally not being transcribed, meaning that the repressor is active and binds to the operator, inhibiting transcription. Transcription takes place when the inducer binds to the repressor, making it unable to bind to the operator. Constitutive mutations cause transcription to take place at all times, whether the inducer is present or not. Constitutive

mutations might occur in the regulator gene, altering the repressor so that it is never able to bind to the operator. Alternatively, constitutive mutations might occur in the operator, altering the binding site for the repressor so that the repressor is unable to bind under any conditions.

2. The *fox* operon, which has sequences A, B, C, and D (which may represent either structural genes or regulatory sequences), encodes enzymes 1 and 2. Mutations in sequences A, B, C, and D have the following effects, where a plus sign (+) indicates that the enzyme is synthesized and a minus sign (−) indicates that the enzyme is not synthesized.

Mutation in sequence	Fox absent		Fox present	
	Enzyme 1	Enzyme 2	Enzyme 1	Enzyme 2
No mutation	−	−	+	+
A	−	−	−	+
B	−	−	−	−
C	−	−	+	−
D	+	+	+	+

a. Is the *fox* operon inducible or repressible?

b. Indicate which sequence (*A*, *B*, *C*, or *D*) is part of the following components of the operon:

Regulator gene _____

Promoter _____

Structural gene for enzyme 1 _____

Structural gene for enzyme 2 _____

• Solution

Because the structural genes in an operon are coordinately expressed, mutations that affect only one enzyme are likely to occur in the structural genes; mutations that affect both enzymes must occur in the promoter or regulator.

a. When no mutations are present, enzymes 1 and 2 are produced in the presence of Fox but not in its absence, indicating that the operon is inducible and Fox is the inducer.

b. The mutation in *A* allows the production of enzyme 2 in the presence of Fox, but enzyme 1 is not produced in the presence or absence of Fox, and so *A* must have a mutation in the structural gene for enzyme 1. With the mutation in *B*, neither enzyme is produced under any conditions, and so this mutation

most likely occurs in the promoter and prevents RNA polymerase from binding. The mutation in *C* affects only enzyme 2, which is not produced in the presence or absence of Fox; enzyme 1 is produced normally (only in the presence of Fox), and so the mutation in *C* most likely occurs in the structural gene for enzyme 2. The mutation in *D* is constitutive, allowing the production of enzymes 1 and 2 whether or not Fox is present. This mutation most likely occurs in the regulator gene, producing a defective repressor that is unable to bind to the operator under any conditions.

Regulator gene	*D*
Promoter	*B*
Structural gene for enzyme 1	*A*
Structural gene for enzyme 2	*C*

3. A mutation occurs in the 5′ UTR of the *trp* operon that reduces the ability of region 2 to pair with region 3. What will the effect of this mutation be when the tryptophan level is high? When the tryptophan level is low?

• Solution

When the tryptophan level is high, regions 2 and 3 do not normally pair, and therefore the mutation will have no effect. When the tryptophan level is low, however, the ribosome normally stalls at the tryptophan codons in region 1 and does not cover region 2, and so regions 2 and 3 are free to pair, which prevents regions 3 and 4 from pairing and forming a terminator, which would end transcription. If regions 2 and 3 cannot pair, then regions 3 and 4 will pair even when tryptophan is low and attenuation will always take place. Therefore, no more tryptophan will be synthesized even in the absence of tryptophan.

COMPREHENSION QUESTIONS

Section 16.1

1. Why is gene regulation important for bacterial cells?

*2. Name six different levels at which gene expression might be controlled.

Section 16.2

*3. Draw a picture illustrating the general structure of an operon and identify its parts.

4. What is the difference between positive and negative control? What is the difference between inducible and repressible operons?

*5. Briefly describe the *lac* operon and how it controls the metabolism of lactose.

6. What is catabolite repression? How does it allow a bacterial cell to use glucose in preference to other sugars?

Section 16.3

*7. What is attenuation? What are the mechanisms by which the attenuator forms when tryptophan levels are high and the antiterminator forms when tryptophan levels are low?

Section 16.4

*8. What is antisense RNA? How does it control gene expression?

9. What are riboswitches? How do they control gene expression? How do riboswitches differ from RNA-mediated repression?

APPLICATION QUESTIONS AND PROBLEMS

Introduction

10. The introduction to this chapter discussed how transformation, in which bacteria take up foreign DNA, has allowed *Streptococcus pneumoniae* to develop resistance to antibiotics. Transformation is brought about by proteins that are encoded in the *com* regulon, which is activated by CSP protein. CSP protein is produced by bacteria and exported into the surrounding medium. When a sufficient

quantity of CSP accumulates, it stimulates the *com* regulon. In addition to stimulating transformation, CSP has another effect—it kills and lyses other cells. Explain how this second effect of CSP—killing and lysing other cells—enhances the ability of *S. pneumoniae* to develop resistance to antibiotics.

Section 16.2

*11. For each of the following types of transcriptional control, indicate whether the protein produced by the regulator gene will be synthesized initially as an active repressor, inactive repressor, active activator, or inactive activator.

a. Negative control in a repressible operon

b. Positive control in a repressible operon

c. Negative control in an inducible operon

d. Positive control in an inducible operon

*12. A mutation at the operator site prevents the regulator protein from binding. What effect will this mutation have in the following types of operons?

a. Regulator protein is a repressor in a repressible operon.

b. Regulator protein is a repressor in an inducible operon.

13. The *blob* operon produces enzymes that convert compound A into compound B. The operon is controlled by a regulatory gene S. Normally, the enzymes are synthesized only in the absence of compound B. If gene S is mutated, the enzymes are synthesized in the presence *and* in the absence of compound B. Does gene S produce a repressor or an activator? Is this operon inducible or repressible?

*14. A mutation prevents the catabolite activator protein (CAP) from binding to the promoter in the *lac* operon. What will the effect of this mutation be on the transcription of the operon?

15. On the basis of the information provided in the introduction, the *com* regulon in *Streptococcus pneumoniae* appears to be controlled through which type of gene regulation? Explain your answer.

a. Negative inducible

b. Negative repressible

c. Postive inducible

d. Positive repressible

16. Under which of the following conditions would a *lac* operon produce the greatest amount of β-glactosidase? The least? Explain your reasoning.

	Lactose present	Glucose present
Condition 1	Yes	No
Condition 2	No	Yes
Condition 3	Yes	Yes
Condition 4	No	No

17. A mutant strain of *E. coli* produces β-galactosidase in the presence *and* in the absence of lactose. Where in the operon might the mutation in this strain occur?

*18. For *E. coli* strains with the *lac* genotypes shown below use a plus sign (+) to indicate the synthesis of β-galactosidase and permease and a minus sign (−) to indicate no synthesis of the proteins.

	Lactose absent		Lactose present	
Genotype of strain	**β-Galactosidase**	**Permease**	**β-Galactosidase**	**Permease**
$lacI^+$ $lacP^+$ $lacO^+$ $lacZ^+$ $lacY^+$				
$lacI^-$ $lacP^+$ $lacO^+$ $lacZ^+$ $lacY^+$				
$lacI^+$ $lacP^+$ $lacO^c$ $lacZ^+$ $lacY^+$				
$lacI^-$ $lacP^+$ $lacO^+$ $lacZ^+$ $lacY^-$				
$lacI^-$ $lacP^-$ $lacO^+$ $lacZ^+$ $lacY^+$				
$lacI^+$ $lacP^+$ $lacO^+$ $lacZ^-$ $lacY^+$ /				
$lacI^-$ $lacP^+$ $lacO^+$ $lacZ^+$ $lacY^-$				
$lacI^-$ $lacP^+$ $lacO^c$ $lacZ^+$ $lacY^+$ /				
$lacI^+$ $lacP^+$ $lacO^+$ $lacZ^-$ $lacY^-$				
$lacI^-$ $lacP^+$ $lacO^+$ $lacZ^+$ $lacY^-$ /				
$lacI^+$ $lacP^-$ $lacO^+$ $lacZ^-$ $lacY^+$				
$lacI^+$ $lacP^-$ $lacO^c$ $lacZ^-$ $lacY^+$ /				
$lacI^-$ $lacP^+$ $lacO^+$ $lacZ^+$ $lacY^-$				
$lacI^+$ $lacP^+$ $lacO^+$ $lacZ^+$ $lacY^+$ /				
$lacI^+$ $lacP^+$ $lacO^+$ $lacZ^+$ $lacY^+$				
$lacI^s$ $lacP^+$ $lacO^+$ $lacZ^+$ $lacY^-$ /				
$lacI^+$ $lacP^+$ $lacO^+$ $lacZ^-$ $lacY^+$				
$lacI^s$ $lacP^-$ $lacO^+$ $lacZ^-$ $lacY^+$ /				
$lacI^+$ $lacP^+$ $lacO^+$ $lacZ^+$ $lacY^+$				

19. Give all possible genotypes of a *lac* operon that produces β-galactosidase and permease under the following conditions. Do not give partial diploid genotypes.

	Lactose absent		Lactose present	
	β-Galactosidase	Permease	β-Galactosidase	Permease
a.	−	−	+	+
b.	−	−	−	+
c.	−	−	+	−
d.	+	+	+	+
e.	−	−	−	−
f.	+	−	+	−
g.	−	+	−	+

*20. Explain why mutations in the *lacI* gene are trans in their effects, but mutations in the *lacO* gene are cis in their effects.

*21. The *mmm* operon, which has sequences *A*, *B*, *C*, and *D* (which may be structural genes or regulatory sequences), encodes enzymes 1 and 2. Mutations in sequences *A*, *B*, *C*, and *D* have the following effects, where a plus sign (+) indicates that the enzyme is synthesized and a minus sign (−) indicates that the enzyme is not synthesized.

Mutation	Mmm absent		Mmm present	
in sequence	Enzyme 1	Enzyme 2	Enzyme 1	Enzyme 2
No mutation	+	+	−	−
A	−	+	−	−
B	+	+	+	+
C	+	−	−	−
D	−	−	−	−

a. Is the *mmm* operon inducible or repressible?

b. Indicate which sequence (*A*, *B*, *C*, or *D*) is part of the following components of the operon:

Regulator gene _____
Promoter _____
Structural gene for enzyme 1 _____
Structural gene for enzyme 2 _____

22. Ellis Engelsberg and his coworkers examined the regulation of genes taking part in the metabolism of arabinose, a sugar (E. Engelsberg et al. 1965. *Journal of Bacteriology* 90:946–957). Four structural genes encode enzymes that help metabolize arabinose (genes *A*, *B*, *D*, and *E*). An additional gene *C* is linked to genes *A*, *B*, and *D*. These genes are in the order *D-A-B-C*. Gene *E* is distant from the other genes. Engelsberg and his colleagues isolated mutations at the *C* gene that affected the expression of structural genes *A*, *B*, *D*, and *E*. In one set of experiments, they created various genotypes at the *A* and *C* loci and determined whether arabinose isomerase (the enzyme encoded by gene *A*) was produced in the presence or absence of arabinose (the substrate of arabinose isomerase). Results from this experiment are shown in the following table, where a plus sign (+) indicates that the

arabinose isomerase was synthesized and a minus sign (−) indicates that the enzyme was not synthesized.

Genotype	Arabinose absent	Arabinose present
1. $C^+ A^+$	−	+
2. $C^- A^+$	−	−
3. $C^- A^+ / C^+ A^-$	−	+
4. $C^c A^- / C^- A^+$	+	+

a. On the basis of the results of these experiments, is the *C* gene an operator or a regulator gene? Explain your reasoning.

b. Do these experiments suggest that the arabinose operon is negatively or positively controlled? Explain your reasoning.

c. What type of mutation is C^c?

23. In *E. coli*, three structural genes (*A*, *D*, and *E*) encode enzymes A, D and E respectively. Gene *O* is an operator. The genes are in the order *O-A-D-E* on the chromosome. These enzymes catalyze the biosynthesis of valine. Mutations were isolated at the *A*, *D*, *E*, and *O* genes to study the production of enzymes A, D, and E when cellular levels of valine were low (T. Ramakrishnan and E. A. Adelberg. 1965. *Journal of Bacteriology* 89:654–660). Levels of the enzymes produced by partial-diploid *E. coli* with various combinations of mutations are shown in the following table.

	Genotype	Amount of enzyme produced		
		E	D	A
1.	$E^+ D^+ A^+ O^+$ / $E^+ D^+ A^+ O^+$	2.40	2.00	3.50
2.	$E^+ D^+ A^+ O^-$ / $E^+ D^+ A^+ O^+$	35.80	38.60	46.80
3.	$E^+ D^- A^+ O^-$ / $E^+ D^+ A^- O^+$	1.80	1.00	47.00
4.	$E^+ D^+ A^- O^-$ / $E^+ D^- A^+ O^+$	35.30	38.00	1.70
5.	$E^- D^+ A^+ O^-$ / $E^+ D^- A^+ O^+$	2.38	38.00	46.70

a. Is the regulator protein that binds to the operator of this operon a repressor (negative control) or an activator (positive control)? Explain your reasoning.

b. Are genes *A*, *D*, and *E* all under the control of operator *O*? Explain your reasoning.

c. Propose an explanation for the low level of enzyme E produced in genotype 3.

Section 16.3

*24. Listed in parts *a* through *g* are some mutations that were found in the 5′ UTR region of the *trp* operon of *E. coli*. What will the most likely effect of each of these mutations be on the transcription of the *trp* structural genes?

a. A mutation that prevents the binding of the ribosome to the 5′ end of the mRNA 5′ UTR

b. A mutation that changes the tryptophan codons in region 1 of the mRNA 5′ UTR into codons for alanine

c. A mutation that creates a stop codon early in region 1 of the mRNA 5′ UTR

d. Deletions in region 2 of the mRNA 5′ UTR

e. Deletions in region 3 of the mRNA 5′ UTR

f. Deletions in region 4 of the mRNA 5′ UTR

g. Deletion of the string of adenine nucleotides that follows region 4 in the 5′ UTR

25. Some mutations in the *trp* 5′ UTR region increase termination by the attenuator. Where might these mutations occur and how might they affect the attenuator?

26. Some of the mutations mentioned in Problem 25 have an interesting property. They prevent the formation of the antiterminator that normally takes place when the tryptophan level is low. In one of the mutations, the AUG start codon for the 5′ UTR peptide has been deleted. How might this mutation prevent antitermination from taking place?

Section 16.4

27. Several examples of antisense RNA regulating translation in bacterial cells have been discovered. Molecular geneticists have also used antisense RNA to artificially control transcription in both bacterial and eukaryotic genes. If you wanted to inhibit the transcription of a bacterial gene with antisense RNA, what sequences might the antisense RNA contain?

CHALLENGE QUESTION

Section 16.3

28. Would you expect to see attenuation in the *lac* operon and other operons that control the metabolism of sugars? Why or why not?

17 Control of Gene Expression in Eukaryotes

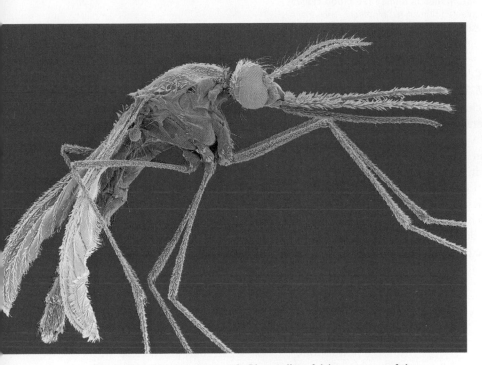

Anopheles mosquitoes transmit Plasmodium falciparum, one of the protozoan parasites that causes malaria, to humans. The parasite survives in humans by altering the expression of its surface proteins through gene regulation, thereby evading the human immune defenses. [Dr. Dennis Kunkel/Visuals Unlimited.]

HOW A PARASITE CHANGES ITS SPOTS

Malaria is one of the most pervasive and destructive of all infectious diseases, infecting about 300 million people worldwide and killing upward of a million people every year. The causative agents of malaria are protozoan parasites of the genus *Plasmodium*; the most deadly forms of the disease are caused by *P. falciparum* and *P. vivax*, which are widespread in many tropical regions of the world. These parasites are transmitted from person to person with deadly precision by mosquitoes. In humans, *Plasmodium* takes up residence in red blood cells, where they produce proteins (antigens) that appear on the surfaces of the host cells (**Figure 17.1**).

Humans are blessed with a powerful immune system, which is capable of detecting foreign antigens and destroying cells that carry them. So why does a malaria parasite survive in its human host, reproduce, and eventually cause malaria? The malaria parasite survives because it continually alters the expression of its surface proteins in a process called antigenic variation and therefore avoids destruction by the immune system.

The surface antigens expressed by *Plasmodium* are encoded by a family of genes that exist in multiple copies. In *P. falciparum*, the antigens are encoded by the *var* gene family, which consists of approximately 60 genes that vary slightly in DNA sequence and encode different surface antigens. In an amazing feat of deception, only one gene of the family is expressed at any one time, and the gene that is expressed changes continuously. As the immune system detects and mounts an attack on the surface antigens of parasitized cells, the antigen that is expressed changes; so that the parasite stays one step ahead of the immune system and is able to survive within the host.

The key to *Plasmodium*'s ability to avoid immune surveillance is gene regulation, the selective and changing expression of its *var* genes. As will be discussed in this chapter, gene regulation in eukaryotic cells can potentially be accomplished by a number of different mechanisms; *Plasmodium* regulates its *var* genes by altering chromatin structure.

The *var* genes of *Plasmodium* are located at the ends of its chromosomes, near the telomeres. These genes are normally silenced in *Plasmodium*, because chromatin in the telomeric regions is highly condensed, with the DNA tightly bound to histone proteins and inaccessible to proteins required for transcription. In 2006, researchers at the Pasteur Institute in Paris showed that individual *var* genes are expressed when chromatin structure

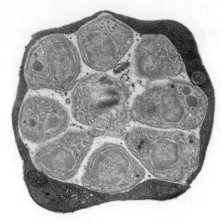

is disrupted by chemical changes in the histone proteins. Enzymes that modify the histones are recruited to the promoter of a specific gene, where they bring about chemical changes in histone proteins, loosening their grip on DNA and providing access to transcription factors, RNA polymerase, and other proteins required for transcription. The mystery of how gene expression is transferred from one *var* gene to another is not yet known but may entail some type of physical interaction between genes.

17.1 *Plasmodium falciparum* **infects human red blood cells.** A cross section through an infected human red blood cell, showing single-celled malaria parasites (green). [Omikron/Photo Researchers.]

This chapter is about gene regulation in eukaryotic cells, which typically takes place at multiple levels. We begin by considering how gene expression is influenced by changes in chromatin structure, which can be altered by several different mechanisms, including the modification of histones as used by *Plasmodium* in controlling the expression of its *var* genes. We next consider the initiation of transcription in eukaryotes, which is controlled by a complex interaction of several types of proteins and the DNA regulatory elements to which they bind. We examine several ways in which gene expression is controlled through the processing, degradation, and translation of mRNA. We end by revisiting some of the similarities in gene regulation in bacteria and eukaryotes.

17.1 Eukaryotic Cells and Bacteria Have Many Features of Gene Regulation in Common, but They Differ in Several Important Ways

As discussed in Chapter 16, many features of gene regulation are common to both bacterial and eukaryotic cells. For example, in both types of cells, DNA-binding proteins influence the ability of RNA polymerase to initiate transcription. However, there are also some differences. First, most eukaryotic genes are not organized into operons and are rarely transcribed together into a single mRNA molecule, although some operon-like gene clusters have been discovered in eukaryotes. In eukaryotic cells, each structural gene typically has its own promoter and is transcribed separately. Second, chromatin structure affects gene expression in eukaryotic cells; DNA must unwind from the histone proteins before transcription can take place. Third, the presence of the nuclear membrane in eukaryotic cells separates transcription and translation in time and space. Therefore, the regulation of gene expression in eukaryotic cells is characterized by a greater diversity of mechanisms that act at different points in the transfer of information from DNA to protein.

Eukaryotic gene regulation is less well understood than bacterial regulation, partly owing to the larger genomes in eukaryotes, their greater sequence complexity, and the difficulty of isolating and manipulating mutations that can be used in the study of gene regulation. Nevertheless, great advances in our understanding of the regulation of eukaryotic genes have been made in recent years, and eukaryotic regulation continues to be a cutting-edge area of research in genetics.

17.2 Changes in Chromatin Structure Affect the Expression of Genes

One type of gene control in eukaryotic cells is accomplished through the modification of chromatin structure. In the nucleus, histone proteins associate to form octamers, around which helical DNA tightly coils to create chromatin (see Figure 11.4). In a general sense, this chromatin structure represses gene expression. For a gene to be transcribed, transcription factors, activators, and RNA polymerase must bind to the DNA. How can these events take place with DNA wrapped tightly around histone proteins? The answer is that, before transcription, chromatin structure changes and the DNA becomes more accessible to the transcriptional machinery.

DNase I Hypersensitivity

Several types of changes are observed in chromatin structure when genes become transcriptionally *active*. As genes become transcriptionally active, regions around the genes become highly sensitive to the action of DNase I (see Chapter 11). These regions, called **DNase I hypersensitive sites**, frequently develop about 1000 nucleotides upstream of the start site of transcription, suggesting that the chromatin in these regions adopts a more open configuration during transcription. This relaxation of the chromatin structure allows regulatory proteins access to binding sites on the DNA. Indeed, many DNase I hypersensitive sites correspond to known binding sites for regulatory proteins. At least three different processes affect gene regulation by altering chromatin structure: (1) the modification of histone proteins;

(2) chromatin remodeling; and (3) DNA methylation. Each of these mechanisms will be discussed in the sections that follow.

Histone Modification

Histones in the octamer core of the nucleosome have two domains: (1) a globular domain that associates with other histones and the DNA and (2) a positively charged tail domain that probably interacts with the negatively charged phosphate groups on the backbone of DNA. The tails of histone proteins are often modified by the addition or removal of phosphate groups, methyl groups, or acetyl groups. These modifications have sometimes been called the **histone code**, because they encode information that affects how genes are expressed.

Methylation of histones One type of histone modification is the addition of methyl groups to the tails of histone proteins. These modifications can bring about either the activation or the repression of transcription, depending on which particular amino acids in the histone tail are methylated. A common modification is the addition of three methyl groups to lysine 4 in the tail of the H3 histone protein, abbreviated H3K4me3 (K is the abbreviation for lysine). The H3K4me3 modification is frequently found in promoters of transcriptionally active genes in eukaryotes. Recent studies have identified proteins that recognize and bind to H3K4me3, including a protein called nucleosome remodeling factor (NURF). NURF and other proteins that recognize H3K4me3 have a common protein-binding domain that binds to the H3 histone tail and then alters chromatin packing, allowing transcription to take place. Research has also demonstrated that some transcription factors, which are necessary for the initiation of transcription (see Chapter 14 and Section 17.4), directly bind to H3K4me3.

Acetylation of histones Another type of histone modification that affects chromatin structure is acetylation, the addition of acetyl groups (CH_3CO) to histone proteins (**Figure 17.2**). The acetylation of histones usually stimulates transcription. For example, the addition of a single acetyl group to lysine 16 in the tail of the H4 histone prevents the formation of the 30-nm chromatin fiber (see Figure 11.4), causing the chromatin to be in an open configuration and available for transcription. In general, acetyl groups destabilize chromatin structure, allowing transcription to take place. Acetyl groups are added to histone proteins by acetyltransferase enzymes; other enzymes called deacetylases strip acetyl groups from histones and restore chromatin structure, which represses transcription. Certain transcription factors (see Chapter 13) and other proteins that regulate transcription either have acetyltransferase activity or attract acetyltransferases to DNA.

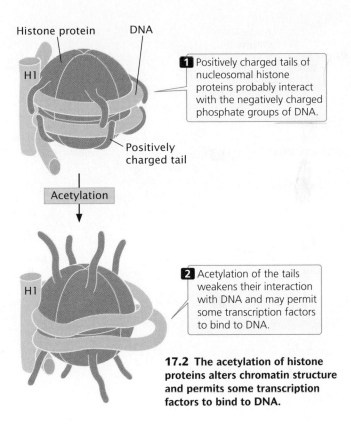

1 Positively charged tails of nucleosomal histone proteins probably interact with the negatively charged phosphate groups of DNA.

2 Acetylation of the tails weakens their interaction with DNA and may permit some transcription factors to bind to DNA.

17.2 The acetylation of histone proteins alters chromatin structure and permits some transcription factors to bind to DNA.

Acetylation of histones controls flowering in *Arabidopsis* The importance of histone acetylation in gene regulation is demonstrated by the control of flowering in *Arabidopsis*, a plant with a number of characteristics that make it an excellent genetic model for plant systems (see material on the model genetic organism *Arabidopsis thaliana* in the Refernce Guide to Model Genetic Organisms at the end of the book). The time at which flowering takes place is critical to the life of a plant; if flowering is initiated at the wrong time of year, pollinators may not be available to fertilize the flowers or environmental conditions may be unsuitable for the survival and germination of the seeds. Consequently, flowering time in most plants is carefully regulated in response to multiple internal and external cues, such as plant size, photoperiod, and temperature.

Among the many genes that control flowering in *Arabidopsis* is *flowering locus C* (*FLC*), which plays an important role in suppressing flowering until after an extended period of coldness (a process called vernalization). The *FLC* gene encodes a regulator protein that represses the activity of other genes that affect flowering (**Figure 17.3**). As long as *FLC* is active, flowering remains suppressed. The activity of *FLC* is controlled by another locus called *flowering locus D* (*FLD*), the key role of which is to stimulate flowering by repressing the action of *FLC*. In essense, flowering is stimulated because *FLD* represses the repressor. How does *FLD* repress *FLC*?

FLD encodes a deacetylase enzyme, which removes acetyl groups from histone proteins in the chromatin surrounding *FLC* (see Figure 17.3). The removal of acetyl groups from

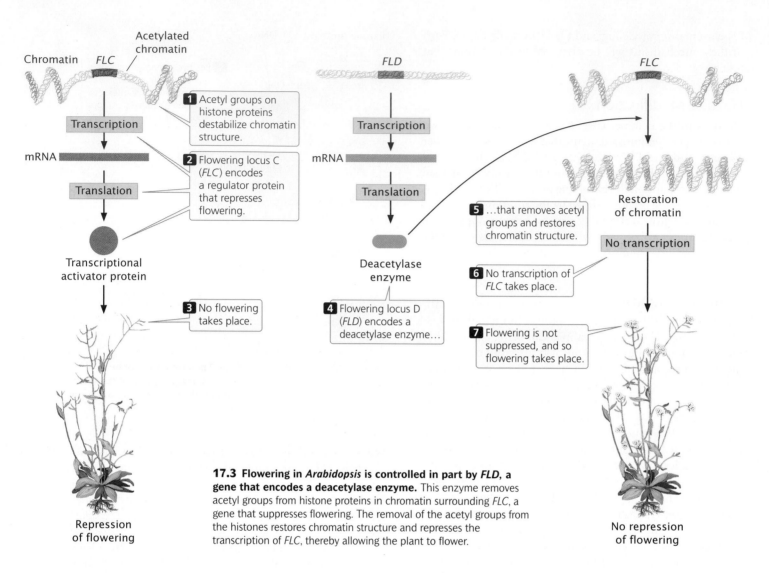

17.3 Flowering in *Arabidopsis* is controlled in part by *FLD*, a gene that encodes a deacetylase enzyme. This enzyme removes acetyl groups from histone proteins in chromatin surrounding *FLC*, a gene that suppresses flowering. The removal of the acetyl groups from the histones restores chromatin structure and represses the transcription of *FLC*, thereby allowing the plant to flower.

histones alters chromatin structure and inhibits transcription. The inhibition of transcription prevents *FLC* from being transcribed and removes its repression on flowering. In short, *FLD* stimulates flowering in *Arabidopsis* by deacetylating the chromatin that surrounds *FLC*, thereby removing its inhibitory effect on flowering. **TRY PROBLEM 19 →**

Chromatin Remodeling

The changes to chromatin structure discussed so far have been through alteration of the structure of the histone proteins. Some transcription factors and other regulatory proteins alter chromatin structure without altering the chemical structure of the histones directly. These proteins are called **chromatin-remodeling complexes**. They bind directly to particular sites on DNA and reposition the nucleosomes, allowing transcription factors to bind to promoters and initiate transcription.

One of the best-studied examples of a chromatin-remodeling complex is SWI–SNF, which is found in yeast, humans, *Drosophila*, and other organisms. This complex uti-

lizes energy derived from the hydrolysis of ATP to reposition nucleosomes, exposing promoters in the DNA to the action of other regulatory proteins and RNA polymerase. Research has shown that SWI–SNF is able to change the position of nucleosomes along a DNA molecule.

Evidence suggests at least two mechanisms by which remodeling complexes reposition nucleosomes. First, some remodeling complexes cause the nucleosome to slide along the DNA, allowing DNA that was wrapped around the nucleosome to occupy a position in between nucleosomes, where it is more accessible to proteins affecting gene expression. Second, some complexes cause conformational changes in the DNA, in nucleosomes, or in both so that DNA that is bound to the nucleosome assumes a more exposed configuration.

Chromatin-remodeling complexes are targeted to specific DNA sequences by transcriptional activators or repressors that attach to a remodeling complex and then bind to the promoters of specific genes. There is also evidence that chromatin-remodeling complexes work together with enzymes that alter histones, such as acetyltransferase enzymes, to change chromatin structure and expose DNA for transcription.

DNA Methylation

Another change in chromatin structure associated with transcription is the methylation of cytosine bases, which yields 5-methylcytosine (see Figure 10.18). The methylation of cytosine in DNA is distinct from the methylation of histone proteins mentioned earlier. Heavily methylated DNA is associated with the repression of transcription in vertebrates and plants, whereas transcriptionally active DNA is usually unmethylated in these organisms. Abnormal patterns of methylation are also associated with some types of cancer.

DNA methylation is most common on cytosine bases adjacent to guanine nucleotides (CpG, where p represents the phosphate group in the DNA backbone); so two methylated cytosines sit diagonally across from each other on opposing strands:

$$5'-\overset{m}{C}\ G-3'$$
$$3'-G\ \underset{m}{C}-5'$$

DNA regions with many CpG sequences are called **CpG islands** and are commonly found near transcription start sites. While genes are not being transcribed, these CpG islands are often methylated, but the methyl groups are removed before the initiation of transcription. CpG methylation is also associated with long-term gene repression, such as on the inactivated X chromosome of female mammals (see Chapter 4).

Evidence indicates that an association exists between DNA methylation and the deacetylation of histones, both of which repress transcription. Certain proteins that bind tightly to methylated CpG sequences form complexes with other proteins that act as histone deacetylases. In other words, methylation appears to attract deacetylases, which remove acetyl groups from the histone tails, stabilizing the nucleosome structure and repressing transcription. The demethylation of DNA allows acetyltransferases to add acetyl groups, disrupting nucleosome structure and permitting transcription. TRY PROBLEM 20 →

CONCEPTS

Sensitivity to DNase I digestion indicates that transcribed DNA assumes an open configuration before transcription. Chromatin structure can be altered by modifications of histone proteins, by chromatin-remodeling complexes that reposition nucleosomes, and by the methylation of DNA.

✔ CONCEPT CHECK 1

What are some of different processes that affect gene regulation by altering chromatin structure?

17.3 Epigenetic Effects Often Result from Alterations in Chromatin Structure

Changes in chromatin structure that affect gene expression, such as the mechanisms just described, are examples of the phenomena of epigenetics. In Chapters 5 and 11, we considered the concept of epigenetics, which was defined as alterations to DNA and chromatin structure that affect traits and are passed on to other cells or future generations but are not caused by changes in the DNA base sequence.

In the past, the term epigenetics was applied to various genetic phenomena that could not be easily explained by traditional Mendelian principles. For example, in the 1950s, Alexander Brink described **paramutation** in corn, in which one allele of a genotype altered the expression of another allele. In corn, alleles at the R locus help to determine pigmentation; the R^r allele is normally dominant and encodes purple kernels. The R^{st} allele encodes spotted kernels. Brink observed that, when the R^r allele was present in a genotype with R^{st}, the effect of the R^r allele was altered so that, in later generations, it encoded reduced levels of purple pigment. In this example of paramutation, the R^{st} allele altered the expression of the R^r allele; this diminished effect of the R^r allele on pigmentation persisted for several generations, even in the absence of the R^{st} allele. Paramutation violates Mendel's principle of segregation, which states that, when gametes are formed, each allele separates and is transmitted independently to the next generation; paramutation violates this rule because the effect of the R^r allele remains associated with the R^{st} allele even after the two alleles have separated. Epigenetics has been called "inheritance, but not as we know it."

In recent years, scientists have discovered that many of these types of phenomena result from changes in chromatin structure that affect the expression of genetic information. Epigenetic changes are heritable in the sense that they are passed from cell to cell and sometimes across generations, but they differ from genetic traits in that they are not encoded by the DNA base sequence and are reversible.

Epigenetic Effects

A number of examples of epigenetic effects have been documented. In some cases, the molecular basis of these effects is at least partly understood.

Epigenetic changes induced by maternal behavior A fascinating example of epigenetics is seen in the long-lasting effects of maternal behavior in rats. Mother rats lick and groom their offspring, usually while the mother arches her back and nurses the offspring. The offspring of mothers who display more licking and grooming behavior are, as adults, less fearful and show reduced hormonal responses to stress compared with the offspring of mothers who lick and groom less. These long-lasting differences in the offspring are not due to genetic differences inherited from their mothers—at least not genetic differences in the base sequences of the DNA. Research has demonstrated that offspring exposed to more licking and grooming from the mother develop a different pattern of DNA methylation compared with offspring exposed to less licking and grooming. These differences in DNA methylation affect the acetylation of histone proteins (discussed earlier) that persist into adulthood and alter the expression of the glucocorticoid receptor gene, a gene that plays a role in hormonal responses to stress. The expression of other stress-response genes also is affected.

To demonstrate the effect of altered chromatin structure on the stress response of the offspring, researchers infused the brains of young rats with a deacetylase inhibitor, which prevents the removal of acetyl groups from the histone proteins. After infusion of the deacetylase inhibitor, differences in DNA methylation and histone acetylation associated with grooming behavior disappeared, as did the difference in responses to fear and stress in the adults. These results demonstrate that the mother rat's behavior (licking and grooming) brings about epigenetic changes in chromatin of the offspring, which cause long-lasting differences in their behavior.

Epigenetic effects caused by prenatal exposure In another study, researchers found that the exposure of embryonic rats to the fungicide vinclozolin, which reduces sperm production, led to reduced sperm production not only in the treated animals (when they reached puberty), but also in several subsequent generations. Increased DNA methylation was seen in sperm of the males that were exposed to vinclozolin and these patterns of methylation were inherited. This study and others have raised concerns that, through epigenetic changes, environmental exposure to some chemicals might have long-lasting effects on the health of future generations.

Epigenetic effects in monozygotic twins Monozygotic (identical) twins develop from a single egg fertilized by a single sperm that divides and gives rise to two zygotes (see Chapter 6). Monozygotic twins are genetically identical, in the sense that they possess identical DNA sequences, but they often differ somewhat in appearance, health, and behavior. The nature of these differences in the phenotypes of identical twins is poorly understood, but recent evidence suggests that at least some of these differences may be due to epigenetic changes. In one study, Mario Fraga and his colleagues examined 80 identical twins and compared the degree and location of their DNA methylation and histone acetylation. They found that DNA methylation and histone acetylation of the two members of an identical twin pair were similar early in life, but older twins had remarkable differences in their overall content and distribution of DNA methylation and histone acetylation. Furthermore, these differences affected gene expression in the twins. This research suggests that identical twins do differ epigenetically and that phenotypic differences between them may be caused by differential gene expression. Additional examples of epigenetic effects include genomic imprinting (discussed in Chapter 5) and X-chromosome inactivation (discussed in Chapter 4).

Molecular Mechanisms of Epigenetic Changes

Epigenetics alters gene expression, alteration that is stable enough to be transmitted through mitosis (and sometimes meiosis) but that can also be changed. Most evidence suggests that epigenetic effects are brought about by physical changes in chromatin structure. Earlier, we considered a number of chemical changes in DNA and histone proteins that potentially alter chromatin structure, including DNA methylation, modification of histone proteins, and repositioning of nucleosomes. There is also growing evidence that small RNA molecules (siRNAs, miRNAs, and piRNAs; see Section 17.6) take part in some types of epigenetic control of gene expression.

The fact that epigenetic marks are passed on to other cells and (sometimes) to future generations means that changes in chromatin structure associated with epigenetic phenotypes must be faithfully maintained when chromosomes replicate. How are epigenetic changes maintained through the process of cell division? Evidence suggests that DNA methylation is often stably maintained through replication. Recall that cytosine bases of CpG sequences are often methylated, which means that two methylated cytosine bases sit diagonally across from each other on opposing strands (see section on DNA methylation). Before replication, cytosine bases on both strands are methylated (**Figure 17.4**). Immediately after semiconservative replication, the cytosine base on one strand (the template strand) will be methylated, but the cytosine base on the other strand (the newly replicated strand) will be unmethylated. Special methyltransferase enzymes recognize the hemimethylated state of CpG dinucleotides and add methyl groups to the unmethylated cytosine bases, creating two new DNA molecules that are fully methylated.

How histone modifications, nucleosome positioning, and other types of epigenetic marks might be maintained across replication is less clear. One possibility, discussed in Chapter 12, is that, after replication, the epigenetic marks remain on the original histones, and these marks recruit enzymes that make similar changes to the new histones.

The Epigenome

The overall pattern of chromatin modifications possessed by each individual organism has been termed the **epigenome**. The epigenome helps to explain stable patterns of gene expression. For example, all cells in the human body are

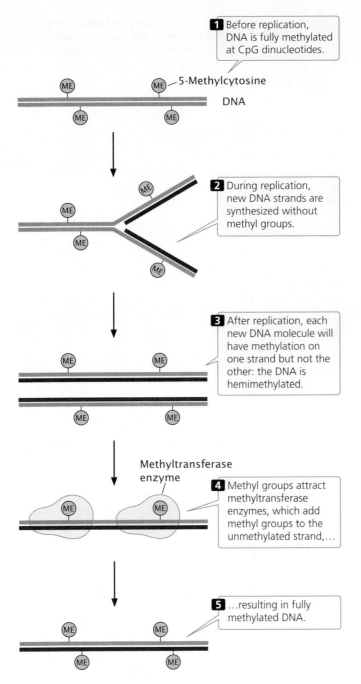

1 Before replication, DNA is fully methylated at CpG dinucleotides.

ME ME 5-Methylcytosine

DNA

ME ME

2 During replication, new DNA strands are synthesized without methyl groups.

ME

ME

ME

ME

3 After replication, each new DNA molecule will have methylation on one strand but not the other: the DNA is hemimethylated.

ME ME

ME ME

Methyltransferase enzyme

4 Methyl groups attract methyltransferase enzymes, which add methyl groups to the unmethylated strand,…

ME ME

5 …resulting in fully methylated DNA.

ME ME

ME ME

17.4 DNA methylation is stably maintained through DNA replication.

genetically identical, and yet different cell types exhibit remarkably different phenotypes—a nerve cell is quite distinct in its shape, size, and function from an intestinal cell. These differences in phenotypes are stable and passed from one cell to another, in spite of the fact that the DNA sequences of all the cells are identical.

Embryonic stem cells are undifferentiated cells that are capable of forming every type of cell in an organism, a property referred to as **pluripotency**. As a stem cell divides and gives rise to a more specialized type of cell, the gene-expression program becomes progressively fixed so that each particular cell type expresses only those genes necessary to carry out the functions of that cell type. Although the control of these cell-specific expression programs is still poorly understood, changes in DNA methylation and chromatin structure clearly play important roles in silencing some genes and activating others.

Recently, researchers have determined the genome-wide distribution of DNA methylation. In mammals, most methylation is at cytosine bases, converting cytosine into 5-methylcytosine. The researchers identified the location of 5-methylcytosine across the entire genome in two cell types: (1) an undifferentiated human stem cell; and (2) a fibroblast, a fully differentiated connective tissue cell. The researchers found widespread differences in the methylation patterns of these two types of cells. The finding that patterns of DNA methylation vary among cell types has led to speculation that cytosine methylation may provide the means by which cell types stably maintain their differences during development.

CONCEPTS

Epigenetic effects—stable alterations to gene expression that are not caused by changes in the DNA base sequence—are frequently associated with changes in chromatin structure. The epigenome is the genomewide pattern of epigenetic changes possessed by an individual organism.

✔ CONCEPT CHECK 2

Which is *not* a mechanism of epigenetic change?

a. DNA methylation

b. Alteration of a DNA base sequence in a promoter

c. Histone actylation

d. Nucleosome repositioning

17.4 The Initiation of Transcription Is Regulated by Transcription Factors and Transcriptional Regulator Proteins

We just considered one level at which gene expression is controlled—the alteration of chromatin and DNA structure. We now turn to another important level of control—that is, control through the binding of proteins to DNA sequences that affect transcription. Transcription is an important level of control in eukaryotic cells, and this control requires

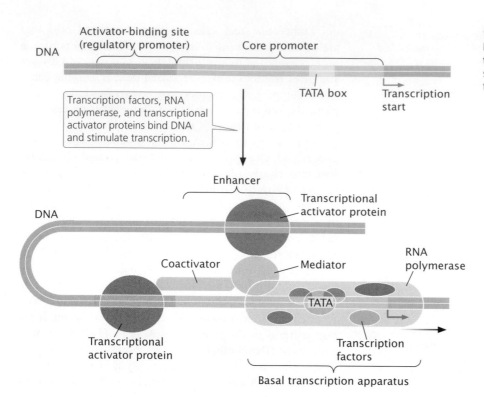

17.5 Transcriptional activator proteins bind to sites on DNA and stimulate transcription. Most act by stimulating or stabilizing the assembly of the basal transcription apparatus.

a number of different types of proteins and regulatory elements. The initiation of eukaryotic transcription was discussed in detail in Chapter 13. Recall that general transcription factors and RNA polymerase assemble into a *basal transcription apparatus*, which binds to a *core promoter* located immediately upstream of a gene. The basal transcription apparatus is capable of minimal levels of transcription; *transcriptional regulator proteins* are required to bring about normal levels of transcription. These proteins bind to a regulatory promoter, which is located upstream of the core promoter (**Figure 17.5**), and to *enhancers*, which may be located some distance from the gene. Some transcriptional regulator proteins are activators, stimulating transcription. Others are represssors, inhibiting transcription.

Transcriptional Activators and Coactivators

Transcriptional activator proteins stimulate and stabilize the basal transcription apparatus at the core promoter. The activators may interact directly with the basal transcription apparatus or indirectly through protein coactivators. Some activators and coactivators, as well as the general transcription factors, also have acetyltransferase activity and so further stimulate transcription by altering chromatin structure.

Transcriptional activator proteins have two distinct functions (see Figure 17.5). First, they are capable of binding DNA at a specific base sequence, usually a consensus sequence in a regulatory promoter or enhancer; for this function, most transcriptional activator proteins contain one or more DNA-binding motifs, such as the helix-turn-helix, zinc finger, and leucine zipper (see Chapter 16). A second function is the ability to interact with other components of the transcriptional apparatus and influence the rate of transcription.

Within the regulatory promoter are typically several different consensus sequences to which different transcriptional activators can bind. Among different promoters, activator-binding sites are mixed and matched in different combinations (**Figure 17.6**), and so each promoter is regulated by a unique combination of transcriptional activator proteins.

Transcriptional activator proteins bind to the consensus sequences in the regulatory promoter and affect the assembly or stability of the basal transcription apparatus at the core promoter. One of the components of the basal transcription apparatus is a complex of proteins called the **mediator** (see Figure 17.5). Transcriptional activator proteins binding to sequences in the regulatory promoter (or enhancer, see next section) make contract with the mediator and affect the rate at which transcription is initiated. Some regulatory promoters also contain sequences that are bound by proteins that lower the rate of transcription through inhibitory interactions with the mediator.

Regulation of galactose metabolism through GAL4

An example of a transcriptional activator protein is GAL4, which regulates the transcription of several yeast genes whose products metabolize galactose. Like the genes in the *lac* operon, the genes that control galactose metabolism are inducible: when galactose is absent, these genes are not

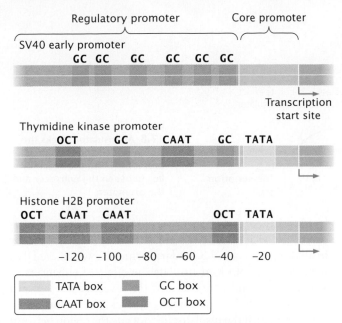

Regulatory promoter | Core promoter

SV40 early promoter
GC GC GC GC GC GC

Transcription start site

Thymidine kinase promoter
OCT GC CAAT GC TATA

Histone H2B promoter
OCT CAAT CAAT OCT TATA

−120 −100 −80 −60 −40 −20

▢ TATA box	▢ GC box	
▢ CAAT box	▢ OCT box	

17.6 The consensus sequences in the promoters of three eukaryotic genes illustrate the principle that different sequences can be mixed and matched in different combinations. A different transcriptional activator protein binds to each consensus sequence, and so each promoter responds to a unique combination of activator proteins.

transcribed and the proteins that break down galactose are not produced; when galactose is present, the genes are transcribed and the enzymes are synthesized. GAL4 contains several zinc fingers and binds to a DNA sequence called UAS$_G$ (upstream activating sequence for GAL4). UAS$_G$ exhibits the properties of an enhancer—a regulatory sequence that may be some distance from the regulated gene and is independent of the gene in position and orientation (see Chapter 13). When bound to UAS$_G$, GAL4 activates the transcription of yeast genes needed for metabolizing galactose.

A particular region of GAL4 binds another protein called GAL80, which regulates the activity of GAL4 in the presence of galactose. When galactose is absent, GAL80 binds to GAL4, preventing GAL4 from activating transcription (**Figure 17.7**). When galactose is present, however, it binds to another protein called GAL3, which interacts with GAL80, causing a conformational change in GAL80 so that it can no longer bind GAL4. The GAL4 protein is then free to activate the transcription of the genes, whose products metabolize galactose.

GAL4 and a number of other transcriptional activator proteins contain multiple amino acids with negative charges that form an *acidic activation domain*. These acidic activators stimulate transcription by enhancing the assembly of the basal transcription apparatus.

Transcriptional Repressors

Some regulatory proteins in eukaryotic cells act as repressors, inhibiting transcription. These repressors bind to sequences in the regulatory promoter or to distant sequences called *silencers*, which, like enhancers, are position and orientation

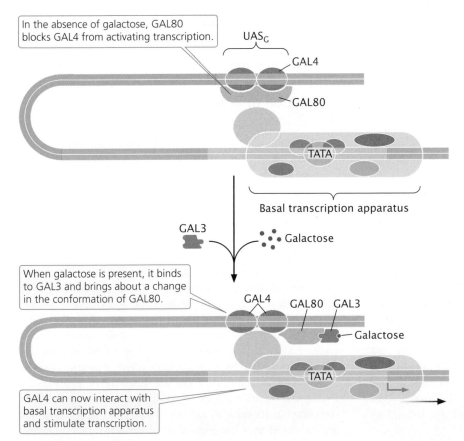

In the absence of galactose, GAL80 blocks GAL4 from activating transcription.

UAS$_G$

GAL4

GAL80

TATA

Basal transcription apparatus

GAL3

Galactose

When galactose is present, it binds to GAL3 and brings about a change in the conformation of GAL80.

GAL4 GAL80 GAL3

Galactose

TATA

GAL4 can now interact with basal transcription apparatus and stimulate transcription.

17.7 Transcription is activated by GAL4 in response to galactose. GAL4 binds to the UAS$_G$ site and controls the transcription of genes in galactose metabolism.

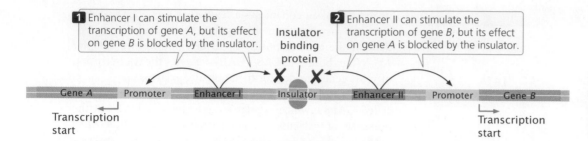

1 Enhancer I can stimulate the transcription of gene *A*, but its effect on gene *B* is blocked by the insulator.

2 Enhancer II can stimulate the transcription of gene *B*, but its effect on gene *A* is blocked by the insulator.

Insulator-binding protein

Gene *A* Promoter Enhancer I Insulator Enhancer II Promoter Gene *B*

Transcription start

Transcription start

17.8 An insulator blocks the action of an enhancer on a promoter when the insulator lies between the enhancer and the promoter.

independent. Unlike repressors in bacteria, most eukaryotic repressors do not directly block RNA polymerase. These repressors may compete with activators for DNA binding sites: when a site is occupied by an activator, transcription is activated, but, if a repressor occupies that site, there is no activation. Alternatively, a repressor may bind to sites near an activator site and prevent the activator from contacting the basal transcription apparatus. A third possible mechanism of repressor action is direct interference with the assembly of the basal transcription apparatus, thereby blocking the initiation of transcription.

CONCEPTS

Transcriptional regulatory proteins in eukaryotic cells can influence the initiation of transcription by affecting the stability or assembly of the basal transcription apparatus. Some regulatory proteins are activators and stimulate transcription; others are repressors and inhibit initiation of transcription.

✔ CONCEPT CHECK 3

Most transcriptional activator proteins affect transcription by interacting with

a. introns.

b. the basal transcription apparatus.

c. DNA polymerase.

d. nucleosomes.

Enhancers and Insulators

Enhancers are capable of affecting transcription at distant promoters. For example, an enhancer that regulates the gene encoding the alpha chain of the T-cell receptor is located 69,000 bp downstream of the gene's promoter. Furthermore, the exact position and orientation of an enhancer relative to the promoter can vary. How can an enhancer affect the initiation of transcription taking place at a promoter that is tens of thousands of base pairs away? In many cases, regulator proteins bind to the enhancer and cause the DNA between the enhancer and the promoter to loop out, bringing the promoter and enhancer close to each other, and so the transcriptional regulator proteins are able to directly interact with the basal transcription apparatus at the core promoter (see Figure 17.5). Some enhancers may be attracted to promoters by proteins that bind to sequences in the regulatory promoter and "tether" the enhancer close to the core promoter. A typical enhancer is some 500 bp in length and contains 10 binding sites for proteins that regulate transcription.

Most enhancers are capable of stimulating any promoter in their vicinities. Their effects are limited, however, by **insulators** (also called boundary elements), which are DNA sequences that block or insulate the effect of enhancers in a position-dependent manner. If the insulator lies between the enhancer and the promoter, it blocks the action of the enhancer; but, if the insulator lies outside the region between the two, it has no effect (**Figure 17.8**). Specific proteins bind to insulators and play a role in their blocking activity. Some insulators also limit the spread of changes in chromatin structure that affect transcription. TRY PROBLEM 24 →

CONCEPTS

Some regulatory proteins bind to enhancers, which are regulatory elements that are distant from the gene whose transcription they stimulate. Insulators are DNA sequences that block the action of enhancers.

✔ CONCEPT CHECK 4

How does the binding of regulatory proteins to enhancers affect transcription at genes that are thousands of base pairs away?

Regulation of Transcriptional Stalling and Elongation

Transcription in eukaryotes is often regulated through factors that affect the initiation of transcription, including changes in chromatin structure, transcription factors, and transcriptional regulatory proteins. Research indicates that transcription may also be controlled through factors that affect stalling and elongation of RNA polymerase after transcription has been initiated.

The basal transcription apparatus—consisting of RNA polymerase, transcription factors, and other proteins—assembles at the core promoter. When the initiation of transcription has taken place, RNA polymerase moves downstream, transcribing the structural gene and producing an RNA product. At some genes, RNA polymerase initiates transcription and transcribes from 24 to 50 nucleotides of RNA but then pauses or stalls. For example, stalling is observed at genes that encode **heat-shock proteins** in *Drosophila*—proteins that help to prevent damage from stressing agents such as extreme heat. Heat-shock proteins are produced by a large number of different genes. During

Table 17.1	A few response elements found in eukaryotic cells	
Response Element	**Responds to**	**Consensus Sequence**
Heat-shock element	Heat and other stress	CNNGAANNTCCNNG
Glucocorticoid response element	Glucocorticoids	TGGTACAAATGTTCT
Phorbol ester response element	Phorbal esters	TGACTCA
Serum response element	Serum	CCATATTAGG

Source: After B. Lewin, *Genes IV* (Oxford University Press, 1994), p. 880.

times of environmental stress, the transcription of all the heat-shock genes is greatly elevated. RNA polymerase initiates transcription at heat-shock genes in *Drosophila* but, in the absence of stress, stalls downstream of the transcription initiation site. Stalled polymerases are released when stress is encountered, allowing rapid transcription of the genes and the production of heat-shock proteins that facilitate adaptation to the stressful environment.

Stalling was formerly thought to take place at only a small number of genes, but recent research indicates that stalling is widespread and common throughout eukaryotic genomes. For example, stalled RNA polymerases were found at hundreds of genes in *Drosophila*. Factors that promote stalling have been identified; a protein called negative elongation factor (NELF) causes RNA polymerase to stall after initiation. Reducing the amount of NELF in cultured cells stimulates the elongation of RNA molecules at many genes with stalled RNA polymerases.

CONCEPTS

At some genes, RNA polymerase may pause or stall downstream of the promoter. Regulatory factors affect stalling and the elongation of transcription.

Coordinated Gene Regulation

Although most eukaryotic cells do not possess operons, several eukaryotic genes may be activated by the same stimulus. Groups of bacterial genes are often coordinately expressed (turned on and off together) because they are physically clustered as an operon and have the same promoter, but coordinately expressed genes in eukaryotic cells are not clustered. How, then, is the transcription of eukaryotic genes coordinately controlled if they are not organized into an operon?

Genes that are coordinately expressed in eukaryotic cells are able to respond to the same stimulus because they have short regulatory sequences in common in their promoters or enhancers. For example, different eukaryotic heat-shock genes possess a common regulatory element upstream of their start sites. Such DNA regulatory sequences are called **response elements**; they are short sequences that typically contain consensus sequences (**Table 17.1**) at varying distances from the gene being regulated. The response elements are binding sites for transcriptional activators. A transcriptional activator protein binds to the response element and elevates transcription. The same response element may be present in different genes, allowing multiple genes to be activated by the same stimulus.

A single eukaryotic gene may be regulated by several different response elements. For example, the metallothionein gene protects cells from the toxicity of heavy metals by encoding a protein that binds to heavy metals and removes them from cells. The basal transcription apparatus assembles around the TATA box, just upstream of the transcription start site for the metallothionein gene, but the apparatus alone is capable of only low rates of transcription.

Other response elements found upstream of the metallothionein gene contribute to increasing its rate of transcription. For example, several copies of a metal response element (MRE) are upstream of the metallothionein gene (**Figure 17.9**). Heavy metals stimulate the binding of activator proteins to MREs, which elevates the rate of transcription of the metallothionein gene. Because there are multiple copies of the MRE, high rates of transcription are induced by metals. Two enhancers also are located in the upstream region of the metallothionein gene; one enhancer contains a response element known as TRE, which stimulates transcription in the presence of an activated protein called AP1. A third response element called GRE is located approximately 250 nucleotides upstream of the metallothionein gene and stimulates transcription in response to certain hormones.

This example illustrates a common feature of eukaryotic transcriptional control: a single gene may be activated by several different response elements found in both promoters and enhancers. Multiple response elements allow the same gene to be activated by different stimuli. At the same time, the presence of the same response element in different genes allows a single stimulus to activate multiple genes. In this way, response elements allow complex biochemical responses in eukaryotic cells.

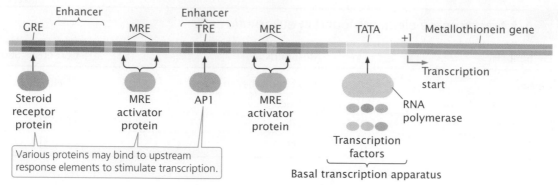

17.9 Multiple response elements (MREs) are found in the upstream region of the metallothionein gene. The basal transcription apparatus binds near the TATA box. In response to heavy metals, activator proteins bind to several MREs and stimulate transcription. The TRE response element is the binding site for transcription factor AP1, which is stimulated by phorbol esters. In response to glucocorticoid hormones, steroid-receptor proteins bind to the GRE response element located approximately 250 nucleotides upstream of the metallothionein gene and stimulate transcription.

17.5 Some Genes Are Regulated by RNA Processing and Degradation

In bacteria, transcription and translation take place simultaneously. In eukaryotes, transcription takes place in the nucleus and the pre-mRNAs are then processed before moving to the cytoplasm for translation, allowing opportunities for gene control after transcription. Consequently, post-transcriptional gene regulation assumes an important role in eukaryotic cells. A common level of gene regulation in eukaryotes is RNA processing and degradation.

Gene Regulation Through RNA Splicing

Alternative splicing allows a pre-mRNA to be spliced in multiple ways, generating different proteins in different tissues or at different times in development (see Chapter 14). Many eukaryotic genes undergo alternative splicing, and the regulation of splicing is probably an important means of controlling gene expression in eukaryotic cells.

Alternative splicing in the T-antigen gene The T-antigen gene of the mammalian virus SV40 is a well-studied example of alternative splicing. This gene is capable of encoding two different proteins, the large T and small t antigens. Which of the two proteins is produced depends on which of two alternative 5′ splice sites is used in RNA splicing (**Figure 17.10**). The use of one 5′ splice site produces mRNA that encodes the large T antigen, whereas the use of the other 5′ splice site (which is farther downstream) produces an mRNA encoding the small t antigen.

A protein called splicing factor 2 (SF2) enhances the production of mRNA encoding the small t antigen (see Figure 17.10). Splicing factor 2 has two binding domains: one domain is an RNA-binding region and the other has alternating serine and arginine amino acids. These two domains are typical of **SR** (serine- and arginine-rich) **proteins**, which often play a role in regulating splicing. Splicing factor 2

stimulates the binding of small nuclear ribonucleoproteins (snRNPs) to the 5′ splice site, one of the earliest steps in RNA splicing (see Chapter 14). The precise mechanism by which SR proteins influence the choice of splice sites is poorly understood. One model suggests that SR proteins bind to specific splice sites on mRNA and stimulate the attachment of snRNPs, which then commit the site to splicing.

Alternative splicing in *Drosophila* sexual development Another example of alternative mRNA splicing that regulates the expression of genes controls whether a fruit fly

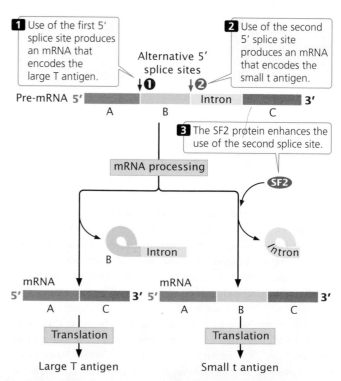

17.10 Alternative splicing leads to the production of the small t antigen and the large T antigen in the mammalian virus SV40.

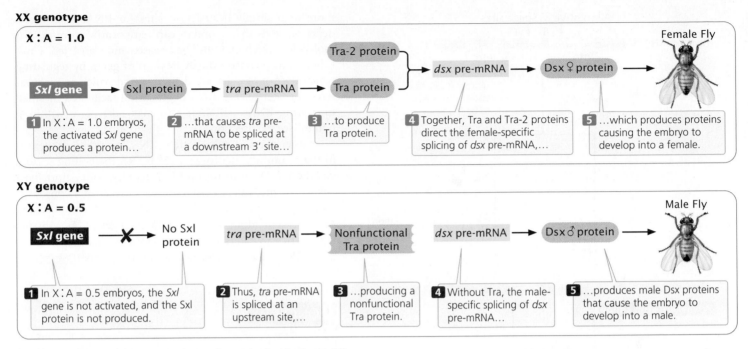

17.11 Alternative splicing controls sex determination in *Drosophila*.

develops as male or female. Sex differentiation in *Drosophila* arises from a cascade of gene regulation. When the ratio of X chromosomes to the number of haploid sets of autosomes (the X : A ratio; see Chapter 4) is 1, a female-specific promoter is activated early in development and stimulates the transcription of the *sex-lethal* (*Sxl*) gene (**Figure 17.11**). The protein encoded by *Sxl* regulates the splicing of the pre-mRNA transcribed from another gene called *transformer* (*tra*). The splicing of *tra* pre-mRNA results in the production of the Tra protein (see Figure 17.11). Together with another protein (Tra-2), Tra stimulates the female-specific splicing of pre-mRNA from yet another gene called *double-sex* (*dsx*). This event produces a female-specific Dsx protein, which causes the embryo to develop female characteristics.

In male embryos, which have an X : A ratio of 0.5 (see Figure 17.11), the promoter that transcribes the *Sxl* gene in females is inactive; so no Sxl protein is produced. In the absence of Sxl protein, *tra* pre-mRNA is spliced at a different 3′ splice site to produce a nonfunctional form of Tra protein (**Figure 17.12**). In turn, the presence of this nonfunctional Tra in males causes *dsx* pre-mRNAs to be spliced differently from that in females, and a male-specific Dsx protein is produced (see Figure 17.11). This event causes the development of male-specific traits.

In summary, the Tra, Tra-2, and Sxl proteins regulate alternative splicing that produces male and female phenotypes in *Drosophila*. Exactly how these proteins regulate alternative splicing is not yet known, but the Sxl protein (produced only in females) possibly blocks the upstream splice site on the *tra* pre-mRNA. This blockage would force the spliceosome to use the downstream 3′ splice site, which

causes the production of Tra protein and eventually results in female traits (see Figure 17.12). **TRY PROBLEM 25 →**

CONCEPTS

Eukaryotic genes can be regulated through the control of mRNA processing. The selection of alternative splice sites leads to the production of different proteins.

The Degradation of RNA

The amount of a protein that is synthesized depends on the amount of corresponding mRNA available for translation. The amount of available mRNA, in turn, depends on both the rate of mRNA synthesis and the rate of mRNA degradation. Eukaryotic mRNAs are generally more stable than bacterial mRNAs, which typically last only a few minutes before being degraded. Nonetheless, there is great variability in the stability of eukaryotic mRNA: some mRNAs persist for only a few minutes; others last for hours, days, or even months. These variations can result in large differences in the amount of protein that is synthesized.

Cellular RNA is degraded by ribonucleases, enzymes that specifically break down RNA. Most eukaryotic cells contain 10 or more types of ribonucleases, and there are several different pathways of mRNA degradation. In one pathway, the 5′ cap is first removed, followed by 5′→3′ removal of nucleotides. A second pathway begins at the 3′ end of the mRNA and removes nucleotides in the 3′→5′ direction. In a third pathway, the mRNA is cleaved at internal sites.

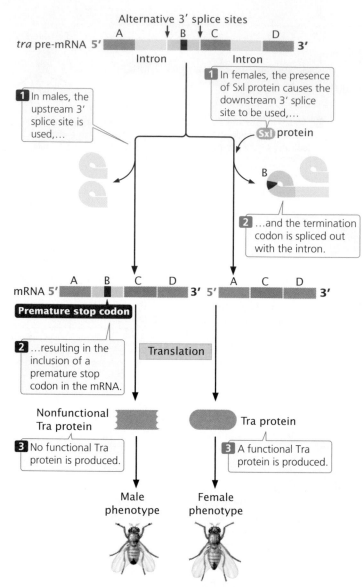

Alternative 3′ splice sites

tra pre-mRNA 5′ ─── 3′

Intron   Intron

1 In males, the upstream 3′ splice site is used,…

1 In females, the presence of Sxl protein causes the downstream 3′ splice site to be used,…

Sxl protein

B

2 …and the termination codon is spliced out with the intron.

mRNA 5′ ──── 3′ 5′ ──── 3′

Premature stop codon

2 …resulting in the inclusion of a premature stop codon in the mRNA.

Translation

Nonfunctional Tra protein   Tra protein

3 No functional Tra protein is produced.

3 A functional Tra protein is produced.

Male phenotype   Female phenotype

17.12 Alternative splicing of *tra* pre-mRNA. Two alternative 3′ splice sites are present.

Messenger RNA degradation from the 5′ end is most common and begins with the removal of the 5′ cap. This pathway is usually preceded by the shortening of the poly(A) tail. Poly(A)-binding proteins (PABPs) normally bind to the poly(A) tail and contribute to its stability-enhancing effect. The presence of these proteins at the 3′ end of the mRNA protects the 5′ cap. When the poly(A) tail has been shortened below a critical limit, the 5′ cap is removed, and nucleases then degrade the mRNA by removing nucleotides from the 5′ end. These observations suggest that the 5′ cap and the 3′ poly(A) tail of eukaryotic mRNA physically interact with each other, most likely by the poly(A) tail bending around so that the PABPs make contact with the 5′ cap (see Figure 15.18).

Much of RNA degradation takes place in specialized complexes called P bodies. However, P bodies appear to be more than simply destruction sites for RNA. Recent evidence suggests that P bodies can temporarily store mRNA molecules, which may later be released and translated. Thus, P bodies help control the expression of genes by regulating which RNA molecules are degraded and which are sequestered for later release. RNA degradation facilitated by small interfering RNAs (siRNAs) also may take place within P bodies (see next section).

Other parts of eukaryotic mRNA, including sequences in the 5′ untranslated region (5′ UTR), the coding region, and the 3′ UTR, also affect mRNA stability. Some short-lived eukaryotic mRNAs have one or more copies of a consensus sequence consisting of 5′-AUUUAUAA-3′, referred to as the AU-rich element, in the 3′ UTR. Messenger RNAs containing AU-rich elements are degraded by a mechanism in which microRNAs take part (see next section). **TRY PROBLEM 26 →**

> **CONCEPTS**
>
> The stability of mRNA influences gene expression by affecting the amount of mRNA available to be translated. The stability of mRNA is affected by the 5′ cap, the poly(A) tail, the 5′ UTR, the coding section, and sequences in 3′ UTR.
>
> ✔ **CONCEPT CHECK 5**
>
> How does the poly(A) tail affect stability?

17.6 RNA Interference Is an Important Mechanism of Gene Regulation

The expression of a number of eukaryotic genes is controlled through RNA interference, also known as RNA silencing and posttranscriptional gene silencing (see Chapter 14). Recent research suggests that as much as 30% of human genes are regulated by RNA interference. Although many of the details of this mechanism are still being investigated, RNA interference is widespread in eukaryotes, existing in fungi, plants, and animals. This mechanism is also widely used as a powerful technique for artificially regulating gene expression in genetically engineered organisms (see Chapter 19).

Small Interfering RNAs and MicroRNAs

RNA interference is triggered by microRNAs (miRNAs) and small interfering RNAs (siRNAs), depending on their origin and mode of action (see Chapter 14). An enzyme called Dicer cleaves and processes double-stranded RNA to produce siRNAs or miRNAs that are from 21 to 25 nucleotides in length (**Figure 17.13**) and pair with proteins to form an RNA-induced silencing complex (RISC). The RNA component of RISC then pairs with complementary base sequences of specific mRNA molecules, most often with sequences in the 3′ UTR of the mRNA. Small interfering RNAs tend to

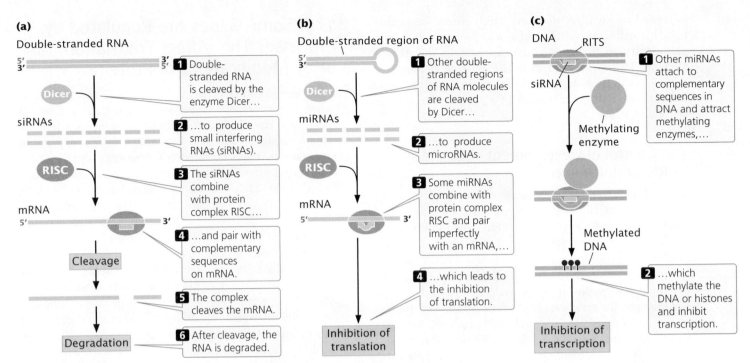

17.13 RNA silencing leads to the degradation of mRNA or to the inhibition of translation or transcription. (a) Small interfering RNAs (siRNAs) degrade mRNA by cleavage. (b) MicroRNAs (miRNAs) lead to the inhibition of translation. (c) Some small interfering RNAs (siRNAs) methylate histone proteins or DNA, inhibiting transcription.

base pair perfectly with the mRNAs, whereas miRNAs often form less-than-perfect pairings.

Mechanisms of Gene Regulation by RNA Interference

Small interfering RNAs and microRNAs regulate gene expression through at least four distinct mechanisms: (1) cleavage of mRNA, (2) inhibition of translation, (3) transcriptional silencing, or (4) degradation of mRNA.

RNA cleavage   RISCs that contain an siRNA (and some that contain an miRNA) pair with mRNA molecules and cleave the mRNA near the middle of the bound siRNA (see Figure 17.13a). This cleavage is sometimes referred to as "Slicer activity." After cleavage, the mRNA is further degraded. Thus, the presence of siRNAs and miRNAs increase the rate at which mRNAs are broken down and decrease the amount of protein produced.

Inhibition of translation   Some miRNAs regulate genes by inhibiting the translation of their complementary mRNAs (see Figure 17.13b). For example, an important gene in flower development in *Arabidopsis thaliana* is *APETALA2*. The expression of this gene is regulated by an miRNA that base pairs with nucleotides in the coding region of *APETALA2* mRNA and inhibits its translation.

The exact mechanism by which miRNAs repress translation is not known, but some research suggests that it can inhibit both the initiation step of translation and steps after translation initiation such as those causing premature termination. Many mRNAs have multiple miRNA-binding sites, and translation is most efficiently inhibited when multiple miRNAs are bound to the mRNA.

Transcriptional silencing   Other siRNAs silence transcription by altering chromatin structure. These siRNAs combine with proteins to form a complex called RITS (for RNA transcriptional silencing; see Figure 17.13c), which is analogous to RISC. The siRNA component of RITS then binds to its complementary sequence in DNA or an RNA molecule in the process of being transcribed and represses transcription by attracting enzymes that methylate the tails of histone proteins. The addition of methyl groups to the histones causes them to bind DNA more tightly, restricting the access of proteins and enzymes necessary to carry out transcription (see earlier section on histone modification). Some miRNAs bind to complementary sequences in DNA and attract enzymes that methylate the DNA directly, which also leads to the suppression of transcription (see earlier section on DNA methylation).

Slicer-independent degradation of mRNA   A final mechanism by which miRNAs regulate gene expression is by triggering the decay of mRNA in a process that does not require Slicer activity. For example, a short-lived mRNA with an AU-rich element in its 3′ UTR is degraded by an RNA-silencing mechanism. Researchers have identified an miRNA with a sequence that is complementary to the consensus sequence in the AU-rich element. This miRNA binds to the AU-rich element and, in a way that is not yet fully understood, brings about the degradation of the mRNA in a process that requires Dicer and RISC.

The Control of Development by RNA Interference

Much of development in multicellular eukaryotes is through gene regulation: different genes are turned on and off at specific times (see Chapter 22). In fact, when miRNAs were first discovered, researchers noticed that a mutation in an miRNA in *C. elegans* caused a developmental defect. Research now demonstrates that miRNA molecules are key factors in controlling development in plants, animals, and humans. For example, the vertebrate heart develops through the programmed differentiation and proliferation of cardiomyocytes, which are controlled by a specific miRNA termed miR-1-1.

Recent studies demonstrate that, through their effects on gene expression, miRNAs play important roles in many diseases and disorders. For example, a genetic form of hearing loss has been associated with a mutation in the gene that encodes an miRNA. Other miRNAs are associated with heart disease. One miRNA called miR-1-2 is highly expressed in heart muscle. Mice genetically engineered to express only 50% of the normal amount of miR-1-2 frequently have holes in the wall that separates their left and right ventricles, a common congenital heart defect seen in newborn humans. Overexpression of another miRNA called miR-1 in the hearts of adult mice causes cardiac arrhythmia—irregular electrical activity of the heart that can be life-threatening in humans. Changes in the expression of other miRNAs have been associated with cancer.

CONCEPTS

RNA silencing is initiated by double-stranded RNA molecules that are cleaved and processed. The resulting siRNAs or miRNAs combine with proteins to form complexes that bind to complementary sequences in mRNA or DNA. The siRNAs and miRNAs affect gene expression by cleaving mRNA, inhibiting translation, altering chromatin structure, or triggering RNA degradation.

✔ CONCEPT CHECK 6

In RNA silencing, siRNAs and miRNAs usually bind to which part of the mRNA molecules that they control?

a. 5′ UTR

b. Segment that encodes amino acids

c. 3′ poly(A) tail

d. 3′ UTR

17.7 Some Genes Are Regulated by Processes That Affect Translation or by Modifications of Proteins

Ribosomes, aminoacyl tRNAs, initiation factors, and elongation factors are all required for the translation of mRNA molecules. The availability of these components affects the rate of translation and therefore influences gene expression. For example, the activation of T lymphocytes (T cells) is critical to the development of immune responses to viruses (see Chapter 22). T cells are normally in the G_0 stage of the cell cycle and not actively dividing. On exposure to viral antigens, however, specific T cells become activated and undergo rapid proliferation (**Figure 17.14**). Activation includes a 7- to 10-fold increase in protein synthesis that causes cells to enter the cell cycle and proliferate. This burst of protein synthesis does not require an increase in mRNA synthesis. Instead, a global increase in protein synthesis is due to the increased availability of initiation factors taking part in translation—initiation factors that allow ribosomes to bind to mRNA and begin translation. Similarly, insulin stimulates the initiation of overall protein synthesis by increasing the availability of initiation factors. Initiation factors exist in inactive forms and, in response to various cell signals, can be activated by chemical modifications of their structure, such as phosphorylation.

Mechanisms also exist for the regulation of translation of specific mRNAs. The initiation of translation in some mRNAs is regulated by proteins that bind to an mRNA's 5′ UTR and inhibit the binding of ribosomes, in a fashion similar to the way in which repressor proteins bind to operators and prevent the transcription of structural genes. The translation of some mRNAs is affected by the binding of proteins to sequences in the 3′ UTR.

Many eukaryotic proteins are extensively modified after translation by the selective cleavage and trimming of amino acids from the ends, by acetylation, or by the addition of phosphate groups, carboxyl groups, methyl groups, or carbohydrates to the protein. These modifications affect the transport, function, and activity of the proteins and have the capacity to affect gene expression.

CONCEPTS

The initiation of translation may be affected by proteins that bind to specific sequences at the 5′ end of mRNA. The availability of ribosomes, tRNAs, initiation and elongation factors, and other components of the translational apparatus may affect the rate of translation.

CONNECTING CONCEPTS

A Comparison of Bacterial and Eukaryotic Gene Control

Now that we have considered the major types of gene regulation in bacteria (see Chapter 16) and eukaryotes (this chapter), let's pause

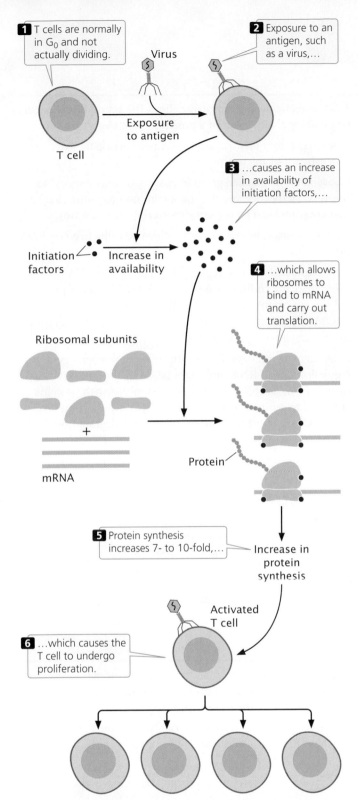

17.14 The expression of some eukaryotic genes is regulated by the availability of components required for translation. In this example, exposure to an antigen stimulates an increased availability of initiation factors and a subsequent increase in protein synthesis, leading to T-cell proliferation.

to consider some of the similarities and differences in bacterial and eukaryotic gene control.

1. Much of gene regulation in bacterial cells is at the level of transcription (although it does exist at other levels). Gene

regulation in eukaryotic cells takes place at multiple levels, including chromatin structure, transcription, mRNA processing, RNA stability, RNA interference, and posttranslational control.

2. Complex biochemical and developmental events in bacterial and eukaryotic cells may require a cascade of gene regulation, in which the activation of one set of genes stimulates the activation of another set.

3. Much of gene regulation in both bacterial and eukaryotic cells is accomplished through proteins that bind to specific sequences in DNA. Regulatory proteins come in a variety of types, but most can be characterized according to a small set of DNA-binding motifs.

4. Chromatin structure plays a role in eukaryotic (but not bacterial) gene regulation. In general, condensed chromatin represses gene expression; chromatin structure must be altered before transcription can take place. Chromatin structure is altered by the modification of histone proteins, chromatin-remodeling proteins, and DNA methylation.

5. Modifications to chromatin structure in eukaryotes may lead to epigenetic changes, which are changes that affect gene expression and are passed on to other cells or future generations.

6. In bacterial cells, genes are often clustered in operons and are coordinately expressed by transcription into a single mRNA molecule. In contrast, most eukaryotic genes typically have their own promoters and are transcribed independently. Coordinate regulation in eukaryotic cells takes place through common response elements, present in the promoters and enhancers of the genes. Different genes that have the same response element in common are influenced by the same regulatory protein.

7. Regulatory proteins that affect transcription exhibit two basic types of control: *repressors* inhibit transcription (negative control); *activators* stimulate transcription (positive control). Both negative control and positive control are found in bacterial and eukaryotic cells.

8. The initiation of transcription is a relatively simple process in bacterial cells, and regulatory proteins function by blocking or stimulating the binding of RNA polymerase to DNA. In contrast, eukaryotic transcription requires complex machinery that includes RNA polymerase, general transcription factors, and transcriptional activators, which allows transcription to be influenced by multiple factors.

9. Some eukaryotic transcriptional regulatory proteins function at a distance from the gene by binding to enhancers, causing the formation of a loop in the DNA, which brings the promoter and enhancer into close proximity. Some distant-acting sequences analogous to enhancers have been described in bacterial cells, but they appear to be less common.

10. The greater time lag between transcription and translation in eukaryotic cells compared with that in bacterial cells allows mRNA stability and mRNA processing to play larger roles in eukaryotic gene regulation.

11. RNA molecules (antisense RNA) may act as regulators of gene expression in bacteria. Regulation by siRNAs and miRNAs, which is extensive in eukaryotes, is absent from bacterial cells.

CONCEPTS SUMMARY

- Eukaryotic cells differ from bacteria in several ways that affect gene regulation, including, in eukaryotes, the absence of operons, the presence of chromatin, and the presence of a nuclear membrane.

- In eukaryotic cells, chromatin structure represses gene expression. In transcription, chromatin structure may be altered by the modification of histone proteins, including acetylation, phosphorylation, and methylation. The repositioning of nucleosomes and the methylation of DNA also affect transcription.

- Epigenetic effects—inherited changes in gene expression that are not caused by changes in the base sequence of DNA—are frequently caused by DNA methylation and alterations to chromatin structure.

- The initiation of eukaryotic transcription is controlled by general transcription factors that assemble into the basal transcription apparatus and by transcriptional activator proteins that stimulate normal levels of transcription by binding to regulatory promoters and enhancers.

- Enhancers affect the transcription of distant genes. Regulatory proteins bind to enhancers and interact with the basal transcription apparatus by causing the intervening DNA to loop out.

- DNA sequences called insulators limit the action of enhancers by blocking their action in a position-dependent manner.

- Some regulatory factors cause RNA polymerase to stall downstream of the promoter.

- Coordinately controlled genes in eukaryotic cells respond to the same factors because they have common response elements that are stimulated by the same transcriptional activator.

- Gene expression in eukaryotic cells can be influenced by RNA processing.

- Gene expression can be regulated by changes in RNA stability. The 5′ cap, the coding sequence, the 3′ UTR, and the poly(A) tail are important in controlling the stability of eukaryotic mRNAs. Proteins binding to the 5′ and 3′ ends of eukaryotic mRNA can affect its translation.

- RNA silencing plays an important role in eukaryotic gene regulation. Small RNA molecules (siRNAs and miRNAs) cleaved from double-stranded DNA combine with proteins and bind to sequences on mRNA or DNA. These complexes cleave RNA, inhibit translation, affect RNA degradation, and silence transcription.

- Control of the posttranslational modification of proteins may play a role in gene expression.

IMPORTANT TERMS

DNase I hypersensitive site (p. 460)	paramutation (p. 463)	insulator (p. 468)
histone code (p. 461)	epigenome (p. 464)	heat-shock protein (p. 468)
chromatin-remodeling complex (p. 462)	pluripotency (p. 465)	response element (p. 469)
CpG island (p. 463)	mediator (p. 466)	SR protein (p. 470)

ANSWERS TO CONCEPT CHECKS

1. Three general processes are the modification of histone proteins (e.g. methylation and acetylation of histones), chromatin remodeling, and DNA methylation.

2. b

3. b

4. The DNA between the enhancer and the promoter loops out, and so regulatory proteins bound to the enhancer are able to interact directly with the transcription apparatus.

5. The poly(A) tail stabilizes the 5′ cap, which must be removed before the mRNA molecule can be degraded from the 5′ end.

6. d

WORKED PROBLEM

What would be the effect of a mutation that caused poly(A)-binding proteins to be nonfunctional?

• Solution

Messenger RNA can be degraded from the 5′ end, from the 3′ end, or through internal cleavage. Degradation from the 5′ end requires the removal of the 5′ cap and is usually preceded by the shortening of the poly(A) tail. Poly(A)-binding proteins bind to the poly(A) tail and prevent it from being shortened. Thus, the presence of these proteins on the poly(A) tail protects the 5′ cap, which prevents RNA degradation. If the gene for poly(A)-binding proteins were mutated in such a way that nonfunctional poly(A) proteins were produced, the proteins would not bind to the poly(A) tail. The tail would be shortened prematurely, the 5′ cap removed, and mRNA degraded more easily. The end result would be less mRNA and thus less protein synthesis.

COMPREHENSION QUESTIONS

Section 17.1

*1. List some important differences between bacterial and eukaryotic cells that affect the way in which genes are regulated.

Section 17.2

2. Where are DNase I hypersensitivity sites found and what do they indicate about the nature of chromatin?

*3. What changes take place in chromatin structure and what role do these changes play in eukaryotic gene regulation?

4. What is the histone code?

Section 17.3

5. What are epigenetic effects? How do they differ from other genetic traits?

*6. What types of changes are thought to be responsible for epigenetic traits?

7. How are patterns of DNA methylation maintained across cell divisions?

Section 17.4

8. Briefly explain how transcriptional activator proteins and repressors affect the level of transcription of eukaryotic genes.

*9. What is an enhancer? How does it affect the transcription of distant genes?

10. What is an insulator?

11. What is a response element? How do response elements bring about the coordinated expression of eukaryotic genes?

Section 17.4

12. Outline the role of alternative splicing in the control of sex differentiation in *Drosophila*.

*13. What role does RNA stability play in gene regulation? What controls RNA stability in eukaryotic cells?

Section 17.5

*14. Briefly list some of the ways in which siRNAs and miRNAs regulate genes.

Section 17.6

*15. How does bacterial gene regulation differ from eukaryotic gene regulation? How are they similar?

APPLICATION QUESTIONS AND PROBLEMS

Introduction

16. The introduction to this chapter described how *Plasmodium* parasites that cause malaria are able to evade the host immune system by constantly altering the expression of their *var* genes, which encode *Plasmodium* surface antigens (L. H. Freitas-Junior et al. 2005. *Cell* 121:25–36). Individual *var* genes are expressed when chromatin structure is disrupted by chemical changes in histone proteins. What type of chemical changes in the histone proteins might be responsible for these changes in gene expression?

Section 17.2

17. A geneticist is trying to determine how many genes are found in a 300,000-bp region of DNA. Analysis shows that four H3K4me3 modifications are found in this piece of DNA. What might their presence suggest about the number of genes located there?

18. In a line of human cells grown in culture, a geneticist isolates a temperature-sensitive mutation at a locus that encodes an acetyltransferase enzyme; at temperatures above 38°C, the mutant cells produce a nonfunctional form of the enzyme. What would be the most likely effect of this mutation when the cells are raised at 40°C?

19. What would be the most likely effect of deleting *flowering locus D (FLD)* in *Arabidopsis thalania*?

20. X31b is an experimental compound that is taken up by rapidly dividing cells. Research has shown that X31b stimulates the methylation of DNA. Some cancer researchers are interested in testing X31b as a possible drug for treating prostate cancer. Offer a possible explanation for why X31b might be an effective anticancer drug.

Section 17.3

21. What would be required to prove that a phenotype is caused by an epigenetic change?

22. Pregnant female rats were exposed to a daily dose of 100 or 200 mg/kg of vinclozolin, a fungicide commonly used in the wine industry (M. D. Anway et al. 2005. *Science* 308:1466–1469). The F_1 offspring from the exposed female rates were interbred, producing F_2, F_3, and F_4 rats. None of the F_2, F_3, or F_4 rats were exposed to vinclozolin.

Testes from the F_1–F_4 male rats were examined and compared with those of control rats from females that were not exposed to vinclozolin (see graphs a–c on the next page). These effects were seen in more than 90% of the F_1–F_4 male offspring. Furthermore, 8% of the F_1–F_4 males from vinclozolin-exposed females developed complete infertility, compared with 0% of the F_1–F_4 males of control females.

Molecular analysis of the testes demonstrated that DNA methylation patterns differed between offspring of vinclozolin-exposed females and offspring of control females. Provide an explanation for the transgenerational effects of vinclozolin on male fertility.

(a)

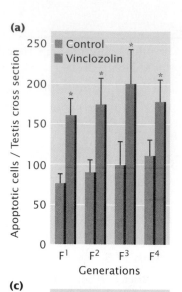

(b)

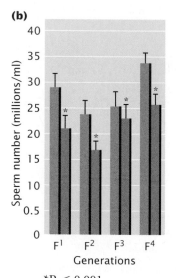

(c)

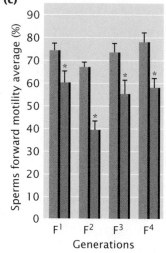

*P < 0.001

[Graphs after M. D. Anway et al., *Science* 308:1466–1469, 2005.]

Section 17.4

23. How do repressors that bind to silencers in eukaryotes differ from repressors that bind to operators in bacteria?

24. An enhancer is surrounded by four genes (*A*, *B*, *C*, and *D*), as shown in the adjoining diagram. An insulator lies between gene *C* and gene *D*. On the basis of the positions of the genes, the enhancer, and the insulator, the transcription of which genes is most likely to be stimulated by the enhancer? Explain your reasoning.

Section 17.5

*25. What will be the effect on sexual development in newly fertilized *Drosophila* embryos if the following genes are deleted?
   a. *sex lethal*
   b. *transformer*
   c. *double-sex*

26. Some eukaryotic mRNAs have an AU-rich element in the 3′ untranslated region. What would be the effect on gene expression if this element were mutated or deleted?

27. A strain of *Arabidopsis thaliana* possesses a mutation in the *APETALA2* gene, in which much of the 3′ untranslated region of mRNA transcribed from the gene is deleted. What is the most likely effect of this mutation on the expression of the *APETALA2* gene?

28. What will be the effect of a mutation that destroys the ability of poly(A)-binding protein (PABP) to attach to a poly(A) tail?

CHALLENGE QUESTIONS

Section 17.2

29. In the fungus *Neurospora*, from about 2% to 3% of cytosine bases are methylated. Researchers isolated *Neurospora* DNA sequences that contained 5-methylcytosine and found that almost all methylated sequences were located in inactive copies of transposable genetic elements (see Chapter 11). On the basis of these observations, propose a possible explanation for why *Neurospora* methylates its DNA and why DNA methylation in this species is associated with transposable genetic elements.

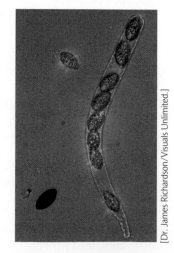

[Dr. James Richardson/Visuals Unlimited.]

Section 17.3

30. In recent years, techniques have been developed to clone mammals through a process called nuclear transfer, in which the nucleus of a somatic cell is transferred to an egg cell from which the nuclear material has been removed (see Chapter 22). Research has demonstrated that, when a nucleus from a differentiated somatic cell is transferred to an egg cell, only a few percent of the resulting embryos complete development and many of those that do complete development die shortly after birth. In contrast, when a nucleus from an undifferentiated embryonic stem cell is transferred into an egg cell, the percentage of embryos that complete development is significantly higher (W. M. Rideout, K. Eggan, and R. Jaenisch. 2001. *Science* 293:1095–1098). Propose a possible reason for why successful development of cloned embryos is higher when the nucleus transferred comes from an undifferentiated embryonic stem cell.

Section 17.4

31. A yeast gene termed *SER3*, which has a role in serine biosynthesis, is repressed during growth in nutrient-rich medium, and so little transcription takes place and little SER3 enzyme is produced. In an investigation of the nature of the repression of the *SER3* gene, a region of DNA upstream of the *SER3* gene was found to be heavily transcribed when the *SER3* gene is repressed (J. A. Martens, L. Laprade, and F. Winston. 2004. *Nature* 429:571–574). Within this upstream region is a promoter that stimulates the transcription of an RNA molecule called *SRG1* (for *SER3* regulatory gene). This RNA molecule has none of the sequences necessary for translation. Mutations in the promoter for *SRG1* result in the disappearance of *SRG1* RNA, and these mutations remove the repression of *SER3*. When RNA polymerase binds to the *SRG1* promoter, the polymerase is found to travel downstream, transcribing the *SGR1* RNA, and to pass through and transcribe the promoter for *SER3*. This activity leads to the repression of *SER3*. Propose a possible explanation for how the transcription of *SGR1* might repress the transcription of *SER3*. (Hint: Remember that the *SGR1* RNA does not encode a protein.)

Section 17.6

32. A common feature of many eukaryotic mRNAs is the presence of a rather long 3′ UTR, which often contains consensus sequences. Creatine kinase B (CK-B) is an enzyme important in cellular metabolism. Certain cells—termed U937D cells—have lots of CK-B mRNA, but no CK-B enzyme is present. In these cells, the 5′ end of the CK-B mRNA is bound to ribosomes, but the mRNA is apparently not translated. Something inhibits the translation of the CK-B mRNA in these cells.

Researchers introduced numerous short segments of RNA containing only 3′ UTR sequences into U937D cells. As a result, the U937D cells began to synthesize the CK-B enzyme, but the total amount of CK-B mRNA did not increase. The introduction of short segments of other RNA sequences did not stimulate the synthesis of CK-B; only the 3′ UTR sequences turned on the translation of the enzyme.

On the basis of these results, propose a mechanism for how CK-B translation is inhibited in the U937D cells. Explain how the introduction of short segments of RNA containing the 3′ UTR sequences might remove the inhibition.

18 Gene Mutations and DNA Repair

People with a mutation in the *tinman* gene, named after Tin Man in *The Wizard of Oz*, often have congenital heart defects. [Mary Evans Picture Library/Alamy.]

A FLY WITHOUT A HEART

The heart of a fruit fly is a simple organ, an open-ended tube that rhythmically contracts, pumping fluid—rather inefficiently—around the body of the fly. Although simple and inelegant, the fruit fly's heart is nevertheless essential, at least for a fruit fly. Remarkably, a few rare mutant fruit flies never develop a heart and die (not surprisingly) at an early embryonic stage. Geneticist Rolf Bodmer analyzed these mutants in the 1980s and made an important discovery—a gene that specifies the development of a heart. He named the gene *tinman*, after the character in *The Wizard of Oz* who also lacked a heart. Bodmer's research revealed that *tinman* encodes a transcription factor, which binds to DNA and turns on other genes that are essential for the normal development of a heart. In the mutant flies studied by Bodmer, this gene was lacking, the transcription factor was never produced, and the heart never developed. Findings from subsequent research revealed the existence of a human gene (called *Nkx2.5*) with a sequence similar to that of *tinman*, but the function of the human gene was unknown.

Then, in the 1990s, physicians Jonathan and Christine Seidman began studying people born with abnormal hearts, such as those with a hole in the septum that separates the chambers on the left and right sides of the heart. Such defects result in abnormal blood flow through the heart, causing the heart to work harder than normally and the mixing of oxygenated and deoxygenated blood. Congenital heart defects are not uncommon; they're found in about 1 of every 125 babies. Some of the defects heal on their own, but others require corrective surgery. Although surgery is often successful in reversing congenital heart problems, many of these patients begin to have irregular heartbeats in their 20s and 30s. The Seidmans and their colleagues found several families in which congenital heart defects and irregular heart beats were inherited together in an autosomal dominant fashion. Detailed molecular analysis of one of these families revealed that the gene responsible for the heart problems was located on chromosome 5, at a spot where the human *tinman* gene (*Nkx2.5*) had been previously mapped. All members of this family who inherited the heart defects also inherited a mutation in the *tinman* gene. Subsequent studies with additional patients found that many people with congenital heart defects have a mutation in the *tinman* gene. The human version of this gene, like its counterpart in flies, encodes a transcription factor that controls heart development. Despite tremendous differences in size, anatomy, and physiology, humans and flies use the same gene to make a heart.

The story of *tinman* illustrates the central importance of studying mutations: the analysis of mutants is often a source of key insights into important biological processes. This chapter focuses on gene mutations—how errors arise in genetic instructions and how those errors are studied. We begin with a brief examination of the different types of mutations, including their phenotypic effects, how they can be suppressed, and mutation rates. The next section explores how mutations spontaneously arise in the course of DNA replication and afterward, as well as how chemicals and radiation induce mutations. We then consider the analysis of mutations. Finally, we take a look at DNA repair and some of the diseases that arise when DNA repair is defective.

18.1 Mutations Are Inherited Alterations in the DNA Sequence

DNA is a highly stable molecule that is replicated with amazing accuracy (see Chapters 10 and 12), but changes in DNA structure and errors of replication do occur. A **mutation** is defined as an inherited change in genetic information; the descendants may be cells or organisms.

The Importance of Mutations

Mutations are both the sustainer of life and the cause of great suffering. On the one hand, mutation is the source of all genetic variation, the raw material of evolution. The ability of organisms to adapt to environmental change depends critically on the presence of genetic variation in natural populations, and genetic variation is produced—at least partly—by mutation. On the other hand, many mutations have detrimental effects, and mutation is the source of many diseases and disorders.

Much of the study of genetics focuses on how variants produced by mutation are inherited; genetic crosses are meaningless if all individual members of a species are identically homozygous for the same alleles. Much of Gregor Mendel's success in unraveling the principles of inheritance can be traced to his use of carefully selected variants of the garden pea; similarly, Thomas Hunt Morgan and his students discovered many basic principles of genetics by analyzing mutant fruit flies.

Mutations are also useful for probing fundamental biological processes. Finding or creating mutations that affect different components of a biological system and studying their effects can often lead to an understanding of the system. This method, referred to as genetic dissection, is analogous to figuring out how an automobile works by breaking different parts of a car and observing the effects; for example, smash the radiator and the engine overheats, revealing that the radiator cools the engine. Using mutations to disrupt function can likewise be a source of insight into biological processes. For example, geneticists have begun to unravel the molecular details of development by studying mutations, such as *tinman*, that interrupt various embryonic stages in *Drosophila* (see Chapter 22). Although this method of breaking "parts" to determine their function might seem like a crude approach to understanding a system, it is actually very powerful and has been used extensively in biochemistry, developmental biology, physiology, and behavioral science (but this method is *not* recommended for learning how your car works).

CONCEPTS

Mutations are heritable changes in DNA. They are essential to the study of genetics and are useful in many other biological fields.

✔ CONCEPT CHECK 1

How are mutations used to help in understanding basic biological processes?

Categories of Mutations

In multicellular organisms, we can distinguish between two broad categories of mutations: somatic mutations and germ-line mutations. **Somatic mutations** arise in somatic tissues, which do not produce gametes (**Figure 18.1**). When

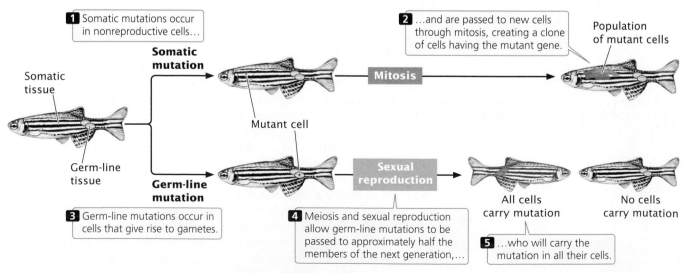

1 Somatic mutations occur in nonreproductive cells…

2 …and are passed to new cells through mitosis, creating a clone of cells having the mutant gene.

Population of mutant cells

Somatic tissue

Somatic mutation

Mitosis

Mutant cell

Germ-line tissue

Germ-line mutation

Sexual reproduction

All cells carry mutation

No cells carry mutation

3 Germ-line mutations occur in cells that give rise to gametes.

4 Meiosis and sexual reproduction allow germ-line mutations to be passed to approximately half the members of the next generation,…

5 …who will carry the mutation in all their cells.

18.1 The two basic classes of mutations are somatic mutations and germ-line mutations.

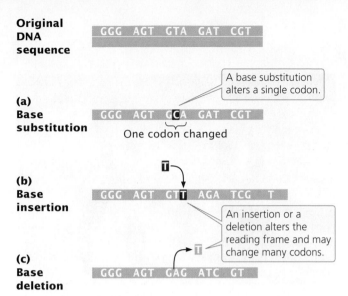

Original DNA sequence

```
GGG AGT GTA GAT CGT
```

(a) Base substitution

A base substitution alters a single codon.

```
GGG AGT GCA GAT CGT
```

One codon changed

(b) Base insertion

```
GGG AGT GTT AGA TCG T
```

An insertion or a deletion alters the reading frame and may change many codons.

(c) Base deletion

```
GGG AGT GAG ATC GT
```

18.2 Three basic types of gene mutations are base substitutions, insertions, and deletions.

a somatic cell with a mutation divides (mitosis), the mutation is passed on to the daughter cells, leading to a population of genetically identical cells (a clone). The earlier in development that a somatic mutation occurs, the larger the clone of cells within that individual organism that will contain the mutation.

Because of the huge number of cells present in a typical eukaryotic organism, somatic mutations are numerous. For example, there are about 10^{14} cells in the human body. Typically, a mutation arises once in every million cell divisions, and so hundreds of millions of somatic mutations must arise in each person. Many somatic mutations have no obvious effect on the phenotype of the organism, because the function of the mutant cell (even the cell itself) is replaced by that of normal cells. However, cells with a somatic mutation that stimulates cell division can increase in number and spread; this type of mutation can give rise to cells with a selective advantage and is the basis for cancers (see Chapter 23).

Germ-line mutations arise in cells that ultimately produce gametes. A germ-line mutation can be passed to future generations, producing individual organisms that carry the mutation in all their somatic and germ-line cells

(see Figure 18.1). When we speak of mutations in multicellular organisms, we're usually talking about germ-line mutations.

Historically, mutations have been partitioned into those that affect a single gene, called *gene mutations*, and those that affect the number or structure of chromosomes, called *chromosome mutations*. This distinction arose because chromosome mutations could be observed directly, by looking at chromosomes with a microscope, whereas gene mutations could be detected only by observing their phenotypic effects. Now, DNA sequencing allows direct observation of gene mutations, and chromosome mutations are distinguished from gene mutations somewhat arbitrarily on the basis of the size of the DNA lesion. Nevertheless, it is practical to use the term *chromosome mutation* for a large-scale genetic alteration that affects chromosome structure or the number of chromosomes and to use the term **gene mutation** for a relatively small DNA lesion that affects a single gene. This chapter focuses on gene mutations; chromosome mutations were discussed in Chapter 9.

Types of Gene Mutations

There are a number of ways to classify gene mutations. Some classification schemes are based on the nature of the phenotypic effect, others are based on the causative agent of the mutation, and still others focus on the molecular nature of the defect. Here, we will categorize mutations primarily on the basis of their molecular nature, but we will also encounter some terms that relate the causes and the phenotypic effects of mutations.

Base substitutions The simplest type of gene mutation is a **base substitution**, the alteration of a single nucleotide in the DNA (**Figure 18.2a**). Base substitutions are of two types. In a **transition**, a purine is replaced by a different purine or, alternatively, a pyrimidine is replaced by a different pyrimidine (**Figure 18.3**). In a **transversion**, a purine is replaced by a pyrimidine or a pyrimidine is replaced by a purine. The number of possible transversions (see Figure 18.3) is twice the number of possible transitions, but transitions arise more frequently. TRY PROBLEM 18 →

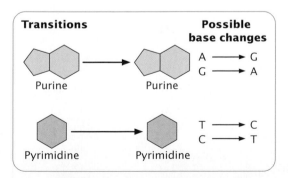

Transitions **Possible base changes**

Purine → Purine

A → G
G → A

Pyrimidine → Pyrimidine

T → C
C → T

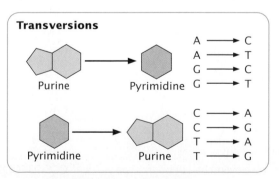

Transversions

Purine → Pyrimidine

A → C
A → T
G → C
G → T

Pyrimidine → Purine

C → A
C → G
T → A
T → G

18.3 A transition is the substitution of a purine for a purine or of a pyrimidine for a pyrimidine; a transversion is the substitution of a pyrimidine for a purine or of a purine for a pyrimidine.

Insertions and deletions The second major class of gene mutations contains **insertions** and **deletions**—the addition or the removal, respectively, of one or more nucleotide pairs (collectively called indels, **Figure 18.2b** and **c**). Although base substitutions are often assumed to be the most common type of mutation, molecular analysis has revealed that insertions and deletions are often more frequent. Insertions and deletions within sequences that encode proteins may lead to **frameshift mutations**, changes in the reading frame (see p. 411 in Chapter 15) of the gene. Frameshift mutations usually alter all amino acids encoded by nucleotides following the mutation, and so they generally have drastic effects on the phenotype. Not all insertions and deletions lead to frameshifts, however; insertions and deletions consisting of any multiple of three nucleotides will leave the reading frame intact, although the addition or removal of one or more amino acids may still affect the phenotype. Mutations not affecting the reading frame are called **in-frame insertions** and **deletions**, respectively.

18.4 The fragile-X chromosome is associated with a characteristic constriction (fragile site) on the long arm. [Visuals Unlimited.]

CONCEPTS

Gene mutations consist of changes in a single gene and can be base substitutions (a single pair of nucleotides is altered) or insertions or deletions (nucleotides are added or removed). A base substitution can be a transition (substitution of like bases) or a transversion (substitution of unlike bases). Insertions and deletions often lead to a change in the reading frame of a gene.

✔ CONCEPT CHECK 2

Which of the following changes is a transition base substitution?

a. Adenine is replaced by thymine.

b. Cytosine is replaced by adenine.

c. Guanine is replaced by adenine.

d. Three nucleotide pairs are inserted into DNA.

Expanding nucleotide repeats Mutations in which the number of copies of a set of nucleotides increases in number are called **expanding nucleotide repeats**. This type of mutation was first observed in 1991 in a gene called *FMR-1*, which causes fragile-X syndrome, the most common hereditary cause of mental retardation. The disorder is so named because, in specially treated cells from persons having the condition, the tip of each long arm of the X chromosome is attached by only a slender thread (**Figure 18.4**). The normal *FMR-1* allele (not containing the mutation) has 60 or fewer copies of CGG but, in persons with fragile-X syndrome, the allele may harbor hundreds or even thousands of copies.

Expanding nucleotide repeats have been found in almost 30 human diseases, several of which are listed in **Table 18.1**. Most of these diseases are caused by the expansion of a set of three nucleotides (called a trinucleotide), most often CNG, where N can be any nucleotide. However,

Table 18.1	Examples of genetic diseases caused by expanding nucleotide repeats			
			Number of Copies of Repeat	
Disease	Repeated Sequence		Normal Range	Disease Range
Spinal and bulbar muscular atrophy	CAG		11–33	40–62
Fragile-X syndrome	CGG		6–54	50–1500
Jacobsen syndrome	CGG		11	100–1000
Spinocerebellar ataxia (several types)	CAG		4–44	21–130
Autosomal dominant cerebellar ataxia	CAG		7–19	37–220
Myotonic dystrophy	CTG		5–37	44–3000
Huntington disease	CAG		9–37	37–121
Friedreich ataxia	GAA		6–29	200–900
Dentatorubral-pallidoluysian atrophy	CAG		7–25	49–75
Myoclonus epilepsy of the Unverricht–Lundborg type	CCCCGCCCCGCG		2–3	12–13

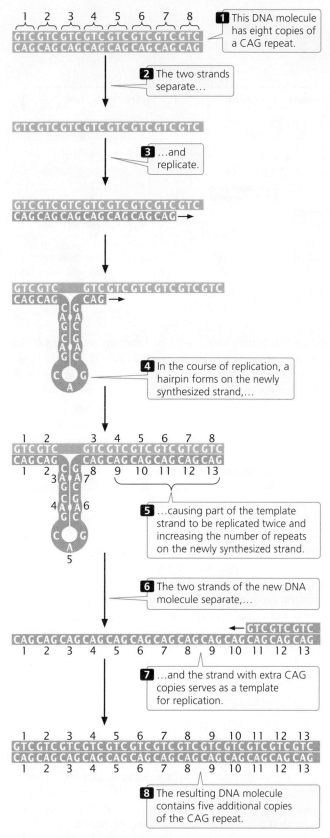

18.5 A model of how the number of copies of a nucleotide repeat may increase in replication.

Try Animation 18.1 to see how copies of a nucleotide repeat can increase.

www.whfreeman.com/pierce4e

some diseases are caused by repeats of four, five, and even twelve nucleotides. The number of copies of the nucleotide repeat often correlates with the severity or age of onset of the disease. The number of copies of the repeat also corresponds to the instability of nucleotide repeats: when more repeats are present, the probability of expansion to even more repeats increases. This association between the number of copies of nucleotide repeats, the severity of the disease, and the probability of expansion leads to a phenomenon known as anticipation (see pp. 122–123 in Chapter 5), in which diseases caused by nucleotide-repeat expansions become more severe in each generation. Less commonly, the number of nucleotide repeats may decrease within a family. Expanding nucleotide repeats have also been observed in some microbes and plants.

Increases in the number of nucleotide repeats can produce disease symptoms in different ways. In several of the diseases (e.g., Huntington disease), the nucleotide expands within the coding part of a gene, producing a toxic protein that has extra glutamine residues (the amino acid encoded by CAG). In other diseases, the repeat is outside the coding region of a gene and affects its expression. In fragile-X syndrome, additional copies of the nucleotide repeat cause the DNA to become methylated, which turns off the transcription of an essential gene.

The mechanism that leads to the expansion of nucleotide repeats is not completely understood. Evidence suggests that expansion occurs in the course of DNA replication and appears to be related to the formation of hairpins and other special DNA structures that form in single-stranded DNA consisting of nucleotide repeats. Such structures may interfere with normal replication by causing strand slippage, misalignment of the sequences, or stalling of replication. One model of how repeat hairpins might result in repeat expansion is shown in **Figure 18.5**.

CONCEPTS

Expanding nucleotide repeats are regions of DNA that consist of repeated copies of sets of nucleotides. Increased numbers of nucleotide repeats are associated with several genetic diseases.

Phenotypic Effects of Mutations

Another way that mutations are classified is on the basis of their phenotypic effects. At the most general level, we can distinguish a mutation on the basis of its phenotype compared with the wild-type phenotype. A mutation that alters the wild-type phenotype is called a **forward mutation**, whereas a **reverse mutation** (a *reversion*) changes a mutant phenotype back into the wild type.

Geneticists use other terms to describe the effects of mutations on protein structure. A base substitution that results in a different amino acid in the protein is referred to as a **missense mutation** (**Figure 18.6a**). A **nonsense mutation**

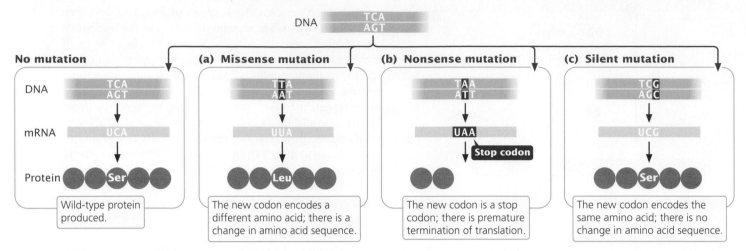

18.6 Base substitutions can cause (a) missense, (b) nonsense, and (c) silent mutations.

changes a sense codon (one that specifies an amino acid) into a nonsense codon (one that terminates translation), as shown in **Figure 18.6b**. If a nonsense mutation occurs early in the mRNA sequence, the protein will be greatly shortened and will usually be nonfunctional.

Because of the redundancy of the genetic code, some different codons specify the same amino acid. A **silent mutation** changes a codon to a synonymous codon that specifies the same amino acid (**Figure 18.6c**), altering the DNA sequence without changing the amino acid sequence of the protein. Not all silent mutations, however, are truly silent: some do have phenotypic effects. For example, silent mutations may have phenotypic effects when different tRNAs (called isoaccepting tRNAs, see Chapter 15) are used for different synonymous codons. Because some isoaccepting tRNAs are more abundant than others, which synonymous codon is used may affect the rate of protein synthesis. The rate of protein synthesis can influence the phenotype by affecting the amount of protein present in the cell and, in a few cases, the folding of the protein. Other silent mutations may alter sequences near the exon–intron junctions that affect splicing (see Chapter 14). Still other silent mutations may influence the folding of the mRNA, affecting its stability.

A **neutral mutation** is a missense mutation that alters the amino acid sequence of the protein but does not change its function. Neutral mutations occur when one amino acid is replaced by another that is chemically similar or when the affected amino acid has little influence on protein function. For example, neutral mutations occur in the genes that encode hemoglobin; although these mutations alter the amino acid sequence of hemoglobin, they do not affect its ability to transport oxygen.

Loss-of-function mutations cause the complete or partial absence of normal protein function. A loss-of-function mutation so alters the structure of the protein that the protein no longer works correctly or the mutation can occur in regulatory regions that affect the transcription, translation, or splicing of the protein. Loss-of-function mutations are frequently recessive, and an individual diploid organism must be homozygous for a loss-of-function mutation before the effects of the loss of the functional protein can be exhibited. The mutations that cause cystic fibrosis are loss-of-function mutations: these mutations produce a nonfunctional form of the cystic fibrosis transmembrane conductance regulator protein, which normally regulates the movement of chloride ions into and out of the cell (see Chapter 5).

In contrast, a **gain-of-function mutation** produces an entirely new trait or it causes a trait to appear in an inappropriate tissue or at an inappropriate time in development. For example, a mutation in a gene that encodes a receptor for a growth factor might cause the mutated receptor to stimulate growth all the time, even in the absence of the growth factor. Gain-of-function mutations are frequently dominant in their expression. Still other types of mutations are **conditional mutations**, which are expressed only under certain conditions. For example, some conditional mutations affect the phenotype only at elevated temperatures. Others are **lethal mutations**, causing premature death (Chapter 5).

TRY PROBLEM 21 →

Suppressor Mutations

A **suppressor mutation** is a genetic change that hides or suppresses the effect of another mutation. This type of mutation is distinct from a reverse mutation, in which the mutated site changes back into the original wild-type sequence (**Figure 18.7**). A suppressor mutation occurs at a site that is distinct from the site of the original mutation; thus, an individual with a suppressor mutation is a double mutant, possessing both the original mutation and the suppressor mutation but exhibiting the phenotype of an unmutated wild type. Like other mutations, suppressors arise randomly. Geneticists distinguish between two classes of suppressor mutations: intragenic and intergenic.

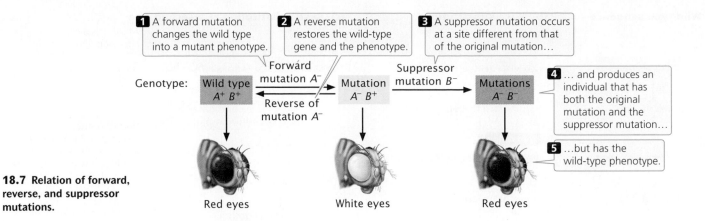

18.7 Relation of forward, reverse, and suppressor mutations.

Red eyes   White eyes   Red eyes

Intragenic suppressor mutation An **intragenic suppressor mutation** is in the same gene as that containing the mutation being suppressed and may work in several ways. The suppressor may change a second nucleotide in the same codon altered by the original mutation, producing a codon that specifies the same amino acid as that specified by the original, unmutated codon (**Figure 18.8**). Intragenic suppressors may also work by suppressing a frameshift mutation. If the original mutation is a one-base deletion, then the addition of a single base elsewhere in the gene will restore the former reading frame. Consider the following nucleotide sequence in DNA and the amino acids that it encodes:

DNA   AAA TCA CTT GGC GTA CAA
Amino acids Phe Ser Glu Pro His Val

Suppose a one-base deletion occurs in the first nucleotide of the second codon. This deletion shifts the reading frame by one nucleotide and alters all the amino acids that follow the mutation.

One-nucleotide deletion

AAA ⅩCAC TTG GCG TAC AA
Phe   Val Asn Arg Met

If a single nucleotide is added to the third codon (the suppressor mutation), the reading frame is restored, although two of the amino acids differ from those specified by the original sequence.

One-nucleotide insertion

AAA CAC TTT GGC GTA CAA
Phe Val Lys Pro His Val

Similarly, a mutation due to an insertion may be suppressed by a subsequent deletion in the same gene.

A third way in which an intragenic suppressor may work is by making compensatory changes in the protein. A first missense mutation may alter the folding of a polypeptide chain by changing the way in which amino acids in the protein interact with one another. A second missense mutation at a different site (the suppressor) may recreate the original folding pattern by restoring interactions between the amino acids.

Intergenic suppressors mutations An **intergenic suppressor mutation**, in contrast, occurs in a gene other than the one bearing the original mutation. These suppressors sometimes work by changing the way that the mRNA is translated. In the example illustrated in **Figure 18.9a**, the

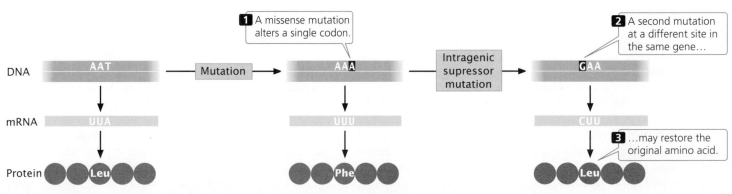

18.8 An intragenic suppressor mutation occurs in the gene containing the mutation being suppressed.

(a) Wild-type sequence   **(b) Base substitution**   **(c) Base substitution at a second site**

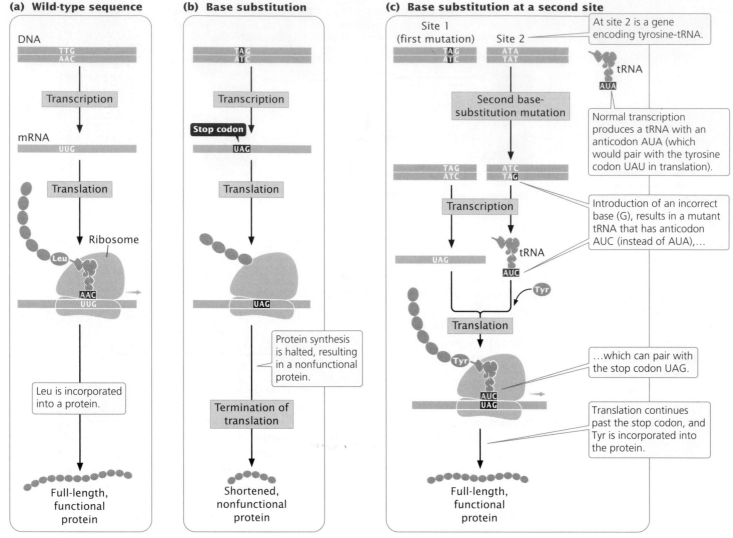

18.9 An intergenic suppressor mutation occurs in a gene other than the one bearing the original mutation. (a) The wild-type sequence produces a full-length, functional protein. (b) A base substitution at a site in the same gene produces a premature stop codon, resulting in a shortened, nonfunctional protein. (c) A base substitution at a site in another gene, which in this case encodes tRNA, alters the anticodon of tRNA^Tyr; tRNA^Tyr can pair with the stop codon produced by the original mutation, allowing tyrosine to be incorporated into the protein and translation to continue.

original DNA sequence is AAC (UUG in the mRNA) and specifies leucine. This sequence mutates to ATC (UAG in mRNA), a termination codon (**Figure 18.9b**). The ATC nonsense mutation could be suppressed by a second mutation in a different gene that encodes a tRNA; this second mutation would result in a codon capable of pairing with the UAG termination codon (**Figure 18.9c**). For example, the gene that encodes the tRNA for tyrosine (tRNA^Tyr), which has the anticodon AUA, might be mutated to have the anticodon AUC, which will then pair with the UAG stop codon. Instead of translation terminating at the UAG codon, tyrosine would be inserted into the protein and a full-length protein would be produced, although tyrosine would now substitute for leucine. The effect of this change would depend on the role of this amino acid in the overall structure of the protein,

but the effect of the suppressor mutation is likely to be less detrimental than the effect of the nonsense mutation, which would halt translation prematurely.

Because cells in many organisms have multiple copies of tRNA genes, other unmutated copies of tRNA^Tyr would remain available to recognize tyrosine codons in the transcripts of the mutant gene in question and other genes being expressed concurrently. We might expect that the tRNAs that have undergone the suppressor mutation described would also suppress the normal termination codons at the ends of other coding sequences, resulting in the production of longer-than-normal proteins, but this event does not usually take place.

Intergenic suppressors can also work through genic interactions (see p. 106 in Chapter 5). Polypeptide chains that are produced by two genes may interact to produce a

Table 18.2	Characteristics of different types of mutations
Type of Mutation	**Definition**
Base substitution	Changes the base of a single DNA nucleotide
Transition	Base substitution in which a purine replaces a purine or a pyrimidine replaces a pyrimidine
Transversion	Base substitution in which a purine replaces a pyrimidine or a pyrimidine replaces a purine
Insertion	Addition of one or more nucleotides
Deletion	Deletion of one or more nucleotides
Frameshift mutation	Insertion or deletion that alters the reading frame of a gene
In-frame deletion or insertion	Deletion or insertion of a multiple of three nucleotides that does not alter the reading frame
Expanding nucleotide repeats	Repeated sequence of a set of nucleotides in which the number of copies of the sequence increases
Forward mutation	Changes the wild-type phenotype to a mutant phenotype
Reverse mutation	Changes a mutant phenotype back to the wild-type phenotype
Missense mutation	Changes a sense codon into a different sense codon, resulting in the incorporation of a different amino acid in the protein
Nonsense mutation	Changes a sense codon into a nonsense codon, causing premature termination of translation
Silent mutation	Changes a sense codon into a synonymous codon, leaving unchanged the amino acid sequence of the protein
Neutral mutation	Changes the amino acid sequence of a protein without altering its ability to function
Loss-of-function mutation	Causes a complete or partial loss of function
Gain-of-function mutation	Causes the appearance of a new trait or function or causes the appearance of a trait in inappropriate tissue or at an inappropriate time
Lethal mutation	Causes premature death
Suppressor mutation	Suppresses the effect of an earlier mutation at a different site
Intragenic suppressor mutation	Suppresses the effect of an earlier mutation within the same gene
Intergenic suppressor mutation	Suppresses the effect of an earlier mutation in another gene

functional protein. A mutation in one gene may alter the encoded polypeptide such that the interaction between the two polypeptides is destroyed, in which case a functional protein is not produced. A suppressor mutation in the second gene may produce a compensatory change in its polypeptide, therefore restoring the original interaction. Characteristics of some of the different types of mutations are summarized in **Table 18.2**. TRY PROBLEM 22 →

CONCEPTS

A suppressor mutation overrides the effect of an earlier mutation at a different site. An intragenic suppressor mutation occurs within the *same* gene as that containing the original mutation, whereas an intergenic suppressor mutation occurs in a *different* gene.

✔ CONCEPT CHECK 3

How is a suppressor mutation different from a reverse mutation?

Worked Problem

A gene encodes a protein with the following amino acid sequence:

Met-Arg-Cys-Ile-Lys-Arg

A mutation of a single nucleotide alters the amino acid sequence to:

Met-Asp-Ala-Tyr-Lys-Gly-Glu-Ala-Pro-Val

A second single-nucleotide mutation occurs in the same gene and suppresses the effects of the first mutation (an intragenic suppressor). With the original mutation and the intragenic suppressor present, the protein has the following amino acid sequence:

Met-Asp-Gly-Ile-Lys-Arg

What is the nature and location of the first mutation and of the intragenic suppressor mutation?

• Solution

The first mutation alters the reading frame, because all amino acids after Met are changed. Insertions and deletions affect the reading frame; the original mutation consists of a single-nucleotide insertion or deletion in the second codon. The intragenic suppressor restores the reading frame; the intragenic suppressor also is most likely a single-nucleotide insertion or deletion. If the first mutation is an insertion, the suppressor must be a deletion; if the first mutation is a deletion, then the suppressor must be an insertion. Notice that the protein produced by the suppressor still differs from the original protein at the second and third amino acids, but the second amino acid produced by the suppressor is the same as that in the protein produced by the original mutation. Thus, the suppressor mutation must have occurred in the third codon, because the suppressor does not alter the second amino acid.

> → For more practice with analyzing mutations, try working Problem 23 at the end of the chapter.

Mutation Rates

The frequency with which a wild-type allele at a locus changes into a mutant allele is referred to as the **mutation rate** and is generally expressed as the number of mutations per biological unit, which may be mutations per cell division, per gamete, or per round of replication. For example, achondroplasia is a type of hereditary dwarfism in humans that results from a dominant mutation. On average, about four achondroplasia mutations arise in every 100,000 gametes, and so the mutation rate is $^4/_{100,000}$, or 0.00004 mutations per gamete. The mutation rate provides information about how often a mutation arises.

Factors affecting mutation rates   Calculations of mutation rates are affected by three factors. First, they depend on the frequency with which a change takes place in DNA. A change in the DNA can arise from spontaneous molecular changes in DNA or it can be induced by chemical, biological, or physical agents in the environment.

The second factor influencing the mutation rate is the probability that, when a change takes place, it will be repaired. Most cells possess a number of mechanisms for repairing altered DNA, and so most alterations are corrected before they are replicated. If these repair systems are effective, mutation rates will be low; if they are faulty, mutation rates will be elevated. Some mutations increase the overall rate of mutation at other genes; these mutations usually occur in genes that encode components of the replication machinery or DNA-repair enzymes.

The third factor is the probability that a mutation will be recognized and recorded. When DNA is sequenced, all mutations are potentially detectable. In practice, however, mutations are usually detected by their phenotypic effects. Some mutations may appear to arise at a higher rate simply because they are easier to detect.

Variation in mutation rates   Mutation rates vary among genes and species (**Table 18.3**), but we can draw several general conclusions about mutation rates. First, spontaneous mutation rates are low for all organisms studied. Typical mutation rates for bacterial genes range from about 1 to 100 mutations per 10 billion cells (from 1×10^{-8} to 1×10^{-10}). The mutation rates for most eukaryotic genes are a bit higher, from about 1 to 10 mutations per million gametes (from 1×10^{-5} to 1×10^{-6}). These higher values in eukaryotes may be due to the fact that the rates are calculated per gamete, and several cell divisions are required to produce a gamete, whereas mutation rates in prokaryotic cells are calculated *per cell division*.

The differences in mutation rates among species may be due to differing abilities to repair mutations, unequal exposures to mutagens, or biological differences in rates of spontaneously arising mutations. Even within a single species, spontaneous rates of mutation vary among genes. The reason for this variation is not entirely understood, but some regions of DNA are known to be more susceptible to mutation than others.

Several recent studies have measured mutation rates directly by sequencing genes of organisms before and after a number of generations. These new studies suggest that mutation rates are often higher than those previously measured on the basis of changes in phenotype. In one study, geneticists sequenced randomly chosen stretches of DNA in the nematode worm *Caenorhabditis elegans* and found about 2.1 mutations per genome per generation, which was 10 times as high as previous estimates based on phenotypic changes. The researchers found that about half of the mutations were insertions and deletions.

Adaptive mutation   As will be discussed in Chapters 24 through 26, evolutionary change that brings about adaptation to new environments depends critically on the presence of genetic variation. New genetic variants arise primarily through mutation. For many years, genetic variation was assumed to arise randomly and at rates that are independent of the need for adaptation. However, some evidence now suggests that stressful environments—where adaptation may be necessary to survive—can induce more mutations in bacteria, a process that has been termed **adaptive mutation**. The idea of adaptive mutation has been intensely debated; critics counter that most mutations are expected to be deleterious, and so increased mutagenesis would likely be harmful most of the time.

Research findings have shown that mutation rates in bacteria collected from the wild do increase in stressful environments, such as those in which nutrients are limited. There

Table 18.3 Mutation rates of different genes in different organisms

Organism	Mutation	Rate	Unit
Bacteriophage T2	Lysis inhibition	1×10^{-8}	Per replication
	Host range	3×10^{-9}	
Escherichia coli	Lactose fermentation	2×10^{-7}	Per cell division
	Histidine requirement	2×10^{-8}	
Neurospora crassa	Inositol requirement	8×10^{-8}	Per asexual spore
	Adenine requirement	4×10^{-8}	
Corn	Kernel color	2.2×10^{-6}	Per gamete
Drosophila	Eye color	4×10^{-5}	Per gamete
	Allozymes	5.14×10^{-6}	
Mouse	Albino coat color	4.5×10^{-5}	Per gamete
	Dilution coat color	3×10^{-5}	
Human	Huntington disease	1×10^{-6}	Per gamete
	Achondroplasia	1×10^{-5}	
	Neurofibromatosis (Michigan)	1×10^{-4}	
	Hemophilia A (Finland)	3.2×10^{-5}	
	Duchenne muscular dystrophy (Wisconsin)	9.2×10^{-5}	

is considerable disagreement about whether (1) the increased mutagenesis is a genetic strategy selected to increase the probability of evolving beneficial traits in the stressful environment or (2) mutagenesis is merely an accidental consequence of error-prone DNA-repair mechanisms that are more active in stressful environments (see Section 18.4 on DNA repair).

CONCEPTS

Mutation rate is the frequency with which a specific mutation arises. Rates of mutations are generally low and are affected by environmental and genetic factors.

✔ CONCEPT CHECK 4

What three factors affect mutation rates?

18.2 Mutations Are Potentially Caused by a Number of Different Natural and Unnatural Factors

Mutations result from both internal and external factors. Those that occur under normal conditions are termed **spontaneous mutations**, whereas those that result from changes caused by environmental chemicals or radiation are **induced mutations**.

Spontaneous Replication Errors

Replication is amazingly accurate: less than one error in a billion nucleotides arises in the course of DNA synthesis (see Chapter 12). However, spontaneous replication errors do occasionally occur.

Tautomeric shifts The primary cause of spontaneous replication errors was formerly thought to be tautomeric shifts, in which the positions of protons in the DNA bases change. Purine and pyrimidine bases exist in different chemical forms called tautomers (**Figure 18.10a**). The two tautomeric forms of each base are in dynamic equilibrium, although one form is more common than the other. The standard Watson-and-Crick base pairings—adenine with thymine, and cytosine with guanine—are between the common forms of the bases, but, if the bases are in their rare tautomeric forms, other base pairings are possible (**Figure 18.10b**).

Watson and Crick proposed that tautomeric shifts might produce mutations, and, for many years, their proposal was the accepted model for spontaneous replication errors. However, there has never been convincing evidence that the rare tautomers are the cause of spontaneous mutations. Furthermore, research now shows little evidence of tautomers in DNA.

Mispairing due to other structures Mispairing can also occur through wobble, in which normal, protonated,

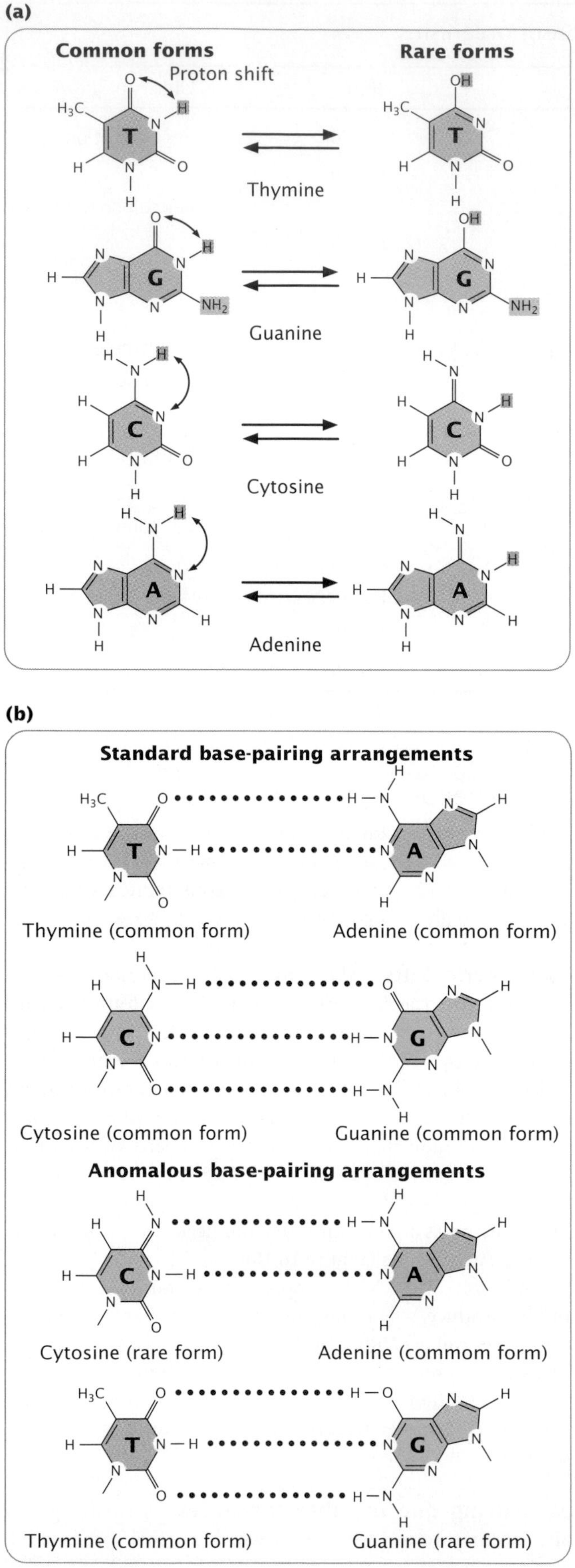

(a)

18.10 Purine and pyrimidine bases exist in different forms called tautomers. (a) A tautomeric shift takes place when a proton changes its position, resulting in a rare tautomeric form. (b) Standard and anomalous base-pairing arrangements that arise if bases are in the rare tautomeric forms.

(b)

and other forms of the bases are able to pair because of flexibility in the DNA helical structure (**Figure 18.11**). These structures have been detected in DNA molecules and are now thought to be responsible for many of the mispairings in replication.

Incorporation errors and replication errors   When a mismatched base has been incorporated into a newly synthesized nucleotide chain, an **incorporated error** is said to have occurred. Suppose that, in replication, thymine (which normally pairs with adenine) mispairs with guanine through wobble (**Figure 18.12**). In the next round of replication, the two mismatched bases separate, and each serves as template for the synthesis of a new nucleotide strand. This time, thymine pairs with adenine, producing another copy of the original DNA sequence. On the other strand, however, the incorrectly incorporated guanine serves as the template and pairs with cytosine, producing a new DNA molecule that has an error—a C·G pair in place of the original T·A pair (a T·A→C·G base substitution). The original incorporated error leads to a **replication error**, which creates a permanent mutation, because all the base pairings are correct and there is no mechanism for repair systems to detect the error.

Causes of deletions and insertions   Mutations due to small insertions and deletions also arise spontaneously in

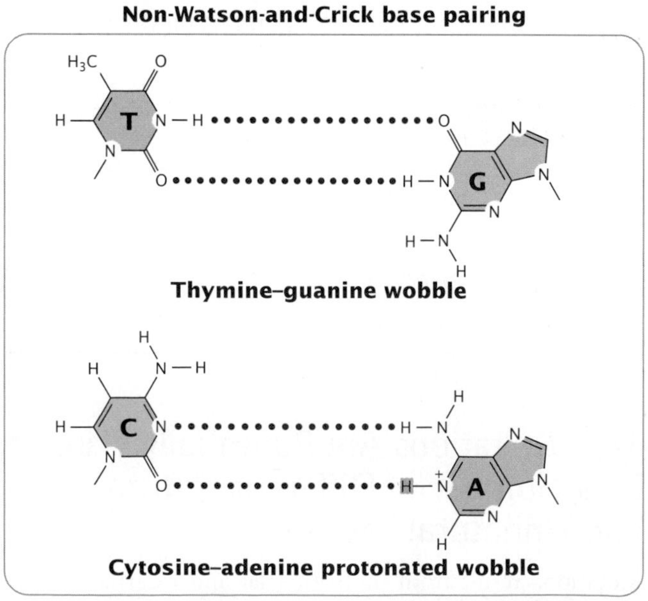

18.11 Nonstandard base pairings can occur as a result of the flexibility in DNA structure. Thymine and guanine can pair through wobble between normal bases. Cytosine and adenine can pair through wobble when adenine is protonated (has an extra hydrogen atom).

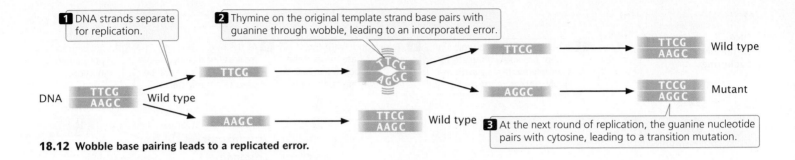

1 DNA strands separate for replication.

2 Thymine on the original template strand base pairs with guanine through wobble, leading to an incorporated error.

DNA

TTCG
AAGC   Wild type

TTCG

AAGC

TTCG
AGGC

TTCG
AAGC   Wild type

TTCG

TTCG
AAGC   Wild type

AGGC

TTCG
AAGC   Wild type

TCCG
AGGC   Mutant

3 At the next round of replication, the guanine nucleotide pairs with cytosine, leading to a transition mutation.

18.12 Wobble base pairing leads to a replicated error.

replication and crossing over. **Strand slippage** can occur when one nucleotide strand forms a small loop (**Figure 18.13**). If the looped-out nucleotides are on the newly synthesized strand, an insertion results. At the next round of replication, the insertion will be replicated and both strands will contain the insertion. If the looped-out nucleotides are on the template strand, then the newly replicated strand has a deletion, and this deletion will be perpetuated in subsequent rounds of replication.

Another process that produces insertions and deletions is unequal crossing over. In normal crossing over, the homologous sequences of the two DNA molecules align, and crossing over produces no net change in the number of nucleotides in either molecule. Misaligned pairing can cause **unequal crossing over**, which results in one DNA molecule with an insertion and the other with a deletion (**Figure 18.14**). Some DNA sequences are more likely than others to undergo strand slippage or unequal crossing over. Stretches of repeated sequences, such as nucleotide repeats or homopolymeric repeats (more than five repeats of the same base in a row), are prone to strand slippage. Duplicated or repetitive sequences may misalign during pairing, leading to unequal crossing over. Both strand slippage and unequal crossing over produce duplicated copies of sequences, which in turn promote further strand slippage and unequal cross-

ing over. This chain of events may explain the phenomenon of anticipation often observed for expanding nucleotide repeats.

CONCEPTS

Spontaneous replication errors arise from altered base structures and from wobble base pairing. Small insertions and deletions can occur through strand slippage in replication and through unequal crossing over.

✔ **CONCEPT CHECK 5**

How does an incorporated error differ from a replicated error?

Spontaneous Chemical Changes

In addition to spontaneous mutations that arise in replication, mutations also result from spontaneous chemical changes in DNA. One such change is **depurination**, the loss of a purine base from a nucleotide. Depurination results when the covalent bond connecting the purine to the 1'-carbon atom of the deoxyribose sugar breaks (**Figure 18.15a**), producing an apurinic site, a nucleotide that lacks its purine base. An apurinic site cannot act as a template for a complementary base in replication. In the absence of base-pairing constraints, an

Newly synthesized strand 5' TACGGACTGAAAA 3'
        Template strand 3' ATGCCTGACTTTTTGCGAAG 5'

1 Newly synthesized strand loops out,…

3 Template strand loops out,…

5' ACGGACTGAA A 3'
3' TGCCTGACTTTTTGCGAA 5'

5' ACGGACTGAAAA 3'
3' TGCCTGACTT T TTGCGAA 5'

2 …resulting in the addition of one nucleotide on the new strand.

4 …resulting in the omission of one nucleotide on the new strand.

5' ACGGACTGAA A AAACGCTTT 3'
3' TGCCTGACTT TTTTGCGAAG 5'

5' ACGGACTGAAAACGCTTT 3'
3' TGCCTGACTT T TTGCGAAG 5'

18.13 Insertions and deletions may result from strand slippage.

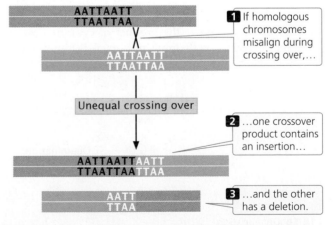

AATTAATT
TTAATTAA

1 If homologous chromosomes misalign during crossing over,…

AATTAATT
TTAATTAA

Unequal crossing over

2 …one crossover product contains an insertion…

AATTAATTAATT
TTAATTAATTAA

AATT
TTAA

3 …and the other has a deletion.

18.14 Unequal crossing over produces insertions and deletions.

(a)

DNA sugar–phosphate backbone

5'

Bases

T

Pyrimidine

G

Purine

G

OH

3'

(b)

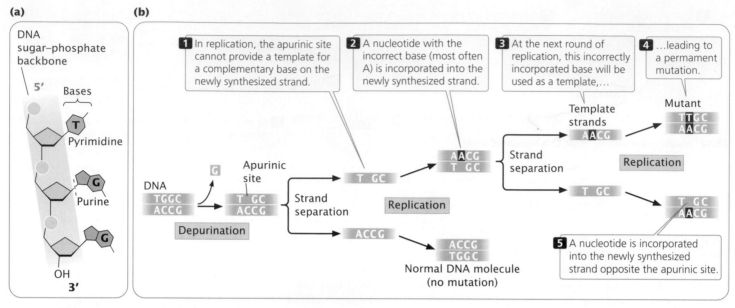

1 In replication, the apurinic site cannot provide a template for a complementary base on the newly synthesized strand.

2 A nucleotide with the incorrect base (most often A) is incorporated into the newly synthesized strand.

3 At the next round of replication, this incorrectly incorporated base will be used as a template,...

4 ...leading to a permanent mutation.

Mutant

Template strands

DNA

Apurinic site

Depuration

Strand separation

Replication

Strand separation

Replication

Normal DNA molecule (no mutation)

5 A nucleotide is incorporated into the newly synthesized strand opposite the apurinic site.

18.15 Depurination (the loss of a purine base from a nucleotide) produces an apurinic site.

incorrect nucleotide (most often adenine) is incorporated into the newly synthesized DNA strand opposite the apurinic site (**Figure 18.15b**), frequently leading to an incorporated error. The incorporated error is then transformed into a replication error at the next round of replication. Depurination is a common cause of spontaneous mutation; a mammalian cell in culture loses approximately 10,000 purines every day.

Another spontaneously occurring chemical change that takes place in DNA is **deamination**, the loss of an amino group (NH_2) from a base. Deamination can be spontaneous or be induced by mutagenic chemicals.

Deamination can alter the pairing properties of a base: the deamination of cytosine, for example, produces uracil (**Figure 18.16a**), which pairs with adenine in replication. After another round of replication, the adenine will pair with thymine, creating a T·A pair in place of the original C·G pair (C·G→U·A→T·A); this chemical change is a transition mutation. This type of mutation is usually repaired by enzymes that remove uracil whenever it is found in DNA. The ability to recognize the product of cytosine deamination may explain why thymine, not uracil, is found in DNA.

In mammals, including humans, some cytosine bases in DNA are naturally methylated and exist in the form of 5-methylcytosine (5mC; see p. 285 in Chapter 10 and Figure 10.18). When deaminated, 5mC becomes thymine (**Figure 18.16b**). Because thymine pairs with adenine in replication, the deamination of 5-methylcytosine changes an original C·G pair to T·A (C·G→5mC·G→T·G→T·A). Consequently, C·G→ T·A transitions are frequent in mammalian cells, and 5mC sites are mutation hotspots in humans. [TRY PROBLEM 27 →]

CONCEPTS

Some mutations arise from spontaneous alterations to DNA structure, such as depurination and deamination, which may alter the pairing properties of the bases and cause errors in subsequent rounds of replication.

Chemically Induced Mutations

Although many mutations arise spontaneously, a number of environmental agents are capable of damaging DNA, including certain chemicals and radiation. Any environmental agent

(a)

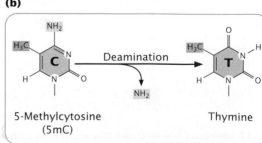

Cytosine → Deamination → Uracil

(b)

5-Methylcytosine (5mC) → Deamination → Thymine

18.16 Deamination alters DNA bases.

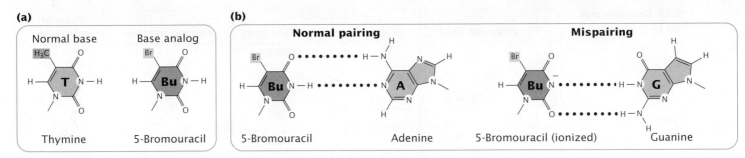

(a)

Normal base   Base analog

Thymine   5-Bromouracil

(b)

Normal pairing   **Mispairing**

5-Bromouracil   Adenine   5-Bromouracil (ionized)   Guanine

18.17 5-Bromouracil (a base analog) resembles thymine, except that it has a bromine atom in place of a methyl group on the 5-carbon atom. Because of the similarity in their structures, 5-bromouracil may be incorporated into DNA in place of thymine. Like thymine, 5-bromouracil normally pairs with adenine but, when ionized, it may pair with guanine through wobble.

that significantly increases the rate of mutation above the spontaneous rate is called a **mutagen.**

The first discovery of a chemical mutagen was made by Charlotte Auerbach, who started her career in Berlin research-ing the development of mutants in *Drosophila*. Faced with increasing anti-Semitism in Nazi Germany, Auerbach emigrat-ed to Britain. There she continued her research on *Drosophila* and collaborated with pharmacologist John Robson on the mutagenic effects of mustard gas, which had been used as a chemical weapon in World War I. The experimental conditions were crude. They heated liquid mustard gas over a Bunsen burner on the roof of the pharmacology building, and the flies were exposed to the gas in a large chamber. After developing serious burns on her hands from the gas, Auerbach let others carry out the exposures, and she analyzed the flies. Auerbach and Robson showed that mustard gas is indeed a powerful mutagen, reducing the viability of gametes and increasing the numbers of mutations seen in the offspring of exposed flies. Because the research was part of the secret war effort, publica-tion of their findings was delayed until 1947.

Base analogs One class of chemical mutagens consists of **base analogs**, chemicals with structures similar to that of any

of the four standard bases of DNA. DNA polymerases cannot distinguish these analogs from the standard bases; so, if base analogs are present during replication, they may be incorpo-rated into newly synthesized DNA molecules. For example, 5-bromouracil (5BU) is an analog of thymine; it has the same structure as that of thymine except that it has a bromine (Br) atom on the 5-carbon atom instead of a methyl group (**Fig-ure 18.17a**). Normally, 5-bromouracil pairs with adenine just as thymine does, but it occasionally mispairs with guanine (**Figure 18.17b**), leading to a transition (T·A→5BU·A→ 5BU·G→C·G), as shown in **Figure 18.18**. Through mispair-ing, 5-bromouracil can also be incorporated into a newly synthesized DNA strand opposite guanine. In the next round of replication 5-bromouracil pairs with adenine, leading to another transition (G·C→G·5BU→A·5BU→A·T).

Another mutagenic chemical is 2-aminopurine (2AP), which is a base analog of adenine. Normally, 2-aminopurine base pairs with thymine, but it may mispair with cytosine, causing a transition mutation (T·A→T·2AP→C·2AP→ C·G). Alternatively, 2-aminopurine may be incorporated through mispairing into the newly synthesized DNA oppo-site cytosine and then later pair with thymine, leading to a C·G→C·2AP→T· 2AP→T·A transition.

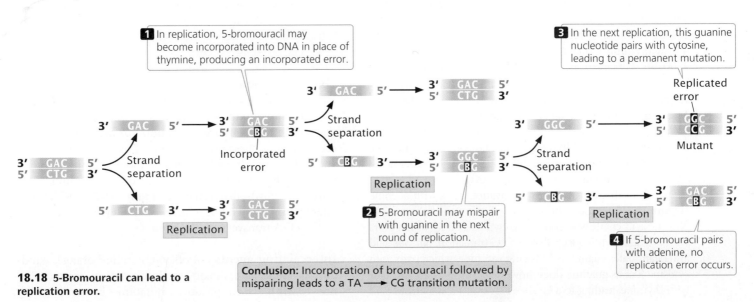

1 In replication, 5-bromouracil may become incorporated into DNA in place of thymine, producing an incorporated error.

3 In the next replication, this guanine nucleotide pairs with cytosine, leading to a permanent mutation.

Replicated error

Mutant

2 5-Bromouracil may mispair with guanine in the next round of replication.

4 If 5-bromouracil pairs with adenine, no replication error occurs.

18.18 5-Bromouracil can lead to a replication error.

Conclusion: Incorporation of bromouracil followed by mispairing leads to a TA ⟶ CG transition mutation.

18.19 Chemicals may alter DNA bases.

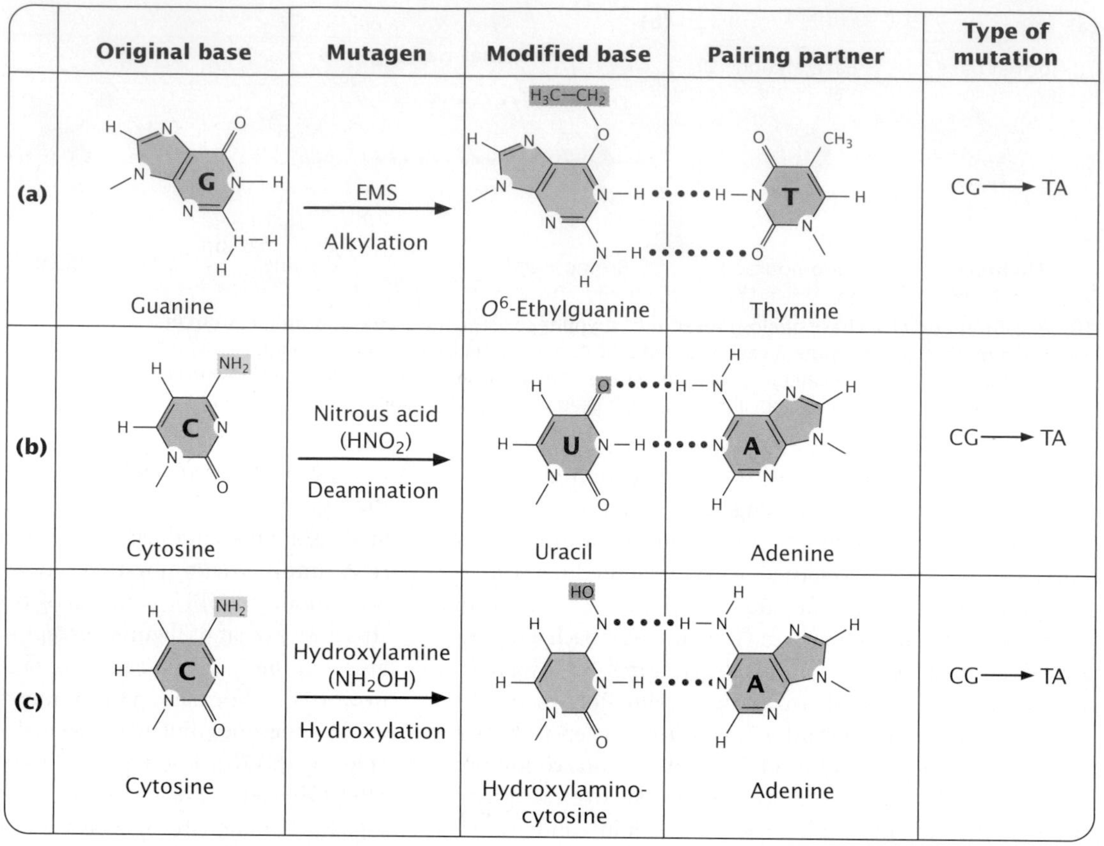

	Original base	Mutagen	Modified base	Pairing partner	Type of mutation
(a)	Guanine	EMS Alkylation	O^6-Ethylguanine	Thymine	CG ⟶ TA
(b)	Cytosine	Nitrous acid (HNO_2) Deamination	Uracil	Adenine	CG ⟶ TA
(c)	Cytosine	Hydroxylamine (NH_2OH) Hydroxylation	Hydroxylamino-cytosine	Adenine	CG ⟶ TA

Thus, both 5-bromouracil and 2-aminopurine can produce transition mutations. In the laboratory, mutations caused by base analogs can be reversed by treatment with the same analog or by treatment with a different analog.

Alkylating agents Alkylating agents are chemicals that donate alkyl groups, such as methyl (CH_3) and ethyl (CH_3–CH_2) groups, to nucleotide bases. For example, ethylmethylsulfonate (EMS) adds an ethyl group to guanine, producing O^6-ethylguanine, which pairs with thymine (**Figure 18.19a**). Thus, EMS produces $C \cdot G \rightarrow T \cdot A$ transitions. EMS is also capable of adding an ethyl group to thymine, producing 4-ethylthymine, which then pairs with guanine, leading to a $T \cdot A \rightarrow C \cdot G$ transition. Because EMS produces both $C \cdot G \rightarrow T \cdot A$ and $T \cdot A \rightarrow C \cdot G$ transitions, mutations produced by EMS can be reversed by additional treatment with EMS. Mustard gas is another alkylating agent.

Deamination In addition to its spontaneous occurrence (see Figure 18.16), deamination can be induced by some chemicals. For instance, nitrous acid deaminates cytosine, creating uracil, which in the next round of replication pairs with adenine (**Figure 18.19b**), producing a $C \cdot G \rightarrow T \cdot A$ transition mutation. Nitrous acid changes adenine into hypoxanthine, which pairs with cytosine, leading to a $T \cdot A \rightarrow C \cdot G$ transition. Nitrous acid also deaminates guanine, producing xanthine, which pairs with cytosine just as guanine does; however, xanthine can also pair with thymine, leading to a $C \cdot G \rightarrow T \cdot A$ transition. Nitrous acid

produces exclusively transition mutations and, because both $C \cdot G \rightarrow T \cdot A$ and $T \cdot A \rightarrow C \cdot G$ transitions are produced, these mutations can be reversed with nitrous acid.

Hydroxylamine Hydroxylamine is a very specific base-modifying mutagen that adds a hydroxyl group to cytosine, converting it into hydroxylaminocytosine (**Figure 18.19c**). This conversion increases the frequency of a rare tautomer that pairs with adenine instead of guanine and leads to $C \cdot G \rightarrow T \cdot A$ transitions. Because hydroxylamine acts only on cytosine, it will not generate $T \cdot A \rightarrow C \cdot G$ transitions; thus, hydroxylamine will not reverse the mutations that it produces. TRY PROBLEM 26 →

Oxidative reactions Reactive forms of oxygen (including superoxide radicals, hydrogen peroxide, and hydroxyl radicals) are produced in the course of normal aerobic metabolism, as well as by radiation, ozone, peroxides, and certain drugs. These reactive forms of oxygen damage DNA and induce mutations by bringing about chemical changes in DNA. For example, oxidation converts guanine into 8-oxy-7,8-dihydrodeoxyguanine (**Figure 18.20**), which frequently mispairs with adenine instead of cytosine, causing a $G \cdot C \rightarrow T \cdot A$ transversion mutation.

Intercalating agents Proflavin, acridine orange, ethidium bromide, and dioxin are **intercalating agents** (**Figure 18.21a**), which produce mutations by sandwiching

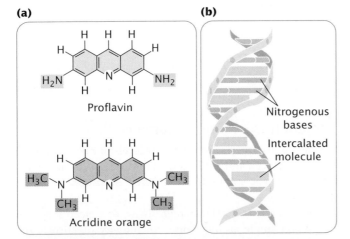

18.20 Oxidative radicals convert guanine into 8-oxy-7, 8-dihydrodeoxyguanine.

themselves (intercalating) between adjacent bases in DNA, distorting the three-dimensional structure of the helix and causing single-nucleotide insertions and deletions in replication (**Figure 18.21b**). These insertions and deletions frequently produce frameshift mutations, and so the mutagenic effects of intercalating agents are often severe. Because intercalating agents generate both additions and deletions, they can reverse the effects of their own mutations.

CONCEPTS

Chemicals can produce mutations by a number of mechanisms. Base analogs are inserted into DNA and frequently pair with the wrong base. Alkylating agents, deaminating chemicals, hydroxylamine, and oxidative radicals change the structure of DNA bases, thereby altering their pairing properties. Intercalating agents wedge between the bases and cause single-base insertions and deletions in replication.

✔ CONCEPT CHECK 6

Base analogs are mutagenic because of which characteristic?

a. They produce changes in DNA polymerase that cause it to malfunction.

b. They distort the structure of DNA.

c. They are similar in structure to the normal bases.

d. They chemically modify the normal bases.

Radiation

In 1927, Hermann Muller demonstrated that mutations in fruit flies could be induced by X-rays. The results of subsequent studies showed that X-rays greatly increase mutation rates in all organisms. Because of their high energies, X-rays, gamma rays, and cosmic rays are all capable of penetrating tissues and damaging DNA. These forms of radiation, called ionizing radiation, dislodge electrons from the atoms that they encounter, changing stable molecules into free radicals and reactive ions, which then alter the structures of bases and break phosphodiester bonds in DNA. Ionizing radiation also frequently results in double-strand breaks in DNA. Attempts to repair these breaks can produce chromosome mutations (discussed in Chapter 9).

Ultraviolet (UV) light has less energy than that of ionizing radiation and does not eject electrons but is nevertheless highly mutagenic. Purine and pyrimidine bases readily absorb UV light, resulting in the formation of chemical bonds between adjacent pyrimidine molecules on the same strand of DNA and in the creation of **pyrimidine dimers** (**Figure 18.22a**). Pyrimidine dimers consisting of two thymine bases (called thymine dimers) are most frequent, but cytosine dimers and thymine cytosine dimers also can form. Dimers distort the configuration of DNA (**Figure 18.22b**) and often block replication. Most pyrimidine dimers are immediately repaired by mechanisms discussed later in this chapter, but some escape repair and inhibit replication and transcription.

When pyrimidine dimers block replication, cell division is inhibited and the cell usually dies; for this reason, UV light kills bacteria and is an effective sterilizing agent. For a mutation—a hereditary error in the genetic instructions—to occur, the replication block must be overcome. Bacteria can sometimes circumvent replication blocks produced by pyrimidine dimers and other types of DNA damage by means of the **SOS system**. This system allows replication blocks to be overcome but, in the process, makes numerous mistakes and greatly increases the rate of mutation. Indeed, the very reason that replication can proceed in the presence of a block is that the enzymes in the SOS system do not strictly adhere to the base-pairing rules. The trade-off is that

18.21 Intercalating agents. Proflavin and acridine orange (a) insert themselves between adjacent bases in DNA, distorting the three-dimensional structure of the helix (b).

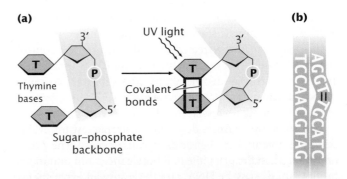

18.22 Pyrimidine dimers result from ultraviolet light.
(a) Formation of thymine dimer. (b) Distorted DNA.

replication may continue and the cell survives, but only by sacrificing the normal accuracy of DNA synthesis.

The SOS system is complex, including the products of at least 25 genes. A protein called RecA binds to the damaged DNA at the blocked replication fork and becomes activated. This activation promotes the binding of a protein called LexA, which is a repressor of the SOS system. The activated RecA complex induces LexA to undergo self-cleavage, destroying its repressive activity. This inactivation enables other SOS genes to be expressed, and the products of these genes allow replication of the damaged DNA to proceed. The SOS system allows bases to be inserted into a new DNA strand in the absence of bases on the template strand, but these insertions result in numerous errors in the base sequence.

> **CONCEPTS**
>
> Ionizing radiation such as X-rays and gamma rays damage DNA by dislodging electrons from atoms; these electrons then break phosphodiester bonds and alter the structure of bases. Ultraviolet light causes mutations primarily by producing pyrimidine dimers that disrupt replication and transcription. The SOS system enables bacteria to overcome replication blocks but introduces mistakes in replication.

18.3 Mutations Are the Focus of Intense Study by Geneticists

Because mutations often have detrimental effects, they are often studied by geneticists. These studies have included the development of tests to determine the mutagenic properties of chemical compounds and the investigation of human populations tragically exposed to high levels of radiation.

Detecting Mutations with the Ames Test

People in industrial societies are surrounded by a multitude of artificially produced chemicals: more than 50,000 different chemicals are in commercial and industrial use today, and from 500 to 1000 new chemicals are introduced each year. Some of these chemicals are potential carcinogens and may cause harm to humans. One method for testing the cancer-causing potential of chemicals is to administer them to laboratory animals (rats or mice) and compare the incidence of cancer in the treated animals with that of control animals. These tests are unfortunately time consuming and expensive. Furthermore, the ability of a substance to cause cancer in rodents is not always indicative of its effect on humans. After all, we aren't rats!

In 1974, Bruce Ames developed a simple test for evaluating the potential of chemicals to cause cancer. The **Ames test** is based on the principle that both cancer and mutations result from damage to DNA, and the results of experiments have demonstrated that 90% of known carcinogens are also

mutagens. Ames proposed that mutagenesis in bacteria could serve as an indicator of carcinogenesis in humans.

The Ames test uses different auxotrophic strains of the bacterium *Salmonella typhimurium* that have defects in the lipopolysaccharide coat, which normally protects the bacteria from chemicals in the environment. Furthermore, the DNA-repair system in these strains has been inactivated, enhancing their susceptibility to mutagens.

The most recent version of the test (called Ames II) uses several auxotrophic strains that detect different types of base-pair substitutions. Other strains detect different types of frameshift mutations. Each strain carries a *his*⁻ mutation, which renders it unable to synthesize the amino acid histidine, and the bacteria are plated onto medium that lacks histidine (**Figure 18.23**). Only bacteria that have undergone a reverse mutation of the histidine gene (*his*⁻ → *his*⁺) are able to synthesize histidine and grow on the medium. Different dilutions of a chemical to be tested are added to plates inoculated with the bacteria, and the number of mutant bacterial colonies that appear on each plate is compared with the number that appear on control plates with no chemical (i.e., that arose through spontaneous mutation). Any chemical that significantly increases the number of colonies appearing on a treated plate is mutagenic and is probably also carcinogenic.

Some compounds are not active carcinogens but can be converted into cancer-causing compounds in the body. To make the Ames test sensitive for such *potential* carcinogens, a compound to be tested is first incubated in mammalian liver extract that contains metabolic enzymes.

The Ames test has been applied to thousands of chemicals and commercial products. An early demonstration of its usefulness was the discovery, in 1975, that many hair dyes sold in the United States contained compounds that were mutagenic to bacteria. These compounds were then removed from most hair dyes.

> **CONCEPTS**
>
> The Ames test uses *his*⁻ strains of bacteria to test chemicals for their ability to produce *his*⁻ → *his*⁺ mutations. Because mutagenic activity and carcinogenic potential are closely correlated, the Ames test is widely used to screen chemicals for their cancer-causing potential.

Radiation Exposure in Humans

People are routinely exposed to low levels of radiation from cosmic, medical, and environmental sources, but there have also been tragic events that produced exposures of much higher degree.

On August 6, 1945, a high-flying American airplane dropped a single atomic bomb on the city of Hiroshima, Japan. The explosion devastated an area of the city measuring 4.5 square miles, killed from 90,000 to 140,000 people,

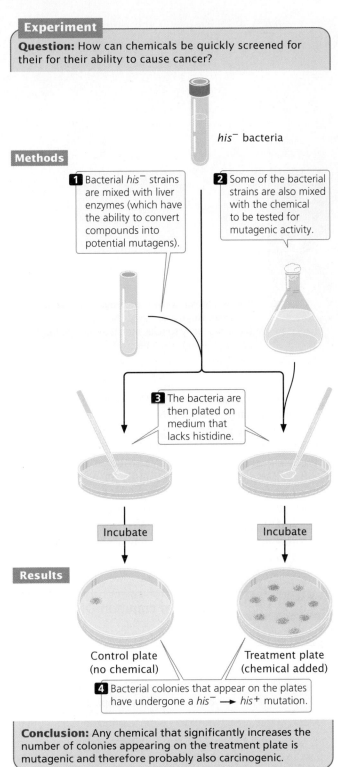

Question: How can chemicals be quickly screened for their for their ability to cause cancer?

his⁻ bacteria

Methods

1 Bacterial *his⁻* strains are mixed with liver enzymes (which have the ability to convert compounds into potential mutagens).

2 Some of the bacterial strains are also mixed with the chemical to be tested for mutagenic activity.

3 The bacteria are then plated on medium that lacks histidine.

Incubate

Incubate

Results

Control plate (no chemical)

Treatment plate (chemical added)

4 Bacterial colonies that appear on the plates have undergone a *his⁻* → *his⁺* mutation.

Conclusion: Any chemical that significantly increases the number of colonies appearing on the treatment plate is mutagenic and therefore probably also carcinogenic.

18.23 The Ames test is used to identify chemical mutagens.

and injured almost as many (**Figure 18.24**). Three days later, the United States dropped an atomic bomb on the city of Nagasaki, this time destroying an area measuring 1.5 square miles and killing between 60,000 and 80,000 people. Huge amounts of radiation were released during these explosions, and many people were exposed.

After the war, a joint Japanese–U.S. effort was made to study the biological effects of radiation exposure on the survivors of the atomic blasts and their children. Somatic mutations were examined by studying radiation sickness and cancer among the survivors; germ-line mutations were assessed by looking at birth defects, chromosome abnormalities, and gene mutations in children born to people that had been exposed to radiation.

Geneticist James Neel and his colleagues examined almost 19,000 children of people who were within 2000 meters (1.2 miles) of the center of the atomic blast at Hiroshima or Nagasaki, along with a similar number of children whose parents did not receive radiation exposure. Radiation doses were estimated for a child's parents on the basis of careful assessment of the parents' location, posture, and position at the time of the blast. A blood sample was collected from each child, and gel electrophoresis was used to investigate amino acid substitutions in 28 proteins. When rare variants were detected, blood samples from the child's parents also were analyzed to establish whether the variant was inherited or a new mutation.

Of a total of 289,868 genes examined by Neel and his colleagues, only one mutation was found in the children of exposed parents; no mutations were found in the control group. From these findings, a mutation rate of 3.4×10^{-6} was estimated for the children whose parents were exposed to the blast, which is within the range of spontaneous mutation rates observed for other eukaryotes. Neel and his colleagues also examined the frequency of chromosome mutations, sex ratios of children born to exposed parents, and frequencies of chromosome aneuploidy. There was no evidence in any of these assays for increased mutations among the children of the people who were exposed to radiation from the atomic explosions, suggesting that germ-line mutations were not elevated.

Animal studies clearly show that radiation causes germ-line mutations; so why was there no apparent increase in germ-line mutations among the inhabitants of Hiroshima and Nagasaki? The exposed parents did exhibit an increased incidence of leukemia and other types of cancers; so somatic mutations were clearly induced. The answer to the question is not known, but the lack of germ-line mutations may be due to the fact that those persons who received the largest radiation doses died soon after the blasts.

The Techa River in southern Russia is another place where people have been tragically exposed to high levels of radiation. The Mayak nuclear facility produced plutonium for nuclear warheads in the early days of the Cold War. Between 1949 and 1956, this plant dumped some 76 million cubic meters of radioactive sludge into the Techa River. People downstream used the river for drinking water and crop irrigation; some received radiation doses 1700 times the annual amount considered safe by today's standards. Radiation in the area was further elevated by a series of nuclear accidents at the Mayak plant; the worst was an explosion of a radioactive liquid storage tank in 1957, which

18.24 Hiroshima was destroyed by an atomic bomb on August 6, 1945. The atomic explosion produced many somatic mutations among the survivors. [Stanley Troutman/AP.]

showered radiation over a 27,000-square-kilometer (10,425-square-mile) area.

Although Soviet authorities suppressed information about the radiation problems along the Techa until the 1990s, Russian physicians led by Mira Kossenko quietly began studying cancer and other radiation-related illnesses among the inhabitants in the 1960s. They found that the overall incidence of cancer was elevated among people who lived on the banks of the Techa River.

Most data on radiation exposure in humans are from the intensive study of the survivors of the atomic bombing of Hiroshima and Nagasaki. However, the inhabitants of Hiroshima and Nagasaki were exposed in one intense burst of radiation, and these data may not be appropriate for understanding the effects of long-term low-dose radiation. Today, U.S. and Russian scientists are studying the people of the Techa River region in an attempt to better understand the effects of chronic radiation exposure on human populations.

18.4 A Number of Pathways Repair Changes in DNA

The integrity of DNA is under constant assault from radiation, chemical mutagens, and spontaneously arising changes. In spite of this onslaught of damaging agents, the rate of mutation remains remarkably low, thanks to the efficiency with which DNA is repaired.

There are a number of complex pathways for repairing DNA, but several general statements can be made about DNA repair. First, most DNA-repair mechanisms require two nucleotide strands of DNA because most replace whole nucleotides, and a template strand is needed to specify the base sequence.

A second general feature of DNA repair is redundancy, meaning that many types of DNA damage can be corrected by more than one pathway of repair. This redundancy testifies to the extreme importance of DNA repair to the survival of the cell: it ensures that almost all mistakes are corrected. If a mistake escapes one repair system, it's likely to be repaired by another system.

We will consider several general mechanisms of DNA repair: mismatch repair, direct repair, base-excision repair, nucleotide-excision repair, and repair of double-strand breaks (**Table 18.4**).

Table 18.4	Summary of common DNA repair mechanisms
Repair System	Type of Damage Repaired
Mismatch	Replication errors, including mispaired bases and strand slippage
Direct	Pyrimidine dimers; other specific types of alterations
Base excision	Abnormal bases, modified bases, and pyrimidine dimers
Nucleotide excision	DNA damage that distorts the double helix, including abnormal bases, modified bases, and pyrimidine dimers
Homologous recombination	Double-strand breaks
Nonhomologous end joining	Double-strand breaks

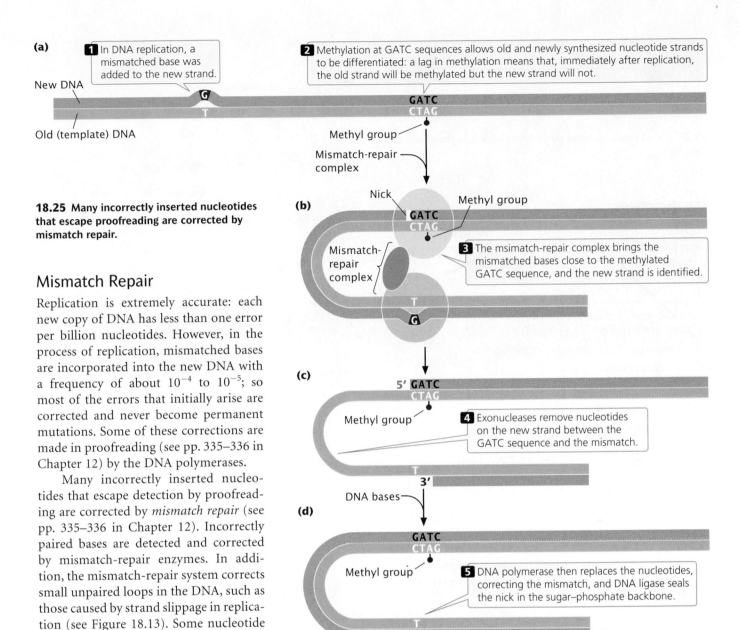

(a)

1 In DNA replication, a mismatched base was added to the new strand.

New DNA

Old (template) DNA

2 Methylation at GATC sequences allows old and newly synthesized nucleotide strands to be differentiated: a lag in methylation means that, immediately after replication, the old strand will be methylated but the new strand will not.

GATC
CTAG

Methyl group

Mismatch-repair complex

(b)

Nick

GATC
CTAG

Methyl group

Mismatch-repair complex

3 The msimatch-repair complex brings the mismatched bases close to the methylated GATC sequence, and the new strand is identified.

(c)

5' GATC
CTAG

Methyl group

4 Exonucleases remove nucleotides on the new strand between the GATC sequence and the mismatch.

3'

DNA bases

(d)

GATC
CTAG

Methyl group

5 DNA polymerase then replaces the nucleotides, correcting the mismatch, and DNA ligase seals the nick in the sugar–phosphate backbone.

A

18.25 Many incorrectly inserted nucleotides that escape proofreading are corrected by mismatch repair.

Mismatch Repair

Replication is extremely accurate: each new copy of DNA has less than one error per billion nucleotides. However, in the process of replication, mismatched bases are incorporated into the new DNA with a frequency of about 10^{-4} to 10^{-5}; so most of the errors that initially arise are corrected and never become permanent mutations. Some of these corrections are made in proofreading (see pp. 335–336 in Chapter 12) by the DNA polymerases.

Many incorrectly inserted nucleotides that escape detection by proofreading are corrected by *mismatch repair* (see pp. 335–336 in Chapter 12). Incorrectly paired bases are detected and corrected by mismatch-repair enzymes. In addition, the mismatch-repair system corrects small unpaired loops in the DNA, such as those caused by strand slippage in replication (see Figure 18.13). Some nucleotide repeats may form secondary structures on the unpaired strand (see Figure 18.5), allowing them to escape detection by the mismatch-repair system.

After the incorporation error has been recognized, mismatch-repair enzymes cut out a section of the newly synthesized strand and fill the gap with new nucleotides by using the original DNA strand as a template. For this strategy to work, mismatch repair must have some way of distinguishing between the old and the new strands of the DNA so that the incorporation error, but not part of the original strand, is removed.

The proteins that carry out mismatch repair in *E. coli* differentiate between old and new strands by the presence of methyl groups on special sequences of the old strand. After replication, adenine nucleotides in the sequence GATC are methylated. The process of methylation is delayed and so, immediately after replication, the old strand is methylated and the new strand is not (**Figure 18.25a**). The mismatch-repair complex brings an unmethylated GATC sequence in close proximity to the mismatched bases. It nicks the unmethylated strand at the GATC site (**Figure 18.25b**), and degrades the strand between the nick and the mismatched bases (**Figure 18.25c**). DNA polymerase and DNA ligase fill in the gap on the unmethylated strand with correctly paired nucleotides (**Figure 18.25d**).

Mismatch repair in eukaryotic cells is similar to that in *E. coli*, but how the old and new strands are recognized in eukaryotic cells is not known. In some eukaryotes, such as yeast and fruit flies, there is no detectable methylation of DNA, and yet mismatch repair still takes place. Humans who possess mutations in mismatch-repair genes often exhibit elevated somatic mutations and are frequently susceptible to colon cancer.

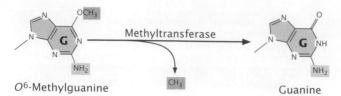

18.26 Direct repair changes nucleotides back into their original structures.

Direct Repair

Direct repair does not replace altered nucleotides but, instead, changes them back into their original (correct) structures. One of the most well understood direct-repair mechanisms is the photoreactivation of UV-induced pyrimidine dimers. *E. coli* and some eukaryotic cells possess an enzyme called photolyase, which uses energy captured from light to break the covalent bonds that link the pyrimidines in a dimer.

Direct repair also corrects O^6-methylguanine, an alkylation product of guanine that pairs with adenine, producing $G \cdot C \rightarrow T \cdot A$ transversions. An enzyme called O^6-methylguanine-DNA methyltransferase removes the methyl group from O^6-methylguanine, restoring the base to guanine (**Figure 18.26**).

Base-Excision Repair

In **base-excision repair**, a modified base is first excised and then the entire nucleotide is replaced. The excision of modified bases is catalyzed by a set of enzymes called DNA glycosylases, each of which recognizes and removes a specific type of modified base by cleaving the bond that links that base to the 1′-carbon atom of deoxyribose sugar (**Figure 18.27a**). Uracil glycosylase, for example, recognizes and removes uracil produced by the deamination of cytosine. Other glycosylases recognize hypoxanthine, 3-methyladenine, 7-methylguanine, and other modified bases.

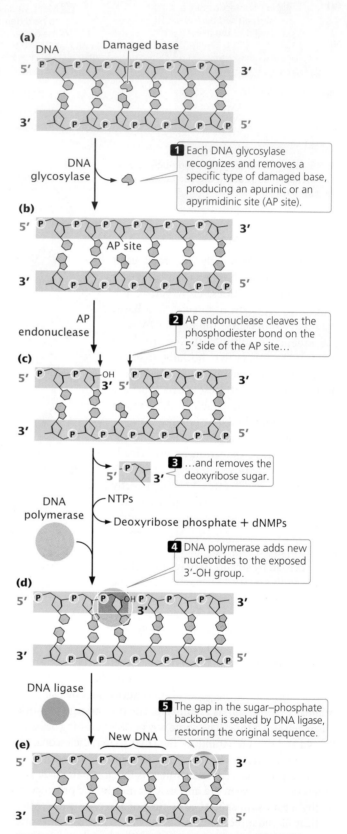

18.27 Base-excision repair excises modified bases and then replaces one or more nucleotides.

After the base has been removed, an enzyme called AP (apurinic or apyrimidinic) endonuclease cuts the phosphodiester bond, and other enzymes remove the deoxyribose sugar (**Figure 18.27b**). DNA polymerase then adds one or more new nucleotides to the exposed 3′-OH group (**Figure 18.27c**), replacing a section of nucleotides on the damaged strand. The nick in the phosphodiester backbone is sealed by DNA ligase (**Figure 18.27d**), and the original intact sequence is restored (**Figure 18.27e**).

Bacteria use DNA polymerase I to replace excised nucleotides, but eukaryotes use DNA polymerase β, which has no proofreading ability and tends to make mistakes. On average, DNA polymerase β makes one mistake per 4000 nucleotides inserted. About 20,000 to 40,000 base modifications per day are repaired by base excision, and so DNA polymerase β may introduce as many as 10 mutations per day into the human genome. How are these errors corrected? Recent research results show that some AP endonucleases have the ability to proofread. When DNA polymerase β inserts a nucleotide with the wrong base into the DNA, DNA ligase cannot seal the nick in the sugar–phosphate backbone, because the 3′-OH and 5′-P groups of adjacent nucleotides are not in the correct orientation for ligase to connect them. In this case, AP endonuclease 1 detects the mispairing and uses its 3′→5′ exonuclease activity to excise the incorrectly paired base. DNA polymerase β then uses its polymerase activity to fill in the missing nucleotide. In this way, the fidelity of base-excision repair is maintained.

CONCEPTS

Direct-repair mechanisms change altered nucleotides back into their correct structures. In base-excision repair, glycosylase enzymes recognize and remove specific types of modified bases. The entire nucleotide is then removed and a section of the polynucleotide strand is replaced.

✔ CONCEPT CHECK 8

How do direct-repair mechanisms differ from mismatch repair and base-excision repair?

Nucleotide-Excision Repair

Another repair pathway is **nucleotide-excision repair**, which removes bulky DNA lesions (such as pyrimidine dimers) that distort the double helix. Nucleotide-excision repair is quite versatile and can repair many different types of DNA damage. It is found in cells of all organisms from bacteria to humans and is among the most important of all repair mechanisms.

The process of nucleotide excision is complex; in humans, a large number of genes take part. First, a complex of enzymes scans DNA, looking for distortions of its

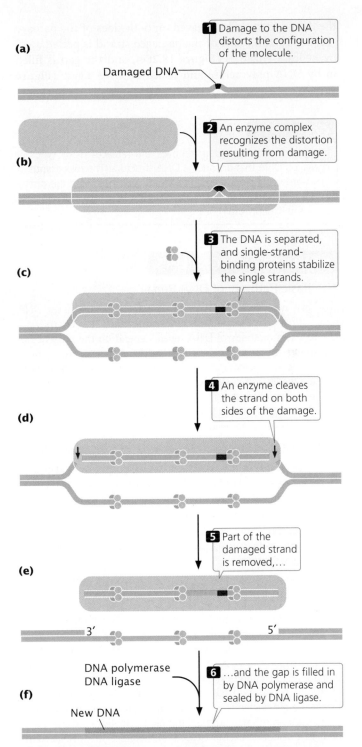

(a)

Damaged DNA

1 Damage to the DNA distorts the configuration of the molecule.

(b)

2 An enzyme complex recognizes the distortion resulting from damage.

(c)

3 The DNA is separated, and single-strand-binding proteins stabilize the single strands.

(d)

4 An enzyme cleaves the strand on both sides of the damage.

(e)

5 Part of the damaged strand is removed,…

3′   5′

DNA polymerase
DNA ligase

(f)

New DNA

6 …and the gap is filled in by DNA polymerase and sealed by DNA ligase.

18.28 Nucleotide-excision repair removes bulky DNA lesions that distort the double helix.

three-dimensional configuration (**Figure 18.28a** and **b**). When a distortion is detected, additional enzymes separate the two nucleotide strands at the damaged region, and single-strand-binding proteins stabilize the separated strands (**Figure 18.28c**). Next, the sugar–phosphate backbone of the damaged strand is cleaved on both sides of the damage (**Figure 18.28d**). Part of the damaged strand is peeled away by helicase enzymes (**Figure 18.28e**), and the gap is filled in by DNA polymerase and sealed by DNA ligase (**Figure 18.28f**).

CONCEPTS

Nucleotide-excision repair removes and replaces many types of damaged DNA that distort the DNA structure. The two strands of DNA are separated, a section of the DNA containing the distortion is removed, DNA polymerase fills in the gap, and DNA ligase seals the filled-in gap.

CONNECTING CONCEPTS

The Basic Pathway of DNA Repair

We have now examined several different mechanisms of DNA repair. What do these methods have in common? How are they different? Most methods of DNA repair depend on the presence of two strands, because nucleotides in the damaged area are removed and replaced. Nucleotides are replaced in mismatch repair, base-excision repair, and nucleotide-excision repair but are not replaced by direct-repair mechanisms.

Repair mechanisms that include nucleotide removal utilize a common four-step pathway:

1. Detection: The damaged section of the DNA is recognized.

2. Excision: DNA-repair endonucleases nick the phosphodiester backbone on one or both sides of the DNA damage and one or more nucleotides are removed.

3. Polymerization: DNA polymerase adds nucleotides to the newly exposed 3'-OH group by using the other strand as a template and replacing damaged (and frequently some undamaged) nucleotides.

4. Ligation: DNA ligase seals the nicks in the sugar–phosphate backbone.

The primary differences in the mechanisms of mismatch, base-excision, and nucleotide-excision repair are in the details of detection and excision. In base-excision and mismatch repair, a single nick is made in the sugar–phosphate backbone on one side of the damage; in nucleotide-excision repair, nicks are made on both sides of the DNA lesion. In base-excision repair, DNA polymerase displaces the old nucleotides as it adds new nucleotides to the 3' end of the nick; in mismatch repair, the old nucleotides are degraded; and, in nucleotide-excision repair, nucleotides are displaced by helicase enzymes. All three mechanisms use DNA polymerase and ligase to fill in the gap produced by the excision and removal of damaged nucleotides.

Repair of Double-Strand Breaks

A common type of DNA damage is a double-strand break, in which both strands of the DNA helix are broken. Double-strand breaks are caused by ionizing radiation, oxidative free radicals, and other DNA-damaging agents. These types of breaks are particularly detrimental to the cell because they stall DNA replication and may lead to chromosome rearrangements, such as deletions, duplications, inversions, and translocations. There are two major pathways for repairing double-strand breaks: homologous recombination and nonhomologous end joining.

Homologous recombination Homologous recombination repairs a broken DNA molecule by using the identical or nearly identical genetic information contained in another DNA molecule, usually a sister chromatid. DNA repair through homologous recombination uses the same mechanism employed in the process of homologous recombination that is responsible for crossing over (see Chapter 12). Homologous recombination begins with the removal of some nucleotides at the broken ends, followed by strand invasion, displacement, and replication (see Figure 12.21). Many of the same enzymes that carry out crossing over are utilized in the repair of double-strand breaks by homologous recombination. Two enzymes that play a role in homologous recombination are BRCA1 and BRCA2. The genes for these proteins are frequently mutated in breast cancer cells.

Nonhomologous end joining Nonhomologous end joining repairs double-strand breaks without using a homologous template. This pathway is often used when the cell is in G_1 and a sister chromatid is not available for repair through homologous recombination. Nonhomologous end joining uses proteins that recognize the broken ends of DNA, bind to the ends, and then joins them together. Nonhomologous end joining is more error prone than homologous recombination and often leads to deletions, insertions, and translocations.

Translesion DNA Polymerases

As discussed in Chapter 12, the high-fidelity DNA polymerases that normally carry out replication operate at high speed and, like a high-speed train, require a smooth track—an undistorted template. Some mutations, such as pyrimidine dimers, produce distortions in the three-dimensional structure of the DNA helix, blocking replication by the high-speed polymerases. When distortions of the template are encountered, specialized translesion DNA polymerases take over replication and bypass the lesion.

The translesion polymerases are able to bypass bulky lesions but, in the process, often make errors. Thus, the translesion polymerases allow replication to proceed past bulky lesion but at the cost of introducing mutations into the sequence. Some of these mutations are corrected by DNA-repair systems, but others escape detection.

An example of a translesion DNA polymerase is polymerase η (eta), which bypasses pyrimidine dimers in eukaryotes.

Polymerase η inserts AA opposite a pyrimidine dimer. This strategy seems to be reasonable because about two-thirds of pyrimidine dimers are thymine dimers. However, the insertion of AA opposite a CT dimer results in a $C \cdot G \rightarrow A \cdot T$ transversion. Polymerase η therefore tends to introduce mutations into the DNA sequence.

CONCEPTS

Two major pathways exist for the repair of double-strand breaks in DNA: homologous recombination and nonhomologous end joining. Special translesion DNA polymerases allow replication to proceed past bulky distortions in the DNA but often introduce errors as they bypass the distorted region.

Genetic Diseases and Faulty DNA Repair

Several human diseases are connected to defects in DNA repair. These diseases are often associated with high incidences of specific cancers, because defects in DNA repair lead to increased rates of mutation. This concept is discussed further in Chapter 23.

Among the best studied of the human DNA-repair diseases is xeroderma pigmentosum (**Figure 18.29**), a rare autosomal recessive condition that includes abnormal skin pigmentation and acute sensitivity to sunlight. Persons who have this disease also have a strong predisposition to skin cancer, with an incidence ranging from 1000 to 2000 times that found in unaffected people.

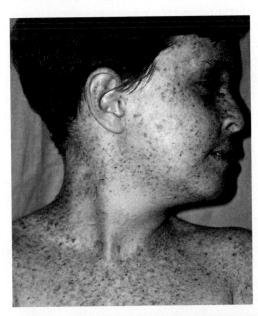

18.29 Xeroderma pigmentosum results from defects in DNA repair. The disease is characterized by frecklelike spots on the skin (shown here) and a predisposition to skin cancer. [Ken Greer/Visuals Unlimited.]

Sunlight includes a strong UV component; so exposure to sunlight produces pyrimidine dimers in the DNA of skin cells. Although human cells lack photolyase (the enzyme that repairs pyrimidine dimers in bacteria), most pyrimidine dimers in humans can be corrected by nucleotide-excision repair (see Figure 18.28). However, the cells of most people with xeroderma pigmentosum are defective in nucleotide-excision repair, and many of their pyrimidine dimers go uncorrected and may lead to cancer.

Xeroderma pigmentosum can result from defects in several different genes. Some persons with xeroderma pigmentosum have mutations in a gene encoding the protein that recognizes and binds to damaged DNA; others have mutations in a gene encoding helicase. Still others have defects in the genes that play a role in cutting the damaged strand on the 5′ or 3′ sides of the pyrimidine dimer. Some persons have a slightly different form of the disease (xeroderma pigmentosum variant) owing to mutations in the gene encoding polymerase η, a translesion DNA polymerase that bypasses pyrimidine dimers.

Two other genetic diseases due to defects in nucleotide-excision repair are Cockayne syndrome and trichothiodystrophy (also known as brittle-hair syndrome). Persons who have either of these diseases do not have an increased risk of cancer but do exhibit multiple developmental and neurological problems. Both diseases result from mutations in some of the same genes that cause xeroderma pigmentosum. Several of the genes taking part in nucleotide-excision repair produce proteins that also play a role in recombination and the initiation of transcription. These other functions may account for the developmental symptoms seen in Cockayne syndrome and trichothiodystrophy.

Another genetic disease caused by faulty DNA repair is an inherited form of colon cancer called hereditary nonpolyposis colon cancer (HNPCC). This cancer is one of the most common hereditary cancers, accounting for about 15% of colon cancers. Research findings indicate that HNPCC arises from mutations in the proteins that carry out mismatch repair (see Figure 18.25). Some additional genetic diseases associated with defective DNA repair are summarized in **Table 18.5**. TRY PROBLEM 32 →

CONCEPTS

Defects in DNA repair are the underlying cause of several genetic diseases. Many of these diseases are characterized by a predisposition to cancer.

✔ CONCEPT CHECK 9

Why are defects in DNA repair often associated with increases in cancer?

Table 18.5 Genetic diseases associated with defects in DNA-repair systems

Disease	Symptoms	Genetic Defect
Xeroderma pigmentosum	Frecklelike spots on skin, sensitivity to sunlight, predisposition to skin cancer	Defects in nucleotide-excision repair
Cockayne syndrome	Dwarfism, sensitivity to sunlight, premature aging, deafness, mental retardation	Defects in nucleotide-excision repair
Trichothiodystrophy	Brittle hair, skin abnormalities, short stature, immature sexual development, characteristic facial features	Defects in nucleotide-excision repair
Hereditary nonpolyposis colon cancer	Predisposition to colon cancer	Defects in mismatch repair
Fanconi anemia	Increased skin pigmentation, abnormalities of skeleton, heart, and kidneys, predisposition to leukemia	Possibly defects in the repair of interstrand cross-links
Li-Fraumeni syndrome	Predisposition to cancer in many different tissues	Defects in DNA damage response
Werner syndrome	Premature aging, predisposition to cancer	Defect in homologous recombination

CONCEPTS SUMMARY

- Mutations are heritable changes in genetic information. Somatic mutations occur in somatic cells; germ-line mutations occur in cells that give rise to gametes.

- The simplest type of mutation is a base substitution, a change in a single base pair of DNA. Transitions are base substitutions in which purines are replaced by purines or pyrimidines are replaced by pyrimidines. Transversions are base substitutions in which a purine replaces a pyrimidine or a pyrimidine replaces a purine.

- Insertions are the addition of nucleotides, and deletions are the removal of nucleotides; these mutations often change the reading frame of the gene.

- Expanding nucleotide repeats are mutations in which the number of copies of a set of nucleotides increases with the passage of time; they are responsible for several human genetic diseases.

- A missense mutation alters the coding sequence so that one amino acid substitutes for another. A nonsense mutation changes a codon that specifies an amino acid into a termination codon. A silent mutation produces a synonymous codon that specifies the same amino acid as the original sequence, whereas a neutral mutation alters the amino acid sequence but does not change the functioning of the protein. A suppressor mutation reverses the effect of a previous mutation at a different site and may be intragenic (within the same gene as the original mutation) or intergenic (within a different gene).

- Mutation rate is the frequency with which a particular mutation arises in a population. Mutation rates are influenced by both genetic and environmental factors.

- Some mutations occur spontaneously. These mutations include the mispairing of bases in replication and spontaneous depurination and deamination.

- Insertions and deletions can arise from strand slippage in replication or from unequal crossing over.

- Base analogs can become incorporated into DNA in the course of replication and pair with the wrong base in subsequent replication events. Alkylating agents and hydroxylamine modify the chemical structure of bases and lead to mutations. Intercalating agents insert into the DNA molecule and cause single-nucleotide additions and deletions. Oxidative reactions alter the chemical structures of bases.

- Ionizing radiation is mutagenic, altering base structures and breaking phosphodiester bonds. Ultraviolet light produces pyrimidine dimers, which block replication. Bacteria use the SOS response to overcome replication blocks produced by pyrimidine dimers and other lesions in DNA.

- The Ames tests uses bacteria to assess the mutagenic potential of chemical substances.

- Most damage to DNA is corrected by DNA-repair mechanisms. These mechanisms include mismatch repair, direct repair, base-excision repair, nucleotide-excision repair, and other repair pathways. Most require two strands of DNA and exhibit some overlap in the types of damage repaired.

- Double-strand beaks are repaired by homologous recombination and nonhomologous end joining. Special translesion DNA polymerases allow replication to proceed past bulky distortions in the DNA.

- Defects in DNA repair are the underlying cause of several genetic diseases.

IMPORTANT TERMS

mutation (p. 482)
somatic mutation (p. 482)
germ-line mutation (p. 483)
gene mutation (p. 483)
base substitution (p. 483)
transition (p. 483)
transversion (p. 483)
insertion (p. 484)
deletion (p. 484)
frameshift mutation (p. 484)
in-frame insertion (p. 484)
in-frame deletion (p. 484)
expanding nucleotide repeat (p. 484)
forward mutation (p. 485)
reverse mutation (reversion) (p. 485)

missense mutation (p. 485)
nonsense mutation (p. 485)
silent mutation (p. 486)
neutral mutation (p. 486)
loss-of-function mutation (p. 486)
gain-of-function mutation (p. 486)
conditional mutation (p. 486)
lethal mutation (p. 486)
suppressor mutation (p. 486)
intragenic suppressor mutation (p. 487)
intergenic suppressor mutation (p. 487)
mutation rate (p. 490)
adaptive mutation (p. 490)
spontaneous mutation (p. 491)
induced mutation (p. 491)

incorporated error (p. 492)
replication error (p. 492)
strand slippage (p. 493)
unequal crossing over (p. 493)
depurination (p. 493)
deamination (p. 494)
mutagen (p. 495)
base analog (p. 495)
intercalating agent (p. 496)
pyrimidine dimer (p. 497)
SOS system (p. 497)
Ames test (p. 498)
direct repair (p. 502)
base-excision repair (p. 502)
nucleotide-excision repair (p. 503)

ANSWERS TO CONCEPT CHECKS

1. Studying mutations that disrupt normal processes often leads to the identification of genes that normally play a role in the process and can help in understanding the molecular details of a process.

2. c

3. A reverse mutation restores the original phenotype by changing the DNA sequence back to the wild-type sequence. A suppressor mutation restores the phenotype by causing an additional change in the DNA at a site that is different from that of the original mutation.

4. The frequency with which changes arise in DNA, how often these changes are repaired by DNA-repair mechanisms, and our ability to detect the mutation

5. An incorporated error is due to a change that takes place in DNA. This change may be corrected by a DNA-repair pathway.

However, if the error has been replicated, it is permanent and cannot be detected by repair pathways.

6. c

7. d

8. Direct-repair mechanisms return an altered base to its correct structure without removing and replacing nucleotides. Mismatch repair and base-excision repair remove and replace nucleotides.

9. Changes in DNA structure do not undergo repair in people with defects in DNA-repair mechanisms. Consequently, increased numbers of mutations occur at all genes, including those that predispose to cancer. This observation indicates that cancer arises from mutations in DNA.

WORKED PROBLEMS

1. A codon that specifies the amino acid Asp undergoes a single-base substitution that yields a codon that specifies Ala. Refer to the genetic code in Figure 15.10 and give all possible DNA sequences for the original and the mutated codon. Is the mutation a transition or a transversion?

• **Solution**

There are two possible RNA codons for Asp: GAU and GAC. The DNA sequences that encode these codons will be complementary to the RNA codons: CTA and CTG. There are four possible RNA codons for Ala: GCU, GCC, GCA, and GCG, which correspond to DNA sequences CGA, CGG, CGT, and CGC. If we organize the original and the mutated sequences as shown in the following table, the type of mutations that may have occurred can be easily seen:

Possible original sequence for Asp	Possible mutated sequence for Ala
CTA	CGA
CTG	CGG
	CGT
	CGC

If the mutation is confined to a single-base substitution, then the only mutations possible are that CTA mutated to CGA or that CTG mutated to CGG. In both, there is a T→G transversion in the middle nucleotide of the codon.

2. The mutations produced by the following compounds are reversed by the substances shown. What conclusions can you make about the nature of the mutations originally produced by these compounds?

Mutations produced by compound	Reversed by			
	5-Bromouracil	EMS	Hydroxyl amine	Acridine orange
a. 1	Yes	Yes	No	No
b. 2	Yes	Yes	Some	No
c. 3	No	No	No	Yes
d. 4	Yes	Yes	Yes	Yes

• Solution

The ability of various compounds to produce reverse mutations reveals important information about the nature of the original mutation.

a. Mutations produced by compound 1 are reversed by 5-bromouracil, which produces both A·T→G·C and G·C→A·T transitions, which tells us that compound 1 produces single-base substitutions that may include the generation of either A·T or G·C pairs. The mutations produced by compound 1 are also reversed by EMS, which, like 5-bromouracil, produces both A·T→G·C and G·C→A·T transitions; so no additional information is provided here. Hydroxylamine does not reverse the mutations produced by compound 1. Because hydroxylamine produces only C·G→T·A transitions, we know

that compound 1 does not generate C·G base pairs. Acridine orange, an intercalating agent that produces frameshift mutations, also does not reverse the mutations, revealing that compound 1 produces only single-base-pair substitutions, not insertions or deletions. In summary, compound 1 appears to cause single-base substitutions that generate T·A but not G·C base pairs.

b. Compound 2 generates mutations that are reversed by 5-bromouracil and EMS, indicating that it may produce G·C or A·T base pairs. Some of these mutations are reversed by hydroxylamine, which produces only C·G→T·A transitions, indicating that some of the mutations produced by compound 2 are C·G base pairs. None of the mutations are reversed by acridine orange; so compound 2 does not induce insertions or deletions. In summary, compound 2 produces single-base substitutions that generate both G·C and A·T base pairs.

c. Compound 3 produces mutations that are reversed only by acridine orange; so compound 3 appears to produce only insertions and deletions.

d. Compound 4 is reversed by 5 bromouracil, EMS, hydroxylamine, and acridine orange, indicating that this compound produces single-base substitutions, which include G·C and A·T base pairs, insertions, and deletions.

COMPREHENSION QUESTIONS

Section 18.1

*1. What is the difference between somatic mutations and germ-line mutations?

*2. What is the difference between a transition and a transversion? Which type of base substitution is usually more common?

*3. Briefly describe expanding nucleotide repeats. How do they account for the phenomenon of anticipation?

4. What is the difference between a missense mutation and a nonsense mutation? Between a silent mutation and a neutral mutation?

5. Briefly describe two different ways that intragenic suppressors may reverse the effects of mutations.

*6. How do intergenic suppressors work?

Section 18.2

*7. What causes errors in DNA replication?

8. How do insertions and deletions arise?

*9. How do base analogs lead to mutations?

10. How do alkylating agents, nitrous acid, and hydroxylamine produce mutations?

11. What types of mutations are produced by ionizing and UV radiation?

*12. What is the SOS system and how does it lead to an increase in mutations?

Section 18.3

13. What is the purpose of the Ames test? How are *his⁻* bacteria used in this test?

Section 18.4

*14. List at least three different types of DNA repair and briefly explain how each is carried out.

15. What features do mismatch repair, base-excision repair, and nucleotide-excision repair have in common?

16. What are two major mechanisms for the repair of double-strand breaks? How do they differ?

APPLICATION QUESTIONS AND PROBLEMS

Section 18.1

*17. A codon that specifies the amino acid Gly undergoes a single-base substitution to become a nonsense mutation. In accord with the genetic code given in Figure 15.10, is this mutation a transition or a transversion? At which position of the codon does the mutation occur?

*18a. If a single transition occurs in a codon that specifies Phe, what amino acids can be specified by the mutated sequence?

b. If a single transversion occurs in a codon that specifies Phe, what amino acids can be specified by the mutated sequence?

c. If a single transition occurs in a codon that specifies Leu, what amino acids can be specified by the mutated sequence?

d. If a single transversion occurs in a codon that specifies Leu, what amino acids can be specified by the mutated sequence?

19. Hemoglobin is a complex protein that contains four polypeptide chains. The normal hemoglobin found in adults—called adult hemoglobin—consists of two alpha and two beta polypeptide chains, which are encoded by different loci. Sickle-cell hemoglobin, which causes sickle-cell anemia, arises from a mutation in the beta chain of adult hemoglobin. Adult hemoglobin and sickle-cell hemoglobin differ in a single amino acid: the sixth amino acid from one end in adult hemoglobin is glutamic acid, whereas sickle-cell hemoglobin has valine at this position. After consulting the genetic code provided in Figure 15.10, indicate the type and location of the mutation that gave rise to sickle-cell anemia.

*20. The following nucleotide sequence is found on the template strand of DNA. First, determine the amino acids of the protein encoded by this sequence by using the genetic code provided in Figure 15.10. Then, give the altered amino acid sequence of the protein that will be found in each of the following mutations:

> Sequence
> of DNA
> template
>   ↳ 3'–TAC TGG CCG TTA GTT GAT ATA ACT–5'
>   ↱ 1   24
> Nucleotide
> number

a. Mutant 1:   A transition at nucleotide 11
b. Mutant 2:   A transition at nucleotide 13
c. Mutant 3:   A one-nucleotide deletion at nucleotide 7
d. Mutant 4:   A T→A transversion at nucleotide 15
e. Mutant 5:   An addition of TGG after nucleotide 6
f. Mutant 6:   A transition at nucleotide 9

21. A polypeptide has the following amino acid sequence:

> Met-Ser-Pro-Arg-Leu-Glu-Gly

The amino acid sequence of this polypeptide was determined in a series of mutants listed in parts *a* through *e*. For each mutant, indicate the type of mutation that occurred in the DNA (single-base substitution, insertion, deletion) and the phenotypic effect of the mutation (nonsense mutation, missense mutation, frameshift, etc.).

a. Mutant 1:   Met-Ser-Ser-Arg-Leu-Glu-Gly
b. Mutant 2:   Met-Ser-Pro
c. Mutant 3:   Met-Ser-Pro-Asp-Trp-Arg-Asp-Lys
d. Mutant 4:   Met-Ser-Pro-Glu-Gly
e. Mutant 5:   Met-Ser-Pro-Arg-Leu-Leu-Glu-Gly

*22. A gene encodes a protein with the following amino acid sequence:

> Met-Trp-His-Arg-Ala-Ser-Phe

A mutation occurs in the gene. The mutant protein has the following amino acid sequence:

> Met-Trp-His-Ser-Ala-Ser-Phe

An intragenic suppressor restores the amino acid sequence to that of the original protein:

> Met-Trp-His-Arg-Ala-Ser-Phe

Give at least one example of base changes that could produce the original mutation and the intragenic suppressor. (Consult the genetic code in Figure 15.10.)

23. A gene encodes a protein with the following amino acid sequence:

> Met-Lys-Ser-Pro-Ala-Thr-Pro

A nonsense mutation from a single-base-pair substitution occurs in this gene, resulting in a protein with the amino acid sequence Met-Lys. An intergenic suppressor mutation allows the gene to produce the full-length protein. With the original mutation and the intergenic suppressor present, the gene now produces a protein with the following amino acid sequence:

> Met-Lys-Cys-Pro-Ala-Thr-Pro

Give the location and nature of the original mutation and of the intergenic suppressor.

24. XG syndrome is a rare genetic disease that is due to an autosomal dominant gene. A complete census of a small European country reveals that 77,536 babies were born in 2004, of whom 3 had XG syndrome. In the same year, this country had a population of 5,964,321 people, and there were 35 living persons with XG syndrome. What is the mutation rate of XG syndrome in this country?

Section 18.2

*25. Can nonsense mutations be reversed by hydroxylamine? Why or why not?

*26. The following nucleotide sequence is found in a short stretch of DNA:

> 5'–ATGT–3'
> 3'–TACA–5'

If this sequence is treated with hydroxylamine, what sequences will result after replication?

27. The following nucleotide sequence is found in a short stretch of DNA:

> 5'–AG–3'
> 3'–TC–5'

a. Give all the mutant sequences that may result from spontaneous depurination in this stretch of DNA.

b. Give all the mutant sequences that may result from spontaneous deamination in this stretch of DNA.

28. In many eukaryotic organisms, a significant proportion of cytosine bases are naturally methylated to 5-methylcytosine. Through evolutionary time, the proportion of AT base pairs in the DNA of these organisms increases. Can you suggest a possible mechanism for this increase?

Section 18.3

*29. A chemist synthesizes four new chemical compounds in the laboratory and names them PFI1, PFI2, PFI3, and PFI4. He gives the PFI compounds to a geneticist friend and asks her to determine their mutagenic potential. The geneticist finds that all four are highly mutagenic. She also tests the capacity of mutations produced by the PFI compounds to be reversed by other known mutagens and obtains the following results. What conclusions can you make about the nature of the mutations produced by these compounds?

| | Reversed by | | | |
Mutations produced by	2-Amino-purine	Nitrous acid	Hydroxyl-amine	Acridine orange
PFI1	Yes	Yes	Some	No
PFI2	No	No	No	No
PFI3	Yes	Yes	No	No
PFI4	No	No	No	Yes

30. Mary Alexander studied the effects of radiation on mutation rates in the sperm of *Drosophila melanogaster*. She irradiated *Drosophila* larvae with either 3000 roentgens (r) or 3975 r, collected the adult males that developed from irradiated larvae, mated them with unirradiated females, and then counted the number of mutant F₁ flies produced by each male. All mutant flies that appeared were used in subsequent crosses to determine if their mutant phenotypes were genetic. She obtained the following results (M. L. Alexander. 1954. *Genetics* 39:409–428):

Group	Number of offspring	Offspring with a genetic mutation
Control (0 r)	45,504	0
Irradiated (3000 r)	49,512	71
Irradiated (3975 r)	50,159	70

a. Calculate the mutation rates of the control group and the two groups of irradiated flies.

b. On the basis of these data, do you think radiation has any effect on mutation? Explain your answer.

31. A genetics instructor designs a laboratory experiment to study the effects of UV radiation on mutation in bacteria. In the experiment, the students expose bacteria plated on petri plates to UV light for different lengths of time, place the plates in an incubator for 48 hours, and then count the number of colonies that appear on each plate. The plates that have received more UV radiation should have more pyrimidine dimers, which block replication; thus, fewer colonies should appear on the plates exposed to UV light for longer periods of time. Before the students carry out the experiment, the instructor warns them that, while the bacteria are in the incubator, the students must not open the incubator door unless the room is darkened. Why should the bacteria not be exposed to light?

Section 18.4

*32. A plant breeder wants to isolate mutants in tomatoes that are defective in DNA repair. However, this breeder does not have the expertise or equipment to study enzymes in DNA-repair systems. How can the breeder identify tomato plants that are deficient in DNA repair? What are the traits to look for?

CHALLENGE QUESTIONS

Section 18.1

33. Robert Bost and Richard Cribbs studied a strain of *E. coli* (*araB14*) that possessed a nonsense mutation in the structural gene that encodes L-ribulokinase, an enzyme that allows the bacteria to metabolize the sugar arabinose (R. Bost and R. Cribbs. 1969. *Genetics* 62:1–8). From the *araB14* strain, they isolated some bacteria that possessed mutations that caused them to revert back to the wild type. Genetic analysis of these revertants showed that they possessed two different suppressor mutations. One suppressor mutation (*R1*) was linked to the original mutation in the L-ribulokinase and probably occurred at the same locus. By itself, this mutation allowed the production of L-ribulokinase, but the enzyme was not as effective in metabolizing arabinose as the enzyme encoded by the wild-type allele. The second suppressor mutation (*Su^B*) was not linked to the original mutation. In conjunction with the *R1* mutation, *Su^B* allowed the production of L-ribulokinase, but *Su^B* by itself was not able to suppress the original mutation.

a. On the basis of this information, are the *R1* and *Su^B* mutations intragenic suppressors or intergenic suppressors? Explain your reasoning.

b. Propose an explanation for how *R1* and *Su^B* restore the ability of *araB14* to metabolize arabinose and why *Su^B* is able to more fully restore the ability to metabolize arabinose.

34. Achondroplasia is an autosomal dominant disorder characterized by disproportionate short stature: the legs and arms are short compared with the head and trunk. The disorder is due to a base substitution in the gene, located on the short arm of chromosome 4, for fibroblast growth factor receptor 3 (FGFR3).

Although achondroplasia is clearly inherited as an autosomal dominant trait, more than 80% of the people who have achondroplasia are born to parents with normal stature. This high percentage indicates that most cases are caused by newly arising mutations; these cases (not inherited from an affected parent) are referred to as sporadic. Findings from molecular studies have demonstrated that sporadic cases of achondroplasia are almost always caused by mutations inherited from the father (paternal mutations). In addition, the occurrence of achondroplasia is higher among older fathers; indeed, approximately 50% of children with achondroplasia are born to fathers older than 35 years of age. There is no association with maternal age. The mutation rate for achondroplasia (about 4×10^{-5} mutations per gamete) is high compared with those for other genetic disorders. Explain why most spontaneous mutations for achondroplasia are paternal in origin and why the occurrence of achondroplasia is higher among older fathers.

A family of three who have achondroplasia. [Gail Burton/AP.]

35. Tay–Sachs disease is a severe, autosomal recessive genetic disease that produces deafness, blindness, seizures, and, eventually, death. The disease results from a defect in the *HEXA* gene, which encodes hexosaminidase A. This enzyme normally degrades G_{M2} gangliosides. In the absence of hexosaminidase A, G_{M2} gangliosides accumulate in the brain. The results of recent molecular studies showed that the most common mutation causing Tay–Sachs disease is a 4-bp insertion that produces a downstream premature stop codon. Results of further studies have revealed that the transcription of the *HEXA* gene is normal in people who have Tay–Sachs disease, but the *HEXA* mRNA is unstable. Propose a mechanism to account for how a premature stop codon could cause mRNA instability.

36. Mutations *ochre* and *amber* are two types of nonsense mutations. Before the genetic code was worked out, Sydney Brenner, Anthony O. Stretton, and Samuel Kaplan applied different types of mutagens to bacteriophages in an attempt to determine the bases present in the codons responsible for *amber* and *ochre* mutations. They knew that *ochre* and *amber* mutants were suppressed by different types of mutations, demonstrating that each is a different termination codon. They obtained the following results:

(1) A single-base substitution could convert an *ochre* mutation into an *amber* mutation.

(2) Hydroxylamine induced both *ochre* and *amber* mutations in wild-type phages.

(3) 2-Aminopurine caused *ochre* to mutate to *amber*.

(4) Hydroxylamine did not cause *ochre* to mutate to *amber*.

These data do not allow the complete nucleotide sequence of the *amber* and *ochre* codons to be worked out, but they do provide some information about the bases found in the nonsense mutations.

a. What conclusions about the bases found in the codons of *amber* and *ochre* mutations can be made from these observations?

b. Of the three nonsense codons (UAA, UAG, UGA), which represents the *ochre* mutation?

Section 18.3

37. To determine whether radiation associated with the atomic bombings of Hiroshima and Nagasaki produced recessive germ-line mutations, scientists examined the sex ratio of the children of the survivors of the blasts. Can you explain why an increase in germ-line mutations might be expected to alter the sex ratio?

Section 18.4

38. Trichothiodystrophy is a human inherited disorder characterized by premature aging, including osteoporosis, osteosclerosis, early graying, infertility, and reduced life span. The results of recent studies showed that the mutation that causes this disorder occurs in a gene that encodes a DNA helicase. Propose a mechanism for how a mutation in a DNA helicase might cause premature aging. Be sure to relate the symptoms of the disorder to possible functions of the helicase enzyme.

19 Molecular Genetic Analysis and Biotechnology

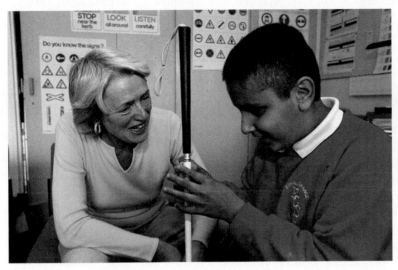

Blindness affects 45 million people throughout the world. Genetic engineering is now being used to treat patients with Leber congenital amaurosis, a genetic form of blindness. [Paul Doyle/Alamy Images.]

HELPING THE BLIND TO SEE

Genetic engineering, the manipulation and transfer of genes from one organism to another, has revolutionized the study of genetics and has been used successfully to create a plethora of new products, including bacteria that generate chemicals, pest-resistant crops, and farm animals that secrete pharmaceutical products in their milk. In perhaps its ultimate application, genetic engineering is being used to treat disease in humans, a process known as gene therapy. In 2007, researchers transferred genes to four blind people, partly restoring their sight—a dramatic example of gene therapy.

Eyesight is the most precious of human senses, enabling us to read, navigate physical obstacles, recognize friends, and enjoy the stunning visual beauty of the natural world. Tragically, millions of people are blind or unable to see well. Eyesight often deteriorates with age; indeed, most blindness is found among the elderly. However, some children are born without sight, and others lose their eyesight at an early age. Research suggests that heredity is responsible for about half of the cases of blindness before the age of 45. Because of the complexity of the eye, its associated nerves, and those parts of the brain taking part in visual perception, defects in a large number of genes may lead to blindness.

Gene therapy was applied to a rare genetic form of blindness known as Leber congenital amaurosis (LCA), in which there is a mutation in one of a number of genes responsible for the development or function of light receptors in the retina. LCA is inherited as an autosomal recessive disorder; in people with two defective copies of a gene, the light-sensing cells die. Children with LCA begin losing sight at birth, and most are completely blind by age 40.

One type of LCA is caused by a defect in the *RPE65* gene, which encodes an enzyme that helps convert vitamin A into rhodopsin, a pigment that absorbs light. Without the enzyme, rhodopsin is not produced and the photoreceptor cells atrophy with the passage of time. In 1998, researchers discovered that some Swedish Briard dogs also have a defect in *RPE65* and become blind. Research on these dogs led to a better understanding of how defects in *RPE65* cause blindness and suggested a possible treatment—gene therapy. In 2001, researchers treated three young dogs with gene therapy. The dogs' vision improved enough that they could avoid objects and negotiate a maze.

In 2007, researchers in Pennsylvania and London used gene therapy to treat four young adults who have LCA. Each patient was injected with a genetically modified virus carrying a functional copy of *RPE65*. The results were dramatic: all the patients showed significant improvement in visual perception. Some who had formerly been able to detect only hand motions were able to read several lines on an eye chart. One patient who had not been able to negotiate an obstacle course was, after treatment, able to make his way through it. All four patients are still legally blind, but the results suggest that gene therapy may be an effective treatment for LCA. Researchers predict that even more dramatic results may be obtained in younger patients who have not yet lost as much of their vision.

This chapter introduces some of the techniques used in genetic engineering. We begin by considering molecular genetic technology and some of its effects. We then examine a number of methods used to isolate, study, alter, and recombine DNA sequences and place them back into cells. Finally, we explore some of the applications of molecular genetic analysis.

19.1 Techniques of Molecular Genetics Have Revolutionized Biology

In 1973, a group of scientists produced the first organisms with recombinant DNA molecules. Stanley Cohen at Stanford University and Herbert Boyer at the University of California School of Medicine at San Francisco and their colleagues inserted a piece of DNA from one plasmid into another, creating an entirely new, recombinant DNA molecule. They then introduced the recombinant plasmid into *E. coli* cells. These experiments ushered in one of the most momentous revolutions in the history of science.

Recombinant DNA technology is a set of molecular techniques for locating, isolating, altering, and studying DNA segments. The term *recombinant* is used because, frequently, the goal is to combine DNA from two distinct sources. Genes from two different bacteria might be joined, for example, or a human gene might be inserted into a viral chromosome. Commonly called **genetic engineering**, recombinant DNA technology now encompasses many molecular techniques that can be used to analyze, alter, and recombine virtually any DNA sequences from any number of sources.

The Molecular Genetics Revolution

The techniques of recombinant DNA technology are just a part of a vast array of molecular methods that are now available for the study of genetics. These molecular techniques have drastically altered the way that genes are studied. Previously, information about the structure and organization of genes was gained by examining their phenotypic effects, but molecular genetic analysis allows the nucleotide sequences themselves to be read. Methods in molecular genetics have provided new information about the structure and function of genes and have altered many fundamental concepts of genetics. Our detailed understanding of genetic processes

such as replication, transcription, translation, RNA processing, and gene regulation has been obtained through the use of molecular genetic techniques. These techniques are used in many other fields as well, including biochemistry, microbiology, developmental biology, neurobiology, evolution, and ecology.

Recombinant DNA technology and other molecular techniques are also being used to create a number of commercial products, including drugs, hormones, enzymes, and crops (**Figure 19.1**). A complete industry—**biotechnology**—has grown up around the use of these techniques to develop new products. In medicine, molecular genetics is being used to probe the nature of cancer, diagnose genetic and infectious diseases, produce drugs, and treat hereditary disorders.

> **CONCEPTS**
>
> Molecular genetics and recombinant DNA technology are used to locate, analyze, alter, study, and recombine DNA sequences. These techniques are used to probe the structure and function of genes, address questions in many areas of biology, create commercial products, and diagnose and treat diseases.

Working at the Molecular Level

The manipulation of genes at the molecular level presents a serious challenge, often requiring strategies that may not, at first, seem obvious. The basic problem is that genes are minute and every cell contains thousands of them. Individual nucleotides cannot be seen, and no physical features mark the beginning or the end of a gene.

Let's consider a typical situation faced by a molecular geneticist. Suppose we wanted to use bacteria to produce large quantities of a human protein. The first and most formidable problem is to find the gene that encodes the desired protein. A haploid human genome consists of 3.2 billion base pairs of DNA. Let's assume that the gene that we want to isolate is 3000 bp long. Our target gene occupies only one-millionth of the genome; so searching for our gene in the huge expanse of genomic DNA is more difficult than looking for a needle in the proverbial haystack. But, even if we are able to locate the gene, how are we to separate it from the rest of the DNA?

19.1 Recombinant DNA technology has been used to create genetically modified crops. Genetically engineered corn, now constitutes 85% of all corn grown in the United States. [Chris Knapton/Photo Researchers.]

If we did succeed in locating and isolating the desired gene, we would next need to insert it into a bacterial cell. Linear fragments of DNA are quickly degraded by bacteria; so the gene must be inserted in a stable form. It must also be able to successfully replicate or it will not be passed on when the cell divides. If we succeed in transferring our gene to bacteria in a stable form, we must still ensure that the gene is properly transcribed and translated.

Finally, the methods used to isolate and transfer genes are inefficient and, of a million cells that are subjected to these procedures, only *one* cell might successfully take up and express the human gene. So we must search through many bacterial cells to find the one containing the recombinant DNA. We are back to the problem of the needle in the haystack.

Although these problems might seem insurmountable, molecular techniques have been developed to overcome all of them. Human genes are routinely transferred to bacterial cells, where the genes are expressed.

CONCEPTS

Molecular genetic analyses require special methods because individual genes make up a tiny fraction of the cellular DNA and they cannot be seen.

19.2 Molecular Techniques Are Used to Isolate, Recombine, and Amplify Genes

A first step in the molecular analysis of a DNA segment or gene is to isolate it from the remainder of the DNA and to make many copies of it so that further analyses can be car-

ried out. The isolation and amplification of DNA frequently requires that it be recombined with other DNA molecules. In the sections that follow, we will examine some of the molecular techniques that are used to isolate, recombine, and amplify DNA segments.

Cutting and Joining DNA Fragments

A key event in the development of molecular genetic methods was the discovery in the late 1960s of **restriction enzymes** (also called **restriction endonucleases**) that recognize and make double-stranded cuts in DNA at specific nucleotide sequences. These enzymes are produced naturally by bacteria, where they are used in defense against viruses. A bacterium protects its own DNA from a restriction enzyme by modifying the recognition sequence, usually by adding methyl groups to its DNA.

Several different types of restriction enzymes have been isolated from bacteria. Type II restriction enzymes recognize specific sequences and cut the DNA at defined sites within or near the recognition sequence. Virtually all molecular genetics work is done with type II restriction enzymes; discussions of restriction enzymes throughout this book refer to type II enzymes.

More than 800 different restriction enzymes that recognize and cut DNA at more than 100 different sequences have been isolated from bacteria. Many of these enzymes are commercially available; examples of some commonly used restriction enzymes are given in **Table 19.1**. The name of each restriction enzyme begins with an abbreviation that signifies its bacterial origin.

The sequences recognized by restriction enzymes are usually from 4 to 8 bp long; most enzymes recognize a sequence of 4 or 6 bp. Most recognition sequences are palindromic—sequences that read the same forward and backward.

Some of the enzymes make staggered cuts in the DNA. For example, *Hin*dIII recognizes the following sequence:

$$\downarrow$$
$$5'\text{--AAGCTT--}3'$$
$$3'\text{--TTCGAA--}5'$$
$$\uparrow$$

*Hin*dIII cuts the sugar–phosphate backbone of each strand at the point indicated by the arrow, generating fragments with short, single-stranded overhanging ends:

$$5'\text{--A} \qquad \text{AGCTT--}3'$$
$$3'\text{--TTCGA} \qquad \text{A--}5'$$

Such ends are called **cohesive ends** or sticky ends because they are complementary to each other and can spontaneously pair to connect the fragments. Thus, DNA fragments can be "glued" together: any two fragments cleaved by the

Table 19.1	Characteristics of some common type II restriction enzymes used in recombinant DNA technology		
Enzyme	**Microorganism from Which Enzyme Is Produced**	**Recognition Sequence**	**Type of Fragment End Produced**
BamHI	Bacillus amyloliquefaciens	↓ 5′–GGATCC–3′ 3′–CCTAGG–3′ ↑	Cohesive
CofI	Clostridium formicoaceticum	↓ 5′–GCGC–3′ 3′–CGCG–5′ ↑	Cohesive
EcoRI	Escherichia coli	↓ 5′–GAATTC–3′ 3′–CTTAAG–5′ ↑	Cohesive
EcoRII	Escherichia coli	↓ 5′–CCAGG–3′ 3′–GGTCC–5′ ↑	Cohesive
HaeIII	Haemophilus aegyptius	↓ 5′–GGCC–3′ 3′–CCGG–5′ ↑	Blunt
HindIII	Haemophilus influenzae	↓ 5′–AAGCTT–3′ 3′–TTCGAA–5′ ↑	Cohesive
PvuII	Proteus vulgaris	↓ 5′–CAGCTG–3′ 3′–GTCGAC–5′ ↑	Blunt

Note: The first three letters of the abbreviation for each restriction enzyme refer to the bacterial species from which the enzyme was isolated (e.g., Eco refers to E. coli). A fourth letter may refer to the strain of bacteria from which the enzyme was isolated (the "R" in EcoRI indicates that this enzyme was isolated from the RY13 strain of E. coli). Roman numerals that follow the letters allow different enzymes from the same species to be identified.

same enzyme will have complementary ends and will pair (**Figure 19.2**). When their cohesive ends have paired, two DNA fragments can be joined together permanently by DNA ligase, which seals nicks between the sugar–phosphate groups of the fragments.

Not all restriction enzymes produce staggered cuts and sticky ends. PvuII cuts in the middle of its recognition site, producing blunt-ended fragments:

$$\downarrow$$
5′–CAGCTG–3′
3′–GTCGAC–5′
$$\uparrow$$

$$\downarrow$$

5′–CAG   CTG–3′
3′–GTC   GAC–5′

Fragments that have blunt ends are usually joined together in other ways.

The sequences recognized by a restriction enzyme are located randomly within the genome. Accordingly, there is a relation between the length of the recognition sequence and the number of times that it is present in a genome: there will be fewer longer recognition sequences than shorter recognition sequences because the probability of the occurrence of a particular sequence consisting of, say, six specific bases is less than the probability of the occurrence of a particular sequence of four specific bases. Consequently, restriction enzymes that recognize longer sequences will cut a given piece of DNA into fewer and longer fragments than will restriction enzymes that recognize shorter sequences.

Restriction enzymes are used whenever DNA fragments must be cut or joined. In a typical restriction reaction, a

(a)

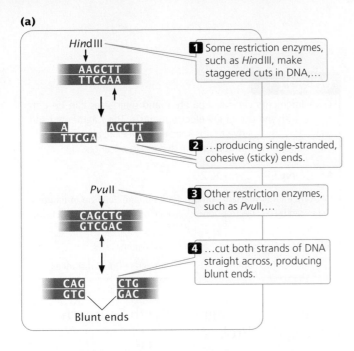

1. Some restriction enzymes, such as *Hind*III, make staggered cuts in DNA,...

2. ...producing single-stranded, cohesive (sticky) ends.

3. Other restriction enzymes, such as *Pvu*II,...

4. ...cut both strands of DNA straight across, producing blunt ends.

Blunt ends

(b)

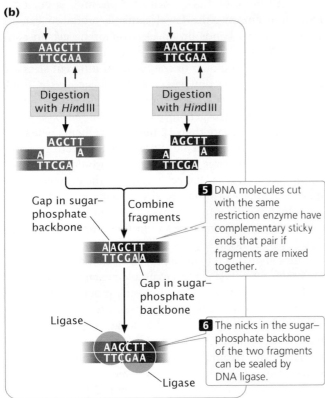

Digestion with *Hind*III

Digestion with *Hind*III

Gap in sugar–phosphate backbone

Combine fragments

5. DNA molecules cut with the same restriction enzyme have complementary sticky ends that pair if fragments are mixed together.

Gap in sugar–phosphate backbone

Ligase

6. The nicks in the sugar–phosphate backbone of the two fragments can be sealed by DNA ligase.

Ligase

19.2 Restriction enzymes make double-stranded cuts in DNA, producing cohesive, or sticky, ends.

concentrated solution of purified DNA is placed in a small tube with a buffer solution and a small amount of restriction enzyme. The reaction mixture is then heated at the optimal temperature for the enzyme, usually 37°C. Within a few hours, the enzyme cuts all the appropriate restriction sites in the DNA, producing a set of DNA fragments.

TRY PROBLEM 29 →

CONCEPTS

Type II restriction enzymes cut DNA at specific base sequences that are palindromic. Some restriction enzymes make staggered cuts, producing DNA fragments with cohesive ends; others cut both strands straight across, producing blunt-ended fragments. There are fewer long recognition sequences in DNA than short sequences.

✔ **CONCEPT CHECK 1**

Where do restriction enzymes come from?

Viewing DNA Fragments

After the completion of a restriction reaction, a number of questions arise. Did the restriction enzyme cut the DNA? Into how many fragments was the DNA cut? What are the sizes of the resulting fragments? Gel electrophoresis provides us with a means of answering these questions.

Electrophoresis is a standard biochemical technique for separating molecules on the basis of their size and electrical charge. There are a number of different types of electrophoresis; to separate DNA molecules, **gel electrophoresis** is used. A porous gel is often made from agarose (a polysaccharide isolated from seaweed), which is melted in a buffer solution and poured into a plastic mold. As it cools, the agarose solidifies, making a gel that looks something like stiff gelatin.

Small wells are made at one end of the gel to hold solutions of DNA fragments (**Figure 19.3a**), and an electrical current is passed through the gel. Because the phosphate group of each DNA nucleotide carries a negative charge, the DNA fragments migrate toward the positive end of the gel. In this migration, the porous gel acts as a sieve, separating the DNA fragments by size. Small DNA fragments migrate more rapidly than do large ones and, with the passage of time, the fragments separate on the basis of their size. Typically, DNA fragments of known length (size standards) are placed in another well. By comparing the migration distance of the unknown fragments with the distance traveled by the size standards, a researcher can determine the approximate size of the unknown fragments (**Figure 19.3b**).

The DNA fragments are still too small to see; so the problem of visualizing the DNA needs to be addressed. Visualization can be accomplished in several ways. The simplest procedure is to stain the gel with a dye specific for nucleic acids, such as ethidium bromide, which wedges itself tightly (intercalates) between the bases of DNA and fluoresces orange when exposed to UV light, producing brilliant orange bands on the gel (**Figure 19.3c**).

Alternatively, DNA fragments can be visualized by adding a radioactive or chemical label to the DNA before it is placed in the gel. Nucleotides with radioactively labeled phosphate

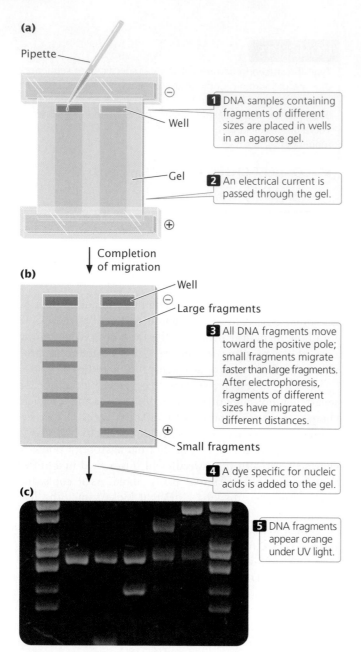

(a)

Pipette

Well

Gel

1 DNA samples containing fragments of different sizes are placed in wells in an agarose gel.

2 An electrical current is passed through the gel.

Completion of migration

(b)

Well

Large fragments

3 All DNA fragments move toward the positive pole; small fragments migrate faster than large fragments. After electrophoresis, fragments of different sizes have migrated different distances.

Small fragments

4 A dye specific for nucleic acids is added to the gel.

(c)

5 DNA fragments appear orange under UV light.

19.3 Gel electrophoresis can be used to separate DNA molecules on the basis of their size and electrical charge.
[Photograph courtesy of Carol Eng.]

(^{32}P) can be used as the substrate for DNA synthesis and will be incorporated into the newly synthesized DNA strand. Radioactively labeled DNA can be detected with a technique called **autoradiography** in which a piece of X-ray film is placed on top of the gel. Radiation from the labeled DNA exposes the film, just as light exposes photographic film in a camera. The developed autoradiograph gives a picture of the fragments in the gel, with each DNA fragment appearing as a dark band on the film. Chemical labels can be detected by adding antibodies or other substances that carry a dye and will attach to the relevant DNA, which can be visualized directly. **TRY PROBLEM 30**

Locating DNA Fragments with Southern Blotting and Probes

If a small piece of DNA, such as a plasmid, is cut by a restriction enzyme, the few fragments produced can be seen as distinct bands on an electrophoretic gel. In contrast, if genomic DNA from a cell is cut by a restriction enzyme, a large number of fragments of different sizes are produced. A restriction enzyme that recognizes a four-base sequence would theoretically cut about once every 256 bp. The human genome, with 3.2 billion base pairs, would generate more than 12 million fragments when cut by this restriction enzyme. Separated by electrophoresis and stained, this large set of fragments would appear as a continuous smear on the gel because of the presence of so many fragments of differing size. Usually, researchers are interested in only a few of these fragments, perhaps those carrying a specific gene. How do they locate the desired fragments in such a large pool of DNA?

One approach is to use a **probe**, which is a DNA or RNA molecule with a base sequence complementary to a sequence in the gene of interest. The bases on a probe will pair only with the bases on a complementary sequence and, if suitably labeled, the probe can be used to locate a specific gene or other DNA sequence. To use a probe, a researcher first cuts the DNA into fragments by using one or more restriction enzymes and then separates the fragments with gel electrophoresis (**Figure 19.4**). Next, the separated fragments must be denatured and transferred to a permanent solid medium (such as nitrocellulose or nylon membrane). **Southern blotting** (named after Edwin M. Southern) is one technique used to transfer the denatured, single-stranded fragments from a gel to a permanent solid medium.

After the single-stranded DNA fragments have been transferred, the membrane is placed in a hybridization solution of a radioactively or chemically labeled probe (see Figure 19.4). The probe will bind to any DNA fragments on the membrane that bear complementary sequences. Often, a probe binds to only a part of the DNA fragment, and so

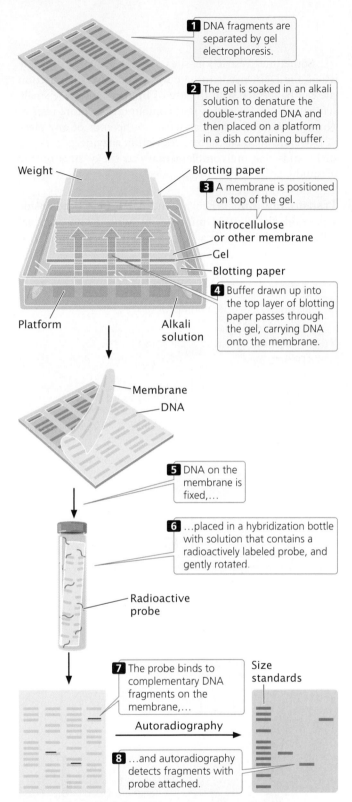

1 DNA fragments are separated by gel electrophoresis.

2 The gel is soaked in an alkali solution to denature the double-stranded DNA and then placed on a platform in a dish containing buffer.

Weight

Blotting paper

3 A membrane is positioned on top of the gel.

Nitrocellulose or other membrane

Gel

Blotting paper

4 Buffer drawn up into the top layer of blotting paper passes through the gel, carrying DNA onto the membrane.

Platform

Alkali solution

Membrane

DNA

5 DNA on the membrane is fixed,…

6 …placed in a hybridization bottle with solution that contains a radioactively labeled probe, and gently rotated.

Radioactive probe

7 The probe binds to complementary DNA fragments on the membrane,…

Size standards

Autoradiography

8 …and autoradiography detects fragments with probe attached.

19.4 Southern blotting and hybridization with probes can locate a few specific fragments in a large pool of DNA.

the DNA fragment may contain sequences not found in the probe. The membrane is then washed to remove any unbound probe; bound probe is detected by autoradiography or another method for chemically labeled probes.

RNA can be transferred from a gel to a solid support by a related procedure called **Northern blotting** (not named after anyone but capitalized to match Southern). The hybridization of a probe can reveal the size of a particular mRNA molecule, its relative abundance, or the tissues in which the mRNA is transcribed. **Western blotting** is the transfer of protein from a gel to a membrane. Here, the probe is usually an antibody, used to determine the size of a particular protein and the pattern of the protein's expression.

CONCEPTS

Labeled probes, which are sequences of RNA or DNA that are complementary to the sequence of interest, can be used to locate individual genes or DNA sequences. Southern blotting can be used to transfer DNA fragments from a gel to a membrane such as nitrocellulose.

✔ CONCEPT CHECK 3

How do Northern and Western blotting differ from Southern blotting?

Cloning Genes

Many recombinant DNA methods require numerous copies of a specific DNA fragment. One way to amplify a specific piece of DNA is to place the fragment in a bacterial cell and allow the cell to replicate the DNA. This procedure is termed **gene cloning** because identical copies (clones) of the original piece of DNA are produced.

A **cloning vector** is a stable, replicating DNA molecule to which a foreign DNA fragment can be attached for introduction into a cell. An effective cloning vector has three important characteristics (**Figure 19.5**): (1) an origin of replication, which ensures that the vector is replicated within

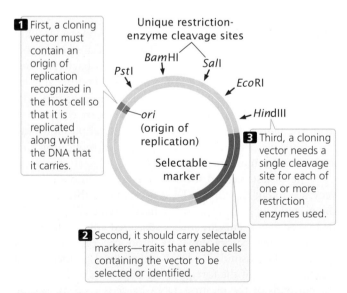

1 First, a cloning vector must contain an origin of replication recognized in the host cell so that it is replicated along with the DNA that it carries.

Unique restriction-enzyme cleavage sites

*Bam*HI

*Sal*I

*Pst*I

*Eco*RI

ori (origin of replication)

*Hind*III

3 Third, a cloning vector needs a single cleavage site for each of one or more restriction enzymes used.

Selectable marker

2 Second, it should carry selectable markers—traits that enable cells containing the vector to be selected or identified.

19.5 An idealized cloning vector has an origin of replication, one or more selectable markers, and one or more unique restriction sites.

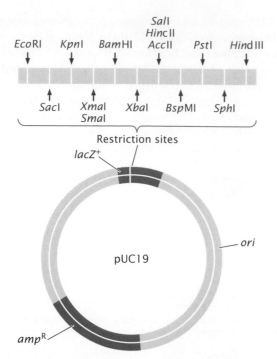

19.6 The pUC19 plasmid is a typical cloning vector. It contains a cluster of unique restriction sites, an origin of replication, and two selectable markers—an ampicillin-resistance gene and a *lacZ* gene.

the cell; (2) selectable markers, which enable any cells containing the vector to be selected or identified; and (3) one or more unique restriction sites into which a DNA fragment can be inserted. The restriction sites used for cloning must be unique; if a vector is cut at multiple recognition sites, generating several pieces of DNA, there will be no way to get the pieces back together in the correct order.

Plasmid vectors Plasmids, circular DNA molecules that exist naturally in bacteria (see Chapter 8), are commonly used vectors for cloning DNA fragments in bacteria. They contain origins of replication and are therefore able to replicate independently of the bacterial chromosome. The plasmids typically used in cloning have been constructed from the larger, naturally occurring bacterial plasmids and have multiple restriction sites, an origin of replication site, and selectable markers (**Figure 19.6**).

The easiest method for inserting a gene into a plasmid vector is to cut the foreign DNA (containing the gene) and the plasmid with the same restriction enzyme (**Figure 19.7**). If the restriction enzyme makes staggered cuts in the DNA, complementary sticky ends are produced on the foreign DNA and the plasmid. The DNA and plasmid are then mixed together; some of the foreign DNA fragments will pair with the cut ends of the plasmid. DNA ligase is used to seal the nicks in the sugar–phosphate backbone, creating a recombinant plasmid that contains the foreign DNA fragment.

Sometimes restriction sites are not available at a site where the DNA needs to be cut. In that case, a restriction site can be created with the use of **linkers**, which are small, synthetic DNA fragments that contain one or more restriction sites. Linkers can be attached to the ends of any piece of DNA and are then cut by a restriction enzyme, generating sticky ends that are complementary to sticky ends on the plasmid.

Transformation When a gene has been placed inside a plasmid, the plasmid must be introduced into bacterial cells. This task is usually accomplished by *transformation*, in

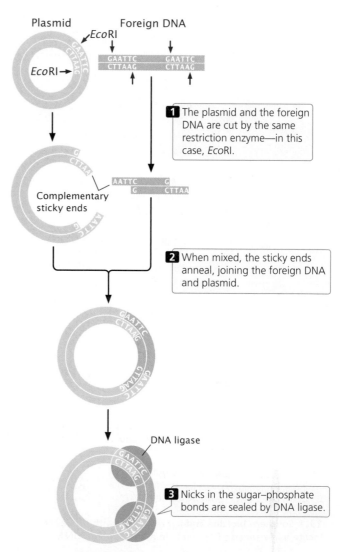

1 The plasmid and the foreign DNA are cut by the same restriction enzyme—in this case, *Eco*RI.

Complementary sticky ends

2 When mixed, the sticky ends anneal, joining the foreign DNA and plasmid.

DNA ligase

3 Nicks in the sugar–phosphate bonds are sealed by DNA ligase.

19.7 A foreign DNA fragment can be inserted into a plasmid with the use of restriction enzymes.

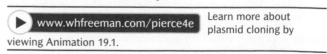
www.whfreeman.com/pierce4e — Learn more about plasmid cloning by viewing Animation 19.1.

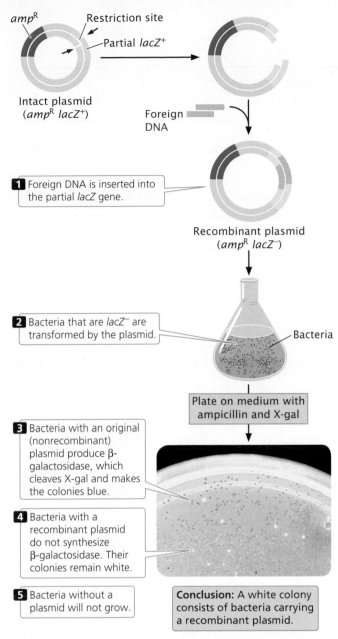

1 Foreign DNA is inserted into the partial *lacZ* gene.

Recombinant plasmid (*amp*^{R} *lacZ*$^-$)

2 Bacteria that are *lacZ*$^-$ are transformed by the plasmid.

Bacteria

Plate on medium with ampicillin and X-gal

3 Bacteria with an original (nonrecombinant) plasmid produce β-galactosidase, which cleaves X-gal and makes the colonies blue.

4 Bacteria with a recombinant plasmid do not synthesize β-galactosidase. Their colonies remain white.

5 Bacteria without a plasmid will not grow.

Conclusion: A white colony consists of bacteria carrying a recombinant plasmid.

19.8 The *lacZ* gene can be used to screen bacteria containing recombinant plasmids. A special plasmid carries a fragment of the *lacZ* gene and an ampicillin-resistance gene. [Photograph: Cytographics/Visuals Unlimited.]

which bacterial cells take up DNA from the external environment (see Chapter 8). Some types of cells undergo transformation naturally; others must be treated chemically or physically before they will undergo transformation. Inside the cell, the plasmids replicate and multiply.

Screening cells for recombinant plasmids Cells bearing recombinant plasmids can be detected with the use of the selectable markers on the plasmid. Common selectable markers are genes that confer resistance to an antibiotic: any cell that contains such a plasmid will be able to live in the presence of the antibiotic, which normally kills bacterial cells.

A common way to screen cells for the presence of a plasmid is to use a plasmid that contains a fragment of the *lacZ* gene—a small part of the front end of the gene (**Figure 19.8**). This partial *lacZ* gene contains a series of unique restriction sites into which can be inserted a piece of DNA to be cloned. The bacteria that will do the work of cloning have special features that make the presence of the recombinant plasmid evident. The bacteria that are transformed by the plasmid are *lacZ*$^-$, missing the front end of the *lacZ* gene but containing its back end. If no foreign DNA has been inserted into the partial *lacZ* gene of the plasmid, the front and back ends of the gene work together within the bacterial cell to produce β-galactosidase. If foreign DNA has been successfully inserted into the restriction site, it disrupts the front end of the *lacZ* gene, and β-galactosidase is *not* produced. The plasmid also usually contains a selectable marker, which may be a gene that confers resistance to an antibiotic such as ampicillin.

Bacteria that are *lacZ*$^-$ are transformed by the plasmids and plated on medium that contains ampicillin. Only cells that have been successfully transformed and contain a plasmid with the ampicillin-resistance gene will survive and grow. Some of these cells will contain an intact plasmid, whereas others possess a recombinant plasmid. The medium also contains the chemical X-gal, which produces a blue substance when cleaved. Bacterial cells with an intact original plasmid—without an inserted fragment—have a functional *lacZ* gene and can synthesize β-galactosidase, which cleaves X-gal and turns the bacteria blue. Bacterial cells with a recombinant plasmid, however, have a β-galactosidase gene that is disrupted by the inserted DNA; they do not synthesize β-galactosidase and remain white. (In these experiments, the bacterium's own β-galactosidase gene has been inactivated, and so only bacteria with the plasmid turn blue.) Thus, the color of the colony allows quick determination of whether a recombinant or intact plasmid is present in the cell. After cells with the recombinant plasmid have been identified, they can be grown in large number, replicating the inserted fragment of DNA.

Plasmids make ideal cloning vectors but can hold only DNA less than about 15 kb in size. When large DNA fragments are inserted into a plasmid vector, the plasmid becomes unstable.

Other gene vectors A number of other vectors have been developed for cloning larger pieces of DNA in bacteria. For example, bacteriophage λ, which infects *E. coli*, can be used to clone as much as about 23,000 bp of foreign DNA; it transfers DNA into bacterial cells with high efficiency. **Cosmids** are plasmids that are packaged into empty viral protein coats and transferred to bacteria by viral infection. They can carry more than twice as much foreign DNA as can a phage vector. **Bacterial artificial chromosomes** (BACs) are vectors originally constructed from the F plasmid (a special

Table 19.2 Comparison of plasmids, phage lambda vectors, cosmids, and bacterial artificial chromosomes

Cloning Vector	Size of DNA That Can Be Cloned	Method of Propagation	Introduction to Bacteria
Plasmid	As large as 15 kb	Plasmid replication	Transformation
Phage lambda	As large as 23 kb	Phage reproduction	Phage infection
Cosmid	As large as 44 kb	Plasmid reproduction	Phage infection
Bacterial artificial chromosome	As large as 300 kb	Plasmid reproduction	Electroporation

Note: 1 kb = 1000 bp. Electroporation consists of electrical pulses that increase the permeability of a membrane.

plasmid that controls mating and the transfer of genetic material in some bacteria; see Chapter 8) and can hold very large fragments of DNA that can be as long as 300,000 bp. **Table 19.2** compares the properties of plasmids, phage λ vectors, cosmids, and BACs.

Sometimes the goal in gene cloning is not just to replicate the gene, but also to produce the protein that it encodes. To ensure transcription and translation, a foreign gene is usually inserted into an **expression vector**, which, in addition to the usual origin of replication, restriction sites, and selectable markers, contains sequences required for transcription and translation in bacterial cells (**Figure 19.9**).

Although manipulating genes in bacteria is simple and efficient, the goal may be to transfer a gene into eukaryotic cells. For example, it might be desirable to transfer a gene conferring herbicide resistance into a crop plant or to transfer a gene for clotting factor into a person suffering from hemophilia. Many eukaryotic proteins are modified

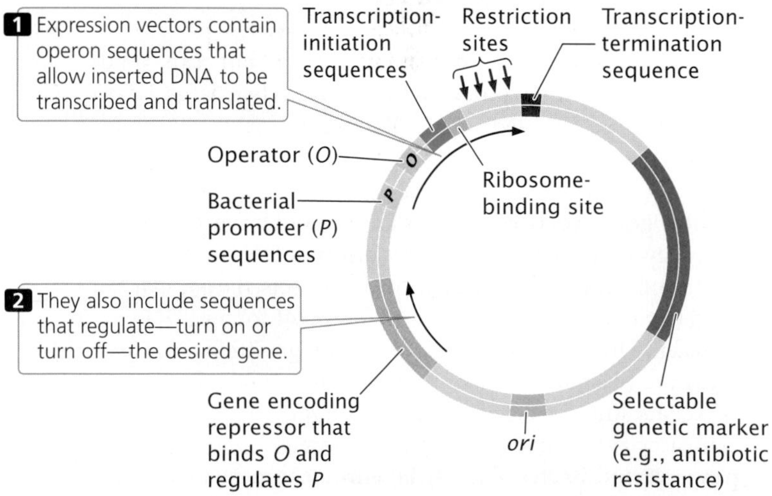

after translation (e.g., sugar groups may be added). Such modifications are essential for proper function, but bacteria do not have the capacity to carry out the modification; thus a functional protein can be produced only in a eukaryotic cell.

A number of cloning vectors have been developed that allow the insertion of genes into eukaryotic cells. Special plasmids have been developed for cloning in yeast, and retroviral vectors have been developed for cloning in mammals. A **yeast artificial chromosome** (YAC) is a DNA molecule that has a yeast origin of replication, a pair of telomeres, and a centromere. These features ensure that YACs are stable, replicate, and segregate in the same way as yeast chromosomes. YACs are particularly useful because they can carry DNA fragments as large as 600 kb, and some special YACs can carry inserts of more than 1000 kb. YACs have been modified so that they can be used in eukaryotic organisms other than yeast.

The soil bacterium *Agrobacterium tumefaciens*, which invades plants through wounds and induces crown galls (tumors), has been used to transfer genes to plants. This bacterium contains a large plasmid called the **Ti plasmid**, part of which is transferred to a plant cell when *A. tumefaciens* infects a plant. In the plant, part of the Ti plasmid DNA integrates into one of the plant chromosomes, where it is transcribed and translated to produce several enzymes that help support the bacterium (**Figure 19.10a**).

Geneticists have engineered a vector that contains the flanking sequences required to transfer DNA (called TL and TR, see Figure 19.10a), a selectable marker, and restriction sites into which foreign DNA can be inserted (**Figure 19.10b**). When placed in *A. tumefaciens* with a helper Ti plasmid (which carries sequences allowing the bacterium to infect the plant cell, which the artificial vector lacks), this vector will transfer the foreign DNA that it carries into a plant cell, where it will integrate into a plant chromosome. This vector has been used to transfer genes that confer economically significant attributes such as resistances to herbicides, plant viruses, and insect pests.

TRY PROBLEM 31 →

1 Expression vectors contain operon sequences that allow inserted DNA to be transcribed and translated.

Transcription-initiation sequences

Restriction sites

Transcription-termination sequence

Operator (*O*)

Bacterial promoter (*P*) sequences

Ribosome-binding site

2 They also include sequences that regulate—turn on or turn off—the desired gene.

Gene encoding repressor that binds *O* and regulates *P*

ori

Selectable genetic marker (e.g., antibiotic resistance)

19.9 To ensure transcription and translation, a foreign gene may be inserted into an expression vector—in this example, an *E. coli* expression vector.

(a)

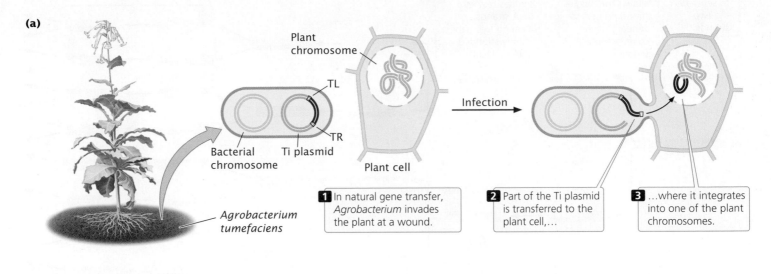

Plant chromosome

TL

TR

Bacterial chromosome

Ti plasmid

Plant cell

Agrobacterium tumefaciens

Infection

1 In natural gene transfer, *Agrobacterium* invades the plant at a wound.

2 Part of the Ti plasmid is transferred to the plant cell,…

3 …where it integrates into one of the plant chromosomes.

(b)

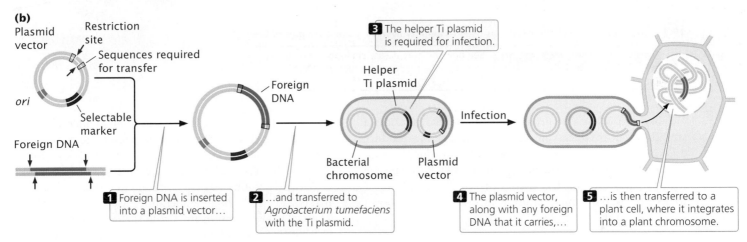

Plasmid vector

Restriction site

Sequences required for transfer

ori

Selectable marker

Foreign DNA

Foreign DNA

3 The helper Ti plasmid is required for infection.

Helper Ti plasmid

Bacterial chromosome

Plasmid vector

Infection

1 Foreign DNA is inserted into a plasmid vector…

2 …and transferred to *Agrobacterium tumefaciens* with the Ti plasmid.

4 The plasmid vector, along with any foreign DNA that it carries,…

5 …is then transferred to a plant cell, where it integrates into a plant chromosome.

19.10 The Ti plasmid can be used to transfer genes into plants. Flanking sequences TL and TR are required for the transfer of the DNA segment from bacteria to the plant cell.

CONCEPTS

DNA fragments can be inserted into cloning vectors, stable pieces of DNA that will replicate within a cell. A cloning vector must have an origin of replication, one or more unique restriction sites, and selectable markers. An expression vector contains sequences that allow a cloned gene to be transcribed and translated. Special cloning vectors have been developed for introducing genes into eukaryotic cells.

✔ CONCEPT CHECK 4

How is a gene inserted into a plasmid cloning vector?

Amplifying DNA Fragments with the Polymerase Chain Reaction

The manipulation and analysis of genes require multiple copies of the DNA sequences used. A major problem in working at the molecular level is that each gene is a tiny fraction of the total cellular DNA. Because each gene is rare, it must be isolated and amplified before it can be studied. As already stated, one way to amplify a DNA fragment is to clone it in bacterial cells. Indeed, for many years, gene cloning was the only way to quickly amplify a DNA fragment, and cloning was a prerequisite for many other molecular methods. Cloning is labor intensive and requires at least several days to grow the bacteria. Today, the polymerase chain reaction makes the amplification of short DNA fragments possible without cloning, but cloning is still widely used for amplifying large DNA fragments and for other manipulations of DNA sequences.

The **polymerase chain reaction** (PCR), first developed in 1983 by Kary Mullis, allows DNA fragments to be amplified a billionfold within just a few hours. It can be used with extremely small amounts of original DNA, even a single molecule. The polymerase chain reaction has revolutionized molecular biology and is now one of the most widely used of all molecular techniques. The basis of PCR is replication catalyzed by a DNA polymerase. Replication in this case has two essential requirements: (1) a single-stranded DNA template from which a new DNA strand can be copied and (2) a pair of primers with a 3′-OH group to which new nucleotides can be added.

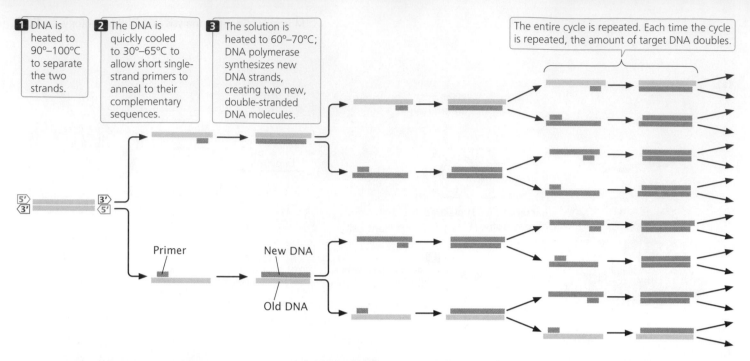

1. DNA is heated to 90°–100°C to separate the two strands.

2. The DNA is quickly cooled to 30°–65°C to allow short single-strand primers to anneal to their complementary sequences.

3. The solution is heated to 60°–70°C; DNA polymerase synthesizes new DNA strands, creating two new, double-stranded DNA molecules.

The entire cycle is repeated. Each time the cycle is repeated, the amount of target DNA doubles.

Primer

New DNA

Old DNA

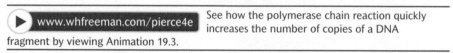

19.11 The polymerase chain reaction is used to amplify even very small samples of DNA.

▶ www.whfreeman.com/pierce4e   See how the polymerase chain reaction quickly increases the number of copies of a DNA fragment by viewing Animation 19.3.

Because a DNA molecule consists of two nucleotide strands, each of which can serve as a template to produce a new molecule of DNA, the amount of DNA doubles with each replication event. The primers used in PCR to replicate the templates are short fragments of DNA, typically from 17 to 25 nucleotides long, that are complementary to known sequences on the template.

The PCR reaction To carry out PCR, researchers begin with a solution that includes the target DNA, DNA polymerase, all four deoxyribonucleoside triphosphates (dNTPs—the substrates for DNA polymerase), primers, and magnesium ions and other salts that are necessary for the reaction to proceed. A typical polymerase chain reaction includes three steps (**Figure 19.11**). In step 1, a starting solution of DNA is heated to between 90° and 100°C to break the hydrogen bonds between the strands and thus produce the necessary single-stranded templates. The reaction mixture is held at this temperature for only a minute or two. In step 2, the DNA solution is cooled quickly to between 30° and 65°C and held at this temperature for a minute or less. During this short interval, the DNA strands will not have a chance to reanneal, but the primers will be able to attach to the template strands. In step 3, the solution is heated for a minute or less to between 60° and 70°C, the temperature at which DNA polymerase can synthesize new DNA strands. At the end of the cycle, two new double-stranded DNA molecules are produced for each original molecule of target DNA.

The whole cycle is then repeated. With each cycle, the amount of target DNA doubles; so the target DNA increases geometrically. One molecule of DNA increases to more than 1000 molecules in 10 PCR cycles, to more than 1 million mol-

ecules in 20 cycles, and to more than 1 billion molecules in 30 cycles. Each cycle is completed within a few minutes; so a large amplification of DNA can be achieved within a few hours.

Two key innovations facilitated the use of PCR in the laboratory. The first was the discovery of a DNA polymerase that is stable at the high temperatures used in step 1 of PCR. The DNA polymerase from *E. coli* that was originally used in PCR denatures at 90°C. For this reason, fresh enzyme had to be added to the reaction mixture in *each* cycle, slowing the process considerably. This obstacle was overcome when DNA polymerase was isolated from the bacterium *Thermus aquaticus*, which lives in the boiling springs of Yellowstone National Park. This enzyme, dubbed **Taq polymerase**, is remarkably stable at high temperatures and is not denatured in the strand-separation step of PCR; so it can be added to the reaction mixture at the beginning of the PCR process and will continue to function through many cycles.

The second key innovation was the development of automated thermal cyclers—machines that bring about the rapid temperature changes necessary for the different steps of PCR. Originally, tubes containing reaction mixtures were moved by hand among water baths set at the different temperatures required for the three steps of each cycle. In automated thermal cyclers, the reaction tubes are placed in a metal block that changes temperature rapidly according to a computer program.

In addition to amplifying DNA, PCR can be used to amplify sequences corresponding to RNA. This amplification is accomplished by first converting RNA into complementary DNA (cDNA) with the use of the enzyme reverse transcriptase. The cDNA can then be amplified by the usual PCR reaction. This technique is referred to as **reverse-transcription PCR**.

Limitations of PCR   The polymerase chain reaction is now often used in place of gene cloning, but it has several limitations. First, the use of PCR requires prior knowledge of at least part of the sequence of the target DNA to allow the construction of the primers. Second, the capacity of PCR to amplify extremely small amounts of DNA makes contamination a significant problem. Minute amounts of DNA from the skin of laboratory workers and even in small particles in the air can enter a reaction tube and be amplified along with the target DNA. Careful laboratory technique and the use of controls are necessary to circumvent this problem.

A third limitation of PCR is accuracy. Unlike other DNA polymerases, *Taq* polymerase does not have the capacity to proofread (see pp. 335–336 in Chapter 12) and, under standard PCR conditions, it incorporates an incorrect nucleotide about once every 20,000 bp. New heat-stable DNA polymerases with proofreading capacity have been isolated, giving more accurate PCR results.

A fourth limitation of PCR is that the size of the fragments that can be amplified by standard *Taq* polymerase is usually less than 2000 bp. By using a combination of *Taq* polymerase and a DNA polymerase with proofreading capacity and by modifying the reaction conditions, investigators have been successful in extending PCR amplification to larger fragments, but even these larger fragments are limited in length to 50,000 bp or smaller. In spite of its limitations, PCR is used routinely in a wide array of molecular applications.

Applications of PCR   The polymerase chain reaction has become one of the most widely used tools within molecular biology. The primers used in PCR are specific for known DNA sequences, and so PCR can be used to detect the presence of a particular DNA sequence in a sample. For example, PCR is often used to detect the presence of viruses in blood samples by adding primers complementary to known viral DNA sequences to the reaction. If viral DNA is present, the primers will attach to it, and PCR will amplify a DNA fragment of known length. The presence of a DNA fragment of the proper length on a gel indicates the presence of viral DNA in the blood sample. Modern diagnostic tests for infection with HIV, the causative agent of AIDS (see Chapter 8), use this type of PCR amplification of HIV sequences.

Another common application of PCR is to identify genetic variation. It allows the identification of restriction sites, restriction fragment length polymorphisms, and microsatellite variations that are used in gene cloning and genetic mapping. Primers specific to a particular mutation can be used to determine whether that mutation is present in the genome of an individual organism.

PCR has also allowed the isolation of DNA from ancient sources, such as DNA from Neanderthals (discussed in the introduction to Chapter 10). It is commonly used to amplify small amounts of DNA from crime scenes and identify the source of a DNA sample through detection of microsatellite variation (see section on DNA fingerprinting).

PCR can be used to introduce new sequences into a fragment of DNA. Although the 3′ end of the primers used in PCR must be complementary to the template, there can be flexibility in the sequences at the 5′ end of the primer. Primers can be designed so that they contain new restriction sites or other desirable sequences at their 5′ ends. These sequences will then be copied by the PCR reaction and thus added to the ends of the DNA amplified by the PCR reaction.

A modification of PCR, known as **real-time PCR**, can be used to quantitatively determine the amount of starting nucleic acid. In this procedure, the usual PCR reaction is used to amplify a specific DNA fragment, and a sensitive instrument is used to accurately determine the amount of DNA that is present in solution after each PCR cycle. A probe that fluoresces and is specific to the DNA sequence of interest is frequently used in the reaction, and so only the DNA of interest is measured. The technique is called real-time PCR because the amount of DNA amplified is measured as the reaction proceeds.

Often, real-time PCR is combined with reverse-transcription PCR to measure the amount of mRNA in a sample, allowing biologists to determine the level of gene expression in different cells and under different conditions. For example, researchers interested in how gene expression changes in response to the administration of a drug often use real-time PCR to quantitatively measure the amount of mRNA produced by specific genes in cells exposed to the drug and compare it with the amount of mRNA produced by the genes in control cells with no drug exposure.

CONCEPTS

The polymerase chain reaction is an enzymatic in vitro method for rapidly amplifying DNA. In this process, DNA is heated to separate the two strands, short primers attach to the target DNA, and DNA polymerase synthesizes new DNA strands from the primers. Each cycle of PCR doubles the amount of DNA. PCR has a number of important applications in molecular biology.

✔ CONCEPT CHECK 5

Why is the use of a heat-stable DNA polymerase important to the success of PCR?

Application: The Genetic Engineering of Plants with Pesticides

An early and economically important use of molecular techniques for transferring genes to other organisms was the genetic engineering of plants that produce their own insecticides. The bacterium *Bacillus thuringiensis* naturally produces a protein, known as the Bt toxin, that is lethal to insects. The Bt toxin is particularly attractive as an insecticide because it is specific to particular insects, breaks down quickly in the environment, and is not toxic to humans and other animals.

In 1987, Mark Vaeck and his colleagues at Plant Genetic Systems in Belgium genetically engineered tobacco plants to

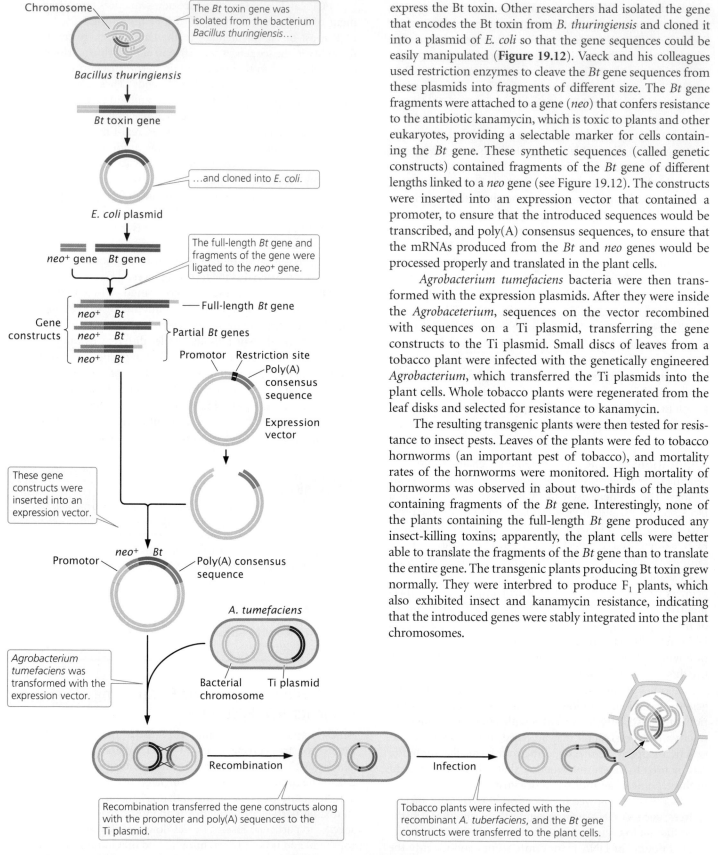

Chromosome

The *Bt* toxin gene was isolated from the bacterium *Bacillus thuringiensis*...

Bacillus thuringiensis

Bt toxin gene

...and cloned into *E. coli.*

E. coli plasmid

neo⁺ gene   *Bt* gene

The full-length *Bt* gene and fragments of the gene were ligated to the *neo⁺* gene.

Gene constructs

neo⁺   Bt   Full-length *Bt* gene

neo⁺   Bt   Partial *Bt* genes

neo⁺   Bt

Promotor   Restriction site

Poly(A) consensus sequence

Expression vector

These gene constructs were inserted into an expression vector.

neo⁺   Bt

Promotor   Poly(A) consensus sequence

A. tumefaciens

Agrobacterium tumefaciens was transformed with the expression vector.

Bacterial chromosome   Ti plasmid

Recombination

Infection

Recombination transferred the gene constructs along with the promoter and poly(A) sequences to the Ti plasmid.

Tobacco plants were infected with the recombinant *A. tuberfaciens,* and the *Bt* gene constructs were transferred to the plant cells.

19.12 The *Bt* toxin gene, which encodes an insecticide, was isolated from bacteria and transferred to tobacco plants.

express the Bt toxin. Other researchers had isolated the gene that encodes the Bt toxin from *B. thuringiensis* and cloned it into a plasmid of *E. coli* so that the gene sequences could be easily manipulated (**Figure 19.12**). Vaeck and his colleagues used restriction enzymes to cleave the *Bt* gene sequences from these plasmids into fragments of different size. The *Bt* gene fragments were attached to a gene (*neo*) that confers resistance to the antibiotic kanamycin, which is toxic to plants and other eukaryotes, providing a selectable marker for cells containing the *Bt* gene. These synthetic sequences (called genetic constructs) contained fragments of the *Bt* gene of different lengths linked to a *neo* gene (see Figure 19.12). The constructs were inserted into an expression vector that contained a promoter, to ensure that the introduced sequences would be transcribed, and poly(A) consensus sequences, to ensure that the mRNAs produced from the *Bt* and *neo* genes would be processed properly and translated in the plant cells.

Agrobacterium tumefaciens bacteria were then transformed with the expression plasmids. After they were inside the *Agrobaceterium*, sequences on the vector recombined with sequences on a Ti plasmid, transferring the gene constructs to the Ti plasmid. Small discs of leaves from a tobacco plant were infected with the genetically engineered *Agrobacterium*, which transferred the Ti plasmids into the plant cells. Whole tobacco plants were regenerated from the leaf disks and selected for resistance to kanamycin.

The resulting transgenic plants were then tested for resistance to insect pests. Leaves of the plants were fed to tobacco hornworms (an important pest of tobacco), and mortality rates of the hornworms were monitored. High mortality of hornworms was observed in about two-thirds of the plants containing fragments of the *Bt* gene. Interestingly, none of the plants containing the full-length *Bt* gene produced any insect-killing toxins; apparently, the plant cells were better able to translate the fragments of the *Bt* gene than to translate the entire gene. The transgenic plants producing Bt toxin grew normally. They were interbred to produce F_1 plants, which also exhibited insect and kanamycin resistance, indicating that the introduced genes were stably integrated into the plant chromosomes.

19.13 Transgenic crops are broadly cultivated throughout the world. The rapeseed shown here has been genetically modified to resist a specific herbicide, which can then be sprayed on the field to kill weeds. [Chris Knapton/Photo Researchers.]

Other researchers using similar methods subsequently introduced the gene for Bt toxin into cotton, tomatoes, corn, and other plants (**Figure 19.13**). Today, transgenic plants expressing the Bt toxin are used broadly throughout the world.

19.3 Molecular Techniques Can Be Used to Find Genes of Interest

To analyze a gene or to transfer it to another organism, the gene must first be located and isolated. For instance, if we wanted to transfer a human gene for growth hormone to bacteria, we must first find the human gene that encodes growth hormone and separate it from the 3.2 billion bp of human DNA. In our consideration of gene cloning, we've glossed over the problem of finding the DNA sequence to be cloned. A discussion of how to solve this problem has been purposely delayed until now because, paradoxically, researchers must often clone a gene to find it.

This approach—to clone first and search later—is called "shotgun cloning," because it is like hunting with a shotgun: a hunter sprays shots widely in the general direction of the quarry, knowing that there is a good chance that one or more of the pellets will hit the intended target. In shotgun cloning, a researcher first clones a large number of DNA fragments, knowing that one or more contains the DNA of interest, and then searches for the fragment of interest among the clones.

Gene Libraries

A collection of clones containing all the DNA fragments from one source is called a **DNA library**. For example, we might isolate genomic DNA from human cells, break it into fragments, and clone all of them in bacterial cells or phages. The set of bacterial colonies or phages containing these fragments is a human **genomic library**, containing all the DNA sequences found in the human genome.

Creating a genomic library To create a genomic library, cells are collected and disrupted, which causes them to release their DNA and other cellular contents into an aqueous solution, and the DNA is extracted from the solution. After the DNA has been isolated, it is cut into fragments by

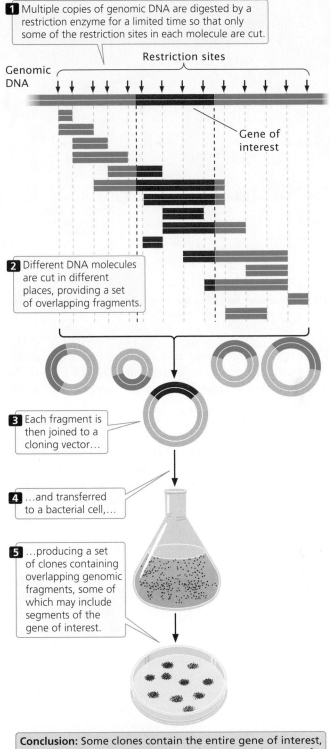

Conclusion: Some clones contain the entire gene of interest, others include part of the gene, and most contain none of the gene of interest.

19.14 A genomic library contains all of the DNA sequences found in an organism's genome.

a restriction enzyme for a limited amount of time only (a partial digestion) so that only *some* of the restriction sites in each DNA molecule are cut. Because which sites are cut is random, different DNA molecules will be cut in different places, and a set of overlapping fragments will be produced (**Figure 19.14**). The fragments are then joined to vectors,

527

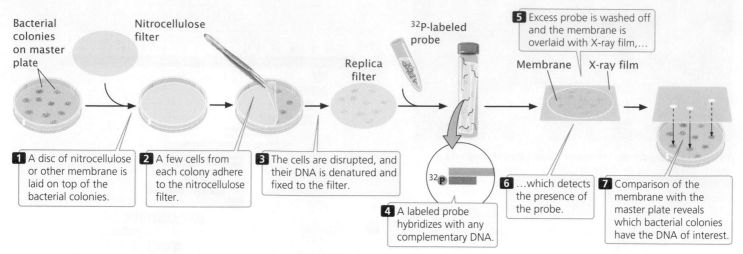

19.15 Genomic and cDNA libraries can be screened with a probe to find the gene of interest.

which can be transferred to bacteria. A few of the clones contain the entire gene of interest, a few contain parts of the gene, but most contain fragments that have no part of the gene of interest.

A genomic library must contain a large number of clones to ensure that all DNA sequences in the genome are represented in the library. A library of the human genome formed by using cosmids, each carrying a random DNA fragment from 35,000 to 44,000 bp long, would require about 350,000 cosmid clones to provide a 99% chance that every sequence is included in the library.

cDNA libraries An alternative to creating a genomic library is to create a library consisting only of those DNA sequences that are transcribed into mRNA (called a **cDNA library** because all the DNA in this library is *complementary* to mRNA). Much of eukaryotic DNA consists of repetitive (and other DNA) sequences that are not transcribed into mRNA (see p. 302 in Chapter 11), and these sequences are not represented in a cDNA library.

A cDNA library has two additional advantages. First, it is enriched with fragments from actively transcribed genes. Second, introns do not interrupt the cloned sequences; introns would pose a problem when the goal is to produce a eukaryotic protein in bacteria, because most bacteria have no means of removing the introns.

The disadvantage of a cDNA library is that it contains only sequences that are present in mature mRNA. Sometimes, researchers are interested in sequences that are not transcribed, such as those in promoters and enhancers, which are important for transcription but are not themselves transcribed. These sequences are not present in a cDNA library. Furthermore, a cDNA library contains only those gene sequences expressed in the tissue from which the RNA was isolated, and the frequency of a particular DNA sequence in a cDNA library depends on the abundance of the corresponding mRNA in the given tissue. So, if a particular gene is not

expressed or is expressed only at low frequency in a particular tissue, it may be absent in a cDNA library prepared from that tissue. In contrast, almost all genes are present at the same frequency in a genomic DNA library. **TRY PROBLEM 34 →**

CONCEPTS

One method of finding a gene is to create and screen a DNA library. A genomic library is created by cutting genomic DNA into overlapping fragments and cloning each fragment in a separate bacterial cell. A cDNA library is created from mRNA that is converted into cDNA and cloned in bacteria.

Screening DNA libraries Creating a genomic or cDNA library is relatively easy compared with screening the library to find clones that contain the gene of interest. The screening procedure used depends on what is known about the gene.

The first step in screening is to plate the clones of the library. If a plasmid or cosmid vector was used to construct the library, the cells are diluted and plated so that each bacterium grows into a distinct colony. If a phage vector was used, the phages are allowed to infect a lawn of bacteria on a petri plate. Each plaque or bacterial colony contains a single, cloned DNA fragment that must be screened for the gene of interest.

A common way to screen libraries is with probes. To use a probe, replicas of the plated colonies or plaques in the library must first be made. **Figure 19.15** illustrates this procedure for a cosmid library.

How is a probe obtained when the gene has not yet been isolated? One option is to use a similar gene from another organism as the probe. For example, if we wanted to screen a human genomic library for the growth-hormone gene and the gene had already been isolated from rats, we could use a purified rat-gene sequence as the probe to find the human gene for growth hormone. Successful hybridization does not require perfect complementarity between the probe and the target sequence; so a related sequence can often be used as a probe.

(a)

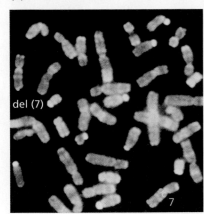

(b)

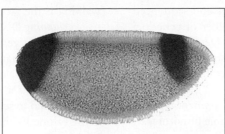

19.16 With in situ hybridization, DNA probes are used to determine the cellular or chromosomal location of a gene or its product. (a) A probe with green fluorescence is specific to chromosome 7, revealing a deletion on one copy of chromosome 7.
 (b) In situ hybridization is used to detect the presence of mRNA from the *tailless* gene in a *Drosophila* embryo. [Part a: Addenbrookes Hospital/Photo Researchers. Part b: Courtesy of L. Tsuda.]

Alternatively, synthetic probes can be created if the protein produced by the gene has been isolated and its amino acid sequence has been determined. With the use of the genetic code and the amino acid sequence of the protein, possible nucleotide sequences of a small region of the gene can be deduced. Although only one sequence in the gene encodes a particular protein, the presence of synonymous codons means that the same protein could be produced by several different nucleotide sequences, and it is impossible to know which is correct. To overcome this problem, a mixture of all the possible nucleotide sequences is used as a probe. To minimize the number of sequences required in the mixture, a region of the protein is selected with relatively little degeneracy in its codons.

When part of the DNA sequence of the gene has been determined, a set of DNA probes can be synthesized chemically by using an automated machine known as an oligonucleotide synthesizer. The resulting probes can be used to screen a library for a gene of interest.

Yet another method of screening a library is to look for the protein product of a gene. This method requires that the DNA library be cloned in an expression vector. The clones can be tested for the presence of the protein by using an antibody that recognizes the protein or by using a chemical test for the protein product. This method depends on the existence of a test for the protein produced by the gene.

Almost any method used to screen a library will identify several clones, some of which will be false positives that do not contain the gene of interest; several screening methods may be needed to determine which clones actually contain the gene.

CONCEPTS

A DNA library can be screened for a specific gene with the use of complementary probes that hybridize to the gene. Alternatively, the library can be cloned into an expression vector, and the gene can be located by examining the clones for the protein product of the gene.

✔ CONCEPT CHECK 6

Briefly explain how synthetic probes are created to screen a DNA library when the protein encoded by the gene is known.

In Situ Hybridization

DNA probes can be used to determine the chromosomal location of a gene in a process called **in situ hybridization**. The name is derived from the fact that DNA (or RNA) is visualized while it is in the cell (in situ). This technique requires that the cells be fixed and the chromosomes be spread on a microscope slide and denatured. A labeled probe is then applied to the slide, just as it is can be applied to a gel. Many probes carry attached fluorescent dyes that can be seen directly with the microscope (**Figure 19.16a**). Several probes with different colored dyes can be used simultaneously to investigate different sequences or chromosomes. Fluorescence in situ hybridization (FISH) has been widely used to identify the chromosomal location of human genes.

In situ hybridization can also be used to determine the tissue distribution of specific mRNA molecules, serving as a source of insight into how gene expression differs among cell types (**Figure 19.16b**). A labeled DNA probe complementary to a specific mRNA molecule is added to tissue, and the location of the probe is determined with the use of either autoradiography or fluorescent tags. Determining where a gene is expressed often helps define its function. For example, finding that a gene is highly expressed only in brain tissue might suggest that the gene has a role in neural function.

Positional Cloning

For many genes with important functions, no associated protein product is yet known. The biochemical bases of many human genetic diseases, for example, are still unknown. How can these genes be isolated? One approach is to first determine the general location of the gene on the chromosome by using recombination frequencies derived from crosses or pedigrees (see Chapter 7). After the chromosomal region where the gene is found has been identified, any genes in this region can be cloned and identified. Then other techniques can be used to identify which of the "candidate" genes might be the one that causes the disease. This approach—to isolate genes on the basis of their position on a gene map—is called **positional cloning**.

The development of molecular techniques for identifying gene sequences, such as restriction sites and sequenced DNA fragments, has provided a large number of gene loci that can be used as markers in mapping. Detailed genetic maps of these markers have been created so that their precise location is known. In the first step of positional cloning, geneticists use mapping studies (see Chapter 7) to establish linkage between molecular markers and a phenotype of interest, such as a human disease or a desirable physical trait in a plant or animal. In regard to a human disease, members of one or more families with the disease would be genotyped for a large number of variable genetic markers from each of the human chromosomes. Demonstration of linkage between the disease phenotype and one or more molecular markers would provide information about which chromosome carries the disease locus and its general location on that chromosome.

The next step is to more precisely locate the locus by using additional molecular markers clustered in the chromosomal region where the locus resides. The goal is to identify molecular markers that are within 1 map unit (1% recombination) of the disease locus. In humans, 1% recombination represents about 1,000,000 bp of DNA; so the region in which the gene is located is still large.

After the gene has been placed on a chromosome map, clones that cover the region can be isolated from a genomic library. With the use of a technique called **chromosome walking** (**Figure 19.17**), it is possible to progress from neighboring genes to linked clones, one of which might contain the gene of interest. The basis of chromosome walking is the fact that a genomic library consists of a set of *overlapping* DNA fragments (see Figure 19.14). We start with a cloned gene marker that is close to the new gene of interest so that the "walk" will be as short as possible. One end of the clone of a neighboring marker (clone A in Figure 19.17) is used to make a complementary probe. This probe is used to screen the genomic library to find a second clone (clone B) that overlaps with the first and extends in the direction of the gene of interest. This second clone is isolated and purified and a probe is prepared from its end. The second probe is used to screen the library for a third clone (clone C) that overlaps with the second. In this way, one can systematically "walk" toward the gene of interest, one clone at a time.

If the region in which the gene of interest lies is very large, chromosome walking can be a laborious task. To cover the region more quickly, geneticists sometimes turn to **chromosome jumping**. In this technique, genomic DNA is cut into fragments by partial digestion with a restriction enzyme, and large fragments between 80 kb and 150 kb are isolated (**Figure 19.18**). The ends of the resulting DNA fragments are ligated together so that they create a circular DNA molecule. In each circle, the beginning and end of the long sequence will be brought close together. The circle is then cut with another restriction enzyme that cuts each circle into smaller fragments. Some of these fragments contain the beginnings and the ends of the original large fragments without the intervening DNA. These fragments are used to create a DNA library. Because the cloned fragments begin and end with sequences that are distantly linked, the DNA at the end of one fragment can then be used to make a probe that can find clones that are distantly linked. New probes are then made to the linked clone and used to jump again. In this way, one is able to "jump" to clones that are more distantly linked to a marker gene.

After clones that cover the delineated region have been obtained by chromosome walking or jumping, all genes located within the region are identified. Genes can be distinguished from other sequences by the presence of characteristic features, such as consensus sequences in the promoter and start codon and stop codons within the same reading frame. After potential "candidate" genes have been identified, they can be evaluated to determine which is most likely to be the gene of interest. The expression pattern of the gene—where and when it is transcribed—can often provide clues about its function. For example, genes for neurological disease would likely be expressed in the brain. Geneticists often look in the coding region of the gene for mutations among people with the disease. More will be said about determining the function of genes in sections that follow and in Chapter 20.

Through positional cloning, researchers can identify genes that encode a phenotype without having a detailed understanding of the underlying biochemical nature of the phenotype. A number of important human diseases have been identified through positional cloning.

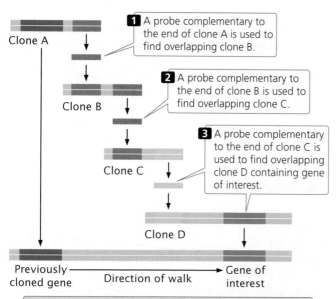

1 A probe complementary to the end of clone A is used to find overlapping clone B.

2 A probe complementary to the end of clone B is used to find overlapping clone C.

3 A probe complementary to the end of clone C is used to find overlapping clone D containing gene of interest.

Clone A

Clone B

Clone C

Clone D

Previously cloned gene — Direction of walk → Gene of interest

Conclusion: By making probes complementary to areas of overlap between cloned fragments in a genomic library, we can connect a gene of interest to a previously mapped, linked gene.

19.17 In chromosome walking, neighboring genes are used to locate a gene of interest.

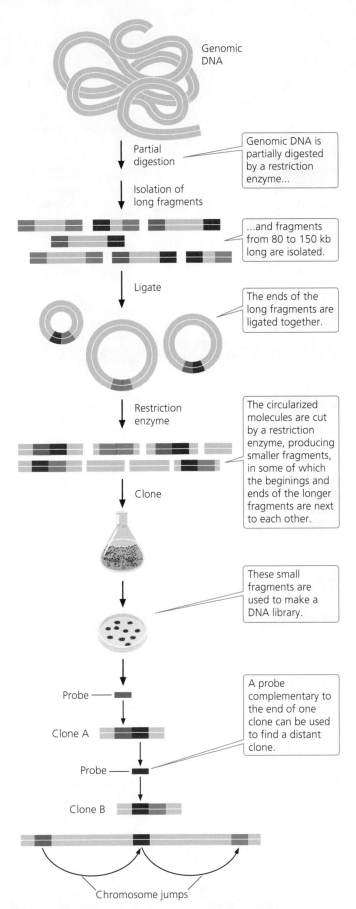

Genomic
DNA

Partial
digestion

Genomic DNA is
partially digested
by a restriction
enzyme...

Isolation of
long fragments

...and fragments
from 80 to 150 kb
long are isolated.

Ligate

The ends of the
long fragments are
ligated together.

Restriction
enzyme

The circularized
molecules are cut
by a restriction
enzyme, producing
smaller fragments,
in some of which
the beginnings and
ends of the longer
fragments are next
to each other.

Clone

These small
fragments are
used to make a
DNA library.

Probe

A probe
complementary to
the end of one
clone can be used
to find a distant
clone.

Clone A

Probe

Clone B

Chromosome jumps

**19.18 Chromosome jumping is used to located distantly
linked clones.**

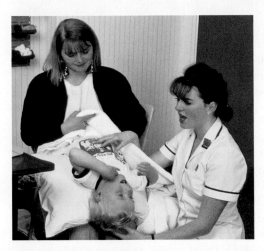

**19.19 Cystic fibrosis was the first genetic disease for which the
causative gene was isolated entirely by positional cloning.** [Simon
Fraser/Science Photo Library.]

CONCEPTS

Positional cloning allows researchers to isolate a gene without having
knowledge of its biochemical basis. Linkage studies are used to map
the locus producing a phenotype of interest to a particular chromosome region. Chromosome walking and jumping are used to progress
from molecular makers to clones containing sequences that cover
the chromosome region. Candidate genes within the region are then
evaluated to determine if they encode the phenotype of interest.

✔ CONCEPT CHECK 7

How are candidate genes that are identified by positional cloning evaluated to determine whether they encode the phenotype of interest?

In Silico Gene Discovery

The complete genomes of more and more species are
sequenced each year (see Chapter 20), and partial sequences
of many other organisms are continually being added to
DNA databases. With this growth in sequence information, the task of finding genes is now often carried out by
using high-speed computers that analyze and search DNA
databases rather than using gene cloning and DNA libraries.
Sometimes called "in silico," these methods of locating and
characterizing genes rely on the identification of characteristic sequences associated with genes and on comparisons
with sequences of known genes in DNA databases. These
methods will be discussed in more detail in Chapter 20.

Application: Isolating the Gene
for Cystic Fibrosis

The first gene responsible for a human genetic disease that
was isolated entirely by positional cloning was the gene for
cystic fibrosis (CF). Cystic fibrosis is an autosomal recessive
disorder characterized by chronic lung infections, insufficient production of pancreatic enzymes that are necessary
for digestion, and increased salt concentration in sweat (**Figure 19.19**). It is among the most common genetic diseases

in Caucasians, occurring with a frequency of about 1 in 2000 live births. About 5% of all Caucasians are carriers of the CF mutation.

Geneticists attempting to isolate the gene for CF faced a formidable task. The symptoms of the disease, especially the elevated salt concentration in sweat, suggested that the gene for CF somehow takes part in the movement of ions into and out of the cell, but no information was available about the protein encoded by the gene. Analyses of pedigrees showed that CF is inherited as an autosomal recessive trait, and so it might be located on any one of the 22 pairs of autosomal chromosomes. Thus, geneticists were seeking an unknown gene—probably encompassing a few thousand or tens of thousands of base pairs—among the 3.2 billion base pairs of the human genome.

Researchers began by looking for associations between the inheritance of CF and that of other traits (**Figure 19.20**). Early studies were limited by the paucity of genetic traits that varied and could be used for gene-mapping studies; but, in the 1980s, advances in molecular biology provided a large number of molecular markers that could be used for linkage analysis (see p. 186 in Chapter 7). Geneticists collected pedigrees of families in which several members had CF. They compared the inheritance of CF with that of molecular markers among the members of these families, looking for evidence of linkage. These studies paid off: the gene for CF was found to be closely linked to two markers, MET and D7S8, located on the long arm of chromosome 7. MET and D7S8 are separated by about 1.5 map units (see Chapter 7). In the human genome, each map unit roughly corresponds to 1 million base pairs; so the gene for CF is located somewhere within a stretch of 1.5 million base pairs of DNA, a huge expanse of sequence.

The next step was to carry out further linkage studies with additional markers, to more precisely delineate where in the 1.5-million-base-pair region the CF gene lies. Researchers selected additional molecular markers from the region surrounding MET and D7S8 and performed linkage studies between these new markers and CF (see Figure 19.20). These studies identified two additional markers, D7S122 and D7S340, that are closely linked to CF. Furthermore, they showed that the order of the four markers is MET-D7S340-D7S122-D7S8 and that the CF gene lies very close to D7S122 and D7S340. This finding narrowed the region in which the gene for CF lies to about 500,000 bp.

At this stage, geneticists began isolating clones of sequences from the delineated region. Starting from the molecular markers, they used a combina-

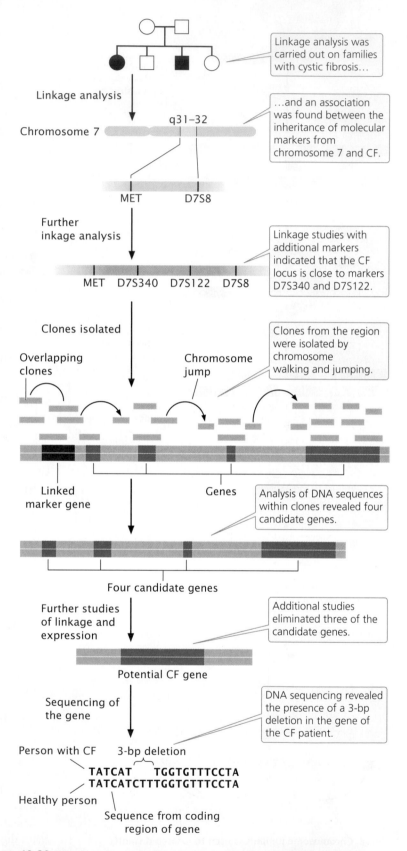

19.20 The gene for cystic fibrosis was located by positional cloning.

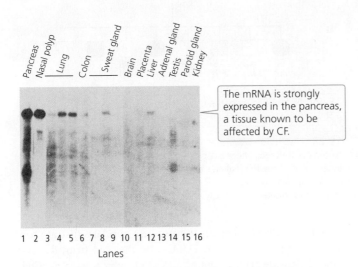

The mRNA is strongly expressed in the pancreas, a tissue known to be affected by CF.

1 2 3 4 5 6 7 8 9 10 11 12 13 14 15 16
Lanes

19.21 A candidate for the cystic fibrosis gene is expressed in pancreatic, respiratory, and sweat-gland tissues—tissues that are affected by the disease. Shown is a Northern blot of mRNA produced by the candidate gene in different tissues. These data provided evidence that the candidate gene is in fact that the gene that causes cystic fibrosis. [After J. R. Riordan et al., *Science* 245:1066–1073, 1989.]

tion of chromosome walking and chromosome jumping to identify clones from human genomic libraries that completely covered the region of interest (see Figure 19.20). An examination of sequences within these clones revealed the presence of four genes in the region encompassed by the linked makers (see Figure 19.20). Additional studies were then carried out to better characterize these candidate genes. Three of the candidate genes were eventually eliminated, either because linkage analysis suggested that they were not closely linked with the inheritance of CF or because characterization of their DNA sequences suggested they were not the gene for CF.

Hybridization studies were carried out with the one remaining gene to determine where it was expressed. Messenger RNA was isolated from different organ tissues, converted into cDNA by reverse transcription, and probed with sequences from the candidate gene. The gene showed high levels of expression in the pancreas, in the lungs, and in sweat glands (**Figure 19.21**), tissues known to be affected by CF.

Copies of the candidate gene from a healthy person and from a person with CF were then sequenced, and the sequence data were examined for differences that might be a mutation causing CF. The findings revealed that the person with CF had a 3-bp deletion in the coding region of the gene; the healthy person did not have this deletion. The deletion resulted in the absence of a phenylalanine amino acid from the protein encoded by the candidate gene. Then, for a large number of patients with CF, geneticists used PCR to amplify the region of the gene where the deletion was found; 68% of the CF patients had this deletion. Subsequent studies demonstrated that the remaining CF patients possessed mutations at other locations within the

candidate gene, thus proving that the candidate gene was indeed the locus that caused CF.

Researchers eventually demonstrated that the gene for CF encodes a membrane protein that controls the movement of chloride into and out of cells and is known today as the *cystic fibrosis transmembrane conductance regulator* (*CFTR*) gene. Patients with CF have two mutated forms of *CFTR*, which cause the chloride channels to remain closed. Chloride ions build up in the cell, leading to the formation of thick mucus and the symptoms of the disease.

19.4 DNA Sequences Can Be Determined and Analyzed

In addition to cloning and amplifying DNA, molecular techniques are used to analyze DNA molecules through the study and determination of their sequences.

Restriction Fragment Length Polymorphisms

A significant contribution of molecular genetics has been to provide numerous genetic markers that can be used in gene mapping. We learned how these markers are essential to the success of positional cloning. One group of such markers comprises **restriction fragment length polymorphisms** (RFLPs), which are variations (polymorphisms) in the patterns of fragments produced when DNA molecules are cut with the same restriction enzyme (**Figure 19.22**). These differences are inherited and can be used in mapping, similar to the way in which allelic differences are used to map conventional genes.

To illustrate mapping with RFLPs, consider Huntington disease, an autosomal dominant disorder. In the family shown in **Figure 19.23**, the father is heterozygous both for Huntington disease (*Hh*) and for a restriction pattern (*AC*). From the father, each child inherits either a Huntington-disease allele (*H*) or a normal allele (*h*); any child inheriting the Huntington-disease allele develops the disease, because it is an autosomal dominant disorder. The child also inherits one of the two RFLP alleles from the father, either *A* or *C*, which produces the corresponding RFLP pattern. In Figure 19.23, every child who inherits the *C* pattern from the father also inherits Huntington disease (and therefore the *H* allele), because the locus for the RFLP is closely linked to the locus for the disease-causing gene. If we had observed no correspondence between the inheritance of the RFLP pattern and the inheritance of the disease, it would indicate that the genes encoding the RFLP and Huntington disease are assorting independently and are not linked. **TRY PROBLEM 39 →**

CONCEPTS

Restriction fragment length polymorphisms are variations in the pattern of fragments produced by restriction enzymes, which reveal variations in DNA sequences. They are used extensively in gene mapping.

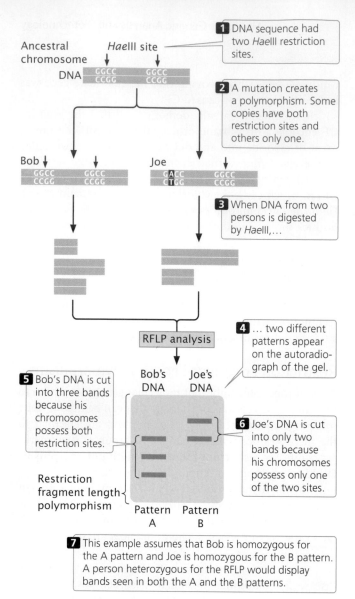

1 DNA sequence had two *HaeIII* restriction sites.

Ancestral chromosome DNA

HaeIII site

2 A mutation creates a polymorphism. Some copies have both restriction sites and others only one.

Bob

Joe

3 When DNA from two persons is digested by *HaeIII*,…

RFLP analysis

4 … two different patterns appear on the autoradiograph of the gel.

Bob's DNA

Joe's DNA

5 Bob's DNA is cut into three bands because his chromosomes possess both restriction sites.

6 Joe's DNA is cut into only two bands because his chromosomes possess only one of the two sites.

Restriction fragment length polymorphism

Pattern A

Pattern B

7 This example assumes that Bob is homozygous for the A pattern and Joe is homozygous for the B pattern. A person heterozygous for the RFLP would display bands seen in both the A and the B patterns.

19.22 Restriction fragment length polymorphisms are genetic markers that can be used in mapping.

DNA Sequencing

A powerful molecular method for analyzing DNA is a technique known as **DNA sequencing**, which quickly determines the sequence of bases in DNA. Sequencing allows the genetic information in DNA to be read, providing an enormous amount of information about gene structure and function. In the mid-1970s, Frederick Sanger and his colleagues created the dideoxy-sequencing method based on the elongation of DNA by DNA polymerase, and it quickly became the standard procedure for sequencing any purified fragment of DNA.

The Sanger, or dideoxy, method of DNA sequencing is based on replication. The fragment to be sequenced is used as a template to make a series of new DNA molecules. In the process, replication is sometimes (but not always) terminated when a specific base is encountered, producing DNA strands of different lengths, each of which ends in the same base.

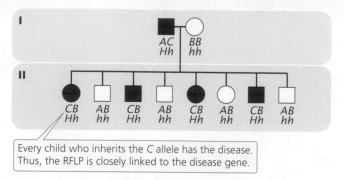

Every child who inherits the *C* allele has the disease. Thus, the RFLP is closely linked to the disease gene.

19.23 Restriction fragment length polymorphisms can be used to detect linkage. There is a close correspondence between the inheritance of the RFLP alleles and the presence of Huntington disease, indicating that the genes that encode the RFLP and Huntington disease are closely linked.

The method relies on the use of a special substrate for DNA synthesis. Normally, DNA is synthesized from deoxyribonucleoside triphosphates (dNTPs), which have an OH group on the 3′-carbon atom (**Figure 19.24a**). In the Sanger method, a special nucleotide, called a **dideoxyribonucleoside triphosphate** (ddNTP; **Figure 19.24b**), is used as one of the substrates. The ddNTPs are identical with dNTPs, except that they lack a 3′-OH group. In the course of DNA synthesis, ddNTPs are incorporated into a growing DNA strand. However, after a ddNTP has been incorporated into the DNA strand, no more nucleotides can be added, because there is no 3′-OH group to form a phosphodiester bond with an incoming nucleotide. Thus, ddNTPs terminate DNA synthesis.

Although the sequencing of a single DNA molecule is technically possible, most sequencing procedures in use today require a considerable amount of DNA; so any DNA fragment to be sequenced must first be amplified by PCR or

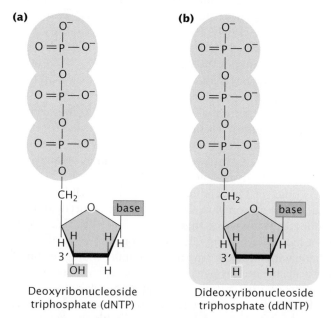

(a) Deoxyribonucleoside triphosphate (dNTP)

(b) Dideoxyribonucleoside triphosphate (ddNTP)

19.24 The dideoxy sequencing reaction requires a special substrate for DNA synthesis. (a) Structure of deoxyribonucleoside triphosphate, the normal substrate for DNA synthesis. (b) Structure of dideoxyribonucleoside triphosphate, which lacks an OH group on the 3′-carbon atom.

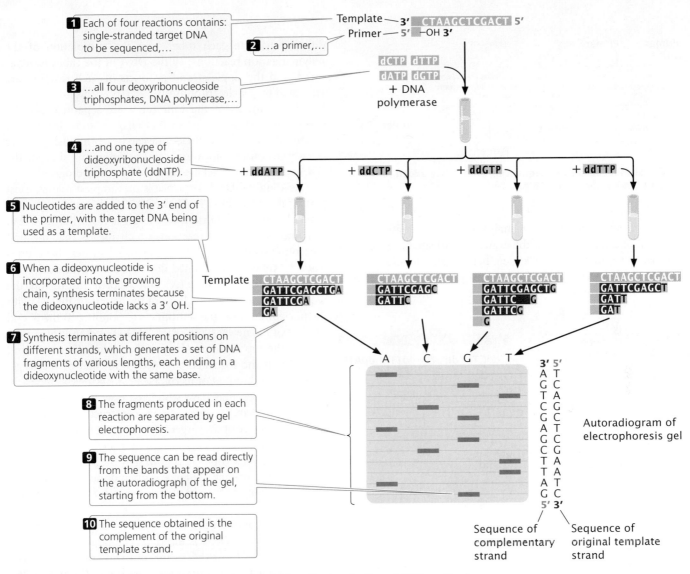

1 Each of four reactions contains: single-stranded target DNA to be sequenced,...

Template — **3'** CTAAGCTCGACT **5'**
Primer — **5'** ▭—OH **3'**

2 ...a primer,...

dCTP dTTP
dATP dGTP
+ DNA polymerase

3 ...all four deoxyribonucleoside triphosphates, DNA polymerase,...

4 ...and one type of dideoxyribonucleoside triphosphate (ddNTP).

+ **ddATP**   + **ddCTP**   + **ddGTP**   + **ddTTP**

5 Nucleotides are added to the 3' end of the primer, with the target DNA being used as a template.

6 When a dideoxynucleotide is incorporated into the growing chain, synthesis terminates because the dideoxynucleotide lacks a 3' OH.

Template CTAAGCTCGACT   CTAAGCTCGACT   CTAAGCTCGACT   CTAAGCTCGACT
          GATTCGAGCTGA   GATTCGAGC   GATTCGAGCTG   GATTCGAGCT
          GATTCGA   GATTC   GATTC G   GATT
          GA   GATTCG   GAT
                                          G

7 Synthesis terminates at different positions on different strands, which generates a set of DNA fragments of various lengths, each ending in a dideoxynucleotide with the same base.

A C G T

8 The fragments produced in each reaction are separated by gel electrophoresis.

9 The sequence can be read directly from the bands that appear on the autoradiograph of the gel, starting from the bottom.

10 The sequence obtained is the complement of the original template strand.

Autoradiogram of electrophoresis gel

3' 5'
A T
G C
T A
C G
G C
A T
G C
C G
T A
T A
A T
G C
5' 3'

Sequence of complementary strand

Sequence of original template strand

19.25 The dideoxy method of DNA sequencing is based on the termination of DNA synthesis.

▶ www.whfreeman.com/pierce4e   View Animation 19.2 to see how the dideoxy sequencing reaction works.

by cloning in bacteria. Copies of the target DNA are isolated and split into four parts (**Figure 19.25**). Each part is placed in a different tube, to which are added:

1. many copies of a primer that is complementary to one end of the target DNA strand;
2. all four types of deoxyribonucleoside triphosphates, the normal precursors of DNA synthesis;
3. a small amount of one of the four types of dideoxyribonucleoside triphosphates, which will terminate DNA synthesis as soon as it is incorporated into any growing chain (each of the four tubes received a different ddNTP); and
4. DNA polymerase.

Either the primer or one of the dNTPs is radioactively or chemically labeled so that newly produced DNA can be detected.

Within each of the four tubes, the DNA polymerase enzyme synthesizes DNA. Let's consider the reaction in one of the four tubes; the one that received ddATP. Within this tube, each of the single strands of target DNA serves as a template for DNA synthesis. The primer pairs to its complementary sequence at one end of each template strand, providing a 3'-OH group for the initiation of DNA synthesis. DNA polymerase elongates a new strand of DNA from this primer. Wherever DNA polymerase encounters a T on the template strand, it uses at random either a dATP or a ddATP to introduce an A in the newly synthesized strand. Because there is more dATP than ddATP in the reaction mixture, dATP is incorporated most often, allowing DNA synthesis to continue. Occasionally, however, ddATP is incorporated into the strand and synthesis terminates. The incorporation of ddA into the new strand occurs randomly at different positions in different

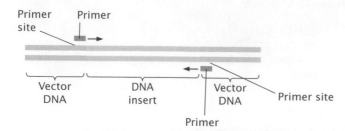

19.26 Sites recognized by sequencing primers are added to the target DNA by cloning the DNA.

copies, producing a set of DNA chains of different lengths (12, 7, and 2 nucleotides long in the example illustrated in Figure 19.25), each ending in a nucleotide that contains adenine.

Equivalent reactions take place in the other three tubes, except that synthesis is terminated at nucleotides with a

different base in each tube. After the completion of the polymerization reactions, all the DNA in the tubes is denatured, and the single-strand products of each reaction are separated by gel electrophoresis.

The contents of the four tubes are separated side by side on an acrylamide gel so that DNA strands differing in length by only a single nucleotide can be distinguished. After electrophoresis, the locations, and therefore the sizes, of the DNA strands in the gel are revealed by autoradiography.

Reading the DNA sequence is the simplest and shortest part of the procedure. In Figure 19.25, you can see that the band closest to the bottom of the gel is from the tube that contained the ddGTP reaction, which means that the first nucleotide synthesized had guanine (G). The next band up is from the tube that contained ddATP; so the next nucleotide in the sequence is adenine (A), and so forth. In this way, the sequence is read from the bottom to the top of the gel, with the nucleotides near the bottom corresponding to the 5′ end of the newly synthesized DNA strand and those near the top corresponding to the 3′ end. Keep in mind that the sequence obtained is not that of the target DNA but that of its *complement*. **TRY PROBLEM 37 →**

You may have wondered how the primers used in dideoxy sequencing are constructed, because the sequence of the target DNA may not be known ahead of time. The trick is to insert a sequence into the target DNA that will be recognized by the primer. Insertion of the primer sequence is often done by first cloning the target DNA in a vector that contains sequences (called universal sequencing primer sites) recognized by a common primer on either side of the site where the target DNA is inserted. The target DNA is then isolated from the vector and will contain universal sequencing primer sites at each end (**Figure 19.26**). Fragments amplified by PCR also can be sequenced. In this case, the primers used for PCR amplification can serve as primers for the sequencing reactions.

For many years, DNA sequencing was done largely by hand and was laborious and expensive. Today, sequencing is usually carried out by automated machines that use fluorescent dyes and laser scanners to sequence thousands of base pairs in a few hours (**Figure 19.27**). The dideoxy reaction is also used here, but the ddNTPs used in the reaction are labeled with fluorescent dyes, and a different colored dye is used for each type of dideoxynucleotide. In this case, the four sequencing reactions can take place in the same test tube and can be placed in the same well during electrophoresis. The most recently developed sequencing machines carry out electrophoresis in gel-containing capillary tubes. The different-size fragments produced by the sequencing reaction separate within a tube and migrate past a laser beam and detector. As the fragments pass the laser, their fluorescent dyes are activated and the resulting fluorescence is detected by an optical

1 A single-stranded DNA fragment whose base sequence is to be determined (the template) is isolated.

2 Each of the four ddNTPs is tagged with a different fluorescent dye, and the Sanger sequencing reaction is carried out.

3 The fragments that end in the same base have the same colored dye attached.

4 The products are denatured, and the DNA fragments produced by the four reactions are mixed and loaded into a single well on an electrophoresis gel. The fragments migrate through the gel according to size,…

5 …and the fluorescent dye on the DNA is detected by a laser beam.

6 Each fragment appears as a peak on the computer printout; the color of the peak indicates which base is present.

7 The sequence information is read directly into the computer, which converts it into the complementary target sequence.

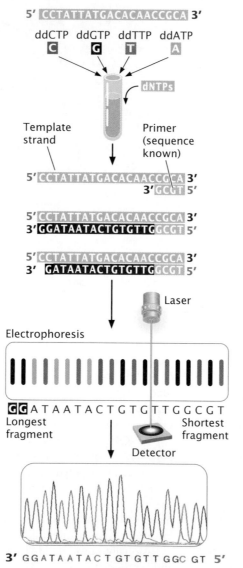

19.27 The dideoxy sequencing method can be automated.

scanner. Each colored dye emits fluorescence of a characteristic wavelength, which is read by the optical scanner. The information is fed into a computer for interpretation, and the results are printed out as a set of peaks on a graph (see Figure 19.27). Automated sequencing machines may contain 96 or more capillary tubes, allowing from 50,000 to 60,000 bp of sequence to be read in a few hours.

CONCEPTS

The dideoxy sequencing method uses ddNTPs, which terminate DNA synthesis at specific bases.

✔ CONCEPT CHECK 8

In the dideoxy sequencing reaction, what terminates DNA synthesis at a particular base?

a. The absence of a base on the ddNTP halts the DNA polymerase.
b. The ddNTP causes a break in the sugar–phosphate backbone.
c. DNA polymerase will not incorporate a ddNTP into the growing DNA strand.
d. The absence of a 3′-OH group on the ddNTP prevents the addition of another nucleotide.

Next-Generation Sequencing Technologies

New methods, called **next-generation sequencing technologies**, have made sequencing hundreds of times faster and less expensive than the traditional Sanger sequencing method. Most next-generation sequencing technologies do sequencing in parallel, which means that hundreds of thousands or even millions of DNA fragments are simultaneously sequenced. One type, called pyrosequencing, is based on DNA synthesis: nucleotides are added one at a time in the order specified by template DNA and the addition of a particular nucleotide is detected with a flash of light, which is generated as the nucleotide is added.

To carry out pyrosequencing, DNA to be sequenced is first fragmented (**Figure 19.28a**). Adaptors, consisting of a short string of nucleotides, are added to each fragment. The adaptor provides a known sequence to prime a PCR reaction. The DNA fragments are then made single stranded. In one version of pyrosequencing, each fragment is then attached to a separate bead and surrounded by a droplet of solution (**Figure 19.28b**). The bead is used to hold the DNA and later deposit it on a plate for the sequencing reaction (see below). Within the droplet, the fragment is then amplified by PCR and the copies of DNA remain attached to the bead. After amplification by PCR, the beads are mixed with DNA polymerase and are deposited on a plate containing more than a million wells (small holes). Each bead is deposited into a different well on the plate.

The sequencing reaction takes place in each well and is based on DNA synthesis. Recall from Chapter 12 that the substrate for DNA synthesis is a deoxynucleoside triphosphate, consisting of a deoxyribose sugar attached to a base and three phosphates. In the process of DNA synthesis, two phosphates (pyrophosphate) are cleaved off, and the resulting nucleotide is attached to the 3′ end of the growing DNA chain. A solution containing one particular type of deoxynucleoside triphosphate—say, deoxyadenosine triphosphate—is passed across the wells (**Figure 19.28c**). If the template within a particular well specifies an adenine nucleotide in the next position of the growing chain, then pyrophosphate is cleaved from the nucleoside triphosphate and the adenine nucleotide is added. A chemical reaction uses the pyrophosphate produced in the reaction to generate a flash of light, which is measured by an optical detector. The amount of light emitted in each well is proportional to the number of nucleotides added: if the template in a well specifies three successive adenine nucleotides (As), then three nucleotides are added and three times more light is emitted than if a single A were added. If the position in the template specifies a base other than adenine, no nucleotide is added, no pyrophosphate is produced, and no light is emitted.

As mentioned, the first solution passed over the plate contains adenosine triphosphate and allows adenine nucleotides to be added to the template. Each well with a template that specifies adenine in the next position will generate a flash of light. Then, a different type of nucleoside triphosphate—say, deoxyguanosine triphosphate—is passed across the wells. Any fragment that specifies a G in the next position of its growing chain will add a guanine nucleotide and emit a flash of light. The nucleotide triphosphates are passed across the well in a predetermined order, and the light emitted by each well is measured. In this way, hundreds of thousands or millions of fragments of DNA are sequenced simultaneously on the basis of the order in which nucleotides are added to the 3′ end of the growing chain. Pyrosequencing determines only the sequence of the fragments in each well; it does not, by itself, allow the sequences of these fragments to be reassembled into the sequence of the entire original piece of DNA. The reassembly of sequenced fragments into a continuous stretch of DNA is a general problem in the sequencing of genomes and will be explained in Chapter 20.

Most next-generation sequencing techniques read shorter DNA fragments than do the Sanger sequencing reactions but, because hundreds of thousands or millions of fragments are sequenced simultaneously, these methods are much faster than traditional Sanger sequencing technology. Even more rapid sequencing technologies, typically called third-generation sequencing, are currently under development.

CONCEPTS

New next-generation sequencing methods sequence many DNA fragments simultaneously and provide a much faster and less-expensive determination of a DNA base sequence than does the Sanger sequencing method.

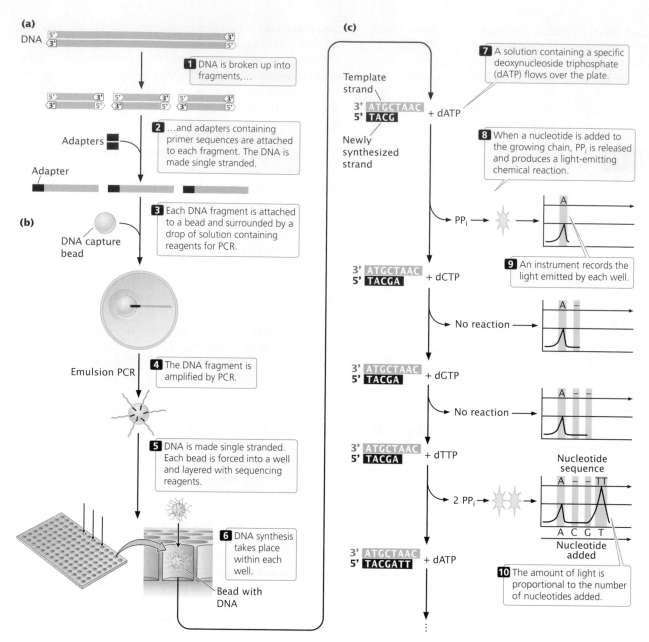

(a)

DNA

1 DNA is broken up into fragments,…

2 …and adapters containing primer sequences are attached to each fragment. The DNA is made single stranded.

Adapters

Adapter

3 Each DNA fragment is attached to a bead and surrounded by a drop of solution containing reagents for PCR.

(b)

DNA capture bead

Emulsion PCR

4 The DNA fragment is amplified by PCR.

5 DNA is made single stranded. Each bead is forced into a well and layered with sequencing reagents.

6 DNA synthesis takes place within each well.

Bead with DNA

(c)

Template strand

Newly synthesized strand

3' ATGCTAAC
5' TACG   + dATP

7 A solution containing a specific deoxynucleoside triphosphate (dATP) flows over the plate.

8 When a nucleotide is added to the growing chain, PP_i is released and produces a light-emitting chemical reaction.

PP_i →

A

9 An instrument records the light emitted by each well.

3' ATGCTAAC
5' TACGA   + dCTP

No reaction →

A –

3' ATGCTAAC
5' TACGA   + dGTP

No reaction →

A – –

3' ATGCTAAC
5' TACGA   + dTTP

2 PP_i →

Nucleotide sequence

A – – TT

A C G T
Nucleotide added

3' ATGCTAAC
5' TACGATT   + dATP

10 The amount of light is proportional to the number of nucleotides added.

19.28 Next-generation sequencing methods are able to simultaneously determine the sequence of hundreds of thousands or millions of DNA fragments. Pyrosequencing is illustrated here.

DNA Fingerprinting

The use of DNA sequences to identify individual persons is called **DNA fingerprinting** or DNA profiling. Because some parts of the genome are highly variable, each person's DNA sequence is unique and, like a traditional fingerprint, provides a distinctive characteristic that allows identification.

Today, most DNA fingerprinting utilizes **microsatellites**, or **short tandem repeats** (STRs), which are very short DNA sequences repeated in tandem (see Chapter 11). Microsatellites are found at many loci throughout the human genome. People vary in the number of copies of repeat sequences that they possess at each locus. Microsatellites are typically detected with PCR, with the use of primers flanking the microsatellite repeats,

so that a DNA fragment containing the repeated sequences is amplified. The length of the amplified segment depends on the number of repeats; DNA from a person with more repeats will produce a longer amplified segment than will DNA from a person with fewer repeats. After PCR has been completed, the amplified fragments are separated by gel electrophoresis and stained, producing a series of bands on a gel (**Figure 19.29**). When several different microsatellite loci are examined, the probability that two people have the same set of patterns becomes vanishingly small, unless they are identical twins.

In a typical application, DNA fingerprinting might be used to confirm that a suspect was present at the scene of a crime (**Figure 19.30**). A sample of DNA from blood, semen,

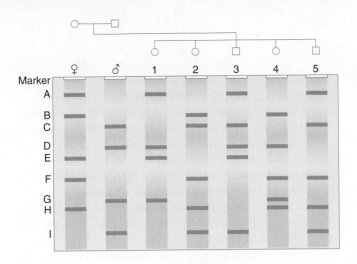

19.29 Variation in banding patterns reveals inherited differences in microsatellite sequences. Microsatellite variation within a family. All bands found in the children are present in the parents. [From A. Griffiths, S. Wessler, R. Lewontin, W. Gelbart, D. Suzuki, and J. Miller, *Introduction to Genetic Analysis*, 8th ed. © 2005 by W. H. Freeman and Company.]

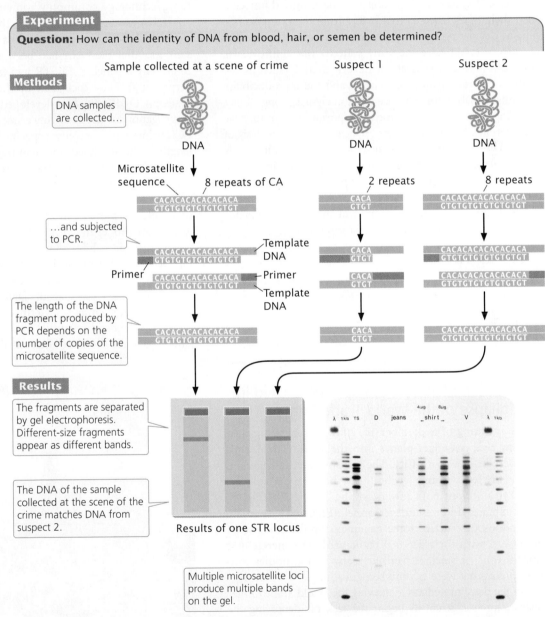

Experiment

Question: How can the identity of DNA from blood, hair, or semen be determined?

Methods

Sample collected at a scene of crime

Suspect 1

Suspect 2

DNA samples are collected...

DNA

DNA

DNA

Microsatellite sequence

8 repeats of CA

2 repeats

8 repeats

CACACACACACACACA
GTGTGTGTGTGTGTGT

CACA
GTGT

CACACACACACACACA
GTGTGTGTGTGTGTGT

...and subjected to PCR.

Template DNA

CACACACACACACACA
GTGTGTGTGTGTGTGT

Primer

CACACACACACACACA
GTGTGTGTGTGTGTGT

Primer

Template DNA

CACA
GTGT

CACA
GTGT

CACACACACACACACA
GTGTGTGTGTGTGTGT

CACACACACACACACA
GTGTGTGTGTGTGTGT

The length of the DNA fragment produced by PCR depends on the number of copies of the microsatellite sequence.

CACACACACACACACA
GTGTGTGTGTGTGTGT

CACA
GTGT

CACACACACACACACA
GTGTGTGTGTGTGTGT

Results

The fragments are separated by gel electrophoresis. Different-size fragments appear as different bands.

The DNA of the sample collected at the scene of the crime matches DNA from suspect 2.

Results of one STR locus

Multiple microsatellite loci produce multiple bands on the gel.

Conclusion: The patterns of bands produced by different samples are compared. The DNA sample collected at the crime scene matches DNA from suspect 2.

19.30 DNA fingerprinting can be used to identify a person. [Gel courtesy of Orchard Cellmark, Germantown, Maryland.]

hair, or other body tissue is collected from the crime scene. If the sample is very small, PCR can be used to amplify it so that enough DNA is available for testing. Additional DNA samples are collected from one or more suspects. The pattern of bands produced by DNA fingerprinting from the sample collected at the crime scene is then compared with the patterns produced by DNA fingerprinting of the DNA from the suspects. A match between the sample from the crime scene and one from a suspect can provide evidence that the suspect was present at the scene of the crime.

Since its introduction in the 1980s, DNA fingerprinting has helped convict a number of suspects in murder and rape cases. Suspects in other cases have been proved innocent when their DNA failed to match that from the crime scenes. Initially, calculating the odds of a match (the probability that two people could have the same pattern) was controversial, and there were concerns about quality control (such as the accidental contamination of samples and the reproducibility of results) in laboratories where DNA analysis is done. Today, DNA fingerprinting has become an important tool in forensic investigations. In addition to its application in the analysis of crimes, DNA fingerprinting is used to assess paternity, study genetic relationships among individual organisms in natural populations, identify specific strains of pathogenic bacteria, and identify human remains. For example, DNA fingerprinting was used to determine that several samples of anthrax mailed to different people in 2001 were all from the same source.

CONCEPTS

DNA fingerprinting detects genetic differences among people by using probes for highly variable regions of chromosomes.

✔ **CONCEPT CHECK 9**

How are microsatellites detected?

Application: Identifying People Who Died in the Collapse of the World Trade Center

On the morning of September 11, 2001, terrorists hijacked and flew two passenger planes into the World Trade Center towers in New York City. The catastrophic damage and ensuing fire led, within a few hours, to the complete collapse of all 110 floors of both towers, killing almost 3000 building occupants and rescue personnel (**Figure 19.31**). The tremendous destructive force generated by the collapse of the towers, with their 425,000 cubic yards of concrete and 200,000 tons of steel, pulverized many of the bodies into small pieces that were beyond recognition.

In the days immediately following the World Trade Center collapse, forensic scientists began the task of identifying the remains of those who perished in the disaster. The goal was to provide evidence for the ongoing criminal investigation of the attack and to identify the remains at the site for families and friends of the victims. This task was unprecedented in scope and difficulty. There was no complete list of victims (such as a passen-

ger list in an airline crash) with which investigators could match the remains. In all, almost 20,000 individual remains were found, varying from whole bodies to tiny fragments of charred bone. The remains were subjected to fires with temperatures exceeding 1000°C that burned for more than 3 months. The collapse of the buildings intermixed many victims' remains, and many body fragments were not recovered for months, during which time they were exposed to dust, water, bacteria, and decay.

The usual means of victim identification—personal items, fingerprints, dental records—were of little use for most of the World Trade Center remains. Identification of the majority of the remains was made with the use of DNA profiling technology, commonly known as DNA fingerprinting.

DNA was first extracted from the tissue samples by using sterile techniques to prevent cross-contamination between samples. After the DNA had been extracted, PCR was used to amplify a panel of 13 STR loci (see section on DNA fingerprinting). These loci make up the Combined DNA Index System (CODIS), a system developed by the Federal Bureau of Investigation for use in solving crimes. For the CODIS system, PCR primers that amplify a specific STR locus are used. The length of the amplified fragment depends on the number of repeats. After PCR, the fragments are separated on a gel or by a capillary electrophoresis machine. Each STR locus in the CODIS system has a large number of alleles and is located on a different human chromosome, and so variation at each locus assorts independently. When all 13 CODIS loci are used together, the probability of two randomly selected people having the same DNA profile is less than 1 in 10 billion.

The DNA fingerprint generated from each body sample was compared with that of DNA extracted from reference samples, such as the victims' toothbrushes and blood samples, provided by families and friends. If the DNA from a body part had the same alleles at all 13 loci as the reference

19.31 DNA fingerprinting was used to identify the remains of people who died in the collapse of the World Trade Center. Ground zero after the catastrophic collapse of the 110-story twin towers. [Michael Reigner/Mai/Mai/Time Life Pictures/Getty Images.]

sample, then a positive identification was made. When no reference sample was available, investigators collected DNA from family members and tried to match the DNA profiles of remains to those of relatives; in this case, some, but not all STR alleles would match.

Unfortunately, many of the remains were so badly degraded that little DNA remained and one or more of the STR loci could not be amplified. For these remains, DNA fingerprinting was also carried out on mitochondrial DNA (see Chapter 21). Because there are many mitochondria per cell and each mitochondrion contains multiple DNA molecules, there are many more copies of mitochondrial DNA per cell than nuclear DNA; mitochondrial DNA has been successfully extracted and analyzed from ancient remains, such as Neanderthals (see introduction to Chapter 10). Alone, fingerprinting from mitochondrial DNA was insufficient to provide identification with a high degree of confidence (because there are not as many sequences that vary among people as in the CODIS loci) but, when it was used in conjunction with data from at least some STR loci, a positive identification could often be made.

Used in combination, these techniques allowed the remains of many victims to be positively identified. Despite the heroic efforts of hundreds of molecular geneticists, forensic anthropologists, and medical examiners to identify the remains, no positively identified remains were recovered for almost half of the people who are thought to have died in the disaster.

19.5 Molecular Techniques Are Increasingly Used to Analyze Gene Function

In the preceding sections, we have learned about powerful molecular techniques for isolating, recombining, and analyzing DNA sequences. Although these techniques provide a great deal of information about the organization and nature of gene sequences, the ultimate goal of many molecular studies is to better understand the function of the sequences. In this section, we will explore some advanced molecular techniques that are frequently used to determine gene function and to better understand the genetic processes that these sequences undergo.

Forward and Reverse Genetics

The traditional approach to the study of gene function begins with the isolation of mutant organisms. For example, suppose a geneticist is interested in genes that affect cardiac function in mammals. A first step would be to find individuals—perhaps mice—that have hereditary defects in heart function. The mutations causing the cardiac problems in the mice could then be mapped, and the implicated genes could be isolated, cloned, and sequenced. The proteins produced by the genes could then be predicted from the gene sequences and isolated. Finally, the biochemistry of the

proteins could be studied and their role in heart function discerned. This approach, which begins with a phenotype (a mutant individual) and proceeds to a gene that encodes the phenotype, is called **forward genetics**.

An alternative approach, made possible by advances in molecular genetics, is to begin with a genotype—a DNA sequence—and proceed to the phenotype by altering the sequence or inhibiting its expression. A geneticist might begin with a gene of unknown function, induce mutations in it, and then look to see what effect these mutations have on the phenotype of the organism. This approach is called **reverse genetics**. Today, both forward and reverse genetic approaches are widely used in analyses of gene function. This section introduces some of the molecular techniques that are used in forward and reverse genetics.

Creating Random Mutations

Forward genetics depends on the identification and isolation of random mutations that affect a phenotype of interest. Early in the study of genetics, geneticists were forced to rely on naturally occurring mutations, which are usually rare and can be detected only if large numbers of organisms are examined. The discovery of mutagenic agents—environmental factors that increase the rate of mutation (see Chapter 18)—provided a means increasing the number of mutants in experimental populations of organisms. One of the first examples of experimentally created mutations was Hermann Muller's use of X-rays in 1927 to induce X-linked mutations in *Drosophila melanogaster*.

Today, radiation, chemical mutagens, and transposable elements are all used to create mutations for genetic analysis. The choice of mutagen is an important consideration because different mutagens produce different types of mutations. For example, some mutagens predominately produce base-pair substitutions, whereas other mutagens produce insertions and deletions. The dose of mutagen also is critical. The mutagen needs to increase the number of mutations so that enough mutants are obtained for analysis. However, too high a mutation rate may cause multiple mutations to appear in the same individual organism, making identification of the gene responsible for a mutant phenotype difficult, or may even kill the organism before the mutant phenotype can be observed.

To determine all genes that might affect a phenotype, the creation of mutations in as many genes as possible is desirable—to saturate the genome with mutations. This procedure is done with a mutagenic screen, which will be described in Chapter 20.

Site-Directed Mutagensis

Reverse genetics depends on the ability to create mutations, not at random, but in particular DNA sequences, and then to study the effects of these mutations on the organism. Mutations are induced at specific locations through a process called **site-directed mutagenesis**.

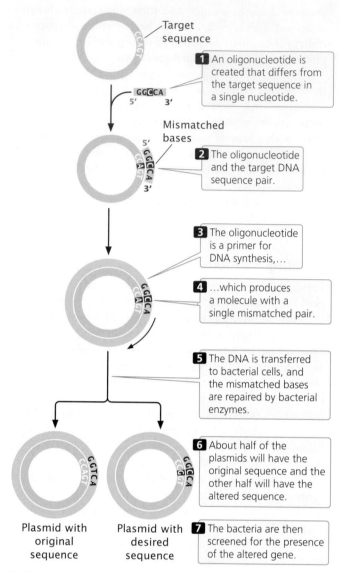

Target sequence

1 An oligonucleotide is created that differs from the target sequence in a single nucleotide.

GGCCA
5'   3'

Mismatched bases

2 The oligonucleotide and the target DNA sequence pair.

3 The oligonucleotide is a primer for DNA synthesis,…

4 …which produces a molecule with a single mismatched pair.

5 The DNA is transferred to bacterial cells, and the mismatched bases are repaired by bacterial enzymes.

6 About half of the plasmids will have the original sequence and the other half will have the altered sequence.

Plasmid with original sequence

Plasmid with desired sequence

7 The bacteria are then screened for the presence of the altered gene.

19.32 Oligonucleotide-directed mutagenesis is used to study gene function when appropriate restriction sites are not available.

A number of different strategies have been developed for site-directed mutagenesis. One strategy that is often used in bacteria is to cut out a short sequence of nucleotides with restriction enzymes and replace it with a short, synthetic oligonucleotide that contains the desired mutated sequence. The success of this method depends on the availability of restriction sites flanking the sequence to be altered.

If appropriate restriction sites are not available, **oligonucleotide-directed mutagenesis** can be used (**Figure 19.32**). In this method, a single-stranded oligonucleotide is produced that differs from the target sequence by one or a few bases. Because they differ in only a few bases, the target DNA and the oligonucleotide will pair. When successfully paired with

the target DNA, the oligonucleotide can act as a primer to initiate DNA synthesis, which produces a double-stranded molecule with a mismatch in the primer region. When this DNA is transferred to bacterial cells, the mismatched bases will be repaired by bacterial enzymes. About half of the time the normal bases will be changed into mutant bases, and about half of the time the mutant bases will be changed into normal bases. The bacteria are then screened for the presence of the mutant gene.

Transgenic Animals

Another way that gene function can be analyzed is by adding DNA sequences of interest to the genome of an organism that normally lacks such sequences and then seeing the effect that the introduced sequence has on the organism's phenotype. This method is a form of reverse genetics. An organism that has been permanently altered by the addition of a DNA sequence to its genome is said to be *transgenic*, and the foreign DNA that it carries is called a **transgene**. Here, we consider techniques for the creation of transgenic mice, which are often used in the study of the function of human genes because they can be genetically manipulated in ways that are impossible with humans and, as mammals, they are more similar to humans than are fruit flies, fish, and other model genetic organisms.

The oocytes of mice and other mammals are large enough that DNA can be injected into them directly. Immediately after penetration by a sperm, a fertilized mouse egg contains two pronuclei, one from the sperm and one from the egg; these pronuclei later fuse to form the nucleus of the embryo. Mechanical devices can manipulate extremely fine, hollow glass needles to inject DNA directly into one of the pronuclei of a fertilized egg (**Figure 19.33**). Typically, a few hundred copies of cloned, linear DNA are injected into a pronucleus, and, in a few of the injected eggs, copies of the cloned DNA integrate randomly into one of the chromosomes through a process called nonhomologous recombination. After injection, the embryos are implanted in a pseudopregnant female—a surrogate mother that has been physiologically prepared for pregnancy by mating with a vasectomized male.

Only about 10% to 30% of the eggs survive and, of those that do survive, only a few have a copy of the cloned DNA stably integrated into a chromosome. Nevertheless, if several hundred embryos are injected and implanted, there is a good chance that one or more mice whose chromosomes contain the foreign DNA will be born. Moreover, because the DNA was injected at the one-cell stage of the embryo, these mice usually carry the cloned DNA in every cell of their bodies, including their reproductive cells, and will therefore pass the foreign DNA on to their progeny. Through interbreeding, a strain of mice that is homozygous for the foreign gene can be created.

Transgenic mice have proved useful in the study of gene function. For example, proof that the *SRY* gene (see Chapter 4) is the male-determining gene in mice was obtained by injecting a copy of the *SRY* gene into XX embryos and observing that these mice developed as males. In addition, researchers have created a number of transgenic mouse strains that serve as experimental models for human genetic diseases.

Knockout Mice

A useful variant of the transgenic approach is to produce mice in which a normal gene has been not just mutated, but fully disabled. These animals, called **knockout mice**, are particularly helpful in determining the function of a gene: the phenotype of the knockout mouse often gives a good indication of the function of the missing gene.

The creation of knockout mice begins when a normal gene is cloned in bacteria and then "knocked out," or disabled. There are a number of ways to disable a gene, but a common method is to insert a gene called *neo*, which confers resistance to the antibiotic G418, into the middle of the target gene (**Figure 19.34**). The insertion of *neo* both disrupts (knocks out) the target gene and provides a convenient marker for finding copies of the disabled gene. In addition, a second gene, usually the herpes simplex viral thymidine kinase (*tk*) gene, is linked to the disrupted gene. The disabled gene is then transferred to cultured embryonic mouse cells, where it may exchange places with the normal copy on the mouse chromosome through homologous recombination.

After the disabled gene has been transferred to the embryonic cells, the cells are screened by adding the antibiotic G418 to the medium. Only cells with the disabled gene containing the *neo* insert will survive. Because the frequency of nonhomologous recombination is higher than that of homologous recombination and because the intact target gene is replaced by the disabled copy only through homologous recombination, a means to select for the rarer

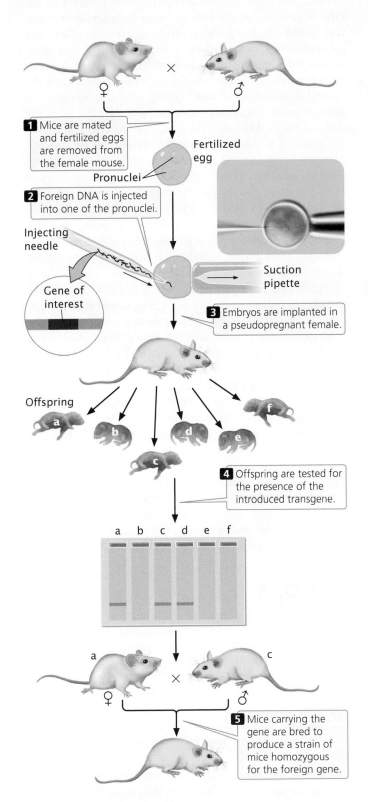

19.33 Transgenic animals have genomes that have been permanently altered through recombinant DNA technology. In the photograph, a mouse embryo is being injected with DNA. [Photograph: Chad Davis/PhotoDisc.]

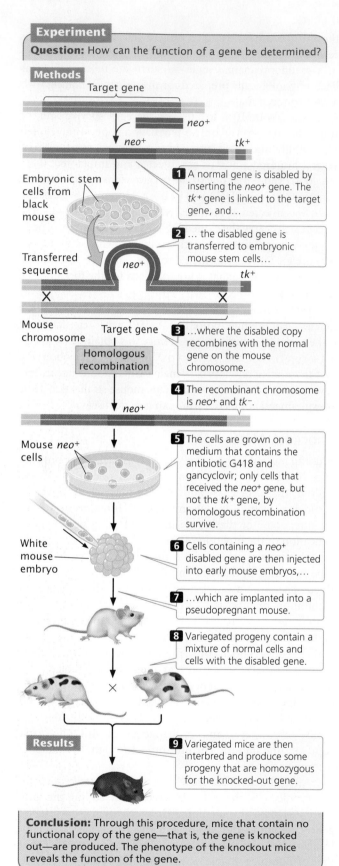

Methods

Target gene

neo⁺

neo⁺   tk⁺

Embryonic stem cells from black mouse

1 A normal gene is disabled by inserting the *neo*⁺ gene. The *tk*⁺ gene is linked to the target gene, and…

2 … the disabled gene is transferred to embryonic mouse stem cells…

Transferred sequence

neo⁺

tk⁺

Mouse chromosome

Target gene

3 …where the disabled copy recombines with the normal gene on the mouse chromosome.

Homologous recombination

4 The recombinant chromosome is *neo*⁺ and *tk*⁻.

neo⁺

Mouse *neo*⁺ cells

5 The cells are grown on a medium that contains the antibiotic G418 and gancyclovir; only cells that received the *neo*⁺ gene, but not the *tk*⁺ gene, by homologous recombination survive.

White mouse embryo

6 Cells containing a *neo*⁺ disabled gene are then injected into early mouse embryos,…

7 …which are implanted into a pseudopregnant mouse.

8 Variegated progeny contain a mixture of normal cells and cells with the disabled gene.

✕

Results

9 Variegated mice are then interbred and produce some progeny that are homozygous for the knocked-out gene.

Conclusion: Through this procedure, mice that contain no functional copy of the gene—that is, the gene is knocked out—are produced. The phenotype of the knockout mice reveals the function of the gene.

19.34 Knockout mice possess a genome in which a gene has been disabled.

homologous recombinants is required. The presence of the viral *tk* gene makes the cells sensitive to gancyclovir. Thus, transfected cells that grow on medium containing G418 and gancyclovir will contain the *neo* gene (disabled target gene) but not the adjacent *tk* gene, because the *tk* gene will be eliminated in the double-crossover event. These cells contain the desired homologous recombinants. The nonhomologous recombinants (random insertions) will contain both the *neo* and the *tk* genes, and these transfected cells will die on the selection medium owing to the presence of gancyclovir. The surviving cells are injected into an early-stage mouse embryo, which is then implanted into a pseudopregnant mouse. Cells in the embryo carrying the disabled gene and normal embryonic cells carrying the wild-type gene will develop together, producing a chimera—a mouse that is a genetic mixture of the two cell types. The production of chimeric mice is not itself desirable, but replacing all the cells of the embryo with injected cells is impossible.

Chimeric mice can be identified easily if the injected embryonic cells came from a black mouse and the embryos into which they are injected came from a white mouse; the resulting chimeras will have variegated black and white fur. The chimeras can then be interbred to produce some progeny that are homozygous for the knockout gene. The effects of disabling a particular gene can be observed in these homozygous mice. Three scientists who helped develop the techniques for creating knockout mice—Mario Capecchi, Oliver Smithies, and Martin Evans—were awarded the Nobel Prize in physiology or medicine in 2007 for this work.

A variant of the knockout procedure is to insert in mice a particular DNA sequence into a known chromosome location. For example, researchers might insert the sequence of a human disease-causing allele into the same locus in mice, creating a precise mouse model of the human disease. Mice that carry inserted sequences at specific locations are called **knock-in mice**.

CONCEPTS

A transgenic mouse is produced by the injection of cloned DNA into the pronucleus of a fertilized egg, followed by implantation of the egg into a female mouse. In knockout mice, the injected DNA contains a mutation that disables a gene. Inside the mouse embryo, the disabled copy of the gene can exchange with the normal copy of the gene through homologous recombination.

✔ CONCEPT CHECK 11

What is the advantage of using the *neo* gene to disrupt the function of a gene in knockout mice?

a. The *neo* gene produces an antibiotic that kills unwanted cells.

b. The *neo* gene is the right size for disabling other genes.

c. The *neo* gene provides a selectable marker for finding cells that contain the disabled gene.

d. The *neo* gene produces a toxin that inhibits transcription of the target gene.

Silencing Genes with RNAi

In the preceding sections, we considered the analysis of gene function by introducing mutations or new DNA sequences into the genome and analyzing the resulting phenotype to provide information about the function of the altered or introduced DNA. We could also analyze gene function by temporarily turning a gene off and seeing what effect the absence of the gene product has on the phenotype. For many years, there was no method for selectively affecting gene expression. However, the discoveries of siRNAs (small interfering RNAs) and miRNAs (microRNAs; see Chapters 14 and 17) provided powerful tools for controlling the expression of individual genes.

Recall that siRNAs and miRNAs are small RNA molecules that combine with proteins to form the RNA-induced silencing complex (RISC). The RISC pairs with complementary sequences on mRNA and either cleaves the mRNA or prevents the mRNA from being translated. Molecular geneticists have exploited this natural machinery (called RNA interference or RNAi) for turning off the expression of specific genes. Studying the effect of silencing a gene with siRNA can often be a source of insight into the gene's function.

The first step in using RNAi technology is to design the siRNAs such that they will be recognized and cleaved by Dicer (the protein that processes siRNAs; see Chapter 14). The complementary sequence must be unique to the target mRNA and not be found on other mRNAs so that the siRNA will not inhibit nontarget mRNAs. Computer programs are often used to design optimal siRNAs.

After the siRNA sequence has been designed, it must be synthesized. One way to do so is to use an oligonucleotide synthesizer to synthesize a DNA fragment corresponding to the siRNA sequence. The synthesized oligonucleotide can be cloned into a plasmid expression vector between two strong promoters (**Figure 19.35**). *Escherichia coli* are then transformed with the plasmid. Within the bacteria, transcription

from the two promoters will proceed in both directions, producing two complementary RNA molecules that will pair to form a double-stranded RNA molecule recognized by Dicer. Alternatively, double-stranded RNA sequences can be synthesized directly with a gene synthesizer.

The next task is to deliver the double-stranded siRNA to the cells. Delivery can be done in a variety of ways, depending on the cell type. *Caenorhabditis elegans* worms can be fed *E. coli* (their natural food) containing the expression vector. Transcription within the bacteria produces double-stranded RNA, which the worms ingest and incorporate into their cells. Alternatively, double-stranded siRNA can be injected directly into cells or the body cavity. Yet another approach is to synthesize a short sequence of DNA that has internal complementarity so that, when transcribed, it folds up into a short hairpin RNA (shRNA) with a double-stranded section. Within a cell, the shRNAs are processed by Dicer to produce siRNAs that bring about gene silencing. DNA sequences containing siRNAs can be introduced into a vector with the use of standard cloning techniques, and the vector can be used to deliver the DNA into a cell. An advantage of this approach is that, with the addition of a DNA sequence, the RNAi sequence has the potential to become a permanent part of the cell's genome and be passed on to progeny.

The use of RNAi for probing gene function is illustrated by a study of genes that affect male fertility. On the basis of chemical analysis of the chromatin structure in sperm, researchers had identified a group of 132 proteins that might have roles in spermatogenesis and fertility. They created siRNAs that targeted the mRNAs for these 132 proteins and injected the siRNAs into *C. elegans* worms. They then measured the fertility of the treated animals. Of the 132 candidate genes, RNAi treatment in 50 of these genes caused sterility or embryonic death. Many of these same genes are also found in humans, and this study identified a number of genes in humans that potentially take part in human fertility.

TRY PROBLEM 41 →

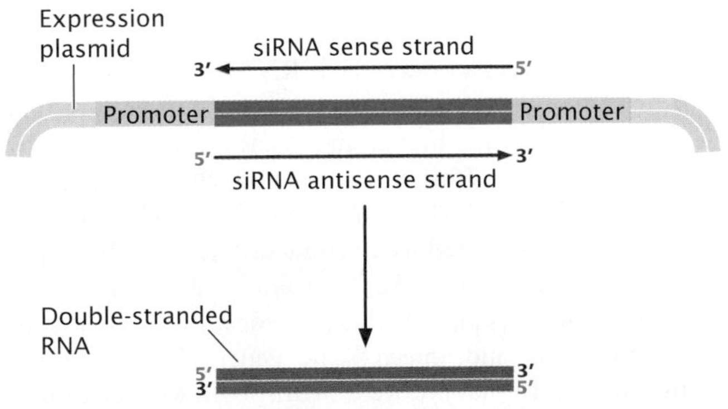

Application: Using RNAi for the Treatment of Human Disease

In addition to its value in determining gene function, RNAi holds great potential as a therapeutic agent for the future treatment of human diseases. This potential includes using siRNAs against RNA viruses, such as HIV, as well treating genetic diseases and cancer.

Treatment of high cholesterol with RNAi Recent research has examined the potential of RNAi for the treatment of disorders of cholesterol metabolism. Although cholesterol is essential for life, too much cholesterol is unhealthy: high blood cholesterol is a major contributor to heart disease, the leading cause of death in the United States. Cholesterol is normally transported throughout the body in the form of small particles called lipoproteins, which consist of a core of lipids surrounded by a shell of phospholipids and proteins

19.35 Small interfering RNAs can be produced by cloning DNA sequences corresponding to the siRNAs between two strong promoters. When cloned into an expression vector, both DNA strands will be transcribed and the complementary RNA molecules will anneal to form double-stranded RNA that will be processed into siRNA by Dicer.

(see Chapter 6). The ApoB protein is an essential part of lipoproteins. Some people possess genetic mutations that cause elevated levels of ApoB, which predisposes them to coronary artery disease. Findings from studies suggest that lowering the amount of ApoB can reduce the number of lipoproteins and lower blood cholesterol in these people, as well as in people who have elevated cholesterol for other reasons.

In 2006, Tracy Zimmermann and her colleagues at Alnylam Pharmaceuticals and Protiva Biotherapeutics demonstrated that RNAi could be used to reduce the levels of ApoB and blood cholesterol in nonhuman primates. The investigators first created siRNAs that targeted *apoB* gene expression (apoB-siRNAs) on the basis of the known sequence of the gene. The apoB-siRNAs were synthesized in the laboratory and consisted of 21 nucleotides on the sense strand (the strand that was complementary to the *apoB* mRNA) and 23 nucleotides on the complementary antisense strand, with a two-nucleotide overhang.

The next task was to get the apoB-siRNA into the cell. Although siRNAs are readily taken up by the cells of invertebrates such as *C. elegans*, most siRNAs will not readily pass through the membranes of mammalian cells in a form that is still effective in gene silencing. In addition, siRNAs are rapidly removed from circulation. To overcome these problems, Zimmermann and her colleagues encapsulated the apoB-siRNAs in lipids, creating stable nucleic-acid–lipid particles (SNALPs). The SNALPs greatly increased the time spent by the siRNAs in circulation and enhanced their uptake by the cell.

The researchers then tested the effects of the apoB-siRNAs on reducing the synthesis of the ApoB protein and on cholesterol levels. They injected cynomolgus monkeys (**Figure 19.36**) with SNALPs containing apoB-siRNA at two

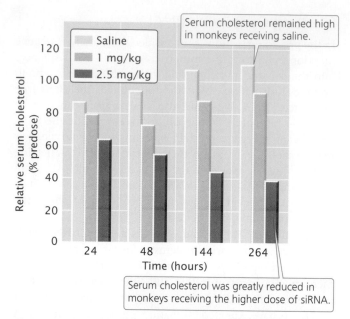

19.37 Treatment of cynomolgus monkeys with siRNAs for the ApoB protein significantly lowered blood-cholesterol levels. Different groups of monkeys were given saline (control), a 1-mg/kg dose, or a 2.5-mg/kg dose of siRNAs. [T. S. Zimmermann, *Nature* 441:112, 2006, Figure 3b.]

doses: 1 mg/kg and 2.5 mg/kg. In addition, they injected a third group of monkeys with saline as a control. They found that the apoB-siRNA clearly silenced the *apoB* gene: 48 hours after treatment, *apoB* mRNA in the liver was reduced by 68% for monkeys receiving the 1-mg dose and 90% for monkeys receiving the 2.5-mg dose.

When the researchers examined serum cholesterol levels in the monkeys, they found that monkeys receiving the apoB-siRNA had a significant reduction in blood-cholesterol levels (**Figure 19.37**). Importantly, they observed no negative effects of the siRNA treatment. Although preliminary, this study suggests that siRNAs have great potential for future treatment of human diseases.

Treament of liver cancer with RNAi Another potential application of RNAi is the treatment of cancer. Research has demonstrated that human liver-cancer cells consistently express a microRNA (miRNA) called miR-26a at lower levels than do normal cells. This miRNA normally helps to regulate the cell cycle; its reduced production leads to impaired cell-cycle control, abnormal cell division, and cancer. Janaiah Kota and colleagues placed sequences for miR-26a in adeno-associated virus and injected the genetically engineered viruses into mice that are were predisposed to liver cancer. Only 2 out of 10 mice that received miR-26a sequences developed cancer, whereas 6 out of 10 control mice that received adeno-associated virus without the added miR-26a sequences developed cancer. These results suggest that miR-NAs might one day be used to effectively treat cancer.

19.36 Transfer of siRNAs for the ApoB protein into cynomolgus monkeys demonstrated the potential use of siRNAs for the treatment of hypercholesterolemia. [Ken Lucas/Visuals Unlimited.]

19.6 Biotechnology Harnesses the Power of Molecular Genetics

In addition to providing valuable new information about the nature and function of genes, techniques of molecular genetics have many practical applications. These applications include the production of pharmaceutical products and other chemicals, specialized bacteria, agriculturally important plants, and genetically engineered farm animals. The technology is also used extensively in medical testing and, in a few cases, is even used to correct human genetic defects. Hundreds of companies now specialize in developing products through genetic engineering, and many large multinational corporations have invested enormous sums of money in molecular genetics research. As discussed earlier, the analysis of DNA is also used in criminal investigations and for the identification of human remains.

Pharmaceutical Products

The first commercial products to be developed with the use of genetic engineering were pharmaceutical products used in the treatment of human diseases and disorders. In 1979, the Eli Lilly corporation began selling human insulin produced with the use of recombinant DNA technology. The gene for human insulin was inserted into plasmids and transferred to bacteria that then produced human insulin. Pharmaceutical products produced through recombinant DNA technology include human growth hormone (for children with growth deficiencies), clotting factors (for hemophiliacs), and tissue plasminogen activator (used to dissolve blood clots in heart-attack patients).

Specialized Bacteria

Bacteria play an important role in many industrial processes, including the production of ethanol from plant material, the leaching of minerals from ore, and the treatment of sewage and other wastes. The bacteria used in these processes are being modified by genetic engineering so that they work more efficiently. New strains of technologically useful bacteria are being developed that will break down toxic chemicals and pollutants, enhance oil recovery, increase nitrogen uptake by plants, and inhibit the growth of pathogenic bacteria and fungi.

Agricultural Products

Recombinant DNA technology has had a major effect on agriculture, where it is now used to create crop plants and domestic animals with valuable traits. For many years, plant pathologists had recognized that plants infected with mild strains of viruses are resistant to infection by virulent strains. Using this knowledge, geneticists have created viral resistance in plants by transferring genes for viral proteins to the plant cells. A genetically engineered squash, called Freedom II, carries genes from the watermelon mosaic virus 2 and the zucchini yellow mosaic virus that protect the squash against viral infections.

Another objective has been to genetically engineer pest resistance into plants to reduce dependence on chemical pesticides. As discussed earlier in the chapter, a gene from the bacterium *Bacillus thuringiensis* that produces an insecticidal toxin has been transferred into corn, tomato, potato, and cotton plants. These Bt crops are now grown worldwide. Other genes that confer resistance to viruses and herbicides have been introduced into a number of crop plants. In 2009, 69.3 million hectares of genetically engineered soybeans, 16.17 million hectares of genetically engineered cotton, and 41.08 million hectares of genetically engineered corn were grown throughout the world. In the United States, 85% of all corn and 91% of all soybeans grown were genetically engineered.

Recombinant DNA techniques are also applied to domestic animals. For example, the gene for growth hormone was isolated from cattle and cloned in *E. coli*; these bacteria produce large quantities of bovine growth hormone, which is administered to dairy cattle to increase milk production. Transgenic animals are being developed to carry genes that encode pharmaceutical products; some eukaryotic proteins must be modified after translation, and only other eukaryotes (but not bacteria) are capable of carrying out the modifications. For example, a gene for human clotting factor VIII has been attached to the regulatory region of the sheep gene for β-lactoglobulin, a milk protein. The fused gene was injected in sheep embryos, creating transgenic sheep that produce in their milk the human clotting factor, which is used to treat hemophiliacs.

The genetic engineering of agricultural products is controversial. One area of concern focuses on the potential effects of releasing novel organisms produced by genetic engineering into the environment. There are many examples in which nonnative organisms released into a new environment have caused ecological disruption because they are free of predators and other natural control mechanisms. Genetic engineering normally transfers only small sequences of DNA, relative to the large genetic differences that often exist between species, but even small genetic differences may alter ecologically important traits that might affect the ecosystem.

Another area of concern is the effect of genetically engineered crops on biodiversity. In the largest field test of genetically engineered plants ever conducted, scientists cultivated beets, corn, and oilseed rape that were genetically engineered to resist herbicide along with traditional crops on 200 test plots throughout the United Kingdom. They then measured the biodiversity of native plants and animals in the agricultural fields. They found that the genetically engineered plants were highly successful in the suppression of weeds; however, plots with genetically engineered beets and oilseed rape have significantly fewer insects that feed on weeds. For example, plots with genetically engineered oilseed rape had 24% fewer butterflies than did plots with traditional crops.

There is also concern that transgenic organisms may hybridize with native organisms and transfer their genetically engineered traits. For example, herbicide resistance engineered into crop plants might be transferred to weeds, which would then be resistant to the herbicides that are now used for their control. Some studies have detected hybridization between genetically engineered crops and wild populations of plants. For example, evidence suggests that transgenic oilseed rape (*Brassica napus*) has hybridized with the weed *Brassica rapa* in Canada. Other concerns focus on health-safety matters associated with the presence of engineered products in natural foods; some critics have advocated required labeling of all genetically engineered foods that contain transgenic DNA or protein. Such labeling is required in countries of the European Union but not in the United States.

On the other hand, the use of genetically engineered crops and domestic animals has potential benefits. Genetically engineered crops that are pest resistant have the potential to reduce the use of environmentally harmful chemicals, and research findings indicate that lower amounts of pesticides are used in the United States as a result of the adoption of transgenic plants. Transgenic crops also increase yields, providing more food per acre, which reduces the amount of land that must be used for agriculture. Genetically engineered plants offer the potential for greater yields that may be necessary to feed the world's future population.

CONCEPTS

Recombinant DNA technology is used to create a wide range of commercial products, including pharmaceutical products, specialized bacteria, genetically engineered crops, and transgenic domestic animals.

✔ **CONCEPT CHECK 12**

What are some matters of concern about the use of genetically engineered crops?

Genetic Testing

The identification and cloning of many important disease-causing human genes have allowed the development of probes for detecting disease-causing mutations. Prenatal testing is already available for several hundred genetic disorders (see Chapter 6). Additionally, presymptomatic genetic tests for adults and children are available for an increasing number of disorders.

The growing availability of genetic tests raises a number of ethical and social questions. For example, is it ethical to test for genetic diseases for which there is no cure or treatment? Other ethical and legal questions concern the confidentiality of test results. Who should have access to the results of genetic testing? Should relatives who also might be at risk be informed of the results of genetic testing?

Another set of concerns is related to the accuracy of genetic tests. For many genetic diseases, the only predictive tests available are those that identify a *predisposing* mutation in DNA, but many genetic diseases may be caused by dozens or hundreds of different mutations. Probes that detect common mutations can be developed, but they won't detect rare mutations and will give a false negative result. Short of sequencing the entire gene—which is expensive and time consuming—there is no way to identify all predisposed persons. These questions and concerns are currently the focus of intense debate by ethicists, physicians, scientists, and patients.

Gene Therapy

Perhaps the ultimate application of recombinant DNA technology is **gene therapy**, the direct transfer of genes into humans to treat disease. In 1990, gene therapy became reality. William French Anderson and his colleagues at the U.S. National Institutes of Health (NIH) transferred a functional gene for adenosine deaminase to a young girl with severe combined immunodeficiency disease (SCID), an autosomal recessive condition that produces impaired immune function. The researchers removed white blood cells from the girl, and used a retrovirus to transfer a recombinant gene for adenosine deaminase into the cells. The cells were cultured in the laboratory and then implanted back into the patient, where they produced adenosine deaminase and helped alleviate the symptoms of the disease.

Today, thousands of patients have received gene therapy, and many clinical trials are underway. Gene therapy has been used as an experimental treatment for genetic diseases, cancer, heart disease, and even some infectious diseases such as AIDS. A number of different methods for transferring genes into human cells are currently under development. Commonly used vectors include genetically modified retroviruses, adenoviruses, and adeno-associated viruses (**Table 19.3**).

In spite of the growing number of clinical trials for gene therapy, significant problems remain in transferring foreign genes into human cells, getting them expressed, and limiting immune responses to the gene products and the vectors used to transfer the genes to the cells. There are also heightened concerns about safety. In 1999, a patient participating in a gene-therapy trial had a fatal immune reaction after he was injected with a viral vector carrying a gene to treat his metabolic disorder. And five children who have undergone gene therapy for severe combined immunodeficiency disease developed leukemia that appeared to be directly related to the insertion of the retroviral gene vectors into cancer-causing genes. Despite these setbacks, gene-therapy research has moved ahead. Unequivocal results demonstrating positive benefits from gene therapy for severe combined immunodeficiency disease and for head and neck cancer were announced in 2000. Gene therapy has been used to successfully treat several children with a genetic disorder called X-linked adrenoleukodystropy, a fatal disease of the central nervous system. Recently, gene therapy was used to partly restore sight in four

Table 19.3	Vectors used in gene therapy	
Vector	**Advantages**	**Disadvantages**
Retrovirus	Efficient transfer	Transfers DNA only to dividing cells, inserts randomly; risk of producing wild-type viruses
Adenovirus	Transfers to nondividing cells	Causes immune reaction
Adeno-associated virus	Does not cause immune reaction	Holds small amount of DNA; hard to produce
Herpes virus	Can insert into cells of nervous system; does not cause immune reaction	Hard to produce in large quantities
Lentivirus	Can accommodate large genes	Safety concerns
Liposomes and other lipid-coated vectors	No replication; does not stimulate immune reaction	Low efficiency
Direct injection	No replication; directed toward specific tissues	Low efficiency; does not work well within some tissues
Pressure treatment	Safe, because tissues are treated outside the body and then transplanted into the patient	Most efficient with small DNA molecules
Gene gun (DNA coated on small gold particles and shot into tissue)	No vector required	Low efficiency

Source: After E. Marshall, Gene therapy's growing pains, *Science* 269:1050–1055, 1995.

young adults with a genetic disorder that causes severe blindness (see the introduction to this chapter).

Gene therapy conducted to date has targeted only nonreproductive, or somatic, cells. Correcting a genetic defect in these cells (termed *somatic gene therapy*) may provide positive benefits to patients but will not affect the genes of future generations. Gene therapy that alters reproductive, or germ-line, cells (termed *germ-line gene therapy*) is technically possible but raises a number of significant ethical issues, because it has the capacity to alter the gene pool of future generations.

CONCEPTS

Gene therapy is the direct transfer of genes into humans to treat disease. Gene therapy was first successfully implemented in 1990 and is now being used to treat genetic diseases, cancer, and infectious diseases.

✔ **CONCEPT CHECK 13**

What is the difference between somatic gene therapy and germ-line gene therapy?

CONCEPTS SUMMARY

- Restriction endonucleases are enzymes that make double-stranded cuts in DNA at specific base sequences.

- DNA fragments can be separated with the use of gel electrophoresis and visualized by staining the gel with a dye that is specific for nucleic acids or by labeling the fragments with a radioactive or chemical tag.

- In gene cloning, a gene or a DNA fragment is placed into a bacterial cell, where it will be multiplied as the cell divides.

- Plasmids, small circular pieces of DNA, are often used as vectors to ensure that a cloned gene is stable and replicated within the recipient cells. Expression vectors contain sequences necessary for foreign DNA to be transcribed and translated.

- The polymerase chain reaction is a method for amplifying DNA enzymatically without cloning. A solution containing DNA is heated, so that the two DNA strands separate, and then quickly cooled, allowing primers to attach to the template DNA. The solution is then heated again, and DNA polymerase synthesizes new strands from the primers. Each time the cycle is repeated, the amount of DNA doubles.

- Genes can be isolated by creating a DNA library—a set of bacterial colonies or viral plaques that each contain a different cloned fragment of DNA. A genomic library contains the entire genome of an organism; a cDNA library contains DNA fragments complementary to all the different mRNAs in a cell.

- In situ hybridization can be used to determine the chromosomal location of a gene and the distribution of the mRNA produced by a gene.

- Positional cloning uses linkage relations to determine the location of genes without any knowledge of their products.

- The Sanger (dideoxy) method of DNA sequencing uses special substrates for DNA synthesis (dideoxynucleoside triphosphates, ddNTPs) that terminate synthesis after they are incorporated into the newly made DNA. Next-generation sequencing methods sequence many DNA fragments simultaneously, providing a much faster and less-expensive determination of a DNA sequence.

- Short tandem repeats (STRs) and microsatellites are used to identify people by their DNA sequences (DNA fingerprinting).

- Forward genetics begins with a phenotype and conducts analyses to locate the responsible genes. Reverse genetics starts with a DNA sequence and conducts analyses to determine its phenotypic effect.

- Site-directed mutagenesis can be used to produce mutations at specific sites in DNA, allowing genes to be tailored for a particular purpose.

- Transgenic animals, produced by injecting DNA into fertilized eggs, contain foreign DNA that is integrated into a chromosome. Knockout mice are transgenic mice in which a normal gene is disabled.

- RNA interference is used to silence the expression of specific genes.

- Techniques of molecular genetics are being used to create products of commercial importance, to develop diagnostic tests, and to treat diseases.

- In gene therapy, diseases are being treated by altering the genes of human cells.

IMPORTANT TERMS

recombinant DNA technology (p. 514)
genetic engineering (p. 514)
biotechnology (p. 514)
restriction enzyme (p. 515)
restriction endonuclease (p. 515)
cohesive end (p. 515)
gel electrophoresis (p. 517)
autoradiography (p. 518)
probe (p. 518)
Southern blotting (p. 518)
Northern blotting (p. 519)
Western blotting (p. 519)
gene cloning (p. 519)
cloning vector (p. 519)
linker (p. 520)
cosmid (p. 521)
bacterial artificial chromosome (BAC)
  (p. 521)

expression vector (p. 522)
yeast artificial chromosome (YAC)
  (p. 522)
Ti plasmid (p. 522)
polymerase chain reaction (PCR) (p. 523)
Taq polymerase (p. 524)
reverse-transcription PCR (p. 524)
real-time PCR (p. 525)
DNA library (p. 527)
genomic library (p. 527)
cDNA library (p. 528)
in situ hybridization (p. 529)
positional cloning (p. 529)
chromosome walking (p. 530)
chromosome jumping (p. 530)
restriction fragment length
  polymorphism (RFLP) (p. 533)
DNA sequencing (p. 534)

dideoxyribonucleoside triphosphate
  (ddNTP) (p. 534)
next-generation sequencing technologies
  (p. 537)
DNA fingerprinting (p. 538)
microsatellite (p. 538)
short tandem repeat (STR) (p. 538)
forward genetics (p. 541)
reverse genetics (p. 541)
site-directed mutagenesis (p. 541)
oligonucleotide-directed mutagenesis
  (p. 542)
transgene (p. 542)
knockout mice (p. 543)
knock-in mice (p. 544)
gene therapy (p. 548)

ANSWERS TO CONCEPT CHECKS

1. Restriction enzymes exist naturally in bacteria, which use them to prevent the entry of viral DNA.

2. c

3. Southern blotting is used to transfer DNA from a gel to a solid medium. Northern blotting is used to transfer RNA from a gel to a solid medium, and Western blotting is used to transfer protein from a gel to a solid medium.

4. The gene and plasmid are cut with the same restriction enzyme and mixed together. DNA ligase is used to seal nicks in the sugar–phosphate backbone.

5. A heat-stable DNA polymerase enzyme is important to the success of PCR because the first step of the PCR reaction requires that the solution be heated to between 90° and 100°C to separate the two DNA strands. Most enzymes are denatured at this temperature. Without a heat-stable DNA polymerase enzyme, new enzyme would have to be added for each PCR cycle. With the use of a heat-stable polymerase, the enzyme can be added at the beginning of the reaction and will function throughout multiple cycles.

6. With the use of the genetic code and the amino acid sequence of the protein, possible nucleotide sequences that cover a small region of the gene can be deduced. A mixture of all the possible nucleotide sequences that might encode the protein, taking into consideration synonymous codons, are used as probes. To minimize the number of sequences required, a region of the protein that has relatively little degeneracy in its codons is selected.

7. The expression pattern of the gene can be examined, and the coding region of copies of the gene from individuals with the mutant phenotype can be compared with the coding region of wild-type individuals.

8. d

9. By using PCR with primers that flank the region containing tandem repeats.

10. b

11. c

12. Possible concerns include: (a) ecological damage caused by introducing novel organisms into the environment; (b) negative effects of transgenic organisms on biodiversity; (c) possible spread of transgenes to native organisms by hybridization; and (d) health effects of eating genetically modified foods.

13. Somatic gene therapy modifies genes only in somatic tissue, and these modifications cannot be inherited. Germ-line gene therapy alters genes in germ-line cells and will be inherited.

WORKED PROBLEMS

1. A molecule of double-stranded DNA that is 5 million base pairs long has a base composition that is 62% G + C. How many times, on average, are the following restriction sites likely to be present in this DNA molecule?

a. *Bam*HI (recognition sequence is GGATCC)

b. *Hin*dIII (recognition sequence is AAGCTT)

c. *Hpa*II (recognition sequence is CCGG)

• Solution

The percentages of G and C are equal in double-stranded DNA; so, if G + C = 62%, then %G = %C = 62%/2 = 31%. The percentage of A + T = (100% – G + C) = 38%, and %A = %T = 38%/2 = 19%. To determine the probability of finding a particular base sequence, we use the multiplication rule, multiplying together the probably of finding each base at a particular site.

a. The probability of finding the sequence GGATCC = 0.31 × 0.31 × 0.19 × 0.19 × 0.31 × 0.31 = 0.0003333. To determine the average number of recognition sequences in a 5-million-base-pair piece of DNA, we multiply 5,000,000 bp × 0.00033 = 1666.5 recognition sequences.

b. The number of AAGCTT recognition sequences is 0.19 × 0.19 × 0.31 × 0.31 × 0.19 × 0.19 × 5,000,000 = 626 recognition sequences.

c. The number of CCGG recognition sequences is 0.31 × 0.31 × 0.31 × 0.31 × 5,000,000 = 46,176 recognition sequences.

2. You are given the following DNA fragment to sequence: 5′–GCTTAGCATC–3′. You first clone the fragment in bacterial cells to produce sufficient DNA for sequencing. You isolate the DNA from the bacterial cells and carry out the dideoxy-sequencing method. You then separate the products of the polymerization reactions by gel electrophoresis. Draw the bands that should appear on the gel from the four sequencing reactions.

• Solution

In the dideoxy-sequencing reaction, the original fragment is used as a template for the synthesis of a new DNA strand; the sequence of the new strand is the sequence that is actually determined. The first task, therefore, is to write out the sequence of the newly synthesized fragment, which will be complementary and antiparallel to the original fragment. The sequence of the newly synthesized strand, written 5′→3′ is: 5′–GATGCTAAGC–3′. Bands representing this sequence will appear on the gel, with the bands representing nucleotides near the 5′ end of the molecule at the bottom of the gel.

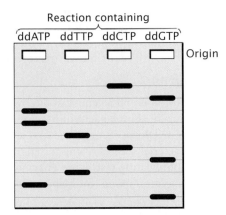

COMPREHENSION QUESTIONS

Section 19.1

1. List some of the effects and practical applications of molecular genetic analyses.

Section 19.2

2. What common feature is seen in the sequences recognized by type II restriction enzymes?

3. What role do restriction enzymes play in bacteria? How do bacteria protect their own DNA from the action of restriction enzymes?

*****4.** Explain how gel electrophoresis is used to separate DNA fragments of different lengths.

*****5.** After DNA fragments have been separated by gel electrophoresis, how can they be visualized?

6. What is the purpose of Southern blotting? How is it carried out?

*7. Give three important characteristics of cloning vectors.

8. Briefly describe two different methods for inserting foreign DNA into plasmids, giving the strengths and weaknesses of each method.

*9. Briefly explain how an antibiotic-resistance gene and the *lacZ* gene can be used to determine which cells contain a particular plasmid.

*10. Briefly explain how the polymerase chain reaction is used to amplify a specific DNA sequence. What are some of the limitations of PCR?

11. What is real-time PCR?

Section 19.3

*12. How does a genomic library differ from a cDNA library? How is a cDNA library created?

13. How are probes used to screen DNA libraries? Explain how a synthetic probe can be prepared when the protein product of a gene is known.

*14. Briefly explain in situ hybridization, giving some applications of this technique.

15. Briefly explain how a gene can be isolated through positional cloning.

16. Explain how chromosome walking can be used to find a gene.

Section 19.4

17. What is the purpose of the dideoxynucleoside triphosphate in the dideoxy sequencing reaction?

*18. What is DNA fingerprinting? What types of sequences are examined in DNA fingerprinting?

Section 19.5

19. How does a reverse genetics approach differ from a forward genetics approach?

20. Briefly explain how site-directed mutagenesis is carried out.

*21. What are knockout mice, how are they produced, and for what are they used?

22. How is RNA interference used in the analysis of gene function?

Section 19.6

23. What is gene therapy?

APPLICATION QUESTIONS AND PROBLEMS

Introduction

24. The introduction to this chapter explains how gene therapy is used to treat Leber congenital amaurosis (LCA, a type of severe blindness). What characteristics of blindness that affects the retina (such LCA) make it an attractive candidate for treatment by gene therapy?

Section 19.2

*25. Suppose that a geneticist discovers a new restriction enzyme in the bacterium *Aeromonas ranidae*. This restriction enzyme is the first to be isolated from this bacterial species. Using the standard convention for abbreviating restriction enzymes, give this new restriction enzyme a name (for help, see footnote to Table 19.1).

26. How often, on average, would you expect a type II restriction endonuclease to cut a DNA molecule if the recognition sequence for the enzyme had 5 bp? (Assume that the four types of bases are equally likely to be found in the DNA and that the bases in a recognition sequence are independent.) How often would the endonuclease cut the DNA if the recognition sequence had 8 bp?

*27. A microbiologist discovers a new type II restriction endonuclease. When DNA is digested by this enzyme, fragments that average 1,048,500 bp in length are produced. What is the most likely number of base pairs in the recognition sequence of this enzyme?

28. Will restriction sites for an enzyme that has 4 bp in its restriction site be closer together, farther apart, or similarly spaced, on average, compared with those of an enzyme that has 6 bp in its restriction site? Explain your reasoning.

*29. About 60% of the base pairs in a human DNA molecule are AT. If the human genome has 3.2 billion base pairs of DNA, about how many times will the following restriction sites be present?

  a. *Bam*HI (restriction site is 5′–GGATCC–3′)

  b. *Eco*RI (restriction site is 5′–GAATTC–3′)

  c. *Hae*III (restriction site is 5′–GGCC–3′)

*30. Restriction mapping of a linear piece of DNA reveals the following *Eco*RI restriction sites.

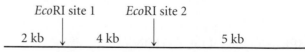

  a. This piece of DNA is cut by *Eco*RI, the resulting fragments are separated by gel electrophoresis, and the gel is stained with ethidium bromide. Draw a picture of the bands that will appear on the gel.

  b. If a mutation that alters *Eco*RI site 1 occurs in this piece of DNA, how will the banding pattern on the gel differ from the one that you drew in part *a*?

c. If mutations that alter *Eco*RI sites 1 and 2 occur in this piece of DNA, how will the banding pattern on the gel differ from the one that you drew in part *a*?

d. If 1000 bp of DNA were inserted between the two restriction sites, how would the banding pattern on the gel differ from the one that you drew in part *a*?

e. If 500 bp of DNA between the two restriction sites were deleted, how would the banding pattern on the gel differ from the one that you drew in part *a*?

*31. Which vectors (plasmid, phage λ, cosmid, bacterial artificial chromosome) can be used to clone a continuous fragment of DNA with the following lengths?

a. 4 kb   **c.** 35 kb

b. 20 kb   **d.** 100 kb

32. A geneticist uses a plasmid for cloning that has the *lacZ* gene and a gene that confers resistance to penicillin. The geneticist inserts a piece of foreign DNA into a restriction site that is located within the *lacZ* gene and uses the plasmid to transform bacteria. Explain how the geneticist can identify bacteria that contain a copy of a plasmid with the foreign DNA.

Section 19.3

*33. Suppose that you have just graduated from college and have started working at a biotechnology firm. Your first job assignment is to clone the pig gene for the hormone prolactin. Assume that the pig gene for prolactin has not yet been isolated, sequenced, or mapped; however, the mouse gene for prolactin has beeen cloned, and the amino acid sequence of mouse prolactin is known. Briefly explain two different strategies that you might use to find and clone the pig gene for prolactin.

34. A genetic engineer wants to isolate a gene from a scorpion that encodes the deadly toxin found in its stinger, with the  ultimate purpose of transferring this gene to bacteria and producing the toxin for use as a commercial pesticide. Isolating the gene requires a DNA library. Should the genetic engineer create a genomic library or a cDNA library? Explain your reasoning.

[Sahara Nature/Fotalia.]

*35. A protein has the following amino acid sequence:

Met-Tyr-Asn-Val-Arg-Val-Tyr-Lys-Ala-Lys-
                                    Trp-Leu-Ile-His-Thr-Pro

You wish to make a set of probes to screen a cDNA library for the sequence that encodes this protein. Your probes should be at least 18 nucleotides in length.

a. Which amino acids in the protein should be used to construct the probes so that the least degeneracy results? (Consult the genetic code in Figure 15.10.)

b. How many different probes must be synthesized to be certain that you will find the correct cDNA sequence that specifies the protein?

Section 19.4

36. Suppose that you want to sequence the following DNA fragment:

$$5'-\text{TCCCGGGAAA-primer site}-3'$$

You first clone the fragment in bacterial cells to produce sufficient DNA for sequencing. You isolate the DNA from the bacterial cells and apply the dideoxy-sequencing method. You then separate the products of the polymerization reactions by gel electrophoresis. Draw the bands that should appear on the gel from the four sequencing reactions.

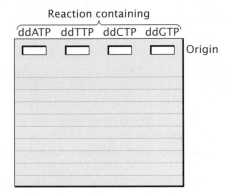

*37. Suppose that you are given a short fragment of DNA to sequence. You clone the fragment, isolate the cloned DNA fragment, and set up a series of four dideoxy reactions. You then separate the products of the reactions by gel electrophoresis and obtain the following banding pattern:

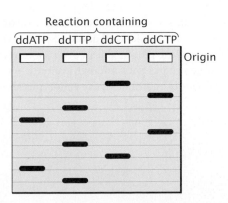

Write out the base sequence of the original fragment that you were given.

Original sequence: 5′– _____ –3′

38. The autoradiograph below is from the original study that first sequenced the cystic fibrosis gene (J. R. Riordan et al. 1989. *Science* 245:1066–1073). From the autoradiograph, determine the sequence of the normal copy of the gene and the sequence of the mutated copy of the gene. Identify the location of the mutation that causes cystic fibrosis (CF).

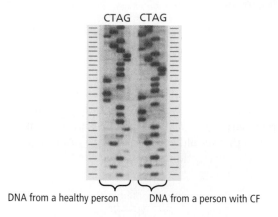

CTAG CTAG

DNA from a healthy person   DNA from a person with CF

*39. A hypothetical disorder called G syndrome is an autosomal dominant disease characterized by visual, skeletal, and cardiovascular defects. The disorder appears in middle age. Because its symptoms are variable, the disorder is difficult to diagnose. Early diagnosis is important, however, because the cardiovascular defects can be treated if the disorder is recognized early. The gene for G syndrome is known to reside on chromosome 7, and it is closely linked to two RFLPs on the same chromosome, one at the *A* locus and one at the *C* locus. The *G*, *A*, and *C* loci are very close together, and there is little crossing over between them. The following RFLP alleles are found at the *A* and *C* loci:

> *A* locus: *A1, A2, A3*
> *C* locus: *C1, C2, C3*

Sally, shown in the following pedigree, is concerned that she might have G syndrome. Her deceased mother had G syndrome, and she has a brother with the disorder. Her other brother is middle-aged and does not have the disease; so assume that he does not carry genes for it. A geneticist genotypes Sally and her immediate family for the *A* and *C* loci and obtains the genotypes shown on the pedigree.

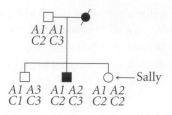

A1 A1
C2 C3

A1 A3   A1 A2   A1 A2 ←— Sally
C1 C3   C2 C3   C2 C2

a. Assume that there is no crossing over between the *A*, *C*, and *G* loci. Does Sally carry the gene that causes G syndrome? Explain why or why not?

b. Draw the arrangement of the *A*, *C*, and *G* alleles on the chromosomes for all members of the family.

Section 19.5

*40. You have discovered a gene in mice that is similar to a gene in yeast. How might you determine whether this gene is essential for development in mice?

41. Andrew Fire, Craig Mello, and their colleagues were among the first to examine the effects of double-stranded RNA on gene expression (A. Fire et al. 1998. *Nature* 391:806–811). In one experiment, they used a transgenic strain of *C. elegans* into which a gene (*gfp*) for a green fluorescent pigment had been introduced. They injected some worms with double-stranded RNA complementary to coding sequences of the *gfp* gene and injected other worms with double-stranded RNA complementary to the coding region of a different gene (*unc22A*) that encodes a muscle protein. The illustration below includes photographs of larvae and adult progeny of the injected worms. Green fluorescent pigment appears as bright spots in the photographs.

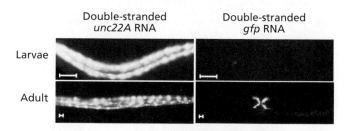

	Double-stranded *unc22A* RNA	Double-stranded *gfp* RNA
Larvae		
Adult		

a. Explain these results.

b. Fire and Mello conducted another experiment in which they injected double-stranded RNA complementary to the introns and promoter sequences of the *gfp* gene. What results would you expect with this experiment? Explain your answer.

CHALLENGE QUESTIONS

Section 19.5

42. Suppose that you are hired by a biotechnology firm to produce a giant strain of fruit flies by using recombinant DNA technology so that genetics students will not be forced to strain their eyes when looking at tiny flies. You go to the library and learn that growth in fruit flies is normally inhibited by a hormone called shorty substance P (SSP). You decide that you can produce giant fruit flies if you can somehow turn off the production of SSP. Shorty substance P is synthesized from a compound called XSP in a single-step reaction catalyzed by the enzyme runtase:

$$XSP \xrightarrow[\text{Runtase}]{} SSP$$

A researcher has already isolated cDNA for runtase and has sequenced it, but the location of the runtase gene in the *Drosophila* genome is unknown. In attempting to devise a strategy for turning off the production of SSP and producing giant flies by using standard recombinant DNA techniques, you discover that deleting, inactivating, or otherwise mutating this DNA sequence in *Drosophila* turns out to be extremely difficult. Therefore you must restrict your genetic engineering to gene augmentation (adding new genes to cells). Describe the methods that you will use to turn off SSP and produce giant flies by using recombinant DNA technology.

Section 19.6

43. Much of the controversy over genetically engineered foods has centered on whether special labeling should be required on all products made from genetically modified crops. Some people have advocated labeling that identifies the product as having been made from genetically modified plants. Others have argued that labeling should be required only to identify the ingredients, not the process by which they were produced. Take one side in this issue and justify your stand.

20 Genomics and Proteomics

The genome of the honeybee, *Apis mellifera*, was seqeunced in 2006, providing new information about honeybee communication, behavior, ecology, and evolution. [Donald Specker/Animals Animals—Earth Scenes.]

DECODING THE WAGGLE DANCE: THE GENOME OF THE HONEYBEE

The honeybee, *Apis mellifera*, is one of the world's most amazing animals. Honeybees are highly social, living in complex colonies in which individual bees cooperate and assume responsibility for specialized tasks that benefit all. In spite of a tiny brain that contains only a million neurons (the human brain contains 100 billion neurons), bees are able to perform intricate behaviors and communicate effectively with thousands of other bees that make up the hive. Each bee recognizes the odor of its own hive, as well as odors that signal alarm, indicate tasks to be performed, and distinguish the caste, age, and sex of other bees. Worker bees possess an extraordinary memory. On finding a clump of flowers that provide a rich supply of nectar, a worker bee returns to the hive and performs the waggle dance, in which it runs a figure eight on the side of the comb. The dance conveys information to other bees about the location of the food source, including the direction in reference to the angle of the sun and the distance from the hive to the source. A bee is able to recall as many as five different flower locations, including the direction and distance of each from the hive, landmarks along the way, and the time of day when the flower produces the most nectar.

Honeybees are not just objects of biological curiosity; they play a critical role in pollinating many important food crops, as well as providing honey and wax. Because of its agricultural and scientific importance, the honeybee was selected for sequencing by the National Genome Research Institute of the U.S. National Institutes of Health. The complete genomic sequence was published in 2006 and is already providing a wealth of new information about the behavior, neural function, ecology, and evolution of honeybees.

The honeybee genome consists of 10,157 genes that encompass about 236 million base pairs of DNA. Compared with other insects, honeybees have more genes that encode olfactory receptors, which is consistent with their sophisticated chemical communication system. Information from the honeybee genome project is being used to better understand the behavior of these remarkable animals. On the basis of the genomic information, 36 genes have been identified that encode neuropeptides—brain proteins that affect behavior and memory. More than 3000 genes are active in a bee's brain, and researchers have located regulatory regions that control the expression of many of these genes, including some that are important in the develop of foraging behavior.

The honeybee has long been associated with humans, but the precise evolutionary origin of honeybees was, until recently, unknown. With information from the genomic sequence, honeybee geneticists used single-nucleotide polymorphisms (SNPs, sites that vary in the base present at a single nucleotide) to study the evolutionary relationships between bees from different parts of the world. This study revealed that *A. mellifera* originated, not in Asia as formerly thought, but in Africa. African populations subsequently expanded into Europe and Asia in two waves, one wave populating western Europe and another wave expanding into Asia and eastern Europe.

European honeybees were introduced into North America by early European settlers, but some have been replaced in recent years by descendants of African "killer bees," which are known for their aggressive stinging behavior. These bees were introduced into Brazil in 1956 and have subsequently spread northward to the southern United States. Analysis of SNPs of bees collected in the United States before 1990 showed that they possessed only European genes but, from 1993 to 2001, bees from southern Texas showed a transition to genes of predominately African ancestry.

The sequencing of the honeybee genome illustrates the diverse ways that genomic information is being used today. **Genomics** is the field of genetics that attempts to understand the content, organization, function, and evolution of genetic information contained in whole genomes. The field of genomics is at the cutting edge of modern biology; information resulting from research in this field has made significant contributions to human health, agriculture, and numerous other areas. It has provided gene sequences necessary for producing medically important proteins through recombinant DNA technology, and comparisons of genome sequences from different organisms are leading to a better understanding of evolution and the history of life.

We begin this chapter by examining genetic and physical maps and methods for sequencing entire genomes. Next, we explore functional genomics—how genes are identified in genomic sequences and how their functions are defined. The sequence of a genome, by itself, is of limited use, and, now that sequencing genomes has become routine, much of genomics is currently focused on deciphering the function of the sequences that are obtained. In the final part of the chapter, we consider proteomics, the study of the complete set of proteins found in a cell.

20.1 Structural Genomics Determines the DNA Sequences of Entire Genomes

Structural genomics concerns the organization and sequence of genetic information contained within a genome. Often, an early step in characterizing a genome is to prepare genetic and physical maps of its chromosomes. These maps provide information about the relative locations of genes, molecular markers, and chromosome segments, which are often essential for positioning chromosome segments and aligning stretches of sequenced DNA into a whole-genome sequence.

Genetic Maps

Everyone has used a map at one time or another. Maps are indispensable for finding a new friend's house, the way to an unfamiliar city in your state, or the location of a country. Each of these examples requires a map with a different scale. To find a friend's house, you would probably use a city street map; to find your way to an unknown city, you might pick up a state highway map; to find a country such as Kazakhstan, you would need a world atlas. Similarly, navigating a genome requires maps of different types and scales.

Construction of genetic maps Genetic maps (also called linkage maps) provide a rough approximation of the locations of genes relative to the locations of other known genes (**Figure 20.1**). These maps are based on the genetic function of recombination (hence the name genetic map). The basic principles of constructing genetic maps are discussed in detail in Chapter 7. In short, individual organisms of known genotype are crossed, and the frequency of recombination between loci is determined by examining the progeny. If the recombination frequency between two loci is 50%, then the loci are located on different chromosomes or are far apart on the same chromosome. If the recombination frequency is less than 50%, the loci are located close together on the same chromosome (they belong to the same linkage group). For linked genes, the rate of recombination is proportional to the physical distance between the loci. Distances on genetic maps are measured in percent recombination (centimorgans, cM), or map units (m.u.). Data from multiple two-point or three-point crosses can be integrated into linkage maps for whole chromosomes.

For many years, genes could be detected only by observing their influence on a trait (the phenotype), and the construction of genetic maps was limited by the availability of single-locus traits that could be examined for evidence of recombination. Eventually, this limitation was overcome by

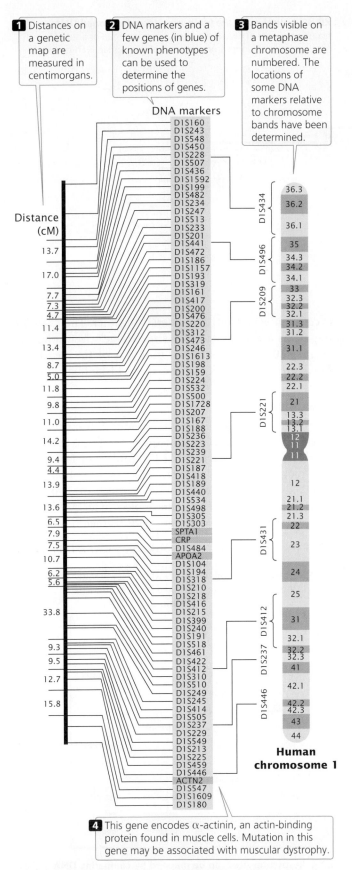

1 Distances on a genetic map are measured in centimorgans.

2 DNA markers and a few genes (in blue) of known phenotypes can be used to determine the positions of genes.

3 Bands visible on a metaphase chromosome are numbered. The locations of some DNA markers relative to chromosome bands have been determined.

Distance (cM)

Human chromosome 1

4 This gene encodes α-actinin, an actin-binding protein found in muscle cells. Mutation in this gene may be associated with muscular dystrophy.

20.1 Genetic maps are based on rates of recombination. Shown here is a genetic map of human chromosome 1.

the development of molecular techniques, such as the analysis of restriction fragment length polymorphisms, the polymerase chain reaction, and DNA sequencing (see Chapter 19), that are able to provide molecular markers that can be used to construct and refine genetic maps.

Limitations of genetic maps Genetic maps have several limitations, the first of which is resolution, or detail. The human genome includes 3.2 billion base pairs of DNA and has a total genetic distance of about 4000 cM, an average of 800,000 bp/cM. Even if a marker were present every centimorgan (which is unrealistic), the resolution in regard to the physical structure of the DNA would still be quite low. In other words, the detail of the map is very limited. A second problem with genetic maps is that they do not always accurately correspond to physical distances between genes. Genetic maps are based on rates of crossing over, which vary somewhat from one part of a chromosome to another; so the distances on a genetic map are only approximations of real physical distances along a chromosome. **Figure 20.2** compares the genetic map of chromosome III of yeast with a physical map determined by DNA sequencing. There are some discrepancies between the distances and even among the positions of some genes. In spite of these limitations, genetic maps have been critical to the development of physical maps and the sequencing of whole genomes.

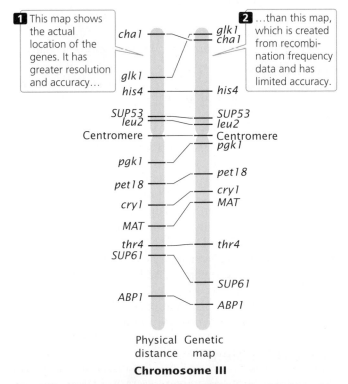

1 This map shows the actual location of the genes. It has greater resolution and accuracy…

2 …than this map, which is created from recombination frequency data and has limited accuracy.

Physical distance Genetic map

Chromosome III

20.2 Genetic and physical maps may differ in relative distances and even in the position of genes on a chromosome. Genetic and physical maps of yeast chromosome III reveal such differences.

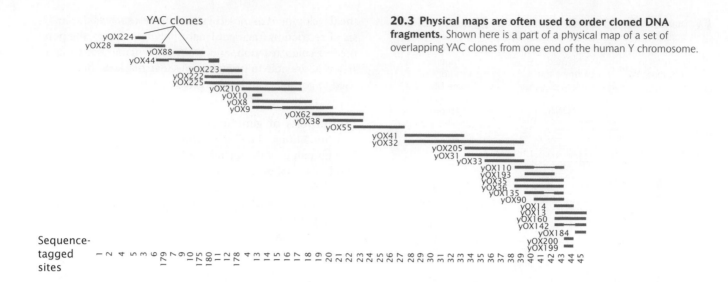

20.3 Physical maps are often used to order cloned DNA fragments. Shown here is a part of a physical map of a set of overlapping YAC clones from one end of the human Y chromosome.

Physical Maps

Physical maps are based on the direct analysis of DNA, and they place genes in relation to distances measured in number of base pairs, kilobases, or megabases. A common type of physical map is one that connects isolated pieces of genomic DNA that have been cloned in bacteria or yeast (**Figure 20.3**). Physical maps generally have higher resolution and are more accurate than genetic maps. A physical map is analogous to a neighborhood map that shows the location of every house along a street, whereas a genetic map is analogous to a highway map that shows the locations of major towns and cities.

One of the techniques for creating physical maps is restriction mapping, which determines the position of restriction sites on DNA. When a piece of DNA is cut with a restriction enzyme and the fragments are separated by gel electrophoresis, the number of restriction sites in the DNA and the distances between them can be determined by the number and positions of bands on the gel (see p. 553 in Chapter 19), but this information does not tell us the order or the precise location of the restriction sites. To map restriction sites, a sample of the DNA is cut with one restriction enzyme, and another sample is cut with a different restriction enzyme. A third sample is cut with both restriction enzymes together (a double digest). The DNA fragments produced by these restriction digests are then separated by gel electrophoresis, and their sizes are compared. Overlap in size of fragments produced by the digests can be used to position the restriction sites on the original DNA molecule. This process is illustrated in the following Worked Problem.

Worked Problem

One sample of a linear 13,000-bp (13-kb) DNA fragment is cut with the restriction enzyme *Eco*RI; a second sample of the same DNA is cut with *Bam*HI; and a third sample is cut with *both Eco*RI and *Bam*HI together. The resulting fragments are separated and sized by gel electrophoresis (**Figure 20.4**). Determine the positions of the *Eco*RI and *Bam*HI restriction sites on the original 13-kb fragment.

• Solution

Using the sizes of the fragments produced from the three digests in Figure 20.4, we can order the positions of the restriction sites on the original 13-kb piece of DNA. First, note that digestion with *Eco*RI alone produced 8-kb, 3-kb, and 2-kb fragments, indicating that there are two *Eco*RI restriction sites in the original linear piece of DNA. Digestion with *Bam*HI produced 9-kb and 4-kb fragments, indicating

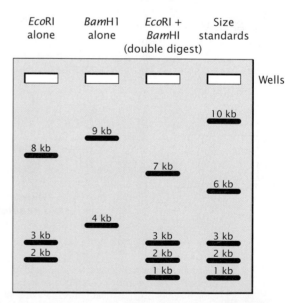

20.4 Restriction sites can be mapped by comparing DNA fragments produced by digestion with restriction enzymes used alone and in various combinations.

that there is only one *Bam*HI site. The *Bam*HI restriction site must be 9 kb from one end and 4 kb from the other end.

The double digest produced four pieces of DNA: 7-kb, 3-kb, 2-kb, and 1-kb fragments. Neither of the fragments generated by *Bam*HI alone is present in the double digest, and so *Eco*RI must have cut both of the *Bam*HI fragments. Consider the 9-kb fragment. How could this fragment be cut by *Eco*RI to produce the fragments found in the double digest? Two of the fragments produced by the double digest, the 7-kb and 2-kb fragments, add up to 9 kb, the length of one fragment produced by digestion by *Bam*HI alone. Similarly, the 3-kb fragment and the 1-kb fragment of the double digest add up to 4 kb, the length of the other fragment produced by *Bam*HI alone. Therefore, *Eco*RI cut the first *Bam*HI fragment into 7-kb and 2-kb fragments and cut the second *Bam*HI fragment into 3-kb and 1-kb fragments:

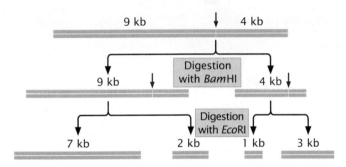

We now know that the *Bam*HI site lies between these two *Eco*RI sites. In regard to the four fragments produced by the double digest, there are several possible arrangements by which a *Bam*HI site could fit in between the two *Eco*RI sites. To determine which of the arrangements is correct, compare the results of the *Eco*RI digestion with the that of double digest. When the original 13-kb DNA fragment was cut by *Eco*RI alone, the three fragments produced were 8 kb, 3 kb, and 2 kb in length. The 2-kb and 3-kb bands are also present in the double digest, indicating that these fragments do not contain a *Bam*HI site. The 8-kb fragment present in the *Eco*RI digest disappears in the double digest and is replaced by the 7-kb fragment and the 1-kb fragment, indicating that the 8-kb fragment has the *Bam*HI site. Thus, the 7-kb and 1-kb fragments must lie next to each other, and the 2-kb and 3-kb fragments are on the ends. Thus, the correct arrangement of the restriction sites is:

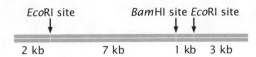

→ Now try your hand at restriction mapping by working Problem 33 at the end of the chapter.

In the Worked Problem, we can map the restriction sites in our heads or with a few simple sketches. Most restriction mapping is done with several restriction enzymes, used alone and in various combinations, producing many restriction fragments. With long pieces of DNA (greater than 30 kb), computer programs are used to determine the restriction maps, and restriction mapping may be facilitated by tagging one end of a large DNA fragment with radioactivity or by identifying the end with the use of a probe.

Physical maps, such as restriction maps of DNA fragments or even whole chromosomes, are often created for genomic analysis. These lengthy maps are put together by combining maps of shorter, overlapping genomic fragments. A number of additional techniques exist for creating physical maps including: sequence-tagged site (STS) mapping, which locates the positions of short unique sequences of DNA on a chromosome; in situ hybridization, by which markers can be visually mapped to locations on chromosomes; and DNA sequencing.

CONCEPTS

Both genetic and physical maps provide information about the relative positions and distances between genes, molecular markers, and chromosome segments. Genetic maps are based on rates of recombination and are measured in percent recombination, or centimorgans. Physical maps are based on physical distances and are measured in base pairs.

✔ **CONCEPT CHECK 1**

What are some of the limitations of genetic maps?

Sequencing an Entire Genome

The first genomes to be sequenced were small genomes of some viruses. The genome of bacteriophage λ, consisting of 49,000 bp, was completed in 1982. In 1995, the first genome of a living organism (*Haemophilus influenzae*) was sequenced by Craig Venter and Claire Fraser of The Institute for Genomic Research (TIGR) and Hamilton Smith of Johns Hopkins University. This bacterium has a small genome of 1.8 million base pairs (**Figure 20.5**). By 1996, the genome of the first eukaryotic organism (yeast) had been determined, followed by the genomes of *Escherichia coli* (1997), *Caenorhabditis elegans* (1998), *Drosophila melanogaster* (2000), and *Arabidopsis thaliana* (2000). The first draft of the human genome was completed in June 2000.

The ultimate goal of structural genomics is to determine the ordered nucleotide sequences of entire genomes of organisms. In Chapter 19, we considered some of the methods used to sequence small fragments of DNA. The main obstacle to sequencing a whole genome is the immense size of most genomes. Bacterial genomes are usually at least several million base pairs long; many eukaryotic genomes are billions

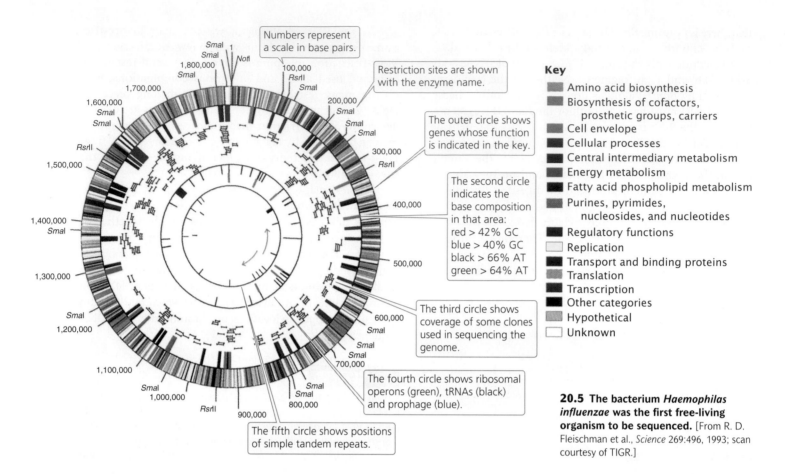

Numbers represent a scale in base pairs.

Restriction sites are shown with the enzyme name.

The outer circle shows genes whose function is indicated in the key.

The second circle indicates the base composition in that area:
red > 42% GC
blue > 40% GC
black > 66% AT
green > 64% AT

The third circle shows coverage of some clones used in sequencing the genome.

The fourth circle shows ribosomal operons (green), tRNAs (black) and prophage (blue).

The fifth circle shows positions of simple tandem repeats.

Key
- Amino acid biosynthesis
- Biosynthesis of cofactors, prosthetic groups, carriers
- Cell envelope
- Cellular processes
- Central intermediary metabolism
- Energy metabolism
- Fatty acid phospholipid metabolism
- Purines, pyrimides, nucleosides, and nucleotides
- Regulatory functions
- Replication
- Transport and binding proteins
- Translation
- Transcription
- Other categories
- Hypothetical
- Unknown

20.5 The bacterium *Haemophilas influenzae* was the first free-living organism to be sequenced. [From R. D. Fleischman et al., *Science* 269:496, 1993; scan courtesy of TIGR.]

of base pairs long and are distributed among dozens of chromosomes. Furthermore, for technical reasons, sequencing cannot begin at one end of a chromosome and continue straight through to the other end; only small fragments of DNA—usually no more than 500 to 700 nucleotides—can be sequenced at one time. Therefore, determining the sequence for an entire genome requires that the DNA be broken into thousands or millions of smaller fragments that can then be sequenced. The difficulty lies in putting these short sequences back together in the correct order. Two different approaches have been used to assemble the short sequenced fragments into a complete genome: map-based sequencing and whole-genome shotgun sequencing. We will consider these two approaches in the context of the Human Genome Project.

The Human Genome Project

By 1980, methods for mapping and sequencing DNA fragments had been sufficiently developed that geneticists began seriously proposing that the entire human genome could be sequenced. An international collaboration was planned to undertake the Human Genome Project (**Figure 20.6**); initial estimates suggested that 15 years and $3 billion would be required to accomplish the task.

The Human Genome Project officially got underway in October 1990. Initial efforts focused on developing new and automated methods for cloning and sequencing DNA and on generating detailed physical and genetic maps of the human genome. The effort was a public project consisting of the international collaboration of 20 research groups and hundreds of individual researchers who formed the International Human Genome Sequencing Consortium. This group used a map-based strategy for sequencing the human genome.

20.6 Craig Venter (left), President of Celera Genomics, and Francis Collins (right), Director of the National Human Genome Research Institute, NIH, announce the completion of a rough draft of the human genome at a press conference in Washington, D.C., on June 26, 2000. [Alex Wong/Newsmakers/Getty Images.]

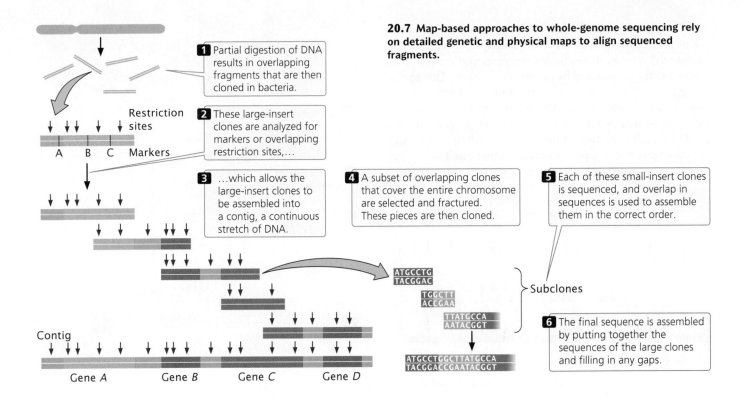

20.7 Map-based approaches to whole-genome sequencing rely on detailed genetic and physical maps to align sequenced fragments.

1 Partial digestion of DNA results in overlapping fragments that are then cloned in bacteria.

2 These large-insert clones are analyzed for markers or overlapping restriction sites,…

3 …which allows the large-insert clones to be assembled into a contig, a continuous stretch of DNA.

4 A subset of overlapping clones that cover the entire chromosome are selected and fractured. These pieces are then cloned.

5 Each of these small-insert clones is sequenced, and overlap in sequences is used to assemble them in the correct order.

6 The final sequence is assembled by putting together the sequences of the large clones and filling in any gaps.

Restriction sites

A   B   C   Markers

Contig

Gene *A*   Gene *B*   Gene *C*   Gene *D*

Subclones

ATGCCTG
TACGGAC

TGGCTT
ACCGAA

TTATGCCA
AATACGGT

ATGCCTGGCTTATGCCA
TACGGACCGAATACGGT

Map-based sequencing In **map-based sequencing**, short sequenced fragments are assembled into a whole-genome sequence by first creating detailed genetic and physical maps of the genome, which provide known locations of genetic markers (restriction sites, other genes, or known DNA sequences) at regularly spaced intervals along each chromosome. These markers are later used to help align the short sequenced fragments into their correct order.

After the genetic and physical maps are available, chromosomes or large pieces of chromosomes are separated by pulsed-field gel electrophoresis (PFGE) or by flow cytometry. Standard gel electrophoresis cannot separate very large pieces of DNA, such as whole chromosomes, but PFGE can separate large molecules of DNA or whole chromosomes in a gel by periodically alternating the orientation of an electrical current. In flow cytometry, chromosomes are sorted optically by size.

Each chromosome (or sometimes the entire genome) is then cut up by partial digestion with restriction enzymes (**Figure 20.7**). Partial digestion means that the restriction enzymes are allowed to act for only a limited time so that not all restriction sites in every DNA molecule are cut. Thus, partial digestion produces a set of large overlapping DNA fragments, which are then cloned with the use of cosmids, yeast artificial chromosomes (YACs), or bacterial artificial chromosomes (BACs; see Chapter 19).

Next, these large-insert clones are put together in their correct order on the chromosome (see Figure 20.7). This assembly can be done in several ways. One method relies on the presence of a high-density map of genetic markers. A complementary DNA probe is made for each genetic marker, and a library of the large-insert clones is screened with the probe, which will hybridize to any colony containing a clone with the marker. The library is then screened for neighboring markers. Because the clones are much larger than the markers used as probes, some clones will have more than one marker. For example, clone A might have markers M1 and M2, clone B markers M2, M3, and M4, and clone C markers M4 and M5. Such a result would indicate that these clones contain areas of overlap, as shown here:

Clone A
M1   M2
M4   M5
Clone C

M2   M3   M4
Clone B

Contig
M1   M2   M3   M4   M5

A set of two or more overlapping DNA fragments that form a contiguous stretch of DNA is called a **contig**. This approach was used in 1993 to create a contig consisting of 196 overlapping YAC clones (see Figure 20.3) of the human Y chromosome.

The order of clones can also be determined without the use of preexisting genetic maps. For example, each clone can be cut with a series of restriction enzymes and the resulting fragments then separated by gel electrophoresis. This method generates a unique set of restriction fragments, called a fingerprint, for each clone. The restriction patterns for the clones are stored in a database. A computer program is then used to examine the restriction patterns of all the clones and look for areas of overlap. The overlap is then used to arrange the clones in order, as shown here:

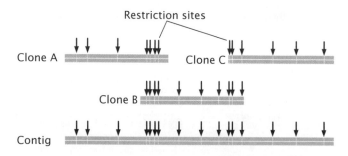

Other genetic markers can be used to help position contigs along the chromosome. TRY PROBLEM 34 →

When the large-insert clones have been assembled into the correct order on the chromosome, a subset of overlapping clones that efficiently cover the entire chromosome can be chosen for sequencing. Each of the selected large-insert clones is fractured into smaller overlapping fragments, which are themselves cloned (see Figure 20.7). These smaller clones (called small-insert clones) are then sequenced. The sequences of the small-insert clones are examined for overlap, which allows them to be correctly assembled to give the sequence of the larger insert clones. Enough overlapping small-insert clones are usually sequenced to ensure that the entire genome is sequenced several times. Finally, the whole genome is assembled by putting together the sequences of all overlapping contigs (see Figure 20.7). Often, gaps in the genome map still exist and must be filled in by using other methods.

The International Human Genome Sequencing Consortium used a similar map-based approach to sequencing the human genome. Many copies of the human genome were cut up into fragments of about 150,000 bp each, which were inserted into bacterial artificial chromosomes. Yeast artificial chromosomes and cosmids had been used in early stages of the project but did not prove to be as stable as the BAC clones, although YAC clones were instrumental in putting together some of the larger contigs. Restriction fingerprints were used to assemble the BAC clones into contigs, which were positioned on the chromosomes with the use of genetic markers and probes. The individual BAC clones were sheared into smaller overlapping fragments and sequenced, and the whole genome was assembled by putting together the sequence of the BAC clones.

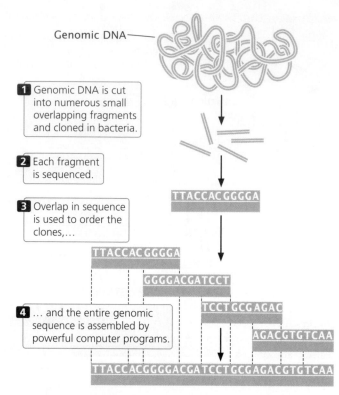

20.8 Whole-genome shotgun sequencing utilizes sequence overlap to align sequenced fragments.

Whole-genome shotgun sequencing

In 1998, Craig Venter announced that he would lead a company called Celera Genomics in a private effort to sequence the human genome. He proposed using a shotgun sequencing approach, which he suggested would be quicker than the map-based approach employed by the International Human Genome Sequencing Consortium.

In **whole-genome shotgun sequencing** (**Figure 20.8**), small-insert clones are prepared directly from genomic DNA and sequenced. Powerful computer programs then assemble the entire genome by examining overlap among the small-insert clones. One advantage of shotgun sequencing is that the small-insert clones can be placed into plasmids, which are simple and easy to manipulate. The requirement for overlap means that most of the genome will be sequenced multiple (often from 10 to 15) times. Shotgun sequencing can be carried out in a highly automated way, with few decisions to be made by the researcher, because the computer assembles the final draft of the sequence.

CONCEPTS

Sequencing a genome requires breaking it up into small overlapping fragments of which the DNA sequences can be determined in a sequencing reaction. In map-based sequencing, sequenced fragments are ordered into the final genome sequence with the use of genetic and physical maps. In whole-genome shotgun sequencing, the genome is assembled by comparing overlap in the sequences of small fragments.

✔ CONCEPT CHECK 2

A contig is

a. a set of molecular markers used in genetic mapping.

b. a set of overlapping fragments that form a continuous stretch of DNA.

c. a set of fragments generated by a restriction enzyme.

d. a small DNA fragment used in sequencing.

Shotgun sequencing was initially used for assembling small genomes such as those of bacteria. When Venter proposed the use of this approach for sequencing the human genome, it was not at all clear that the approach could successfully assemble a complex genome consisting of billions of base pairs such as the human genome.

In the summer of 2000, both public and private sequencing projects announced the completion of a rough draft that included most of the sequence of the human genome, 5 years ahead of schedule. The human genome sequence was declared completed in the spring of 2003, although some gaps still remain. For most chromosomes, the finished sequence is 99.999% accurate, with less than one base-pair error per 100,000 bp, an accuracy rate 10 times that of the initial goal.

Results and implications of the Human Genome Project With the first human genome determined, sequencing additional genomes is an easier but still formidable task. Genomes from several different people have now been completely sequenced. The cost of sequencing a complete human genome has dropped dramatically and will continue to fall as sequencing technology improves. A number of commercial companies are racing to develop the technology to sequence a complete human genome for $1,000 or less.

The availability of the complete sequence of the human genome is proving to be of great benefit. The sequence has provided tools for detecting and mapping genetic variants across the human genome, which has greatly facilitated gene mapping in humans. For example, several million sites where people differ in a single nucleotide (called single-nucleotide polymorphisms; see next section) have now been identified, and these sites are being used in genomewide association studies to locate genes that affect diseases and traits in humans. The sequence is also providing important information about development and many basic cellular processes. Comparisons of the human genome with those of other organisms are adding to our understanding of evolution and the history of life.

Along with the many potential benefits of having complete sequence information are concerns about the misuse of this information. With the knowledge gained from genomic sequencing, many more genes for diseases, disorders, and behavioral and physical traits will be identified, increasing the number of genetic tests that can be performed to make predictions about the future phenotype and health of a person. There is concern that information from genetic testing might be used to discriminate against people who are carriers of disease-causing genes or who might be at risk for some future disease. This matter has been addressed to some extent in the United States with the passage of the Genetic Information Nondiscrimination Act (see Chapter 6), which prohibits health insurers and employers from using genetic information to make decisions about health-insurance coverage and employment. Questions also arise about who should have access to a person's genome sequence. What about relatives, who have similar genomes and who also might be at risk for some of the same diseases? There are also questions about the use of this information to select for specific traits in future offspring. All of these concerns are legitimate and must be addressed if we are to use the information from genome sequencing responsibly.

CONCEPTS

The Human Genome Project was an effort to sequence the entire human genome. A rough draft of the sequence was completed by two competing teams, both of which finished a rough draft of the genome sequence in 2000. The entire sequence was completed in 2003. The ability to rapidly sequence human genomes raises a number of ethical questions.

Single-Nucleotide Polymorphisms

Since the completion of the sequencing of the human genome, sequencers have focused much of their effort on mapping differences among people in their genomic sequences.

Imagine that you are riding in an elevator with a random stranger. How much of your genome do you have in common with this person? Studies of variation in the human genome indicate that you and the stranger will be identical at about 99.9% of your DNA sequences. The difference between you and the stranger is very small in *relative* terms but, because the human genome is so large (3.2 billion base pairs), you and the stranger will be different at more than 3 million base pairs of your genomic DNA. These differences are what makes each of us unique, and they greatly affect our physical features, our health, and possibly even our intelligence and personality.

A site in the genome where individual members of a species differ in a single base pair is called a **single-nucleotide polymorphism** (SNP, pronounced "snip"). Arising through mutation, SNPs are inherited as allelic variants (just as are alleles that produce phenotypic differences, such as blood types), although SNPs do not usually produce a phenotypic difference. Single-nucleotide polymorphisms are numerous and are present throughout genomes. In a comparison of the same chromosome from two different people, a SNP can be found approximately every 1000 bp.

Haplotypes and linkage disequilibrium Most SNPs present within a population arose once from a single mutation that occurred on a particular chromosome and subsequently spread throughout the population. Thus, each SNP is initially associated with other SNPs (as well as other types of genetic variants or alleles) that were present on the particular chromosome on which the mutation arose. The specific set of SNPs and other genetic variants observed on a single chromosome or part of a chromosome is called a **haplotype** (**Figure 20.9**). SNPs within a haplotype are physically linked and therefore tend to be inherited together. New haplotypes can arise through mutation or crossing over, which breaks up the particular set of SNPs in a haplotype. Because the rate of crossing over is proportional to the physical distances between genes, SNPs and other genetic variants that are located close together on the chromosome will be strongly associated as haplotypes. The nonrandom association between genetic variants within a haplotype is called **linkage disequilibrium**. Because the SNPs in a haplotype are inherited together, a haplotype consisting of thousands of SNPs can be identified with the use of only a few SNPs. The few SNPs used to identify a haplotype are called **tag SNPs**. There are about 10 million SNPs in the human population but, because of linkage disequilibrium, these SNPs are present in a much smaller number of haplotypes. Therefore, a relatively small number of SNPs—perhaps only 100,000—can be used to identify most of the haplotypes in humans.

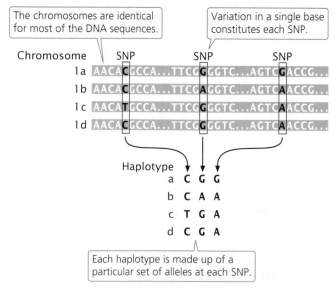

20.9 A haplotype is a specific set of SNPs and other genetic variants observed on a single chromosome or part of a chromosome. Chromosomes 1a, 1b, 1c, and 1d represent different copies of a chromosome that might be found in a population.

The use of single-nucleotide polymorphisms Because of their variability and widespread occurrence throughout the genome, SNPs are valuable as markers in linkage studies. When a SNP is physically close to a disease-causing locus, it will tend to be inherited along with the disease-causing allele. People with the disease will tend to have different SNPs from those of healthy people. A comparison of SNP haplotypes in people with a disease and in healthy people can reveal the presence of genes that affect the disease; because the disease gene and the SNP are closely linked, the location of the disease-causing gene can be determined from the location of associated SNPs. This approach is the same as that used in gene mapping with RFLPs (see Chapter 19), but there are many more SNPs than RFLPs, providing a dense set of variable markers covering the entire genome that can be used more effectively in mapping.

An international effort, called the International HapMap Project, was begun in 2002 with the goal of cataloging and mapping SNPs and other genetic variants that could be used to identify common haplotypes in human populations for use in linkage and family studies. Phase I of the project, completed in 2005, cataloged more than 1 million SNPs in the genomes of 269 people from four diverse human populations (African, Japanese, Chinese, and European). These SNPs are spread over all 23 human chromosomes at a distance of about one SNP per 5000 bp. Phase II of the project, completed in 2006, cataloged a total of 4.6 million SNPs.

Alleles from most of the common SNPs are found in all four human ethnic groups, although the frequencies of the alleles vary considerably among human populations. The greatest genetic diversity of SNPs is found within Africans, which is consistent with many other studies that suggest that humans first evolved in Africa.

Data from the HapMap Project have provided important information about the function and evolution of the human genome. For example, studies of linkage disequilibrium with the use of SNPs have determined that recombination does not take place randomly across the chromosomes: there are numerous recombination hotspots, where more recombination takes place than expected merely by chance. The distribution of SNPs is also providing information about regions of the human genome that have evolved quickly in the recent past, providing clues to how humans have responded to natural selection.

Genomewide association studies As expected, the availability of SNPs has greatly facilitated the search for genes that cause human diseases. The use of numerous SNPs scattered across the genome to find genes of interest is termed a **genome-wide association study**. Soon after the completion of phase I of the HapMap Project, researchers used the SNP data to conduct a genomewide association study for the presence of a major gene that causes age-related macular degeneration, one of the leading causes of blindness among the elderly. In this study, researchers genotyped 96 people with macular degeneration and 50 healthy people for more than 100,000 SNPs that blanketed the genome. The study revealed a strong association between the disease and a gene on chromosome 1 that encodes complement factor H, which has a role in immune function. The finding suggests that macular degeneration might be treated with drugs that affect complement. Another study uncovered a major gene associated with Crohn disease, a common inflammatory disorder of the gastrointestinal tract. Other studies using genomewide scans of SNPs have identified genes that contribute to heart disease, bone density, prostate cancer, and diabetes.

In one of the most successful applications of SNPs for finding disease associations, a consortium of 50 researchers genotyped each of 17,000 people in the United Kingdom for 500,000 SNPs in 2007. They detected strong associations between 24 genes and chromosome segments and the incidence of seven common diseases, including coronary artery disease, Crohn disease, rheumatoid arthritis, bipolar disorder, hypertension, and two types of diabetes. The importance of this study is its demonstration that genomewide association studies utilizing SNPs can successfully locate genes that contribute to complex diseases caused by multiple genetic and environmental factors.

Within the past few years, SNPs have been used in numerous genomewide association studies to successfully locate genes that influence the age of puberty and menopause in women, height, skin pigmentation, eye color, body weight, bone density, and glaucoma, in addition to those identified in the U.K. study. Unfortunately, the genes identified often explain only a small proportion of the genetic influence on the trait. For example, human height is a highly heritable trait, yet the top 20 genetic variants identified in genomewide association studies explain only 3% of the total differences in human height. The low percentage of variation explained means that the genes identified are not, by themselves, useful predictors of the risk of inheriting the diease or trait. Neverthelss the identification of specific genes that influence a disease or trait can lead to a better understanding of the biological processes that produce the phenotype.

Copy-Number Variations

Each diploid person normally possesses two copies of every gene, one inherited from the mother and one inherited from the father. Nevertheless, studies of the human genome have revealed differences among people in the number of copies of large DNA sequences (greater than 1000 bp); these variations are called **copy-number variations** (CNVs). Copy-number variations may include deletions, causing some people to have only a single copy of a sequence, or duplications, causing some people to have more than two copies. A study of CNVs in 270 people from four populations (those studied in the HapMap Project) identified a surprising number of these types of variants: more than 1447 genomic regions varied in copy number, encompassing 12% of the human genome. But only a small part of the

human genome was surveyed in this study, and so the CNVs detected are only a small subset of all that exist. Many of the CNVs encompassed large regions of DNA sequence, often several hundred thousand base pairs in length. Thus, people differ not only at millions of individual SNPs, but also in the number of copies of many larger segments of the genome.

Most CNVs contain multiple genes and potentially affect the phenotype by altering gene dosage and by changing the position of sequences, which may affect the regulation of nearby genes. Indeed, several studies have found associations between CNVs and disease and even between CNVs and normal phenotypic variability in human populations. For example, variations in the number of copies of the *UGT2B17* gene contribute to differences in testosterone metabolism among individuals and affect the risk of prostate cancer. Copy-number variations have been associated with psoriasis, schizophrenia, autism, and mental retardation.

Expressed-Sequence Tags

Another type of data identified by sequencing projects is the **expressed-sequence tag** (EST). In most eukaryotic organisms, only a small percentage of the DNA encodes proteins; in humans, less than 2% of the DNA encodes the amino acids of proteins. If only protein-encoding genes are of interest, an examination of mRNA rather than the entire DNA genomic sequence is often more efficient. RNA can be examined by using ESTs—markers associated with DNA sequences that are expressed as RNA. Expressed-sequence tags are obtained by isolating RNA from a cell and subjecting it to reverse transcription, producing a set of cDNA fragments that correspond to RNA molecules from the cell. Short stretches from the ends of these cDNA fragments are then sequenced, and the sequence obtained (called a tag) provides a marker that identifies the DNA fragment. Expressed-sequence tags can be used to find active genes in a particular tissue or at a particular point in development.

CONCEPTS

In addition to collecting genomic-sequence data, genomic projects are collecting databases of nucleotides that vary among individual organisms (single-nucleotide polymorphisms, SNPs), variations in the number of copies of sequences (copy-number variations), and markers associated with transcribed sequences (expressed-sequence tags, ETSs).

✔ **CONCEPT CHECK 3**

What was the goal of the HapMap Project?

Bioinformatics

Complete genome sequences have now been determined for more than 1000 organisms, with many additional projects underway. These studies are producing tremendous quantities of sequence data. Cataloging, storing, retrieving, and analyzing this huge data set are major challenges of modern genetics. **Bioinformatics** is an emerging field consisting of molecular biology and computer science that centers on the development of databases, computer-search algorithms, gene-prediction software, and other analytical tools that are used to make sense of DNA-, RNA-, and protein-sequence data. Bioinformatics develops and applies these tools to "mine the data," extracting the useful information from sequencing projects. The development and use of algorithms and computer software for analyzing DNA- and protein-sequence data have helped to make molecular biology a more-quantitative field. Sequence data in publicly available databases, freely searchable with an Internet connection, enable scientists and students throughout the world to access this tremendous resource.

A number of databases have been established for the collection and analysis of DNA- and protein-sequence information. Primary databases contain the sequence information, along with information that describes the source of the sequence and its determination. Secondary databases contain the results of analyses carried out on the primary sequence data, such as information about particular sequence patterns, variations, mutations, and evolutionary relationships. Some widely used bioinformatics databases are listed in **Table 20.1**.

After a genome has been sequenced, one of the first tasks is to identify potential genes within the sequence. Unfortunately, there are no universal characteristics that mark the beginning and end of a gene. Before being sequenced, most genomes contain just a few known genes of which the locations have already been determined. The enormous amount of DNA in a genome and the complexities of gene structure make finding genes within the sequence a difficult task. Computer programs have been developed to look for specific sequences in DNA that are associated with certain genes. There are two general approaches to finding genes. The *ab initio* approach scans the sequence looking for features that are usually within a gene. For example, protein-encoding genes are characterized by an open reading frame, which includes a start codon and a stop codon in the same reading frame. Specific sequences mark the splice sites at the beginning and end of introns; other specific sequences are present in promoters immediately upstream of start codons. The comparative approach looks for similarity between a new sequence and sequences of all known genes. If a match is found, then the new sequence is assumed to be a similar gene. Some of these computer programs are capable of examining databases of EST and protein sequences to see if there is evidence that a potential gene is expressed.

Importantly, the programs that have been developed to identify genes on the basis of DNA sequence are not perfect. Therefore, the numbers of genes reported in most genome projects are estimates. The presence of multiple introns, alternative splicing, multiple copies of some genes, and much noncoding DNA between genes makes accurate identification and counting of genes difficult.

After a gene has been identified, it must be **annotated**, which means linking its sequence information to other information about its function and expression, the protein that it encodes, and information on similar genes in other species. There are a number of methods of probing a gene's function, which will be discussed in the Section 20.2 on functional genomics. Computer programs are available for determining whether similar sequences have already been found, either in the same species or in different species. The most widely used of these programs is the Basic Local Alignment Search Tool (BLAST). To conduct a BLAST search, a researcher submits

Table 20.1	Common bioinformatics databases	
Name	**Description**	**URL**
GenBank	Primary DNA-sequence information maintained by the U.S. National Institutes of Health	www.ncbi.nlm.nih.gov/Genbank
EMBL-Bank	Primary DNA-sequence information maintained by European researchers	www.ebi.ac.uk/embl
dbEST	Database of ESTs from many organisms	www.ncbi.nlm.nih.gov/dbEST
MMDB	Molecular Modeling Database, three-dimensional macromolecular structures of proteins and nucleotides	www.ncbi.nlm.nih.gov/Structure/MMDB/mmdb.shtml
FlyBase	Diverse genetic information on *Drosophila melanogaster*	flybase.bio.indiana.edu
UniProt	Protein-sequence data and other information about proteins from a variety of organisms	www.ebi.ac.uk/uniprot

a query sequence and the program searches the database for any other sequences that have regions of high similarity to the query sequence. The program returns all sequences in the database that are similar, along with information about the degree of similarity and the significance of the match (how likely the similarity would be to occur by chance alone).

Additional databases contain information on sequence diversity—how and where a genome varies among individual organisms. The human genome has now been mapped for millions of SNPs, and information about the frequency of these variants in different populations is currently being collected. Additional databases have been developed to catalog all known mutations causing particular diseases; the Human Variome Project is an effort to collect and make available all genetic variations that affect human health. Important information about the expression patterns of the thousands of genes found in genomes also is being compiled.

TRY PROBLEM 35 →

CONCEPTS

Bioinformatics is a interdisciplinary field that combines molecular biology and computer science. It develops databases of DNA and protein sequences and tools for analyzing those sequences.

✔ CONCEPT CHECK 4

The *ab initio* approach finds genes by looking for

a. common sequences found in most genes.

b. similarity in sequence with known genes.

c. mRNA with the use of in situ hybridization.

d. mutant phenotypes.

Metagenomics

Advances in sequencing technology, which have made sequencing faster and less expensive, now provide the possibility of sequencing not just the genomes of individual species but the genomes of entire communities of organisms. **Metagenomics** is an emerging field in which the genome sequences of an entire group of organisms that inhabit a common environment are sampled and determined.

Thus far, metagenomics has been applied largely to microbial communities. Traditionally, bacteria have been studied by growing and analyzing them in the laboratory. However, many bacteria cannot be cultured with the use of laboratory techniques. Metagenomics analyzes microbial communities by extracting DNA from the environment, determining its sequences, and reconstructing community composition and function on the basis of genome sequences. This technique allows the identification and genetic analysis of species that have never been studied by traditional microbiological methods. The entire genomes of some dominant species have been reconstructed from environmental samples, providing scientists with a great deal of information on the biology of these microbes.

An early metagenomic study analyzed the community composition of acid drainage from a mine and determined that this community consisted of only a few dominant bacterial species. Another study, called the Global Ocean Sampling Expedition, followed the route of Darwin's voyage on the H.M.S. *Beagle* in the 1800s. Scientists collected ocean samples and used metagenomic methods to determine their microbial communities. In this study, scientists cataloged sequences for more than 6 million proteins, including more than 1700 new protein families. Other studies have examined the composition of the bacterial community that inhabits the human gut.

Some important results have already emerged from metagenomic studies. Analyses of bacterial genomes found in ocean samples led to the discovery of proteorhodopsin proteins, which are light-driven proton pumps. Subsequent research demonstrated that these proteins are found in diverse microbial groups and are a major source of energy flux in the world's oceans.

Another study examined the gut microflora of obese and lean people. Two groups of bacteria are common in the human gut: Bacteroidetes and Firmicutes. Researchers discovered that obese people have relatively more Firmicutes than do lean people and that the proportion of Firmicutes decreases in obese people who lose weight on a low-calorie diet. These same results were observed in obese and lean mice. In an elegant experiment, researchers transferred bacteria from obese to lean mice. The mice that received bacteria from obese mice extracted more calories from their food and stored more fat, suggesting that gut microflora might play some role in obesity.

Ecological communities have traditionally been characterized on the basis of the species that they contain. Metagenomics affords the possibility of characterizing communities on the basis of their component genes, an approach that has been termed gene-centric. A gene-centric approach leads to new questions: Are certain types of genes more common in some communities than others? Are some genes essential for energy flow and nutrient recycling within a community? Because of the larger sizes of their genomes, eukaryotic communities have not yet been the focus of these approaches, but many researchers predict that, in the future, metagenomic studies will be applied to eukaryotic communities as well as to those of prokaryotes.

CONCEPTS

Metagenomic studies examine the genomes of communities of organisms that inhabit a common environment. This approach has been applied to microbial communities and allows the composition and genetic makeup of the community to be determined without cultivating and isolating individual species. Metagenomic studies are a source of important new insights into microbial communities.

Synthetic Biology

The ability to sequence and study whole genomes, coupled with an increased understanding of the genetic information required for basic biological processes, now provides the possibility of creating—entirely from scratch—novel organisms that have never before existed. Synthetic biology is a new field that seeks to design organisms that might provide useful functions, such as microbes that provide clean energy or break down toxic wastes.

Synthetic biologists have already mixed and matched parts from different organisms to synthesize microbes. In 2002, geneticists recreated the poliovirus by joining together pieces of DNA that were synthesized in the laboratory. Even more impressively, in 2010, Daniel Gibson and his colleagues synthesized from scratch the complete 1.08-million-base-pair genome of the bacterium *Mycoplasma mycoides*. They started with a thousand pieces of DNA that were synthesized in the laboratory and then joined them together in successively larger pieces until they had assembled a complete copy of the genome. Within their synthetic genome they included a set of DNA sequences that spelled out—in code—an email address, the names of the researchers who participated in the project, and several famous quotations. Finally, the researchers transplanted the artificial genome into a cell of a different bacterial species, *M. capricolum*, whose original genome had been removed. The new cell then began expressing the traits specified by the synthetic genome. This experiment constitutes the creation of the first synthetic cell.

These types of experiments have raised a number of concerns about this technology. The ability to tailor-make novel genomes and mix and match parts from different organisms creates the potential for the synthesis of dangerous microbes that might create ecological havoc if they escape from the laboratory or that could be used in biological warfare or bioterrorism.

20.2 Functional Genomics Determines the Function of Genes by Using Genomic-Based Approaches

A genomic sequence is, by itself, of limited use. Merely knowing the sequence would be like having a huge set of encyclopedias without being able to read: you could recognize the different letters but the text would be meaningless. **Functional genomics** characterizes what the sequences do—their function. The goals of functional genomics include the identification of all the RNA molecules transcribed from a genome, called the **transcriptome** of that genome, and all the proteins encoded by the genome, called the **proteome**. Functional genomics exploits both bioinformatics and laboratory-based experimental approaches in its search to define the function of DNA sequences.

Chapter 19 considered several methods for identifying genes and assessing their functions, including in situ hybrid-

ization, experimental mutagenesis, and the use of transgenic animals and knockouts. These methods can be applied to individual genes and can provide important information about the locations and functions of genetic information. In this section, we will focus primarily on methods that rely on knowing the sequences of other genes or that can be applied to large numbers of genes simultaneously.

Predicting Function from Sequence

The nucleotide sequence of a gene can be used to predict the amino acid sequence of the protein that it encodes. The protein can then be synthesized or isolated and its properties studied to determine its function. However, this biochemical approach to understanding gene function is both time consuming and expensive. A major goal of functional genomics has been to develop computational methods that allow gene function to be identified from DNA sequence alone, bypassing the laborious process of isolating and characterizing individual proteins.

Homology searches   One computational method (often the first employed) for determining gene function is to conduct a homology search, which relies on comparisons of DNA and protein sequences from the same organism and from different organisms. Genes that are evolutionarily related are said to be **homologous**. Homologous genes found in different species that evolved from the same gene in a common ancestor are called **orthologs** (**Figure 20.10**). For example, both mouse and human genomes contain a gene that encodes the alpha subunit of hemoglobin; the mouse and human alpha-hemoglobin genes are said to be orthologs, because both genes evolved from an alpha-hemoglobin gene in a mammalian ancestor common to mice and humans. Homologous genes in the same organism (arising by duplication of a single gene in the evolutionary past) are called **paralogs** (see Figure 20.10). Within the human genome is a gene that encodes the alpha subunit of hemoglobin and another, homologous gene that encodes the beta subunit of hemoglobin. These two genes arose because an ancestral gene underwent duplication and the resulting two genes diverged through evolutionary time, giving rise to the alpha- and beta-subunit genes (see Figure 26.18); these two genes are paralogs. Homologous genes (both orthologs and paralogs) often have the same or related functions; so, after a function has been assigned to a particular gene, it can provide a clue to the function of a homologous gene.

Databases containing genes and proteins found in a wide array of organisms are available for homology searches. Powerful computer programs, such as BLAST, have been developed for scanning these databases to look for particular sequences. Suppose a geneticist sequences a genome and locates a gene that encodes a protein of unknown function. A homology search conducted on databases containing the DNA or protein sequences of other organisms may identify

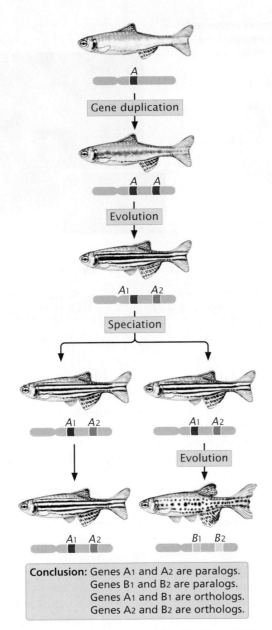

20.10 Homologous sequences are evolutionarily related.
Orthologs are homologous sequences found in different species;
paralogs are homologous genes in the same species and arise from
gene duplication.

Conclusion: Genes A1 and A2 are paralogs.
Genes B1 and B2 are paralogs.
Genes A1 and B1 are orthologs.
Genes A2 and B2 are orthologs.

number of protein domains, which have mixed and matched
through evolutionary time to yield the protein diversity seen
in present-day organisms.

Many protein domains have been characterized, and
their molecular functions have been determined. The
sequence from a newly identified gene can be scanned
against a database of known domains. If the gene sequence
encodes one or more domains whose functions have been
previously determined, the function of the domain can
provide important information about a possible function of
the new gene.

CONCEPTS

The function of an unknown gene can sometimes be determined
by finding genes with similar sequence whose function is known.
A gene's function may also be determined by identifying functional domains in the protein that it encodes.

✔ **CONCEPT CHECK 5**

What is the difference between orthologs and paralogs?

a. Orthologs are homologous sequences; paralogs are analogous
sequences.

b. Orthologs are more similar than paralogs.

c. Orthologs are in the same organism; paralogs are in different
organisms.

d. Orthologs are in different organisms; paralogs are in the same
organism.

Gene Expression and Microarrays

Many important clues about gene function come from
knowing when and where the genes are expressed. The
development of microarrays has allowed the expression of
thousand of genes to be monitored simultaneously.

Microarrays rely on nucleic acid hybridization (see
Chapter 19), in which a known DNA fragment is used as
a probe to find complementary sequences (**Figure 20.11**).
Numerous known DNA fragments are fixed to a solid support in an orderly pattern, or array, usually as a series of dots.
These DNA fragments (the probes) usually correspond to
known genes.

After the microarray has been constructed, mRNA, DNA,
or cDNA isolated from experimental cells is labeled with fluorescent nucleotides and applied to the array. Any of the DNA
or RNA molecules that are complementary to probes on the
array will hybridize with them and emit fluorescence, which
can be detected by an automated scanner. An array containing
tens of thousands of probes can be applied to a glass slide or
silicon wafer just a few square centimeters in size.

The use of microarrays   Used with cDNA, microarrays
can provide information about the expression of thousands
of genes, enabling scientists to study which genes are active

one or more orthologous sequences. If a function is known
for a protein encoded by one of these sequences, that function may provide information about the function of the
newly discovered protein.

Protein domains   Complex proteins often contain
regions that have specific shapes or functions called **protein
domains**. For example, certain DNA-binding proteins attach
to DNA in the same way; these proteins have in common a
domain that provides the DNA-binding function. Each protein domain has an arrangement of amino acids common
to that domain. There are probably a limited, though large,

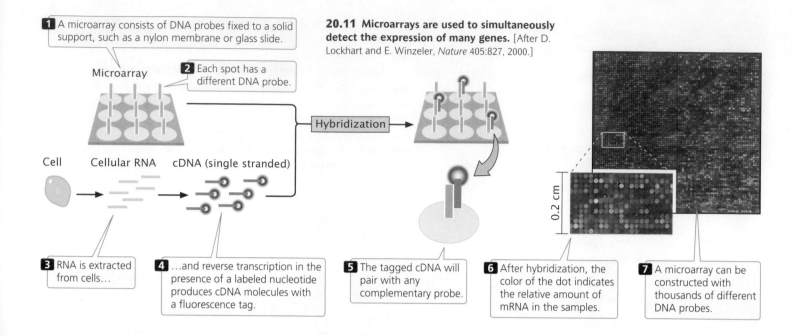

1 A microarray consists of DNA probes fixed to a solid support, such as a nylon membrane or glass slide.

20.11 Microarrays are used to simultaneously detect the expression of many genes. [After D. Lockhart and E. Winzeler, *Nature* 405:827, 2000.]

Microarray

2 Each spot has a different DNA probe.

Cell   Cellular RNA   cDNA (single stranded)

Hybridization

0.2 cm

3 RNA is extracted from cells…

4 …and reverse transcription in the presence of a labeled nucleotide produces cDNA molecules with a fluorescence tag.

5 The tagged cDNA will pair with any complementary probe.

6 After hybridization, the color of the dot indicates the relative amount of mRNA in the samples.

7 A microarray can be constructed with thousands of different DNA probes.

in particular tissues. They can also be used to investigate how gene expression changes in the course of biological processes such as development or disease progression. For example, breast cancer is a disease that affects 1 of 10 women in the United States, and half of those women die from it. Current treatment depends on a number of factors, including a woman's age, the size of the tumor, the characteristics of tumor cells, and whether the cancer has already spread to nearby lymph nodes. Many women whose cancer has not spread are treated by removal of the tumor and radiation therapy, yet the cancer later reappears in some of the women thus treated. These women might benefit from more-aggressive treatment when the cancer is first detected.

Using microarrays, researchers examined the expression patterns of 25,000 genes from primary tumors of 78 young women who had breast cancer (**Figure 20.12**). Messenger RNA from cancer cells and noncancer cells was converted into cDNA and labeled with red fluorescent nucleotides and with green fluorescent nucleotides, respectively. The labeled cDNAs were mixed and hybridized to a DNA chip, which contains DNA probes from different genes. Hybridization of the red (cancer) and green (noncancer) cDNAs is proportional to the relative amounts of mRNA in the samples. The fluorescence of each spot is assessed with microscopic scanning and appears as a single color. Red indicates the overexpression of a gene in the cancer cells relative to that in the noncancer cells (more red-labeled cDNA hybridizes), whereas green indicates the underexpression of a gene in the cancer cells relative to that in the noncancer cells (more green-labeled cDNA hybridizes). Yellow indicates equal expression in both types of cells (equal hybridization of red- and green-labeled cDNAs), and no color indicates no expression in either type of cell.

In 34 of the 78 patients, the cancer later spread to other sites; the other 44 patients remained free of breast cancer for 5 years after their initial diagnoses. The researchers identified a subset of 70 genes whose expression patterns in the initial tumors accurately predicted whether the cancer would later spread (see Figure 20.12). This degree of prediction was much higher than that of traditional predictive measures, which are based on the size and histology of the tumor. These results, though preliminary and confined to a small sample of cancer patients, suggest that gene-expression data obtained from microarrays can be a powerful tool in determining the nature of cancer treatment. **TRY PROBLEM 38** →

In another application of microarrays, geneticists examined differences in the gene-expression patterns of patients having different forms of leprosy, a debilitating disease caused by the bacterium *Mycobacterium leprae*. Some people infected with *M. leprae* have an active form of leprosy, called lepromatous leprosy. They have typical symptoms and lesions with high numbers of bacteria. Other people infected with *M. leprae* have limited symptoms and lesions with few bacteria that spontaneously heal. This form of the disease is called tuberculoid leprosy. Why do some people become lepromatous when infected with *M. leprae*, whereas other are tuberculoid?

To answer this question, geneticists used microarrays to study the expression patterns of genes in the two groups of patients. Biopsies were conducted on skin lesions from five lepromatous patients and six tuberculoid patients. RNA was isolated from samples, converted into cDNA, labeled, and hybridized to a microarray containing 12,000 human genes. The expression patterns of a number of human genes differed in lepromatous and tuberculoid patients. Many of

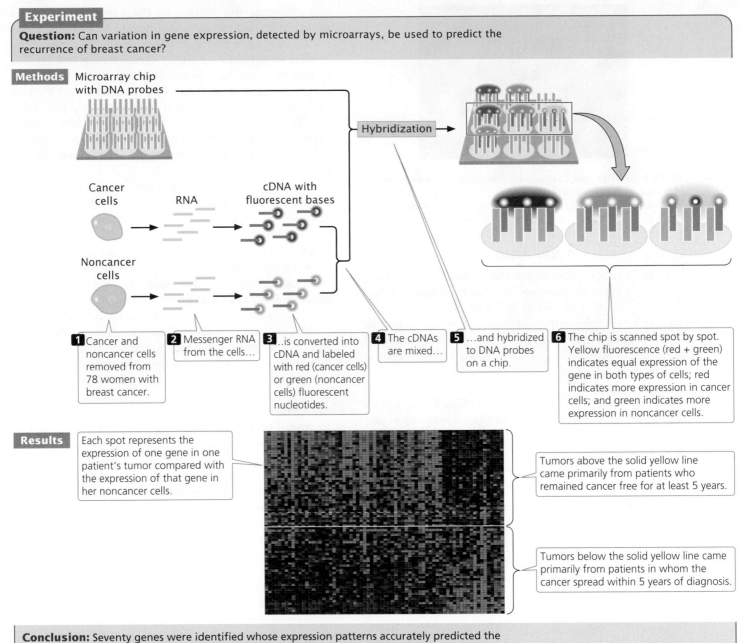

Experiment

Question: Can variation in gene expression, detected by microarrays, be used to predict the recurrence of breast cancer?

Methods Microarray chip with DNA probes

Hybridization

Cancer cells

RNA

cDNA with fluorescent bases

Noncancer cells

1 Cancer and noncancer cells removed from 78 women with breast cancer.

2 Messenger RNA from the cells…

3 …is converted into cDNA and labeled with red (cancer cells) or green (noncancer cells) fluorescent nucleotides.

4 The cDNAs are mixed…

5 …and hybridized to DNA probes on a chip.

6 The chip is scanned spot by spot. Yellow fluorescence (red + green) indicates equal expression of the gene in both types of cells; red indicates more expression in cancer cells; and green indicates more expression in noncancer cells.

Results Each spot represents the expression of one gene in one patient's tumor compared with the expression of that gene in her noncancer cells.

Tumors above the solid yellow line came primarily from patients who remained cancer free for at least 5 years.

Tumors below the solid yellow line came primarily from patients in whom the cancer spread within 5 years of diagnosis.

Conclusion: Seventy genes were identified whose expression patterns accurately predicted the recurrence of breast cancer within 5 years of treatment.

20.12 Microarrays can be used to examine gene expression associated with disease progression.
[After L. J. van't Veer, *Nature* 405:532, 2002.]

these genes participate in the immune response, suggesting that differences in immune function are critical in the progression of active leprosy. The products of the genes that show differences in expression are being examined as possible targets for drug therapy.

Microarrays that allow the detection of specific alleles, SNPs, and even particular proteins also have been created. Importantly, not all DNA molecules bind equally to microarrays, and so microarrays may sometimes overestimate or underestimate the expression of specific genes. Thus, verifi-

cation of the results of microarrays through other methods is desirable.

CONCEPTS

Microarrays, consisting of DNA probes attached to a solid support, can be used to determine which RNA and DNA sequences are present in a mixture of nucleic acids. They are capable of determining which RNA molecules are being synthesized and can thus be used to examine changes in gene expression.

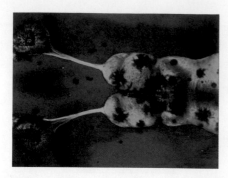

20.13 A reporter sequence can be used to examine expression of a gene. Expression of the neural-specific β-tubulin gene in the brain of a tadpole is revealed by fluorescent green pigment. [Miranda Gomperts, Courtesy Enrique Amaya, The Amaya Lab.]

Gene Expression and Reporter Sequences

Patterns of gene expression can also be determined visually by using a reporter sequence. In this approach, genomic fragments are first cloned in BACs or other vectors that are capable of holding the coding region of a gene plus its regulatory sequences. The coding region of a gene whose expression is to be studied is then replaced with a reporter sequence, which encodes an easily observed product. For example, a commonly used reporter sequence encodes a green fluorescent protein (GFP) from jellyfish. The BAC is then inserted into an embryo, creating a transgenic organism. The regulatory sequence of the cloned gene ensures that it is expressed at the appropriate time and in the appropriate tissue within the transgenic organism. The product of the gene expression is a fluorescent green pigment, which is easily observed (**Figure 20.13**).

This technique is being used to study the expression patterns of genes that affect brain function. In the Gene Expression Nervous System Atlas (GENSAT) project, scientists are systematically replacing the coding regions of hundreds of genes with the GFP reporter sequence and observing their patterns of expression in transgenic mice. The goal is to produce a comprehensive atlas of gene expression in the mouse brain. This project has already shed light on where in the brain several genes having roles in inherited neurological disorders are expressed. Another functional-genomics project, the Allen Brain Atlas, has compiled information on the expression patterns of 20,000 genes in the mouse brain.

Genomewide Mutagenesis

As discussed in Chapter 18, one of the best methods for determining the function of a gene is to examine the phenotypes of individual organisms that possess a mutation in the gene. Traditionally, genes encoding naturally occurring variations in a phenotype were mapped, the causative genes were isolated, and their products were studied. But

this procedure was limited by the number of naturally occurring mutations and the difficulty of mapping genes with a limited number of chromosomal markers. The number of naturally occurring mutations can be increased by exposure to mutagenic agents, and the accuracy of mapping is increased dramatically by the availability of mapped molecular markers, such as RFLPs, microsatellites, STSs, ETSs, and SNPs. These two methods—random inducement of mutations on a genomewide basis and mapping with molecular markers—are coupled and automated in a **mutagenesis screen**.

Genomewide mutagenesis screens can be used to search for all genes affecting a particular function or trait. For example, mutagenesis screens of mice are being used to identify genes having roles in cardiovascular function. When genes that affect cardiovascular function are located in mice, homology searches are carried out to determine if similar genes exist in humans. These genes can then be studied to better understand cardiac disease in humans.

To conduct a mutagenesis screen, random mutations are induced in a population of organisms, creating new phenotypes. The mutations are induced by exposing the organisms to radiation, a chemical mutagen (see Chapter 18), or transposable elements (DNA sequences that insert randomly into the DNA; see Chapter 11). The procedure for a typical mutagenesis screen is illustrated in **Figure 20.14**. Here, male zebrafish are treated with ethylmethylsulfonate (EMS), a chemical that induces germ-line mutations in their sperm. The treated males are mated with wild-type female fish. A few of the offspring will be heterozygous for mutations induced by EMS; so the offspring are screened for any mutant phenotypes that might be the products of dominant mutations expressed in these heterozygous fish. Recessive mutations will not be expressed in the F_1 progeny but can be revealed with further breeding.

The fish with variant phenotypes undergo further breeding experiments to verify that each variant phenotype is, in fact, due to a single-gene mutation. Although these breeding studies may indicate that the mutant phenotype is due to a single-gene mutation, the gene in which the mutation occurs is not known. The gene can be located, however, by positional cloning (see Chapter 19).

Mutagenesis screens have been used to study genes that control vertebrate development. A team of developmental geneticists have produced thousands of mutations in the zebrafish that affect development and are systematically locating and characterizing the loci where the mutations occur. Zebrafish are ideal genetic models for this type of study because they reproduce quickly, are easily reared in the laboratory, and have transparent embryos in which developmental deformities are easy to spot. This research has already identified a number of genes that are important in embryonic development, many of which have counterparts in humans.

Experiment

Question: How can genes encoding a particular trait or function be identified?

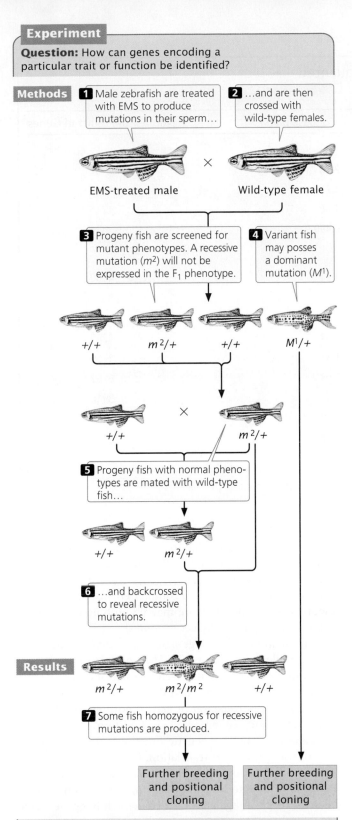

Methods

1 Male zebrafish are treated with EMS to produce mutations in their sperm...

2 ...and are then crossed with wild-type females.

EMS-treated male × Wild-type female

3 Progeny fish are screened for mutant phenotypes. A recessive mutation (m^2) will not be expressed in the F_1 phenotype.

4 Variant fish may posses a dominant mutation (M^1).

$+/+$ $m^2/+$ $+/+$ $M^1/+$

$+/+$ × $m^2/+$

5 Progeny fish with normal phenotypes are mated with wild-type fish...

$+/+$ $m^2/+$

6 ...and backcrossed to reveal recessive mutations.

Results

$m^2/+$ m^2/m^2 $+/+$

7 Some fish homozygous for recessive mutations are produced.

Further breeding and positional cloning

Further breeding and positional cloning

Conclusion: The mutagenesis screen produces fish with a mutation affecting the trait. These fish are further analyzed to identify the mutation.

20.14 Genes affecting a particular characteristic or function can be identified by a genomewide mutagenesis screen. In this illustration, M^1 represents a dominant mutation and m^2 represents a recessive mutation.

CONCEPTS

Genomewide mutagenesis screening coupled with positional cloning can be used to identify genes that affect a specific characteristic or function.

✔ CONCEPT CHECK 6

Which is the correct order of steps in a mutagenesis screen?

a. Positional cloning, mutagenesis, identification of mutants, verification of genetic basis.

b. Mutagenesis, positional cloning, identification of mutants, verification of genetic basis.

c. Mutagenesis, identification of mutants, verification of genetic basis, positional cloning.

d. Identification of mutants, positional cloning, mutagenesis, verification of genetic basis.

20.3 Comparative Genomics Studies How Genomes Evolve

Genome-sequencing projects provide detailed information about gene content and organization in different species and even in different members of the same species, allowing inferences about how genes function and genomes evolve. They also provide important information about evolutionary relationships among organisms and about factors that influence the speed and direction of evolution. **Comparative genomics** is the field of genomics that compares similarities and differences in gene content, function, and organization among genomes of different organisms.

Prokaryotic Genomes

More than 1000 bacterial genomes have now been sequenced. Most prokaryotic genomes consist of a single circular chromosome. However, there are exceptions, such as that of *Vibrio cholerae*, the bacterium that causes cholera, which has two circular chromosomes, and that of *Borrelia burgdorferi*, which has one large linear chromosome and 21 smaller chromosomes.

Genome size and number of genes The total amount of DNA in prokaryotic genomes ranges from 490,885 bp in *Nanoarchaeum equitans*, an archaea that lives entirely within another archaea, to 9,105,828 bp in *Bradyrhizobium japonicum*, a soil bacterium (**Table 20.2**). Although this range in genome size might seem extensive, it is much less than the enormous range of genome sizes seen in eukaryotes, which can vary from a few million base pairs to hundreds of billions of base pairs. *Escherichia coli*, the most widely used bacterium for genetic studies, has a fairly typically genome size at 4.6 million base pairs. Archaea and bacteria are similar

Table 20.2 Characteristics of some completely sequenced representative prokaryotic genomes

Species	Size (millions of base pairs)	Number of Predicted Genes
Archaea		
Archaeoglobus fulgidus	2.18	2407
Methanobacterium thermoautotrophicum	1.75	1869
Methanococcus jannaschii	1.66	1715
Nanoarchaeum equitans	0.490	536
Eubacteria		
Bacillus subtilis	4.21	4100
Bradyrhizobium japonicum	9.11	8317
Buchnera species	0.64	564
Escherichia coli	4.64	4289
Haemophilus influenzae	1.83	1709
Mesorhizobium loti	7.04	6752
Mycobacterium tuberculosis	4.41	3918
Mycoplasma genitalium	0.58	480
Staphylococcus aureus	2.88	2697
Vibrio cholerae	4.03	3828

Source: Data from the Genome Atlas of the Center for Biological Sequence Analysis, www.cbs.dtu.dk/services/GenomeAtlas.

in their ranges of genome size. Surprisingly, genome size also varies extensively within some species; for example, different strains of *E. coli* vary in genome size by more than 1 million base pairs.

Among prokaryotes, the number of genes typically varies from 1000 to 2000, but some species have as many as 6700 and others as few as 480. Interestingly, the density of genes is rather constant across all species, with an average of about one gene per 1000 bp. Thus, prokaryotes with larger genomes will have more genes, in contrast with eukaryotes, for which there is little association between genome size and number of genes (see the section on eukaryotic genomes). The evolutionary factors that determine the size of prokaryotic genomes (as well as eukaryotic genomes) are still largely unknown. However, the time required for cell division is frequently longer in organisms with larger genomes because of the time required to replicate the DNA before division. Thus, selection may favor smaller genomes in organisms that occupy environments where rapid reproduction is advantageous.

Prokaryotes with the smallest genomes tend to be in species that occupy restricted habitats, such as bacteria that live inside other organisms. The constant environment and metabolic functions supplied by the host organism may allow these bacteria to survive with fewer genes. Bacteria with the larger genomes tend to occupy highly complex and variable environments, such as the soil or the root nodules of plants. In these environments, genes that are needed only occasionally may be useful. In complex environments where resources are abundant, there may be little need for rapid division (which favors smaller genomes).

An example of a large-genome bacterium occupying a complex environment is *Streptomyces coeliocolor*, a filamentous bacterium with a genome size of 8,677,507 bp. This bacterium has been called the "Boy Scout" bacterium because it has a diverse set of genes and therefore follows the Scout motto: be prepared. In addition to the usual housekeeping genes for genetic functions such as replication, transcription, and translation, its genome contains a number of genes for the breakdown of complex carbohydrates, allowing it to consume decaying matter from plants, animals, insects, fungi, and other bacteria. It has genes for a large number of proteins that produce secondary metabolites (breakdown products) that can function as antibiotics and protect against desiccation and low temperatures. In this species, a large genome with lots of genes appears to be beneficial in the complex environment that it inhabits.

Horizontal gene transfer Bacteria possess a number of mechanisms by which they can gain and lose DNA. DNA

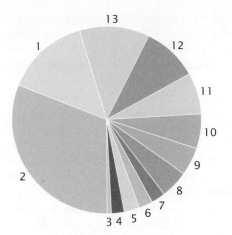

1. Metabolism
2. Unknown
3. Ionic homeostasis
4. Protein synthesis
5. Energy
6. Transport facilitation
7. Cellular biogenesis
8. Intracellular transport
9. Protein destination
10. Cellular communication and signal transduction
11. Cell rescue, defense, cell death, and aging
12. Cell growth, cell division, and DNA synthesis
13. Transcription

20.15 The functions of many genes in prokaryotes cannot be determined by comparison with genes in other prokaryotes. Proportion of the circle occupied by each color represents the proportion of genes affecting various known and unknown functions in *E. coli*.

can be lost through simple deletion; DNA can be gained by gene duplication and through the insertion of transposable genetic elements. Another mechanism for gaining new genetic information is horizontal gene transfer, a process by which both closely and distantly related bacterial species periodically exchange genetic information over evolutionary time. Such exchange may take place through the bacterial uptake of DNA in the environment (transformation), through the exchange of plasmids, and through viral vectors (see Chapter 8). Horizontal gene transfer has been recognized for some time, but analyses of many microbial genomes now indicate that it is more extensive than was formerly recognized. The widespread occurrence of horizontal transfer has caused some biologists to question whether distinct species even exist in bacteria (see Chapter 8 for more discussion of this matter).

Function of genes   Only about half of the genes identified in prokaryotic genomes can be assigned a function. Almost a quarter of the genes have no significant sequence similarity to any known genes in other bacteria. The number of genes that encode biological functions such as transcription and translation tends to be similar among species, even when their genomes differ greatly in size. This similarity suggests that these functions are encoded by a basic set of genes that does not vary among species. In contrast, the number of genes taking part in biosynthesis, energy metabolism, transport, and regulatory functions varies greatly among species and tends to be higher in species with larger genomes. The functions of predicted genes (i.e., genes identified by computer programs) and known genes in *E. coli* are presented in **Figure 20.15**. A substantial part of the "extra" DNA found in the larger bacterial genomes is made up of paralogous genes that have arisen by duplication.

CONCEPTS

Comparative genomics compares the content and organization of whole genomic sequences from different organisms. Prokaryotic genomes are small, usually ranging from 1 million to 3 million base pairs of DNA, with several thousand genes. Prokaryotes with the smallest genomes tend to occupy restricted habitats, whereas those with the largest genomes are usually found in complex environments. Horizontal gene transfer has played a major role in bacterial genome evolution.

✔ **CONCEPT CHECK 7**

What is the relation between genome size and gene number in prokaryotes?

Eukaryotic Genomes

The genomes of more than 100 eukaryotic organisms have been completely sequenced, including a number of fungi and protists, several insects, a number of plants, and numerous vertebrates. Sequenced eukaryotes include papaya, corn, rice, sorghum, grapevine, silkworms, fruit flies, aphids, mosquitoes, anemone, mouse, rat, dog, cow, horse, chimpanzee, and human. Even the genomes of extinct organisms have now been sequenced, including those of the wooly mammoth and Neanderthals. Hundreds of additional eukaryotic genomes are in the process of being sequenced. It is important to note that, even though the genomes of these organisms have been "completely sequenced," many of the final assembled sequences contain gaps, and regions of heterochromatin may not have been sequenced at all. Thus, the sizes of eukaryotic genomes are often estimates, and the number of base pairs given for the genome size of a particular species may vary. Predicting the number of genes that are present in a genome also is difficult and may vary, depending on the assumptions made and the particular gene-finding software used.

Genome size and number of genes The genomes of eukaryotic organisms (**Table 20.3**) are larger than those of prokaryotes, and, in general, multicellular eukaryotes have more DNA than do simple, single-celled eukaryotes such as yeast (see p. 301 in Chapter 11). However, there is no close relation between genome size and complexity among the multicellular eukaryotes. For example, the roundworm *Caenorhabditis elegans* is structurally more complex than the plant *Arabidopsis thaliana* but has considerably less DNA.

In general, eukaryotic genomes also contain more genes than do prokaryotes (but some large bacteria have more genes than single-celled yeasts have), and the genomes of multicellular eukaryotes have more genes than do the genomes of single-celled eukaryotes. In contrast with bacteria, there is no correlation between genome size and number of genes in eukaryotes. The *number* of genes among multicellular eukaryotes also is not obviously related to phenotypic complexity: humans have more genes than do invertebrates but only twice as many as fruit flies and only slightly more than the plant *A. thaliana*. The nematode *C. elegans* has more genes than does *Drosophila melanogaster* but is less complex. Additionally, the pufferfish has only about one-tenth the amount of DNA present in humans and mice but has almost as many genes. Eukaryotic genomes contain multiple copies of many genes, indicating that gene duplication has been an important process in genome evolution.

Segmental duplications and multigene families
Many eukaryotic genomes, especially those of multicellular organisms, are filled with **segmental duplications**, regions greater than 1000 bp that are almost identical in sequence. For example, about 4% of the human genome consists of segmental duplications. In most segmental duplications, the two copies are found on the same chromosome (an intrachromosomal duplication), but, in others, the two copies are found on different chromosomes (an interchromosomal duplication). In the human genome, the average size of segmental duplications is 15,000 bp.

Segmental duplications arise from processes that generate chromosome duplications, such as unequal crossing over (see Chapter 9). After a segmental duplication arises, it promotes further duplication by the occurrence of misalignment among the duplicated regions. Segmental duplication plays an important role in evolution by giving rise to new genes. After a segmental duplication arises, the original copy of the gene can continue its function while the new copy undergoes mutation. These changes may eventually lead to new function. The importance of gene duplication in genome evolution is demonstrated by the large number of multigene families that exist in many eukaryotic genomes. A **multigene family** is a group of evolutionarily related genes that arose through repeated evolution of an ancestral gene. For example, the globin gene family in humans consists of 13 genes that encode globinlike molecules, most of which produce proteins that carry oxygen. An even more spectacular example is the human olfactory multigene family, which consists of about 1000 genes that encode olfactory receptor molecules used in our sense of smell.

Gene deserts The density of genes in a typical eukaryotic genome varies greatly, with some chromosomes having a high density of genes and others being relatively gene poor. In some areas of the genome, long stretches of DNA, often consisting of hundreds of thousands to millions of base pairs are completely devoid of any known genes or other functional sequences; these regions are known as **gene**

Table 20.3	Characteristics of some eukaryotic genomes that have been completely sequenced	
Species	Genome Size (millions of base pairs)	Number of Predicted Genes
Saccharomyces cerevisiae (yeast)	12	6,144
Arabidopsis thaliana (plant)	125	25,706
Caenorhabditis elegans (roundworm)	103	20,598
Drosophila melanogaster (fruit fly)	170	13,525
Anopheles gambiae (mosquito)	278	14,707
Danio rerio (zebrafish)	1,465	22,409
Takifugu rubripes (tiger pufferfish)	329	22,089
Mus musculus (mouse)	2,627	26,762
Rattus novergicus (Norway rat)	2,571	23,761
Pan troglodytes (chimpanzee)	2,733	22,524
Homo sapiens (human)	3,223	~24,000

Source: Ensembl Web site: www.ensembl.org/index.html and plants.ensembl.org/index.html

deserts and are surprisingly common in eukaryotes. The human genome contains about 500 gene deserts or more, making up approximately 25% of the total euchromatin in the human genome. Gene deserts are particularly common on human chromosomes 4, 5, and 13, where they cover as much as 40% of the entire chromosome.

What is the purpose of a gene desert? Why does it exist? A possible answer is that it contains DNA sequences that have a functional role—perhaps in regulating genes or in the overall architecture of the chromosome—but we are unable to recognize the function of the sequences that it contains. To address the functional significance of gene deserts, Marcelo Nobrega and his colleagues deleted gene deserts from mice. They created transgenic mice that were missing either a 1,500,000-bp gene desert from mouse chromosome 3 or a 845,000-bp gene desert from chromosome 19. Through interbreeding, mice were created that were homozygous for the deletions, meaning that their genomes were completely missing the DNA in these gene deserts. The researchers carefully monitored the resulting transgenic mice, looking at survival rate, weight gain, and blood chemistry. They conducted visual and pathological examinations at various ages on multiple organs, including brain, thymus, heart, lungs, spleen, stomach, intestine, kidney, bladder, and reproductive organs. Remarkably, the mice with the deleted gene deserts appeared healthy and were indistinguishable from control mice. Although the mice carrying the deletions could have had abnormalities that went undetected, these results suggest that large regions of the mammalian genome can be deleted without major phenotypic effects and may, in fact, be superfluous.

Transposable elements A substantial part of the genomes of most multicellular organisms consists of moderately and highly repetitive sequences (see Chapter 11), and the percentage of repetitive sequences is usually higher in those species with larger genomes (**Table 20.4**). Most of these repetitive sequences appear to have arisen through transposition and are particularly evident in the human genome: 45% of the DNA in the human genome is derived from transposable elements, many of which are defective and no longer able to move. Most of the DNA in multicellular organisms is noncoding, and many genes are interrupted by introns. In the more complex eukaryotes, both the number and the length of the introns are greater.

Protein diversity In spite of only a modest increase in gene number, vertebrates have considerably more protein diversity than do invertebrates. One way to measure the amount of protein diversity is by counting the number of protein domains, which are characteristic parts of proteins that are often associated with a function. Vertebrate genomes do not encode significantly more protein domains than do invertebrate genomes; for example, there are 1262 domains in humans compared with 1035 in fruit flies (**Table 20.5**). However, the existing domains in humans are

Table 20.4 Percentage of genome consisting of interspersed repeats derived from transposable elements

Organism	Percentage of Genome
Arabidopsis thaliana (plant)	10.5
Caenorhabditis elegans (worm)	6.5
Drosophila melanogaster (fly)	3.1
Takifugu rubripes (tiger pufferfish)	2.7
Homo sapiens (human)	44.4

assembled into more combinations, leading to many more types of proteins. For example, the human genome contains almost twice as many arrangements of protein domains as do worms or flies and almost six times as many as does yeast.

Homologous genes An obvious and remarkable trend seen in eukaryotic genomes is the degree of homology among genes found in even distantly related species. For example, mice and humans have about 99% of their genes in common. About 50% of the genes in fruit flies are homologous to genes in humans, and, even in plants, about 18% of the genes are homologous to those found in humans.

Colinearity between related genomes One of the features of genome evolution revealed by comparing the gene sequences of different organisms is that many genes are present in the same order in related genomes, a phenomenon that is sometimes termed colinearity. The reason for colinearity among genomes is that they are descended from a common ancestral genome, and evolutionary forces have

Table 20.5 Number of estimated protein domains encoded by some eukaryotic genomes

Species	Number of Predicted Protein Domains
Saccharomyces cerevisiae (yeast)	851
Arabidopsis thaliana (plant)	1012
Caenorhabditis elegans (roundworm)	1014
Drosophila melanogaster (fruit fly)	1035
Homo sapiens (human)	1262

Source: Number of genes and protein-domain families from the International Human Genome Sequencing Consortium, Initial sequencing and analysis of the human genome, *Nature* 409:860–921, Table 23, 2001.

Rice

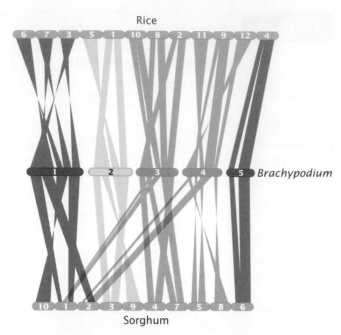

20.16 Colinear relationships among blocks of genes found in rice, sorghum, and *Brachypodium* (wild grass). Each colored band represents a block of genes that is collinear between chromosomes of the three species.

maintained the same gene order in the genomes of descendants. Genomic studies of grasses—plants in the family Poaceae—illustrate the principle of colinearity.

The grasses comprise more than 10,000 species, including economically important crops such as rice, wheat, barley, corn, millet, oat, and sorghum. Taken together, grasses make up about 60% of the world's food production. The genomes of these species vary greatly in size and chromosome number. For example, chromosome number in grasses ranges from 4 to 266; the rice genome consists of only about 460 million base pairs, whereas the genome of wheat contains 17 billion base pairs. In spite of these large differences in chromosome number and genome size, the position and order of many genes within the genomes are remarkably conserved. Through evolutionary time, regions of DNA between the genes (intergenic regions) have increased, decreased, and undergone rearrangements, whereas the genes themselves have stayed relatively constant in order and content. An example of the collinear relationship of genes between rice, sorghum, and *Brachypodium* (wild grass) is shown in **Figure 20.16**.

TRY PROBLEMS 40 AND 41 ➡

CONCEPTS

Genome size varies greatly among eukaryotic species. For multicellular eukaryotic organisms, there is no clear relation between organismal complexity and amount of DNA or gene number. A substantial part of the genome in eukaryotic organisms consists of

repetitive DNA, much of which is derived from transposable elements. Vast regions of DNA may contain no genes or other functional sequences. Many eukaryotic genomes have homologous genes in common, and genes are often in the same order in the genomes of related organisms.

✔ CONCEPT CHECK 8

Segmental duplications play an important role in evolution by

a. giving rise to new genes and multigene families.

b. keeping the number of genes in a genome constant.

c. eliminating repetitive sequences produced by transposition.

d. controlling the base content of the genome.

Comparative *Drosophila* Genomics

The fruit fly *Drosophila melanogaster* is one of the workhorses of genetics. Its genome was sequenced in 2000, the second animal genome to be deciphered. In 2007, a consortium of 250 researchers from different parts of the world published genomic sequences of 10 additional species of *Drosophila*. Combined with the already sequenced genomes of *D. melanogaster* and *D. pseudoobscura*, this effort provided genomic data for 12 species of *Drosophila*, allowing detailed evolutionary analysis of this group of insects.

The 12 *Drosophila* species that were sequenced are found throughout the world, and all diverged from a common ancestor some 60 million years ago (**Figure 20.17**). Genome size in the group varies from 130 million base pairs in *D. mojavensis* to 364 million base pairs in *D. virilis* (**Table 20.6**). The number of protein-encoding genes varies less widely, from 13,733 to 17,325. Many of the genes are found in all 12 of the species; for example, 77% of the 13,733 genes found in *D. melanogaster* have homologs in all of the other species.

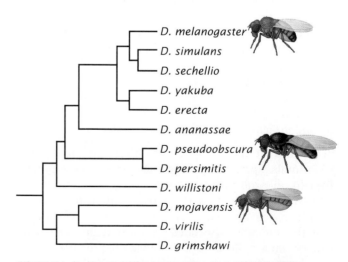

20.17 Twelve *Drosophila* species whose genomes have been determined and analyzed diverged from a common ancestor approximately 60 million years ago.

Table 20.6	Characteristics of the genomes of 12 *Drosophila* species	
Species	Genome Size (millions of base pairs)	Number of Protein-Encoding Genes
D. melanogaster	200	13,733
D. simulans	162	15,983
D. sechellia	171	16,884
D. yakuba	190	16,423
D. erecta	135	15,324
D. ananassae	217	15,276
D. pseudoobscura	193	16,363
D. persimilis	193	17,325
D. willistoni	222	15,816
D. virilis	364	14,680
D. mojavensis	130	14,849
D. grimshawi	231	15,270

Source: *Drosophila* 12 Genome Consortium, Evolution of genes and genomes on the *Drosophila* phylogeny, *Nature* 450:203–218, 2007.

Transposable elements have played an important role in the evolution of *Drosophila* genomes. The extent of DNA that consists of the remnants of transposable elements ranges from 2.7% of the genome in *D. simulans* to 25% of the genome of *D. ananassae*. The amount of DNA contributed by transposable elements accounts for much of the difference in genome size among the species. Genomic rearrangements (inversions and translocations) were frequent in the evolution of this group. The rate of evolution varies among different classes of genes: genes controlling olfaction, immunity, and insecticide resistance have evolved at a faster rate compared with other genes.

The Human Genome

The human genome, which is fairly typical of mammalian genomes, has been extensively studied and analyzed because of its importance to human health and evolution. It is 3.2 billion base pairs in length, but only about 25% of the DNA is transcribed into RNA, and less than 2% encodes proteins. Active genes are often separated by vast regions of noncoding DNA, much of which consists of repeated sequences derived from transposable elements.

The average gene in the human genome is approximately 27,000 bp in length, with about 9 exons (**Table 20.7**). (One exceptional gene has 234 exons.) The introns of human genes are much longer, and there are more of them than in other genomes (**Figure 20.18**). The human genome does not encode substantially more protein domains (see Table 20.5), but the domains are combined in more ways to produce

Table 20.7	Average characteristics of genes in the human genome
Characteristic	Average
Number of exons	8.8
Size of internal exon	145 bp
Size of intron	3,365 bp
Size of 5′ untranslated region	300 bp
Size of 3′ untranslated region	770 bp
Size of coding region	1,340 bp
Total length of gene	27,000 bp

a relatively diverse proteome. Gene functions encoded by the human genome are presented in **Figure 20.19**. A single gene often encodes multiple proteins through alternative splicing; each gene encodes, on the average, two or three different mRNAs, meaning that the human genome, with approximately 24,000 genes, might encode 72,000 proteins or more.

Gene density varies among human chromosomes; chromosomes 17, 19, and 22 have the highest densities and chromosomes X, Y, 4, 13, and 18 have the lowest densities. Some proteins encoded by the human genome that are not found in other animals include: those affecting immune function; neural development, structure, and function; intercellular and intracellular signaling pathways in development; hemostasis; and apoptosis.

Transposable elements are much more common in the human genome than in worm, plant, and fruit-fly genomes (see Table 20.4). The density of transposable elements varies, depending on chromosome location. In one region of the X

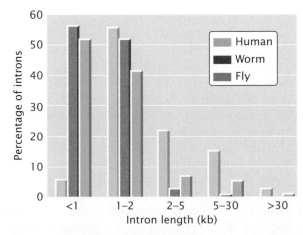

20.18 The introns of genes in humans are generally longer than the introns of genes in worms and flies.

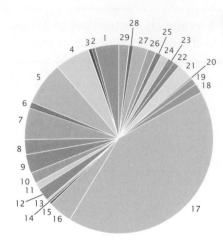

1. Miscellaneous
2. Viral protein
3. Transfer or carrier protein
4. Transcription factor
5. Nucleic acid enzyme
6. Signaling molecule
7. Receptor
8. Kinase
9. Select regulatory molecule
10. Transferase

11. Synthase and synthetase
12. Oxidoreductase
13. Lyase
14. Ligase
15. Isomerase
16. Hydrolase
17. Molecular function unknown
18. Transporter
19. Intracellular transporter
20. Select calcium-binding protein

21. Protooncogene
22. Structural protein of muscle
23. Motor
24. Ion channel
25. Immunoglobulin
26. Extracellular matrix
27. Cytoskeletal structural protein
28. Chaperone
29. Cell adhesion

20.19 Functions for many human genes have yet to be determined. Proportion of the circle occupied by each color represents the proportion of genes affecting various known and unknown functions.

chromosome, 89% of the DNA is made up of transposable elements, whereas other regions are largely devoid of these elements. The human genome contains a variety of types of transposable elements, including LINEs, SINEs, retrotransposons, and DNA transposons (see Chapter 11). Most appear to be evolutionarily old and are defective, containing mutations and deletions making them no longer capable of transposition.

20.4 Proteomics Analyzes the Complete Set of Proteins Found in a Cell

DNA sequence data are tremendous sources of insight into the biology of an organism, but they are not the whole story. Many genes encode proteins, and proteins carry out the vast majority of the biochemical reactions that shape the phenotype of an organism. Although proteins are encoded by DNA, many proteins undergo modifications after translation and, in more-complex eukaryotes, there are many more proteins than genes. Thus, in recent years, molecular biologists have turned their attention to analysis of the protein content of cells. The ultimate goal is to determine the proteome, the complete set of proteins found in a given cell, and the study of the proteome is termed **proteomics**.

Plans are underway to identify and characterize all proteins in the human body, an effort that has been called the **Human Proteome Project**. The project would catalog which proteins are present in which cell types, where each protein is located within the cell, and which other proteins each interacts with. Many researchers feel that this information will be of immense benefit in identifying drug targets, understanding the biological basis of disease, and understanding the molecular basis of many biological processes.

Deciphering the proteome of even a single cell is a challenging task. Every cell contains a complete sequence of genes, but different cells express vastly different proteins. Each gene may produce a number of different proteins through alternative processing (see Chapter 14) and post-translational protein processing (see Chapter 15). A typical human cell contains as many as 100,000 different proteins that vary greatly in abundance, and no technique such as PCR can be used to easily amplify proteins. Many technical advances in analyzing proteins have been made in recent years, but the complete proteome of even a single species has yet to be determined.

Determination of Cellular Proteins

The basic procedure for characterizing the proteome is first to separate the proteins found in a cell and then to identify and quantify the individual proteins. One method for separating proteins is **two-dimensional polyacrylamide gel electrophoresis** (2D-PAGE), in which the proteins are separated in one dimension by charge, separated in a second dimension by mass, and then stained (**Figure 20.20a**). This procedure separates the different proteins into spots, with the size of each spot proportional to the amount of protein present. A typical 2D-PAGE gel may contain several hundred to several thousand spots (**Figure 20.20b**).

Because 2D-PAGE does not detect some proteins in low abundance and is difficult to automate, researchers have recently turned to liquid chromatography for separating proteins. In liquid chromatography, a mixture of molecules is dissolved in a liquid and passed through a column packed with solid particles. Different affinities for the liquid and solid phases cause some components of the mixture to travel through the column more slowly than others, resulting in separation of the components of the mixture.

The traditional method for identifying a protein is to remove its amino acids one at a time and determine the identity of each amino acid removed. This method is far too slow and labor intensive for analyzing the thousands of

(a)

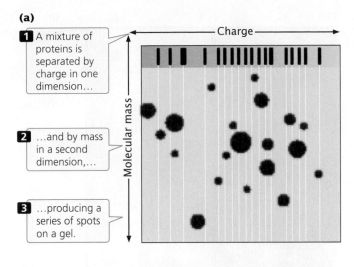

1 A mixture of proteins is separated by charge in one dimension…

2 …and by mass in a second dimension,…

3 …producing a series of spots on a gel.

← Charge →

Molecular mass

(b)

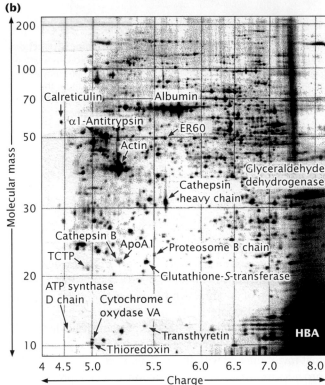

Molecular mass

200

100

70 — Calreticulin — Albumin

α1-Antitrypsin

50 — ER60

Actin

Glyceraldehyde dehydrogenase

Cathepsin heavy chain

30

Cathepsin B — ApoA1 — Proteosome B chain

TCTP — Glutathione-S-transferase

ATP synthase D chain — Cytochrome c oxydase VA

20

Transthyretin

10 — Thioredoxin

HBA

4 4.5 5.0 5.5 6.0 6.5 7.0 8.0

← Charge →

20.20 Two-dimensional acrylamide gel electrophoresis can be used to separate cellular proteins. [After G. Gibson and S. Muse, *A Primer of Genome Science*, 2d ed. (Sinauer, 2004), Figure 5.4.]

proteins present in a typical cell. Today, researchers use **mass spectrometry**, which is a method for precisely determining the molecular mass of a molecule. In mass spectrometry, a molecule is ionized and its migration rate in an electrical field is determined. Because small molecules migrate more rapidly than larger molecules, the migration rate can accurately determine the mass of the molecule.

To analyze proteins with mass spectrometry, a protein is first digested with the enzyme trypsin, which cleaves the protein into smaller peptide fragments, each of which contains several amino acids (**Figure 20.21a**). Mass spectrometry is then used to separate the peptides on the basis of their mass-to-charge ratio (**Figure 20.21b**). This separation

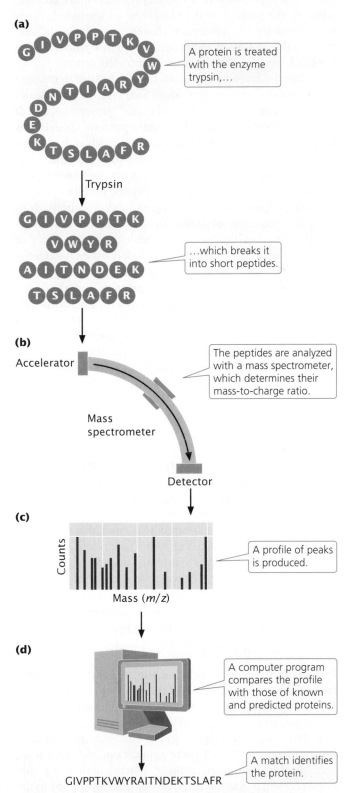

(a)

GIVPPTKVW Y R A I T N D E K T S L A F R

A protein is treated with the enzyme trypsin,…

Trypsin

GIVPPTK

VWYR

AITNDEK

TSLAFR

…which breaks it into short peptides.

(b)

Accelerator

Mass spectrometer

The peptides are analyzed with a mass spectrometer, which determines their mass-to-charge ratio.

Detector

(c)

Counts

Mass (*m/z*)

A profile of peaks is produced.

(d)

A computer program compares the profile with those of known and predicted proteins.

GIVPPTKVWYRAITNDEKTSLAFR

A match identifies the protein.

20.21 Mass spectrometry is used to identify proteins.

produces a profile of peaks, in which each peak corresponds to the mass-to-charge ratio of one peptide (**Figure 20.21c**). A computer program then searches through a database of proteins to find a match between the profile generated and the profile expected with a known protein (**Figure 20.21d**), allowing the protein in the sample to be identified. Using bioinformatics, the computer creates "virtual digests" and predicts the profiles of all proteins found in a genome, given the DNA sequences of the protein-encoding genes.

Mass spectrometric methods can also be used to measure the amount of each protein identified. With recent advances, researchers now carry out "shotgun" proteomics, which eliminates most of the initial protein-separation stage. In this procedure, a complex mixture of proteins (such as those from a tissue sample) is digested and analyzed with mass spectrometry. The computer program then sorts out the proteins present in the original sample from the peptide profiles.

Mary Lipton and her colleagues used this approach to study the proteome of *Deinococcus radiodurans*, an exceptional bacterium that is able to withstand high doses of ionizing radiation that are lethal to all other organisms. The genome of *D. radiodurans* had already been sequenced. Lipton and her colleagues extracted proteins from the bacteria, digested them with trypsin, separated the fragments with liquid chromatography, and then determined the proteins from the peptide fragments with mass spectrometry. They were able to identify 1910 proteins, more than 60% of the proteins predicted on the basis of the genome sequence.

Affinity Capture

Proteomics concerns not just the identification of all proteins in a cell, but also an understanding of how these proteins interact and how their expression varies with the passage of time. Researchers have developed a number of techniques for identifying proteins that interact within the cell. In **affinity capture**, an antibody to a specific protein is used to capture one protein from a complex mixture of proteins. The protein captured will "pull down" with it any proteins with which the captured protein physically interacts. The pulled-down mixture of proteins can then be analyzed by mass spectrometry to identify the proteins. Various modifications of affinity capture and other techniques can be used to determine the complete set of protein interactions in a cell, termed the **interactome**.

Protein Microarrays

Protein–protein interactions can also be analyzed with **protein microarrays** (**Figure 20.22**), which are similar to the microarrays used for examining gene expression. With this technique, a large number of different proteins are applied to a glass slide as a series of spots, with each spot containing a different protein. In one application, each spot is an antibody for a different protein, labeled with a tag that fluoresces when bound. An extract of tissue is applied to the protein microarray. A spot of fluorescence appears when a

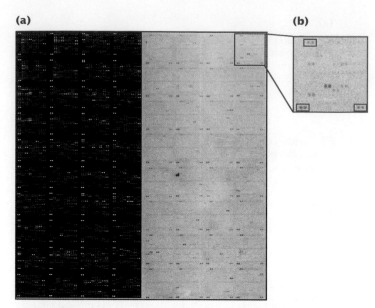

20.22 Protein microarrays can be used to examine interactions among proteins. (a) A microarray containing 4400 proteins found in yeast. (b) The array was probed with an enzyme that phosphylates proteins to determine which proteins serve as substrate for the enzyme. Dark spots represent proteins that were phosphylated by the enzyme. Proteins that phosphylate themselves (autophosphorylate) are included in each block of the microarray (shown in blue boxes) to serve as reference points. [D. Hall, J. Ptacek, and M. Snyder, *Mechanisms of Ageing and Development* 128:161–167, 2007.]

protein in the extract binds to antibody, indicating the presence of that particular protein in the tissue.

Structural Proteomics

The high-resolution structure of a protein provides a great deal of useful information. It is often a source of insight into the function of an unknown protein; it may also suggest the location of active sites and provide information about other molecules that interact with the protein. Knowledge of a protein's structure often suggests targets for potential drugs that might interact with the protein. Because structure often provides information about function, a goal of proteomics is to determine the structure of every protein found in a cell.

Currently, determining the structure for even a single protein is a major task, often requiring several years of intensive labor by a team of researchers. Two procedures are currently used to solve the structures of complex proteins: (1) X-ray crystallography, in which crystals of the protein are bombarded with X-rays and the diffraction patterns of the X-rays are used to determine the structure (see Chapter 10), and (2) nuclear magnetic resonance (NMR), which provides information on the position of specific atoms within a molecule by using the magnetic properties of nuclei.

Both X-ray crystallography and NMR require human intervention at many stages and are too slow for determining the structure of thousands of proteins that may exist within a cell. Because the structures of hundreds of thousands of proteins are required for studies of the proteome, researchers ultimately hope to be able to predict the structure of a protein from its amino acid sequence. This method is not possible at the present time, but the hope is that, if enough high-resolution structures are solved, it may be possible in the future to model the structure from the amino acid sequence alone. As scientists work on automated methods that will speed the structural determination of proteins, bioinformaticists are developing better computer programs for predicting protein structure from sequence.

CONCEPTS

The proteome is the complete set of proteins found in a cell. Techniques of protein separation and mass spectrometry are used to identify the proteins present within a cell. Affinity capture and microarrays are used to determine sets of interacting proteins. Structural proteomics attempts to determine the structure of all proteins.

✔ **CONCEPT CHECK 9**

Why is knowledge of a protein's structure important?

CONCEPTS SUMMARY

- Genomics is the field of genetics that attempts to understand the content, organization, and function of genetic information contained in whole genomes.

- Genetic maps position genes relative to other genes by determining rates of recombination and are measured in percent recombination. Physical maps are based on the physical distances between genes and are measured in base pairs.

- The Human Genome Project is an effort to determine the entire sequence of the human genome. The project began officially in 1990; rough drafts of the human genome sequence were completed in 2000. The final draft of the human genome sequence was completed in 2003.

- Sequencing a whole genome requires breaking the genome into small overlapping fragments whose DNA sequences can be determined in sequencing reactions. The individual sequences can be ordered into a whole-genome sequence with the use of a map-based approach, in which fragments are assembled in order by using previously created genetic and physical maps, or with the use of a whole-genome shotgun approach, in which overlap between fragments is used to assemble them into a whole-genome sequence.

- Single-nucleotide polymorphisms are single-base differences in DNA between individual organisms and are valuable as markers in linkage studies. Individual organisms may also differ in the number of copies of DNA sequences, called copy-number variations.

- Expressed-sequence tags are markers associated with expressed (transcribed) DNA sequences and can be used to find the genes expressed in a genome.

- Bioinformatics is a synthesis of molecular biology and computer science that develops tools to store, retrieve, and analyze DNA-, cDNA-, and protein-sequence data.

- Metagenomics studies the genomes of entire groups of organims. Synthetic biology is developing techniques for creating genomes and organisms.

- Homologous genes are evolutionarily related. Orthologs are homologous sequences found in different organisms, whereas paralogs are homologous sequences found in the same organism. Gene function may be determined by looking for homologous sequences (both orthologs and paralogs) whose function has been previously determined.

- A microarray consists of DNA fragments fixed in an orderly pattern to a solid support, such as a nylon filter or glass slide. When a solution containing a mixture of DNA or RNA is applied to the array, any nucleic acid that is complementary to the probe being used will bind to the probe. Microarrays can be used to monitor the expression of thousands of genes simultaneously.

- By linking a reporter sequence with the regulatory sequences of a gene, the expression pattern of the gene can be observed by looking for the product of the reporter sequence. Genes affecting a particular function can also be identified through whole-genome mutagenesis screens.

- Most prokaryotic species have between 1 million and 3 million base pairs of DNA and from 1000 to 2000 genes. Compared with that of eukaryotic genomes, the density of genes in prokaryotic genomes is relatively uniform, with about one gene per 1000 bp. There is relatively little noncoding DNA between prokaryotic genes. Horizontal gene transfer (the movement of genes between different species) has been an important evolutionary process in prokaryotes.

- Eukaryotic genomes are larger and more variable in size than prokaryotic genomes. There is no clear relation between organismal complexity and the amount of DNA or number of genes among multicellular organisms. Much of the genomes of eukaryotic organisms consists of repetitive DNA. Transposable elements are very common in most eukaryotic genomes.

- Proteomics determines the protein content of a cell and the functions of those proteins. Proteins within a cell can be separated and identified with the use of mass spectrometry. Structural proteomics attempts to determine the three-dimensional shape of proteins.

IMPORTANT TERMS

genomics (p. 558)
structural genomics (p. 558)
genetic (linkage) map (p. 558)
physical map (p. 560)
map-based sequencing (p. 563)
contig (p. 563)
whole-genome shotgun sequencing
  (p. 564)
single-nucleotide polymorphism (SNP)
  (p. 565)
haplotype (p. 566)
linkage disequilibrium (p. 566)
tag single-nucleotide polymorphism
  (tag-SNP) (p. 566)

genomewide association study (p. 566)
copy-number variation (CNV) (p. 567)
expressed-sequence tag (EST) (p. 567)
bioinformatics (p. 568)
annotation (p. 568)
metagenomics (p. 569)
functional genomics (p. 570)
transcriptome (p. 570)
proteome (p. 570)
homologous genes (p. 570)
orthologous genes (p. 570)
paralogous genes (p. 570)
protein domain (p. 571)
microarray (p. 571)

mutagenesis screen (p. 574)
comparative genomics (p. 575)
segmental duplication (p. 578)
multigene family (p. 578)
gene desert (p. 578)
proteomics (p. 582)
Human Proteome Project (p. 582)
two-dimensional polyacrylamide gel
  electrophoresis (2D-PAGE) (p. 582)
mass spectrometry (p. 583)
affinity capture (p. 584)
interactome (p. 584)
protein microarray (p. 584)

ANSWERS TO CONCEPT CHECKS

1. Accuracy and resolution

2. b

3. To catalog and map SNPs and other human genetic variants

4. a

5. d

6. c

7. Species with larger genomes generally have more genes than species with smaller genomes, and so gene density is quite constant.

8. a

9. Structure often provides important information about how a protein functions and the types of proteins with which it is likely to interact.

WORKED PROBLEM

A linear piece of DNA that is 30 kb long is first cut with *Bam*HI, then with *Hpa*II, and, finally, with both *Bam*HI and *Hpa*II together. Fragments of the following sizes were obtained from this reaction:

  *Bam*HI: 20-kb, 6-kb, and 4-kb fragments

  *Hpa*II: 21-kb and 9-kb fragments

  *Bam*HI and HpaII: 20-kb, 5-kb, 4-kb, and 1-kb fragments

Draw a restriction map of the 30-kb piece of DNA, indicating the locations of the *Bam*HI and *Hpa*II restriction sites.

• Solution

This problem can be solved correctly through a variety of approaches; this solution applies one possible approach.

When cut by *Bam*HI alone, the linear piece of DNA is cleaved into three fragments; so there must be two *Bam*HI restriction sites. When cut with *Hpa*II alone, a clone of the same piece of DNA is cleaved into only two fragments; so there is a single *Hpa*II site.

Let's begin to determine the location of these sites by examining the *Hpa*II fragments. Notice that the 21-kb fragment produced when the DNA is cut by *Hpa*II is not present in the fragments produced when the DNA is cut by *Bam*HI and *Hpa*II together (the double digest); this result

indicates that the 21-kb *Hpa*II fragment has within it a *Bam*HI site. If we examine the fragments produced by the double digest, we see that the 20-kb and 1-kb fragments sum to 21 kb; so a *Bam*HI site must be 20 kb from one end of the fragment and 1 kb from the other end.

Similarly, we see that the 9-kb *Hpa*II fragment does not appear in the double digest and that the 5-kb and 4-kb fragments in the double digest add up to 9 kb; so another *Bam*HI site must be 5 kb from one end of this fragment and 4 kb from the other end.

Now, let's examine the fragments produced when the DNA is cut by *Bam*HI alone. The 20-kb and 4-kb fragments are also present in the double digest; so neither of these fragments contains an *Hpa*II site. The 6-kb fragment, however, is not present in the double digest, and the 5-kb and 1-kb fragments in the double digest sum to 6 kb; so this fragment contains an *Hpa*II site that is 5 kb from one end and 1 kb from the other end.

We have accounted for all the restriction sites, but we must still determine the order of the sites on the original 30-kb fragment.

Notice that the 5-kb fragment must be adjacent to both the 1-kb and the 4-kb fragments; so it must be in between these two fragments.

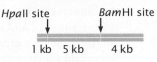

We have also established that the 1-kb and 20-kb fragments are adjacent; because the 5-kb fragment is on one side, the 20-kb fragment must be on the other, completing the restriction map:

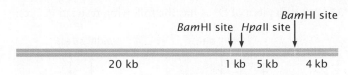

COMPREHENSION QUESTIONS

Section 20.1

*1. What is the difference between a genetic map and a physical map? Which generally has higher resolution and accuracy and why?

*2. What is the difference between a map-based approach to sequencing a whole genome and a whole-genome shotgun approach?

3. How are DNA fragments ordered into a contig with the use of restriction sites?

*4. Describe the different approaches to sequencing the human genome that were taken by the international collaboration and by Celera Genomics.

5. What are some of the ethical concerns arising out of the information produced by the Human Genome Project?

6. What is a single-nucleotide polymorphism (SNP)? How are SNPs used in genomic studies?

7. What is a haplotype? How do different haplotypes arise?

8. What is linkage disequilibrium? How does it result in haplotypes?

9. How is a genomewide association carried out?

10. What is copy-number variation? How does it arise?

11. (a) What is an expressed-sequence tag (EST)? (b) How are ESTs created? (c) How are ESTs used in genomics studies?

12. Give some examples of important findings from metagenomic studies.

13. How are genes recognized within genomic sequences?

Section 20.2

*14. What are homologous sequences? What is the difference between orthologs and paralogs?

15. Describe several different methods for inferring the function of a gene by examining its DNA sequence.

16. What is a microarray? How can it be used to obtain information about gene function?

17. Explain how a reporter sequence can be used to provide information about the expression pattern of a gene.

*18. Briefly outline how a mutagenesis screen is carried out.

Section 20.3

19. What is the relation between genome size and gene number in prokaryotes?

20. Eukaryotic genomes are typically much larger than prokaryotic genomes. What accounts for the increased amount of DNA in eukaryotic genomes?

*21. What is horizontal gene transfer? How might it take place between different species of bacteria?

22. DNA content varies considerably among different multicellular organisms. Is this variation closely related to the number of genes and the complexity of the organism? If not, what accounts for the differences?

*23. More than half of the genome of *Arabidopsis thaliana* consists of duplicated sequences. What mechanisms are thought to have been responsible for these extensive duplications?

24. What is a segmental duplication?

25. What is a gene desert?

26. What accounts for much of the differences in genome size among species of *Drosophila*?

27. The human genome does not encode substantially more protein domains than do invertebrate genomes, yet it encodes many more proteins. How are more proteins encoded when the number of domains does not differ substantially?

28. (a) What is genomics and how does structural genomics differ from functional genomics? (b) What is comparative genomics?

Section 20.4

29. How does proteomics differ from genomics?

30. How is mass spectrometry used to identify proteins in a cell?

31. What are the goals of the Human Proteome Project?

APPLICATION QUESTIONS AND PROBLEMS

Section 20.1

*32. A 22-kb piece of DNA has the following restriction sites:

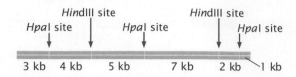

A batch of this DNA is first fully digested by *Hpa*I alone, then another batch is fully digested by *Hind*III alone, and, finally, a third batch is fully digested by both *Hpa*I and *Hind*III together. The fragments resulting from each of the three digestions are placed in separate wells of an agarose gel, separated by gel electrophoresis, and stained by ethidium bromide. Draw the bands as they would appear on the gel.

*33. A piece of DNA that is 14 kb long is cut first by *Eco*RI alone, then by *Sma*I alone, and, finally, by both *Eco*RI and *Sma*I together. The following results are obtained:

Digestion by *Eco*RI alone	Digestion by *Sma*I alone	Digestion by both *Eco*RI and *Sma*I
3-kb fragment	7-kb fragment	2-kb fragment
5-kb fragment	7-kb fragment	3-kb fragment
6-kb fragment		4-kb fragment
		5-kb fragment

Draw a map of the *Eco*RI and *Sma*I restriction sites on this 14-kb piece of DNA, indicating the relative positions of the restriction sites and the distances between them.

34. The presence (+) or absence (−) of six sequence-tagged sites (STSs) in each of five bacterial artificial chromosome (BAC) clones (A–E) is indicated in the following table. Using these markers, put the BAC clones in their correct order and indicate the locations of the STS sites within them.

	STSs					
BAC clone	1	2	3	4	5	6
A	+	−	−	−	+	−
B	−	−	−	+	−	+
C	−	+	+	−	−	−
D	−	−	+	−	+	−
E	+	−	−	+	−	−

35. How does the density of genes found on chromosome 22 compare with the density of genes found on chromosome 21, two similar-sized chromosomes? How does the number of genes on chromosome 22 compare with the number found on the Y chromosome?

To answer these questions, go to www.ensembl.org. Under the heading *Species*, select *Human*. On the left-hand side of the next page click on *Karyotype*. Pictures of the human chromosomes will appear. Click on chromosome 22 and select *Chromosome Summary*. You will be shown a picture of this chromosome and a histogram illustrating the densities of total genes (uncolored bars) and of known genes (colored bars). The total numbers of genes, along with the chromosome length in base pairs are given at the bottom of the diagram. Write down the total length of the chromosome and the number of protein-coding genes.

Now go to chromosome 21 by pulling down the *Jump to Chromosome* menu and selecting chromosome 21. Examine the total length and total number of protein-coding genes for chromosome 21. Now do the same for the Y chromosome. Calculate the gene density (number of genes/length) for chromosomes 22, 21, and Y.

a. Which chromosome has the highest density and greatest number of genes? Which has the fewest?

b. Examine in more detail the genes at the tip of the short arm of the Y chromosome by clicking on the top bar in the histogram of genes. Jump to location view. A more detailed view will be shown. What known genes are found in this region? How many protein-coding genes are there in this region?

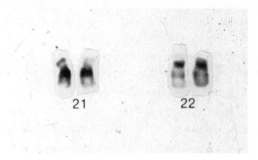

Human chromosomes 21 and 22. [Leonard Lessin/ Photolibrary/Peter Arnold Images.]

36. In recent years, honeybee colonies throughout North American have been decimated by the rapid death of worker bees, a disorder termed colony collapse disorder (CCD). First noticed by beekeepers in 2004, the disorder has been responsible for the loss of 50% to 90% of beekeeping operations in the United States. Evidence suggests that CCD is caused by a pathogen. Diana Cox-Foster and her colleagues (2007. *Science* 318:283–287) used a metagenomic approach to try to identify the causative agent of CCD. They isolated DNA from normal honeybee hives and from hives that had experienced CCD. A number of different bacteria, fungi, and viruses

were identified in the metagenomic analysis. The following table gives the percentage of CCD hives and non-CCD hives that tested positive for four potential pathogens identified in the metagenomic analysis. On the basis of these data, which potential pathogen appears most likely to be responsible for CCD? Explain your reasoning. Do these data prove that this pathogen is the cause of CCD? Explain.

Virus	CCD hives infected (*n* = 30)	Non-CCD hives infected (*n* = 21)
Israeli acute paralysis virus	83.3%	4.8%
Kashmir bee virus	100%	76.2%
Nosema apis	90%	47.6%
Nosema cernae	100%	80.8%

37. James Noonan and his colleagues (2005. *Science* 309:597–599) set out to study genome sequences of an extinct species of cave bear. They extracted DNA from 40,000-year-old bones from a cave bear. They then used a metagenomic approach to isolate, identify, and sequence the cave-bear DNA. Why did they use a metagenomics approach when their objective was to sequence the genome of one species (the cave bear)?

Section 20.2

38. Microarrays can be used to determine levels of gene expression. In one type of microarray, hybridization of the red (experimental) and green (control) cDNAs is proportional to the relative amounts of mRNA in the samples. Red indicates the overexpression of a gene and green indicates the underexpression of a gene in the experimental cells relative to the control cells, yellow indicates equal expression in experimental and control cells, and no color indicates no expression in either experimental or control cells.

In one experiment, mRNA from a strain of antibiotic-resistant bacteria (experimental cells) is converted into cDNA and labeled with red fluorescent nucleotides; mRNA from a nonresistant strain of the same bacteria (control cells) is converted into cDNA and labeled with green fluorescent nucleotides. The cDNAs from the resistant and nonresistant cells are mixed and hybridized to a chip containing spots of DNA from genes 1 through 25. The results are shown in the adjoining illustration. What conclusions can you make about which genes might be implicated in antibiotic resistance in these bacteria? How might this information be used to design new antibiotics that are less vulnerable to resistance?

Section 20.3

39. *Dictyostelium discoideum* is a soil-dwelling, social ameba: much of the time, the organism consists of single, solitary cells, but, during times of starvation, individual amebae come together to form aggregates that have many characteristics of multicellular organisms. Biologists have long debated whether *D. discoideum* is a unicellular or multicellular organism. In 2005, the genome of *D. discoideum* was completely sequenced. The table below lists some genomic characteristics of *D. discoideum* and other eukaryotes (L. Eichinger et al. 2005. *Nature* 435:43–57).

Table for Problem 39

Feature	*D. discoideum*	*P. falciparium*	*S. cerevisiae*	*A. thaliana*	*D. melanogaster*	*C. elegans*	*H. sapiens*
Organism	ameba	malaria parasite	yeast	plant	fruit fly	worm	human
Cellularity	?	uni	uni	multi	multi	multi	multi
Genome size (millions of base pairs)	34	23	13	125	180	103	2,851
Number of genes	12,500	5,268	5,538	25,498	13,676	19,893	22,287
Average gene length (bp)	1,756	2,534	1,428	2,036	1,997	2,991	27,000
Genes with introns (%)	69	54	5	79	38	5	85
Mean number of introns	1.9	2.6	1.0	5.4	4.0	5.0	8.1
Mean intron size (bp)	146	179	nd*	170	nd*	270	3,365
Mean G + C (exons)	27%	24%	28%	28%	55%	42%	45%

*nd = not determined

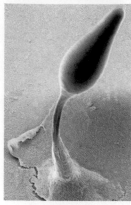

Dictyostelium discoideum.
[Dr. Richard Kessel & Dr. Gene Shih/Visuals Unlimited.]

a. On the basis of the organisms listed in the table other than *D. discoideum*, what are some differences in genome characteristics between unicellular and multicellular organisms?

b. On the basis of these data, do you think the genome of *D. discoideum* is more like those of other unicellular eukaryotes or more like those of multicellular eukaryotes? Explain your answer.

40. What are some of the major differences in the ways in which genetic information is organized in the genomes of prokaryotes compared with eukaryotes?

41. How do the following genomic features of prokaryotic organisms compare with those of eukaryotic organisms? How do they compare among eukaryotes?

a. Genome size   c. Gene density (bp/gene)

b. Number of genes   d. Number of exons

42. A group of 250 scientists sequenced and analyzed the genomes of 12 species of *Drosophila* (*Drosophila* 12 Genomes Consortium. 2007. *Nature* 450:203–218). Data on genome size and number of protein-encoding genes from this study are given in Table 20.6 (see p. 581). Using data in from this table, plot the number of protein-encoding genes as a function of genome size for 12 species of *Drosophila*. Is there a relation between genome size and number of genes in fruit flies? How does this result compare with the relation between genome size and number of genes across all eukaryotes?

Section 20.4

43. A scientist determines the complete genomes and proteomes of a live cell and a muscle cell from the same person. Would you expect bigger differences in the genomes or proteomes of these two cell types? Explain your answer.

CHALLENGE QUESTIONS

Section 20.1

44. The genome of *Drosophila melanogaster,* a fruit fly, was sequenced in 2000. However, this "completed" sequence did not include most heterochromatin regions. The heterochromatin was not sequenced until 2007 (R. A. Hoskins et al. 2007. *Science* 316:1625–1628). Most completed genome sequences do not include heterochromatin. Why is heterochromatin usually not sequenced in genomic projects? (Hint: See Chapter 11 for a more detailed discussion of heterochromatin.)

45. In metagenomic studies, a comparison of ribosomal RNA sequences is often used to determine the number of different species present. What are some characteristics of ribosomal sequences that make them useful for determining what species are present?

*46. Some synthetic biologists have proposed creating an entirely new, free-living organism with a minimal genome, the smallest set of genes that allows for replication of the organism in a particular environment. This genome could be used to design and create, from "scratch," novel organisms that might perform specific tasks such as the breakdown of toxic materials in the environment.

a. How might the minimal genome required for life be determined?

b. What, if any, social and ethical concerns might be associated with the construction of an entirely new organism with a minimal genome?

21 Organelle DNA

Analysis of DNA sequences from the mitochondria of asses indicates that the domesticated donkey (top photograph) is descended from African wild asses (bottom photograph). Mitochondrial DNA has a number of characteristics that make it well suited for the study of evolutionary relationships. [Top: PhotoDisc; bottom: Joanna Van Gruisen/Ardea London.]

THE DONKEY: A WILD ASS OR A HALF ASS?

The donkey (*Equus asinus*) has a long and distinguished past, serving as a bearer of people, possessions, and produce in many cultures and throughout much of human history. In spite of its long association with human culture, the origins of domesticated donkeys have been uncertain. What little archeological evidence is available suggests that donkeys were domesticated approximately 5000 years ago—about the same time as the domestication of horses.

Domesticated donkeys are clearly related to other asses, which include two subspecies of African wild asses, the Nubian wild ass (*E. africanus africanus*) and the Somalian wild ass (*E. a. somaliensis*), and two species of Asian half asses (*E. hemionus and E. kiang*). Which of these asses gave rise to donkeys and where domestication took place have, until recently, been unclear.

To address these questions, a multinational team of scientists lead by Albano Beja-Pereira turned to DNA in the mitochondria. As will be discussed in this chapter, mitochondria, along with chloroplasts and a few other cell organelles, possess their own DNA, which encodes some of the proteins and RNA molecules found in the organelles. **Mitochondrial DNA** (mtDNA) has a number of advantages for the study of evolutionary relationships, which is why Beja-Pereira and his colleagues used it in their study of donkeys. First, the DNA molecule found in mitochondria is shorter than DNA in the chromosomes found in the nuclei of eukaryotic cells, which makes it easier to analyze. Second, mtDNA is abundant because each cell typically has numerous mitochondria, each with several copies of the mitochondrial chromosome. Thus, mtDNA is easier to isolate and study than is nuclear DNA. Third, mtDNA in animals tends to evolve more rapidly than nuclear DNA, and so it's useful for looking at relationships among closely related organisms. Finally, mtDNA is typically inherited from only one parent (usually the mother), and so its genes are not reshuffled every generation by recombination, which tends to obscure genetic relationships.

To study the genetic origin of donkeys, Beja-Pereira and his colleagues obtained tissue from domestic donkeys in 52 countries throughout the Old World and from African wild asses and Asian half asses. After isolating DNA from the samples, they determined the nucleotide sequence of 479 bp of the mitochondrial chromosome. The DNA sequences were then used to construct an evolutionary tree (**Figure 21.1**) depicting the relationships among the different groups of donkeys and asses.

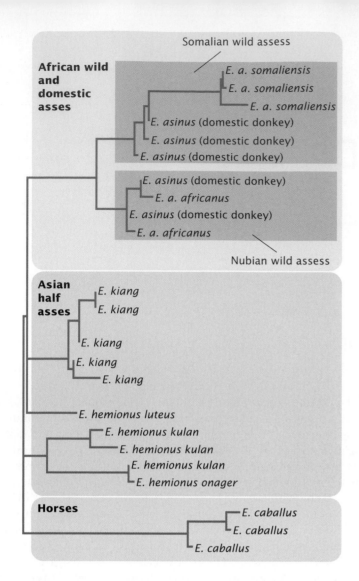

21.1 Evolutionary relationship of domestic donkeys, African wild asses, and Asian half asses. [From A. Beja-Pereira et al., *Science* 304:1781, 2004.]

The analysis revealed that African wild asses and Asian half asses are genetically distinct: they formed two completely separate groups in the evolutionary tree (see Figure 21.1). All the domestic donkeys clustered within the African wild-ass group, indicating that donkeys evolved from the wild asses rather than from the half asses. Another interesting feature indicated by the analysis was that donkeys appear to have at least two distinct origins from African wild asses, as revealed by the fact that some donkeys cluster with the Somalian wild asses and other donkeys cluster with the Nubian wild asses. This finding suggests that at least two independent domestication events took place. There was also significantly more genetic diversity in the domestic donkeys of North Africa than in the donkeys of other regions of the world.

The clear affinity between domestic donkeys and African wild asses, coupled with the finding of greater diversity among North African donkeys, suggests that donkey domestication took place at least twice in Africa. Much evidence indicates that the first farm animals—sheep and goats—were domesticated in the Middle East. Donkeys are the only ungulate known to have been domesticated solely in Africa, highlighting the important role North Africa played in early population expansion and trade throughout the Old World. **TRY PROBLEM 12 →**

DNA sequences found in mitochondria and other organelles possess unique properties that make these sequences useful in the fields of conservation biology, evolution, and medical genetics. Uniparental inheritance exhibited by genes found in mitochondria and chloroplasts was discussed in Chapter 5; the present chapter examines molecular aspects of organelle DNA. We begin by briefly considering the structures of mitochondria and chloroplasts, the inheritance of traits encoded by their genes, and the evolutionary origin of these organelles. We then examine the general characteristics of mtDNA, followed by a discussion of the organization and function of different types of mitochondrial genomes. Finally, we turn to **chloroplast DNA** (cpDNA) and examine its characteristics, organization, and function.

21.1 Mitochondria and Chloroplasts Are Eukaryotic Cytoplasmic Organelles

Mitochondria and chloroplasts are membrane-bounded organelles located in the cytoplasm of eukaryotic cells (**Figure 21.2**). Mitochondria are present in almost all eukaryotic cells, whereas chloroplasts are found in plants and some

protists. Both organelles generate ATP, the universal energy carrier of cells.

Mitochondrion and Chloroplast Structure

Mitochondria are tubular structures that are from 0.5 to 1.0 micrometer (μm) in diameter, about the size of a typical bacterium, whereas chloroplasts are typically from about 4 to 6 μm in diameter. Both are surrounded by two membranes enclosing a region (called the matrix in mitochondria and the stroma in chloroplasts) that contains enzymes, ribosomes, RNA, and DNA. In mitochondria, the inner membrane is highly folded; embedded within it are the enzymes that catalyze electron transport and oxidative phosphorylation. Chloroplasts have a third membrane, called the thylakoid membrane, which is highly folded and stacked to form aggregates called grana. This membrane bears the pigments and enzymes required for photophosphorylation. New mitochondria and chloroplasts arise by the division of existing organelles—divisions that take place throughout the cell cycle and are independent of mitosis and meiosis. Mitochondria and chloroplasts possess DNA that encodes polypeptides used by the organelle, as well as ribosomal RNAs and transfer RNAs needed for the translation of these proteins.

Mitochondrion   **Chloroplast**

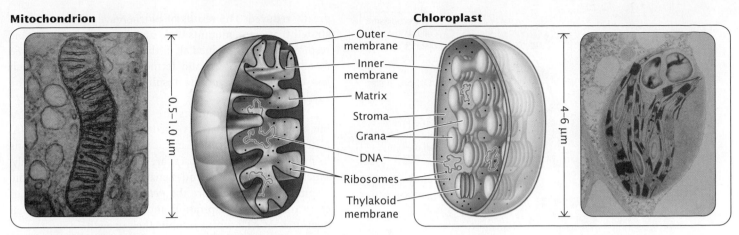

Outer membrane
Inner membrane
Matrix
Stroma
Grana
DNA
Ribosomes
Thylakoid membrane

0.5–1.0 μm

4–6 μm

21.2 Comparison of the structures of mitochondria and chloroplasts. [Left: Don Fawcett/Visuals Unlimited. Right: Biophoto Associates/Photo Researchers.]

The Genetics of Organelle-Encoded Traits

Mitochondria and chloroplasts are present in the cytoplasm, as already stated, and are usually inherited from a single parent. Thus, traits encoded by mtDNA and cpDNA exhibit uniparental inheritance. In animals, mtDNA is inherited almost exclusively from the female parent, although occasional male transmission of mtDNA has been documented. Paternal inheritance of organelles is common in gymnosperms (conifers) and in a few angiosperms (flowering plants). Some plants even exhibit biparental inheritance of mtDNA and cpDNA.

Replicative segregation   Individual cells may contain from dozens to hundreds of organelles, each with numerous copies of the organelle genome; so each cell typically possesses from hundreds to thousands of copies of mitochondrial and chloroplast genomes (**Figure 21.3**). A mutation arising within one organelle DNA molecule generates a mixture of mutant and wild-type DNA sequences within that cell. The occurrence of two distinct varieties of DNA within the cytoplasm of a single cell is termed **heteroplasmy**. When a heteroplasmic cell divides, the organelles segregate randomly into the two progeny cells in a process called **replicative segregation** (**Figure 21.4**), and chance determines the proportion of mutant organelles in each cell. Although most progeny cells will inherit a mixture of mutant and normal organelles, just by chance some cells may receive organelles with only mutant or only wild-type sequences; this situation, in which all organelles are genetically identical, is known as **homoplasmy**. Fusion of mitochondria also takes place frequently.

When replicative segregation takes place in somatic cells, it may create phenotypic variation within a single organism; different cells of the organism may possess different proportions of mutant and wild-type sequences, resulting in different degrees of phenotypic expression in different tissues. When replicative segregation takes place in the germ cells of a heteroplasmic cytoplasmic donor there may be different phenotypes among the offspring.

The disease known as myoclonic epilepsy and ragged-red fiber syndrome (MERRF) is caused by a mutation in an mtDNA gene. A 20-year-old person who carried this mutation in 85% of his mtDNAs displayed a normal phenotype, whereas a cousin who had the mutation in 96% of his mtDNAs was severely affected. In diseases caused by mutations in mtDNA, the severity of the disease is frequently related to the proportion of mutant mtDNA sequences inherited at birth. **TRY PROBLEM 13** →

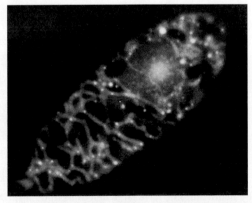

21.3 Individual cells may contain many mitochondria, each with several copies of the mitochondrial genome. Shown is a cell of *Euglena gracilis*, a protist, stained so that the nucleus appears red, mitochondria green, and mtDNA yellow. [From Y. Huyashi and K. Veda, *Journal of Cell Sciences* 93:565, 1989.]

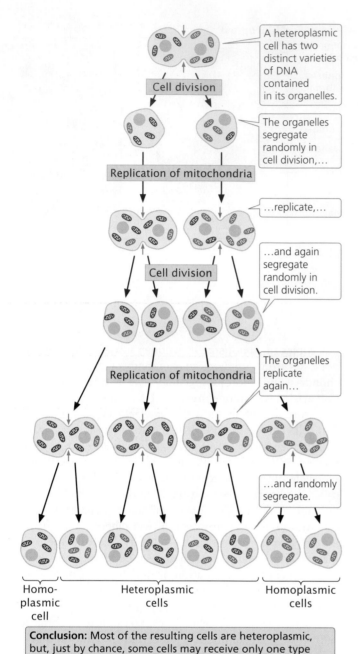

A heteroplasmic cell has two distinct varieties of DNA contained in its organelles.

Cell division

The organelles segregate randomly in cell division,…

Replication of mitochondria

…replicate,…

…and again segregate randomly in cell division.

Cell division

Replication of mitochondria

The organelles replicate again…

…and randomly segregate.

Homo-plasmic cell   Heteroplasmic cells   Homoplasmic cells

Conclusion: Most of the resulting cells are heteroplasmic, but, just by chance, some cells may receive only one type of organelle (e.g., they may receive all normal or all mutant).

21.4 Organelles in a heteroplasmic cell divide randomly into the progeny cells. This diagram illustrates replicative segregation in mitosis; the same process also takes place in meiosis.

Traits encoded by mtDNA

Traits encoded by mtDNA A number of traits encoded by organelle DNA have been studied. One of the first to be examined in detail was the phenotype produced by *petite* mutations in yeast (**Figure 21.5**). In the late 1940s, Boris Ephrussi and his colleagues noticed that, when grown on solid medium, some colonies of yeast were much smaller than normal. Examination of these *petite* colonies revealed that the growth rates of the cells within the colonies were

greatly reduced. The results of biochemical studies demonstrated that *petite* mutants were unable to carry out aerobic respiration; they obtained all of their energy from anaerobic metabolism (glycolysis and fermentation), which is much less efficient than aerobic respiration and results in the smaller colony size.

Some *petite* mutations are defects in nuclear DNA, but most *petite* mutations occur in mitochondrial DNA. Mitochondrial *petite* mutants often have large deletions in mtDNA or, in some cases, are missing mtDNA entirely. Much of the mtDNA encodes enzymes that catalyze aerobic respiration, and therefore the *petite* mutants are unable to carry out aerobic respiration and cannot produce normal quantities of ATP, which inhibits their growth.

Another known mtDNA mutation occurs in *Neurospora* (see pp. 402–405 in Chapter 15). Isolated by Mary Mitchell in 1952, *poky* mutants grow slowly, display cytoplasmic inheritance, and have abnormal amounts of cytochromes. Cytochromes are protein components of the electron-transport chain of the mitochondria and play an integral role in the production of ATP. Most organisms have three primary types of cytochromes: cytochrome *a*, cytochrome *b*, and cytochrome *c*. *Poky* mutants have cytochrome *c* but no cytochrome *a* or *b*. Like *petite* mutants, *poky* mutants are defective in ATP synthesis and therefore grow more slowly than do normal, wild-type cells. **TRY PROBLEM 16 →**

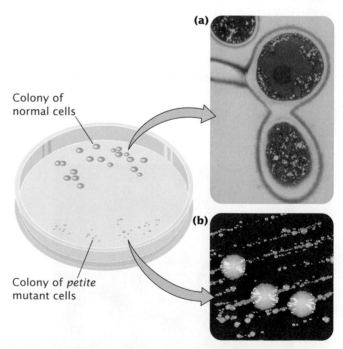

Colony of normal cells

Colony of *petite* mutant cells

21.5 The *petite* mutants have large deletions in their mtDNA and are unable to carry out oxidative phosphorylation. (a) A normal yeast cell and (b) a *petite* mutant. [Part a: David M. Phillips/Visuals Unlimited. Part b: Courtesy of Dr. Des Clark-Walker, Research School of Biological Sciences, the Australian National University.]

In recent years, a number of genetic diseases that result from mutations in mtDNA have been identified in humans. Leber hereditary optic neuropathy (LHON), which typically leads to sudden loss of vision in middle age, results from mutations in the mtDNA genes that encode electron-transport proteins. Another disease caused by mtDNA mutations is neurogenic muscle weakness, ataxia, and retinitis pigmentosa (NARP), which is characterized by seizures, dementia, and developmental delay. Other mitochondrial diseases include Kearns–Sayre syndrome (KSS) and chronic external opthalmoplegia (CEOP), both of which result in paralysis of the eye muscles, droopy eyelids, and, in severe cases, vision loss, deafness, and dementia. All of these diseases exhibit cytoplasmic inheritance and variable expression (see Chapter 5).

A trait in plants that is produced by mutations in mitochondrial genes is cytoplasmic male sterility, a mutant phenotype found in more than 140 different plant species and inherited only from the maternal parent. These mutations inhibit pollen development but do not affect female fertility.

A number of cpDNA mutants also have been discovered. One of the first to be recognized was leaf variegation in the four o'clock plant *Mirabilis jalapa*, which was studied by Carl Correns in 1909 (see pp. 118–119 in Chapter 5). In the green alga *Chlamydomonas*, streptomycin-resistant mutations occur in cpDNA, and a number of mutants exhibiting altered pigmentation and growth in higher plants have been traced to defects in cpDNA.

Worked Problem

To illustrate the inheritance of a trait encoded by organelle DNA, consider the following problem. A physician examines a young man who has a progressive muscle disorder and visual abnormalities. A number of the patient's relatives have the same condition, as shown in the adjoining pedigree. The degree of expression of the trait is highly variable among members of the family: some are only slightly affected, whereas others developed severe symptoms at an early age. The physician concludes that this disorder is due to a mutation in the mitochondrial genome. Do you agree with the physician's conclusion? Why or why not? Could the disorder be due to a mutation in a nuclear gene? Explain your reasoning.

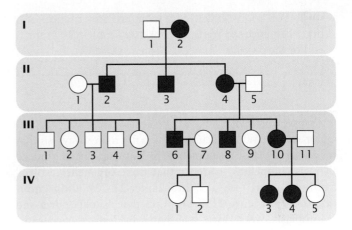

• Solution

The conclusion that the disorder is caused by a mutation in the mitochondrial genome is supported by the pedigree and the observation of variable expression in affected members of the same family. The disorder is passed only from affected mothers to both male and female offspring; when fathers are affected, none of their children have the trait (as seen in the children of II-2 and III-6). This outcome is expected of traits determined by mutations in mtDNA, because mitochondria are in the cytoplasm and usually inherited only from a single (in humans, the maternal) parent. The trait cannot be X-linked recessive, because a cross between a female with the trait (X^aX^a) and a male without the trait (X^+Y) would *not* produce daughters with the trait (X^aX^a), which we see in III-10, IV-3, and IV-4.

The facts that some offspring of affected mothers do not show the trait (III-9 and IV-5) and that expression varies from one person to another suggest that affected persons are heteroplasmic, with both mutant and wild-type mitochondria. Random segregation of mitochondria in meiosis may produce gametes having different proportions of mutant and wild-type sequences, resulting in different degrees of phenotypic expression among the offspring. Most likely, symptoms of the disorder develop when some minimum proportion of the mitochondria are mutant. Just by chance, some of the gametes produced by an affected mother contain few mutant mitochondria and result in offspring that lack the disorder.

Another possible explanation for the disorder is that it results from an autosomal dominant gene. When an affected (heterozygous) person mates with an unaffected (homozygous) person, about half of the offspring are expected to have the trait, but, just by chance, some affected parents will have no affected offspring. Individuals II-2 and III-6 in the pedigree could just have happened to be male and their sex could be unrelated to the mode of transmission. The variable expression could be explained by variable expressivity (see Chapter 5).

> → For more experience with the inheritance of organelle-encoded traits, try working Problem 14 at the end of the chapter.

The Endosymbiotic Theory

Chloroplasts and mitochondria are in many ways similar to bacteria. This resemblance is not superficial; indeed there is compelling evidence that these organelles evolved from eubacteria (see p. 17 in Chapter 2). The **endosymbiotic theory** (**Figure 21.6**) proposes that mitochondria and chloroplasts were once free-living bacteria that became internal inhabitants (endosymbionts) of early eukaryotic cells. According to this theory, between 1 billion and 1.5 billion years ago, a large, anaerobic eukaryotic cell engulfed an aerobic eubacterium, one that possessed the enzymes necessary for oxidative phosphorylation. The eubacterium provided the formerly anaerobic cell with the capacity for oxidative phosphorylation and allowed it to produce more ATP for each organic molecule digested. With time, the endosymbiont became an integral part of the eukaryotic host cell, and its descendants evolved into present-day mitochondria. Sometime later, a similar relation arose between photosynthesizing eubacteria and eukaryotic cells, leading to the evolution of chloroplasts.

A great deal of evidence supports the idea that mitochondria and chloroplasts originated as eubacterial cells. Many modern single-celled eukaryotes (protists) are hosts to endosymbiotic bacteria. Mitochondria and chloroplasts are similar in size to present-day eubacteria and possess their own DNA, which has many characteristics in common with eubacterial DNA. Mitochondria and chloroplasts possess ribosomes, some of which are similar in size and structure to eubacterial ribosomes. Finally, antibiotics that inhibit protein synthesis in eubacteria but do not affect protein synthesis in eukaryotic cells also inhibit protein synthesis in these organelles.

The strongest evidence for the endosymbiotic theory comes from the study of DNA sequences in organelle DNA. Ribosomal RNA and protein-encoding gene sequences in mitochondria and chloroplasts have been found to be more closely related to sequences in the genes of eubacteria than they are to those found in the eukaryotic nucleus. Mitochondrial DNA sequences are most similar to sequences found in a group of eubacteria called the α-proteobacteria, suggesting that the original bacterial endosymbiont came from this group. Chloroplast DNA sequences are most closely related to sequences found in cyanobacteria, a group of photosynthesizing eubacteria. All of this evidence indicates that mitochondria and chloroplasts are more closely related to eubacterial cells than they are to the eukaryotic cells in which they are now found.

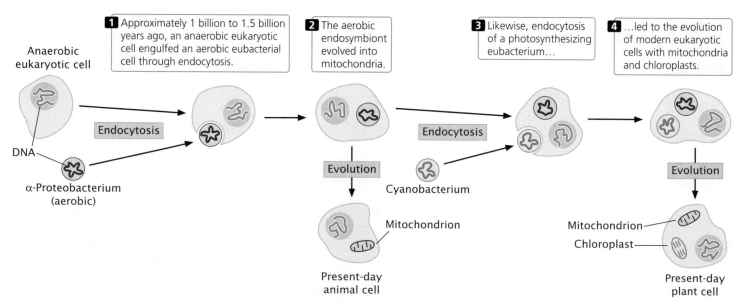

21.6 The endosymbiotic theory proposes that mitochondria and chloroplasts in eukaryotic cells arose from eubacteria.

21.2 Mitochondrial DNA Varies Widely in Size and Organization

In most animals and fungi, the mitochondrial genome consists of a single, highly coiled, circular DNA molecule, which is similar in structure to a eubacterial chromosome. Plant mitochondrial genomes often exist as a complex collection of multiple circular DNA molecules. In some species, the mitochondrial genome consists of a single, linear DNA molecule. Each mitochondrion contains multiple copies of the mitochondrial genome, and a cell may contain many mitochondria. A typical rat liver cell, for example, has from 5 to 10 mtDNA molecules in each of about 1000 mitochondria; so each cell possesses from 5000 to 10,000 copies of the mitochondrial genome. Mitochondrial DNA constitutes about 1% of the total cellular DNA in a rat liver cell. Like eubacterial chromosomes, mtDNA lacks the histone proteins normally associated with eukaryotic nuclear DNA, although it is complexed with other proteins that have some histone-like properties. The guanine–cytosine (GC) content of mtDNA is often sufficiently different from that of nuclear DNA that mtDNA can be separated from nuclear DNA by density-gradient centrifugation.

Mitochondrial genomes are small compared with nuclear genomes and vary greatly in size among different organisms (**Table 21.1**). The sizes of mitochondrial genomes of most species range from 15,000 bp to 65,000 bp, but those of a few species are much smaller (e.g., the genome of *Plasmodium falciparum*, the parasite that causes malaria, is only 6,000 bp) and those of some plants are several million base pairs. Although the amount of DNA in mitochondrial genomes varies widely, there is no correlation between genome size and number of genes. The number of genes is more constant than genome size; most species have only from 40 to 50 genes. These genes encode five basic functions: respiration and oxidative phosphorylation, translation, transcription, RNA processing, and the import of proteins into the cell. Most of the variation in size of mitochondrial genomes is due to differences in noncoding sequences such as introns and intergenic regions.

Table 21.1	Sizes of mitochondrial genomes in selected organisms
Organism	**Size of mtDNA (bp)**
Pichia canadensis (fungus)	27,694
Podospora anserina (fungus)	100,314
Saccharomyces cerevisiae (fungus)	85,779*
Drosophila melanogaster (fruit fly)	19,517
Lumbricus terrestris (earthworm)	14,998
Xenopus laevis (frog)	17,553
Mus musculus (house mouse)	16,295
Homo sapiens (human)	16,569
Chlamydomonas reinhardtii (green alga)	15,758
Plasmodium falciparum (protist)	5,966
Paramecium aurelia (protist)	40,469
Arabidopsis thaliana (plant)	166,924
Cucumis melo (plant)	2,400,000

*Size varies among strains.

The Gene Structure and Organization of Mitochondrial DNA

The nucleotide sequence of the mitochondrial genome has been determined for a variety of different organisms, including protists, fungi, plants, and animals. The genes for most of the 900 or so structural proteins and enzymes found in mitochondria are actually encoded by nuclear DNA, translated on cytoplasmic ribosomes, and then transported into the mitochondria; the mitochondrial genome typically encodes only a few proteins and a few rRNA and tRNA molecules needed for mitochondrial protein synthesis. The organization of these mitochondrial genes and how they are expressed is extremely diverse across organisms.

Ancestral and derived mitochondrial genomes Mitochondrial genomes can be divided in two basic types—ancestral genomes and derived genomes—although there is much variation within each type and the mtDNA of some organisms does not fit well into either category. Ancestral mitochondrial genomes are found in some plants and protists. These genomes contain more genes than do derived genomes, have rRNA genes that encode eubacterial-like ribosomes, and have a complete or almost complete set of tRNA genes. They possess few introns and little noncoding DNA between genes, generally use universal codons, and have their genes organized into clusters similar to those found in eubacteria. In summary, ancestral mitochondrial genomes retain many characteristics of their eubacterial ancestors.

(a)

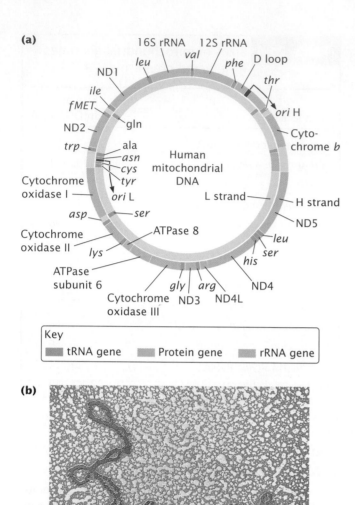

(b)

21.7 The human mitochondrial genome, consisting of 16,569 bp, is highly economical in its organization. (a) The outer circle represents the heavy (H) strand, and the inner circle represents the light (L) strand. The origins of replication for the H and L strands are *ori* H and *ori* L, respectively. ND identifies genes that encode subunits of NADH dehydrogenase. (b) Electron micrograph of isolated mtDNA. [Part b: CNRI/Photo Researchers.]

Derived mitochondrial genomes, in contrast, are usually smaller than ancestral genomes and contain fewer genes. Their rRNA genes and ribosomes differ from those found in eubacteria. The DNA sequences found in derived mitochondrial genomes differ more from typical eubacterial sequences than do ancestral genomes, and they contain nonuniversal codons. Most animal and fungal mitochondrial genomes fit into this category. Derived mitochondrial genomes have characteristics that differ substantially from those found in typical eubacteria.

Human mtDNA Human mtDNA is a circular molecule encompassing 16,569 bp that encode two rRNAs, 22 tRNAs, and 13 proteins. The two nucleotide strands of the molecule

differ in their base composition: the heavy (H) strand has more guanine nucleotides, whereas the light (L) strand has more cytosine nucleotides. The H strand is the template for both rRNAs, 14 of the 22 tRNAs, and 12 of the 13 proteins, whereas the L strand serves as template for only 8 of the tRNAs and 1 protein.

The origin of replication for the H strand is within a region known as the **D loop** (**Figure 21.7**), which also contains promoters for both the H and the L strands. Human mtDNA is highly economical in its organization: there are few noncoding nucleotides between the genes; almost all the messenger RNA is translated (there are no 5′ and 3′ untranslated regions); and there are no introns. Each strand has only a single promoter; so transcription produces two very large RNA precursors that are later cleaved into individual RNA molecules. Many of the genes that encode polypeptides even lack a complete termination codon, ending in either U or UA; the addition of a poly(A) tail to the 3′ end of the mRNA provides a UAA termination codon that halts translation. Human mtDNA also contains very little repetitive DNA. The one region of the human mtDNA that does contain some noncoding nucleotides is the D loop. **TRY PROBLEM 19 →**

Yeast mtDNA The organization of yeast mtDNA is quite different from that of human mtDNA. Although the yeast mitochondrial genome with 78,000 bp is nearly five times as large, it encodes only six additional genes, for a total of 2 rRNAs, 25 tRNAs, and 16 polypeptides (**Figure 21.8**). Most of the extra DNA in the yeast mitochondrial genome consists

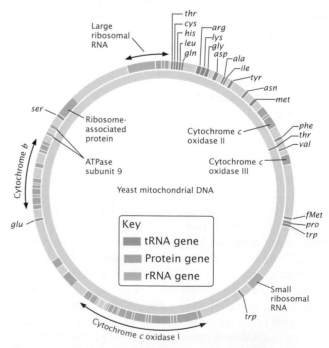

21.8 The yeast mitochondrial genome, consisting of 78,000 bp, contains much noncoding DNA.

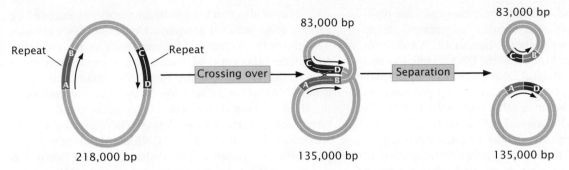

21.9 Size variation in plant mtDNA can be generated through recombination between direct repeats. In turnips, the mitochondrial genome consists of a "master circle" of 218,000 bp; crossing over between the direct repeats produces two smaller circles of 135,000 bp and 83,000 nucleotide pairs.

of introns and noncoding sequences. Yeast mitochondrial genes are separated by long intergenic spacer regions that have no known functions. The genes encoding polypeptides often include regions that encode 5′ and 3′ untranslated regions of the mRNA; there are also short repetitive sequences and some duplications.

Flowering-plant mtDNA Flowering plants (angiosperms) have the largest and most-complex mitochondrial genomes known; their mitochondrial genomes range in size from 186,000 bp in white mustard to 2,400,000 bp in muskmelon. Even closely related plant species may differ greatly in the sizes of their mtDNA.

Part of the extensive size variation in the mtDNA of flowering plants can be explained by the presence of large direct repeats, which constitute large parts of the mitochondrial genome. Crossing over between these repeats can generate multiple circular chromosomes of different sizes. The mitochondrial genome in turnips, for example, consists of a "master circle" consisting of 218,000 bp that has direct repeats. Homologous recombination between the repeats can generate two smaller circles of 135,000 bp and 83,000 bp (**Figure 21.9**). Other species contain several direct repeats, providing possibilities for complex crossing-over events that may increase or decrease the number and sizes of the circles.

Nonuniversal Codons in Mitochondrial DNA

In most bacterial and eukaryotic DNA, the same codons specify the same amino acids (see p. 412 in Chapter 15). However, there are exceptions to this universal code, and many of these exceptions are in mtDNA (**Table 21.2**). There is no "mitochondrial code"; rather, exceptions to the universal code exist in mitochondria, and these exceptions often differ among organisms. For example, AGA specifies arginine in the universal code, but AGA encodes serine in *Drosophila* mtDNA and is a stop codon in mammalian mtDNA.

CONCEPTS

The mitochondrial genome consists of circular DNA with no associated histone proteins, although it is complexed with other proteins that have some histonelike properties. The sizes and structures of mtDNA differ greatly among organisms. Human mtDNA exhibits extreme economy, but mtDNAs found in yeast and flowering plants contain many noncoding nucleotides and repetitive sequences. Mitochondrial DNA in most flowering plants is large and typically has one or more large direct repeats that can recombine to generate smaller or larger molecules.

✔ CONCEPT CHECK 3

Which of the following characteristics applies to ancestral mitochondrial genomes?

a. Have genes with many introns

b. Contain more genes than derived genomes do

c. Possess much noncoding DNA between genes

d. Have nonuniversal codons

The Replication, Transcription, and Translation of Mitochondrial DNA

Mitochondrial DNA does not replicate in the orderly, regulated manner of nuclear DNA. Mitochondrial DNA is synthesized throughout the cell cycle and is not coordinated with the synthesis of nuclear DNA. Which mtDNA molecules are replicated at any particular moment appears to be random;

		mtDNA		
Table 21.2	Nonuniversal codons found in mitochondrial DNA			
Codon	Universal Code	Vertebrate	*Drosophila*	Yeast
UGA	Stop	Tryptophan	Tryptophan	Tryptophan
AUA	Isoleucine	Methionine	Methionine	Methionine
AGA	Arginine	Stop	Serine	Arginine

Source: After T. D. Fox, *Annual Review of Genetics* 21:69, 1987.

within the same mitochondrion, some molecules are replicated two or three times, whereas others are not replicated at all. Furthermore, the two strands in human mtDNA may not replicate synchronously. Mitochondrial DNA is replicated by a special DNA polymerase called DNA polymerase γ (gamma). Single-strand-binding proteins, helicases, and topoisomerases are required for mitochondrial DNA replication, just as they are in eubacterial and nuclear DNA replication. In many organisms, these proteins are encoded by nuclear genes.

The processes of the transcription and translation of mitochondrial genes exhibit extensive variation among different organisms. In human mtDNA, eubacterial-like operons are absent, and there are two promoters, one for each nucleotide strand, within the D loop. Transcription of the two strands proceeds in opposite directions, generating two giant precursor RNAs that are then cleaved to yield individual rRNAs, tRNAs, and mRNAs. As the tRNAs are transcribed, they fold up into three-dimensional configurations. These configurations are recognized and cut out by enzymes. The tRNA genes generally flank the protein and rRNA genes; so cleavage of the tRNAs releases mRNAs and rRNAs. In the mitochondrial genomes of fungi, plants, and protists, there are multiple promoters, although genes are occasionally arranged and transcribed in operons.

Most mRNA molecules produced by the transcription of mtDNA are not capped at their 5′ ends, unlike mRNA transcribed from nuclear genes (see Figure 14.6). Poly(A) tails are added to the 3′ ends of some mRNAs encoded by animal mtDNA, but poly(A) tails are missing from those encoded by mtDNA in fungi, plants, and protists. The poly(A) tails added to animal mitochondrial mRNAs are shorter than those attached to nuclear-encoded mRNA and are probably added by an entirely different mechanism.

Some of the genes in yeast and plant mtDNA contain introns, many of which are self-splicing. RNA encoded by some mitochondrial genomes undergoes extensive editing (see pp. 387–388 in Chapter 14).

Translation in mitochondria has some similarities to eubacterial translation, but there are also important differences. In mitochondria, protein synthesis is initiated at AUG start codons by N-formylmethionine, just as in the eubacterial initiation of translation. Mitochondrial translation also employs elongation factors similar to those seen in eubacteria, and the same antibiotics that inhibit translation in eubacteria also inhibit translation in mitochondria. However, mitochondrial ribosomes are variable in structure and are often different from those seen in both eubacterial and eukaryotic cells. Additionally, the initiation of translation in mitochondria must be different from that of both eubacterial and eukaryotic cells, because animal mitochondrial mRNA contains no Shine–Dalgarno ribosome-binding site and no 5′ cap. (A Shine–Dalgarno sequence has been observed in mitochondrial mRNA of the protozoan *Reclinomonas americana*, which has a very primitive, eubacterial-like mitochondrion.)

There is also much diversity in the tRNAs encoded by various mitochondrial genomes. Human mtDNA encodes 22 of the 32 tRNAs required for translation in the cytoplasm. In human mitochondrial translation, there is even more wobble than in cytoplasmic translation; many mitochondrial tRNAs will recognize any of the four nucleotides in the third position of the codon, permitting translation to take place with even fewer tRNAs. The increased wobble also means that any change in a DNA nucleotide at the third position of the codon will be a silent mutation (see p. 486 in Chapter 18) and will not alter the amino acid sequence of the protein. Thus, more of the changes due to wobble in mtDNA are silent and accumulate over time, contributing to a higher rate of evolution. In some organisms, fewer than 22 tRNAs are encoded by mtDNA; in these organisms, nuclear-encoded tRNAs are imported from the cytoplasm to help carry out translation in the mitochondria. In other organisms, the mitochondrial genome encodes a complete set of all 32 tRNAs.

CONCEPTS

The processes of replication, transcription, and translation vary widely among mitochondrial genomes and exhibit a curious mix of eubacterial, eukaryotic, and unique characteristics.

✔ **CONCEPT CHECK 4**

Which of the following statements applies to mitochondrial translation?

a. Elongation factors are similar to those in eukaryotic translation.

b. There is no wobble.

c. Antibiotics that inhibit bacterial translation have no effect on mitochondrial translation.

d. The initiation codon specifies N-formylmethionine.

The Evolution of Mitochondrial DNA

As already mentioned, comparisons of mitochondrial DNA sequences with DNA sequences in bacteria strongly support a common eubacterial origin for all mtDNA. Nevertheless, patterns of evolution seen in mtDNA vary greatly among different groups of organisms.

The sequences of vertebrate mtDNA exhibit an accelerated rate of evolution: the sequences in mammalian mtDNA, for example, typically change from 5 to 10 times as fast as those in mammalian nuclear DNA. The number of genes present and the organization of vertebrate mitochondrial genomes, however, are relatively constant. In contrast, sequences of plant mtDNA evolve slowly at a rate only $1/_{10}$th that of the nuclear genome, but their gene content and organization change rapidly. The reason for these basic differences in rates of evolution is not yet known.

A possible reason for the accelerated rate of evolution seen in vertebrate mtDNA is its high mutation rate, which would allow DNA sequences to change quickly. Increased errors associated with replication, the absence of DNA-repair functions, and the frequent replication of mtDNA may increase the number of mutations. The large amount of wobble in mitochondrial translation may allow mutations to accumulate over time, as discussed earlier. The use of mtDNA in evolutionary studies will be described in more detail in Chapter 26.

At conception, a mammalian zygote inherits approximately 100,000 copies of mtDNA inherited from the egg. Because of the large number of mtDNA molecules in each cell and the high rate of mutation in mtDNA, most cells would be expected to contain a mixture of wild-type and mutant mtDNA molecules (heteroplasmy). However, heteroplasmy is rarely present: in most organisms, the copies of mtDNA are genetically identical (homoplasmy). To account for the uniformity of mtDNA within individual organisms, geneticists hypothesize that, in early development or gamete formation, mtDNA goes through some type of bottleneck, during which the mtDNAs within a cell are reduced to just a few copies, which then replicate and give rise to all subsequent copies of mtDNA. Through this process, genetic variation in mtDNA within a cell is eliminated and most copies of mtDNA are identical. Recent studies have provided evidence that a bottleneck does exist, but there is contradictory evidence concerning where in development the mtDNA bottleneck arises.

> **CONCEPTS**
>
> All mtDNA appears to have evolved from a common eubacterial ancestor, but the patterns of evolution seen in different mitochondrial genomes vary greatly. Vertebrate mtDNA exhibits rapid change in sequence but little change in gene content and organization, whereas the mtDNA of plants exhibits little change in sequence but much variation in gene content and organization.

Mitochondrial DNA Variation and Human History

Mitochondrial DNA has been studied extensively to reconstruct the evolution and migration patterns of humans. Some of the advantages of mtDNA for studying evolution were mentioned in the introduction to the chapter: (1) the small size and abundance of mtDNA in the cell; (2) the rapid evolution of mtDNA sequences, facilitating study of closely related groups; and (3) the maternal inheritance of mtDNA. Samples of mtDNA have been analyzed from thousands of people belonging to hundreds of different ethnic groups throughout the world and are helping to unravel many aspects of human

evolution and history. For example, initial studies on mtDNA sequences led to the proposal that a small groups of humans migrated out of Africa about 85,000 years ago and populated the rest of the world (called the Out of Africa hypothesis or the African Replacement hypothesis), which has now gained wide acceptance and is supported by additional studies of DNA sequences from the Y chromosome and nuclear genes.

Sequences from mtDNA have also been used to study the spread of agriculture in Europe. Agriculture first arose about 10,000 years ago in the Middle East (see Chapter 1) and then spread to many parts of the world. Arriving in Europe about 7500 years ago, agriculture replaced the hunting-and-gathering life style of early inhabitants, who had lived there for at least 30,000 years. Two rival theories have been proposed to explain how agriculture spread into Europe: (1) the demic diffusion theory, which proposes that farmers from the Near East migrated into Europe and replaced the original hunter–gatherers; and (2) the cultural diffusion theory, which proposes that the indigenous hunter–gatherers adopted the practice of farming. Thus, the demic diffusion theory suggests that the modern European gene pool is derived largely from people who migrated from the Near East less than 10,000 years ago, and the cultural diffusion theory suggests that modern Europeans derive their genes largely from the ancient hunter–gatherers. Although early studies of languages and blood types suggested that agriculture might have spread through demic diffusion, analyses of the mtDNA of living Europeans indicate that less than 20% of the modern European gene pool is descended from immigrating Near Eastern farmers, supporting the cultural diffusion theory.

In 2009, scientists succeeded in extracting DNA from the bones and teeth of 20 early European hunter–gatherers from sites in Lithuania, Poland, Russia, and Germany. Using PCR, they amplified and sequenced mtDNA from these remains. The scientists then compared the hunter–gatherer mtDNA sequences with those from 25 early European farmers and with those from a group of 484 modern Europeans. They found large genetic differences in all three groups, with less than 20% of the hunter–gatherer mtDNA types also present in modern Europeans. These results suggest that the first farmers in Europe were not descendants of the early hunter–gatherers, supporting the demic diffusion model. However, the genetic relationship of the early farmers to modern Europeans is unclear, and these conclusions remain controversial. **TRY PROBLEM 20 →**

21.3 Chloroplast DNA Exhibits Many Properties of Eubacterial DNA

Geneticists have long recognized that many traits associated with chloroplasts exhibit cytoplasmic inheritance, indicating that these traits are not encoded by nuclear

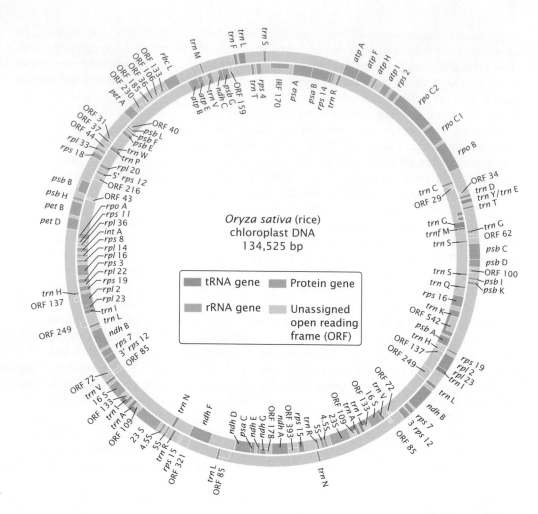

21.10 Chloroplast DNA of rice.

genes. In 1963, chloroplasts were shown to have their own DNA (**Figure 21.10**).

Among different plants, the chloroplast genome ranges in size from 80,000 to 600,000 bp, but most chloroplast genomes range from 120,000 to 160,000 bp (**Table 21.3**). Chloroplast DNA is usually a single, double-stranded DNA molecule that is circular, is highly coiled, and lacks associ-

ated histone proteins. As in mtDNA, multiple copies of the chloroplast genome are found in each chloroplast, and there are multiple organelles per cell; so there are several hundred to several thousand copies of cpDNA in a typical plant cell.

The Gene Structure and Organization of Chloroplast DNA

The chloroplast genomes from a number of plant and algal species have been sequenced, and cpDNA is now recognized to be basically eubacterial in its organization: the order of some groups of genes is the same as that observed in *E. coli*, and many chloroplast genes are organized into operon-like clusters.

Among vascular plants, chloroplast chromosomes are similar in gene content and gene order. A typical chloroplast genome encodes 4 rRNA genes, from 30 to 35 tRNA genes, a number of ribosomal proteins, many proteins engaged in photosynthesis, and several proteins having roles in nonphotosynthesis processes. A key protein encoded by cpDNA is ribulose-1,5-bisphosphate carboxylase-oxygenase (abbreviated RuBisCO), which participates in the fixation of carbon in photosynthesis. RuBisCO makes up about 50% of the protein found in green plants and is therefore considered the most abundant protein on Earth. It is a complex protein consist-

Table 21.3	Sizes of chloroplast genomes in selected organisms
Organism	**Size of cpDNA (bp)**
Euglena gracilis (protist)	143,172
Porphyra purpurea (red alga)	191,028
Chlorella vulgaris (green alga)	150,613
Marchantia polymorpha (liverwort)	121,024
Nicotiana tabacum (tobacco)	155,939
Zea mays (corn)	140,387
Pinus thunbergii (black pine)	119,707

ing of eight identical large subunits and eight identical small subunits. The large subunit is encoded by chloroplast DNA, whereas the small subunit is encoded by nuclear DNA.

The circular chloroplast genome has genes on both of its strands. Some chloroplast genes have been identified on the basis of the presence of an open reading frame. An open reading frame is a sequence of nucleotides that contains a start codon and a stop codon in the same reading frame and is, most likely, a gene that encodes a protein, although, in many cases, no protein products have yet been isolated for these seqeunces. A prominent feature of most chloroplast genomes is the presence of a large inverted repeat. In rice, this repeat includes genes for 23S rRNA, 4.5S rRNA, and 5S rRNA, as well as several genes for tRNAs and proteins (see Figure 21.10). In some plants, these repeats constitute most of the genome, whereas, in others, the repeats are absent entirely. Much of cpDNA consists of noncoding sequences, and introns are found in many chloroplast genes. Finally, many of the sequences in cpDNA are quite similar to those found in equivalent eubacterial genes.

CONCEPTS

Most chloroplast genomes consist of a single circular DNA molecule not complexed with histone proteins. Although there is considerable size variation among species, the cpDNAs found in most vascular plants are about 150,000 bp. Genes are scattered in the circular chloroplast genome, and many contain introns. Most cpDNAs contain a large inverted repeat.

✔ CONCEPT CHECK 5

In its organization, chloroplast DNA is most similar to

a. eubacteria.

b. archaea.

c. nuclear DNA of plants.

d. nuclear DNA of primitive eukaryotes.

The Replication, Transcription, and Translation of Chloroplast DNA

Little is known about the process of cpDNA replication. The results of studies viewing cpDNA replication with electron microscopy suggest that replication begins within two D loops and spreads outward to form a theta structure (see Figure 12.4). After an initial round of replication, DNA synthesis may switch to a rolling-circle-type mechanism (see Figure 12.5).

The transcription and translation of chloroplast genes are similar in many respects to these processes in eubacteria. For example, promoters found in cpDNA are virtually identical with those found in eubacteria and possess sequences similar to the −10 and −35 consensus sequences of eubacterial promoters. The same antibiotics that inhibit protein synthesis in eubacteria (as well as in mitochondria) inhibit protein synthesis in chloroplasts, indicating that protein synthesis in eubacteria and chloroplasts is similar.

Chloroplast translation is initiated by *N*-formylmethionine, just as it is in eubacteria.

Most genes in cpDNA are transcribed in groups; only a few genes have their own promoters and are transcribed as separate mRNA molecules. The RNA polymerase that transcribes cpDNA is more similar to eubacterial RNA polymerase than to any of the RNA polymerases that transcribe eukaryotic nuclear genes. Like eubacterial mRNAs, chloroplast mRNAs are not capped at the 5′ ends, and poly(A) tails are not added to the 3′ ends. However, introns are removed from some RNA molecules after transcription, and the 5′ and 3′ ends may undergo some additional processing before the molecules are translated. Like eubacterial mRNAs, many chloroplast mRNAs have a Shine–Dalgarno sequence in the 5′ untranslated region, which may serve as a ribosome-binding site.

Chloroplasts, like eubacteria, contain 70S ribosomes that consist of two subunits, a large 50S subunit and a smaller 30S subunit. The small subunit includes a single RNA molecule that is 16S in size, similar to that found in the small subunit of eubacterial ribosomes. The larger 50S subunit includes three rRNA molecules: a 23S rRNA, a 5S rRNA, and a 4.5S rRNA. In eubacterial ribosomes, the large subunit possesses only two rRNA molecules, which are 23S and 5S in size. The 4.5S rRNA molecule found in the large subunit of chloroplast ribosomes is homologous to the 3′ end of the 23S rRNA found in eubacteria; so the structure of the chloroplast ribosome is very similar to that of ribosomes found in eubacteria.

Initiation factors, elongation factors, and termination factors function in chloroplast translation and eubacterial translation in similar ways. Most chloroplast chromosomes encode from 30 to 35 different tRNAs, suggesting that the expanded wobble seen in mitochondria does not exist in chloroplast translation. Only universal codons have been found in cpDNA.

The Evolution of Chloroplast DNA

The DNA sequences of chloroplasts are very similar to those found in cyanobacteria; so chloroplast genomes clearly have a eubacterial ancestry. Overall, cpDNA sequences evolve slowly compared with sequences in nuclear DNA and some mtDNA. For most chloroplast genomes, size and gene organization are similar, although there are some notable exceptions. Because they evolve slowly and, like mtDNA, are inherited from only one parent, cpDNA is often useful for determining the evolutionary relationships among different plant species.

CONCEPTS

Many aspects of the transcription and translation of cpDNA are similar to those of eubacteria. Chloroplast DNA sequences are most similar to DNA sequences in cyanobacteria, which supports the endosymbiotic theory. Most cpDNA evolves slowly in sequence and structure.

✔ CONCEPT CHECK 6

Which characteristic does not apply to chloroplast DNA or mRNA?

a. Promoters are similar to eubacterial promoters.

b. Genes in cpDNA are transcribed in groups.

c. Many chloroplast mRNAs have Shine–Dalgarno sequences.

d. Poly(A) tails are added to chloroplast mRNAs.

CONNECTING CONCEPTS

Genome Comparisons

A theme running through the preceding discussions of mitochondrial and chloroplast genomes has been a comparison of these genomes with those found in eubacterial and eukaryotic cells (**Table 21.4**). The endosymbiotic theory indicates that mitochondria and chloroplasts evolved from eubacterial ancestors, and therefore mtDNA and cpDNA might be assumed to be similar to DNA found in eubacterial cells. The actual situation is more complex: mitochondrial DNA and chloroplast DNA possess a mixture of eubacterial, eukaryotic, and unique characteristics.

The mitochondrial and chloroplast genomes are similar to those of eubacterial cells in that they are small, lack histone proteins, and are usually on circular DNA molecules. The gene organization and the expression of organelle genomes, however, display some similarities to those of eubacterial genomes and some similarities to those of eukaryotic genomes. For example, different types of introns are present in some organelle genomes but are absent from others (Table 21.4). Human mtDNA, which has little noncoding DNA between genes and little repetitive DNA, is similar in organization to that of typical eubacterial chromosomes, but other mitochondrial and chloroplast genomes possess long noncoding sequences between genes. Shine–Dalgarno sequences, the ribosome-binding sites characteristic of eubacterial DNA, are present in some cpDNA but are absent in mtDNA. Some mitochondrial genomes use nonuniversal codons and have extended wobble, which is rare in both eubacterial and eukaryotic DNA. Comparison of additional characteristics in eukaryotic, eubacterial, and organelle genomes is shown in Table 21.4.

What conclusions can we draw from these comparisons? Clearly, the genomes of mitochondria and chloroplasts are not typical of the nuclear genomes of the eukaryotic cells in which they reside. In sequence, organelle DNA is most similar to eubacterial DNA, but many aspects of the organization and expression of organelle genomes are unique. It is important to remember that the endosymbiotic theory does not propose that mitochondria and chloroplasts are eubacterial in nature but that they arose from eubacterial ancestors more than a billion years ago. Through time, the genomes of the endosymbionts have undergone considerable evolutionary change and have evolved characteristics that distinguish them from contemporary eubacterial and eukaryotic genomes.

Table 21.4 Comparison of nuclear eukaryotic, eubacterial, mitochondrial, and chloroplast genomes

Characteristic	Eukaryotic Genome	Eubacterial Genome	Mitochondrial Genome	Chloroplast Genome
Genome consists of double-stranded DNA	Yes	Yes	Yes	Yes
Circular	No	Yes	Most	Yes
Histone proteins	Yes	No	No	No
Size	Large	Small	Small	Small
Number of molecules per genome	Several	One	One in animals; several in some plants	One
Pre-mRNA introns	Common	Absent	Absent	Absent
Group I introns	Present	Present	Present	Present
Group II introns	Absent	Present	Present	Present
Polycistronic mRNA	Uncommon	Common	Present	Common
5′ cap added to mRNA	Yes	No	No	No
3′ poly(A) tail added to mRNA	Yes	No	Some in animals	No
Shine–Dalgarno sequence in 5′ untranslated region of mRNA	No	Yes	Rare	Some
Nonuniversal codons	Rare	Rare	Yes	No
Extended wobble	No	No	Yes	No
Translation inhibited by tetracycline	No	Yes	Yes	Yes

21.4 Through Evolutionary Time, Genetic Information Has Moved Between Nuclear, Mitochondrial, and Chloroplast Genomes

Many proteins found in modern mitochondria and chloroplasts are encoded by nuclear genes, which suggests that much of the original genetic material in the endosymbiont has probably been transferred to the nucleus. This assumption is supported by the observation that some DNA sequences normally found in mtDNA have been detected in the nuclear DNA of some strains of yeast and maize. Likewise, chloroplast sequences have been found in the nuclear DNA of spinach. Furthermore, the sequences of nuclear genes that encode organelle proteins are most similar to their eubacterial counterparts.

There is also evidence that genetic material has moved from chloroplasts to mitochondria. For example, DNA fragments from the 16S rRNA gene and two tRNA genes that are normally encoded by cpDNA have been found in the mtDNA of maize. Sequences from the gene that encodes the large subunit of RuBisCO, which is normally encoded by cpDNA, are duplicated in maize mtDNA. And there is even evidence that some nuclear genes have moved into mitochondrial genomes. The exchange of genetic material between the nuclear, mitochondrial, and chloroplast genomes has given rise to the term "promiscuous DNA" to describe this phenomenon. The mechanism by which this exchange takes place is not entirely clear.

21.5 Damage to Mitochondrial DNA Is Associated with Aging

The symptoms of many human genetic diseases caused by defects in mtDNA first appear in middle age or later and increase in severity as people age. One hypothesis to explain the late onset and progressive worsening of mitochondrial diseases is related to the decline in oxidative phosphorylation with aging.

Oxidative phosphorylation is the process that generates ATP, the primary carrier of energy in the cell. This process takes place on the inner membrane of the mitochondrion and requires a number of different proteins, some encoded by mtDNA and others encoded by nuclear genes. Oxidative phosphorylation normally declines with age and, if it falls below some critical threshold, tissues do not make enough ATP to sustain vital functions and disease symptoms appear. Most people start life with an excess capacity for oxidative phosphorylation; this capacity decreases with age, but most people reach old age or die before the critical threshold is passed. Persons born with mitochondrial diseases carry mutations in their mtDNA that lower their oxidative phosphorylation capacity. At birth, their capacity may be sufficient to support their ATP needs but, as their oxidative phosphorylation capacity declines with age, they cross the critical threshold and begin to experience symptoms. These symptoms usually first appear in tissues that are most critically dependent on mitochondrial energy: the central nervous system, heart and skeletal muscle, pancreatic islets, kidneys, and the liver.

Why does oxidative phosphorylation capacity decline with age? A possible explanation is that damage to mtDNA accumulates with age: deletions and base substitutions in mtDNA increase with age. For example, a common 5000-bp deletion in mtDNA is absent in normal heart muscle cells before the age of 40, but, afterward, this deletion is present with increasing frequency. The same deletion is found at a low frequency in normal brain tissue before age 75 but is found in 11% to 12% of mtDNAs in the basal ganglia by age 80. People with mtDNA genetic diseases may age prematurely because they begin life with damaged mtDNA.

The mechanism of age-related increases in mtDNA damage is not yet known. Oxygen radicals—highly reactive compounds that are natural by-products of oxidative phosphorylation—are known to damage DNA (see p. 496 in Chapter 18). Because mtDNA is physically close to the enzymes taking part in oxidative phosphorylation, mtDNA may be more prone to oxidative damage than is nuclear DNA. When mtDNA has been damaged, the cell's capacity to produce ATP drops. To produce sufficient ATP to meet the cell's energy needs, even more oxidative phosphorylation must take place, which in turn may stimulate further production of oxygen radicals, leading to a vicious cycle.

Evidence for the hypothesis that mtDNA damage is associated with aging comes from a study in which geneticists increased the rate of mutations in the mtDNA of mice. In mice, as in humans, DNA polymerase γ catalyzes the replication of mtDNA. Polymerase γ both synthesizes DNA and proofreads (see Chapter 12). Geneticists created transgenic mice in which the proofreading activity of the enzyme was defective, though the polymerase activity was unaffected. The somatic tissues of these transgenic mice accumulated extensive mutations in their mtDNA. The mice showed symptoms of premature aging, including weight and hair loss, reductions in fertility, curvature of the spine, and reduced life span. These results support the hypothesis that mutations in mtDNA can lead to at least some features of aging. However, the cause of mtDNA damage in the course of normal aging is not clear. Although the transgenic mice had increased mutations in their mtDNA, they did not show evidence of increased oxidative damage, as would be expected under the vicious-cycle hypothesis. Furthermore, subsequent studies found that some of the genetically engineered mice had greatly increased numbers of mutations but did not show signs of increased aging, so the association between aging and damage to mtDNA remains controversial. It is important to recognize that normal aging is a complex process and is unlikely to be caused by a single factor.

Significantly elevated levels of mtDNA defects have been observed in some patients with late-onset degenerative diseases, such as diabetes mellitus, ischemic heart disease, Parkinson disease, Alzheimer disease, and Huntington disease. All of these diseases appear in middle to old age and have symptoms associated with tissues that critically depend on oxidative phosphorylation for ATP production.

However, because Huntington disease and some cases of Alzheimer disease are inherited as autosomal dominant conditions, mtDNA defects cannot be the primary cause of these diseases, although they may contribute to their progression.

CONCEPTS SUMMARY

- Mitochondria and chloroplasts are eukaryotic organelles that possess their own DNA. Traits encoded by mtDNA and cpDNA are usually inherited from a single parent, most often the mother. Random segregation of organelles in cell division may produce phenotypic variation among cells within an individual organism and among the offspring of a single female.

- The endosymbiotic theory proposes that mitochondria and chloroplasts originated as free-living prokaryotic (specifically eubacterial) organisms that entered into a beneficial association with eukaryotic cells.

- The mitochondrial genome usually consists of a single circular DNA molecule that lacks histone proteins. Mitochondrial DNA varies in size among different groups of organisms.

- Ancestral mitochondrial genomes typically have characteristics of eubacterial genomes. Derived mitochondrial genomes are smaller and contain fewer genes. Their rRNA genes and ribosomes differ from those found in eubacteria, and they use some nonuniversal codons.

- Human mtDNA is highly economical, with few noncoding nucleotides. Fungal and plant mtDNAs contain much noncoding DNA between genes, introns within genes, and extensive 5′ and 3′ untranslated regions.

- Mitochondrial DNA is synthesized throughout the cell cycle, and its synthesis is not coordinated with the replication of nuclear DNA.

- The transcription of mitochondrial genes varies among different organisms.

- Antibiotics that inhibit eubacterial ribosomes also inhibit mitochondrial ribosomes. Many mitochondrial genomes encode a limited number of tRNAs, with relaxed codon–anticodon pairing rules and extended wobble.

- Comparisons of mtDNA sequences suggest that mitochondria evolved from a eubacterial ancestor. Vertebrate mtDNA exhibits rapid change in sequence but little change in gene content and organization. Plant mtDNA exhibits little change in sequence but much variation in gene content and organization.

- Mitochondrial DNA sequences have been widely used to study patterns of human evolution.

- Chloroplast genomes consist of a single circular DNA molecule that varies little in size and lacks histone proteins. Each plant cell contains multiple copies of cpDNA.

- Transcription and translation are similar in chloroplasts and eubacteria.

- Chloroplast DNA sequences are most similar to those in cyanobacteria and tend to evolve slowly.

- Through evolutionary time, many mitochondrial and chloroplast genes have moved to nuclear chromosomes. In some plants, there is evidence that copies of chloroplast genes have moved to the mitochondrial genome.

IMPORTANT TERMS

mitochondrial DNA (mtDNA) (p. 591)
chloroplast DNA (cpDNA) (p. 592)
heteroplasmy (p. 593)

replicative segregation (p. 593)
homoplasmy (p. 593)

endosymbiotic theory (p. 596)
D loop (p. 598)

ANSWERS TO CONCEPT CHECKS

1. c
2. Many modern protists are host to endosymbiotic bacteria. Mitochondria and chloroplasts are similar in size to eubacteria and have their own DNA, as well as ribosomes that are similar in size and shape to eubacterial ribosomes. Antibiotics that inhibit protein synthesis in eubacteria also inhibit protein synthesis in mitochondria and chloroplasts. Gene sequences in

mtDNA and cpDNA are most similar to eubacterial DNA sequences.

3. b
4. d
5. a
6. d

WORKED PROBLEM

1. Suppose that a new organelle is discovered in an obscure group of protists. This organelle contains a small DNA genome, and some scientists are arguing that, like chloroplasts and mitochondria, this organelle originated as a free-living eubacterium that entered into an endosymbiotic relation with the protist. Outline a research plan to determine whether the new organelle evolved from a free-living eubacterium. What kinds of data would you collect and what predictions would you make if the theory were correct?

• Solution

We could examine the structure, organization, and sequences of the organelle genome. If the organelle shows only characteristics of eukaryotic DNA, then it most likely has a eukaryotic origin but, if it displays some characteristics of eubacterial DNA, then this finding supports the theory of a eubacterial origin. However, on the basis of our knowledge of mitochondrial and chloroplast genomes, we should not expect the organelle genome to be entirely eubacterial in its characteristics.

We could start by examining the overall characteristics of the organelle DNA. If it has a eubacterial origin, we might

expect that the organelle genome will consist of a circular molecule and will lack histone proteins. We might then sequence the organelle DNA to determine its gene content and organization. The presence of any group II introns would suggest a eubacterial origin, because these introns have been found only in eubacterial genomes and genomes derived from eubacteria. The presence of any pre-mRNA introns, on the other hand, would suggest a eukaryotic origin, because these introns have been found only in nuclear eukaryotic genomes. If the organelle genome has a eubacterial origin, we might expect to see polycistronic mRNA, the absence of a 5′ cap, and the inhibition of translation by those antibiotics that typically inhibit eubacterial translation.

Finally, we could compare the DNA sequences found in the organelle genome with homologous sequences from eubacteria and eukaryotic genomes. If the theory of an endosymbiotic origin is correct, then the organelle sequences should be most similar to homologous sequences found in eubacteria.

COMPREHENSION QUESTIONS

Section 21.1

1. Explain why many traits encoded by mtDNA and cpDNA exhibit considerable variation in their expression, even among members of the same family.

*2. What is the endosymbiotic theory? How does it help to explain some of the characteristics of mitochondria and chloroplasts?

3. What evidence supports the endosymbiotic theory?

Section 21.2

4. How are genes organized in the mitochondrial genome? How does this organization differ between ancestral and derived mitochondrial genomes?

*5. What are nonuniversal codons? Where are they found?

6. How does the replication of mtDNA differ from the replication of eukaryotic nuclear DNA.

*7. The human mitochondrial genome encodes only 22 tRNAs, whereas at least 32 tRNAs are required for cytoplasmic translation. Why are fewer tRNAs needed in mitochondria?

8. What are some possible explanations for an accelerated rate of evolution in the sequences of vertebrate mtDNA?

Section 21.3

*9. Briefly describe the organization of genes on the chloroplast genome.

*10. Briefly describe the general structures of mtDNA and cpDNA. How are they similar? How do they differ? How do their structures compare with the structures of eubacterial and eukaryotic (nuclear) DNA?

Section 21.4

11. What is meant by the term "promiscuous DNA"?

APPLICATION QUESTIONS AND PROBLEMS

Introduction

12. The Asian half asses consist of two species *Equus kiang* and *E. hemionus*, with *E. hemionus* divided into three subspeces (*E. h. luteus*, *E. h. kulan*, and *E. h. onager*). Using mtDNA sequences, Albano Beja-Pereira and colleagues (2004. *Science* 304:1781) created an evolutionary tree for these groups of Asian half asses, as well as African wild asses and domestic asses. Using the results of Beja-Pereira et al.

shown in Figure 21.1, describe the evolutionary relationships among the different species and subspecies of the Asian half asses.

Section 21.1

13. A wheat plant that is light green in color is found growing in a field. Biochemical analysis reveals that chloroplasts in this plant produce only 50% of the chlorophyll normally found in wheat chloroplasts. Propose a set of crosses to determine whether the light-green phenotype is caused by a mutation in a nuclear gene or in a chloroplast gene.

*14. A rare neurological disease is found in the family illustrated in the following pedigree. What is the most likely mode of inheritance for this disorder? Explain your reasoning.

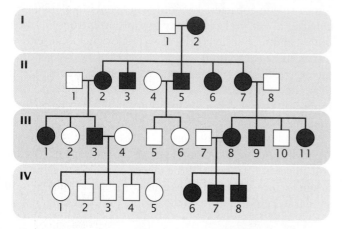

15. Fredrick Wilson and his colleagues studied members of a large family who had low levels of magnesium in their blood (see the adjoining pedigree). They argued that this disorder of magnesium (and associated high blood pressure and high cholesterol) is caused by a mutation in mtDNA (F. H. Wilson et al. 2004. *Science* 306:1190–1194).

 a. What evidence suggests that a gene in the mtDNA is causing this disorder?

 b. Could this disorder be caused by an autosomal dominant gene? Why or why not?

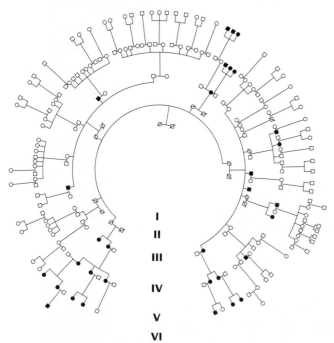

[After F. H. Wilson et al. 2004. *Science* 306:1190–1194.]

16. In a particular strain of *Neurospora*, a *poky* mutation exhibits biparental inheritance, whereas *poky* mutations in other strains are inherited only from the maternal parent. Explain these results.

Section 21.2

*17. A scientist collects cells at various points in the cell cycle and isolates DNA from them. Using density-gradient centrifugation, she separates the nuclear and mtDNA. She then measures the amount of mtDNA and nuclear DNA present at different points in the cell cycle. On the following graph, draw a line to represent the relative amounts of nuclear DNA that you expect her to find per cell throughout the cell cycle. Then, draw a dotted line on the same graph to indicate the relative amount of mtDNA that you would expect to see at different points throughout the cell cycle.

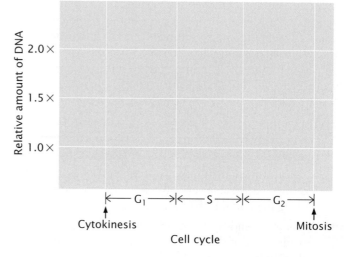

18. In 1979, bones found outside Ekaterinburg, Russia, were shown to be those of Tsar Nicholas and his family, who were executed in 1918 by a Bolshevik firing squad in the Russian Revolution. To prove that the skeletons were those of the royal family, mtDNA was extracted from the bone samples, amplified by PCR, and compared with mtDNA from living relatives of the tsar's family. Why was DNA from the mitochondria analyzed instead of nuclear DNA? What are some of the advantages of using mtDNA for this type of study?

19. From Figure 21.7, determine as best you can the percentage of human mtDNA that is coding (transcribed into RNA) and the percentage that is noncoding (not transcribed).

20. Richard Cordaux and his colleagues analyzed mtDNA sequences to determine whether the recent spread of agriculture into new areas in India was by demic diffusion (traditional farmers replacing hunter–gatherers) or cultural diffusion (the practice of farming spreading among hunter–gatherers; see p. 601). They examined sequences of mtDNA from (1) 229 hunter–gatherers from

southern India, (2) 201 recent farmers from southern India, (3) 140 traditional farmers from southern India, and (4) 62 traditional farmers from northern India (R. Cordaux et al. 2004. *Science* 304:1125). On the basis of the mtDNA sequences, they computed *Fst* values, which represent the overall degree of genetic difference between groups. The *Fst* values for differences between the four groups are listed in the following table.

	Hunter–gatherers from south	Traditional farmers from south	Traditional farmers from north
Recent farmers from south	0.038	0.028	0.036
Hunter–gatherers from south		0.039	0.048
Traditional farmers from south			0.017

a. Do these data support demic diffusion or cultural diffusion as the mechanism for the spread of farming to the new areas of India? Explain your reasoning.

b. These researchers also looked at sequences on the Y chromosome and found the same patterns as those shown by the mtDNA sequences. Why did they use sequences on the Y chromosome?

Section 21.3

21. Antibiotics such as chloramphenicol, tetracycline, and erythromycin inhibit protein synthesis in eubacteria but have no effect on protein synthesis encoded by nuclear genes. Cycloheximide inhibits protein synthesis encoded by nuclear genes but has no effect on eubacterial protein synthesis. How might these compounds be used to determine which proteins are encoded by mitochondrial and chloroplast genomes?

CHALLENGE QUESTIONS

Section 21.2

22. Steven Frank and Laurence Hurst argued that a cytoplasmically inherited mutation in humans that has severe effects in males but no effect in females will not be eliminated from a population by natural selection, because only females pass on mtDNA (S. A. Frank and L. D. Hurst. 1996. *Nature* 383:224). Using this argument, explain why males with Leber hereditary optic neuropathy are more severely affected than females.

23. In a study of a myopathy, several families exhibited vision problems, muscle weakness, and deafness (M. Zeviani et al. 1990. *American Journal of Human Genetics* 47:904–914). Analysis of the mtDNA from affected persons in these families revealed that large numbers of their mtDNA possessed deletions of varying length. Different members of the same family and even different mitochondria from the same person possessed deletions of different sizes; so the underlying defect appeared to be a tendency for the mtDNA of affected persons to have deletions. A pedigree of one of the families studied is shown here. The researchers concluded that this disorder is inherited as an autosomal dominant trait and they mapped the disease-causing gene to a position on chromosome 10 in the nucleus.

a. What characteristics of the pedigree rule out inheritance of a trait encoded by a gene in the mtDNA?

b. Explain how a mutation in a nuclear gene might lead to deletions in mtDNA.

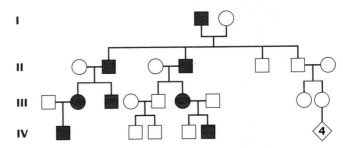

[After M. Zeviani et al. 1990. *American Journal of Human Genetics* 47:904–914.]

Section 21.4

24. Mitochondrial DNA sequences have been detected in the nuclear genomes of many organisms, and cpDNA sequences are sometimes found in the mitochondrial genome. Propose a mechanism for how such "promiscuous DNA" might move between nuclear, mitochondrial, and chloroplast genomes.

22 Developmental Genetics and Immunogenetics

Mexican tetras that live in caves have lost their eyes through a developmental change in gene expression. [Breck P. Kent/Animals Animals—Earth Scenes.]

HOW A CAVEFISH LOST ITS EYES

The vertebrate eye is an exquisite organ. It bends and focuses light through the lens onto the retina at the back of the eye, activating millions of photoreceptors that send information to the brain about wavelength and intensity. Within the brain, this information is reconstructed into the image that we see. Charles Darwin once called the eye the "organ of perfection."

Early vertebrates evolved complex eyes, and all vertebrates have a complete set of genes for eye development. However, not all vertebrates develop eyes. The Mexican tetra, *Astyanax mexicanus*, normally swims the surface waters of streams and rivers in Texas and northern Mexico. Some 10,000 years ago, a few tetras migrated into caves and set up residence. In the total darkness of their cave environment, eyes were not needed and, with the passage of time, these tetras lost their eyes. Today, some 30 distinct populations of Mexican tetras are totally blind and eyeless, whereas surface-dwelling populations of the same species have retained normal eye development.

So how does a cavefish lose its eyes? Amazingly, Mexican tetra zygotes begin to develop eyes, just like their surface-dwelling cousins; but, about 24 hours after fertilization, eye development aborts, and the cells that were destined to become the lens spontaneously die. The absence of the lens prevents other components of the eye, such as the cornea and iris, from developing. The optic cup and retina form, but their growth is retarded and photoreceptor cells never differentiate. The degenerate eye sinks into the orbit and is eventually covered by a flap of skin, forever obliterating all sight.

Blind Mexican tetras have the same genes as surface-dwelling Mexican tetra; how they differ is in their gene expression. Geneticist William Jeffrey and his colleagues demonstrated that the expression of two genes named *sonic hedgehog* (*shh*) and *tiggy-winkle hedgehog* (*twhh*) are more widely expressed in the eye primordium of blind cavefish than in surface fish. The expanded expression of *shh* and *twhh* activates the transcription of other genes, which cause lens cells to undergo a type of spontaneous suicide called apoptosis (discussed in more detail later in this chapter), and the lens degenerates.

When Jeffrey and his colleagues injected *twhh* or *shh* mRNA into the embryos of surface fish, the development of the lens aborted, and adults that developed from these embryos were missing eyes. When the scientists used drugs to inhibit the expression of *twhh* and *shh* in cavefish embryos, eye development in these fish was partly restored. These results demonstrate that eye development in Mexican tetras is regulated by the precise expression of *shh* and *twhh*. Overexpression of one or both of these genes in the cavefish induces the death of lens cells and aborts normal eye development. Incredibly, a small increase in the transcription of either gene during embryonic development results in a major anatomical change in adult fish, change that has allowed the fish to adapt to the darkness of the cave environment.

The story of the blind Mexican cavefish illustrates how the degree of gene expression regulates key developmental events, a theme that permeates this chapter on the genetic control of development. The chapter begins with a discussion of the genetic control of early development of *Drosophila* embryos, one of the best-understood developmental systems. We next consider the genetic control of floral structure in plants, another model system that has been well studied. We take a more detailed look at two topics illustrated in the story of blind Mexican cavefish: (1) programmed cell death and (2) the use of development for understanding evolution. At the end of the chapter, we turn to the development of immunity and its genetic control.

TRY PROBLEM 10 →

22.1 Development Takes Place Through Cell Determination

Every multicellular organism begins life as a unicellular, fertilized egg. This single-celled zygote undergoes repeated cell divisions, eventually producing millions or trillions of cells that constitute a complete adult organism. Initially, each cell in the embryo is **totipotent**: it has the potential to develop into any cell type. Many cells in plants and fungi remain totipotent, but animal cells usually become committed to developing into specific types of cells after just a few early embryonic divisions. This commitment often comes well before a cell begins to exhibit any characteristics of a particular cell type; when the cell becomes committed, it cannot

reverse its fate and develop into a different cell type. A cell becomes committed by a process called **determination**.

For many years, the work of developmental biologists was limited to describing the changes that take place in the course of development, because techniques for probing the intracellular processes behind these changes were unavailable. But, in recent years, powerful genetic and molecular techniques have had a tremendous influence on the study of development. In a few model systems such as *Drosophila* and *Arabidopsis*, the molecular mechanisms underlying developmental change are now beginning to be understood.

If all cells in a multicellular organism are derived from the same original cell, how do different cell types arise? Before the 1950s, two hypotheses were considered. One possibility is that, throughout development, genes might be selectively lost or altered, causing different cell types to have different genomes. Alternatively, each cell might contain the same genetic information, but different genes might be expressed in each cell type. The results of early cloning experiments helped settle this issue.

Cloning Experiments on Plants

In the 1950s, Frederick Steward developed methods for cloning plants. He disrupted phloem tissue from the root of a carrot by separating and isolating individual cells. He then placed individual cells in a sterile medium that contained nutrients. Steward was successful in getting the cells to grow and divide, and, eventually, he obtained whole edible carrots from single cells (**Figure 22.1**). Because all parts of the plant

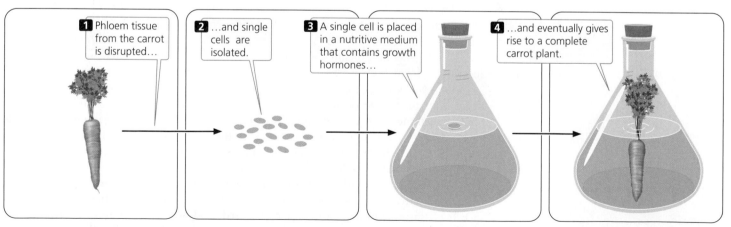

1 Phloem tissue from the carrot is disrupted...

2 ...and single cells are isolated.

3 A single cell is placed in a nutritive medium that contains growth hormones...

4 ...and eventually gives rise to a complete carrot plant.

22.1 Many kinds of plants can be cloned from isolated single cells. Thus none of the original genetic material is lost in development.

were regenerated from a specialized phloem cell, Steward concluded that each phloem cell contained the genetic potential for a whole plant; none of the original genetic material was lost during determination.

Cloning Experiments on Animals

The results of other studies demonstrated that most animal cells also retain a complete set of genetic information during development. In 1952, Robert Briggs and Thomas King removed the nuclei from unfertilized oocytes of the frog *Rana pipiens.* They then isolated nuclei from frog blastulas (an early embryonic stage) and injected these nuclei individually into the oocytes. The eggs were then pricked with a needle to stimulate them to divide. Although most were damaged in the process, a few eggs developed into complete tadpoles that eventually metamorphosed into frogs.

In the late 1960s, John Gurdon used these methods to successfully clone a few frogs with nuclei isolated from the intestinal cells of tadpoles. This accomplishment suggested that the differentiated intestinal cells carried the genetic information necessary to encode traits found in all other cells. However, Gurdon's successful clonings may have resulted from the presence of a few undifferentiated stem cells in the intestinal tissue, which were inadvertently used as the nuclei donors.

In 1997, researchers at the Roslin Institute of Scotland announced that they had successfully cloned a sheep by using the genetic material from a differentiated cell of an adult animal. To perform this experiment, they fused an udder cell from a white-faced Finn Dorset ewe with an enucleated egg cell and stimulated the egg electrically to initiate development. After growing the embryo in the laboratory for a week, they implanted it into a Scottish black-faced surrogate mother. Dolly, the first mammal cloned from an adult cell, was born on July 5, 1996 (**Figure 22.2**). Since the cloning of Dolly, a number of other animals including sheep, goats,

mice, rabbits, cows, pigs, horses, mules, dogs, and cats have been cloned from differentiated adult cells. Importantly, although Dolly and other mammals that have been cloned contain the same *nuclear* genetic material as that of their cloned parent, they are not identical for *cytoplasmic* genes, such as those on the mitochondrial chromosome, because the cytoplasm is donated by both the donor cell and the enucleated egg cell.

The cloning experiments demonstrated that genetic material is not lost or permanently altered during development: development must require the selective expression of genes. But how do cells regulate their gene expression in a coordinated manner to give rise to a complex, multicellular organism? Research has now begun to provide some answers to this important question.

CONCEPTS

The ability to clone plants and animals from single specialized cells demonstrates that genes are not lost or permanently altered during development.

✔ CONCEPT CHECK 1

Scientists have cloned some animals by injecting a nucleus from an early embryo into an enucleated egg cell. The resulting animals are genetically identical with the donor of the nucleus. Does this outcome prove that genetic material is not lost during development? Why or why not?

22.2 Pattern Formation in *Drosophila* Serves As a Model for the Genetic Control of Development

Pattern formation consists of the developmental processes that lead to the shape and structure of complex, multicellular organisms. One of the best-studied systems for the genetic control of pattern formation is the early embryonic development of *Drosophila melanogaster*. Geneticists have isolated a large number of mutations in fruit flies that influence all aspects of their development, and these mutations have been subjected to molecular analysis, providing much information about how genes control early development in *Drosophila*.

The Development of the Fruit Fly

An adult fruit fly possesses three basic body parts: head, thorax, and abdomen (**Figure 22.3**). The thorax consists of three segments: the first thoracic segment carries a pair of legs; the second thoracic segment carries a pair of legs and a pair of wings; and the third thoracic segment carries a pair of legs and the halteres (rudiments of the second pair of wings found in most other insects). The abdomen consists of a number of segments.

22.2 In 1996, researchers at the Roslin Institute of Scotland successfully cloned a sheep named Dolly. They used the genetic material from a differentiated cell of an adult animal. [Paul Clements/AP.]

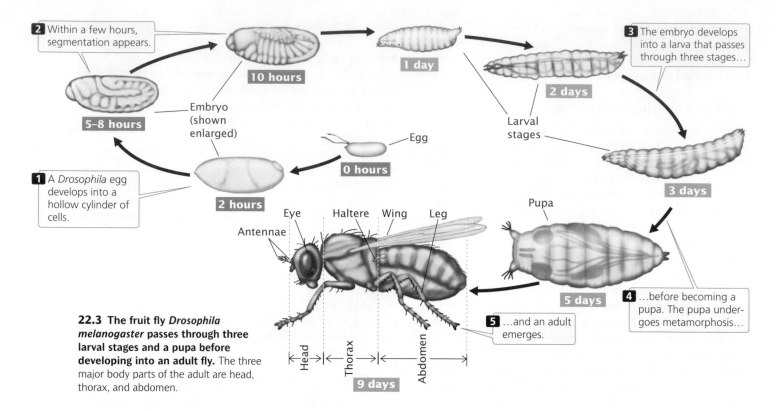

2 Within a few hours, segmentation appears.

10 hours

1 day

3 The embryo develops into a larva that passes through three stages…

5–8 hours

2 days

Embryo (shown enlarged)

Larval stages

Egg

0 hours

1 A *Drosophila* egg develops into a hollow cylinder of cells.

2 hours

3 days

Antennae

Eye   Haltere   Wing   Leg

Pupa

4 …before becoming a pupa. The pupa undergoes metamorphosis…

5 days

5 …and an adult emerges.

22.3 The fruit fly *Drosophila melanogaster* passes through three larval stages and a pupa before developing into an adult fly. The three major body parts of the adult are head, thorax, and abdomen.

Head   Thorax   Abdomen

9 days

When a *Drosophila* egg has been fertilized, its diploid nucleus (**Figure 22.4a**) immediately divides nine times without division of the cytoplasm, creating a single, multinucleate cell (**Figure 22.4b**). These nuclei are scattered throughout the cytoplasm but later migrate toward the periphery of the embryo and divide several more times (**Figure 22.4c**). Next, the cell membrane grows inward and around each nucleus, creating a layer of approximately 6000 cells at the outer surface of the embryo (**Figure 22.4d**). Nuclei at one end of the embryo develop into pole cells, which eventually give rise to germ cells. The early embryo then undergoes further development in three distinct stages: (1) the anterior–posterior axis and the dorsal–ventral axis of the embryo are established (**Figure 22.5a**); (2) the number and orientation of the body segments are determined (**Figure 22.5b**); and (3) the identity of each individual segment is established (**Figure 22.5c**). Different sets of genes control each of these three stages (**Table 22.1**).

Egg-Polarity Genes

The **egg-polarity genes** play a crucial role in establishing the two main axes of development in fruit flies. You can think of these axes as the longitude and latitude of development: any location in the *Drosophila* embryo can be defined in relation to these two axes.

The egg-polarity genes are transcribed into mRNAs in the course of egg formation in the maternal parent, and these mRNAs become incorporated into the cytoplasm of the egg. After fertilization, the mRNAs are translated into proteins that play an important role in determining the anterior–posterior and dorsal–ventral axes of the embryo. Because the mRNAs of the polarity genes are produced by the female parent and influence the phenotype of the offspring, the traits encoded by them are examples of genetic maternal effects (see p. 119–120 in Chapter 5).

There are two sets of egg-polarity genes: one set determines the anterior–posterior axis, and the other determines the dorsal–ventral axis. These genes work by setting up concentration gradients of morphogens within the developing embryo. A **morphogen** is a protein that varies in concentration and elicits different developmental responses at different concentrations. Egg-polarity genes function by producing proteins that become asymmetrically distributed in the cytoplasm, giving the egg polarity, or direction. This asymmetrical distribution may take place in a couple of ways. An mRNA may be localized to particular regions of the egg

Table 22.1	Stages in the early development of fruit flies and the genes that control each stage
Developmental Stage	**Genes**
Establishment of main body axes	Egg-polarity genes
Determination of number and polarity of body segments	Segmentation genes
Establishment of identity of each segment	Homeotic genes

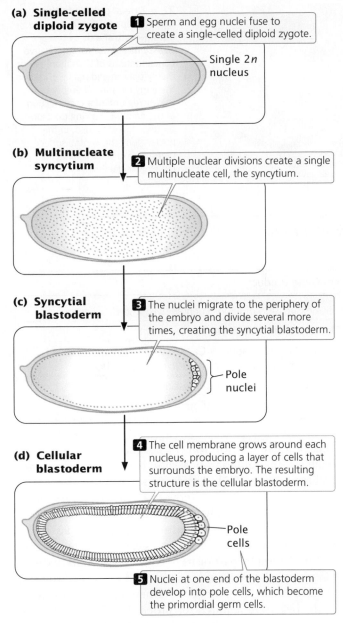

(a) Single-celled diploid zygote

1 Sperm and egg nuclei fuse to create a single-celled diploid zygote.

Single 2*n* nucleus

(b) Multinucleate syncytium

2 Multiple nuclear divisions create a single multinucleate cell, the syncytium.

(c) Syncytial blastoderm

3 The nuclei migrate to the periphery of the embryo and divide several more times, creating the syncytial blastoderm.

Pole nuclei

4 The cell membrane grows around each nucleus, producing a layer of cells that surrounds the embryo. The resulting structure is the cellular blastoderm.

(d) Cellular blastoderm

Pole cells

5 Nuclei at one end of the blastoderm develop into pole cells, which become the primordial germ cells.

22.4 Early development of a *Drosophila* embryo.

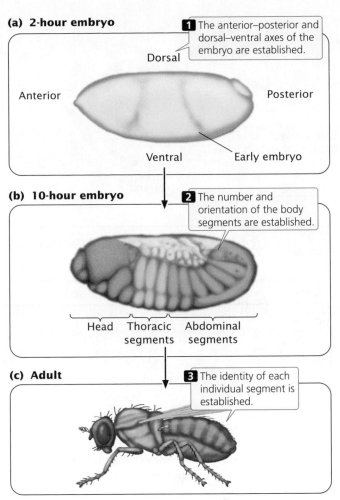

(a) 2-hour embryo

1 The anterior–posterior and dorsal–ventral axes of the embryo are established.

Dorsal

Anterior — Posterior

Ventral — Early embryo

(b) 10-hour embryo

2 The number and orientation of the body segments are established.

Head — Thoracic segments — Abdominal segments

(c) Adult

3 The identity of each individual segment is established.

22.5 In an early *Drosophila* embryo, the major body axes are established, the number and orientation of the body segments are determined, and the identity of each individual segment is established. Different sets of genes control each of these three stages.

cell, leading to an abundance of the protein in those regions when the mRNA is translated. Alternatively, the mRNA may be randomly distributed, but the protein that it encodes may become asymmetrically distributed by a transport system that delivers it to particular regions of the cell, by regulation of its translation, or by its removal from particular regions by selective degradation.

Determination of the dorsal–ventral axis

The dorsal–ventral axis defines the back (dorsum) and belly (ventrum) of a fly (see Figure 22.5). At least 12 different genes determine this axis, one of the most important being a gene called *dorsal*. The *dorsal* gene is transcribed and translated in the maternal ovary, and the resulting mRNA and protein are transferred to the egg during oogenesis. In a newly laid egg, mRNA and protein encoded by the *dorsal* gene are uniformly distributed throughout the cytoplasm but, after the nuclei have migrated to the periphery of the embryo (see Figure 22.4c), Dorsal protein becomes redistributed. Along one side of the embryo, Dorsal protein remains in the cytoplasm; this side will become the dorsal surface. Along the other side, Dorsal protein is taken up into the nuclei; this side will become the ventral surface. At this point, there is a smooth gradient of increasing nuclear Dorsal concentration from the dorsal to the ventral side (**Figure 22.6**).

The nuclear uptake of Dorsal protein is thought to be governed by a protein called Cactus, which binds to Dorsal protein and traps it in the cytoplasm. The presence of yet another protein, called Toll, leads to the phosphylation of Cactus, causing it to be degraded. When Cactus is degraded, Dorsal is released and can move into the nucleus. Together,

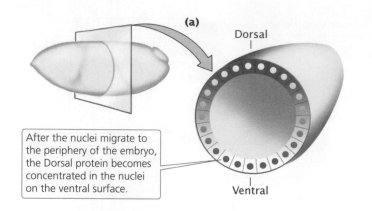

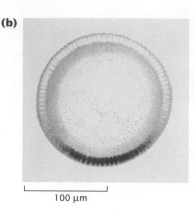

22.6 Dorsal protein in the nuclei helps to determine the dorsal–ventral axis of the *Drosophila* embryo. (a) Relative concentrations of Dorsal protein in the cytoplasm and nuclei of cells in the early *Drosophila* embryo. (b) Micrograph of a cross section of the embryo showing the Dorsal protein, darkly stained, in the nuclei along the ventral surface. [Part b: Max Planck Institute for Developmental Biology.]

Dorsal

After the nuclei migrate to the periphery of the embryo, the Dorsal protein becomes concentrated in the nuclei on the ventral surface.

Ventral

100 μm

Table 22.2	Key genes that control the development of the dorsal–ventral axis in fruit flies and their action	
Gene	Where Expressed	Action of Gene Product
dorsal	Ovary	Affects the expression of genes such as *twist* and *decapentaplegic*
cactus	Ovary	Traps Dorsal protein in the cytoplasm
toll	Ovary	Leads to the phosphorylation of Cactus, which is then degraded, releasing Dorsal to move into the nuclei of ventral cells
twist	Embryo	Takes part in the development of mesodermal tissues
decapentaplegic	Embryo	Takes part in the development of gut structures

Cactus and Toll regulate the nuclear distribution of Dorsal protein, which in turn determines the dorsal–ventral axis of the embryo.

Inside the nucleus, Dorsal protein acts as a transcription factor, binding to regulatory sites on the DNA and activating or repressing the expression of other genes (**Table 22.2**). High nuclear concentration of Dorsal protein (as in cells on the ventral side of the embryo) activates a gene called *twist*, which causes ventral tissues to develop. Low nuclear concentrations of Dorsal protein (as in cells on the dorsal side of the embryo), activate a gene called *decapentaplegic*, which specifies dorsal structures. In this way, the ventral and dorsal sides of the embryo are determined.

Determination of the anterior–posterior axis One of the most important early developmental events is the determination of the anterior (head) and posterior (butt)

ends of an animal. We will consider several key genes that establish this anterior–posterior axis of the *Drosophila* embryo (**Table 22.3**). An important gene in this regard is *bicoid*, which is first transcribed in the ovary of an adult female during oogenesis. The *bicoid* mRNA becomes incorporated into the cytoplasm of the egg; as it passes into the egg, *bicoid* mRNA becomes anchored to the anterior end of the egg by part of its 3′ end. This anchoring causes *bicoid* mRNA to become concentrated at the anterior end (**Figure 22.7a**). (A number of other genes that are active in the ovary are required for proper localization of *bicoid* mRNA in the egg.) When the egg has been laid, *bicoid* mRNA is translated into Bicoid protein. Because most of the mRNA is at the anterior end of the egg, Bicoid protein is synthesized there and forms a concentration gradient along the anterior–posterior axis of the embryo, with a high concentration at the anterior end and a low concen-

Table 22.3	Some key genes that determine the anterior–posterior axis in fruit flies	
Gene	Where Expressed	Action
bicoid	Ovary	Regulates expression of genes responsible for anterior structures; stimulates *hunchback*
nanos	Ovary	Regulates expression of genes responsible for posterior structures; inhibits translation of *hunchback* mRNA
hunchback	Embryo	Regulates transcription of genes responsible for anterior structures

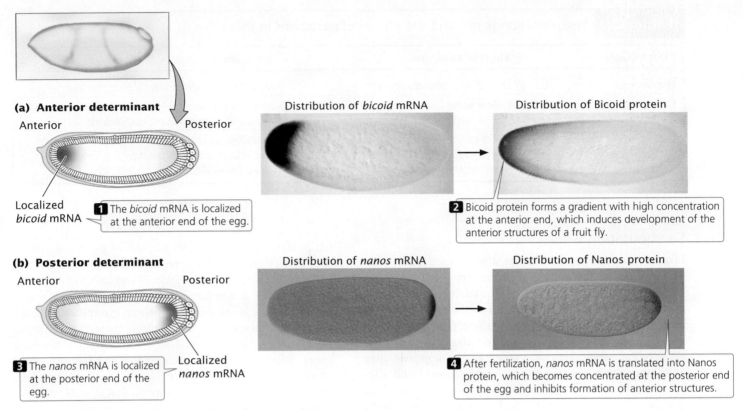

(a) Anterior determinant

Anterior   Posterior

Localized
bicoid mRNA

1 The *bicoid* mRNA is localized at the anterior end of the egg.

Distribution of *bicoid* mRNA

Distribution of Bicoid protein

2 Bicoid protein forms a gradient with high concentration at the anterior end, which induces development of the anterior structures of a fruit fly.

(b) Posterior determinant

Anterior   Posterior

Localized
nanos mRNA

3 The *nanos* mRNA is localized at the posterior end of the egg.

Distribution of *nanos* mRNA

Distribution of Nanos protein

4 After fertilization, *nanos* mRNA is translated into Nanos protein, which becomes concentrated at the posterior end of the egg and inhibits formation of anterior structures.

22.7 The anterior–posterior axis in a *Drosophila* embryo is determined by concentrations of Bicoid and Nanos proteins. [Part a: From Christiane Nüsslein-Volhard, Determination of the embryonic axes of *Drosophila*, *Development* (Suppl.) 1:1, 1991. Part b: Courtesy of E. R. Gavis, L. K. Dickinson, and R. Lehmann, Massachusetts Institute of Technology.]

tration at the posterior end. This gradient is maintained by the continuous synthesis of Bicoid protein and its short half-life.

The high concentration of Bicoid protein at the anterior end induces the development of anterior structures such as the head of the fruit fly. Bicoid—like Dorsal—is a morphogen. It stimulates the development of anterior structures by binding to regulatory sequences in the DNA and influencing the expression of other genes. One of the most important of the genes stimulated by Bicoid protein is *hunchback*, which is required for the development of the head and thoracic structures of the fruit fly.

The development of the anterior–posterior axis is also greatly influenced by a gene called *nanos*, an egg-polarity gene that acts at the posterior end of the axis. The *nanos* gene is transcribed in the adult female, and the resulting mRNA becomes localized at the posterior end of the egg (**Figure 22.7b**). After fertilization, *nanos* mRNA is translated into Nanos protein, which diffuses slowly toward the anterior end. The Nanos protein gradient is opposite that of the Bicoid protein: Nanos is most concentrated at the posterior end of the embryo and is least concentrated at the anterior end. Nanos protein inhibits the formation of anterior structures by repressing the translation of *hunchback* mRNA. The

synthesis of the Hunchback protein is therefore stimulated at the anterior end of the embryo by Bicoid protein and is repressed at the posterior end by Nanos protein. This combined stimulation and repression results in a Hunchback protein concentration gradient along the anterior–posterior axis that, in turn, affects the expression of other genes and helps determine the anterior and posterior structures.

CONCEPTS

The major axes of development in early fruit-fly embryos are established as a result of initial differences in the distribution of specific mRNAs and proteins encoded by genes in the female parent (genetic maternal effect). These differences in distribution establish concentration gradients of morphogens, which cause different genes to be activated in different parts of the embryo.

✔ CONCEPT CHECK 2

High concentration of which protein stimulates the development of anterior structures?

a. Dorsal   c. Bicoid

b. Toll   d. Nanos

Table 22.4   Segmentation genes and the effects of mutations in them

Class of Gene	Effect of Mutations	Examples of Genes
Gap genes	Delete adjacent segments	*hunchback, Krüppel, knirps, giant, tailless*
Pair-rule genes	Delete same part of pattern in every other segment	*runt, hairy, fushi tarazu, even-paired, odd-paired, skipped, sloppy, paired, odd-skipped*
Segment-polarity genes	Affect polarity of segment; part of segment replaced by mirror image of part of another segment	*engrailed, wingless, gooseberry, cubitus interruptus, patched, hedgehog, disheveled, costal-2, fused*

Segmentation Genes

Like all insects, the fruit fly has a segmented body plan. When the basic dorsal–ventral and anterior–posterior axes of the fruit-fly embryo have been established, **segmentation genes** control the differentiation of the embryo into individual segments. These genes affect the number and organization of the segments, and mutations in them usually disrupt whole sets of segments. The approximately 25 segmentation genes in *Drosophila* are transcribed after fertilization; so they don't exhibit a genetic maternal effect, and their expression is regulated by the Bicoid and Nanos protein gradients.

The segmentation genes fall into three groups as shown in **Table 22.4** and **Figure 22.8**. The three groups of genes act sequentially, affecting progressively smaller regions of the embryo. First, the products of the egg-polarity genes activate or repress **gap genes**, which divide the embryo into broad regions. The gap genes, in turn, regulate **pair-rule genes**, which affect the development of pairs of segments. Finally, the pair-rule genes influence **segment-polarity genes**, which guide the development of individual segments.

Gap genes define large sections of the embryo; mutations in these genes eliminate whole groups of adjacent segments. Mutations in the *Krüppel* gene, for example, cause the absence of several adjacent segments. Pair-rule genes define regional sections of the embryo and affect alternate segments. Mutations in the *even-skipped* gene cause the deletion of even-numbered segments, whereas mutations in the *fushi tarazu* gene cause the absence of odd-numbered segments. Segment-polarity genes affect the organization of segments. Mutations in these genes cause part of each segment to be deleted and replaced by a mirror image of part or all of an adjacent segment. For example, mutations in the *gooseberry* gene cause the posterior half of each segment to be replaced by the anterior half of an adjacent segment.

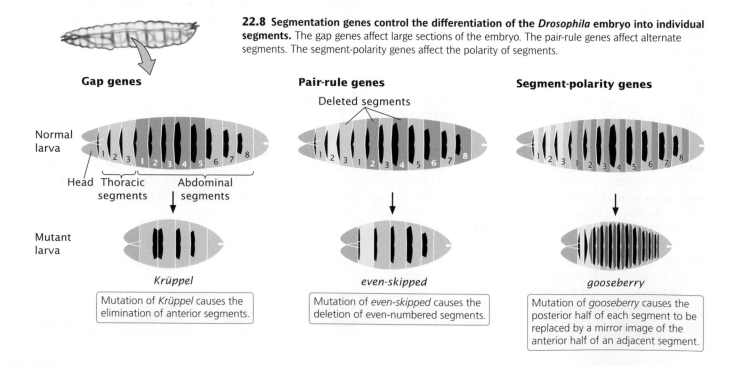

22.8 Segmentation genes control the differentiation of the *Drosophila* embryo into individual segments. The gap genes affect large sections of the embryo. The pair-rule genes affect alternate segments. The segment-polarity genes affect the polarity of segments.

Gap genes

Normal larva

Head   Thoracic segments   Abdominal segments

Mutant larva

Krüppel

Mutation of *Krüppel* causes the elimination of anterior segments.

Pair-rule genes

Deleted segments

even-skipped

Mutation of *even-skipped* causes the deletion of even-numbered segments.

Segment-polarity genes

gooseberry

Mutation of *gooseberry* causes the posterior half of each segment to be replaced by a mirror image of the anterior half of an adjacent segment.

When the major axes of the fruit-fly embryo have been established, segmentation genes determine the number, orientation, and basic organization of the body segments.

✔ CONCEPT CHECK 3

The correct sequence in which the segmentation genes act is:

a. segment-polarity genes ⟶ gap genes ⟶ pair-rule genes.

b. gap genes ⟶ pair-rule genes ⟶ segment-polarity genes.

c. segment-polarity genes ⟶ pair-rule genes ⟶ gap genes.

d. gap genes ⟶ segment-polarity genes ⟶ pair-rule genes.

Homeotic Genes in *Drosophila*

After the segmentation genes have established the number and orientation of the segments, **homeotic genes** become active and determine the *identity* of individual segments. Eyes normally arise only on the head segment, whereas legs develop only on the thoracic segments. The products of homeotic genes activate other genes that encode these segment-specific characteristics. Mutations in the homeotic genes cause body parts to appear in the wrong segments.

In the late 1940s, Edward Lewis began to study homeotic mutations in *Drosophila*—mutations that cause bizarre rearrangements of body parts. Mutations in the *Antennapedia* gene, for example, cause legs to develop on the head of a fly in place of the antenna (**Figure 22.9**). Homeotic genes create addresses for the cells of particular segments, telling the cells where they are within the regions defined by the segmentation genes. When a homeotic gene is mutated, the address is wrong and cells in the segment develop as though they were somewhere else in the embryo.

Homeotic genes in *Drosophila* are expressed after fertilization and are activated by specific concentrations of the proteins produced by the gap, pair-rule, and segment-polarity genes. The homeotic gene *Ultrabithorax* (*Ubx*), for example, is activated when the concentration of Hunchback protein (a product of a gap gene) is within certain values. These concentrations exist only in the middle region of the embryo; so *Ubx* is expressed only in these segments.

The homeotic genes in animals encode regulatory proteins that bind to DNA; each gene contains a subset of nucleotides, called a **homeobox**, that are similar in all homeotic genes. The homeobox encodes 60 amino acids that serve as a DNA-binding domain; this domain is related to the helix-turn-helix motif (see Figure 16.2a). Homeoboxes are also present in segmentation genes and other genes that play a role in spatial development.

There are two major clusters of homeotic genes in *Drosophila*. One cluster, the ***Antennapedia* complex**, affects the development of the adult fly's head and anterior thoracic segments. The other cluster consists of the ***bithorax* complex** and includes genes that influence the adult fly's posterior thoracic and abdominal segments. Together, the *bithorax* and *Antennapedia* genes are termed the **homeotic complex** (HOM-C). In *Drosophila*, the *bithorax* complex comprises three genes, and the *Antennapedia* complex has five; all are located on the same chromosome (**Figure 22.10**). In addition to these eight genes, HOM-C contains many sequences that regulate the homeotic genes. Remarkably, the order of the genes in the HOM-C is the same as the order in which the genes are expressed along the anterior–posterior axis of the body. The genes that are expressed in the more-anterior segments are found at one end of the complex, whereas those expressed in the more-posterior end of the embryo are found at the other end of the complex (see Figure 22.10).

Homeotic genes help determine the identity of individual segments in *Drosophila* embryos by producing DNA-binding proteins that activate other genes. Each homeotic gene contains a consensus sequence called a homeobox, which encodes the DNA-binding domain.

✔ CONCEPT CHECK 4

Mutations in homeotic genes often cause

a. the deletion of segments. b. the absence of structures.

c. too many segments. d. structures to appear in the wrong place.

(a) 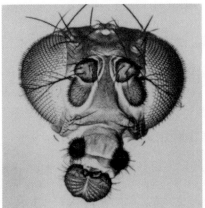 **(b)**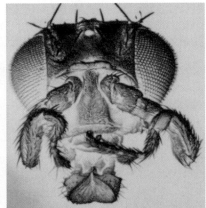

22.9 The homeotic mutation *Antennapedia* substitutes legs for the antennae of a fruit fly. (a) Normal, wild-type antennae. (b) *Antennapedia* mutant. [F. R. Turner/Biological Photo Services.]

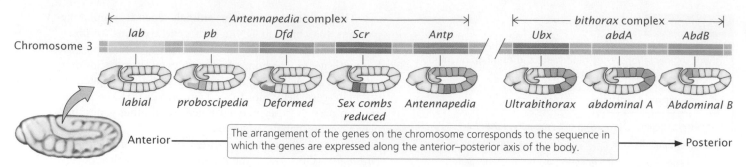

22.10 Homeotic genes, which determine the identity of individual segments in *Drosophila*, are present in two complexes. The *Antennapedia* complex has five genes, and the *bithorax* complex has three genes.

Homeobox Genes in Other Organisms

After homeotic genes in *Drosophila* had been isolated and cloned, molecular geneticists set out to determine if similar genes exist in other animals; probes complementary to the homeobox of *Drosophila* genes were used to search for homologous genes that might play a role in the development of other animals. The search was hugely successful: homeobox-containing genes have been found in all animals, including nematodes, beetles, sea urchins, frogs, birds, and mammals. Genes with homeoboxes have even been discovered in fungi and plants, indicating that the homeobox arose early in the evolution of eukaryotes. One group of homeobox genes comprises the **Hox genes**, which include the homeotic complex of *Drosophila* described in the preceding section. *Hox* genes are found in all animals except sponges.

In vertebrates, there are usually four clusters of *Hox* genes, each of which contains from 9 to 11 genes. Mammalian *Hox* genes, like those in *Drosophila*, encode transcription factors that help determine the identity of body regions along an anterior–posterior axis. The *Hox* genes of other organisms often exhibit the same relation between order on the chromosome and order of their expression along the anterior–posterior axis of the embryo as that of *Drosophila* (**Figure 22.11**), but this pattern is not universal. For exam-

ple, the tunicate *Oikopleura duiuca* (a primitive relative of vertebrates) has 9 *Hox* genes, but the genes are scattered throughout the genome, unlike the clustered arrangement seen in most animals. Despite a lack of physical clustering, the *Hox* genes in *O. duiuca* are expressed in the same anterior–posterior order as that seen in vertebrates.

The *Hox* genes of vertebrates also exhibit a relation between their order on the chromosome and the timing of their expression: genes at one end of the complex (those expressed at the anterior end) are expressed early in development, whereas genes at the other end (those expressed at the posterior end) are expressed later. If a *Hox* gene is experimentally moved to a new location within the *Hox*-gene complex, it is expressed in the appropriate tissue but the timing of its expression is altered, suggesting that the timing of gene expression is controlled by its physical location within the complex. Although the mechanism of this sequential control is not well understood, evidence suggests that, in mice, it is controlled by a progressive change in the methylation patterns of histone proteins, an epigenetic change that alters chromatin structure and affects transcription. Recent studies have also identified microRNA genes (see Chapter 14) within the *Hox*-gene clusters, and evidence suggests that miRNAs play a role in controlling the expression of some *Hox* genes.

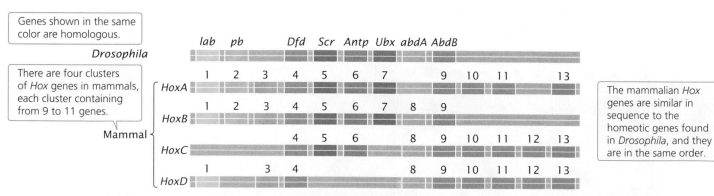

22.11 Homeotic genes in mammals are similar to those found in *Drosophila*. The complexes are arranged so that genes with similar sequences lie in the same column. See Figure 22.10 for the full names of the *Drosophila* genes.

Hox genes and their expression are often correlated with anatomical changes in animals, and *Hox* genes have been hypothesized to play an important role in the evolution of animals. For example, the lancet *Branchiostoma* (another primitive relative of vertebrates) has a simple body form and possesses only 14 *Hox* genes in a single cluster, whereas some fishes, with a much more complex body architecture, have as many as 48 *Hox* genes in seven clusters.

CONNECTING CONCEPTS

The Control of Development

Development is a complex process consisting of numerous events that must take place in a highly specific sequence. The results of studies in fruit flies and other organisms reveal that this process is regulated by a large number of genes. In *Drosophila,* the dorsal–ventral axis and the anterior–posterior axis are established by maternal genes; these genes encode mRNAs and proteins that are localized to specific regions within the egg and cause specific genes to be expressed in different regions of the embryo. The proteins of these genes then stimulate other genes, which in turn stimulate yet other genes in a cascade of control. As might be expected, most of the gene products in the cascade are regulatory proteins, which bind to DNA and activate other genes.

In the course of development, successively smaller regions of the embryo are determined (**Figure 22.12**). In *Drosophila,* first, the major

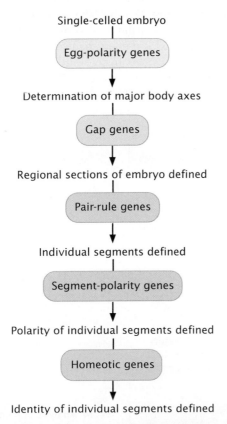

22.12 A cascade of gene regulation establishes the polarity and identity of individual segments of *Drosophila*. In development, successively smaller regions of the embryo are determined.

axes and regions of the embryo are established by egg-polarity genes. Next, patterns within each region are determined by the action of segmentation genes: the gap genes define large sections, the pair-rule genes define regional sections of the embryo and affect alternate segments, and the segment-polarity genes affect individual segments. Finally, the homeotic genes provide each segment with a unique identity. Initial gradients in proteins and mRNA stimulate localized gene expression, which produces more finely located gradients that stimulate even more localized gene expression. Developmental regulation thus becomes more and more narrowly defined.

The processes by which limbs, organs, and tissues form (called morphogenesis) are less well understood, although this pattern of generalized-to-localized gene expression is encountered frequently.
TRY PROBLEM 20 →

Epigenetic Changes in Development

Early development of the fruit fly is controlled in large part by the products of certain key genes selectively activating or repressing the expression of other genes. As discussed in Chapter 17, gene expression in eukaryotes is affected by epigenetic changes, and indeed epigenetics also plays an important role in development.

In Chapter 17, epigenetic changes were defined as heritable alterations to DNA and chromatin structure—alterations that affect gene expression and are passed on to other cells but are not changes to the DNA base sequence. In the course of development, a single-cell zygote divides and gives rise to many cells, which differentiate and acquire the characteristics of specific organs and tissues. Each type of cell eventually expresses a different subset of genes, producing the proteins needed for that cell type, and this program of gene expression is passed on when the differentiated cell divides.

The gene-expression program of cells that make up a particular organ or tissue type is often defined by epigenetic marks. As development and differentiation proceed, cells acquire epigenetic changes that turn specific sets of genes on and off. In Chapter 17, we considered several types of epigenetic marks, including DNA methylation, the modification of histone proteins, and chromatin remodeling. These epigenetic changes help determine which genes are expressed by a cell. In early stages of development, genes that may be required at later stages of development are often held in a transiently silent state by histone modifications. Histone modifications are generally flexible and easily reversed, and so these genes can be activated later in developmental stages. In the course of development, other genes are permanently silenced; this more-long-term silencing is often accomplished by DNA methylation.

22.3 Genes Control the Development of Flowers in Plants

One of the most important developmental events in the life of a plant is the switch from vegetative growth to flowering. The precise timing of this switch is affected by season,

22.13 The flower produced by *Arabidopsis thaliana* has four sepals, four white petals, six stamens, and two carpels. [Darwin Dale/Photo Researchers.]

day length, plant size, and a number of other factors and is under the control of a large number of different genes. The development of the flower itself also is under genetic control, and homeotic genes play a crucial role in the determination of the floral structures.

Flower Anatomy

A flower is made up of four concentric rings of modified leaves, called whorls. The outermost whorl (whorl 1) consists of the green leaflike sepals. The next whorl (whorl 2) consists of the petals, which typically lack chlorophyll. Whorl 3 consists of the stamens, which bear pollen; and whorl 4 consists of carpels that are often fused to form the stigma bearing the ovules. In wild-type *Arabidopsis*, a model genetic plant (see the Reference Guide to Model Genetic Organisms and **Figure 22.13**), there are four sepals, four white petals, six stamens (four long and two short), and two carpels (**Figure 22.14a**).

Genetic Control of Flower Development

Elliot Meyerowitz and his colleagues conducted a series of experiments to examine the genetic basis of flower development in *Arabidopsis*. They began by isolating and analyzing homeotic mutations in *Arabidopsis*. Homeotic mutations were actually first identified in plants, in 1894, when William Bateson noticed that the floral parts of plants occasionally appeared in the wrong place: he found, for example, flowers in which stamens grew in the place where petals normally grow.

Meyerowitz and his coworkers used these types of mutants to reveal the genes that control flower development. They were able to place the homeotic mutations that they isolated into three groups on the basis of their effect on floral structure. Class A mutants had carpels instead of sepals in the first whorl and stamens instead of petals in the second whorl (**Figure 22.14b**). The third whorl consisted of stamens, and the fourth whorl consisted of carpels, the normal pattern. Class B mutants had sepals in the first and second whorls and carpels in the third and fourth whorls (**Figure 22.14c**). The final group, class C mutants, had sepals and petals in the first and second whorls, respectively, as is

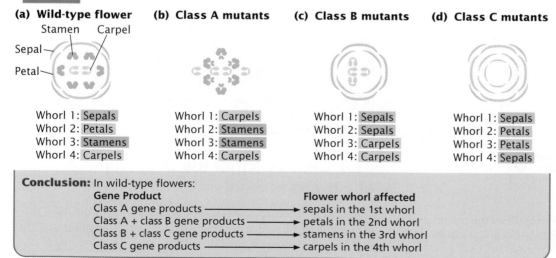

Experiment

Question: How do genes control the development of flower structures?

Methods Isolate and analyze homeotic mutants that affect flower development.

Results

(a) Wild-type flower

Stamen   Carpel

Sepal

Petal

Whorl 1: Sepals
Whorl 2: Petals
Whorl 3: Stamens
Whorl 4: Carpels

(b) Class A mutants

Whorl 1: Carpels
Whorl 2: Stamens
Whorl 3: Stamens
Whorl 4: Carpels

(c) Class B mutants

Whorl 1: Sepals
Whorl 2: Sepals
Whorl 3: Carpels
Whorl 4: Carpels

(d) Class C mutants

Whorl 1: Sepals
Whorl 2: Petals
Whorl 3: Petals
Whorl 4: Sepals

Conclusion: In wild-type flowers:

Gene Product	Flower whorl affected
Class A gene products	→ sepals in the 1st whorl
Class A + class B gene products	→ petals in the 2nd whorl
Class B + class C gene products	→ stamens in the 3rd whorl
Class C gene products	→ carpels in the 4th whorl

22.14 Analysis of homeotic mutants in *Arabidopsis thaliana* led to an understanding of the genes that determine floral structures in plants.

normal, but had petals in the third whorl and sepals in the fourth whorl (**Figure 22.14d**).

Meyerowitz and his colleagues concluded that each class of mutants was missing the product of a gene or the products of a set of genes that are critical to proper flower development: class A mutants were missing gene A activity, class B mutants were missing gene B activity, and class C mutants were missing gene C activity. They hypothesized that the class A genes are active in the first and second whorls. Class A gene products alone cause the first whorl to differentiate into sepals. Class A gene products together with Class B gene products cause the second whorl to develop into petals. Class C gene products together with Class B gene products induce the third whorl to develop into stamens. Class C genes alone cause the fourth whorl to become carpels. The products of the different gene classes and their effects are summarized in the conclusion of Figure 22.14.

To explain the results, they also proposed that the genes of some classes affect the activities of others. Where class A is active, class C is repressed, and where class C is active, class A is repressed. Additionally, if a mutation inactivates A, then C becomes active and vice versa. Class A genes are normally expressed in whorls 1 and 2, class B genes are expressed in whorls 2 and 3, and class C genes are expressed in whorls 3 and 4 (**Figure 22.15**).

The interaction of these three classes of genes explains the different classes of mutants in Figure 22.14. For example, class A mutants are lacking class A gene products, and therefore class C genes are active in all tissues because, when A is inactivated, C becomes active. Therefore whorl 1, with only class C gene products, will consist of carpels; whorl 2, with class C and class B gene products, will produce stamens; whorl 3, with class B and class C gene products, will produce stamens; and whorl 4, with only class C gene activity, will produce carpels (see Figure 22.14b):

Class C (in the absence of class A)
    gene products   → carpels (1st whorl)
Class B + class C (in the absence
    of class A) gene products   → stamens (2nd whorl)
Class B + class C gene products   → stamens (3rd whorl)
Class C gene products   → carpels (4th whorl)

To confirm this explanation, Meyerowitz and his colleagues bred double and triple mutants and predicted the outcome. The resulting flower structures fit their predictions. In subsequent

studies, they isolated the genes of each class. There are two class A genes, termed *APETALA1* (*AP1*) and *APETALA2* (*AP2*); two class B genes, termed *APETALA3* (*AP3*) and *PISTILLATA* (*PI*); and one class C gene termed *AGAMOUS* (*AG*). The cloning and sequencing of these genes revealed that all are MADS-box genes—genes that function as transcription factors and affect the expression of other genes. MADS-box genes in plants play a similar role to that of homeobox genes in animals, although MADS-box genes and homeobox genes are not homologous.

The results of other studies have demonstrated the presence of an additional group of genes called *SEPALLATA* (*SEP*) that are expressed in whorls 2, 3, and 4, and they, too, are required for normal floral development. If the *SEP* genes are defective, the flower consists entirely of sepals. Findings from studies of other species have demonstrated that this system of flower development exists not only in *Arabidopsis* but also in other flowering plants. It is important to note that these genes are necessary but not sufficient for proper flower development; other genes also take part in the identity of the different parts of flowers. TRY PROBLEM 23 →

22.4 Programmed Cell Death Is an Integral Part of Development

Cell death is an important part of multicellular life. Cells in many tissues have a limited life span, and they die and are replaced continually by new cells. Cell death shapes many body parts in the course of development: it is responsible for the disappearance of a tadpole's tail during metamorphosis and causes the removal of tissue between the digits to produce the human hand. As discussed in the introduction to this chapter, the death of lens cells causes the absence of eyes in blind cavefish. Cell death is also used to eliminate dangerous cells that have escaped normal controls (see section on mutations in cell-cycle control and cancer in Chapter 23).

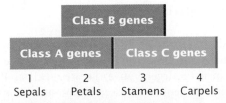

	Class B genes		
Class A genes		**Class C genes**	
1	2	3	4
Sepals	Petals	Stamens	Carpels

22.15 Expression of class A, B, and C genes varies among the structures of a flower.

Apoptosis Cell death in animals is often initiated by the cell itself in **apoptosis**, or cellular suicide. In this process, a

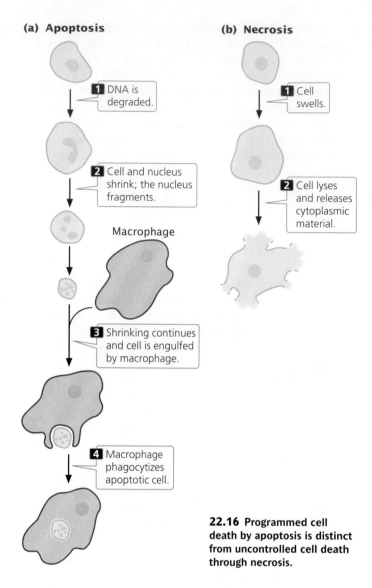

(a) Apoptosis

1 DNA is degraded.

2 Cell and nucleus shrink; the nucleus fragments.

Macrophage

3 Shrinking continues and cell is engulfed by macrophage.

4 Macrophage phagocytizes apoptotic cell.

(b) Necrosis

1 Cell swells.

2 Cell lyses and releases cytoplasmic material.

22.16 Programmed cell death by apoptosis is distinct from uncontrolled cell death through necrosis.

cell's DNA is degraded, its nucleus and cytoplasm shrink, and the cell undergoes phagocytosis by other cells without any leakage of its contents (**Figure 22.16a**). Cells that are injured, on the other hand, die in a relatively uncontrolled manner called *necrosis*. In this process, a cell swells and bursts, spilling its contents over neighboring cells and eliciting an inflammatory response (**Figure 22.16b**). Apoptosis is essential to embryogenesis; most multicellular animals cannot complete development if the process is inhibited.

Regulation of apoptosis Surprisingly, most cells are programmed to undergo apoptosis and will survive only if the internal death program is continually held in check. The process of apoptosis is highly regulated and depends on numerous signals inside and outside the cell. Geneticists have identified a number of genes having roles in various stages of the regulation of apoptosis. Some of these genes encode enzymes called **caspases**, which cleave other proteins at specific sites (after aspartic acid). Each caspase is

synthesized as a large, inactive precursor (a procaspase) that is activated by cleavage, often by another caspase. When one caspase is activated, it cleaves other procaspases that trigger even more caspase activity. The resulting cascade of caspase activity eventually cleaves proteins essential to cell function, such as those supporting the nuclear membrane and cytoskeleton. Caspases also cleave a protein that normally keeps an enzyme that degrades DNA (DNase) in an inactive form. Cleavage of this protein activates DNase and leads to the breakdown of cellular DNA, which eventually leads to cell death.

Procaspases and other proteins required for cell death are continuously produced by healthy cells, and so the potential for cell suicide is always present. A number of different signals can trigger apoptosis; for instance, infection by a virus can activate immune cells to secrete substances onto an infected cell, causing that cell to undergo apoptosis. This process is believed to be a defense mechanism designed to prevent the reproduction and spread of viruses. Similarly, DNA damage can induce apoptosis and thus prevent the replication of mutated sequences. Damage to mitochondria and the accumulation of a misfolded protein in the endoplasmic reticulum also stimulate programmed cell death.

Apoptosis in development Apoptosis plays a critical role in development. As animals develop, excess cells are often produced and then later culled by apoptosis to produce the proper number of cells required for an organ or a tissue. In some cases, whole structures are created that are later removed by apoptosis. For example, early mammalian embryos develop both male and female reproductive ducts, but the Wolffian ducts degenerate in females and the Mullerian ducts degenerate in males.

During embryonic development in *Drosophila*, large numbers of cells die. Three genes in *Drosophila* activate caspases that are essential for apoptosis: *reaper* (*rpg*), *grim*, and *head involution defective* (*hid*). Embryos possessing a deletion that removes all three genes exhibit no apoptosis and die in the course of embryogenesis with an excess of cells. Numerous other genes also affect the process of apoptosis.

Apoptosis is also crucial in metamorphosis, the process by which larval structures are transformed into adult structures. For example, the large salivary glands of larval fruit flies regress during metamorphosis. The hormone ecdysone stimulates metamorphosis, including the onset of apoptosis. Ecdysone induces the expression of *rpg* and *hid* and inhibits the expression of other genes, which then leads to apoptosis of salivary gland cells.

Apoptosis in disease The symptoms of many diseases and disorders are caused by apoptosis or, in some cases, its absence. In neurodegenerative diseases such as Parkinson disease and Alzheimer disease, symptoms are caused by a loss of neurons through apoptosis. In heart attacks and stroke, some cells die through necrosis, but many others undergo apoptosis. Cancer is often stimulated by mutations in genes that regulate apoptosis, leading to a failure of apoptosis that would normally eliminate cancer cells.

22.5 The Study of Development Reveals Patterns and Processes of Evolution

"Ontogeny recapitulates phylogeny" is a familiar phrase that was coined in the 1860s by German zoologist Ernst Haeckel to describe his belief—now considered an oversimplication—that during their development (ontogeny) organisms repeat their evolutionary history (phylogeny). According to Haeckel's belief, a human embryo passes through fish, amphibian, reptilian, and mammalian stages before developing human traits. Scientists have long recognized that organisms do not pass through the adult stages of their ancestors during their development, but the embryos of these related organisms often display similarities.

Common genes in developmental pathways

Although ontogeny does not precisely recapitulate phylogeny, many evolutionary biologists today are turning to the study of development for a better understanding of the processes and patterns of evolution. Sometimes called "evo-devo," the study of evolution through the analysis of development is revealing that the same genes often shape developmental pathways in distantly related organisms. Biologists once thought that segmentation in vertebrates and invertebrates was only superficially similar, but we now know that, in both *Drosophila* and the primitive chordate *Branchiostoma*, the *engrailed* gene divides the embryo into specific segments. A gene called *distalless*, which creates the legs of a fruit fly, plays a role in the development of crustacean branched appendages. This same gene also stimulates body outgrowths of many other organisms, from polychaete worms to starfish.

An amazing example of how the same genes in distantly related organisms can shape similar developmental pathways is seen in the development of eyes in fruit flies, mice, and humans. Walter Gehring and his collaborators examined the effect of the *eyeless* gene in *Drosophila*, which is required for

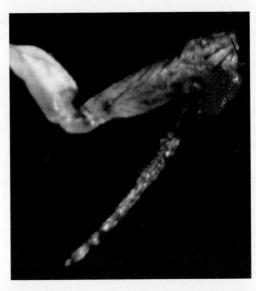

22.17 Expression of the *eyeless* gene causes the development of an eye on the leg of a fruit fly. Genes similar to *eyeless* also control eye development in mice and humans. [U. Kloter and G. Halder/Biozentrum.]

proper development of the fruit-fly eye. Gehring and his coworkers genetically engineered cells that expressed *eyeless* in parts of the fly where the gene is not normally expressed. When these flies hatched, they had eyes on their wings, antennae, and legs (**Figure 22.17**). These structures were not just tissue that resembled eyes, but complete eyes with a cornea, cone cells, and photoreceptors that responded to light, although the flies could not use these eyes to see, because they lacked a connection to the nervous system.

The *eyeless* gene has counterparts in mice and humans that affect the development of mammalian eyes. There is a striking similarity between the *eyeless* gene of *Drosophila* and the *Small eye* gene that exists in mice. In mice, a mutation in one copy of *Small eye* causes small eyes; a mouse that is homozygous for the *Small eye* mutation has no eyes. There is also a similarity between the *eyeless* gene in *Drosophila* and the *Aniridia* gene in humans; a mutation in *Aniridia* produces a severely malformed human eye. Similarities in the sequences of *eyeless*, *Small eye*, and *Aniridia* suggest that all three genes evolved from a common ancestral sequence. This possibility is surprising, because the eyes of insects and mammals were thought to have evolved independently. Similarities among *eyeless*, *Small eye*, and *Aniridia* suggest that a common pathway underlies eye development in flies, mice, and humans.

Similar genes may be part of a developmental pathway common to two different species but have quite different effects. For example, a *Hox* gene called *AbdB* helps define the posterior end of a *Drosophila* embryo; a similar group of genes in birds divides the wing into three segments. In another example, the *sog* gene in fruit flies stimulates cells to assume a ventral orientation in the embryo, but the expression of a similar gene called *chordin* in vertebrates causes cells to assume dorsal orientation, exactly the opposite of the situation in fruit flies. In fruit flies, the *toll* gene encodes a protein that helps determine the dorsal–ventral axis, as mentioned earlier. Similar genes in vertebrates encode proteins called Toll-like receptors that bind to molecules on pathogens and stimulate the immune system. The theme emerging from these studies is that a small, common set of genes may underlie many basic developmental processes in many different organisms.

Evolution through change in gene expression
Another concept revealed by studies in evo-devo is that many major evolutionary adaptations are not through changes in the types of proteins produced but through changes in the expression of genes that encode proteins that regulate development. This concept is seen in the evolution of blind cavefish discussed in the introduction to the chapter, where a small difference in the expression of two transcription-factor-encoding genes, *sonic hedgehog* and *tiggy-winkle hedgehog*, cause the development of the lens to abort and result in eyelessness.

Another example of differences in gene expression bringing about evolutionary change is seen in Darwin's finches, a group of 14 closely related birds found on the Galápagos Islands (see Figure 26.9). The birds differ primarily in the size and shape of their beaks: ground finches have deep and wide beaks, cactus finches have long and pointed beaks, and warbler finches have sharp and thin beaks. These differences are associated with diet, and evolutionary changes in beak shape and size have taken place in the past when climate changes brought about shifts in the abundance of food items.

To investigate the underlying genetic basis of these evolutionary changes, Arkhat Abzhanov and his colleagues used microarrays (see Chapter 20) to examine differences in transcription levels of several thousand genes in five species of Darwin's finches. They found differences in the expression of a gene that encodes a protein called calmodulin (CaM); the gene that encodes CaM was more highly expressed in the long and pointed beak of cactus-finch embryos than in the beaks of the other species. CaM takes part in a process called calcium signaling, which is known to affect many aspects of development. When Abzhanov and his coworkers activated calcium signaling in developing chicken embryos, the chickens had longer beaks, like those of the cactus finch. Thus, these researchers were able to reproduce, at least in part, the evolutionary difference that distinguishes cactus finches. The importance of this experiment is that it shows that changes in the expression of a single gene in the course of development can produce significant anatomical differences in adults.

In these studies and others, the combined efforts of developmental biologists, geneticists, and evolutionary biologists are sources of important insights into how evolution takes place. Although Haeckel's euphonious phrase "ontogeny recapitulates phylogeny" was incorrect, evo-devo is proving that development can reveal much about the process of evolution.

22.6 The Development of Immunity Is Through Genetic Rearrangement

A basic assumption of developmental biology is that every somatic cell carries an identical set of genetic information and that no genes are lost in development. Although this assumption holds for most cells, there are some important exceptions, one of which concerns genes that encode immune function in vertebrates. In the development of immunity, individual segments of certain genes are rearranged into different combinations, producing immune cells that contain different genetic information and that are each adapted to attack one particular foreign substance. This rearrangement and loss of genetic material is key to the power of our immune systems to protect us against almost any conceivable foreign substance.

The immune system provides protection against infection by bacteria, viruses, fungi, and parasites. The focus of an immune response is an **antigen**, defined as any molecule (usually a protein) that elicits an immune reaction. The immune system is remarkable in its ability to recognize an almost unlimited number of potential antigens. The body is full of proteins, and so it is essential that the immune system be able to distinguish between self-antigens and foreign antigens. Occasionally, the ability to make this distinction breaks down, and the body produces an immune reaction to its own antigens, resulting in an **autoimmune disease** (**Table 22.5**).

The Organization of the Immune System
The immune system contains a number of different components and uses several mechanisms to provide protection against pathogens, but most immune responses can be

Table 22.5 Examples of autoimmune diseases

Disease	Tissues Attacked
Graves disease, Hashimoto thyroiditis	Thyroid gland
Rheumatic fever	Heart muscle
Systemic lupus erythematosus	Joints, skin, and other organs
Rheumatoid arthritis	Joints
Insulin-dependent diabetes mellitus	Insulin-producing cells in pancreas
Multiple sclerosis	Myelin sheath around nerve cells

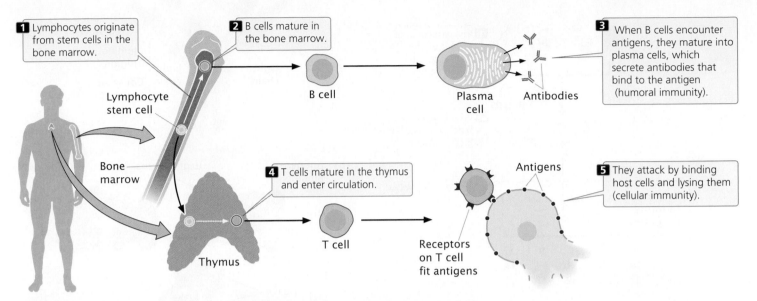

1 Lymphocytes originate from stem cells in the bone marrow.

Lymphocyte stem cell

Bone marrow

Thymus

2 B cells mature in the bone marrow.

B cell

3 When B cells encounter antigens, they mature into plasma cells, which secrete antibodies that bind to the antigen (humoral immunity).

Plasma cell   Antibodies

4 T cells mature in the thymus and enter circulation.

T cell

Receptors on T cell fit antigens

Antigens

5 They attack by binding host cells and lysing them (cellular immunity).

22.18 Immune responses are classified as humoral immunity, in which antibodies are produced by B cells, and cellular immunity, which is produced by T cells.

grouped into two major classes: humoral immunity and cellular immunity. Although it is convenient to think of these classes as separate systems, they interact and influence each other significantly.

Humoral immunity   Immune function is carried out by specialized blood cells called lymphocytes, which are a type of white blood cell. **Humoral immunity** centers on the production of antibodies by specialized lymphocytes called **B cells** (**Figure 22.18**), which mature in the bone marrow. **Antibodies** are proteins that circulate in the blood and other body fluids, binding to specific antigens and marking them for destruction by phagocytic cells. Antibodies also activate a set of proteins called complement that help to lyse cells and attract macrophages.

Cellular immunity   **T cells** (see Figure 22.18), are specialized lymphocytes that mature in the thymus and respond only to antigens found on the surfaces of the body's own cells. These lymphocytes are responsible for the second type of immune response, **cellular immunity**.

After a pathogen such as a virus has infected a host cell, some viral antigens appear on the cell's surface. Proteins, called **T-cell receptors**, on the surfaces of T cells bind to these antigens and mark the infected cell for destruction. T-cell receptors must simultaneously bind a foreign antigen and a self-antigen called a **major histocompatibility complex antigen** (MHC antigen) on the cell surface. Not all T cells attack cells having foreign antigens; some help regulate immune responses, providing communication among different components of the immune system.

Clonal selection   How can the immune system recognize an almost unlimited number of foreign antigens? Remark-

ably, each mature lymphocyte is genetically programmed to attack one and only one specific antigen: each mature B cell produces antibodies against a single antigen, and each T cell is capable of attaching to only one type of foreign antigen.

If each lymphocyte is specific for only one type of antigen, how does an immune response develop? The **theory of clonal selection** states that, initially, there is a large pool of millions of different lymphocytes, each capable of binding only one antigen (**Figure 22.19**); so millions of different foreign antigens can be detected. To illustrate clonal selection, let's imagine that a foreign protein enters the body. Only a few lymphocytes in the pool will be specific for this particular foreign antigen. When one of these lymphocytes encounters the foreign antigen and binds to it, that lymphocyte is stimulated to divide. The lymphocyte proliferates rapidly, producing a large population of genetically identical cells—a clone—each of which is specific for that particular antigen.

This initial proliferation of antigen-specific B and T cells is known as a **primary immune response** (see Figure 22.19); in most cases, the primary response destroys the foreign antigen. Following the primary immune response, most of the lymphocytes in the clone die, but a few continue to circulate in the body. These **memory cells** may remain in circulation for years or even for the rest of a person's life. Should the same antigen reappear at some time in the future, memory cells specific to that antigen become activated and quickly give rise to another clone of cells capable of binding the antigen. The rise of this second clone is termed a **secondary immune response** (see Figure 22.19). The ability to quickly produce a second clone of antigen-specific cells permits the long-lasting immunity that often follows recovery from a disease. For example, people who have chicken pox usually have life-long immunity to the disease. The secondary immune response is also the basis for vaccination, which

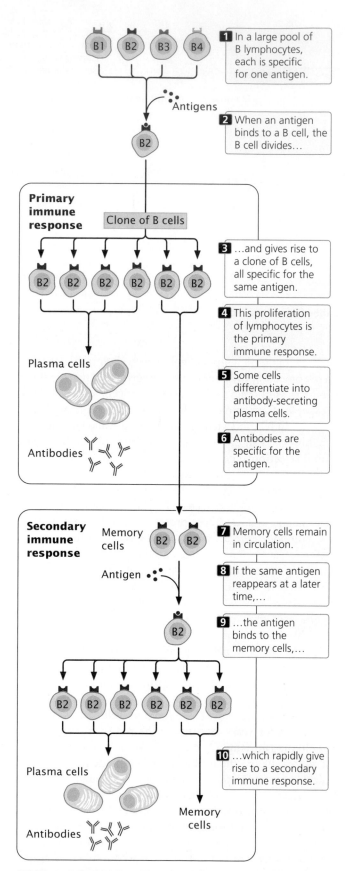

22.19 An immune response to a specific antigen is produced through clonal selection.

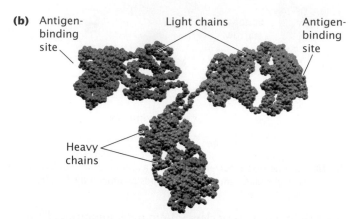

22.20 Each immunoglobulin molecule consists of four polypeptide chains—two light chains and two heavy chains—that combine to form a Y-shaped structure. (a) Structure of an immunoglobulin. (b) Folded, space-filling model.

stimulates a primary immune response to an antigen and results in memory cells that can quickly produce a secondary response if that same antigen appears in the future.

Three sets of proteins are used in immune responses: antibodies, T-cell receptors, and the major histocompatibility antigens. The next section explores how the enormous diversity in these proteins is generated.

CONCEPTS

Each B cell and T cell of the immune system is genetically capable of binding one type of foreign antigen. When a lymphocyte binds to an antigen, the lymphocyte undergoes repeated division, giving rise to a clone of genetically identical lymphocytes (the primary response), all of which are specific for that same antigen. Memory cells remain in circulation for long periods of time; if the antigen reappears, the memory cells undergo rapid proliferation and generate a secondary immune response.

Immunoglobulin Structure

The principal products of the humoral immune response are antibodies—also called immunoglobulins. Each immunoglobulin (Ig) molecule consists of four polypeptide chains—two identical light chains and two identical heavy chains—that form a Y-shaped structure (**Figure 22.20**). Disulfide bonds link the two heavy chains in the stem of the Y and attach a light chain to a heavy chain in each arm of the Y. Binding sites for antigens are at the ends of the two arms.

The light chains of an immunoglobulin are of two basic types, called kappa chains and lambda chains. An immunoglobulin molecule can have two kappa chains or two lambda chains, but it cannot have one of each type. Both the light and the heavy chains have a variable region at one end and a constant region at the other end; the variable regions of different immunoglobulin molecules vary in amino acid sequence, whereas the constant regions of different immunoglobulins are similar in sequence. The variable regions of both light and heavy chains make up the antigen-binding regions and specify the type of antigen to which the antibody can bind.

The Generation of Antibody Diversity

The immune system is capable of making antibodies against virtually any antigen that might be encountered in a person's lifetime: each person is capable of making about 10^{15} different antibody molecules. Antibodies are proteins; so the amino acid sequences of all 10^{15} potential antibodies must be encoded in the human genome. However, there are fewer than 1×10^5 genes in the human genome and, in fact, only 3×10^9 total base pairs; so how can this huge diversity of antibodies be encoded?

The answer lies in the fact that antibody genes are composed of segments. There are a number of copies of each type of segment, each differing slightly from the others. In the maturation of a lymphocyte, the segments are joined together to create an immunoglobulin gene. The particular copy of each segment used is random and, because there are multiple copies of each type, there are many possible combinations of the segments. A limited number of segments can therefore encode a huge diversity of antibodies.

To illustrate this process of antibody assembly, let's consider the immunoglobulin light chains. Kappa and lambda chains are encoded by separate genes on different chromosomes. Each gene is composed of three types of segments: V, for variable; J, for joining; and C, for constant. The V segments encode most of the variable region of the light chains, the C segment encodes the constant region of the chain, and the J segments encode a short set of nucleotides that join the V segment and the C segments together. The number of V, J, and C segments differs among species. For the human kappa gene, there are from 30 to 35 different functional V gene segments, 5 different J genes, and a single C gene segment, all of which are present in the germline DNA (**Figure 22.21a**). The V gene segments, which are about 400 bp in length, are located on the same chromosome and are separated from one another by about 7000 bp. The

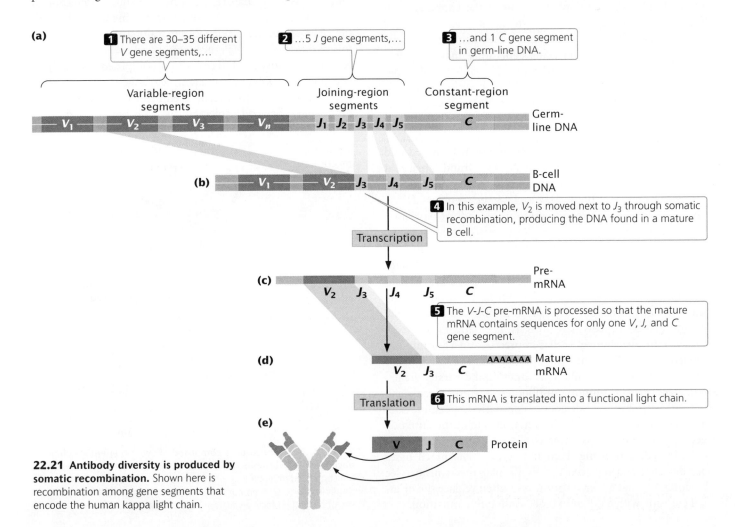

22.21 Antibody diversity is produced by somatic recombination. Shown here is recombination among gene segments that encode the human kappa light chain.

J gene segments are about 30 bp in length and all together encompass about 1400 bp.

Initially, an immature lymphocyte inherits all of the *V* gene segments and all of the *J* gene segments present in the germ line. In the maturation of the lymphocyte, **somatic recombination** within a single chromosome moves one of the *V* gene segments to a position next to one of the *J* gene segments (**Figure 22.21b**). In Figure 22.21b, V_2 (the second of approximately 35 different *V* gene segments) undergoes somatic recombination, which places it next to J_3 (the third of 5 *J* gene segments); the intervening segments are lost.

After somatic recombination has taken place, the B-cell DNA is transcribed into pre-mRNA that contains one *V* gene and several *J* genes, along with the *C* gene (**Figure 22.21c**). The resulting pre-mRNA is processed (**Figure 22.21d**) to produce a mature mRNA that contains only sequences for a single *V*, *J*, and *C* segment; this mRNA is translated into a functional light chain (**Figure 22.21e**). In this way, each mature human B cell produces a unique type of kappa light chain, and different B cells produce slightly different kappa chains, depending on the combination of *V* and *J* segments that are joined together.

The gene that encodes the lambda light chain is organized in a similar way but differs from the kappa gene in the number of copies of the different segments. Somatic recombination takes place among the segments in the same way as that in the kappa gene, generating many possible combinations of lambda light chains. The gene that encodes the immunoglobulin heavy chain also is arranged in *V*, *J*, and *C* segments, but this gene possesses *D* (for diversity) segments as well. Thus, many different types of light and heavy chains are possible.

Somatic recombination is brought about by RAG1 and RAG2 proteins, which generate double-strand breaks at specific nucleotide sequences called recombination signal sequences that flank the *V*, *D*, *J*, and *C* gene segments. DNA-repair proteins then process and join the ends of particular segments together.

In addition to somatic recombination, other mechanisms add to antibody diversity. First, each type of light chain can potentially combine with each type of heavy chain to make a functional immunoglobulin molecule, increasing the amount of possible variation in antibodies. Second, the recombination process that joins *V*, *J*, *D*, and *C* gene segments in the developing B cell is imprecise, and a few random nucleotides are frequently lost or gained at the junctions of the recombining segments. This **junctional diversity** greatly enhances variation among antibodies. A third mechanism that adds to antibody diversity is **somatic hypermutation**, a process that leads to a high mutation rate in the antibody genes. This process is initiated when cytosine bases are deaminated, converting them into uracil. The uracil bases are detected and replaced by DNA-repair mechanisms (see Chapter 18) that are error prone and often replace the original cytosine with a different base, leading to a mutation.

Through the processes of somatic recombination, junctional diversity, and somatic hypermutation, each lymphocyte comes to possess a unique set of genetic information (different from that in other lymphocytes) that encodes an antibody specific to a particular antigen. **TRY PROBLEM 25** →

CONCEPTS

The genes encoding the antibody chains are organized in segments, and germ-line DNA contains multiple versions of each segment. The many possible combinations of *V*, *J*, and *D* segments permit an immense variety of different antibodies to be generated. This diversity is augmented by the different combinations of light and heavy chains, the random addition and deletion of nucleotides at the junctions of the segments, and the high mutation rates in the immunoglobulin genes.

✔ CONCEPT CHECK 7

How does somatic recombination differ from alternative splicing of RNA?

T-Cell-Receptor Diversity

Like B cells, each mature T cell has genetically determined specificity for one type of antigen that is mediated through the cell's receptors. T-cell receptors are structurally similar to immunoglobulins (**Figure 22.22**) and are located on the cell surface; most T-cell receptors are composed of one alpha and one beta polypeptide chain held together by disulfide bonds. One end of each chain is embedded in the cell membrane; the other end projects away from the cell and binds

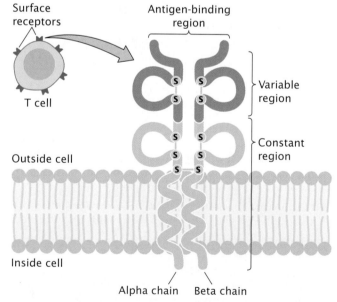

22.22 A T-cell receptor is composed of two polypeptide chains, each having a variable and constant region. Most T-cell receptors are composed of alpha and beta polypeptide chains held together by disulfide bonds. One end of each chain traverses the cell membrane; the other end projects away from the cell and binds antigens.

antigens. Like the immunoglobulin chains, each chain of the T-cell receptor possesses a constant region and a variable region (see Figure 22.22); the variable regions of the two chains provide the antigen-binding site.

The genes that encode the alpha and beta chains of the T-cell receptor are organized much like those that encode the heavy and light chains of immunoglobulins: each gene is made up of segments that undergo somatic recombination before the gene is transcribed. For example, the human gene for the alpha chain initially consists of 44 to 46 *V* gene segments, 50 *J* gene segments, and a single *C* gene segment. The organization of the gene for the beta chain is similar, except that it also contains *D* gene segments. Alpha and beta chains combine randomly and there is junctional diversity, but there is no evidence for somatic hypermutation in T-cell-receptor genes.

CONCEPTS

Like the genes that encode antibodies, the genes for the T-cell-receptor chains consist of segments that undergo somatic recombination, generating an enormous diversity of antigen-binding sites.

Major Histocompatibility Complex Genes

When tissues are transferred from one species to another or even from one member to another within a species, the transplanted tissues are usually rejected by the host animal. The results of early studies demonstrated that this graft rejection is due to an immune response that takes place when antigens on the surface of cells of the grafted tissue are detected and attacked by T cells in the host organism. The antigens that elicit graft rejection are referred to as histocompatibility antigens, and they are encoded by a cluster of genes called the major histocompatibility complex (MHC).

T cells are activated only when the T-cell receptor simultaneously binds both a foreign antigen and the host cell's own histocompatibility antigen. The reason for this requirement is not clear; it may reserve T cells for action against pathogens that have invaded cells. When a foreign body, such as a virus, is ingested by a macrophage or other cell, partly digested pieces of the foreign body containing antigens are displayed on the cell's surface (**Figure 22.23**). A cell infected with a virus may also express viral antigens on its cell surface. Through their T-cell receptors, T cells bind to both the histocompatibility protein and the foreign antigen and secrete substances that either destroy the antigen-containing cell or activate other B and T cells or do both.

The MHC genes are among the most variable genes known: there are more than 100 different alleles for some MHC loci. Because each person possesses five or more MHC loci and because many alleles are possible at each locus, no two people (with the exception of identical twins) produce the same set of histocompatibility antigens. The variation

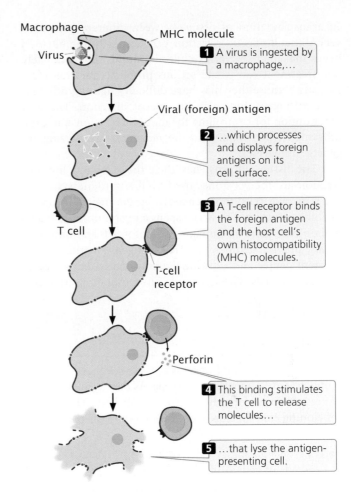

22.23 T cells are activated by binding both to a foreign antigen and to a histocompatibility antigen on the surface of a self-cell.

Labels in figure:
- Macrophage
- Virus
- MHC molecule
- **1** A virus is ingested by a macrophage,…
- Viral (foreign) antigen
- **2** …which processes and displays foreign antigens on its cell surface.
- T cell
- **3** A T-cell receptor binds the foreign antigen and the host cell's own histocompatibility (MHC) molecules.
- T-cell receptor
- Perforin
- **4** This binding stimulates the T cell to release molecules…
- **5** …that lyse the antigen-presenting cell.

in histocompatibility antigens provides each of us with a unique identity for our own cells, which allows our immune systems to distinguish self from nonself. This variation is also the cause of rejection in organ transplants.

CONCEPTS

The MHC genes encode proteins that provide identity to the cells of each individual organism. To bring about an immune response, a T-cell receptor must simultaneously bind both a histocompatibility (self) antigen and a specific foreign antigen.

Genes and Organ Transplants

For a person with a seriously impaired organ, a transplant operation may offer the only hope of survival. Successful transplantation requires more than the skills of a surgeon; it also requires a genetic match between the patient and the person donating the organ.

The fate of transplanted tissue depends largely on the type of antigens present on the surface of its cells. Because foreign tissues are usually rejected by the host, the successful transplantation of tissues between different persons is very difficult. Tissue rejection can be partly inhibited by drugs that interfere with cellular immunity. Unfortunately, this treatment can create serious problems for transplant patients because they may have difficulty fighting off common pathogens and may thus die of infection. The only other option for controlling the immune reaction is to carefully match the donor and the recipient, maximizing the genetic similarities.

The tissue antigens that elicit the strongest immune reaction are the very ones used by the immune system to mark its own cells, those encoded by the major histocompatibility complex. The MHC spans a region of more than 3 million base pairs on human chromosome 6 and has many alleles, providing different MHC antigens on the cells of different people and allowing the immune system to recognize foreign cells. The severity of an immune rejection of a transplanted organ depends on the number of mismatched MHC antigens on the cells of the transplanted tissue. The ABO red-blood-cell antigens also are important because they elicit a strong immune reaction. The ideal donor is the patient's own identical twin, who will have exactly the same MHC and ABO antigens. Unfortunately, most patients don't have an identical twin. The next-best donor is a sibling with the same major MHC and ABO antigens. If a sibling is not available, donors from the general population are considered. An attempt is made to match as many of the MHC antigens of the donor and recipient as possible, and immunosuppressive drugs are used to control rejection due to the mismatches. The long-term success of transplants depends on the closeness of the match. Survival rates after kidney transplants (the most successful of the major organ transplants) increase from 63% with zero or one MHC match to 90% with four matches.

CONCEPTS SUMMARY

- Each multicellular organism begins as a single cell that has the potential to develop into any cell type. As development proceeds, cells become committed to particular fates. The results of cloning experiments demonstrated that this process arises from differential gene expression.

- In the early *Drosophila* embryo, determination is effected through a cascade of gene control.

- The dorsal–ventral and anterior–posterior axes of the *Drosophila* embryo are established by egg-polarity genes, which are expressed in the female parent and produce RNA and proteins that are deposited in the egg cytoplasm. Initial differences in the distribution of these molecules regulate gene expression in various parts of the embryo. The dorsal–ventral axis is defined by a concentration gradient of the Dorsal protein, and the anterior–posterior axis is defined by concentration gradients of the Bicoid and Nanos proteins.

- Three types of segmentation genes act sequentially to determine the number and organization of the embryonic segments in *Drosophila*. The gap genes establish large sections of the embryo, the pair-rule genes affect alternate segments, and the segment-polarity genes affect the organization of individual segments. Homeotic genes then define the identity of individual *Drosophila* segments.

- Homeotic genes control the development of flower structure. Three sets of genes interact to determine the identity of the four whorls found in a complete flower.

- Apoptosis, or programmed cell death, plays an important role in the development of many animals. Apoptosis is a highly regulated process that depends on caspases—proteins that cleave proteins. Apoptosis plays an important role in animal development.

- The immune system is the primary defense network in vertebrates. In humoral immunity, B cells produce antibodies that bind foreign antigens; in cellular immunity, T cells attack cells carrying foreign antigens.

- Each B and T cell is capable of binding only one type of foreign antigen. When a lymphocyte binds to an antigen, the lymphocyte divides and gives rise to a clone of cells, each specific for that same antigen—the primary immune response. A few memory cells remain in circulation for long periods of time and, on exposure to that same antigen, can proliferate rapidly and generate a secondary immune response.

- Immunoglobulins (antibodies) are encoded by genes that consist of several types of gene segments; germ-line DNA contains multiple copies of these gene segments, which differ slightly in sequence. Somatic recombination randomly brings together one version of each segment to produce a single complete gene, allowing many combinations. Diversity is further increased by the random addition and deletion of nucleotides at the junctions of the segments and by a high mutation rate.

- The germ-line genes for T-cell receptors consist of segments with multiple varying copies. Somatic recombination generates many different types of T-cell receptors in different cells. Junctional diversity also adds to T-cell-receptor variability.

- The major histocompatibility complex encodes a number of histocompatibility antigens. The MHC antigen allows the immune system to distinguish self from nonself. Each locus for the MHC contains many alleles.

IMPORTANT TERMS

totipotency (p. 612)
determination (p. 612)
egg-polarity gene (p. 614)
morphogen (p. 614)
segmentation gene (p. 618)
gap gene (p. 618)
pair-rule gene (p. 618)
segment-polarity gene (p. 618)
homeotic gene (p. 619)
homeobox (p. 619)
Antennapedia complex (p. 619)

bithorax complex (p. 619)
homeotic complex (HOM-C) (p. 619)
Hox gene (p. 620)
apoptosis (p. 623)
caspase (p. 624)
antigen (p. 626)
autoimmune disease (p. 626)
humoral immunity (p. 627)
B cell (p. 627)
antibody (p. 627)
T cell (p. 627)

cellular immunity (p. 627)
T-cell receptor (p. 627)
major histocompatibility complex
  (MHC) antigen (p. 627)
theory of clonal selection (p. 627)
primary immune response (p. 627)
memory cell (p. 627)
secondary immune response (p. 627)
somatic recombination (p. 630)
junctional diversity (p. 630)
somatic hypermutation (p. 630)

ANSWERS TO CONCEPT CHECKS

1. No, it does *not* prove that genetic material is not lost during development, because differentiation has not yet taken place in an early embryo. The early embryo would likely still contain all its genes and could therefore give rise to a complete animal. The use of specialized cells, such as a cell from an udder, does prove that genes are not lost during development because, if they were lost, there would be no cloned animal.

2. c

3. b

4. d

5. d

6. In cell death from necrosis, the cell swells and bursts, causing an inflammatory reponse. In cell death through apoptosis, the cell's DNA is degraded, its nucleus and cytoplasm shrink, and the cell is phagocytized, without leakage of cellular contents.

7. Somatic recombination takes place through the rearrangement of DNA segments, and so each lymphocyte has a different sequence of nucleotides in its DNA. Alternative splicing (discussed in Chapter 14) takes place through the rearrangement of RNA sequences in pre-mRNA; there is no change in the DNA that encodes the pre-mRNA. The generation of antibody diversity requires both the somatic recombination of DNA sequences and the alternative splicing of the pre-mRNA sequences.

WORKED PROBLEMS

1. If a fertilized *Drosophila* egg is punctured at the anterior end and a small amount of cytoplasm is allowed to leak out, what will be the most likely effect on the development of the fly embryo?

• **Solution**

The egg-polarity genes determine the major axes of development in the *Drosophila* embryo. One of these genes is *bicoid*, which is transcribed in the maternal parent. As *bicoid* mRNA passes into the egg, the mRNA becomes anchored to the anterior end of the egg. After the egg has been laid, *bicoid* mRNA is translated into Bicoid protein, which forms a concentration gradient along the anterior–posterior axis of the embryo. The high concentration of Bicoid protein at the anterior end induces the development of anterior structures such as the head of the fruit fly. If the anterior end of the egg is punctured, cytoplasm containing high concentrations of Bicoid protein will leak out, reducing the concentration of Bicoid protein at the anterior end. The result will be that the embryo fails to develop head and thoracic structures at the anterior end.

2. The immunoglobulin molecules of a particular mammalian species have kappa and lambda light chains and heavy chains. The kappa gene consists of 250 V and 8 J segments. The lambda gene contains 200 V and 4 J segments. The gene for the heavy chain consists of 300 V, 8 J, and 4 D segments. If just somatic recombination and random combinations of light and heavy chains are taken into consideration, how many different types of antibodies can be produced by this species?

• **Solution**

For the kappa light chain, there are $250 \times 8 = 2000$ combinations; for the lambda light chain, there are $200 \times 4 = 800$ combinations; so a total of 2800 different types of light chains are possible. For the heavy chains, there are $300 \times 8 \times 4 = 9600$ possible types. Any of the 2800 light chains can combine with any of the 9600 heavy chains; so there are $2800 \times 9600 = 26,880,000$ different types of antibodies possible from somatic recombination and random combination alone. Junctional diversity and somatic hypermutation would greatly increase this diversity.

COMPREHENSION QUESTIONS

Section 22.1

*1. What experiments suggested that genes are not lost or permanently altered in development?

Section 22.2

2. Briefly explain how the Dorsal protein is redistributed in the formation of the *Drosophila* embryo and how this redistribution helps to establish the dorsal–ventral axis of the early embryo.

*3. Briefly describe how the *bicoid* and *nanos* genes help to determine the anterior–posterior axis of the fruit fly.

*4. List the three major classes of segmentation genes and outline the function of each.

5. What role do homeotic genes play in the development of fruit flies?

Section 22.3

6. How do class A, B, and C genes in plants work together to determine the structures of the flower?

Section 22.4

*7. What is apoptosis and how is it regulated?

Section 22.6

*8. Explain how each of the following processes contributes to antibody diversity.

a. Somatic recombination

b. Junctional diversity

c. Hypermutation

9. What is the function of the MHC antigens? Why are the genes that encode these antigens so variable?

APPLICATION QUESTIONS AND PROBLEMS

Introduction

10. William Jeffrey and his colleagues crossed surface-dwelling Mexican tetra that had fully developed eyes with blind Mexican tetras from caves (see the introduction to this chapter). The progeny from this cross had uniformly small eyes compared with those of surface fish (Y. Yamamoto, D. W. Stock, and W. R. Jeffrey. 2004. *Nature* 431:844–847). What prediction can you make about the expression of *shh* in the embryos of these progeny relative to its expression in the embryos of surface fish?

Section 22.1

11. If telomeres are normally shortened after each round of replication in somatic cells (see Chapter 12), what prediction would you make about the length of telomeres in Dolly, the first cloned sheep?

Section 22.2

12. Why do mutations in *bicoid* and *nanos* exhibit genetic maternal effects (a mutation in the maternal parent produces a phenotype that shows up in the offspring, see Chapter 5), but mutations in *runt* and *gooseberry* do not? (Hint: See Tables 22.3 and 22.4.)

*13. Give examples of genes that affect development in fruit flies by regulating gene expression at the level of (a) transcription and (b) translation.

14. What would be the most likely effect on development of puncturing the posterior end of a *Drosophila* egg, allowing a small amount of cytoplasm to leak out, and then injecting that cytoplasm into the anterior end of another egg?

15. Christiane Nüsslein-Volhard and her colleagues carried out several experiments in an attempt to understand what determines the anterior and posterior ends of a *Drosophila* larva (reviewed in C. Nüsslein-Volhard, H. G. Frohnhofer, and R. Lehmann. 1987. *Science* 238:1675–1681). They isolated fruit flies with mutations in the *bicoid* gene (*bcd⁻*). These flies produced embryos that lacked a head and thorax. When they transplanted cytoplasm from the anterior end of an egg from a wild-type female into the anterior end of an egg from a mutant *bicoid* female, normal head and thorax development took place in the embryo. However, transplanting cytoplasm from the posterior end of an egg from a wild-type female into the anterior end of an egg from a *bicoid* female had no effect. Explain these results in regard to what you know about proteins that control the determination of the anterior–posterior axis.

*16. What would be the most likely result of injecting *bicoid* mRNA into the posterior end of a *Drosophila* embryo and inhibiting the translation of *nanos* mRNA?

17. What would be the most likely effect of inhibiting the translation of *hunchback* mRNA throughout the embryo?

*18. Molecular geneticists have performed experiments in which they altered the number of copies of the *bicoid* gene in flies, affecting the amount of Bicoid protein produced.

a. What would be the effect on development of an increased number of copies of the *bicoid* gene?

b. What would be the effect of a decreased number of copies of *bicoid*? Justify your answers.

19. What would be the most likely effect on fruit-fly development of a deletion in the *nanos* gene?

20. Give an example of a gene found in each of the categories of genes (egg-polarity, gap, pair-rule, and so forth) listed in Figure 22.12.

21. In Chapter 1, we considered an early idea about heredity called preformationism, which suggested that inside the egg or sperm is a tiny adult called a homunculus, with all the features of an adult human in miniature. According to this idea, the homunculus simply enlarges during development. What types of evidence presented in this chapter prove that preformationism is false?

Section 22.3

22. Explain how (a) the absence of class B gene expression produces the flower structures seen in class B mutants (see Figure 22.14c) and (b) the absence of class C gene product produces the structures seen in class C mutants (see Figure 22.14d).

23. What would you expect a flower to look like in a plant that lacked both class A and class B genes? In a plant that lacked both class B and class C genes?

24. What will be the flower structure of a plant in which expression of the following genes is inhibited:

a. Expression of class B genes is inhibited in the second whorl, but not in the third whorl.

b. Expression of class C genes is inhibited in the third whorl, but not in the fourth whorl.

c. Expression of class A genes is inhibited in the first whorl, but not in the second whorl.

d. Expression of class A genes is inhibited in the second whorl, but not in the first whorl.

Section 22.6

*25. In a particular species, the gene for the kappa light chain has 200 *V* gene segments and 4 *J* segments. In the gene for the lambda light chain, this species has 300 *V* segments and 6 *J* segments. If only the variability arising from somatic recombination is taken into consideration, how many different types of light chains are possible?

26. In the fictional book *Chromosome 6* by Robin Cook, a biotechnology company genetically engineers individual bonobos (a type of chimpanzee) to serve as future organ donors for clients. The genes of the bonobos are altered so that no tissue rejection takes place when their organs are transplanted into a client. What genes would need to be altered for this scenario to work? Explain your answer.

CHALLENGE QUESTIONS

Section 22.2

27. As we have learned in this chapter, the Nanos protein inhibits the translation of *hunchback* mRNA, thus lowering the concentration of Hunchback protein at the posterior end of a fruit-fly embryo and stimulating the differentiation of posterior characteristics. The results of experiments have demonstrated that the action of Nanos on *hunchback* mRNA depends on the presence of an 11-base sequence that is located in the 3′ untranslated region (3′ UTR) of *hunchback* mRNA. This sequence has been termed the Nanos response element (NRE). There are two copies of NRE in the 3′ UTR of *hunchback* mRNA. If a copy of NRE is added to the 3′ UTR of another mRNA produced by a different gene, the mRNA now becomes repressed by Nanos. The repression is greater if several NREs are added. On the basis of these observations, propose a mechanism for how Nanos inhibits Hunchback translation.

28. Given the distribution of *Hox* genes among animals, what would you predict about the number and type of *Hox* genes in the common ancestor of all animals?

Section 22.6

29. Ataxia-telangiectasis (ATM) is a rare genetic neurodegenerative disease. About 20% of people with ATM develop acute lymphocytic leukemia or lymphoma, cancers of the immune cells. Cells in many of these cancers exhibit chromosome rearrangements, with chromosome breaks occurring at antibody and T-cell-receptor genes (A. L. Bredemeyer et al. 2006. *Nature* 442:466–470). Many people with ATM also have a weakened immune system, making them susceptible to respiratory infections. Research has shown that the locus that causes ATM has a role in the repair of double-strand breaks. Explain why people who have a genetic defect in the repair of double-strand breaks might have a high incidence of chromosome rearrangements in their immune cells and why their immune systems might be weakened.

23 Cancer Genetics

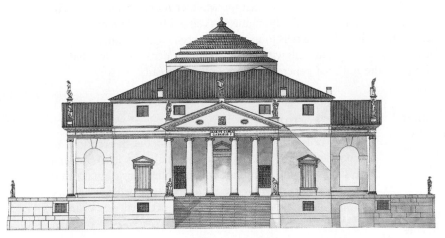

Villa designed by Renaissance architect Andrea Palladio, for whom the *palladin* gene is named. *Palladin* encodes an essential component of a cell's cytoskeleton; when mutated, *palladin* contributes to the spread of pancreatic cancer. [© *La Rotunda a Vicenzo* by Giovanni Giaconi.]

PALLADIN AND THE SPREAD OF CANCER

Pancreatic cancer is among the most serious of all cancers. Although only the eleventh most common form of cancer, with about 43,000 new cases each year in the United States, pancreatic cancer is the fourth leading cause of death due to cancer, killing more than 36,000 people each year. Most people with pancreatic cancer survive less than 6 months after the cancer is diagnosed; only 5% survive more than 5 years. A primary reason for pancreatic cancer's lethality is its propensity to spread rapidly to the lymph nodes and other organs. Most symptoms don't appear until the disease is advanced and the cancer has invaded other organs. So what makes pancreatic cancer so likely to spread?

In 2006, researchers identified a key gene that contributes to the development of pancreatic cancer—an important source of insight into pancreatic cancer's aggressive nature. Geneticists at the University of Washington in Seattle had found a unique family in which nine members over three generations were diagnosed with pancreatic cancer (**Figure 23.1**). Nine additional family members had precancerous growths that were likely to develop into pancreatic cancer. In this family, pancreatic cancer was inherited as an autosomal dominant trait.

Using gene-mapping techniques, the geneticists determined that the gene causing pancreatic cancer in the family was located within a region on the long arm of chromosome 4. Unfortunately, this region encompasses 16 million base pairs and includes 250 genes.

To determine which of the 250 genes in the delineated region might be responsible for cancer in the family, researchers designed a unique microarray (see Chapter 20) that contained sequences from the region. They used this microarray to examine gene expression in pancreatic tumors and precancerous growths in family members, as well as in sporadic pancreatic tumors in other people and in normal pancreatic tissue from unaffected people. The researchers reasoned that the cancer gene might be overexpressed or underexpressed in the tumors relative to normal tissue. Data from the microarray revealed that the most overexpressed gene in the pancreatic tumors and precancerous growths was a gene encoding a critical component of the cytoskeleton—a gene called *palladin*. Sequencing demonstrated that all members of the family with pancreatic cancer had an identical mutation in exon 2 of the *palladin* gene.

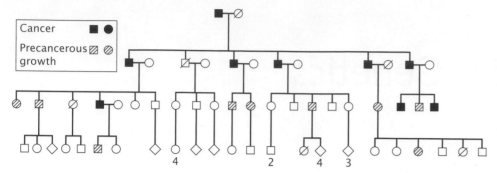

23.1 Pancreatic cancer is inherited as an autosomal dominant trait in a family that possesses a mutant *palladin* gene. [After K. L. Pogue et al., *Plos Medicine* 3:2216–2228, 2006.]

The *palladin* gene is named for Renaissance architect Andrea Palladio because *palladin* plays a central role in the architecture of the cell. Palladin protein functions as a scaffold for the binding of the other cytoskeleton proteins that are necessary for maintaining cell shape, movement, and differentiation. The ability of a cancer cell to spread is directly related to its cytoskeleton; cells that spread typically have poor cytoskeleton architecture, enabling them to detach easily from a primary tumor mass and migrate through other tissues. To determine whether mutations in the *palladin* gene affect cell mobility, researchers genetically engineered cells with a mutant copy of the *palladin* gene and tested the ability of these cells to migrate. The cells with mutated *palladin* were 33% more efficient at migrating than cells with normal *palladin,* demonstrating that the *palladin* gene contributes to the ability of pancreatic cancer cells to spread.

The discovery of *palladin*'s link to pancreatic cancer illustrates the power of modern molecular genetics for unraveling the biological nature of cancer. In this chapter, we examine the genetic nature of cancer, a peculiar disease that is fundamentally genetic but is often not inherited. We begin by considering the nature of cancer and how multiple genetic alterations are required to transform a normal cell into a cancerous one. We then consider some of the types of genes that contribute to cancer, including oncogenes and tumor-suppressor genes, genes that control the cell cycle, genes in signal-transduction pathways, genes encoding DNA-repair systems and telomerase, and genes that, like *palladin*, contribute to the spread of cancer. Next, we take a look at chromosome mutations associated with cancer and genomic instability. We examine the role that viruses play in some cancers and epigenetic changes associated with cancer. Finally, we take a detailed look at how specific genes contribute to the progression of colon cancer.

23.1 Cancer Is a Group of Diseases Characterized by Cell Proliferation

Cancer kills one of every five people in the United States, and cancer treatments cost billions of dollars per year. Cancer is not a single disease; rather, it is a heterogeneous group of disorders characterized by the presence of cells that do not respond to the normal controls on division. Cancer cells divide rapidly and continuously, creating tumors that crowd out normal cells and eventually rob healthy tissues of nutrients (**Figure 23.2**). The cells of an advanced tumor can separate from the tumor and travel to distant sites in the body, where they may take up residence and develop into new tumors. The most common cancers in the United States are those of the lung, prostate gland, breast, colon and rectum, and blood (**Table 23.1**).

Tumor Formation

Normal cells grow, divide, mature, and die in response to a complex set of internal and external signals. A normal cell receives both stimulatory and inhibitory signals, and its growth and division are regulated by a delicate balance between these opposing forces. In a cancer cell, one or more

(a)

(b)

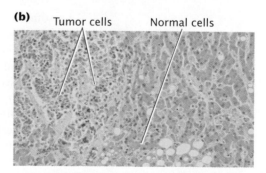

23.2 Abnormal proliferation of cancer cells produces a tumor that crowds out normal cells. (a) Metastatic lung-tumor masses (white protrusions) growing on a human liver. (b) A light micrograph of the section in part *a* showing areas of small, dark tumor cells invading a region of larger, lighter normal liver cells. [Courtesy of J. Braun.]

Table 23.1	Estimated incidences of various cancers and cancer mortality in the United States in 2010

Type of Cancer	New Cases per Year	Deaths per Year
Lung and bronchus	222,520	157,300
Prostate	217,730	32,050
Breast	209,060	40,230
Colon and rectum	142,570	51,370
Lymphoma	74,030	21,530
Bladder	70,530	14,680
Melanoma	68,130	8,700
Kidney and renal pelvis	58,240	13,040
Thyroid	44,670	1,690
Uterus	43,470	7,950
Pancreas	43,140	36,800
Leukemias	43,050	21,840
Oral cavity and pharynx	36,540	7,880
Liver	24,120	18,910
Brain and nervous system	22,020	13,140
Ovary	21,880	13,850
Stomach	21,000	10,570
Myeloma	20,180	10,650
Larynx	12,720	3,600
Uterine cervix	12,200	4,210
Cancers of soft tissues including heart	10,520	3,920
All cancers	1,529,560	569,490

Source: American Cancer Society, *Cancer Facts and Figures, 2007* (Atlanta: American Cancer Society, 2010), p. 4.

of the signals has been disrupted, which causes the cell to proliferate at an abnormally high rate. As they lose their response to the normal controls, cancer cells gradually lose their regular shape and boundaries, eventually forming a distinct mass of abnormal cells—a tumor. If the cells of the tumor remain localized, the tumor is said to be benign; if the cells invade other tissues, the tumor is said to be **malignant**. Cells that travel to other sites in the body, where they establish secondary tumors, have undergone **metastasis**.

Cancer As a Genetic Disease

Cancer arises as a result of fundamental defects in the regulation of cell division, and its study therefore has significance not only for public health, but also for our basic understanding of cell biology. Through the years, a large number of theories have been put forth to explain cancer, but we now recognize that most, if not all, cancers arise from defects in DNA.

Genetic evidence for cancer Early observations suggested that cancer might result from genetic damage. First, many agents, such as ionizing radiation and chemicals, that cause mutations also cause cancer (are carcinogens, see Chapter 18). Second, some cancers are consistently associated with particular chromosome abnormalities. About 90% of people with chronic myeloid leukemia, for example, have a reciprocal translocation between chromosome 22 and chromosome 9. Third, some specific types of cancers tend to run in families. Retinoblastoma, a rare childhood cancer of the retina, appears with high frequency in a few families and is inherited as an autosomal dominant trait, suggesting that a single gene is responsible for these cases of the disease.

Although these observations hinted that genes play some role in cancer, the theory of cancer as a genetic disease had several significant problems. If cancer is inherited, every cell in the body should receive the cancer-causing gene, and therefore every cell should become cancerous. In those types of cancer that run in families, however, tumors typically appear only in certain tissues and often only when the person reaches an advanced age. Finally, many cancers do not run in families at all and, even in regard to those cancers that generally do, isolated cases crop up in families with no history of the disease.

Knudson's multistep model of cancer In 1971, Alfred Knudson proposed a model to explain the genetic basis of cancer. Knudson was studying retinoblastoma, which usually develops in only one eye but occasionally appears in both. Knudson found that, when retinoblastoma appears in both eyes, onset is at an early age, and affected children often have close relatives who also have retinoblastoma.

Knudson proposed that retinoblastoma results from two separate genetic defects, both of which are necessary for cancer to develop (**Figure 23.3**). He suggested that, in the cases in which the disease affects just one eye, a single cell in one eye undergoes two successive mutations. Because the chance of these two mutations occurring in a single cell is remote, retinoblastoma is rare and typically develops in only one eye. For the bilateral case, Knudson proposed that the child inherited one of the two mutations required for the cancer, and so every cell contains this initial mutation. In these cases, all that is required for cancer to develop is for one eye cell to undergo the second mutation. Because each eye possesses millions of cells, the probability that the second mutation will occur in at least one cell of each eye is high, producing tumors in both eyes at an early age.

Knudson's proposal suggests that cancer is the result of a multistep process that requires several mutations. If one or more of the required mutations is inherited, fewer additional

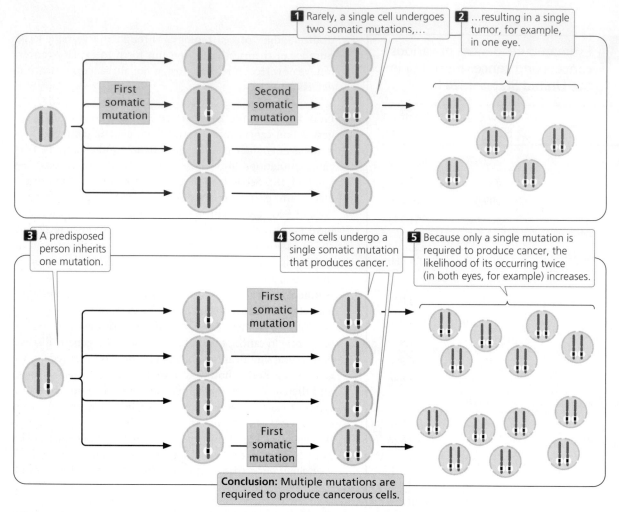

23.3 Alfred Knudson proposed that retinoblastoma results from two separate genetic defects, both of which are necessary for cancer to develop.

mutations are required to produce cancer, and the cancer will tend to run in families. The idea that cancer results from multiple mutations turns out to be correct for most cancers.

Knudson's genetic theory for cancer has been confirmed by the identification of genes that, when mutated, cause cancer. Today, we recognize that cancer is fundamentally a genetic disease, although few cancers are actually inherited. Most tumors arise from somatic mutations that accumulate in a person's life span, either through spontaneous mutation or in response to environmental mutagens.

TRY PROBLEM 23 →

The clonal evolution of tumors Cancer begins when a single cell undergoes a mutation that causes the cell to divide at an abnormally rapid rate. The cell proliferates, giving rise to a clone of cells, each of which carries the same mutation. Because the cells of the clone divide more rapidly than normal, they soon outgrow other cells. An additional mutation that arises in some of the clone's cells may further enhance the ability of those cells to proliferate, and cells carrying both mutations soon become the most common cells in the clone. Eventually, they may be overtaken by cells that contain yet

more mutations that enhance proliferation. In this process, called **clonal evolution**, the tumor cells acquire more mutations that allow them to become increasingly more aggressive in their proliferative properties (**Figure 23.4**).

The rate of clonal evolution depends on the frequency with which new mutations arise. Any genetic defect that allows more mutations to arise will accelerate cancer progression. Genes that regulate DNA repair are often found to have been mutated in the cells of advanced cancers, and inherited disorders of DNA repair are usually characterized by increased incidences of cancer. Because DNA-repair mechanisms normally eliminate many of the mutations that arise, cells with defective DNA-repair systems are more likely to retain mutations than are normal cells, including mutations in genes that regulate cell division. Xeroderma pigmentosum, for example, is a rare disorder caused by a defect in DNA repair (see p. 505 in Chapter 18). People with this condition have elevated rates of skin cancer when exposed to sunlight (which induces mutation).

Mutations in genes that affect chromosome segregation also may contribute to the clonal evolution of tumors. Many cancer cells are aneuploid, and chromosome mutations

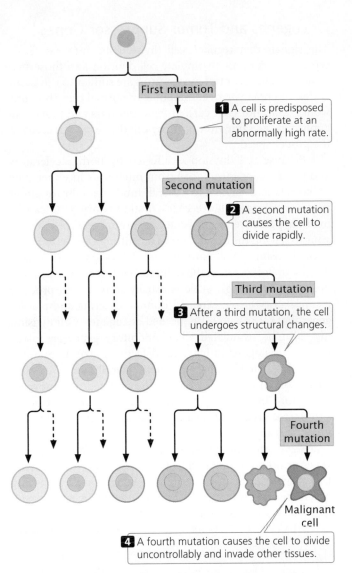

First mutation

1 A cell is predisposed to proliferate at an abnormally high rate.

Second mutation

2 A second mutation causes the cell to divide rapidly.

Third mutation

3 After a third mutation, the cell undergoes structural changes.

Fourth mutation

Malignant cell

4 A fourth mutation causes the cell to divide uncontrollably and invade other tissues.

23.4 Through clonal evolution, tumor cells acquire multiple mutations that allow them to become increasingly more aggressive and proliferative. To conserve space, a dashed arrow is used to represent a second cell of the same type in each case.

clearly contribute to cancer progression by duplicating some genes (those on extra chromosomes) and eliminating others (those on deleted chromosomes). Cellular defects that interfere with chromosome separation increase aneuploidy and may therefore accelerate cancer progression.

CONCEPTS

Cancer is fundamentally a genetic disease. Mutations in several genes are usually required to produce cancer. If one of these mutations is inherited, fewer somatic mutations are necessary for cancer to develop, and the person may have a predisposition to cancer. Clonal evolution is the accumulation of mutations in a clone of cells.

✔ CONCEPT CHECK 1

How does the multistep model of cancer explain the observation that sporadic cases of retinoblastoma usually appear in only one eye, whereas inherited forms of the cancer appear in both eyes?

The Role of Environmental Factors in Cancer

Although cancer is fundamentally a genetic disease, most cancers are not inherited, and there is little doubt that many cancers are influenced by environmental factors. The role of environmental factors in cancer is suggested by differences in the incidence of specific cancers throughout the world (**Table 23.2**). The results of studies show that migrant populations typically take on the cancer incidence of their host country. For example, the overall rates of cancer are considerably lower in Japan than in Hawaii. However, within a single generation after migration to Hawaii, Japanese people develop cancer at rates similar to those of native Hawaiians.

Smoking is a good example of an environmental factor that is strongly associated with cancer. Other environmental factors that induce cancer are certain types of chemicals, such as benzene (used as an industrial solvent), benzo[a]pyrene (found in cigarette smoke), and polychlori-

Table 23.2	Examples of geographic variation in the incidence of cancer	
Type of Cancer	Location	Incidence Rate*
Lip	Canada (Newfoundland)	15.1
	Brazil (Fortaleza)	1.2
Nasopharynx	Hong Kong	30.0
	United States (Utah)	0.5
Colon	United States (Iowa)	30.1
	India (Mumbai)	3.4
Lung	United States (New Orleans, African Americans)	110.0
	Costa Rica	17.8
Prostate	United States (Utah)	70.2
	China (Shanghai)	1.8
Bladder	United States (Connecticut, Whites)	25.2
	Philippines (Rizal)	2.8
All cancer	Switzerland (Basel)	383.3
	Kuwait	76.3

Source: C. Muir et al., *Cancer Incidence in Five Continents*, vol. 5 (Lyon: International Agency for Research on Cancer, 1987), Table 12-2.
*The incidence rate is the age-standardized rate in males per 100,000 population.

nated biphenyls (PCBs; used in industrial transformers and capacitors). Ultraviolet light, ionizing radiation, and viruses are other known carcinogens and are associated with variation in the incidence of many cancers. Most environmental factors associated with cancer cause somatic mutations that stimulate cell division or otherwise affect the process of cancer progression.

Environmental factors may interact with genetic predispositions to cancer. For example, lung cancer is clearly associated with smoking, an environmental factor. Recent genomewide association studies (see Chapter 20) have discovered that variation at several genes predisposes some people to smoking-induced lung cancer. Variants at some of these genes cause people to be more likely to become addicted to smoking. Other predisposing genes encode receptors that bind potential carcinogens in cigarette smoke.

 TRY PROBLEM 22 →

23.2 Mutations in a Number of Different Types of Genes Contribute to Cancer

As we have learned, cancer is a disease caused by alterations in the DNA. There are, however, many different types of genetic alterations that may contribute to cancer. More than 350 different human genes have been identified that contribute to cancer; the actual number is probably much higher. Research on mice suggests that more than 2000 genes can, when mutated, contribute to the development of cancer. In the next several sections, we will consider some of the different types of genes that frequently have roles in cancer.

Oncogenes and Tumor-Suppressor Genes

The signals that regulate cell division fall into two basic types: molecules that stimulate cell division and those that inhibit it. These control mechanisms are similar to the accelerator and brake of an automobile. In normal cells (but, one would hope, not your car), both accelerators and brakes are applied at the same time, causing cell division to proceed at the proper speed.

Because cell division is affected by both accelerators and brakes, cancer can arise from mutations in either type of signal, and there are several fundamentally different routes to cancer (**Figure 23.5**). A stimulatory gene can be made hyperactive or active at inappropriate times, analogous to having the accelerator of an automobile stuck in the floored position. Mutations in stimulatory genes are usually dominant because even the reduced amount of gene product produced by a single allele is usually sufficient to produce a stimulatory effect. Mutated dominant-acting stimulatory genes that cause cancer are termed **oncogenes**. Cell division may also be stimulated when inhibitory genes are made *inactive*, analogous to having a defective brake in an automobile. Mutated inhibitory genes generally have recessive effects, because both copies must be mutated to remove all inhibition. Inhibitory genes in cancer are termed **tumor-suppressor genes**. Many cancer cells have mutations in both oncogenes and tumor-suppressor genes.

Although oncogenes or mutated tumor-suppressor genes or both are required to produce cancer, mutations in DNA-repair genes can increase the likelihood of acquiring mutations in these genes. Having mutated DNA-repair genes is analogous to having a lousy car mechanic who does not make the necessary repairs on a broken accelerator or brake.

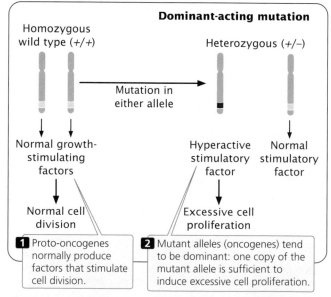

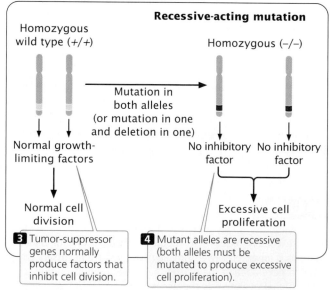

23.5 Both oncogenes and tumor-suppressor genes contribute to cancer but differ in their modes of action and dominance.

Oncogenes Oncogenes were the first cancer-causing genes to be identified. In 1909, a farmer brought physician Peyton Rous a hen with a large connective-tissue tumor (sarcoma) growing on its breast. When Rous injected pieces of this tumor into other hens, they also developed sarcomas. Rous conducted experiments that demonstrated that the tumors were being transmitted by a virus, which became known as the Rous sarcoma virus. A number of other cancer-causing viruses were subsequently isolated from various animal tissues. These viruses were generally assumed to carry a cancer-causing gene that was transferred to the host cell. The first oncogene, called *src*, was isolated from the Rous sarcoma virus in 1970.

In 1975, Michael Bishop, Harold Varmus, and their colleagues began to use probes for viral oncogenes to search for related sequences in normal cells. They discovered that the genomes of all normal cells carry DNA sequences that are closely related to viral oncogenes. These cellular genes are called **proto-oncogenes**. They are responsible for basic cellular functions in normal cells but, when mutated, they become oncogenes that contribute to the development of cancer. When a virus infects a cell, a proto-oncogene may become incorporated into the viral genome through recombination. Within the viral genome, the proto-oncogene may mutate to an oncogene that, when inserted back into a cell, causes rapid cell division and cancer. Because the proto-oncogenes are more likely to undergo mutation or recombination within a virus, viral infection is often associated with the cancer.

Proto-oncogenes can be converted into oncogenes in viruses by several different ways. The sequence of the proto-oncogene may be altered or truncated as it is being incorporated into the viral genome. This mutated copy of the gene may then produce an altered protein that causes uncontrolled cell proliferation. Alternatively, through recombination, a proto-oncogene may end up next to a viral promoter or enhancer, which then causes the gene to be overexpressed. Finally, sometimes the function of a proto-oncogene in the host cell may be altered when a virus inserts its own DNA into the gene, disrupting its normal function.

Many oncogenes have been identified by experiments in which selected fragments of DNA are added to cells in culture. Some of the cells take up the DNA and, if these cells become cancerous, then the DNA fragment that was added to the culture must contain an oncogene. The fragments can then be sequenced, and the oncogene can be identified. A large number of oncogenes have now been discovered (**Table 23.3**). About 90% of all cancer genes are thought to be dominant oncogenes.

Tumor-suppressor genes Tumor-suppressor genes are more difficult than oncogenes to identify because they *inhibit* cancer and are recessive; both alleles must be mutated before the inhibition of cell division is removed. Because the *failure* of their function promotes cell proliferation, tumor-suppressor genes cannot be identified by adding them to cells and looking for cancer. About 10% of cancer-causing genes are thought to be tumor-suppressor genes.

Table 23.3	Some oncogenes and functions of their corresponding proto-oncogenes	
Oncogene	Cellular Location of Product	Function of Proto-oncogene
sis	Secreted	Growth factor
erbB	Cell membrane	Part of growth-factor receptor
erbA	Cytoplasm	Thyroid-hormone receptor
src	Cell membrane	Protein tyrosine kinase
ras	Cell membrane	GTP binding and GTPase
myc	Nucleus	Transcription factor
fos	Nucleus	Transcription factor
jun	Nucleus	Transcription factor
bcl-1	Nucleus	Cell cycle

Defects in both copies of a tumor-suppressor gene are usually required to cause cancer; an organism can inherit one defective copy of the tumor-suppressor gene (is heterozygous for the cancer-causing mutation) and not have cancer, because the remaining normal allele produces the tumor-suppressing product. However, these heterozygotes are often predisposed to cancer, because inactivation or loss of the one remaining allele is all that is required to completely eliminate the tumor-suppressor product. Inactivation of the remaining wild-type allele in heterozygotes is referred to as **loss of heterozygosity**. A common mechanism for loss of heterozygosity is a deletion on the chromosome that carried the normal copy of the tumor-suppressor gene (**Figure 23.6**).

Among the first tumor-suppressor genes to be identified was the retinoblastoma gene. In 1985, Raymond White and Webster Cavenne showed that large segments of chromosome 13 were missing in cells of retinoblastoma tumors, and, later, the tumor-suppressor gene was isolated from these segments. A number of tumor-suppressor genes have now been discovered (**Table 23.4**).

Sometimes the mutation or loss of a single allele of a recessive tumor-suppressor gene is sufficient to cause cancer. This effect—the appearance of the trait in an individual cell or organism that is heterozygous for a normally recessive trait—is called **haploinsufficiency**. This phenomenon is thought to be due to dosage effects: the heterozygote produces only half as much of the product encoded by the tumor-suppressing gene. Normally, this amount is sufficient for the cellular processes that prevent tumor formation, but it is less than the optimal amount, and other factors may sometimes combine with the lowered tumor-suppressor product to cause cancer.

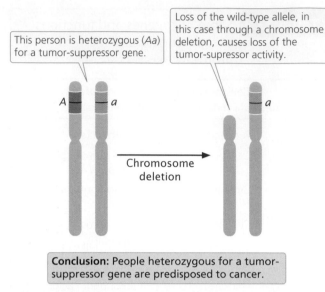

This person is heterozygous (*Aa*) for a tumor-suppressor gene.

Loss of the wild-type allele, in this case through a chromosome deletion, causes loss of the tumor-supressor activity.

Chromosome deletion

Conclusion: People heterozygous for a tumor-suppressor gene are predisposed to cancer.

23.6 Loss of heterozygosity often leads to cancer in a person heterozygous for a tumor-suppressor gene.

Haploinsufficiency is seen in some inherited predispositions to cancer. Bloom syndrome is an autosomal recessive disease characterized by short stature, male infertility, and a predisposition to cancers of many types. The disease results from a defect in the *BLM* locus, which encodes a DNA helicase enzyme that plays a key role in the repair of double-strand breaks. Persons homozygous for mutations at the *BLM* locus have a greatly elevated risk of cancer. Persons heterozygous for mutations at the *BLM* locus were thought to be unaffected. However, a recent survey of Ashkenazi Jews (who have a high frequency of Bloom syndrome) showed that heterozygous carriers of a *BLM* mutation were at increased risk of colorectal cancer. Similarly, mice with one mutated copy of the *BLM* gene are more than twice as likely to develop intestinal tumors as are mice with no *BLM* mutations. **TRY PROBLEM 24 →**

Table 23.4	Some tumor-suppressor genes and their functions	
Gene	Cellular Location of Product	Function
NF1	Cytoplasm	GTPase activator
p53	Nucleus	Transcription factor, regulates apoptosis
RB	Nucleus	Transcription factor
WT-1	Nucleus	Transcription factor

Source: J. Marx, Learning how to suppress cancer, *Science* 261:1385, 1993.

CONCEPTS

Proto-oncogenes are genes that control normal cellular functions. When mutated, proto-oncogenes become oncogenes that stimulate cell proliferation. They tend to be dominant in their action. Tumor-suppressor genes normally inhibit cell proliferation; when mutated, they allow cells to proliferate. Tumor-suppressor genes tend to be recessive in their action. Individual organisms that are heterozygous for tumor-suppressor genes are often predisposed to cancer.

✔ CONCEPT CHECK 2

Why are oncogenes usually dominant in the their action, whereas tumor-suppressor genes are recessive?

Genes That Control the Cycle of Cell Division

The cell cycle is the normal process by which cells undergo growth and division. Normally, progression through the cell cycle is tightly regulated so that cells divide only when additional cells are needed, when all the components necessary for division are present, and when the DNA has been replicated without damage. Sometimes, however, errors arise in one or more of the components that regulate the cell cycle. These errors often cause cells to divide at inappropriate times or rates, leading to cancer. Indeed, many proto-oncogenes and tumor-suppressor genes function normally by helping to control the cell cycle. Before considering how errors in this system contribute to cancer, we must first understand how the cell cycle is usually regulated.

Control of the cell cycle As discussed in Chapter 2, the cell cycle consists of the period from one cell division to the next. Cells that are actively dividing pass through the G_1, S, and G_2 phases of interphase and then move directly into the M phase, when cell division takes place. Nondividing cells exit from G_1 into the G_0 stage, in which they are functional but not actively growing or dividing. Progression from one stage of the cell cycle to another is influenced by a number of internal and external signals and is regulated at key points in the cycle called checkpoints.

For many years, the biochemical events that control the progression of cells through the cell cycle were completely unknown, but research findings have now revealed many of the details of this process. Key events of the cell cycle are controlled by **cyclin-dependent kinases** (CDKs), which are enzymes that add phosphate groups to other proteins. Sometimes, phosphorylation activates the other protein and, other times, it inactivates the protein. As their name implies, CDKs are functional only when they associate with another protein called a **cyclin**. The level of cyclin oscillates in the course of the cell cycle; when bound to a CDK, cyclin specifies which proteins the CDK will phosphorylate. Each cyclin appears at a specific point in the cell cycle, usually because its synthesis and destruction

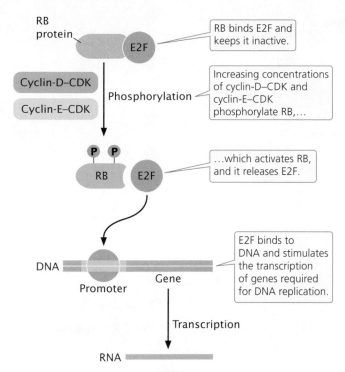

RB protein

RB binds E2F and keeps it inactive.

Cyclin-D–CDK

Cyclin-E–CDK

Phosphorylation

Increasing concentrations of cyclin-D–CDK and cyclin-E–CDK phosphorylate RB,…

RB   E2F

…which activates RB, and it releases E2F.

DNA

Promoter   Gene

E2F binds to DNA and stimulates the transcription of genes required for DNA replication.

Transcription

RNA

23.7 The RB protein helps control the progression through the G$_1$/S checkpoint by binding transcription factor E2F.

are regulated by another cyclin. Cyclins and CDKs are called by different names in different organisms; here, we will use the terms applied to these molecules in mammals.

G$_1$-to-S transition Let's begin by looking at the G$_1$-to-S transition. As mentioned earlier, progression through the cell cycle is regulated at several checkpoints, which ensure that all cellular components are present and in good working order before the cell proceeds to the next stage. The G$_1$/S checkpoint is in G$_1$, just before the cell enters into the S phase and replicates its DNA. The cell is prevented from passing through the G$_1$/S checkpoint by a molecule called the retinoblastoma (RB) protein (**Figure 23.7**), which binds to another molecule called E2F and keeps it inactive. In G$_1$, cyclin D and cyclin E continuously increase in concentration and combine with their associated CDKs. Cyclin-D–CDK and cyclin-E–CDK both phosphorylate molecules of RB. Late in G$_1$, the phosphorylation of RB is completed, which inactivates RB. Without the inhibitory effects of RB, the E2F protein is released. E2F is a transcription factor that stimulates the transcription of genes that produce enzymes necessary for the replication of DNA, and the cell moves into the S stage of the cell cycle.

G$_2$-to-M transition Regulation of the G$_2$-to-M transition is similar to that of the G$_1$-to-S transition. In the G$_2$-to-M transition, cyclin B combines with a CDK to form an inactive complex called *mitosis-promoting factor* (MPF). After MPF has been formed, it must be activated by the

removal of a phosphate group (**Figure 23.8a**). During G$_1$, cyclin B levels are low; so the amount of MPF also is low. As more cyclin B is produced, it combines with CDK to form increasing amounts of MPF. Near the end of G$_2$, the amount of active MPF reaches a critical level, which commits the cell to divide. The MPF concentration continues to increase, reaching a peak in mitosis.

The active form of MPF phosphorylates other proteins, which then bring about many of the events associated with mitosis, such as nuclear-membrane breakdown, spindle formation, and chromosome condensation. At the end of metaphase, cyclin B is abruptly degraded, which lowers the amount of MPF and, initiating anaphase, sets in motion a chain of events that ultimately brings mitosis to a close (**Figure 23.8b**). In brief, high levels of active MPF stimulate mitosis, and low levels of MPF bring a return to interphase conditions.

The G$_2$/M checkpoint is at the end of G$_2$, before the cell enters mitosis. A number of factors stimulate the synthesis of cyclin B and the activation of MPF, whereas other factors inhibit MPF. Together, these factors ensure that mitosis is not initiated until conditions are appropriate for cell division. For example, DNA damage inhibits the activation of MPF; consequently, the cell is arrested in G$_2$ and does not undergo division.

Spindle-assembly checkpoint Yet another checkpoint, called the spindle-assembly checkpoint, is in metaphase. This checkpoint delays the onset of anaphase until all chromosomes are aligned on the metaphase plate and sister kinetochores are attached to spindle fibers from opposite poles. If all chromosomes are not properly aligned, the checkpoint blocks the destruction of cyclin B. The persistence of cyclin B keeps MPF active and maintains the cell in a mitotic state. An additional checkpoint controls the cell's exit from mitosis.

Mutations in cell-cycle control and cancer Many cancers are caused by defects in the cell cycle's regulatory machinery. For example, mutations in the gene that encodes the RB protein—which normally holds the cell in G$_1$ until the DNA is ready to be replicated—are associated with many cancers, including retinoblastoma (from which the RB protein gets it name). When the *RB* gene is mutated, cells pass through the G$_1$/S checkpoint without the normal controls that prevent cell proliferation. The gene that encodes cyclin D (thus stimulating the passage of cells through the G$_1$/S checkpoint) is overexpressed in about 50% of all breast cancers, as well as some cases of esophageal and skin cancer. Likewise, the tumor-suppressor gene *p53*, which is mutated in about 75% of all colon cancers, regulates a potent inhibitor of CDK activity.

Some proto-oncogenes and tumor-suppressor genes have roles in apoptosis. Cells have the ability to assess themselves and, when they are abnormal or damaged, they normally undergo apoptosis (see pp. 623–625 in Chapter 22). Cancer

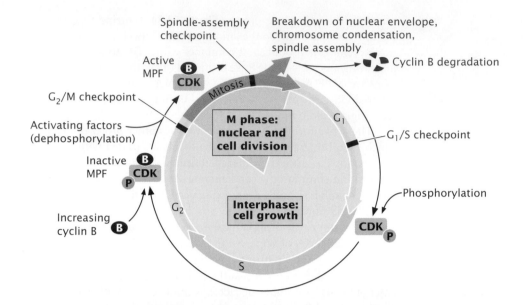

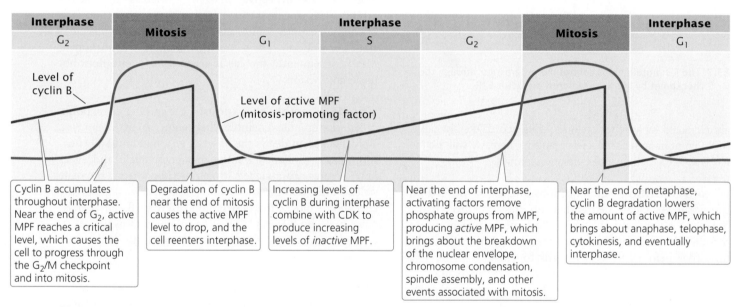

23.8 Progression through the G₂/M checkpoint is regulated by cyclin B.

cells frequently have chromosome mutations, DNA damage, and other cellular anomalies that would normally stimulate apoptosis and prevent their proliferation. Often, these cells have mutations in genes that regulate apoptosis, and therefore they do not undergo programmed cell death. The ability of a cell to initiate apoptosis in response to DNA damage, for example, depends on *p53*, which is inactive in many human cancers.

CONCEPTS

Progression through the cell cycle is controlled at checkpoints, which are regulated by interactions between cyclins and cyclin-dependent kinases. Genes that control the cell cycle are frequently mutated in cancer cells.

✔ CONCEPT CHECK 3

What would be the most likely effect of a mutation that causes cyclin B to be unable to bind to CDK?

a. Cells pass through the G₂/M checkpoint and enter mitosis even when DNA has not been replicated.

b. Cells never pass through the G₁/S checkpoint.

c. Cells pass through mitosis more quickly than unmutated cells.

d. Cells fail to pass the G₂/M checkpoint and do not enter into mitosis.

Signal-transduction pathways Whether cells pass through the cell cycle and continue to divide is influenced by a large number of internal and external signals. External signals are initiated by hormones and growth factors. These molecules are often unable to pass through the cell membrane because of their size or charge; they exert their effects by binding to receptors on the cell surface, which triggers a series of intracellular reactions that then carry the message to the nucleus or other site within the cell. This type of system, in which an external signal triggers a cascade of intracellular reactions that ultimately produce a specific response, is called a **signal-transduction pathway**. Defects in signal-transduction pathways are often associated with cancer.

A signal-transduction pathway begins with the binding of an external signaling molecule to a specific receptor that is embedded in the cell membrane. Receptors in signal-transduction pathways usually have three parts: (1) an extracellular domain that protrudes from the cell and binds the signaling molecule; (2) a transmembrane domain that passes across the membrane and conducts the signal to the interior of the cell; (3) an intracellular domain that extends into the cytoplasm and, on the binding of the signaling molecule, undergoes a chemical or conformational change that is transmitted to molecules of the signal-transduction pathway in the cytoplasm. The binding of a signaling molecule to the membrane-bound receptor activates a protein in the pathway. On activation, this protein activates the next molecule in the pathway, often by adding or removing phosphate groups or causing changes in the conformation of the protein. The newly activated protein activates the next molecule in the pathway, and in this way, the signal is passed along through a cascade of reactions and ultimately produces the response, such as stimulating or inhibiting the cell cycle.

In the past decade, much research has been conducted to determine the pathways by which various signals influence the cell cycle. To illustrate signal transduction, let's consider the Ras signal-transduction pathway, a pathway that plays an important role in control of the cell cycle. Each Ras protein cycles between an active form and an inactive form. In the inactive form, the Ras protein is bound to guanosine diphosphate (GDP); in the active form, it is bound to guanosine triphosphate (GTP).

The Ras signal-transduction pathway is activated when a growth factor, such as epidermal growth factor (EGF), binds to a receptor on the cell membrane (**Figure 23.9**). The binding of EGF causes a conformational change in the receptor and the addition of phosphate groups to it. The addition of the phosphate groups allows adaptor molecules to bind to the receptor. These adaptor molecules link the receptor with an inactive molecule of Ras protein. The adaptor molecules stimulate Ras to release GDP and bind GTP, activating Ras. The newly activated Ras protein then binds to an inactive form of another protein called Raf and activates it. After activating Raf, Ras hydrolyzes GTP to GDP, which converts Ras back into the inactive form.

Activated Raf then sets in motion a cascade of reactions, ending in the activation of a protein called MAP kinase. Activated MAP kinase moves into the nucleus and activates a number of transcription factors that stimulate the transcription of genes taking part in the cell cycle. In this way, the

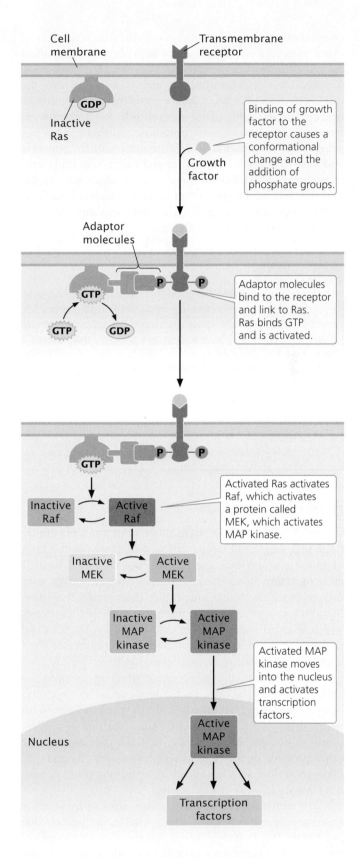

23.9 The Ras signal-transduction pathway conducts signals from growth factors and hormones to the nucleus and stimulates the cell cycle. Mutations in this pathway often contribute to cancer.

original external signal promotes cell division. A number of other transduction pathways have been identified that affect the cell cycle and cell proliferation.

Because signal-transduction pathways help control the cell cycle, defects in their components often contribute to cancer. For example, genes that encode Ras proteins are frequently oncogenes, and mutations in these genes are often found in cancer cells: 75% of tumors of the pancreas and 50% of those in the thyroid and colon have mutations in *ras* genes. Mutations in these genes produce mutant Ras proteins that can bind GTP but cannot hydrolyze it to GDP, and so they are stuck in the activated form and therefore continuously stimulate cell division.

CONCEPTS

Molecules outside the cell often bring about intracellular responses by binding to a membrane receptor and stimulating a cascade of intracellular reactions, known as a signal-transduction pathway. Many molecules in the pathway are proteins that alternate between active and inactive forms. Defects in signal-transduction pathways are often associated with cancer.

✔ CONCEPT CHECK 4

Ras proteins are activated when they

a. bind GTP.   c. bind GDP.

b. release GTP.   d. undergo acetylation.

DNA-Repair Genes

Cancer arises from the accumulation of multiple mutations in a single cell. Some cancer cells have normal rates of mutation, and multiple mutations accumulate because each mutation gives the cell a replicative advantage over cells without the mutations. Other cancer cells may have higher-than-normal rates of mutation in all of their genes, which leads to more-frequent mutation of oncogenes and tumor-suppressor genes. What might be the source of these high rates of mutation in some cancer cells?

Two processes control the rate at which mutations arise within a cell: (1) the rate at which errors arise in the course of replication and afterward and (2) the efficiency with which these errors are corrected. The error rate in replication is controlled by the fidelity of DNA polymerases and other proteins in the replication process (see Chapter 12). However, defects in genes encoding replication proteins have not been strongly linked to cancer.

The mutation rate is also strongly affected by whether errors are corrected by DNA-repair systems (see pp. 500–506 in Chapter 18). Defects in genes that encode components of these repair systems have been consistently associated with a number of cancers. People with xeroderma pigmentosum, for example, are defective in nucleotide-excision repair, an important cellular repair system that normally corrects

DNA damage caused by a number of mutagens, including ultraviolet light. Likewise, about 13% of colorectal, endometrial, and stomach cancers have cells that are defective in mismatch repair, another major repair system in the cell.

A particular type of colon cancer called nonpolyposis colorectal cancer is inherited as an autosomal dominant trait. In families with this condition, a person can inherit one mutated and one normal allele of a gene that controls mismatch repair. The normal allele provides sufficient levels of the protein for mismatch repair to function, but it is highly likely that this normal allele will become mutated or lost in at least a few cells. If it does so, there is no mismatch repair, and these cells undergo higher-than-normal rates of mutation, leading to defects in oncogenes and tumor-suppressor genes that cause the cells to proliferate.

Defects in DNA-repair systems may also contribute to the generation of chromosome rearrangements and genomic instability. Many DNA-repair systems make single- and double-strand breaks in the DNA. If these breaks are not repaired properly, then chromosome rearrangements often result.

Genes That Regulate Telomerase

Another factor that may contribute to the progression of cancer is the inappropriate activation of an enzyme called telomerase. Telomeres are special sequences at the ends of eukaryotic chromosomes. Recall that the ends of chromosomes cannot be replicated, and telomeres become shorter with each cell division. This shortening eventually leads to the destruction of the chromosome and cell death; so somatic cells are capable of a limited number of cell divisions.

In germ cells, telomerase replicates the chromosome ends (see pp. 340–342 in Chapter 12), thereby maintaining the telomeres, but this enzyme is not normally expressed in somatic cells. In many tumor cells, however, sequences that regulate the expression of the telomerase gene are mutated, allowing the enzyme to be expressed, and the cell is capable of unlimited cell division. Although the expression of telomerase appears to contribute to the development of many cancers, its precise role in tumor progression is unknown and under investigation.

Genes That Promote Vascularization and the Spread of Tumors

A final set of factors that contribute to the progression of cancer includes genes that affect the growth and spread of tumors. Oxygen and nutrients, which are essential to the survival and growth of tumors, are supplied by blood vessels, and the growth of new blood vessels (angiogenesis) is important to tumor progression. Angiogenesis is stimulated by growth factors and others proteins encoded by genes whose expression is carefully regulated in normal cells. In tumor cells, genes encoding these proteins are often overexpressed compared with normal cells, and inhibitors of angiogenesis-promoting factors may be inactivated or underexpressed. At least one inherited cancer syndrome—van Hippel–Lindau disease, in which people develop multiple types of tumors—is caused by the mutation of a gene that affects angiogenesis.

In the development of many cancers, the primary tumor gives rise to cells that spread to distant sites, producing secondary tumors. This process of metastasis is the cause of death in 90% of human cancer cases; it is influenced by cellular changes induced by somatic mutation. As discussed in the introduction to this chapter, the *palladin* gene, when mutated, contributes to the metastasis of pancreatic tumors. By using microarrays to measure levels of gene expression, researchers have identified other genes that are transcribed at a significantly higher rate in metastatic cells compared with nonmetastatic cells. For example, one study detected a set of 95 genes that were overexpressed or underexpressed in a population of metastatic breast-cancer cells that were strongly metastatic to the lung, compared with a population of cells that were only weakly metastatic to the lung. Genes that contribute to metastasis often encode components of the extracellular matrix and the cytoskeleton. Others encode adhesion proteins, which help hold cells together.

Advances in sequencing technology have now made possible the complete sequencing of the DNA of tumor cells to see how their genomes differ from those of normal cells. In one experiment, researchers sequenced the entire genome of cells from a metastasized breast-cancer tumor and compared it with the genome of noncancer cells from the same person. They also compared the genome of the metastasized tumor with the genome of the primary tumor (from which the metastasis originated), which had been removed from the patient 9 years earlier. The researchers found 32 different somatic mutations in the coding regions of genes from the tumor cells, 19 of which were not detected in the primary tumor. This finding suggests that the metastasized tumor underwent considerable genetic changes in its 9-year evolution from the primary tumor. In contrast, another study of a breast-cancer metastasis found only 2 mutations that were not present in the primary tumor but, in this case, the metastasis had evolved in only 1 year.

CONCEPTS

Mutations in genes that encode components of DNA-repair systems are often associated with cancer; these mutations increase the rate at which mutations are retained and result in an increased number of mutations in proto-oncogenes, tumor-suppressor genes, and other genes that contribute to cell proliferation. Mutations that allow telomerase to be expressed in somatic cells and those that affect vascularization and metastasis also may contribute to cancer progression.

✔ **CONCEPT CHECK 5**

Which type of mutation in telomerase could be associated with cancer cells?

---

a. Mutations that produce an inactive form of telomerase

b. Mutations that decrease the expression of telomerase

c. Mutations that increase the expression of telomerase

d. All of the above

MicroRNAs and Cancer

MicroRNAs (miRNAs) are a class of small RNA molecules that pair with complementary sequences on mRNA and degrade the mRNA or inhibit its translation (see Chapter 14). Given the fact that miRNAs are important in controlling gene expression and development, it is not surprising that researchers are finding that they are also associated with tumor development. Many tumor cells exhibit widespread reduction in the expression of many miRNAs. Recently, researchers genetically engineered mouse tumor cells that lacked the machinery to generate miRNAs and found that these cells showed enhanced tumor progression when implanted into mice. Interestingly, this effect was seen only in cells that had already initiated tumor development, suggesting that miRNAs play a role in later stages of tumor progression.

Lowered levels of miRNAs may contribute to cancer by allowing oncogenes that are normally controlled by the miRNAs to be expressed at high levels. For example, let-7 miRNA normally controls the expression of the *ras* oncogene, probably by binding to complementary sequences in the 3' untranslated region of mRNA and inhibiting translation. In lung-cancer cells, levels of let-7 miRNA are often low, allowing the Ras protein to be highly expressed, which then leads to the development of lung cancer.

A transcription factor called c-MYC is often expressed at high levels in cancer cells. Evidence suggests that c-MYC helps to drive cell proliferation and the development of cancer. Among other effects, c-MYC binds to the promoters of miRNA genes and decreases their transcription, decreasing the abundance of the miRNAs. Some of these miRNAs are known to suppress tumor development. Research has shown that, if, through genetic manipulation, the miRNAs are expressed at high levels, the development of tumors decreases. All of these findings suggest that altered expression of miRNAs plays an important role in cancer.

Several miRNAs have been implicated in the process of metastasis. A particular miRNA called miR-10b has been associated with the formation of metastatic breast tumors. In one experiment, investigators manipulated a line of breast-cancer cells so that miR-10b was overexpressed. When the manipulated cells were injected into mice, many of the mice developed metastatic tumors. In contrast with the preceding examples in which the lower expression of miRNA is associated with cancer, here high levels of miRNA appear to promote the spread of cancer cells. Further study revealed that, in humans, the levels of miR-10b are elevated in metastatic tumors compared with tumors in metastasis-free patients. miR-10b regulates the expression of a number of other genes, including some that are known to suppress the spread of tumor cells. Other miRNAs are known to inhibit metastasis.

The Cancer Genome Project

Formed in 2008, the International Cancer Genome Consortium coordinates efforts to determine the genomic sequences of tumors. A goal of the consortium is to completely sequence 500 tumors from each of 50 different types of cancer, along with the genomes of normal tissues from the same persons. This effort is already producing important results, revealing the numbers and types of mutations that are associated with particular cancers. The hope is that new cancer-causing genes will be identified, which will lead to a better understanding of the nature of cancer and suggest new targets for cancer treatment. The genomes of a number of tumors have been sequenced and many more are currently being sequenced. For example, the entire genome of a small-cell lung carcinoma (a type of lung cancer) was sequenced in 2010 and compared with the genome of normal cells from the same person. More than 22,000 base-pair mutations were identified in the tumor, of which 134 were within protein-encoding genes. The tumor also possessed 58 chromosome rearrangements and 334 copy-number variations (see Chapter 20).

The large number of mutations found in cancer genomes can be divided into two types, called drivers and passengers. Drivers are mutations that drive the cancer process: they directly contribute to the development of cancer. Drivers include mutations in oncogenes, tumor-suppressor genes, DNA-repair genes, and the other types of cancer genes discussed in this chapter. Passengers are mutations that arise randomly in the process of tumor development and do not contribute to the cancer process. Many of the passengers are in introns (regions between genes) and other DNA that is not transcribed and translated, but they can also arise within protein-encoding genes. A major challenge is to determine which of the numerous mutations found in tumors are drivers and actually contribute to the development of cancer and which are passengers with no effect.

23.3 Changes in Chromosome Number and Structure Are Often Associated with Cancer

Most tumors contain cells with chromosome mutations. For many years, geneticists argued about whether these chromosome mutations were the cause or the result of cancer. Some types of tumors are consistently associated with *specific* chromosome mutations; for example, most cases of chronic myelogenous leukemia are associated with a reciprocal translocation between chromosomes 22 and 9. These types of associations suggest that chromosome mutations contribute

to the cause of the cancer. Yet many cancers are not associated with specific types of chromosome abnormalities, and individual *gene* mutations are now known to contribute to many types of cancer. Nevertheless, chromosome instability is a general feature of cancer cells, causing them to accumulate chromosome mutations, which then affect individual genes that may contribute to the cancer process. Thus, chromosome mutations appear to both *cause* cancer and *result* from it.

At least three types of chromosome rearrangements—deletions, inversions, and translocations—are associated with certain types of cancer. Deletions may result in the loss of one or more tumor-suppressor genes. Inversions and translocations contribute to cancer in several ways. First, the chromosomal breakpoints that accompany these mutations may lie within tumor-suppressor genes, disrupting their function and leading to cell proliferation. Second, translocations and inversions may bring together sequences from two different genes, generating a fused protein that stimulates some aspect of the cancer process.

Fusion proteins are seen in most cases of chronic myelogenous leukemia, a form of leukemia affecting bone-marrow cells. As mentioned earlier, most patients with chronic myelogenous leukemia have a reciprocal translocation between the long arm of chromosome 22 and the tip of the long arm of chromosome 9 (**Figure 23.10**). This translocation produces a shortened chromosome 22, called the Philadelphia chromosome because it was first discovered in Philadelphia. At the end of a normal chromosome 9 is

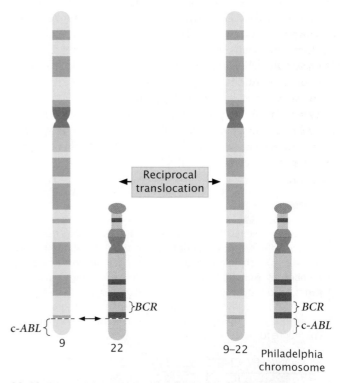

23.10 A reciprocal translocation between chromosomes 9 and 22 causes chronic myelogenous leukemia.

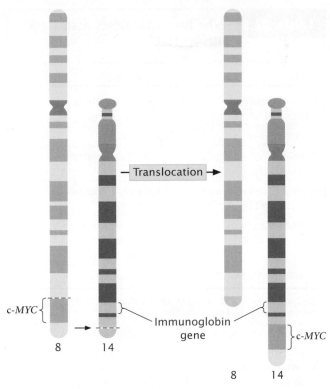

23.11 A reciprocal translocation between chromosomes 8 and 14 causes Burkitt lymphoma.

Many tumors contain a variety of types of chromosome mutations. Some tumors are associated with specific deletions, inversions, and translocations. Deletions can eliminate or inactivate genes that control the cell cycle; inversions and translocations can cause breaks in genes that suppress tumors, fuse genes to produce cancer-causing proteins, or move genes to new locations, where they are under the influence of different regulatory sequences.

✔ **CONCEPT CHECK 6**

Chronic myelogenous leukemia is usually associated with which type of chromosome rearrangement?

a. Duplication   c. Inversion

b. Deletion   d. Translocation

Most advanced tumors contain cells that exhibit a dramatic variety of chromosome anomalies, including extra chromosomes, missing chromosomes, and chromosome rearrangements (**Figure 23.12**). Some cancer researchers believe that cancer is initiated when genetic changes take place that cause the genome to become unstable, generating numerous chromosome abnormalities that then alter the expression of oncogenes and tumor-suppressor genes.

A number of genes that contribute to genomic instability and lead to missing or extra chromosomes (aneuploidy) have now been identified. Aneuploidy in somatic cells usually arises when chromosomes do not segregate properly in mitosis. Normal cells have a spindle-assembly checkpoint

a potential cancer-causing gene called c-*ABL*. As a result of the translocation, part of the c-*ABL* gene is fused with the *BCR* gene from chromosome 22. The protein produced by this *BCR*–c-*ABL* fusion gene is much more active than the protein produced by the normal c-*ABL* gene; the fusion protein stimulates increased, unregulated cell division and eventually leads to leukemia.

A third mechanism by which chromosome rearrangements may produce cancer is by the transfer of a potential cancer-causing gene to a new location, where it is activated by different regulatory sequences. Burkitt lymphoma is a cancer of the B cells, the lymphocytes that produce antibodies. Many people with Burkitt lymphoma possess a reciprocal translocation between chromosome 8 and chromosome 2, 14, or 22 (**Figure 23.11**). This translocation relocates a gene called c-*MYC* from the tip of chromosome 8 to a position in chromosome 2, 14, or 22 that is next to a gene that encodes an immunoglobulin protein. At this new location, c-*MYC*, a cancer-causing gene, comes under the control of regulatory sequences that normally activate the production of immunoglobulins, and c-*MYC* is expressed in B cells. The c-*MYC* protein stimulates the division of the B cells and leads to Burkitt lymphoma.

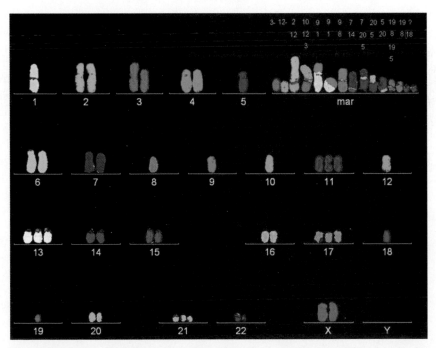

23.12 Cancer cells often possess chromosome abnormalities, including extra chromosomes, missing chromosomes, and chromosome rearrangements. Shown here are chromosomes from a colon-cancer cell, which has numerous chromosome abnormalities. [Courtesy of Dr. Peter Duesberg, University of California at Berkeley.]

that monitors the proper assembly of the mitotic spindle (see p. 645); if chromosomes are not properly attached to the microtubules at metaphase, the onset of anaphase is blocked. Some aneuploid cancer cells contain mutant alleles for genes that encode proteins having roles in this checkpoint; in these cells, anaphase is entered despite the improper assembly of the spindle or lack of it, and chromosome abnormalities result. For example, mutations in *RB* increase aneuploidy by increasing the expression of a protein called Mad2, which is a critical component of the spindle-assembly checkpoint.

Mutations in genes that encode parts of the spindle apparatus also may contribute to abnormal segregation and lead to chromosome abnormalities. *APC* is a tumor-suppressor gene that is often mutated in colon-cancer cells. *APC* has several functions, one of which is to interact with the ends of the microtubules that associate with the kinetochore. Dividing mouse cells that have defective copies of the *APC* gene give rise to cells with many chromosome defects.

The tumor-suppressor gene *p53*, in addition to controlling apoptosis, plays a role in the duplication of the centrosome, which is required for proper formation of the spindle and for chromosome segregation. Normally, the centrosome duplicates once per cell cycle. If *p53* is mutated or missing, however, the centrosome may undergo extra duplications, resulting in the unequal segregation of chromosomes. In this way, mutation of the *p53* gene may generate chromosome mutations that contribute to cancer. The *p53* gene is also a tumor-suppressor gene that prevents cell division when the DNA is damaged. **TRY PROBLEM 30 →**

23.4 Viruses Are Associated with Some Cancers

As mentioned earlier in the chapter, viruses are responsible for a number of cancers in animals, and there is evidence that some viruses contribute to at least a few cancers in humans (**Table 23.5**). For example, about 95% of all women with cervical cancer are infected with **human papilloma viruses** (HPVs). Similarly, infection with the virus that causes hepatitis B increases the risk of liver cancer in some people. The Epstein–Barr virus, which is responsible for mononucleosis, has been linked to several types of cancer that are prevalent in parts of Africa, including Burkitt lymphoma.

Retroviruses and cancer Many of the viruses that cause cancer in animals are retroviruses; earlier, we saw how studies of the Rous sarcoma retrovirus in chickens led to the identification of oncogenes in humans. Retroviruses sometimes cause cancer by mutating and rearranging host genes, converting proto-oncogenes into oncogenes (**Figure 23.13a**). Another way that viruses may contribute to cancer is by altering the expression of host genes (**Figure 23.13b**). Retroviruses often contain strong promoters to ensure that their own genetic material is transcribed by the host cell. If

Table 23.5	Some human cancers associated with viruses
Virus	**Cancer**
Human papilloma viruses (HPVs)	Cervical, penile, and vulvar cancers
Hepatitis B virus	Liver cancer
Human T-cell leukemia virus 1 (HTLV-1)	Adult T-cell leukemia
Human T-cell leukemia virus 2 (HTLV-2)	Hairy-cell leukemia
Epstein–Barr virus	Burkitt lymphoma, nasopharyngeal cancer, Hodgkin lymphoma
Human herpes virus	Kaposi sacrcoma

Note: Some of these associations between cancer and viruses exist only in certain populations and geographic areas.

the provirus inserts near a proto-oncogene, viral promoters may stimulate high levels of expression of the proto-oncogene, leading to cell proliferation.

There are only a few retroviruses that cause cancer in humans. HTLV-1, the first human retrovirus discovered, is associated with human adult T-cell leukemia. Other human cancers are associated with DNA viruses, which, like retroviruses, integrate into the host chromosome but, unlike retroviruses, do not utilize reverse transcription. Human papilloma virus, for example, is a DNA virus that is strongly associated with cervical cancer, as mentioned earlier.

Human papilloma virus and cervical cancer Human papilloma virus causes warts and other types of benign tumors of epithelial cells. More than 100 different types of HPV are known, about 30 of which are sexually transmitted and cause genital warts. A few of them are associated with cervical cancer. In the United States, 70% of the cases of cervical cancer are caused by HPV-16 and HPV-18. These viruses cause cervical cancer by producing a protein that attaches to and inactivates RB and p53, two proteins that play key roles in the regulation of the cell cycle. When these proteins are inactivated, cells are stimulated to progress through the cell cycle and divide without the normal controls that prevent cell proliferation.

About 75% of sexually active women in the United States are infected with HPV, but only a small number of these women will ever develop cervical cancer. The risk of infection by HPV can be reduced by limiting the number of sexual partners and by a new vaccine that was approved by the U.S. Food and Drug Administration in 2006. This vaccine is highly effective against four of the most common HPVs associated with cervical cancer.

(a)

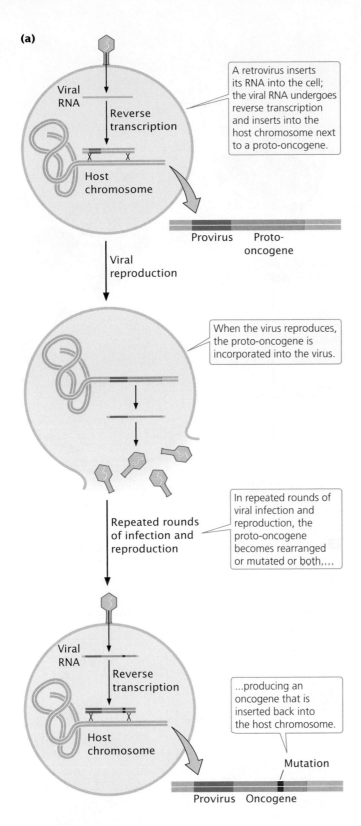

(b)

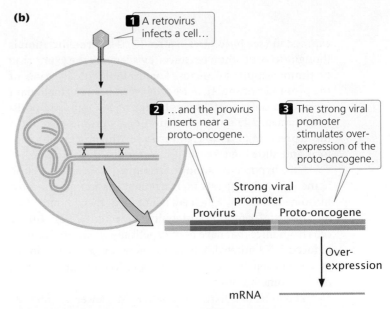

23.13 Retroviruses cause cancer by (a) mutating and rearranging proto-oncogenes and (b) inserting strong promoters near proto-oncogenes.

Although rare in the United States, cervical cancer is the second most common cause of cancer in women worldwide, with high incidences in many developing countries such as those of Sub-Saharan Africa, South Asia, and South and Central America. An estimated 510,000 women throughout the world develop cervical cancer each year, and 288,000 die from the disease. The primary reason for the high incidence and death rate in developing regions is lack of access to the cervical-cancer screening procedures such as the Pap test. The availability of the vaccine against cervical cancer in these developing countries could greatly reduce the incidence of cervical cancer.

> **CONCEPTS**
>
> Viruses contribute to a few cancers in humans by mutating and rearranging host genes that then contribute to cell proliferation and by altering the expression of host genes.

23.5 Epigenetic Changes Are Often Associated with Cancer

Epigeneic changes—alterations to DNA methylation and chromatin structure (see p. 285 in Chapter 10)—are seen in many cancer cells. Abnormal patterns of DNA methylation are particulary prevalent in many cancers. In some cases, the DNA of cancer cells is overmethylated (hypermethylated); in other cases, the DNA of the cancer cells is undermethylated (hypomethylated). Methylation of DNA affects gene expression, and many genes in cancer cells show abnormal patterns of expression. In general, DNA methylation is associated with the repression of gene

In spite of the large number of women infected with HPVs, the incidence of cervical cancer in the United States has declined 75% in the past 40 years. The primary reason for this decline is widespread use of the Pap test, which detects early stages of cervical cancer, as well as cervical dysplasia, a precancerous growth that can be removed before it develops into cancer.

expression (see p. 463 in Chapter 17). Hypermethylation is thought to contribute to cancer by silencing the expression of tumor-suppressor genes. For example, methylation of the promoter of the *Apaf-1* gene is seen in many malignant melanoma cells. *Apaf-1* helps bring about apoptosis of cells with damaged DNA; methylation of its promoter reduces the expression of *Apaf-1*, interrupting the process of apoptosis and allowing abnormal cancer cells to survive.

How hypomethylation contributes to cancer is less clear. Some evidence suggests that hypomethylation causes chromosome instability, which is a hallmark of many tumors (see discussion of genomic instability in Section 23.2). Tumor cells from mice that have been genetically engineered to have reduced DNA methylation show increased gains and losses of chromosomes, but how hypomethylation might cause chromosome instability is unclear.

The role of DNA methylation in cancer is interesting because, unlike the other genetic changes discussed so far, DNA methylation is reversible and is not a mutation. Epigenetic processes are receiving increasing attention by cancer researchers because they may be particularly amenable to drug therapy.

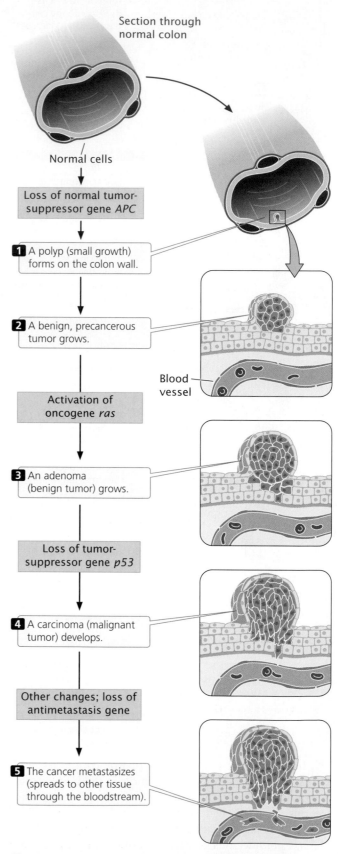

CONCEPTS

Changes in patterns of methylation of DNA are often associated with cancer. Changes in DNA methylation potentially affect the expression of tumor-suppressor genes and may contribute to chromosome instability.

✔ CONCEPT CHECK 7

Hypermethylation is thought to contribute to cancer by

a. inhibiting DNA replication.

b. inhibiting the expression of tumor-suppressor genes.

c. stimulating the translation of ongenes.

d. stimulating telomerase.

23.6 Colorectal Cancer Arises Through the Sequential Mutation of a Number of Genes

Mutations that contribute to colorectal cancer have received extensive study. This cancer is an excellent example of how cancer often arises through the accumulation of successive genetic defects.

Colorectal cancer arises in the cells lining the colon and rectum. More than 142,000 new cases of colorectal cancer are diagnosed in the United States each year, where this cancer is responsible for more than 51,000 deaths annually. If detected early, colorectal cancer can be treated successfully; consequently, there has been much interest in identifying the molecular events responsible for the initial stages of this cancer.

Colorectal cancer is thought to originate as benign tumors called adenomatous polyps (**Figure 23.14**). Initially,

23.14 Mutations in multiple genes contribute to the progression of colorectal cancer.

these polyps are microscopic but, in time, they enlarge and the cells of the polyp acquire the abnormal characteristics of cancer cells. In the later stages of the disease, the tumor may invade the muscle layer surrounding the gut and metastasize. The progression of the disease is slow; from 10 to 35 years may be required for a benign tumor to develop into a malignant tumor.

Most cases of colorectal cancer are sporadic, developing in people with no family history of the disease, but a few families display a clear genetic predisposition to this disease. In one form of hereditary colon cancer, known as familial adenomatous polyposis coli, hundreds or thousands of polyps develop in the colon and rectum; if these polyps are not removed, one or more almost invariably become malignant.

Because polyps and tumors of the colon and rectum can be easily observed and removed with a colonoscope (a fiber-optic instrument used to view the interior of the rectum and colon), much is known about the progression of colorectal cancer, and some of the genes responsible for its clonal evolution have been identified. Mutations in these genes are responsible for the different steps of colorectal-cancer progression. One of the earliest steps is a mutation that inactivates the *APC* gene, which increases the rate of cell division, leading to polyp formation (see Figure 23.14). A person with familial adenomatous polyposis coli inherits one defective copy of the *APC* gene, and defects in this gene are associated with the numerous polyps that appear in those who have this disorder. Mutations in *APC* are also found in the polyps that develop in people who do not have adenomatous polyposis coli.

Mutations of the *ras* oncogene usually occur later, in larger polyps consisting of cells that have acquired some genetic mutations. As discussed earlier in the chapter, the normal *ras* proto-oncogene is a key player in a signal-transduction pathway that relays signals from growth factors to the nucleus, where the signal stimulates cell division. When *ras* is mutated, the protein that it encodes continually relays a stimulatory signal for cell division, even when growth factor is absent.

Mutations in *p53* and other genes appear still later in tumor progression; these mutations are rare in polyps but common in malignant cells. About 75% of colorectal cancers have mutations in tumor-suppressor gene *p53*. Because *p53* prevents the replication of cells with genetic damage and controls proper chromosome segregation, mutations in *p53* may allow a cell to rapidly acquire further gene and chromosome mutations, which then contribute to further proliferation and invasion into surrounding tissues.

The sequence of steps just outlined is not the only route to colorectal cancer, and the mutations need not occur in the order presented here, but this sequence is a common pathway by which colon and rectal cells become cancerous.

CONCEPTS SUMMARY

- Cancer is fundamentally a genetic disorder, arising from somatic mutations in multiple genes that affect cell division and proliferation. If one or more mutations are inherited, then fewer additional mutations are required for cancer to develop.

- A mutation that allows a cell to divide rapidly provides the cell with a growth advantage; this cell gives rise to a clone of cells having the same mutation. Within this clone, other mutations occur that provide additional growth advantages, and cells with these additional mutations become dominant in the clone. In this way, the clone evolves.

- Environmental factors play an important role in the development of many cancers by increasing the rate of somatic mutations.

- Oncogenes are dominant mutated copies of normal genes (proto-oncogenes) that stimulate cell division. Tumor-suppressor genes normally inhibit cell division; recessive mutations in these genes may contribute to cancer. Sometimes, the mutation of a single allele of a tumor-suppressor gene is sufficient to cause cancer, a phenomenon known as haploinsufficiency.

- The cell cycle is controlled by cyclins and cyclin-dependent kinases. Mutations in genes that control the cell cycle are often associated with cancer.

- Signal-transduction pathways conduct external signals to intracellular responses. These pathways consist of a series of proteins that are activated and inactivated in a cascade of reactions. Mutations in the components of signal-transduction pathways disrupt the cell cycle and may contribute to cancer.

- Defects in DNA-repair genes often increase the overall mutation rate of other genes, leading to defects in proto-oncogenes and tumor-suppressor genes that may contribute to cancer progression.

- Mutations in sequences that regulate telomerase allow cells to divide indefinitely, contributing to cancer progression.

- Tumor progression is also affected by mutations in genes that promote vascularization and the spread of tumors.

- Many tumor cells exhibit a widespread reduction in the expression of many miRNAs, suggesting that a reduction in miRNA control of gene expression may play a role in tumor progression.

- Some cancers are associated with specific chromosome mutations, including chromosome deletions, inversions, and translocations. Deletions may cause cancer by removing or disrupting genes that suppress tumors; inversions and translocations may break tumor-suppressing genes or they may move genes to positions next to different regulatory sequences, which alters their expression.

- Mutations in some genes cause or allow the missegregation of chromosomes, leading to aneuploidy that may contribute to cancer.

- Changes in patterns of DNA methylation are seen in many cancer cells. DNA in cancer cells may be over- or undermethylated.

- Viruses are associated with some cancers; they contribute to cell proliferation by mutating and rearranging host genes and by altering the expression of host genes.

- Colorectal cancer offers a model system for understanding tumor progression in humans. Initial mutations stimulate cell division, leading to a small benign polyp. Additional mutations allow the polyp to enlarge, invade the muscle layer of the gut, and eventually spread to other sites. Mutations in particular genes affect different stages of this progression.

IMPORTANT TERMS

malignant tumor (p. 639)
metastasis (p. 639)
clonal evolution (p. 640)
oncogene (p. 642)

tumor-suppressor gene (p. 642)
proto-oncogene (p. 643)
loss of heterozygosity (p. 643)
haploinsufficiency (p. 643)

cyclin-dependent kinase (CDK) (p. 644)
cyclin (p. 644)
signal-transduction pathway (p. 647)
human papilloma virus (HPV) (p. 652)

ANSWERS TO CONCEPT CHECKS

1. Retinoblastoma results from at least two separate genetic defects, both of which are necessary for cancer to develop. In sporadic cases, two successive mutations must occur in a single cell, which is unlikely and therefore typically affects only one eye. In people who have inherited one of the two required mutations, every cell contains this mutation, and so a single additional mutation is all that is required for cancer to develop. Given the millions of cells in each eye, there is a high probability that the second mutation will occur in at least one cell of each eye, producing tumors in both eyes and the inheritance of this type of retinoblastoma.

2. Oncogenes have a stimulatory effect on cell proliferation. Mutations in oncogenes are usually dominant because a

mutation in a single copy of the gene is usually sufficient to produce a stimulatory effect. Tumor-suppressor genes inhibit cell proliferation. Mutations in tumor-suppressor genes are generally recessive, because both copies must be mutated to remove all inhibition.

3. d

4. a

5. c

6. d

7. b

WORKED PROBLEM

1. In some cancer cells, a specific gene has become duplicated many times. Is this gene likely to be an oncogene or a tumor-suppressor gene? Explain your reasoning.

• Solution

The gene is likely to be an oncogene. Oncogenes stimulate cell proliferation and act in a dominant manner. Therefore, extra

copies of an oncogene will result in cell proliferation and cancer. Tumor-suppressor genes, on the other hand, suppress cell proliferation and act in a recessive manner; a single copy of a tumor-suppressor gene is sufficient to prevent cell proliferation. Therefore, extra copies of the tumor-suppressor gene will not lead to cancer.

COMPREHENSION QUESTIONS

Section 23.1

*1. What types of evidence indicate that cancer arises from genetic changes?

2. How is cancer different from most other types of genetic diseases?

Section 23.2

*3. Outline Knudson's multistage theory of cancer and describe how it helps to explain unilateral and bilateral cases of retinoblastoma.

4. Briefly explain how cancer arises through clonal evolution.

*5. What is the difference between an oncogene and a tumor-suppressor gene? Give some examples of the functions of proto-oncogenes and tumor suppressers in normal cells.

6. What is haploinsufficiency? How might it affect cancer risk?

*7. How do cyclins and CDKs differ? How do they interact in controlling the cell cycle.

8. Briefly outline the events that control the progression of cells through the G_1/S checkpoint in the cell cycle.

9. Briefly outline the events that control the progression of cells through the G_2/M checkpoint of the cell cycle.

*10. What is a signal-transduction pathway? Why are mutations in components of signal-transduction pathways often associated with cancer?

11. How is the Ras protein activated and inactivated?

12. Why do mutations in genes that encode DNA-repair enzymes often produce a predisposition to cancer?

*13. What role do telomeres and telomerase play in cancer progression?

Section 23.3

*14. Explain how chromosome deletions, inversions, and translocations may cause cancer.

15. Briefly outline how the Philadelphia chromosome leads to chronic myelogenous cancer.

16. What is genomic instability? Give some ways in which genomic instability may arise.

Section 23.4

*17. How do viruses contribute to cancer?

Section 23.5

18. How is an epigenetic change different from a mutation?

19. How is DNA methylation related to cancer?

Section 23.6

20. Briefly outline some of the genetic changes that are commonly associated with the progression of colorectal cancer.

APPLICATION QUESTIONS AND PROBLEMS

Section 23.1

21. If cancer is fundamentally a genetic disease, how might an environmental factor such as smoking cause cancer?

22. Both genes and environmental factors contribute to cancer. Table 23.2 shows that prostate cancer is 30 times as common among Caucasians from Utah as among Chinese from Shanghai. Briefly outline how you might go about determining if these differences in the incidence of prostate cancer are due to differences in the genetic makeup of two populations or to differences in their environments.

*23. A couple has one child with bilateral retinoblastoma. The mother is free from cancer, but the father has unilateral retinoblastoma and he has a brother who has bilateral retinoblastoma.

   a. If the couple has another child, what is the probability that this next child will have retinoblastoma?

   b. If the next child has retinoblastoma, is it likely to be bilateral or unilateral?

   c. Explain why the father's case of retinoblastoma is unilateral, whereas his son's and brother's cases are bilateral.

Section 23.2

*24. The *palladin* gene, which plays a role in pancreatic cancer (see the introduction to this chapter), is said to be an oncogene. Which of its characteristics suggest that it is an oncogene rather than a tumor-suppressor gene?

25. Mutations in the *RB* gene are often associated with cancer. Explain how a mutation that results in a nonfunctional RB protein contributes to cancer.

26. Cells in a tumor contain mutated copies of a particular gene that promotes tumor growth. Gene therapy can be used to introduce a normal copy of this gene into the tumor cells. Would you expect this therapy to be effective if the mutated gene were an oncogene? A tumor-suppressor gene? Explain your reasoning.

*27. Genes in cancer cells are frequently amplified, meaning that the gene exists in many copies. Would you expect to see gene amplification in oncogenes, tumor-suppresor genes, or both? Explain your answer.

28. David Seligson and his colleagues examined levels of histone protein modification in prostate tumors and their association with clinical outcomes (D. B. Seligson et al. 2005. *Nature* 435:1262–1266). They used antibodies to

stain for acetylation at three different sites and for methylation at two different sites on histone proteins. They found that the degree of histone acetylation and methylation helped predict whether prostate cancer would return within 10 years in the patients who had a prostate tumor removed. Explain how acetylation and methylation might be associated with tumor recurrence in prostate cancer. (Hint: See Chapter 17.)

29. Radiation is known to cause cancer, yet radiation is often used as treatment for some types of cancer. How can radiation be a contributor to both the cause and the treatment of cancer?

Section 23.3

30. Some cancers are consistently associated with the deletion of a particular part of a chromosome. Does the deleted region contain an oncogene or a tumor-suppressor gene? Explain.

Section 23.5

31. Some cancers have been treated with drugs that demethylate DNA. Explain how these drugs might work. Do you think the cancer-causing genes that respond to the demethylation are likely to be oncogenes or tumor-suppressor genes? Explain your reasoning.

CHALLENGE QUESTIONS

Section 23.2

32. Many cancer cells are immortal (will divide indefinitely) because they have mutations that allow telomerase to be expressed. How might this knowledge be used to design anticancer drugs?

33. Bloom syndrome is an autosomal recessive disease that exhibits haploinsufficiency. As described on page 644, a recent survey showed that people heterozygous for mutations at the *BLM* locus are at increased risk of colon cancer. Suppose you are a genetic counselor. A young woman is referred to you whose mother has Bloom syndrome; the young woman's father has no family history of Bloom syndrome. The young woman asks whether she is likely to experience any other health problems associated with her family history of Bloom syndrome. What advice would you give her?

34. Imagine that you discover a large family in which bladder cancer is inherited as an autosomal dominant trait. Briefly outline a series of studies that you might conduct to identify the gene that causes bladder cancer in this family.

24 QUANTITATIVE GENETICS

Methods of quantitative genetics coupled with molecular techniques have been used to identify a gene that determines oil content in corn. [Walter Bibikow/Getty Images.]

CORN OIL AND QUANTITATIVE GENETICS

In the year 2010, the world's population was estimated to be more than 6.8 billion people. The United Nations projects that the world population will surpass 7 billion in 2012 and reach 9 billion by 2050. Feeding those billions of additional people will be a major challenge for agriculture in next few decades. Crop plants will have to provide most of the calories and nutrients required for world's future population. Because of dwindling petroleum supplies and concerns about global warming, plants are also increasingly being utilized as sources of biofuels, placing additional demands on crop production. How will agriculture provide sufficient crop yields to supply the food and energy requirements of the future?

To help meet the need for increased crop production, plant breeders are using the latest techniques of genetics in their quest to develop higher-yielding, more-efficient crop plants. The power of this approach is demonstrated by research aimed at increasing the oil content of corn. The oil content of corn is inherited, but it is not a simple single-gene characteristic such as seed shape in peas. Numerous genes and environmental factors contribute to the oil content of corn, and the inheritance of oil content is more complex than that of the characteristics that we have studied so far. For traits such as oil content, several loci frequently interact and their expression is affected by environmental factors. Can the inheritance of a complex characteristic such as oil content be studied? Is it possible to predict the oil content of a plant on the basis of its breeding? The answers are yes—at least in part—but these questions cannot be addressed with the methods that we used for simple genetic characteristics. Instead, we must use statistical procedures that have been developed for analyzing complex characteristics. The genetic analysis of complex characteristics such as the oil content of corn is known as **quantitative genetics**.

In 2008, geneticists used a combination of quantitative genetics and molecular techniques to identify a key gene that controls oil content in corn. First, they conducted crosses between high-oil corn plants and low-oil plants to identify chromosomal regions that play an important role in determining oil production. Chromosome regions containing genes that influence a quantitative trait are termed **quantitative trait loci** (QTLs).

Through these crosses, the geneticists located a QTL that affected oil content on corn chromosome 6.

Fine-scale genetic mapping further narrowed the QTL down to a small region of 4.2 centiMorgans on chromosome 6. Researchers sequenced DNA from the region and found that it contained five genes, one of which was *DGAT1-2*, a gene known to produce an enzyme that catalyzes the final step in a pathway for triacyglycerol biosynthesis. DNA from the *DGAT1-2* gene in a high-oil-producing strain contained an insertion of a codon that added phenylalanine to the enzyme, an insertion that was missing from a low-oil strain. The researchers confirmed the effect of the additional phenylalanine codon on oil production by producing transgenic corn that contained the extra codon; oil production of these transgenic strains increased compared with that of transgenic strains without the extra codon. Interestingly, the extra phenylalanine codon is present in wild relatives of corn, suggesting that the codon was lost in the process of domestication or subsequent breeding of modern varieties.

This research suggests that the oil content of corn and other plants might be increased by genetically modifying their *DGAT1-2* genes to contain the extra codon for phenylalanine. Other studies that similarly combine quantitative and molecular analyses have recently led to the identification of genes that increase the vitamin A content of rice and increase sugar production by tomatoes.

This chapter is about the genetic analysis of complex characteristics such as oil content in corn. We begin by considering the differences between quantitative and qualitative characteristics and why the expression of some characteristics varies continuously. We'll see how quantitative characteristics are often influenced by many genes, each of which has a small effect on the phenotype. Next, we will examine statistical procedures for describing and analyzing quantitative characteristics. We will consider how much of phenotypic variation can be attributed to genetic and environmental influences and will conclude by looking at the effects of selection on quantitative characteristics.

24.1 Quantitative Characteristics Vary Continuously and Many Are Influenced by Alleles at Multiple Loci

Qualitative, or discontinuous, characteristics possess only a few distinct phenotypes (**Figure 24.1a**); these characteristics are the types studied by Mendel and have been the focus of our attention thus far. However, many characteristics vary continuously along a scale of measurement with many overlapping phenotypes (**Figure 24.1b**). They are referred to as *continuous characteristics*; they are also called *quantitative characteristics* because any individual's phenotype must be described with a quantitative measurement. Quantitative characteristics might include height, weight, and blood pressure in humans, growth rate in mice, seed weight in plants, and milk production in cattle.

Quantitative characteristics arise from two phenomena. First, many are polygenic: they are influenced by genes at many loci. If many loci take part, many genotypes are possible, each producing a slightly different phenotype. Second,

(a) Discontinuous characteristic

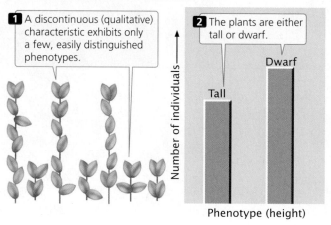

1 A discontinuous (qualitative) characteristic exhibits only a few, easily distinguished phenotypes.

2 The plants are either tall or dwarf.

Number of individuals →

Tall

Dwarf

Phenotype (height)

(b) Continuous characteristic

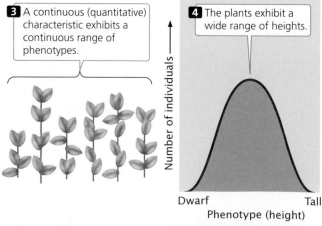

3 A continuous (quantitative) characteristic exhibits a continuous range of phenotypes.

4 The plants exhibit a wide range of heights.

Number of individuals →

Dwarf   Tall
Phenotype (height)

24.1 Discontinuous and continuous characteristics differ in the number of phenotypes exhibited.

Table 24.1 Hypothetical example of plant height determined by pairs of alleles at each of three loci

Plant Genotype	Doses of Hormone	Height (cm)
A^-A^- B^-B^- C^-C^-	0	10
A^+A^- B^-B^- C^-C^-	1	11
A^-A^- B^+B^- C^-C^-		
A^-A^- B^-B^- C^-C^+		
A^+A^+ B^-B^- C^-C^-	2	12
A^-A^- B^+B^+ C^-C^-		
A^-A^- B^-B^- C^+C^+		
A^+A^- B^+B^- C^-C^-		
A^+A^- B^-B^- C^+C^-		
A^-A^- B^+B^- C^+C^-		
A^+A^+ B^+B^- C^-C^-	3	13
A^+A^+ B^-B^- C^+C^-		
A^+A^- B^+B^+ C^-C^-		
A^-A^- B^+B^+ C^+C^-		
A^+A^- B^-B^- C^+C^+		
A^-A^- B^+B^- C^+C^+		
A^+A^- B^+B^- C^+C^-		
A^+A^+ B^+B^+ C^-C^-	4	14
A^+A^+ B^+B^- C^+C^-		
A^+A^- B^+B^+ C^+C^-		
A^-A^- B^+B^+ C^+C^+		
A^+A^+ B^-B^- C^+C^+		
A^+A^- B^+B^- C^+C^+		
A^+A^+ B^+B^+ C^-C^+	5	15
A^+A^- B^+B^+ C^+C^+		
A^+A^+ B^+B^- C^+C^+		
A^+A^+ B^+B^+ C^+C^+	6	16

Note: Each + allele contributes 1 cm in height above a baseline of 10 cm.

quantitative characteristics often arise when environmental factors affect the phenotype because environmental differences result in a single genotype producing a range of phenotypes. Most continuously varying characteristics are *both* polygenic *and* influenced by environmental factors, and these characteristics are said to be multifactorial.

The Relation Between Genotype and Phenotype

For many discontinuous characteristics, the relation between genotype and phenotype is straightforward. Each genotype produces a single phenotype, and most phenotypes are encoded by a single genotype. Dominance and epistasis may allow two or three genotypes to produce the same phenotype, but the relation remains simple. This simple relation between genotype and phenotype allowed Mendel to decipher the basic rules of inheritance from his crosses with pea plants; it also permits us both to predict the outcome of genetic crosses and to assign genotypes to individuals.

For quantitative characteristics, the relation between genotype and phenotype is often more complex. If the characteristic is polygenic, many different genotypes are possible, several of which may produce the same phenotype. For instance, consider a plant whose height is determined by three loci (A, B, and C), each of which has two alleles. Assume that one allele at each locus (A^+, B^+, and C^+) encodes a plant hormone that causes the plant to grow 1 cm above its baseline height of 10 cm. The other allele at each locus (A^-, B^-, and C^-) does not encode a plant hormone and thus does not contribute to additional height. If we consider only the two alleles at a single locus, 3 genotypes are possible (A^+A^+, A^+A^-, and A^-A^-). If all three loci are taken into account, there are a total of $3^3 = 27$ possible multilocus genotypes (A^+A^+ B^+B^+ C^+C^+, A^+A^- B^+B^+ C^+C^+, etc.). Although there are 27 genotypes, they produce only seven phenotypes (10 cm, 11 cm, 12 cm, 13 cm, 14 cm, 15 cm, and 16 cm in height). Some of the genotypes produce the same phenotype (**Table 24.1**); for example, genotypes A^+A^- B^-B^- C^-C^-, A^-A^- B^+B^- C^-C^-, and A^-A^- B^-B^- C^+C^- all have one gene that encodes a plant hormone. These genotypes produce one dose of the hormone and a plant that is 11 cm tall. Even in this simple example of only three loci, the relation between genotype and phenotype is quite complex. The more loci encoding a characteristic, the greater the complexity.

The influence of environment on a characteristic also can complicate the relation between genotype and phenotype. Because of environmental effects, the same genotype may produce a range of potential phenotypes. The phenotypic ranges of different genotypes may overlap, making it difficult to know whether individuals differ in phenotype because of genetic or environmental differences (**Figure 24.2**).

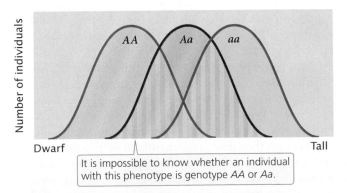

It is impossible to know whether an individual with this phenotype is genotype *AA* or *Aa*.

24.2 For a quantitative characteristic, each genotype may produce a range of possible phenotypes. In this hypothetical example, the phenotypes produced by genotypes *AA*, *Aa*, and *aa* overlap.

In summary, the simple relation between genotype and phenotype that exists for many qualitative (discontinuous) characteristics is absent for quantitative characteristics, and it is impossible to assign a genotype to an individual on the basis of its phenotype alone. The methods used for analyzing qualitative characteristics (examining the phenotypic ratios of progeny from a genetic cross) will not work for quantitative characteristics. Our goal remains the same: we wish to make predictions about the phenotypes of offspring produced in a genetic cross. We may also want to know how much of the variation in a characteristic results from genetic differences and how much results from environmental differences. To answer these questions, we must turn to statistical methods that allow us to make predictions about the inheritance of phenotypes in the absence of information about the underlying genotypes. **TRY PROBLEM 16 →**

Types of Quantitative Characteristics

Before we look more closely at polygenic characteristics and relevant statistical methods, we need to more clearly define what is meant by a quantitative characteristic. Thus far, we have considered only quantitative characteristics that vary continuously in a population. A *continuous characteristic* can theoretically assume any value between two extremes; the number of phenotypes is limited only by our ability to precisely measure the phenotype. Human height is a continuous characteristic because, within certain limits, people can theoretically have any height. Although the number of phenotypes possible with a continuous characteristic is infinite, we often group similar phenotypes together for convenience; we may say that two people are both 5 feet 11 inches tall, but careful measurement may show that one is slightly taller than the other.

Some characteristics are not continuous but are nevertheless considered quantitative because they are determined by multiple genetic and environmental factors. **Meristic characteristics**, for instance, are measured in whole numbers. An example is litter size: a female mouse may have 4, 5, or 6 pups but not 4.13 pups. A meristic characteristic has a limited number of distinct phenotypes, but the underlying determination of the characteristic may still be quantitative. These characteristics must therefore be analyzed with the same techniques that we use to study continuous quantitative characteristics.

Another type of quantitative characteristic is a **threshold characteristic**, which is simply present or absent. For example, the presence of some diseases can be considered a threshold characteristic. Although threshold characteristics exhibit only two phenotypes, they are considered quantitative because they, too, are determined by multiple genetic and environmental factors. The expression of the characteristic depends on an underlying susceptibility (usually referred to as liability or risk) that varies continuously. When the susceptibility is larger than a threshold value, a specific trait is expressed (**Figure 24.3**). Diseases are often threshold

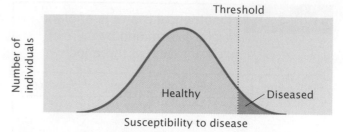

24.3 Threshold characteristics display only two possible phenotypes—the trait is either present or absent—but they are quantitative because the underlying susceptibility to the characteristic varies continuously. When the susceptibility exceeds a threshold value, the characteristic is expressed.

characteristics because many factors, both genetic and environmental, contribute to disease susceptibility. If enough of the susceptibility factors are present, the disease develops; otherwise, it is absent. Although we focus on the genetics of continuous characteristics in this chapter, the same principles apply to many meristic and threshold characteristics.

Just because a characteristic can be measured on a continuous scale does not mean that it exhibits quantitative variation. One of the characteristics studied by Mendel was height of the pea plant, which can be described by measuring the length of the plant's stem. However, Mendel's particular plants exhibited only two distinct phenotypes (some were tall and others short), and these differences were determined by alleles at a single locus. The differences that Mendel studied were therefore discontinuous in nature.

CONCEPTS

Characteristics whose phenotypes vary continuously are called quantitative characteristics. For most quantitative characteristics, the relation between genotype and phenotype is complex. Some characteristics whose phenotypes do not vary continuously also are considered quantitative because they are influenced by multiple genes and environmental factors.

Polygenic Inheritance

After the rediscovery of Mendel's work in 1900, questions soon arose about the inheritance of continuously varying characteristics. These characteristics had already been the focus of a group of biologists and statisticians, led by Francis Galton, who used statistical procedures to examine the inheritance of quantitative characteristics such as human height and intelligence. The results of these studies showed that quantitative characteristics are inherited, although the mechanism of inheritance was not yet known. Some biometricians argued that the inheritance of quantitative characteristics could not be explained by Mendelian principles, whereas others felt that Mendel's principles acting on numerous genes (polygenes) could adequately account for the inheritance of quantitative characteristics.

This conflict began to be resolved by the work of Wilhelm Johannsen, who showed that continuous variation in the weight of beans was influenced by both genetic and environmental factors. George Udny Yule, a mathematician, proposed in 1906 that several genes acting together could produce continuous characteristics. This hypothesis was later confirmed by Herman Nilsson-Ehle, working on wheat and tobacco, and by Edward East, working on corn. The argument was finally laid to rest in 1918, when Ronald Fisher demonstrated that the inheritance of quantitative characteristics could indeed be explained by the cumulative effects of many genes, each following Mendel's rules.

Kernel Color in Wheat

To illustrate how multiple genes acting on a characteristic can produce a continuous range of phenotypes, let us examine one of the first demonstrations of polygenic inheritance. Nilsson-Ehle studied kernel color in wheat and found that the intensity of red pigmentation was determined by three unlinked loci, each of which had two alleles.

Nilsson-Ehle's cross Nilsson-Ehle obtained several homozygous varieties of wheat that differed in color. Like Mendel, he performed crosses between these homozygous varieties and studied the ratios of phenotypes in the progeny. In one experiment, he crossed a variety of wheat that possessed white kernels with a variety that possessed purple (very dark red) kernels and obtained the following results:

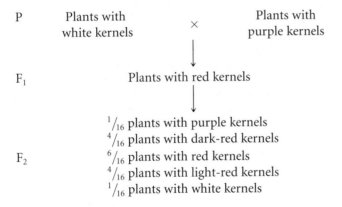

| P | Plants with white kernels | × | Plants with purple kernels |

F$_1$ — Plants with red kernels

F$_2$:
$^1/_{16}$ plants with purple kernels
$^4/_{16}$ plants with dark-red kernels
$^6/_{16}$ plants with red kernels
$^4/_{16}$ plants with light-red kernels
$^1/_{16}$ plants with white kernels

Interpretation of the cross Nilsson-Ehle interpreted this phenotypic ratio as the result of the segregation of alleles at two loci. (Although he found alleles at three loci that affect kernel color, the two varieties used in this cross differed at only two of the loci.) He proposed that there were two alleles at each locus: one that produced red pigment and another that produced no pigment. We'll designate the alleles that encoded pigment A^+ and B^+ and the alleles that encoded no pigment A^- and B^-. Nilsson-Ehle recognized that the effects of the genes were additive. Each gene seemed to contribute equally to color; so the overall phenotype could be determined by adding the effects of all the genes, as shown in the following table.

Genotype	Doses of pigment	Phenotype
$A^+A^+\ B^+B^+$	4	Purple
$A^+A^+\ B^+B^-$ $A^+A^-\ B^+B^+$	3	Dark red
$A^+A^+\ B^-B^-$ $A^-A^-\ B^+B^+$ $A^+A^-\ B^+B^-$	2	Red
$A^+A^-\ B^-B^-$ $A^-A^-\ B^+B^-$	1	Light red
$A^-A^-\ B^-B^-$	0	White

Notice that the purple and white phenotypes are each encoded by a single genotype, but other phenotypes may result from several different genotypes.

From these results, we see that five phenotypes are possible when alleles at two loci influence the phenotype and the effects of the genes are additive. When alleles at more than two loci influence the phenotype, more phenotypes are possible, and the color would appear to vary continuously between white and purple. If environmental factors had influenced the characteristic, individuals of the same genotype would vary somewhat in color, making it even more difficult to distinguish between discrete phenotypic classes. Luckily, environment played little role in determining kernel color in Nilsson-Ehle's crosses, and only a few loci encoded color; so Nilsson-Ehle was able to distinguish among the different phenotypic classes. This ability allowed him to see the Mendelian nature of the characteristic.

Let's now see how Mendel's principles explain the ratio obtained by Nilsson-Ehle in his F$_2$ progeny. Remember that Nilsson-Ehle crossed the homozygous purple variety (A^+A^+ B^+B^+) with the homozygous white variety ($A^-A^-\ B^-B^-$), producing F$_1$ progeny that were heterozygous at both loci ($A^+A^-\ B^+B^-$). This is a dihybrid cross, like those that we worked in Chapter 3, except that both loci encode the same trait. All the F$_1$ plants possessed two pigment-producing alleles that allowed two doses of color to make red kernels. The types and proportions of progeny expected in the F$_2$ can be found by applying Mendel's principles of segregation and independent assortment.

Let's first examine the effects of each locus separately. At the first locus, two heterozygous F$_1$s are crossed ($A^+A^- \times A^+A^-$). As we learned in Chapter 3, when two heterozygotes are crossed, we expect progeny in the proportions $^1/_4\ A^+A^+$, $^1/_2\ A^+A^-$, and $^1/_4\ A^-A^-$. At the second locus, two heterozygotes also are crossed, and, again, we expect progeny in the proportions $^1/_4\ B^+B^+$, $^1/_2\ B^+B^-$, and $^1/_4\ B^-B^-$.

To obtain the probability of combinations of genes at both loci, we must use the multiplication rule of probability (see Chapter 3), which is based on Mendel's principle of independent assortment. The expected proportion of F$_2$ progeny with genotype $A^+A^+\ B^+B^+$ is the product of the probability of obtaining genotype A^+A^+ ($^1/_4$) and the probability of obtaining genotype B^+B^+ ($^1/_4$), or $^1/_4 \times ^1/_4 = ^1/_{16}$

(**Figure 24.4**). The probabilities of each of the phenotypes can then be obtained by adding the probabilities of all the genotypes that produce that phenotype. For example, the red phenotype is produced by three genotypes:

Genotype	Probability
$A^+A^+ \, B^-B^-$	$^1/_{16}$
$A^-A^- \, B^+B^+$	$^1/_{16}$
$A^+A^- \, B^+B^-$	$^1/_4$

Thus, the overall probability of obtaining red kernels in the F_2 progeny is $^1/_{16} + ^1/_{16} + ^1/_4 = ^6/_{16}$. Figure 24.4 shows that the phenotypic ratio expected in the F_2 is $^1/_{16}$ purple, $^4/_{16}$ dark red, $^6/_{16}$ red, $^4/_{16}$ light red, and $^1/_{16}$ white. This phenotypic ratio is precisely what Nilsson-Ehle observed in his F_2 progeny, demonstrating that the inheritance of a continuously varying characteristic such as kernel color is indeed according to Mendel's basic principles.

Conclusions and implications Nilsson-Ehle's crosses demonstrated that the difference between the inheritance of genes influencing quantitative characteristics and the inheritance of genes influencing discontinuous characteristics is in the *number* of loci that determine the characteristic. When multiple loci affect a character, more genotypes are possible; so the relation between the genotype and the phenotype is less obvious. As the number of loci affecting a character increases, the number of phenotypic classes in the F_2 increases (**Figure 24.5**).

Several conditions of Nilsson-Ehle's crosses greatly simplified the polygenic inheritance of kernel color and made it possible for him to recognize the Mendelian nature of the characteristic. First, genes affecting color segregated at only two or three loci. If genes at many loci had been segregating, he would have had difficulty in distinguishing the phenotypic classes. Second, the genes affecting kernel color had strictly additive effects, making the relation between genotype and phenotype simple. Third, environment played almost no role in the phenotype; had environmental factors modified the phenotypes, distinguishing between the five phenotypic classes would have been difficult. Finally, the loci that Nilsson-Ehle studied were not linked; so the genes assorted independently. Nilsson-Ehle was fortunate: for many polygenic characteristics, these simplifying conditions are not present and Mendelian inheritance of these characteristics is not obvious. TRY PROBLEM 17 ➔

Determining Gene Number for a Polygenic Characteristic

The proportion of F_2 individuals that resemble one of the original parents can be used to estimate the number of genes affecting a polygenic trait. When two individuals homozygous for different alleles at a single locus are crossed ($A^1A^1 \times A^2A^2$) and the resulting F_1 are interbred ($A^1A^2 \times A^1A^2$), one-fourth of the F_2 should be homozygous like each of the original

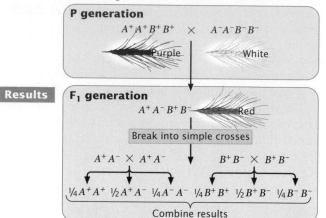

Experiment

Question: How is a continous trait, such as kernel color in wheat, inherited?

Methods Cross wheat having white kernels and wheat having purple kernels. Intercross the F_1 to produce F_2.

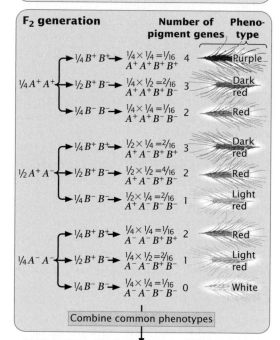

Conclusion: Kernel color in wheat is inherited according to Mendel's principles acting on alleles at two loci.

24.4 Nilsson-Ehle demonstrated that kernel color in wheat is inherited according to Mendelian principles. The ratio of phenotypes in the F_2 can be determined by breaking the dihybrid cross into two simple single-locus crosses and combining the results by using the multiplication rule.

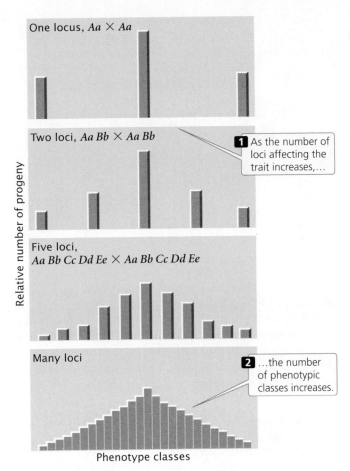

24.5 The results of crossing individuals heterozygous for different numbers of loci affecting a characteristic.

parents. If the original parents are homozygous for different alleles at two loci, as are those in Nilsson-Ehle's crosses, then $\frac{1}{4} \times \frac{1}{4} = \frac{1}{16}$ of the F_2 should resemble one of the original homozygous parents. Generally, $(\frac{1}{4})^n$ will be the number of individuals in the F_2 progeny that should resemble each of the original homozygous parents, where n equals the number of loci with a segregating pair of alleles that affects the characteristic. This equation provides us with a possible means of determining the number of loci influencing a quantitative characteristic.

To illustrate the use of this equation, assume that we cross two different homozygous varieties of pea plants that differ in height by 16 cm, interbreed the F_1, and find that approximately $\frac{1}{256}$ of the F_2 are similar to one of the original homozygous parental varieties. This outcome would suggest that four loci with segregating pairs of alleles ($\frac{1}{256} = (\frac{1}{4})^4$) are responsible for the height difference between the two varieties. Because the two homozygous strains differ in height by 16 cm and there are four loci each of which has two alleles (eight alleles in all), each of the alleles contributes 16 cm$/8 = 2$ cm in height.

This method for determining the number of loci affecting phenotypic differences requires the use of homozygous strains, which may be difficult to obtain in some organisms.

It also assumes that all the genes influencing the characteristic have equal effects, that their effects are additive, and that the loci are unlinked. For many polygenic characteristics, these assumptions are not valid, and so this method of determining the number of genes affecting a characteristic has limited application. **TRY PROBLEM 19 →**

CONCEPTS

The principles that determine the inheritance of quantitative characteristics are the same as the principles that determine the inheritance of discontinuous characteristics, but more genes take part in the determination of quantitative characteristics.

✔ CONCEPT CHECK 1

Briefly explain how the number of genes influencing a polygenic trait can be determined?

24.2 Statistical Methods Are Required for Analyzing Quantitative Characteristics

Because quantitative characteristics are described by a measurement and are influenced by multiple factors, their inheritance must be analyzed statistically. This section will explain the basic concepts of statistics that are used to analyze quantitative characteristics.

Distributions

Understanding the genetic basis of any characteristic begins with a description of the numbers and kinds of phenotypes present in a group of individuals. Phenotypic variation in a group can be conveniently represented by a **frequency distribution**, which is a graph of the frequencies (numbers or proportions) of the different phenotypes (**Figure 24.6**). In a typical frequency distribution, the phenotypic classes

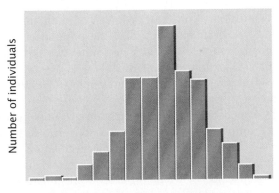

24.6 A frequency distribution is a graph that displays the number or proportion of different phenotypes. Phenotypic values are plotted on the horizontal axis, and the numbers (or proportions) of individuals in each class are plotted on the vertical axis.

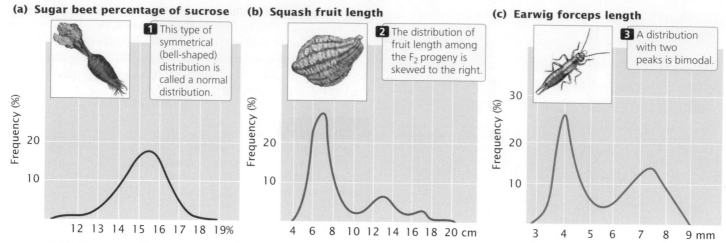

24.7 Distributions of phenotypes can assume several different shapes.

are plotted on the horizontal (x) axis, and the numbers (or proportions) of individuals in each class are plotted on the vertical (y) axis. A frequency distribution is a concise method of summarizing all phenotypes of a quantitative characteristic.

Connecting the points of a frequency distribution with a line creates a curve that is characteristic of the distribution. Many quantitative characteristics exhibit a symmetrical (bell-shaped) curve called a **normal distribution** (**Figure 24.7a**). Normal distributions arise when a large number of independent factors contribute to a measurement, as is often the case in quantitative characteristics. Two other common types of distributions (skewed and bimodal) are illustrated in **Figure 24.7b** and **c**.

Samples and Populations

Biologists frequently need to describe the distribution of phenotypes exhibited by a group of individuals. We might want to describe the height of students at the University of Texas (UT), but there are more than 40,000 students at UT, and measuring every one of them would be impractical. Scientists are constantly confronted with this problem: the group of interest, called the **population**, is too large for a complete census. One solution is to measure a smaller collection of individuals, called a **sample**, and use measurements made on the sample to describe the population.

To provide an accurate description of the population, a good sample must have several characteristics. First, it must be representative of the whole population. If our sample consisted entirely of members of the UT basketball team, for instance, we would probably overestimate the true height of the students. One way to ensure that a sample is representative of the population is to select the members of the sample randomly. Second, the sample must be large enough that chance differences between individuals in the sample and the overall population do not distort the

estimate of the population measurements. If we measured only three students at UT and just by chance all three were short, we would underestimate the true height of the student population. Statistics can provide information about how much confidence to expect from estimates based on random samples.

✔ CONCEPT CHECK 2

A geneticist is interested in whether asthma is caused by a mutation in the *DS112* gene. The geneticist collects DNA from 120 people with asthma and 100 healthy people and sequences the DNA. She finds that 35 of the people with asthma and none of the healthy people have a mutation in the *DS112* gene. What is the population in this study?

a. The 120 people with asthma

b. The 100 healthy people

c. The 35 people with a mutation in their gene

d. All people with asthma

The Mean

The **mean**, also called the average, provides information about the center of the distribution. If we measured the heights of 10-year-old and 18-year-old boys and plotted a frequency distribution for each group, we would find that both distributions are normal, but the two distributions would be centered at different heights, and this difference would be indicated in their different means (**Figure 24.8**).

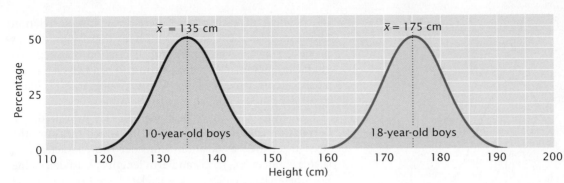

24.8 The mean provides information about the center of a distribution. Both distributions of heights of 10-year-old and 18-year-old boys are normal, but they have different locations along a continuum of height, which makes their means different.

Suppose we have five measurements of height in centimeters: 160, 161, 167, 164, and 165. If we represent a group of measurements as x_1, x_2, x_3, and so forth, then the mean ($\bar{x}$) is calculated by adding all the individual measurements and dividing by the total number of measurements in the sample (n):

$$\bar{x} = \frac{x_1 + x_2 + x_3 + \ldots + x_n}{n} \quad (24.1)$$

In our example, $x_1 = 160$, $x_2 = 161$, $x_3 = 167$, and so forth. The mean height ($\bar{x}$) equals:

$$\bar{x} = \frac{160 + 161 + 167 + 164 + 165}{5} = \frac{817}{5} = 163.4$$

A shorthand way to represent this formula is

$$\bar{x} = \frac{\sum x_i}{n} \quad (24.2)$$

or

$$\bar{x} = \frac{1}{n} \sum x_i \quad (24.3)$$

where the symbol $\sum$ means "the summation of" and x_i represents individual x values.

The Variance and Standard Deviation

A statistic that provides key information about a distribution is the **variance**, which indicates the variability of a group of measurements, or how spread out the distribution is. Distributions may have the same mean but different variances (**Figure 24.9**). The larger the variance, the greater the spread of measurements in a distribution about its mean.

The variance (s^2) is defined as the average squared deviation from the mean:

$$s^2 = \frac{\sum (x_i - \bar{x})^2}{n - 1} \quad (24.4)$$

To calculate the variance, we (1) subtract the mean from each measurement and square the value obtained, (2) add all the squared deviations, and (3) divide this sum by the number of original measurements minus 1.

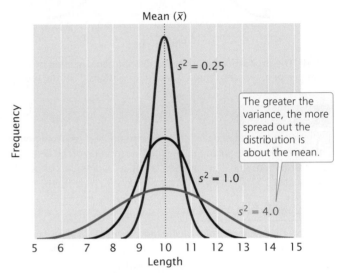

24.9 The variance provides information about the variability of a group of phenotypes. Shown here are three distributions with the same mean but different variances.

Another statistic that is closely related to the variance is the **standard deviation** (s), which is defined as the square root of the variance:

$$s = \sqrt{s^2} \quad (24.5)$$

Whereas the variance is expressed in units squared, the standard deviation is in the same units as the original measurements; so the standard deviation is often preferred for describing the variability of a measurement.

A normal distribution is symmetrical; so the mean and standard deviation are sufficient to describe its shape. The mean plus or minus one standard deviation ($\bar{x} \pm s$) includes approximately 66% of the measurements in a normal distribution; the mean plus or minus two standard deviations ($\bar{x} \pm 2s$) includes approximately 95% of the measurements, and the mean plus or minus three standard deviations ($\bar{x} \pm 3s$) includes approximately 99% of the measurements (**Figure 24.10**). Thus, only 1% of a normally distributed population lies outside the range of ($\bar{x} \pm 3s$).

TRY PROBLEM 22 →

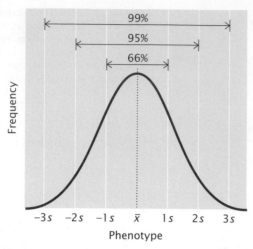

24.10 The proportions of a normal distribution occupied by plus or minus one, two, and three standard deviations from the mean.

CONCEPTS

The mean and variance describe a distribution of measurements: the mean provides information about the location of the center of a distribution, and the variance provides information about its variability.

✔ CONCEPT CHECK 3

The measurements of a distribution with a higher _____ will be more spread out.

a. mean   c. standard deviation

b. variance   d. variance and standard deviation

Correlation

The mean and the variance can be used to describe an individual characteristic, but geneticists are frequently interested in more than one characteristic. Often, two or more characteristics vary together. For instance, both the number and the weight of eggs produced by hens are important to the poultry industry. These two characteristics are not independent of each other. There is an inverse relation between egg number and weight: hens that lay more eggs produce smaller eggs. This kind of relation between two characteristics is called a **correlation**. When two characteristics are correlated, a change in one characteristic is likely to be associated with a change in the other.

Correlations between characteristics are measured by a **correlation coefficient** (designated r), which measures the strength of their association. Consider two characteristics, such as human height (x) and arm length (y). To determine how these characteristics are correlated, we first obtain the covariance (cov) of x and y:

$$\text{cov}_{xy} = \frac{\sum (x_i - \bar{x})(y_i - \bar{y})}{n-1} \qquad (24.6)$$

The covariance is computed by (1) taking an x value for an individual and subtracting it from the mean of x ($\bar{x}$); (2) taking the y value for the same individual and subtracting it from the mean of y ($\bar{y}$); (3) multiplying the results of these two subtractions; (4) adding the results for all the xy pairs; and (5) dividing this sum by $n - 1$ (where n equals the number of xy pairs).

The correlation coefficient (r) is obtained by dividing the covariance of x and y by the product of the standard deviations of x and y:

$$r = \frac{\text{cov}_{xy}}{s_x s_y} \qquad (24.7)$$

A correlation coefficient can theoretically range from -1 to $+1$. A positive value indicates that there is a direct association between the variables (**Figure 24.11a**): as one variable increases, the other variable also tends to increase. A positive correlation exists for human height and weight: tall people tend to weigh more. A negative correlation coefficient indicates that there is an inverse relation between the two variables (**Figure 24.11b**): as one variable increases, the other tends to decrease (as is the case for egg number and egg weight in chickens).

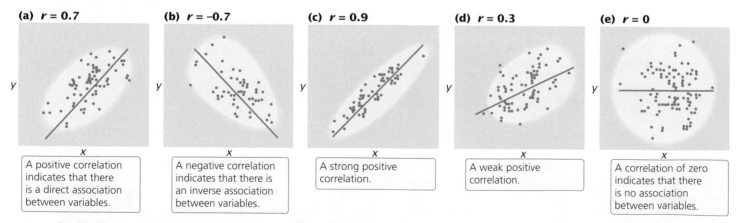

(a) $r = 0.7$

A positive correlation indicates that there is a direct association between variables.

(b) $r = -0.7$

A negative correlation indicates that there is an inverse association between variables.

(c) $r = 0.9$

A strong positive correlation.

(d) $r = 0.3$

A weak positive correlation.

(e) $r = 0$

A correlation of zero indicates that there is no association between variables.

24.11 The correlation coefficient describes the relation between two or more variables.

The absolute value of the correlation coefficient (the size of the coefficient, ignoring its sign) provides information about the strength of association between the variables. A coefficient of −1 or +1 indicates a perfect correlation between the variables, meaning that a change in x is always accompanied by a proportional change in y. Correlation coefficients close to −1 or close to +1 indicate a strong association between the variables: a change in x is almost always associated with a proportional increase in y, as seen in **Figure 24.11c**. On the other hand, a correlation coefficient closer to 0 indicates a weak correlation: a change in x is associated with a change in y but not always (**Figure 24.11d**). A correlation of 0 indicates that there is no association between variables (**Figure 24.11e**).

A correlation coefficient can be computed for two variables measured for the same individual, such as height (x) and weight (y). A correlation coefficient can also be computed for a single variable measured for pairs of individuals. For example, we can calculate for fish the correlation between the number of vertebrae of a parent (x) and the number of vertebrae of its offspring (y), as shown in **Figure 24.12**. This approach is often used in quantitative genetics.

A correlation between two variables indicates only that the variables are associated; it does not imply a cause-and-effect relation. Correlation also does not mean that the values of two variables are the same; it means only that a change in one variable is associated with a proportional change in the other variable. For example, the x and y variables in the following list are almost perfectly correlated, with a correlation coefficient of 0.99.

	x value	y value
	12	123
	14	140
	10	110
	6	61
	3	32
Average:	9	90

A high correlation is found between these x and y variables; larger values of x are always associated with larger values of y. Note that the y values are about 10 times as large as the corresponding x values; so, although x and y are correlated, they are not identical. The distinction between correlation and identity becomes important when we consider the effects of heredity and environment on the correlation of characteristics. TRY PROBLEM 25 →

Regression

Correlation provides information only about the strength and direction of association between variables. However, we often want to know more than just whether two variables are associated; we want to be able to predict the value of one variable, given a value of the other.

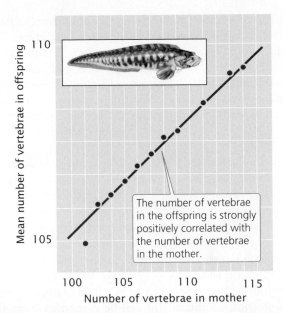

24.12 **A correlation coefficient can be computed for a single variable measured for pairs of individuals.** Here, the numbers of vertebrae in mothers and offspring of the fish *Zoarces viviparus* are compared.

> The number of vertebrae in the offspring is strongly positively correlated with the number of vertebrae in the mother.

A positive correlation exists between the body weight of parents and the body weight of their offspring; this correlation exists in part because genes influence body weight, and parents and children have genes in common. Because of this association between parental and offspring phenotypes, we can predict the weight of an individual on the basis of the weights of its parents. This type of statistical prediction is called **regression**. This technique plays an important role in quantitative genetics because it allows us to predict the characteristics of offspring from a given mating, even without knowledge of the genotypes that encode the characteristics.

Regression can be understood by plotting a series of x and y values. **Figure 24.13** illustrates the relation between

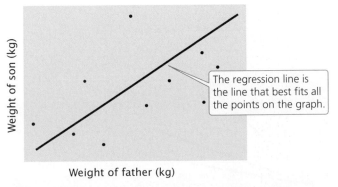

> The regression line is the line that best fits all the points on the graph.

24.13 **A regression line defines the relation between two variables.** Illustrated here is a regression of the weights of fathers against the weights of sons. Each father–son pair is represented by a point on the graph: the x value of a point is the father's weight and the y value of the point is the son's weight.

the weight of a father (x) and the weight of his son (y). Each father–son pair is represented by a point on the graph. The overall relation between these two variables is depicted by the regression line, which is the line that best fits all the points on the graph (deviations of the points from the line are minimized). The regression line defines the relation between the x and y variables and can be represented by

$$y = a + bx \qquad (24.8)$$

In Equation 24.8, x and y represent the x and y variables (in this case, the father's weight and the son's weight, respectively). The variable a is the y intercept of the line, which is the expected value of y when x is 0. Variable b is the slope of the regression line, also called the **regression coefficient**.

Trying to position a regression line by eye is not only very difficult but also inaccurate when there are many points scattered over a wide area. Fortunately, the regression coefficient and the y intercept can be obtained mathematically. The regression coefficient (b) can be computed from the covariance of x and y (cov_{xy}) and the variance of x (s) by

$$b = \frac{\text{cov}_{xy}}{s_x^2} \qquad (24.9)$$

The regression coefficient indicates how much y increases, on average, per increase in x. Several regression lines with different regression coefficients are illustrated in **Figure 24.14**. Notice that as the regression coefficient increases, the slope of the regression line increases.

After the regression coefficient has been calculated, the y intercept can be calculated by substituting the regression coefficient and the mean values of x and y into the following equation:

$$a = \bar{y} - b\bar{x} \qquad (24.10)$$

The regression equation ($y = a + bx$, Equation 24.8) can then be used to predict the value of any y given the value of x.

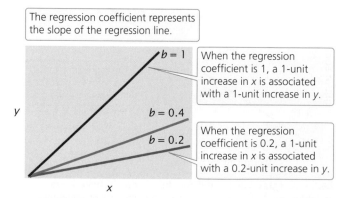

The regression coefficient represents the slope of the regression line.

$b = 1$ — When the regression coefficient is 1, a 1-unit increase in x is associated with a 1-unit increase in y.

$b = 0.4$

$b = 0.2$ — When the regression coefficient is 0.2, a 1-unit increase in x is associated with a 0.2-unit increase in y.

24.14 The regression coefficient, b, represents the change in y per unit change in x. Shown here are regression lines with different regression coefficients.

Worked Problem

Body weights of 11 female fishes and the numbers of eggs that they produce are:

Weight (mg)	Eggs (thousands)
x	y
14	61
17	37
24	65
25	69
27	54
33	93
34	87
37	89
40	100
41	90
42	97

What are the correlation coefficient and the regression coefficient for body weight and egg number in these 11 fishes?

• Solution

The computations needed to answer this question are given in the table on the facing page. To calculate the correlation and regression coefficients, we first obtain the sum of all the x_i values ($\sum x_i$) and the sum of all the y_i values ($\sum y_i$); these sums are shown in the last row of the table. We can calculate the means of the two variables by dividing the sums by the number of measurements, which is 11:

$$\bar{x} = \frac{\sum x_i}{n} = \frac{334}{11} = 30.36$$

$$\bar{y} = \frac{\sum y_i}{n} = \frac{842}{11} = 76.55$$

After the means have been calculated, the deviations of each value from the mean are computed; these deviations are shown in columns B and E of the table.

A Weight (mg) x	B $x_i - \bar{x}$	C $(x_i - \bar{x})^2$	D Eggs (thousands) y	E $y_i - \bar{y}$	F $(y_i - \bar{y})^2$	G $(x_i - \bar{x})(y_i - \bar{y})$
14	−16.36	267.65	61	−15.55	241.80	254.40
17	−13.36	178.49	37	−39.55	1564.20	528.39
24	−6.36	40.45	65	−11.55	133.40	73.46
25	−5.36	28.73	69	−7.55	57.00	40.47
27	−3.36	11.29	54	−22.55	508.50	75.77
33	2.64	6.97	93	16.45	270.60	43.43
34	3.64	13.25	87	10.45	109.20	38.04
37	6.64	44.09	89	12.45	155.00	82.67
40	9.64	92.93	100	23.45	549.90	226.06
41	10.64	113.21	90	13.45	180.90	143.11
42	11.64	135.49	97	20.45	418.20	238.04
$\sum x_i = 334$		$\sum(x_i - \bar{x})^2 = 932.55$	$\sum y_i = 842$		$\sum(y_i - \bar{y})^2 = 4188.70$	$\sum(x_i - \bar{x})(y_i - \bar{y}) = 1743.84$

Source: R. R. Sokal and F. J. Rohlf, *Biometry*, 2d ed. (San Francisco: W. H. Freeman and Company, 1981).

The deviations are then squared (columns C and F) and summed (last row of columns C and F). Next, the products of the deviation of the x values and the deviation of the y values ($(x_i - \bar{x})(y_i - \bar{y})$) are calculated; these products are shown in column G, and their sum is shown in the last row of column G.

To calculate the covariance, we use Equation 24.6:

$$\text{cov}_{xy} = \frac{\sum(x_i - \bar{x})(y_i - \bar{y})}{n-1} = \frac{1743.84}{10} = 174.38$$

To calculate the covariance and the regression requires the variances and standard deviations of x and y:

$$s_x^2 = \frac{\sum(x_i - \bar{x})^2}{n-1} = \frac{932.55}{10} = 93.26$$

$$s_x = \sqrt{s_x^2} = \sqrt{93.26} = 9.66$$

$$s_y^2 = \frac{\sum(y_i - \bar{y})^2}{n-1} = \frac{4188.70}{10} = 418.87$$

$$s_y = \sqrt{s_y^2} = \sqrt{418.87} = 20.47$$

We can now compute the correlation and regression coefficients as shown here.

Correlation coefficient:

$$r = \frac{\text{cov}_{xy}}{s_x s_y} = \frac{174.38}{9.66 \times 20.47} = 0.88$$

Regression coefficient:

$$b = \frac{\text{cov}_{xy}}{s_x^2} = \frac{174.38}{93.26} = 1.87$$

→ Practice your understanding of correlation and regression by working Problem 26 at the end of the chapter.

Applying Statistics to the Study of a Polygenic Characteristic

Edward East carried out one early statistical study of polygenic inheritance on the length of flowers in tobacco (*Nicotiana longiflora*). He obtained two varieties of tobacco that differed in flower length: one variety had a mean flower length of 40.5 mm, and the other had a mean flower length of 93.3 mm (**Figure 24.15**). These two varieties had been inbred for many generations and were homozygous at all loci contributing to flower length. Thus, there was no genetic variation in the original parental strains; the small differences in flower length within each strain were due to environmental effects on flower length.

When East crossed the two strains, he found that flower length in the F_1 was about halfway between that in the two parents (see Figure 24.15), as would be expected if the genes determining the differences in the two strains were additive in their effects. The variance of flower length in the F_1 was similar to that seen in the parents, because all the F_1 had the same genotype, as did each parental strain (the F_1 were all heterozygous at the genes that differed between the two parental varieties).

East then interbred the F_1 to produce F_2 progeny. The mean flower length of the F_2 was similar to that of the F_1, but the variance of the F_2 was much greater (see Figure 24.15).

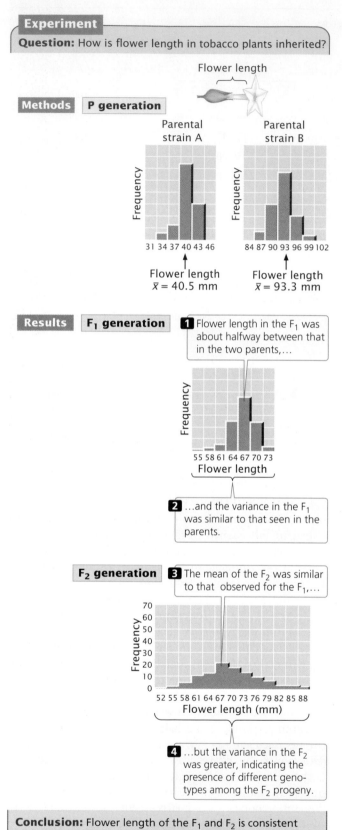

Experiment

Question: How is flower length in tobacco plants inherited?

Methods   **P generation**

Flower length

Parental strain A

Parental strain B

Frequency

Frequency

31 34 37 40 43 46

84 87 90 93 96 99 102

Flower length
$\bar{x} = 40.5$ mm

Flower length
$\bar{x} = 93.3$ mm

Results   **F₁ generation**

1 Flower length in the F₁ was about halfway between that in the two parents,...

Frequency

55 58 61 64 67 70 73
Flower length

2 ...and the variance in the F₁ was similar to that seen in the parents.

F₂ generation

3 The mean of the F₂ was similar to that observed for the F₁,...

70
60
50
40
30
20
10
0

Frequency

52 55 58 61 64 67 70 73 76 79 82 85 88
Flower length (mm)

4 ...but the variance in the F₂ was greater, indicating the presence of different genotypes among the F₂ progeny.

Conclusion: Flower length of the F₁ and F₂ is consistent with the hypothesis that the characteristic is determined by several genes that are additive in their effects.

24.15 Edward East conducted an early statistical study of the inheritance of flower length in tobacco.

This greater variability indicates that not all of the F₂ progeny had the same genotype.

East selected some F₂ plants and interbred them to produce F₃ progeny. He found that flower length of the F₃ depended on flower length in the plants selected as their parents. This finding demonstrated that flower-length differences in the F₂ were partly genetic and were therefore passed to the next generation. None of the 444 F₂ plants raised by East exhibited flower lengths similar to those of the two parental strains. This result suggested that more than four loci with pairs of alleles affected flower length in his varieties, because four allelic pairs are expected to produce 1 of 256 progeny $[(^1/_4)^4 = {}^1/_{256}]$ having one or the other of the original parental phenotypes.

24.3 Heritability Is Used to Estimate the Proportion of Variation in a Trait That Is Genetic

In addition to being polygenic, quantitative characteristics are frequently influenced by environmental factors. Knowing how much of the variation in a quantitative characteristic is due to genetic differences and how much is due to environmental differences is often useful. The proportion of the total phenotypic variation that is due to genetic differences is known as the **heritability**.

Consider a dairy farmer who owns several hundred milk cows. The farmer notices that some cows consistently produce more milk than others. The nature of these differences is important to the profitability of his dairy operation. If the differences in milk production are largely genetic in origin, then the farmer may be able to boost milk production by selectively breeding the cows that produce the most milk. On the other hand, if the differences are largely environmental in origin, selective breeding will have little effect on milk production, and the farmer might better boost milk production by adjusting the environmental factors associated with higher milk production. To determine the extent of genetic and environmental influences on variation in a characteristic, phenotypic variation in the characteristic must be partitioned into components attributable to different factors.

Phenotypic Variance

To determine how much of phenotypic differences in a population is due to genetic and environmental factors, we must first have some quantitative measure of the phenotype under consideration. Consider a population of wild plants that differ in size. We could collect a representative sample of plants from the population, weigh each plant in the sample, and calculate the mean and variance of plant weight. This **phenotypic variance** is represented by V_P.

Components of phenotypic variance First, some of the phenotypic variance may be due to differences in genotypes among individual members of the population. These differences are termed the **genetic variance** and are represented by V_G.

Second, some of the differences in phenotype may be due to environmental differences among the plants; these differences are termed the **environmental variance**, V_E. Environmental variance includes differences in environmental factors such as the amount of light or water that the plant receives; it also includes random differences in development that cannot be attributed to any specific factor. Any variation in phenotype that is not inherited is, by definition, a part of the environmental variance.

Third, **genetic—environmental interaction variance** (V_{GE}) arises when the effect of a gene depends on the specific environment in which it is found. An example is shown in **Figure 24.16**. In a dry environment, genotype *AA* produces a plant that averages 12 g in weight, and genotype *aa* produces a smaller plant that averages 10 g. In a wet environment, genotype *aa* produces the larger plant, averaging 24 g in weight, whereas genotype *AA* produces a plant that averages 20 g. In this example, there are clearly differences in the two environments: both genotypes produce heavier plants in the wet environment. There are also differences in the weights of the two genotypes, but the relative performances of the genotypes depend on whether the plants are grown in a wet or a dry environment. In this case, the influences on phenotype cannot be neatly allocated into genetic and environmental components, because the expression of the genotype depends on the environment in which the plant grows. The phenotypic variance must therefore include a component that accounts for the way in which genetic and environmental factors interact.

In summary, the total phenotypic variance can be apportioned into three components:

$$V_P = V_G + V_E + V_{GE} \qquad (24.11)$$

Components of genetic variance Genetic variance can be further subdivided into components consisting of different types of genetic effects. First, **additive genetic variance** (V_A) comprises the additive effects of genes on the phenotype, which can be summed to determine the overall effect on the phenotype. For example, suppose that, in a plant, allele A^1 contributes 2 g in weight and allele A^2 contributes 4 g. If the alleles are strictly additive, then the genotypes would have the following weights:

$$A^1A^1 = 2 + 2 = 4 \text{ g}$$
$$A^1A^2 = 2 + 4 = 6 \text{ g}$$
$$A^2A^2 = 4 + 4 = 8 \text{ g}$$

The genes that Nilsson-Ehle studied, which affect kernel color in wheat, were additive in this way. The additive genetic variance primarily determines the resemblance between parents and offspring. For example, if all of the phenotypic variance is due to additive genetic variance, then the phenotypes of the offspring will be exactly intermediate between those of the parents, but, if some genes have dominance, then offspring may be phenotypically different from both parents (i.e., *Aa* × *Aa* → *aa* offspring).

Second, there is **dominance genetic variance** (V_D) when some genes have a dominance component. In this case, the alleles at a locus are not additive; rather, the effect of an allele depends on the identity of the other allele at that locus. For example, with a dominant allele (*T*), genotypes *TT* and *Tt* have the same phenotype. Here, we cannot simply add the effects of the alleles together, because the effect of the small *t* allele is masked by the presence of the large *T* allele. Instead, we must add a component (V_D) to the genetic variance to account for the way in which alleles interact.

Third, genes at different loci may interact in the same way that alleles at the same locus interact. When this genic interaction takes place, the effects of the genes are not additive. For example, Chapter 5 described how coat color in Labrador retrievers exhibits genic interaction; genotypes *BB ee* and *bb ee* both produce yellow dogs, because the effect of alleles at the *B* locus are masked when *ee* alleles are present at the *E* locus. With genic interaction, we must add

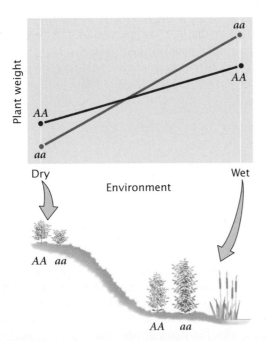

24.16 Genetic–environmental interaction variance is obtained when the effect of a gene depends on the specific environment in which it is found. In this example, the genotype affects plant weight, but the environmental conditions determine which genotype produces the heavier plant.

a third component, called **genic interaction variance** (V_I), to the genetic variance:

$$V_G = V_A + V_D + V_I \qquad (24.12)$$

Summary equation We can now integrate these components into one equation to represent all the potential contributions to the phenotypic variance:

$$V_P = V_A + V_D + V_I + V_E + V_{GE} \qquad (24.13)$$

This equation provides us with a model that describes the potential causes of differences that we observe among individual phenotypes. It's important to note that this model deals strictly with the observable *differences* (variance) in phenotypes among individual members of a population; it says nothing about the absolute value of the characteristic or about the underlying genotypes that produce these differences.

Types of Heritability

The model of phenotypic variance that we've just developed can be used to determine how much of the phenotypic variance in a characteristic is due to genetic differences. **Broad-sense heritability** (H^2) represents the proportion of phenotypic variance that is due to genetic variance and is calculated by dividing the genetic variance by the phenotypic variance:

$$\text{broad-sense heritability} = H^2 = \frac{V_G}{V_P} \qquad (24.14)$$

The symbol H^2 represents broad-sense heritability because it is a measure of variance, which is in units squared.

Broad-sense heritability can potentially range from 0 to 1. A value of 0 indicates that none of the phenotypic variance results from differences in genotype and all of the differences in phenotype result from environmental variation. A value of 1 indicates that all of the phenotypic variance results from differences in genotypes. A heritability value between 0 and 1 indicates that both genetic and environmental factors influence the phenotypic variance.

Often, we are more interested in the proportion of the phenotypic variance that results from the additive genetic variance because, as mentioned earlier, the additive genetic variance primarily determines the resemblance between parents and offspring. **Narrow-sense heritability** (h^2) is equal to the additive genetic variance divided by the phenotypic variance:

$$\text{narrow-sense heritability} = h^2 = \frac{V_A}{V_P} \qquad (24.15)$$

TRY PROBLEM 27 →

Calculating Heritability

Having considered the components that contribute to phenotypic variance and having developed a general concept of heritability, we can ask, How do we go about estimating these different components and calculating heritability? There are several ways to measure the heritability of a characteristic. They include eliminating one or more variance components, comparing the resemblance of parents and offspring, comparing the phenotypic variances of individuals with different degrees of relatedness, and measuring the response to selection. The mathematical theory that underlies these calculations of heritability is complex and beyond the scope of this book. Nevertheless, we can develop a general understanding of how heritability is measured.

Heritability by elimination of variance components One way of calculating the broad-sense heritability is to eliminate one of the variance components. We have seen that $V_P = V_G + V_E + V_{GE}$. If we eliminate all environmental variance ($V_E = 0$), then $V_{GE} = 0$ (because, if either V_G or V_E is zero, no genetic–environmental interaction can take place), and $V_P = V_G$. In theory, we might make V_E equal to 0 by ensuring that all individuals were raised in exactly the same environment but, in practice, it is virtually impossible. Instead, we could make V_G equal to 0 by raising genetically identical individuals, causing V_P to be equal to V_E. In a typical experiment, we might raise cloned or highly inbred, identically homozygous individuals in a defined environment and measure their phenotypic variance to estimate V_E. We could then raise a group of genetically variable individuals and measure their phenotypic variance (V_P). Using V_E calculated on the genetically identical individuals, we could obtain the genetic variance of the variable individuals by subtraction:

$$V_{G(\text{of genetically varying individuals})}$$
$$= V_{P\,(\text{of genetically varying individuals})} - V_{E\,(\text{of genetically identical individuals})} \qquad (24.16)$$

The broad-sense heritability of the genetically variable individuals would then be calculated as follows:

$$H^2 = \frac{V_{G\,(\text{of genetically varying individuals})}}{V_{P\,(\text{of genetically varying individuals})}} \qquad (24.17)$$

Sewall Wright used this method to estimate the heritability of white spotting in guinea pigs. He first measured the phenotypic variance for white spotting in a genetically variable population and found that $V_P = 573$. Then he inbred the guinea pigs for many generations so that they were essentially homozygous and genetically identical. When he measured their phenotypic variance in white spotting, he obtained V_P equal to 340. Because $V_G = 0$ in this group, their $V_P = V_E$. Wright assumed this value of environmental vari-

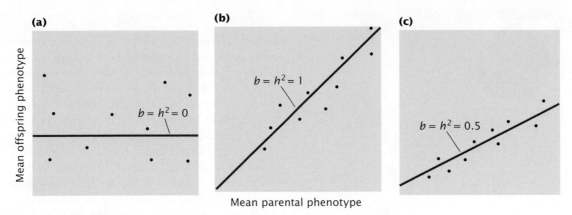

24.17 The narrow-sense heritability, h^2, equals the regression coefficient, b, in a regression of the mean phenotype of the offspring against the mean phenotype of the parents. (a) There is no relation between the parental phenotype and the offspring phenotype. (b) The offspring phenotype is the same as the parental phenotypes. (c) Both genes and environment contribute to the differences in phenotype.

ance for the original (genetically variable) population and estimated their genetic variance:

$$V_P - V_E = V_G$$

$$573 - 340 = 233$$

He then estimated the broad-sense heritability from the genetic and phenotypic variance:

$$H^2 = \frac{V_G}{V_P}$$

$$H^2 = \frac{233}{573} = 0.41$$

This value implies that 41% of the variation in spotting of guinea pigs in Wright's population was due to differences in genotype.

Estimating heritability by using this method assumes that the environmental variance of genetically identical individuals is the same as the environmental variance of the genetically variable individuals, which may not be true. Additionally, this approach can be applied only to organisms for which it is possible to create genetically identical individuals. **TRY PROBLEM 32** →

Heritability by parent–offspring regression Another method for estimating heritability is to compare the phenotypes of parents and offspring. When genetic differences are responsible for phenotypic variance, offspring should resemble their parents more than they resemble unrelated individuals, because offspring and parents have some genes in common that help determine their phenotype. Correlation and regression can be used to analyze the association of phenotypes in different individuals.

To calculate the narrow-sense heritability in this way, we first measure the characteristic on a series of parents and offspring. The data are arranged into families, and the mean parental phenotype is plotted against the mean offspring phenotype (**Figure 24.17**). Each data point in the graph represents one family; the value on the x (horizontal) axis is the mean phenotypic value of the parents in a family, and the value on the y (vertical) axis is the mean phenotypic value of the offspring for the family.

Let's assume that there is no narrow-sense heritability for the characteristic ($h^2 = 0$), meaning that genetic differences do not contribute to the phenotypic differences among individuals. In this case, offspring will be no more similar to their parents than they are to unrelated individuals, and the data points will be scattered randomly, generating a regression coefficient of zero (see Figure 24.17a). Next, let's assume that all of the phenotypic differences are due to additive genetic differences ($h^2 = 1.0$). In this case, the mean phenotype of the offspring will be equal to the mean phenotype of the parents, and the regression coefficient will be 1 (see Figure 24.17b). If genes and environment both contribute to the differences in phenotype, both heritability and the regression coefficient will lie between 0 and 1 (see Figure 24.17c). The regression coefficient therefore provides information about the magnitude of the heritability.

A complex mathematical proof (which we will not go into here) demonstrates that, in a regression of the mean phenotype of the offspring against the mean phenotype of the parents, narrow-sense heritability (h^2) equals the regression coefficient (b):

$$h^2 = b_{\text{(regression of mean offspring against mean of both parents)}} \qquad (24.18)$$

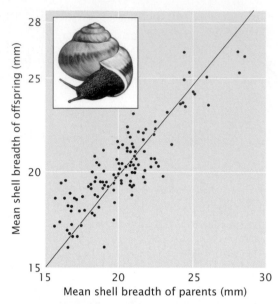

24.18 The heritability of shell breadth in snails can be determined by regression of the phenotype of offspring against the mean phenotype of the parents. The regression coefficient, which equals the heritability, is 0.70. [From L. M. Cook, *Evolution* 19:86–94, 1965.]

An example of calculating heritability by regression of the phenotypes of parents and offspring is illustrated in **Figure 24.18**.

Sometimes, only the phenotype of one parent is known. In a regression of the mean offspring phenotype against the phenotype of only one parent, the narrow-sense heritability equals twice the regression coefficient:

$$h^2 = 2b_{(\text{regression of mean offspring against mean of one parent})} \qquad (24.19)$$

With only one parent, the heritability is twice the regression coefficient because only half the genes of the offspring come from one parent; thus, we must double the regression coefficient to obtain the full heritability.

Heritability and degrees of relatedness   A third method for calculating heritability is to compare the phenotypes of individuals having different degrees of relatedness. This method is based on the concept that the more closely related two individuals are, the more genes they have in common.

Monozygotic (identical) twins have 100% of their genes in common, whereas dizygotic (nonidentical) twins have, on average, 50% of their genes in common. If genes are important in determining variability in a characteristic, then monozygotic twins should be more similar in a particular characteristic than dizygotic twins. By using correlation to compare the phenotypes of monozygotic and dizygotic twins, we can estimate broad-sense heritability. A rough estimate of the broad-sense heritability can be obtained by taking twice the difference of the correlation coefficients for a quantitative characteristic in monozygotic and dizygotic twins:

$$H^2 = 2(r_{\text{MZ}} - r_{\text{DZ}}) \qquad (24.20)$$

where r_{MZ} equals the correlation coefficient among monozygotic twins and r_{DZ} equals the correlation coefficient among dizygotic twins. For example, suppose we found the correlation of height among the two members of monozygotic twin pairs (r_{MZ}) to be 0.9 and the correlation of height among the two members of dizygotic twins (r_{DZ}) to be 0.5. The broad-sense heritability for height would be $H^2 = 2(0.9 - 0.5) = 2(0.4) = 0.8$. This calculation assumes that the two individuals of a monozygotic twin pair experience environments that are no more similar to each other than those experienced by the two individuals of a dizygotic twin pair. This assumption is often not met when twins have been reared together.

Narrow-sense heritability also can be estimated by comparing the phenotypic variances for a characteristic in full sibs (siblings who have both parents in common as well as 50% of their genes on the average) and half sibs (who have only one parent in common and thus 25% of their genes on the average).

All estimates of heritability depend on the assumption that the environments of related individuals are not more similar than those of unrelated individuals. This assumption is difficult to meet in human studies, because related people are usually reared together. Heritability estimates for humans should therefore always be viewed with caution.

TRY PROBLEM 35 ➡

CONCEPTS

Broad-sense heritability is the proportion of phenotypic variance that is due to genetic variance. Narrow-sense heritability is the proportion of phenotypic variance that is due to additive genetic variance. Heritability can be measured by eliminating one of the variance components, by analyzing parent–offspring regression, or by comparing individuals having different degrees of relatedness.

✔ **CONCEPT CHECK 5**

If the environmental variance (V_E) increases and all other variance components remain the same, what will the effect be?

a. Broad-sense heritability will decrease.

b. Broad-sense heritability will increase.

c. Narrow-sense heritability will increase.

d. Broad-sense heritability will increase, but narrow-sense heritability will decrease.

The Limitations of Heritability

Knowledge of heritability has great practical value because it allows us to statistically predict the phenotypes of offspring on the basis of their parent's phenotype. It also provides useful information about how characteristics will respond to selection (see Section 24.4). In spite of its importance, heritability is frequently misunderstood. Heritability does not provide information about an individual's genes or the environmental factors that control the development of a characteristic, and it says nothing about the nature of differences between groups. This section outlines some limitations and common misconceptions concerning broad- and narrow-sense heritability.

Heritability does not indicate the degree to which a characteristic is genetically determined Heritability is the proportion of the phenotypic variance that is due to genetic variance; it says nothing about the degree to which genes determine a characteristic. Heritability indicates only the degree to which genes determine *variation* in a characteristic. The determination of a characteristic and the determination of variation in a characteristic are two very different things.

Consider polydactyly (the presence of extra digits) in rabbits, which can be caused either by environmental factors or by a dominant gene. Suppose we have a group of rabbits all homozygous for a gene that produces the usual numbers of digits. None of the rabbits in this group carries a gene for polydactyly, but a few of the rabbits are polydactylous because of environmental factors. Broad-sense heritability for polydactyly in this group is zero, because there is no genetic variation for polydactyly; all of the variation is due to environmental factors. However, it would be incorrect for us to conclude that genes play no role in determining the number of digits in rabbits. Indeed, we know that there are specific genes that can produce extra digits. Heritability indicates nothing about whether genes control the development of a characteristic; it provides information only about causes of the variation in a characteristic within a defined group.

An individual does not have heritability Broad- and narrow-sense heritabilities are statistical values based on the genetic and phenotypic variances found in a *group* of individuals. Heritability cannot be calculated for an individual, and heritability has no meaning for a specific individual. Suppose we calculate the narrow-sense heritability of adult body weight for the students in a biology class and obtain a value of 0.6. We could conclude that 60% of the variation in adult body weight among the students in this class is determined by additive genetic variation. We should not, however, conclude that 60% of any particular student's body weight is due to additive genes.

There is no universal heritability for a characteristic The value of heritability for a characteristic is specific for a given population in a given environment. Recall that broad-sense heritability is genetic variance divided by phenotypic variance. Genetic variance depends on which genes are present, which often differs between populations. In the example of polydactyly in rabbits, there were no genes for polydactyly in the group; so the heritability of the characteristic was zero. A different group of rabbits might contain many genes

for polydactyly, and the heritability of the characteristic might then be high.

Environmental differences also may affect heritability because V_P is composed of both genetic and environmental variance. When the environmental differences that affect a characteristic differ between two groups, the heritabilities for the two groups also often differ.

Because heritability is specific to a defined population in a given environment, it is important not to extrapolate heritabilities from one population to another. For example, human height is determined by environmental factors (such as nutrition and health) and by genes. If we measured the heritability of height in a developed country, we might obtain a value of 0.8, indicating that the variation in height in this population is largely genetic. This population has a high heritability because most people have adequate nutrition and health care (V_E is low); so most of the phenotypic variation in height is genetically determined. It would be incorrect for us to assume that height has a high heritability in all human populations. In developing countries, there may be more variation in a range of environmental factors; some people may enjoy good nutrition and health, whereas others may have a diet deficient in protein and suffer from diseases that affect stature. If we measured the heritability of height in such a country, we would undoubtedly obtain a lower value than we observed in the developed country because there is more environmental variation and the genetic variance in height constitutes a smaller proportion of the phenotypic variation, making the heritability lower. The important point to remember is that heritability must be calculated separately for each population and each environment.

Even when heritability is high, environmental factors may influence a characteristic High heritability does not mean that environmental factors cannot influence the expression of a characteristic. High heritability indicates only that the environmental variation to which the population is *currently* exposed is not responsible for variation in the characteristic. Let's look again at human height. In most developed countries, heritability of human height is high, indicating that genetic differences are responsible for most of the variation in height. It would be wrong for us to conclude that human height cannot be changed by alteration of the environment. Indeed, height decreased in several European cities during World War II owing to hunger and disease, and height can be increased dramatically by the administration of growth hormone to children. The absence of environmental variation in a characteristic does not mean that the characteristic will not respond to environmental change.

Heritabilities indicate nothing about the nature of population differences in a characteristic A common misconception about heritability is that it provides information about population differences in a characteristic.

Heritability is specific for a given population in a given environment, and so it cannot be used to draw conclusions about why populations differ in a characteristic.

Suppose we measured heritability for human height in two groups. One group is from a small town in a developed country, where everyone consumes a high-protein diet. Because there is little variation in the environmental factors that affect human height and there is some genetic variation, the heritability of height in this group is high. The second group comprises the inhabitants of a single village in a developing country. The consumption of protein by these people is only 25% of that consumed by those in the first group; so their average adult height is several centimeters less than that in the developed country. Again, there is little variation in the environmental factors that determine height in this group, because everyone in the village eats the same types of food and is exposed to the same diseases. Because there is little environmental variation and there is some genetic variation, the heritability of height in this group also is high.

Thus, the heritability of height in both groups is high, and the average height in the two groups is considerably different. We might be tempted to conclude that the difference in height between the two groups is genetically based—that the people in the developed country are genetically taller than the people in the developing country. This conclusion is obviously wrong, however, because the differences in height are due largely to diet—an environmental factor. Heritability provides no information about the causes of differences between populations.

These limitations of heritability have often been ignored, particularly in arguments about possible social implications of genetic differences between humans. Soon after Mendel's principles of heredity were rediscovered, some geneticists began to claim that many human behavioral characteristics are determined entirely by genes. This claim led to debates about whether characteristics such as human intelligence are determined by genes or environment. Many of the early claims of genetically based human behavior were based on poor research; unfortunately, the results of these studies were often accepted at face value and led to a number of eugenic laws that discriminated against certain groups of people. Today, geneticists recognize that many behavioral characteristics are influenced by a complex interaction of genes and environment, and separating genetic effects from those of the environment is very difficult.

The results of a number of modern studies indicate that human intelligence as measured by IQ and other intelligence tests has a moderately high heritability (usually from 0.4 to 0.8). On the basis of this observation, some people have argued that intelligence is innate and that enhanced educational opportunities cannot boost intelligence. This argument is based on the misconception that, when heritability is high, changing the environment will not alter the characteristic. In addition, because heritabilities of intelligence range from 0.4 to 0.8, a considerable amount of the variance in intelligence originates from environmental differences.

Another argument based on a misconception about heritability is that ethnic differences in measures of intelligence are genetically based. Because the results of some genetic studies show that IQ has moderately high heritability and because other studies find differences in the average IQ of ethnic groups, some people have suggested that ethnic differences in IQ are genetically based. As in the example of the effects of diet on human height, heritability provides no information about causes of differences among groups; it indicates only the degree to which phenotypic variance within a single group is genetically based. High heritability for a characteristic does not mean that phenotypic differences between ethnic groups are genetic. We should also remember that separating genetic and environmental effects in humans is very difficult; so heritability estimates themselves may be unreliable.

TRY PROBLEM 34 →

CONCEPTS

Heritability provides information only about the degree to which *variation* in a characteristic is genetically determined. There is no universal heritability for a characteristic; heritability is specific for a given population in a specific environment. Environmental factors can potentially affect characteristics with high heritability, and heritability says nothing about the nature of population differences in a characteristic.

✔ CONCEPT CHECK 6

Suppose that you just learned that the narrow-sense heritability of blood pressure measured among a group of African Americans in Detroit, Michigan, is 0.40. What does this heritability tell us about genetic and environmental contributions to blood pressure?

Locating Genes That Affect Quantitative Characteristics

The statistical methods described for use in analyzing quantitative characteristics can be used both to make predictions about the average phenotype expected in offspring and to estimate the overall contribution of genes to variation in the characteristic. These methods do not, however, allow us to identify and determine the influence of individual genes that affect quantitative characteristics. As discussed in the introduction to this chapter, chromosome regions with genes that control polygenic characteristics are referred to as quantitative trait loci. Although quantitative genetics has made important contributions to basic biology and to plant and animal breeding, the past inability to identify QTLs and measure their individual effects severely limited the application of quantitative genetic methods.

24.19 The availability of molecular markers makes the mapping of QTLs possible in many organisms. QTL mapping is used to identify genes that determine muscle mass in pigs. [USDA.]

Mapping QTLs In recent years, numerous genetic markers have been identified and mapped with the use of molecular techniques, making it possible to identify QTLs by linkage analysis. The underlying idea is simple: if the inheritance of a genetic marker is associated consistently with the inheritance of a particular characteristic (such as increased height), then that marker must be linked to a QTL that affects height. The key is to have enough genetic markers so that QTLs can be detected throughout the genome. With the introduction of restriction fragment length polymorphisms, microsatellite variations, and single-nucleotide polymorphisms (SNPs; see Chapters 19 and 20), variable markers are now available for mapping QTLs in a number of different organisms (**Figure 24.19**).

A common procedure for mapping QTLs is to cross two homozygous strains that differ in alleles at many loci. The resulting F_1 progeny are then intercrossed or backcrossed to allow the genes to recombine through independent assortment and crossing over. Genes on different chromosomes and genes that are far apart on the same chromosome will recombine freely; genes that are closely linked will be inherited together. The offspring are measured for one or more quantitative characteristics; at the same time, they are genotyped for numerous genetic markers that span the genome. Any correlation between the inheritance of a particular marker allele and a quantitative phenotype indicates that a QTL is linked to that marker. If enough markers are used, the detection of all the QTLs affecting a characteristic is theoretically possible. It is important to recognize that a QTL is not a gene; rather, it is a map location for a chromosome region that is associated with that trait. After a QTL has been identified, it can be studied for the presence

of specific genes that influence the quantitative trait. The introduction to the chapter describes how this approach was used to identify a major gene that affects oil production in corn. QTL mapping has been used to detect genes affecting a variety of characteristics in plant and animal species (**Table 24.2**).

Genomewide association studies The traditional method of identifying QTLs is to carry out crosses between varieties that differ in a quantitative trait and then genotype numerous progeny for many markers. Although effective, this method is slow and labor intensive.

An alternative technique for identifying genes that affect quantitative traits is to conduct genomewide association studies, which were introduced in Chapter 7. Unlike traditional linkage analysis, which examines the association of a trait and gene markers among the *progeny of a cross*, genomewide association studies look for associations between traits and genetic markers in a *biological popula-*

Table 24.2	Quantitative characteristics for which QTLs have been detected	
Organism	Quantitative Characteristic	Number of QTLs Detected
Tomato	Soluble solids	7
	Fruit mass	13
	Fruit pH	9
	Growth	5
	Leaflet shape	9
	Height	9
Corn	Height	11
	Leaf length	7
	Tiller number	1
	Glume hardness	5
	Grain yield	18
	Number of ears	9
	Thermotolerance	6
Common bean	Number of nodules	4
Mung bean	Seed weight	4
Cow pea	Seed weight	2
Wheat	Preharvest sprout	4
Pig	Growth	2
	Length of small intestine	1
	Average back fat	1
	Abdominal fat	1
Mouse	Epilepsy	2
Rat	Hypertension	2

Source: After S. D. Tanksley, Mapping polygenes, *Annual Review of Genetics* 27:218, 1993.

tion, a group of interbreeding individuals. The presence of an association between genetic markers and a trait indicate that the genetic markers are closely linked to one or more genes that affect variation in the trait. Genomewide association studies have been facilitated by the identification of single-nucleotide polymorphisms, which are positions in the genome where individual organisms vary in a single nucleotide base (see Chapter 20). Many SNPs have been identified in some organisms through genomic sequencing. Individual organisms can often be quickly and inexpensively genotyped for numerous SNPs, providing the genetic markers necessary to conduct genomewide association studies.

Genomewide association studies have been widely used to locate genes that affect quantitative traits in humans, including disease susceptibility, obesity, intelligence, and height. A number of quantitative traits in plants have been studied with association studies, including kernel composition, size, color and taste, disease resistance, and starch quality. Genomewide association studies in domestic animals have identified chromosomal segments affecting body weight, body composition, reproductive traits, hormone levels, hair characteristics, and behaviors.

CONCEPTS

The availability of numerous genetic markers revealed by molecular methods makes it possible to map chromosome segments containing genes that contribute to polygenic characteristics. Genomewide association studies locate genes that affect quantitative traits by detecting associations between genetic markers and a trait within a population of individuals.

24.4 Genetically Variable Traits Change in Response to Selection

Evolution is genetic change among members of a population. Several different forces are potentially capable of producing evolution, and we will explore these forces and the process of evolution more fully in Chapter 25. Here, we consider how one of these forces—natural selection—may bring about genetic change in a quantitative characteristic.

Charles Darwin proposed the idea of natural selection in his book *On the Origin of Species* in 1859. **Natural selection** arises through the differential reproduction of individuals with different genotypes, allowing individuals with certain genotypes to produce more offspring than others. Natural selection is one of the most important of the forces that brings about evolutionary change. Through natural selection, organisms become genetically suited to their environments; as environments change, groups of organisms change in ways that make them better able to survive and reproduce.

For thousands of years, humans have practiced a form of selection by promoting the reproduction of organisms with traits perceived as desirable. This form of selection is **artificial selection**, and it has produced the domestic plants and animals that make modern agriculture possible. The power of artificial selection, the first application of genetic principles by humans, is illustrated by the tremendous diversity of shapes, colors, and behaviors of modern domesticated dogs (**Figure 24.20**).

Predicting the Response to Selection

When a quantitative characteristic is subjected to natural or artificial selection, it will frequently change with the passage of time, provided that there is genetic variation for that characteristic in the population. Suppose that a dairy farmer breeds only those cows in his herd that have the highest milk production. If there is genetic variation in milk production, the mean milk production in the offspring of the selected cows should be higher than the mean milk production of the original herd. This increased production is due to the fact that the selected cows possess more genes for high milk production than does the average cow, and these genes are passed on to the offspring. The offspring of the selected cows possess a higher proportion of genes for greater milk yield and therefore produce more milk than the average cow in the initial herd.

The extent to which a characteristic subjected to selection changes in one generation is termed the **response to selection**. Suppose that the average cow in a dairy herd produces 80 liters of milk per week. A farmer selects for increased milk production by breeding the highest milk producers, and the progeny of these selected cows produce 100 liters of milk per week on average. The response to selection is calculated by subtracting the mean phenotype of the original population (80 liters) from the mean phenotype of the offspring (100 liters), obtaining a response to selection of $100 - 80 = 20$ liters per week.

Factors influencing response to selection The response to selection is determined primarily by two factors. First, it is affected by narrow-sense heritability, which largely determines the degree of resemblance between parents and offspring. When the narrow-sense heritability is high, offspring will tend to resemble their parents; conversely, when the narrow-sense heritability is low, there will be little resemblance between parents and offspring.

The second factor that determines the response to selection is how much selection there is. If the farmer is very stringent in the choice of parents and breeds only the highest milk producers in the herd (say, the top 2 cows), then all the offspring will receive genes for high-quality milk production. If the farmer is less selective and breeds the top 20 milk producers in the herd, then the offspring will not carry as many superior genes for high milk production, and they will not, on average, produce as much milk as the offspring of the top 2 producers. The response to selection depends on the

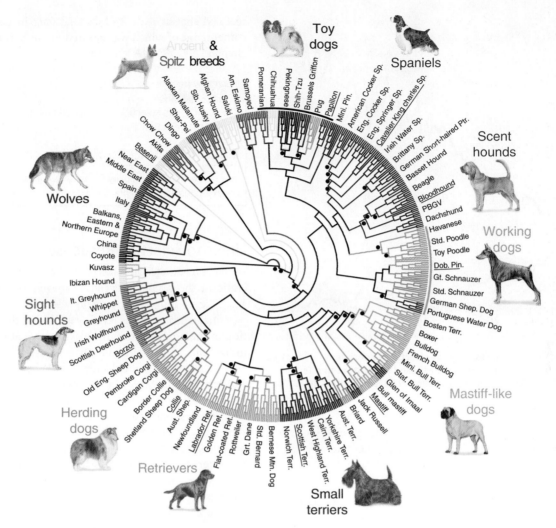

24.20 Artificial selection has produced the tremendous diversity of shape, size, color, and behavior seen today among breeds of domestic dogs. This diagram depicts the evolutionary relationships among wolves and different breeds of dogs from analyses of DNA sequences. [J. & C. Sohns/AgeFotostock.]

phenotypic difference of the individuals that are selected as parents; this phenotypic difference is measured by the **selection differential**, defined as the difference between the mean phenotype of the selected parents and the mean phenotype of the original population. If the average milk production of the original herd is 80 liters and the farmer breeds cows with an average milk production of 120 liters, then the selection differential is 120 − 80 = 40 liters.

Calculation of response to selection The response to selection (*R*) depends on the narrow-sense heritability (h^2) and the selection differential (*S*):

$$R = h^2 \times S \qquad (24.21)$$

This equation can be used to predict the magnitude of change in a characteristic when a given selection differential is applied. G. A. Clayton and his colleagues estimated the response to selection that would take place in abdomi-

nal bristle number of *Drosophila melanogaster*. Using several different methods, including parent−offspring regression, they first estimated the narrow-sense heritability of abdominal bristle number in one population of fruit flies to be 0.52. The mean number of bristles in the original population was 35.3. They selected individual flies with a mean bristle number of 40.6 and intercrossed them to produce the next generation. The selection differential was 40.6 − 35.3 = 5.3; so they predicted a response to selection to be

$$R = 0.52 \times 5.3 = 2.8$$

The response to selection of 2.8 is the expected increase in the characteristic of the offspring above the mean of the original population. They therefore expected the average number of abdominal bristles in the offspring of their selected flies to be 35.3 + 2.8 = 38.1. Indeed, they found an average bristle number of 37.9 in these flies.

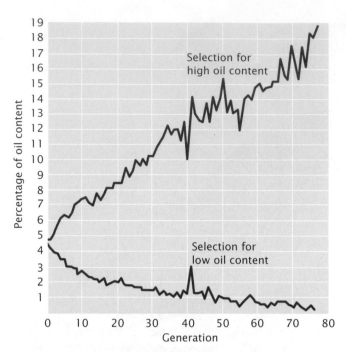

24.21 In a long-term response-to-selection experiment, selection for oil content in corn increased oil content in one line to about 20%, whereas it almost eliminated it in another line.

Estimating heritability from response to selection

Rearranging Equation 24.21 provides another way to calculate narrow-sense heritability:

$$h^2 = \frac{R}{S} \qquad (24.22)$$

In this way, h^2 can be calculated by conducting a response-to-selection experiment. First, the selection differential is obtained by subtracting the population mean from the mean of selected parents. The selected parents are then interbred, and the mean phenotype of their offspring is measured. The difference between the mean of the offspring and that of the initial population is the response to selection, which can be used with the selection differential to estimate heritability. Heritability determined by a response-to-selection experiment is usually termed the **realized heritability**. If certain assumptions are met, the realized heritability is identical with the narrow-sense heritability.

One of the longest selection experiments is a study of oil and protein content in corn seeds (**Figure 24.21**). This experiment began at the University of Illinois on 163 ears of corn with an oil content ranging from 4% to 6%. Corn plants having high oil content and those having low oil content were selected and interbred. Response to selection for increased oil content (the upper line in Figure 24.21) reached about 20%, whereas response to selection for decreased oil content reached a lower limit near zero. Genetic analyses of the high- and low-oil-content strains

revealed that at least 20 loci take part in determining oil content, one of which we explored in the introduction to the chapter. TRY PROBLEM 38 →

Limits to Selection Response

When a characteristic has been selected for many generations, the response eventually levels off, and the characteristic no longer responds to selection (**Figure 24.22**). A potential reason for this leveling off is that the genetic variation in the population may be exhausted; at some point, all individuals in the population have become homozygous for alleles that encode the selected trait. When there is no more additive genetic variation, heritability equals zero, and no further response to selection can take place.

The response to selection may level off even while some genetic variation remains in the population, however, because natural selection opposes further change in the characteristic. Response to selection for small body size in mice, for example, eventually levels off because the smallest animals are sterile and cannot pass on their genes for small body size. In this case, artificial selection for small size is

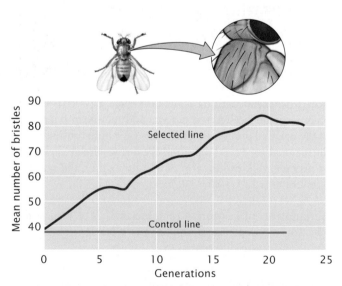

24.22 The response of a population to selection often levels off at some point in time. In a response-to-selection experiment for increased abdominal chaetae bristle number in female fruit flies, the number of bristles increased steadily for about 20 generations and then leveled off.

opposed by natural selection for fertility, and the population can no longer respond to the artificial selection.

Correlated Responses

Often when a specific trait is selected, other traits change at the same time. This type of associated response is due to the fact that the traits are encoded by the same genes.

Two or more characteristics are often correlated. Human height and weight exhibit a positive correlation: tall people, on the average, weigh more than short people. This correlation is a **phenotypic correlation** because the association is between two phenotypes of the same person. Phenotypic correlations may be due to environmental or genetic correlations. Environmental correlations refer to two or more characteristics that are influenced by the same environmental factor. Moisture availability, for example, may affect both the size of a plant and the number of seeds produced by the plant. Plants growing in environments with lots of water are large and produce many seeds, whereas plants growing in environments with limited water are small and have few seeds.

Alternatively, a phenotypic correlation may result from a **genetic correlation**, which means that the genes affecting two characteristics are associated. The primary genetic cause of phenotypic correlations is pleiotropy, which is due to the effect of one gene on two or more characteristics (see Chapter 5). In humans, for example, many body structures respond to growth hormone, and there are genes that affect the amount of growth hormone secreted by the pituitary gland. People with certain genes produce high levels of growth hormone, which increases both height and hand size. Others possess genes that produce lower levels of growth hormone, which leads to both short stature and small hands. Height and hand size are therefore phenotypically correlated in humans, and this correlation is due to a genetic correlation—the fact that both characteristics are affected by the same genes that control the amount of growth hormone. Genetically speaking, height and hand size are the same characteristic because they are the phenotypic manifestation of a single set of genes. When two characteristics are influenced by the same genes they are genetically correlated.

Genetic correlations are quite common (**Table 24.3**) and may be positive or negative. A positive genetic correlation between two characteristics means that genes that cause an increase in one characteristic also produce an increase in the other characteristic. Thorax length and wing length in *Drosophila* are positively correlated because the genes that increase thorax length also increase wing length. A negative genetic correlation means that genes that cause an increase in one characteristic produce a decrease in the other characteristic. Milk yield and percentage of butterfat are negatively correlated in cattle: genes that cause higher milk production result in milk with a lower percentage of butterfat.

Genetic correlations are important in animal and plant breeding because they produce a correlated response to

Table 24.3 Genetic correlations in various organisms

Organism	Characteristics	Genetic Correlation
Cattle	Milk yield and percentage of butterfat	0.38
Pig	Weight gain and back-fat thickness	0.13
	Weight gain and efficiency	0.69
Chicken	Body weight and egg weight	0.42
	Body weight and egg production	−0.17
	Egg weight and egg production	−0.31
Mouse	Body weight and tail length	0.29
Fruit fly	Abdominal bristle number and sternopleural bristle number	0.41

Source: After D. S. Falconer, *Introduction to Quantitative Genetics* (London: Longman, 1981), p. 284.

selection, which means that, when one characteristic is selected, genetically correlated characteristics also change. Correlated responses to selection are due to the fact that both characteristics are influenced by the same genes; selection for one characteristic causes a change in the genes affecting that characteristic, and these genes also affect the second characteristic, causing it to change at the same time. Correlated responses may well be undesirable and may limit the ability to alter a characteristic by selection. From 1944 to 1964, domestic turkeys were subjected to intense selection for growth rate and body size. At the same time, fertility, egg production, and egg hatchability all declined. These correlated responses were due to negative genetic correlations between body size and fertility; eventually, these genetic correlations limited the extent to which the growth rate of turkeys could respond to selection. Genetic correlations may also limit the ability of natural populations to respond to selection in the wild and adapt to their environments.

TRY PROBLEM 42 →

CONCEPTS

Genetic correlations result from pleiotropy. When two characteristics are genetically correlated, selection for one characteristic will produce a correlated response in the other characteristic.

✔ CONCEPT CHECK 8

In a herd of dairy cattle, milk yield and the percentage of butterfat exhibit a genetic correlation of −0.38. If greater milk yield is selected in this herd, what will be the effect on the percentage of butterfat?

CONCEPTS SUMMARY

- Quantitative genetics focuses on the inheritance of complex characteristics whose phenotypes vary continuously. For many quantitative characteristics, the relation between genotype and phenotype is complex because many genes and environmental factors influence a characteristic.

- The individual genes that influence a polygenic characteristic follow the same Mendelian principles that govern discontinuous characteristics, but, because many genes participate, the expected ratios of phenotypes are obscured.

- A population is the group of interest, and a sample is a subset of the population used to describe it.

- A frequency distribution, in which the phenotypes are represented on one axis and the number of individuals possessing each phenotype is represented on the other, is a convenient means of summarizing phenotypes found in a group of individuals.

- The mean and variance provide key information about a distribution: the mean gives the central location of the distribution, and the variance provides information about how the phenotype varies within a group.

- The correlation coefficient measures the direction and strength of association between two variables. Regression can be used to predict the value of one variable on the basis of the value of a correlated variable.

- Phenotypic variance in a characteristic can be divided into components that are due to additive genetic variance, dominance genetic variance, genic interaction variance, environmental variance, and genetic–environmental interaction variance.

- Broad-sense heritability is the proportion of the phenotypic variance due to genetic variance; narrow-sense heritability is the proportion of the phenotypic variance due to additive genetic variance.

- Heritability provides information only about the degree to which variation in a characteristic results from genetic differences. Heritability is based on the variances present within a group of individuals, and an individual does not have heritability. The heritability of a characteristic varies among populations and among environments. Even if the heritability for a characteristic is high, the characteristic may still be altered by changes in the environment. Heritabilities provide no information about the nature of population differences in a characteristic.

- Quantitative trait loci are chromosome segments containing genes that control polygenic characteristics. QTLs can be mapped by examining the association between the inheritance of a quantitative characteristic and the inheritance of genetic markers. Genes influencing quantitative traits can also be located with the use of genomewide association studies.

- The amount that a quantitative characteristic changes in a single generation when subjected to selection (the response to selection) is directly related to the selection differential and narrow-sense heritability.

- A genetic correlation may be present when the same gene affects two or more characteristics (pleiotropy). Genetic correlations produce correlated responses to selection.

IMPORTANT TERMS

quantitative genetics (p. 659)
quantitative trait locus (QTL) (p. 659)
meristic characteristic (p. 662)
threshold characteristic (p. 662)
frequency distribution (p. 665)
normal distribution (p. 666)
population (p. 666)
sample (p. 666)
mean (p. 666)
variance (p. 667)
standard deviation (p. 667)

correlation (p. 668)
correlation coefficient (p. 668)
regression (p. 669)
regression coefficient (p. 670)
heritability (p. 672)
phenotypic variance (p. 672)
genetic variance (p. 673)
environmental variance (p. 673)
genetic–environmental interaction
    variance (p. 673)
additive genetic variance (p. 673)

dominance genetic variance (p. 673)
genic interaction variance (p. 674)
broad-sense heritability (p. 674)
narrow-sense heritability (p. 674)
natural selection (p. 680)
artificial selection (p. 680)
response to selection (p. 680)
selection differential (p. 681)
realized heritability (p. 682)
phenotypic correlation (p. 683)
genetic correlation (p. 683)

ANSWERS TO CONCEPT CHECKS

1. Cross two individuals that are each homozygous for different genes affecting the traits and then intercross the resulting F_1 progeny to produce the F_2. Determine what proportion of the F_2 progeny resembles one of the original homozygotes in the P generation. This proportion should be $(1/4)^n$, where n equals the number of loci with a segregating pair of alleles that affect the characteristic.

2. d

3. d

4. b

5. a

6. It indicates that about 40% of the differences in blood pressure among African Americans in Detroit are due to additive genetic differences. It neither provides information about the heritability of blood pressure in other groups of people nor indicates anything about the nature of differences in blood pressure between African Americans in Detroit and people in other groups.

7. 0.2

8. The percentage of butterfat will decrease.

WORKED PROBLEMS

1. Seed weight in a particular plant species is determined by pairs of alleles at two loci (a^+a^- and b^+b^-) that are additive and equal in their effects. Plants with genotype $a^-a^-\ b^-b^-$ have seeds that average 1 g in weight, whereas plants with genotype $a^+a^+\ b^+b^+$ have seeds that average 3.4 g in weight. A plant with genotype $a^-a^-\ b^-b^-$ is crossed with a plant of genotype $a^+a^+\ b^+b^+$.

a. What is the predicted weight of seeds from the F_1 progeny of this cross?

b. If the F_1 plants are intercrossed, what are the expected seed weights and proportions of the F_2 plants?

• Solution

The difference in average seed weight of the two parental genotypes is 3.4 g − 1 g = 2.4 g. These two genotypes differ in four genes; so, if the genes have equal and additive effects, each gene difference contributes an additional 2.4 g/4 = 0.6 g of weight to the 1-g weight of a plant ($a^-a^-\ b^-b^-$), which has none of the contributing genes.

The cross between the two homozygous genotypes produces the F_1 and F_2 progeny shown in the table below.

a. The F_1 are heterozygous at both loci ($a^+a^-\ b^+b^-$) and possess two genes that contribute an additional 0.6 g each to the 1-g weight of a plant that has no contributing genes. Therefore, the seeds of the F_1 should average 1 g + 2(0.6 g) = 2.2 g.

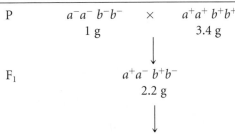

P $\quad a^-a^-\ b^-b^- \quad \times \quad a^+a^+\ b^+b^+$
$\quad\quad$ 1 g $\quad\quad\quad\quad$ 3.4 g

$F_1 \quad\quad\quad a^+a^-\ b^+b^-$
$\quad\quad\quad\quad$ 2.2 g

	Genotype	Probability	Number of contributing genes	Average seed weight
	$a^-a^-\ b^-b^-$	$^1/_4 \times ^1/_4 = ^1/_{16}$	0	1 g + (0 × 0.6 g) = 1 g
	$a^+a^-\ b^-b^-$	$^1/_2 \times ^1/_4 = ^1/_8$ $\Big\}\,^2/_8 = ^4/_{16}$	1	1 g + (1 × 0.6 g) = 1.6 g
	$a^-a^-\ b^+b^-$	$^1/_4 \times ^1/_2 = ^1/_8$		
F_2	$a^+a^+\ b^-b^-$	$^1/_4 \times ^1/_4 = ^1/_{16}$	2	1 g + (2 × 0.6 g) = 2.2 g
	$a^-a^-\ b^+b^+$	$^1/_4 \times ^1/_4 = ^1/_{16}$ $\Big\}\,^2/_{16} + ^1/_4 + = ^6/_{16}$		
	$a^+a^-\ b^+b^-$	$^1/_2 \times ^1/_2 = ^1/_4$		
	$a^+a^+\ b^+b^-$	$^1/_4 \times ^1/_2 = ^1/_8$ $\Big\}\,^2/_8 = ^4/_{16}$	3	1 g + (3 × 0.6 g) = 2.8 g
	$a^+a^-\ b^+b^+$	$^1/_4 \times ^1/_2 = ^1/_8$		
	$a^+a^+\ b^+b^+$	$^1/_4 \times ^1/_4 = ^1/_{16}$	4	1 g + (4 × 0.6 g) = 3.4 g

b. The F_2 will have the following phenotypes and proportions: $^1/_{16}$ 1 g; $^4/_{16}$ 1.6 g; $^6/_{16}$ 2.2 g; $^4/_{16}$ 2.8 g; and $^1/_{16}$ 3.4 g.

2. Phenotypic variation is analyzed for milk production in a herd of dairy cattle and the following variance components are obtained:

Additive genetic variance (V_A)	$= 0.4$
Dominance genetic variance (V_D)	$= 0.1$
Genic interaction variance (V_I)	$= 0.2$
Environmental variance (V_E)	$= 0.5$
Genetic–environmental interaction variance (V_{GE}) $= 0.0$	

a. What is the narrow-sense heritability of milk production?

b. What is the broad-sense heritability of milk production?

• Solution

To determine the heritabilities, we first need to calculate V_P and V_G:

$$V_P = V_A + V_D + V_I + V_E + V_{GE}$$

$$= 0.4 + 0.1 + 0.2 + 0.5 + 0$$

$$= 1.2$$

$$V_G = V_A + V_D + V_I$$

$$= 0.7$$

a. The narrow-sense heritability is:

$$h^2 = \frac{V_A}{V_P} = \frac{0.4}{1.2} = 0.33$$

b. The broad-sense heritability is:

$$H^2 = \frac{V_G}{V_P} = \frac{0.7}{1.2} = 0.58$$

3. The heights of parents and their offspring are measured for 10 families:

Mean height of parents (cm)	Mean height of offspring (cm)
150	152
157	163
188	193
165	163
160	152
142	157
170	183
183	175
152	163
173	180

From these data, determine

a. the mean, variance, and standard deviation of the height of parents and of offspring;

b. the correlation and regression coefficients for a regression of mean offspring height against mean parental height; and

c. the narrow-sense heritability of height in these families.

d. What conclusions can be drawn from the heritability value determined in part c?

• Solution

a. The best way to begin is by constructing a table, as shown on the facing page. To calculate the means, we need to sum the values of x and y, which are shown in the last rows of columns A and D of the table. For the mean of parental height,

$$\bar{x} = \frac{\sum x_i}{n} = \frac{1640}{10} = 164 \text{ cm}$$

For the mean of the offspring height,

$$\bar{y} = \frac{\sum y_i}{n} = \frac{1681}{10} = 168.1 \text{ cm}$$

For the variance, we subtract each x and y value from its mean (columns B and E) and square these differences (columns C and F). The sums of the these squared deviations are shown in the last row of columns C and F. For the regression, we need the covariance, which requires that we take the difference between each x value and its mean, multiply it by the difference between each y value and its mean [$(x_i - \bar{x})(y_i - \bar{y})$, column G], and then sum these products (last row of column G).

The variance is the sum of the squared deviations from the mean divided by $n - 1$, where n is the number of measurements:

$$s_x^{\ 2} = \frac{\sum(x_i - \bar{x})^2}{n-1} = \frac{1944}{9} = 216$$

$$s_y^{\ 2} = \frac{\sum(y_i - \bar{y})^2}{n-1} = \frac{1750.9}{9} = 194.54$$

The standard deviation is the square root of the variance:

$$s_x = \sqrt{s_x^{\ 2}} = \sqrt{216} = 14.70$$

$$s_y = \sqrt{s_y^{\ 2}} = \sqrt{194.54} = 13.95$$

b. To calculate the correlation coefficient and regression coefficient, we need the covariance:

$$\text{cov}_{xy} = \frac{\sum(x_i - \bar{x})(y_i - \bar{y})}{n-1} = \frac{1551}{9} = 172.33 \text{ cm}$$

The correlation coefficient is the covariance divided by the standard deviation of x and the standard deviation of y:

$$r = \frac{\text{cov}_{xy}}{s_x s_y} = \frac{172.33}{(14.70)(13.95)} = 0.84$$

The regression coefficient is the covariance divided by the variance of x:

$$r = \frac{\text{cov}_{xy}}{s_x^{\ 2}} = \frac{172.33}{216} = 0.80$$

A Mean height parents (cm)	B	C	D Mean height offspring (cm)	E	F	G
x_i	$x_i - \bar{x}$	$(x_i - \bar{x})^2$	y_i	$y_i - \bar{y}$	$(y_i - \bar{y})^2$	$(x_i - \bar{x})(y_i - \bar{y})$
150	−14	196	152	−16.1	259.21	225.4
157	−7	49	163	−5.1	26.01	35.7
188	24	576	193	24.9	620.01	597.6
165	1	1	163	−5.1	26.01	−5.1
160	−4	16	152	−16.1	259.21	64.4
142	−22	484	157	−11.1	123.21	244.2
170	6	36	183	14.9	222.01	89.4
183	19	361	175	6.9	47.61	131.1
152	−12	144	163	−5.1	26.01	61.2
173	9	81	180	11.9	141.61	107.1
$\Sigma x_i = 1640$		$\Sigma (x_i - \bar{x})^2 = 1944$	$\Sigma y_i = 1681$		$\Sigma(y_i - \bar{y})^2 = 1750.9$	$\Sigma (x_i - \bar{x})(y_i - \bar{y}) = 1551$

c. In a regression of the mean phenotype of the offspring against the mean phenotype of the parents, the regression coefficient equals the narrow-sense heritability, which is 0.80.

d. We conclude that 80% of the variance in height among the members of these families results from additive genetic variance.

4. A farmer is raising rabbits. The average body weight in his population of rabbits is 3 kg. The farmer selects the 10 largest rabbits in his population, whose average body weight is 4 kg, and interbreeds them. If the heritability of body weight in the rabbit population is 0.7, what is the expected body weight among offspring of the selected rabbits?

• Solution

The farmer has carried out a response-to-selection experiment, in which the response to selection will equal the selection differential times the narrow-sense heritability. The selection differential equals the difference in average weights of the selected rabbits and the entire population: 4 kg − 3 kg = 1 kg. The narrow-sense heritability is given as 0.7; so the expected response to selection is: $R = h^2 \times S = 0.7 \times 1$ kb = 0.7 kg. This value is the increase in weight that is expected in the offspring of the selected parents; so the average weight of the offspring is expected to be: 3 kg + 0.7 kg = 3.7 kg.

COMPREHENSION QUESTIONS

Section 24.1

*1. How does a quantitative characteristic differ from a discontinuous characteristic?

2. Briefly explain why the relation between genotype and phenotype is frequently complex for quantitative characteristics.

*3. Why do polygenic characteristics have many phenotypes?

Section 24.2

*4. Explain the relation between a population and a sample. What characteristics should a sample have to be representative of the population?

5. What information do the mean and variance provide about a distribution?

6. How is the standard deviation related to the variance?

*7. What information does the correlation coefficient provide about the association between two variables?

8. What is regression? How is it used?

Section 24.3

*9. List all the components that contribute to the phenotypic variance and define each component.

*10. How do broad-sense and narrow-sense heritabilities differ?

11. Briefly outline some of the ways in which heritability can be calculated.

12. Briefly describe common misunderstandings or misapplications of the concept of heritability.

13. Briefly explain how genes affecting a polygenic characteristic are located with the use of QTL mapping.

Section 24.4

*14. How is the response to selection related to the narrow-sense heritability and the selection differential? What information does the response to selection provide?

15. Why does the response to selection often level off after many generations of selection?

APPLICATION QUESTIONS AND PROBLEMS

Section 24.1

*16. For each of the following characteristics, indicate whether it would be considered a discontinuous characteristic or a quantitative characteristic. Briefly justify your answer.

a. Kernel color in a strain of wheat, in which two codominant alleles segregating at a single locus determine the color. Thus, there are three phenotypes present in this strain: white, light red, and medium red.

b. Body weight in a family of Labrador retrievers. An autosomal recessive allele that causes dwarfism is present in this family. Two phenotypes are recognized: dwarf (less than 13 kg) and normal (greater than 13 kg).

c. Presence or absence of leprosy. Susceptibility to leprosy is determined by multiple genes and numerous environmental factors.

d. Number of toes in guinea pigs, which is influenced by genes at many loci.

e. Number of fingers in humans. Extra (more than five) fingers are caused by the presence of an autosomal dominant allele.

*17. Assume that plant weight is determined by a pair of alleles at each of two independently assorting loci (*A* and *a*, *B* and *b*) that are additive in their effects. Further assume that each allele represented by an uppercase letter contributes 4 g to weight and that each allele represented by a lowercase letter contributes 1 g to weight.

a. If a plant with genotype *AA BB* is crossed with a plant with genotype *aa bb*, what weights are expected in the F_1 progeny?

b. What is the distribution of weight expected in the F_2 progeny?

*18. Assume that three loci, each with two alleles (*A* and *a*, *B* and *b*, *C* and *c*), determine the differences in height between two homozygous strains of a plant. These genes are additive and equal in their effects on plant height. One strain (*aa bb cc*) is 10 cm in height. The other strain (*AA BB CC*) is 22 cm in height. The two strains are crossed, and the resulting F_1 are interbred to produce F_2 progeny. Give the phenotypes and the expected proportions of the F_2 progeny.

*19. A farmer has two homozygous varieties of tomatoes. One variety, called Little Pete, has fruits that average only 2 cm in diameter. The other variety, Big Boy, has fruits that average a whopping 14 cm in diameter. The farmer crosses Little Pete and Big Boy; he then intercrosses the F_1 to produce F_2 progeny. He grows 2000 F_2 tomato plants and doesn't find any F_2 offspring that produce fruits as small as Little Pete or as large as Big Boy. If we assume that the differences in fruit size of these varieties are produced by genes with equal and additive effects, what conclusion can we make about the minimum number of loci with pairs of alleles determining the differences in fruit size of the two varieties?

20. Seed size in a plant is a polygenic characteristic. A grower crosses two pure-breeding varieties of the plant and measures seed size in the F_1 progeny. She then backcrosses the F_1 plants to one of the parental varieties and measures seed size in the backcross progeny. The grower finds that seed size in the backcross progeny has a higher variance than does seed size in the F_1 progeny. Explain why the backcross progeny are more variable.

Section 24.2

*21. The following data are the numbers of digits per foot in 25 guinea pigs. Construct a frequency distribution for these data.

4, 4, 4, 5, 3, 4, 3, 4, 4, 5, 4, 4, 3, 2, 4, 4, 5, 6, 4, 4, 3, 4, 4, 4, 5

22. Ten male Harvard students were weighed in 1916. Their weights are given here in kilograms. Calculate the mean, variance, and standard deviation for these weights.

51, 69, 69, 57, 61, 57, 75, 105, 69, 63

23. Among a population of tadpoles, the correlation coefficient for size at metamorphosis and time required for metamorphosis is −0.74. On the basis of this correlation, what conclusions can you make about the relative sizes of tadpoles that metamorphose quickly and those that metamorphose more slowly?

*24. A researcher studying alcohol consumption in North American cities finds a significant, positive correlation between the number of Baptist preachers and alcohol consumption. Is it reasonable for the researcher to conclude that the Baptist preachers are consuming most of the alcohol? Why or why not?

25. Body weight and length were measured on six mosquito fish; these measurements are given in the following table. Calculate the correlation coefficient for weight and length in these fish.

[A. Hartl/AgeFotostock.]

Wet weight (g)	Length (mm)
115	18
130	19
210	22
110	17
140	20
185	21

*26. The heights of mothers and daughters are given in the following table:

Height of mother (in)	Height of daughter (in)
64	66
65	66
66	68
64	65
63	65
63	62
59	62
62	64
61	63
60	62

a. Calculate the correlation coefficient for the heights of the mothers and daughters.

b. Using regression, predict the expected height of a daughter whose mother is 67 inches tall.

Section 24.3

*27. Phenotypic variation in tail length of mice has the following components:

Additive genetic variance (V_A)	= 0.5
Dominance genetic variance (V_D)	= 0.3
Genic interaction variance (V_I)	= 0.1
Environmental variance (V_E)	= 0.4
Genetic–environmental interaction variance (V_{GE})	= 0.0

a. What is the narrow-sense heritability of tail length?

b. What is the broad-sense heritability of tail length?

28. The narrow-sense heritability of ear length in Reno rabbits is 0.4. The phenotypic variance (V_P) is 0.8, and the environmental variance (V_E) is 0.2. What is the additive genetic variance (V_A) for ear length in these rabbits?

*29. Assume that human ear length is influenced by multiple genetic and environmental factors. Suppose you measured ear length on three groups of people, in which group A consists of five unrelated persons, group B consists of five siblings, and group C consists of five first cousins.

a. With the assumption that the environment for each group is similar, which group should have the highest phenotypic variance? Explain why.

b. Is it realistic to assume that the environmental variance for each group is similar? Explain your answer.

30. A characteristic has a narrow-sense heritability of 0.6.

a. If the dominance variance (V_D) increases and all other variance components remain the same, what will happen to the narrow-sense heritability? Will it increase, decrease, or remain the same? Explain.

b. What will happen to the broad-sense heritability? Explain.

c. If the environmental variance (V_E) increases and all other variance components remain the same, what will happen to the narrow-sense heritability? Explain.

d. What will happen to the broad-sense heritability? Explain.

31. Flower color in the varieties of pea plants studied by Mendel is controlled by alleles at a single locus. A group of peas homozygous for purple flowers is grown in a garden. Careful study of the plants reveals that all their flowers are purple, but there is some variability in the intensity of the purple color. If heritability were estimated for this variation in flower color, what would it be? Explain your answer.

*32. A graduate student is studying a population of bluebonnets along a roadside. The plants in this population are genetically variable. She counts the seeds produced by 100 plants and measures the mean and variance of seed number. The variance is 20. Selecting one plant, the graduate student takes cuttings from it and cultivates these cuttings in the greenhouse, eventually producing many genetically identical clones of the same plant. She then transplants these clones into the roadside population, allows them to grow for 1 year, and then counts the number of seeds produced by each of the cloned plants. The graduate student finds that the variance in seed number among these cloned plants is 5. From the phenotypic variance of the genetically variable and genetically identical plants, she calculates the broad-sense heritability.

[Amy Lenise/Stock.xchng.]

a. What is the broad-sense heritability of seed number for the roadside population of bluebonnets?

b. What might cause this estimate of heritability to be inaccurate?

33. Many researchers have estimated the heritability of human traits by comparing the correlation coefficients of monozygotic and dizygotic twins (see p. 676). One of the assumptions in using this method is that two monozygotic twins experience environments that are no more similar to each other than those experienced by two dizygotic twins. How might this assumption be violated? Give some specific examples of ways in which the environments of two monozygotic twins might be more similar than the environments of two dizyotic twins.

34. A genetics researcher determines that the broad-sense heritability of height among Southwestern University undergraduate students is 0.90. Which of the following conclusions would be resonable? Explain your answer.

a. Because Sally is a Southwestern University undergraduate student, 10% of her height is determined by nongenetic factors.

b. Ninety percent of variation in height among all undergraduate students in the United States is due to genetic differences.

c. Ninety percent of the height of Southwestern University undergraduate students is determined by genes.

d. Ten percent of the variation in height among Southwestern University undergraduate students is determined by variation in nongenetic factors.

e. Because the heritability of height among Southwestern Unversity students is so high, any change in the students' environment will have minimal effect on their height.

*35. The length of the middle joint of the right index finger was measured on 10 sets of parents and their adult offspring. The mean parental lengths and the mean offspring lengths for each family are listed in the following table. Calculate the regression coefficient for regression of mean offspring length against mean parental length and estimate the narrow-sense heritability for this characteristic.

Mean parental length (mm)	Mean offspring length (mm)
30	31
35	36
28	31
33	35
26	27
32	30
31	34
29	28
40	38
33	34

36. *Drosophila buzzatii* is a fruit fly that feeds on the rotting fruits of cacti in Australia. Timothy Prout and Stuart Barker calculated the heritabilities of body size, as measured by thorax length, for a natural population of *D. buzzatii* raised in the wild and for a population of *D. buzzatii* collected in the wild but raised in the laboratory (T. Prout and J. S. F. Barker. 1989. *Genetics* 123:803−813). They found the following heritabilities.

Population	Heritability of body size (± standard error)
Wild population	0.0595 ± 0.0123
Laboratory-reared population	0.3770 ± 0.0203

Why do you think the heritability measured in the laboratory-reared population is higher than that measured in the natural population raised in the wild?

*37. Mr. Jones is a pig farmer. For many years, he has fed his pigs the food left over from the local university cafeteria, which is known to be low in protein, deficient in vitamins, and downright untasty. However, the food is free, and his pigs don't complain. One day a salesman from a feed company visits Mr. Jones. The salesman claims that his company sells a new, high-protein, vitamin-enriched feed that enhances weight gain in pigs. Although the food is expensive, the salesman claims that the increased weight gain of the pigs will more than pay for the cost of the feed, increasing Mr. Jones's profit. Mr. Jones responds that he took a genetics class when he went to the university and that he has conducted some genetic experiments on his pigs; specifically, he has calculated the narrow-sense heritability of weight gain for his pigs and found it to be 0.98. Mr. Jones says that this heritability value indicates that 98% of the variance in weight gain among his pigs is determined by genetic differences, and therefore the new pig feed can have little effect on the growth of his pigs. He concludes that the feed would be a waste of his money. The salesman doesn't dispute Mr. Jones' heritability estimate, but he still claims that the new feed can significantly increase weight gain in Mr. Jones' pigs. Who is correct and why?

Section 24.4

38. Joe is breeding cockroaches in his dorm room. He finds that the average wing length in his population of cockroaches is 4 cm. He chooses six cockroaches that have the largest wings; the average wing length among these selected cockroaches is 10 cm. Joe interbreeds these selected cockroaches. From earlier studies, he knows that the narrow-sense heritability for wing length in his population of cockroaches is 0.6.

a. Calculate the selection differential and expected response to selection for wing length in these cockroaches.

b. What should be the average wing length of the progeny of the selected cockroaches?

39. Three characteristics in beef cattle—body weight, fat content, and tenderness—are measured, and the following variance components are estimated:

	Body weight	Fat content	Tenderness
V_A	22	45	12
V_D	10	25	5
V_I	3	8	2
V_E	42	64	8
V_{GE}	0	0	1

In this population, which characteristic would respond best to selection? Explain your reasoning.

*40. A rancher determines that the average amount of wool produced by a sheep in her flock is 22 kg per year. In an attempt to increase the wool production of her flock, the rancher picks five male and five female sheep with the greatest wool production; the average amount of wool produced per sheep by those selected is 30 kg. She interbreeds these selected sheep and finds that the average wool production among the progeny of the selected sheep is 28 kg. What is the narrow-sense heritability for wool production among the sheep in the rancher's flock?

41. A strawberry farmer determines that the average weight of individual strawberries produced by plants in his garden is 2 g. He selects the 10 plants that produce the largest strawberries; the average weight of strawberries among these selected plants is 6 g. He interbreeds these selected strawberry plants. The progeny of these selected plants produce strawberries that weigh 5 g. If the farmer were to select plants that produce an average strawberry weight of 4 g, what would be the predicted weight of strawberries produced by the progeny of these selected plants?

42. The narrow-sense heritability of wing length in a population of *Drosophila melanogaster* is 0.8. The narrow-sense heritability of head width in the same population is 0.9. The genetic correlation between wing length and head width is −0.86. If a geneticist selects for increased wing length in these flies, what will happen to head width?

43. Pigs have been domesticated from wild boars. Would you expect to find higher heritability for weight among domestic pigs or among wild boars? Explain your answer.

CHALLENGE QUESTIONS

Section 24.1

44. Bipolar illness is a psychiatric disorder that has a strong hereditary basis, but the exact mode of inheritance is not known. Research has shown that siblings of patients with bipolar illness are more likely to develop the disorder than are siblings of unaffected persons. Findings from one study demonstrated that the ratio of bipolar brothers to bipolar sisters is higher when the patient is male than when the patient is female. In other words, relatively more brothers of bipolar patients also have the disease when the patient is male than when the patient is female. What does this observation suggest about the inheritance of bipolar illness?

Section 24.3

45. We have explored some of the difficulties in separating the genetic and environmental components of human behavioral characteristics. Considering these difficulties and what you know about calculating heritability, propose an experimental design for accurately measuring the heritability of musical ability.

46. A student who has just learned about quantitative genetics says, "Heritability estimates are worthless! They don't tell you anything about the genes that affect a characteristic. They don't provide any information about the types of offspring to expect from a cross. Heritability estimates measured in one population can't be used for other populations; so they don't even give you any general information about how much of a characteristic is genetically determined. I can't see that heritabilities do anything other than make undergraduate students sweat during tests." How would you respond to this statement? Is the student correct? What good are heritabilities, and why do geneticists bother to calculate them?

Section 24.4

47. Eugene Eisen selected for increased 12-day litter weight (total weight of a litter of offspring 12 days after birth) in a population of mice (E. J. Eisen. 1972. *Genetics* 72:129–142). The 12-day litter weight of the population steadily increased but then leveled off after about 17 generations. At generation 17, Eisen took one family of mice from the selected population and reversed the selection procedure: in this group, he selected for *decreased* 12-day litter size. This group immediately responded to decreased selection; the 12-day litter weight dropped 4.8 g within one generation and dropped 7.3 g after 5 generations. On the basis of the results of the reverse selection, what is the most likely explanation for the leveling off of 12-day litter weight in the original population?

[J & C Sohns/AgeFotostock.]

25 Population Genetics

Rocky Mountain bighorn sheep (*Ovis canadensis*). A population of bighorn sheep at the National Bison Range suffered the loss of genetic variation due to genetic drift; the introduction of sheep from other populations dramatically increased genetic variation and the fitness of the sheep. [Tom J. Ulrich/Visuals Unlimited.]

GENETIC RESCUE OF BIGHORN SHEEP

Rocky Mountain bighorn sheep (*Ovis canadensis*) are among North America's most spectacular animals, characterized by the male's magnificent horns that curve gracefully back over the ears, spiraling down and back up beside the face. Two hundred years ago, bighorn sheep were numerous throughout western North America, ranging from Mexico to southern Alberta and from Colorado to California. Meriwether Lewis and William Clark reported numerous sightings of these beautiful animals in their expedition across the western United States from 1804 to 1806. Before 1900, the number of bighorn sheep in North America was about 2 million.

Unfortunately, settlement of the west by Europeans was not kind to the bighorns. Beginning in the late 1800s, hunting, loss of habitat, competition from livestock, and diseases carried by domestic sheep decimated the bighorns. Today, fewer than 70,000 bighorn sheep remain, scattered across North America in fragmented and isolated populations.

In 1922, wildlife biologists established a population of bighorn sheep at the National Bison Range, an isolated tract of 18,000 acres nestled between the mountains of northwestern Montana. In that year, 12 bighorn sheep—4 males (rams) and 8 females (ewes)—were trapped at Banff National Park in Canada and transported to the National Bison Range. No additional animals were introduced to this population for the next 60 years.

At first, the population of bighorns at the National Bison Range flourished, protected there from hunting and livestock. Within 8 years, the population had grown to 90 sheep but then began to slowly decrease in size. Population size waxed and waned through the years, but the number of sheep had dropped to about 50 animals by 1985, and the population was in trouble. The amount of genetic variation was low compared with other native populations of bighorn sheep. The reproductive rate of both male and female sheep had dropped, and the size and survival of the sheep were lower than in more healthy populations. The population at the National Bison Range was suffering from genetic drift, an evolutionary force operating in small populations that causes random changes in the gene pool and the loss of genetic variation.

To counteract the negative effects of genetic drift, biologists added 5 new rams from other herds in Montana and Wyoming in 1985, mimicking the effects of natural migration among herds. Another 10 sheep were introduced between 1990 and 1994. This influx of new genes had a dramatic effect on the genetic health of the population. Genetic variation among individual sheep increased significantly. Outbred rams (those containing the new genes) were more dominant, were more likely to copulate, and were more likely to produce offspring. Outbred ewes had more than twice the annual reproductive success of inbred females. Adult survival increased after the introduction of new genes, and, slowly, the population grew in size, reaching 69 sheep by 2003.

The bighorn sheep at the National Bison Range illustrate an important principle of genetics: small populations lose genetic variation with the passage of time through genetic drift, often with catastrophic consequences for survival and reproduction. The introduction of new genetic variation into an inbred population, called **genetic rescue**, often dramatically improves the health of the population and can better ensure its long-term survival. These effects have important implications for wildlife management, as well as for how organisms evolve in the natural world.

This chapter introduces *population genetics*, the branch of genetics that studies the genetic makeup of groups of individuals and how a *group's* genetic composition changes with time. Population geneticists usually focus their attention on a **Mendelian population**, which is a group of interbreeding, sexually reproducing individuals that have a common set of genes, the **gene pool**. A population evolves through changes in its gene pool; therefore, population genetics is also the study of evolution. Population geneticists study the variation in alleles within and between groups and the evolutionary forces responsible for shaping the patterns of genetic variation found in nature. In this chapter, we will learn how the gene pool of a population is measured and what factors are responsible for shaping it.

25.1 All organisms exhibit genetic variation. (a) Extensive variation among humans. (b) Variation in the spotting patterns of Asian lady beetles. [Part a: Paul Warner/AP.]

25.1 Genotypic and Allelic Frequencies Are Used to Describe the Gene Pool of a Population

An obvious and pervasive feature of life is variability. Consider a group of students in a typical college class, the members of which vary in eye color, hair color, skin pigmentation, height, weight, facial features, blood type, and susceptibility to numerous diseases and disorders. No two students in the class are likely to be even remotely similar in appearance.

Humans are not unique in their extensive variability (**Figure 25.1a**); almost all organisms exhibit variation in phenotype. For instance, lady beetles are highly variable in their patterns of spots (**Figure 25.1b**), mice vary in body size, snails have different numbers of stripes on their shells, and plants vary in their susceptibility to pests. Much of this phenotypic variation is hereditary. Recognition of the extent of phenotypic variation led Charles Darwin to the idea of evolution

through natural selection. Genetic variation is the basis of all evolution, and the extent of genetic variation within a population affects its potential to adapt to environmental change.

In fact, even more genetic variation exists in populations than is visible in the phenotype. Much variation exists at the molecular level owing, in part, to the redundancy of the genetic code, which allows different codons to specify the same amino acid. Thus, two members of a population can produce the same protein even if their DNA sequences are different. DNA sequences between the genes and introns within genes do not encode proteins; much of the variation in these sequences also has little effect on the phenotype.

An important, but frequently misunderstood, tool used in population genetics is the mathematical model. Let's take a moment to consider what a model is and how it can be used. A mathematical model usually describes a process as an equation. Factors that may influence the process are represented by variables in the equation; the equation defines the way in which the variables influence the process. Most models are simplified representations of a process, because the simultaneous consideration of all of the influencing factors is impossible; some factors must be ignored in order to examine the effects of others. At first, a model might consider only one factor or a few factors, but, after their effects are understood, the model can be improved by the addition of more details. Importantly, even a simple model can be a source of valuable insight into how a process is influenced by key variables.

Before we can explore the evolutionary processes that shape genetic variation, we must be able to describe the genetic structure of a population. The usual way of describing this structure is to enumerate the types and frequencies of genotypes and alleles in a population.

Calculating Genotypic Frequencies

A frequency is simply a proportion or a percentage, usually expressed as a decimal fraction. For example, if 20% of the alleles at a particular locus in a population are A, we would say that the frequency of the A allele in the population is 0.20. For large populations, for which a determination of the genes of all individual members is impractical, a sample of the population is usually taken and the genotypic and allelic frequencies are calculated for this sample (see Chapter 24 for a discussion of samples). The genotypic and allelic frequencies of the sample are then used to represent the gene pool of the population.

To calculate a **genotypic frequency**, we simply add up the number of individuals possessing the genotype and divide by the total number of individuals in the sample (N). For a locus with three genotypes AA, Aa, and aa, the frequency (f) of each genotype is

$$f(AA) = \frac{\text{number of } AA \text{ individuals}}{N}$$

$$f(Aa) = \frac{\text{number of } Aa \text{ individuals}}{N} \quad (25.1)$$

$$f(aa) = \frac{\text{number of } aa \text{ individuals}}{N}$$

The sum of all the genotypic frequencies always equals 1.

Calculating Allelic Frequencies

The gene pool of a population can also be described in terms of the allelic frequencies. There are always fewer alleles than genotypes; so the gene pool of a population can be described in fewer terms when the allelic frequencies are used. In a sexually reproducing population, the genotypes are only temporary assemblages of the alleles: the genotypes break down each generation when individual alleles are passed to the next generation through the gametes, and so the types and numbers of alleles, rather than genotypes, have real continuity from one generation to the next and make up the gene pool of a population.

Allelic frequencies can be calculated from (1) the numbers or (2) the frequencies of the genotypes. To calculate the **allelic frequency** from the numbers of genotypes, we count the number of copies of a particular allele present in a sample and divide by the total number of all alleles in the sample:

$$\text{frequency of an allele} = \frac{\begin{array}{c}\text{number of copies}\\\text{of the allele}\end{array}}{\begin{array}{c}\text{number of copies of all}\\\text{alleles at the locus}\end{array}} \quad (25.2)$$

For a locus with only two alleles (A and a), the frequencies of the alleles are usually represented by the symbols p and q, and can be calculated as follows:

$$p = f(A) = \frac{2n_{AA} + n_{Aa}}{2N}$$
$$q = f(a) = \frac{2n_{aa} + n_{Aa}}{2N} \quad (25.3)$$

where n_{AA}, n_{Aa}, and n_{aa} represent the numbers of AA, Aa, and aa individuals, and N represents the total number of individuals in the sample. We divide by $2N$ because each diploid individual has two alleles at a locus. The sum of the allelic frequencies always equals 1 ($p + q = 1$); so, after p has been obtained, q can be determined by subtraction: $q = 1 - p$.

Alternatively, allelic frequencies can be calculated from the genotypic frequencies. To calculate an allelic frequency from genotypic frequencies, we add the frequency of the homozygote for each allele to half the frequency of the heterozygote (because half of the heterozygote's alleles are of each type):

$$p = f(A) = f(AA) + \frac{1}{2}f(Aa)$$
$$q = f(a) = f(aa) + \frac{1}{2}f(Aa) \quad (25.4)$$

We obtain the same values of p and q whether we calculate the allelic frequencies from the numbers of genotypes (see Equation 25.3) or from the genotypic frequencies (see Equation 25.4). A sample calculation of allelic frequencies is provided in the next Worked Problem.

TRY PROBLEM 16 →

Loci with multiple alleles   We can use the same principles to determine the frequencies of alleles for loci with more than two alleles. To calculate the allelic frequencies from the numbers of genotypes, we count up the number of copies of an allele by adding twice the number of homozygotes to the number of heterozygotes that possess the allele and divide this sum by twice the number of individuals in the sample. For a locus with three alleles (A^1, A^2, and A^3) and six genotypes (A^1A^1, A^1A^2, A^2A^2, A^1A^3, A^2A^3, and A^3A^3), the frequencies (p, q, and r) of the alleles are

$$p = f(A^1) = \frac{2n_{A^1A^1} + n_{A^1A^2} + n_{A^1A^3}}{2N}$$

$$q = f(A^2) = \frac{2n_{A^2A^2} + n_{A^1A^2} + n_{A^2A^3}}{2N} \qquad (25.5)$$

$$r = f(A^3) = \frac{2n_{A_3A_3} + n_{A^1A^3} + n_{A^2A^3}}{2N}$$

Alternatively, we can calculate the frequencies of multiple alleles from the genotypic frequencies by extending Equation 25.4. Once again, we add the frequency of the homozygote to half the frequency of each heterozygous genotype that possesses the allele:

$$p = f(A^1A^1) + \tfrac{1}{2} f(A^1A^2) + \tfrac{1}{2} f(A^1A^3)$$

$$q = f(A^2A^2) + \tfrac{1}{2} f(A^1A^2) + \tfrac{1}{2} f(A^2A^3) \qquad (25.6)$$

$$r = f(A^3A^3) + \tfrac{1}{2} f(A^1A^3) + \tfrac{1}{2} f(A^2A^3)$$

X-linked loci   To calculate allelic frequencies for genes at X-linked loci, we apply these same principles. However, we must remember that a female possesses two X chromosomes and therefore has two X-linked alleles, whereas a male has only a single X chromosome and has one X-linked allele.

Suppose there are two alleles at an X-linked locus, X^A and X^a. Females may be either homozygous (X^AX^A or X^aX^a) or heterozygous (X^AX^a). All males are hemizygous (X^AY or X^aY). To determine the frequency of the X^A allele (p), we first count the number of copies of X^A: we multiply the number of X^AX^A females by two and add the number of X^AX^a females and the number of X^AY males. We then divide the sum by the total number of alleles at the locus, which is twice the total number of females plus the number of males:

$$p = f(X^A) = \frac{2n_{X^AX^A} + n_{X^AX^a} + n_{X^AY}}{2n_{females} + n_{males}} \qquad (25.7a)$$

Similarly, the frequency of the X^a allele is

$$q = f(X^a) = \frac{2n_{X^aX^a} + n_{X^AX^a} + n_{X^aY}}{2n_{females} + n_{males}} \qquad (25.7b)$$

The frequencies of X-linked alleles can also be calculated from genotypic frequencies by adding the frequency of the females that are homozygous for the allele, half the fre-

quency of the females that are heterozygous for the allele, and the frequency of males hemizygous for the allele:

$$p = f(X^A) = f(X^AX^A) + \tfrac{1}{2} f(X^AX^a) + f(X^AY)$$

$$q = f(X^a) = f(X^aX^a) + \tfrac{1}{2} f(X^AX^a) + f(X^aY) \qquad (25.8)$$

If you remember the logic behind all of these calculations, you can determine allelic frequencies for any set of genotypes, and it will not be necessary to memorize the formulas.   **TRY PROBLEM 18** →

CONCEPTS

Population genetics concerns the genetic composition of a population and how it changes with time. The gene pool of a population can be described by the frequencies of genotypes and alleles in the population.

✔ CONCEPT CHECK 1

What are some advantages of using allelic frequencies to describe the gene pool of a population instead of using genotypic frequencies?

Worked Problem

The human MN blood-type antigens are determined by two codominant alleles, L^M and L^N (see p. 102 in Chapter 5). The MN blood types and corresponding genotypes of 398 Finns from Karjala are tabulated here.

Phenotype	Genotype	Number
MM	L^ML^M	182
MN	L^ML^N	172
NN	L^NL^N	44

Source: W. C. Boyd, *Genetics and the Races of Man* (Boston: Little, Brown, 1950.)

Calculate the genotypic and allelic frequencies at the MN locus for the Karjala population.

• Solution

The genotypic frequencies for the population are calculated with the following formula:

genotypic frequency

$$= \frac{\text{number of individuals with genotype}}{\text{total number of individual in sample } (N)}$$

$$f(L^ML^M) = \frac{\text{number of } L^ML^M \text{ individuals}}{N} = \frac{182}{398} = 0.457$$

$$f(L^ML^N) = \frac{\text{number of } L^ML^M \text{ individuals}}{N} = \frac{172}{398} = 0.432$$

$$f(L^NL^N) = \frac{\text{number of } L^NL^N \text{ individuals}}{N} = \frac{44}{398} = 0.111$$

The allelic frequencies can be calculated from either the numbers or the frequencies of the genotypes. To calculate allelic frequencies from numbers of genotypes, we add the number of copies of the allele and divide by the number of copies of all alleles at that locus.

$$\text{frequency of an allele} = \frac{\text{number of copies of the allele}}{\text{number of copies all alleles}}$$

$$p = f(L^M) = \frac{(2n_{L^M L^M}) + (n_{L^M L^N})}{2N}$$

$$= \frac{2(182) + 172}{2(398)} = \frac{536}{796} = 0.673$$

$$q = f(L^N) = \frac{(2n_{L^N L^N}) + (n_{L^M L^N})}{2N}$$

$$= \frac{2(44) + 172}{2(398)} = \frac{260}{796} = 0.327$$

To calculate the allelic frequencies from genotypic frequencies, we add the frequency of the homozygote for that genotype to half the frequency of each heterozygote possessing that allele:

$$p = f(L^M) = f(L^M L^M) + \tfrac{1}{2} f(L^M L^N)$$
$$= 0.457 + \tfrac{1}{2} (0.432) = 0.673$$

$$q = f(L^N) = f(L^N L^N) + \tfrac{1}{2} f(L^M L^N)$$
$$= 0.111 + \tfrac{1}{2} (0.432) = 0.327$$

> → Now try your hand at calculating genotypic and allelic frequencies by working Problem 17 at the end of the chapter.

25.2 The Hardy–Weinberg Law Describes the Effect of Reproduction on Genotypic and Allelic Frequencies

The primary goal of population genetics is to understand the processes that shape a population's gene pool. First, we must ask what effects reproduction and Mendelian principles have on the genotypic and allelic frequencies: How do the segregation of alleles in gamete formation and the combining of alleles in fertilization influence the gene pool? The answer to this question lies in the **Hardy–Weinberg law**, among the most important principles of population genetics.

The Hardy–Weinberg law was formulated independently by both G. H. Hardy and Wilhelm Weinberg in 1908. (Similar conclusions were reached by several other geneticists at about the same time.) The law is actually a mathematical model that evaluates the effect of reproduction on the genotypic and allelic frequencies of a population. It makes several simplifying assumptions about the population and provides two key predictions if these assumptions are met. For an autosomal locus with two alleles, the Hardy–Weinberg law can be stated as follows:

Assumptions If a population is large, randomly mating, and not affected by mutation, migration, or natural selection, then:

Prediction 1 the allelic frequencies of a population do not change; and

Prediction 2 the genotypic frequencies stabilize (will not change) after one generation in the proportions p^2 (the frequency of AA), $2pq$ (the frequency of Aa), and q^2 (the frequency of aa), where p equals the frequency of allele A and q equals the frequency of allele a.

The Hardy–Weinberg law indicates that, when the assumptions are met, reproduction alone does not alter allelic or genotypic frequencies and the allelic frequencies determine the frequencies of genotypes.

The statement that genotypic frequencies stabilize after one generation means that they may change in the first generation after random mating, because one generation of random mating is required to produce Hardy–Weinberg proportions of the genotypes. Afterward, the genotypic frequencies, like allelic frequencies, do not change as long as the population continues to meet the assumptions of the Hardy–Weinberg law. When genotypes are in the expected proportions of p^2, $2pq$, and q^2, the population is said to be in **Hardy–Weinberg equilibrium**.

CONCEPTS

The Hardy–Weinberg law describes how reproduction and Mendelian principles affect the allelic and genotypic frequencies of a population.

✔ CONCEPT CHECK 2

Which statement is not an assumption of the Hardy–Weinberg law?

a. The allelic frequencies (p and q) are equal.

b. The population is randomly mating.

c. The population is large.

d. Natural selection has no effect.

Genotypic Frequencies at Hardy–Weinberg Equilibrium

How do the conditions of the Hardy–Weinberg law lead to genotypic proportions of p^2, $2pq$, and q^2? Mendel's principle of segregation says that each individual organism possesses two alleles at a locus and that each of the two alleles has an equal probability of passing into a gamete. Thus, the frequencies of alleles in gametes will be the same as the frequencies of alleles in the parents. Suppose we have a Mendelian population in which the frequencies of alleles A and a are p and q, respectively. These frequencies will also be those in the gametes. If mating is random (one of the assumptions

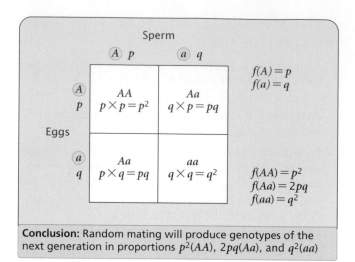

Sperm

	$\textcircled{A}\; p$	$\textcircled{a}\; q$
$\textcircled{A}\; p$	AA $p \times p = p^2$	Aa $q \times p = pq$
$\textcircled{a}\; q$	Aa $p \times q = pq$	aa $q \times q = q^2$

Eggs

$f(A) = p$
$f(a) = q$

$f(AA) = p^2$
$f(Aa) = 2pq$
$f(aa) = q^2$

Conclusion: Random mating will produce genotypes of the next generation in proportions $p^2(AA)$, $2pq(Aa)$, and $q^2(aa)$

25.2 Random mating produces genotypes in the proportions p^2, $2pq$, and q^2.

of the Hardy–Weinberg law), the gametes will come together in random combinations, which can be represented by a Punnett square as shown in **Figure 25.2**.

The multiplication rule of probability can be used to determine the probability of various gametes pairing. For example, the probability of a sperm containing allele A is p and the probability of an egg containing allele A is p. Applying the multiplication rule, we find that the probability that these two gametes will combine to produce an AA homozygote is $p \times p = p^2$. Similarly, the probability of a sperm containing allele a combining with an egg containing allele a to produce an aa homozygote is $q \times q = q^2$. An Aa heterozygote can be produced in one of two ways: (1) a sperm containing allele A may combine with an egg containing allele a ($p \times q$) or (2) an egg containing allele A may combine with a sperm containing allele a ($p \times q$). Thus, the probability of alleles A and a combining to produce an Aa heterozygote is $2pq$. In summary, whenever the frequencies of alleles in a population are p and q, the frequencies of the genotypes in the next generation will be p^2, $2pq$, and q^2.

Closer Examination of the Assumptions of the Hardy–Weinberg Law

Before we consider the implications of the Hardy–Weinberg law, we need to take a closer look at the three assumptions that it makes about a population. First, it assumes that the population is large. How big is "large"? Theoretically, the Hardy–Weinberg law requires that a population be infinitely large in size, but this requirement is obviously unrealistic. In practice, many large populations are in the predicated Hardy–Weinberg proportions, and significant deviations arise only when population size is rather small. Later in the chapter, we will examine the effects of small population size on allelic frequencies.

The second assumption of the Hardy–Weinberg law is that members of the population mate randomly, which means that each genotype mates relative to its frequency. For example, suppose that three genotypes are present in a population in the following proportions: $f(AA) = 0.6$, $f(Aa) = 0.3$, and $f(aa) = 0.1$. With random mating, the frequency of mating between two AA homozygotes ($AA \times AA$) will be equal to the multiplication of their frequencies: $0.6 \times 0.6 = 0.36$, whereas the frequency of mating between two aa homozygotes ($aa \times aa$) will be only $0.1 \times 0.1 = 0.01$.

The third assumption of the Hardy–Weinberg law is that the allelic frequencies of the population are not affected by natural selection, migration, and mutation. Although mutation occurs in every population, its rate is so low that it has little short-term effect on the predictions of the Hardy–Weinberg law (although it may largely shape allelic frequencies over long periods of time when no other forces are acting). Although natural selection and migration are significant factors in real populations, we must remember that the purpose of the Hardy–Weinberg law is to examine only the effect of reproduction on the gene pool. When this effect is known, the effects of other factors (such as migration and natural selection) can be examined.

A final point is that the assumptions of the Hardy–Weinberg law apply to a single locus. No real population mates randomly for all traits; and a population is not completely free of natural selection for all traits. The Hardy–Weinberg law, however, does not require random mating and the absence of selection, migration, and mutation for all traits; it requires these conditions only for the locus under consideration. A population may be in Hardy–Weinberg equilibrium for one locus but not for others.

Implications of the Hardy–Weinberg Law

The Hardy–Weinberg law has several important implications for the genetic structure of a population. One implication is that a population cannot evolve if it meets the Hardy–Weinberg assumptions, because evolution consists of change in the allelic frequencies of a population. Therefore the Hardy–Weinberg law tells us that reproduction alone will not bring about evolution. Other processes such as natural selection, mutation, migration, or chance are required for populations to evolve.

A second important implication is that, when a population is in Hardy–Weinberg equilibrium, the genotypic frequencies are determined by the allelic frequencies. When a population is not in Hardy–Weinberg equilibrium, we have no basis for predicting the genotypic frequenices. Although we can always determine the allelic frequencies from the genotypic frequencies (see Equation 25.3), the reverse (determining the genotypic frequencies from the allelic frequencies) is possible only when the population is in Hardy–Weinberg equilibrium.

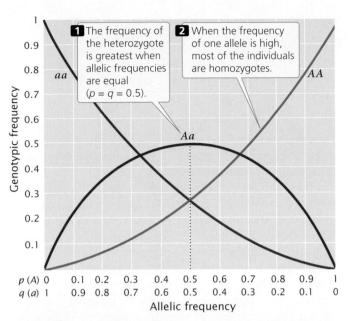

25.3 When a population is in Hardy–Weinberg equilibrium, the proportions of genotypes are determined by the frequencies of alleles.

▶ www.whfreeman.com/pierce4e   Animation 25.1 examines the effect of changing allelic frequencies on genotypic frequencies when a population is in Hardy–Weinberg equilibrium.

Table 25.1 Extensions of the Hardy–Weinberg law

Situation	Allelic Frequencies	Genotypic Frequencies
Three alleles	$f(A^1) = p$ $f(A^2) = q$ $f(A^3) = r$	$f(A^1A^1) = p^2$ $f(A^1A^2) = 2pq$ $f(A^2A^2) = q^2$ $f(A^1A^3) = 2pr$ $f(A^2A^3) = 2qr$ $f(A^3A^3) = r^2$
X-linked alleles	$f(X^1) = p$ $f(X^2) = q$	$f(X^1X^1 \text{ female}) = p^2$ $f(X^1X^2 \text{ female}) = 2pq$ $f(X^2X^2 \text{ female}) = q^2$ $f(X^1Y \text{ male}) = p$ $f(X^2Y \text{ male}) = q$

Note: For X-linked female genotypes, the frequencies are the proportions among all females; for X-linked male genotypes, the frequencies are the proportions among all males.

For a locus with two alleles, the frequency of the heterozygote is greatest when allelic frequencies are between 0.33 and 0.66 and is at a maximum when allelic frequencies are each 0.5 (**Figure 25.3**). The heterozygote frequency also never exceeds 0.5 when the population is in Hardy–Weinberg equilibrium. Furthermore, when the frequency of one allele is low, homozygotes for that allele will be rare, and most of the copies of a rare allele will be present in heterozygotes. As you can see from Figure 25.3, when the frequency of allele a is 0.2, the frequency of the aa homozygote is only 0.04 (q^2), but the frequency of Aa heterozygotes is 0.32 ($2pq$); 80% of the a alleles are in heterozygotes.

A third implication of the Hardy–Weinberg law is that a single generation of random mating produces the equilibrium frequencies of p^2, $2pq$, and q^2. The fact that genotypes are in Hardy–Weinberg proportions does not prove that the population is free from natural selection, mutation, and migration. It means only that these forces have not acted since the last time random mating took place.

Extensions of the Hardy–Weinberg Law

The Hardy–Weinberg expected proportions can also be applied to multiple alleles and X-linked alleles (**Table 25.1**). With multiple alleles, the genotypic frequencies expected at equilibrium are the square of the allelic fre-

quencies. For an autosomal locus with three alleles, the equilibrium genotypic frequencies will $(p + q + r)^2 = p^2 + 2pq + q^2 + 2pr + 2qr + r^2$. For an X-linked locus with two alleles, X^A and X^a, the equilibrium frequencies of the female genotypes are $(p + q)^2 = p^2 + 2pq + q^2$, where p^2 is the frequency of X^AX^A, $2pq$ is the frequency of X^AX^a, and q^2 is the frequency of X^aX^a. Males have only a single X-linked allele, and so the frequencies of the male genotypes are p (frequency of X^AY) and q (frequency of X^aY). These proportions are those of the genotypes among males and females rather than the proportions among the entire population. Thus, p^2 is the expected proportion of females with the genotype X^AX^A; if females make up 50% of the population, then the expected proportion of this genotype in the entire population is $0.5 \times p^2$.

The frequency of an X-linked recessive trait among males is q, whereas the frequency among females is q^2. When an X-linked allele is uncommon, the trait will therefore be much more frequent in males than in females. Consider hemophilia A, a clotting disorder caused by an X-linked recessive allele with a frequency (q) of approximately 1 in 10,000, or 0.0001. At Hardy–Weinberg equilibrium, this frequency will also be the frequency of the disease among males. The frequency of the disease among females, however, will be $q^2 = (0.0001)^2 = 0.00000001$, which is only 1 in 10 million. Hemophilia is 1000 times as frequent in males as in females.

Testing for Hardy–Weinberg Proportions

To determine if a population's genotypes are in Hardy–Weinberg equilibrium, the genotypic proportions expected under the Hardy–Weinberg law must be compared with the observed genotypic frequencies. To do so, we first calculate the allelic frequencies, then find the expected genotypic frequencies by using the square of the allelic frequencies, and, finally, compare the observed and expected genotypic frequencies by using a chi-square test.

Worked Problem

Jeffrey Mitton and his colleagues found three genotypes (R^2R^2, R^2R^3, and R^3R^3) at a locus encoding the enzyme peroxidase in ponderosa pine trees growing at Glacier Lake, Colorado. The observed numbers of these genotypes were:

Genotypes	Number observed
R^2R^2	135
R^2R^3	44
R^3R^3	11

Are the ponderosa pine trees at Glacier Lake in Hardy–Weinberg equilibrium at the peroxidase locus?

• Solution

If the frequency of the R^2 allele equals p and the frequency of the R^3 allele equals q, the frequency of the R^2 allele is

$$p = f(R^2) = \frac{(2n_{R^2R^2}) + (n_{R^2R^3})}{2N} = \frac{2(135) + 44}{2(190)} = 0.826$$

The frequency of the R^3 allele is obtained by subtraction:

$$q = f(R^3) = 1 - p = 0.174$$

The frequencies of the genotypes expected under Hardy–Weinberg equilibrium are then calculated separately by using p^2, $2pq$, and q^2:

$$R^2R^2 = p^2 = (0.826)^2 = 0.683$$

$$R^2R^3 = 2pq = 2(0.826)(0.174) = 0.287$$

$$R^3R^3 = q^2 = (0.174)^2 = 0.03$$

Multiplying each of these expected genotypic frequencies by the total number of observed genotypes in the sample (190), we obtain the numbers expected for each genotype:

$$R^2R^2 = 0.683 \times 190 = 129.8$$

$$R^2R^3 = 0.287 \times 190 = 54.5$$

$$R^3R^3 = 0.03 \times 190 = 5.7$$

By comparing these expected numbers with the observed numbers of each genotype, we see that there are more R^2R^2 homozygotes and fewer R^2R^3 heterozygotes and R^3R^3 homozygotes in the population than we expect at equilibrium.

A goodness-of-fit chi-square test is used to determine whether the differences between the observed and the

expected numbers of each genotype are due to chance:

$$X^2 = \sum \frac{(\text{observerd} - \text{expected})^2}{\text{expected}}$$

$$= \frac{(135 - 129.8)^2}{129.8} + \frac{(44 - 54.5)^2}{54.5} + \frac{(11 - 5.7)^2}{5.7}$$

$$= 0.21 + 2.02 + 4.93 = 7.16$$

The calculated chi-square value is 7.16; to obtain the probability associated with this chi-square value, we determine the appropriate degrees of freedom.

Up to this point, the chi-square test for assessing Hardy–Weinberg equilibrium has been identical with the chi-square tests that we used in Chapter 3 to assess progeny ratios in a genetic cross, where the degrees of freedom were $n - 1$ and n equaled the number of expected genotypes. For the Hardy–Weinberg test, however, we must subtract an additional degree of freedom, because the expected numbers are based on the observed allelic frequencies; therefore, the observed numbers are not completely free to vary. In general, the degrees of freedom for a chi-square test of Hardy–Weinberg equilibrium equal the number of expected genotypic classes minus the number of associated alleles. For this particular Hardy–Weinberg test, the degree of freedom is $3 - 2 = 1$.

After we have calculated both the chi-square value and the degrees of freedom, the probability associated with this value can be sought in a chi-square table (see Table 3.5). With one degree of freedom, a chi-square value of 7.16 has a probability between 0.01 and 0.001. It is very unlikely that the peroxidase genotypes observed at Glacier Lake are in Hardy–Weinberg proportions.

→ For additional practice, determine whether the genotypic frequencies in Problem 20 at the end of the chapter are in Hardy–Weinberg equilibrium.

CONCEPTS

The observed number of genotypes in a population can be compared with the Hardy–Weinberg expected proportions by using a goodness-of-fit chi-square test.

✔ CONCEPT CHECK 3

What is the expected frequency of heterozygotes in a population with allelic frequencies x and y that is in Hardy–Weinberg equilibrium?

a. $x + y$ c. $2xy$

b. xy d. $(x - y)^2$

Estimating Allelic Frequencies with the Hardy–Weinberg Law

A practical use of the Hardy–Weinberg law is that it allows us to calculate allelic frequencies when dominance is present. For example, cystic fibrosis is an autosomal recessive disorder characterized by respiratory infections, incomplete digestion, and abnormal sweating (see p. 102 in Chapter 5). Among North American Caucasians, the incidence of the disease is approximately 1 person in 2000. The formula for calculating allelic frequency (see Equation 25.3) requires that we know the numbers of homozygotes and heterozygotes, but cystic fibrosis is a recessive disease and so we cannot easily distinguish between homozygous normal persons and heterozygous carriers. Although molecular tests are available for identifying heterozygous carriers of the cystic fibrosis gene, the low frequency of the disease makes widespread screening impractical. In such situations, the Hardy–Weinberg law can be used to estimate the allelic frequencies.

If we assume that a population is in Hardy–Weinberg equilibrium with regard to this locus, then the frequency of the recessive genotype (aa) will be q^2, and the allelic frequency is the square root of the genotypic frequency:

$$q = \sqrt{f(aa)} \qquad (25.9)$$

If the frequency of cystic fibrosis in North American Caucasians is approximately 1 in 2000, or 0.0005, then $q = \sqrt{0.0005} = 0.02$. Thus, about 2% of the alleles in the Caucasian population encode cystic fibrosis. We can calculate the frequency of the normal allele by subtracting: $p = 1 - q = 1 - 0.02 = 0.98$. After we have calculated p and q, we can use the Hardy–Weinberg law to determine the frequencies of homozygous normal people and heterozygous carriers of the gene:

$$f(AA) = p^2 = (0.98)^2 = 0.960$$
$$f(Aa) = 2pq = 2(0.02)(0.98) = 0.0392$$

Thus, about 4% (1 of 25) of Caucasians are heterozygous carriers of the allele that causes cystic fibrosis. $\boxed{\text{TRY PROBLEM 23} \rightarrow}$

CONCEPTS

Although allelic frequencies cannot be calculated directly for traits that exhibit dominance, the Hardy–Weinberg law can be used to estimate the allelic frequencies if the population is in Hardy–Weinberg equilibrium for that locus. The frequency of the recessive allele will be equal to the square root of the frequency of the recessive trait.

✔ CONCEPT CHECK 4

In cats, all-white color is dominant over not all-white. In a population of 100 cats, 19 are all-white cats. Assuming that the population is in Hardy–Weinberg equilibrium, what is the frequency of the all-white allele in this population?

25.3 Nonrandom Mating Affects the Genotypic Frequencies of a Population

An assumption of the Hardy–Weinberg law is that mating is random with respect to genotype. Nonrandom mating affects the way in which alleles combine to form genotypes and alters the genotypic frequencies of a population.

We can distinguish between two types of nonrandom mating. **Positive assortative mating** refers to a tendency for like individuals to mate. For example, humans exhibit positive assortative mating for height: tall people mate preferentially with other tall people; short people mate preferentially with other short people. **Negative assortative mating** refers to a tendency for unlike individuals to mate. If people engaged in negative assortative mating for height, tall and short people would preferentially mate. Assortative mating is usually for a particular trait and will affect only those genes that encode the trait (and genes closely linked to them).

One form of nonrandom mating is **inbreeding**, which is preferential mating between related individuals. Inbreeding is actually positive assortative mating for relatedness, but it differs from other types of assortative mating because it affects all genes, not just those that determine the trait for which the mating preference exists. Inbreeding causes a departure from the Hardy–Weinberg equilibrium frequencies of p^2, $2pq$, and q^2. More specifically, it leads to an increase in the proportion of homozygotes and a decrease in the proportion of heterozygotes in a population. **Outcrossing** is the avoidance of mating between related individuals.

In a diploid organism, a homozygous individual has two copies of the same allele. These two copies may be the same in *state*, which means that the two alleles are alike in structure and function but do not have a common origin. Alternatively, the two alleles in a homozygous individual may be the same because they are identical by *descent*; that is, the copies are descended from a single allele that was present in an ancestor (**Figure 25.4**). If we go back far enough in time, many alleles are likely to be identical by descent but, for calculating the effects of inbreeding, we consider identity by descent by going back only a few generations.

Inbreeding is usually measured by the **inbreeding coefficient**, designated F, which is a measure of the probability that two alleles are identical by descent. Inbreeding coefficients can range from 0 to 1. A value of 0 indicates that mating in a large population is random; a value of 1 indicates that all alleles are identical by descent. Inbreeding coefficients can be calculated from analyses of pedigrees or they can be determined from the reduction in the heterozygosity of a population. Although we will not go into the details of how F is calculated, an understanding of how inbreeding affects genotypic frequencies is important.

When inbreeding takes place, the proportion of heterozygotes decreases by $2Fpq$, and half of this value (Fpq) is *added* to the proportion of each homozygote each generation. The frequencies of the genotypes will then be

$$f(AA) = p^2 + Fpq$$
$$f(Aa) = 2pq - 2Fpq \qquad (25.10)$$
$$f(aa) = q^2 + Fpq$$

Consider a population that reproduces by self-fertilization (so $F = 1$). We will assume that this population begins

(a) Alleles identical by descent

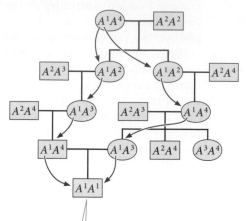

These two copies of the A^1 allele are descended from the same copy in a common ancestor; so they are identical by descent.

(b) Alleles identical by state

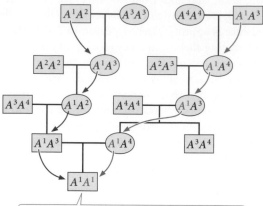

These two copies of the A^1 allele are the same in structure and function, but are descended from two different copies in ancestors; so they are identical in state.

25.4 Individuals may be homozygous by descent or by state. Inbreeding is a measure of the probability that two alleles are identical by descent.

with genotypic frequencies in Hardy–Weinberg proportions (p^2, $2pq$, and q^2). With selfing, each homozygote produces progeny only of the same homozygous genotype ($AA \times AA$ produces all AA; and $aa \times aa$ produces all aa), whereas only half the progeny of a heterozygote will be like the parent ($Aa \times Aa$ produces $^1/_4\ AA$, $^1/_2\ Aa$, and $^1/_4\ aa$). Selfing therefore reduces the proportion of heterozygotes in the population by half with each generation, until all genotypes in the population are homozygous (**Table 25.2** and **Figure 25.5**).

TRY PROBLEM 26 →

For most outcrossing species, close inbreeding is harmful because it increases the proportion of homozygotes and thereby boosts the probability that deleterious and lethal recessive alleles will combine to produce homozygotes with

a harmful trait. Assume that a recessive allele (a) that causes a genetic disease has a frequency (q) of 0.01. If the population mates randomly ($F = 0$), the frequency of individuals affected with the disease (aa) will be $q^2 = 0.01^2 = 0.0001$; so only 1 in 10,000 individuals will have the disease. However, if $F = 0.25$ (the equivalent of brother–sister mating), then the expected frequency of the homozygote genotype is $q^2 + Fpq = (0.01)^2 + (0.25)(0.99)(0.01) = 0.0026$; thus, the genetic

Table 25.2	Generational increase in frequency of homozygotes in a self-fertilizing population starting with $p = q = 0.5$		
	Genotypic Frequencies		
Generation	AA	Aa	aa
1	$^1/_4$	$^1/_2$	$^1/_4$
2	$^1/_4 + ^1/_8 = ^3/_8$	$^1/_4$	$^1/_4 + ^1/_8 = ^3/_8$
3	$^3/_8 + ^1/_{16} = ^7/_{16}$	$^1/_8$	$^3/_8 + ^1/_{16} = ^7/_{16}$
4	$^7/_{16} + ^1/_{32} = ^{15}/_{32}$	$^1/_{16}$	$^7/_{16} + ^1/_{32} = ^{15}/_{32}$
n	$\dfrac{1 - (^1/_2)^n}{2}$	$(^1/_2)^n$	$\dfrac{1 - (^1/_2)^n}{2}$
	$^1/_2$	0	$^1/_2$

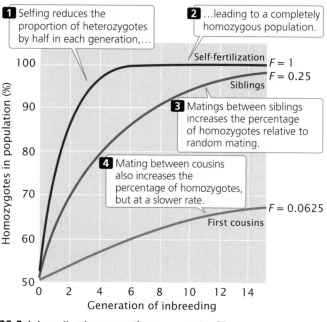

1 Selfing reduces the proportion of heterozygotes by half in each generation,…

2 …leading to a completely homozygous population.

Self-fertilization $F = 1$
$F = 0.25$
Siblings

3 Matings between siblings increases the percentage of homozygotes relative to random mating.

4 Mating between cousins also increases the percentage of homozygotes, but at a slower rate.

$F = 0.0625$
First cousins

25.5 Inbreeding increases the percentage of homozygous individuals in a population.

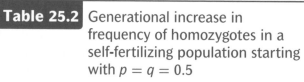

www.whfreeman.com/pierce4e

To see the effects of inbreeding on genotypic frequencies, view the Mini-Tutorial in Animation 25.1.

Table 25.3	Effects of inbreeding on Japanese children	
Genetic Relationship of Parents	*F*	Mortality of Children (through 12 years of age)
Unrelated	0	0.082
Second cousins	0.016 ($^1/_{64}$)	0.108
First cousins	0.0625 ($^1/_{16}$)	0.114

Source: After D. L. Hartl and A. G. Clark, *Principles of Population Genetics*, 2d ed. (Sunderland, Mass.: Sinauer, 1989), Table 2. Original data from W. J. Schull and J. V. Neel, *The Effects of Inbreeding on Japanese Children* (New York: Harper & Row, 1965).

25.7 Although inbreeding is generally harmful, a number of inbreeding organisms are successful. Shown here is the terrestrial slug *Arion circumscriptos*, an inbreeding species that causes damage in greenhouses and flower gardens. [William Leonard/DRK Photo.]

disease is 26 times as frequent at this level of inbreeding. This increased appearance of lethal and deleterious traits with inbreeding is termed **inbreeding depression**; the more intense the inbreeding, the more severe the inbreeding depression.

The harmful effects of inbreeding have been recognized by people for thousands of years and may be the basis of cultural taboos against mating between close relatives. William Schull and James Neel found that, for each 10% increase in *F*, the mean IQ of Japanese children dropped six points. Child mortality also increases with close inbreeding (**Table 25.3**); children of first cousins have a 40% increase in mortality over that seen among the children of unrelated people. Inbreeding also has deleterious effects on crops (**Figure 25.6**) and domestic animals.

Inbreeding depression is most often studied in humans, as well as in plants and animals reared in captivity, but the negative effects of inbreeding may be more severe in natural populations. Julie Jimenez and her colleagues collected wild

mice from a natural population in Illinois and bred them in the laboratory for three to four generations. Laboratory matings were chosen so that some mice had no inbreeding, whereas others had an inbreeding coefficient of 0.25. When both types of mice were released back into the wild, the weekly survival of the inbred mice was only 56% of that of the noninbred mice. Inbred male mice also continously lost body weight after release into the wild, whereas noninbred male mice initially lost weight but then regained it within a few days after release.

In spite of the fact that inbreeding is generally harmful for outcrossing species, a number of plants and animals regularly inbreed and are successful (**Figure 25.7**). Inbreeding is commonly used to produce domesticated plants and animals having desirable traits. As stated earlier, inbreeding increases homozygosity, and eventually all individuals in the population become homozygous for the same allele. If a species undergoes inbreeding for a number of generations, many deleterious recessive alleles are weeded out by natural or artificial selection so that the population becomes homozygous for beneficial alleles. In this way, the harmful effects of inbreeding may eventually be eliminated, leaving a population that is homozygous for beneficial traits.

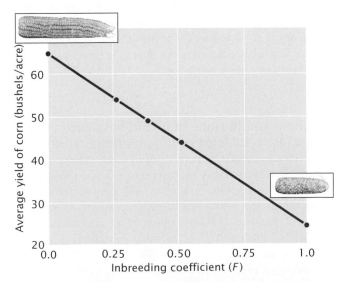

25.6 Inbreeding often has deleterious effects on crops. As inbreeding increases, the average yield of corn, for example, decreases.

CONCEPTS

Nonrandom mating alters the frequencies of the genotypes but not the frequencies of the alleles. Inbreeding is preferential mating between related individuals. With inbreeding, the frequency of homozygotes increases, whereas the frequency of heterozygotes decreases.

✔ CONCEPT CHECK 5

What is the effect of outcrossing on a population?

a. Allelic frequencies change.

b. There will be more heterozygotes than predicted by the Hardy–Weinberg law.

c. There will be fewer heterozygotes than predicted by the Hardy–Weinberg law.

d. Genotypic frequencies will equal those predicted by the Hardy–Weinberg law.

25.4 Several Evolutionary Forces Potentially Cause Changes in Allelic Frequencies

The Hardy–Weinberg law indicates that allelic frequencies do not change as a result of reproduction; thus, other processes must cause alleles to increase or decrease in frequency. Processes that bring about change in allelic frequency include mutation, migration, genetic drift (random effects due to small population size), and natural selection.

Mutation

Before evolution can take place, genetic variation must exist within a population; consequently, all evolution depends on processes that generate genetic variation. Although new *combinations* of existing genes may arise through recombination in meiosis, all genetic variants ultimately arise through mutation.

The effect of mutation on allelic frequencies Mutation can influence the rate at which one genetic variant increases at the expense of another. Consider a single locus in a population of 25 diploid individuals. Each individual possesses two alleles at the locus under consideration; so the gene pool of the population consists of 50 allelic copies. Let us assume that there are two different alleles, designated G^1 and G^2 with frequencies p and q, respectively. If there are 45 copies of G^1 and 5 copies of G^2 in the population, $p = 0.90$ and $q = 0.10$. Now suppose that a mutation changes a G^1 allele into a G^2 allele. After this mutation, there are 44 copies of G^1 and 6 copies of G^2, and the frequency of G^2 has increased from 0.10 to 0.12. Mutation has changed the allelic frequency.

If copies of G^1 continue to mutate to G^2, the frequency of G^2 will increase and the frequency of G^1 will decrease (**Figure 25.8**). The amount that G^2 will change (Δq) as a result of mutation depends on: (1) the rate of G^1-to-G^2 mutation (μ) and (2) p, the frequency of G^1 in the population. When p is large, there are many copies of G^1 available to mutate to G^2, and the amount of change will be relatively large. As more mutations occur and p decreases, there will be fewer copies of G^1 available to mutate to G^2. The change in G^2 as a result of mutation equals the mutation rate times the allelic frequency:

$$\Delta q = \mu p \qquad (25.11)$$

As the frequency of p decreases as a result of mutation, the change in frequency due to mutation will be less and less.

So far, we have considered only the effects of $G^1 \rightarrow G^2$ forward mutations. Reverse $G^2 \rightarrow G^1$ mutations also occur at rate ν, which will probably be different from the forward mutation rate, μ. Whenever a reverse mutation occurs, the frequency of G^2 decreases and the frequency of G^1 increases

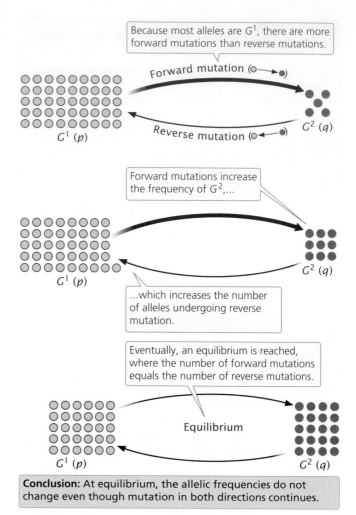

Because most alleles are G^1, there are more forward mutations than reverse mutations.

Forward mutation ($\circ \longrightarrow \bullet$)

Reverse mutation ($\circ \longleftarrow \bullet$)

$G^1 (p)$ $G^2 (q)$

Forward mutations increase the frequency of G^2,...

...which increases the number of alleles undergoing reverse mutation.

Eventually, an equilibrium is reached, where the number of forward mutations equals the number of reverse mutations.

Equilibrium

$G^1 (p)$ $G^2 (q)$

Conclusion: At equilibrium, the allelic frequencies do not change even though mutation in both directions continues.

25.8 Recurrent mutation changes allelic frequencies. Forward and reserve mutations eventually lead to a stable equilibrium.

(see Figure 25.8). The rate of change due to reverse mutations equals the reverse mutation rate times the allelic frequency of G^2 ($\Delta q = \nu q$). The overall change in allelic frequency is a balance between the opposing forces of forward mutation and reverse mutation:

$$\Delta q = \mu p - \nu q \qquad (25.12)$$

Reaching equilibrium of allelic frequencies Consider an allele that begins with a high frequency of G^1 and a low frequency of G^2. In this population, many copies of G^1 are initially available to mutate to G^2, and the increase in G^2 due to forward mutation will be relatively large. However, as the frequency of G^2 increases as a result of forward mutations, fewer copies of G^1 are available to mutate; so the number of forward mutations decreases. On the other hand, few copies of G^2 are initially available to undergo a reverse mutation to G^1 but, as the frequency of G^2 increases, the number of copies of G^2 available to

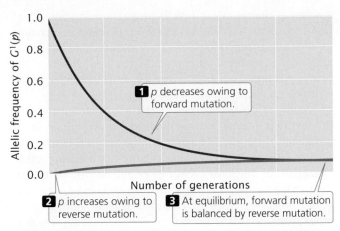

1 p decreases owing to forward mutation.

2 p increases owing to reverse mutation.

3 At equilibrium, forward mutation is balanced by reverse mutation.

25.9 Change due to recurrent mutation slows as the frequency of p drops. Allelic frequencies are approaching mutational equilibrium at typical low mutation rates. The allelic frequency of G^1 decreases as a result of forward ($G^1 \rightarrow G^2$) mutation at rate m (0.0001) and increases as a result of reverse ($G^2 \rightarrow G^1$) mutation at rate n (0.00001). Owing to the low rate of mutations, eventual equilibrium takes many generations to be reached.

undergo reverse mutation to G^1 increases; so the number of genes undergoing reverse mutation will increase. Eventually, the number of genes undergoing forward mutation will be counterbalanced by the number of genes undergoing reverse mutation. At this point, the increase in q due to forward mutation will be equal to the decrease in q due to reverse mutation, and there will be no net change in allelic frequency ($\Delta q = 0$), in spite of the fact that forward and reverse mutations continue to occur. The point at which there is no change in the allelic frequency of a population is referred to as **equilibrium** (see Figure 25.8). At equilibrium, the frequency of G^2 ($\hat{q}$) will be

$$\hat{q} = \frac{\mu}{\mu + \nu} \qquad (25.13)$$

This final equation tells us that the allelic frequency at equilibrium is determined solely by the forward (μ) and reverse (ν) mutation rates. **TRY PROBLEM 28** →

Summary of effects When the only evolutionary force acting on a population is mutation, allelic frequencies change with the passage of time because some alleles mutate into others. Eventually, these allelic frequencies reach equilibrium and are determined only by the forward and reverse mutation rates. When the allelic frequencies reach equilibrium, the Hardy–Weinberg law tells us that genotypic frequencies also will remain the same.

The mutation rates for most genes are low; so change in allelic frequency due to mutation in one generation is very small, and long periods of time are required for a population to reach mutational equilibrium. For example, if the

forward and reverse mutation rates for alleles at a locus are 1×10^{-5} and 0.3×10^{-5} per generation, respectively (rates that have actually been measured at several loci in mice), and the allelic frequencies are $p = 0.9$ and $q = 0.1$, then the net change in allelic frequency per generation due to mutation is

$$\begin{aligned}
\Delta q &= \mu p - \nu q \\
&= (1 \times 10^{-5})(0.9) - (0.3 \times 10^{-5})(0.1) \\
&= 8.7 \times 10^{-6} = 0.0000087
\end{aligned}$$

Therefore, change due to mutation in a single generation is extremely small and, as the frequency of p drops as a result of mutation, the amount of change will become even smaller (**Figure 25.9**). The effect of typical mutation rates on Hardy–Weinberg equilibrium is negligible, and many generations are required for a population to reach mutational equilibrium. Nevertheless, if mutation is the only force acting on a population for long periods of time, mutation rates will determine allelic frequencies.

CONCEPTS

Recurrent mutation causes changes in the frequencies of alleles. At equilibrium, the allelic frequencies are determined by the forward and reverse mutation rates. Because mutation rates are low, the effect of mutation per generation is very small.

✔ **CONCEPT CHECK 6**

When a population is in equilibrium for forward and reverse mutation rates, which of the following is true:

a. The number of forward muations is greater than the number of reverse mutations.

b. No forward or reverse mutations occur.

c. The number of forward mutations is equal to the number of reverse mutations.

d. The population is in Hardy–Weinberg equilibrium.

Migration

Another process that may bring about change in allelic frequencies is the influx of genes from other populations, commonly called **migration** or **gene flow**. One of the assumptions of the Hardy–Weinberg law is that migration does not take place, but many natural populations do experience migration from other populations. The overall effect of migration is twofold: (1) it prevents populations from becoming genetically different from one another and (2) it increases genetic variation within populations.

The effect of migration on allelic frequencies Let us consider the effects of migration by looking at a simple, unidirectional model of migration between two populations that differ in the frequency of an allele a. Say that the frequency of this allele in population I is q_I and in population

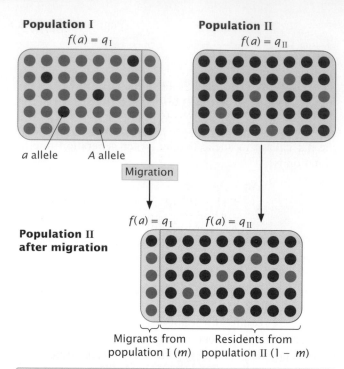

Population I
$f(a) = q_I$

Population II
$f(a) = q_{II}$

a allele   *A* allele

Migration

$f(a) = q_I$ $f(a) = q_{II}$

Population II after migration

Migrants from population I (*m*)   Residents from population II (1 − *m*)

Conclusion: The frequency of allele *a* in population II after migration is $q'_{II} = q_1 m + q_{II}(1 − m)$.

25.10 The amount of change in allelic frequency due to migration between populations depends on the difference in allelic frequency and the extent of migration. Shown here is a model of the effect of unidirectional migration on allelic frequencies. The frequency of allele *a* in the source population (population I) is q_I. The frequency of this allele in the recipient population (population II) is q_{II}.

II is q_{II} (**Figure 25.10**). In each generation, a representative sample of the individuals in population I migrates to population II and reproduces, adding its genes to population II's gene pool. Migration is only from population I to population II (is unidirectional), and all the conditions of the Hardy–Weinberg law apply, except the absence of migration.

After migration, population II consists of two types of individuals. Some are migrants; they make up proportion *m* of population II, and they carry genes from population I; so the frequency of allele *a* in the migrants is q_I. The other individuals in population II are the original residents. If the migrants make up proportion *m* of population II, then the residents make up 1 − *m*; because the residents originated in population II, the frequency of allele *a* in this group is q_{II}. After migration, the frequency of allele *a* in the merged population II (q'_{II}) is

$$q'_{II} = q_I(m) + q_{II}(1 − m) \qquad (25.14)$$

where $q_I(m)$ is the contribution to *q* made by the copies of allele *a* in the migrants and $q_{II}(1 − m)$ is the contribution to *q* made by copies of allele *a* in the residents. The change in the allelic frequency due to migration (Δq) will be

$$\Delta q = m(q_I − q_{II}) \qquad (25.15)$$

Equation 25.15 summarizes the factors that determine the amount of change in allelic frequency due to migration. The

amount of change in *q* is directly proportional to the migration (*m*); as the amount of migration increases, the change in allelic frequency increases. The magnitude of change is also affected by the differences in allelic frequencies of the two populations ($q_I − q_{II}$); when the difference is large, the change in allelic frequency will be large.

With each generation of migration, the frequencies of the two populations become more and more similar until, eventually, the allelic frequency of population II equals that of population I. When $q_I − q_{II} = 0$, there will be no further change in the allelic frequency of population II, in spite of the fact that migration continues. If migration between two populations takes place for a number of generations with no other evolutionary forces present, an equilibrium is reached at which the allelic frequency of the recipient population equals that of the source population.

The simple model of unidirectional migration between two populations just outlined can be expanded to accommodate multidirectional migration between several populations.

The overall effect of migration   Migration has two major effects. First, it causes the gene pools of populations to become more similar. Later, we will see how genetic drift and natural selection lead to genetic differences between populations; migration counteracts this tendency and tends to keep populations homogeneous in their allelic frequencies. Second, migration adds genetic variation to populations. Different alleles may arise in different populations owing to rare mutational events, and these alleles can be spread to new populations by migration, increasing the genetic variation within the recipient population.   TRY PROBLEM 31 →

CONCEPTS

Migration causes changes in the allelic frequency of a population by introducing alleles from other populations. The magnitude of change due to migration depends on both the extent of migration and the difference in allelic frequencies between the source and the recipient populations. Migration decreases genetic differences between populations and increases genetic variation within populations.

✔ CONCEPT CHECK 7

Each generation, 10 random individuals migrate from population A to population B. What will happen to allelic frequency *q* as a result of migration when *q* is equal in populations A and B?

a. *q* in A will decrease.   c. *q* will not change in either A or B.

b. *q* in B will increase.   d. *q* in B will become q^2.

Genetic Drift

The Hardy–Weinberg law assumes random mating in an infinitely large population; only when population size is infinite will the gametes carry genes that perfectly represent the parental gene pool. But no real population is infinitely large, and when population size is limited, the gametes that unite to form individuals of the next generation carry a sample of alleles present in the parental gene pool. Just by chance, the composition of this sample will often deviate from that of the parental gene pool, and this deviation may cause allelic frequencies to change. The smaller the gametic sample, the greater the chance that its composition will deviate from that of the entire gene pool.

The role of chance in altering allelic frequencies is analogous to the flip of a coin. Each time we flip a coin, we have a 50% chance of getting a head and a 50% chance of getting a tail. If we flip a coin 1000 times, the observed ratio of heads to tails will be very close to the expected 50 : 50 ratio. If, however, we flip a coin only 10 times, there is a good chance that we will obtain not exactly 5 heads and 5 tails, but rather maybe 7 heads and 3 tails or 8 tails and 2 heads. This kind of deviation from an expected ratio due to limited sample size is referred to as **sampling error**.

Sampling error arises when gametes unite to produce progeny. Many organisms produce a large number of gametes but, when population size is small, a limited number of gametes unite to produce the individuals of the next generation. Chance influences which alleles are present in this limited sample and, in this way, sampling error may lead to **genetic drift**, or changes in allelic frequency. Because the deviations from the expected ratios are random, the direction of change is unpredictable. We can nevertheless predict the magnitude of the changes.

The magnitude of genetic drift The effect of genetic drift can be viewed in two ways. First, we can see how it influences the change in allelic frequencies of a single population with the passage of time. Second, we can see how it affects differences that accumulate among series of populations. Imagine that we have 10 small populations, all beginning with the exact same allelic frequencies of $p = 0.5$ and $q = 0.5$. When genetic drift occurs in a population, allelic frequencies within the population will change but, because drift is random, the way in which allelic frequencies change in each population will not be the same. In some populations, p may increase as a result of chance. In other populations, p may decrease as a result of chance. In time, the allelic frequencies in the 10 populations will become different: the populations will genetically diverge. As time passes, the change in allelic frequency within a population and the genetic divergence among populations are due to the same force—the random change in allelic frequencies. The magnitude of genetic drift can be assessed either by examining the change in allelic frequency within a single population or by examining the magnitude of genetic differences that accumulate among populations.

The amount of genetic drift can be estimated from the variance in allelic frequency. Variance, s^2, is a statistical measure that describes the degree of variability in a trait

(see p. 667 in Chapter 24). Suppose that we observe a large number of separate populations, each with N individuals and allelic frequencies of p and q. After one generation of random mating, genetic drift expressed in terms of the variance in allelic frequency among the populations (s_p^2) will be

$$s_p^{\,2} = \frac{pq}{2N} \qquad (25.16)$$

The amount of change resulting from genetic drift (the variance in allelic frequency) is determined by two parameters: the allelic frequencies (p and q) and the population size (N). Genetic drift will be maximal when p and q are equal (each 0.5). For example, assume that a population consists of 50 individuals. When the allelic frequencies are equal ($p = q = 0.5$), the variance in allelic frequency (s_p^2) will be $(0.5 \times 0.5)/(2 \times 50) = 0.0025$. In contrast, when $p = 0.9$ and $q = 0.1$, the variance in allelic frequency will be only 0.0009. Genetic drift will also be higher when the population size is small. If $p = q = 0.5$, but the population size is only 10 instead of 50, then the variance in allelic frequency becomes $(0.5 \times 0.5)/(2 \times 10) = 0.0125$, which is five times as great as when population size is 50.

This divergence of populations through genetic drift is strikingly illustrated in the results of an experiment carried out by Peter Buri on fruit flies (**Figure 25.11**). Buri examined the frequencies of two alleles (bw^{75} and bw) that affect eye color in the flies. He set up 107 populations, each consisting of 8 males and 8 females. He began each population with a frequency of bw^{75} equal to 0.5. He allowed the fruit flies within each replicate to mate randomly and, each generation, he randomly selected 8 male and 8 female flies to be the parents of the next generation. He followed the changes in the frequencies of the two alleles over 19 generations. In one population, the average frequency of bw^{75} (p) over the 19 generations was 0.5312. We can use Equation 25.16 to calculate the expected variance in allelic frequency due to genetic drift. The frequency of the bw allele (q) will be $1 - p = 1 - 0.53125 = 0.46875$. The population size ($N$) equals 16. The expected variance in allelic frequency will be

$$\frac{pq}{2N} = \frac{0.53125 \times 0.46875}{2 \times 16} = 0.0156$$

which was very close to the actual observed variance of 0.0151.

The effect of population size on genetic drift is illustrated by a study conducted by Luca Cavalli-Sforza and his colleagues. They studied variation in blood types among villagers in the Parma Valley of Italy, where the amount of migration between villages was limited. They found that variation in allelic frequency was greatest between small isolated villages in the upper valley but decreased between larger villages and towns farther down the valley. This result is exactly what we expect with genetic drift: there should be more genetic drift and thus more variation among villages when population size is small.

Experiment

Question: What effect does genetic drift have on the genetic composition of populations?

Methods Buri examined the frequencies of two alleles (bw^{75} and *bw*) that affect *Drosophila* eye color in 107 replicate small populations over 19 generations.

Results

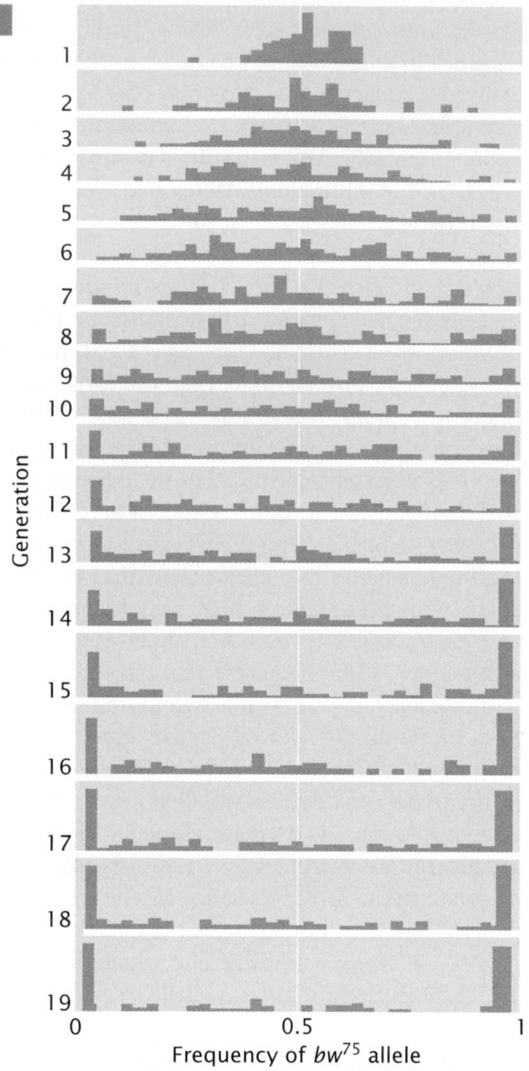

Conclusion: As a result of genetic drift, allelic frequencies in the different populations diverged and often became fixed for one allele or the other.

25.11 Populations diverge in allelic frequency and become fixed for one allele as a result of genetic drift. In Buri's study of two eye-color alleles (bw^{75} and *bw*) in *Drosophila*, each population consisted of 8 males and 8 females and began with the frequency of bw^{75} equal to 0.5.

For ecological and demographic studies, population size is usually defined as the number of individuals in a group. The evolution of a gene pool depends, however, only on those individuals who contribute genes to the next generation. Population geneticists usually define population size as the equivalent number of breeding adults, the **effective population size** (N_e). Several factors determine the equivalent number of breeding adults, including the sex ratio, variation between individuals in reproductive success, fluctuations in population size, the age structure of the population, and whether mating is random.

CONCEPTS

Genetic drift is change in allelic frequency due to chance factors. The amount of change in allelic frequency due to genetic drift is inversely related to the effective population size (the equivalent number of breeding adults in a population).

✔ CONCEPT CHECK 8

Which of the following statements is an example of genetic drift?

a. Allele *g* for fat production increases in a small population because birds with more body fat have higher survivorship in a harsh winter.

b. Random mutation increases the frequency of allele *A* in one population but not in another.

c. Allele *R* reaches a frequency of 1.0 because individuals with genotype *rr* are sterile.

d. Allele *m* is lost when a virus kills all but a few individuals and just by chance none of the survivors possess allele *m*.

Causes of genetic drift All genetic drift arises from sampling error, but there are several different ways in which sampling error can arise. First, a population may be reduced in size for a number of generations because of limitations in space, food, or some other critical resource. Genetic drift in a small population for multiple generations can significantly affect the composition of a population's gene pool.

A second way that sampling error can arise is through the **founder effect**, which is due to the establishment of a population by a small number of individuals; the population of bighorn sheep at the National Bison Range, discussed in the introduction to this chapter, underwent a founder effect. Although a population may increase and become quite large, the genes carried by all its members are derived from the few genes originally present in the founders (assuming no migration or mutation). Chance events affecting which genes were present in the founders will have an important influence on the makeup of the entire population.

A third way in which genetic drift arises is through a **genetic bottleneck**, which develops when a population undergoes a drastic reduction in population size. A genetic bottleneck developed in northern elephant seals (**Figure 25.12**). Before 1800, thousands of elephant seals were found along the California coast, but the population was devastated by hunting between 1820 and 1880. By 1884, as few as 20 seals survived on a remote beach of Isla de Guadelupe west of Baja, California. Restrictions on hunting enacted by the United States and Mexico allowed the seals to recover, and there are now more

25.12 Northern elephant seals underwent a severe genetic bottleneck between 1820 and 1880. Today, these seals have low levels of genetic variation. [PhotoDisc/Getty Images.]

than 30,000 seals in the population. All seals in the population today are genetically similar, because they have genes that were carried by the few survivors of the population bottleneck.

The effects of genetic drift Genetic drift has several important effects on the genetic composition of a population. First, it produces change in allelic frequencies within a population. Because drift is random, allelic frequency is just as likely to increase as it is to decrease and will wander with the passage of time (hence the name genetic drift). **Figure 25.13** illustrates a computer simulation of genetic drift in five populations over 30 generations, starting with $q = 0.5$ and maintaining a constant population size of 10 males and 10 females. These allelic frequencies change randomly from generation to generation.

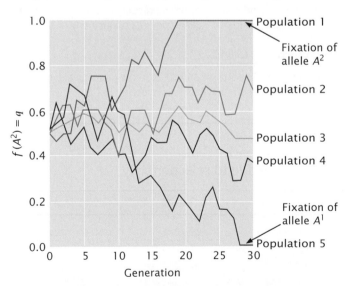

25.13 Genetic drift changes allelic frequencies within populations, leading to a reduction in genetic variation through fixation and genetic divergence among populations. Shown here is a computer simulation of changes in the frequency of allele A^2 (q) in five different populations due to random genetic drift. Each population consists of 10 males and 10 females and begins with $q = 0.5$.

A second effect of genetic drift is to reduce genetic variation within populations. Through random change, an allele may eventually reach a frequency of either 1 or 0, at which point all individuals in the population are homozygous for one allele. When an allele has reached a frequency of 1, we say that it has reached **fixation**. Other alleles are lost (reach a frequency of 0) and can be restored only by migration from another population or by mutation. Fixation, then, leads to a loss of genetic variation within a population. This loss can be seen in the northern elephant seals described earlier. Today, these seals have low levels of genetic variation; a study of 24 protein-encoding genes found no individual or population differences in these genes. A subsequent study of sequence variation in mitochondrial DNA also revealed low levels of genetic variation. In contrast, the southern elephant seals had much higher levels of mitochondrial DNA variation. The southern elephant seals also were hunted, but their population size never dropped below 1000; therefore, unlike the northern elephant seals, they did not experience a genetic bottleneck.

Given enough time, all small populations will become fixed for one allele or the other. Which allele becomes fixed is random and is determined by the initial frequency of the allele. If a population begins with two alleles, each with a frequency of 0.5, both alleles have an equal probability of fixation. However, if one allele is initially common, it is more likely to become fixed.

A third effect of genetic drift is that different populations diverge genetically with time. In Figure 25.13, all five populations begin with the same allelic frequency ($q = 0.5$) but, because drift is random, the frequencies in different populations do not change in the same way, and so populations gradually acquire genetic differences. Eventually, all the populations reach fixation; some will become fixed for one allele, and others will become fixed for the alternative allele.

The three results of genetic drift (allelic frequency change, loss of variation within populations, and genetic divergence between populations) take place simultaneously, and all result from sampling error. The first two results take place *within* populations, whereas the third takes place *between* populations. TRY PROBLEM 32 →

CONCEPTS

Genetic drift results from continuous small population size, the founder effect (establishment of a population by a few founders), and the bottleneck effect (population reduction). Genetic drift causes a change in allelic frequencies within a population, a loss of genetic variation through the fixation of alleles, and genetic divergence between populations.

Natural Selection

A final process that brings about changes in allelic frequencies is natural selection, the differential reproduction of genotypes (see p. 680 in Chapter 24). Natural selection

25.14 Natural selection produces adaptations, such as those seen in the polar bears that inhabit the extreme Arctic environment. These bears blend into the snowy background, which helps them in hunting seals. The hairs of their fur stay erect even when wet, and thick layers of blubber provide insulation, which protects against subzero temperatures. Their digestive tracts are adapted to a seal-based carnivorous diet. [Digital Vision.]

takes place when individuals with adaptive traits produce a greater number of offspring than that produced by others in the population. If the adaptive traits have a genetic basis, they are inherited by the offspring and appear with greater frequency in the next generation. A trait that provides a reproductive advantage thereby increases with the passage of time, enabling populations to become better suited to their environments—to become better adapted. Natural selection is unique among evolutionary forces in that it promotes adaptation (**Figure 25.14**).

Fitness and the selection coefficient The effect of natural selection on the gene pool of a population depends on the fitness values of the genotypes in the population. **Fitness** is defined as the relative reproductive success of a genotype. Here, the term *relative* is critical: fitness is the reproductive success of one genotype compared with the reproductive successes of other genotypes in the population.

Fitness (W) ranges from 0 to 1. Suppose the average number of viable offspring produced by three genotypes is

Genotypes:	A^1A^1	A^1A^2	A^2A^2
Mean number of offspring produced:	10	5	2

To calculate fitness for each genotype, we take the mean number of offspring produced by a genotype and divide it by the mean number of offspring produced by the most prolific genotype:

Fitness (W):

$$A^1A^1 \qquad\qquad A^1A^2 \qquad\qquad A^2A^2$$

$$W_{11} = \frac{10}{10} = 1.0 \quad W_{12} = \frac{5}{10} = 0.5 \quad W_{22} = \frac{2}{10} = 0.2 \quad (25.17)$$

The fitness of genotype A^1A^1 is designated W_{11}, that of A^1A^2 is W_{12}, and that of A^2A^2 is W_{22}. A related variable is the **selection coefficient** (s), which is the relative intensity of selection against a genotype. We usually speak of selection for a particular genotype, but keep in mind that, when selection is *for* one genotype, selection is automatically *against* at least one other genotype. The selection coefficient is equal to $1 - W$; so the selection coefficients for the preceding three genotypes are

$$\qquad\qquad\qquad A^1A^1 \qquad A^1A^2 \qquad A^2A^2$$
Selection coefficient $(1 - W)$: $\quad s_{11} = 0 \quad s_{12} = 0.5 \quad s_{22} = 0.8$

CONCEPTS

Natural selection is the differential reproduction of genotypes. It is measured as fitness, which is the reproductive success of a genotype compared with other genotypes in a population.

✔ **CONCEPT CHECK 9**

The average numbers of offspring produced by three genotypes are: $GG = 6$; $Gg = 3$, $gg = 2$. What is the fitness of Gg?

a. 3   b. 0.5   c. 0.3   d. 0.27

Table 25.4 Method for determining changes in allelic frequency due to selection

	A^1A^1	A^1A^2	A^2A^2
Initial genotypic frequencies	p^2	$2pq$	q^2
Fitnesses	W_{11}	W_{12}	W_{22}
Proportionate contribution of genotypes to population	p^2W_{11}	$2pqW_{12}$	q^2W_{22}
Relative genotypic frequency after selection	$\dfrac{p^2W_{11}}{\bar{w}}$	$\dfrac{2pqW_{12}}{\bar{w}}$	$\dfrac{q^2W_{22}}{\bar{w}}$

Note: $\bar{w} = p^2W_{11} + 2pqW_{12} + q^2W_{22}$

Allelic frequencies after selection: $p' = f(A^1) = f(A^1A^1) + \frac{1}{2}f(A^1A^2)$; $q' = 1 - p$.

Table 25.5 Formulas for calculating change in allelic frequencies with different types of selection

Type of Selection	Fitness Values			
	A^1A^1	A^1A^2	A^2A^2	Change in q
Selection against a recessive trait	1	1	$1-s$	$\dfrac{-spq^2}{1-sq^2}$
Selection against a dominant trait	1	$1-s$	$1-s$	$\dfrac{-spq^2}{1-s+sq^2}$
Selection against a trait with no dominance	1	$1-\frac{1}{2}s$	$1-s$	$\dfrac{-\frac{1}{2}spq}{1-sa}$
Selection against both homozygotes (overdominance)	$1-s_{11}$	1	$1-s_{22}$	$\dfrac{pq(s_{11}p-s_{22}q)}{1-s_{11}p^2-s_{22}q^2}$

The general selection model   With selection, differential fitness among genotypes leads to changes in the frequencies of the genotypes over time which, in turn, lead to changes in the frequencies of the alleles that make up the genotypes. We can predict the effect of natural selection on allelic frequencies by using a general selection model, which is outlined in **Table 25.4**. Use of this model requires knowledge of both the initial allelic frequencies and the fitness values of the genotypes. It assumes that mating is random and that the only force acting on a population is natural selection. The general selection model can be used to calculate the allelic frequencies after any type of selection. It is also possible to work out formulas for determining the change in allelic frequency when selection is against recessive, dominant, and codominant traits, as well as traits in which the heterozygote has highest fitness (**Table 25.5**).

CONCEPTS

The change in allelic frequency due to selection can be determined for any type of genetic trait by using the general selection model.

Worked Problem

Let's apply the general selection model in Table 25.4 to a real example. Alcohol is a common substance in rotting fruit, where fruit-fly larvae grow and develop; larvae use the enzyme alcohol dehydrogenase (ADH) to detoxify the effects of the alcohol. In some fruit-fly populations, two alleles are present at the locus that encodes ADH: Adh^F, which encodes a form of the enzyme that migrates rapidly (fast) on an electrophoretic gel; and Adh^S, which encodes a form of the enzyme that migrates slowly on an electro-

phoretic gel. Female fruit flies with different Adh genotypes produce the following numbers of offspring when alcohol is present:

Genotype	Mean number of offspring
Adh^F/Adh^F	120
Adh^F/Adh^S	60
Adh^S/Adh^S	30

a. Calculate the relative fitnesses of females having these genotypes.

b. If a population of fruit flies has an initial frequency of Adh^F equal to 0.2, what will the frequency be in the next generation when alcohol is present?

• **Solution**

a. First, we must calculate the fitnesses of the three genotypes. Fitness is the relative reproductive output of a genotype and is calculated by dividing the mean number of offspring produced by that genotype by the mean number of offspring produced by the most prolific genotype. The fitnesses of the three Adh genotypes therefore are:

Genotype	Mean number of offspring	Fitness
Adh^F/Adh^F	120	$W_{FF}=\dfrac{120}{120}=1$
Adh^F/Adh^S	60	$W_{FS}=\dfrac{60}{120}=0.5$
Adh^S/Adh^S	30	$W_{SS}=\dfrac{30}{120}=0.25$

	$Adh^F Adh^F$	$Adh^F Adh^S$	$Adh^S Adh^S$
Initial genotypic frequencies:	$p^2 = (0.2)^2 = 0.04$	$2pq = 2(0.2)(0.8) = 0.32$	$q^2 = (0.8)^2 = 0.64$
Fitnesses:	$W_{FF} = 1$	$W_{FS} = 0.5$	$W_{SS} = 0.25$
Proportionate contribution of genotypes to population:	$p^2 W_{FF} = 0.04(1) = 0.04$	$2pq W_{FS} = (0.32)(0.5) = 0.16$	$q^2 W_{SS} = (0.64)(0.25) = 0.16$

b. To calculate the frequency of the Adh^F allele after selection, we can apply the table method. In the first row of the table above, we record the initial genotypic frequencies before selection has acted. If mating has been random (an assumption of the model), the genotypes will have the Hardy–Weinberg equilibrium frequencies of p^2, $2pq$, and q^2. In the second row of the table above, we put the fitness values of the corresponding genotypes. The proportion of the population represented by each genotype after selection is obtained by multiplying the initial genotypic frequency times its fitness (third row of Table 25.4). Now the genotypes are no longer in Hardy–Weinberg equilibrium.

The mean fitness ($\bar{w}$) of the population is the sum of the proportionate contributions of the three genotypes: $\bar{w} = p^2 W_{11} + 2pq W_{12} + q^2 W_{22} = 0.04 + 0.16 + 0.16 = 0.36$. The mean fitness $\bar{w}$ is the average fitness of all individuals in the population and allows the frequencies of the genotypes after selection to be obtained.

The frequency of a genotype after selection will be equal to its proportionate contribution divided by the mean fitness of the population ($p^2 W_{11}/\bar{w}$ for genotype A^1A^1, $2pq W_{12}/\bar{w}$ for genotype A^1A^2, and $q^2 W_{22}/\bar{w}$ for genotype A^2A^2) as shown in the fourth line of Table 25.4. We can now add these values to our table as shown below:

	$Adh^F Adh^F$	$Adh^F Adh^S$	$Adh^S Adh^S$
Initial genotypic frequencies:	$p^2 = (0.2)^2 = 0.04$	$2pq = 2(0.2)(0.8) = 0.32$	$q^2 = (0.8)^2 = 0.64$
Fitnesses:	$W_{FF} = 1$	$W_{FS} = 0.5$	$W_{SS} = 0.25$
Proportionate contribution of genotypes to population:	$p^2 W_{FF} = 0.04(1) = 0.04$	$2pq W_{FS} = (0.32)(0.5) = 0.16$	$q^2 W_{SS} = (0.64)(0.25) = 0.16$
Relative genotypic frequency after selection:	$\dfrac{p^2 w_{FF}}{\bar{w}} = \dfrac{0.04}{0.36} = 0.11$	$\dfrac{2pq w_{FS}}{\bar{w}} = \dfrac{0.16}{0.36} = 0.44$	$\dfrac{q^2 w_{ss}}{\bar{w}} = \dfrac{0.16}{0.36} = 0.44$

After the new genotypic frequencies have been calculated, the new allelic frequency of Adh^F (p') can be determined by using the now-familiar formula of Equation 25.4:

$$p' = f(Adh^F) = f(Adh^F/Adh^F) + \tfrac{1}{2} f(Adh^F/Adh^S)$$
$$= 0.11 + \tfrac{1}{2}(0.44) = 0.33$$

and that of q' can be obtained by subtraction:

$$q' = 1 - p'$$
$$= 1 - 0.33 = 0.67$$

We predict that the frequency of Adh^F will increase from 0.2 to 0.33.

> → For more practice with the selection model, try Problem 33 at the end of the chapter.

The results of selection   The results of selection depend on the relative fitnesses of the genotypes. If we have three genotypes (A^1A^1, A^1A^2, and A^2A^2) with fitnesses W_{11}, W_{12}, and W_{22}, we can identify six different types of natural selection (**Table 25.6**). In type 1 selection, a dominant allele A^1

confers a fitness advantage; in this case, the fitnesses of genotypes A^1A^1 and A^1A^2 are equal and higher than the fitness of A^2A^2 ($W_{11} = W_{12} > W_{22}$). Because both the heterozygote and the A^1A^1 homozygote have copies of the A^1 allele and produce more offspring than the A^2A^2 homozygote does, the frequency of the A^1 allele will increase with time, and the frequency of the A^2 allele will decrease. This form of selection, in which one allele or trait is favored over another, is termed **directional selection**.

Type 2 selection (see Table 25.6) is directional selection against a dominant allele A^1 ($W_{11} = W_{12} < W_{22}$). In this case, the A^2 allele increases and the A^1 allele decreases. Type 3 and type 4 selection also are directional selection but, in these cases, there is incomplete dominance and the heterozygote has a fitness that is intermediate between the two homozygotes ($W_{11} > W_{12} > W_{22}$ for type 3; $W_{11} < W_{12} < W_{22}$ for type 4). When A^1A^1 has the highest fitness (type 3), the A^1 allele increases and the A^2 allele decreases with the passage of time. When A^2A^2 has the highest fitness (type 4), the A^2 allele increases and the A^1 allele decreases with time. Eventually, directional selection leads to fixation of the favored allele and elimination of the other allele, as long as no other evolutionary forces act on the population.

Table 25.6 Types of natural selection

Type	Fitness Relation	Form of Selection	Result
1	$W_{11} = W_{12} > W_{22}$	Directional selection against recessive allele A^2	A^1 increases, A^2 decreases
2	$W_{11} = W_{12} < W_{22}$	Directional selection against dominant allele A^1	A^2 increases, A^1 decreases
3	$W_{11} > W_{12} > W_{22}$	Directional selection against incompletely dominant allele A^2	A^1 increases, A^2 decreases
4	$W_{11} < W_{12} < W_{22}$	Directional selection against incompletely dominant allele A^1	A^2 increases, A^1 decreases
5	$W_{11} < W_{12} > W_{22}$	Overdominance	Stable equilibrium, both alleles maintained
6	$W_{11} > W_{12} < W_{22}$	Underdominance	Unstable equilibrium

Note: W_{11}, W_{12}, and W_{22} represent the fitnesses of genotypes A^1A^1, A^1A^2, and A^2A^2, respectively.

 **www.whfreeman.com/pierce4e**   To see the effects of natural selection on allelic and genotypic frequencies, view the Mini-Tutorial in Animation 25.1.

Two types of selection (types 5 and 6) are special situations that lead to equilibrium, where there is no further change in allelic frequency. Type 5 selection is referred to as **overdominance** or heterozygote advantage. Here, the heterozygote has higher fitness than the fitnesses of the two homozygotes ($W_{11} < W_{12} > W_{22}$). With overdominance, both alleles are favored in the heterozygote, and neither allele is eliminated from the population. Initially, the allelic frequencies may change because one homozygote has higher fitness than the other; the direction of change will depend on the relative fitness values of the two homozygotes. The allelic frequencies change with overdominant selection until a stable equilibrium is reached, at which point there is no further change. The allelic frequency at equilibrium ($\hat{q}$) depends on the relative fitnesses (usually expressed as selection coefficients) of the two homozygotes:

$$\hat{q} = f(A^2) = \frac{s_{11}}{s_{11} + s_{22}} \qquad (25.18)$$

where s_{11} represents the selection coefficient of the A^1A^1 homozygote and s_{22} represents the selection coefficient of the A^2A^2 homozygote.

The last type of selection (type 6) is **underdominance**, in which the heterozygote has lower fitness than both homozygotes ($W_{11} > W_{12} < W_{22}$). Underdominance leads to an unstable equilibrium; here, allelic frequencies will not change as long as they are at equilibrium but, if they are disturbed from the equilibrium point by some other evolutionary force, they will move away from equilibrium until one allele eventually becomes fixed. **TRY PROBLEM 34** →

CONCEPTS

Natural selection changes allelic frequencies; the direction and magnitude of change depend on the intensity of selection, the dominance relations of the alleles, and the allelic frequencies. Directional selection favors one allele over another and eventually leads to fixation of the favored allele. Overdominance leads to a stable equilibrium with maintenance of both alleles in the population. Underdominance produces an unstable equilibrium because the heterozygote has lower fitness than those of the two homozygotes.

✔ **CONCEPT CHECK 10**

How does overdominance differ from directional selection?

Change in the allelic frequency of a recessive allele due to natural selection The rate at which selection changes allelic frequencies depends on the allelic frequency itself. If an allele (A^2) is lethal and recessive, $W_{11} = W_{12} = 1$, whereas $W_{22} = 0$. The frequency of the A^2 allele will decrease with time (because the A^2A^2 homozygote produces no offspring), and the rate of decrease will be proportional to the frequency of the recessive allele. When the frequency of the allele is high, the change in each generation is relatively large but, as the frequency of the allele drops, a higher proportion of the alleles are in the heterozygous genotypes, where they are immune to the action of natural selection (the heterozygotes have the same phenotype as the favored homozygote). Thus, selection against a rare recessive allele is very inefficient and its removal from the population is slow.

The relation between the frequency of a recessive allele and its rate of change under natural selection has an important implication. Some people believe that the medical treatment of patients with rare recessive diseases will cause the disease gene to increase, eventually leading to degeneration of the human gene pool. This mistaken belief was the basis of eugenic laws that were passed in the early part of the twentieth century prohibiting the marriage of persons with certain genetic conditions and allowing the involuntary sterilization of others. However, most copies of rare recessive alleles are present in heterozygotes, and selection against the homozygotes will have little effect on the frequency of a recessive allele. Thus, whether the homozygotes for a recessive trait reproduce or not has little effect on the frequency of the disorder.

Mutation and natural selection Recurrent mutation and natural selection act as opposing forces on detrimental alleles; mutation increases their frequency and natural selection decreases their frequency. Eventually, these two forces reach an equilibrium, in which the number of alleles added by mutation is balanced by the number of alleles removed by selection.

The frequency of a recessive allele at equilibrium ($\hat{q}$) is equal to the square root of the mutation rate divided by the selection coefficient:

$$\hat{q} = \sqrt{\frac{\mu}{s}} \qquad (25.19)$$

For selection acting on a dominant allele, the frequency of the dominant allele at equilibrium can be shown to be

$$\hat{q} = \frac{\mu}{s} \qquad (25.20)$$

Achondroplasia is a common type of human dwarfism that results from a dominant gene. People with this condition are fertile, although they produce only about 74% as many children as are produced by people without achondroplasia. The fitness of people with achondroplasia therefore averages 0.74, and the selection coefficient (s) is $1 - W$, or 0.26. If we assume that the mutation rate for achondroplasia is about 3×10^{-5} (a typical mutation rate in humans), then

we can predict that the equilibrium frequency for the achondroplasia allele will be

$$\hat{q} = \frac{0.00003}{0.26} = 0.0001153$$

This frequency is close to the actual frequency of the disease.

TRY PROBLEM 37 →

CONCEPTS

Mutation and natural selection act as opposing forces on detrimental alleles: mutation tends to increase their frequency and natural selection tends to decrease their frequency, eventually producing an equilibrium.

CONNECTING CONCEPTS

The General Effects of Forces That Change Allelic Frequencies

You now know that four processes bring about change in the allelic frequencies of a population: mutation, migration, genetic drift, and natural selection. Their short- and long-term effects on allelic frequencies are summarized in **Table 25.7**. In some cases, these changes continue until one allele is eliminated and the other becomes fixed in the population. Genetic drift and directional selection will eventually result in fixation, provided these forces are the only ones acting on a population. With the other evolutionary forces, allelic frequencies change until an equilibrium point is reached, and then there is no additional change in allelic frequency. Mutation, migration, and some forms of natural selection can lead to stable equilibria (see Table 25.7).

The different evolutionary forces affect both genetic variation within populations and genetic divergence between populations. Evolutionary forces that maintain or increase genetic variation within populations are listed in the upper-left quadrant of **Figure 25.15**. These forces include some types of natural selection, such as overdominance, in which both alleles are favored. Mutation and migration also increase genetic variation within populations because they introduce new alleles to the population. Evolutionary forces that decrease genetic variation within populations are listed in the lower-left quadrant of Figure 25.15. These forces include genetic drift, which decreases variation through the fixation of

Table 25.7	Effects of different evolutionary forces on allelic frequencies within populations	
Force	**Short-Term Effect**	**Long-Term Effect**
Mutation	Change in allelic frequency	Equilibrium reached between forward and reverse mutations
Migration	Change in allelic frequency	Equilibrium reached when allelic frequencies of source and recipient population are equal
Genetic drift	Change in allelic frequency	Fixation of one allele
Natural selection	Change in allelic frequency	Directional selection: fixation of one allele Overdominant selection: equilibrium reached

	Within populations	Between populations
Increase genetic variation	**Mutation Migration Some types of natural selection**	**Mutation Genetic drift Some types of natural selection**
Decrease genetic variation	**Genetic drift Some types of natural selection**	**Migration Some types of natural selection**

25.15 Mutation, migration, genetic drift, and natural selection have different effects on genetic variation within populations and on genetic divergence between populations.

alleles, and some forms of natural selection such as directional selection.

The various evolutionary forces also affect the amount of genetic divergence between populations. Natural selection increases divergence between populations if different alleles are favored in the different populations, but it can also decrease divergence between populations by favoring the same allele in the different populations. Mutation almost always increases diver-

gence between populations because different mutations arise in each population. Genetic drift also increases divergence between populations because changes in allelic frequencies due to drift are random and are likely to change in different directions in separate populations. Migration, on the other hand, decreases divergence between populations because it makes populations similar in their genetic composition.

Migration and genetic drift act in opposite directions: migration increases genetic variation within populations and decreases divergence between populations, whereas genetic drift decreases genetic variation within populations and increases divergence among populations. Mutation increases both variation within populations and divergence between populations. Natural selection can either increase or decrease variation within populations, and it can increase or decrease divergence between populations.

An important point to keep in mind is that real populations are simultaneously affected by many evolutionary forces. We have examined the effects of mutation, migration, genetic drift, and natural selection in isolation so that the influence of each process would be clear. However, in the real world, populations are commonly affected by several evolutionary forces at the same time, and evolution results from the complex interplay of numerous processes.

CONCEPTS SUMMARY

- Population genetics examines the genetic composition of groups of individuals and how this composition changes with time.

- A Mendelian population is a group of interbreeding, sexually reproducing individuals, whose set of genes constitutes the population's gene pool. Evolution takes place through changes in this gene pool.

- A population's genetic composition can be described by its genotypic and allelic frequencies.

- The Hardy–Weinberg law describes the effects of reproduction and Mendel's laws on the allelic and genotypic frequencies of a population. It assumes that a population is large, randomly mating, and free from the effects of mutation, migration, and natural selection. When these conditions are met, the allelic frequencies do not change and the genotypic frequencies stabilize after one generation in the Hardy–Weinberg equilibrium proportions p^2, $2pq$, and q^2, where p and q equal the frequencies of the alleles.

- Nonrandom mating affects the frequencies of genotypes but not those of alleles.

- Inbreeding, a type of positive assortative mating, increases the frequency of homozygotes while decreasing the frequency of heterozygotes. Inbreeding is frequently detrimental because it increases the appearance of lethal and deleterious recessive traits.

- Mutation, migration, genetic drift, and natural selection can change allelic frequencies.

- Recurrent mutation eventually leads to an equilibrium, with the allelic frequencies being determined by the relative rates of forward and reverse mutation. Change due to mutation in a single generation is usually very small because mutation rates are low.

- Migration, the movement of genes between populations, increases the amount of genetic variation within populations and decreases the number of differences between populations.

- Genetic drift is change in allelic frequencies due to chance factors. Genetic drift arises when a population consists of a small number of individuals, is established by a small number of founders, or undergoes a major reduction in size. Genetic drift changes allelic frequencies, reduces genetic variation within populations, and causes genetic divergence among populations.

- Natural selection is the differential reproduction of genotypes; it is measured by the relative reproductive successes (fitnesses) of genotypes. The effects of natural selection on allelic frequency can be determined by applying the general selection model. Directional selection leads to the fixation of one allele. The rate of change in allelic frequency due to selection depends on the intensity of selection, the dominance relations, and the initial frequencies of the alleles.

- Mutation and natural selection can produce an equilibrium, in which the number of new alleles introduced by mutation is balanced by the elimination of alleles through natural selection.

IMPORTANT TERMS

genetic rescue (p. 694)
Mendelian population (p. 694)
gene pool (p. 694)
genotypic frequency (p. 695)
allelic frequency (p. 695)
Hardy–Weinberg law (p. 697)
Hardy–Weinberg equilibrium (p. 697)
positive assortative mating (p. 701)
negative assortative mating (p. 701)

inbreeding (p. 701)
outcrossing (p. 701)
inbreeding coefficient (p. 701)
inbreeding depression (p. 703)
equilibrium (p. 705)
migration (gene flow) (p. 705)
sampling error (p. 707)
genetic drift (p. 707)
effective population size (p. 708)

founder effect (p. 708)
genetic bottleneck (p. 708)
fixation (p. 709)
fitness (p. 710)
selection coefficient (p. 710)
directional selection (p. 712)
overdominance (p. 713)
underdominance (p. 713)

ANSWERS TO CONCEPT CHECKS

1. There are fewer alleles than genotypes, and so the gene pool can be described by fewer parameters when allelic frequencies are used. Additionally, the genotypes are temporary assemblages of alleles that break down each generation; rather than the genotypes, the alleles are passed from generation to generation in sexually reproducing organisms.

2. a

3. c

4. 0.10

5. b

6. c

7. c

8. d

9. b

10. In overdominance, the heterozygote has the highest fitness. In directional selection, one allele or one trait is favored over another.

WORKED PROBLEM

1. A recessive allele for red hair (r) has a frequency of 0.2 in population I and a frequency of 0.01 in population II. A famine in population I causes a number of people in population I to migrate to population II, where they reproduce randomly with the members of population II. Geneticists estimate that, after migration, 15% of the people in population II consist of people who migrated from population I. What will be the frequency of red hair in population II after the migration?

• Solution

From Equation 25.14, the allelic frequency in a population after migration (q'_{II}) is

$$q'_{II} = q_I(m) + q_{II}(1 - m)$$

where q_I and q_{II} are the allelic frequencies in population I (migrants) and population II (residents), respectively, and m is the proportion of population II that consist of migrants. In this problem, the frequency of red hair is 0.2 in population I and 0.01 in population II. Because 15% of population II consists of migrants, $m = 0.15$. Substituting these values into Equation 25.14, we obtain

$$q'_{II} = 0.2(0.15) + (0.01)(1 - 0.15) = 0.03 + 0.0085 = 0.0385$$

which is the expected frequency of the allele for red hair in population II after migration. Red hair is a recessive trait; if mating is random for hair color, the frequency of red hair in population II after migration will be

$$f(rr) = q^2 = (0.0385)^2 = 0.0015$$

COMPREHENSION QUESTIONS

Section 25.1

1. What is a Mendelian population? How is the gene pool of a Mendelian population usually described?

Section 25.2

2. What are the predictions given by the Hardy–Weinberg law?

*3. What assumptions must be met for a population to be in Hardy–Weinberg equilibrium?

4. What is random mating?

*5. Give the Hardy–Weinberg expected genotypic frequencies for (a) an autosomal locus with three alleles, and (b) an X-linked locus with two alleles.

Section 25.3

6. Define inbreeding and briefly describe its effects on a population.

Section 25.4

7. What determines the allelic frequencies at mutational equilibrium?

*8. What factors affect the magnitude of change in allelic frequencies due to migration?

9. Define genetic drift and give three ways in which it can arise. What effect does genetic drift have on a population?

*10. What is effective population size? How does it affect the amount of genetic drift?

11. Define natural selection and fitness.

12. Briefly describe the differences between directional selection, overdominance, and underdominance. Describe the effect of each type of selection on the allelic frequencies of a population.

13. What factors affect the rate of change in allelic frequency due to natural selection?

*14. Compare and contrast the effects of mutation, migration, genetic drift, and natural selection on genetic variation within populations and on genetic divergence between populations.

APPLICATION QUESTIONS AND PROBLEMS

Section 25.1

15. How would you respond to someone who said that models are useless in studying population genetics because they represent oversimplifications of the real world?

*16. Voles (*Microtus ochrogaster*) were trapped in old fields in southern Indiana and were genotyped for a transferrin locus. The following numbers of genotypes were recorded, where T^E and T^F represent different alleles.

[Tom McHugh/Photo Reseachers]

$T^E T^E$	$T^E T^F$	$T^F T^F$
407	170	17

Calculate the genotypic and allelic frequencies of the transferrin locus for this population.

17. Jean Manning, Charles Kerfoot, and Edward Berger 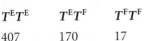 studied genotypic frequencies at the phosphoglucose isomerase (GPI) locus in the cladoceran *Bosmina longirostris*. At one location, they collected 176 animals from Union Bay in Seattle, Washington, and determined their GPI genotypes by using electrophoresis (J. Manning, W. C. Kerfoot, and E. M. Berger. 1978. *Evolution* 32:365–374).

Genotype	Number
$S^1 S^1$	4
$S^1 S^2$	38
$S^2 S^2$	134

Determine the genotypic and allelic frequencies for this population.

18. Orange coat color of cats is due to an X-linked allele (X^O) that is codominant with the allele for black (X^+). Genotypes of the orange locus of cats in Minneapolis and St. Paul, Minnesota, were determined, and the following data were obtained:

$X^O X^O$ females	11
$X^O X^+$ females	70
$X^+ X^+$ females	94
$X^O Y$ males	36
$X^+ Y$ males	112

Calculate the frequencies of the X^O and X^+ alleles for this population.

Section 25.2

19. A total of 6129 North American Caucasians were blood typed for the MN locus, which is determined by two codominant alleles, L^M and L^N. The following data were obtained:

Blood type	Number
M	1787
MN	3039
N	1303

Carry out a chi-square test to determine whether this population is in Hardy–Weinberg equilibrium at the MN locus.

20. Most black bears (*Ursus americanus*) are black or brown in color. However, occasional white bears of this species appear in some populations along the coast of British Columbia. Kermit Ritland and his colleagues determined that white coat color in these bears results from a recessive mutation (*G*)

[Marc Muench/Alamy.]

caused by a single nucleotide replacement in which guanine substitutes for adenine at the melanocortin-1 receptor locus (*mcr1*), the same locus responsible for red hair in humans (K. Ritland, C. Newton, and H. D. Marshall. 2001. *Current Biology* 11:1468–1472). The wild-type allele at this locus (*A*) encodes black or brown color. Ritland and his colleagues collected samples from bears on three islands and determined their genotypes at the *mcr1* locus.

Genotype	Number
AA	42
AG	24
GG	21

a. What are the frequencies of the *A* and *G* alleles in these bears?

b. Give the genotypic frequencies expected if the population is in Hardy–Weinberg equilibrium.

c. Use a chi-square test to compare the number of observed genotypes with the number expected under Hardy–Weinberg equilibrium. Is this population in Hardy–Weinberg equilibrium? Explain your reasoning.

21. Genotypes of leopard frogs from a population in central Kansas were determined for a locus (*M*) that encodes the enzyme malate dehydrogenase. The following numbers of genotypes were observed:

Genotype	Number
M^1M^1	20
M^1M^2	45
M^2M^2	42
M^1M^3	4
M^2M^3	8
M^3M^3	6
Total	125

a. Calculate the genotypic and allelic frequencies for this population.

b. What would the expected numbers of genotypes be if the population were in Hardy–Weinberg equilibrium?

22. Full color (*D*) in domestic cats is dominant over dilute color (*d*). Of 325 cats observed, 194 have full color and 131 have dilute color.

a. If these cats are in Hardy–Weinberg equilibrium for the dilution locus, what is the frequency of the dilute allele?

b. How many of the 194 cats with full color are likely to be heterozygous?

23. Tay–Sachs disease is an autosomal recessive disorder. Among Ashkenazi Jews, the frequency of Tay–Sachs disease is 1 in 3600. If the Ashkenazi population is mating randomly for the Tay–Sachs gene, what proportion of the population consists of heterozygous carriers of the Tay–Sachs allele?

24. In the plant *Lotus corniculatus*, cyanogenic glycoside protects the plant against insect pests and even grazing by cattle. This glycoside is due to a simple dominant allele. A population of *L. corniculatus* consists of 77 plants that possess cyanogenic glycoside and 56 that lack the compound. What is the frequency of the dominant allele responsible for the presence of cyanogenic glycoside in this population?

*25. Color blindness in humans is an X-linked recessive trait. Approximately 10% of the men in a particular population are color blind.

a. If mating is random for the color-blind locus, what is the frequency of the color-blind allele in this population?

b. What proportion of the women in this population are expected to be color blind?

c. What proportion of the women in the population are expected to be heterozygous carriers of the color-blind allele?

Section 25.3

*26. The human MN blood type is determined by two codominant alleles, L^M and L^N. The frequency of L^M in Eskimos on a small Arctic island is 0.80. If the inbreeding coefficient for this population is 0.05, what are the expected frequencies of the M, MN, and N blood types on the island?

27. Demonstrate mathematically that full-sib mating ($F = {}^1\!/_4$) reduces the heterozygosity by ${}^1\!/_4$ with each generation.

Section 25.4

28. The forward mutation rate for piebald spotting in guinea pigs is 8×10^{-5}; the reverse mutation rate is 2×10^{-6}. If no other evolutionary forces are assumed to be present, what is the expected frequency of the allele for piebald spotting in a population that is in mutational equilibrium?

29. For a period of 3 years, Gunther Schlager and Margaret Dickie estimated the forward and reverse mutation rates for five loci in mice that encode various aspects of coat color by examining more than 5 million mice for spontaneous mutations (G. Schlager and M. M. Dickie. 1966. *Science* 151:205–206). The numbers of mutations detected at the dilute locus are as follows:

	Number of gametes examined	Number of mutations detected
Forward mutations	260,675	5
Reverse mutations	583,360	2

Calculate the forward and reverse mutation rates at this locus. If these mutations rates are representative of rates in natural populations of mice, what would the expected equilibrium frequency of dilute mutations be?

*30. In German cockroaches, curved wing (*cv*) is recessive to normal wing (*cv⁺*). Bill, who is raising cockroaches in his dorm room, finds that the frequency of the gene for curved wings in his cockroach population is 0.6. In the apartment of his friend Joe, the frequency of the gene for curved wings is 0.2. One day Joe visits Bill in his dorm room, and several cockroaches jump out of Joe's hair and join the population in Bill's room. Bill estimates that, now, 10% of the cockroaches in his dorm room are individual roaches that jumped out of Joe's hair. What is the new frequency of curved wings among cockroaches in Bill's room?

31. A population of water snakes is found on an island in Lake Erie. Some of the snakes are banded and some are unbanded; banding is caused by an autosomal allele that is recessive to an allele for no bands. The frequency of banded snakes on the island is 0.4, whereas the frequency of banded snakes on the mainland is 0.81. One summer, a large number of snakes migrate from the mainland to the island. After this migration, 20% of the island population consists of snakes that came from the mainland.

a. If both the mainland population and the island population are assumed to be in Hardy–Weinberg equilibrium for the alleles that affect banding, what is the frequency of the allele for bands on the island and on the mainland before migration?

b. After migration has taken place, what is the frequency of the banded allele on the island?

32. Pikas are small mammals that live at high elevation in the talus slopes of mountains. Most populations located on mountain tops in Colorado and Montana in North America are isolated from one another, because the pikas don't occupy the low-elevation habitats that separate the mountain tops and don't venture far from the talus slopes. Thus, there is little gene flow between populations. Furthermore, each population is small in size and was founded by a small number of pikas.

A group of population geneticists propose to study the amount of genetic variation in a series of pika populations and to compare the allelic frequencies in different populations. On the bases of the biology and the distribution of pikas, predict what the population geneticists will find concerning the within- and between-population genetic variation.

33. In a large, randomly mating population, the frequency of the allele (*s*) for sickle-cell hemoglobin is 0.028. The results of studies have shown that people with the following genotypes at the beta-chain locus produce the average numbers of offspring given:

Genotype	Average number of offspring produced
SS	5
Ss	6
ss	0

a. What will the frequency of the sickle-cell allele (*s*) be in the next generation?

b. What will the frequency of the sickle-cell allele be at equilibrium?

34. Two chromosomal inversions are commonly found in populations of *Drosophila pseudoobscura*: Standard (*ST*) and Arrowhead (*AR*). When treated with the insecticide DDT, the genotypes for these inversions exhibit overdominance, with the following fitnesses:

Genotype	Fitness
ST/ST	0.47
ST/AR	1
AR/AR	0.62

What will the frequencies of *ST* and *AR* be after equilibrium has been reached?

*35. In a large, randomly mating population, the frequency of an autosomal recessive lethal allele is 0.20. What will the frequency of this allele be in the next generation if the lethality takes place before reproduction?

36. The fruit fly *Drosophila melanogaster* normally feeds on rotting fruit, which may ferment and contain high levels of alcohol. Douglas Cavener and Michael Clegg studied allelic frequencies at the locus for alcohol dehydrogenase (*Adh*) in experimental populations of *D. melanogaster* (D. R. Cavener and M. T. Clegg. 1981. *Evolution* 35:1–10). The experimental populations were established from wild-caught flies and were raised in cages in the laboratory. Two control populations (C1 and C2) were raised on a standard cornmeal–molasses–agar diet. Two ethanol populations (E1 and E2) were raised on a cornmeal–molasses–agar diet to which was added 10% ethanol. The four populations were periodically sampled to determine the allelic frequencies of two alleles at the alcohol dehydrogenase locus, *Adh*^S and *Adh*^F. The frequencies of these alleles in the experimental populations are shown in the graph.

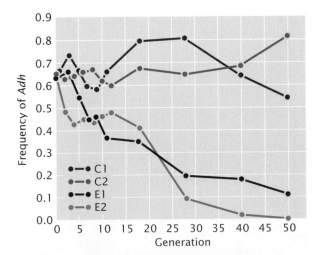

a. On the basis of these data, what conclusion might you draw about the evolutionary forces that are affecting the *Adh* alleles in these populations?

b. Cavener and Clegg measured the viability of the different *Adh* genotypes in the alcohol environment and obtained the following values:

Genotype	Relative viability
Adh^F/Adh^F	0.932
Adh^F/Adh^S	1.288
Adh^S/Adh^S	0.596

Using these relative viabilities, calculate relative fitnesses for the three genotypes. If a population has an initial frequency of $p = f(Adh^F) = 0.5$, what will the expected frequency of Adh^F be in the next generation on the basis of these fitness values?

37. A certain form of congenital glaucoma is caused by an autosomal recessive allele. Assume that the mutation rate is 10^{-5} and that persons having this condition produce, on the average, only about 80% of the offspring produced by persons who do not have glaucoma.

a. At equilibrium between mutation and selection, what will the frequency of the gene for congenital glaucoma be?

b. What will the frequency of the disease be in a randomly mating population that is at equilibrium?

CHALLENGE QUESTION

Section 25.4

38. The Barton Springs salamander is an endangered species found only in a single spring in the city of Austin, Texas. There is growing concern that a chemical spill on a nearby freeway could pollute the spring and wipe out the species. To provide a source of salamanders to repopulate the spring in the event of such a catastrophe, a proposal has been made to establish a captive breeding population of the salamander in a local zoo. You are asked to provide a plan for the establishment of this captive breeding population, with the goal of maintaining as much of the genetic variation of the species as possible in the captive population. What factors might cause loss of genetic variation in the establishment of the captive population? How could loss of such variation be prevented? With the assumption that only a limited number of salamanders can be maintained in captivity, what procedures should be instituted to ensure the long-term maintenance of as much of the variation as possible?

[Danté B. Fenolio.]

26 Evolutionary Genetics

Some chimpanzees, like humans, have the ability to taste phenylthiocarbamide (PTC), whereas others cannot taste it. Recent research indicates that the PTC taste polymorphism evolved independently in humans and chimpanzees. [Jong & Petra Wegner/Animals Animals-Earth Scenes.]

TASTER GENES IN SPITTING APES

Almost every student of biology knows about the taster test. The teacher passes out small pieces of paper impregnated with a compound called phenylthiocarbamide (PTC), and the students, following the teacher's instructions, put the paper in their mouths. The reaction is always the same: a number of the students immediately spit the paper out, repelled by the bitter taste of PTC. A few students, however, can't taste the PTC and continue to suck on the paper, wondering what all the spitting is about. Variation among individuals in a trait such as the ability to taste PTC is termed a polymorphism.

The ability to taste PTC is inherited as an autosomal dominant trait in humans. The frequencies of taster and nontaster alleles have been estimated in hundreds of human populations worldwide. Almost all populations have both tasters and nontasters; the frequency of the two alleles varies widely.

PTC is not found in nature, but the ability to taste it is strongly correlated with the ability to taste other naturally occurring bitter compounds, some of which are toxic. The ability to taste PTC has also been linked to dietary preferences and may be associated with susceptibility to certain diseases, such as thyroid deficiency. These observations suggest that natural selection has played a role in the evolution of the taster trait. Some understanding of the mechanism of the evolution of the taster trait was gained when well-known population geneticist Ronald A. Fisher and his colleagues took a trip to the zoo.

Fisher wondered whether other primates also might have the ability to taste PTC. To answer this question, he prepared some drinks with different levels of PTC and set off for the zoo with his friends, fellow biologists Edmund (E. B.) Ford and Julian Huxley. At the zoo, the PTC-laced drinks were offered to eight chimpanzees and one orangutan. Fisher and his friends were initially concerned that they might not be able to tell whether the apes could taste the PTC. That concern disappeared, however, when the first one sampled the drink and immediately spat on Fisher. Of the eight chimpanzees tested, six were tasters and two were nontasters.

The observation that chimpanzees and humans both have the PTC taste polymorphism led Fisher and his friends to assume that the polymorphism arose in a common

721

ancestor of humans and chimpanzees, which passed it on to both species. However, they had no way to test their hypothesis. Sixty-five years later, geneticists armed with the latest techniques of modern molecular biology were able to determine the origin of the PTC taste polymorphism and test the hypothesis of Fisher and his friends.

Molecular studies reveal that our ability to taste PTC is controlled by alleles at the *TAS2R38* locus, a 1000-bp gene found on chromosome 7. This locus encodes receptors for bitter compounds and is expressed in the cells of our taste buds. One common allele encodes a receptor that allows the ability to taste PTC; an alternative allele encodes a receptor that does not respond to PTC.

Recent research has demonstrated that PTC taste sensitivity in chimpanzees also is controlled by alleles at the *TAS2R38* locus. However, much to the surprise of the investigators, the taster alleles in humans and chimpanzees are not the same at the molecular level. In the human taster and nontaster alleles, nucleotide differences at three positions affect which amino acids are present in the taste-receptor protein. In chimpanzees, none of these nucleotide differences are present. Instead, a mutation in the initiation codon produces the nontaster allele. This substitution eliminates the normal initiation codon, and the ribosome initiates translation at an alternative downstream initiation codon, resulting in the production of a shortened protein receptor that fails to respond to PTC.

What these findings mean is that Fisher and his friends were correct in their observation that humans and chimpanzees both have PTC taste polymorphism but were incorrect in their hypothesis about its origin: humans and chimpanzees independently evolved the PTC taste polymorphism.

This chapter is about the genetic basis of evolution. As illustrated by the story of the PTC taster polymorphism, evolutionary genetics has a long history but has been transformed in recent years by the application of powerful techniques of molecular genetics. In Chapter 25, we considered the evolutionary forces that bring about change in the allelic frequencies of a population: mutation, migration, genetic drift, and selection. In this chapter, we examine some specific ways that these forces shape the genetic makeup of populations and bring about long-term evolution. We begin by looking at how the process of evolution depends on genetic variation and how much genetic variation is found in natural populations. We then turn to the evolutionary changes that bring about the appearance of new species and how evolutionary histories (phylogenies) are constructed. We end the chapter by taking a look at patterns of evolutionary change at the molecular level.

26.1 Organisms Evolve Through Genetic Change Taking Place Within Populations

Evolution is one of the foundational principles of all of biology. Theodosius Dobzhansky, an important early leader in the field of evolutionary genetics, once remarked "Nothing in biology makes sense except in the light of evolution." Indeed, evolution is an all-encompassing theory that helps to make sense of much of the natural world, from the sequences of DNA found in our cells to the types of plants and animals that surround us. The evidence for evolution is overwhelm-

ing. Evolution has been directly observed numerous times; for example, hundreds of different insects evolved resistance to common pesticides following their introduction after World War II. Evolution is supported by the fossil record, comparative anatomy, embryology, the distribution of plants and animals (biogeography), and molecular genetics.

Biological evolution   In spite of its vast importance to all fields of biology, evolution is often misunderstood and misinterpreted. In our society, the term *evolution* frequently refers to any type of change. However, biological **evolution** refers only to a specific type of change—genetic change taking place in a group of organisms. Two aspects of this definition should be emphasized. First, evolution includes genetic change only. Many nongenetic changes take place in living organisms, such as the development of a complex intelligent person from a single-celled zygote. Although remarkable, this change isn't evolution, because it does not include changes in genes. The second aspect to emphasize is that evolution takes place in *groups* of organisms. An individual organism does not evolve; what evolves is the gene pool common to a group of organisms.

Evolution as a two-step process   Evolution can be thought of as a two-step process. First, genetic variation arises. Genetic variation has its origin in the processes of mutation, which produces new alleles, and recombination, which shuffles alleles into new combinations. Both of these processes are random and produce genetic variation continually, regardless of evolution's need for it. The second step in the process of evolution is the increase and decrease in the frequencies of

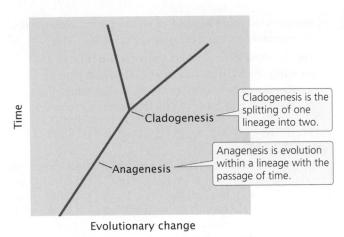

26.1 Anagenesis and cladogenesis are two different types of evolutionary change. Anagenesis is change within an evolutionary lineage; cladogenesis is the splitting of lineages (speciation).

Cladogenesis is the splitting of one lineage into two.

Anagenesis is evolution within a lineage with the passage of time.

genetic variants. Various evolutionary forces discussed in Chapter 25 cause some alleles in the gene pool to increase in frequency and other alleles to decrease in frequency. This shift in the composition of the gene pool common to a group of organisms constitutes evolutionary change.

Types of evolution We can differentiate between two types of evolution that take place within a group of organisms connected by reproduction. **Anagenesis** refers to evolution taking place in a single group (a lineage) with the passage of time (**Figure 26.1**). Another type of evolution is **cladogenesis**, the splitting of one lineage into two. When a lineage splits, the two branches no longer have a common gene pool and evolve independently of one another. New species arise through cladogenesis. TRY PROBLEM 22 ➡

CONCEPTS

Biological evolution is genetic change that takes place within a group of organisms. Anagenesis is evolution that takes place within a single lineage; cladogenesis is the splitting of one lineage into two.

✔ CONCEPT CHECK 1

Briefly describe how evolution takes place as a two-step process.

26.2 Many Natural Populations Contain High Levels of Genetic Variation

Because genetic variation must be present for evolution to take place, evolutionary biologists have long been interested in the amounts of genetic variation in natural populations and the forces that control the amount and nature of that variation. For many years, they could not examine genes directly and were limited to studying the phenotypes of organisms. Although genetic variation could not be quantified directly, studies of phenotypes suggested that many populations of organisms harbor considerable genetic variation. Populations of organisms in nature exhibit tremendous amounts of phenotypic variation: frogs vary in color pattern, birds differ in size, butterflies vary in spotting patterns, mice differ in coat color, humans vary in blood types, to mention just a few. Crosses revealed that a few of these traits were inherited as simple genetic traits (**Figure 26.2**) but, for most traits, the precise genetic basis was complex and unknown, preventing early evolutionary geneticists from quantifying the amount of genetic variation in natural populations.

As discussed in Chapter 24, the ability of a population to respond to selection (the response to selection) depends on narrow-sense heritability, which is a measure of the additive genetic variation of a trait within a population. Many organisms respond to artificial selection carried out by humans, suggesting that the populations of these organisms contain much additive genetic variation. For example, humans have used artificial selection to produce numerous breeds of dogs that vary tremendously in size, shape, color, and behavior

Normal homozygotes

Heterozygotes

Recessive bimacula phenotype

26.2 Early evolutionary geneticists were forced to rely on the phenotypic traits that had a simple genetic basis. Variation in the spotting patterns of the butterfly *Panaxia dominula* is an example.

(see Figure 24.20). Early studies of chromosome variations in *Drosophila* and plants also suggested that genetic variation in natural populations is plentiful and widespread.

Molecular Variation

Tremendous advances in molecular genetics have made it possible to investigate evolutionary change directly by analyzing protein and nucleic acid sequences. These molecular data offer a number of advantages for studying the process and pattern of evolution:

Molecular data are genetic. Evolution results from genetic change with time. Many anatomical, behavioral, and physiological traits have a genetic basis, but the relation between the underlying genes and the trait may be complex. Protein- and nucleic-acid-sequence variation has a clear genetic basis that is often easy to interpret.

Molecular methods can be used with all organisms. Early studies of population genetics relied on simple genetic traits such as human blood types, banding patterns in snails, or spotting patterns in butterflies (see Figure 26.2), which are restricted to a small group of organisms. However, all living organisms have proteins and nucleic acids; so molecular data can be collected from any organism.

Molecular methods can be applied to a huge amount of genetic variation. An enormous amount of data can be accessed by molecular methods. The human genome, for example, contains more than 3 billion base pairs of DNA, which constitutes a large pool of information about our evolution.

All organisms can be compared with the use of some molecular data. Trying to assess the evolutionary history of distantly related organisms is often difficult because they have few characteristics in common. The evolutionary relationships between angiosperms were traditionally assessed by comparing floral anatomy, whereas the evolutionary relationships of bacteria were determined by their nutritional and staining properties. Because plants and bacteria have so few structural characteristics in common, evaluating how they are related to one another was difficult in the past. All organisms have certain molecular traits in common, such as ribosomal RNA sequences and some fundamental proteins. These molecules offer a valid basis for comparisons among all organisms.

Molecular data are quantifiable. Protein and nucleic acid sequence data are precise, accurate, and easy to quantify, which facilitates the objective assessment of evolutionary relationships.

Molecular data often provide information about the process of evolution. Molecular data can reveal important clues about the process of evolution. For example, the results of a study of DNA sequences have revealed that one type of insecticide resistance in

mosquitoes probably arose from a single mutation that subsequently spread throughout the world.

The database of molecular information is large and growing. Today, this database of DNA and protein sequences can be used for making evolutionary comparisons and inferring mechanisms of evolution.

CONCEPTS

Molecular techniques and data offer a number of advantages for evolutionary studies. Molecular data are genetic in nature and can be investigated in all organisms; they provide potentially large data sets, allow all organisms to be compared by using the same characteristics, are easily quantifiable, and provide information about the process of evolution.

Protein Variation

The initial breakthrough in quantifying genetic variation in natural populations came with the application of electrophoresis (see Figure 19.3) to population studies. This technique separates macromolecules, such as proteins or nucleic acids, on the basis of their size and charge. In 1966, Richard Lewontin and John Hubby extracted proteins from fruit flies, separated the proteins by electrophoresis, and stained them for specific enzymes. An examination of the pattern of bands on gels enabled them to assign genotypes to individual flies and to quantify the amount of genetic variation in natural populations. In the same year, Harry Harris quantified genetic variation in human populations by using the same technique. Protein variation has now been examined in hundreds of different species by using protein electrophoresis (**Figure 26.3**).

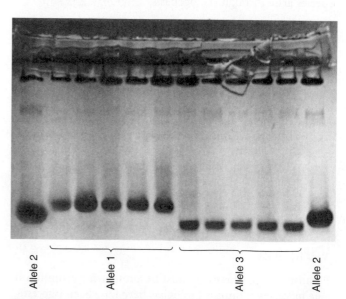

26.3 Molecular variation in proteins is revealed by electrophoresis. Tissue samples from *Drosophila pseudoobscura* were subjected to electrophoresis and stained for esterase. Esterases encoded by different alleles migrate different distances. Shown on the gel are homozygotes for three different alleles.

Table 26.1 Proportion of polymorphic loci and expected heterozygosity for different organisms, as determined by protein electrophoresis

Group	Number of Species	Proportion of Polymorphic Loci		Expected Heterozygosity	
		Mean	SD*	Mean	SD*
Plants	15	0.26	0.17	0.07	0.07
Invertebrates (excluding insects)	28	0.40	0.28	0.10	0.07
Insects (excluding *Drosophila*)	23	0.33	0.20	0.07	0.08
Drosophila	32	0.43	0.13	0.14	0.05
Fish	61	0.15	0.01	0.05	0.04
Amphibians	12	0.27	0.13	0.08	0.04
Reptiles	15	0.22	0.13	0.05	0.02
Birds	10	0.15	0.11	0.05	0.04
Mammals	46	0.15	0.10	0.04	0.02

*SD, standard deviation from the mean.

Source: After L. E. Mettler, T. G. Gregg, and H. E. Schaffer, *Population Genetics and Evolution*, 2d ed. (Englewood Cliffs, N.J.: Prentice Hall, 1988), Table 9.3. Original data from E. Nevo, Genetic variation in natural populations: Patterns and theory, *Theoretical Population Biology* 13:121–177, 1978

Measures of genetic variation The amount of genetic variation in populations is commonly measured by two parameters. The **proportion of polymorphic loci** is the proportion of examined loci in which more than one allele is present in a population. If we examined 30 different loci and found two or more alleles present at 15 of these loci, the percentage of polymorphic loci would be $^{15}/_{30} = 0.5$. The **expected heterozygosity** is the proportion of individuals that are expected to be heterozygous at a locus under the Hardy–Weinberg conditions, which is $2pq$ when there are two alleles present in the population. The expected heterozygosity is often preferred to the observed heterozygosity because expected heterozygosity is independent of the breeding system of an organism. For example, if a species self-fertilizes, it may have little or no heterozygosity but still have considerable genetic variation, which the expected heterozygosity will represent. Expected heterozygosity is typically calculated for a number of loci and is then averaged over all the loci examined.

The proportion of polymorphic loci and the expected heterozygosity (averaged over all loci measured) have been determined by protein electrophoresis for a number of species (**Table 26.1**). About one-third of all protein loci are polymorphic, and expected heterozygosity averages about 10%, although there is considerable diversity among species. These measures actually underestimate the true amount of genetic variation though, because protein electrophoresis does not detect some amino acid substitutions; nor does it detect genetic variation in DNA that does not alter the amino acids of a protein (synonymous codons and variation in noncoding regions of the DNA). **TRY PROBLEM 23** ➔

Explanations for protein variation By the late 1970s, geneticists recognized that most populations possess large amounts of genetic variation, although the evolutionary significance of this fact was not at all clear. Two opposing hypotheses arose to account for the presence of the extensive molecular variation in proteins. The **neutral-mutation hypothesis** proposed that the molecular variation revealed by protein electrophoresis is adaptively neutral; that is, individuals with different molecular variants have equal fitness at realistic population sizes. This hypothesis does not propose that the proteins are functionless; rather, it suggests that most variants are functionally equivalent. Because these variants are functionally equivalent, natural selection does not differentiate between them, and their evolution is shaped largely by the random processes of genetic drift and mutation. The neutral-mutation hypothesis accepts that natural selection is an important force in evolution but views selection as a process that favors the "best" allele while eliminating others. It proposes that, when selection is important, there will be *little* genetic variation.

The **balance hypothesis** proposed, on the other hand, that the genetic variation in natural populations is maintained by selection that favors variation (balancing selection). Overdominance, in which the heterozygote has higher fitness than that of either homozygote, is one type of balancing selection. Under this hypothesis, the molecular variants are not physiologically equivalent and do not have the same fitness. Instead, genetic variation within natural populations is shaped largely by selection, and, when selection is important, there will be *much* variation.

Many attempts to prove one hypothesis or the other failed, because precisely how much variation was actually present was not clear (remember that protein electrophoresis detects only some genetic variation) and because both hypotheses are capable of explaining many different patterns of genetic variation. The results of recent studies that provide direct information about DNA sequence variation demonstrate that much variation at the level of DNA has little obvious effect on the phenotype and is therefore likely to be neutral.

CONCEPTS

The application of electrophoresis to the study of protein variation in natural populations revealed that most organisms possess large amounts of genetic variation. The neutral-mutation hypothesis proposes that most molecular variation is neutral with regard to natural selection and is shaped largely by mutation and genetic drift. The balance hypothesis proposes that genetic variation is maintained by balancing selection.

✔ CONCEPT CHECK 2

Which statement is true of the neutral-mutation hypothesis?

a. All proteins are functionless.

b. Natural selection plays no role in evolution.

c. Most molecular variants are functionally equivalent.

d. All of the above.

DNA Sequence Variation

The development of techniques for isolating, cutting, and sequencing DNA in the past 30 years has provided powerful tools for detecting, quantifying, and investigating genetic variation. The application of these techniques has provided a detailed view of genetic variation at the molecular level.

Restriction-site variation Among the first techniques for detecting and analyzing genetic variation in DNA sequences was the use of restriction enzymes. Each restriction enzyme recognizes and cuts a particular sequence of DNA nucleotides known as that enzyme's restriction site (see Chapter 19). Variation in the presence of a restriction site is called a restriction fragment length polymorphism (RFLP; see Figure 19.22). Each restriction enzyme recognizes a limited number of nucleotide sites in a particular piece of DNA but, if a number of different restriction enzymes are used and the sites recognized by the enzymes are assumed to be random sequences, RFLPs can be used to estimate the amount of variation in the DNA and the proportion of nucleotides that differ between organisms. RFLPs can also be used to analyze the genetic structure of populations and to assess evolutionary relationships among organisms. RFLPs were widely used in evolutionary studies before the development of rapid and inexpensive methods for directly sequencing DNA, and restriction analysis is still employed

today in studies of molecular evolution. However, the use of restriction enzymes to analyze DNA sequence variation gives an incomplete picture of the underlying variation, because it detects variation only at restriction sites.

In an evolutionary application of RFLPs, Nicholas Georgiadis and his colleagues studied genetic relationships among African elephants (*Loxodonta africana*; **Figure 26.4a**) from 10 protected areas of Africa. Mitochondrial DNA (mtDNA) was extracted from samples collected from 270 elephants and amplified with the polymerase chain reaction (PCR; see pp. 523–524 in Chapter 19). The mtDNA was then cleaved with 10 different restriction enzymes, and RFLPs were detected with gel electrophoresis. The degree of genetic differences among elephants from different sites was measured from the sequence variation revealed by the RFLPs. The results of the study showed that the elephant populations are genetically differentiated across Africa, but there are no major differences in their genetic structure among regions (**Figure 26.4b**). From the patterns of variation, the researchers concluded that the elephants have a complex population history, with subdivided populations that exhibit intermittent gene flow.

(a)

(b)

Population	Haplotype									
Location	1	2	3	4	5	6	7	8	9	10
Kenya	▦	▦	▦							
Kenya	▦	▦	▦							
Tanzania		▦					▦			
Zimbabwe			▦							
Zimbabwe			▦					▦	▦	
Zimbabwe			▦	▦						
Botswana			▦							▦
Botswana			▦					▦		▦
Botswana			▦	▦						
South Africa			▦							

26.4 Restriction fragment length polymorphisms have been used to study population structure and gene flow among populations of African elephants. (a) African elephant. (b) Geographic distribution of 10 mtDNA haplotypes determined by restriction fragment length analysis. Color indicates the presence of the restriction haplotype. [(a) Digital Vision; (b) data from N. Georgiadis et al., *Journal of Heredity* 85:100–104, 1994.]

Microsatellite variation   Microsatellites are short DNA sequences that exist in multiple copies repeated in tandem (see pp. 538–539 in Chapter 19). Variation in the number of copies of the repeats is common, with individual organisms often differing in the number of repeat copies. Microsatellites can be detected by using PCR. Pairs of primers are used that flank a region of repeated copies of the sequence. The DNA fragments that are synthesized in the PCR reaction vary in length, depending on the number of tandem repeats present. DNA from an individual organism with more repeats will produce a longer amplified segment. After PCR has been completed, the amplified fragments are separated with the use of gel electrophoresis and stained, producing a series of bands on a gel (see Figure 19.29). The banding patterns that result represent different alleles (variants in the DNA sequence) and can be used to quantify genetic variation, assess genetic relationships among individual organisms, and quantify genetic differences in a population. An advantage of using microsatellites is that the PCR reaction can be used on very small amounts of DNA and is rapid. The amplified fragments can be fluorescently labeled and detected by a laser, allowing the process to be automated.

David Coltman and his colleagues used microsatellite variation to study paternity in bighorn sheep (**Figure 26.5**; also see the introduction to Chapter 25) and showed that sport hunting of trophy rams has reduced the weight and horn size of the animals. Samples of blood, hair, and ear tissue were collected from bighorn sheep at Ram Mountain in Alberta, Canada—a population that has been monitored since 1971. DNA was extracted from the tissue samples and amplified with PCR, revealing variation at 20 microsatellite loci. On the basis of the microsatellite variation, paternity was assigned to 241 rams, and the family relationships of

the sheep were worked out. Using these family relationships and the quantitative genetic techniques described in Chapter 24, the geneticists were able to show that ram weight and horn size had high heritability and exhibited a strong positive genetic correlation (see p. 683 in Chapter 24). Trophy hunters selectively shoot rams with large horns, often before they are able to reproduce. This selective pressure has produced a response to selection: the rams are evolving smaller horns. Between 1971 and 2002, horn size in the population decreased by about one-quarter. Because of the positive genetic correlation between horn size and body size, the body size of rams also is decreasing. Unfortunately, the killing of trophy rams with large horns has led to a decrease in the very traits that are prized by the hunters. This research illustrates the use of microsatellites in evolutionary analysis that has practical application.

Variation detected by DNA sequencing   The development of techniques for rapidly and inexpensively sequencing DNA (see pp. 534–537 in Chapter 19) have made this type of data an important tool in population and evolutionary studies. DNA sequence data often reveal processes that influence evolution and are invaluable for determining the evolutionary relationships of different organisms. The use of PCR for producing the DNA used in the sequencing reactions means that data can be obtained from a very small initial sample of DNA.

DNA sequencing was used to reassess the genetic relationships among African elephants. Recall that data from RFLPs found little evidence of regional subdivision among elephants in Africa. Alfred Roca and his colleagues obtained tissue samples from 195 elephants by shooting them with needlelike darts, which fell to the ground after hitting an elephant but retained a small plug of skin. From the skin samples, the scientists sequenced 1732 base pairs of DNA from four nuclear genes. Their analysis revealed large genetic differences between forest elephants and savannah elephants, suggesting that there is limited gene flow (migration) between these two groups of elephants. On the basis of these results, the scientists proposed that two different species of elephant exist in Africa.

Another example of the use of DNA sequence data to decipher evolutionary relationships is the unusual case of HIV infection in a dental practice in Florida. In July 1990, the U.S. Centers for Disease Control and Prevention (CDC) reported that a young woman in Florida (later identified as Kimberly Bergalis) had become HIV positive after undergoing an invasive dental procedure performed by a dentist who had AIDS. Bergalis had no known risk factors for HIV infection and no known contact with other HIV-positive persons. The CDC acknowledged that Bergalis might have acquired the infection from her dentist. Subsequently, the dentist wrote to all of his patients, suggesting that they be tested for HIV infection. By 1992, seven of the dentist's patients had tested positive for HIV, and this number eventually increased to ten.

26.5 Microsatellite variation has been used to study the response of bighorn sheep to selective pressure on horn size due to trophy hunting. [Eyewire.]

Table 26.2 HIV-positive persons included in study of HIV isolates from a Florida dental practice

Person	Sex	Known Risk Factors	Average Differences in DNA Sequences (%)	
			From HIV from Dentist	From HIV from Controls
Dentist	M	Yes		11.0
Patient A	F	No	3.4	10.9
Patient B	F	No	4.4	11.2
Patient C	M	No	3.4	11.1
Patient E	F	No	3.4	10.8
Patient G	M	No	4.9	11.8
Patient D	M	Yes	13.6	13.1
Patient F	M	Yes	10.7	11.9

Source: After C. Y. Ou et al., *Science* 256:1165–1171, 1992, Table 1.

Originally diagnosed with HIV infection in 1986, the dentist began to develop symptoms of AIDS in 1987 but continued to practice dentistry for another 2 years. All of his HIV-positive patients had received invasive dental procedures, such as root canals and tooth extractions, in the period when the dentist was infected. Among the seven patients originally studied by the CDC (patients A–G, **Table 26.2**), two had known risk factors for HIV infection (intravenous drug use, homosexual behavior, or sexual relations with HIV-infected persons), and a third had possible but unconfirmed risk factors.

To determine whether the dentist had infected his patients, the CDC conducted a study of the molecular evolution of HIV isolates from the dentist and from the patients. HIV undergoes rapid evolution, making it possible to trace the path of its transmission. This rapid evolution also allows HIV to develop drug resistance quickly, making the development of a treatment for AIDS difficult.

Blood specimens were collected from the dentist, the patients, and a group of 35 local controls (other HIV-infected people who lived within 90 miles of the dental practice but who had no known contact with the dentist). DNA was extracted from white blood cells, and a 680-bp fragment of the *envelope* gene of the virus was amplified by PCR. The fragments from the dentist, the patients, and the local controls were then sequenced and compared.

The divergence between the viral sequences taken from the dentist, the seven patients, and the controls is shown in Table 26.2. Viral DNA taken from patients with no confirmed risk factors (patients A, B, C, E, and G) differed from the dentist's viral DNA by 3.4% to 4.9%, whereas the viral DNA from the controls differed from the dentist's by an average of 11%. The viral sequences collected from five patients (A, B, C, E, and G) were more closely related to the viral sequences collected from the dentist than to viral sequences from the general population, strongly suggesting

that these patients acquired their HIV infection from the dentist. The viral isolates from patients D and F (patients with confirmed risk factors), however, differed from that of the dentist by 10.7% and 13.6%, suggesting that these two patients did not acquire their infection from the dentist.

An analysis of the evolutionary relationships of the viral sequences (**Figure 26.6**) confirmed that the virus taken from the dentist had a close evolutionary relationship to viruses taken from patients A, B, C, E, and G. The viruses from patients D and F, with known risk factors, were no more similar to the virus from the dentist than to viruses from local controls, indicating that the dentist most likely infected five of his patients, whereas the other two patients probably acquired their infections elsewhere. Of three additional HIV-positive patients that have been identified since 1992, only one has viral sequences that are closely related to those from the dentist.

The study of HIV isolates from the dentist and his patients provides an excellent example of the relevance of evolutionary studies to real-world problems. How the dentist infected his patients during their visits to his office remains a mystery, but this case is clearly unusual. A study of almost 16,000 patients treated by HIV-positive health-care workers failed to find a single case of confirmed transmission of HIV from the health-care worker to the patient.

CONCEPTS

Variation in DNA nucleotide sequence can be analyzed by using restriction fragment length polymorphisms, microsatellites, and data from direct DNA sequencing.

✔ **CONCEPT CHECK 3**

What are some of the advantages of using microsatellites for evolutionary studies?

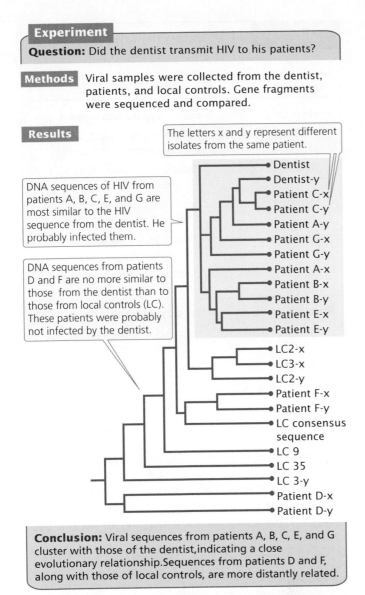

Question: Did the dentist transmit HIV to his patients?

Methods Viral samples were collected from the dentist, patients, and local controls. Gene fragments were sequenced and compared.

Results

The letters x and y represent different isolates from the same patient.

DNA sequences of HIV from patients A, B, C, E, and G are most similar to the HIV sequence from the dentist. He probably infected them.

DNA sequences from patients D and F are no more similar to those from the dentist than to those from local controls (LC). These patients were probably not infected by the dentist.

Dentist
Dentist-y
Patient C-x
Patient C-y
Patient A-y
Patient G-x
Patient G-y
Patient A-x
Patient B-x
Patient B-y
Patient E-x
Patient E-y
LC2-x
LC3-x
LC2-y
Patient F-x
Patient F-y
LC consensus sequence
LC 9
LC 35
LC 3-y
Patient D-x
Patient D-y

Conclusion: Viral sequences from patients A, B, C, E, and G cluster with those of the dentist, indicating a close evolutionary relationship. Sequences from patients D and F, along with those of local controls, are more distantly related.

26.6 Evolutionary tree showing the relationships of HIV isolates from a dentist, seven of his patients (A through G), and other HIV-positive persons from the same region (local controls, LC). The phylogeny is based on DNA sequences taken from the *envelope* gene of the virus. [After C. Ou et al., Molecular epidemiology of HIV transmission in a dental practice, *Science* 256:1167, 1992.]

26.3 New Species Arise Through the Evolution of Reproductive Isolation

The term *species* literally means kind or appearance; **species** are different kinds or types of living organisms. In many cases species differences are easy to recognize: a horse is clearly a different species from a chicken. Sometimes, however, species differences are not so clear cut. Some species of *Plethodon* salamanders are so similar in appearance that they can be distinguished only by looking at their proteins or genes.

The concept of a species has two primary uses in biology. First, a species is a name given to a particular type of organism. For effective communication, biologists must use a standard set of names for the organisms that they study, and species names serve that purpose. When a geneticist talks about conducting crosses with *Drosophila melanogaster*, other biologists immediately understand which organism was used. The second use of the term species is in an evolutionary context: a species is considered an evolutionarily independent group of organisms.

The Biological Species Concept

What kinds of differences are required to consider two organisms different species? A widely used definition of species is the **biological species concept**, first fully developed by evolutionary biologist Ernst Mayr in 1942. Mayr was primarily interested in the biological characteristics that are responsible for separating organisms into independently evolving units. He defined a species as a group of organisms whose members are capable of interbreeding with one another but are reproductively isolated from the members of other species. In other words, members of the same species have the biological potential to exchange genes, and members of different species cannot exchange genes. Because different species do not exchange genes, each species evolves independently.

Not all biologists adhere to the biological species concept, and there are several problems associated with it. For example, reproductive isolation, on which the biological species concept is based, cannot be determined of fossils and, in practice, it is often difficult to determine even whether living species are biologically capable of exchanging genes. Furthermore, the biological species concept cannot be applied to asexually reproducing organisms, such as bacteria. In practice, most species are distinguished on the basis of phenotypic (usually anatomical) differences. Biologists often assume that phenotypic differences represent underlying genetic differences; if the phenotypes of two organisms are quite different, then they probably cannot and do not interbreed in nature.

Reproductive Isolating Mechanisms

The key to species differences under the biological species concept is reproductive isolation—biological characteristics that prevent genes from being exchanged between different species. Any biological factor or mechanism that prevents gene exchange is termed a **reproductive isolating mechanism**.

Prezygotic reproductive isolating mechanisms **Prezygotic reproductive isolating mechanisms** prevent gametes from two different species from fusing and forming a hybrid zygote. In **ecological isolation**, members of two species do not encounter one another and therefore do not reproduce with one another, because they have different ecological niches, living in different habitats and interacting with the environment in different ways. For example, some species of forest-dwelling birds feed and nest in the forest canopy, whereas other species confine their activities to the forest floor. Because they never come into contact, these birds are reproductively isolated from one another. Other species are separated by **behavioral isolation**, differences in behavior that prevent interbreeding. Many male frogs attract females of the same species by using a unique, species-specific call. Two closely related frogs may use the same pond but never interbreed, because females are attracted only to the call of their own species.

Another type of prezygotic reproductive isolation is **temporal isolation**, in which reproduction takes place at different times of the year. Some species of plants do not exchange genes, because they flower at different times of the year. **Mechanical isolation** results from anatomical differences that prevent successful copulation. This type of isolation is seen in many insects, in which closely related species differ in their male and female genitalia, and so copulation is physically impossible. Finally, some species are separated by **gametic isolation**, in which mating between individuals of different species takes place, but the gametes do not form zygotes. Male gametes may not survive in the female reproductive tract or may not be attracted to female gametes. In other cases, male and female gametes meet but are too incompatible to fuse to form a zygote. Gametic isolation is seen in many plants, where pollen from one species cannot fertilize the ovules of another species.

Postzygotic reproductive isolating mechanisms

Other species are separated by **postzygotic reproductive isolating mechanisms**, in which gametes of two species fuse and form a zygote, but there is no gene flow between the two species, either because the resulting hybrids are inviable or sterile or because reproduction breaks down in subsequent generations.

If prezygotic reproductive isolating mechanisms fail or have not yet evolved, mating between two organisms of different species may take place, with the formation of a hybrid zygote containing genes from two different species. In many cases, such species are still separated by **hybrid inviability**, in which incompatibility between genomes of the two species prevents the hybrid zygote from developing. Hybrid inviability is seen in some groups of frogs, in which mating between different species and fertilization take place, but the resulting embryos never complete development.

Other species are separated by **hybrid sterility**, in which hybrid embryos complete development but are sterile, so that genes are not passed between species. Donkeys and horses frequently mate and produce a viable offspring—a mule—but most mules are sterile; thus, there is no gene flow between donkeys and horses (but see Problem 44 at the end of Chapter 9). Finally, some closely related species

Table 26.3 Types of reproductive isolating mechanisms

Type	Characteristics
Prezygotic	**Before a zygote has formed**
Ecological	Differences in habitat; individuals do not meet
Temporal	Reproduction takes place at different times
Mechanical	Anatomical differences prevent copulation
Behavioral	Differences in mating behavior prevent mating
Gametic	Gametes incompatible or not attracted to each other
Postzygotic	**After a zygote has formed**
Hybrid inviability	Hybrid zygote does not survive to reproduction
Hybrid sterility	Hybrid is sterile
Hybrid breakdown	F_1 hybrids are viable and fertile, but F_2 are inviable or sterile

are capable of mating and producing viable and fertile F_1 progeny. However, genes do not flow between the two species because of **hybrid breakdown**, in which further crossing of the hybrids produces inviable or sterile offspring. The different types of reproductive isolating mechanisms are summarized in **Table 26.3**.

CONCEPTS

The biological species concept defines a species as a group of potentially interbreeding organisms that are reproductively isolated from the members of other species. Under this concept, species are separated by reproductive isolating mechanisms, which may intervene before a zygote is formed (prezygotic reproductive isolating mechanisms) or after a zygote is formed (postzygotic reproductive isolating mechanisms).

✔ CONCEPT CHECK 4

Which statement is an example of postzygotic reproductive isolation?

a. Sperm of species A dies in the oviduct of species B before fertilization can take place.

b. Hybrid zygotes between species A and B are spontaneously aborted early in development.

c. The mating seasons of species A and B do not overlap.

d. Males of species A are not attracted to the pheromones produced by the females of species B.

Modes of Speciation

Speciation is the process by which new species arise. In regard to the biological species concept, speciation comes about through the evolution of reproductive isolating mechanisms—mechanisms that prevent the exchange of genes between groups of organisms.

There are two principle ways in which new species arise. **Allopatric speciation** arises when a geographic barrier first splits a population into two groups and blocks the exchange of genes between them. The interruption of gene flow then leads to the evolution of genetic differences that result in reproductive isolation. **Sympatric speciation** arises in the absence of any external barrier to gene flow; reproductive isolating mechanisms evolve within a single population. We will take a more-detailed look at both of these mechanisms next.

Allopatric speciation Allopatric speciation is initiated when a geographic barrier splits a population into two or more groups and prevents gene flow between the isolated groups (**Figure 26.7a**). Geographic barriers can take a number of forms. Uplifting of a mountain range may split a population of lowland plants into separate groups on each side of the mountains. Oceans serve as effective barriers for many types of terrestrial organisms, separating individuals on different islands from one another and from those on the mainland. Rivers often separate populations of fish located in separate drainages. The erosion of mountains may leave populations of alpine plants isolated on separate mountain peaks.

After two populations have been separated by a geographic barrier that prevents gene flow between them, they evolve independently (**Figure 26.7b**). The genetic isolation allows each population to accumulate genetic differences that are not found in the other population; genetic differences arise through natural selection, unique mutations, and genetic drift (if the populations are small). Genetic differentiation eventually leads to prezygotic and postzygotic isolation. It is important to note that prezygotic isolation and postzygotic isolation arise simply as a consequence of genetic divergence.

If the geographic barrier that once separated the two populations disappears or individuals are able to disperse over it, the populations come into secondary contact (**Figure 26.7c**). At this point, several outcomes are possible. If limited genetic differentiation has taken place during the separation of the populations, reproductive isolating mechanisms may not have evolved or may be incomplete. Genes will flow between the two populations, eliminating any genetic differences that did arise, and the populations will remain a single species.

A second possible outcome is that genetic differentiation during separation leads to prezygotic reproductive isolating mechanisms; in this case, the two populations are different

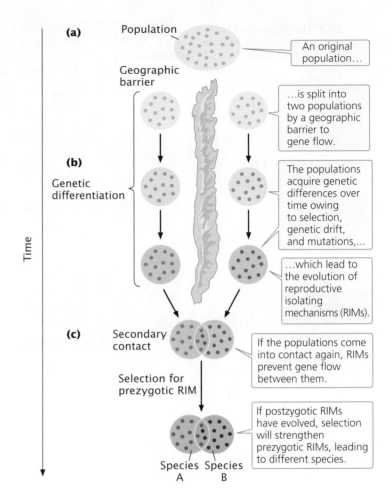

26.7 Allopatric speciation is initiated by a geographic barrier to gene flow between two populations.

species. A third possible outcome is that, during their time apart, some genetic differentiation took place between the populations, leading to incompatibility in their genomes and postzygotic isolation. If postzygotic isolating mechanisms have evolved, any mating between individuals from the different populations will produce hybrid offspring that are inviable or sterile. Individuals that mate only with members of the same population will have higher fitness than that of individuals that mate with members of the other population; so natural selection will increase the frequency of any trait that prevents interbreeding between members of the different populations. With the passage of time, prezygotic reproductive isolating mechanisms will evolve. In short, if some postzygotic reproductive isolation exists, natural selection will favor the evolution of prezygotic reproductive isolating mechanisms to prevent wasted reproduction by individuals mating with members of the other population. This process of postzygotic reproductive isolation leading to the evolution of prezygyotic isolating mechanisms is termed *reinforcement*.

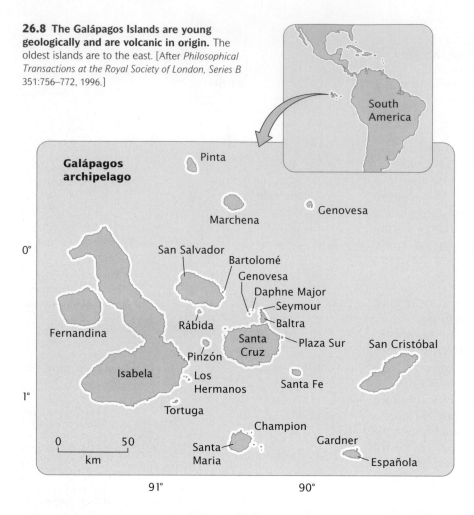

26.8 The Galápagos Islands are young geologically and are volcanic in origin. The oldest islands are to the east. [After *Philosophical Transactions at the Royal Society of London, Series B* 351:756–772, 1996.]

to the west (Isabela and Fernandina). With the passage of time, the number of islands in the archipelago increased as new volcanoes arose.

Darwin's finches consist of 14 species that are found on various islands in the Galápagos archipelago (**Figure 26.9**). The birds vary in the shape and sizes of their beaks, which are adapted for eating different types of food items. Recent studies of the development of finch embryos have helped to reveal some of the molecular details of how differences in beak shapes have evolved (see p. 626 in Chapter 22). Genetic studies have demonstrated that all the birds are closely related and evolved from a single ancestral species that migrated to the islands from the coast of South America some 2 million to 3 million years ago. The evolutionary relationships among the 14 species, based on studies of microsatellite data, are depicted in the evolutionary tree shown in **Figure 26.9**. Most of the species are separated by a behavioral isolating mechanism (song in particular), but some of the species can and occasionally do hybridize in nature.

The first finches to arrive in the Galápagos probably colonized one of the larger eastern islands. A breeding population became established and increased with time. At some point, a few birds dispersed to other islands, where they were effectively isolated from the original population, and established a new population. This population underwent genetic differentiation owing to genetic drift and adaptation to the local conditions of the island. It eventually became reproductively isolated from the original population. Individual birds from the new population then dispersed to other islands and gave rise to additional species. This process was repeated many times. Occasionally, newly evolved birds dispersed to an island where another species was already present, giving rise to secondary contact between the species. Today, many of the islands have more than one resident finch.

The age of the 14 species has been estimated with data from mitochondrial DNA. **Figure 26.10** shows that there is a strong correspondence between the number of bird species present at various times in the past and the number of islands in the archipelago. This correspondence is one of the most compelling pieces of evidence for the theory that the different species of finches arose through allopatric speciation.

A number of variations in this general model of allopatric speciation are possible. Many new species probably arise when a small group of individuals becomes geographically isolated from the main population; for example, a few individuals of a mainland population might migrate to a geographically isolated island. In this situation, founder effect and genetic drift play a larger role in the evolution of genetic differences between the populations.

An excellent example of allopatric speciation can be found in Darwin's finches, a group of birds on the Galápagos Islands; these finches were discovered by Charles Darwin in his voyage aboard the *Beagle*. The Galápagos are an archipelago of islands located some 900 km off the coast of South America (**Figure 26.8**). Consisting of more than a dozen large islands and many smaller ones, the Galápagos formed from volcanoes that erupted over a geological hot spot that has remained stationary while the geological plate over it moved eastward in the past 3 million years. The movement of the geological plate pulled newly formed islands eastward, and so the islands to the east (San Cristóbal and Española) are older than those

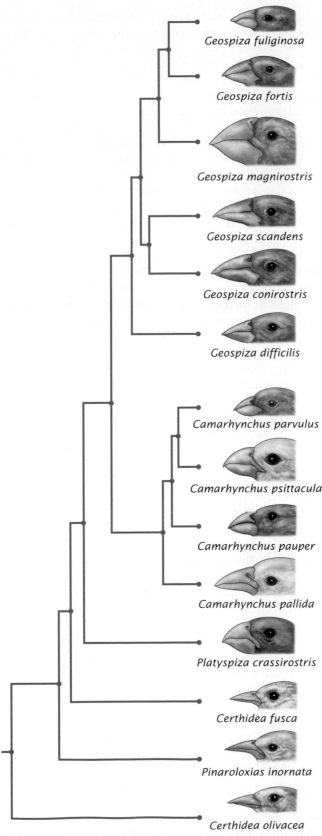

26.9 Darwin's finches consist of 14 species that evolved from a single ancestral species that migrated to the Galápagos Islands and underwent repeated allopatric speciation. [After B. R. Grant and P. R. Grant, *Bioscience* 53:965–975, 2003.]

Geospiza fuliginosa

Geospiza fortis

Geospiza magnirostris

Geospiza scandens

Geospiza conirostris

Geospiza difficilis

Camarhynchus parvulus

Camarhynchus psittacula

Camarhynchus pauper

Camarhynchus pallida

Platyspiza crassirostris

Certhidea fusca

Pinaroloxias inornata

Certhidea olivacea

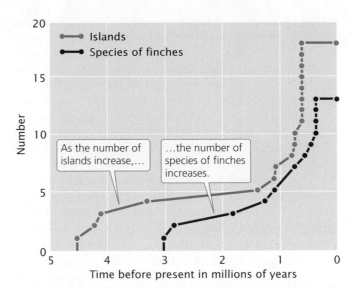

As the number of islands increase,...

...the number of species of finches increases.

26.10 The number of species of Darwin's finches present at various times in the past corresponds with the number of islands in the Galápagos archipelago. [Data from P. R. Grant, B. R. Grant, and J. C. Deutsch, Speciation and hybridization in island birds, *Philosophical Transactions of the Royal Society of London, Series B* 351:765–772, 1996.]

CONCEPTS

Allopatric speciation is initiated by a geographic barrier to gene flow. A single population is split into two or more populations by a geographic barrier. With the passage of time, the populations evolve genetic differences, which bring about reproductive isolation. After postzygotic reproductive isolating mechanisms have evolved, selection favors the evolution of prezygotic reproductive isolating mechanisms.

✔ CONCEPT CHECK 5

What role does genetic drift play in allopatric speciation?

Sympatric speciation Sympatric speciation arises in the absence of any geographic barrier to gene flow; reproductive isolating mechanisms evolve within a single interbreeding population. Sympatric speciation has long been controversial within evolutionary biology. Ernst Mayr believed that sympatric speciation was impossible, and he demonstrated that many apparent cases of sympatric speciation could be explained by allopatric speciation. More recently, however, evidence has accumulated that sympatric speciation can and has arisen under special circumstances. The difficulty with sympatric speciation is that isolating mechanisms arise as a *consequence* of genetic differentiation, which takes place only if gene flow between groups is interrupted. But, without reproductive isolation (or some external barrier), how can gene flow be interrupted? How can genetic differentiation arise within a single group that is freely exchanging genes?

Most models of sympatric speciation assume that genetic differentiation is initiated by strong disruptive selection taking place within a single population. One homozygote (A^1A^1) is strongly favored on one resource (perhaps the plant species that is host to an insect) and the other homozygote (A^2A^2) is favored on a different resource (perhaps a different host plant). Heterozygotes (A^1A^2) have low fitness on both resources. In this situation, natural selection will favor genotypes at other loci that cause assortative mating (matings between like individuals, see Chapter 25), and so no matings take place between A^1A^1 and A^2A^2, which would produce A^1A^2 offspring with low fitness.

Now imagine that alleles at a second locus affect mating behavior, such that C^1C^1 individuals prefer mating only with other C^1C^1 individuals, and C^2C^2 individuals prefer mating with other C^2C^2 individuals. If alleles at the A locus are non-randomly associated with alleles at the C locus so that only $A^1A^1\,C^1C^1$ individuals and $A^2A^2\,C^2C^2$ individuals exist, then gene flow will be restricted between individuals using the different resources, allowing the two groups to evolve further genetic differences that might lead to reproductive isolation and sympatric speciation.

The difficulty with this model is that recombination quickly breaks up the nonrandom associations between genotypes at the two loci, producing individuals such as $A^1A^1\,C^2C^2$, which would prefer to mate with $A^2A^2\,C^2C^2$ individuals. This mating would produce all A^1A^2 offspring, which do poorly on both resources. Thus, even limited recombination will prevent the evolution of the mating-preference genes.

Sympatric speciation is more probable if the genes that affect resource utilization also affect mating preferences. Genes that affect both resource utilization and mating preference are indeed present in some host races—populations of specialized insects that feed on different host plants. Guy Bush studied what appeared to be initial stages of speciation in host races of the apple maggot fly (*Rhagoletis pomonella*, **Figure 26.11**). The flies of this species feed on the fruits of a specific host tree. Mating takes place near the fruits, and the flies lay their eggs on the ripened fruits, where their larvae grow and develop. *R. pomnella* originally existed only on the fruits of hawthorn trees, which are native to North America; 150 years ago, *R. pomnella* was first observed on cultivated apples, which are related to hawthorns but a different species. Infestations of apples by this new apple host race of *R. pomnella* quickly spread, and, today, many apple trees throughout North America are infested with the flies.

The apple host race of *R. pomnella* probably originated when a few flies acquired a mutation that allowed them to feed on apples instead of the hawthorn fruits. Because mating takes place on and near the fruits, flies that use apples are more likely to mate with other flies using apples, leading to genetic isolation between flies using hawthorns and those using apples. Indeed, Bush found that some genetic differentiation has already taken place between the two host races. Flies lay their eggs on ripening fruit, and there has been strong selection for the flies to synchronize their reproduction with the period when their host species has ripening fruit. Apples ripen several weeks earlier than hawthorns. Correspondingly, the peak mating period of the apple host races is 3 weeks earlier than that of the hawthorn race. These differences in the timing of reproduction between apple and hawthorn races have further reduced gene flow—to about 2%—between the two host races and have led to significant genetic differentiation between them. These differences have evolved in the past 150 years and evolution appears to be ongoing. Although genetic differentiation has taken place between apple and hawthorn host races of *R. pomnella* and some degree of reproductive isolation has evolved between them, reproductive isolation is not yet complete and speciation has not fully taken place.

Speciation through polyploidy A special type of sympatric speciation takes place through polyploidy (see Chapter 9). Polyploid organisms have more than two genomes ($3n$, $4n$, $5n$, etc.). As discussed in Chapter 9, allopolyploidy often arises when two diploid species hybridize, producing $2n$ hybrid offspring. Nondisjunction in one of the hybrid offspring produces a $4n$ tetraploid. Because this tetraploid contains exactly two copies of each chromosome, it is usually fertile and will be reproductively isolated from the two parental species by differences in chromosome number (see Figure 9.28).

Numerous species of flowering plants are allopolyploids. Speciation through polyploidy was observed when it led to a new species of salt-marsh grass that arose along the coast of England about 1870. This polyploid contains genomes of the European salt grass *Spartina maritima* ($2n = 60$) and the American salt grass *S. alternaflora* ($2n = 62$; **Figure 26.12**). Seeds from the American salt grass were probably transported to England in the ballast of a ship. Regardless of how it got there, *S. alternaflora* grew in an English marsh and eventually crossed with *S. maritima*, producing a hybrid with $2n = 61$. Nondisjunction in the hybrid then led to chro-

26.11 Host races of the apple maggot fly, *Rhagoletis pomenella*, have evolved some reproductive isolation without any geographic barrier to gene flow. [Tom Murray.]

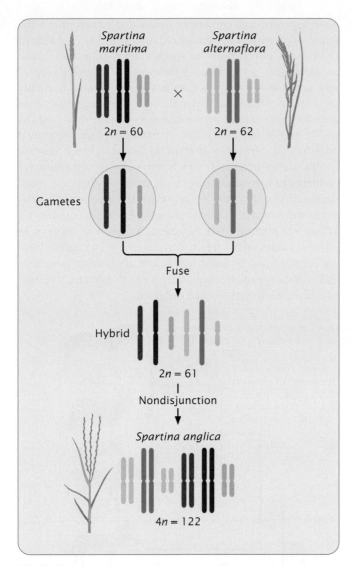

26.12 *Spartina anglica* **arose sympatrically through allopolyploidy.**

mosome doubling, producing a new species *S. anglica* with $4n = 122$ (see Figure 26.12). This new species subsequently spread along the coast of England.

CONCEPTS

Sympatric speciation arises within a single interbreeding population without any geographic barrier to gene flow. Sympatric speciation may arise under special circumstances, such as when resource use is linked to mating preference (in host races) or when species hybridization leads to allopolyploidy.

Genetic Differentiation Associated with Speciation

As we have seen, genetic differentiation leads to the evolution of reproductive isolating mechanisms, which restrict gene flow between populations and lead to speciation. How

Table 26.4	Genetic similarity in groups of the *Drosophila willistoni* complex
Group	Mean Genetic Similarity
Geographic populations	0.970
Subspecies	0.795
Sibling species	0.517
Nonsibling species	0.352

much genetic differentiation is required for reproductive isolation to take place? This question has received considerable study by evolutionary geneticists, but, unfortunately, there is no universal answer. Some newly formed species differ in many genes, whereas others appear to have undergone divergence in just a few genes.

One group of organisms that have been extensively studied for genetic differences associated with speciation is the genus *Drosophila*. The *Drosophila willistoni* group consists of at least 12 species found in Central and South America in various stages in the process of speciation. Using protein electrophoresis, Francisco Ayala and his colleagues genotyped flies from different geographic populations (populations with limited genetic differences), subspecies (populations with considerable genetic differences), sibling species (newly arisen species), and nonsibling species (older species). For each group, they computed a measure of genetic similarity, which ranges from 1 to 0 and represents the overall level of genetic differentiation (**Table 26.4**). They found that there was a general decrease in genetic similarity as flies evolve from geographic populations to subspecies to sibling species to nonsibling species. These data suggest that considerable genetic differentiation at many loci is required for speciation to arise. A study of *D. simulans* and *D. melanogaster*, two species that produce inviable hybrids when crossed, suggested that at least 200 genes contribute to the inviability of hybrids between the two species.

However, other studies suggest that speciation may have arisen through changes in just a few genes. For example, *D. heteroneura* and *D. silvestris* are two species of Hawaiian fruit flies that exhibit behavioral reproductive isolation. The isolation is determined largely by differences in head shape; *D. heteroneura* has a hammer-shaped head with widely separated eyes that is recognized by females of the same species but rejected by *D. silvestris* females. Genetic studies indicate that only a few loci (about 10) determine the differences in head shape.

CONCEPTS

Some newly arising species have a considerable number of genetic differences; others have few genetic differences.

26.4 The Evolutionary History of a Group of Organisms Can Be Reconstructed by Studying Changes in Homologous Characteristics

The evolutionary relationships among a group of organisms are termed a **phylogeny**. Because most evolution takes place over long periods of time and is not amenable to direct observation, biologists must reconstruct phylogenies by inferring the evolutionary relationships among present-day organisms. The discovery of fossils of ancestral organisms can aid in the reconstruction of phylogenies, but the fossil record is often too poor to be of much help. Thus, biologists are often restricted to analyses of characteristics in present-day organisms to determine their evolutionary relationships. In the past, phylogenetic relationships were reconstructed on the basis of phenotypic characteristics—often, anatomical traits. Today, molecular data, including protein and DNA sequences, are frequently used to construct phylogenetic trees.

Phylogenies are reconstructed by inferring changes that have taken place in homologous characteristics. Such characteristics evolved from the same character in a common ancestor. For example, the front leg of a mouse and the wing of a bat are homologous structures, because both evolved from the forelimb of an early mammal that was an ancestor to both mouse and bat. Although these two anatomical features look different and have different functions, close examination of their structure and development reveals that they are indeed homologous. And, because mouse and bat have these homologous features and others in common, we know that they are both mammals. Similarly, DNA sequences are homologous if two present-day sequences evolved from a single sequence found in an ancestor. For example, all eukaryotic organisms have a gene for cytochrome *c*, an enzyme that helps carry out oxidative respiration. This gene is assumed to have arisen in a single organism in the distant past and was then passed down to descendants of that early ancestor. Today, all copies of the cytochrome *c* gene are homologous, because they all evolved from the same original copy in the distant ancestor of all organisms that possess this gene.

A graphical representation of a phylogeny is called a **phylogenetic tree**. As shown in **Figure 26.13**, a phylogenetic tree depicts the evolutionary relationships among different organisms, similarly to the way in which a pedigree represents the genealogical relationships among family members.

A phylogenetic tree consists of **nodes** that represent the different organisms being compared, which might be different individuals, populations, or species. Terminal nodes (those at the end of the outermost branches of the tree) represent organisms for which data have been obtained, usually present-day organisms. Internal nodes represent common ancestors that existed before divergence between organisms took place. In most cases, the internal nodes represent past ancestors that are inferred from the analysis. The nodes are connected by **branches**, which represent the evolutionary connections between organisms. In many phylogenetic trees, the lengths of the branches represent the amount of evolutionary divergence that has taken place between organisms. When one internal node represents a common ancestor to all other nodes on the tree, the tree is said to be **rooted**. Trees are often rooted by including in the analysis an organism that is distantly related to all the others; this distantly related organism is referred to as an outgroup.

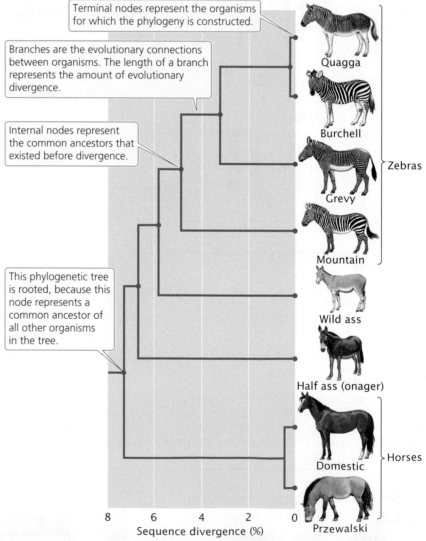

Terminal nodes represent the organisms for which the phylogeny is constructed.

Branches are the evolutionary connections between organisms. The length of a branch represents the amount of evolutionary divergence.

Internal nodes represent the common ancestors that existed before divergence.

This phylogenetic tree is rooted, because this node represents a common ancestor of all other organisms in the tree.

Quagga

Burchell

Grevy — Zebras

Mountain

Wild ass

Half ass (onager)

Domestic — Horses

Przewalski

8   6   4   2   0
Sequence divergence (%)

26.13 A phylogenetic tree is a graphical representation of the evolutionary relationships among a group of organisms.

Phylogenetic trees are created to depict the evolutionary relationships among organisms; they are also created to depict the evolutionary relationships among DNA sequences. The latter type of phylogenetic tree is termed a **gene tree** (**Figure 26.14**). TRY PROBLEM 28 →

CONCEPTS

A phylogeny represents the evolutionary relationships among a group of organisms and is often depicted graphically by a phylogenetic tree, which consists of nodes representing the organisms and branches representing their evolutionary connections.

✔ CONCEPT CHECK 6

Which feature is found in a rooted tree but not in an unrooted tree?

a. Terminal nodes

b. Internal nodes

c. A common ancestor to all other nodes

d. Branch lengths that represent the amount of evolutionary divergence between nodes

---

The Alignment of Homologous Sequences

Today, phylogenetic trees are often constructed from DNA sequence da-ta. This construction requires that homologous sequences be compared. Thus, a first step in constructing phylogenetic trees from DNA sequence data is to identify homologous genes and properly align their nucleotide bases. Consider the following sequences that might be found in two different organisms:

Nucleotide position	1 2 3 4 5 6 7 8
Gene *X* from species A	5′–A T T G C G A A–3′
Gene *X* from species B	5′–A T G C C A A C–3′

These two sequences can be aligned in several possible ways. We might assume that there have been base substitutions at positions 3, 4, 6, and 8:

Nucleotide position	1 2 3 4 5 6 7 8
Gene *X* from species A	5′–A T T G C G A A–3′
Gene *X* from species B	5′–A T G C C A A C–3′

Alternatively, we might assume that a nucleotide at position 3 has been inserted or deleted, generating a gap in the sequence of species B, and that there has been a single nucleotide substitution at position 6:

Nucleotide position	1 2 3 4 5 6 7 8
Gene *X* from species A	5′–A T T G C G A A–3′
Gene *X* from species B	5′–A T – G C C A A C–3′

The second alignment requires fewer evolutionary steps (a deletion or insertion plus one base substitution) than does the first alignment (four base substitutions). Sequence alignments are usually made by computer programs that include

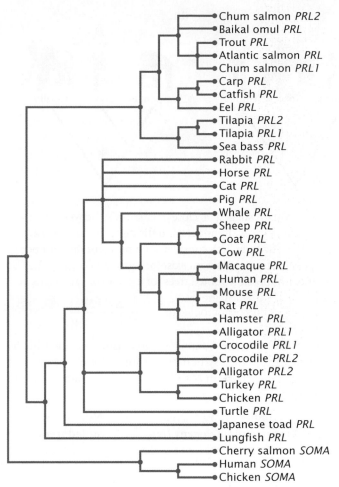

26.14 A gene tree can be used to represent the evolutionary relationships among a group of genes. This gene tree is a rooted tree, in which *PRL* represents a prolactin gene; *PRL1* and *PRL2* are two different prolactin genes found in the same organism; and *SOMA* represents a somatropin gene, which is related to prolactin genes. [After M. P. Simmons and J. V. Freudestein, Uninode coding vs. gene tree parsimony for phylogenetic reconstruction using duplicate genes, *Molecular Phylogenetics and Evolution* 23:488, 2002.]

assumptions about which types of change are more likely to take place. If two sequences have undergone much divergence, then generating alignments can be difficult. Most phylogenetic comparisons are based on alignments requiring the insertion of gaps into the sequence to maximize their similarity.

The Construction of Phylogenetic Trees

Consider a simple phylogeny that depicts the evolutionary relationships among three organisms—humans, chimpanzees, and gorillas. Charles Darwin originally proposed that chimpanzees and gorillas were closely related to humans and modern research supports a close relationship between these three species. There are three possible phylogenetic trees for humans, chimpanzees, and gorillas (**Figure 26.15**). The goal of the evolutionary biologist is to determine which of the trees is correct. Molecular data have been applied to this question;

(a)

(b)

(c)

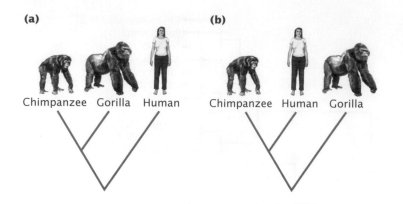

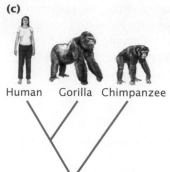

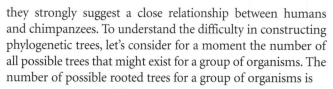

26.15 There are three potential phylogenetic trees for a group of three organisms.

they strongly suggest a close relationship between humans and chimpanzees. To understand the difficulty in constructing phylogenetic trees, let's consider for a moment the number of all possible trees that might exist for a group of organisms. The number of possible rooted trees for a group of organisms is

$$\text{number of rooted trees} = \frac{(2N-3)!}{2^{N-2}(N-2)!}$$

where N equals the number of organisms included in the phylogeny, and the ! symbol stands for factorial, the product of all the integers from N to 1. Substituting values of N into this equation, we find:

Number of organisms included in phylogeny (N)	Number of rooted trees
2	1
3	3
4	15
5	105
10	34,459,425
20	8.2×10^{21}

As the number of organisms in the phylogeny increases beyond just a few, the number of possible rooted trees becomes astronomically large. Clearly, choosing the best tree by directly comparing all the possibilities is impossible.

There are several different approaches to inferring evolutionary relationships and constructing phylogenetic trees. In one approach, termed the *distance approach*, evolutionary relationships are inferred on the basis of the overall degree of similarity between organisms. Typically, a number of different phenotypic characteristics or gene sequences are examined and the organisms are grouped on the basis of their overall similarity, taking into consideration all the examined characteristics and sequences. A second approach, called the *maximum parsimony approach*, infers phylogenetic relationships on the basis of the fewest number of evolutionary changes that must have taken place since the organisms last had an ancestor in common. A third approach, called *maximum likelihood* and *Baysian methods*, infers phylogenetic relationships on the basis of which phylogeny maximizes the probability of obtaining the set of characteristics exhibited by the organisms. In this approach, a phylogeny with a higher probability of

producing the observed characters in the organisms studied is preferred over a phylogeny with a lower probability.

With all three approaches to constructing phylogenies, several different numerical methods are available for the construction of phylogenetic trees. These methods are beyond the scope of this book. All include certain assumptions that help limit the number of different trees that must be considered; most rely on computer programs that compare phenotypic characteristics or sequence data to sequentially group organisms in the construction of the tree.

CONCEPTS

Molecular data can be used to infer phylogenies (evolutionary histories) of groups of living organisms. The construction of phylogenies requires the proper alignment of homologous DNA sequences. Several different approaches are used to reconstruct phylogenies, including distance methods, maximum parsimony methods, and maximum likelihood and Baysian methods.

26.5 Patterns of Evolution Are Revealed by Changes at the Molecular Level

In recent years, the ability to analyze genetic variation at the molecular level has revealed a number of evolutionary processes and features that were formerly unsuspected. This section considers several aspects of evolution at the molecular level.

Rates of Molecular Evolution

Findings from molecular studies of numerous genes have demonstrated that different genes and different parts of the same gene often evolve at different rates.

Rates of nucleotide substituion Rates of evolutionary change in nucleotide sequences are usually measured as the rate of nucleotide substitution, which is the number of substitutions taking place per nucleotide site per year. To calculate the rate of nucleotide substitution, we begin by looking at homologous sequences from different organisms. We first align the homologous sequences and then compare the sequences and determine the number of nucleotides that differ between the two sequences. We might compare the growth-hormone sequences for mice and rats, which diverged from a common ancestor some 15 million years ago. From the number of different nucleotides in their growth-hormone genes, we compute the number of nucleotide substitutions that must have taken place since they diverged. Because the same site may have mutated more than once, the number of nucleotide substitutions is larger than the number of nucleotide differences in two sequences; so special mathematical methods have been developed for inferring the actual number of substitutions likely to have taken place.

When we have the number of nucleotide substitutions per nucleotide site, we divide by the amount of evolutionary time that separates the two organisms (usually obtained from the fossil record) to obtain an overall rate of nucleotide substitution. For the mouse and rat growth-hormone gene, the overall rate of nucleotide substitution is approximately 8×10^{-9} substitutions per site per year.

Nonsynonymous and synonymous rates of substituion

Nucleotide changes in a gene that alter the amino acid sequence of a protein are referred to as nonsynonymous substitutions. Nucleotide changes, particularly those at the third position of a codon, that do not alter the amino acid sequence are called synonymous substitutions. The rate of nonsynonymous substitution varies widely among mammalian genes. The rate for the α-actin protein is only 0.01×10^{-9} substitutions per site per year, whereas the rate for interferon γ is 2.79×10^{-9}, about 1000 times as high. The rate of synonymous substitution also varies among genes, but not to the extent of variation in the nonsynonymous rate. For most protein-encoding genes, the synonymous rate of change is considerably higher than the nonsynonymous rate because synonymous mutations are tolerated by natural selection (**Table 26.5**). Nonsynonymous mutations, on the other hand, alter the amino acid sequence of the protein and, in many cases, are detrimental to the fitness of the organism; so most of these mutations are eliminated by natural selection.

Substituion rates for different parts of a gene

Different parts of a gene also evolve at different rates, with the highest rates of substitutions in regions of the gene that have the least effect on function, such as the third position of a codon, flanking regions, and introns (**Figure 26.16**). The 5′ and 3′ flanking regions of genes are not transcribed into RNA; therefore, substitutions in these regions do not alter the amino acid sequence of the protein, although they may affect gene expression (see Chapters 16 and 17). Rates of substitution in introns are nearly as high. Although these nucleotides do not encode amino acids, introns must be spliced out of the pre-mRNA for a functional protein to be produced, and particular sequences are required at the 5′ splice site, 3′ splice site, and branch point for correct splicing (see Chapter 14).

Table 26.5	Rates of nonsynonymous and synonymous substitutions in mammalian genes based on human–rodent comparisons

Gene	Nonsynonymous Rate (per site per 10^9 years)	Synonymous Rate (per site per 10^9 years)
α-Actin	0.01	3.68
β-Actin	0.03	3.13
Albumin	0.91	6.63
Aldolase A	0.07	3.59
Apoprotein E	0.98	4.04
Creatine kinase	0.15	3.08
Erythropoietin	0.72	4.34
α-Globin	0.55	5.14
β-Globin	0.80	3.05
Growth hormone	1.23	4.95
Histone 3	0.00	6.38
Immunoglobulin heavy chain (variable region)	1.07	5.66
Insulin	0.13	4.02
Interferon α 1	1.41	3.53
Interferon γ	2.79	8.59
Luteinizing hormone	1.02	3.29
Somatostatin-28	0.00	3.97

Source: After W. Li and D. Graur, *Fundamentals of Molecular Evolution* (Sunderland, Mass.: Sinauer, 1991), p. 69, Table 1.

Substitution rates are somewhat lower in the 5′ and 3′ untranslated regions of a gene. These regions are transcribed into RNA but do not encode amino acids. The 5′ untranslated region contains the ribosome-binding site, which is essential for translation, and the 3′ untranslated region contains sequences that may function in regulating mRNA stability and translation; so substitutions in these regions may have deleterious effects on organismal fitness and may not be tolerated.

The lowest rates of substitution are seen in nonsynonymous changes in the coding region, because these substitutions always alter the amino acid sequence of the protein and are often deleterious. High rates of substitution occur in pseudogenes, which are duplicated nonfunctional copies of genes that have acquired mutations. Such genes no longer produce a functional product; so mutations in pseudogenes have no effect on the fitness of the organism. The highest rates of substitution are seen in noncoding regions of DNA found between genes.

In summary, there is a relation between the function of a sequence and its rate of evolution; higher rates are found

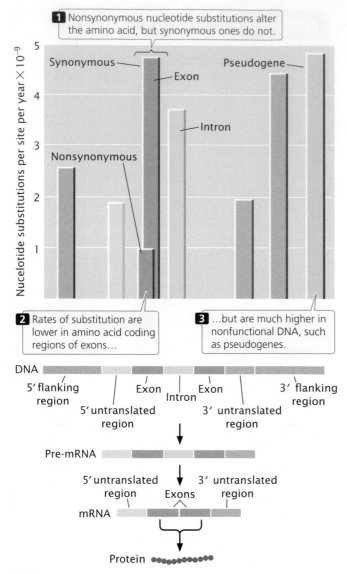

1 Nonsynonymous nucleotide substitutions alter the amino acid, but synonymous ones do not.

2 Rates of substitution are lower in amino acid coding regions of exons…

3 …but are much higher in nonfunctional DNA, such as pseudogenes.

26.16 Different parts of genes evolve at different rates. The highest rates of nucleotide substitution are in sequences that have the least effect on protein function.

where they have the least effect on function. This observation fits with the neutral-mutation hypothesis, which predicts that molecular variation is not affected by natural selection.

The Molecular Clock

The neutral-mutation hypothesis proposes that evolutionary change at the molecular level takes place primarily through the fixation of neutral mutations by genetic drift. The rate at which one neutral mutation replaces another depends only on the mutation rate, which should be fairly constant for any particular gene. If the rate at which a protein evolves is roughly constant over time, the amount of molecular change that a protein has undergone can be used as a **molecular clock** to date evolutionary events.

For example, the enzyme cytochrome *c* could be examined in two organisms known from fossil evidence to have had a common ancestor 400 million years ago. By determining the number of differences in the cytochrome *c* amino acid sequences in each organism, we could calculate the number of substitutions that have occurred per amino acid site. The occurrence of 20 amino acid substitutions since the two organisms diverged indicates an average rate of 5 substitutions per 100 million years. Knowing how fast the molecular clock ticks allows us to use molecular changes in cytochrome *c* to date other evolutionary events: if we found that cytochrome *c* in two organisms differed by 15 amino acid substitutions, our molecular clock would suggest that they diverged some 300 million years ago. If we assumed some error in our estimate of the rate of amino acid substitution, statistical analysis would show that the true divergence time might range from 160 million to 440 million years. The molecular clock is analogous to geological dating based on the radioactive decay of elements.

The molecular clock was proposed by Emile Zuckerandl and Linus Pauling in 1965 as a possible means of dating evolutionary events on the basis of molecules in present-day organisms. A number of studies have examined the rate of evolutionary change in proteins (**Figure 26.17**), and the molecular clock has been widely used to date evolutionary events when the fossil record is absent or ambiguous. However, the results of several studies have shown that the molecular clock does not always tick at a constant rate, particularly over shorter time periods, and this method remains controversial.

CONCEPTS

Different genes and different parts of the same gene evolve at different rates. Those parts of genes that have the least effect on function tend to evolve at the highest rates. The idea of the molecular clock is that individual proteins and genes evolve at a constant rate and that the differences in the sequences of present-day organisms can be used to date past evolutionary events.

✔ CONCEPT CHECK 7

In general, which types of sequences are expected to exhibit the slowest evolutionary change?

a.   Synonymous changes in amino acid coding regions of exons

b.   Nonsynonymous changes in amino acid coding regions of exons

c.   Introns

d.   Pseudogenes

Genome Evolution

The rapid growth of sequence data available in DNA databases has been a source of insight into evolutionary processes. Whole-genome sequences also are providing new information about how genomes evolve and the processes that shape the size, complexity, and organization of genomes.

(a)

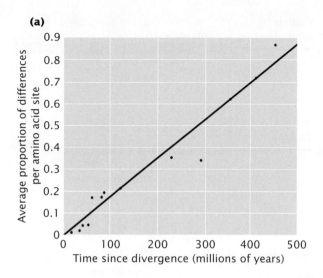

(b)

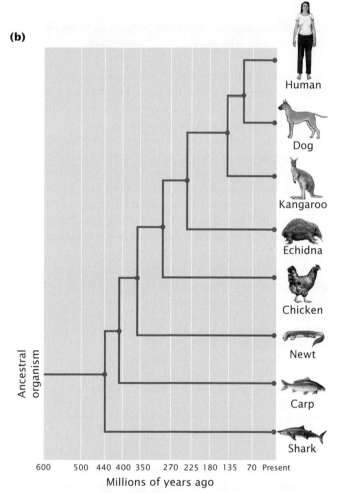

26.17 The molecular clock is based on the assumption of a constant rate of change in protein or DNA sequence. (a) Relation between the rate of amino acid substitution and time since divergence, based in part on amino acid sequences of alpha hemoglobin from the eight species shown in part *b*. The constant rate of evolution in protein and DNA sequences has been used as a molecular clock to date past evolutionary events. (b) Phylogeny of eight of the species that are plotted in part *a* and their approximate times of divergence based on the fossil record.

Exon shuffling Many proteins are composed of groups of amino acids, called domains, that specify discrete functions or contribute to the molecular structure of a protein. For example, in Chapter 16, we considered the DNA-binding domains of proteins that regulate gene expression. Analyses of gene sequences from eukaryotic organisms indicate that exons often encode discrete functional domains of proteins.

Some genes elongated and evolved new functions when one or more exons duplicated and underwent divergence. For example, the human serum-albumin gene is made up of three copies of a sequence that encodes a protein domain consisting of 195 amino acids. Additionally, the genes that encode human immunoglobulins have undergone repeated tandem duplications, creating many similar *V, J, D,* and *C* segments (see pp. 629–630 in Chapter 22) that enable the immune system to respond to almost any foreign substance that enters the body.

A comparison of DNA sequences from different genes reveals that new genes have repeatedly evolved through a process called **exon shuffling**. In this process, exons of different genes are exchanged, creating genes that are mosaics of other genes. For example, tissue plasminogen activator (TPA) is an enzyme that contains four domains of three different types, called kringle, growth factor, and finger. Each domain is encoded by a different exon. The gene for TPA is believed to have acquired its exons from other genes that encode different proteins: the kringle exon came from the plasminogen gene, the growth-factor exon came from the epidermal growth-factor gene; and the finger exon came from the fibronectin gene. The mechanism by which exon shuffling takes place is poorly known, but new proteins with different combinations of functions encoded by other genes apparently have repeatedly evolved by this mechanism.

Gene duplication New genes have also evolved through the duplication of whole genes and their subsequent divergence. This process creates **multigene families**, sets of genes that are similar in sequence but encode different products. For example, humans possess 13 different genes found on chromosomes 11 and 16 that encode globinlike molecules, which take part in oxygen transport (**Figure 26.18**). All of these genes have a similar structure, with three exons separated by two introns, and are assumed to have evolved through repeated duplication and divergence from a single globin gene in a distant ancestor. This ancestral gene is thought to have been most similar to the present-day myoglobin gene and first duplicated to produce an α/β-globin precursor gene and the myoglobin gene. The α/β-globin gene then underwent another duplication to give rise to a primordial α-globin gene and a primordial β-globin gene. Subsequent duplications led to multiple α-globin and β-globin genes. Similarly, vertebrates contain four clusters of *Hox* genes, each cluster comprising from 9 to 11 genes. *Hox* genes play an important role in development (see pp. 620–621 in Chapter 22).

Some gene families include genes that are arrayed in tandem on the same chromosome; others are dispersed

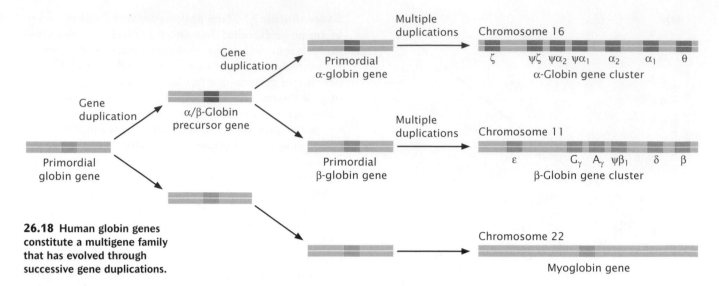

26.18 Human globin genes constitute a multigene family that has evolved through successive gene duplications.

among different chromosomes. Gene duplication is a common occurrence in eukaryotic genomes; for example, about 5% of the human genome consists of duplicated segments.

Gene duplication provides a mechanism for the addition of new genes with novel functions; after a gene duplicates, there are two copies of the sequence, one of which is free to change and potentially take on a new function. The extra copy of the gene may, for example, become active at a different time in development or be expressed in a different tissue or even diverge and encode a protein having different amino acids. However, the most common fate of gene duplication is that one copy acquires a mutation that renders it nonfunctional, giving rise to a pseudogene. Pseudogenes are common in the genomes of complex eukaryotes; the human genome is estimated to contain as many as 20,000 pseudogenes.

Whole-genome duplication In addition to the duplication of individual genes, whole genomes of some organisms have apparently duplicated in the past. For example, a comparison of the genome of the yeast *Saccharomyces cerevisiae* with the genomes of other fungi reveals that *S. cerevisiae* or one of its immediate ancestors underwent a whole-genome duplication, generating two copies of every gene. Many of the copies subsequently acquired new functions; others acquired mutations that destroyed the original function and then diverged into random DNA sequences. Whole-genome duplication can take place through polyploidy.

Horizontal gene transfer Most organisms acquire their genomes through vertical transmission—transfer through the reproduction of genetic information from parents to offspring. Most phylogenetic trees assume vertical transmission of genetic information. Findings from DNA sequence studies reveal that DNA sequences are sometimes exchanged by a horizontal gene transfer, in which DNA is transferred between individuals of different species (see Chapter 8). This process is especially common among bacteria, and there are a number of documented cases in which genes are transferred from bacteria to eukaryotes. The extent of horizontal gene transfer among eukaryotic organisms is controversial, with few well-documented cases. Horizontal gene transfer can obscure phylogenetic relationships and make the reconstruction of phylogenetic trees difficult.

> **CONCEPTS**
>
> New genes may evolve through the duplication of exons, shuffling of exons, duplication of genes, and duplication of whole genomes. Genes can be passed among distantly related organisms through horizontal gene transfer.

CONCEPTS SUMMARY

- Evolution is genetic change taking place within a group of organisms. It is a two-step process: (1) genetic variation arises and (2) genetic variants change in frequency.
- Anagenesis refers to change within a single lineage; cladogenesis is the splitting of one lineage into two.
- Molecular methods offer a number of advantages for the study of evolution.
- The use of protein electrophoresis to study genetic variation in natural populations showed that most natural populations have large amounts of genetic variation in their proteins. The neutral-mutation hypothesis proposes that molecular variation is selectively neutral and is shaped largely by mutation and genetic drift. The balance hypothesis proposes that molecular variation is maintained largely by balancing selection.
- Variation in DNA sequences can be assessed by analyzing restriction fragment length polymorphisms, microsatellites, and data from direct sequencing.

- A species can be defined as a group of organisms that are capable of interbreeding with one another and are reproductively isolated from the members of other species.

- Species are prevented from exchanging genes by reproductive isolating mechanisms, either before a zygote has formed (prezygotic reproductive isolation) or after a zygote has formed (postzygotic reproductive isolation).

- Allopatric speciation arises when a geographic barrier prevents gene flow between two populations. With the passage of time, the two populations acquire genetic differences that may lead to reproductive isolating mechanisms.

- Sympatric speciation arises when reproductive isolation exists in the absence of any geographic barrier. It may arise under special circumstances.

- Some species arise only after populations have undergone considerable genetic differences; others arise after changes have taken place in only a few genes.

- Evolutionary relationships (a phylogeny) can be represented by a phylogenetic tree, consisting of nodes that represent organisms and branches that represent their evolutionary connections.

- Approaches to constructing phylogenetic trees include the distance approach, the maximum parsimony approach, and the maximum likelihood (Baysian methods) approach.

- Different parts of the genome show different amounts of genetic variation. In general, those parts that have the least effect on function evolve at the highest rates.

- The molecular-clock hypothesis proposes a constant rate of nucleotide substitution, providing a means of dating evolutionary events by looking at nucleotide differences between organisms.

- Genome evolution takes place through the duplication and shuffling of exons, the duplication of genes to form gene families, whole-genome duplication, and the horizontal transfer of genes between organisms.

IMPORTANT TERMS

evolution (p. 727)
anagenesis (p. 723)
cladogenesis (p. 723)
proportion of polymorphic loci (p. 725)
expected heterozygosity (p. 725)
neutral-mutation hypothesis (p. 725)
balance hypothesis (p. 725)
species (p. 729)
biological species concept (p. 729)
reproductive isolating mechanism (p. 729)
prezygotic reproductive isolating mechanism (p. 729)

ecological isolation (p. 729)
behavioral isolation (p. 730)
temporal isolation (p. 730)
mechanical isolation (p. 730)
gametic isolation (p. 730)
postzygotic reproductive isolating mechanism (p. 730)
hybrid inviability (p. 730)
hybrid sterility (p. 730)
hybrid breakdown (p. 730)
speciation (p. 731)
allopatric speciation (p. 731)
sympatric speciation (p. 731)

phylogeny (p. 736)
phylogenetic tree (p. 736)
node (p. 736)
branch (p. 736)
rooted tree (p. 736)
gene tree (p. 737)
molecular clock (p. 740)
exon shuffling (p. 741)
multigene family (p. 741)

ANSWERS TO CONCEPT CHECKS

1. First genetic variation arises. Then various evolutionary forces cause changes in the frequency of genetic variants.

2. c

3. Microsatellites are often highly variable among individuals. They can be amplified with the use of PCR, and so they can be detected with a small amount of starting DNA. Finally, the detection and analysis of microsatellites can be automated.

4. b

5. Genetic drift can bring about changes in the allelic frequencies of populations and lead to genetic differences among populations. Genetic differentiation is the cause of postzygotic and prezygotic reproductive isolation between populations that leads to speciation.

6. c

7. b

COMPREHENSION QUESTIONS

Section 26.1

1. How is biological evolution defined?
*2. What are the two steps in the process of evolution?
3. How is anagenesis different from cladogenesis?

Section 26.2

*4. As a measure of genetic variation, why is the expected heterozygosity often preferred to the observed heterozygosity?

5. Why does protein variation, as revealed by electrophoresis, underestimate the amount of true genetic variation?

6. What are some of the advantages of using molecular data in evolutionary studies?

*7. What is the key difference between the neutral-mutation hypothesis and the balance hypothesis?

8. Describe some of the methods that have been used to study variation in DNA.

Section 26.3

*9. What is the biological species concept?

10. What is the difference between prezygotic and postzygotic reproductive isolating mechanisms. List the different types of each.

11. What is the basic difference between allopatric and sympatric modes of speciation?

*12. Briefly outline the process of allopatric speciation.

13. What are some of the difficulties with sympatric speciation?

*14. Briefly explain how switching from hawthorn fruits to apples has led to genetic differentiation and partial reproductive isolation in *Rhagoletis pomonella*.

Section 26.4

15. Draw a simple phylogenetic tree and identify a node, a branch, and an outgroup.

*16. Briefly describe differences among the distance approach, the maximum parsimony approach, and the maximum likelihood approach to the reconstruction of phylogenetic trees.

Section 26.5

17. Outline the different rates of evolution that are typically seen in different parts of a protein-encoding gene. What might account for these differences?

*18. What is the molecular clock?

19. What is exon shuffling? How can it lead to the evolution of new genes?

20. What is a multigene family? What processes produce multigene families?

*21. Define horizontal gene transfer. What problems does it cause for evolutionary biologists?

APPLICATION QUESTIONS AND PROBLEMS

Section 26.1

22. The following illustrations represent two different patterns of evolution. Briefly discuss the differences in these two patterns, particularly in regard to the role of cladogenesis in evolutionary change.

 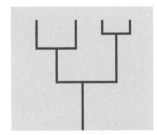

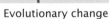

Evolutionary change   Evolutionary change

Section 26. 2

*23. Donald Levin used protein electrophoresis to study genetic variation in two species of wildflowers in Texas, *Phlox drummondi* and *P. cuspidate* (D. Levin. 1978. *Evolution* 32:245–263). *P. drummondi* reproduces only by cross fertilization, whereas *P. cuspidate* is partly self-fertilizing. The table below gives allelic frequencies for several loci and populations studied by Levin.

a. Calculate the percentage of polymorphic loci and the expected heterozygosity for each species. For the expected heterozygosity, calculate the mean for all loci in a population, and then calculate the mean for all populations.

b. What tentative conclusions can you draw about the effect of self-fertilization on genetic variation in these flowers?

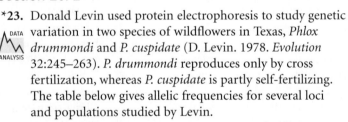

| | | Allelic frequencies | | | | | | | |
| | | Adh | | Got-2 | | Pgi-2 | | Pgm-2 | |
Species	Population	a	b	a	b	a	b	a	b
P. drummondii	1	0.10	0.90	0.02	0.98	0.04	0.96	1.0	0.0
P. drummondii	2	0.11	0.89	0.0	1.0	0.31	0.69	0.83	0.17
P. drummondii	3	0.26	0.74	1.0	0.0	0.14	0.86	1.0	0.0
P. cuspidate	1	0.0	1.0	0.0	1.0	0.0	1.0	0.0	1.0
P. cuspidate	2	0.0	1.0	0.0	1.0	0.0	1.0	0.0	1.0
P. cuspidate	3	0.91	0.09	0.0	1.0	0.0	1.0	0.0	1.0

Note: *a* and *b* represent different alleles at each locus.

Section 26.3

24. Which of the isolating mechanisms listed in Table 26.3 have partly evolved between apple and hawthorn host races of *Rhagoletis pomenella*, the apple maggot fly?

25. In Section 26.3, we considered the sympatric evolution of reproductive isolating mechanisms in host races of *Rhagoletis pomenella*, the apple maggot fly. The wasp *Diachasma alloeum* parasitizes apple maggot flies, laying its eggs on the larvae of the flies. Immature wasps hatch from the eggs and feed on the fly larvae. Research by Andrew Forbes and his colleagues (Forbes et al. 2009. *Science* 323:776–779) demonstrated that wasps that parasitize apple races of *R. pomenella* are genetically differentiated from those that parasitize hawthorn races of *R. pomenella*. They also found that wasps that prey on the apple race of the flies are attracted to odors from apples, whereas wasps that prey on the hawthorn race are attracted to odors from hawthorn fruit. (a) Propose an explanation for how genetic differences might have evolved between the wasps that parasitize the two races of *R. pomenella*. (b) How might these differences lead to speciation in the wasps?

Section 26.4

*****26.** How many rooted trees are theoretically possible for a group of 7 organisms? How many for 12 organisms?

27. Align the following sequence so as to maximize their similarity. What is the minimum number of evolutionary steps that that separate these two sequences?

TTGCAAAC

TGAAACTG

28. Michael Bunce and his colleagues in England, Canada, and the United States extracted and sequenced mitochondrial DNA from fossils of Haast's eagle, a gigantic eagle that was driven to extinction 700 years ago when humans first arrived in New Zealand (M. Bunce et al. 2005. *Plos Biology* 3:44–46). Using mitochondrial DNA sequences from living eagles and those from Haast eagle fossils, they created the phylogenetic tree at the right.
On this phylogenetic tree, identify (a) all terminal nodes; (b) all internal nodes; (c) one example of a branch; and (d) the outgroup.

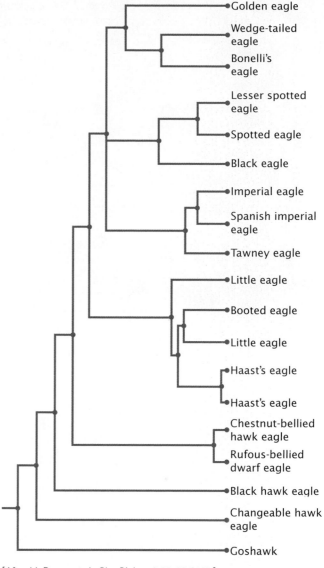

[After M. Bunce et al., *Plos Biology* 3:44–46, 2005.]

29. On the basis of the phylogeny of Darwin's finches shown in Figure 26.9, predict which two species in each of the following groups will be the most similar genetically.

a. *Camarhynchus parvulus, Camarhynchus psittacula, Camarhynchus pallida*

b. *Camarhynchus parvulus, Camarhynchus pallida, P. crassirostris*

c. *Geospiza difficilis, Geospiza conirostris, Geospiza scandens*

d. *Camarhynchus parvulus, Certhidea fusca, Pinaroloxias inornata*

CHALLENGE QUESTIONS

Section 26.3

30. Explain why natural selection may cause prezygotic reproductive isolating mechanisms to evolve if postzygotic reproductive isolating mechanisms are already present but natural selection can never cause the evolution of postzygotic reproductive isolating mechanisms.

31. Polyploidy is very common in flowering plants: approximately 40% of all flowering plant species are polyploids. Although polyploidy exists in many different animal groups, it is much less common. Why is polyploidy more common in plants than in animals? Give one or more possible reasons.

[Wally Eberhart/Visuals Unlimited.]

[Sinclair Stammers/Photo Researchers.]

[Eye of Science/Photo Researchers.]

[Steve Gschmeissner/Science Photo Library/Alamy.]

[Biophoto Associates/ Photo Researchers.]

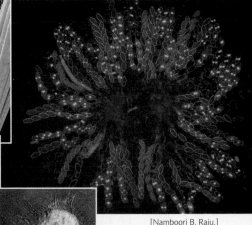
[Namboori B. Raju.]

Model genetic organisms possess characteristics that make them useful for genetic studies. Shown are several organisms commonly used in genetic studies.

Reference Guide to Model Genetic Organisms

What do Lou Gehrig, the finest first baseman in major league history, and Stephen Hawking, the world's most famous theoretical physicist, have in common? They both suffered or suffer from amyotrophic lateral sclerosis (ALS, also known as Lou Gehrig disease), a degenerative neurological disease that leads to progressive weakness and wasting of skeletal muscles. Interestingly, recent research findings reveal that some athletes with ALS symptoms in fact suffer from brain trauma due to sports injuries; so Lou Gehrig might not have had the disease named for him.

Most cases of ALS are sporadic, appearing in people who have no family history of the disease. However, about 10% of the cases run in families and are inherited. In 2004, geneticists discovered a large Brazilian family with multiple cases of ALS. Genetic analysis revealed that ALS in this family is due to a mutation in a gene called *VABP*, which encodes a vesicle-associated membrane protein.

To better understand how mutations in *VABP* lead to the symptoms of ALS, geneticists turned to an unlikely subject—the fruit fly *Drosophila melanogaster*. Fruit flies don't have ALS, but they do possess a gene very similar to *VABP*. Using a wide array of techniques that have been developed for genetically manipulating fruit flies, geneticists created transgenic flies with the mutant sequence of the *VABP* gene that causes ALS in humans. These flies are a disease model for ALS and are being used to better understand what the gene *VABP* does normally and how its disruption can lead to ALS.

The field of genetics has been greatly influenced and shaped by a few key organisms—called model genetic organisms—whose characteristics make them particularly amenable to genetic studies. The use of *Drosophila* for studying ALS in humans illustrates the power of this approach. Because features of genetic systems are common to many organisms, research conducted on one species can often be a source of insight into the genetic systems of other species. This commonality of genetic function means that geneticists can focus their efforts on model organisms that are easy to work with and likely to yield results.

Model genetic organisms possess life cycles and genomic features that make them well suited to genetic study and analysis. Some key features possessed by many model genetic organisms include:

- a short generation time, and so several generations of genetic crosses can be examined in reasonable time;

- the production of numerous progeny, and so genetic ratios can be easily observed;

- the ability to carry out and control genetic crosses in the organism;

- the ability to be reared in a laboratory environment, requiring little space and few resources to maintain;

- the availability of numerous genetic variants; and

- an accumulated body of knowledge about their genetic systems.

In recent years, the genomes of many model genetic organisms have been completely sequenced, greatly facilitating their use in genetic research.

Not all model organisms possess all of these characteristics. However, each model genetic organism has one or more features that make it useful for genetic analysis. For example, corn cannot be easily grown in the laboratory (and usually isn't) and it has a relatively long generation time, but it produces numerous progeny and there are many genetic variants of corn available for study.

This reference guide highlights six model genetic organisms with important roles in the development of genetics: the fruit fly (*Drosophila melanogaster*), bacterium (*Escherichia coli*), roundworm (*Caenorhabditis elegans*), thale cress plant (*Arabidopsis thaliana*), house mouse (*Mus musculus*), and yeast (*Saccharomyces cerevisiae*). These six organisms have been widely used in genetic research and instruction. A number of other organisms also are used as model systems in genetics, including corn (*Zea mays*), zebrafish (*Danio rerio*), clawed frog (*Xenopus lavis*), bread mold (*Neurospora crassa*), rat (*Rattus norvegicus*), and Rhesus macaque (*Macaca mulatta*), just to mention a few.

A1

The Fruit Fly *Drosophila melanogaster*

*D*rosophila melanogaster, a fruit fly, was among the first organisms used for genetic analysis and, today, it is one of the most widely used and best known genetically of all organisms. It has played an important role in studies of linkage, epistasis, chromosome genetics, development, behavior, and evolution. Because all organisms use a common genetic system, understanding a process such as replication or transcription in fruit flies helps us to understand these same processes in humans and other eukaryotes.

Drosophila is a genus of more than 1000 described species of small flies (about 1 to 2 mm in length) that frequently feed and reproduce on fruit, although they rarely cause damage and are not considered economic pests. The best known and most widely studied of the fruit flies is *D. melanogaster*, but genetic studies have been extended to many other species of the genus as well. *D. melanogaster* first began to appear in biological laboratories about 1900. After first taking up breeding experiments with mice and rats, Thomas Hunt Morgan began using fruit flies in experimental studies of heredity at Columbia University. Morgan's laboratory, located on the top floor of Schermerhorn Hall, became known as the Fly Room (see Figure 4.11b). To say that the Fly Room was unimpressive is an understatement. The cramped room, only about 16 by 23 feet, was filled with eight desks, each occupied by a student and his experiments. The primitive laboratory equipment consisted of little more than milk bottles for rearing the flies and hand-held lenses for observing their traits. Later, microscopes replaced the hand-held lenses, and crude incubators were added to maintain the fly cultures, but even these additions did little to increase the physical sophistication of the laboratory. Morgan and his students were not tidy: cockroaches were abundant (living off spilled *Drosophila* food), dirty milk bottles filled the sink, ripe bananas—food for the flies—hung from the ceiling, and escaped fruit flies hovered everywhere. In spite of its physical limitations, the Fly Room was the source of some of the most important research in the history of biology. There was daily excitement among the students, some of whom initially came to the laboratory as undergraduates. The close quarters facilitated informality and the free flow of ideas. Morgan and the Fly Room illustrate the tremendous importance of "atmosphere" in producing good science. Morgan and his students eventually used *Drosophila* to elucidate many basic principles of heredity, including sex-linked inheritance, epistasis, multiple alleles, and gene mapping.

Advantages of *D. melanogaster* as a model genetic organism
Drosophila's widespread use in genetic studies is no accident. The fruit fly has a number of characteristics that make it an ideal subject for genetic investigations. Compared with other organisms, it has a relatively short generation time; fruit flies will complete an entire generation in about 10 days at room temperature, and so several generations can be studied within a few weeks. Although *D. melanogaster* has a short generation time, it possesses a complex life cycle, passing through several different developmental stages, including egg, larva, pupa, and adult. A female fruit fly is capable of mating within 8 hours of emergence and typically begins to lay eggs after about 2 days. Fruit flies also produce a large number of offspring, laying as many as 400 to 500 eggs in a 10-day period. Thus, large numbers of progeny can be obtained from a single genetic cross.

Another advantage is that fruit flies are easy to culture in the laboratory. They are usually raised in small glass vials or bottles and are fed easily prepared, pastelike food consisting of bananas or corn meal and molasses. Males and females are readily distinguished and virgin females are easily isolated, facilitating genetic crosses. The flies are small, requiring little space—several hundred can be raised in a half-pint bottle—but they are large enough for many mutations to be easily observed with a hand lens or a dissecting microscope.

Finally, *D. melanogaster* is the organism of choice for many geneticists because it has a relatively small genome consisting of 175 million base pairs of DNA, which is only about 5% of the size of the human genome. It has four pairs of chromosomes: three pairs of autosomes and one pair of sex chromosomes. The X chromosome (designated chromosome 1) is large and acrocentric, whereas the Y chromosome is large and submetacentric, although it contains very little genetic information. Chromosomes 2 and 3 are large and metacentric; chromosome 4 is a very small acrocentric chromosome. In the salivary glands, the chromosomes are very large (see p. 297 in Chapter 11), making *Drosophila* an excellent subject for chromosome studies. In 2000, the complete genome of *D. melanogaster* was sequenced, followed by the sequencing of the genome of *D. pseudoobscura* in 2005 and the genomes of 10 additional *Drosophila* genomes in 2007. *Drosophila* continues today to be one of the most versatile and powerful of all genetic model organisms. ■

The Fruit Fly
Drosophila melanogaster

ADVANTAGES

- Small size
- Short generation time of 10 days at room temperature
- Each female lays 400–500 eggs
- Easy to culture in laboratory
- Small genome
- Large chromosomes
- Many mutations available

STATS

Taxonomy:	Insect
Size:	2–3 mm in length
Anatomy:	3 body segments, 6 legs, 1 pair of wings
Habitat:	Feeds and reproduces on fruit

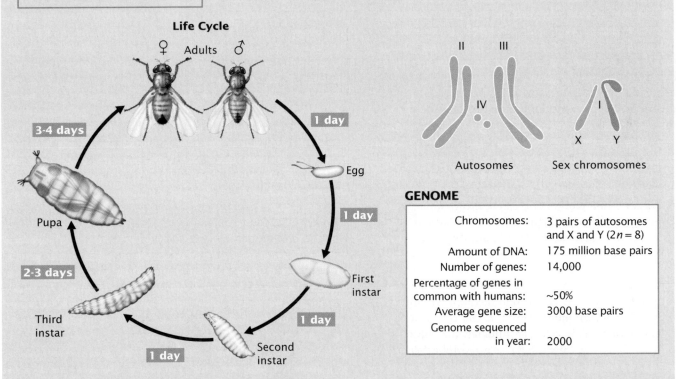

Life Cycle

♀ Adults ♂

3-4 days

1 day

Egg

1 day

First instar

1 day

Second instar

1 day

Third instar

2-3 days

Pupa

II   III

IV

Autosomes

I

X   Y

Sex chromosomes

GENOME

Chromosomes:	3 pairs of autosomes and X and Y (2*n* = 8)
Amount of DNA:	175 million base pairs
Number of genes:	14,000
Percentage of genes in common with humans:	~50%
Average gene size:	3000 base pairs
Genome sequenced in year:	2000

CONTRIBUTIONS TO GENETICS

- Basic principles of heredity including sex-linked inheritance, multiple alleles, epistatsis, gene mapping, etc.
- Mutation research
- Chromosome variation and behavior
- Population genetics
- Genetic control of pattern formation
- Behavioral genetics

The Bacterium *Escherichia coli*

The most widely studied prokaryotic organism and one of the best genetically characterized of all species is the bacterium *Escherichia coli*. Although some strains of *E. coli* are toxic and cause disease, most are benign and reside naturally in the intestinal tracts of humans and other warm-blooded animals. *E. coli* was first described by Theodore Escherich in 1885 but, for many years, the assumption was that all bacteria reproduced only asexually and that genetic crosses were impossible. In 1946, Joshua Lederberg and Edward Tatum demonstrated that *E. coli* undergoes a type of sexual reproduction; their finding initiated the use of *E. coli* as a model genetic organism. A year later, Lederberg published the first genetic map of *E. coli* based on recombination frequencies and, in 1952, William Hays showed that mating between bacteria is asymmetrical, with one bacterium serving as genetic donor and the other as genetic recipient.

Advantages of *E. coli* as a model genetic organism

Escherichia coli is one of the true workhorses of genetics; its twofold advantage is rapid reproduction and small size. Under optimal conditions, this organism can reproduce every 20 minutes and, in a mere 7 hours, a single bacterial cell can give rise to more than 2 million descendants. One of the values of rapid reproduction is that enormous numbers of cells can be grown quickly, and so even very rare mutations will appear in a short period. Consequently, numerous mutations in *E. coli,* affecting everything from colony appearance to drug resistance, have been isolated and characterized.

Escherichia coli is easy to culture in the laboratory in liquid medium (see Figure 8.1a) or on solid medium within petri plates (see Figure 8.1b). In liquid culture, *E. coli* cells will grow to a concentration of a billion cells per milliliter, and trillions of bacterial cells can be easily grown in a single test tube. When *E. coli* cells are diluted and spread onto the solid medium of a petri dish, individual bacteria reproduce asexually, giving rise to a concentrated clump of 10 million to 100 million genetically identical cells, called a colony. This colony formation makes it easy to isolate genetically pure strains of the bacteria.

The *E. coli* genome

The *E. coli* genome is on a single chromosome and—compared with those of humans, mice, plants, and other multicellular organisms—is relatively small, consisting of 4,638,858 base pairs. If stretched out straight, the DNA molecule in the single *E. coli* chromosome would be 1.6 mm long, almost a thousand times as long as the *E. coli* cell within which it resides (see Figure 11.1). To accommodate this huge amount of DNA within the confines of a single cell, the *E. coli* chromosome is highly coiled and condensed. The information within the *E. coli* chromosome also is compact, having little noncoding DNA between and within the genes and having few sequences for which there is more than one copy. The *E. coli* genome contains an estimated 4300 genes, more than half of which have no known function. These "orphan genes" may play important roles in adapting to unusual environments, coordinating metabolic pathways, organizing the chromosome, or communicating with other bacterial cells. The haploid genome of *E. coli* makes it easy to isolate mutations because there are no dominant genes at the same loci to suppress and mask recessive mutations.

Life cycle of *E. coli*

Wild-type *E. coli* is prototrophic and can grow on minimal medium that contains only glucose and some inorganic salts. Under most conditions, *E. coli* divides about once an hour, although, in a richer medium containing sugars and amino acids, it will divide every 20 minutes. It normally reproduces through simple binary fission, in which the single chromosome of a bacterium replicates and migrates to opposite sides of the cell, followed by cell division, giving rise to two identical daughter cells (see Figure 2.5). Mating between bacteria, called conjugation, is controlled by fertility genes normally located on the F plasmid (see p. 211). In conjugation, one bacterium donates genetic material to another bacterium, followed by genetic recombination that integrates new alleles into the bacterial chromosome. Genetic material can also be exchanged between strains of *E. coli* through transformation and transduction (see Figure 8.7).

Genetic techniques with *E. coli*

Escherichia coli is used in a number of experimental systems in which fundamental genetic processes are studied in detail. For example, in vitro translation systems contain within a test tube all the components necessary to translate the genetic information of a messenger RNA molecule into a polypeptide chain. Similarly, in vitro systems containing components from *E. coli* cells allow transcription, replication, gene expression, and many other important genetic functions to be studied and analyzed under controlled laboratory conditions.

Escherichia coli is also widely used in genetic engineering (recombinant DNA; see Chapter 19). Plasmids have been isolated from *E. coli* and genetically modified to create effective vectors for transferring genes into bacteria and eukaryotic cells. Often, new genetic constructs (DNA sequences created in the laboratory) are assembled and cloned in *E. coli* before transfer to other organisms. Methods have been developed to introduce specific mutations within *E. coli* genes, and so genetic analysis no longer depends on the isolation of randomly occurring mutations. New DNA sequences produced by recombinant DNA can be introduced by transformation into special strains of *E. coli* that are particularly efficient (competent) at taking up DNA.

Because of its powerful advantages as a model genetic organism, *E. coli* has played a leading role in many fundamental discoveries in genetics, including elucidation of the genetic code, probing the nature of replication, and working out the basic mechanisms of gene regulation. ■

Bacterium
Escherichia coli

ADVANTAGES

- Small size
- Rapid reproduction, dividing every 20 minutes under optimal conditions
- Easy to culture in liquid medium or on petri plates
- Small genome
- Many mutants available
- Numerous methods available for genetic engineering

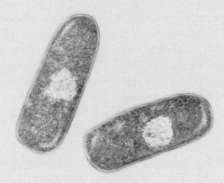

STATS

Taxonomy:	Eubacteria
Size:	1–2 μm in length
Anatomy:	Single cell surrounded by cell wall with nucleoid region
Habitat:	Intestinal tract of warm-blooded animals

Life Cycle

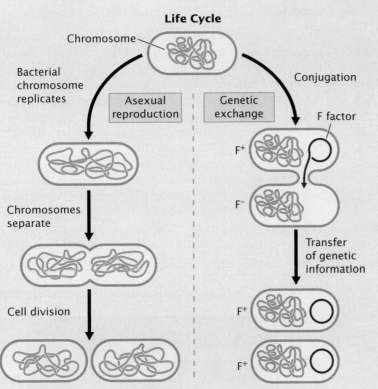

Chromosome

Bacterial chromosome replicates

Asexual reproduction

Genetic exchange

Conjugation

F factor

F⁺

F⁻

Chromosomes separate

Transfer of genetic information

Cell division

F⁺

F⁺

Chromosome

GENOME

Chromosomes:	1 circular chromosome
Amount of DNA:	4.64 million base pairs
Number of genes:	4300
Percentage of genes in common with humans:	8%
Average gene size:	1000 base pairs
Genome sequenced in:	1997

CONTRIBUTIONS TO GENETICS

- Gene regulation
- Molecular biology and biochemistry of genetic processes, such as replication, transcription, translation, recombination
- Gene structure and organization in bacteria
- Workhorse of recombinant DNA
- Gene mutations

The Nematode Worm *Caenorhabditis elegans*

You may be asking, What is a nematode, and why is it a model genetic organism? Although rarely seen, nematodes are among the most abundant organisms on Earth, inhabiting soils throughout the world. Most are free living and cause no harm, but a few are important parasites of plants and animals, including humans. Although *Caenorhabditis elegans* has no economic importance, it has become widely used in genetic studies because of its simple body plan, ease of culture, and high reproductive capacity. First introduced to the study of genetics by Sydney Brenner, who formulated plans in 1962 to use *C. elegans* for the genetic dissection of behavior, this species has made important contributions to the study of development, cell death, aging, and behavior.

Advantages of *C. elegans* as a model genetic organism An ideal genetic organism, *C. elegans* is small, easy to culture, and produces large numbers of offspring. The adult *C. elegans* is about 1 mm in length. Most investigators grow *C. elegans* on agar-filled petri plates that are covered with a lawn of bacteria, which the nematodes devour. Thousands of worms can be easily cultured in a single laboratory. Compared with most multicellular animals, they have a very short generation time, about 3 days at room temperature. And they are prolific reproducers, with a single female producing from 250 to 1000 fertilized eggs in 3 to 4 days.

Another advantage of *C. elegans*, particularly for developmental studies, is that the worm is transparent, allowing easy observation of internal development at all stages. It has a simple body structure, with a small, invariant number of somatic cells: 959 cells in a mature hermaphroditic female and 1031 cells in a mature male.

Life cycle of *C. elegans* Most mature adults are hermaphrodites, with the ability to produce both eggs and sperm and undergo self-fertilization. A few are male, which produce only sperm and mate with hermaphrodites. The hermaphrodites have two sex chromosomes (XX); the males possess a single sex chromosome (XO). Thus, hermaphrodites that self-fertilize produce only hermaphrodites (with the exception of a few males that result from nondisjunction of the X chromosomes). When hermaphrodites mate with males, half of the progeny are XX hermaphrodites and half are XO males.

Eggs are fertilized internally, either from sperm produced by the hermaphrodite or from sperm contributed by a male. The eggs are then laid, and development is completed externally. Approximately 14 hours after fertilization, a larva hatches from the egg and goes through four larval stages—termed L1, L2, L3, and L4—that are separated by molts. The L4 larva undergoes a final molt to produce the adult worm. Under normal laboratory conditions, worms will live for 2 to 3 weeks.

The *C. elegans* genome Geneticists began developing plans in 1989 to sequence the genome of *C. elegans*, and the complete genome sequence was obtained in 1998. Compared with the genomes of most multicellular animals, that of *C. elegans*, at 103 million base pairs of DNA, is small, which facilitates genomic analysis. The availability of the complete genome sequence provides a great deal of information about gene structure, function, and organization in this species. For example, the process of programmed cell death (apoptosis, see Chapter 22) plays an important role in development and in the suppression of cancer. Apoptosis in *C. elegans* is remarkably similar to that in humans. Having the complete genome sequence of *C. elegans*, and given its ease of genetic manipulation, geneticists have identified genes that participate in apoptosis, which has increased our understanding of apoptosis in humans and its role in cancer.

Genetic techniques with *C. elegans* Chemical mutagens are routinely used to generate mutations in *C. elegans*—mutations that are easy to identify and isolate. The ability of hermaphrodites to self-fertilize means that progeny homozygous for recessive mutations can be obtained in a single generation; the existence of males means that genetic crosses can be carried out.

Developmental studies are facilitated by the transparent body of the worms. As stated earlier, *C. elegans* has a small and exact number of somatic cells. Researchers studying the development of *C. elegans* have meticulously mapped the entire cell lineage of the species, and so the developmental fate of every cell in the adult body can be traced to the original single-celled fertilized egg. Developmental biologists often use lasers to destroy (ablate) specific cells in a developing worm and then study the effects on physiology, development, and behavior.

RNA interference has proved to be an effective tool for turning off genes in *C. elegans*. Geneticists inject double-stranded

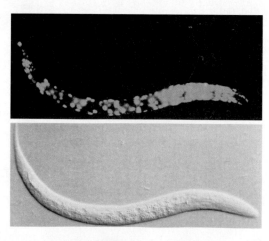

Figure 1 A sequence for the green fluorescent protein (GFP) has been used to visually determine the expression of genes inserted into *C. elegans* (lower photograph). The gene for GFP is injected into the ovary of a worm and becomes incorporated into the worm genome. The expression of this transgene produces GFP, which fluoresces green (upper photograph). [From Huaqi Jiang et al., *PNAS* 98:7916–7921, 2001. Copyright 2001 National Academy of Sciences, U.S.A.]

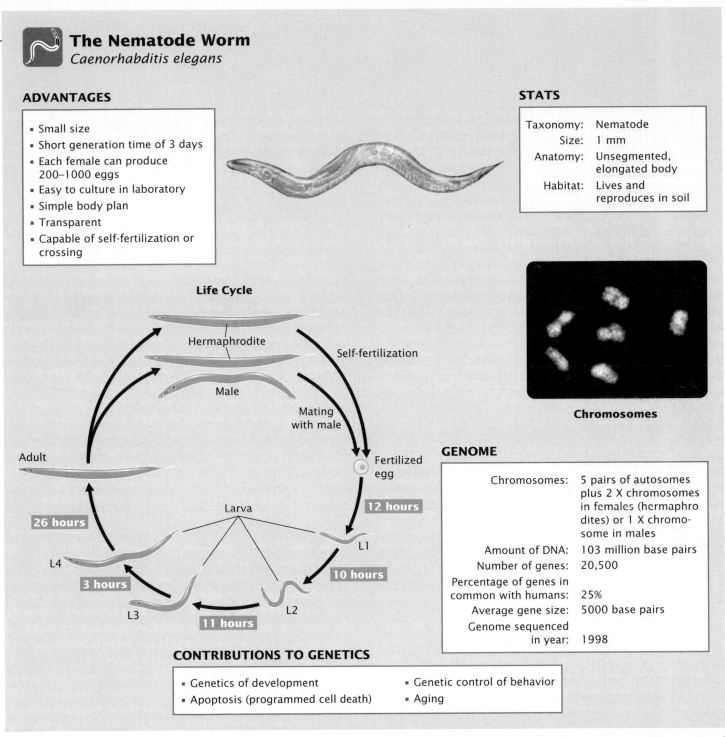

The Nematode Worm
Caenorhabditis elegans

ADVANTAGES

- Small size
- Short generation time of 3 days
- Each female can produce 200–1000 eggs
- Easy to culture in laboratory
- Simple body plan
- Transparent
- Capable of self-fertilization or crossing

STATS

Taxonomy:	Nematode
Size:	1 mm
Anatomy:	Unsegmented, elongated body
Habitat:	Lives and reproduces in soil

Life Cycle

Hermaphrodite

Self-fertilization

Male

Mating with male

Fertilized egg

Adult

Larva

12 hours

26 hours

L1

10 hours

L4

3 hours

L3

11 hours

L2

Chromosomes

GENOME

Chromosomes:	5 pairs of autosomes plus 2 X chromosomes in females (hermaphrodites) or 1 X chromosome in males
Amount of DNA:	103 million base pairs
Number of genes:	20,500
Percentage of genes in common with humans:	25%
Average gene size:	5000 base pairs
Genome sequenced in year:	1998

CONTRIBUTIONS TO GENETICS

- Genetics of development
- Apoptosis (programmed cell death)
- Genetic control of behavior
- Aging

[Photography courtesy of William Goodyer and Monique Zetka.]

copies of RNA that is complementary to specific genes; the double-stranded RNA then silences the expression of these genes through the RNAi process. The worms can even be fed bacteria that have been genetically engineered to express the double-stranded RNA, thus avoiding the difficulties of microinjection.

Transgenic worms can be produced by injecting DNA into the ovary, where the DNA becomes incorporated into the oocytes. Geneticists have created a special reporter gene that produces the jellyfish green fluorescent protein (GFP). When this reporter gene is injected into the ovary and becomes inserted into the worm genome, its expression produces GFP, which fluoresces green, allowing the expression of the gene to be easily observed (**Figure 1**). ■

The Plant *Arabidopsis thaliana*

Much of the early work in genetics was carried out on plants, including Mendel's seminal discoveries in pea plants as well as important aspects of heredity, gene mapping, chromosome genetics, and quantitative inheritance in corn, wheat, beans, and other plants. However, by the mid-twentieth century, many geneticists had turned to bacteria, viruses, yeast, *Drosophila,* and mouse genetic models. Because a good genetic plant model did not exist, plants were relatively neglected, particularly for the study of molecular genetic processes.

This changed in the last part of the twentieth century with the widespread introduction of a new genetic model organism, the plant *Arabidopsis thaliana. A. thaliana* was identified in the sixteenth century, and the first mutant was reported in 1873; but this species was not commonly studied until the first detailed genetic maps appeared in the early 1980s. Today, *Arabidopsis* figures prominently in the study of genome structure, gene regulation, development, and evolution in plants, and it provides important basic information about plant genetics that is applied to economically important plant species.

Advantages of *Arabidopsis* as a model genetic organism
The thale cress *Arabidopsis thaliana* is a member of the Brassicaceae family and grows as a weed in many parts of the world. Except in its role as a model genetic organism, *Arabidopsis* has no economic importance, but it has a number of characteristics that make it well suited to the study of genetics. As an angiosperm, it has features in common with other flowering plants, some of which play critical roles in the ecosystem or are important sources of food, fiber, building materials, and pharmaceutical agents. *Arabidopsis*'s chief advantages are its small size (maximum height of 10–20 cm), prolific reproduction, and small genome.

Arabidopsis thaliana completes development—from seed germination to seed production—in about 6 weeks. Its small size and ability to grow under low illumination make it ideal for laboratory culture. Each plant is capable of producing from 10,000 to 40,000 seeds, and the seeds typically have a high rate of germination; so large numbers of progeny can be obtained from single genetic crosses.

The *Arabidoposis* genome
A key advantage for molecular studies is *Arabidopsis*'s small genome, which consists of only 125 million base pairs of DNA on five pairs of chromosomes, compared with 2.5 billion base pairs of DNA in the maize genome and 16 billion base pairs in the wheat genome. The genome of *A. thaliana* was completely sequenced in 2000, providing detailed information about gene structure and organization in this species. A number of variants of *A. thaliana*—called ecotypes—that vary in shape, size, physiological characteristics, and DNA sequence are available for study.

Life cycle of *Arabidopsis*
The *Arabidopsis* life cycle is fairly typical of most flowering plants (see Figure 2.22). The main, vegetative part of the plant is diploid; haploid gametes are produced in the pollen and ovaries. When a pollen grain lands on the stigma of a flower, a pollen tube grows into the pistil and ovary. Two haploid sperm nuclei contained in each pollen grain travel down the pollen tube and enter the embryo sac. There, one of the haploid sperm cells fertilizes the haploid egg cell to produce a diploid zygote. The other haploid sperm cell fuses with two haploid nuclei to form the $3n$ endosperm, which provides tissue that will nourish the growing embryonic plant. The zygotes develop within the seeds, which are produced in a long pod.

Under appropriate conditions, the embryo germinates and begins to grow into a plant. The shoot grows upward and the roots downward, a compact rosette of leaves is produced and, under the right conditions, the shoot enlarges and differentiates into flower structures. At maturity, *A. thaliana* is a low-growing plant with roots, a main shoot with branches that bear mature leaves, and small white flowers at the tips of the branches.

Genetic techniques with *Arabidopsis*
A number of traditional and modern molecular techniques are commonly used with *Arabidopsis* and provide it with special advantages for genetic studies. *Arabidopsis* can self-fertilize, which means that any recessive mutation appearing in the germ line can be recovered in the immediate progeny. Cross-fertilization also is possible by removing the anther from one plant and dusting pollen on the stigma of another plant—essentially the same technique used by Gregor Mendel with pea plants (see Figure 3.4).

As already mentioned, many naturally occurring variants of *Arabidopsis* are available for study, and new mutations can be produced by exposing its seeds to chemical mutagens, radiation, or transposable elements that randomly insert into genes. The large number of offspring produced by *Arabidopsis* facilitates screening for rare mutations.

Genes from other organisms can be transferred to *Arabidopsis* by the Ti plasmid from the bacterium *Agrobacterium tumefaciens,* which naturally infects plants and transfers the Ti plasmid to plant cells (see Chapter 19). After transfer to a plant cell, the

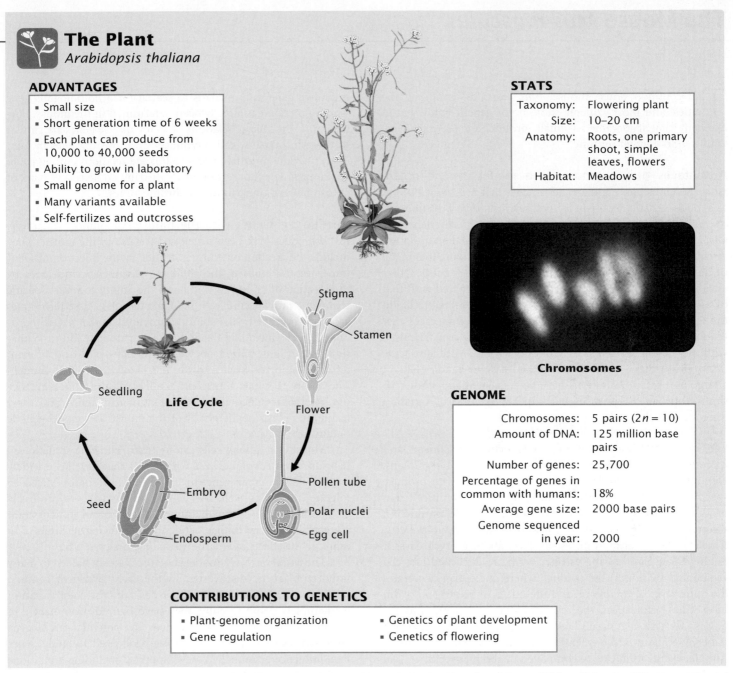

The Plant
Arabidopsis thaliana

ADVANTAGES

- Small size
- Short generation time of 6 weeks
- Each plant can produce from 10,000 to 40,000 seeds
- Ability to grow in laboratory
- Small genome for a plant
- Many variants available
- Self-fertilizes and outcrosses

STATS

Taxonomy:	Flowering plant
Size:	10–20 cm
Anatomy:	Roots, one primary shoot, simple leaves, flowers
Habitat:	Meadows

Chromosomes

GENOME

Chromosomes:	5 pairs ($2n = 10$)
Amount of DNA:	125 million base pairs
Number of genes:	25,700
Percentage of genes in common with humans:	18%
Average gene size:	2000 base pairs
Genome sequenced in year:	2000

Life Cycle

Seedling

Stigma

Stamen

Flower

Pollen tube

Polar nuclei

Egg cell

Embryo

Endosperm

Seed

CONTRIBUTIONS TO GENETICS

- Plant-genome organization
- Gene regulation
- Genetics of plant development
- Genetics of flowering

[Photography courtesy of Anand P. Tyazi and Luca Comai, Dept. of Biology, University of Washington, Seattle.]

Ti plasmid randomly inserts into the DNA of the plant that it infects, thereby generating mutations in the plant DNA in a process called *insertional mutagenesis*. Geneticists have modified the Ti plasmid to carry a *GUS* gene, which has no promoter of its own. The *GUS* gene encodes an enzyme that converts a colorless compound (X-Glu) into a blue dye. Because the *GUS* gene has no promoter, it is expressed only when inserted into the coding sequence of a plant gene. When that happens, the enzyme encoded by *GUS* is synthesized and converts X-Glu into a blue dye that stains the cell. This dye provides a means to visually determine the expression pattern of a gene that has been interrupted by Ti DNA, producing information about the expression of genes that are mutated by insertional mutagenesis. ■

The Mouse *Mus musculus*

The common house mouse, *Mus musculus*, is among the oldest and most valuable subjects for genetic study. It's an excellent genetic organism—small, prolific, and easy to keep, with a short generation time.

Advantages of the mouse as a model genetic organism   Foremost among many advantages that *Mus musculus* has as a model genetic organism is its close evolutionary relationship to humans. Being a mammal, the mouse is genetically, behaviorally, and physiologically more similar to humans than are other organisms used in genetics studies, making the mouse the model of choice for many studies of human and medical genetics. Other advantages include a short generation time compared with that of most other mammals. *Mus musculus* is well adapted to life in the laboratory and can be easily raised and bred in cages that require little space; thus several thousand mice can be raised within the confines of a small laboratory room. Mice have large litters (8–10 pups), and are docile and easy to handle. Finally, a large number of mutations have been isolated and studied in captive-bred mice, providing an important source of variation for genetic analysis.

Life cycle of the mouse   The production of gametes and reproduction in the mouse are very similar to those in humans. Diploid germ cells in the gonads undergo meiosis to produce sperm and oocytes, as outlined in Chapter 2. Male mice begin producing sperm at puberty and continue sperm production throughout the remainder of their lives. Starting at puberty, female mice go through an estrus cycle about every 4 days. If mating takes place during estrus, sperm are deposited into the vagina and swim into the oviduct, where one sperm penetrates the outer layer of the ovum, and the nuclei of sperm and ovum fuse. After fertilization, the diploid embryo implants into the uterus. Gestation typically takes about 21 days. Mice reach puberty in about 5 to 6 weeks and will live for about 2 years. A complete generation can be completed in about 8 weeks.

The mouse genome   The mouse genome contains about 2.6 billion base pairs of DNA, which is similar in size to the human genome. Mice and humans also have similar numbers of genes. For most human genes, there are homologous genes in the mouse. An important tool for determining the function of an unknown gene in humans is to search for a homologous gene whose function has already been determined in the mouse. Furthermore, the linkage relations of many mouse genes are similar to those in humans, and the linkage relations of genes in mice often provide important clues to linkage relations among genes in humans. The mouse genome is distributed across 19 pairs of autosomes and one pair of sex chromosomes.

Genetic techniques with the mouse   A number of powerful techniques have been developed for use in the mouse. They include the creation of transgenic mice by the injection of DNA into a mouse embryo, the ability to disrupt specific genes by the creation of knockout mice, and the ability to insert specific sequences into specific loci (see Chapter 19). These techniques are made possible by the ability to manipulate the mouse reproductive cycle, including the ability to hormonally induce ovulation, isolate unfertilized oocytes from the ovary, and implant fertilized embryos back into the uterus of a surrogate mother. The ability to create transgenic, knockout, and knock-in mice has greatly facilitated the study of human genetics, and these techniques illustrate the power of the mouse as a model genetic organism.

Mouse and human cells can be fused, allowing somatic-cell hybridization techniques (see Chapter 7) that have been widely used to assign human genes to specific chromosomes. Mice also tolerate inbreeding well, and inbred strains of mice are easily created by brother–sister mating. Members of an inbred strain are genetically very similar or identical, allowing researchers to examine the effects of environmental factors on a trait.

The use of mice as a model genetic organism has led to many important genetic discoveries. In the early twentieth century, mice were used to study the genetic basis of coat-color variation in mammals. More recently, they have figured prominently in research on the genetic basis of cancer, and potential carcinogens are often tested in mice. Mice have been used to study genes that influence mammalian development, including mutations that produce birth defects in humans. A large number of mouse models of specific human diseases have been created—in some cases, by isolating and inbreeding mice with naturally occurring mutations and, in other cases, by using knockout and knock-in techniques to disable and modify specific genes. ∎

The Mouse
Mus musculus

ADVANTAGES

- Closely related to humans
- Small size
- Rapid reproduction
- Easy to rear in the laboratory
- Tolerates inbreeding

STATS

Taxonomy:	Mammal
Size:	2–3 inches 20 grams
Anatomy:	Typical rodent body plan
Habitat:	Fields, houses, and other human structures

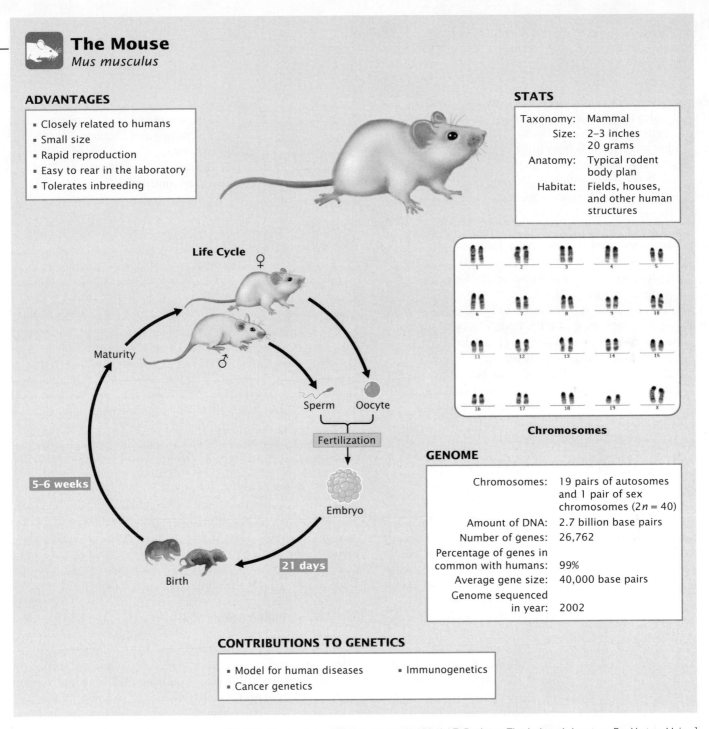

Life Cycle

Maturity

5–6 weeks

Birth

21 days

Embryo

Fertilization

Sperm   Oocyte

Chromosomes

GENOME

Chromosomes:	19 pairs of autosomes and 1 pair of sex chromosomes ($2n = 40$)
Amount of DNA:	2.7 billion base pairs
Number of genes:	26,762
Percentage of genes in common with humans:	99%
Average gene size:	40,000 base pairs
Genome sequenced in year:	2002

CONTRIBUTIONS TO GENETICS

- Model for human diseases
- Cancer genetics
- Immunogenetics

[Photography courtesy of Ellen C. Akeson and Muriel T. Davisson, The Jackson Laboratory, Bar Harton, Maine.]

The Yeast *Saccharomyces cerevisiae*

Common baker's yeast *(Saccharomyces cerevisiae)* has been widely adopted as a simple model system for the study of eukaryotic genetics. Long used for baking bread and making beer, yeast has more recently been utilized for the production of biofuels. Louis Pasteur identified *S. cerevisiae* as the microorganism responsible for fermentation in 1857, and its use in genetic analysis began about 1935. Having been the subject of extensive studies in classical genetics for many years, yeast genes are well known and characterized. At the same time, yeast's unicellular nature makes it amenable to molecular techniques developed for bacteria. Thus, yeast combines both classical genetics and molecular biology to provide a powerful model for the study of eukaryotic genetic systems.

Advantages of yeast as a model genetic organism The great advantage of yeast is that it not only is a eukaryotic organism, with genetic and cellular systems similar to those of other, more-complex eukaryotes such as humans, but also is unicellular, with many of the advantages of manipulation found with bacterial systems. Like bacteria, yeast cells require little space, and large numbers of cells can be grown easily and inexpensively in the laboratory.

Yeast exists in both diploid and haploid forms. When haploid, the cells possess only a single allele at each locus, which means that the allele will be expressed in the phenotype; unlike the situation in diploids, there is no dominance through which some alleles mask the expression of others. Therefore, recessive alleles can be easily identified in haploid cells, and then the interactions between alleles can be examined in the diploid cells.

Another feature that makes yeast a powerful genetic model system is that, subsequent to meiosis, all of the products of a meiotic division are present in a single structure called an ascus (see the next subsection) and remain separate from the products of other meiotic divisions. The four cells produced by a single meiotic division are termed a tetrad. In most organisms, the products of different meiotic divisions mix, and so identification of the results of a single meiotic division is impossible. For example, if we were to isolate four sperm cells from the testes of a mouse, it is extremely unlikely that all four would have been produced by the same meiotic division. Having tetrads separate in yeast allows us to directly observe the effects of individual meiotic divisions on the types of gametes produced and to more easily identify crossover events. The genetic analysis of a tetrad is termed tetrad analysis.

Yeast has been subjected to extensive genetic analysis, and thousands of mutants have been identified. In addition, many powerful molecular techniques developed for manipulating genetic sequences in bacteria have been adapted for use in yeast. Yet, in spite of a unicellular structure and ease of manipulation, yeast cells possess many of the same genes found in humans and other complex multicellular eukaryotes, and many of these genes have identical or similar functions in these eukaryotes. Thus, the genetic study of yeast cells often contributes to our understanding of other, more-complex eukaryotic organisms, including humans.

Life cycle of yeast As stated earlier, *Saccharomyces cerevisiae* can exist as either haploid or diploid cells. Haploid cells usually exist when yeast is starved for nutrients and reproduce mitotically, producing identical, haploid daughter cells through budding. Yeast cells can also undergo sexual reproduction. There are two mating types, **a** and α; haploid cells of different mating types fuse and then undergo nuclear fusion to create a diploid cell. The diploid cell is capable of budding mitotically to produce genetically identical diploid cells. Starvation induces the diploid cells to undergo meiosis, resulting in four haploid nuclei, which become separated into different cells, producing haploid spores. Because the four products of meiosis (a tetrad) are enclosed in a common structure, the ascus, all the products of a single meiosis can be isolated (tetrad analysis).

The yeast genome *Saccharomyces cerevisiae* has 16 pairs of typical eukaryotic chromosomes. The rate of recombination is high, giving yeast a relatively long genetic map compared with those of other organisms. The genome of *S. cerevisiae* contains 12 million base pairs, plus the 2 million to 3 million base pairs of rRNA genes. In 1996, *S. cerevisiae* was the first eukaryotic organism whose genome was completely sequenced.

Genetic techniques with yeast One advantage of yeast to researchers is the use of plasmids to transfer genes or DNA sequences of interest into cells. Yeast cells naturally possesses a circular plasmid, named 2μ, that is 6300 bp long and is transmitted to daughter cells in mitosis and meiosis. This plasmid has an origin of replication recognized by the yeast replication system, and so it replicates autonomously in the cell. The 2μ plasmid has been engineered to provide an efficient vector for transferring genes into yeast. In other cases, bacterial plasmids have been adapted for use in yeast. Some of them undergo homologous recombination with the yeast chromosome, transferring their sequences to the yeast chromosome. Shuttle vectors, which can be propagated in both bacteria and yeast, are particularly effective. Such vectors make it possible to construct and manipulate gene sequences in bacteria, where often more-powerful tech-

Yeast
Saccharomyces cerevisiae

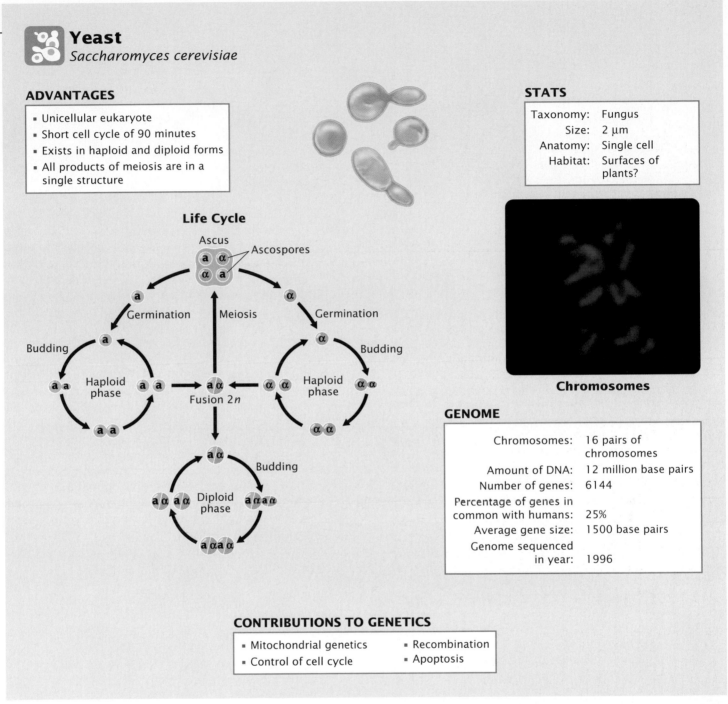

ADVANTAGES

- Unicellular eukaryote
- Short cell cycle of 90 minutes
- Exists in haploid and diploid forms
- All products of meiosis are in a single structure

STATS

Taxonomy:	Fungus
Size:	2 μm
Anatomy:	Single cell
Habitat:	Surfaces of plants?

Chromosomes

Life Cycle

Ascus

Ascospores

Germination Meiosis Germination

Budding Budding

Haploid phase Haploid phase

Fusion 2*n*

Budding

Diploid phase

GENOME

Chromosomes:	16 pairs of chromosomes
Amount of DNA:	12 million base pairs
Number of genes:	6144
Percentage of genes in common with humans:	25%
Average gene size:	1500 base pairs
Genome sequenced in year:	1996

CONTRIBUTIONS TO GENETICS

- Mitochondrial genetics
- Control of cell cycle
- Recombination
- Apoptosis

[Photography courtesy of Mara Stewart and Dean Dawson, Dept. of Microbiology and Molecular Biology, Sackler School of Biomedical Sciences, Tufts Univeristy.]

niques are available for genetic manipulation and selection, and then transfer the gene sequences into yeast cells, where their function can be tested.

Plasmids are limited in the size of DNA fragments that they can carry (see p. 521 in Chapter 19). Yeast artificial chromosomes (YACs) overcome this limitation; YACs can hold DNA fragments as large as several hundred thousand base pairs (as described in Chapter 11. Yeast artificial chromosomes are engineered DNA fragments that contain centromeric and telomeric sequences and segregate like chromosomes in meiosis and mitosis. ∎

Glossary

acceptor arm The arm in tRNA to which an amino acid attaches.

acentric chromatid Lacks a centromere; produced when crossing over takes place within a paracentric inversion. The acentric chromatid does not attach to a spindle fiber and does not segregate in meiosis or mitosis; so it is usually lost after one or more rounds of cell division.

acidic activation domain Commonly found in some transcriptional activator proteins, a domain that contains multiple amino acids with negative charges and stimulates the transcription of certain genes.

acrocentric chromosome Chromosome in which the centromere is near one end, producing a long arm at one end and a knob, or satellite, at the other end.

activator *See* **transcriptional activator protein**.

adaptive mutation Process by which a specific environment induces mutations that enable organisms to adapt to the environment.

addition rule States that the probability of any of two or more mutually exclusive events occurring is calculated by adding the probabilities of the individual events.

additive genetic variance Component of the genetic variance that can be attributed to the additive effect of different genotypes.

adenine (A) Purine base in DNA and RNA.

adenosine-3′,5′-cyclic monophosphate (cAMP) Modified nucleotide that functions in catabolite repression. Low levels of glucose stimulate high levels of cAMP; cAMP then attaches to CAP, which binds to the promoter of certain operons and stimulates transcription.

adjacent-1 segregation Type of segregation that takes place in a heterozygote for a translocation. If the original, nontranslocated chromosomes are N_1 and N_2 and the chromosomes containing the translocated segments are T_1 and T_2, then adjacent-1 segregation takes place when N_1 and T_2 move toward one pole and T_1 and N_2 move toward the opposite pole.

adjacent-2 segregation Type of segregation that takes place in a heterozygote for a translocation. If the original, nontranslocated chromosomes are N_1 and N_2 and the chromosomes containing the translocated segments are T_1 and T_2, then adjacent-2 segregation takes place when N_1 and T_1 move toward one pole and T_2 and N_2 move toward the opposite pole.

A-DNA Right-handed helical structure of DNA that exists when little water is present.

affinity capture Use of an antibody to capture one protein from a complex mixture of proteins. The captured protein will "pull down" with it any proteins with which it interacts, which can then be analyzed by mass spectrometry to identify these proteins.

allele One of two or more alternate forms of a gene.

allelic frequency Proportion of a particular allele in a population.

allopatric speciation Speciation that arises when a geographic barrier first splits a population into two groups and blocks the exchange of genes between them.

allopolyploidy Condition in which the sets of chromosomes of a polyploid individual possessing more than two haploid sets are derived from two or more species.

allosteric protein Protein that changes its conformation on binding with another molecule.

alternate segregation Type of segregation that takes place in a heterozygote for a translocation. If the original, nontranslocated chromosomes are N_1 and N_2 and the chromosomes containing the translocated segments are T_1 and T_2, then alternate segregation takes place when N_1 and N_2 move toward one pole and T_1 and T_2 move toward the opposite pole.

alternation of generations Complex life cycle in plants that alternates between the diploid sporophyte stage and the haploid gametophyte stage.

alternative processing pathway One of several pathways by which a single pre-mRNA can be processed in different ways to produce alternative types of mRNA.

alternative splicing Process by which a single pre-mRNA can be spliced in more than one way to produce different types of mRNA.

Ames test Test in which special strains of bacteria are used to evaluate the potential of chemicals to cause cancer.

amino acid Repeating unit of proteins; consists of an amino group, a carboxyl group, a hydrogen atom, and a variable R group.

aminoacyl (A) site One of three sites in a ribosome occupied by a tRNA in translation. All charged tRNAs (with the exception of the initiator tRNA) first enter the A site in translation.

aminoacyl-tRNA synthetase Enzyme that attaches an amino acid to a tRNA. Each aminoacyl-tRNA synthetase is specific for a particular amino acid.

amniocentesis Procedure used for prenatal genetic testing to obtain a sample of amniotic fluid from a pregnant woman. A long sterile needle is inserted through the abdominal wall into the amniotic sac to obtain the fluid.

amphidiploidy Type of allopolyploidy in which two different diploid genomes are combined such that every chromosome has one and only one homologous partner and the genome is functionally diploid.

anagenesis Evolutionary change within a single lineage.

anaphase Stage of mitosis in which chromatids separate and move toward the spindle poles.

anaphase I Stage of meiosis I. In anaphase I, homologous chromosomes separate and move toward the spindle poles.

anaphase II Stage of meiosis II. In anaphase II, chromatids separate and move toward the spindle poles.

aneuploidy Change from the wild type in the number of chromosomes; most often an increase or decrease of one or two chromosomes.

annotation Linking the sequence information of a gene that has been identified to other information about the gene's function and expression, about the protein encoded by the gene, and about similar genes in other species.

Antennapedia **complex** Cluster of five homeotic genes in fruit flies that affects the development of the adult fly's head and anterior thoracic segments.

antibody Produced by a B cell, a protein that circulates in the blood and other body fluids. An antibody binds to a specific antigen and marks the antigen for destruction by a phagocytic cell.

anticipation Increasing severity or earlier age of onset of a genetic trait in succeeding generations. For example, symptoms of a genetic disease may become more severe as the trait is passed from generation to generation.

anticodon Sequence of three nucleotides in transfer RNA that pairs with the corresponding codon in messenger RNA in translation.

antigen Substance that is recognized by the immune system and elicits an immune response.

antigenic drift Used in reference to a rapidly evolving virus, with new strains appearing frequently because of mutations.

antigenic shift Used in reference to a virus that has undergone major changes in its genome through the reassortment of genetic material from two different strains of the virus.

antiparallel Refers to a characteristic of the DNA double helix in which the two polynucleotide strands run in opposite directions.

antisense RNA Small RNA molecule that base pairs with a complementary DNA or RNA sequence and affects its functioning.

antiterminator Protein or DNA sequence that inhibits the termination of transcription.

apoptosis Programmed cell death, in which a cell degrades its own DNA, the nucleus and cytoplasm shrink, and the cell undergoes phagocytosis by other cells without leakage of its contents.

archaea One of the three primary divisions of life. Archaea consist of unicellular organisms with prokaryotic cells.

artificial selection Selection practiced by humans.

attenuation Type of gene regulation in some bacterial operons, in which transcription is initiated but terminates prematurely before the transcription of the structural genes.

attenuator Secondary structure that forms in the 5′ untranslated region of some operons and causes the premature termination of transcription.

autoimmune disease Characterized by an abnormal immune response to a person's own (self) antigen.

autonomous element Transposable element that is fully functional and able to transpose on its own.

autonomously replicating sequence (ARS) DNA sequence that confers the ability to replicate; contains an origin of replication.

autopolyploidy Condition in which all the sets of chromsomes of a polyploid individual possessing more than two haploid sets are derived from a single species.

autoradiography Method for visualizing DNA or RNA molecules labeled with radioactive substances. A piece of X-ray film is placed on top of a slide, gel, or other substance that contains DNA labeled with radioactive chemicals. Radiation from the labeled DNA exposes the film, providing a picture of the labeled molecules.

autosome Chromosome that is the same in males and females; nonsex chromosome.

auxotrophic bacterium A bacterium or fungus that possesses a nutritional mutation that disrupts its ability to synthesize an essential biological molecule; cannot grow on minimal medium but can grow on minimal medium to which has been added the biological molecule that it cannot synthesize.

backcross Cross between an F_1 individual and one of the parental (P) genotypes.

bacterial artificial chromosome (BAC) Cloning vector used in bacteria that is capable of carrying DNA fragments as large as 500 kb.

bacterial colony Clump of genetically identical bacteria derived from a single bacterial cell that undergoes repeated rounds of division.

bacteriophage Virus that infects bacterial cells.

balance hypothesis Proposes that much of the molecular variation seen in natural populations is maintained by balancing selection that favors genetic variation.

Barr body Condensed, darkly staining structure that is found in most cells of female placental mammals and is an inactivated X chromosome.

basal transcription apparatus Complex of transcription factors, RNA polymerase, and other proteins that assemble on the promoter and are capable of initiating minimal levels of transcription.

base *See* **nitrogenous base.**

base analog Chemical substance that has a structure similar to that of one of the four standard bases of DNA and may be incorporated into newly synthesized DNA molecules in replication.

base-excision repair DNA repair that first excises modified bases and then replaces the entire nucleotide.

base substitution Mutation in which a single pair of bases in DNA is altered.

B cell Particular type of lymphocyte that produces humoral immunity; matures in the bone marrow and produces antibodies.

B-DNA Right-handed helical structure of DNA that exists when water is abundant; the secondary structure described by Watson and Crick and probably the most common DNA structure in cells.

behavioral isolation Reproductive isolation due to differences in behavior that prevent interbreeding.

bidirectional replication Replication at both ends of a replication bubble.

bioinformatics Synthesis of molecular biology and computer science that develops databases and computational tools to store, retrieve, and analyze nucleic-acid- and protein-sequence data.

biological species concept Defines a species as a group of organisms whose members are capable of interbreeding with one another but are reproductively isolated from the members of other species. Because different species do not exchange genes, each species evolves independently. Not all biologists adhere to this concept.

biotechnology Use of biological processes, particularly molecular genetics and recombinant DNA technology, to produce products of commercial value.

***bithorax* complex** Cluster of three homeotic genes in fruit flies that influences the adult fly's posterior thoracic and abdominal segments.

bivalent Refers to a synapsed pair of homologous chromosomes.

blending inheritance Early concept of heredity proposing that offspring possess a mixture of the traits from both parents.

branch Evolutionary connections between organisms in a phylogenetic tree.

branch migration Movement of a cross bridge along two DNA molecules.

branch point Adenine nucleotide in nuclear pre-mRNA introns that lies from 18 to 40 nucleotides upstream of the 3′ splice site.

broad-sense heritability Proportion of the phenotypic variance that can be attributed to genetic variance.

caspase Enzyme that cleaves other proteins and regulates apoptosis. Each caspase is synthesized as a large, inactive precursor (a procaspase) that is activated by cleavage, often by another caspase.

catabolite activator protein (CAP) Protein that functions in catabolite repression. When bound with cAMP, CAP binds to the promoter of certain operons and stimulates transcription.

catabolite repression System of gene control in some bacterial operons in which glucose is used preferentially and the metabolism of other sugars is repressed in the presence of glucose.

cDNA (complementary DNA) library Collection of bacterial colonies or phage colonies containing DNA fragments that have been produced by reverse transcription of cellular mRNA.

cell cycle Stages through which a cell passes from one cell division to the next.

cell line Genetically identical cells that divide indefinitely and can be cultured in the laboratory.

cell theory States that all life is composed of cells, that cells arise only from other cells, and that the cell is the fundamental unit of structure and function in living organisms.

cellular immunity Type of immunity resulting from T cells, which recognize antigens found on the surfaces of self cells.

centiMorgan Another name for map unit.

central dogma Concept that genetic information passes from DNA to RNA to protein in a one-way information pathway.

centriole Cytoplasmic organelle consisting of microtubules; present at each pole of the spindle apparatus in animal cells.

centromere Constricted region on a chromosome that stains less strongly than the rest of the chromosome; region where spindle microtubules attach to a chromosome.

centromeric sequence DNA sequence found in functional centromeres.

centrosome Structure from which the spindle apparatus develops; contains the centriole.

Chargaff's rules Rules developed by Erwin Chargaff and his colleagues concerning the ratios of bases in DNA.

checkpoint A key transition point at which progression to the next stage in the cell cycle is regulated.

chiasma (pl., **chiasmata**) Point of attachment between homologous chromosomes at which crossing over took place.

chloroplast DNA (cpDNA) DNA in chloroplasts; has many characteristics in common with eubacterial DNA and typically consists of a circular molecule that lacks histone proteins and encodes some of the rRNAs, tRNAs, and proteins found in chloroplasts.

chorionic villus sampling (CVS) Procedure used for prenatal genetic testing in which a small piece of the chorion (the outer layer of the placenta) is removed from a pregnant woman. A catheter is inserted through the vagina and cervix into the uterus. Suction is then applied to remove the sample.

chromatin Material found in the eukaryotic nucleus; consists of DNA and proteins.

chromatin-remodeling complex Complex of proteins that alters chromatin structure without acetylating histone proteins.

chromatin-remodeling protein Binds to a DNA sequence and disrupts chromatin structure, causing the DNA to become more accessible to RNA polymerase and other proteins.

chromatosome Consists of a nucleosome and an H1 histone protein.

chromosomal puff Localized swelling of a polytene chromosome; a region of chromatin in which DNA has unwound and is undergoing transcription.

chromosome Structure consisting of DNA and associated proteins that carries and transmits genetic information.

chromosome deletion Loss of a chromosome segment.

chromosome duplication Mutation that doubles a segment of a chromosome.

chromosome inversion Rearrangement in which a segment of a chromosome has been inverted 180 degrees.

chromosome jumping Method of moving from a gene on a cloned fragment to sequences on distantly linked fragments.

chromosome mutation Difference from the wild type in the number or structure of one or more chromosomes; often affects many genes and has large phenotypic effects.

chromosome rearrangement Change from the wild type in the structure of one or more chromosomes.

chromosome theory of heredity States that genes are located on chromosomes.

chromosome walking Method of locating a gene by using partly overlapping genomic clones to move in steps from a previously cloned, linked gene to the gene of interest.

cis configuration Arrangement in which two or more wild-type genes are on one chromosome and their mutant alleles are on the homologous chromosome; also called coupling configuration.

cladogenesis Evolution in which one lineage is split into two.

clonal evolution Process by which mutations that enhance the ability of cells to proliferate predominate in a clone of cells, allowing the clone to become increasingly rapid in growth and increasingly aggressive in proliferation properties.

cloning vector Stable, replicating DNA molecule to which a foreign DNA fragment can be attached and transferred to a host cell.

cloverleaf structure Secondary structure common to all tRNAs.

coactivator Protein that cooperates with an activator of transcription. In eukaryotic transcriptional control, coactivators often physically interact with transcriptional activators and the basal transcription apparatus.

codominance Type of allelic interaction in which the heterozygote simultaneously expresses traits of both homozygotes.

codon Sequence of three nucleotides that encodes one amino acid in a protein.

coefficient of coincidence Ratio of observed double crossovers to expected double crossovers.

cohesin Molecule that holds the two sister chromatids of a chromosome together. The breakdown of cohesin at the centromeres enables the chromatids to separate in anaphase of mitosis and anaphase II of meiosis.

cohesive end Short, single-stranded overhanging end on a DNA molecule produced when the DNA is cut by certain restriction enzymes. Cohesive ends are complementary and can spontaneously pair to rejoin DNA fragments that have been cut with the same restriction enzyme.

cointegrate structure Produced in replicative transposition, an intermediate structure in which two DNA molecules with two copies of the transposable element are fused.

colinearity Concept that there is a direct correspondence between the nucleotide sequence of a gene and the continuous sequence of amino acids in a protein.

colony *See* **bacterial colony.**

comparative genomics Comparative studies of the genomes of different organisms.

competent cell Capable of taking up DNA from its environment (capable of being transformed).

complementary DNA strands The relation between the two nucleotide strands of DNA in which each purine on one strand pairs with a specific pyrimidine on the opposite strand (A pairs with T, and G pairs with C).

complementation Two different mutations in the heterozygous condition are exhibited as the wild-type phenotype; indicates that the mutations are at different loci.

complementation test Test designed to determine whether two different mutations are at the same locus (are allelic) or at different loci (are nonallelic). Two individuals that are homozygous for two independently derived mutations are crossed, producing F_1 progeny that are heterozygous for the mutations. If the mutations are at the same locus, the F_1 will have a mutant phenotype. If the mutations are at different loci, the F_1 will have a wild-type phenotype.

complete dominance Refers to an allele or a phenotype that is expressed in homozygotes (*AA*) and in heterozygotes (*Aa*); only the dominant allele is expressed in a heterozygote phenotype.

complete linkage Linkage between genes that are located close together on the same chromosome with no crossing over between them.

complete medium Used to culture bacteria or some other microorganism; contains all the nutrients required for growth and synthesis, including those normally synthesized by the organism. Nutritional mutants can grow on complete medium.

composite transposon Type of transposable element in bacteria that consists of two insertion sequences flanking a segment of DNA.

concept of dominance Principle of heredity discovered by Mendel stating that, when two different alleles are present in a genotype, only one allele may be expressed in the phenotype. The dominant allele is the allele that is expressed, and the recessive allele is the allele that is not expressed.

concordance Percentage of twin pairs in which both twins have a particular trait.

concordant Refers to a pair of twins both of whom have the trait under consideration.

conditional mutation Expressed only under certain conditions.

conjugation Mechanism by which genetic material can be exchanged between bacterial cells. In conjugation, two bacteria lie close together and a cytoplasmic connection forms between them. A plasmid or sometimes a part of the bacterial chromosome passes through this connection from one cell to the other.

consanguinity Mating between related individuals.

consensus sequence Comprises the most commonly encountered nucleotides found at a specific location in DNA or RNA.

−10 consensus sequence (Pribnow box) Consensus sequence (TATAAT) found in most bacterial promoters approximately 10 bp upstream of the transcription start site.

−35 consensus sequence Consensus sequence (TTGACA) found in many bacterial promoters approximately 35 bp upstream of the transcription start site.

constitutive gene A gene that is not regulated and is expressed continually.

constitutive mutation Causes the continuous transcription of one or more structural genes.

contig Set of overlapping DNA fragments that have been assembled in the correct order to form a continuous stretch of DNA sequence.

continuous characteristic Displays a large number of possible phenotypes that are not easily distinguished, such as human height.

continuous replication Replication of the leading strand in the same direction as that of unwinding, allowing new nucleotides to be added continuously to the 3′ end of the new strand as the template is exposed.

coordinate induction Simultaneous synthesis of several enzymes that is stimulated by a single environmental factor.

copy-number variation (CNV) Difference among individual organisms in the number of copies of any large DNA sequence (larger than 1000 bp).

core enzyme Part of bacterial RNA polymerase that, during transcription, catalyzes the elongation of the RNA molecule by the addition of RNA nucleotides; consists of four subunits: two copies of alpha (α), a single copy of beta (β), and a single copy of beta prime (β′).

corepressor Substance that inhibits transcription in a repressible system of gene regulation; usually a small molecule that binds to a repressor protein and alters it so that the repressor is able to bind to DNA and inhibit transcription.

core promoter Located immediately upstream of the eukaryotic promoter, DNA sequences to which the basal transcription apparatus binds.

correlation Degree of association between two or more variables.

correlation coefficient Statistic that measures the degree of association between two or more variables. A correlation coefficient can range from −1 to +1. A positive value indicates a direct relation between the variables; a negative correlation indicates an inverse relation. The absolute value of the correlation coefficient provides information about the strength of association between the variables.

cosmid Cloning vector that combines the properties of plasmids and phage vectors and is used to clone large pieces of DNA in bacteria. A cosmid is a small plasmid that carries a λ *cos* site, allowing the plasmid to be packaged into a viral coat.

cotransduction Process in which two or more genes are transferred together from one bacterial cell to another. Only genes located close together on a bacterial chromosome will be cotransduced.

cotransformation Process in which two or more genes are transferred together during cell transformation.

coupling configuration *See* **cis configuration.**

CpG island DNA region that contains many copies of a cytosine base followed by a guanine base; often found near transcription start sites in eukaryotic DNA. The cytosine bases in CpG islands are commonly methylated when genes are inactive but are demethylated before the initiation of transcription.

cross bridge In a heteroduplex DNA molecule, the point at which each nucleotide strand passes from one DNA molecule to the other.

crossing over Exchange of genetic material between homologous but nonsister chromatids.

cruciform Structure formed by the pairing of inverted repeats on both strands of double-stranded DNA.

C value Haploid amount of DNA found in a cell of an organism.

cyclin A key protein in the control of the cell cycle; combines with a cyclin-dependent kinase (CDK). The levels of cyclin rise and fall in the course of the cell cycle.

cyclin-dependent kinase (CDK) A key protein in the control of the cell cycle; combines with cyclin.

cytokinesis Process by which the cytoplasm of a cell divides.

cytoplasmic inheritance Inheritance of characteristics encoded by genes located in the cytoplasm. Because the cytoplasm is usually contributed entirely by only one parent, most cytoplasmically inherited characteristics are inherited from a single parent.

cytosine (C) Pyrimidine base in DNA and RNA.

deamination Loss of an amino group (NH_2) from a base.

degenerate genetic code Refers to the fact that the genetic code contains more information than is needed to specify all 20 common amino acids.

deletion Mutation in which one or more nucleotides are deleted from a DNA sequence.

deletion mapping Technique for determining the chromsomal location of a gene by studying the association of its phenotype or product with particular chromosome deletions.

delta sequence Long terminal repeat in *Ty* elements of yeast.

denaturation (melting) Process that separates the strands of double-stranded DNA when DNA is heated.

deoxyribonucleotide Basic building block of DNA, consisting of deoxyribose, a phosphate, and a nitrogenous base.

deoxyribose Five-carbon sugar in DNA; lacks a hydroxyl group on the 2′-carbon atom.

depurination Break in the covalent bond connecting a purine base to the 1′-carbon atom of deoxyribose, resulting in the loss of the purine base.

determination Process by which a cell becomes committed to developing into a particular cell type.

diakinesis Fifth substage of prophase I in meiosis. In diakinesis, chromosomes contract, the nuclear membrane breaks down, and the spindle forms.

dicentric bridge Structure produced when the two centromeres of a dicentric chromatid are pulled toward opposite poles, stretching the dicentric chromosome across the center of the nucleus. Eventually, the dicentric bridge breaks as the two centromeres are pulled apart.

dicentric chromatid Chromatid that has two centromeres; produced when crossing over takes place within a paracentric inversion. The two centromeres of the dicentric chromatid are frequently pulled toward opposite poles in mitosis or meiosis, breaking the chromosome.

dideoxyribonucleoside triphosphate (ddNTP) Special substrate for DNA synthesis used in the Sanger dideoxy sequencing method; identical with dNTP (the usual substrate for DNA synthesis) except that it lacks a 3′-OH group. The incorporation of a ddNTP into DNA terminates DNA synthesis.

dihybrid cross A cross between two individuals that differ in two characteristics—more specifically, a cross between individuals that are homozygous for different alleles at the two loci (*AA BB* × *aa bb*); also refers to a cross between two individuals that are both heterozygous at two loci (*Aa Bb* × *Aa Bb*).

dioecious organism Belongs to a species whose members have either male or female reproductive structures.

diploid Possessing two sets of chromosomes (two genomes).

diplotene Fourth substage of prophase I in meiosis. In diplotene, centromeres of homologous chromosomes move apart, but the homologs remain attached at chiasmata.

directional selection Selection in which one trait or allele is favored over another.

direct repair DNA repair in which modified bases are changed back into their original structures.

direct-to-consumer genetic test Test for a genetic condition; the test can be purchased directly by a consumer, without the involvement of a physician or other health care provider.

discontinuous characteristic Exhibits only a few, easily distinguished phenotypes. An example is seed shape in which seeds are either round or wrinkled.

discontinuous replication Replication of the lagging strand in the direction opposite that of unwinding, which means that DNA must be synthesized in short stretches (Okazaki fragments).

discordant Refers to a pair of twins of whom one twin has the trait under consideration and the other does not.

displaced duplication Chromosome rearrangement in which the duplicated segment is some distance from the original segment, either on the same chromosome or on a different one.

dizygotic twins Nonidentical twins that arise when two different eggs are fertilized by two different sperm; also called fraternal twins.

D loop Region of mitochondrial DNA that contains an origin of replication and promoters; it is displaced during the initiation of replication, leading to the name displacement, or D, loop.

DNA fingerprinting Technique used to identify individuals by examining their DNA sequences.

DNA gyrase *E. coli* topoisomerase enzyme that relieves the torsional strain that builds up ahead of the replication fork.

DNA helicase Enzyme that unwinds double-stranded DNA by breaking hydrogen bonds.

DNA library Collection of bacterial colonies containing all the DNA fragments from one source.

DNA ligase Enzyme that catalyzes the formation of a phosphodiester bond between adjacent 3′-OH and 5′-phosphate groups in a DNA molecule.

DNA methylation Modification of DNA by the addition of methyl groups to certain positions on the bases.

DNA polymerase Enzyme that synthesizes DNA.

DNA polymerase I Bacterial DNA polymerase that removes and replaces RNA primers with DNA nucleotides.

DNA polymerase II Bacterial DNA polymerase that takes part in DNA repair; restarts replication after synthesis has halted because of DNA damage.

DNA polymerase III Bacterial DNA polymerase that synthesizes new nucleotide strands by using the 3′-OH group provided by the primer.

DNA polymerase IV Bacterial DNA polymerase; probably takes part in DNA repair.

DNA polymerase V Bacterial DNA polymerase; probably takes part in DNA repair.

DNA polymerase α Eukaryotic DNA polymerase that initiates replication.

DNA polymerase β Eukaryotic DNA polymerase that participates in DNA repair.

DNA polymerase δ Eukaryotic DNA polymerase that replicates the lagging strand during DNA synthesis; also carries out DNA repair and translesion in DNA synthesis.

DNA polymerase ε Eukaryotic DNA polymerase that replicates the leading strand during DNA synthesis.

DNA polymerase γ Eukaryotic DNA polymerase that replicates mitochondrial DNA. A γ-like DNA polymerase replicates chloroplast DNA.

DNase I hypersensitive site Chromatin region that becomes sensitive to digestion by the enzyme DNase I.

DNA sequencing Process of determining the sequence of bases along a DNA molecule.

DNA transposon *See* **transposable element.**

domain Functional part of a protein.

dominance genetic variance Component of the genetic variance that can be attributed to dominance (interaction between genes at the same locus).

dominant Refers to an allele or a phenotype that is expressed in homozygotes (*AA*) and in heterozygotes (*Aa*); only the dominant allele is expressed in a heterozygote phenotype.

dosage compensation Equalization in males and females of the amount of protein produced by X-linked genes. In placental mammals, dosage compensation is accomplished by the random inactivation of one X chromosome in the cells of females.

double fertilization Fertilization in plants; includes the fusion of a sperm cell with an egg cell to form a zygote and the fusion of a second sperm cell with the polar nuclei to form an endosperm.

double-strand-break model Model of homologous recombination in which a DNA molecule undergoes double-strand breaks.

down mutation Decreases the rate of transcription.

downstream core promoter element Consensus sequence [RG(A or T)CGTG] found in some eukaryotic RNA polymerase II core promoters; usually located approximately 30 bp downstream of the transcription start site.

Down syndrome (trisomy 21) Characterized by variable degrees of mental retardation, characteristic facial features, some retardation of growth and development, and an increased incidence of heart defects, leukemia, and other abnormalities; caused by the duplication of all or part of chromosome 21.

ecological isolation Reproductive isolation in which different species live in different habitats and interact with the environment in different ways. Thus, their members do not encounter one another and do not reproduce with one another.

Edward syndrome (trisomy 18) Characterized by severe retardation, low-set ears, a short neck, deformed feet, clenched fingers, heart problems, and other disabilities; results from the presence of three copies of chromosome 18.

effective population size Effective number of breeding adults in a population; influenced by the number of individuals contributing genes to the next generation, their sex ratio, variation between individuals in reproductive success, fluctuations in population size, the age structure of the population, and whether mating is random.

egg Female gamete.

egg-polarity gene Determines the major axes of development in an early fruit-fly embryo. One set of egg-polarity genes determines the anterior–posterior axis and another determines the dorsal–ventral axis.

elongation factor G (EF-G) Protein that combines with GTP and is required for movement of the ribosome along the mRNA during translation.

elongation factor Ts (EF-Ts) Protein that regenerates elongation factor Tu in the elongation stage of protein synthesis.

elongation factor Tu (EF-Tu) Protein taking part in the elongation stage of protein synthesis; forms a complex with GTP and a charged amino acid and then delivers the charged tRNA to the ribosome.

endosymbiotic theory States that some membrane-bounded organelles, such as mitochondria and chloroplasts, in eukaryotic cells originated as free-living eubacterial cells that entered into an endosymbiotic relation with a eukaryotic host cell and evolved into the present-day organelles; supported by a number of similarities in structure and sequence between organelle and eubacterial DNAs.

enhancer Sequence that stimulates maximal transcription of distant genes; affects only genes on the same DNA molecule (is cis acting), contains short consensus sequences, is not fixed in relation to the transcription start site, can stimulate almost any promoter in its vicinity, and may be upstream or downstream of the gene. The function of an enhancer is independent of sequence orientation.

environmental variance Component of the phenotypic variance that is due to environmental differences among individual members of a population.

epigenetic change Stable alteration of chromatin structure that may be passed on to other cells or to an individual organism. *See also* **epigenetics.**

epigenetics Phenonmena due to alterations to DNA that do not include changes in the base sequence; often affect the way in which the DNA sequences are expressed. Such alterations are often stable and heritable in the sense that they are passed from one cell to another.

epigenome All epigenetic modifications within the genome of an individual organism.

episome Plasmid capable of integrating into a bacterial chromosome.

epistasis Type of gene interaction in which a gene at one locus masks or suppresses the effects of a gene at a different locus.

epistatic gene Masks or suppresses the effect of a gene at a different locus.

equilibrium Situation in which no further change takes place; in population genetics, refers to a population in which allelic frequencies do not change.

equilibrium density gradient centrifugation Method used to separate molecules or organelles of different density by centrifugation.

eubacteria One of the three primary divisions of life. Eubacteria consist of unicellular organisms with prokaryotic cells and include most of the common bacteria.

euchromatin Chromatin that undergoes condensation and decondensation in the course of the cell cycle.

eukaryote Organism with a complex cell structure including a nuclear envelope and membrane-bounded organelles. One of the three primary divisions of life, eukaryotes include unicellular and multicellular forms.

evolution Genetic change taking place in a group of organisms.

exit (E) site One of three sites in a ribosome occupied by a tRNA. In the elongation stage of translation, the tRNA moves from the peptidyl (P) site to the E site from which it then exits the ribosome.

exon Coding region of a split gene (a gene that is interrupted by introns). After processing, the exons remain in messenger RNA.

exon shuffling Process, important in the evolution of eukaryotic genes, by which exons of different genes are exchanged and mixed into new combinations, creating new genes that are mosaics of other preexisting genes.

expanding trinucleotide repeat Mutation in which the number of copies of a set of nucleotides (most often three nucleotides) increases in succeeding generations.

expected heterozygosity Proportion of individuals that are expected to be heterozygous at a locus when the Hardy–Weinberg assumptions are met.

expressed-sequence tag (EST) Unique fragment of DNA from the coding region of a gene, produced by the reverse transcription of cellular RNA. Parts of the fragments are sequenced so that they can be identified.

expression vector Cloning vector containing DNA sequences such as a promoter, a ribosome-binding site, and transcription initiation and termination sites that allow DNA fragments inserted into the vector to be transcribed and translated.

expressivity Degree to which a trait is expressed.

familial Down syndrome Caused by a Robertsonian translocation in which the long arm of chromosome 21 is translocated to another chromosome; tends to run in families.

fertilization Fusion of gametes (sex cells) to form a zygote.

fetal cell sorting Separation of fetal cells from maternal blood. Genetic testing on the fetal cells can provide information about genetic diseases and disorders in the fetus.

F factor Episome of *E. coli* that controls conjugation and gene exchange between *E. coli* cells. The F factor contains an origin of replication and genes that enable the bacterium to undergo conjugation.

F_1 (first filial) generation Offspring of the initial parents (P) in a genetic cross.

F_2 (second filial) generation Offspring of the F_1 generation in a genetic cross; the third generation of a genetic cross.

first polar body One of the products of meiosis I in oogenesis; contains half the chromosomes but little of the cytoplasm.

fitness Reproductive success of a genotype compared with that of other genotypes in a population.

5′ cap Modified 5′ end of eukaryotic mRNA, consisting of an extra nucleotide (methylated) and methylation of the 2′ position of the ribose sugar in one or more subsequent nucleotides; plays a role in the binding of the ribosome to mRNA and affects mRNA stability and the removal of introns.

5′ end End of the polynucleotide chain where a phosphate is attached to the 5′-carbon atom of the nucleotide.

5′ splice site The 5′ end of an intron where cleavage takes place in RNA splicing.

5′ untranslated (5′ UTR) region Sequence of nucleotides at the 5′ end of mRNA; does not encode the amino acids of a protein.

fixation Point at which one allele reaches a frequency of 1. At this point, all members of the population are homozygous for the same allele.

flanking direct repeat Short, directly repeated sequence produced on either side of a transposable element when the element inserts into DNA.

forward genetics Traditional approach to the study of gene function that begins with a phenotype (a mutant organism) and proceeds to a gene that encodes the phenotype.

forward mutation Alters a wild-type phenotype.

founder effect Sampling error that arises when a population is established by a small number of individuals; leads to genetic drift.

fragile site Constriction or gap that appears at a particular location on a chromosome when cells are cultured under special conditions. One fragile site on the human X chromosome is associated with mental retardation (fragile-X syndrome) and results from an expanding trinucleotide repeat.

fragile-X syndrome A form of X-linked mental retardation that appears primarily in males; results from an increase in the number of repeats of a CGG trinucleotide.

frameshift mutation Alters the reading frame of a gene.

fraternal twins Nonidentical twins that arise when two different eggs are fertilized by two different sperm; also called dizygotic twins.

frequency distribution Graphical way of representing values. In genetics, usually the phenotypes found in a group of individuals are displayed as a frequency distribution. Typically, the phenotypes are plotted on the horizontal (x) axis and the numbers (or proportions) of individuals with each phenotype are plotted on the vertical (y) axis.

functional genomics Area of genomics that studies the functions of genetic information contained within genomes.

G_0 (gap 0) Nondividing stage of the cell cycle.

G_1 (gap 1) Stage in interphase of the cell cycle in which the cell grows and develops.

G_2 (gap 2) Stage of interphase in the cell cycle that follows DNA replication. In G_2, the cell prepares for division.

gain-of-function mutation Produces a new trait or causes a trait to appear in inappropriate tissues or at inappropriate times in development.

gametic isolation Reproductive isolation due to the incompatibility of gametes. Mating between members of different species takes place, but the gametes do not form zygotes. Seen in many plants, where pollen from one species cannot fertilize the ovules of another species.

gametophyte Haploid phase of the life cycle in plants.

gap genes In fruit flies, set of segmentation genes that define large sections of the embryo. Mutations in these genes usually eliminate whole groups of adjacent segments.

gel electrophoresis Technique for separating charged molecules (such as proteins or nucleic acids) on the basis of molecular size or charge or both.

gene Genetic factor that helps determine a trait; often defined at the molecular level as a DNA sequence that is transcribed into an RNA molecule.

gene cloning Insertion of DNA fragments into bacteria in such a way that the fragments will be stable and copied by the bacteria.

gene conversion Process of nonreciprocal genetic exchange that can produce abnormal ratios of gametes following meiosis.

gene desert In reference to the density of genes in the genome, a region that is gene poor—that is, a long stretch of DNA possibly consisting of hundreds of thousands to millions of base pairs completely devoid of any known genes or other functional sequences.

gene family *See* **multigene family.**

gene flow Movement of genes from one population to another; also called migration.

gene interaction Interaction between genes at different loci that affect the same characteristic.

gene mutation Affects a single gene or locus.

gene pool Total of all genes in a population.

generalized transduction Transduction in which any gene can be transferred from one bacterial cell to another by a virus.

general transcription factor Protein that binds to eukaryotic promoters near the start site and is a part of the basal transcription apparatus that initiates transcription.

gene regulation Mechanisms and processes that control the phenotypic expression of genes.

gene therapy Use of recombinant DNA to treat a disease or disorder by altering the genetic makeup of the patient's cells.

genetic bottleneck Sampling error that arises when a population undergoes a drastic reduction in population size; leads to genetic drift.

genetic conflict hypothesis Suggests that genomic imprinting evolved because different and conflicting pressures act on maternal and paternal alleles for genes that affect fetal growth. For example, paternally derived alleles often favor maximum fetal growth, whereas maternally derived alleles favor less than maximum fetal growth because of the high cost of fetal growth to the mother.

genetic correlation Phenotypic correlation due to the same genes affecting two or more characteristics.

genetic counseling Educational process that attempts to help patients and family members deal with all aspects of a genetic condition.

genetic drift Change in allelic frequency due to sampling error.

genetic engineering Common term for recombinant DNA technology.

genetic–environmental interaction variance Component of the phenotypic variance that results from an interaction between genotype and environment. Genotypes are expressed differently in different environments.

Genetic Information Nondiscrimination Act (GINA) U.S. law prohibiting health insurers from using genetic information to make decisions about health-insurance coverage and rates; prevents employers from using genetic information in employment decisions; also prevents health insurers and employers from asking or requiring a person to take a genetic test.

genetic (linkage) map Map of the relative distances between genetic loci, markers, or other chromosome regions determined by rates of recombination; measured in percent recombination or map units.

genetic marker Any gene or DNA sequence used to identify a location on a genetic or physical map.

genetic maternal effect Determines the phenotype of an offspring. With genetic maternal effect, an offspring inherits genes for the characteristics from both parents, but the offspring's phenotype is determined not by its own genotype but by the nuclear genotype of its mother.

genetic rescue Introduction of new genetic variation into an inbred population that often dramatically improves the health of the population in an effort to increase its chances of long-term survival.

genetic variance Component of the phenotypic variance that is due to genetic differences among individual members of a population.

gene tree Phylogenetic tree representing the evolutionary relationships among a set of genes.

genic balance system Sex-determining system in which sexual phenotype is controlled by a balance between genes on the X chromosome and genes on the autosomes.

genic interaction variance Component of the genetic variance that can be attributed to genic interaction (interaction between genes at different loci).

genic sex determination Sex determination in which the sexual phenotype is specified by genes at one or more loci, but there are no obvious differences in the chromosomes of males and females.

genome Complete set of genetic instructions for an organism.

genomewide association studies Studies that look for nonrandom associations between the presence of a trait and alleles at many different loci scattered across a genome—that is, for associations between traits and particular suites of alleles in a population.

genomic imprinting Differential expression of a gene that depends on the sex of the parent that transmitted the gene. If the gene is inherited from the father, its expression differs from that if the gene is inherited from the mother.

genomic library Collection of bacterial or phage colonies containing DNA fragments that consist of the entire genome of an organism.

genomics Study of the content, organization, and function of genetic information in whole genomes.

genotype The set of genes possessed by an individual organism.

genotypic frequency Proportion of a particular genotype within a population.

germ-line mutation Mutation in a germ-line cell (one that gives rise to gametes).

germ-plasm theory States that cells in the reproductive organs carry a complete set of genetic information.

G$_2$/M checkpoint Important point in the cell cycle near the end of G$_2$. After this checkpoint has been passed, the cell undergoes mitosis.

goodness-of-fit chi-square test Statistical test used to evaluate how well a set of observed values fit the expected values. The probability associated with a calculated chi-square value is the probability that the differences between the observed and the expected values may be due to chance.

G overhang A guanine-rich sequence of nucleotides that protrudes beyond the complementary C-rich strand at the end of a chromosome.

group I intron Belongs to a class of introns in some ribosomal RNA genes that are capable of self-splicing.

group II intron Belongs to a class of introns in some protein-encoding genes that are capable of self-splicing and are found in mitochondria, chloroplasts, and a few eubacteria.

G$_1$/S checkpoint Important point in the cell cycle. After the G$_1$/S checkpoint has been passed, DNA replicates and the cell is committed to dividing.

guanine (G) Purine base in DNA and RNA.

guide RNA (gRNA) RNA molecule that serves as a template for an alteration made in mRNA during RNA editing.

gynandromorph Individual organism that is a mosaic for the sex chromosomes, possessing tissues with different sex-chromosome constitutions.

gyrase *See* **DNA gyrase.**

hairpin Secondary structure formed when sequences of nucleotides on the same strand are complementary and pair with each other.

haploid Possessing a single set of chromosomes (one genome).

haploinsufficiency The appearance of a mutant phenotype in an individual cell or organism that is heterozygous for a normally recessive trait.

haploinsufficient gene Must be present in two copies for normal function. If one copy of the gene is missing, a mutant phenotype is produced.

haplotype A specific set of linked genetic variants or alleles on a single chromosome or on part of a chromosome.

Hardy–Weinberg equilibrium Frequencies of genotypes when the conditions of the Hardy–Weinberg law are met.

Hardy–Weinberg law Important principle of population genetics stating that, in a large, randomly mating population not affected by mutation, migration, or natural selection, allelic frequencies will not change and genotypic frequencies stabilize after one generation in the proportions p^2 (the frequency of *AA*), $2pq$ (the frequency of *Aa*), and q^2 (the frequency of *aa*), where p equals the frequency of allele *A* and q equals the frequency of allele *a*.

heat-shock protein Produced by many cells in response to extreme heat and other stresses; helps cells prevent damage from such stressing agents.

helicase *See* **DNA helicase.**

hemizygosity Possession of a single allele at a locus. Males of organisms with XX-XY sex determination are hemizygous for X-linked loci because their cells possess a single X chromosome.

heritability Proportion of phenotypic variation due to genetic differences. *See also* **broad-sense heritability** and **narrow-sense heritability.**

hermaphroditism Condition in which an individual organism possesses both male and female reproductive structures. True hermaphrodites produce both male and female gametes.

heterochromatin Chromatin that remains in a highly condensed state throughout the cell cycle; found at the centromeres and telomeres of most chromosomes.

heteroduplex DNA DNA consisting of two strands, each of which is from a different chromosome.

heterogametic sex The sex (male or female) that produces two types of gametes with respect to sex chromosomes. For example, in the XX-XY sex-determining system, the male produces both X-bearing and Y-bearing gametes.

heterokaryon Cell possessing two nuclei derived from different cells through cell fusion.

heteroplasmy Presence of two or more distinct variants of DNA within the cytoplasm of a single cell.

heterozygote screening Tests members of a population to identify heterozygous carriers of a disease-causing allele who are healthy but have the potential to produce children who have the disease.

heterozygous Refers to an individual organism that possesses two different alleles at a locus.

highly repetitive DNA DNA that consists of short sequences that are present in hundreds of thousands to millions of copies; clustered in certain regions of chromosomes.

histone Low-molecular-weight protein found in eukaryotes that complexes with DNA to form chromosomes.

histone code Modification of histone proteins, such as the addition or removal of phosphate groups, methyl groups, or acetyl groups, that encode information affecting how genes are expressed.

Holliday intermediate Structure that forms in homologous recombination; consists of two duplex molecules connected by a cross bridge.

Holliday junction Model of homologous recombination that is initiated by single-strand breaks in a DNA molecule.

holoenzyme Complex of an enzyme and other protein factors necessary for complete function.

homeobox Conserved subset of nucleotides in homeotic genes. In *Drosophila*, it consists of 180 nucleotides that encode 60 amino acids of a DNA-binding domain related to the helix-turn-helix motif.

homeotic complex (HOM-C) Major cluster of homeotic genes in fruit flies; consists of the *Antennapedia* complex, which affects the development of the adult fly's head and anterior segments, and the *bithorax* complex, which affects the adult fly's posterior thoracic and abdominal segments.

homeotic gene Determines the identity of individual segments or parts in an early embryo. Mutations in such genes cause body parts to appear in the wrong places.

homogametic sex The sex (male or female) that produces gametes that are all alike with regard to sex chromosomes. For example, in the XX-XY sex-determining system, the female produces only X-bearing gametes.

homologous genes Evolutionarily related genes, having descended from a gene in a common ancestor.

homologous pair of chromosomes Two chromosomes that are alike in structure and size and that carry genetic information for the same set of hereditary characteristics. One chromosome of a homologous pair is inherited from the male parent and the other is inherited from the female parent.

homologous recombination Exchange of genetic information between homologous DNA molecules.

homoplasmy Presence of only one version of DNA within the cytoplasm of a single cell.

homozygous Refers to an individual organism that possesses two identical alleles at a locus.

horizontal gene transfer Transfer of genes from one organism to another by a mechanism other than reproduction.

***Hox* gene** Gene that contains a homeobox.

human papilloma virus (HPV) Virus associated with cervical cancer.

Human Proteome Project Project with the goal of identifying and characterizing all proteins in the human body.

humoral immunity Type of immunity resulting from antibodies produced by B cells.

hybrid breakdown Reproductive isolating mechanism in which closely related species are capable of mating and producing viable and fertile F_1 progeny, but genes do not flow between the two species, because further crossing of the hybrids produces inviable or sterile offspring.

hybrid dysgenesis Sudden appearance of numerous mutations, chromosome aberrations, and sterility in the offspring of a cross between a male fly that possesses *P* elements and a female fly that lacks them.

hybrid inviability Reproductive isolating mechanism in which mating between two organisms of different species take place and hybrid offspring are produced but are not viable.

hybridization Pairing of two partly or fully complementary single-stranded nucleotide chains.

hybrid sterility Hybrid embryos complete development but are sterile; exemplified by mating between donkeys and horses to produce a mule, a viable but usually sterile offspring.

hypostatic gene Gene that is masked or suppressed by the action of a gene at a different locus.

identical twins Twins that arise when a single egg fertilized by a single sperm splits into two separate embryos; also called monozygotic twins.

inbreeding Mating between related individuals that takes place more frequently than expected on the basis of chance.

inbreeding coefficient Measure of inbreeding; the probability (ranging from 0 to 1) that two alleles are identical by descent.

inbreeding depression Decreased fitness arising from inbreeding; often due to the increased expression of lethal or deleterious recessive traits.

incomplete dominance Refers to the phenotype of a heterozygote that is intermediate between the phenotypes of the two homozygotes.

incomplete linkage Linkage between genes that exhibit some crossing over; intermediate in its effects between independent assortment and complete linkage.

incomplete penetrance Refers to a genotype that does not always express the expected phenotype. Some individuals possess the genotype for a trait but do not express the phenotype.

incorporated error Incorporation of a damaged nucleotide or mismatched base pair into a DNA molecule.

independent assortment Independent separation of chromosome pairs in anaphase I of meiosis; contributes to genetic variation.

induced mutation Results from environmental agents, such as chemicals or radiation.

inducer Substance that stimulates transcription in an inducible system of gene regulation; usually a small molecule that binds to a repressor

protein and alters that repressor so that it can no longer bind to DNA and inhibit transcription.

inducible operon Operon or other system of gene regulation in which transcription is normally off. Something must take place for transcription to be induced, or turned on.

induction Stimulation of the synthesis of an enzyme by an environmental factor, often the presence of a particular substrate.

in-frame deletion Deletion of some multiple of three nucleotides, which does not alter the reading frame of the gene.

in-frame insertion Insertion of some multiple of three nucleotides, which does not alter the reading frame of the gene.

inheritance of acquired characteristics Early notion of inheritance proposing that acquired traits are passed to descendants.

initiation codon The codon in mRNA that specifies the first amino acid (fMet in bacterial cells; Met in eukaryotic cells) of a protein; most commonly AUG.

initiation factor 1 (IF-1) Protein required for the initiation of translation in bacterial cells; enhances the dissociation of the large and small subunits of the ribosome.

initiation factor 2 (IF-2) Protein required for the initiation of translation in bacterial cells; forms a complex with GTP and the charged initiator tRNA and then delivers the charged tRNA to the initiation complex.

initiation factor 3 (IF-3) Protein required for the initiation of translation in bacterial cells; binds to the small subunit of the ribosome and prevents the large subunit from binding during initiation.

initiator protein Binds to an origin of replication and unwinds a short stretch of DNA, allowing helicase and other single-strand-binding proteins to bind and initiate replication.

insertion Mutation in which nucleotides are added to a DNA sequence.

insertion sequence Simple type of transposable element found in bacteria and their plasmids that contains only the information necessary for its own movement.

in situ hybridization Method used to determine the chromosomal location of a gene or other specific DNA fragment or the tissue distribution of an mRNA by using a labeled probe that is complementary to the sequence of interest.

insulator DNA sequence that blocks or insulates the effect of an enhancer; must be located between the enhancer and the promoter to have blocking activity; also may limit the spread of changes in chromatin structure.

integrase Enzyme that inserts prophage, or proviral, DNA into a chromosome.

interactome Complete set of protein interactions in a cell.

intercalating agent Chemical substance that is about the same size as a nucleotide and may become sandwiched between adjacent bases in DNA, distorting the three-dimensional structure of the helix and causing single-nucleotide insertions and deletions in replication.

interchromosomal recombination Recombination among genes on different chromosomes.

interference Degree to which one crossover interferes with additional crossovers.

intergenic suppressor mutation Suppressor mutation that occurs in a gene (locus) that is different from the gene containing the original mutation.

interkinesis Period between meiosis I and meiosis II.

internal promoter Promoter located within the sequences of DNA that are transcribed into RNA.

interphase Period in the cell cycle between the cell divisions. In interphase, the cell grows, develops, and prepares for cell division.

interspersed repeat sequence Repeated sequence at multiple locations throughout the genome.

intrachromosomal recombination Recombination among genes located on the same chromosome.

intragenic mapping Maps the locations of mutations within a single locus.

intragenic suppressor mutation Suppressor mutation that occurs in the same gene (locus) as the mutation that it suppresses.

intron Intervening sequence in a split gene; removed from the RNA after transcription.

inverted repeats Sequences on the same strand that are inverted and complementary.

isoaccepting tRNAs Different tRNAs with different anticodons that specify the same amino acid.

isotopes Different forms of an element that have the same number of protons and electrons but differ in the number of neutrons in the nucleus.

junctional diversity Addition or deletion of nucleotides at the junctions of gene segments brought together in the somatic recombination of genes that encode antibodies and T-cell receptors.

karyotype Picture of an individual organism's complete set of metaphase chromosomes.

kinetochore Set of proteins that assemble on the centromere, providing the point of attachment for spindle microtubules.

Klinefelter syndrome Human condition in which cells contain one or more Y chromosomes along with multiple X chromosomes (most commonly XXY but may also be XXXY, XXXXY, or XXYY). Persons with Klinefelter syndrome are male in appearance but frequently possess small testes, some breast enlargement, and reduced facial and pubic hair; often taller than normal and sterile, most have normal intelligence.

knock-in mouse Mouse that carries a foreign sequence inserted at a specific chromosome location.

knockout mouse Mouse in which a normal gene has been disabled ("knocked out").

labeling Method for adding a radioactive or chemical label to a molecule such as DNA or RNA.

lagging strand DNA strand that is replicated discontinuously.

large ribosomal subunit The larger of the two subunits of a functional ribosome.

lariat Looplike structure created in the splicing of nuclear pre-mRNA in which the 5′ end of an intron is attached to a branch point in pre-mRNA.

leading strand DNA strand that is replicated continuously.

leptotene First substage of prophase I in meiosis. In leptotene, chromosomes contract and become visible.

lethal allele Causes the death of an individual organism, often early in development, and so the organism does not appear in the progeny of a genetic cross. Recessive lethal alleles kill individual organisms that are homozygous for the allele; dominant lethals kill both heterozygotes and homozygotes.

lethal mutation Causes premature death.

LINE *See* **long interspersed element.**

linkage analysis Gene mapping based on the detection of physical linkage between genes, as measured by the rate of recombination, in progeny from a cross.

linkage disequilibrium Nonrandom association between genetic variants within a haplotype.

linkage group Genes located together on the same chromosome.

linked genes Genes located on the same chromosome.

linker Small, synthetic DNA fragment that contains one or more restriction sites. Can be attached to the ends of any piece of DNA and used to insert a gene into a plasmid vector when restrictions sites are not available.

linker DNA Stretch of DNA separating two nucleosomes.

locus Position on a chromosome where a specific gene is located.

lod (logarithm of odds) score Logarithm of the ratio of the probability of obtaining a set of observations, assuming a specified degree of linkage, to the probability of obtaining the same set of observations with independent assortment; used to assess the likelihood of linkage between genes from pedigree data.

long interspersed element (LINE) Long DNA sequence repeated many times and interspersed throughout the genome.

loss-of-function mutation Causes the complete or partial absence of normal function.

loss of heterozygosity At a locus having a normal allele and a mutant allele, inactivation or loss of the normal allele.

Lyon hypothesis Proposed by Mary Lyon in 1961, this hypothesis proposes that one X chromosome in each female cell becomes inactivated (a Barr body) and suggests that which of the X chromosomes becomes inactivated is random and varies from cell to cell.

lysogenic cycle Life cycle of a bacteriophage in which phage genes first integrate into the bacterial chromosome and are not immediately transcribed and translated.

lytic cycle Life cycle of a bacteriophage in which phage genes are transcribed and translated, new phage particles are produced, and the host cell is lysed.

major histocompatibility complex (MHC) antigen Belongs to a large and diverse group of antigens found on the surfaces of cells that mark those cells as self; encoded by a large cluster of genes known as the major histocompatibility complex. T cells simultaneously bind to foreign and MHC antigens.

malignant tumor Consists of cells that are capable of invading other tissues.

map-based sequencing Method of sequencing a genome in which sequenced fragments are ordered into contigs with the use of genetic or physical maps.

mapping function Relates recombination frequencies to actual physical distances between genes.

map unit (m.u.) Unit of measure for distances on a genetic map; 1 map unit equals 1% recombination.

mass spectrometry Method for precisely determining the molecular mass of a molecule by using the migration rate of an ionized molecule in an electrical field.

maternal blood screening test Tests for genetic conditions in a fetus by analyzing the blood of the mother. For example, the level of α-fetoprotein in maternal blood provides information about the probability that a fetus has a neural-tube defect.

mean Statistic that describes the center of a distribution of measurements; calculated by dividing the sum of all measurements by the number of measurements; also called the average.

mechanical isolation Reproductive isolation resulting from anatomical differences that prevent successful copulation.

mediator Complex of proteins that is one of the components of the basal transcription apparatus.

megaspore One of the four products of meiosis in plants.

megasporocyte In the ovary of a plant, a diploid reproductive cell that undergoes meiosis to produce haploid macrospores.

meiosis Process in which chromosomes of a eukaryotic cell divide to give rise to haploid reproductive cells. Consists of two divisions: meiosis I and meiosis II.

meiosis I First phase of meiosis. In meiosis I, chromosome number is reduced by half.

meiosis II Second phase of meiosis. Events in meiosis II are essentially the same as those in mitosis.

melting *See* **denaturation.**

melting temperature Midpoint of the melting range of DNA.

memory cell Long-lived lymphocyte among the clone of cells generated when a foreign antigen is encountered. If the same antigen is encountered again, the memory cells quickly divide and give rise to another clone of cells specific for that particular antigen.

Mendelian population Group of interbreeding, sexually reproducing individuals.

meristic characteristic Characteristic whose phenotype varies in whole numbers, such as number of vertebrae.

merozygote Bacterial cell that has two copies of some genes—one copy on the bacterial chromosome and a second copy on an introduced F plasmid; also called partial diploid.

messenger RNA (mRNA) RNA molecule that carries genetic information for the amino acid sequence of a protein.

metacentric chromosome Chromosome in which the two chromosome arms are approximately the same length.

metagenomics An emerging field of sequencing technology in which the genome sequences of an entire group of organisms inhabiting a common environment are sampled and determined.

metaphase Stage of mitosis. In metaphase, chromosomes align in the center of the cell.

metaphase I Stage of meiosis I. In metaphase I, homologous pairs of chromosomes align in the center of the cell.

metaphase II Stage of meiosis II. In metaphase II, individual chromosomes align on the metaphase plate.

metaphase plate Plane in a cell between two spindle poles. In metaphase, chromosomes align on the metaphase plate.

metastasis Refers to cells that separate from malignant tumors and travel to other sites, where they establish secondary tumors.

5′-methylcytosine Modified nucleotide, consisting of cytosine to which a methyl group has been added; predominate form of methylation in eukaryotic DNA.

microarray Ordered array of DNA fragments fixed to a solid support, which serve as probes to detect the presence of complementary sequences; often used to assess the expression of genes in various tissues and under different conditions.

microRNA (miRNA) Small RNA molecule, typically 21 or 22 bp in length, produced by cleavage of double-stranded RNA arising from small hairpins within RNA that is mostly single stranded. The miRNAs combine with proteins to form a complex that binds (imperfectly) to mRNA molecules and inhibits their translation.

microsatellite *See* **variable number of tandem repeats.**

microspore Haploid product of meiosis in plants.

microsporocyte Diploid reproductive cell in the stamen of a plant; undergoes meiosis to produce four haploid microspores.

microtubule Long fiber composed of the protein tubulin; plays an important role in the movement of chromosomes in mitosis and meiosis.

migration Movement of genes from one population to another; also called gene flow.

minimal medium Used to culture bacteria or some other microorganism; contains only the nutrients required by prototrophic (wild-type) cells—typically, a carbon source, essential elements such as nitrogen and phosphorus, certain vitamins, and other required ions and nutrients.

mismatch repair Process that corrects mismatched nucleotides in DNA after replication has been completed. Enzymes excise incorrectly paired nucleotides from the newly synthesized strand and use the original nucleotide strand as a template when replacing them.

missense mutation Alters a codon in mRNA, resulting in a different amino acid in the protein encoded.

mitochondrial DNA (mtDNA) DNA in mitochondria; has some characteristics in common with eubacterial DNA and typically consists of a circular molecule that lacks histone proteins and encodes some of the rRNAs, tRNAs, and proteins found in mitochondria.

mitosis Process by which the nucleus of a eukaryotic cell divides.

mitosis-promoting factor (MPF) Protein functioning in the control of the cell cycle; consists of a cyclin combined with cyclin-dependent kinase (CDK). Active MPF stimulates mitosis.

mitotic spindle Array of microtubules that radiate from two poles; moves chromosomes in mitosis and meiosis.

model genetic organism An organism that is widely used in genetic studies because it has characteristics, such as short generation time and large numbers of progeny, that make it particularly useful for genetic analysis.

moderately repetitive DNA DNA consisting of sequences that are from 150 to 300 bp in length and are repeated thousands of times.

modified base Rare base found in some RNA molecules. Such bases are modified forms of the standard bases (adenine, guanine, cytosine, and uracil).

molecular chaperone Molecule that assists in the proper folding of another molecule.

molecular clock Refers to the use of molecular differences to estimate the time of divergence between organisms; assumes a roughly constant rate at which one neutral mutation replaces another.

molecular genetics Study of the chemical nature of genetic information and how it is encoded, replicated, and expressed.

molecular motor Specialized protein that moves cellular components.

monoecious organism Individual organism that has both male and female reproductive structures.

monohybrid cross A cross between two individuals that differ in a single characteristic—more specifically, a cross between individuals that are homozygous for different alleles at the same locus ($AA \times aa$); also refers to a cross between two individuals that are heterozygous for two alleles at a single locus ($Aa \times Aa$).

monosomy Absence of one of the chromosomes of a homologous pair.

monozygotic twins Identical twins that arise when a single egg fertilized by a single sperm splits into two separate embryos.

Morgan 100 map units.

morphogen Molecule whose concentration gradient affects the developmental fate of surrounding cells.

mosaicism Condition in which regions of tissue within a single individual have different chromosome constitutions.

M (mitotic) phase Period of active cell division; includes mitosis (nuclear division) and cytokinesis (cytoplasmic division).

mRNA surveillance Mechanisms for the detection and elimination of mRNAs that contain errors that may create problems in the course of translation.

multifactorial characteristic Determined by multiple genes and environmental factors.

multigene family Set of genes similar in sequence that arose through repeated duplication events; often encode different proteins.

multiple alleles Presence in a group of individuals of more than two alleles at a locus. Although, for the group, the locus has more than two alleles, each member of the group has only two of the possible alleles.

multiple 3′ cleavage sites Refers to the presence of more than one 3′ cleavage site on a single pre-mRNA, which allows cleavage and polyadenylation to take place at different sites, producing mRNAs of different lengths.

multiplication rule States that the probability of two or more independent events occurring together is calculated by multiplying the probabilities of each of the individual events.

mutagen Any environmental agent that significantly increases the rate of mutation above the spontaneous rate.

mutagenesis screen Method for identifying genes that influence a specific phenotype. Random mutations are induced in a population of organisms, and individual organisms with mutant phenotypes are identified. These individual organisms are crossed to determine the genetic basis of the phenotype and to map the location of mutations that cause the phenotype.

mutation Heritable change in genetic information.

mutation rate Frequency with which a gene changes from the wild type to a specific mutant; generally expressed as the number of mutations per biological unit (that is, mutations per cell division, per gamete, or per round of replication).

narrow-sense heritability Proportion of the phenotypic variance that can be attributed to additive genetic variance.

natural selection Differential reproduction of genotypes.

negative assortative mating Mating between unlike individuals that is more frequent than would be expected on the basis of chance.

negative control Gene regulation in which the binding of a regulatory protein to DNA inhibits transcription (the regulatory protein is a repressor).

negative supercoiling *See* **supercoiling.**

neutral mutation Changes the amino acid sequence of a protein but does not alter the function of the protein.

neutral-mutation hypothesis Proposes that much of the molecular variation seen in natural populations is adaptively neutral and unaffected by natural selection. Under this hypothesis, individuals with different molecular variants have equal fitnesses.

newborn screening Tests newborn infants for certain genetic disorders.

next-generation sequencing technologies Sequencing methods, such as pyrosequencing, that are capable of simultaneously determining the sequences of many DNA fragments; the technologies are much faster and less expensive than the Sanger method.

nitrogenous base Nitrogen-containing base that is one of the three parts of a nucleotide.

node Point in a phylogenetic tree that represents an organism. Terminal nodes are those that are at the outermost branches of the tree and represent organisms for which data have been obtained. Internal nodes represent ancestors common to organisms on different branches of the tree.

nonautonomous element Transposable element that cannot transpose on its own but can transpose in the presence of an autonomous element of the same family.

nondisjunction Failure of homologous chromosomes or sister chromatids to separate in meiosis or mitosis.

nonhistone chromosomal protein One of a heterogeneous assortment of nonhistone proteins in chromatin.

nonidentical twins Twins that arise when two different eggs are fertilized by two different sperm; also called dizygotic twins or fraternal twins.

nonoverlapping genetic code Refers to the fact that, generally, each nucleotide is a part of only one codon and encodes only one amino acid in a protein.

nonreciprocal translocation Movement of a chromosome segment to a nonhomologous chromosome or region without any (or with unequal) reciprocal exchange of segments.

nonrecombinant (parental) gamete Contains only the original combinations of genes present in the parents.

nonrecombinant (parental) progeny Possesses the original combinations of traits possessed by the parents.

nonreplicative transposition Type of transposition in which a transposable element excises from an old site and moves to a new site, resulting in no net increase in the number of copies of the transposable element.

nonsense codon Codon in mRNA that signals the end of translation; also called a stop codon or termination codon. The three common nonsense codons are UAA, UAG, and UGA.

nonsense-mediated mRNA decay (NMD) Process that brings about the rapid elimination of mRNA that has a premature stop codon.

nonsense mutation Changes a sense codon (one that specifies an amino acid) into a stop codon.

nonstop RNA decay Mechanism in eukaryotic cells for dealing with ribosomes stalled at the 3′ end of an mRNA that lacks a termination codon. A protein binds to the A site of the stalled ribosome and recruits other proteins that degrade the mRNA from the 3′ end.

nontemplate strand The DNA strand that is complementary to the template strand; not ordinarily used as a template during transcription.

normal distribution Common type of frequency distribution that exhibits a symmetrical, bell-shaped curve; usually arises when a large number of independent factors contribute to the measurement.

Northern blotting Process by which RNA is transferred from a gel to a solid support such as a nitrocellulose or nylon filter.

nuclear envelope Membrane that surrounds the genetic material in eukaryotic cells to form a nucleus; segregates the DNA from other cellular contents.

nuclear matrix Network of protein fibers in the nucleus; holds the nuclear contents in place.

nuclear pre-mRNA intron Belongs to a class of introns in protein-encoding genes that reside in the nuclei of eukaryotic cells; removed by spliceosomal-mediated splicing.

nucleoid Bacterial DNA confined to a definite region of the cytoplasm.

nucleoside Ribose or deoxyribose bonded to a base.

nucleosome Basic repeating unit of chromatin, consisting of a core of eight histone proteins (two each of H2A, H2B, H3, and H4) and about 146 bp of DNA that wraps around the core about two times.

nucleotide Repeating unit of DNA or RNA made up of a sugar, a phosphate, and a base.

nucleotide-excision repair DNA repair that removes bulky DNA lesions and other types of DNA damage.

nucleus Space in eukaryotic cells that is enclosed by the nuclear envelope and contains the chromosomes.

nullisomy Absence of both chromosomes of a homologous pair ($2n - 2$).

Okazaki fragment Short stretch of newly synthesized DNA. Produced by discontinuous replication on the lagging strand, these fragments are eventually joined together.

oligonucleotide-directed mutagenesis Method of site-directed mutagenesis that uses an oligonucleotide to introduce a mutant sequence into a DNA molecule.

oncogene Dominant-acting gene that stimulates cell division, leading to the formation of tumors and contributing to cancer; arises from mutated copies of a normal cellular gene (proto-oncogene).

one gene, one enzyme hypothesis Idea proposed by Beadle and Tatum that each gene encodes a separate enzyme.

one gene, one polypeptide hypothesis Modification of the one gene, one enzyme hypothesis; proposes that each gene encodes a separate polypeptide chain.

oogenesis Egg production in animals.

oogonium Diploid cell in the ovary; capable of undergoing meiosis to produce an egg cell.

open reading frame (ORF) Continuous sequence of DNA nucleotides that contains a start codon and a stop codon in the same reading frame; is assumed to be a gene that encodes a protein but, in many cases, the protein has not yet been identified.

operator DNA sequence in the operon of a bacterial cell. A regulator protein binds to the operator and affects the rate of transcription of structural genes.

operon Set of structural genes in a bacterial cell along with a common promoter and other sequences (such as an operator) that control the transcription of the structural genes.

origin of replication Site where DNA synthesis is initiated.

orthologous genes Homologous genes found in different species because the two species have a common ancestor that also possessed the gene.

outcrossing Mating between unrelated individuals that is more frequent than would be expected on the basis of chance.

overdominance Selection in which the heterozygote has higher fitness than that of either homozygote; also called heterozygote advantage.

ovum Final product of oogenesis.

pachytene Third substage of prophase I in meiosis. The synaptonemal complex forms during pachytene.

pair-rule genes Set of segmentation genes in fruit flies that define regional sections of the embryo and affect alternate segments. Mutations in these genes often cause the deletion of every other segment.

palindrome Sequence of nucleotides that reads the same on complementary strands; inverted repeats.

pangenesis Early concept of heredity proposing that particles carry genetic information from different parts of the body to the reproductive organs.

paracentric inversion Chromosome inversion that does not include the centromere in the inverted region.

paralogous genes Homologous genes in the same species that arose through the duplication of a single ancestral gene.

paramutation Epigenetic change in which one allele of a genotype alters the expression of another allele; the altered expression persists for several generations, even after the altering allele is no longer present.

parental gamete See **nonrecombinant (parental) gamete**.

parental progeny See **nonrecombinant (parental) progeny.**

partial diploid Bacterial cell that possesses two copies of genes, including one copy on the bacterial chromosome and the other on an extra piece of DNA (usually a plasmid); also called merozygote.

Patau syndrome (trisomy 13) Characterized by severe mental retardation, a small head, sloping forehead, small eyes, cleft lip and palate, extra fingers and toes, and other disabilities; results from the presence of three copies of chromosome 13.

pedigree Pictorial representation of a family history outlining the inheritance of one or more traits or diseases.

penetrance Percentage of individuals with a particular genotype that express the phenotype expected of that genotype.

pentaploidy Possession of five haploid sets of chromosomes ($5n$).

peptide bond Chemical bond that connects amino acids in a protein.

peptidyl (P) site One of three sites in a ribosome occupied by a tRNA in translation. In the elongation stage of protein synthesis, tRNAs move from the aminoacyl (A) site into the P site.

peptidyl transferase Activity in the ribosome that creates a peptide bond between two amino acids. Evidence suggests that this activity is carried out by one of the RNA components of the ribosome.

pericentric inversion Chromosome inversion that includes the centromere in the inverted region.

P (parental) generation First set of parents in a genetic cross.

phage See **bacteriophage.**

phenocopy Phenotype that is produced by environmental effects and is the same as the phenotype produced by a genotype.

phenotype Appearance or manifestation of a characteristic.

phenotypic correlation Correlation between two or more phenotypes in the same individual.

phenotypic variance Measures the degree of phenotypic differences among a group of individuals; composed of genetic, environmental, and genetic–environmental interaction variances.

phenylketonuria (PKU) Genetic disease characterized by mental retardation, light skin, and eczema; caused by mutations in the gene that encodes phenylalanine hydroxylase (PAH), a liver enzyme that normally metabolizes the amino acid phenylalanine. When the enzyme is defective, phenylalanine is not metabolized and builds up to high levels in the body, eventually causing mental retardation and other characteristics of the disease. The disease is inherited as an autosomal recessive disorder and can be effectively treated by limiting phenylalanine in the diet.

phosphate group A phosphorus atom attached to four oxygen atoms; one of the three components of a nucleotide.

phosphodiester Molecule containing R–O–P–O–R, where R is a carbon-containing group, O is oxygen, and P is phosphorus.

phosphodiester linkage Phosphodiester bond connecting two nucleotides in a polynucleotide strand.

phylogenetic tree Graphical represenation of the evolutionary connections between organisms or genes.

phylogeny Evolutionary relationships among a group of organisms or genes, usually depicted as a family tree or branching diagram.

physical map Map of physical distances between loci, genetic markers, or other chromosome segments; measured in base pairs.

pilus (pl., **pili**) Extension of the surface of some bacteria that allows conjugation to take place. When a pilus on one cell makes contact with a receptor on another cell, the pilus contracts and pulls the two cells together.

Piwi-interacting RNA (piRNA) Small RNA molecule belonging to a class named after Piwi proteins with which these molecules interact; similar to microRNA and small interfering RNA and thought to have a role in the regulation of sperm development.

plaque Clear patch of lysed cells on a continuous layer of bacteria on the agar surface of a petri plate. Each plaque represents a single original phage that multiplied and lysed many cells.

plasmid Small, circular DNA molecule found in bacterial cells that is capable of replicating independently from the bacterial chromosome.

pleiotropy A single genotype influences multiple phenotypes.

pluripotency In embryonic stem cells, the property of being undifferentiated, with the capacity to form every type of cell in an organism.

poly(A)-binding protein (PABP) Binds to the poly(A) tail of eukaryotic mRNA and makes the mRNA more stable. There are several types of PABPs, one of which is PABII.

poly(A) tail String of adenine nucleotides added to the 3′ end of a eukaryotic mRNA after transcription.

polycistronic mRNA Single bacterial RNA molecule that encodes more than one polypeptide chain; uncommon in eukaryotes.

polygenic characteristic Encoded by genes at many loci.

polymerase chain reaction (PCR) Method of enzymatically amplifying DNA fragments.

polynucleotide strand Series of nucleotides linked together by phosphodiester bonds.

polypeptide Chain of amino acids linked by peptide bonds; also called a protein.

polyploidy Possession of more than two haploid sets of chromosomes.

polyribosome Messenger RNA molecule with several ribosomes attached to it.

polytene chromosome Giant chromosome in the salivary glands of *Drosophila melanogaster*. Each polytene chromosome consists of a number of DNA molecules lying side by side.

population The group of interest; often represented by a subset called a sample. Also, a group of members of the same species.

population genetics Study of the genetic composition of populations (groups of members of the same species) and how a population's collective group of genes changes with the passage of time.

positional cloning Method that allows for the isolation and identification of a gene by examining the cosegregation of a phenotype with previously mapped genetic markers.

position effect Dependence of the expression of a gene on the gene's location in the genome.

positive assortative mating Mating between like individuals that is more frequent than would be expected on the basis of chance.

positive control Gene regulation in which the binding of a regulatory protein to DNA stimulates transcription (the regulatory protein is an activator).

positive supercoiling *See* **supercoiling.**

posttranslational modification Alteration of a protein after translation; may include cleavage from a larger precursor protein, the removal of amino acids, and the attachment of other molecules to the protein.

postzygotic reproductive isolating mechanism Reproductive isolation that arises after a zygote is formed, either because the resulting hybrids are inviable or sterile or because reproduction breaks down in subsequent generations.

preformationism Early concept of inheritance proposing that a miniature adult (homunculus) resides in either the egg or the sperm and increases in size in development, with all traits being inherited from the parent that contributes the homunculus.

preimplantation genetic diagnosis (PGD) Genetic testing on an embryo produced by in vitro fertilization before implantation of the embryo in the uterus.

pre-messenger RNA (pre-mRNA) Eukaryotic RNA molecule that is modified after transcription to become mRNA.

presymptomatic genetic testing Tests people to determine whether they have inherited a disease-causing gene before the symptoms of the disease have appeared.

prezygotic reproductive isolating mechanism Reproductive isolation in which gametes from two different species are prevented from fusing and forming a hybrid zygote.

primary Down syndrome Caused by the presence of three copies of chromosome 21.

primary immune response Initial clone of cells specific for a particular antigen and generated when the antigen is first encountered by the immune system.

primary oocyte Oogonium that has entered prophase I.

primary spermatocyte Spermatogonium that has entered prophase I.

primary structure of a protein The amino acid sequence of a protein.

primase Enzyme that synthesizes a short stretch of RNA on a DNA template; functions in replication to provide a 3′-OH group for the attachment of a DNA nucleotide.

primer Short stretch of RNA on a DNA template; provides a 3′-OH group for the attachment of a DNA nucleotide at the initiation of replication.

principle of independent assortment (Mendel's second law) Important principle of heredity discovered by Mendel that states that genes encoding different characteristics (genes at different loci) separate independently; applies only to genes located on different chromosomes or to genes far apart on the same chromosome.

principle of segregation (Mendel's first law) Important principle of heredity discovered by Mendel that states that each diploid individual possesses two alleles at a locus and that these two alleles separate when gametes are formed, one allele going into each gamete.

probability Likelihood of the occurrence of a particular event; more formally, the number of times that a particular event occurs divided by the number of all possible outcomes. Probability values range from 0 to 1.

proband A person having a trait or disease for whom a pedigree is constructed.

probe Known sequence of DNA or RNA that is complementary to a sequence of interest and will pair with it; used to find specific DNA sequences.

prokaryote Unicellular organism with a simple cell structure. Prokaryotes include eubacteria and archaea.

prometaphase Stage of mitosis. In prometaphase, the nuclear membrane breaks down and the spindle microtubules attach to the chromosomes.

promoter DNA sequence to which the transcription apparatus binds so as to initiate transcription; indicates the direction of transcription, which of the two DNA strands is to be read as the template, and the starting point of transcription.

proofreading Ability of DNA polymerases to remove and replace incorrectly paired nucleotides in the course of replication.

prophage Phage genome that is integrated into a bacterial chromosome.

prophase Stage of mitosis. In prophase, the chromosomes contract and become visible, the cytoskeleton breaks down, and the mitotic spindle begins to form.

prophase I Stage of meiosis I. In prophase I, chromosomes condense and pair, crossing over takes place, the nuclear membrane breaks down, and the spindle forms.

prophase II Stage of meiosis after interkinesis. In prophase II, chromosomes condense, the nuclear membrane breaks down, and the spindle forms. Some cells skip this stage.

proportion of polymorphic loci Percentage of loci in which more than one allele is present in a population.

protein-coding region The part of mRNA consisting of the nucleotides that specify the amino acid sequence of a protein.

protein domain Region of a protein that has a specific shape or function.

protein kinase Enzyme that adds phosphate groups to other proteins.

protein microarray Large number of different proteins applied to a glass slide as a series of spots, each spot containing a different protein; used to analyze protein–protein interactions.

proteome Set of all proteins encoded by a genome.

proteomics Study of the proteome, the complete set of proteins found in a given cell.

proto-oncogene Normal cellular gene that controls cell division. When mutated, it may become an oncogene and contribute to cancer progression.

prototrophic bacterium A wild-type bacterium that can use a carbon source, essential elements such as nitrogen and phosphorus, certain vitamins, and other required ions and nutrients to synthesize all the compounds that they need for growth and reproduction.

provirus DNA copy of viral DNA or viral RNA; integrated into the host chromosome and replicated along with the host chromosome.

pseudoautosomal region Small region of the X and Y chromosomes that contains homologous gene sequences.

pseudodominance Expression of a normally recessive allele owing to a deletion on the homologous chromosome.

Punnett square Shorthand method of determining the outcome of a genetic cross. On a grid, the gametes of one parent are written along the upper edge and the gametes of the other parent are written along the left-hand edge. Within the cells of the grid, the alleles in the gametes are combined to form the genotypes of the offspring.

purine Type of nitrogenous base in DNA and RNA. Adenine and guanine are purines.

pyrimidine Type of nitrogenous base in DNA and RNA. Cytosine, thymine, and uracil are pyrimidines.

pyrimidine dimer Structure in which a bond forms between two adjacent pyrimidine molecules on the same strand of DNA; disrupts normal hydrogen bonding between complementary bases and distorts the normal configuration of the DNA molecule.

quantitative characteristic Continuous characteristic; displays a large number of possible phenotypes, which must be described by a quantitative measurement.

quantitative genetics Genetic analysis of complex characteristics or characteristics influenced by multiple genetic factors.

quantitative trait locus (QTL) A gene or chromosomal region that contributes to the expression of quantitative characteristics.

quaternary structure of a protein Interaction of two or more polypeptides to form a functional protein.

reading frame Particular way in which a nucleotide sequence is read in groups of three nucleotides (codons) in translation. Each reading frame begins with a start codon and ends with a stop codon.

realized heritability Narrow-sense heritability measured from a response-to-selection experiment.

real-time PCR Modification of the polymerase chain reaction that quantitatively determines the amount of starting nucleic acid; the amount of DNA amplified is measured as the reaction proceeds.

reannealing *See* **renaturation.**

recessive Refers to an allele or phenotype that is expressed only when the recessive allele is present in two copies (homozygous). The recessive allele is not expressed in the heterozygote phenotype.

reciprocal crosses Crosses in which the phenotypes of the male and female parents are reversed. For example, in one cross, a tall male is crossed with a short female and, in the other cross, a short male is crossed with a tall female.

reciprocal translocation Reciprocal exchange of segments between two nonhomologous chromosomes.

recombinant DNA technology Set of molecular techniques for locating, isolating, altering, combining, and studying DNA segments.

recombinant gamete Possesses new combinations of genes.

recombinant progeny Possesses new combinations of traits formed from recombinant gametes.

recombination Process that produces new combinations of alleles.

recombination frequency Proportion of recombinant progeny produced in a cross.

regression Analysis of how one variable changes in response to another variable.

regression coefficient Statistic that measures how much one variable changes, on average, with a unit change in another variable.

regulator gene Gene associated with an operon in bacterial cells that encodes a protein or RNA molecule that functions in controlling the transcription of one or more structural genes.

regulator protein Produced by a regulator gene, a protein that binds to another DNA sequence and controls the transcription of one or more structural genes.

regulatory element DNA sequence that affects the transcription of other DNA sequences to which it is physically linked.

regulatory gene DNA sequence that encodes a protein or RNA molecule that interacts with DNA sequences and affects their transcription or translation or both.

regulatory promoter DNA sequence located immediately upstream of the core promoter that affects transcription; contains consensus sequences to which transcriptional regulator proteins bind.

relaxed state of DNA Energy state of a DNA molecule when there is no structural strain on the molecule.

release factor Protein required for the termination of translation; binds to a ribosome when a stop codon is reached and stimulates the release of the polypeptide chain, the tRNA, and the mRNA from the ribosome. Eukaryotic cells require two release factors (eRF_1 and eRF_2), whereas *E. coli* requires three (RF_1, RF_2, and RF_3).

renaturation The process by which two complementary single-stranded DNA molecules pair; also called reannealing.

repetitive DNA Sequences that exist in multiple copies in a genome.

replication Process by which DNA is synthesized from a single-stranded nucleotide template.

replication bubble Segment of a DNA molecule that is unwinding and undergoing replication.

replication error Replication of an incorporated error in which a change in the DNA sequence has been replicated and all base pairings in the new DNA molecule are correct.

replication fork Point at which a double-stranded DNA molecule separates into two single strands that serve as templates for replication.

replication licensing factor Protein that ensures that replication takes place only once at each origin; required at the origin before replication can be initiated and removed after the DNA has been replicated.

replication origin Sequence of nucleotides where replication is initiated.

replication terminus Point at which replication stops.

replicative segregation Random segregation of organelles into progeny cells in cell division. If two or more versions of an organelle are present in the original cell, chance determines the proportion of each type that will segregate into each progeny cell.

replicative transposition Type of transposition in which a copy of the transposable element moves to a new site while the original copy remains at the old site; increases the number of copies of the transposable element.

replicon Unit of replication, consisting of DNA from the origin of replication to the point at which replication on either side of the origin ends.

repressible operon Operon or other system of gene regulation in which transcription is normally on. Something must take place for transcription to be repressed, or turned off.

repressor Regulatory protein that binds to a DNA sequence and inhibits transcription.

reproductive isolating mechanism Any biological factor or mechanism that prevents gene exchange.

repulsion *See* **trans configuration.**

resolvase Enzyme required for some types of transposition; brings about resolution—that is, crossing over between sites located within the transposable element. Resolvase may be encoded by the transposable element or by a cellular enzyme that normally functions in homologous recombination.

response element Common DNA sequence found upstream of some groups of eukaryotic genes. A regulatory protein binds to a response element and stimulates the transcription of a gene. The presence of the same response element in several promoters or enhancers allows a single factor to simultaneously stimulate the transcription of several genes.

response to selection The amount that a characteristic changes in one generation owing to selection; equals the selection differential times the narrow-sense heritability.

restriction endonuclease Technical term for a restriction enzyme, which recognizes particular base sequences in DNA and makes double-stranded cuts nearby.

restriction enzyme Recognizes particular base sequences in DNA and makes double-stranded cuts nearby; also called restriction endonuclease.

restriction fragment length polymorphism (RFLP) Variation in the pattern of fragments produced when DNA molecules are cut with the same restriction enzyme; represents a heritable difference in DNA sequences and can be used in genetic analyses.

restriction mapping Determines the locations of sites cut by restriction enzymes in a piece of DNA.

retrotransposon Type of transposable element in eukaryotic cells that possesses some characteristics of retroviruses and transposes through an RNA intermediate.

retrovirus RNA virus capable of integrating its genetic material into the genome of its host. The virus injects its RNA genome into the host cell, where reverse transcription produces a complementary, double-stranded DNA molecule from the RNA template. The DNA copy then integrates into the host chromosome to form a provirus.

reverse duplication Duplication of a chromosome segment in which the sequence of the duplicated segment is inverted relative to the sequence of the original segment.

reverse genetics A molecular approach that begins with a genotype (a DNA sequence) and proceeds to the phenotype by altering the sequence or by inhibiting its expression.

reverse mutation (reversion) Mutation that changes a mutant phenotype back into the wild type.

reverse transcriptase Enzyme capable of synthesizing complementary DNA from an RNA template.

reverse transcription Synthesis of DNA from an RNA template.

reverse-transcription PCR Amplifies sequences corresponding to RNA. Reverse transcriptase is used to convert RNA into complementary DNA, which can then be amplified by the usual polymerase chain reaction.

rho-dependent terminator Sequence in bacterial DNA that requires the presence of the rho subunit of RNA polymerase to terminate transcription.

rho factor Subunit of bacterial RNA polymerase that facilitates the termination of transcription of some genes.

rho-independent terminator Sequence in bacterial DNA that does not require the presence of the rho subunit of RNA polymerase to terminate transcription.

ribonucleoside triphosphate (rNTP) Substrate of RNA synthesis; consists of ribose, a nitrogenous base, and three phosphates linked to the 5′-carbon atom of the ribose. In transcription, two of the phosphates are cleaved, producing an RNA nucleotide.

ribonucleotide Nucleotide containing ribose; present in RNA.

ribose Five-carbon sugar in RNA.

ribosomal RNA (rRNA) RNA molecule that is a structural component of the ribosome.

riboswitch Regulatory sequences in an RNA molecule. When an inducer molecule binds to the riboswitch, the binding changes the configuration of the RNA molecule and alters the expression of the RNA, usually by affecting the termination of transcription or affecting translation.

ribozyme RNA molecule that can act as a biological catalyst.

RNA-coding region Sequence of DNA nucleotides that encodes an RNA molecule.

RNA editing Process in which the protein-coding sequence of an mRNA is altered after transcription. The amino acids specified by the altered mRNA are different from those predicted from the nucleotide sequence of the gene encoding the protein.

RNA-induced silencing complex (RISC) Combination of a small interfering RNA (siRNA) molecule or a microRNA (miRNA) molecule and proteins that can cleave mRNA, leading to the degradation of the mRNA, or affect transcription or repress translation of the mRNA.

RNA interference (RNAi) Process in which cleavage of double-stranded RNA produces small interfering RNAs (siRNAs) that bind to mRNAs containing complementary sequences and bring about their cleavage and degradation.

RNA polymerase Enzyme that synthesizes RNA from a DNA template during transcription.

RNA polymerase I Eukaryotic RNA polymerase that transcribes large ribosomal RNA molecules (18 S rRNA and 28 S rRNA).

RNA polymerase II Eukaryotic RNA polymerase that transcribes pre-messenger RNA, some small nuclear RNAs, and some microRNAs.

RNA polymerase III Eukaryotic RNA polymerase that transcribes transfer RNA, small ribosomal RNAs (5 S rRNA), some small nuclear RNAs, and some microRNAs.

RNA polymerase IV Transcribes small interfering RNAs in plants.

RNA polymerase V Transcribes RNA that has a role in heterochromatin formation in plants.

RNA replication Process in some viruses by which RNA is synthesized from an RNA template.

RNA silencing Mechanism by which double-stranded RNA is cleaved and processed to yield small single-stranded interfering RNAs (siRNAs), which bind to complementary sequences in mRNA and bring about the cleavage and degradation of mRNA; also known as RNA interference and posttranscriptional RNA gene silencing. Some siRNAs also bind to complementary sequences in DNA and guide enzymes to methylate the DNA.

RNA splicing Process by which introns are removed and exons are joined together.

Robertsonian translocation Translocation in which the long arms of two acrocentric chromosomes become joined to a common centromere, resulting in a chromosome with two long arms and usually another chromosome with two short arms.

rolling-circle replication Replication of circular DNA that is initiated by a break in one of the nucleotide strands, producing a double-stranded circular DNA molecule and a single-stranded linear DNA molecule, the latter of which may circularize and serve as a template for the synthesis of a complementary strand.

rooted tree Phylogenetic tree in which one internal node represents the common ancestor of all other organisms (nodes) on the tree. In a rooted tree, all the organisms depicted have a common ancestor.

R plasmid (R factor) Plasmid having genes that confer antibiotic resistance to any cell that contains the plasmid.

sample Subset used to describe a population.

sampling error Deviations from expected ratios due to chance occurrences when the number of events is small.

secondary immune response Clone of cells generated when a memory cell encounters an antigen; provides long-lasting immunity.

secondary oocyte One of the products of meiosis I in female animals; receives most of the cytoplasm.

secondary spermatocyte Product of meiosis I in male animals.

secondary structure of a protein Regular folding arrangement of amino acids in a protein. Common secondary structures found in proteins include the alpha helix and the beta pleated sheet.

second polar body One of the products of meiosis II in oogenesis; contains a set of chromosomes but little of the cytoplasm.

segmental duplications Regions larger than 1000 bp that are almost identical in sequence in eukaryotic genomes.

segmentation genes Set of about 25 genes in fruit flies that control the differentiation of the embryo into individual segments, affecting the number and organization of the segments. Mutations in these genes usually disrupt whole sets of segments.

segment-polarity genes Set of segmentation genes in fruit flies that affect the organization of segments. Mutations in these genes cause part of each segment to be deleted and replaced by a mirror image of part or all of an adjacent segment.

selection coefficient Measure of the relative intensity of selection against a genotype; equals 1 minus fitness.

selection differential Difference in phenotype between the selected individuals and the average of the entire population.

semiconservative replication Replication in which the two nucleotide strands of DNA separate, each serving as a template for the synthesis of a new strand. All DNA replication is semiconservative.

sense codon Codon that specifies an amino acid in a protein.

separase Molecule that cleaves cohesin molecules, which hold the sister chromatids together.

sequential hermaphroditism Phenomenon in which the sex of an individual organism changes in the course of its lifetime; the organism is male at one age or developmental stage and female at a different age or stage.

70S initiation complex Final complex formed in the initiation of translation in bacterial cells; consists of the small and large subunits of the ribosome, mRNA, and initiator tRNA charged with fMet.

sex Male or female.

sex chromosomes Chromosomes that differ morphologically or in number in males and females.

sex determination Specification of sex (male or female). Sex-determining mechanisms include chromosomal, genic, and environmental sex-determining systems.

sex-determining region Y (*SRY*) gene On the Y chromosome, a gene that triggers male development.

sex-influenced characteristic Characteristic encoded by autosomal genes that are more readily expressed in one sex. For example, an autosomal dominant gene may have higher penetrance in males than in females or an autosomal gene may be dominant in males but recessive in females.

sex-limited characteristic Characteristic encoded by autosomal genes and expressed in only one sex. Both males and females carry genes for sex-limited characteristics, but the characteristics appear in only one of the sexes.

sex-linked characteristic Characteristic determined by a gene or genes on sex chromosomes.

shelterin A multiprotein complex that binds to mammalian telomeres and protects the ends of the DNA from being inadvertently repaired as a double-stranded break in the DNA.

Shine–Dalgarno sequence Consensus sequence found in the bacterial 5′ untranslated region of mRNA; contains the ribosome-binding site.

short interspersed element (SINE) Short DNA sequence repeated many times and interspersed throughout the genome.

short tandem repeat (STR) Very short DNA sequence repeated in tandem and found widely in the human genome.

shuttle vector Cloning vector that allows DNA to be transferred to more than one type of host cell.

sigma factor Subunit of bacterial RNA polymerase that allows the RNA polymerase to recognize a promoter and initiate transcription.

signal sequence From 15 to 30 amino acids that are found at the amino end of some eukaryotic proteins and direct the protein to specific locations in the cell; usually cleaved from the protein.

signal-transduction pathway System in which an external signal (initiated by a hormone or growth factor) triggers a cascade of intracellular reactions that ultimately produce a specific response.

silencer Sequence that has many of the properties possessed by an enhancer but represses transcription.

silent mutation Change in the nucleotide sequence of DNA that does not alter the amino acid sequence of a protein.

single-nucleotide polymorphism (SNP) Single-base-pair differences in DNA sequence between individual members of a species.

single-strand-binding (SSB) protein Binds to single-stranded DNA in replication and prevents it from annealing with a complementary strand and forming secondary structures.

sister chromatids Two copies of a chromosome that are held together at the centromere. Each chromatid consists of a single DNA molecule.

site-directed mutagenesis Produces specific nucleotide changes at selected sites in a DNA molecule.

small cytoplasmic RNA (scRNA) Small RNA molecule found in the cytoplasm of eukaryotic cells.

small interfering RNA (siRNA) Single-stranded RNA molecule (usually from 21 to 25 nucleotides in length) produced by the cleavage and processing of double-stranded RNA; binds to complementary sequences in mRNA and brings about the cleavage and degradation of the mRNA. Some siRNAs bind to complementary sequences in DNA and bring about their methylation.

small nuclear ribonucleoprotein (snRNP) Structure found in the nuclei of eukaryotic cells that consists of small nuclear RNA (snRNA) and protein; functions in the processing of pre-mRNA.

small nuclear RNA (snRNA) Small RNA molecule found in the nuclei of eukaryotic cells; functions in the processing of pre-mRNA.

small nucleolar RNA (snoRNA) Small RNA molecule found in the nuclei of eukaryotic cells; functions in the processing of rRNA and in the assembly of ribosomes.

small ribosomal subunit The smaller of the two subunits of a functional ribosome.

somatic-cell hybridization Fusion of different cell types.

somatic hypermutation High rate of somatic mutation such as that in genes encoding antibodies.

somatic mutation Mutation in a cell that does not give rise to a gamete.

somatic recombination Recombination in somatic cells, such as maturing lymphocytes, among segments of genes that encode antibodies and T-cell receptors.

SOS system System of proteins and enzymes that allow a cell to replicate its DNA in the presence of a distortion in DNA structure; makes numerous mistakes in replication and increases the rate of mutation.

Southern blotting Process by which DNA is transferred from a gel to a solid support such as a nitrocellulose or nylon filter.

specialized transduction Transduction in which genes near special sites on the bacterial chromosome are transferred from one bacterium to another; requires lysogenic bacteriophages.

speciation Process by which new species arise. *See also* **biological species concept, allopatric speciation,** and **sympatric speciation.**

species Term applied to different kinds or types of living organisms.

spermatid Immediate product of meiosis II in spermatogenesis; matures to sperm.

spermatogenesis Sperm production in animals.

spermatogonium Diploid cell in the testis; capable of undergoing meiosis to produce a sperm.

S (synthesis) phase Stage of interphase in the cell cycle. In S phase, DNA replicates.

spindle microtubule Microtubule that moves chromosomes in mitosis and meiosis.

spindle pole Point from which spindle microtubules radiate.

spliceosome Large complex consisting of several RNAs and many proteins that splices protein-encoding pre-mRNA; contains five small ribonucleoprotein particles (U1, U2, U4, U5, and U6).

spontaneous mutation Arises from natural changes in DNA structure or from errors in replication.

sporophyte Diploid phase of the life cycle in plants.

SR proteins Group of serine- and arginine-rich proteins that regulate alternative splicing of pre-mRNA.

standard deviation Statistic that describes the variability of a group of measurements; the square root of the variance.

stop (termination or nonsense) codon Codon in mRNA that signals the end of translation. The three common stop codons are UAA, UAG, and UGA.

strand slippage Slipping of the template and newly synthesized strands in replication in which one of the strands loops out from the other and nucleotides are inserted or deleted on the newly synthesized strand.

structural gene DNA sequence that encodes a protein that functions in metabolism or biosynthesis or that has a structural role in the cell.

structural genomics Area of genomics that studies the organization and sequence of information contained within genomes; sometimes used by protein chemists to refer to the determination of the three-dimensional structure of proteins.

submetacentric chromsome Chromosome in which the centromere is displaced toward one end, producing a short arm and a long arm.

supercoiling Coiled tertiary structure that forms when strain is placed on a DNA helix by overwinding or underwinding of the helix. An overwound DNA exhibits positive supercoiling; an underwound DNA exhibits negative supercoiling.

suppressor mutation Mutation that hides or suppresses the effect of another mutation at a site that is distinct from the site of the original mutation.

sympatric speciation Speciation arising in the absence of any geographic barrier to gene flow; reproductive isolating mechanisms evolve with a single interbreeding population.

synapsis Close pairing of homologous chromosomes.

synaptonemal complex Three-part structure that develops between synapsed homologous chromosomes.

synonymous codons Different codons that specify the same amino acid.

tag single-nucleotide polymorphism (tag-SNP) Used to identify a haplotype.

tandem duplication Chromosome rearrangement in which a duplicated chromosome segment is adjacent to the original segment.

tandem repeat sequences DNA sequences repeated one after another; tend to be clustered at specific locations on a chromosome.

***Taq* polymerase** DNA polymerase commonly used in PCR reactions. Isolated from the bacterium *Thermus aquaticus*, the enzyme is stable at high temperatures, and so it is not denatured during the strand-separation step of the cycle.

TATA-binding protein (TBP) Polypeptide chain found in several different transcription factors that recognizes and binds to sequences in eukaryotic promoters.

TATA box Consensus sequence (TATAAAA) commonly found in eukaryotic RNA polymerase II promoters; usually located from 25 to 30 bp upstream of the transcription start site. The TATA box determines the start point for transcription.

T cell Particular type of lymphocyte that produces cellular immunity; originates in the bone marrow and matures in the thymus.

T-cell receptor Found on the surface of a T cell, a receptor that simultaneously binds a foreign and a self-antigen on the surface of a cell.

telocentric chromosome Chromosome in which the centromere is at or very near one end.

telomerase Enzyme that is made up of both protein and RNA and replicates the ends (telomeres) of eukaryotic chromosomes. The RNA part of the enzyme has a template that is complementary to repeated sequences in the telomere and pairs with them, providing a template for the synthesis of additional copies of the repeats.

telomere Stable end of a chromosome.

telomere-associated sequence Sequence found at the ends of a chromosome next to the telomeric sequence; consists of long, complex repeated sequences.

telomeric sequence Sequence found at the ends of a chromosome; consists of many copies of short, simple sequences repeated one after the other.

telophase Stage of mitosis. In telophase, the chromosomes arrive at the spindle poles, the nuclear membrane re-forms, and the chromosomes relax and lengthen.

telophase I Stage of meiosis I. In telophase I, chromosomes arrive at the spindle poles.

telophase II Stage of meiosis II. In telophase II, chromosomes arrive at the spindle poles.

temperate phage Bacteriophage that utilizes the lysogenic cycle, in which the phage DNA integrates into the bacterial chromosome and remains in an inactive state.

temperature-sensitive allele Expressed only at certain temperatures.

template strand The strand of DNA that is used as a template during transcription. The RNA synthesized during transcription is complementary and antiparallel to the template strand.

temporal isolation Reproductive isolation in which the reproduction of different groups takes place at different times of the year, and so there is no gene flow between groups; exemplified by species of plants that flower at different times of the year and thus do not exchange genes.

terminal inverted repeats Sequences found at both ends of a transposable element that are inverted complements of one another.

termination codon Codon in mRNA that signals the end of translation; also called nonsense codon or stop codon. The three common termination codons are UAA, UAG, and UGA.

terminator Sequence of DNA nucleotides that causes the termination of transcription.

tertiary structure of a protein Higher-order folding of amino acids in a protein to form the overall three-dimensional shape of the molecule.

testcross A cross between an individual with an unknown genotype and an individual with the homozygous recessive genotype.

tetrad The four products of meiosis; all four chromatids of a homologous pair of chromosomes.

tetrad analysis Genetic analysis of a tetrad, the products of a single meiosis.

tetraploidy Possession of four haploid sets of chromosomes ($4n$).

tetrasomy Presence of two extra copies of a chromosome ($2n + 2$).

TFIIB recognition element (BRE) Consensus sequence [(G or C)(G or C)(G or C)CGCC] found in some RNA polymerase II core promoters; usually located from 32 to 38 bp upstream of the transcription start site.

theory of clonal selection Explains the generation of primary and secondary immune responses. The binding of a B cell to an antigen stimulates the cell to divide, giving rise to a clone of genetically identical cells, all of which are specific for the antigen.

theta replication Replication of circular DNA that is initiated by the unwinding of the two nucleotide strands, producing a replication bubble. Unwinding continues at one or both ends of the bubble, making it progressively larger. DNA replication on both of the template strands is simultaneous with unwinding until the two replication forks meet.

30S initiation complex Initial complex formed in the initiation of translation in bacterial cells; consists of the small subunit of the ribosome, mRNA, initiator tRNA charged with fMet, GTP, and initiation factors 1, 2, and 3.

three-point testcross Cross between an individual heterozygous at three loci and an individual homozygous for recessive alleles at those loci.

3′ end End of a polynucleotide chain where an OH group is attached to the 3′-carbon atom of the nucleotide.

3′ splice site The 3′ end of an intron where cleavage takes place in RNA splicing.

3′ untranslated (3′ UTR) region Sequence of nucleotides at the 3′ end of mRNA; does not encode the amino acids of a protein but affects both the stability of the mRNA and its translation.

threshold characteristic Discontinuous characteristic whose expression depends on an underlying susceptibility that varies continuously.

thymine (T) Pyrimidine base in DNA but not in RNA.

Ti plasmid Large plasmid isolated from the bacterium *Agrobacterium tumefaciens* and used to transfer genes to plant cells.

topoisomerase Enzyme that adds or removes rotations in a DNA helix by temporarily breaking nucleotide strands; controls the degree of DNA supercoiling.

totipotency The potential of a cell to develop into any other cell type.

trans configuration Arrangement in which each chromosome contains one wild-type (dominant) gene and one mutant (recessive) gene; also called repulsion.

transcription Process by which RNA is synthesized from a DNA template.

transcriptional activator protein Protein in eukaryotic cells that binds to consensus sequences in regulatory promoters or enhancers and affects transcription initiation by stimulating the assembly of the basal transcription apparatus.

transcription bubble Region of a DNA molecule that has unwound to expose a single-stranded template, which is being transcribed into RNA.

transcription factor Protein that binds to DNA sequences in eukaryotic cells and affects transcription.

transcription start site The first DNA nucleotide that is transcribed into an RNA molecule.

transcription unit Sequence of nucleotides in DNA that encodes a single RNA molecule, along with the sequences necessary for its transcription; normally contains a promoter, an RNA-coding sequence, and a terminator.

transcriptome Set of all RNA molecules transcribed from a genome.

transducing phage Contains a piece of the bacterial chromosome inside the phage coat. *See also* **generalized transduction.**

transductant Bacterial cell that has received genes from another bacterium through transduction.

transduction Type of gene exchange that takes place when a virus carries genes from one bacterium to another. After it is inside the cell, the newly introduced DNA may undergo recombination with the bacterial chromosome.

transesterification Chemical reaction in some RNA-splicing reactions.

transfer–messenger RNA (tmRNA) An RNA molecule that has properties of both mRNA and tRNA; functions in rescuing ribosomes that are stalled at the end of mRNA.

transfer RNA (tRNA) RNA molecule that carries an amino acid to the ribosome and transfers it to a growing polypeptide chain in translation.

transfer RNA intron Belongs to a class of introns in tRNA genes. The splicing of these genes relies on enzymes.

transformant Cell that has received genetic material through transformation.

transformation Mechanism by which DNA found in the medium is taken up by the cell. After transformation, recombination may take place between the introduced genes and the cellular chromosome.

transforming principle Substance responsible for transformation. DNA is the transforming principle.

transgene Foreign gene or other DNA fragment carried in germ-line DNA.

transgenic mouse Mouse whose genome contains a foreign gene or genes added by employing recombinant DNA methods.

transition Base substitution in which a purine is replaced by a different purine or a pyrimidine is replaced by a different pyrimidine.

translation Process by which a protein is assembled from information contained in messenger RNA.

translocation Movement of a chromosome segment to a nonhomologous chromosome or to a region within the same chromosome. Also, movement of a ribosome along mRNA in the course of translation.

translocation carrier Individual organism heterozygous for a chromosome translocation.

transmission genetics Field of genetics that encompasses the basic principles of genetics and how traits are inherited.

transposable element DNA sequence capable of moving from one site to another within the genome through a mechanism that differs from that of homologous recombination.

transposase Enzyme encoded by many types of transposable elements that is required for their transposition. The enzyme makes single-strand breaks at each end of the transposable element and on either side of the target sequence where the element inserts.

transposition Movement of a transposable genetic element from one site to another. Replicative transposition increases the number of copies of the transposable element; nonreplicative transposition does not increase the number of copies.

trans-splicing The process of splicing together exons from two or more pre-mRNAs.

transversion Base substitution in which a purine is replaced by a pyrimidine or a pyrimidine is replaced by a purine.

trihybrid cross A cross between two individuals that differ in three characteristics (*AA BB CC* × *aa bb cc*); also refers to a cross between two individuals that are both heterozygous at three loci (*Aa Bb Cc* × *Aa Bb Cc*).

triplet code Refers to the fact that three nucleotides encode each amino acid in a protein.

triploidy Possession of three haploid sets of chromosomes ($3n$).

triplo-X syndrome Human condition in which cells contain three X chromosomes. A person with triplo-X syndrome has a female phenotype without distinctive features other than a tendency to be tall and thin; a few such women are sterile, but many menstruate regularly and are fertile.

trisomy Presence of an additional copy of a chromosome ($2n + 1$).

trisomy 8 Presence of three copies of chromosome 8; in humans, results in mental retardation, contracted fingers and toes, low-set malformed ears, and a prominent forehead.

trisomy 13 Presence of three copies of chromosome 13; in humans, results in Patau syndrome.

trisomy 18 Presence of three copies of chromosome 18; in humans, results in Edward syndrome.

trisomy 21 Presence of three copies of chromosome 21; in humans, results in Down syndrome.

tRNA charging Chemical reaction in which an aminoacyl-tRNA synthetase attaches an amino acid to its corresponding tRNA.

tRNA-modifying enzyme Creates a modified base in RNA by catalyzing a chemical change in the standard base.

tubulin Protein found in microtubules.

tumor-suppressor gene Gene that normally inhibits cell division. Recessive mutations in such genes often contribute to cancer.

Turner syndrome Human condition in which cells contain a single X chromosome and no Y chromosome (XO). Persons with Turner syndrome are female in appearance but do not undergo puberty and have poorly developed female secondary sex characteristics; most are sterile but have normal intelligence.

two-dimensional polyacrylamide gel electrophorsis (2D-PAGE) Method for separating proteins into spots in which the proteins are separated in one dimension by charge, separated in a second dimension by mass, and then stained. Each spot is proportional to the amount of protein present.

two-point testcross Cross between an individual heterozygous at two loci and an individual homozygous for recessive alleles at those loci.

ultrasonography Procedure for visualizing the fetus. High-frequency sound is beamed into the uterus. Sound waves that encounter dense tissue bounce back and are transformed into a picture of the fetus.

unbalanced gamete Gamete that has a variable number of chromosomes; some chromosomes may be missing and others may be present in more than one copy.

underdominance Selection in which the heterozygote has lower fitness than that of either homozygote.

unequal crossing over Misalignment of the two DNA molecules during crossing over, resulting in one DNA molecule with an insertion and the other with a deletion.

uniparental disomy Inheritance of both chromosomes of a homologous pair from a single parent.

unique-sequence DNA Sequence present only once or a few times in a genome.

universal genetic code Refers to the fact that particular codons specify the same amino acids in almost all organisms.

up mutation Mutation that increases the rate of transcription.

upstream element Consensus sequence found in some bacterial promoters that contains a number of A–T pairs and is found about 40 to 60 bp upstream of the transcription start site.

uracil (U) Pyrimidine base in RNA but not normally in DNA.

variable number of tandem repeats (VNTRs) Short sequences repeated in tandem that vary greatly in number among individuals; also called microsatellites. Because they are quite variable, VNTRs are commonly used in DNA fingerprinting.

variance Statistic that describes the variability of a group of measurements.

virulent phage Bacteriophage that reproduces only through the lytic cycle and kills its host cell.

virus Noncellular replicating agent consisting of nucleic acid surrounded by a protein coat; can replicate only within its host cell.

Western blotting Process by which protein is transferred from a gel to a solid support such as a nitrocellulose or nylon filter.

whole-genome shotgun sequencing Method of sequencing a genome in which sequenced fragments are assembled into the correct sequence in contigs by using only the overlaps in sequence.

wild type The trait or allele that is most commonly found in natural (wild) populations.

wobble Base pairing between codon and anticodon in which there is nonstandard pairing, usually at the third ($3'$) position of the codon; allows more than one codon to pair with the same anticodon.

X : A : ratio Ratio of the number of X chromosomes to the number of haploid autosomal sets of chromosomes; determines sex in fruit flies.

X-linked characteristic Characteristic determined by a gene or genes on the X chromosome.

X-ray diffraction Method for analyzing the three-dimensional shape and structure of chemical substances. Crystals of a substance are bombarded with X-rays, which hit the crystals, bounce off, and produce a diffraction pattern on a detector. The pattern of the spots produced on the detector provides information about the molecular structure.

yeast artificial chromosome (YAC) Cloning vector consisting of a DNA molecule with a yeast origin of replication, a pair of telomeres, and a centromere. YACs can carry very large pieces of DNA (as large as several hundred thousand base pairs) and replicate and segregate as yeast chromosomes do.

Y-linked characteristic Characteristic determined by a gene or genes on the Y chromosome.

Z-DNA Secondary structure of DNA characterized by 12 bases per turn, a left-handed helix, and a sugar–phosphate backbone that zigzags back and forth.

zygotene Second substage of prophase I in meiosis. In zygotene, chromosomes enter into synapsis.

Answers to Selected Questions and Problems

Chapter 1

1. In the Hopi culture, albinos were considered special and given special status. Because extensive exposure to sunlight could be damaging or deadly, Hopi male albinos did no agricultural work. Albinism was a considered a positive trait rather than a negative physical condition, allowing albinos to have more children and thus increasing the frequency of the allele. Finally, the small population size of the Hopi tribe may have helped increase the allele frequency of the albino gene owing to chance.

3. Genetics plays important roles in the diagnosis and treatment of hereditary diseases, in breeding plants and animals for improved production and disease resistance, and in producing pharmaceuticals and novel crops through genetic engineering.

5. Transmission genetics: The inheritance of genes from one generation to the next, gene mapping, characterization of the phenotypes produced by mutations.

   Molecular genetics: The structure, organization, and function of genes at the molecular level.

   Population genetics: Genes and changes in genes in populations.

8. Pangenesis proposes that information for creating each part of an offspring's body originates in each part of the parent's body and is passed through the reproductive organs to the embryo at conception. Pangenesis suggests that changes in parts of the parent's body may be passed to the offspring's body. The germ-plasm theory, in contrast, states that the reproductive cells possess all of the information required to make the complete body; the rest of the body contributes no information to the next generation.

10. Preformationism is the idea that an offspring results from a miniature adult form that is already preformed in the sperm or the egg. All traits are thus inherited from only one parent, either the father or the mother, depending on whether the homunculus (the preformed miniature adult) resides in the sperm or the egg.

13. Gregor Mendel

16. Genes are composed of DNA nucleotide sequences and are located at specific positions in chromosomes.

18. (a) Transmission genetics; (b) population genetics; (c) population genetics; (d) molecular genetics; (e) molecular genetics; (f) transmission genetics.

21. Genetics is old in the sense that humans have been aware of hereditary principles for thousands of years and have applied them since the beginning of agriculture and the domestication of plants and animals. It is very young in the sense that the fundamental principles were not uncovered until Mendel's time and the structure of DNA and the principles of recombinant DNA were discovered within the past 60 years.

23. (a) Pangenesis postulates that specific particles (called gemmules) carry genetic information from all parts of the body to the reproductive organs, and then the genetic information is conveyed to the embryo where each unit directs the formation of its own specific part of the body. According to the germ-plasm theory, gamete-producing cells found within the reproductive organs contain a complete set of genetic information that is passed to the gametes. The two concepts are similar in that both propose that genetic information is contained in discrete units that are passed on to offspring. They differ in where that genetic information resides. In pangenesis, it resides in different parts of the body and must travel to the reproductive organs. In the germ-plasm theory, all the genetic information is already in the reproductive cells.
    (b) Preformationism holds that the sperm or egg contains a miniature preformed adult called a homunculus. In development, the homunculus grows to produce an offspring. Only one parent contributes genetic traits to the offspring. Blending inheritance requires contributions of genetic material from both parents. The genetic contributions from the parents blend to produce the genetic material of the offspring. Once blended, the genetic material cannot be separated for future generations.
    (c) The inheritance of acquired characteristics postulates that traits acquired in a person's lifetime alter the genetic material and can be transmitted to offspring. Our modern theory of heredity indicates that offspring inherit genes located on chromosomes passed from their parents. These chromosomes segregate in meiosis in the parent's germ cells and are passed into the gametes.

24. (a) Both cell types have lipid bilayer membranes, DNA genomes, and machinery for DNA replication, transcription, translation, energy metabolism, response to stimuli, growth, and reproduction. Eukaryotic cells have a nucleus containing chromosomal DNA and possess internal membrane-bounded organelles.
    (b) A gene is a basic unit of hereditary information, usually encoding a functional RNA or polypeptide. Alleles are variant forms of a gene, arising through mutation.
    (c) The genotype is the set of genes or alleles inherited by an organism from its parent(s). The expression of the genes of a particular genotype, through interaction with environmental factors, produces the phenotype, the observable trait.
    (d) Both are nucleic acid polymers. RNA contains a ribose sugar, whereas DNA contains a deoxyribose sugar. RNA also contains uracil as one of the four bases, whereas DNA contains thymine. The other three bases are common to both DNA and RNA. Finally, DNA is usually double stranded, consisting of two complementary strands, whereas RNA is single stranded.
    (e) Chromosomes are structures consisting of DNA and associated proteins. The DNA contains the genetic information.

27. All genomes must have the ability to store complex information and to vary. The blueprint for an entire organism must be contained within the genome of each reproductive cell. The information has to be in the form of a code that can be used as a set of instructions for assembling the components of the cells. The genetic material of any organism must be stable, be replicated precisely, and be transmitted faithfully to the progeny but must be capable of mutating.

29. Legally, the woman is not required to inform her children or other relatives, but people may have different opinions about her moral and parental responsibilities. On the one hand, she has the legal right to keep private the results of any medical information,

including the results of genetic testing. On the other hand, her children may be at increased risk of developing these disorders and might benefit from that information. For example, the risk of colon cancer can be reduced by regular exams, in which any tumors can be detected and removed before they become cancerous. Some people might argue that her parental responsibilities include providing her children with information about possible medical problems. Another issue to consider is the possibility that her children or other relatives might not want to know their genetic risk, particularly for a disorder such as Alzheimer disease, for which there is no cure.

30. (a) Having the genetic test removes doubt about the potential for the disorder: you are either susceptible or not. Knowing about the potential of a genetic disorder enables you to make life-style changes that might lessen the effect of the disease or lessen the risk. The types and nature of future medical tests could be positively affected by the genetic testing, thus allowing for early warning and screening for the disease. The knowledge could also enable you to make informed decisions regarding offspring and the potential of passing the trait to your offspring. Additionally, by knowing what to expect, you could plan your life accordingly.

   Reasons for not having the test typically concern the potential for testing positive for the susceptibility to the genetic disease. If the susceptibility is detected, there is potential for discrimination. For example, your employer (or possibly future employer) might consider you a long-term liability, thus affecting employment options. Insurance companies may not want to insure you for the condition or its symptoms, and social stigmatization regarding the disease could be a factor. Knowledge of the potential future condition could lead to psychological difficulties in coping with the anxiety of waiting for the disease to manifest.

   (b) There is no "correct" answer, but some of the reasons for wanting to be tested are: the test would remove doubt about the susceptibility, particularly if family members have had the genetic disease; either a positive or a negative result would allow for informed planning of life style, medical testing, and family choices in the future.

Chapter 2

1.

Prokaryotic cell	Eukaryotic cell
No nucleus	Nucleus present
No paired chromosomes	Paired chromosomes common
Typically a single circular chromosome containing a single origin of replication	Typically multiple linear chromosomes containing centromeres, telomeres, and multiple origins of replication
Single chromosome is replicated, with each copy moving to opposite sides of the cell	Chromosomes require mitosis or meiosis to segregate to the proper location
No histone proteins complexed to DNA	Histone proteins are complexed to DNA

3. The events are (i) a cell's genetic information must be copied; (ii) the copies of the genetic information must be separated from each other; and (iii) the cell must divide.

6.

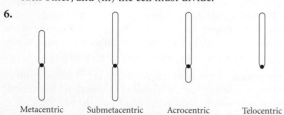

Metacentric   Submetacentric   Acrocentric   Telocentric

8. Prophase: The chromosomes condense and become visible, the centrosomes move apart, and microtubule fibers form from the centrosomes.

Prometaphase: The nucleoli disappear and the nuclear envelope begins to disintegrate, allowing the cytoplasm and nucleoplasm to join. The sister chromatids of each chromosome are attached to microtubules from the opposite centrosomes.

Metaphase: The spindle microtubules are clearly visible, and the chromosomes arrange themselves on the metaphase plate of the cell.

Anaphase: The sister chromatids separate at the centromeres after the breakdown of cohesin protein, and the newly formed daughter chromosomes move to the opposite poles of the cell.

Telophase: The nuclear envelope re-forms around each set of daughter chromosomes. Nucleoli reappear. Spindle microtubules disintegrate.

10. In the mitotic cell cycle, the genetic material is precisely copied, and so the two resulting cells contain the same genetic information. In other words, the daughter cells have genomes identical with each other and with that of the mother cell.

14. Meiosis includes two cell divisions, thus producing four new cells (in many species). The chromosome number of a haploid cell produced by meiosis I is half the chromosome number of the original diploid cell. Finally, the cells produced by meiosis are genetically different from the original cell and genetically different from each other.

16.

Mitosis	Meiosis
A single cell division produces two genetically identical progeny cells.	Two cell divisions usually result in four progeny cells that are not genetically identical.
Chromosome number in progeny cells and the original cell remains the same.	Daughter cells are haploid and have half the chromosomal complement of the original diploid cell as a result of the separation of homologous pairs in anaphase I.
Daughter cells and the original cell are genetically identical. No crossing over or separation of homologous chromosomes takes place.	Crossing over in prophase I and separation of homologous pairs in anaphase I produce daughter cells that are genetically different from each other and from the original cell.
Homologous chromosomes do not synapse.	Homologous chromosomes synapse in prophase I.
In metaphase, individual chromosomes line up on the metaphase plate.	In metaphase I, homologous pairs of chromosomes line up on the metaphase plate.
	Individual chromosomes line up in metaphase II.
In anaphase, sister chromatids separate.	In anaphase I, homologous chromosomes separate. Sister chromatids separate in anaphase II.

The most important difference is that mitosis produces cells genetically identical with each other and with the original cell, resulting in the orderly passage of information from one cell to its progeny. In contrast, by producing progeny that do not contain pairs of homologous chromosomes, meiosis results in the reduction of chromosome number from that of the original cell. Meiosis also allows for genetic variation through crossing over and the random assortment of homologous chromosomes.

20. (a) The two chromatids of a chromosome
   (b) The two chromosomes of a homologous pair
   (c) Cohesin
   (d) The enzyme separase
   (e) The hands of the two blind men
   (f) If one man failed to grasp his sock, use of the knife to cut the string holding them together would be difficult. The two socks of a

pair would not be separated, and both would end up in one man's bag. Similarly, if each chromatid is not attached to spindle fibers and pulled in opposite directions, the two chromatids will not separate and both will migrate to the same cell. This cell will have two copies of one chromosome.

23. **(a)** 12 chromosomes and 24 DNA molecules; **(b)** 12 chromosomes and 24 DNA molecules; **(c)** 12 chromosomes and 24 DNA molecules; **(d)** 12 chromosomes and 24 DNA molecules; **(e)** 12 chromosomes and 12 DNA molecules; **(f)** 6 chromosomes and 12 DNA molecules; **(g)** 12 chromosomes and 12 DNA molecules; **(h)** 6 chromosomes and 6 DNA molecules.

25. The diploid number of chromosomes is six. The left-hand cell is in anaphase I of meiosis; the middle cell is in anaphase of mitosis; the right-hand cell is in anaphase II of meiosis.

27. **(a)** Cohesin is needed to hold the sister chromatids together until anaphase of mitosis. If cohesin fails to form early in mitosis, the sister chromatids could separate before anaphase. As a result, chromosomes would not segregate properly to daughter cells.
(b) Shugoshin protects cohesin proteins from degradation at the centromere during meiosis I. Cohesin at the arms of the homologous chromosomes is not protected by shugoshin and is broken in anaphase I, enabling the two homologs to separate. If shugoshin is absent during meiosis, then the cohesin at the centromere may be broken, allowing for the separation of sister chromatids along with the homologs in anaphase I and leading to the improper segregation of chromosomes to daughter cells.
(c) If shugoshin does not break down, the cohesin proteins at the centromere remain protected from degradation. The intact cohesin prevents the sister chromatids from separating in anaphase II of meiosis; as a result, sister chromatids would not separate, leading to daughter cells that have too many chromosomes or too few.
(d) If separase is defective, homologous chromosomes and sister chromatids would not separate in meiosis and mitosis, resulting in some cells that have too few chromosomes and some cells that have too many chromosomes.

30. The progeny of an organism with the larger number of homologous pairs of chromosomes should be expected to exhibit more variation. The number of different combinations of chromosomes that are possible in the gametes is $2n$, where n is equal to the number of homologous pairs of chromosomes. For the fruit fly, which has four pairs of chromosomes, the number of possible combinations is $2^4 = 16$. For the house fly, which has six pairs of chromosomes, the number of possible combinations is $2^6 = 64$.

31. **(a)** Metaphase I

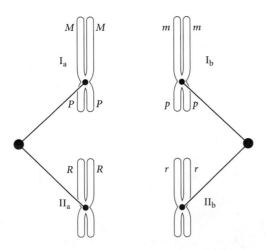

(b) Gametes

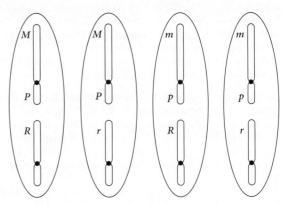

34. **(a)** No. The first polar body and the secondary oocyte are the result of meiosis I. In meiosis I, homologous chromosomes segregate; thus, both the first polar body and the secondary oocyte will contain only one homolog of each original chromosome pair, and the alleles of some genes in these homologs will differ. Additionally, crossing over in prophase I will have generated new and different arrangements of genetic material for each homolog of the pair.
(b) No. The second polar body and the ovum will contain the same members of the homologous pairs of chromosomes that were separated in meiosis I and produced by the separation of sister chromatids in anaphase II. However, the sister chromatids of each pair are no longer identical, because they underwent crossing over in prophase I and thus contain genetic information that is not identical.

37. Most male animals produce sperm by meiosis. Because meiosis takes place only in diploid cells, haploid male bees do not undergo meiosis. Male bees can produce sperm but only through mitosis. Haploid cells that divide mitotically produce haploid cells.

Chapter 3

1. Mendel was successful for several reasons. He chose to work with a plant, *Pisum sativum*, that was easy to cultivate, grew relatively rapidly compared with other plants, and produced many offspring whose phenotypes were easy to determine, which allowed Mendel to detect mathematical ratios of progeny phenotypes. The seven characteristics that he chose to study exhibited only a few distinct phenotypes and did not show a range of variation. Finally, by looking at each trait separately and counting the numbers of the different phenotypes, Mendel adopted a reductionist experimental approach and applied the scientific method. From his observations, he proposed hypotheses that he was then able to test empirically.

3. The principle of segregation states that an organism possesses two alleles for any particular characteristic. These alleles separate in the formation of gametes. In other words, one allele goes into each gamete. The principle of segregation is important because it explains how the genotypic ratios in the haploid gametes are produced.

8. Walter Sutton's chromosome theory of heredity states that genes are located on chromosomes. The independent segregation of pairs of homologous chromosomes in meiosis provides the biological basis for Mendel's two principles of heredity.

9. The principle of independent assortment states that alleles at different loci segregate independently of one another. The principle of independent assortment is an extension of the principle of segregation: the principle of segregation states that the two alleles at a locus separate; according to the principle of independent assortment, when these two alleles separate, their separation is independent of the separation of alleles at other loci.

14. (a) The parents are *RR* (orange fruit) and *rr* (cream fruit). All the F₁ are *Rr* (orange). The F₂ are 1 *RR* : 2 *Rr* : 1 *rr* and have an orange-to-cream phenotypic ratio of 3 : 1.
(b) Half of the progeny are homozygous for orange fruit (*RR*) and half are heterozygous for orange fruit (*Rr*).
(c) Half of the progeny are heterozygous for orange fruit (*Rr*) and half are homozygous for cream fruit (*rr*).

16. (a) Although the white female guinea pig gave birth to the off-spring, her eggs were produced by the ovary from the black female guinea pig. The transplanted ovary produced only eggs containing the allele for black coat color. Like most mammals, guinea pig fe-males produce primary oocytes early in development, and thus the transplanted ovary already contained primary oocytes produced by the black female guinea pig.
(b) The white male guinea pig contributed a *w* allele, and the white female guinea pig contributed the *W* allele from the transplanted ovary. The offspring are thus *Ww*.
(c) The transplant experiment supports the germ-plasm theory. According to the germ-plasm theory, only the genetic information in the germ-line tissue in the reproductive organs is passed to the offspring. The production of black guinea pig offspring suggests that the allele for black coat color was passed to the offspring from the transplanted ovary in agreement with the germ-plasm theory. According to the pangenesis theory, the genetic informa-tion passed to the offspring originates at various parts of the body and travels to the reproductive organs for transfer to the gametes. If pangenesis were correct, then the guinea pig offspring should have been white. The white-coat alleles would have traveled to the transplanted ovary and then into the white female's gametes. The absence of any white offspring indicates that the pangenesis hypothesis is invalid.

17. (a) Female parent is *iᴮiᴮ*; male parent is *IᴬiᴮB*.
(b) Both parents are *iᴮiᴮ*.
(c) Male parent is *iᴮiᴮ*; female parent is *IᴬIᴬ* or, possibly, *Iᴬiᴮ*, but a heterozygous female in this mating is unlikely to have produced eight blood-type-A kittens owing to chance alone.
(d) Both parents are *IᴬiᴮB*.
(e) Either both parents are *IᴬIᴬ* or one parent is *IᴬIᴬ* and the other parent is *IᴬiᴮB*. The blood type of the offspring does not allow a determination of the precise genotype of either parent.
(f) Female parent is *iᴮiᴮB*; male parent is *IᴬiᴮB*.

20. (a) Sally (*Aa*), Sally's mother (*Aa*), Sally's father (*aa*), and Sally's brother (*aa*); (b) $^1/_2$; (c) $^1/_2$.

22. Use *h* for the hairless allele and *H* for the dominant allele for the presence of hair. Because *H* is dominant over *h*, a rat terrier with hair could be either homozygous (*HH*) or heterozygous (*Hh*). To determine which genotype is present in the rat terrier with hair, cross this dog with a hairless rat terrier (*hh*). If the terrier with hair is homozygous (*HH*), then no hairless offspring will be produced by the testcross. However, if the terrier is heterozygous (*Hh*), then $^1/_2$ of the offspring will be hairless.

24. (a) $^1/_{18}$; (b) $^1/_{36}$; (c) $^{11}/_{36}$; (d) $^1/_6$; (e) $^1/_4$; (f) $^3/_4$.

25. (a) $^1/_{128}$; (b) $^1/_{64}$; (c) $^7/_{128}$; (d) $^{35}/_{128}$; (e) $^{35}/_{128}$.

27. Parents:

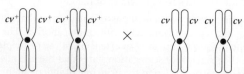

F₁ generation:

F₂ generation:

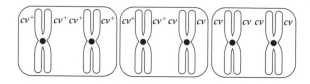

28. (a) In the F₁ black guinea pigs (*Bb*), only one chromosome pos-sesses the black allele, and so the number of copies present at each stage are: G₁, one black allele; G₂, two black alleles; metaphase of mitosis, two black alleles; metaphase I of meiosis, two black alleles; after cytokinesis of meiosis, one black allele but only in half of the cells produced by meiosis. (The remaining half will not contain the black allele.)
(b) In the F₁ brown guinea pigs (*bb*), both homologs possess the brown allele, and so the number of copies present at each stage are: G₁, two brown alleles; G₂, four brown alleles; metaphase of mitosis, four brown alleles; metaphase I of meiosis, four brown alleles; metaphase II, two brown alleles; after cytokinesis of meiosis, one brown allele.

30. (a) $^9/_{16}$ black and curled, $^3/_{16}$ black and normal, $^3/_{16}$ gray and curled, and $^1/_{16}$ gray and normal.
(b) $^1/_4$ black and curled, $^1/_4$ black and normal, $^1/_4$ gray and curled, $^1/_4$ gray and normal.

31. (a) $^1/_2$ (*Aa*) × $^1/_2$ (*Bb*) × $^1/_2$ (*Cc*) × $^1/_2$ (*Dd*) × $^1/_2$ (*Ee*) = $^1/_{32}$
(b) $^1/_2$ (*Aa*) × $^1/_2$ (*bb*) × $^1/_2$ (*Cc*) × $^1/_2$ (*dd*) × $^1/_4$ (*ee*) = $^1/_{64}$
(c) $^1/_4$ (*aa*) × $^1/_2$ (*bb*) × $^1/_4$ (*cc*) × $^1/_2$ (*dd*) × $^1/_4$ (*ee*) = $^1/_{256}$
(d) No offspring having this genotype. The *Aa Bb Cc dd Ee* parent cannot contribute a *D* allele, the *Aa bb Cc Dd Ee* parent cannot contribute a *B* allele. Therefore, their offspring cannot be homozygous for the *BB* and *DD* gene loci.

34. (a) Gametes from *Aa Bb* individual:

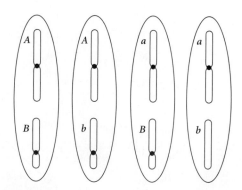

Gametes from *aa bb* individual:

(b) Progeny at G₁:

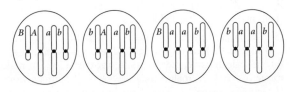

Progeny at G₂:

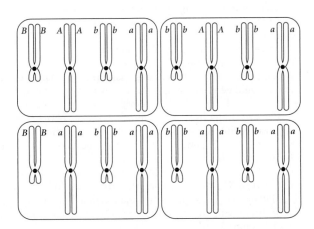

Progeny at prophase of mitosis:

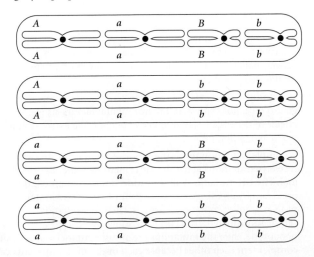

The order of chromosomes on the metaphase plate can vary.

35. (a) The burnsi × burnsi cross produced both burnsi and pipiens offspring, suggesting that the parents were heterozygous with each possessing a burnsi allele and a pipiens allele. The cross also suggests that the burnsi allele is dominant over the pipiens allele. The progeny of the burnsi × pipiens crosses suggest that each of the crosses was between a homozygous recessive frog (pipiens) and a heterozygous dominant frog (burnsi). The results of both crosses are consistent with the burnsi phenotype being recessive to the pipiens phenotype.

(b) Let B represent the burnsi allele and B^+ represent the pipiens allele.

burnsi (BB^+) × burnsi (BB^+)

burnsi (BB^+) × pipiens (B^+B^+)

burnsi (BB^+) × pipiens (B^+B^+)

c. For the burnsi × burnsi ($BB^+ \times BB^+$) cross, we would expect a phenotypic ratio of 3 : 1 in the offspring. A chi-square test to evaluate the fit of the observed numbers of progeny with an expected 3 : 1 ratio gives a chi-square value of 2.706 with 1 degree of freedom. The probability associated with this chi-square value is between 0.1 and 0.05, indicating that the differences between what we expected and what we observed could have been due to chance.

For the first burnsi × pipiens ($BB^+ \times B^+B^+$) cross, we would expect a phenotypic ratio of 1 : 1. A chi-square test comparing observed and expected values yields $\chi^2 = 1.78$, df $= 1$, $P > 0.05$. For the second burnsi × pipiens ($BB^+ \times B^+B^+$) cross, we would expect a phenotypic ratio of 1 : 1. A chi-square test of the fit of the observed numbers with those expected with a 1 : 1 ratio yields $\chi^2 = 0.46$, df $= 1$, $P > 0.05$. Thus, all three crosses are consistent with the predication that the burnsi allele is dominant over the pipiens allele.

38. (a) For the cross between a heterozygous F₁ plant (*Cc Ff*) and a homozygous recessive plant (*cc ff*), we would expect a phenotypic ratio of 1 : 1 : 1 : 1 for the different phenotypic classes. A chi-square test comparing the fit of the observed data with the expected 1 : 1 : 1 : 1 ratio yields a chi-square value of 35 with df $= 3$ and $P < 0.005$.

(b) From the chi-square value, it is unlikely that chance produced the differences between the observed and the expected ratios, indicating that the progeny are not in a 1 : 1 : 1 : 1 ratio.

(c) The number of plants with the *cc ff* genotype is fewer than expected. The *cc ff* genotype is possibly sublethal. In other words, California poppies with the homozygous recessive genotypes are possibly less viable than the other possible genotypes.

40. The first geneticist has identified an allele for obesity that he believes to be recessive. Let's call his allele for obesity o_1 and the normal allele O_1. On the basis of the crosses that the geneticist performed, the allele for obesity appears to be recessive.

Cross 1 with possible genotype:

Obese (o_1o_1) × Normal (O_1O_1)

↓

F₁   All normal (O_1o_1)

Cross 2 with possible genotypes:

F₁   Normal (O_1o_1) × Normal (O_1o_1)

↓

F₂   8 normal (O_1O_1 and O_1o_1)

2 obese (o_1o_1)

Cross 3 with possible genotypes:

Obese (o_1o_1) × Obese (o_1o_1)

↓

F₁   All obese (o_1o_1)

Let's call the second geneticist's allele for obesity o_2 and the normal allele O_2. The cross between obese mice from the two laboratories produced only normal mice. The alleles for obesity from both laboratories are recessive. However, they are located at different gene loci. Essentially, the obese mice from the different laboratories have separate obesity genes that are independent of one another.

The likely genotypes of the obese mice are as follows:

Obese mouse 1 ($o_1o_1\ O_2O_2$) × Obese mouse 2 ($O_1O_1\ o_2o_2$)

↓

F$_1$ All normal ($O_1o_1\ O_2o_2$)

Chapter 4

1. Females produce larger gametes than those produced by males.

5. The pseudoautosomal region is a region of similarity between the X and the Y chromosomes that is responsible for pairing the X and Y chromosomes in meiotic prophase I. Genes in this region are present in two copies in males and females and are thus inherited in the same way as autosomal genes are inherited, whereas other Y-linked genes are passed only from father to son.

9. Males show the phenotypes of all X-linked traits, regardless of whether the X-linked allele is normally recessive or dominant. Males inherit X-linked traits from their mothers, pass X-linked traits to all of their daughters, and, through their daughters, to their daughters' descendants but not to their sons or their sons' descendants.

11. Y-linked traits appear only in males and are always transmitted from fathers to sons, thus following a strict paternal lineage. Autosomal male-limited traits also appear only in males, but they can be transmitted to sons through their mothers.

15. **(a)** Female; **(b)** male; **(c)** male, sterile; **(d)** female; **(e)** male; **(f)** female; **(g)** metafemale; **(h)** male; **(i)** intersex; **(j)** female; **(k)** metamale, sterile; **(l)** metamale; **(m)** intersex.

20. Her father's mother and her father's father, but not her mother's mother or her mother's father.

22. **(a)** Yes; **(b)** yes; **(c)** no; **(d)** no.

23. **(a)** F$_1$: $^1/_2$ X$^+$Y (gray males), $^1/_2$ X$^+$X^y (gray females); F$_2$: $^1/_4$ X$^+$Y (gray males), $^1/_4$ X^{y}Y (yellow males), $^1/_4$ X$^+$X^y (gray females), $^1/_4$ X$^+$X$^+$ (gray females).
(b) F$_1$: $^1/_2$ X^{y}Y (yellow males), $^1/_2$ X$^+$X^y (gray females); F$_2$: $^1/_4$ X$^+$Y (gray males), $^1/_4$ X^{y}Y (yellow males), $^1/_4$ X$^+$X^y (gray females), $^1/_4$ X^{y}X^y (yellow females).
(c) F$_2$: $^1/_4$ X$^+$Y (gray males), $^1/_4$ X^{y}Y (yellow males), $^1/_4$ X$^+$X$^+$ (gray females), $^1/_4$ X$^+$X^y (gray females).
(d) F$_3$: $^1/_8$ gray males, $^3/_8$ yellow males, $^5/_{16}$ gray females, and $^3/_{16}$ yellow females.

25. Because color blindness is a recessive trait, the color-blind daughter must be homozygous recessive. Because the color blindness is X-linked, John has grounds for suspicion. Normally, their daughter would have inherited John's X chromosome. Because John is not color blind, he could not have transmitted an X chromosome with a color-blind allele to his daughter.

A remote alternative possibility is that the daughter is XO, having inherited a recessive color-blind allele from her mother and no sex chromosome from her father. In that case, the daughter would have Turner syndrome. A new X-linked color-blind mutation also is possible, albeit even less likely.

If Cathy had given birth to a color-blind son, then John would have no grounds for suspicion. The son would have inherited John's Y chromosome and the color-blind X chromosome from Cathy.

27. Because Bob must have inherited the Y chromosome from his father and his father has normal color vision, a nondisjunction event

in the paternal lineage cannot account for Bob's genotype. Bob's mother must be heterozygous X$^+$X^c because she has normal color vision, and she must have inherited a color-blind X chromosome from her color-blind father. For Bob to inherit two color-blind X chromosomes from his mother, the egg must have arisen from a nondisjunction in meiosis II. In meiosis I, the homologous X chromosomes separate, and so one cell has the X$^+$ chromosome and the other has X^c. The failure of sister chromatids to separate in meiosis II would then result in an egg with two copies of X^c.

31. **(a)** Male parent is X$^+$Y; female parent is X$^+$X^m.
(b) Male parent is X^{m}Y; female parent is X$^+$X$^+$.
(c) Male parent is X^{m}Y; female parent is X$^+$X^m.
(d) Male parent is X$^+$Y; female parent is X^{m}X^m.
(e) Male parent is X$^+$Y; female parent is X$^+$X$^+$.

32. F$_1$: $^1/_2$ Z^{b}Z$^+$ (normal males); $^1/_2$ Z^{b}W (bald females).
F$_2$: $^1/_4$ Z$^+$Z^b (normal males), $^1/_4$ Z$^+$W (normal females), $^1/_4$ Z^{b}Z^b (bald males), $^1/_4$ Z^{b}W (bald females).

36. **(a)** F$_1$: all males are have miniature wings and red eyes (X^{m}Y s^+s), and all females have long wings and red eyes (X$^+$X^m s^+s).
F$_2$: $^3/_{16}$ male, normal, red; $^1/_{16}$ male, normal, sepia; $^3/_{16}$ male, miniature, red; $^1/_{16}$ male, miniature, sepia; $^3/_{16}$ female, normal, red; $^1/_{16}$ female, normal, sepia; $^3/_{16}$ female, miniature, red; $^1/_{16}$ female, miniature, sepia.
(b) F$_1$: all females have long wings and red eyes (X^{m+} X^m s^+s), and all males have long wings and red eyes (X^{m+}Y s^+s). F$_2$: $^3/_{16}$ males, long wings, red eyes; $^1/_{16}$ males, long wings, sepia eyes; $^3/_{16}$ males, mini wings, red eyes; $^1/_{16}$ males, mini wings, sepia eyes; $^6/_{16}$ females, long wings, red eyes; $^2/_{16}$ females, long wings, sepia eyes.

38. The trivial explanation for these observations is that this form of color blindness is an autosomal recessive trait. In that case, the father would be a heterozygote, and we would expect equal proportions of color-blind and normal children of either sex.

If, on the other hand, we assume that this form of color blindness is an X-linked trait, then the mother is X^{c}X^c and the father must be X$^+$Y. Normally, all the sons would be color blind, and all the daughters should have normal vision. The most likely way to have a daughter who is color blind would be for her not to have inherited an X$^+$ from her father. The observation that the color-blind daughter is short in stature and has failed to undergo puberty is consistent with Turner syndrome (XO). The color-blind daughter would then be X^{c}O.

39. **(a)** 1; **(b)** 0; **(c)** 0; **(d)** 1; **(e)** 1; **(f)** 2; **(g)** 0; **(h)** 2; **(i)** 3.

Chapter 5

1. In incomplete dominance, the phenotype of the heterozygote is intermediate between the phenotypes of the two homozygotes. In codominance, both alleles are expressed and both phenotypes are manifested simultaneously.

2. In incomplete penetrance, the expected phenotype of a particular genotype is not expressed. Environmental factors, as well as the effects of other genes, may alter the phenotypic expression of a particular genotype.

5. A complementation test is used to determine whether two different recessive mutations are at the same locus (are allelic) or at different loci. The two mutations are introduced into the same individual organism by crossing homozygotes for each of the mutants. If the progeny show a mutant phenotype, then the mutations are allelic (at the same locus). If the progeny show a wild-type (dominant) phenotype, then the mutations are at different loci and are said to complement each other because each of the mutant parents can supply a functional copy (or dominant allele) of the gene mutated in the other parent.

6. Cytoplasmically inherited traits are encoded by genes in the cytoplasm. Because the cytoplasm is usually inherited from a single (most often the female) parent, reciprocal crosses do not show the same results. Cytoplasmically inherited traits often show great variability because different egg cells (female gametes) may have differing proportions of cytoplasmic alleles owing to random sorting of mitochondria (or plastids in plants).

11. Continuous characteristics, also called quantitative characteristics, exhibit many phenotypes with a continuous distribution. They result from the interaction of multiple genes (polygenic traits), the influence of environmental factors on the phenotype, or both.

14. (a) The results of the crosses indicate that cremello and chestnut are pure-breeding traits (homozygous). Palomino is a heterozygous trait that produces a 1 : 2 : 1 ratio when palominos are crossed with each other. The simplest hypothesis consistent with these results is incomplete dominance, with palomino as the phenotype of the heterozygotes resulting from chestnuts crossed with cremellos.
(b) Let C^B = chestnut, C^W = cremello. The parents and offspring of these crosses have the following genotypes: chestnut = $C^B C^B$; cremello = $C^W C^W$; palomino = $C^B C^W$.

15. (a) $^1/_2\ L^M L^M$ (type M), $^1/_2\ L^M L^N$ (type MN); (b) all $L^N L^N$ (type N);
(c) $^1/_2\ L^M L^N$ (type MN), $^1/_4\ L^M L^M$ (type M), $^1/_4\ L^N L^N$ (type N);
(d) $^1/_2\ L^M L^N$ (type MN), $^1/_2\ L^N L^N$ (type N); (e) all $L^M L^N$ (type MN).

18. (a) The 2 : 1 ratio in the progeny of two spotted hamsters suggests lethality, and the 1 : 1 ratio in the progeny of a spotted hamster and a hamster without spots indicates that spotted is a heterozygous phenotype. If S and s represent alleles at the locus for white spotting, spotted hamsters are Ss and solid-colored hamsters are ss. One-quarter of the zygotes expected from a mating of two spotted hamsters are SS—embryonic lethal—and missing from the progeny, resulting in the 2 : 1 ratio of spotted to solid progeny.
(b) Because spotting is a heterozygous phenotype, obtaining Chinese hamsters that breed true for spotting is impossible.

24. The child's genotype has an allele for blood-type B and an allele for blood-type N that could not have come from the mother and must have come from the father. Therefore, the child's father must have an allele for type B and an allele for type N. George, Claude, and Henry are eliminated as possible fathers because they lack an allele for either type B or type N.

25. (a) All walnut ($Rr\ Pp$); (b) $^1/_4$ walnut ($Rr\ Pp$), $^1/_4$ rose ($Rr\ pp$), $^1/_4$ pea ($rr\ Pp$), $^1/_4$ single ($rr\ pp$); (c) $^9/_{16}$ walnut ($R\_\ P\_$), $^3/_{16}$ rose ($R\_\ pp$), $^3/_{16}$ pea ($rr\ P\_$), $^1/_{16}$ single ($rr\ pp$); (d) $^3/_4$ rose ($R\_\ pp$), $^1/_4$ single ($rr\ pp$); (e) $^1/_4$ walnut ($Rr\ Pp$), $^1/_4$ rose ($Rr\ pp$), $^1/_4$ pea ($rr\ Pp$), $^1/_4$ single ($rr\ pp$); (f) $^1/_2$ rose ($Rr\ pp$), $^1/_2$ single ($rr\ pp$).

26. (a) Each of the backcross progeny received recessive alleles b and r from the homozygous white parent. Their phenotype is therefore determined by the alleles received from the F_1 parent: brown fish are $Bb\ Rr$; blue fish are $Bb\ rr$; red fish are $bb\ Rr$; and white fish are $bb\ rr$.
(b) We expect a 1 : 1 : 1 : 1 ratio of the four phenotypes

	Observed	Expected	O − E	(O − E)²/E
Brown	228	229.25	−1.25	0.007
Blue	230	229.25	0.75	0.002
Red	237	229.25	7.75	0.262
White	222	229.25	−7.25	0.229
Total	917	917		0.5 = χ^2

df = 4 − 1 = 3; 0.9 < P < 0.975
we cannot reject the hypothesis.

(c) The homozygous red fish would be $bb\ RR$, crossed with $bb\ rr$. All progeny would be $bb\ Rr$, or red fish.
(d) Homozygous red fish $bb\ RR$ × homozygous blue fish $BB\ rr$

F_1 will be all brown: $Bb\ RR$ backcrossed to $bb\ RR$ (homozygous red parent).

Backcross progeny will be in equal proportions $Bb\ RR$ (brown); $Bb\ Rr$ (brown); $bb\ RR$ (red); and $bb\ Rr$ (red). Overall, $^1/_2$ brown and $^1/_2$ red.

29. (a) Labrador retrievers vary in two loci, B and E. Black dogs have dominant alleles at both loci ($B\_\ E\_$), brown dogs have $bb\ E\_$, and yellow dogs have $B\_\ ee$ or $bb\ ee$. Because all the puppies were black, all of them must have inherited a dominant B allele from the yellow parent and a dominant E allele from the brown parent. The brown female parent must have been $bb\ EE$, and the yellow male must have been $BB\ ee$. All the black puppies were $Bb\ Ee$.
(b) Mating two yellow Labradors will produce yellow puppies only. Mating two brown Labradors will produce either brown puppies only, if at least one of the parents is homozygous EE, or $^3/_4$ brown and $^1/_4$ yellow if both parents are heterozygous Ee.

31. Let A and B represent the two loci. The F_1 heterozygotes are $Aa\ Bb$. The F_2 are: $A\_\ B\_$ disc-shaped, $A\_\ bb$ spherical, $aa\ B\_$ spherical, $aa\ bb$ long.

Chapter 6

1. The three factors are: (i) controlled mating experiments are impossible; (ii) humans have a long generation time, and so tracking the inheritance of traits for more than one generation takes a long time; and (iii) the number of progeny per mating is limited, and so phenotypic ratios are uncertain.

3. Autosomal recessive trait: affected males and females arise with equal frequency from unaffected parents; often appears to skip generations; unaffected child of an affected parent is a carrier.

Autosomal dominant trait: affected males and females arise with equal frequency from a single affected parent; does not usually skip generations.

X-linked recessive trait: affects males predominantly and is passed from an affected male through his unaffected daughter to his grandson; not passed from father to son.

X-linked dominant trait: affects males and females; is passed from an affected male to all his daughters but not to his sons; is passed from an affected woman (usually heterozygous for a rare dominant trait) equally to half her daughters and half her sons.

Y-linked trait: affects males exclusively; is passed from father to all sons.

6. Monozygotic and dizygotic. Monozygotic twins arise when a single fertilized egg splits into two embryos in early embryonic development divisions. They are genetically identical. Dizygotic twins arise from two different eggs fertilized by two different sperm. They have, on the average, 50% of their genes in common.

9. Genetic counseling provides assistance to clients by interpreting the results of genetic testing and diagnosis; provides information about relevant disease symptoms, treatment, and progression; assesses and calculates the various genetic risks facing the person or couple; and helps clients and family members cope with the stress of decision-making and facing up to the drastic changes in their lives that may be precipitated by a genetic condition.

21. (a)

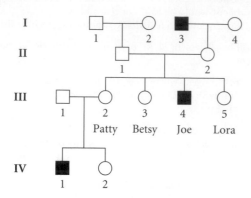

(b) X-linked recessive; **(c)** zero; **(d)** $^1/_4$; **(e)** $^1/_4$.

25. (a) Autosomal dominant. The trait must be autosomal because affected males pass the trait to both sons and daughters. It is dominant because it does not skip generations, all affected individuals have affected parents, and it is extremely unlikely that multiple unrelated individuals mating into the pedigree are carriers of a rare trait.
(b) X-linked dominant. Superficially, this pedigree appears to be similar to the pedigree in part *a* in that both males and females are affected, and it appears to be a dominant trait. However, closer inspection reveals that, whereas affected females can pass the trait to either sons or daughters, affected males pass the trait only to all daughters.
(c) Y-linked. The trait affects only males and is passed from father to son. All sons of an affected male are affected.
(d) X-linked recessive or sex-limited autosomal dominant. Because only males show the trait, the trait could be X-linked recessive, Y-linked, or sex-limited. We can eliminate Y-linkage because affected males do not pass the trait to their sons. X-linked recessive inheritance is consistent with the pattern of unaffected female carriers producing both affected and unaffected sons and affected males producing unaffected female carriers, but no affected sons. Sex-limited autosomal dominant inheritance is also consistent with unaffected heterozygous females producing affected heterozygous sons, unaffected homozygous recessive sons, and unaffected heterozygous or homozygous recessive daughters. The two remaining possibilities of X-linked recessive versus sex-limited autosomal dominant could be distinguished if we had enough data to determine whether affected males have both affected and unaffected sons, as expected from autosomal dominant inheritance, or whether affected males have only unaffected sons, as expected from X-linked recessive inheritance. Unfortunately, this pedigree shows only two sons from affected males. In both cases, the sons are unaffected, consistent with X-linked recessive inheritance, but two male progeny are not enough to conclude that affected males cannot produce affected sons.
(e) Autosomal recessive. Unaffected parents produce affected progeny, and so the trait is recessive. The affected daughter must have inherited recessive alleles from both unaffected parents, and so the trait must be autosomal. If it were X-linked, her father would show the trait.

27. (a) X-linked recessive; **(b)** $^1/_4$; **(c)** $^1/_2$.

31. Migraine headaches: genetic and environmental. Markedly greater concordance in monozygotic twins, who are 100% genetically identical, than in dizygotic twins, who are 50% genetically identical, is indicative of a genetic influence. However, only 60% concordance for monozygotic twins indicates that environmental factors also play a role.

Eye color: genetic. The concordance is greater in monozygotic twins than in dizygotic twins. Moreover, the monozygotic twins have 100% concordance for this trait, indicating that environment has no detectable influence.
Measles: no detectable genetic influence. There is no difference in concordance between monozygotic and dizygotic twins. Some environmental influence can be detected because monozygotic twins show less than 100% concordance.
Clubfoot: genetic and environmental. The reasoning for migraine headaches applies here. A strong environmental influence is indicated by the high discordance in monozygotic twins.
High blood pressure: genetic and environmental. Reasoning is similar to that for clubfoot.
Handedness: no genetic influence. The concordance is the same in monozygotic and dizygotic twins. Environmental influence is indicated by the less than 100% concordance in monozygotic twins.
Tuberculosis: no genetic influence. Concordance is the same in monozygotic and dizygotic twins. The importance of environmental influence is indicated by the very low concordance in monozygotic twins.

Chapter 7

1. Recombination means that meiosis generates gametes with allelic combinations that differ from the original gametes inherited by an organism. Recombination may be caused by the independent assortment of loci on different chromosomes or by a physical crossing over between two loci on the same chromosome.

2. (a) Nonrecombinant gametes; **(b)** 50% recombinant gametes and 50% nonrecombinant; **(c)** more than 50% of the gametes nonrecombinant, and less than 50% recombinant.

5. For genes in coupling configuration, two wild-type alleles are on the same chromosome and the two mutant alleles are on the homologous chromosome. For genes in repulsion, the wild-type allele of one gene and the mutant allele of the other gene are on the same chromosome, and vice versa on the homologous chromosome. The two arrangements have opposite effects on the results of a cross. For genes in coupling configuration, most of the progeny will be either wild type for both genes or mutant for both genes, with relatively few that are wild type for one gene and mutant for the other. For genes in repulsion, most of the progeny will be mutant for only one gene and wild type for the other, with relatively few recombinants that are wild type for both or mutant for both.

8. The farther apart two loci are, the more likely there will be double crossovers between them. The calculated recombination frequency will underestimate the true crossover frequency because the double-crossover progeny are not counted as recombinants.

10. Positive interference indicates that a crossover inhibits or interferes with the occurrence of a second crossover nearby. Negative interference suggests that a crossover event can stimulate additional crossover events in the same region of the chromosome.

16. The genes are linked and have not assorted independently.

18. (a) Both plants are *Dd Pp*.
(b) Yes. Recombination frequency = 3.8%.
(c) The two plants have different coupling configurations. In plant A, the dominant alleles *D* and *P* are coupled; one chromosome is $\underline{\quad D \quad\quad P \quad}$ and the other is $\underline{\quad d \quad\quad p \quad}$. In plant B, they are in repulsion; its chromosomes are $\underline{\quad D \quad\quad p \quad}$ and $\underline{\quad d \quad\quad P \quad}$.

21. The distances between the genes are indicated by the recombination rates. Because loci R and L_2 have the greatest recombination rate, they must be the farthest apart, and W_2 is in the middle. The order of the genes is: R, W_2, L_2.

24.

Genotype	Body color	Eyes	Bristles	Proportion
(a) $e^+\ ro^+\ f^+$	normal	normal	normal	20%
$e^+\ ro^+\ f$	normal	normal	forked	20%
$e\ ro\ f^+$	ebony	rough	normal	20%
$e\ ro\ f$	ebony	rough	forked	20%
$e^+\ ro\ f^+$	normal	rough	normal	5%
$e^+\ ro\ f$	normal	rough	forked	5%
$e\ ro^+\ f^+$	ebony	normal	normal	5%
$e\ ro^+\ f$	ebony	normal	forked	5%
(b) $e^+\ ro^+\ f^+$	normal	normal	normal	5%
$e^+\ ro^+\ f$	normal	normal	forked	5%
$e\ ro\ f^+$	ebony	rough	normal	5%
$e\ ro\ f$	ebony	rough	forked	5%
$e^+\ ro\ f^+$	normal	rough	normal	20%
$e^+\ ro\ f$	normal	rough	forked	20%
$e\ ro^+\ f^+$	ebody	normal	normal	20%
$e\ ro^+\ f$	ebony	normal	forked	20%

26.

Gene *f* is unlinked to either of these groups; it is on a third linkage group.

29. **(a)** *V* is the middle gene.
(b) The *Wx–V* distance = 7 m.u, and the *Sh–V* distance = 30 m.u. The *Wx–Sh* distance is the sum of these two distances, or 37 m.u.
(c) Coefficient of coincidence = 0.80; interference = 0.20

32.

Genotype	Body	Eyes	Wings	Proportion
$b^+\ pr^+\ vg^+$	normal	normal	normal	40.7%
$b\ pr\ vg$	black	purple	vestigial	40.7%
$b^+\ pr^+\ vg$	normal	normal	vestigial	6.3%
$b\ pr\ vg^+$	black	purple	normal	6.3%
$b^+\ pr\ vg$	normal	purple	vestigial	2.8%
$b\ pr^+\ vg^+$	black	normal	normal	2.8%
$b^+\ pr\ vg^+$	normal	purple	normal	0.2%
$b\ pr^+\ vg$	black	normal	vestigial	0.2%

34.

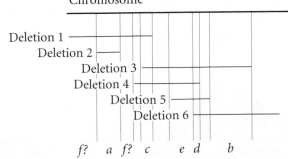

The location of *f* is ambiguous; it could be in either location shown in the deletion map.

36. Enzyme 1 is on chromosome 9; enzyme 2 is on chromosome 4; enzyme 3 is on the X chromosome.

Chapter 8

3.

Types of matings	Outcomes
$F^+ \times F^-$	Two F^+ cells
Hfr $\times F^-$	One F^+ cell and one F^- cell
$F' \times F^-$	Two F' cells

The F factor contains a number of genes that take part in the conjugation process, including genes necessary for the synthesis of the sex pilus. The F factor has an origin of replication that enables the factor to be replicated in the conjugation process and has genes for opening the plasmid and initiating the chromosome transfer.

4. To map genes by conjugation, an Hfr strain is mixed with an F^- strain. The two strains must have different genotypes and must remain in physical contact for the transfer to take place. The conjugation process is interrupted at regular intervals. The chromosomal transfer from the Hfr strain always begins with a part of the integrated F factor and proceeds in a linear fashion. Transfer of the entire chromosome takes approximately 100 minutes. The time required for individual genes to be transferred is relative to their positions on the chromosome and the direction of transfer initiated by the F factor. Gene distances are typically mapped in minutes. The genes that are transferred by conjugation to the recipient must be incorporated into the recipient's chromosome by recombination to be expressed.

In transformation, the relative frequency at which pairs of genes are transferred, or cotransformed, indicates the distance between the two genes. Closer gene pairs are cotransformed more frequently. As in interrupted conjugation, the donor DNA must be incorporated into the recipient cell's chromosome by recombination. Physical contact of the donor and recipient cells is not needed. The recipient cell takes up the DNA directly from the environment, and so the DNA from the donor strain must be isolated and broken up before transformation can take place.

The transfer of DNA by transduction requires a viral vector. DNA from the donor cell is packaged into a viral protein coat. The viral particle containing the bacterial donor DNA then infects a recipient bacterial cell. The donor bacterial DNA is incorporated into the recipient cell's chromosome by recombination. Only genes that are close together on the bacterial chromosome can be cotransduced. Therefore, the rate of cotransduction, like the rate of cotransformation, is an indication of the physical distances between genes on the chromosome.

In all three processes, the recipient cell takes up a piece of the donor chromosome and incorporates some of that piece into its own chromosome by recombination. In all three processes, mapping distance is calculated by measuring the frequency with which recipient cells are transformed. The processes use different methods to incorporate donor DNA in the recipient cell.

6. Reproduction is rapid, asexual, and produces lots of progeny. Their genomes are small. They are easy to grow in the laboratory. Techniques are available for isolating and manipulating their genes. Mutant phenotypes, especially auxotrophic phenotypes, are easy to measure.

10. In generalized transduction, randomly selected bacterial genes are transferred from one bacterial cell to another by a virus. In specialized transduction, only genes from a particular locus on the bacterial chromosome are transferred to another bacterium. The process of specialized transduction requires lysogenic phages that integrate into specific locations on the host cell's chromosome. When the phage DNA excises from the host chromosome and the excision process is imprecise, the phage DNA will carry a small part of the bacterial DNA. The hybrid DNA must be injected by the phage into another bacterial cell in another round of infection.

The transfer of DNA by generalized transduction requires that the host DNA be broken down into smaller pieces and that a piece of the host DNA, instead of phage DNA, be packaged into a phage coat. The defective phage cannot produce new phage particles in a subsequent infection, but it can inject the bacterial DNA into another bacterium or recipient. Through a double-crossover event, the donor DNA can become incorporated into the bacterial recipient's chromosome.

11. To conduct the complementation test, Benzer infected cells of *E. coli* K with large numbers of the two mutant phage types. For successful infection, each mutant phage must supply the gene product or protein missing in the other mutant phage. Complementation will result only if the mutations are at separate loci. If the two mutations are at the same locus, there will be no complementation of gene products, and no plaques will be produced on the *E. coli* K lawns.

12. A retrovirus is able to integrate its RNA genome into its host cell's DNA genome through the action of the enzyme reverse transcriptase, which can synthesize complementary DNA from either an RNA or a DNA template. Reverse transcriptase uses the retroviral single-stranded RNA as a template to synthesize double-stranded DNA. The newly synthesized DNA molecule can then integrate into the host chromosome to form a provirus.

17. By using low doses of antibiotics for 5 years, Farmer Smith selected for bacteria that are resistant to the antibiotics. The doses used killed sensitive bacteria but not moderately sensitive or slightly resistant bacteria. As time passed, only resistant bacteria remained in his pigs because any sensitive bacteria were eliminated by the low doses of antibiotics.

In the future, Farmer Smith should continue to use the vitamins but should use the antibiotics only when a sick pig requires them. In this manner, he will not be selecting for antibiotic-resistant bacteria, and the chances of successful treatment of his sick pigs will be greater.

20. In each of the Hfr strains, the F factor has been inserted into a different location in the chromosome. The orientation of the F factor in the strains varies as well.

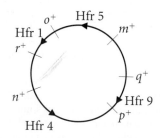

21. Distances between genes are in minutes.

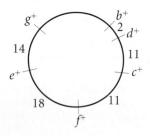

30. Number of plaques produced by $c^+ m^+ = 460$; by $c^- m^- = 460$; by $c^+ m^- = 40$ (recombinant); by $c^+ m^- = 40$ (recombinant).

32. (a) The recombination frequency between r_2 and h is $^{13}/_{180} = 0.072$, or 7.2%. The RF between r_{13} and h is $^3/_{216} = 0.014$, or 1.4%.

(b)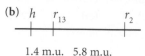

1.4 m.u.   5.8 m.u.

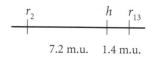

7.2 m.u.   1.4 m.u.

33. (a)

(b)

h^+ $\qquad$ t^+ $\qquad\qquad\qquad\qquad$ c^+

7.1 m.u.   24.1 m.u.

(c) Coefficient of coincidence $= (6 + 5)/(0.071 \times 0.241 \times 942) = 0.68$. Interference $= 1 - 0.68 = 0.32$.

Chapter 9

1. Chromosome rearrangements:

Deletion: Loss of a part of a chromosome.

Duplication: Addition of an extra copy of a part of a chromosome.

Inversion: A part of a chromosome is reversed in orientation.

Translocation: A part of one chromosome becomes incorporated into a different (nonhomologous) chromosome or into the same chromosome.

Aneuploidy: Loss or gain of one or more chromosomes, causing the chromosome number to deviate from $2n$ or the normal euploid complement.

Polyploidy: Gain of entire sets of chromosomes, causing the chromosome number to change from $2n$ to $3n$ (triploid), $4n$ (tetraploid), and so on.

2. The expression of some genes is balanced with the expression of other genes; the ratios of their gene products, usually proteins, must be maintained within a narrow range for proper cell function. Extra copies of one of these genes cause that gene to be expressed at proportionately higher levels, thereby upsetting the balance of gene products.

5. A paracentric inversion does not include the centromere; a pericentric inversion includes the centromere.

9. Translocations can produce phenotypic effects if the translocation breakpoint disrupts a gene or if a gene near the breakpoint is altered in its expression because of relocation to a different chromosomal environment (a position effect).

13. Primary Down syndrome is caused by the spontaneous, random nondisjunction of chromosome 21, leading to trisomy 21. Familial Down syndrome most frequently arises as a result of a Robertsonian translocation of chromosome 21 with another chromosome, usually chromosome 14.

14. Uniparental disomy is the inheritance of both copies of a chromosome from the same parent. It may arise from a trisomy condition

in which the early embryo loses one of the three chromosomes, and the two remaining chromosomes are from the same parent.

16. In autopolyploidy, all sets of chromosomes are derived from a single species. In allopolyploidy, the sets of chromosomes are derived from two or more different species. Autopolyploidy may arise through nondisjunction in an early $2n$ embryo or through meiotic nondisjunction that produces a gamete with extra sets of chromosomes. Allopolyploidy is usually preceded by hybridization between two different species, followed by chromosome doubling.

18. (a) Duplications; **(b)** polyploidy; **(c)** deletions; **(d)** inversions; **(e)** translocations.

19. (a) Tandem duplication of AB; **(b)** displaced duplication of AB; **(c)** paracentric inversion of DEF; **(d)** deletion of B; **(e)** deletion of FG; **(f)** paracentric inversion of CDE; **(g)** pericentric inversion of ABC; **(h)** duplication and inversion of DEF; **(i)** duplication of CDEF, inversion of EF.

22. (a) $^{1}/_{3}$ Notch white-eyed females, $^{1}/_{3}$ wild-type females, and $^{1}/_{3}$ wild-type males; **(b)** $^{1}/_{3}$ Notch red-eyed females, $^{1}/_{3}$ wild-type females, and $^{1}/_{3}$ white-eyed males; **(c)** $^{1}/_{3}$ Notch white-eyed females, $^{1}/_{3}$ white-eyed females, and $^{1}/_{3}$ white-eyed males.

24. (a)

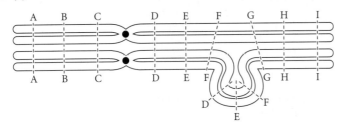

(b)

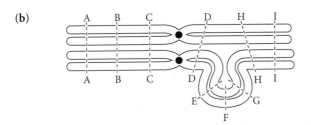

(c)

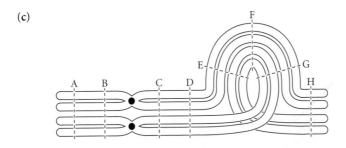

(d)

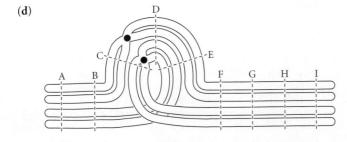

27. (a)

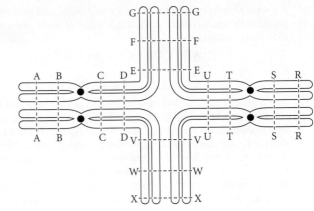

(b) Alternate:

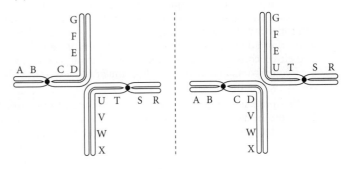

Adjacent-1:

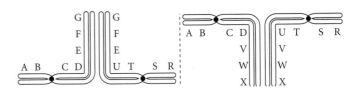

Adjacent-2:

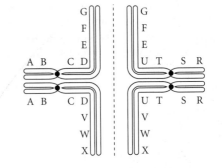

(c) Alternate: Gametes contain either both normal or both translocation chromosomes, and are all viable.

$$\underline{A B \cdot C D E F G} + \underline{R S \cdot T U V W X}$$

and

$$\underline{A B \cdot C D V W X} + \underline{R S \cdot T U E F G}$$

Adjacent-1: Gametes contain one normal and one translocation chromosome, resulting in the duplication of some genes and a deficiency of others.

$$\underline{A B \cdot C D E F G} + \underline{R S \cdot T U E F G}$$

and

$$\underline{A B \cdot C D V W X} + \underline{R S \cdot T U V W X}$$

Adjacent-2 (rare): Gametes contain one normal and one trans-location chromosome, with the duplication of some genes and a deficiency of others.

and

$$A\,B \cdot C\,D\,E\,F\,G + A\,B \cdot C\,D\,V\,W\,X$$

$$R\,S \cdot T\,U\,V\,W\,X + R\,S \cdot T\,U\,E\,F\,G$$

31. The high incidence of Down syndrome in Bill's family and among Bill's relatives is consistent with familial Down syndrome, caused by a Robertsonian translocation of chromosome 21. Bill and his sister, who are unaffected, are phenotypically normal carriers of the translocation and have 45 chromosomes. Their children and Bill's brother, who have Down syndrome, have 46 chromosomes. From the information given, there is no reason to suspect that Bill's wife Betty has any chromosomal abnormalities. Therefore, the statement in part *d* is most likely correct. All other statements are incorrect.

32. In mammals, the higher frequency of sex-chromosome aneuploids compared with that of autosomal aneuploids is due to X-chromosome inactivation and the lack of essential genes on the Y chromosome. If fish do not have X-chromosome inactivation and if both of their sex chromosomes have numerous essential genes, then the frequency of aneuploids should be similar for both sex chromosomes and autosomes.

33. **(a)** Possible gamete types: (i) normal chromosome 13 and normal chromosome 22; (ii) translocated chromosome 13 + 22; (iii) translocated chromosome 13 + 22 and normal chromosome 22; (iv) normal chromosome 13; (v) normal chromosome 13 and translocated chromosome 13 + 22; (vi) normal chromosome 22.
    (b) Zygotes types: (i) 13, 13, 22, 22; normal; (ii) 13, 13 + 22, 22; translocation carrier; (iii) 3, 13 + 22, 22, 22; trisomy 22; (iv) 13, 13, 22; monosomy 22; (v) 13, 13, 13 + 22, 22; trisomy 13; (vi) 13, 22, 22; monosomy 13.
    (c) 50%

35. **(a)** 15; **(b)** 24; **(c)** 32; **(d)** 17; **(e)** 14; **(f)** 14; **(g)** 40; **(h)** 18.

Chapter 10

1. The genetic material must (i) contain complex information, (ii) replicate or be replicated faithfully, and (iii) encode the phenotype.

3. Experiments by Hershey and Chase on the bacteriophage T2 by Avery, Macleod, and McCarty, who demonstrated that the transforming material initially identified by Griffiths is DNA.

5. Hershey and Chase used ^{32}P-labeled phages to infect bacteria. The progeny phage released from these bacteria emitted radioactivity from ^{32}P. The presence of the ^{32}P in the progeny phages indicated that the infecting phages had passed DNA on to the progeny phages.

7. Deoxyguanosine 5′-phosphate (dGMP)

Deoxyribose sugar

10.

14. Replication, transcription, and translation—the components of the central dogma of molecular biology.

20. The IIR strain of *Streptococcus pneumonia* must have been naturally competent—was capable of taking up DNA from the environment. The heat-killed strains of IIIS bacteria lysed, releasing their DNA into the environment and enabling the IIIS chromosomal DNA fragments to come into contact with IIR cells. The IIIS DNA responsible for the virulence of the IIIS strain was taken up by IIR cells and integrated into the IIR chromosome, thus "transforming" the IIR cells into virulent IIIS cells.

24. Approximately 5×10^{20} nucleotide pairs. Stretched end to end, the DNA would reach 1.7×10^8 km.

25. **(a)**

Organism and tissue	(A + G)/(C + T)	(A + T)/(C + G)
Sheep thymus	1.03	1.36
Pig liver	0.99	1.44
Human thymus	1.03	1.52
Rat bone marrow	1.02	1.36
Hen erythrocytes	0.97	1.38
Yeast	1.00	1.80
E. coli	1.04	1.00
Human sperm	1.00	1.67
Salmon sperm	1.02	1.43
Herring sperm	1.04	1.29

(b) The (A + G)/(T + C) ratio of ~1.0 is constant for these organisms. Each of them has a double-stranded genome. The percentage of purines should equal the percentage of pyrimidines in double-stranded DNA, which means that (A + G) = (C + T). The (A + T)/(C + G) ratios are not constant. The number of A−T base pairs relative to the number of G−C base pairs is unique to each organism and can vary among the different organisms.

(c) The $(A + G)/(T + C)$ ratio is about the same for the sperm samples, as should be expected. As in part *b*, the percentage of purines should equal the percentage of pyrimidines.

27. The relations in parts *b, c, e, g,* and *h*.

28. Adenine = 15%; guanine = 35%; cytosine = 35%.

31. Virus I is a double-stranded RNA virus. Uracil is present, indicating an RNA genome. As expected for a double-stranded genome, the percentages of adenine and uracil are equal, as are the percentages of guanine and cytosine.
    Virus II is a doubled-stranded DNA virus. The presence of thymine indicates that the viral genome is DNA. As expected for a double-stranded DNA molecule, the percentages of adenine and thymine are equal, as are the percentages of guanine and cytosine.
    Virus III is a single-stranded DNA virus. The presence of thymine indicates a DNA genome. However, the percentages of thymine and adenine are unequal, as are the percentages of guanine and cytosine, which suggests a single-stranded DNA molecule.
    Virus IV is a single-stranded RNA virus. The presence of uracil indicates an RNA genome. However, the percentage of adenine does not equal that of uracil and the percentage of guanine does not equal that of cytosine, which suggests a single-stranded genome.

32. 100,000

34. No. The flow of information predicted by the central dogma is from DNA to RNA to protein. An exception is reverse transcription, whereby RNA encodes DNA. However, biologists do not currently know of a process in which the flow of information is from proteins to DNA, which is required by the theory of the inheritance of acquired characteristics.

38. Although the chemical composition of the genetic material on the planet may be different from that of DNA, it will more than likely have properties similar to those of DNA. As stated in this chapter, the genetic material must contain complex information, replicate faithfully, and encode the phenotype. Even if the material on the planet is not DNA, it must meet these criteria. Additionally, the genetic material must be stable.

40. The carrier of genetic information must be stable so that the genetic information is faithfully transmitted from one generation to the next. The $3'$-hydroxyl group on the ribose sugar of RNA makes RNA more reactive than DNA, which lacks a free $3'$-hydroxyl group on its deoxyribose sugar. Because it is single stranded, RNA can assume a number of secondary structures and a number of different functions, decreasing its stability. The double-stranded nature of DNA, with its bases hydrogen bonded to other bases, stabilizes DNA and makes it more suitable as the carrier of genetic information.

Chapter 11

1. Supercoiling arises from the overwinding (positive supercoiling) or underwinding (negative supercoiling) of the DNA double helix; from a lack of free ends, as in circular DNA molecules; when the ends of the DNA molecule are bound to proteins that prevent the ends from rotating about each other.

3. The nucleosome consists of a core particle, which contains two molecules each of histones H2A, H2B, H3, and H4, and from 145 to 147 bp of DNA that winds around the core. A chromatosome contains the nucleosome core and a molecule of histone H1.

7. Centromeres are the points at which spindle fibers attach to the chromosome. They are necessary for proper segregation of the chromosomes in mitosis and meiosis. Most eukaryotic centromeres are characterized by heterochromatin consisting of highly repetitive DNA. Centromeres are thought to exist at specific locations on the chromosome because of epigenetic changes to chromatin structure at those locations. For example, nucleosomes at centromeres often possess the variant histone CenH3.

8. Telomeres are the stable ends of the linear chromosomes in eukaryotes. They prevent degradation by exonucleases and prevent joining of the ends. Telomeres also enable the replication of the ends of the chromosome. Telomeric DNA sequences consist of repeats of a simple sequence, usually in the form of $5'-C_n(A \text{ or } T)_m$.

11. Unique-sequence DNA, present in only one or a few copies per haploid genome, constitutes most of the protein-coding sequences as well as a large number of sequences with unknown function.

    Moderately repetitive sequences, ranging from a few hundred to a few thousand base pairs in length, are present in as many as several thousand copies per haploid genome. Some moderately repetitive DNA consists of functional genes that encode rRNAs and tRNAs, but most is made up of transposable elements and remnants of transposable elements.

    Highly repetitive DNA, or satellite DNA, consists of clusters of tandem repeats of short sequences (often less than 10 base pairs in length) present in hundreds of thousands to millions of copies per haploid genome.

12. Most transposable elements have terminal inverted repeats and are flanked by short direct repeats. Replicative transposons use a copy-and-paste mechanism in which the transposon is replicated and inserted in a new location, leaving the original transposon in place. Nonreplicative transposons use a cut-and-paste mechanism in which the original transposon is excised and moved to a new location.

13. A retrotransposon is a transposable element that relocates through an RNA intermediate. First, it is transcribed into RNA. Then, a reverse transcriptase encoded by the retrotransposon reverse transcribes the RNA template into a DNA copy of the transposon, which then integrates into a new location in the host genome.

14. First, a transposase makes single-stranded nicks on either side of the transposon and on either side of the target sequence. Second, a DNA ligase connects the free ends of the transposon to the free ends of the DNA at the target site. Third, the free $3'$ ends of DNA on either side of the transposon are used to replicate the transposon sequence, forming the cointegrate. The enzymes normally required for DNA replication are required for this step, including DNA polymerase. The cointegrate has two copies of the transposon plus the target-site sequence on one side of each copy. Fourth, the cointegrate undergoes resolution, in which resolvase enzymes (such as those used in homologous recombination) catalyze a crossing over within the transposon.

15.

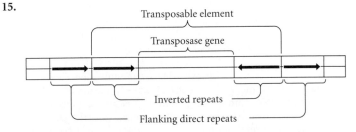

22. Because they do not benefit the cell and may be harmful to it. They exist only because they are efficient at replicating and spreading.

24. Prokaryotic chromosomes are usually circular, whereas eukaryotic chromosomes are linear. Prokaryotic chromosomes generally contain the entire genome, whereas each eukaryotic chromosome has only a part of the genome: the eukaryotic genome is divided into multiple chromosomes. Prokaryotic chromosomes are generally much smaller than eukaryotic chromosomes and have only a single

origin of DNA replication, whereas eukaryotic chromosomes contain multiple origins of DNA replication. Prokaryotic chromosomes are typically condensed into nucleoids, which have loops of DNA compacted into a dense body. Eukaryotic chromosomes contain DNA packaged into nucleosomes, which are further coiled and packaged into structures of successively higher order. The condensation state of eukaryotic chromosomes varies with the cell cycle.

27. (a) 3.2×10^7; (b) 2.9×10^8.

28. More acetylation. Regions of DNase I sensitivity are less condensed than DNA that is not sensitive to DNase I, the sensitive DNA is less tightly associated with nucleosomes, and it is in a more-open state. Such a state is associated with the acetylation of lysine residues in the N-terminal histone tails. Acetylation eliminates the positive charge of the lysine residue and reduces the affinity of the histone for the negatively charged phosphates of the DNA backbone.

31. The upper molecule, with a high percentage of A−T base pairs, will have a lower melting temperature than that of the lower molecule, which has mostly G−C base pairs. A−T base pairs have two hydrogen bonds and are thus less stable than G−C base pairs, which have three hydrogen bonds.

33. The flanking repeat is in boldface type.
   (a) 5′–ATTCGAAC**TGAC**[transposable element]**TGAC**CGATCA–3′
   (b) 5′–ATT**CGAA**[transposable element]**CGAA**CTGACCGATCA–3′

34. A fly that has white eyes with small red spots may be homozygous (female) or hemizygous (male) for an allele of the white-eye locus that contains a transposon insertion. The eye cells in these flies cannot make red pigment. In the course of eye development, the transposon may spontaneously transpose out of the white-eye locus, restoring function to this gene and allowing the cell and its mitotic progeny to make red pigment. The number and size of red spots in the eyes depend on how early in eye development the transposition occurs.

35. The number of base pairs between the staggered single-stranded nicks made at the target site by the transposase.

38. Without a functional transposase gene of its own, the transposon would be able to transpose only if another transposon of the same type were in the cell and able to express a functional transposase enzyme.

Chapter 12

2. Meselson and Stahl grew *E. coli* cells in a medium containing the heavy isotope of nitrogen (^{15}N) for several generations. The ^{15}N was incorporated into the DNA of the *E. coli* cells. The *E. coli* cells were then switched to a medium containing the common form of nitrogen (^{14}N) and allowed to proceed through a few cycles of cellular generations. Samples of the bacteria were removed at each cellular generation. Using equilibrium density gradient centrifugation, Meselson and Stahl were able to distinguish DNA fragments that contained only ^{15}N from those that contained only ^{14}N or a mixture of ^{15}N and ^{14}N because DNA containing the ^{15}N isotope is "heavier." The more ^{15}N in a DNA fragment, the lower the fragment will sink. DNA from cells grown in the ^{15}N medium produced only a single band at the expected position during centrifugation. After one round of replication in the ^{14}N medium, a single band was present after centrifugation, but the band was located at an intermediate position between that of a DNA band containing only ^{15}N and that of a DNA band containing only ^{14}N. After two rounds of replication, two bands of DNA were present. One band was located at an intermediate position between that of a DNA band containing only ^{15}N and that of a DNA band containing only ^{14}N, whereas the other band was at a position expected for DNA containing only ^{14}N. These results were consistent with the predictions of semi-

conservative replication and incompatible with the predictions of conservative and dispersive replication.

3.

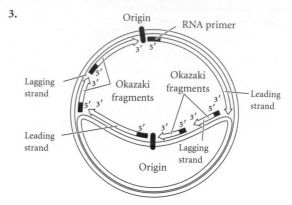

7. Deoxyadenosine triphosphate, deoxyguanosine triphosphate, deoxycytosine triphosphate, and deoxythymidine triphosphate.

10. Primase, an RNA polymerase, synthesizes the short RNA molecules, or primers, that provide a 3′-OH group to which DNA polymerase can attach deoxyribonucleotides in replication initiation.

13. Ways in which DNA replication is similar in eukaryotes and bacteria:

   Replication is semiconservative.

   Replication origins serve as starting points for replication.

   A short segment of RNA called a primer provides a 3′-OH group for DNA polymerases to begin the synthesis of the new strands.

   Synthesis is in the 5′-to-3′ direction.

   The template strand is read in the 3′-to-5′ direction.

   Deoxyribonucleoside triphosphates are the substrates.

   Replication is continuous on the leading strand and discontinuous on the lagging strand.

   Ways in which eukaryotic DNA replication differs from bacterial replication:

   There are multiple origins of replications per chromosome.

   Several different DNA polymerases have different functions.

   Immediately after DNA replication, nucleosomes are assembled.

16. RecBCD protein unwinds double-stranded DNA and can cleave nucleotide strands.

   RecA protein allows a single strand to invade a double-stranded DNA.

   RuvA and RuvB proteins promote branch migration in homologous recombination.

   DNA ligase repairs nicks or cuts in the DNA.

18. The strands of the novel double-stranded nucleic acid must be parallel to each other, unlike antiparallel double-stranded DNA on Earth. Replication by *E. coli* DNA polymerases can proceed only in the 5′-to-3′ direction, which requires the template to be read in the 3′-to-5′ direction. If replication is continuous on both strands, the two strands must have the same direction and be parallel.

19. Immediately after transfer, no ^{32}P incorporation. After one round of replication, one strand of each newly synthesized DNA molecule will contain ^{32}P, whereas the other strand will contain only nonradioactive phosphorous. After two rounds of replication, 50% of the DNA molecules will have ^{32}P in both strands, whereas one strand of the remaining 50% will contain ^{32}P and the other strand will contain nonradioactive phosphorous.

21. Theta replication, 5 minutes; rolling-circle replication, 10 minutes.

23.

Okazaki fragments — Origin — RNA primer

Lagging — Leading

Leading — Lagging

Unwinding — Unwinding

Origin

24. (a) More errors in replication; (b) primers would not be removed; (c) primers that have been removed would not be replaced.

28. The Geminin protein inhibits replication licensing by the mini-chromosome maintenance complex (MCM). Licensing at the origins is necessary for replication to begin. MCM must bind to DNA to initiate replication. After replication has begun at an origin, Geminin prevents MCM from binding to DNA and reinitiating replication at that origin. At the end of mitosis, Geminin protein is degraded. Treatment of the cells with siRNAs prevents the production of Geminin, allowing MCM to bind at the origins without regulation. The continued and repeated licensing of the replication origin by MCM could lead to an increase in chromosome number within the nucleus of each cell. The polyploid nucleus was likely blocked from undergoing mitotic division by cell-cycle checkpoints, generating a single nucleus with the multiple sets of chromosomes.

35. Protein B may be needed for the successful initiation of replication at replication origins. Protein B is present at the beginning of S phase but disappears by the end of it. Protein A may be responsible for removing or inactivating protein B. As levels of protein A increase, the levels of protein B decrease, preventing extra initiation events. When protein A is mutated, it can no longer inactivate protein B; thus successive rounds of replication can begin, owing to the high levels of protein B. When protein B is mutated, it cannot assist initiation and replication ceases.

Chapter 13

1.

Thymine (DNA only)   Uracil (RNA only)

Deoxyribonucleotide   Ribonucleotide

RNA and DNA are polymers of nucleotides that are held together by phosphodiester bonds. An RNA nucleotide contains ribose, whereas a DNA nucleotide contains deoxyribose. RNA contains uracil but not thymine. DNA contains thymine but not uracil. RNA is typically single stranded even though RNA molecules can pair with other complementary sequences. DNA molecules are almost always double stranded.

4.

Promoter → Transcription start site — Terminator

5′ ———————————————— 3′
3′ ———————————————— 5′

RNA coding region

7. RNA polymerase I transcribes rRNA.

RNA polymerase II transcribes pre-mRNA, some snRNAs, snoRNAs and some miRNAs.

RNA polymerase III transcribes small RNA molecules such as 5S rRNA, tRNAs, some snRNAs, and some miRNAs.

RNA polymerases IV and V transcribe some siRNAs in plants.

9.

Transcription start site

5′–〜〜TTGACA〜〜〜TATAAT〜〜–3′
3′–〜〜AACTGT〜〜〜ATATTA〜〜–5′
　　−35　　　　　−10
　　region　　　region

13. Transcription and replication are similar in that both use DNA templates; synthesize molecules in the 5′-to-3′ direction; synthesize molecules that are antiparallel and complementary to their templates; use nucleotide triphosphates as substrates; and use complexes of proteins and enzymes necessary for catalysis.

Transcription differs from replication in its unidirectional synthesis of only a single strand of nucleic acid; its initiation does not require a primer; it is subject to numerous regulatory mechanisms; and each gene is transcribed separately.

Replication differs from transcription in its bidirectional synthesis of two strands of nucleic acid and its use of replication origins.

16. 5′–AUAGGCGAUGCCA–3′

20. Similarities: Both use DNA templates; the DNA template is read in the 3′-to-5′ direction; the complementary strand is synthesized in the 5′-to-3′ direction, which is antiparallel to the template; both use triphosphates as substrates; and their actions are enhanced by accessory proteins.

Differences: RNA polymerases use ribonucleoside triphosphates as substrates, whereas DNA polymerases use deoxyribonucleoside triphophates; DNA polymerases require a primer that provides an available 3′-OH group at which synthesis begins, whereas RNA polymerases do not require primers to begin synthesis; and RNA polymerases synthesize a copy of only one of the DNA strands, whereas DNA polymerases synthesize copies of both strands.

23. (a) Would probably affect the −10 consensus sequence, which would most likely decrease transcription; (b) could affect the binding of the sigma factor to the promoter and downregulate transcription, reducing or inhibiting transcription; (c) unlikely to have any effect on transcription; (d) would have little effect on transcription.

25. (a)

Promoter　　Transcription start site　　Terminator

5′ ———————————————— 3′
3′ ———————————————— 5′

Template strand

(b)

RNA polymerase II promoter　　Transcription start site

5′ ————————————————
3′ ————————————————

Template strand

27. (a) No. Because the rho protein has a role in transcription termination, it should not affect transcription initiation or elongation. So you would expect to see transcription.

(b) No. Without the rho protein, transcription would be expected to continue past the normal termination site of rho-dependent terminators, producing some longer molecules than expected.
(c) No. Only RNA molecules produced from genes using rhodependent termination should be longer. Genes that are terminated through rho-independent termination should remain unaffected.
(d) Yes. You would expect to see some RNA molecules that are longer than normal. Only genes that use rho-dependent termination would be expected to not terminate at the normal termination site, thus producing some RNA molecules that are longer than normal.
(e) No. RNA would be copied from only a single strand because the rho protein does not affect transcription initiation or elongation.

32. If TBP cannot bind to the TATA box, then genes with these promoters will be transcribed at very low levels or not at all. Because the TATA box is the most common promoter element for RNA polymerase II transcription units and is found in some RNA polymerase III promoters, transcription will decline significantly. The lack of proteins encoded by these genes will most likely result in cell death.

36. **(a)** From experiment 3 and its corresponding lane in the gel, pTFIIB and the RNA polymerase from *S. pombe* (pPolII) appear to have been sufficient to determine the start site of transcription. The *S. pombe* start site was used even though many of the other transcription factors were from *S. cerevisiae*.
(b) When TFIIE and TFIIH were individually exchanged, transcription was affected. However, the paired exchange of TFIIE and TFIIH allowed for transcription, suggesting that TFIIE and TFIIH undergo a species-specific interaction essential for transcription or that the absence of this interaction will inhibit transcription. Similar results were observed in the paired exchange of TFIIB and RNA polymerase, which suggests that species-specific interaction between TFIIB and RNA polymerase is needed for transcription.
(c) TFIIB likely interacts both with RNA polymerase and with other transcription factors necessary for initiation at the promoter. The data indicate that TFIIB and RNA polymerase II must be from the same species for transcription to take place. TFIIB potentially assembles downstream of the TATA box but in association with other transcription factors located at or near the TATA box. Possibly, these downstream interactions between TFIIB and the other transcription factors stimulate conformational changes in RNA polymerase and the DNA sequence, enabling transcription to begin at the appropriate start site. In experiment 3, both the RNA polymerase and TFIIB were from *S. pombe*, leading to the positioning of the RNA polymerase at the *S. pombe* transcription start site by pTFIIB.

Chapter 14

1. According to the concept of colinearity, the number of amino acids in a protein should be proportional to the number of nucleotides in the gene encoding the protein. In bacteria, the concept is nearly fulfilled. However, in eukaryotes, long regions of noncoding DNA sequence separate coding regions. So, the concept of colinearity at the DNA level is not fulfilled.

3. Group I introns are found in rRNA genes and some bacteriophage genes. Group II introns are found in the protein-encoding genes of mitochondria, chloroplasts, and a few eubacteria. Nuclear pre-mRNA introns are found in the protein-encoding genes of the nucleus. Transfer RNA introns are found in tRNA genes.

4. The 5′ untranslated region, which contains the Shine–Dalgarno sequence; the protein-encoding region; and the 3′ untranslated region.

6. **(a)** The 5′ end of eukaryotic mRNA is modified by the addition of the 5′ cap. The cap consists of an extra guanine nucleotide (linked 5′ to 5′ to the mRNA molecule) that is methylated at position 7 of the base and adjacent bases whose sugars are methylated at the 2′-OH group.
(b) Initially, the terminal phosphate of the three 5′ phosphates linked to the end of the mRNA molecule is removed. Subsequently, a guanine nucleotide binds to the 5′ end of the mRNA by 5′-to-5′ phosphate linkage. Next, a methyl group binds to position 7 of the guanine base. Ribose sugars of adjacent nucleotides also may undergo methylation at the 2′-OH group.
(c) The 5′ cap binds to the ribosome and to the mRNA molecule. The 5′ cap may increase mRNA stability in the cytoplasm and is needed for the efficient splicing of the intron that is nearest the 5′ end of the pre-mRNA molecule.

8. The spliceosome consists of five snRNPs. The splicing of pre-mRNA nuclear introns takes place within the spliceosome.

11. RNA editing alters the sequence of an RNA molecule after transcription by the insertion, deletion, or modification of nucleotides within the transcript. The guide RNAs provide templates for the alteration of nucleotides in RNA molecules undergoing editing and are complementary to regions within the preedited RNA molecule. At these complementary regions, the gRNAs base pair to the preedited RNA molecule. This binding determines the location of the alteration of nucleotides.

12. The different types of processing are: addition of the 5′ cap to the 5′ end of the pre-mRNA; cleavage of the 3′ end of a site downstream of the AAUAAA consensus sequence of the last exon.; addition of the poly(A) tail to the 3′ end of the mRNA immediately after cleavage; and removal of the introns (splicing).

13. Cleavage of the precursor RNA into smaller molecules; trimming or removal of the nucleotides at both the 5′ and the 3′ ends of the tRNAs; and alteration of standard bases by base-modifying enzymes.

15. Most rRNAs are synthesized as large precursor RNAs that undergo methylation, cleavage, and trimming to produce the mature mRNA molecules.

19. The large size of the dystrophin gene is likely due to the presence of many intervening sequences, or introns, within the coding region of the gene. Excision of the introns through RNA splicing yields the mature mRNA that encodes the dystrophin protein.

21. DNA

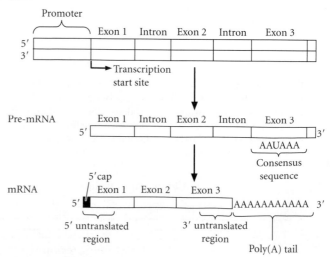

(a) The 5′ untranslated region lies upstream of the translation start site. In bacteria, the ribosome-binding site, or Shine–Dalgarno sequence, is within the 5′ untranslated region. However, eukaryotic mRNA does not have the equivalent sequence, and a eukaryotic ribosome binds at the 5′ cap of the mRNA molecule.
(b) The promoter is the DNA sequence recognized and bound by the transcription apparatus to initiate transcription.
(c) The AAUAAA consensus sequence, which lies downstream of the coding region of the gene, determines the location of the 3′ cleavage site in the pre-mRNA molecule.
(d) The transcription start site begins the coding region of the gene and is located from 25 to 30 nucleotides downstream of the TATA box.
(e) The 3′ untranslated region is a sequence of nucleotides that is located at the 3′ end of the mRNA and is not translated into proteins. However, it does affect the stability and translation of the mRNA.
(f) Introns are noncoding sequences of DNA that intervene within coding regions of a gene.
(g) Exons are coding regions of a gene.
(h) A poly(A) tail is added to the 3′ end of the pre-mRNA; it affects mRNA stability and the binding of the ribosome to the mRNA.
(i) The 5′ cap functions in the initiation of translation and in mRNA stability.

23. (a) It would prevent the cleavage and polyadenylation factor (CPSF) from binding, resulting in no polyadenylation of the pre-mRNA, which would in turn affect the stability and translation of the mRNA.
(b) It would most likely prevent splicing of the intron nearest to the 5′ cap. Ultimately, elimination of the cap will affect the stability of the pre-mRNA and its ability to be translated.
(c) It would cause the mRNA to be degraded quickly by nucleases in the cytoplasm.

26. Because the stability of mRNA depends on the proteins that bind to the poly(A) tail, the mRNAs that contain poly(A) tails will be degraded at a greater-than-normal rate in the cultured cells.

Chapter 15

2. Nirenberg and Matthaei used homopolymer and copolymer methods.

   Advantages: Using a cell-free protein synthesizing system, they were able to determine the amino acid encoded by each homopolymer and thus the amino acids specified by the codons UUU, AAA, CCC, and GGG.

   Disadvantages: The homopolymer method yielded the specificities of only four codons. The copolymer method depended on the random incorporation of the nucleotides, which did not always happen. A further problem was that the base sequence of the codon could not be determined; only the bases contained within the codon could be determined. The redundancy of the code created difficulties because several different codons specified the same amino acid.

   Nirenberg and Leder mixed ribosomes bound to short RNAs of known sequences with charged tRNAs. The mixture was passed through a nitrocellulose filter to which the tRNAs paired to ribosome–RNA complexes adhered. They next determined the amino acids attached to the bound tRNAs.

4. In synonymous codons that differ only at the third nucleotide position, the "wobble" and nonstandard base-pairing with the anticodons will likely result in the correct amino acid being inserted in the protein.

5. (a) The reading frame refers to how the nucleotides in a nucleic acid molecule are grouped into codons, with each codon containing three nucleotides. Any sequence of nucleotides has three potential reading frames that have completely different sets of codons.
(b) In an overlapping code, a single nucleotide is included in more than one codon. The result for a sequence of nucleotides is that more than one type of polypeptide can be encoded within that sequence.
(c) In a nonoverlapping code, a single nucleotide is part of only one codon, which results in the production of a single type of polypeptide from one polynucleotide sequence.
(d) An initiation codon establishes the appropriate reading frame and specifies the first amino acid of the protein chain. Typically, the initiation codon is AUG; however, GUG and UUG also can serve as initiation codons.
(e) The termination codon signals the termination, or end, of translation and the end of the protein molecule. The three types of termination codons—UAA, UAG, and UGA—are also referred to as stop codons or nonsense codons. These codons do not encode amino acids.
(f) A sense codon is a group of three nucleotides that encode an amino acid. Sixty-one sense codons encode the 20 amino acids commonly found in proteins.
(g) A nonsense codon, or termination codon, signals the end of translation. Nonsense codons do not encode amino acids.
(h) In a universal code, each codon specifies the same amino acid in all organisms. The genetic code is nearly universal but note completely. Most of the exceptions are in mitochondrial genes.
(i) Although most codons are universal (or nearly universal) in that they specify the same amino acids in almost all organisms, there are exceptions in which a codon has different meanings in different organisms.

7. For each of the 20 different amino acids commonly found in proteins, a corresponding aminoacyl-tRNA synthetase covalently links the amino acid to the correct tRNA molecule.

10. Three elongation factors have been identified in bacteria: EF-Tu, EF-Ts, and EF-G. EF-Tu joins GTP, which in turn joins a tRNA charged with an amino acid. The charged tRNA is delivered to the ribosome at the A site. In the process of delivery, the GTP joined to EF-Tu is cleaved to form a EF-Tu–GDP complex. EF-Ts regenerates EF-Tu–GTP. The elongation factor EF-G binds GTP and is necessary for the translocation, or movement, of the ribosome along the mRNA during translation.

15. A number of antibiotics bind the ribosome and inhibit protein synthesis at different steps in translation. Some antibiotics, such as streptomycin, bind to the small subunit and inhibit the initiation of translation. Other antibiotics, such as chloramphenicol, bind to the large subunit and block the elongation of the peptide by preventing peptide-bond formation. Antibiotics such as tetracycline and neomycin bind the ribosome near the A site yet have different effects. Tetracyclines block the entry of charged tRNAs to the A site, whereas neomycin induces translational errors. Finally, some antibiotics, such as erythromycin, block the translocation of the ribosome along the mRNA.

16. On the basis of the mutant strain's ability to grow on the given substrates, the mutations can be assembled into four groups:

    Group 1 mutants (*trp-1*, *trp-10*, *trp-11*, *trp-9*, *trp-6*, and *trp-7*) can grow only on minimal medium supplemented with trpytophan.

    Group 2 mutants (*trp-3*) can grow on minimal medium supplemented with either trpytophan or indole.

Group 3 mutants (*trp-2* and *trp-4*) can grow on minimal medium supplemented with tryptophan, indole, or indole glycerol phosphate.

Group 4 mutants (*trp-8*) can grow on minimal medium supplemented with the addition of tryptophan, indole, indole glycerol phosphate, or anthranilic acid.

Group 4 → Group 3 → Group 2 → Group 1

Precursor ⟶ anthranilic acid ⟶ indole glycerol phosphate ⟶ indole ⟶ tryptophan

18. (a) 1; (b) 2; (c) 3; (d) 3; (e) 4.

20. (a) amino–fMet-Phe-Lys-Phe-Lys-Phe–carboxyl
(b) amino–fMet-Tyr-Ile-Tyr-Ile–carboxyl
(c) amino–fMet-Asp-Glu-Arg-Phe-Leu-Ala–carboxyl
(d) amino–fMet-Gly–carboxyl (The stop codon UAG follows the codon for glycine.)

22. There are two possible sequences:

mRNA:	5′–AUGUGGCAU–3′
DNA template:	3′–TACACCGTA–5′
DNA nontemplate:	5′–ATGTGGCAT–3′
mRNA:	5′–AUGUGGCAC–3′
DNA template:	3′–TACACCGTG–5′
DNA nontemplate:	5′–AT GTGGCAC–3′

24. (a) 3′–CCG–5′ or 3′–UCG–5′; (b) 3′–UUC–5′; (c) 3′–AUU–5′ or 3′–UUU–5′ or 3′–CUU–5′; (d) 3′–ACC–5′ or 3′–GCC–5′; (e) 3′–GUC–5′.

30. (a) The lack of IF-1 would decrease the amount of protein synthesized. IF-1 promotes the dissociation of the large and small ribosomal subunits. The initiation of translation requires a free small subunit. The absence of IF-1 would reduce the rate of initiation because more of the small ribosomal subunits would remain bound to the large ribosomal subunits.
(b) No translation would take place. IF-2 is necessary for the initiation of translation. The lack of IF-2 would prevent fMet-tRNA$_f^{met}$ from being delivered to the small ribosomal subunit, thus blocking translation.
(c) Although translation would be initiated by the delivery of methionine to the ribosome–mRNA complex, no other amino acids would be delivered to the ribosome. EF-Tu, which binds to GTP and the charged tRNA, is necessary for elongation. This three-part complex enters the A site of the ribosome. If EF-Tu is not present, the charged tRNA will not enter the A site, thus stopping translation.
(d) EF-G is necessary for the translocation (movement) of the ribosome along the mRNA in a 5′-to-3′ direction. When a peptide bond has formed between Met and Pro, the lack of EF-G would prevent the movement of the ribosome along the mRNA, and so no new codons would be read. The formation of the dipeptide Met-Pro does not not require EF-G.
(e) The release factors RF$_1$ and RF$_2$ recognize the stop codons and bind to the ribosome at the A site. They then interact with RF$_3$ to promote cleavage of the peptide from the tRNA at the P site. The absence of the release factors would prevent the termination of translation at the stop codon.
(f) ATP is required for tRNAs to be charged with amino acids by aminoacyl-tRNA synthetases. Without ATP, the charging would not take place, and no amino acids will be available for protein synthesis.
(g) GTP is required for the initiation, elongation, and termination of translation. If GTP is absent, protein synthesis will will not take place.

36. (a) The results suggest that, to initiate translation, the ribosome is scanned to find the appropriate start sequence.
(b) The initiation of translation in bacteria requires the 16S RNA of the small ribosomal subunit to interact with the Shine–Dalgarno sequence. This interaction serves to line up the ribosome over the start codon. If the anticodon has been changed such that the start codon cannot be recognized, then protein synthesis is not likely to take place.

Chapter 16

2. DNA and chromatin structure, transcription, mRNA processing, mRNA stability, translation, and protein modification.

3.

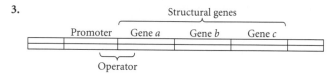

5. The *lac* operon consists of three structural genes—*lacZ*, *lacY*, and *lacA*. The *lacZ* gene encodes the enzyme β-galactosidase, which cleaves the disaccharide lactose into galactose and glucose and converts lactose into allolactose. The *lacY* gene encodes lactose permease, an enzyme necessary for the passage of lactose through the *E. coli* cell membrane. The *lacA* gene encodes the enzyme thiogalactoside transacetylase, whose function in lactose metabolism has not yet been determined. All three genes have an overlapping promoter and operator region in common. Upstream of the lactose operon, the *lacI* gene encodes the *lac* operon repressor, which binds at the operator region and inhibits the transcription of the *lac* operon by preventing RNA polymerase from successfully initiating transcription.

When lactose is present in the cell, the enzyme β-galactosidase converts some of it into allolactose, which binds to the *lac* repressor, altering its shape and reducing the repressor's affinity for the operator. Because the allolactose-bound repressor does not bind to the operator, RNA polymerase can initiate transcription of the *lac* structural genes from the *lac* promoter.

7. Attenuation is the termination of transcription before the structural genes of an operon are transcribed. Attenuation results from the formation of a termination hairpin, or attenuator, in the RNA.

Two types of secondary structures can be formed by the mRNA 5′ UTR of the *trp* operon. One of them allows transcription to proceed; the other one terminates transcription.

When tryptophan levels are high, region 3 pairs with region 4 to form the attenuator hairpin structure, stopping transcription. When tryptophan levels are low, region 2 pairs with region 3 to form the antiterminator hairpin, allowing transcription to proceed through the structural genes.

8. Antisense RNA molecules are small RNA molecules that are complementary to other DNA or RNA sequences. In bacterial cells, antisense RNA can bind to a complementary region in the 5′ UTR of an mRNA molecule, blocking the attachment of the ribosome to the mRNA and thus stopping translation.

11. (a) Inactive repressor; (b) active activator; (c) active repressor; (d) inactive activator.

12. (a) The operon will never be turned off, and transcription will take place all the time.
(b) The result will be constitutive expression, and transcription will take place all the time.

14. RNA polymerase will bind the *lac* promoter poorly, significantly decreasing the transcription of the *lac* structural genes.

18.

Genotype of strain	Lactose absent		Lactose present	
	β-Galactosidase	Permease	β-Galactosidase	Permease
$lacI^+ \ lacP^+ \ lacO^+ \ lacZ^+ \ lacY^+$	−	−	+	+
$lacI^- \ lacP^+ \ lacO^+ \ lacZ^+ \ lacY^+$	+	+	+	+
$lacI^+ \ lacP^+ \ lacO^c \ lacZ^+ \ lacY^+$	+	+	+	+
$lacI^- \ lacP^+ \ lacO^+ \ lacZ^+ \ lacY^-$	+	−	+	−
$lacI^- \ lacP^- \ lacO^+ \ lacZ^+ \ lacY^+$	−	−	−	−
$lacI^+ \ lacP^+ \ lacO^+ \ lacZ^- \ lacY^+/$ $lacI^- \ lacP^+ \ lacO^+ \ lacZ^+ \ lacY^-$	−	−	+	+
$lacI^- \ lacP^+ \ lacO^c \ lacZ^+ \ lacY^+/$ $lacI^+ \ lacP^+ \ lacO^+ \ lacZ^- \ lacY^-$	+	+	+	+
$lacI^- \ lacP^+ \ lacO^+ \ lacZ^+ \ lacY^-/$ $lacI^+ \ lacP^- \ lacO^+ \ lacZ^- \ lacY^+$	−	−	+	−
$lacI^+ \ lacP^- \ lacO^c \ lacZ^- \ lacY^+/$ $lacI^- \ lacP^+ \ lacO^+ \ lacZ^+ \ lacY^-$	−	−	+	−
$lacI^+ \ lacP^+ \ lacO^+ \ lacZ^+ \ lacY^+/$ $lacI^+ \ lacP^+ \ lacO^+ \ lacZ^+ \ lacY^+$	−	−	+	+
$lacI^s \ lacP^+ \ lacO^+ \ lacZ^+ \ lacY^-/$ $lacI^+ \ lacP^+ \ lacO^+ \ lacZ^- \ lacY^+$	−	−	−	−
$lacI^s \ lacP^- \ lacO^+ \ lacZ^- \ lacY^+/$ $lacI^+ \ lacP^+ \ lucO^+ \ lacZ^+ \ lacY^+$	−	−	−	−

20. The *lacI* gene encodes the *lac* repressor protein, which can diffuse within the cell and attach to any operator. It can therefore affect the expression of genes on the same molecule or on different molecules of DNA. The *lacO* gene encodes the operator. It affects the binding of RNA polymerase to DNA and therefore affects the expression of genes only on the same molecule of DNA.

21. (**a**) Repressible; (**b**) B is the regulator gene, D is the promoter, A is the structural gene for enzyme 1, and C is the structural gene for enzyme 2.

24. (**a**) No gene expression; (**b**) transcription of the structural genes only when alanine levels are low; (**c**) no transcription; (**d**) no transcription; (**e**) transcription will proceed; (**f**) transcription will proceed; (**g**) transcription will proceed.

Chapter 17

1. Bacterial genes are often clustered in operons and are coordinately expressed through the synthesis of a single polygenic mRNA. Eukaryotic genes are typically separate, with each containing its own promoter and transcribed on individual mRNAs.

In eukaryotic cells, chromatin must assume a more open configuration, making DNA accessible to transcription factors, activators, and RNA polymerase. Activators and repressors function in both eukaryotic and bacterial cells. However, activators are more common in eukaryotic cells than in bacterial cells.

In bacteria, transcription and translation can take place concurrently. In eukaryotes, the nuclear membrane separates transcription from translation both physically and temporally. This separation results in a greater diversity of regulatory mechanisms at different points in gene expression.

3. Changes in chromatin structure can result in the repression or stimulation of gene expression. The acetylation of histone proteins increases transcription. The reverse reaction, deacetylation, restores repression. Chromatin-remodeling complexes bind directly to the DNA, altering chromatin structure without acetylating histone proteins and allowing transcription to be initiated by making the promoters accessible to transcription factors. The methylation of DNA sequences represses transcription. The demethylation of DNA sequences often increases transcription.

6. Changes in chromatin structure, brought about by DNA methylation, the alteration of histone proteins, and the repositioning of nucleosomes.

9. An enhancer is a DNA sequence that, when bound to transcriptional activator proteins, can affect the transcription of a distant gene. Transcription at a distant gene is affected when the DNA sequence between the gene's promoter and the enhancer loops out, bringing the promotor and the enhancer close together and allowing the transcriptional activator proteins to directly interact with the basal transcription apparatus at the promoter, which stimulates transcription.

13. The total amount of protein synthesized depends on the amount of mRNA available for translation. The amount of available mRNA depends on the rates of mRNA synthesis and degradation. Less-stable mRNAs degrade faster than stable mRNAs, and so fewer copies of the mRNA are available as templates for translation. The 5′ cap, the 3′ poly(A) tail, the 5′ UTR, the 3′ UTR, and the coding region in an mRNA molecule affect its stability. Poly(A)-binding proteins bind at the 3′ poly(A) tail. These proteins contribute to the stability of the tail and protect the 5′ cap through direct interaction. When a critical number of adenine nucleotides have been removed from the tail, the protection is lost and the 5′ cap is removed. The removal of the 5′ cap enables 5′-to-3′ nucleases to degrade the mRNA.

14. Through Slicer activity, which cleaves mRNA sequences; through the binding of miRNAs to complementary regions in mRNA, which prevents translation; through transcriptional silencing, in which siRNAs play a role in altering chromatin structure; and through Slicer-independent mRNA degradation stimulated by miRNA, which binds to complementary regions in the 3′ UTR of the mRNA.

15. In eukaryotic cells, the coding regions of pre-mRNA are interrupted by introns—noncoding regions that may be much longer than the coding regions. The splicing of pre-mRNA to remove these noncoding regions is a point at which gene expression can be regulated. In most prokaryotic cells, coding regions are not interrupted by introns. In eukaryotic cells, chromatin structure plays a role in gene regulation. The process of transcription initiation is more complex in eukaryotic cells than in bacterial cells. In eukaryotes, initiation requires RNA polymerase, general transcription factors, and transcriptional activators. Bacterial RNA polymerase is either blocked or stimulated by the actions of regulatory proteins. Finally, in eukaryotes, activator proteins may bind to enhancers at a great distance from the promoter and structural gene. Distant enhancers are less common in bacterial cells.

    The regulation of both bacterial genes and eukaryotic genes requires the action of protein repressors and protein activators. Cascades of gene regulation in which the activation of one set of genes affects another set of genes are found in both eukaryotes and bacteria. The regulation of gene expression at the transcriptional level is common to both types of cells.

25. The fruit flies will develop (**a**) male characteristics, (**b**) male characteristics, (**c**) both male and female characteristics.

Chapter 18

1. Germ-line mutations are changes in the DNA of reproductive (germ) cells and can be passed to offspring. Somatic mutations are changes in the DNA of an organism's somatic-tissue cells and cannot be passed to offspring.

2. Transition mutations result from purine-to-purine or pyrimidine-to-pyrimidine base substitutions. Transversions result from purine-to-pyrimidine or pyrimidine-to-purine base substitutions. Transition mutations are more common because spontaneous mutations typically result in transition mutations rather than transversions.

3. Expanding nucleotide repeats result when a DNA insertion mutation increases the number of copies of a nucleotide repeat sequence. The increase may be due to errors in replication or to unequal recombination. Within a given family, a particular type of nucleotide repeat may increase in number from generation to subsequent generation, increasing the severity of the mutation in a process called anticipation.

6. Intergenic suppressor mutations restore the wild-type phenotype. However, they do not revert the original mutation. The suppression is the result of a mutation in a gene other than the gene containing the original mutation and reverses the effect of the original mutation. For example, because many proteins interact with other proteins, the original mutation may have altered one protein in a way that disrupts a protein–protein interaction, and the second mutation alters another protein to restore that interaction. Another type of intergenic suppression is due to a mutation within the anticodon region of a tRNA molecule that allows the tRNA to pair with the codon containing the original mutation and results in the substitution of a functional amino acid in the protein.

7. Two types of events have been proposed that could lead to DNA replication errors: (i) mispairing due to tautomeric shifts in nucleotides and (ii) mispairing through wobble or flexibility of the DNA molecule. Mispairing through wobble caused by flexibility in the DNA helix is now thought to be the most likely cause.

9. Base analogs have structures similar to those of the nucleotides and may be incorporated into the DNA in the course of replication. Many analogs tend to mispair, which can lead to mutations. DNA replication is required for the base-analog-induced mutations to be incorporated into the DNA.

12. The SOS system is an error-prone DNA-repair system consisting of at least 25 genes. Induction of the SOS system causes damaged DNA regions to be bypassed, which allows for DNA replication across the damaged regions. However, the insertion of bases into the new DNA strand in the absence of bases on the template strand results in a less-accurate replication process; thus, more mutations will occur.

14. Mismatch repair: Repairs replication errors that result from base-pair mismatches. Mismatch-repair enzymes recognize distortions in the DNA structure due to mispairing and detect the newly synthesized strand by its lack of methylation. The distorted segment is excised, and DNA polymerase and DNA ligase fill in the gap.

    Direct repair: Repairs DNA damage by directly changing the damaged nucleotide back into its original structure.

    Base-excision repair: Excises the damaged base and then replaces the entire nucleotide.

    Nucleotide-excision repair: Relies on repair enzymes to recognize distortions of the DNA double helix. These enzymes excise a damaged region by cleaving phosphodiester bonds on either side of the damaged region. The gap created by the excision is filled in by DNA polymerase.

17. Transversion at the first position: GGA→UGA

18. (**a**) Leucine, serine, or phenylalanine; (**b**) isoleucine, tyrosine, leucine, valine, or cysteine; (**c**) phenylalanine, proline, serine, or leucine; (**d**) methionine, phenylalanine, valine, arginine, tryptophan, leucine, isoleucine, tyrosine, histidine, or glutamine, or a stop codon could result as well.

20. (**a**) amino–Met-Thr-Gly-**Ser**-Gln-Leu-Tyr–carboxyl
    (**b**) amino–Met-Thr-Gly-Asn–carboxyl
    (**c**) amino–Met-Thr-**Ala-Ile-Asn-Tyr-Ile**–carboxyl
    (**d**) amino–Met-Thr-Gly-Asn-**His**-Leu-Tyr–carboxyl
    (**e**) amino–Met-Thr-**Thr**-Gly-Asn-Gln-Leu-Tyr–carboxyl
    (**f**) amino–Met-Thr-**Gly**-Asn-Gln-Leu-Tyr–carboxyl

22. Four of the six Arg codons could be mutated by a single-base substitution to produce a Ser codon. However, only two of the Arg codons mutated to form Ser codons could subsequently be mutated at a second position by a single-base substitution to regenerate the Arg codon. In both events, the mutations are transversions.

Original Arg codon	Ser codon	Restored Arg codon
CGU	AGU	AGG or AGA
CGC	AGC	AGG or AGA

25. No, hydroxylamine cannot reverse nonsense mutations. Hydroxylamine modifies cytosine-containing nucleotides and can result only in GC-to-AT transition mutations. In a stop codon, the GC-to-AT transition will result only in a different stop codon.

26. | Original sequence | Mutated sequence |
    | --- | --- |
    | 5′–ATGT–3′ | 5′–AT**A**T–3′ |
    | 3′–TACA–5′ | 3′–TA**T**A–5′ |

29. PFI1 causes transitions, PFI2 causes transversions or large deletions, PFI3 causes transitions, and PFI4 causes single-base insertions or deletions.

32. By looking for plants that have increased levels of mutations either in their germ-line or somatic tissues. Potentially mutant plants may have been exposed to standard mutagens that damage DNA. If they are defective in DNA repair, they should have higher rates of mutation. For example, tomato plants with defective DNA-repair systems should have an increased mutation rate when exposed to high levels of ultraviolet light. Therefore, they need to be grown in an environment that has lower levels of sunlight.

Chapter 19

4. Gel electrophoresis uses an electric field to drive negatively charged DNA molecules through a gel that acts as a molecular sieve. Shorter DNA molecules are less hindered by the agarose or polyacrylamide matrix and migrate faster than longer DNA molecules.

5. By staining with a fluorescent dye such as ethidium bromide, which intercalates between the stacked bases of the DNA double helix.

7. Cloning vectors should have (i) an origin of DNA replication so they can be maintained in a cell; (ii) a gene, such as an antibiotic-resistance gene, to select for cells that carry the vector; and (iii) a unique restriction site or series of sites at which a foreign DNA molecule can be inserted.

9. Foreign DNAs are inserted into one of the unique restriction sites in the *lacZ* gene of plasmids and transform the plasmids into *E. coli* cells. Transformed cells are plated on a medium containing the appropriate antibiotic to select for cells that carry the plasmid, an inducer of the *lac* operon, and X-gal, a substrate for β-galactosidase that turns blue when cleaved. Colonies that carry the plasmid without foreign DNA inserts will have intact *lacZ* genes, make functional β-galactosidase, cleave X-gal, and turn blue. Colonies that carry the plasmid with foreign DNA inserts will not make functional β-galactosidase and will remain white.

10. First, the double-stranded template DNA is denatured by high temperature. Then, synthetic oligonucleotide primers corresponding to the ends of the DNA sequence to be amplified are annealed to the single-stranded DNA template strands. These primers are extended by a thermostable DNA polymerase so that the target DNA sequence is duplicated. These steps are repeated 30 times or more. Each cycle of denaturation, primer annealing, and extension doubles the number of copies of the target sequence between the primers.

A limitation of PCR amplification is that, to synthesize the PCR primers, the sequence of the gene to be amplified must be known, at least at the ends of the region to be amplified. Another is that the extreme sensitivity of the technique renders it susceptible to contamination. A third limitation is that the most-common thermostable DNA polymerase used for PCR, Taq DNA polymerase, has a high error rate. A fourth limitation is that PCR amplification is usually limited to DNA fragments of a few thousand base pairs or less; optimal DNA polymerase mixtures and reaction conditions extend the amplifiable length to about 20 kb.

12. A genomic library is created by inserting fragments of chromosomal DNA into a cloning vector. Chromosomal DNA is randomly fragmented by shearing or by partial digestion with a restriction enzyme. A cDNA library is made from mRNA sequences. Cellular mRNAs are isolated and then reverse transcriptase is used to convert the mRNA sequences into cDNA, which is cloned into plasmid or phage vectors.

14. In in situ hybridization, radiolabeled or fluorescently labeled DNA or RNA probes are hybridized to DNA or RNA molecules that are still in the cell. This technique can be used to visualize not only the expression of specific mRNAs in different cells and tissues, but also the location of genes on metaphase or polytene chromosomes.

18. DNA fingerprinting detects genetic differences among people by using probes for highly variable regions (usually, microsatellites, or short tandem repeats) of chromosomes. Typically, PCR is used to detect the microsatellites. Because people differ in the number of repeated sequences that they possess, the technique is used in the analysis of crimes, in paternity cases, and in identifying remains.

21. A knockout mouse has a target gene disrupted or deleted ("knocked out"). First, the target gene is cloned. The middle part of the gene is replaced by a selectable marker—typically, by *neo*, a gene that confers resistance to G418. This construct is then introduced back into mouse embryonic stem cells, and cells with G418 resistance are selected. The surviving cells are screened for cells in which the *neo*-containing construct has replaced the chromosomal copy of the target gene through homologous recombination of the flanking sequences. These embryonic stem cells are then injected into mouse blastocyst-stage embryos, and the chimeric embryos are transferred to the uterus of a pseudopregnant female mouse. The knockout cells will participate in the formation of many tissues in the mouse fetus, including germ-line cells. The chimeric offspring are interbred to produce offspring that are homozygous for the knockout allele. The offspring phenotype provides information about the function of the gene.

25. *Ara*I

27. 10

29. (a) 460,800; (b) 1,036,800; (c) 5,120,000.

30.

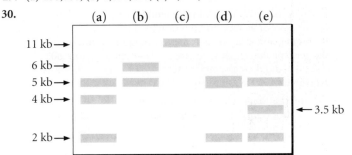

31. (a) Plasmid; (b) phage λ; (c) cosmid; (d) bacterial artificial chromosome.

33. One strategy would be to use the mouse gene for prolactin as a probe to find the homologous pig gene from a pig genomic or cDNA library. A second strategy would be to use the amino acid sequence of mouse prolactin to design degenerate oligonucleotides as hybridization probes to screen a pig DNA library. Yet a third strategy would be to use the amino acid sequence of mouse prolactin to design a pair of degenerate oligonucleotide PCR primers for PCR amplification of the pig gene for prolactin.

35. (a) Val-Tyr-Lys-Ala-Lys-Trp; (b) 128.

37. 5′–NGCATCAGTA–3′

39. (a) Sally's mother must have had *A2A3* and *C2C2*. The linkage relations of these chromosomes are *A1C1* and *A1C3* from the father, and *A2C2* and *A3C2* from the mother. The mother passed an *A2C2* to Sally's brother who has G syndrome; therefore, the G-syndrome allele must be linked with *A2C2*. Because Sally inherited the *A2C2* chromosome from her mother, she also must have inherited the G-syndrome allele, assuming that no crossover took place between the *A*, *C*, and *G* loci.

(b) Father: <u>*A1* *C1* *g*</u> , <u>*A1* *C3* *g*</u>

Mother: <u>*A2* *C2* *G*</u> , <u>*A3* *C2* *g*</u>

Sally's unaffected brother: <u>*A1* *C3* *g*</u>, <u>*A3* *C2* *g*</u>

Sally's affected brother: <u>*A1* *C3* *g*</u> , <u>*A2* *C2* *G*</u>

Sally: <u>*A1* *C1* *g*</u> , <u>*A2* *C2* *G*</u>

40. This gene must first be cloned, possibly by using the yeast gene as a probe to screen a mouse genomic DNA library. The cloned gene is then engineered to replace a substantial part of the protein-coding sequence with the *neo* gene. This construct is then introduced into mouse embryonic stem cells, which are transferred to the uterus of a pseudopregnant mouse. The progeny are tested for the presence of the knockout allele, and those having the knockout allele are interbred.

If the gene is essential for embryonic development, no homozygous knockout mice will be born. The arrested or spontaneously aborted fetuses can then be examined to determine how development has gone awry in fetuses that are homozygous for the knockout allele.

Chapter 20

1. A genetic map locates genes or markers on the basis of genetic recombination frequencies. A physical map locates genes or markers on the basis of the physical lengths of DNA sequences. Because recombination frequencies vary from one region of a chromosome to another, genetic maps are approximate. Genetic maps also have lower resolution because recombination is difficult to observe between loci that are very close to each other. Physical maps based on DNA sequences or restriction maps have much greater accuracy and resolution, down to a single base pair of DNA sequence.

2. The map-based approach first assembles large clones into contigs on the basis of genetic and physical maps and then selects clones for sequencing. The whole-genome shotgun approach breaks the genome into short sequence reads—typically, from 600 to 700 bp—and then assembles them into contigs on the basis of sequence overlap with the use of powerful computers.

4. The international collaboration took the ordered, map-based approach, beginning with the construction of detailed genetic and physical maps. Celera took the whole-genome shotgun approach. Celera did make use of the physical map produced by the international collaboration to order sequences in the assembly phase.

14. Homologous sequences are derived from a common ancestor. Orthologs are sequences in different species that are descended from a sequence in a common ancestral species. Paralogs are sequences in the same species that originated by duplication of an ancestral sequence and subsequently diverged.

18. After random mutagenesis with chemicals or transposons, the mutant progeny population is screened for phenotypes of interest. The mutant gene can be identified by cosegregation with molecular markers or by sequencing the position of transposon insertion. To verify that the mutation identified is truly responsible for the phenotype, a mutation can be introduced into a wild-type copy of the gene and the offspring can be searched for the phenotype.

21. Horizontal gene transfer is the exchange of genetic material between closely or distantly related species over evolutionary time. In bacteria, horizontal gene transfer may take place through uptake of environmental DNA (transformation), through conjugative plasmids with a broad host range, or through transfection with bacteriophages with a broad host range.

23. The *Arabidopsis* genome appears to have undergone at least one round of duplication of the whole genome (tetraploidy) and numerous localized duplications through unequal crossing over.

32.
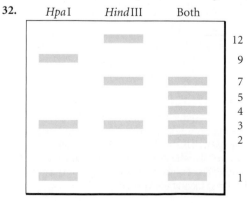

HpaI	HindIII	Both	
			12
			9
			7
			5
			4
			3
			2
			1

33.
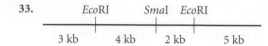

46. **(a)** The minimal genome required might be determined by examining simple free-living organisms having small genomes to determine which genes they have in common. Mutations can then be systematically made to determine which of the common genes are essential for these organisms to survive. The apparently nonessential genes (those genes in which mutations do not affect the viability of the organism) can then be deleted one by one until only the essential genes are left. Elimination of any of the essential genes will result in loss of viability. Alternatively, essential genes could be assembled through genetic engineering, creating an entirely novel organism.
(b) The novel organism would prove that humans have acquired the ability to create a new species or form of life. Humans would then be able to direct evolution as never before. Among the social and ethical concerns would be the question of whether human society has the wisdom to temper its power and whether such novel synthetic organisms can or will be used to develop pathogens for biological warfare or terrorism. After all, no person or animal would have been previously exposed or have acquired immunity to such a novel synthetic organism. There would also be uncertainty about the new organism's effect on ecosystems if it were released or escaped.

Chapter 21

2. The endosymbiotic theory proposes that mitochondria and chloroplasts evolved from formerly free-living bacteria that became endosymbionts within a larger eukaryotic cell. Both chloroplasts and mitochondria contain genomes that encode proteins, tRNAs, and rRNAs. Chloroplasts, mitochondria, and eubacteria are similar in genome size, circular chromosome structure, and other aspects of genome structure. Moreover, the chloroplast and mitochondrial ribosomes are similar in size and function to eubacterial ribosomes. DNA sequences in mitochondrial and chloroplast genomes are similar to those in eubacteria.

5. A nonuniversal codon specifies an amino acid in one organism but not in most other organisms. Typically, nonuniversal codons are found in mitochondrial genomes, and these exceptions vary in the mitochondria of different organisms.

7. In mitochondrial translation, "wobble" pairing at the third position of a codon is more frequent than in the cytoplasmic translation of nuclear genes. Most anticodons of the mitochondrial tRNAs can pair with more than one codon. Essentially, the first position of the anticodon can pair with any of the four nucleotides present at the third position of the codon.

9. The chloroplast genome—typically, a double-stranded circular DNA molecule whose organization resembles that of eubacterial genomes—contains sequences similar to those of eubacteria. There are long noncoding nucleotide sequences between genes and many duplicated sequences. Genes are located on both strands of cpDNA and may contain introns. The chloroplast genome usually encodes ribosomal proteins, 5 rRNAs, from 30 to 35 tRNA genes, and proteins needed for photosynthesis, as well as proteins not needed for photosynthesis. A large inverted repeat is found in the genomes of most chloroplasts.

10. Most mitochondrial and chloroplast chromosomes are small, are circular, and lack histone proteins—characteristics that are similar to those of eubacterial, but not eukaryotic, cells. Chloroplasts and some mitochondria produce polycistronic mRNA, another characteristic common to eubacteria. Chloroplast genes, but not most

mitochondrial genes, typically possess Shine–Dalgarno sequences. Antibiotics that inhibit eubacterial translation inhibit mitochondrial and chloroplast translation.

Eukaryotic nuclear genomes are typically composed of linear chromosomes in which DNA is bound to histone proteins. Eukaryotic nuclear DNA sequences contain pre-mRNA introns, which are found in chloroplast genomes but are usually not found in mitochondrial genomes. Eukaryotic genomes contain large numbers of repeated sequences and pseudogenes, which are usually not found in mitochondrial genomes.

14. The disorder is cytoplasmically inherited. Only females pass the trait to their offspring. It does not appear to be sex-specific in that both males and females can have the disorder. These characteristics are consistent with cytoplasmic inheritance.

17.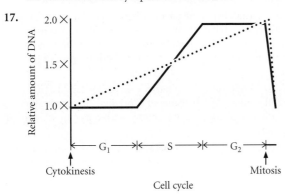

8. (a) Somatic recombination: produces many combinations of variable domains with junction segments and diversity segments.
(b) Junctional diversity: in recombination, the process that joins the V, J, and D gene segments is imprecise, resulting in small deletions or insertions and frameshifts.
(c) Hypermutation: somatic hypermutation (accelerated random mutation) of the V gene segments further diversifies antibodies.

13. (a) The products of *bicoid* and *dorsal* affect embryonic polarity by regulating the transcription of target genes.
(b) The product of *nanos* regulates the translation of *hunchback* mRNA.

16. The presence of *bicoid* mRNA in the posterior end of the embryo would cause transcription of *hunchback* in the posterior regions. Without Nanos protein, the *hunchback* mRNA would be translated to create high levels of Hunchback protein in the posterior end as well as in the anterior end. The result would be an embryo with anterior structures at both ends.

18. (a) Females with an increased number of copies of the *bicoid* gene would have higher levels of maternal *bicoid* mRNA in the anterior cytoplasm of their eggs. After fertilization, embryos would have higher levels of Bicoid protein and thus enlarged anterior and thoracic structures.
(b) A decreased number of copies of the *bicoid* gene would ultimately result in lower levels of Bicoid protein and an embryo with small head structures.

25. 2600

Chapter 22

1. Cloning experiments with plants and animals from differentiated cells showed that the nuclei of these differentiated cells still retained all of the genetic information required for the development of a whole organism.

3. Maternally transcribed *bicoid* and *nanos* mRNAs are localized to the anterior and posterior ends of the egg, respectively. After fertilization, these mRNAs are translated, and the proteins diffuse to form opposing gradients: Bicoid protein concentrations are highest at the anterior end, whereas Nanos protein concentrations are highest at the posterior end. Bicoid protein at the anterior end acts as a transcription factor to activate the transcription of *hunchback*, a gene required for the formation of head and thoracic structures. Nanos protein at the posterior end inhibits the translation of *hunchback* mRNA, thereby preventing the formation of anterior structures in the posterior regions.

4. Gap genes specify broad regions (multiple adjacent segments) along the anterior–posterior axis of the embryo. Interactions among the gap genes regulate the transcription of the pair-rule genes.

Pair-rule genes compartmentalize the embryo into segments and regulate the expression of the segment-polarity genes. Each pair-rule gene is expressed in alternating segments.

Segment-polarity genes specify the anterior and posterior compartments within each segment.

7. Apoptosis is programmed cell death, characterized by nuclear DNA fragmentation, shrinkage of the cytoplasm and nucleus, and phagocytosis of the remnants of the dead cell. Apoptosis is regulated by internal and external signals that regulate the activation of procaspases—cysteine proteases that are activated by proteolytic cleavage. When activated, these caspases activate other caspases in a cascade and degrade key cellular proteins.

Chapter 23

1. Many types of cancer are associated with exposure to radiation and other environmental mutagens. Some types of cancer tend to run in families, and a few cancers are linked to chromosomal abnormalities. Finally, the discovery of oncogenes and specific mutations that cause proto-oncogenes to become oncogenes or that inactivate tumor-suppressor genes proved that cancer has a genetic basis.

3. The multistage theory of cancer states that more than one mutation is required for most cancers to develop. Most retinoblastomas are unilateral because the likelihood of any cell acquiring two rare mutations is very low, and thus retinoblastomas develop in only one eye. Bilateral cases of retinoblastoma develop in people born with a predisposing mutation, and so only one additional mutational event will result in cancer. Thus, the probability of retinoblastoma development is higher and likely to be in both eyes. Because the predisposing mutation is inherited, people with bilateral retinoblastoma have relatives with retinoblastoma.

5. An oncogene stimulates cell division, whereas a tumor-suppressor gene puts the brakes on cell growth. Proto-oncogenes are normal cellular genes that function in cell growth and in the regulation of the cell cycle. Some examples are *erbA*, *erbB*, *myc*, *src*, and *ras*. Tumor suppressors inhibit cell-cycle progression; examples are transcription factors RB and P53 and the GTPase activator NF1.

7. The CDKs, or cyclin-dependent kinases, have enzymatic activity and phosphorylate multiple substrate molecules when activated by binding to the appropriate cyclin. Cyclins are regulators of CDKs and have no enzymatic activity of their own. Each cyclin molecule binds to a single CDK molecule. Whereas CDK levels remain relatively stable, cyclin levels oscillate in the course of the cell cycle.

10. A signal-transduction pathway enables a cell to respond to an external signal. It begins with the binding of the external signal molecule and then proceeds through a cascade of intracellular events that relay and amplify the signaling event to bring about changes in transcription, metabolism, morphology, or other aspects of cell function. Because cell growth and division is regulated by

external signals, mutations in signal-transduction components may cause the cell to grow and divide in the absence of external growth stimuli or cause the cell to stop responding to external signals that inhibit growth.

13. DNA polymerases are unable to replicate the ends of linear DNA molecules. Therefore, the ends of eukaryotic chromosomes shorten with every round of DNA replication unless telomerase uses its RNA component to add back telomeric DNA sequences, which it does in reproductive cells. Normally, somatic cells do not express telomerase; their telomeres progressively shorten with each cell division until vital genes are lost and the cells undergo apoptosis. Transformed cells (cancerous cells) induce the expression of the telomerase gene, thus leading to cell proliferation.

14. Deletions may cause the loss of one or more tumor-suppressor genes. Inversions and translocations may inactivate tumor-suppressor genes if the chromosomal breakpoints are within tumor-suppressor genes. Alternatively, a translocation may place a proto-oncogene in a new location, where it is activated by different regulatory sequences, causing overexpression or unregulated expression of the proto-oncogene. Finally, inversions and translocations may also bring parts of two different genes together, causing the synthesis of a novel protein that is oncogenic.

17. Retroviruses have strong promoters. After its integration into a host genome, a retrovirus promoter may drive the overexpression of a cellular proto-oncogene. Alternatively, the integration of a retrovirus may inactivate a tumor-suppressor gene. A few retroviruses carry oncogenes that are altered versions of host proto-oncogenes. Other viruses, such as human papilloma virus, express gene products (proteins or RNA molecules) that interact with the host cell-cycle machinery and inactivate tumor-suppressor proteins.

23. (a) 50%.
(b) Bilateral.
(c) The father may have unilateral retinoblastoma because of incomplete penetrance of the mutation in the *RB* gene. Alleles at another locus or multiple other loci may have contributed to resistance to retinoblastoma in the father, and so he suffered from retinoblastoma in only one eye. Alternatively, good fortune (random chance) may have spared the other eye from the second mutation event that would have led to retinoblastoma in that eye also.

24. Because oncogenes promote cell proliferation, they act in a dominant manner. In contrast, mutations in tumor-suppressor genes cause loss of function and act in a recessive manner. When introduced into cells, the mutated *palladin* gene increases cell migration. Such a dominant effect suggests that *palladin* is an oncogene.

27. Gene amplification in cancer cells would most likely affect oncogenes. Oncogenes stimulate cell division, and so extra copies of oncogenes would add to the stimulation of cell proliferation. Tumor-suppressor genes, however, tend to inhibit cell proliferation. Extra copies of tumor-suppressor genes would add to the inhibition of cell proliferation and therefore not likely lead to cancer.

Chapter 24

1. Discontinuous characteristics have only a few distinct phenotypes. In contrast, a quantitative characteristic shows a continuous variation in phenotype.

3. Many genotypes are possible with multiple genes. Even for the simplest two-allele loci, the number of possible genotypes is equal to 3^n, where n is the number of loci, or genes. Thus, for three genes, there are 27 genotypes; four genes yield 81 genotypes; and so forth. If each genotype corresponds to a unique phenotype, the the number of phenotypes is the same: 27 for three genes and 81 for four genes. Moreover, the phenotype for a given genotype may be influenced by environmental factors, leading to an even greater array of phenotypes.

4. A sample is a subset of a population. To be representative of the population, the sample should be randomly selected and sufficiently large to minimize random differences between members of the sample and the population.

7. The magnitude or absolute value of the correlation coefficient reports the strength of the association between the two variables. A value close to $+1$ or -1 indicates a strong association; values close to zero indicate weak association.

9. V_G, component of variance due to variation in genotype; V_A, component of variance due to additive genetic variance; V_D, component of variance due to dominance genetic variance; V_I, component of variance due to genic interaction variance; V_E, component of variance due to environmental differences; V_{GE}, component of variance due to interaction between genes and environment.

10. Broad-sense heritability is the part of phenotypic variance that is due to all types of genetic variance, including additive, dominance, and genic-interaction variances. Narrow-sense heritability is just that part of the phenotypic variance due to additive genetic variance.

14. The response to selection (R) = narrow-sense heritability (h^2) × selection differential (S). The value of R predicts how much the mean quantitative phenotype will change with different selection in a single generation.

16. (a) Discontinuous characteristic because only a few distinct phenotypes are present and it is determined by alleles at a single locus.
(b) Discontinuous characteristic because there are only two phenotypes (dwarf and normal) and a single locus determines the characteristic.
(c) Quantitative characteristic because susceptibility is a continuous trait determined by multiple genes and environmental factors (an example of a quantitative phenotype with a threshold effect).
(d) Quantitative characteristic because it is determined by many loci (an example of a meristic characteristic).
(e) Discontinuous characteristic because only a few distinct phenotypes are determined by alleles at a single locus.

17. (a) All weigh 10 grams; (b) $^1/_{16}$ weighing 16 grams, $^4/_{16}$ weighing 13 grams, $^6/_{16}$ weighing 10 grams, $^4/_{16}$ weighing 7 grams, and $^1/_{16}$ weighing 4 grams.

18. $^1/_{64}$ 22 cm tall; $^6/_{64}$ 20 cm tall; $^{15}/_{64}$ 18 cm tall; $^{20}/_{64}$ 16 cm tall; $^{15}/_{64}$ 14 cm tall; $^6/_{64}$ 12 cm tall; $^1/_{64}$ 10 cm tall.

19. That six or more loci take part.

21.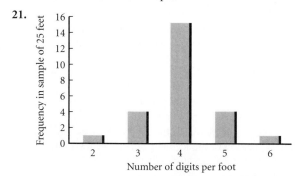

24. No. Correlation does not mean causation. A number of other factors may be responsible for the observed correlation. For example, Baptist preachers may find that their services are needed in areas of high alcohol consumption.

26. (a) $r = 0.80$; (b) 67.8 inches.

27. (a) 0.38; (b) 0.69.

29. (a) Group A, because unrelated people have the greatest genetic variance; (b) no, because siblings raised in the same house should have smaller environmental variance than that of unrelated people.

32. (**a**) 0.75; (**b**) its inaccuracy might be due to a difference between the environmental variance of the genetically identical population and that of the genetically diverse population.

35. 0.8

37. The salesman is correct because Mr. Jones's determination of heritability was conducted for a population of pigs under one environmental condition: low nutrition. His findings do not apply to any other population or even to the same population under different environmental conditions. High heritability for a trait does not mean that environmental changes will have little effect.

40. 0.75

Chapter 25

3. Large population, random mating, and not affected by migration, selection, or mutation.

5. (**a**) If the frequencies of alleles A^1, A^2, and A^3 are defined as p, q, and r, respectively: $f(A^1A^1) = p^2$; $f(A^1A^2) = 2pq$; $f(A^2A^2) = q^2$; $f(A^1A^3) = 2pr$; $f(A^2A^3) = 2qr$; $f(A^3A^3) = r^2$.

(**b**) For an X-linked locus with two alleles:

$f(X^1X^1) = p^2$ among females; $p^2/2$ for the whole population

$f(X^1X^2) = 2pq$ among females; pq for the whole population

$f(X^2X^2) = q^2$ among females; $q^2/2$ for the whole population

$f(X^1Y) = p$ among males; $p/2$ for the whole population

$f(X^2Y) = q$ among males; $q/2$ for the whole population

8. The proportion of the population due to migrants and the difference in allelic frequencies between the migrant population and the original resident population.

10. The effective population size N_E differs from the actual population size if the sex ratio is disproportionate or if a few dominant individuals of either sex contribute disproportionately to the gene pool of the next generation.

14. Mutation increases genetic variation within populations and increases divergence between populations because different mutations arise in each population.

Migration increases genetic variation within a population by introducing new alleles, but it decreases divergence between populations.

Genetic drift decreases genetic variation within populations because it causes alleles to eventually become fixed, but it increases divergence between populations because the effects of drift differ in each population.

Natural selection may either increase or decrease genetic variation, depending on whether the selection is directional or balanced. It may increase or decrease divergence between populations, depending on whether different populations have similar or different selection pressures.

16. $f(T^ET^E) = 0.685$; $f(T^ET^F) = 0.286$; $f(T^FT^F) = 0.029$; $f(T^E) = 0.828$; $f(T^F) = 0.172$.

25. (**a**) 0.1; (**b**) 1%; (**c**) 18%.

26. $f(L^ML^M) = 0.648$; $f(L^ML^N) = 0.304$; $f(L^NL^N) = 0.048$.

30. 0.31

35. 0.17

Chapter 26

2. Mutation and recombination.

4. Because the expected heterozygosity is independent of the breeding system.

7. Neutral-mutation hypothesis: Most molecular variation is adaptively neutral.

Balance hypothesis: Most genetic variation is maintained by balanced selection, favoring heterozygosity at most loci.

9. The biological species concept defines a species as a group of organisms whose members can potentially interbreed with one another other but are reproductively isolated from members of other species. One problem with this concept is that it does not apply to asexually reproducing organisms. Another is that, in practice, most species are defined by phenotypic or anatomical differences, and determining their potential for interbreeding may be difficult or impossible.

12. First, a geographic barrier splits a population into, say, two groups and prevents gene flow between the groups on either side of the barrier. The two groups then evolve independently. Through natural selection, genetic drift, and mutation the two groups become genetically different. These genetic differences lead to reproductive isolation, and the two groups will have undergone allopatric speciation.

14. The flies that feed on apples mate on or near apples, whereas flies that feed on hawthorn fruit mate on or near the hawthorn fruit. Because apples ripen several weeks earlier than do hawthorn fruit, flies that feed on apples mate earlier than flies that feed on hawthorn fruit. These differences in hosts and time of mating lead to reproductive isolation and an accumulation of genetic differences between the two groups of flies.

16. The distance approach relies on the degree of overall similarity between phenotypic characteristics or gene sequences; the most-similar species are grouped together. The parsimony approach tries to reconstruct an evolutionary pathway on the basis of the minimum number of evolutionary changes that must have taken place since the species last had an ancestor in common. Maximum likelihood and Bayesian methods infer phylogenetic relationship on the basis of which phylogeny maximize the probability of obtaining the set of characteristics exhibited by the organisms.

18. The molecular clock estimates the rate at which nucleotide changes take place in a DNA sequence.

21. Horizontal gene transfer, also called lateral gene transfer, is the transmission of genetic information across species boundaries. Horizontal gene transfer takes place frequently in bacteria, through transformation and phage-mediated transduction. In eukaryotes, horizontal gene transfer may take place through endosymbiosis (e.g., mitochondrial and chloroplast genes), viral infections, parasitic infections, and human intervention (genetic engineering). Horizontal gene transfers create problems in the construction of evolutionary relationships with the use of molecular sequence data because the evolutionary history of horizontally transferred genes differs from the evolutionary history of other genes.

23. (**a**) For *P. drummondii*, the percentage of polymorphic loci ranges from 50% in population 3 to 75% in populations 1 and 2. The mean expected heterozygosities in each population are 0.074 for population 1, 0.227 for population 2, and 0.156 for population 3. The mean expected heterozygosity for all populations is 0.15.

For *P. cuspidate*, the percentage of polymorphic loci ranges from 0% in populations 1 and 2 to 25% in population 3. The mean expected heterozygosities are 0.0 for populations 1 and 2 and 0.16 for population 3. The mean expected heterozygosity for all populations is 0.055.

(**b**) Self-fertilization reduces genetic variation.

26. For 7 organisms, 10,395 rooted trees; for 12 organisms, 13,749,310,575 rooted trees.

Index

Note: Page numbers followed by f indicate figures; those followed by t indicate tables. Page numbers preceded by A refer to the *Reference Guide to Model Genetic Organisms*.

A site in ribosome, 410–411, 416–417, 416f–418f, 420f
AbdB gene, 625
ABO blood group antigens, 105, 105f, 108–109
  Bombay phenotype and, 108–109
  nail-patella syndrome and, 185–186, 185f
  in organ transplantation, 632
Abzahanov, Arkaht, 626
Ac elements, 310, 311, 311f, 312f
Acentric chromatids, 247
Acetylation, histone, in gene regulation, 461–462
Achondroplasia, 148t, 714
Acidic activation domains, 467
Acquired immunodeficiency syndrome, 227–229, 727–728
Acquired traits, inheritance of, 47
Acridine orange, as mutagen, 496–497, 497f
Acrocentric chromosomes, 20, 20f, 240
Activator gene, 311
Adaptive mutations, 490–491
Addition rule, 52f, 53
Additive genetic variance, 673
Adenine, 280, 280f, 282, 282f. *See also* Base(s)
Adenosine monophosphate (AMP), in translation, 413, 414f
Adenosine triphosphate. *See* ATP
Adenosine-3′,5′ cyclic monophosphate (cAMP), in catabolite repression, 445
Adjacent-1 segregation, 249, 250f
Adjacent-2 segregation, 249, 250f
A-DNA, 283, 283f
Adoption studies, 146
  of obesity, 146, 146f
Affinity capture, 584
African Replacement hypothesis, 601
African sleeping sickness, 387, 388f
Agar plates, 205, 205f
Age, maternal, aneuploidy and, 256, 257f
Aging
  apoptosis and, 623–624, 624f
  mitochondrial DNA and, 605–606
  premature, 341–342
    in Hutchinson–Gilford progeria syndrome, 135–136, 135f, 136f
    in Werner syndrome, 341–342
  telomerase and, 341–342
  telomere shortening and, 341–342
Agriculture. *See also* Breeding; Crop plants
  genetic engineering in, 525–527, 526f, 547–548
  genetics in, 3, 3f, 7–8, 8f, 9
  origin and spread of, 601
  quantitative genetics in, 659–660
Agrobacterium tumefaciens, in cloning, 522, 523f

AIDS, 227–229, 727–728
Alanine, 406f
Albinism. *See also* Color/pigmentation
  gene interaction and, 109–110
  in Hopi, 1–2, 2f
  inheritance of, 53–55
  in snails, 110, 111f
Alkylating agents, as mutagens, 496
Alleles, 11, 19. *See also* Gene(s)
  in crossing-over, 27, 28f
  definition of, 46, 46t
  dominant, 48f, 49, 100–103
  lethal, 100f, 103–104
  letter notation for, 48, 55
  multiple, 104, 104f, 105f
  recessive, 49
  segregation of, 9, 50f, 57f, 59–60, 162–163
  temperature-sensitive, 123
  wild-type, 55
Allelic fixation, 709
Allelic frequency
  calculation of, 695–696
  at equilibrium, 697–698, 704–705, 704f
  estimation of, 700–701
  fixation and, 709
  genetic drift and, 706–709, 709f, 714–715, 714t
  Hardy–Weinberg law and, 700–701
  migration and, 705–706, 706f, 714–715, 714t
  for multiple alleles, 696
  mutations and, 704–705, 704f, 714–715, 714t
  natural selection and, 709–714, 710t, 711t, 713t–715t
  nonrandom mating and, 701–703
  rate of change for, 713–714, 715f
  for X-linked alleles, 696
Allelic series, 104
Allolactose, 440f, 441
Allopatric speciation, 731, 731f
Allopolyploidy, 260, 260f, 262t
  speciation and, 734–735
Allosteric proteins, 436
Alpha chains, of T-cell receptor, 630–631, 630f
Alpha (α) helix, 283, 283f, 405, 407f
α-Amanitin, 351–352
α-Fetoprotein, 150
Alternate segregation, 249, 250f
Alternation of generations, 35, 35f
Alternative processing pathways, 385–387, 387f, 423
Alternative splicing, 385–387, 387f
  in gene regulation, 470–471
Alu elements, 313
Alzheimer's disease, 606
Ames, Bruce, 498
Ames test, 498, 499f
Amino acids, 405
  assembly of into proteins, 412–420. *See also* Translation
  definition of, 405

  in domains, 434
  in genetic code, 407–412, 410f
  peptide bonds of, 405, 407f, 416, 417f
  proportions of, 409
  sequence of, 407, 407f
  structure of, 405–407, 406f
  triplet code and, 407
  types of, 406f
Aminoacyl (A) site in robosome, 416–417, 416f–418f, 420f
Aminoacyl-tRNA
  synthesis of, 413, 414f
  in translation, 474
Aminoacyl-tRNA synthetases, 413, 414f
2-Aminopurine, as mutagen, 495–496
Amniocentesis, 148–149, 149f
AMP
  cyclic, in catabolite repression, 445
  in translation, 413, 414f
Amphidiploidy, 260
Anagenesis, 723
Anaphase
  in meiosis, 27, 27t, 28f
  in mitosis, 22–24, 23f, 24t, 27
Anderson, W. French, 548
Androgen insensitivity syndrome, 80–81
Anemia, Fanconi, 506t
Aneuploidy, 241, 241f, 252–257, 262t
  autosomal, 254–256, 255f, 257f
  cancer, 256
  causes of, 252, 253f
  definition of, 252
  in Down syndrome, 254–255, 255f, 257f
  in Edward syndrome, 256
  effects of, 252–254
  in evolution, 262–263
  in humans, 254–256
  maternal age and, 256, 257f
  mosaicism and, 257, 258f
  in Patau syndrome, 256
  rate of, 253
  of sex chromosomes, 254
    in Klinefelter syndrome, 79, 79f, 90
    in Turner syndrome, 79, 79f, 90
  in trisomy 8, 256
  types of, 252
  uniparental disomy and, 257
Angelman syndrome, 121
Angiogenesis, in cancer, 648–649
Angiosperms. *See also Arabidopsis thaliana* (mustard plant); Plants
  flower color in
    inheritance of, 100–101, 101f, 102
    lethal alleles and, 103–104
  flower development in, 622–623, 622f
  flower length in, inheritance of, 671–672, 672f
  mitochondrial DNA in, 599
Animals
  breeding of, 7–8, 547–548, 591–592
    artificial selection and, 680, 681f